Learning the Art of Electronics

This introduction to circuit design is unusual in several respects. First, it offers not just explanations, but a full lab course. Each of the labs begins with a discussion of a particular sort of circuit followed by the chance to try it out and see how it actually behaves. Accordingly, students understand the circuit's operation in a way that is deeper and much more satisfying than the manipulation of formulas. Second, it describes circuits that more traditional engineering introductions would postpone: thus, in the third lab, we build a radio receiver; in the fifth, we build an operational amplifier from an array of transistors. The digital part of the course, rebuilt around FPGAs and a powerful ARM microcontroller, allows more sophisticated applications, including a lunar lander, a voice recorder, and a lullaby jukebox. Third, it proceeds at a rapid pace but requires no prior knowledge of electronics. Students gain intuitive understanding through immersion in good circuit design.

- Each session is divided into several parts, including Notes, Labs; many also have Worked Examples, Supplementary Notes, and information about Online Supplements.
- Very little math: focus is on intuition and practical skills.

Thomas C. Hayes worked as a Wall Street lawyer, then moved to Boston where he learned electronics by attending courses at Harvard and M.I.T. He then taught at Harvard and Boston University. His notes, designed for students new to electronics, grew into a succession of books, culminating in this current edition.

David Abrams designed electronic instrumentation after graduating from M.I.T. before co-founding Galactic Industries Corp. In 2001, David negotiated the sale of Galactic and entered Harvard Law School. After working as an Intellectual Property Associate and clerking for a Federal Judge, he ended up teaching circuit design at Harvard, where he updated the Laboratory Electronics course with the new FPGA and microcontroller lessons contained in this second edition.

Paul Horowitz is a Professor of Physics and of Electrical Engineering, emeritus, at Harvard University, where he originated the Laboratory Electronics course in 1974 from which emerged *The Art of Electronics* (1980). He is one of the pioneers of the search of intelligent life beyond Earth. Other research interests have included observational astrophysics, X-ray and particle microscopy, and optical interferometry. He is the author of some 200 scientific articles and reports, has consulted widely for industry and government, and is the designer of numerous electronic and photographic instruments.

Learning the Art of Electronics

A Hands-On Lab Course

Second Edition

Thomas C. Hayes

David Abrams

with the assistance of Paul Horowitz

Shaftesbury Road, Cambridge CB2 8EA, United Kingdom

One Liberty Plaza, 20th Floor, New York, NY 10006, USA

477 Williamstown Road, Port Melbourne, VIC 3207, Australia

314–321, 3rd Floor, Plot 3, Splendor Forum, Jasola District Centre,
New Delhi – 110025, India

103 Penang Road, #05–06/07, Visioncrest Commercial, Singapore 238467

Cambridge University Press is part of Cambridge University Press & Assessment,
a department of the University of Cambridge.

We share the University's mission to contribute to society through the pursuit of
education, learning and research at the highest international levels of excellence.

www.cambridge.org
Information on this title: www.cambridge.org/9781009535182

DOI: 10.1017/9781108946650

First edition © Cambridge University Press 2016
Second edition © David Abrams, Thomas C. Hayes, and Paul Horowitz 2025

This publication is in copyright. Subject to statutory exception and to the provisions
of relevant collective licensing agreements, no reproduction of any part may take
place without the written permission of Cambridge University Press & Assessment.

When citing this work, please include a reference to the DOI 10.1017/9781108946650

First published 2016
5th printing with corrections 2020
Second edition 2025

Printed in Mexico by Litográfica Ingramex, S.A. de C.V.

A catalogue record for this publication is available from the British Library

A Cataloging-in-Publication data record for this book is available from the Library of Congress

ISBN 978-1-009-53518-2 Paperback

Cambridge University Press & Assessment has no responsibility for the persistence
or accuracy of URLs for external or third-party internet websites referred to in this
publication and does not guarantee that any content on such websites is, or will remain,
accurate or appropriate.

For Debbie, Tessa, Turner and Jamie

And in memory of my beloved friend, Jonathan

———————————

For Ak

And my mentors, Jim Williams and J.K. Roberge, who sadly are no longer with us

Contents

Preface to the Second Edition		*page* xxi
Preface to the First Edition		xxvi
Overview, as the Course begins		xxx

Part I Analog: Passive Devices — 1

1N DC Circuits — 3
- 1N.1 Overview — 3
- 1N.2 Three laws — 5
- 1N.3 First application: voltage divider — 12
- 1N.4 Loading, and "output impedance" — 15
- 1N.5 AoE Reading — 24

1L Lab: DC Circuits — 25
- 1L.1 Ohm's law — 25
- 1L.2 Meters, VOM and DVM — 27
- 1L.3 Voltage divider — 29
- 1L.4 Ohms law applied to convert a meter movement into a voltmeter and ammeter — 30
- 1L.5 The diode — 32
- 1L.6 I versus V for some mystery boxes — 33
- 1L.7 Oscilloscope and function generator — 35

1S Supplementary Notes: Resistors, Voltage, Current — 38
- 1S.1 Reading resistors — 38
- 1S.2 Voltage versus current — 41

1W Worked Examples: DC Circuits — 45
- 1W.1 Design a voltmeter, current meter — 45
- 1W.2 Resistor power dissipation — 47
- 1W.3 Working around imperfections of instruments — 48
- 1W.4 Thevenin models — 50
- 1W.5 "Looking through" a circuit fragment, and R_{in}, R_{out} — 51
- 1W.6 Effects of loading — 52

1O Online Content: Breadboarding Hints & Tips — 54
- 1O.1 Solderless breadboards — 54
- 1O.2 Two possible breadboard choices — 54
- 1O.3 Building circuits with the solderless breadboard — 54

2N	**RC Circuits**		55
	2N.1	Capacitors	55
	2N.2	Time-domain view of RCs	57
	2N.3	Frequency domain view of RCs	63
	2N.4	Two unglamorous but important cap applications: "blocking" and "decoupling"	78
	2N.5	A somewhat mathy view of RC filters	80
	2N.6	AoE Reading	81
2L	**Lab: Capacitors**		82
	2L.1	Time-domain view	82
	2L.2	Frequency-domain view	85
2S	**Supplementary Notes: RC Circuits**		89
	2S.1	Reading capacitors	89
	2S.2	C notes: trying for an intuitive grip on capacitors' behavior	94
	2S.3	Sweeping frequencies	97
2W	**Worked Examples: RC Circuits**		104
	2W.1	RC filters	104
	2W.2	RC step response	109
3N	**Diode Circuits**		112
	3N.1	Overloaded filter: another reason to follow our $10\times$ loading rule	112
	3N.2	Scope probe	113
	3N.3	Inductors	116
	3N.4	LC resonant circuit	117
	3N.5	Diode circuits	122
	3N.6	An important diode application: DC from AC	123
	3N.7	Unregulated power supply	127
	3N.8	Radio!	130
	3N.9	AoE Reading	135
3L	**Lab: Resonant and Diode Circuits**		136
	3L.1	LC resonant circuit	136
	3L.2	Diode X–Y plots: a vivid way to display diode I–V behavior	138
	3L.3	Half-wave rectifier	141
	3L.4	Full-wave bridge rectifier	142
	3L.5	Design exercise: AM radio receiver (fun!)	143
	3L.6	Signal diodes	144
3S	**Supplementary Notes and Jargon: Diode Circuits**		146
	3S.1	A puzzle: why LC's ringing dies away despite Fourier	146
	3S.2	Jargon: passive devices	147
3W	**Worked Examples: Diode Circuits**		149
	3W.1	Power supply design	149
	3W.2	Z_{in}	152

		Contents	ix

Part II Analog: Discrete Transistors — 157

4N Transistors I — 159
- 4N.1 Overview of Labs 4 and 5 — 159
- 4N.2 Preliminary: introductory sketch — 162
- 4N.3 The simplest view: forgetting beta — 163
- 4N.4 Add quantitative detail: use beta explicitly — 166
- 4N.5 A strikingly different transistor circuit: the switch — 174
- 4N.6 Recapitulation: the important transistor circuits at a glance — 175
- 4N.7 AoE Reading — 176

4L Lab: Transistors I — 177
- 4L.1 Transistor preliminaries: look at devices out of circuit — 177
- 4L.2 Emitter follower — 178
- 4L.3 Fixing a defect in the follower's R_{out} — 180
- 4L.4 Current source — 181
- 4L.5 Common-emitter amplifier — 182
- 4L.6 Transistor switch — 183
- 4L.7 A note on power-supply noise — 185

4W Worked Examples: Transistors I — 187
- 4W.1 Emitter follower — 187
- 4W.2 Phase splitter: input and output impedances of a transistor circuit — 190
- 4W.3 Transistor switch — 194

5N Transistors II — 197
- 5N.1 Some novelty, but the earlier view of transistors still holds — 198
- 5N.2 Reviewish: phase splitter — 198
- 5N.3 Another view of transistor behavior: Ebers–Moll — 200
- 5N.4 Complication: distortion in a high-gain amplifier — 203
- 5N.5 Complication: temperature instability — 205
- 5N.6 Current mirror — 210
- 5N.7 Reconciling the two views: Ebers–Moll meets $I_C = \beta \times I_B$ — 210
- 5N.8 "Difference" or "differential" amplifier — 211
- 5N.9 Postscript: deriving r_e — 216
- 5N.10 AoE Reading — 217

5L Lab: Transistors II — 219
- 5L.1 Difference or differential amplifier — 219
- 5L.2 AM modulator — 229
- 5L.3 Current mirror — 232

5S Supplementary Notes and Jargon: Transistors II — 237
- 5S.1 Two surprises, perhaps, in behavior of differential amp — 237
- 5S.2 Current mirrors: Early effect — 239
- 5S.3 Transistor summary — 248
- 5S.4 Important circuits — 249

		5S.5 Jargon: bipolar transistors	253
5W		**Worked Examples: Transistors II**	**255**
		5W.1 High-gain amplifiers	255
		5W.2 Differential amplifier	256
		5W.3 Op-amp innards: diff-amp within an IC operational amplifier	257

Part III Analog: Operational Amplifiers and their Applications — 261

6N Op-Amps I — 263
- 6N.1 Overview of feedback — 263
- 6N.2 Preliminary: negative feedback as a general notion — 266
- 6N.3 Feedback in electronics — 267
- 6N.4 The op-amp Golden Rules — 269
- 6N.5 Applications — 270
- 6N.6 Two amplifiers — 271
- 6N.7 Inverting amplifier — 272
- 6N.8 When do the Golden Rules apply? — 274
- 6N.9 Strange things can be put into feedback loop — 277
- 6N.10 AoE Reading — 279

6L Lab: Op-Amps I — 280
- 6L.1 A few preliminaries — 280
- 6L.2 Open-loop test circuit — 281
- 6L.3 Close the loop: follower — 281
- 6L.4 Non-inverting amplifier — 283
- 6L.5 Inverting amplifier — 283
- 6L.6 Summing amplifier — 284
- 6L.7 Design exercise: unity-gain phase shifter — 284
- 6L.8 Push–pull buffer — 286
- 6L.9 Current-to-voltage converter — 287
- 6L.10 Current source — 289

6W Worked Examples: Op-Amps I — 291
- 6W.1 Basic difference amp made with an op-amp — 291
- 6W.2 A more exotic difference amp: IC with wide common-mode input range: INA149 — 294
- 6W.3 Problem: odd summing circuit — 295

7N Op-Amps II: Departures from Ideal — 298
- 7N.1 Old: subtler cases, for analysis — 299
- 7N.2 Op-amp departures from ideal — 302
- 7N.3 Five more applications: integrator, differentiator, difference amplifier, instrumentation amplifier, AC amplifier — 313
- 7N.4 Differentiator — 318
- 7N.5 Op-amp difference amplifier — 318
- 7N.6 Instrumentation amplifier — 320

	7N.7	AC amplifier: an elegant way to minimize effects of op-amp DC errors	323
	7N.8	AoE Reading	324

7L Lab: Op-Amps II   325
- 7L.1 Integrator   325
- 7L.2 Differentiator   328
- 7L.3 Slew rate   330
- 7L.4 Instrumentation amplifier   330
- 7L.5 AC amplifier: microphone amplifier   340

7S Supplementary Notes: Op-Amp Jargon   342

7W Worked Examples: Op-Amps II   343
- 7W.1 The problem   343
- 7W.2 Op-amp millivoltmeter   346

8N Op-Amps III: Nice Positive Feedback   351
- 8N.1 Useful positive feedback   351
- 8N.2 Comparators   352
- 8N.3 *RC* relaxation oscillator   359
- 8N.4 Sinewave oscillator: Wien bridge   363
- 8N.5 AoE Reading   367

8L Lab: Op-Amps III   368
- 8L.1 Two comparators   368
- 8L.2 Op-amp *RC* relaxation oscillator   371
- 8L.3 Easiest *RC* oscillator, using IC Schmitt trigger   371
- 8L.4 Apply the sawtooth: PWM motor drive   372
- 8L.5 IC *RC* relaxation oscillator: '555   373
- 8L.6 '555 for low-frequency frequency modulation ("FM")   374
- 8L.7 Sinewave oscillator: Wien bridge   375

8W Worked Examples: Op-Amp III   377
- 8W.1 Schmitt trigger design tips   377
- 8W.2 Problem: heater controller   380

9N Op-Amps IV: Parasitic Oscillations; Active Filter   385
- 9N.1 Introduction   385
- 9N.2 Active filters   386
- 9N.3 Nasty "parasitic" oscillations: the problem, generally   388
- 9N.4 Parasitic oscillations in op-amp circuits   388
- 9N.5 Op-amp remedies for keeping loops stable   393
- 9N.6 A general criterion for stability: loop gain where phase shift approaches 180°   398
- 9N.7 Parasitic oscillation without op-amps   400
- 9N.8 Remedies for parasitic oscillation   402
- 9N.9 Recapitulation: to keep circuits quiet. . .   403
- 9N.10 AoE Reading   404

9L	**Lab: Op-Amps IV**	405
	9L.1 VCVS active filter	405
	9L.2 Discrete transistor follower	406
	9L.3 Op-amp instability: phase shift can make an op-amp oscillate	408
	9L.4 Op-amp with buffer in feedback loop	410
9S	**Supplementary Notes: Op-Amps IV**	412
	9S.1 Op-amp frequency compensation	412
	9S.2 Active filters: how to improve a simple RC filter	416
	9S.3 Noise: diagnosing fuzz	421
	9S.4 Annotated LF411 op-amp schematic	428
	9S.5 Quantitative effects of feedback	429
9W	**Worked Examples: Op-Amps IV**	433
	9W.1 What all that op-amp gain does for us	433
	9W.2 Stability questions	434
10N	**Op-Amps V: PID Motor Control Loop and Lock-in Amplifier**	439
	10N.1 PID Controller	439
	10N.2 Lock-in Amplifier	453
	10N.3 AoE Reading	457
10L	**Lab: Op-Amps V**	458
	10L.1 PID Controller	458
	10L.2 Lock-in amplifier	469
11N	**Voltage Regulators**	476
	11N.1 Evolving a regulated power supply	477
	11N.2 Easier: 3-terminal IC regulators	482
	11N.3 Thermal design	484
	11N.4 Current sources	486
	11N.5 Crowbar overvoltage protection	487
	11N.6 A different scheme: switching regulators	488
	11N.7 AoE Reading	493
11L	**Lab: Voltage Regulators**	494
	11L.1 Linear voltage regulators	494
	11L.2 A switching voltage regulator	500
11W	**Worked Examples: Voltage Regulators**	505
	11W.1 Choosing a heat sink	505
	11W.2 Applying a current-source IC	506
12N	**MOSFET Switches and an Introduction to JFETs**	508
	12N.1 Why we treat FETs as we do	509
	12N.2 Power switching: turning something ON or OFF	512
	12N.3 A power switch application: audio amplifier	514

	12N.4	Logic gates	516
	12N.5	Analog switches	517
	12N.6	Applications	519
	12N.7	Testing a sample-and-hold circuit	524
	12N.8	AoE Reading	528
12L	**Lab: MOSFET Switches and a JFET AGC**		**529**
	12L.1	Power MOSFET	529
	12L.2	Analog switches	532
	12L.3	Switching audio amplifier	538
	12L.4	A Glimpse of JFETs	539
	12L.5	Applying the voltage-controlled resistance	543
	12L.6	Put the pieces together: AGC	545
12S	**Supplementary Notes: MOSFET Switches**		**546**
	12S.1	A physical picture	546
13N	**Group Audio Project**		**552**
	13N.1	Overview: a day of group effort	552
	13N.2	One concern for everyone: stability	555
	13N.3	Sketchy datasheets for LED and phototransistor	556
13L	**Lab: Group Audio Project**		**557**
	13L.1	Typical waveforms	557
	13L.2	Debugging strategies	558
Part IV	**Digital: Gates, Flip-Flops, Counters, PLD, Memory**		**561**
14N	**Logic Gates**		**563**
	14N.1	Analog versus digital	564
	14N.2	Number codes: Two's-complement	568
	14N.3	Combinational logic	570
	14N.4	Gate types. TTL and CMOS	578
	14N.5	Noise immunity	580
	14N.6	More on gate types	584
	14N.7	AoE Reading	586
14L	**Lab: Logic Gates**		**587**
	14L.1	Set up	587
	14L.2	Input and output characteristics of integrated gates: TTL and CMOS	592
	14L.3	Pathologies	593
	14L.4	Applying IC gates to generate particular logic functions	595
	14L.5	Gate innards; looking within the black box of CMOS logic	597
14S	**Supplementary Notes: Digital Jargon and Logic Interfacing**		**601**
	14S.1	Digital glossary	601
	14S.2	Interfacing among logic families	603

14W	**Worked Examples: Logic Gates**	606
	14W.1 Multiplexing: generic	606
	14W.2 Binary arithmetic	610
14O	**Online Content: Logic Gates**	619
	14O.1 Adding a fixed 3.3 V supply to the PB-503 breadboard	619
15N	**Introduction to Programmable Logic**	620
	15N.1 The hardware	621
	15N.2 Creating the bitstream	623
	15N.3 Introduction to Verilog	624
	15N.4 Digital logic recap	635
	15N.5 AoE Reading	635
15L	**Lab: Programmable Logic**	636
	15L.1 Introduction to FPGA combinational logic	636
	15L.2 Active-low design	639
	15L.3 Solutions	640
15S	**Supplementary Notes: The WebFPGA**	641
	15S.1 Introducing the WebFPGA	641
	15S.2 Verilog resources	647
15W	**Worked Examples: Programmable Logic**	648
	15W.1 Digital comparators	648
15O	**Online Content: Programmable Logic**	650
	15O.1 Online FPGA resources	650
	15O.2 Alternatives to the WebFPGA	650
16N	**Flip-Flops**	651
	16N.1 Sequential circuits generally, and flip-flops	651
	16N.2 Applications: more debouncers	661
	16N.3 Another flop application: shift-register	662
	16N.4 Flip-flops in Verilog	663
	16N.5 AoE Reading	667
16L	**Lab: Flip-Flops**	668
	16L.1 A primitive flip-flop: *SR* latch	668
	16L.2 D type	668
	16L.3 Switch bounce, and three debouncers	671
	16L.4 Shift register	673
	16L.5 That was fun, let's do it again in Verilog	675
	16L.6 Solutions	676
16S	**Supplementary Notes: Flip-Flops**	677
	16S.1 Flip-flop tricks	677
	16S.2 Blocking versus non-blocking assignments	679

17N	**Counters**	681
	17N.1 But first some circuit dangers and anomalies	681
	17N.2 Counters: ripple vs. synchronous	685
	17N.3 Designing a larger, more versatile synchronous counter	689
	17N.4 Some useful counter functions	691
	17N.5 Creating divide-by-N counters	692
	17N.6 Counters in Verilog	695
	17N.7 AoE Reading	698
17L	**Lab: Counters**	699
	17L.1 Counter lab	699
	17L.2 Counters: ripple and synchronous	699
	17L.3 Test a divide-by-3 built two ways	700
	17L.4 8-bit counter	701
	17L.5 Build a selectable divide-by-N counter	703
	17L.6 Counter applications: stopwatch	706
	17L.7 Counters in Verilog	707
17S	**Supplementary Notes: Digital Debugging**	709
	17S.1 Digital debugging tips	709
17W	**Worked Examples: Applications of Counters**	714
	17W.1 Using carry out as a clock signal	714
	17W.2 Counting as a digital design strategy	716
	17W.3 Using a counter to measure period, thus many possible input quantities	716
	17W.4 Bullet timer	722
17O	**Online Content: Counters**	728
	17O.1 74LS469 alternatives	728
18N	**Memory**	729
	18N.1 Buses	729
	18N.2 Memory	730
	18N.3 Memory in Verilog	736
	18N.4 AoE Reading	741
18L	**Lab: Memory**	742
	18L.1 Build a Verilog RAM	742
	18L.2 Build a display decoder ROM	743
18S	**Supplementary Notes: Memory Blocks**	754
	18S.1 Accessing FPGA resource blocks	754
	18S.2 Using WebFPGA EBR memory blocks	755
19N	**Finite State Machines**	759
	19N.1 State machine: new name for old notion	759
	19N.2 Designing any sequential circuit: beyond counters	760
	19N.3 Designing an odd two-bit counter	760

		19N.4 Selecting the FSM clock frequency	769
		19N.5 Recapitulation: several possible sequential machines	770
		19N.6 AoE Reading	770
19L		**Lab: Finite State Machines**	771
		19L.1 Design the control logic for a reaction timer	771
		19L.2 Solutions	775
19W		**Worked Examples: Finite State Machines**	776
		19W.1 Complex blink five ways	776
		19W.2 Takeaway	789
Part V		**Digital: Analog–Digital, PLL, Digital Project Lab**	791
20N		**Analog ↔ Digital; PLL**	793
		20N.1 Digital ⇔ analog conversion, generally	794
		20N.2 Digital to analog (DAC) methods	798
		20N.3 Analog-to-digital conversion	802
		20N.4 Sampling artifacts	814
		20N.5 Dither	816
		20N.6 Phase-locked loop	817
		20N.7 AoE Reading	824
20L		**Lab: Analog ↔ Digital; PLL**	825
		20L.1 Analog-to-digital converter	825
		20L.2 Phase-locked loop: frequency multiplier	831
20S		**Supplementary Notes: Sampling Rules; Sampling Artifacts**	835
		20S.1 What's in this chapter?	835
		20S.2 General notion: sampling produces predictable artifacts in the sampled data	835
		20S.3 Examples: sampling artifacts in time- and frequency-domains	837
		20S.4 Explanation: the images, intuitively	840
20W		**Worked Examples: Analog ↔ Digital**	847
		20W.1 ADC	847
21N		**Digital Project Lab**	851
		21N.1 A digital project	851
21L		**Lab: Digital Project**	854
		21L.1 Multiplex the four-digit LED display	854
		21L.2 Some project ideas	857
Part VI		**Microcontrollers**	867
22N		**Microcontrollers I: Introduction**	869
		22N.1 Microcomputer basics	869

	22N.2	Which microcontroller to use?	876
	22N.3	The ARM Cortex architecture	878
	22N.4	Yikes! Isn't there an easier way?	886
	22N.5	The first day of the microcontroller lab	889
	22N.6	AoE Reading	889
22L		**Lab: Microcontrollers I**	891
	22L.1	Introduction to the SparkFun SAMD21 Mini	891
	22L.2	Install the SparkFun SAMD21 Mini	892
	22L.3	Testing your development environment	893
	22L.4	A debugging primer	895
	22L.5	Blink the external LED	897
	22L.6	Add a delay to permit full speed operation	898
	22L.7	Make the delay a function	899
	22L.8	Use read-modify-write to toggle the LED	900
	22L.9	Use a single register write to toggle the LED	900
	22L.10	Solutions	901
22S		**Supplementary Notes: Microcontrollers I**	902
	22S.1	Preparing the SparkFun SAMD21 Mini Breakout	902
	22S.2	Installing Segger Embedded Studio for ARM	902
	22S.3	Programming in C	905
22O		**Online Content: Microcontrollers I**	907
	22O.1	Alternatives to the SparkFun SAMD21 Mini	907
23N		**Microcontrollers II: Stacks, Timers and Input**	908
	23N.1	The computer stack	908
	23N.2	Subroutine (function) calls	910
	23N.3	An improved delay function	911
	23N.4	SAMD21 port input	913
	23N.5	Intrinsic functions for CPU instructions	916
	23N.6	AoE Reading	916
23L		**Lab: Microcontrollers II**	917
	23L.1	Create a better Delay() function	917
	23L.2	Add a pushbutton to control the LED	918
	23L.3	Scan a matrix keyboard	920
	23L.4	Solutions	923
23S		**Supplementary Notes: Creating Robust, Readable and Maintainable Code**	924
	23S.1	Write functions to isolate hardware from your application	924
	23S.2	Write your application as a finite state machine	927
	23S.3	A deeper dive into the CMSIS structure format	930
24N		**Microcontrollers III: Using Internal Peripherals**	933
	24N.1	Initialize the DAC	934

xviii   Contents

		24N.2 Write the DAC output function	939
		24N.3 Lab preview	939
		24N.4 AoE Reading	939

24L   **Lab: Microcontrollers III**   940
    24L.1   DAC skeleton code   940
    24L.2   Test the SAMD21 DAC   943
    24L.3   Add an audio amp, so you can listen   944
    24L.4   Conclusion   948
    24L.5   Solutions   948

24S   **Supplementary Notes: Waveform Processing**   949
    24S.1   Add the hardware   949
    24S.2   Test your waveform processing system   949
    24S.3   Some suggested lab exercises   950
    24S.4   Skeleton code for waveform processing   955
    24S.5   Solutions   958

24W   **Worked Examples: Speeding Up the SAMD21 CPU Clock**   959
    24W.1   Why does the SAMD21 run so slowly after a reset?   959
    24W.2   Let's look at the CPU clock (literally!)   959
    24W.3   Speeding up the SAMD21   960

25N   **Microcontrollers IV: Timers & Interrupts**   967
    25N.1   The SAMD21 Timer/Counter peripherals   967
    25N.2   Interrupts   970
    25N.3   Lab preview   975
    25N.4   AoE Reading   975

25L   **Lab: Microcontrollers IV**   976
    25L.1   Use Timer/Counter 4 to generate a square wave   976
    25L.2   Use the TC4 interrupt to update the DAC   979
    25L.3   Timer/DAC skeleton code   981
    25L.4   Conclusion   985
    25L.5   Solutions   985

25S   **Supplementary Notes: A Detailed Look at SAMD21 Interrupt Handling**   986
    25S.1   The test program   986
    25S.2   Let's take a look at the foreground program   987
    25S.3   Conclusion   992

25W   **Worked Examples: Using SAMD Interrupts**   993
    25W.1   An interrupt-driven pulse measurement system   993

25O   **Online Content: Creating PlaySong() data**   1003
    25O.1   What is PlaySong?   1003

26N	**Microcontrollers V: Serial Communication**	1004
	26N.1 Some serial buses	1004
	26N.2 The SAMD21 SERCOM peripheral	1010
	26N.3 Other serial protocols...	1011
	26N.4 Readings	1013
26L	**Lab: Microcontrollers V**	1014
	26L.1 Using the SAMD21 SERCOM peripheral	1014
	26L.2 Selecting the SparkFun SAMD21 Port I/O pins	1014
	26L.3 The SparkFun SerLCD display	1016
	26L.4 Program the SERCOM for SPI communications	1017
	26L.5 LCD display tests	1019
	26L.6 The MCP41010 digital potentiometer	1020
	26L.7 SPI digipot test program	1020
	26L.8 Build a theremin	1022
	26L.9 Cleanup	1026
	26L.10 SPI test code	1027
	26L.11 Conclusion	1031
	26L.12 Solutions	1031
26W	**Worked Examples: Interrupt-Driven Serial I/O and SPI Off-chip RAM**	1032
	26W.1 An interrupt-driven version of SPI_SendData()	1032
	26W.2 Adding more RAM	1036
	26W.3 Make a voice recorder	1044
	26W.4 Solutions	1048
27N	**Microcontrollers VI: Using an RTOS**	1049
	27N.1 Real-time operating systems	1049
	27N.2 Segger embOS	1050
	27N.3 Readings	1059
27L	**Lab: Microcontrollers VI**	1060
	27L.1 Complete the hardware build	1060
	27L.2 Install embOS and updated sample code	1060
	27L.3 Get familiar with the RTOS	1062
	27L.4 Build a lullaby jukebox	1066
	27L.5 Testing and debugging an embOS program	1074
	27L.6 Conclusion	1074
	27L.7 Solutions	1074
27S	**Supplementary Notes: Installing embOS**	1075
	27S.1 Installing Segger embOS	1075
27W	**Worked Examples: Adding IR Remote Control to the Jukebox**	1079
	27W.1 Remote control	1079
	27W.2 The hardware	1079
	27W.3 Choosing an IR transmitter	1080
	27W.4 Decoding the demodulated IR signal	1081

Contents

28N **Project Possibilities: Toys in the Attic** — 1087
- 28N.1 Projects: an invitation and a caution — 1088
- 28N.2 Some pretty projects — 1088
- 28N.3 Some other memorable projects — 1093
- 28N.4 Games — 1105
- 28N.5 Sensors, actuators, gadgets — 1107
- 28N.6 Stepper-motor drive — 1107
- 28N.7 Project ideas — 1109
- 28N.8 And many examples are shown in AoE — 1110
- 28N.9 Now go forth — 1110

28O **Online Content: Toys in the Attic Sensors, Actuators, and Gadgets** — 1111
- 28O.1 A few good sources of sensors, actuators, and other devices — 1111
- 28O.2 A source of small parts for mechanical linkages — 1111
- 28O.3 "Servo" motors — 1111
- 28O.4 Nitinol muscle wires — 1111
- 28O.5 Transducers — 1111
- 28O.6 Displays — 1111
- 28O.7 Interface devices — 1111

A **Appendix: Debugging Circuits** — 1112
- A.1 A debugging checklist — 1113
- A.2 David's Axiom — 1113

B **Appendix: Pinouts** — 1114
- B.1 Analog — 1114
- B.2 Digital — 1116

C **Appendix: Transmission Lines** — 1119
- C.1 A topic we have dodged till now — 1119
- C.2 A new case: transmission line — 1120
- C.3 Reflections — 1122
- C.4 But why do we care about reflections? — 1124
- C.5 Transmission-line effects for sinusoidal signals — 1127

D **Appendix: Scope Advice** — 1130
- D.1 What we don't intend to tell you — 1130
- D.2 What we'd like to tell you — 1130

O **Online Appendices** — 1136

Index — 1137

Preface to the Second Edition

This edition of *Learning the Art of Electronics* (LAoE) makes some changes that reflect what we have learned about the way the book has been used since its publication in 2016.

The most important changes are two.

- In our own course, we have begun experimenting with a change that we had resisted for the first few decades of the course: separating, into separate terms, the analog and digital halves of the course.
- In the digital half of the course, we have substituted an ARM microcontroller board for the two-path 8051 microcontroller labs of the first edition of LAoE. In addition, Field Programmable Gate Array (FPGA) exercises replace the PLDs of the first edition (which we here will call "LAoE_1").

The more relaxed pace relieves our sense that we were overdoing the hurry. (We recall the old saw, "We should cover less, uncover more.") The two-term format allows us time to add some new labs.

LAoE_1 envisioned assigning one chapter for each day of the course. This second edition relaxes this expectation: more than one day will be used for nearly every chapter.

The eased pace allows us to pause to discuss and build some circuits that are mentioned but not demonstrated in LAoE_1. For example, on Day One we propose exploring the details of DVM behavior. In Lab 3, we try out a variety of diodes, and show ways to use an oscilloscope to display their behavior. Midway in the operational-amplifier section of the course, we pause to try applying a junction field-effect transistor (JFET), a device we excluded from LAoE_1, though LAoE_1 does use FETs (MOSFETs) as switches.

Analog Additions

We add three labs that ask construction of relatively complex circuits:

- Instrumentation amplifier: this is a special case of an op-amp-based differential amplifier, and thus appears midway in the op-amp section of the book. The designs begin with circuits based literally on operational amplifiers, but then ask you to try two integrated versions of the amplifier, including one that relies on translation into the digital domain. That exercise relies on an Arduino microcontroller. A course (or solo hobbyist!) choosing not to bring in a digital process so early in the course can simply omit the Arduino application.
- Lock-in Amplifier: this method for extracting information from a noisy environment we apply with a home-brew circuit, not a commercial lock-in amp. This lab, too, is placed in the op-amp section of the course.
- Automatic Gain Control: this circuit, a relatively-complex negative feedback loop, requires introducing the JFET (junction field-effect transistor), the FET type omitted from the first edition.

Digital Additions and Changes

Programmable Logic Additions

In LAoE_1 we used a PAL (Programmable Array Logic) to replace discrete devices in the Big Board microcomputer. In this edition, programmable logic, in the form of an FPGA, assumes equal standing with discrete logic.

We use a small FPGA initially to replicate combinational logic, then to create sequential circuits including flip-flops, counters, shift registers, and finite state machines of increasing complexity. In general, we first look at a standard discrete logic device like a gate or counter, then replicate its function in the FPGA.

The digital portion of the course ends with a greatly expanded Digital Project Lab, using the FPGA to drive a multiplexed four-digit LED display and then using that display to build a full, working system of your own design. Suggested projects include a digital clock, a capacitance meter, a digital voltmeter, and a hotel safe, but any design that can make use of the numerical output of the display is possible.

Microcontroller Additions

With the exception of the introduction to Chapter 22N reviewing computer history and introducing microcontrollers, the six chapters dedicated to embedded processing in this edition are entirely new. We've bid a fond farewell to our 1980's Intel 8-bit 8051 microcontroller, assembly language, and crude development tools to move into the modern age of RISC processors.

Our microcontroller labs now develop code for the 32-bit ARM Cortex-M0+ microcontroller using the C language and an Integrated Development Environment incorporating a hardware debugging tool that allows us to stop and peer into the controller innards whenever a voyeuristic impulse strikes us (which is usually when our program is not working correctly). We glory in the [almost] unlimited reach of 32-bit variables and the availability of thirteen general-purpose registers, each of which can be used as a data pointer register into memory.[1]

Of course with great power comes great complexity. Even the low-end ARM microcontroller we use is loaded with peripherals including timer/counters, analog comparators, DAC, ADCs, serial communications units, touch screen interfaces, and more, all with dozens of configuration registers compared to the paltry timer and UART controlled by a handful of registers in the 8051.

We start our journey with small ambition, to blink an LED (the "Hello World" of embedded development). Along the way we tackle the internal peripherals one at a time, scanning a keypad with port I/O commands for user input, using a DAC to output synthetic sine waves, adding a timer interrupt to create musical notes with little CPU intervention, and programming the serial interface to output text on an LCD display. By the last lab, we will have built an entire application, a jukebox that plays children's lullabies using a Real Time Operating System (RTOS), which enables us to reuse the code developed in the previous labs as a set of separate tasks interacting through inter-task mailboxes and events.

---

[1] Those of you who struggled with the limited range of 8-bit registers and the single DPTR register in the 8051 should be shouting for joy about now.

Structural Innovations

Online Content

Alas, with the additional content in the second edition, the book is close to reaching a size that can no longer be printed in a single volume. Rather than forcing you to buy two books (and perhaps saving a few readers from a painful hernia), we have moved some less-used content online to a new website, `https://LAoE.link/Online_Chapters.html`.

The online chapter numbers are included in the Table of Contents with an "O" suffix but you will not find them in the book, you will need to visit the website. However, only optional or rarely used material has been moved online. For example, Chapter 22O, which describes alternatives to the SparkFun SAMD21 Mini breakout we use in the microcontroller labs, is unlikely to be needed more than once, if at all.

Similarly, moving the second half of the previous Chapter 26N online (now renamed to Chapter 28O) allows us to keep the list of interesting Sensors, Actuators, and Gadgets up to date without burdening your book with extra pages.

Maintainable Hyperlinks

We often, particularly in the new microcontroller material, need to point you to web references that are not our own. Unfortunately, the permanency of print does not coexist well with the ephemeral Internet. Law professor Jonathan Zittrain analyzed 2 million links at nytimes.com and found that 25% of them no longer pointed to the original content (what he coins "link rot"). Our solution is to direct you to a link we create prefaced with `https://LAoE.link/` which then redirects you to the actual reference.

For example, the statistic above comes from Prof. Zittrain's article "The Internet is Rotting" in *The Atlantic*, available at `https://www.theatlantic.com/technology/archive/2021/06/the-internet-is-a-collective-hallucination/619320/`. At the time of this edition's publication, our link `https://LAoE.link/Link_Rot.html` pointed to this same page. But should this article move or become unavailable, we will update our link to point to its new location or, at worst, a similar reference.

Should you be concerned about typing a link into your browser that leads to you-know-not-where (or you just object to typing in links in general) you can go to our link redirect web site by typing a single URL, `https://LAoE.link/`, and scrolling to the bottom of the home page. There you will find a list of our links and the sites they redirect to, organized by book chapter. You can then just click on the link directly with no additional typing.

We rely on you, dear reader, to let us know when a link is broken. Do email us at `BadLink@LAoE.link` with the page number and link that does not appear to work and we will endeavor to correct it.

Thanks where thanks are due

Finally, we would like to acknowledge the friends and colleagues who have generously given their time to making sure the new material in this edition not only is correct, but that the lab exercises work as expected. Thanks to Harvard Preceptor Kathryn Ledbetter for testing the FPGA labs without preconceptions – her comments and criticisms were invaluable in helping us make confusing content clearer. Thanks also to University of Maryland Professor Sarah Eno, whose beta testing of the new microcontroller material ensured smoother sailing for you, our readers, in completing those labs. Finally, we once again are indebted to Jim MacArthur, whose extensive knowledge and experience in all things electronic not only caught many of our errors in the content but whose suggestions led to much clearer explanations overall as well. As always, any remaining errors are our own. If

you bring them to our attention, we will endeavor to correct them in a future edition. Contact us at
`authors@LAoE.link`.

Tom Hayes
David Abrams
February 2024

Legal notice

In this book we have attempted to teach the techniques of electronic design, using circuit examples and data that we believe to be accurate. However, the examples, data, and other information are intended solely as teaching aids and should not be used in any particular application without independent testing and verification by the person making the application. Independent testing and verification are especially important in any application in which incorrect functioning could result in personal injury or damage to property.

For these reasons, we make no warranties, express or implied, that the examples, data, or other information in this volume are free of error, that they are consistent with industry standards, or that they will meet the requirements for any particular application. THE AUTHORS AND PUBLISHER EXPRESSLY DISCLAIM THE IMPLIED WARRANTIES OF MERCHANTABILITY AND OF FITNESS FOR ANY PARTICULAR PURPOSE, even if the authors have been advised of a particular purpose, and even if a particular purpose is indicated in the book. The authors and publisher also disclaim all liability for direct, indirect, incidental, or consequential damages that result from any use of the examples, data, or other information in this book.

In addition, we make no representation regarding whether use of the examples, data, or other information in this volume might infringe others' intellectual property rights, including US and foreign patents. It is the reader's sole responsibility to ensure that he or she is not infringing any intellectual property rights, even for use which is considered to be experimental in nature. By using any of the examples, data, or other information in this volume, the reader has agreed to assume all liability for any damages arising from or relating to such use, regardless of whether such liability is based on intellectual property or any other cause of action, and regardless of whether the damages are direct, indirect, incidental, consequential, or any other type of damage. The authors and publisher disclaim any such liability.

Preface to the First Edition

A book and a course

This is a book for the impatient. It's for a person who's eager to get at the fun and fascination of putting electronics to work. The course squeezes what we facetiously call "all of electronics" into about twenty five days of class. Of course, it is nowhere near *all*, but we hope it is enough to get an eager person launched and able to design circuits that do their tasks well.

Our title claims that this volume, which obviously is a *book*, is also a *course*. It is that, because it embodies a class that Paul Horowitz and I taught together at Harvard for more than 25 years. It embodies that course with great specificity, providing what are intended as day-at-a-time doses.

A day at a time: Notes, Lab, Problems, Supplements

Each day's dose includes not only the usual contents of a *book* on electronics – notes describing and explaining new circuits – but also a *lab* exercise, a chance to try out the day's new notions by building circuits that apply these ideas. We think that building the circuits will let you understand them in a way that reading about them cannot.

In addition, nearly every day includes a *worked example* and many days include what we call "supplementary notes." These – for example, early notes on how to read resistors and capacitors – are not for every reader. Some people don't need the note because they already understand the topic. Others will skip the note because they don't want to invest the time on a first pass through the book. That's fine. That's just what we mean by "supplementary:" it's something (like a supplementary vitamin) that may be useful, but that you can quite safely live without.

What's new?

If any reader is acquainted with the *Student Manual...*, published in 1989 to accompany the second edition of *The Art of Electronics*, it may be worth noting principal differences between this book and that one. First, this book means to be self-sufficient, whereas the earlier book was meant to be read alongside the larger work. Second, the most important changes in content are these:

- Analog:
  - we devote a day primarily to the intriguing and difficult topic of *parasitic* oscillations and their cures; and
  - we give a day to building a "PID" circuit, stabilizing a feedback loop that controls a motor's position. We apply signals that form three functions of an *error* signal, the difference between target voltage and output voltage: "Proportional" (P), "Integral" (I), and "Derivative" (D) functions of that difference.

- Digital:
  - application of Programmable Logic Devices (PLDs or "PALs")[2] programmed with the high-level *hardware description language* (HDL), Verilog; and
  - a shift from use of a microprocessor to a *microcontroller*, in the computer section that concludes the course. This microcontroller, unlike a microprocessor, can operate with little or no additional circuitry, so it is well-suited to the construction of useful devices rather than computers.
- Website: The book's website `https://learningtheartofelectronics.com/` has a lot more things, in particular code in machine readable form.

...And the style of this book

A reader will gather early on that this book, like the *Student Manual* is strikingly informal. Many figures are hand-drawn; notation may vary; explanations aim to help intuition rather than to offer a mathematical view of circuits. We emphasize *design* rather than *analysis*. And we try hard to devise applications for circuits that are fun: we like it when our designs make sounds (on a good day they emit *music*), and we like to see motors spin.

Who's likely to enjoy this book and course

You need not resemble the students who take our course at the university, but you may be interested to know who they are, since the course evolved with them in mind. We teach the course in three distinct forms. Most of our students take it during fall and spring daytime classes. There, about half are undergraduates in the sciences and engineering; the other half are graduate students, including a few cross-registered from MIT who need an introduction quicker (and, admittedly, less deep) than electronics courses offered down there. (We don't get EE majors from there; we get people who want a less formal introduction to the subject.)

In the night version of the course, we get mostly older students, many of whom work with technology and who have become curious about what's in the "box" that they work with. Most often the mysterious "box" is simply a computer, and the student is a programmer. Sometimes the "box" is a lab setup (we get students from medical labs), or an industrial control apparatus that the student would like to demystify.

In the summer version of the course, about half our students are rising high school seniors – and the ablest of these prove a point we've seen repeatedly: to learn circuit design you don't need to know any substantial amount of physics or sophisticated math. We see this in the College course, too, where some of our outstanding students have been Freshmen (though most students are at least two or three years older).

And we can't help boasting, as we did in the preface to the 1989 *Student Manual*, that once in a great while a professor takes our course, or at least sits in. One of these buttonholed one of us recently in a hallway, on a visit to the University where he was to give a talk. "Well, Tom," he said, "one of your students finally made good." He was modestly referring to the fact that he'd recently won a Nobel Prize. We wish we could claim that we helped him get it. We can't. But we're happy to have him as an alumnus.[3]

We expect that some of these notes will strike you as elementary, some as excessively dense: your reaction naturally will reflect the uneven experience you have had with the topics we treat. Some of

---

[2] We have switched to a Field Programmable Gate Array, FPGA, in the second edition.
[3] This was Frank Wilczek. He did sit quietly at the back of our class for a while, hoping for some insights into a simulation that he envisioned. If those insights came, they probably didn't come from us.

you are sophisticated programmers, and will sail through the assembly-language programming near the course's end; others will find it heavy going.[4] That's all right. The course out of which this book grew has a reputation as fun, and not difficult in one sense, but difficult in another: the concepts are straightforward; abstractions are few. But we do pass a lot of information to our students in a short time; we do expect them to achieve literacy rather fast. This course is a lot like an introductory language course, and we hope to teach by the method sometimes called immersion. It is the laboratory exercises that do the best teaching; we hope this book will help to make those exercises instructive. I have to add though, in the spirit of modern jurisprudence, a reminder to read the legal notice appended to this Preface.

The mother ship: Horowitz & Hill's *The Art of Electronics*

Paul Horowitz launched this course, 40 odd years ago, and he and Winfield Hill wrote the book that, in its various editions, has served as textbook for this course. That book, now in its third edition and which we will refer to as "AoE," remains the reference work on which we rely. We no longer require that students buy it as they take our course. It is so rich and dense that it might cause intellectual indigestion in a student just beginning their study of electronics. But we know that some of our students and readers will want to look more deeply into topics treated in this book, and to help those people we provide cross-references to AoE throughout this book. The fortunate student who has access to AoE can get more than this book by itself can offer.

Analog and digital: a possible split

In our College course we go through all the book's material in one term of about thirteen weeks. In the night course, which meets just once each week, we do the same material in two terms. The first term treats analog (Days 1–13), the second treats digital (Days 14–26).[5] We know that some other universities use the same split, analog versus digital. It is quite possible to do the digital half before the analog. Only on the first day of digital – when we ask that people build a logic gate from MOSFET switches – would a person without analog training need a little extra guidance. For the most part, the digital half treats its devices as black boxes that one need not crack open and understand. We do need to be aware of input and output properties, but these do not raise any subtle analog questions

It is also possible to pare the course somewhat, if necessary. We don't like to see any of our labs missed, but we know that the summer version of the class, which compresses it all into a bit more than six weeks, makes the tenth lab optional (Chapter 10 presents a "PID" motor controller). And the summer course omits the gratifying but not-essential digital project lab, 21L, in which students build a device of their own design.

Who helped especially with this book

First, and most obviously, comes Paul Horowitz, my teacher long ago, my co-teacher for so many years, and all along a demanding and invaluable critic of the book as it evolved. Most of the book's hand-drawn figures, as well, still are his handiwork. Without Paul and his support, this book would not exist.

Second, I want to acknowledge the several friends and colleagues who have looked closely at parts

---

[4] We have switched to the C programming language in the second edition but the sentiment remains the same.

[5] In the second edition, we have added two chapters to the digital half for the expanded programmable logic material.

of the book and have improved and corrected these parts. Two are friends with whom I once taught, and who thus not only are expert in electronics but also know the course well. These are Steve Morss and Jason Gallicchio. Steve and I taught together nearly thirty years ago. Back then, he helped me to try out and to understand new circuits. He then went off to found a company, but we stayed in touch, and when we began to use a logic compiler in the course (Verilog) I took advantage of his experience. Steve was generous with his advice and then with a close reading of our notes on the subject. As I first met Verilog's daunting range of powers it was very good to be able to consult a patient and experienced practitioner.

Jason helped especially with the notes on sampling. He has the appealing but also intimidating quality of being unable to give half-power, light criticism. I was looking for pointers on details. The draft of my notes came back glowing red with his astute markups. I got more help than I'd hoped for – but, of course, that was good for the notes.

A happy benefit of working where I do is to be able to draw on the extremely knowledgeable people about me, when I'm stumped. Jim MacArthur runs the electronics shop, here, and is always overworked. I could count on finding him in his lab on most weekends, and, if I did, he would accept an interruption for questions either practical or deep. David Abrams is a similarly knowledgeable colleague who twice has helped me to explain to students results that I and the rest of us could not understand. With experience in industry as well as in teaching our course, David is another specially valuable resource.

Curtis Mead, one of Paul Horowitz's graduate students, gave generously of his skill in circuit layout, to help us make the LCD board that we use in the digital parts of this course. Jake Connors, who had served as our teaching assistant, also helped to produce the LCD boards that Curtis had laid out. Randall Briggs, another of our former TAs, helped by giving a keen, close reading.

It probably goes without saying, but let's say it: whatever is wrong in this book, despite the help I've had, is my own responsibility, my own contribution, not that of any wise advisor.

In the laborious process of producing readable versions of the book's thousand-odd diagrams, two people gave essential help. My son, Jamie Hayes, helped first by drawing, and then by improving the digital images of scanned drawings. Ray Craighead, a skilled illustrator whom we found online,[6] made up intelligently rendered computer images from our raggedy hand-drawn originals. He was able to do this in a style that does not jar too strikingly when placed alongside our many hand-drawn figures. We found no one else able to do what Ray did.

Then, when the pieces were approximately assembled, but still very ragged, the dreadfully hard job of putting the pieces together, finding inconsistencies and repetitions, cutting references to figures that had been cut, attempting to impose some consistency, ("Carry_Out" rather than "Carry$_{OUT}$" or "C$_{out}$" – at least on the same page – and so on), in 1000 pages or so, fell to my editor, David Tranah. He put up not only with the initial raggedness, but also with continual small changes, right to the end, and he did this soon after he had completed a similarly exhausting editing of AoE. For this unflagging effort I am both admiring and grateful.

And, finally, I should thank my wife, Debbie Mills, for tolerating the tiresome sight of me sitting, distracted, in many settings – on back porch, vacation terrace in Italy, fireside chair – poking away at revisions. She will be glad that the book, at last, is done.

Tom Hayes

---

[6] See https://LAoE.link/Craighead.com.

Overview, as the Course begins

The circuits of the first three lessons in this course are humbler than what you will see later, and the devices you meet here are probably more familiar to you than, say, transistors, operational amplifiers – or microprocessors: Ohm's Law will surprise none of you; $I = C\frac{dV}{dt}$ probably sounds at least vaguely familiar.

But the circuit elements that this section treats – passive devices – appear over and over in later active circuits. So, if a student happens to tell us, "I'm going to be away on the day you're doing Lab 2," we tell her she will have to make up the lab somehow. We tell her that the second lab, on RC circuits, is the most important in the course. If you do not use that lab to cement your understanding of RC circuits – especially filters – then you will be haunted by muddled thinking for at least the remainder of the analog part of the course.

Resistors will give you no trouble; diodes will seem simple enough, at least in the view that we settle for: they are one-way conductors. Capacitors and inductors behave more strangely. We will see very few circuits that use inductors, but a great many that use capacitors. You are likely to need a good deal of practice before you get comfortable with the central facts of capacitors' behavior – easy to state, hard to get an intuitive grip on: they pass AC, block DC, and only *rarely* cause large phase shifts.

We should also restate a word of reassurance: you can manage this course perfectly even if the "$-j$" in the expression for the capacitor's impedance is completely unfamiliar to you (it's the square root of minus 1). If you consult AoE, and after reading about complex impedances in AoE's spectacularly dense Math Review (Appendix A) you feel that you must be spectacularly dense, don't worry. That is the place in the course where the squeamish may begin to wonder if they ought to retreat to some slower-paced treatment of the subject. Do not give up at this point; hang on until you have seen transistors, at least. One of the most striking qualities of this book is its cheerful evasion of complexity whenever a simpler account can carry you to a good design. The treatment of transistors offers a good example, and you ought to stay with the course long enough to see that: the transistor chapter is difficult, but wonderfully simpler than most other treatments of the subject. You will begin designing useful transistor circuits on your first day with the subject.

It is also in the first three labs that you will get used to the lab instruments – and especially to the most important of these, the oscilloscope. It is a complex machine; only practice will teach you to use it well. Do not make the common mistake of thinking that the person next to you who is turning knobs so confidently, flipping switches and adjusting trigger level – all on the first or second day of the course – is smarter than you are. No, that person has done it before. In two weeks, you too will be making the scope do your bidding – assuming that you don't leave the work to that person next to you, who knew it all from the start.

The images on the scope screen make silent and invisible events visible, though strangely abstracted as well; these scope traces will become your mental images of what happens in your circuits. The scope will serve as a time microscope that will let you see events that last a handful of nanoseconds: the length of time light takes to get from you to the person sitting a little way down the lab bench.

Overview, as the Course begins　　xxxi

You may even find yourself reacting emotionally to shapes on the screen, feeling good when you see a smooth, handsome sinewave, disturbed when you see the peaks of the sine clipped, or its shape warped; annoyed when fuzz grows on your waveforms.

Anticipating some of these experiences, and to get you in the mood to enjoy the coming weeks in which small events will paint their self-portraits on your screen, we offer you a view of some scope traces that never quite occurred, and that nevertheless seem just about right: just what a scope would show if it could. This drawing was posted on my door for years, and students who happened by would pause, peer, hesitate – evidently working a bit to put a mental frame around these not-quite-possible pictures. Sometimes a person would ask if these are scope traces. They are not, of course; the leap beyond what a scope can show was down to the artist: Saul Steinberg. Graciously, he has allowed us to show his drawing here. We hope you enjoy it. Perhaps it will help you to look on your less exotic scope displays with a little of the respect and wonder with which we have to look on the traces below.

Overview, as the Course begins

COUNTRY NOISES

(zigzag)	*Phone rings*
(slashes)	*Phone rings in TV drama*
(slashes spaced)	*Phone rings in house across the road*
(blocks)	*Furnace*
(small dots pattern)	*Refrigerator*
(down-pointing triangles)	*Lawnmower*
(small triangles)	*Small plane*
(fine dotted line)	*T.W.A. from Paris to Kennedy*
(wavy pattern)	*Dishwasher*
(ornamental stars)	*Cricket*
(arrows pattern)	*Electric clock*
(thin line)	*Car*
(line with ornaments)	*Truck*
(decorative pattern)	*Dead leaves cross the road*
(thin line)	*Mosquito*
(floral scroll)	*Paper uncrumples in wastebasket*
(two gradient bars)	*Willow*
(spirals)	*Raccoon?*
(small ornament)	*Chest of drawers creaks*
(diamonds)	*Frog*
(zigzag band)	*Woodpecker*
(beaded line)	*Rain on roof*
(beaded line)	*Rain on deck*
(leaf ornaments)	*Blue jay*
(flourishes)	*Catbird*
(miscellaneous symbols)	*Unidentified*

—Saul Steinberg

Drawing by Saul Steinberg, copyright Saul Steinberg Foundation; originally published in *The New Yorker Magazine*, 1979. reproduced with permission.

Part I

Analog: Passive Devices

1N DC Circuits

Contents

1N.1	**Overview**	**3**
	1N.1.1 Why?	3
	1N.1.2 What is "the art of electronics"?	4
	1N.1.3 What the course is *not* about	4
	1N.1.4 What the course *is* about	5
1N.2	**Three laws**	**5**
	1N.2.1 Ohm's law: $V = IR$	5
	1N.2.2 Kirchhoff's laws: V, I	9
1N.3	**First application: voltage divider**	**12**
	1N.3.1 A voltage divider to analyze	14
1N.4	**Loading, and "output impedance"**	**15**
	1N.4.1 Two possible methods	15
	1N.4.2 Justifying the Thevenin shortcut	16
	1N.4.3 Applying the Thevenin model	17
	1N.4.4 VOM versus DVM: a conclusion?	20
	1N.4.5 Digression on *ground*	20
	1N.4.6 A rule of thumb for relating $R_{out, A}$ to $R_{in, B}$	21
1N.5	**AoE reading**	**24**

1N.1 Overview

We will start by looking at circuits made up entirely of

- DC voltage sources (things whose output voltage is constant over time; things like a battery, or a lab power supply); and . . .
- resistors.

Sounds simple, and it is. We will try to point out quick ways to handle these familiar circuit elements. We will concentrate on one circuit fragment, the voltage divider.

1N.1.1 Why?

In each day's class notes we will sketch the sort of task that the day's material might let us accomplish. We do this to try to head off a challenge likely to occur to any skeptical reader: OK, this is a something-or-other circuit, but what's it for? Why do I need a something-or-other? This is an integrator – but why do I want an integrator? Here is our first try at providing such a sample application:

Problem Given a constant ("DC") voltage source, design a lower voltage source, strong enough to "drive" a particular "load" resistance.

Shorthand version of the problem: Make a voltage divider to deliver a specified voltage. Arrange things so that increasing load current to a maximum causes V_out to vary by no more than a specified percentage.

1N.1.2 What is "the art of electronics?"

Not art that you're likely to find in a museum,[1] but *art* in an older sense: a craft.[2] No doubt the title of *The Art of Electronics* (hereafter referred to as "AoE") was chosen with an awareness of the suggestion that there's something borderline-magical available here: perhaps a hint of "black art"?

AoE §1.1

Here is AoE's formulation of the subject of this course:

> *the laws, rules of thumb, and tricks that constitute the art of electronics as we see it.*

As you may have gathered, if you have looked at the text, this course differs from an engineering electronics course in concentrating on the "rules of thumb" and the bag of "tricks." You will learn to use rules of thumb and reliable "tricks" without apology. With their help you will be able to leave the calculator-bound novice engineer in the dust!

1N.1.3 What the course is *not* about

Wire my basement? Fix my TV?

Alumni of this course sometimes are asked for help that is beyond their capacities, and sometimes below – or beside – what they know. "So, now you can wire some outlets in my basement?" No. This course won't help much with that task, which is easy in a sense but difficult in another, in that it requires a detailed knowledge of electrical codes (required wire gauges; types of jacketing; where ground-fault-interrupters are required). And when your friend's TV quits, you're probably not going to want to fix it: much of the set's circuitry will be embodied in mysterious proprietary integrated circuits; an effective repair – if it were economically worthwhile – would likely amount to ordering a replacement for a substantial module, rather than replacing a burned-out resistor or transistor, as in the good old days of big and fixable devices.

Delivering power

A subtler point is worth making as well: only now and then, in this course, do we undertake to deliver *power* to something (the "something" is conventionally called a "load"). Occasionally, we are interested in doing that: when we want to make a loud sound from a speaker, or want to spin a motor. But much more often, we would like to minimize the flow of power; we are concerned, instead, with the flow of *information*.

On the wall of the lobby of MIT's Electrical Engineering building is a huge blowup of a photo of some MIT engineers standing among what look like large generators or motors, each about the size of a small cow. The photo in Fig. 1N.1 seems to date from the 1930s.

The "Electricals," back then, were concerned mostly with those big machines: with delivering power. It was the power companies that were hiring, when one of our uncles finished at MIT, around 1936.

---

[1] But see, if you find yourself in Munich, a spectacular exception: the world's greatest museum of science and technology, the Deutsches Museum. There you will find wonderful machines demonstrating such arcana as the history of the manufacture of threaded fasteners.

[2] "An industrial pursuit... of a skilled nature; a craft, business, profession." Oxford English Dictionary (1989).

Figure 1N.1 Electronics ca. 1935 [used with permission of MIT.]

Hoover Dam, finished in 1935, was the engineering wonder of the day. Big was beautiful. (Even now, Hoover Dam's website boasts of the dam's *weight*! – 6.6 million tons, in case you were wondering.)

1N.1.4 What the course *is* about: processing *information*

Times have changed, as you may have noticed. Small is beautiful; nano is extra beautiful – and electronics, these days, is concerned mostly with processing information.[3] So, we like circuits that pass and process signals while generating very little heat – using very little power. We like, for example, digital circuits made out of field-effect transistors that form switches; they offer low output impedance, gargantuan input impedance, and quiescent current of approximately zero. To a good approximation, they're not transferring power, not using it, not delivering it. They're dealing in information. That's almost always what we'll be doing in this course.

Obvious, perhaps? Perhaps.

We will postpone till next time – not to overload you, on the first day – discussion of a related topic: just what form the *information* is likely to take, in our circuits: *voltage* versus *current*. The answer may surprise you; or you may be inclined to reject the question as empty, since you know that long ago Ohm taught us that current and voltage in a device can be intimately related. Next time, we'll try to persuade you that you ought not to reject the question; that it's worth considering whether the signal is represented as a *voltage* or as a *current* (and see Note 1S on this topic).

Now on to less abstract topics, and our first useful circuit: a *voltage divider*.

1N.2 Three laws

AoE §1.2.1

A glance at three *laws*: Ohm's law, and Kirchhoff's laws (Voltage – "KVL," and current – "KCL").

We rely on these rules continually, in electronics. Nevertheless, we rarely will mention Kirchhoff again. We use his observations *implicitly*. By contrast, we will see and use Ohm's law a lot; no one has gotten around to doing what's demanded by the bumper sticker one sees around MIT: *Repeal Ohm's Law!*

1N.2.1 Ohm's law: $V = IR$

- V is the analog of water pressure or 'head' of water.
- R describes the restriction of flow.

---

[3] We guess a potentially big exception is the continuing struggle to produce a ever more efficient and economically-viable *electrically-powered* cars. Some glory awaits the *electricals* who succeed at that task.

- I is the rate of flow (volume/unit time).

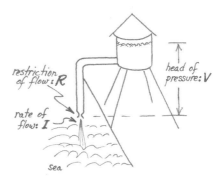

Figure 1N.2 Hydraulic analogy: voltage as head of water, etc. Use it if it helps your intuition.

The homely hydraulic analogy works pretty well, if you don't push it too far – and if you're not too proud to use such an aid to intuition.

What is "voltage," and other deeper questions

For the most part, we will evade such deep questions in this course. We're inclined to say, "Oh, a Volt is what pushes an Amp through an Ohm." But you don't have to be quite so glib (you don't have to sound so much like a Harvard student!). A less circular definition of *voltage* is the potential energy per unit charge. Or, equivalently, it can be defined as the work done to move a unit charge against an electric field (a word that we hope doesn't worry you; we suggest you try to get accustomed to use of the word, even if you have reservations about its usefulness[4]), from one electric potential (analogous to a position on a hillside) to a higher potential (higher on the hillside).

Figure 1N.3 Voltage is work to raise a unit charge from one level to a higher level (or "potential").

The voltage *difference* between two points on the hillside (or staircase, as in Fig. 1N.3) can be described as a difference in electric potential or voltage. The so-called "electric field" will tend to push that charge back down, just as gravity will tend to pull the water down from the tank. You may or may not be interested to know that one volt is the work done as one adds one joule of potential energy to one coulomb of charge.[5] But we'll not again speak in these terms – which sound more like physics than like language for the "art of electronics."

"Ground"

Sometimes we speak of a voltage relative to some absolute reference – perhaps the planet Earth (or, a bit more practically, the potential at the place where a copper spike has been driven into the ground, in the basement of the building where you are doing your electronics). In the hydraulic analogy, that

---

[4] You may be inclined to wonder, as Purcell suggests in his excellent book, "... what is a field? Is it something real, or is it merely a name for a factor in an equation which has to be multiplied by something else to give the numerical value of the force we measure in an experiment?" E.M. Purcell and D.J. Morin, *Electricity and Magnetism*, 3rd ed. (2013), §1.7. Purcell makes a persuasive argument that the concept of "field" is useful.

[5] See Purcell and Morin, §2.2.

absolute *zero*-reference might be sea-level. More often, as we will reiterate below, we are interested only in *relative* voltages: *differences* in potential, measured relative to an arbitrary reference point, not relative to planet Earth.

Ohm's is a very useful rule; but it applies only to things that behave like *resistors*. What are these? They are things that obey Ohm's Law! (Sorry folks: that's as deeply as we'll look at this question, in this course.[6])

Why does Ohm's law hold?

The restriction of current flow that we call "resistance" – which we might contrast with the very-easy flow of current in a piece of wire[7] – occurs because the charge-carrying electrons, accelerated by an electric field, bump into obstacles (vibrations of the atomic lattice) after a short free flight, and then have to be re-accelerated in the direction of the field. Materials that are good conductors – metals – have a substantial population of electrons that are not tightly bound, and consequently are free to travel when pushed. The conductivity of a metal depends on the density of the population of charge carriers (usually, un-bound electrons), and it's kind of reassuring to find that conductivity usually degrades with rising temperature: the free flights become shorter, as electrons bump into the jumpier atoms of the hotter material. This effect you will see confirmed in Lab 1L if things come out right in your experiment (you'll have to do a little reasoning to see this effect confirmed; the notes to Lab 1L do not point out where this occurs). The stronger the field, the faster the drift of the electrons. Field strength goes with *voltage* difference between two points on the conductor; rate of drift of the electrons measures *current*. So, Ohm's Law is pretty plausible.

AoE §C.4

What determines the value of a resistor?

A "resistor" is also, of course, a "conductor"; it may seem a bit perverse to call this thing that is inserted in a circuit to permit current flow a "resistor." But the name comes from the assumption that the resistor is inserted where an excellent conductor – a piece of wire – might have stood, instead. To make a resistor, one can use either of two strategies: to make a "carbon composition" resistor (the sort that we'll use in lab because their values are relatively easy to read), one mixes up a batch of powdered insulator and powdered conductor (carbon), adjusting the proportions to give the material a particular resistivity. To make a "metal film" resistor (much the more common type, these days), one "deposits" a thin film of metal on a ceramic substrate, and then partially cuts away the thin conducting film.

How generally does Ohm's law apply? We begin almost at once to meet devices that do *not* obey Ohm's Law (see Lab 1L: a lamp; a diode). Ohm's Law describes *one* possible relation between V and I in a component; but there are others. As AoE says,

Crudely speaking, the name of the game is to make and use gadgets that have interesting and useful I versus V characteristics.

In a resistor, current and voltage are proportional in a nice, linear way: double the voltage and you get double the current. Ohm's Law holds. Don't expect to use it where it doesn't fit. Even the lamp – whose filament is just a piece of metal that one might expect would behave like a resistor – doesn't follow Ohm's Law, as you'll see in Lab 1L. Why not?[8]

---

[6] If this remark frustrates you, see an ordinary E&M book; for example, see the good discussion of the topic in E.M. Purcell and D.J. Morin, *Electricity & Magnetism*, 3rd ed. (2013), or in S. Burns and P. Bond, *Principles of Electronic Circuits* (1987).

[7] You may prefer the contrast with a superconductor, whose resistance is not just small but is *zero*.

[8] Here's a powerful clue: it *would* follow Ohm's Law if you could hold the filament's temperature constant.

DC Circuits

...But we can extend the reach of Ohm's law? Dynamic resistance: After today, we rarely will limit ourselves to devices that show simply *resistance* – and as we have said, even the resistor-like lamp that you meet in Lab 1L, along with the *diode*, defy Ohm's-Law treatment. But an extended version of Ohm's Law that we'll call *dynamic resistance* will allow us to apply the familiar rule in settings where otherwise it would not work. The idea is just to define a *local* resistance – the tangent to the slope of the device's *V–I* curve:

AoE §1.2.6

$$R_{\text{dynamic}} \equiv \Delta V / \Delta I$$

This redefinition allows us to talk about the effective resistance of a diode, a transistor, or a current source (a circuit that holds current constant). Here is a sketch of a diode's *V–I* curve – oriented so that *V* is the vertical axis. This orientation puts the curves' slopes into the familiar units, *Ohms* (rather than into 1/Ohms,[9] as in the more standard *I–V* plot).

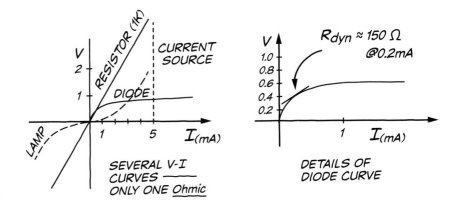

Figure 1N.4 Dynamic resistance illustrated: local slope can be defined for devices that are not Ohmic.

You may like the *resistor's* well-behaved straight line, because it is familiar. But the nice thing about the notion of R_{dynamic} is that is so broad-minded: it is happy to describe the *V–I* curve (or "*I–V* curve," as it is more often called) for any device. It will happily fit a transistor, an exotic current source, or just about anything. The nearly-vertical plot of the *current source*, implying enormous R_{dynamic}, will become important to your understanding of transistors.

AoE §1.2.2C

Power in a resistor: Power is the rate of doing work, as you may recall from a course on mechanics. The concept comes up most often, in electronics, when one tries to specify a component that can safely handle the *power* that it is likely to have to dissipate. High power produces high heat, and calls for a component capable of unloading or dissipating that heat. The three resistors shown below illustrate the rough relation between power rating and size – because large size usually offers large area in contact with the surroundings or "ambient."

The indicated power ratings show the maximum that each can dissipate without damage. The tiny "surface-mount" on the left ("'0805 size," large by surface-mount standards) dissipates more than one might expect if one compares its size to that the of the 1/4 W carbon-comp (the sort we use in the lab); it does better than one might expect because it is soldered directly to a circuit board, whose copper traces help to draw off and dissipate its heat.

In the coming labs you will sometimes run into the question whether your components can handle the power that is expected. Our usual resistors are rated at 1/4 W. You can confirm that such a resistor

---

[9] ...or *Siemens*, the official name for inverse Ohms.

1N.2 Three laws

Figure 1N.5 Three resistors (plus a pretty good copper–nickel alloy conductor).

can handle 15 V (our usual maximum supply voltage) if the resistor's value is at least 1 kΩ. Let's try that calculation: $P = I \times V$.

Thanks to Ohm's Law, the formula for power can be written in any of three ways:

- $P = I \times V$ (as we just said); but since $V = IR$, and
- $P = I^2 R$; and since $I = V/R$,
- $P = V^2/R$.

In the present case, it is the last form that is most useful: $\frac{1}{4}W = 15^2/R_{\min}$. So $R_{\min} = 225/0.25 = 900$. So 1k is close to the minimum safe value, at 15 V (910 would be safe, but let's not be so fussy; call it 1k).

So far, we have mentioned only power in a *resistor*. The notion is more general than that, and the formula

$$P = V \times I$$

holds for any electronic component.

A closer look at what we mean by V and I makes this formula seem almost obvious:

- current measures charge/time; and
- voltage measures work/charge.

So, the product, $V \times I$ = work/charge × charge/time = work/time, and this is *power*.

In this course, the exceptional cases where we do worry about power are those cases in which we either use large voltage swings (for example, the 30 V output swing of the "comparator" in Lab 8L) or want to provide unusually large currents (for example, the speaker drive of Lab 6L, the light-emitting diode drive of the music-transmission lab, 13L, and the voltage regulators of Lab 11L).

1N.2.2 Kirchhoff's laws: V, I

These two 'laws' probably only codify what you think you know through common sense:

- Sum of voltages around the loop (or "circuit") is zero; see Fig. 1N.6, left.
- Sum of currents in and out of a node is zero (algebraic sum, of course); see Fig. 1N.6, right.

Applications of these laws: series and parallel circuits:

Series $I_{\text{total}} = I_1 = I_2$ $V_{\text{total}} = V_1 + V_2$

Parallel $I_{\text{total}} = I_1 + I_2$ $V_{\text{total}} = V_1 = V_2$

DC Circuits

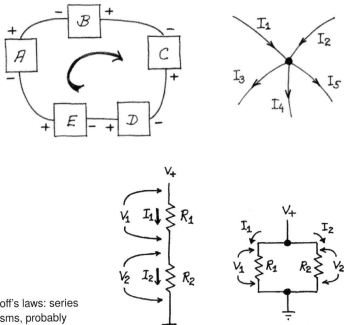

Figure 1N.6 Kirchhoff's two laws. Left: KVL – sum of voltages around a loop is zero; right: KCL – sum of currents in and out of a node is zero.

Figure 1N.7 Applications of Kirchhoff's laws: series and parallel circuits: a couple of truisms, probably familiar to you already.

Query Incidentally, where is the "loop" that Kirchhoff's law refers to? *Answer:* the "loop" (or "circuit," a near-synonym) is apparent if one draws the voltage source as a circuit element, and ties its foot to the foot of the R: see Fig. 1N.8.

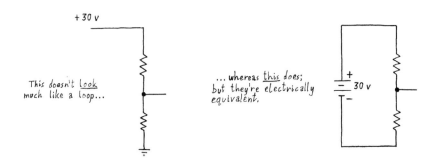

Figure 1N.8 Voltage divider redrawn to look more like a "loop" or "circuit".

Usually we don't bother to draw the voltage source that way; we label points with voltage values, and assume that you can picture the circuit path for yourself, if you choose to.

This is kind of *boring*. So, let's hurry on to less abstract circuits: to applications – and tricks. First, some labor-saving tricks.

Parallel resistances: calculating equivalent R

The *conductances* add:

$$\text{Conductance}_{\text{total}} = \text{Conductance}_1 + \text{Conductance}_2 = 1/R_1 + 1/R_2$$

This is the easy notion to remember, but not usually convenient to apply, for one rarely speaks of conductances. The notion "resistance" is so generally used that you will sometimes want to use the formula for the effective resistance of two parallel resistors:

$$R_{\text{tot}} = (R_1 \cdot R_2)/(R_1 + R_2)$$

Figure 1N.9
Parallel resistors: the *conductances* add; unfortunately, the *resistances* don't.

AoE §1.2.2B

Believe it or not, even this formula is messier than what we like to ask you to work with in this course. So we proceed immediately to some tricks that let you do most work in your head. Consider the easy cases in Fig. 1N.10. The first two are especially important, because they help one to *estimate* the effect of a circuit one can liken to either case. Labor-saving tricks that give you an estimate are not to be scorned: if you see an easy way to an estimate, you're likely to make the estimate. If you have to work too hard to get the answer, you may find yourself simply *not* making the estimate. A deadly trap for the student doing a lab is the thought, "Oh, I'll calculate this later – some time this evening, when I'm comfortable in front of a spreadsheet." This student won't get to that calculation! The leftmost case in Fig. 1N.10 surely doesn't call for pulling out a formula: two equal Rs paralleled behave like $R/2$. The middle case is easier still, given that we're willing to ignore errors under 10%. (On the other hand, when you *do* want to trim an R value by 10% this is an easy way to do just that.) The rightmost calls for slightly more imagination: think of the lone R as a paralleling of two resistors of equal value: $1/R = 2/2R$. Then the whole looks like three paralleled resistors, each of value $2R$. The result then is $2R/3$.

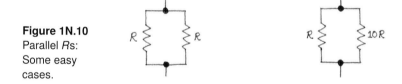

Figure 1N.10
Parallel *R*s: Some easy cases.

In this course we usually are content with answers good to 10%. So, if two parallel resistors differ by a factor of ten or more, then we can ignore the larger of the two.

Let's elevate this observation to a **rule of thumb** (our first). While we're at it, we can state the equivalent rule for resistors in series.

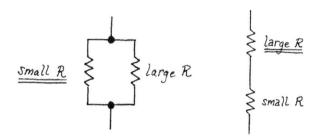

Figure 1N.11 Resistor calculation shortcut: parallel, series. In a parallel circuit, a resistor much *smaller* than others dominates. In a series circuit, the *larger* resistor dominates.

1N.3 First application: voltage divider

AoE §1.2.3

Why dividers? Are dividers necessary? Why not start with the *right* voltage? The answer, as you know, is just that a typical circuit needs several voltages, and building a "power supply" to deliver each voltage is impractical (meaning, mostly, *expensive*). You'll soon be designing power supplies, and certainly at that time will appreciate how much simpler a voltage divider is, compared to a full power supply.

To illustrate the point that voltage dividers are useful, and not just an academic device used to provide an easy introduction to circuitry, we offer here a piece of a fairly complex device, a "function generator" – the box that soon will be providing waveforms to the circuits that you build in Lab 2L. Figure 1N.12 shows part of the circuitry that converts a triangular waveform into a sinusoidal shape.

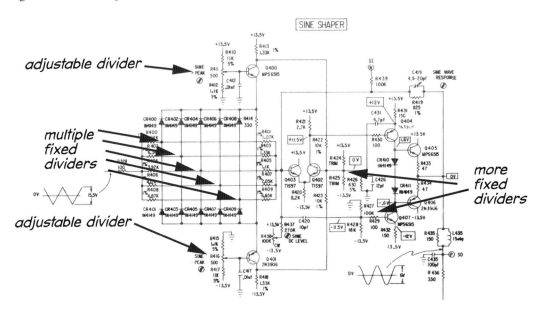

Figure 1N.12 Voltage dividers in a function generator: dividers are not just for beginners [Krohn–Hite 1600 A function generator].

Aside: variable dividers or "potentiometers": Before we look closely at an ordinary divider, let's note a variation that's often useful: a voltage divider that is adjustable. This circuit, available as a ready-made component, is called a "potentiometer."

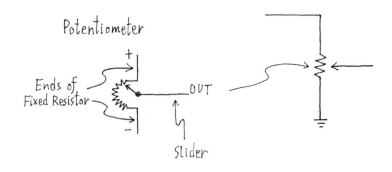

Figure 1N.13 Symbol for potentiometer, and its construction.

1N.3 First application: voltage divider

The name describes the circuit pretty well: the device "meters" or measures out "potential." Recalling this may help you to keep separate the two ways to use the device:

- as a potentiometer versus..., and
- as a variable resistor.

"Variable resistor:" just one way to use a pot: The component is called a potentiometer (a 3-terminal device), but it can be used as a *variable resistor* (a 2-terminal device).

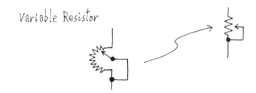

Figure 1N.14 A "pot" can be wired to operate as a variable resistor.

The pot becomes a variable resistor if one uses just one end of the fixed resistor and the slider, or if (somewhat better) one ties the slider to one end.[10]

How the potentiometer is constructed: It helps to see how the thing is constructed. In Fig. 1N.15 are photos of two potentiometers. It is not hard to recognize how the large one on the left works. A (fixed) wire-wound resistor, not insulated, follows most of the way around a circle. A sliding contact presses against this wire-wound resistor. It can be rotated to either extreme.

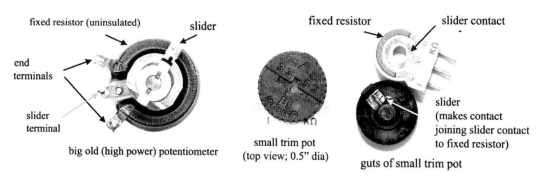

Figure 1N.15 Potentiometer construction details.

At the position shown, the contact seems to be about 70% of the way between lower and upper terminals. If the upper terminal were at 10 V and the lower one at ground, the output at the *slider* terminal would be about 7 V.

The smaller pot shown in middle and right of Fig. 1N.15 (and shown enlarged, relative to the left-hand device) is fundamentally the same, but constructed in a way that makes it compact. Its fixed resistor – of value 1 kΩ – is made not of wound wire but of "cermet."[11]

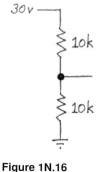

Figure 1N.16 Voltage divider.

[10] The difference between those two options is subtle. If the fixed resistance is, say 100k, the variable resistance range for either arrangement is 0 to 100k. The difference – and the reason to prefer tying slider to an end – appears if the pot becomes dirty with age. If the slider should momentarily lose contact with the fixed resistor, the tied arrangement takes the effective R value to 100k. In contrast, the lazier arrangement takes R to *open* (call it "infinite resistance," if you prefer) when the slider loses contact.

The difference often is not important. But since it costs you nothing to use the "tied" configuration, you might as well make that your habit.

[11] "Cermet" is a composite material formed of ceramic and metal.

1N.3.1 A voltage divider to analyze

Figure 1N.16 is a simple example of the more common *fixed* voltage divider. At last we have reached a circuit that does something useful. It delivers a voltage of the designer's choice; a voltage less than the original or "source" voltage.

First, a note on labeling: we label the resistors "10k"; we omit "Ω." It goes without saying. The "k" means kilo- or 10^3, as you probably know.

One can calculate V_{out} in several ways. We will try to push you toward the way that makes it easy to get an answer in your head.

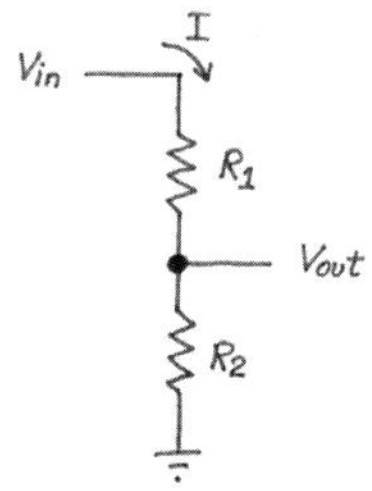

Three ways to analyze this circuit

First method: calculate the current through the series resistance...: Calculate the current (see Fig. 1N.17):

$$I = V_{in}/(R_1 + R_2)$$

That's $30\,\text{V}/20\,\text{k}\Omega = 1.5\,\text{mA}$. After calculating that current, use it to calculate the voltage in the lower leg of the divider:

$$V_{out} = I \cdot R_2$$

Here that product is $1.5\,\text{mA} \cdot 10\text{k} = 15\,\text{V}$. That takes *too long*.

Figure 1N.17
Voltage divider: first method (too hard!): calculate current explicitly.

Second method: rely on the fact that I is the same in top and bottom...: But rely on this equality only *implicitly*.

If you want an algebraic argument, you might say,

$$V_2/(V_1 + V_2) = I \cdot R_2/(I \cdot (R_1 + R_2)) = R_2/(R_1 + R_2)$$

or,

$$V_{out} = V_{in} \cdot R_2/(R_1 + R_2) \qquad (1\text{N}.1)$$

In this case that means

$$V_{out} = V_{in} \cdot (10\text{k}/20\text{k}) = V_{in}/2$$

That's *much better*, and you will use formula (1N.1) fairly often. But we would like to push you not to memorize that equation, but instead to work less formally.

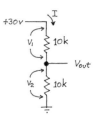

Figure 1N.18
Voltage divider: second method: (a little better): current implicit.

Third method: say to yourself in words how the divider works...: Something like

> Since the currents in top and bottom are equal, the voltage drops are proportional to the resistances (later, *impedances* – a more general notion that covers devices other than resistors).

So in this case, where the lower R makes up half the total resistance, it also will show half the total voltage.

For another example, if the lower leg is 10 times the upper leg, it will show about 90% of the input voltage (10/11, if you're fussy, but 90%, to our usual tolerances).

1N.4 Loading, and "output impedance"

Now – after you've calculated V_{out} for the divider – suppose someone comes along and puts in a third resistor, as in Fig. 1N.19. (*Query*: Are you entitled to be outraged? Is this not fair?[12])

Again there is more than one way to make the new calculation – but one way is tidier than the other.

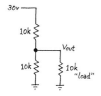

Figure 1N.19 Voltage divider *loaded*.

1N.4.1 Two possible methods

Tedious method: Model the two lower Rs as one R; calculate V_{out} for this new voltage divider: see Fig. 1N.20. The new divider delivers $1/3\ V_{\text{in}}$. That's reasonable, but it requires you to draw a new model to describe each possible loading.

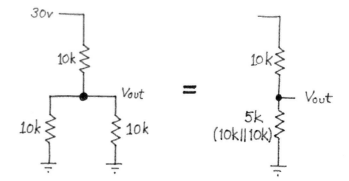

Figure 1N.20 Voltage divider *loaded*: load and lower R combined in model.

AoE §1.2.5

Better method: Thevenin's model: Here's how to calculate the two elements of the Thevenin model:

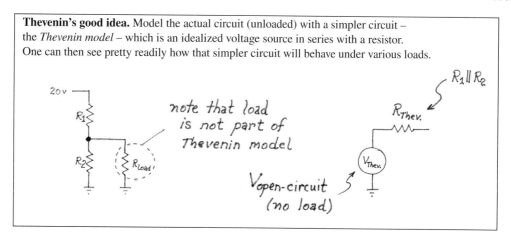

Figure 1N.21 Thevenin model: perfect voltage source in series with output resistance.

V_{Thev} Just $V_{\text{open circuit}}$: the voltage out when nothing is attached ("no load").

---

[12] We don't think you're entitled to be outraged – but perhaps you should be mildly offended. Certainly it's normal to see *something* attached to the output of your circuit (that "something" is called a "load"). You built the divider in order to provide current to something. But, as we'll soon be saying repeatedly, you are entitled to expect loads that are not too "heavy" (that is, don't draw too much current). By the standards of our course, this 10k load *is* too heavy. You'll see why in a moment.

16   DC Circuits

R_Thev Often formulated as the quotient of $V_\text{Thev}/I_\text{short circuit}$, which is the current that flows from the circuit output to ground if you simply *short* the output to ground.

In practice, you are not likely to discover R_Thev by so brutal an experiment. Often, shorting the output to ground is a very *bad* idea: bad for the circuit and sometimes dangerous to you. Imagine the result, for example, if you decided to try this with a 12 V car battery! And if you have a diagram of the circuit to look at, a much faster shortcut is available: see Fig. 1N.22.

Shortcut calculation of R_Thev. Given a circuit diagram, the fastest way to calculate R_Thev is to see it as *the parallel resistance of the several resistances viewed from the output*. (This formulation assumes that the voltage sources are ideal, incidentally; when they are not, we need to include their output resistance. For the moment, let's ignore this complication.)

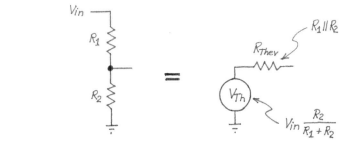

Figure 1N.22 $R_\text{Thev} = R_1$ parallel R_2.

1N.4.2   Justifying the Thevenin shortcut

Our *shortcut*, shown in Fig. 1N.22, asserts that $R_\text{Thev} = R_\text{parallel}$, but the result still may strike you as a little odd: why should R_1, going up to the positive supply, be treated as *parallel* to R_2? Well, suppose the positive supply were set at zero volts. Then surely the two resistances would be in parallel, right?

And try another thought experiment: redefine the positive supply as 0 V. It follows then that the voltage we had been calling "ground" or "zero volts" now is -30 V (to use the numbers of Fig. 1N.16). Now it seems clear that the *upper* resistor, to ground, matters; the lower resistor, R_2, going to -30 V now is the one that seems odd. But, of course, the circuit doesn't know or care how we humans have chosen to define our "zero volts" or "ground."

The point of this playing with "ground" definitions is just that the voltages at the far end of those resistors "seen" from the output do not matter. It matters only that those voltages are *fixed*.[13]

Or suppose a different divider (chosen to make the numbers easy): 20 V divided by two 10k resistors. To discover the *impedance* at the output, do the usual experiment (one that we will speak of again and again):

A general definition and procedure for determining *impedance* at a point:
To discover the *impedance* at a point:
apply a ΔV; find ΔI.
The quotient is the impedance.

---

[13] Later we will relax this requirement, too. Thevenin's model works as well to define the output resistance of a time-varying voltage source as it does for a DC source.

1N.4 Loading, and "output impedance"

You will recognize the circuits in Fig. 1N.23 as just a "small signal" or "dynamic" version of Ohm's Law. In this case 1 mA was flowing before the wiggle. After we force the output up by 1 V, the currents in top and bottom resistors no longer match: upstairs: 0.9 mA; downstairs, 1.1 mA. The difference must come from you, the wiggler.

$$\text{Result: impedance} = \Delta V/\Delta I = 1\,\text{V}/0.2\,\text{mA} = 5\text{k}$$

AoE §1.2.6

And – happily – that is the parallel resistance of the two Rs. Does that argument make the result easier to accept?

You may be wondering why this model is useful. Fig. 1N.23 shows one way to put the answer, though probably you will remain skeptical until you have seen the model at work in several examples.

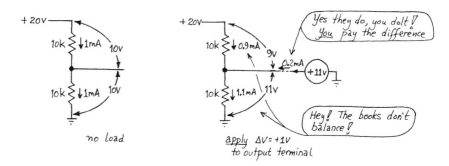

Figure 1N.23 Hypothetical divider: current = 1 mA; apply a wiggle of voltage, ΔV; see what ΔI results.

Rationalizing Thevenin's result: parallel Rs

Any non-ideal voltage source "droops" when loaded. How much it droops depends on its "output impedance." The Thevenin equivalent model, with its R_Thev, describes this property neatly in a single number.

1N.4.3 Applying the Thevenin model

First, let's make sure Thevenin had it right: let's make sure his model behaves the way the original circuit does. We found that the 10k, 10k divider from 30 V, which put out 15 V when not loaded, drooped to 10 V under a 10k load; see Fig. 1N.24. Does the model do the same?

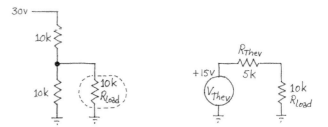

Figure 1N.24 Thevenin model and load: droops as original circuit drooped.

Yes, the model droops to the extent the original did: down to 10 V. What the model provides that the original circuit lacked is that single value, R_Thev, expressing how droopy/stiff the output is.

If someone changed the value of the load, the Thevenin model would help you to see what droop to expect; if, instead, you didn't use the model and had to put the two lower resistors in parallel again and recalculate their parallel resistance, you'd take longer to get each answer, and still you might not get a feel for the circuit's *output impedance*.

DC Circuits

Let's try using the model on a set of voltage sources that differ *only* in R_{Thev}. At the same time we can see the effect of an instrument's *input impedance*.

Suppose we have a set of voltage dividers, but dividing a 20 V input by two; see Fig. 1N.25. Let's assume that we use 1% resistors (value good to ±1%): V_{Thev} is obvious, and is the same in all cases; but R_{Thev} evidently varies from divider to divider.

Suppose now that we try to *measure* V_{out} at the output of each divider. If we measured with a *perfect* voltmeter, the answer in all cases would be 10 V. (*Query*: is it 10.000 V? 10.0 V?[14])

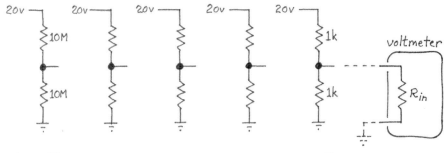

Figure 1N.25 Set of similar voltage dividers: same V_{Thev}, differing R_{Thev}.

But if we actually perform the measurement, we will see the effect of loading by the R_{in} of our imperfect lab voltmeters. Let's try it with a VOM ("volt-ohm-milliammeter," the conventional name for the old-fashioned "analog" meter, which gives its answers by deflecting its needle to a degree that forms an analog to the quantity measured), and then with a DVM ("digital voltmeter," a more recent invention, which usually can measure current and resistance as well as voltage, despite its name; both types sometimes are called simply "multimeters").

Suppose you poke the several divider outputs, beginning from the right side, where the resistors are 1 kΩ. Here's a table showing what we found, at three of the dividers:

R values	measured V_{out}	inference
1k	9.95	within R tolerance
10k	9.76	loading barely apparent
100k	8.05	loading obvious

The 8.05 V reading shows such obvious loading – and such a nice round number, if we treat it as "8 V" – that we can use this to calculate the meter's R_{in} without much effort: see Fig. 1N.26.

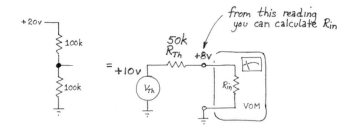

Figure 1N.26 VOM reading departs from ideal; we can infer $R_{in\text{-}VOM}$.

---

[14] It could be 10.0 V – but it could be a bit less than 9.9 V or a bit more than 10.1 V, if the 20 V source were good to 1% and we used 1% resistors. We should not expect perfection.

1N.4 Loading, and "output impedance"

As usual, one has a choice now, whether to pull out a formula and calculator, or whether to try, instead, to do the calculation "back-of-the-envelope" style. Let's try the latter method.

First, we know that R_{Thev} is 100k parallel 100k: 50k. Now let's talk our way to a solution (an *approximate* solution: we'll treat the measured V_{out} as just "8 V"):

> The meter shows us 8 parts in 10; across the divider's R_{Thev} (or call it "R_{out}") we must be dropping the other 2 parts in 10. The relative sizes of the two resistances are in proportion to these two voltage drops: 8 to 2, so $R_{\text{in-VOM}}$ must be $4 \cdot R_{\text{Thev}}$: 200k.

If we look closely at the front of the VOM, we'll find a little notation,

$$20{,}000 \text{ ohms/volt.}$$

That specification means that if we used the meter on its 1 V scale (that is, if we set things so that an input of 1 V would deflect the needle fully), then the meter would show an input resistance of 20k. In fact, it's showing us 200k. Does that make sense? It will when you've figured out what must be inside a VOM to allow it to change voltage ranges: a set of big series resistors.[15] Our answer, 200k, is correct when we have the meter set to the 10 V scale, as we do for this measurement.

This is probably a good time to take a quick look at what's inside a multimeter – VOM or DVM:

How a meter works: some meter types fundamentally sense *current*, others sense *voltage*; see Figs. 1N.27 and 1N.28. Both meter mechanisms, however, can be rigged to measure both quantities, operating as a "multimeter."

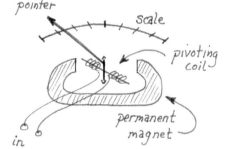

Figure 1N.27 An *analog* meter senses current in its guts.

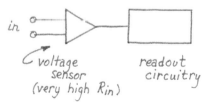

Figure 1N.28 A *digital* meter senses voltage in its innards.

The VOM specification, 20,000 ohms/volt, describes the sensitivity of the meter *movement* – the guts of the instrument. This movement puts a fairly low ceiling on the VOM's input resistance at a given range setting.

Let's try the same experiment with a DVM, and let's suppose we get the following readings:

---

[15] You'll understand this fully when you have done AoE Problem 1.8; for now, take our word for it.

DC Circuits

R values	Measured V_{out}	Inference
100k	9.92	within R tolerance
1M	9.55	loading apparent
10M	6.69	loading obvious

Again let's use the case where the droop is obvious; again let's talk our way to an answer:

This time R_{Thev} is 5M; we're dropping 2/3 of the voltage across $R_{in\text{-}DVM}$, 1/3 across R_{Thev}. So, $R_{in\text{-}DVM}$ must be $2 \times R_{Thev}$, or 10M.

If we check the datasheet for this particular DVM we find that its R_{in} is specified to be "10M, all ranges." Again our readings make sense.

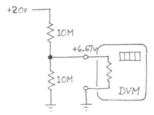

Figure 1N.29 DVM reading departs from ideal; we can infer $R_{in\text{-}DVM}$.

1N.4.4 VOM versus DVM: a conclusion?

Evidently, the DVM is a better voltmeter, at least in its R_{in} – as well as much easier to use. As a current meter, however, it is no better than the VOM: it drops 1/4 V full scale, as the VOM does; it measures current simply by letting it flow through a small resistor; the meter then measures the voltage across that resistor.

1N.4.5 Digression on *ground*

The concept "ground" ("earth," in Britain) sounds solid enough. It turns out to be ambiguous. Try your understanding of the term by looking at some cases: see Fig. 1N.30.

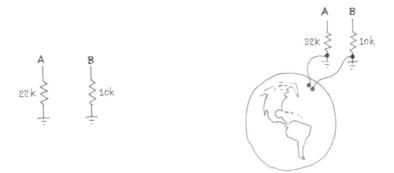

Figure 1N.30 Ground in two senses.

Query What is the resistance between points A and B? (Easy, if you don't think about it too hard.[16])

---

[16] Well, not easy, unless you understand the "ground" indicated in Fig. 1N.30 to be the common zero-reference in the circuit at hand. In that case, the shared "ground" connection means that the 10k and 22k resistances are in series. If you took "ground" to mean "the world, with one connection in New York, the other in Chicago," the answer would not be so easy. Luckily, that would be a rare – and perverse – meaning to attribute to the circuit diagram.

1N.4 Loading, and "output impedance"

We know that the ground symbol means, in any event, that the bottom ends of the two resistors are electrically joined. Does it matter whether that point is also tied to the pretty planet we live on? It turns out that it does not.

And where is "ground" in the circuit in Fig. 1N.31?

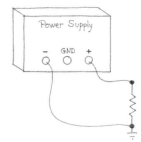

Figure 1N.31 Ground in two senses, revisited.

Two senses for "Ground"

"Local ground" Local ground is what we care about: the common point in our circuit that we arbitrarily choose to call zero volts. Only rarely do we care whether or not that local reference is tied to a spike driven into the earth.

"Earth ground" But, *be warned*, sometimes you are confronted with lines that are tied to world ground – for example, the ground clip on a scope probe, and the "ground" of the breadboards that we use in the lab; then you must take care not to clip the scope ground to, say, +15 on the breadboard.

For a vivid illustration of "earth" grounding, see an image from Wikipedia: https://LAoE.link/Earth_Ground.html.

It might be useful to use different *symbols* for the two senses of ground, and such a convention does exist: see Fig. 1N.32.

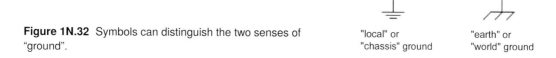

Figure 1N.32 Symbols can distinguish the two senses of "ground".

Unfortunately, this distinction is not widely observed, and in this book as in AoE, we will not maintain this graphical distinction.[17] Ordinarily, we will use the symbol that Fig. 1N.32 suggests for "local ground." In the rare case when we intend a connection to "world" ground, we will say so in words, not graphically.

1N.4.6 A rule of thumb for relating $R_{\text{out, A}}$ to $R_{\text{in, B}}$

The voltage dividers whose outputs we tried to measure introduced us to a problem we will see over and over again: some circuit tries to "drive" a load. To some extent, the load changes the output. We need to be able to predict and control this change. To do that, we need to understand, first, the characteristic we call R_{in} (this rarely troubles anyone) and, second, the one we have called R_{Thev} (this one takes

---

[17] Our symbols, here, do not follow the usual convention, but are consistent with use in *The Art of Electronics*, a consistency that seems necessary.

22 DC Circuits

longer to get used to). Next time, when we meet frequency-dependent circuits, we will generalize both characteristics to "Z_{in}" and "Z_{out}."

Here we will work our way to another rule of thumb; one that will make your life as designer relatively easy. We start with a *Design goal*: When circuit A drives circuit B: arrange things so that B loads A lightly enough to cause only insignificant attenuation of the signal. And this goal leads to the rule of thumb in Fig. 1N.33.

How does this rule get us the desired result? Look at the problem as a familiar voltage divider question. If $R_{out, A}$ is much smaller than $R_{in, B}$, then the divider delivers nearly all of the original signal. If the relation is 1:10, then the divider delivers 10/11 of the signal: attenuation is just under 10%, and that's good enough for our purposes.

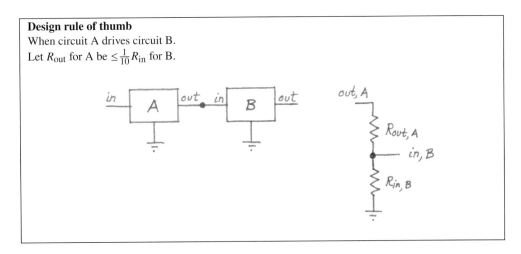

Figure 1N.33 Circuit A drives circuit B.

We like this arrangement not just because we like big signals. (If that were the only concern, we could always boost the output signal, simply amplifying it.) We like this arrangement above all because it *allows us to design circuit-fragments independently*: we can design A, then design B, and so on. We need not consider A, B as a large single circuit. That's good: makes our work of design and analysis lots easier than it would be if we had to treat every large circuit as a unit.

An example, with numbers as in Fig. 1N.34: what R_{Thev} for droop of $\leq 10\%$? What R's, therefore?

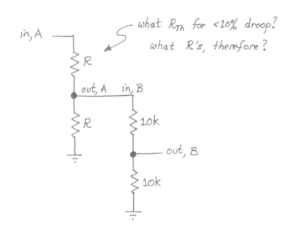

Figure 1N.34 One divider driving another: a chance to apply our rule of thumb.

The effects of this rule of thumb become more interesting if you extend this chain: from A and B, on to C: see Fig. 1N.35.

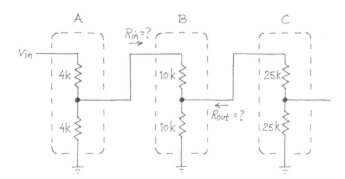

Figure 1N.35 Extending the divider: testing the claim that our rule of thumb lets us consider one circuit fragment at a time.

As we design C, what R_{Thev} should we use for B? Is it just 10k parallel 10k? That's the answer if we can consider B by itself, using the usual simplifying assumptions: source ideal ($R_{out} = 0$) and load ideal (R_{in} infinitely large).

But should we be more precise? Should we admit that the upper branch really looks like 10K+2K: 12k? That's 20% different. Is our whole scheme crumbling? Are we going to have to look all the way back to the start of the chain, in order to design the next link? Must we, in other words, consider the whole circuit at once, not just the fragment B, as we had hoped?

No. Relax. That 20% error gets diluted to half its value: R_{Thev} for B is 10k *parallel* 12k, but that's a shade under 5.5k. So we need not look beyond B. We can, indeed, consider the circuit in the way we had hoped: fragment by fragment.

If this argument has not exhausted you, you might give our claim a further test by looking in the other direction: does C alter B's input resistance appreciably (>10%)? You know the answer, but confirming it would let you check your understanding of our rule of thumb and its effects.

Two important exceptions to our rule of thumb: signals that are currents, and transmission lines

The rule of §1N.4.6 for relating impedances is extremely useful and important. But we should acknowledge two important classes of circuits to which it does *not* apply:

- Signal sources that provide *current* signals rather than voltage signals. At the moment, this distinction may puzzle you. Yet the distinction between *voltage sources* – the sort familiar to most of us – versus *current sources* is substantial and important.[18]

- High-frequency circuits where the signal paths must be treated as "transmission lines" (see Appendix C).

  AoE §H.1

  Only at frequencies above what we will use in this course do such effects become apparent. These are frequencies (or *frequency components*, since "steep" waveform edges include high-frequency components, as Fourier teaches – see Chapter 3N) where the period of a signal is comparable to the time required for that signal to travel to the end of its path. For example, a six-foot cable's propagation time would be about 9 ns, the period of a 110 MHz sinusoid. Square waves are still more troublesome; one at even a few MHz would include quite a strong component

---

[18] You may be inclined to protest that you can convert a current into a voltage and vice versa. You will build your first circuit with an output that is a determined *current* in Lab 4L, and you will meet your first current-source transducer (a photodiode) in Lab 6L.

at that frequency, and this component would be distorted by such a cable if we did not take care to "terminate" it properly. In fact, these calculations understate the difficulties; a round-trip path length greater than about 1/10th wavelength calls for termination.

We don't want you to worry, right now, about these two points. Signals as currents are unusual in this course, and in this course you are not likely to confront transmission-line problems. But be warned that you will meet these later, if you begin to work with signals at higher frequencies.

1N.5 AoE reading

This is the first instance in which we've tried to steer you toward particular sections of *The Art of Electronics*. That book is, of course a sibling (or is it the mother?) of the book you are reading. You should not consider the readings listed below to be *required*. But if you are proceeding by using the two books in parallel, these are the sections we consider most relevant.

Readings:
    Chapter 1, §§1.1–1.3.2.
    Appendix C on resistor types: resistor color code and precision resistors.
    Appendix H on transmission lines.
Problems:
    Problems in text.
    Exercises 1.37, 1.38.

1L   Lab: DC Circuits

1L.1   Ohm's law

A preliminary note on *procedure*

The principal challenge here is simply to get used to the breadboard and the way to connect instruments to it. We do not expect you to find Ohm's law surprising. Try to build your circuit *on the breadboard*, not in the air. Novices often begin by suspending a resistor between the jaws of alligator clips that run to power supply and meters. Try to do better: plug the two leads of the DUT ("Device Under Test") into the plastic breadboard strip. Bring power supply and meters to the breadboard using banana-to-wire leads, if you have these, to go direct from banana jack on the source right into the breadboard. In Fig. 1L.1 is a sketch of the poor way and better way to build a circuit.

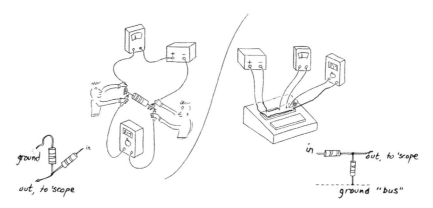

Figure 1L.1 Bad and good breadboarding technique: Left: labor intensive, mid-air method; Right: tidy method, circuit wired in place.

Color coding, and making the circuit look like its diagram

This is also the right time to begin to establish some conventions that will help you keep your circuits intelligible: in Fig. 1L.2 is a photo of one of our powered breadboard boxes, showing the usual power supply connections. The individual breadboard strips make connections by inserting #22 solid wires in small spring-metal troughs: see Fig. 1L.3. Be sure to strip at least three-eights of an inch (.375 in or ≈ 9 mm) of the insulation off the end of each wire before you insert it. Use a variable regulated DC supply, and the hookup shown in Fig. 1L.10.[1]

Caution! You must *not* connect the external supply to the colored banana jacks on the powered breadboard (Red, Yellow, Blue): these are the *outputs* of the internal supplies of the breadboard, and even when the breadboard power is **off**, these connectors tie to internal electronics that can be damaged by an attempt to drive it. The same warning applies to the three white horizontal power supply "buses" at the top of the board, strips that are internally tied to those three internal power supplies. (In case you are curious, Red is +5, while Yellow and Blue are adjustable, and normally are set to about ±15 V.)

---

[1] See online Chapter 1O at https://LAoE.link/LAoE_Chapter_010.pdf for hints & tips on using the breadboard to build the lab circuits.

Lab: DC Circuits

Note by the way, that voltages are measured *between* points in the circuit, while currents are measured *through* a part of a circuit. Therefore you usually have to "break" or interrupt the circuit in order to measure a current.

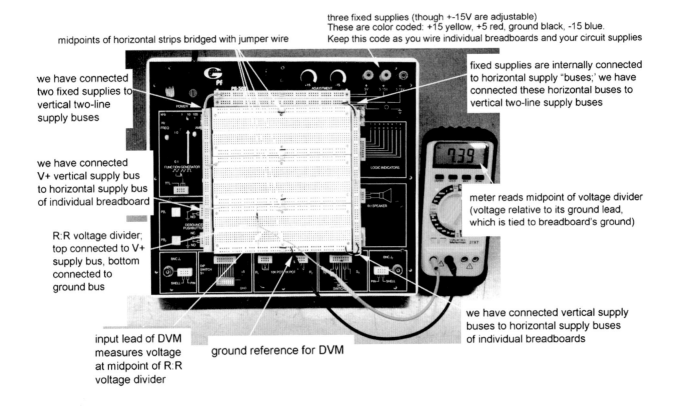

Figure 1L.2 Breadboard power bus connections, again.

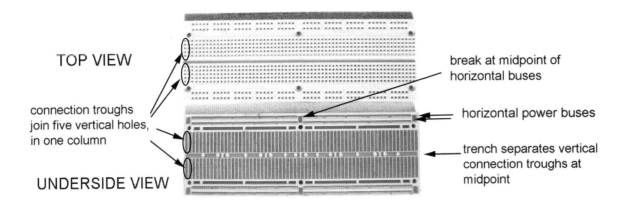

Figure 1L.3 Single breadboard strip: a look at underside reveals the pattern of connecting metal troughs.

1L.2 Meters, VOM and DVM

1L.2.1 Analog "VOM" versus digital "DVM"

First, let's get used to some jargon, and a fundamental difference between the two types of "multimeter" that you will use in this course.

Figure 1L.4 DVM and VOM: digital and analog versions of an instrument to measure voltage and current.

The older *analog* multimeters are called "volt-ohm-milliammeters, VOM".[2] We call them "analog" because they display their result by moving a needle to a degree that is *analogous* to the quantity measured. Double the input voltage, and the needle's displacement from the zero position doubles.

They are *current*-measuring devices, using a meter movement like the one you will meet in §1L.4, with some switching circuitry added on the input side, to make the bare meter movement into a useful instrument. This makes the meter more versatile than it would be if used alone. VOMs are harder to use than DVMs, so after the first day you will probably reach for a DVM, not a VOM.

The newer multimeters are called "*Digital* Voltmeters, DVM," even though they can measure much more than just voltage. We call them "digital" because they display their result as a set of numerals and sometimes characters: as a code, in other words, rather than any movement proportional to the quantity measured.

Our DVMs measure voltage, and also current and resistance. So, they could be called "VOM" as the analog meters are. To avoid confusion with the analog version, they never are called "VOM." Our DVMs also measure some other quantities: frequency, and in most cases also capacitance. Some will measure a transistor's "gain" (amplification of current). This last feature will not interest you until we introduce you to the bipolar transistor in Lab 4.

Whereas a VOM is fundamentally a current-measuring device that one might say can be tricked into measuring voltage (as you will demonstrate in §1L.4), a DVM is a voltage-measuring device that can be tricked into measuring current. Fundamentally a DVM is just an analog-to-digital converter ("ADC" – a device you will look at closely midway through the digital part of this course). In other words, it measures the voltage applied to it, displaying the result in "digital" form – as a set of characters on a

---

[2] Old timers sometimes call these "Simpson meters" because that company dominated the market.

display. Some simple circuitry precedes the ADC's input: a set of voltage dividers that allow the ADC to accept varying full-scale input voltages, and a simple current measurement scheme, described in §1L.2.2.

1L.2.2 Current measurement

To measure current, the DVM does just what you would do, if you had only a voltmeter: it runs the input current through a small-valued resistor, and measures the voltage drop across that resistor.

A particular resistor value is selected automatically for each full-scale current range, and this keeps the resulting voltage under a volt (on some ranges, far under a volt). The small voltage that the DVM drops when measuring current is called the "burden" that the meter imposes.[3] Ideally, it would be zero, so as not to alter the current value from what it would be absent the meter.

A side-effect of the fact that DVM is essentially a voltage-measuring device is the need to switch input leads when measuring current. If you make the mistake of applying a substantial *voltage* to the current input – as all of us are capable of doing, when touching a voltage source, forgetting that we just finished measuring current – you will burn out the meter's protective fuse.

Blown fuse? The need to switch lead placement is awkward, but necessary. If you do burn out the fuse, the DVM gives you no help, no indication that the fuse has blown – apart from a current reading of *zero*. Despite other clever qualities, DVMs are dumb – and dumb in the sense of *mute*. The persistent reading, "current = 0," is a maddeningly ambiguous reading, of course. It is ambiguous in the sense that it could indicate that the fuse has blown – but it doesn't necessarily mean that, of course.

If you suspect your DVM has blown its fuse, test it by measuring a known current. For example, tie a 1k resistor to +5 V, and measure current that flows through the resistor to ground. If the meter reads zero, the fuse has blown and you need to open the meter (a tedious process, sometimes demanding that you remove several screws) in order to replace the fuse.

1L.2.2.1 An experiment: Try your DVM's current-measurement *burden*

Just *how low* is the voltage that the DVM measures, when it is purporting to measure current? With your DVM set to measure current (let's call this DVM "A"), tie a 1k resistor to +5 V as we just suggested, and use a *second* DVM ("B") to measure the *voltage* between the two leads of the DVM A, which is measuring *current*.

This voltage is the one called the meter's "burden" on the circuit. We hope the value is a small fraction of 5 V, since it tends to distort the current relative to what it would have been, absent the current-measuring meter. See if the *burden* changes if you replace the 1k R, tied to +5, with a 1M resistor.

1L.2.3 Resistance measurement

Measuring resistance requires that the meter use an internal battery to apply either voltage or current to the resistance under test. Analog VOMs apply a voltage and measure the resulting current. DVMs do the opposite – appropriately, since the guts of a DVM is a voltage measuring device.

Contemporary DVMs have the good sense to adjust the test current so as to keep the resulting voltage very low. They apply a current selected according to the resistance range, a current that will keep the resulting voltage under about 0.5 V.[4] This value is designed to be too low to cause conduction in any diode and transistor junctions that might be attached to the resistor under test. Nevertheless, you

---

[3] See, for example, the Keysight U1240 series datasheet: https://LAoE.link/DMM_Specs.pdf.
[4] For the Keithley U1240b DVM.

should always be skeptical when measuring resistance *in circuit* as the reading could be compromised if any of the test current gets shunted through other components. It is best to disconnect at least one end of a resistor from a circuit before trying to measure its value with a multimeter.

1L.2.3.1 An experiment: Try your DVM's *resistance* measurement

Rather than take our word for the claim that the DVM's voltage drop is low in *resistance* mode, let's measure it. Use a second DVM to measure the voltage drop between the two leads of the DVM that is measuring *resistance*. The drop should be under 0.5 V, but may be *far* under that value.

1L.2.4 Diode test

The DVM puts out a small current, usually 0.5 mA to 1 mA,[5] and displays the resulting voltage across the diode under test. This function can also be used as a "continuity" test on many DVMs: if the voltage drop that results is very low, the DVM beeps.[6]

1L.2.4.1 Measure your DVM's current output in *diode test* mode

Incidentally, you will notice that *Diode Test* provides a fixed current, something that can useful for a variety of purposes. For example, you can use it to check whether a second DVM's fuse has blown. The process that we here are describing as a way to measure the value of the *Diode Test* current could also be described as an easy way to determine whether a second DVM's current-measurement fuse has blown. No need to rig the 5V source and resistor that we proposed in §1L.2.2.1 on the facing page.

1L.2.5 Capacitance

Today, the concept "capacitance" may be entirely unfamiliar, but you will become familiar with it in Lab 2.

The same current source that is used to measure resistance (above, §1L.2.3) can be used (by clever circuitry internal to the DVM) to charge the capacitance under test. Software can sample the capacitor voltage and infer capacitance. This process, of course, requires considerable brains in the DVM.

You may meet a simpler method in an early digital lab, where capacitance can be inferred from the frequency of an oscillation. If we build an oscillator timed by an *RC* "time-constant" (much more on this topic, soon!) and install a known *R*, the frequency (or period) of oscillation allows one to infer the value of the applied *C*. Take a reading with a particular value of *C*. Try another *C*. If period doubles, for example, then the value of the new *C* apparently is twice the first.

1L.2.6 Transistor β

Another sneak preview – save this note for a later time, when you've met bipolar transistors. Transistor β characterizes it's *current gain*: the ratio of current out ($I_{Collector}$) to current in (I_{Base}). This β is a slippery characteristic, one that varies widely with temperature, with applied current and from one individual transistor to another. Perhaps for this reason, fancy DVMs omit this test, while simpler ones often include it. Our DVMs that offer a β test assume an I_C of about 1mA.

1L.3 Voltage divider

Construct the voltage divider shown in Fig. 1L.5 Apply V_{in}=15 V (use the DC voltages on the breadboard). Measure the (open circuit) output voltage. Then attach a 7.5k load and see what happens.

---

[5] 0.5 mA for the Keithley U1240b.
[6] For Keithley U1240b, beeps if the drop is less than 50 mV.

Now measure the short circuit current. (That means "short the output to ground, but make the current flow through your current meter." Don't let the scary word "short" throw you: the current in this case will be very modest. You may have grown up thinking "a short blows a fuse." That's a good generalization around the house, but it often does not hold in electronics.)

From $I_{\text{Short Circuit}}$ and $V_{\text{Open Circuit}}$ you can calculate the Thevenin equivalent circuit.

Now build the Thevenin equivalent circuit, using the variable regulated DC supply as the voltage source, and check that its open circuit voltage and short circuit current match those of the circuit that it models. Then attach a 7.5k load, just as you did with the original voltage divider, to see if it behaves identically.

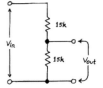

Figure 1L.5
Voltage divider.

> **A note on the practical use of Thevenin models**
>
> You will rarely do again what you just did: *short* the output of a circuit to ground in order to discover its R_{Thev} (or "output impedance," as we soon will begin to call this characteristic). This method is too brutal in the lab, and too slow when you want to calculate R_{Thev} on paper. In the lab, I_{SC} could be too large for the health of your circuit (as in your fuse-blowing experience). You will soon learn a gentler way to get the same information.
>
> On paper, if you are given the circuit diagram the fastest way to get R_{Thev} for a divider is always to take the *parallel* resistance of the several resistances that make up the divider – again assuming R_{source} is ideal: zero ohms. (A hard point to get used to, here: look at all paths in parallel, going to any fixed voltage, not just to ground. See the attempt to rationalize this result in Fig. 1N.23 on p. 17.) The case you just examined is illustrated in Fig. 1L.6.

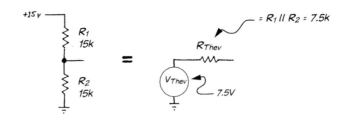

Figure 1L.6 R_{Thev} = parallel resistances as seen from the circuit's output.

1L.4 Ohm's law applied to convert a meter movement into a voltmeter and ammeter

A Thevenin model is extremely useful as a *concept*; but you'll not again *build* such a model. The circuit you just put together is useful only as a device to help you get a grip on the concept. Now comes, instead, a chance to apply Ohm's law to make something almost useful! – a *voltmeter*, and then a *current meter* ("ammeter").

You will start with a bare-bones "meter movement" – a device that lets a current deflect a needle. This mechanism is basically a current-measuring device: the needle's deflection is proportional to the torque developed by a coil that sits in the field of a permanent magnet, and this torque is proportional to the current through the coil. See the sketch in Fig. 1L.7.[7]

An analog "multimeter" or VOM (volt–ohm-milliammeter) is just such a movement, with switchable resistor networks attached. We would now like you to re-invent the multimeter.

---

[7] After Google images file: "analog meter movement." Source unknown.

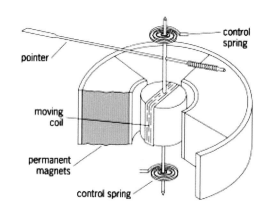

Figure 1L.7 Mechanism of analog current/voltmeter. From *The Free Dictionary*, see www.globalspec.com.

1L.4.1 Internal resistance of the movement

You could design a voltmeter – as you will in the next subsection – simply assuming that the movement's R_{INT} is negligibly small. The resistance of this *movement* is low – well under 1 kΩ. But let's be more careful. Let's first measure the internal resistance. Here, *please note*, we want you to work with a simple *bare meter movement*, not with a multimeter. (It's more fun to start from scratch.)

Do this measuring any way you like, using the *variable power supply*, *DVM* (digital voltmeter – really a *multimeter*, despite its name), and *VOM* if you need it. If you use DVM as current meter, we should warn you that it is easy to blow the meter's internal fuse, because a moment's touch to a power supply will pass an essentially unlimited current. To make things worse, a DVM with blown fuse offers no display to indicate the problem. Instead, it simply persists in reading *zero* current. There are several good ways to do this task.

The meter movements are protected[8]. Don't worry about burning out the movement, even if you make some mistakes. Even "pinning" the needle against the stop will not damage the device, as we found by experiment – though in general you'll want to avoid doing this to analog meters. We have provided diodes that protect the movement against overdrive in both forward and reverse directions. You should *note*, as a result, that these diodes that protect the movement also make it behave very strangely if you overdrive it. (No doubt you can guess what device it behaves like.) So, don't use data gathered while driving the movement beyond full scale.

Sketch the arrangement you use to measure R_{INT} and note the values of R_{INT} and $I_{FULL\text{-}SCALE}$.

1L.4.2 10 V voltmeter

Show how to use the movement, plus whatever else is needed, to form a 10 V-full-scale voltmeter (that just means that 10 V applied to the input of your circuit should deflect the needle fully). Draw your circuit.

Note: this is a good time to start getting used to our canny use of approximations, as we design. As you specify the resistor that you want to add to the bare meter movement, recall that you are limited to "5% values" – resistors whose true value is known only to lie within ±5% of the nominal value.

Does the movement's internal resistance cause a significant error? (What does "significant" mean? Well, you might compare the contribution of R_{INT} against the contribution of error from your uncertainty about the value of the resistor that (we hope!) you have included.)

---

[8] Note to instructors: please add two paralleled and oppositely-oriented silicon diodes such as 1N4004 in parallel with the meter movement. These will conduct at about 0.6 V, protecting the movement during overdrive.

1L.4.3 10 mA current meter ("ammeter")

Show how to use the movement, plus whatever else is needed, to form a 10 mA-full-scale ammeter. (This is a bit trickier than the 10 V voltmeter.) Sketch it, and test your circuit.

1L.5 The diode

Here is another device that does not obey Ohm's law: the **diode**. (We don't expect you to understand how the diode works yet; we just want you to meet it, to get some perspective on Ohm's Law devices: to see that they constitute an important but *special* case.

We need to modify the test setup here, because you can't just stick a voltage across a diode, as you did for the resistor and lamp below (§1L.6).[9] You'll see why after you've measured the diode's V versus I. Do that by wiring up the circuit shown in Fig. 1L.8.

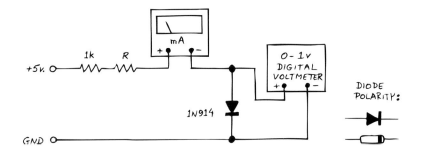

Figure 1L.8 Diode *VI* measuring circuit.

In this circuit you are applying a *current*, and noting the diode voltage that results; earlier, you applied a voltage and read resulting current. The 1k resistor limits the current to safe values. Vary R – use a 100k variable resistor (usually called a *potentiometer* or "pot" even when wired, as here, as a variable resistor), a resistor substitution box, or a selection of various fixed resistors) – and look at I versus V. First, get a feel for the behavior by sweeping the R value by hand, and noticing what happens to the diode current. Then sketch the plot in two forms: linear and lin–log (or semi-log), in Fig. 1L.9.

First, get an impression of the shape of the linear plot; just four or five points should define the shape of the curve. Then draw the same points on a *lin–log* plot, which compresses one of the axes. (Evidently, it is the fast-growing current axis that needs compressing, in this case.) If you have some lin–log paper use it. If you don't have such paper, you can use the small version laid out below. The point is to see the pattern.

See what happens if you reverse the direction of the diode. How would you summarize the V versus I behavior of a diode? Now explain what would happen if you were to put 5 V across the diode (**don't try it!**). Look at a diode datasheet, if you're curious: see what the manufacturer thinks would happen. The datasheet won't say "Boom" or "Pfft," but that is what it will mean.

We'll do lots more with this important device.

---

[9] Well, you *can*; but you can't do it twice with the same diode!

1L.6 I versus V for some mystery boxes

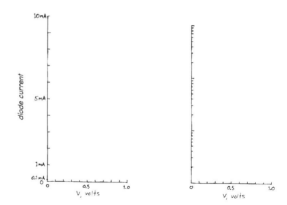

Figure 1L.9 Diode *I* versus *V*: linear plot; lin–log plot.

1L.6 I versus V for some mystery boxes

Find two *mystery* boxes which your instructor[10] will have set up for you (we will call these DUTs: Device Under Test). These are two-terminal devices, one of which is an ordinary resistor, the other an odder thing. The devices are hidden inside black plastic 35 mm film cans. Apply voltages in the range zero to a couple of volts, using a *variable* power supply, and note voltage and current pairs. For the range between 0 and 1 V, measure at increments of 0.1 V (because this is the range where you need a detailed picture).

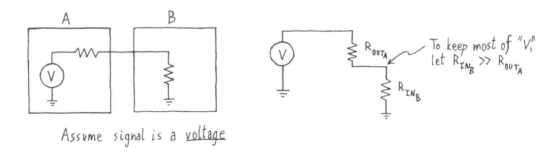

Figure 1L.10 Circuit for measurement of *I* versus *V*.

Sketch a graph of a few points to get the trend. Turning the power-supply voltage knob by hand, you may be able to get a sense of the shape of the curve, or see where more points are needed.

Decide which device is which: which ordinary, which odd. *Warning*: keep the applied voltage below 7 V, or you may destroy one of the DUTs.

To make your task challenging, we ask that you measure *voltage* and *current simultaneously*, as you do this exercise. We ask this so that you will be obliged to consider the effects of the instruments on your measurements (more on this point, below).

Effects of the instruments on your readings: Consider a couple of practical questions that arise in even this simplest of "experiments."

---

[10] If you're working through this book on your own, no one will have set this up of course. But you and instructors can get more information at www.learningtheartofelectronics.com. For now, we reveal that one 'mystery' is a #47 lamp, the other a 33 Ω power resistor.

A qualitative view: Is the voltmeter measuring the voltage at the place you want, namely across the object under test? Or does the voltage reading include the effect of the ammeter, in your arrangement? Does that matter? If you are measuring the object's *voltage* precisely, then are you reading its *current* or are you measuring the current in object plus the current passing through the DVM as well? If you can't have it both ways (as you can't), and must live with one of the two errors, which experimental setup gives you the smaller error?

Figure 1L.11 shows two sketches of the two possible placements of the voltmeter.

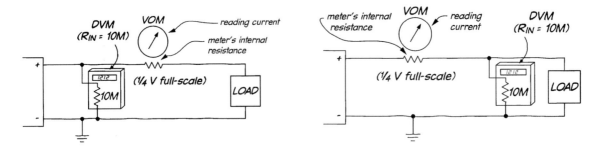

Figure 1L.11 Measuring *I* and *V* simultaneously: only one instrument gets a true reading at DUT.

If you know (and we'll now reveal this fact to you) that the *LOAD* resistance is under 1k ("resistance" or some near-equivalent; we're not promising you *Ohmic* behavior), then you can judge which of the two setups illustrated above gives the truer pair of *I* and *V* readings. Use that preferred setup to get a set of *I* and *V* readings, and try to infer an *R* value, if you can.

1L.6.1 Estimate % error caused by the instruments

Once you have your best estimate of *R* in hand, estimate the errors that the instruments cause. Specifically, *when the ammeter is on the 10mA full-scale range*, what percentage error in *inferred R* results from the presence of...

- ammeter, when you measure *V* where the ammeter causes an error?
- voltmeter, when you measure *V* where the ammeter causes no error?

After getting answers to these questions, probably you can say what an ideal voltmeter (or ammeter) should do to the circuit under test? What does that say about its "internal resistance"? (Perhaps you caught on to this theme, pages earlier.)

Now sketch the curves described by your data points. Don't work too hard: it's the shapes that we're after. The resistor isn't very interesting (what's its value?); the mystery device is a little more intriguing.

What do you suppose the mystery gadget is? We hope you saw its curve. In fact, it's made of material much like what forms the resistor. Why does its *I–V* curve look different? The bend can be useful: much later, in Lab 8L, we will exploit this curvature to make a circuit regulate its own *gain*. Now you have earned the right to open up the film canister and discover what's inside.

1L.6.2 Where should DVM and ammeter go, for large R_{LOAD}?

Now you may have "learned" in the previous exercise where to place the voltmeter in order to get the best simultaneous *I* and *V* readings. Just to make sure you don't go away thinking you've learned something,[11] we'd like you to try *a pencil-and-paper* exercise, this time assuming that the DUT (R_{LOAD}) is a very large resistance: 10 MΩ. Assume that the DVM's R_{in}=10 MΩ. (Note that we're *not* asking you to carry out this experiment; only to predict what you would see if you did do the experiment.)

If you placed your DVM at the load, the DVM would appear in parallel with the load. What percentage error would this evoke in the simultaneous *current* measurement? In contrast, the *voltage* error caused by reading upstream of the ammeter in this case would be no larger than in the case you tried in the lab experiment.

Perhaps now you can begin to generalize about which cases oblige one to worry about the DVM's R_{in}. But if you are not sure – read on.

1L.7 Oscilloscope and function generator

This is just a first view of oscilloscopes to give you a head start (more about oscilloscopes can be found in Appendix D and AoE, Appendix O). You'll get a big (oscillo)scope workout later in Lab 2L.

We'll be using the oscilloscope and function generator in virtually every lab from now on. For today's lab, voltmeters were sufficient – and, in fact, more convenient than an oscilloscope, because our circuits' voltages and currents have been politely sitting still, giving us time to measure them. If you are familiar with the scope, go right on to the first real exercise in Lab 2L – or go home, having earned a rest.

The scope soon will become our favorite instrument as we begin to look at signals that are not static (called DC in electronics jargon): signals that, instead, vary with time. Lab 2L is concerned exclusively with such circuits, and with rare exceptions this will be true of all our work from now on. The scope draws a plot of voltage (on the vertical axis) versus time (on the horizontal axis).

Get familiar with scope and *function generator* (a box that puts out time-varying voltages, or "waveforms:" things like sinewaves, triangle waves and square waves) by generating a 1 kHz sinewave with the function generator and displaying it on the scope. Connect the function generator directly to the scope, using a "BNC cable," not a *probe*.[12]

If it seems to you that both instruments present you with a bewildering array of switches and knobs, don't blame yourself. These front-panels just *are* complicated. You will need several lab sessions to get fully used to them – and at term's end you may still not know all: it may be a long time before you find any occasion for use of the *holdoff* control, for example, or of *single-shot* triggering, if your scope offers this.

Play with the scope's sweep and trigger controls. Most of the discussion, here, applies fully only to an *analog* scope. The digital scope automates some functions, such as triggering and gain setting, if you want it to. We suggest that you ought *not* to begin your scope career using such fancy features: *beware* the mind-stunting button labeled *AUTOSET*!

Specifically, try the following:

- The *vertical gain* switch. This controls "*volts/div*"; note that "div" or "division" refers to the *centimeter* marks, not to the tiny 0.2 cm marks;

---

[11] Just kidding.
[12] BNC (Bayonet Neill–Concelman) names the connectors on the ends of this coaxial cable. (See AoE §1.8). We usually call the *cable* by this name, though strictly the term applies to its connectors.

Lab: DC Circuits

- The *horizontal sweep speed* selector: *time* per division.

    On this knob as on the vertical gain knob, make sure the switch is in its **CAL** position, not **VAR** or "variable." Usually that means that you should turn a small center knob clockwise till you feel the switch *detent* click into place. If you don't do this, you can't trust *any* reading you take.

- The *trigger* controls. Don't feel dumb if you have a hard time getting the scope to trigger properly. Triggering is by far the subtlest part of scope operation. "Trigger" circuits tell the scope when to begin moving the trace across the screen: when to begin *drawing* a waveform. When you think you have triggering under control, invite your partner to prove to you that you *don't*: have your partner randomize some of the scope controls, then see if you can regain a sensible display (don't let your partner overdo it, here).

    Beware the tempting so-called "normal" settings (usually labeled "**NORM**"). They rarely help, and instead cause much misery when misapplied. Think of "normal" here as short for **abnormal**! Save it for the rare occasion when you know you need it. "**AUTO**" is almost always the better choice.

    > The scope waits around till triggered. In AUTO mode, it sweeps either when triggered, or when it has waited so long that it loses patience. Thus you always get at least a trace, in AUTO mode.
    >
    > NORMAL, in contrast, makes the scope infinitely patient: it will wait forever, drawing nothing on the screen, until it sees a valid trigger signal. Meanwhile, you look at a dark screen; usually, that is not helpful!
    >
    > Trigger "source": here's another way to go wrong: trigger can be set to look at either of the scope's two input "channels" (CH1 or CH2), or at the external BNC connector, called EXT.

Switch the function generator to square waves and use the scope to measure the "risetime" of the square wave (defined as time to pass from 10% to 90% of its full amplitude).

At first you may be inclined to despair, saying "Risetime? The square wave rises instantaneously." The scope, properly applied, will show you this is not so.

> **A suggestion on triggering**
>
> It's a good idea to *watch* the waveform edge that *triggers* the scope, rather than trigger on one event and watch another. If you watch the *trigger event*, you will find that you can sweep the scope fast without losing the display off the right side of the screen.
>
> So, to measure *risetime* of a signal connected to Channel One, trigger on *Ch 1, rising-edge*.

What comes out of the function generator's **SYNC OUT** or **TTL**[13] connector?

Look at this on one channel while you watch a triangle or square wave on the other scope channel. To see how SYNC or TTL can be useful, try to trigger the scope on the peak of a sinewave *without* using these aids; then notice how entirely easy it is to trigger so when you *do* use SYNC or TTL to trigger the scope.

Triggering on a well-defined point in a waveform is especially useful when you become interested in measuring a difference in *phase* between two waveforms; this you will do several times in the next lab. You can use a *zero-crossing* (where the waveform crosses the midpoint of its excursion), or, when using SYNC OUT, you can use the peak or trough of the waveform.

---

[13] TTL stands for the equally cryptic phrase, Transistor–Transistor Logic, which is a kind of digital logic gate that you will find described in Chapter 14N. In the present context, TTL should be understood to mean simply Logic-Level square wave, a square wave that swings between ground and approximately 4 V.

1L.7 Oscilloscope and function generator

How about the terminal marked **CALIBRATOR** (or CAL) on the scope's front panel? (We won't ask you to *use* this signal yet; not until Lab 3L do we explain how a scope probe works, and how you "calibrate" it with this signal. For now, just note that this signal is available to you). *Postpone* using scope probes until you understand what is within one of these gadgets. A "10×" scope probe is *not* just a piece of coaxial cable with a grabber on the end.

Put an "offset" onto the signal, if your function generator permits, then see what the **AC/DC** switch (located near the scope inputs) does.

> **Note on AC/DC switch**
>
> Common sense may seem to invite you to use the **AC** position most of the time: after all, are these time-varying signals that you're looking at not "AC" – alternating current (in some sense)? *Eschew* this plausible error. The *AC* setting on the scope puts a capacitor in series with the scope input, and this can produce startling distortions of waveforms if you forget it is there. (See what a 50 Hz square wave looks like on AC, if you need convincing.) Furthermore, the AC setting washes away DC information, as you have seen: it hides from you the fact that a sinewave is sitting on a DC offset, for example. You don't want to wash away information except when you choose to do so knowingly and purposefully. Once in a while you *will* want to look at a little sine with its DC level stripped away; but always you will want to *know* that this DC information has been made invisible.

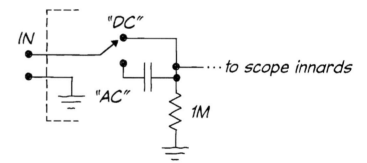

Figure 1L.12 AC/DC scope input select.

Set the function generator to some frequency in the middle of its range, then try to make an accurate frequency measurement with the scope. (Directly, you are obliged to measure *period*, of course, not frequency.[14]) You will do this operation hundreds of times during this course. Soon you will be good at it.

Trust the scope period readings; distrust the analog function generator frequency markings; these are useful only for very *approximate* guidance.

---

[14] Well, if you're using a digital scope you could resort to the shabby trick of asking it to measure frequency for you. But don't start that way today.

1S Supplementary Notes: Resistors, Voltage, Current

1S.1 Reading resistors

In the lab exercises, from now and ever after, you will want to be able to read resistor values without pulling out a meter to *measure* the part's value (we do sometimes find desperate students resorting to such desperate means). The process will seem laborious, at first; but soon, as you get used to at least the common resistance values, you will be able to read many color codes at a glance. We will use resistors whose packages are big enough to be readable. The now-standard *surface-mount* parts are so tiny that the smaller ones normally are not labeled at all. This spares you the trouble (and opportunity) of reading them. If you mix a few surface-mount parts on the bench, you can only sweep them up and start over – unless you are willing to measure each.

1S.1.1 A menagerie of 10k resistors

Fig. 1S.1 shows a handsome collection of 10k resistors in a variety of packages.

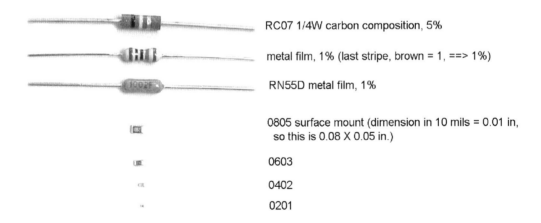

Figure 1S.1 10k resistors.

The type that we use – the carbon composition resistor at the top – is nearly obsolete, and relatively expensive, as a result. But we like them for lab work, because (on a good day) *we can read them*. The others are pretty nasty for breadboarding. The one with the value written out in numerals may appeal to you, if you don't want to learn the color codes – but it really isn't much fun to work with, because if it happens to be mounted with the value label *down*, you're out of luck.

1S.1.2 Resistor values and tolerances

Figure 1S.2 shows a resistor of the sort we use in this course's labs: it is a "5% carbon composition" part. "5%" "tolerance" means that its *actual* value is guaranteed to lie within 5% of its *nominal* value. If it is labeled "100k" (100 000 Ω), its actual value can be expected to lie in the range ~95–105 kΩ.

The first problem one confronts on a first day with resistors, is *which way* to orient the part.

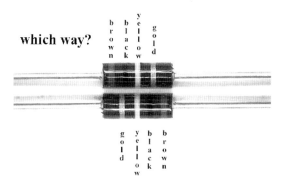

Figure 1S.2 Which way? Put the *tolerance* stripe to the right.

"*Tolerance stripe?*," you protest. "How do I know which that is?" In our labs, the answer is the fourth band, colored silver or gold. If you are using color-coded 1% resistors, the tolerance band will be the *fifth*. Fig. 1S.3 shows such a part.

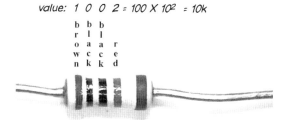

Figure 1S.3 Extra band for color-coded 1% resistor.

The use of four bands to define the value indicates (implicitly) that this is better than the usual 5% part; the color of the fifth band, brown represents "1," for this "1%" resistor. (A red band would indicate 2%, and so on.)

Tolerance: Only two tolerance values are common, for the 3-band carbon composition parts you are likely[1] to meet in the labs: 5% and 10%:

- silver: ±10%
- gold: ±5%

Value: Once you've oriented it properly – tolerance to the right – you can read off the value colors, and then can translate those to numbers. Finally, with the three numbers in hand, you will have enough information to discover the value. The resistor we just looked at is brown-black-yellow: see Fig. 1S.4.

Brown-black-yellow is one-zero-four, and the fourth band, gold, says "±5%."[2] The third band, the "four," is an exponent – a power of ten. So, this resistor's value is 100k.

---

[1] *No* tolerance band, in a 3-stripe part, indicates ±20%. Such resistors are rare.
[2] Sometimes the fourth band is *not* the last band; one more is added, for military components, to indicate the "failure rate."

Supplementary Notes: Resistors, Voltage, Current

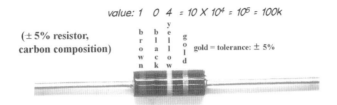

Figure 1S.4 Value stripes reveal what this resistor is.

The color code: The colors represent numbers, as set out below. A variety of mnemonics have arisen, most of them more-or-less offensive, in order to help engineers to memorize this code. One of the blander mnemonics is "Big Boys Race Our Young Girls, But Violet Generally Wins." This mnemonic is not very clever. It fails to distinguish the trio Black, Brown and Blue and also between the pair Green and Gray. *Black* is a plausible representation of the value zero, since black represents the absence of color; the color Brown is close to the color Black; perhaps one might remember Gray's position as "next to White." None of this is very satisfactory. We keep charts of the color codes up on the walls of our own teaching lab.

Here are the colors and the values they represent:

- black: zero
- brown: one
- red: two
- orange: three
- yellow: four
- green: five
- blue: six
- violet: seven
- gray: eight
- white: nine
  the next two are used as multipliers *only*, and are rarely seen:
- gold: 0.1
- silver: 0.01

1S.1.3 The set of "10% values"

It is hard to get used to the strange set of values that are "standard" in the lab. They are not the nice, round values that one might expect. They seem weird and arbitrary at first. But their strangeness does make sense. Because of our uncertainty about the *actual* value of a resistor, it doesn't make sense to specify distinct *nominal* values that are too close together; if we specify nominal values that are too close, their actual values are likely to overlap. To avoid this, the nominal values are placed far enough apart so as to make overlap slight.

A "10%" resistor of nominal value "10 Ω," for example, could be as large as 11 Ω. A "12 Ω" resistor could be 10% smaller – a bit under 11 Ω. So, 12 is about as close as it makes sense to place the next *10%* value after "10." And so on – the steps growing proportionally as the values rise, producing such unfamiliar numbers as *27, 39, 47*, and so on. Here is the 10% set (known as "E12," twelve values per decade. See Appendix C in AoE). Most of our laboratory circuits will use these *10%* values (with appropriate multipliers: we rarely use 10 ohm resistors, but often use 10 kΩ parts, for example). This is so even though our lab resistors are better than 10% parts – normally they are good to ±5%:

$$10, 12, 15, 18, 22, 27, 33, 39, 47, 56, 68, 82, 100$$

1S.1.4 Power

Only now and then are we obliged to consider *power* ratings of the components we use in the labs. That is true because our signal voltages are modest (under ±10 V) and our currents are small (a few tens of milliamps). *Power* in our components, therefore, is modest as well, since power is the product of the two: Voltage *times* Current. 10 V×10 mA, for example, dissipates 100 mW – one tenth of a watt. But our standard resistors cannot handle much power: 1/4 watt is the most they can stand, sustained. So, recall this limitation – or you may be reminded by burned fingertips.

Incidentally, you may need to remind yourself, until your intuition catches up with your book knowledge, that it is *low-valued* resistors that are likely to overheat. 10 V across *1k* is no problem; 10 V across *100 Ω* will hurt, if you touch that quarter-watt resistor, so don't do this.

1S.2 Voltage versus current

Two points to make here:

1. What is it that "flows" in most of our circuits? Current versus voltage.
2. It's usually information, not power, that interests us in this course.

1S.2.1 Is this note necessary?

I hope you don't feel insulted by this note – by its suggestion that you could get *current* and *voltage* muddled. Lest you should feel so insulted, let's adopt the device that we often find useful in this course: assume that we're talking not to you but to your slightly confused lab partner. We're trying to provide you with some arguments that will help you straighten out that intelligent but slightly-addled fellow.

Figure 1S.5 is a piece of evidence to support our claim that intelligent students can find it hard to keep *I* and *V* in their places. This is a drawing from a very good answer to a recent exam question. The answer was good – but includes a startling error that shows the student thinking *current* while handling *voltage*.

Do you see a place where the student seems to have interchanged the two concepts? (We don't think you need to understand this circuit fully to see the problem: you *do* need to understand that two circuits are included to provide a voltage gain of 2; we have indicated those.) We think this drawing shows that a smart student with quite a good grasp of electronics can still get this fundamental point wrong.

The evidence for the misunderstanding is just the presence of the two 2× amplifiers, with the slightly cryptic explanatory note at the left side, "1/2 of V_1 takes path 2× amplify." Notice that "V_1" appears again at the output of the upper 2× amplifier. Apparently, the author thought that V_1 coming from the 10k potentiometer would be cut in half because it goes two places – horizontally to the upper part of the circuit, and vertically to the lower part of the circuit.

This view is wrong: both destinations are very high-impedance inputs (input to triangle labeled "311", and input to lower triangle's "+" input). This *voltage* can go to many high-impedance inputs without suffering attenuation. The thought that "two similar destinations implies it's cut in half" sounds like a correct view of how a *current* would behave. Figure 1S.6 sketches what we guess the student imagined.

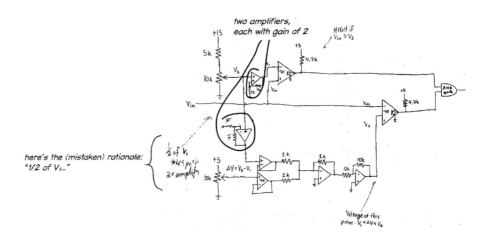

Figure 1S.5 Exam drawing: current and voltage confounded?

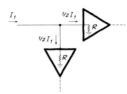

Figure 1S.6 *Currents*, in contrast to voltages, *do* split 50–50 when given two equal resistive paths.

Well, perhaps this example has convinced you that the question can puzzle a good student; perhaps we have not convinced you. Either way, let's carry on with the argument.

1S.2.2 What is it that "flows" in most of our circuits?

Your intuition may be inclined to say, "current flows," so the *signals* that we process and pass through our circuits probably are currents. That argument is plausible – but misleading at best, and usually just *wrong*.

It is *true* that *current* flows, whereas *voltage* emphatically does *not*. That's an important point – and many students need to correct themselves during the first weeks of the course as they find themselves referring to *voltages* flowing through a resistor.

But – and here is the subtler point – though *voltages* don't flow, we often speak of *signals* flowing through a circuit; indeed, all our circuits serve to process *signals* – and nearly always the *signals* that we process are *voltages*, not *currents*.

How can we reconcile these two truths: (1) currents, not voltages, flow; but (2) the signals that interest us as we design and analyze circuits nearly always are voltages, not currents? We can, as we hope some illustrations will persuade you.

And here is a preliminary question that we ought to get out of the way: *is this a question we need worry about?* Aren't voltages and current proportional, one to the other, in any case? So isn't it unimportant which we choose to describe – since one needs only to multiply by a constant in order to convert one into the other? Not a bad question, or challenge, but flawed: the deep flaw is not that some devices are not *ohmic* (though certainly this is true it is a minor point); the deep flaw is that circuit design strategy will head off in opposite directions, depending upon whether the *signals* that one intends to pass are *voltages* or *currents*.

1S.2 Voltage versus current

***Signal as* voltage:** Let's illustrate that point. Suppose one assumes – as we usually do – that our signal encodes its information as a *voltage*, not as a *current*. What follows? It follows that to keep this signal strong and healthy as it passes from one circuit to another – a passage that entails passing a voltage divider – we want $R_{\text{in}_B} \gg R_{\text{out}_A}$. Ideally, we want R_{in_B} infinite.

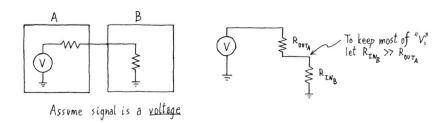

Figure 1S.7 Signal as voltage: preserving it implies impedance relation, *A* versus *B*.

This is the case that we normally assume. It is the case introduced in Chapter 1N, and then used to define a design "rule of thumb."

***Signal as* current:** Notice how different our design rules would be if our signals were *currents*: we would want circuit *B* to interfere as little as possible with the output of *A* – and that means that we would want *B* to look like a *short circuit* to *A*. Ideally, we want R_{in_B} zero!

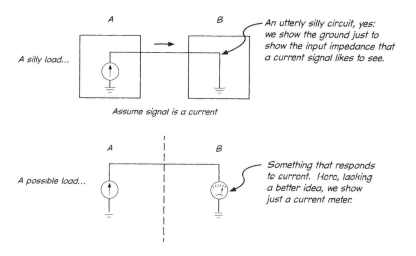

Figure 1S.8 Signal as current: preserving it implies a very different impedance relation, *A* versus *B*.

In our course, this case turns out to be *rare* (but you'll see it now and then: for example, see Lab 6L's photodiode circuit).

AoE Exercise 1.10.

***A hybrid case? Signal as* power:** Sometimes our goal is to transfer, from *A* to *B*, *power*: say for the case where we want to drive a speaker and get a big sound out. In this case we want to deliver *both current and voltage* to the speaker. We will be frustrated if either one is small since the speaker's *power* will be the product of *voltage times current*. For a given R_{out}, the R_{in} value that maximizes power in the load turns out to be neither large nor small relative to R_{out_A}; instead, we want to make the two values, R_{out_A} and R_{in_B}, *equal*.[3]

But this, again, is a case very exceptional in our course. We are concerned almost always with processing and transferring *information*, not power; and that information we encode almost always as a voltage.

An illustrative example: lowpass filter (a preview): Next time you will meet *RC* frequency-dependent "filters." But we think you will be able to follow the argument presented in this example. A lowpass filter illustrates the point that it is a *voltage* signal that we want to *pass*: a wiggle of information. We do not want our filter to pass *current* from input to output. This example is particularly striking because the frequency range in which the *signal* is passed is the frequency range when input *current* is minimal – and vice versa.

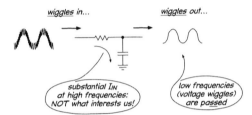

Figure 1S.9 Lowpass illustrates that it's voltage, not current, that we view as *signal*.

Incidental point: the decibel definition that interests us: Because we are concerned with processing *voltages*, we use a definition of the term *decibel* that reflects that interest:

$$dB = 20 \log_{10} A_2/A_1 \qquad (1S.1)$$

We use this in preference to the definition of a ratio of *powers*:

$$dB = 10 \log_{10} P_2/P_1 \qquad (1S.2)$$

The difference between the two forms reflects the fact that power in a resistive load varies with the *square* of voltage.

1S.2.3 Punchline: it's information, not power that interests us in this course

So though the world still needs electrical power (and storage, delivery and control of electrical power for automobiles is a newly hot topic), our interests in this course concentrate rather on the use of electrical signals to convey and process *information*.

---

[3] Note a subtlety here: we have stated a rule (first proposed by Jacobi) for choosing R_{LOAD} for a given R_{out}. The opposite problem, choosing R_{out} for a given R_{in}, yields a different result. For that case, the maximum power to the load is delivered by R_{out} is zero.

1W Worked Examples: DC circuits

1W.1 Design a voltmeter, current meter

<small>AoE 1.2.5, Multimeters, ex. 1.8</small>

A 50 µA meter movement has an internal resistance of 5k. What shunt resistance is needed to convert it to a 0–1 A ammeter? What series resistance will convert it to a 0–10 V voltmeter?

This exercise gives you an insight into the instrument, of course, but it also will give you some practice in judging when to use approximations: how precise to make your calculations, to say this another way.

1A meter: "50 µA meter movement" means that the needle deflects fully when 50 µA flows through the movement (a coil that deflects in the magnetic field of a permanent magnet: see Fig. 1L.7 for a sketch). The remaining current must bypass the movement; but the current through the movement must remain proportional to the whole current.

Such a long sentence makes the design sound complicated. In fact, as probably you have seen all along, the design is extremely simple: just add a resistance in parallel with the movement (this is the shunt mentioned in the problem): see Fig. 1W.1. What value?

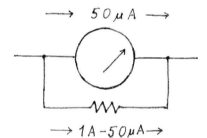

Figure 1W.1 Shunt resistance allows sensitive meter movement to measure a total current of 1 A.

Well, what else do we know? We know the *resistance* of the meter movement. That characteristic plus the full-scale current tell us the *full-scale voltage drop* across the movement: that's

$$V_{\text{movement(full-scale)}} = I_{\text{full-scale}} \times R_{\text{movement}} = 50\,\mu\text{A} \times 5\,\text{k}\Omega = 250\,\text{mV}$$

Now we can choose R_{shunt}, since we know current and voltage that we want to see across the parallel combination. At this point we have a chance to work too hard, or instead to use a sensible approximation. The occasion comes up as we try to answer the question, "How much current should pass through the shunt?"

One possible answer is "1A less 50 µA, or 0.99995 A." Another is "1 A."

Which do you like? If you're fresh from a set of physics courses, you may be inclined toward the first answer. If we take that, then the resistance we need is

$$R = \frac{V_{\text{full-scale}}}{I_{\text{full-scale}}} = \frac{250\,\text{mV}}{0.99995\,\text{A}} = 0.2500125\,\Omega$$

Now in some settings that might be a good answer. In this setting, it is not. It is a very silly answer. That resistor specification claims to be good to a few parts in a million. If that were possible at all, it would be a preposterous goal in an instrument that makes a needle move so we can peer at it.

So we should have chosen the second branch at the outset: seeing that the $50\,\mu$A movement current is small relative to the 1 A total current, we should then ask ourselves, "About how small (in fractional or percentage terms)?" The answer would be 50 parts in a million. And that fraction is so small relative to reasonable resistor and meter tolerances that we should conclude at once that we should neglect the $50\,\mu$A.

Neglecting the movement current, we find the shunt resistance is just $250\,\text{mV}/1\,\text{A} = 250\,\text{m}\Omega$. In short, the problem is very easy if we have the good sense to let it be easy. You will find this happening repeatedly in this course: if you find yourself churning through a lot of math, and especially if you are carrying lots of digits with you, you're probably overlooking an easy way to do the job. There is no sense carrying all those digits and then having to reach for a 5% resistor and 10% capacitor. In this case – designing a meter, we'll do a little better: we'll use 1% resistors – but approximations still make sense.

Voltmeter: Here we want to arrange things so that 10 V applied to the circuit causes a full-scale deflection of the movement. Which way should we think of the cause of that deflection – "$50\,\mu$A flowing," or "250 mV across the movement?"

Either is fine. Thinking in *voltage* terms probably helps one to see that most of the 10 V must be dropped across some element we are to add, since only 0.25 V will be dropped across the meter movement. That should help us sketch the solution: see Fig. 1W.2.

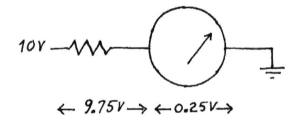

Figure 1W.2 Voltmeter: series resistance needed.

What series resistance should we add? There are two equivalent ways of answering:

1. The resistance must drop 9.75 volts out of 10, when $50\,\mu$A flows; so $R = 9.75\,\text{V}/50\,\mu\text{A} = 195\,\text{k}\Omega$.
2. Total resistance must be $10\,\text{V}/50\,\mu\text{A} = 200\,\text{k}\Omega$. The meter movement looks like 5k, we were told; so we need to add the difference, $195\,\text{k}\Omega$.

If you got stung on the first part of this problem, giving an answer like "$0.2500125\,\Omega$," then you might be inclined to say, "Oh, $50\,\mu$A is very small; the meter is delicate, so I'll neglect it. I'll put in a 200k series resistor, and be just a little off."

Well, just to keep you off-balance, we must now push you the other way: this time, "$50\,\mu$A," though a small current, is not negligibly small because it is not to be compared with some much larger current. On the contrary, it is the crucial characteristic we need to work with: it determines the value of the series resistor. And we should *not* say "200k is close enough," though 195k is the exact answer. The difference is 2.5%: much less than what we ordinarily worry about in this course (because we need to get used to components with 5 and 10% tolerances); but in a meter it's surely worth a few pennies extra to get a 1% resistor: a 196k.

1W.2 Resistor power dissipation

Problem Please specify a *10%* value in each case. Here are the 10% values (of resistance), as a reminder:

10; 12; 15; 18; 22; 27; 33; 39; 47; 56; 68; 82; 100

Problem What is the smallest 1/4 W resistor one should put across a 5 V supply?

Solution
$$P = V^2/R; \qquad R = 25\,V^2/0.25\,\text{W} = 100\,\Omega$$

Problem What is the lowest value surface-mount 1/8 W resistor one should put between −15 V and +15 V?

Solution
$$P = V^2/R; \qquad R = 900\,V^2/0.125\,\text{W} = 7200\,\Omega \approx 7.5\text{k}$$

8.2k is the nearest 10% value that would not overheat.

Problem What is the maximum voltage one ought to put across a 10 W, 10 Ω power resistor?

Solution
$$P = V^2/R; \qquad V = \sqrt{P \times R} = \sqrt{10\,\text{W} \times 10\,\Omega} = 10\,\text{V}$$

Problem (Power transmission. Effect of voltage step-up.) Electric power is transmitted at very high voltages to minimize power losses in the transmission cables. By what factor are long-distance *power line losses* reduced if the power company manages to raise its transmission voltage from 100,000 volts to 1 million volts? (We assume the power company is obliged to deliver a given amount of power to the customer, unchanged between the two cases.)

Solution Power losses in the lines are proportional to $V \times I$ in the line, or, equivalently, to $I^2 \times R$, where R is the resistance of a unit length of the line (a fat cable, no doubt; but its resistance is not zero[1]).

Stepping *up* the voltage by a factor of 10 steps *down* the current by the same factor (while transmitting a given level of power). Since power loss is proportional to the *square* of the current, the power dissipated will fall by $(1/10)^2 = 1/100$. A big reward. The change sounds worthwhile – though extreme high voltages are troublesome to insulate; so, 1 MV is about the practical limit.

Problem (Why AC?) Why does high-voltage transmission militate strongly against Thomas Edison's program of DC power transmission?

Solution AC transmission is favored, for most purposes, because stepping AC voltages up and down is easy, requiring only transformers. Stepping DC voltages up and down requires more complex processes. High-voltage DC transmission is, in fact, used for special cases, where AC losses caused by inductance and capacitance are unusually high, as in undersea cables. For the most part, however, AC transmission still prevails because of its simplicity and the low losses that are achieved in the stepping up/down process.

---

[1] ... not, at least, until all power lines are replaced by superconductors.

1W.3 Working around imperfections of instruments

§1L.6 asks you to go through the chore of confirming Ohm's Law. But it also confronts you at once with the difficulty that you cannot quite do what the experiment asks you to do: measure I and V in the resistor simultaneously. Two placements of the DVM are suggested in Fig. 1W.3 (one is drawn, the other hinted at).

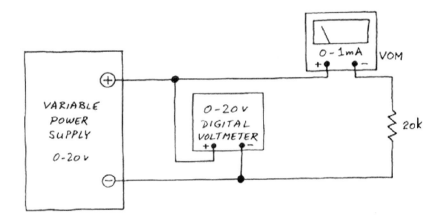

Figure 1W.3 §1L.6 setup: DVM and VOM cannot both measure the relevant quantity.

A qualitative view

Just a few minutes' reflection will tell you that the voltage reading is off, in the circuit as drawn; moving the DVM solves that problem (above), but now makes the current reading inaccurate.

A quantitative view

Here's the problem we want to spend a few minutes on.

Problem (Errors caused by the meters)   If the analog meter movement is as described in §1W.1: what percentage error in the *voltage* reading results, if the voltage probe is connected as shown in the figure for the first Lab 1L experiment, when the measured resistor has the following values.

- $R = 20\,\text{k}\Omega$.
- $R = 200\,\Omega$.
- $R = 2\,\text{M}\Omega$.

Assume that you are applying 20 V, and that you can find a meter setting that lets you get *full-scale deflection* in the current meter.

Solution   This question is easier than it may appear. The error we get results from the voltage drop across the current meter; but we know what that drop is from earlier: full-scale: 0.25 V. So the resistor values do not matter. Our voltage readings always are high by a quarter volt, if we can set the current meter to give full-scale deflection. The value of the resistor being measured does *not* matter.

When the DVM reads 20 V, the true voltage (at the top of the resistor) is 19.75 V. Our voltage reading is high by 0.25 V/19.75 V – about 0.25/20 or 1 part in 80: 1.25% (if we applied a lower voltage, the voltage error would be more important, assuming we still managed to get full-scale deflection from the current meter, as we might be able to by changing ranges).

1W.3 Working around imperfections of instruments

Problem Same question, but concerning *current* measurement error, if the voltmeter probe is moved to connect directly to the top of the resistor, for the same resistor values. Assume the DVM has an input resistance of 20 MΩ.

Solution If we move the DVM to the top of the resistor, then the voltage reading becomes correct: we are measuring what we meant to measure. But now the current meter is reading a little high: it measures not only the resistor current but also the DVM current, which flows parallel to the current in R.

The size of *this* error depends directly on the size of R we are measuring. You don't even need a pencil and paper to calculate how large the errors are:

- If R is 20 kΩ – and the DVM looks like 20 M – then one part in a thousand of the total current flows through the DVM: the current reading will be high by 0.1%.
- If R is 200 Ω, then the current error is minute: 1 part in 100,000: 0.001%.
- If R is 2 MΩ, then the error is large: 1 part in ten.

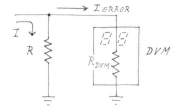

Figure 1W.4 DVM causes current-reading error: how large? % error same as ratio of R to R_{DVM}.

Conclusion? There is *no* general answer to the question "Which is the better way to place the DVM in this circuit?" The answer depends on R, on the applied voltage, and on the consequent ammeter range setting.

And before we leave this question, let's notice the implication of that last phrase: the error depends on the VOM *range* setting. Why? Well, this is our first encounter with the concept we like to call Electronic Justice, or the principle that The Greedy Will Be Punished. No doubt these names mystify you, so we'll be specific: the thought is that if you want good resolution from the VOM, you will pay a price: the meter will alter results more than if you looked for less resolution: see Fig. 1W.5.

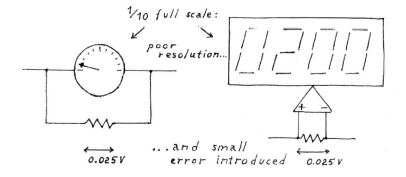

Figure 1W.5 Tradeoffs, or Electronic Justice I. VOM or DVM as *ammeter*: the larger the reading, the larger the voltage error introduced; VOM as *voltmeter*: the larger the deflection at a given V_{in}, the lower the input impedance.

50 Worked Examples: DC circuits

If you want the current meter needle to swing nearly all the way across the dial (giving best resolution: small changes in current cause relatively large needle movement), then you'll get nearly the full-scale 1/4-volt drop across the ammeter. The same goes for the DVM as ammeter, if you understand that 'full scale' for the DVM means filling its digital range: "3 1/2 digits," as the jargon goes (the "half digit" is a character that takes only the values zero or one). So, if you set the DVM current range so that your reading looks like

$$0.093$$

you have poor resolution: about 1%. If you can choose a setting that makes the same current look like 0.930, you've improved resolution tenfold. But you have also increased the voltage drop across the meter by the same factor; for the DVM, like the analog VOM, drops 1/4 V full-scale, and proportionately less for smaller "deflection" (in the VOM) or smaller fractions of the full-scale range (for the DVM).

1W.4 Thevenin models

Problem Draw Thevenin Models for the circuits in Fig. 1W.6. Give answers to 10% and to 1%.

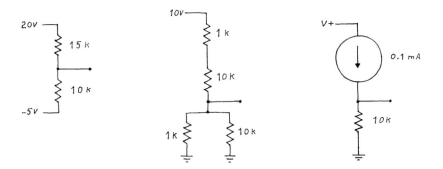

Figure 1W.6 Some circuits to be reduced to Thevenin models.

Some of these examples show typical difficulties that can slow you down until you have done a lot of Thevenin models.

The leftmost circuit is most easily done by temporarily redefining *ground*. That trick puts the circuit into the entirely familiar form in Fig 1W.7.

The only difficulty that the middle circuit presents comes when we try to approximate. The 1% answer is easy, here. The 10% answer is tricky. If you have been paying attention to our exhortations to use 10% approximations, then you may be tempted to model each of the resistor blocks with the dominant R: the small one, in the parallel case, the big one in the series case, see Fig. 1W.8.

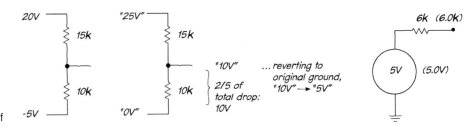

Figure 1W.7 A slightly novel problem reduced to a familiar one by temporary redefinition of ground.

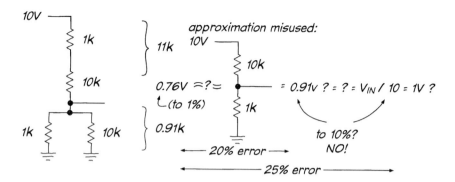

Figure 1W.8 10% approximations: errors can accumulate.

Unfortunately, this is a rare case when the errors gang up on us; we are obliged to carry greater precision for the two elements that make up the divider. For R_{Thev}, too, we need greater precision: $0.91\text{k} \parallel 11\text{k} = 0.84\text{k}$ – not acceptably close to the usual "1k" approximation.

This example is not meant to make you give up approximations. It makes the point that it's the *result* that should be good to the stated precision, not just the intermediate steps.

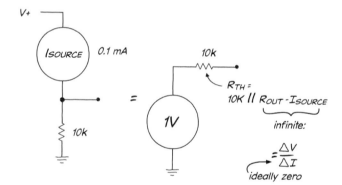

Figure 1W.9 Current source feeding resistor, and equivalent Thevenin model.

The *current source* shown in Fig. 1W.9 probably looks odd to you. But you needn't understand how to make one to see its effect; just take it on faith that it does what's claimed: sources (squirts) a fixed current, down toward the negative supply. The rest follows from Ohm's Law. (In Lab 4L you will learn to design these things – and you will discover that some devices just do behave like current sources without being coaxed into it: transistors behave this way – both bipolar and FET.)

The point that the current source shows a very high output impedance helps to remind us of the definition of impedance – always the same: $\Delta V/\Delta I$. It is better to carry that general notion with you than to memorize a truth like "Current sources show high output impedance." Recalling that definition of impedance, you can always figure out the current source's approximate output impedance (large versus small); soon you will know the particular result for a current source, just because you will have seen this case repeatedly.

1W.5 "Looking through" a circuit fragment, and R_{in}, R_{out}

Problem What are R_{in}, R_{out} at the indicated points in Fig. 1W.10?

Solution A. R_{in}. It's clear what R_{in} the divider should show: just $R_1 + R_2$. But when we say that

52 Worked Examples: DC circuits

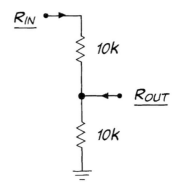

Figure 1W.10 Determining R_{in}, R_{out}; you need to decide what's beyond the circuit to which you're connecting.

are we answering the right question? Isn't the divider surely going to drive something down the line? If not, why was it constructed?

The answer is yes, it *is* going to drive something else – the *load*. But that something else should present an R_{in} high enough so that it does not appreciably alter the result you got when you ignored the load. If we follow our 10× rule of thumb (see §1N.4.6) you won't be far off this idealization: less than 10% off. To put this concisely, you might say simply that we assume an *ideal* load: a load with infinite input impedance.

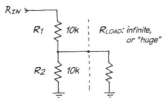

Figure 1W.11 R_{in}: we need an assumption about the load that the circuit drives, if we are to determine R_{in}.

Solution B. R_{out}. Here the same problem arises – and we settle it in the same way: by assuming an ideal source. The difficulty, again, is that we need to make some assumption about what is on the far side of the divider if we are to determine R_{out}: see Fig. 1W.12. The assumption we make in determining R_{out} is familiar to you from Thevenin models: we assume source impedance so low that we can neglect it, calling it zero. If each resistor is 10k, R_{out} is 5k.

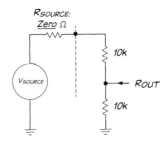

Figure 1W.12 R_{out}: we need an assumption about the source that drives the circuit, if we are to determine R_{out}.

1W.6 Effects of loading

Problem In Fig. 1W.13, what is the voltage at X:

1W.6 Effects of loading

1. with no load attached?
2. when measured with a VOM labeled "10 000 ohms/volt?"
3. when measured with a scope whose input resistance is 1 MΩ?

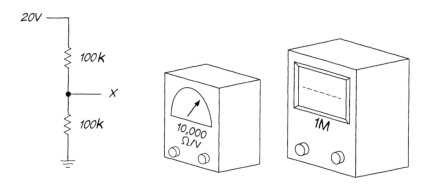

Figure 1W.13 V_{out}: calculated versus measured.

This example recapitulates a point made several times over in Chapter 1N, as you recognize. *Reminder*: The "... ohms/volt" rating implies that on the 1-volt scale (that is, when 1V causes full deflection of the meter) the meter will present that input resistance. What resistance would the meter present when set to the 10-volt scale? (Answer: 10 times the 1-volt R_{in}: 100 kΩ.)

We start, as usual, by trying to reduce the circuit to familiar form. The Thevenin model does that for us. Then we add meter or scope as *load*, forming a voltage divider, and see (Fig. 1W.14) what voltage results.

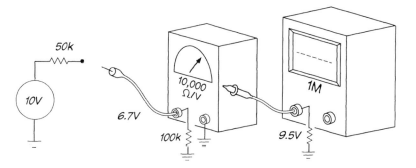

Figure 1W.14 Thevenin model of the circuit under test; and showing the "load" – this time, a meter or scope.

You will go through this general process again and again, in this course: reduce an unfamiliar circuit diagram to one that looks familiar. Sometimes you will do that by merely redrawing or rearranging the circuit; more often you will invoke a model, and often that model will be Thevenin's.

10 Online Content: Breadboarding Hints & Tips

10.1 **Solderless breadboards**

10.2 **Two possible breadboard choices**

10.3 **Building circuits with the solderless breadboard**

Available at `https://LAoE.link/LAoE_Chapter_010.pdf`

2N *RC* Circuits

Contents

2N.1	**Capacitors**	**55**
	2N.1.1 Why?	56
	2N.1.2 Capacitor structure	56
2N.2	**Time-domain view of *RC*s**	**57**
	2N.2.1 Integrators and differentiators	59
2N.3	**Frequency domain view of *RC*s**	**63**
	2N.3.1 The impedance or *reactance* of a cap	63
	2N.3.2 *RC* filters	65
	2N.3.3 Decibels	68
	2N.3.4 Estimating a filter's attenuation	70
	2N.3.5 Input and output impedance of an *RC* circuit	72
	2N.3.6 Phase shift	73
	2N.3.7 Phasor diagrams	74
2N.4	**Two unglamorous but important cap applications: "blocking" and "decoupling"**	**78**
	2N.4.1 *Blocking* capacitor	78
	2N.4.2 Decoupling or bypass capacitor	79
2N.5	**A somewhat mathy view of *RC* filters**	**80**
2N.6	**AoE reading**	**81**

2N.1 Capacitors

Now things get a little more complicated, and more interesting, as we meet frequency-dependent circuits. We rely on the *capacitor* (or just "cap") to implement this new trick, which depends on the capacitor's ability to "remember" its recent history.

That ability allows us to make timing circuits (circuits that let something happen a predetermined time after something else occurs); the most important of such circuits are *oscillators* – circuits that do this timing operation over and over, endlessly, in order to set the frequency of an output waveform. The capacitor's memory also lets us make circuits that respond mostly to changes (*differentiators*) or mostly to averages (*integrators*). And the capacitor's memory implements the *RC* circuit that is by far the most important to us: a circuit that favors one frequency range over another (a *filter*).[1]

All of these circuit fragments will be useful within later, more complicated circuits. The filters, above

---

[1] Incidentally, in case you need to be persuaded that remembering is the essence of the service that capacitors provide, note that much later in this course, partway into the digital material, we will meet large arrays of capacitors used simply and explicitly to remember: several sorts of digital memory (dynamic RAM, Flash, EEPROM and EPROM) use millions to billions of tiny capacitors to store their information, holding that data in some cases for many *years*.

all others, will be with us constantly as we meet other analog circuits. They are nearly as ubiquitous as the (resistive-) *voltage divider* that we met in the first class.

2N.1.1 Why?

We have suggested a collection of applications for *RC* circuits. Here is a shorter answer to what we mean to do with the circuits we meet today:

- generate an output voltage transition that occurs a particular delay time after an input voltage transition; and
- design a circuit that will treat inputs of different frequencies differently: it will pass more in one range of frequencies than in another (this circuit we call a "filter").

2N.1.2 Capacitor structure

The capacitor in Fig. 2N.1 is drawn to look like a ham sandwich: metal plates are the bread, some dielectric is the ham (*ceramic* capacitors really are about as simple as this). More often, capacitors achieve large area (thus large capacitance) by doing something tricky, such as putting the dielectric between two thin layers of metal foil, then rolling the whole thing up like a roll of paper towel (*mylar* capacitors are built this way).

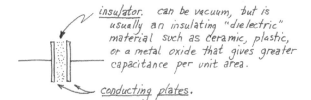

Figure 2N.1 The simplest capacitor configuration: sandwich.

A static description of cap behavior...

AoE §1.4.1

A *static* description of the way a capacitor behaves would say

$$Q = CV$$

where Q is total charge, C is the measure of how big the cap is (how much charge it can store at a given voltage: $C = Q/V$), and V is the voltage across the cap.

This statement just defines the notion of capacitance. It is the way a physicist might describe how a cap behaves, and rarely will we use it again.

...A dynamic description of cap behavior

Instead, we use a dynamic description – a statement of how things change with time:

$$I = C \frac{dV}{dt} \qquad (2\text{N}.1)$$

This is just the time derivative of the "static" description. C is constant with time; I is defined as the rate at which charge flows. This equation isn't hard to grasp. It says "The bigger the current, the faster the cap's voltage changes."

2N.2 Time-domain view of RCs

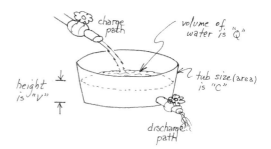

Figure 2N.2 A cap with one end grounded works a lot like a tub of water.

A hydraulic analogy: Again, flowing water helps intuition: think of the cap (with one end grounded) as a tub that can hold charge: see Fig. 2N.2.

A tub of large diameter (cap) holds a lot of water (charge), for a given height or depth (V). If you fill the tub through a thin straw (small *I*), the water level – *V* – will rise slowly; if you fill or drain through a fire hose (big *I*) the tub will fill ("charge") or drain ("discharge") quickly. A tub of large diameter (large capacitor) takes longer to fill or drain than a small tub. Self-evident, isn't it?

2N.2 Time-domain view of RCs

Now let's leave tubs of water, and anticipate what we will see when we watch the voltage on a cap change with time: when we look on a scope screen, as you will do in Lab 2L.

An easy case: constant *I*

This tidy waveform, called a *ramp*, is useful, and you will come to recognize it as the signature of this circuit fragment: capacitor driven by constant current (or "current source").

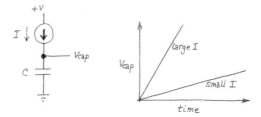

Figure 2N.3 Easy case: constant *I* ⟶ constant dV/dt.

This arrangement is used to generate a triangle waveform, see Fig. 2N.4.

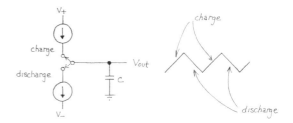

Figure 2N.4 How to use a cap to generate a triangle waveform: ramp up, ramp down.

But the ramp waveform is relatively rare, because current sources are relatively rare. Much more common is the next case.

RC Circuits

A harder case but more common: constant voltage source in series with a resistor ("exponential" charging)

AoE §1.4.2A

Here, the voltage on the cap approaches the applied voltage – but at a rate that diminishes toward zero as V_{cap} approaches its destination. Figure 2N.5 shows V_{cap} starting out bravely, moving fast toward its V_{in} (charging at 10 mA in the example above, thus at 10 V/ms); but as it gets nearer to its goal, it loses its nerve. By the time it is 1V away it has slowed to 1/10 its starting rate.

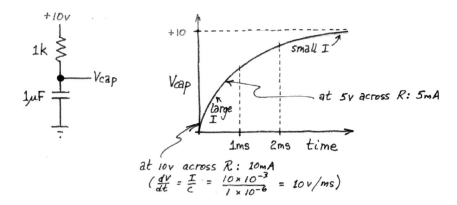

Figure 2N.5 The more usual case: cap charged and discharged from a *voltage* source, through a series resistor.

The cap behaves a lot like the walker in Zeno's stadium paradox (not so famous as Achilles chasing the tortoise). Remember her? According to Plato, Zeno visited Athens and teased his audience by asking a question something like this: "If a walker means to reach the end of the stadium, she must first go halfway, and before that, halfway to the halfway point, and so on. The number of halfway points is unlimited, so she cannot reach the end in a limited time. So, she never gets there."

Here, our cap is a little different, starting boldly and then becoming timid. Does it ever reach its destination?

Exponential charge and discharge

The behavior of *RC* charging and discharging is called "exponential" because it shows the quality common to members of that large class of functions. A function whose slope is proportional to its value is called exponential: the function e^x behaves as the charging *RC* circuit does. Its slope is equal to its value: $(e^x)' = e^x$.

Familiar exponential functions are those describing population growth (the more bacteria, the faster the colony grows) and a draining bathtub (the shallower the water, the more slowly it drains). The capacitor's *discharge* curve follows this rule more obviously than the *charging* curve, so let's consider that case: as the capacitor discharges, the voltage across the *R* diminishes, reducing the rate at which it discharges. This is evident in the *discharge* curve of Fig. 2N.6. The *RC* discharges the way a bathtub drains.

Two numbers in the plot of Fig. 2N.6 are worth remembering:

- in *one RC* (called "one time-constant") V_{cap} goes 63% of the way toward its destination; and
- in *five RC*s, 99% of the way.

If you need an exact solution to such a timing problem:

$$V_{cap} = V_{applied} \cdot (1 - e^{-t/RC})$$

In case you can't see at a glance what this equation is trying to tell you, look at $e^{-t/RC}$ by itself:

2N.2 Time-domain view of RCs

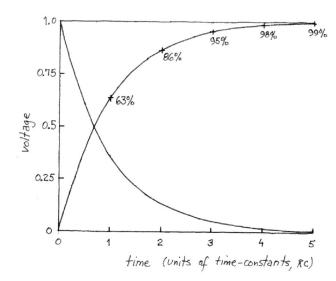

Figure 2N.6 *RC* charge, discharge curves.

- when $t = RC$, this expression is $e^{-1} = 1/e$, or 0.37; and
- when $t \gg RC$, i.e., very large, this expression is tiny, and $V_{cap} \approx V_{applied}$.

> A tip to help you calculate time-constants:
> MΩ and μF give time-constant in seconds; and
> kΩ and μF give time-constant in milliseconds.

(The second case, kΩ and μF, is very common.) In the example above, RC is the product of 1k and 1 μF: 1ms.

The "time constant," *RC*

You may be puzzled, at first, to see frequent reference to a circuit's "time constant," even in settings where no one plans to use the quantity *RC* directly, as one might, say, in an oscillator circuit. The *time constant* serves to give a ballpark measure of how quick or slow a circuit element may be. If you see fuzz on an oscilloscope screen at 1 MHz, for example, you know at once not to blame an *RC* circuit whose time constant is in the millisecond range.

...and a trick for the calculator-less

The exponential expression, $e^{-t/RC}$, which describes the "error", may look pretty formidable to work with by hand. In fact, however, it is not hard to estimate. Just note the following pattern:

- in one time-constant the *error* decreases to about 40% of what it was at the start of that time-constant.

So, after a couple of time constants, the error is down to 0.4×0.4, or about 0.16. One more time constant takes the error down to about 0.06. And so on.

2N.2.1 Integrators and differentiators

The very useful formula, $I = C\, dV/dt$ will let us figure out when these two circuits perform pretty well as integrator and differentiator, respectively.

RC Circuits

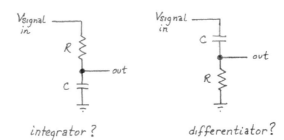

Figure 2N.7 Can we exploit cap's $I = C\, dV/dt$ to make differentiators and integrators?

Differentiator Let's first consider the simpler circuit in Fig. 2N.8. The current that flows in the cap is proportional to dV_{in}/dt; so, in once sense, the circuit differentiates the input signal perfectly: its *current* is proportional to the slope of V_{in}. But the circuit is pretty evidently useless (*perfectly* useless). It gives us no way to measure that current. If we could devise a way to measure the current, we would have a differentiator. Figure 2N.9 shows our earlier proposal, again. Does it work?

Figure 2N.8 Useless "differentiator"?

Answer: Yes and No. Yes, to the extent that $V_{cap} = V_{in}$ (and thus $dV_{cap}/dt = dV_{in}/dt$), it works. But to the extent that V_{out} moves, it is imperfect, because that movement makes V_{cap} differ from V_{in}.

So the circuit errs to the extent that the output moves away from ground; but of course it must move away from ground to give us an output. This differentiator is compromised. So is the *RC* integrator, it turns out. When we meet operational amplifiers, we will manage to make nearly-ideal integrators, and pretty good differentiators.

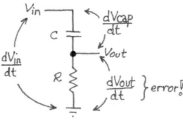

Figure 2N.9 Differentiator? – again.

AoE puts that point this way:

- for the differentiator: "... choose R and C small enough so that $dV_{out}/dt \ll dV_{in}/dt$..."; and
- for the integrator: "... [make sure that] $V_{out} \ll V_{in} \ldots \omega \cdot RC \gg 1$."[2]

AoE §1.4.3

AoE §1.4.4

We can put this simply – perhaps crudely. Assume a sinewave input. Then,

the *RC* differentiator (and integrator, too) works pretty well *if* it is *murdering* the signal (that is, attenuates it severely), so that V_{out} (and dV_{out}) is tiny: hardly moves away from ground.

It follows[3], along the way, that differentiator and integrator will impose a 90° phase shift on a sinusoidal input. This result, obvious here, should help you anticipate how *RC* circuits viewed as "filters" (below) will impose phase shifts.

---

[2] This may be the first appearance of ω (*omega*) in these notes. ω describes frequency not in cycles per second, properly called "hertz," but in radians per second. Since a full cycle includes 2π radians, $\omega = 2\pi f$.

[3] When we say "it follows," we are assuming that you accept the proposition that the derivative or integral of a sinusoid is another sinusoid 90° phase-shifted relative to the original. If this point is new to you, hang on. We'll soon see this happening in today's lab.

2N.2 Time-domain view of RCs

Try out the differentiator: Figure 2N.10 shows what a differentiator like the one you'll build in Lab 2L puts out, given a square wave input. The slow sweep on the left makes the output look pretty much like a spike, responding to each edge of the square wave. The detail on the right shows the decay of the RC that forms the differentiator[4] (the time constant, RC, is $100\,\Omega \times 100\,\text{pF} = 10^4 \times 10^{-12} = 10 \times 10^{-9} = 10\,\text{ns}$).

The square wave produces the most dramatic response from a differentiator. A sinusoid produces a quieter response: another sinusoid, simply phase shifted. Note that the output amplitude is much less than input: see Fig. 2N.11.

The output looks more or less correct: it measures the slope of the input. But it also looks a little funny: it is not a smooth sinusoid. That's the fault not of the differentiator but of the "sinusoid" that goes in – coming from the function generator. That sinusoid is a bit of a phoney, produced by whittling away the points of a triangle! The differentiator exposes this fraud: one can make out at least a constant-slope section as the underlying triangle crosses zero.

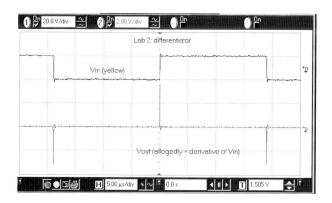
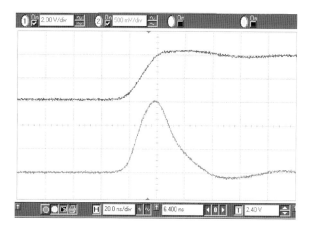

Figure 2N.10 Differentiator, fed a square wave as input; detail shows RC.

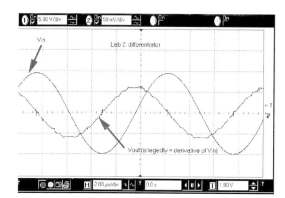

Figure 2N.11 Response of RC differentiator to sinusoid: shows the "sine" isn't clean (note scale change from 'in' to 'out': Ch1 is input).

AoE §1.4.4

Integrator

The circuit on the left in Fig. 2N.12 is a *perfect* integrator – but its input is a *current*. When the input is a voltage, as on the right, the RC integrator becomes imperfect. The limitations of the RC integrator

---

[4] The amplitude is not the same as in the slower sweep, because the detail shows response to a positive 4 V step (the edge of a "TTL" square wave), used because its edge is steeper than the edge of the square wave shown in the slower sweep. The steeper edge produces a more-nearly-ideal response pulse.

62 RC Circuits

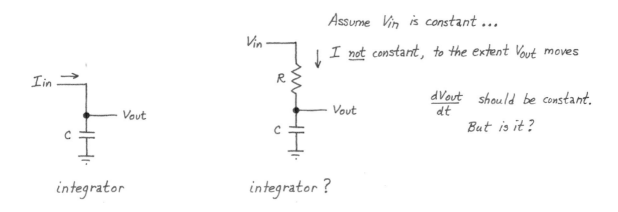

Figure 2N.12 Integrator? – again, only sort-of.

resemble those of the *RC* differentiator. To keep things simple, imagine that you apply a *step* input; ask what waveform out you would like to see out, and what, therefore, you would like the current to do.[5]

The integrator works if $V_{out} \ll V_{in}$ – or, less formally, it works if it's murdering the input. We achieve that by making sure that the cap doesn't have time to charge much between reversals of the input waveform: $RC \gg$ {half-period of a square wave input}, for example.

Try out the integrator: Again a square wave provides the easiest test of the integrator, since it's pretty plausible that a square wave input should charge, then discharge the cap at a nearly-constant rate, producing a ramp up, then a ramp down. Figure 2N.13 shows the lab's *RC* integrator responding to square wave, then sinusoid.

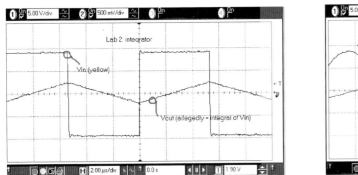 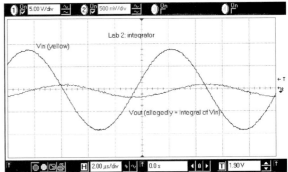

Figure 2N.13 Lab 2L's *RC* integrator responding to square wave and sinusoid (note scale change from 'in' to 'out': Ch1 is input).

Note that this integrator is severely attenuating the signal, as required for an *RC* integrator: the gain on the output channel is *ten times* the gain on the input channel, in Fig. 2N.13.

Incidentally, if you find it hard to get an intuitive grip on what the *integral* of a sinusoid should look

---

[5] What we'd like to see is an output *ramp*: constant dV_{out}/dt for constant V_{in}.

like, try this trick: *reverse the process*. Check that the *input* waveform shows the *slope* of the output; in other words, that the input shows the derivative of the output – the *inverse* of the integral function.

2N.3 Frequency domain view of *RC*s

We'll now switch to speaking of the same old *RC* circuits in a different way, describing not what a scope image might show – the so-called "time-domain" view – but instead how the circuit behaves as frequency changes. This is the "frequency domain" view. Don't let the name scare you: the reference to "domain" (a word that evokes castles and lords and ladies) is just a high-fallutin' way to specify what's on the *x*-axis ("domain" describes the input to a mathematical function, as you may recall).

2N.3.1 The impedance or *reactance* of a cap

A cap's impedance varies with frequency. Impedance is the generalized form of what we called "resistance" for "resistors;" "reactance" is the term reserved for capacitors and inductors. The latter usually are coils of wire, often wound around an iron core.

It's obvious that a cap cannot conduct a DC current: just recall what the cap's insides look like: an insulator separating two plates. That takes care of the cap's impedance at DC: clearly it's infinite (or *huge*, anyway).

A time-varying voltage across a cap can cause a *current* to pass through the capacitor. This is a point we hope we established in discussing differentiators, §2N.2.1.

A related but different issue concerns us when we consider filters, however: we would like to understand why a rapidly-varying *voltage* can pass through the capacitor. But that does happen, and we need to understand the process in order to understand a *highpass* filter. If you're already happy with the result, skip this subsection.

When we say the AC signal passes through a filter, all we mean is that a wiggle on the left causes a wiggle of similar size on the right: see Fig. 2N.14.

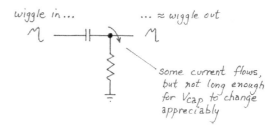

Figure 2N.14 How a cap "passes" a signal.

The wiggle makes it "across" the cap so long as there isn't time for the voltage on the cap to change much before the wiggle has ended – before the voltage heads in the other direction. In other words, quick wiggles pass; slow wiggles don't.

Please *note* that what we just described is the passing of a *voltage* signal through a filter. This is not the same as the passing of a *current* through a capacitor – an effect that cannot be continuous. The passing of current can occur only so long as the voltage difference between the capacitor's plates is *changing*, as it is for a continuous sinusoidal input, for example.

We can stop worrying about our intuition, if you like[6], and give the expression for the cap's *impedance*:

---

[6] In §2S.2 we make another attempt to offer intuitive fingerholds on this idea: ways to grasp intuitively what the math is trying to tell us.

RC Circuits

$$Z_C = -j/\omega C = -j/2\pi f C$$

We can see at once that the cap's impedance falls with frequency and is inversely proportional to C. But what's this "j"? Two answers: symbolically, it's just an electronics convention representing the mathematician's "i," i.e., $\sqrt{-1}$.[7] That answer may not help if you consider $\sqrt{-1}$ to be a pretty weird thing, as most right-minded people do! The second answer is that j happens to provide a mathematically convenient way to talk about phase shifts. But it is a way that we will not rely on in this course. Instead, we will usually sweep j and even phase-shift itself under the rug (see later in this section).

And once we have an expression for the impedance of the cap – an expression that shows it varying smoothly with frequency – we can see how capacitors will perform in voltage dividers.

Digression: deriving Z_C

Skip the following, if you're in a hurry. This is just for the person who feels uncomfortable when handed a formula that has not been justified.

Why does the formula $Z_C = -j/\omega C = -j/2\pi f C$, hold? We will sneak up on the answer, using some lazy approximations (very much in the style of this course).

What we mean by "impedance of the capacitor" is equivalent to what we mean by "impedance of the resistor:" the quotient of V/I. If we plug in expressions for the somewhat novel V and I in the capacitor, we should get the advertised formula for Z_C.

Since we're interested in how the capacitor responds to time-varying signals, and the sinusoid provides the simplest case,[8] we'll assume a V_{cap} that is not DC, but, rather, sinusoidal:

$$V_{cap} = A\sin(\omega t)$$

Because the current through a capacitor is the time-derivative of the voltage across the cap (see §2N.1, above), the expression for the current through a capacitor across which one applies V_{cap} will be

$$I_{cap} = C\left(\frac{dV_{cap}}{dt}\right) = CA\omega \cos \omega t$$

and Z_{cap} is the quotient, V/I, given by

$$\frac{V}{I} = \frac{V_{cap}}{I_{cap}} = \frac{A\sin(\omega t)}{A\omega \cos(\omega t) C}$$

To evaluate this exactly one is obliged to confront the nasty fact that sine and cosine are out of phase, a fact that is handily expressed – but cryptically, to those of us not yet accustomed to complex quantities – in the "$-j$" that appears in the expression for Z_{cap}. Just now, we choose *not* to confront that nasty fact, and choose instead to sweep $-j$ and phase shift under the rug.

Hiding from phase shift

If we neglect phase shift, then we can simplify this quotient, $\frac{A\sin(\omega t)}{A\omega \cos(\omega t) C}$. We can consider just the maximum values of the sine and cosine functions, i.e., 1, and then divide out the amplitude, A, along with the values of sin and cos which we take to be *one*, so as to get the very simple expression:

AoE §1.7.1A

$$X_{cap} = |Z_{cap}| = \frac{1}{\omega C} \quad \text{or} \quad \frac{1}{2\pi f C}$$

---

[7] Electrical engineers avoid saying "i" because it sounds like *current*.

[8] You'll recall that Fourier teaches that any waveform other than sinusoid – say, square wave, or triangle wave – can be formed as a sum of sinusoids. It follows that we need only consider how a circuit responds to a sinusoid, in order to gain a fully-general understanding of the circuit's behavior.

2N.3 Frequency domain view of RCs

The "|" indicates "magnitude, ignoring phase". *Reactance* thus describes the ratio of current and voltage *magnitudes* only. This is not quite the whole story, evidently, since this account ignores phase shift. But this expression provides a pretty good rough description of the way a capacitor's impedance behaves, and it is easy to understand. The expression tells us two important truths, as we noticed when we first met Z_C:

(1) the capacitor's impedance is *inversely* proportional to C; and
(2) the capacitor's impedance varies inversely with *frequency*.

The second point, of course, is the exciting one: here's a device that will let us build a new sort of voltage divider, more intriguing than the ones built with resistors. These dividers can be *frequency-dependent*. That sounds interesting.

Later, we will return to the topic of phase shift. For the moment, we won't worry about it.

2N.3.2 RC filters

These are the most important applications of capacitors. These circuits are just voltage dividers, essentially like the resistive dividers you have met already. The *resistive* dividers treated DC and time-varying signals alike. To some of you that seemed obvious and inevitable (and maybe you felt insulted by the exercise at the end of Lab 1L that asked you to confirm that AC was like DC to the divider). Resistive dividers treat AC and DC alike because resistors can't remember what happened even an instant ago. They're little existentialists, living in the present. (We're talking about ideal resistors, of course.)

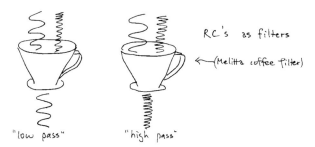

Figure 2N.15 *RCs* are most important as filters – more or less like coffee filters.

A rare but simple case: C:C divider

You know how a resistive divider works on a sine. How would you expect a divider made of capacitors to treat a time-varying signal? The C:C divider is not quite a realistic circuit, since we're normally stuck with substantial stray R values that complicate it. But it is useful to get us started in considering phase shifts.

If this case worries you, good: you're probably worrying about phase shifts. Turns out they cause no trouble here: output is in phase with input. (If you can handle the complex notation, write $Z_C = -j/\omega C$, and you'll see the js wash out.)

But what happens in the combined case, where the divider is made up of both R and C? This turns out to be the case that usually concerns us. This problem is harder, but still fits the voltage-divider model. Let's generalize that model a bit in Fig. 2N.17.

A qualitative view of the filter's frequency response

The behavior of these voltage dividers – which we call *filters* when we speak of them in frequency terms, because each favors either high or low frequencies – is easy to analyze, at least roughly:

RC Circuits

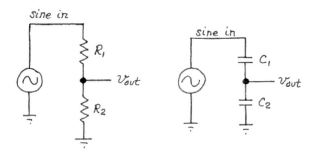

Figure 2N.16 Two dividers that deliver 1/2 of V_{in} – with no phase shift.

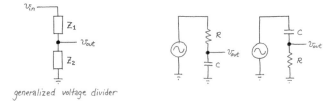

Figure 2N.17 Generalized voltage divider; and voltage dividers made up of R paired with C.

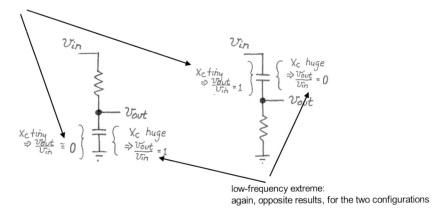

Figure 2N.18 Establish the endpoints of each filter's frequency response curve, by trying f = tiny, f = huge.

(1) See what the filter does at the two frequency extremes. (This looking at extremes is a useful trick; you will soon use it again to find the filters' worst-case Z_{in} and Z_{out}.)

At one end of the frequency range, one configuration of filter passes almost none of an input signal, while the other passes all, and vice versa:

At $f = 0$, what fraction out

At very high f, what fraction out?

As the annotations in Fig. 2N.18 say, each filter can deliver all of its input to the output at one frequency extreme, none at the other extreme; for the two filters those extremes occur, of course, at opposite ends of the frequency spectrum.

(2) Determine where the output "turns the corner" (corner defined arbitrarily[9]) as the frequency where the output is 3 dB less than the input (always called just "the 3 dB point"; the "minus" is understood).

---

[9] Well, not quite arbitrarily: a signal reduced by 3 dB delivers half its original power.

2N.3 Frequency domain view of RCs

Knowing the endpoints, which tell us whether the filter is *highpass* or *lowpass*, and knowing the 3 dB point, we can draw the full frequency-response curve; see Fig. 2N.19.

The "3 dB point," the frequency where the filter "turns the corner" is

$$f_{3dB} = \frac{1}{2\pi RC}$$

Beware the more elegant formulation that uses ω:

$$\omega_{3dB} = \frac{1}{RC}$$

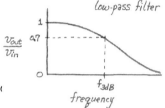 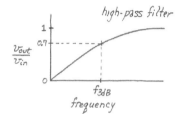

Figure 2N.19 *RC* filter's frequency curve.

That is tidy, but is very likely to give you answers off by about a factor of 6, since you will be measuring *period* and its inverse in the lab: frequency in *hertz* (or "cycles-per-second," as it used to be called), *not* in radians. So avoid using ω as you evaluate frequency.

Two asides

Caution! Do not confuse these *frequency-domain* pictures with the earlier *RC* step-response picture, (which speaks in the *time-domain*). Both sorts of plots are sketched in Fig. 2N.20.

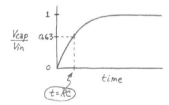 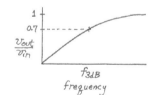

Figure 2N.20 Deceptive similarity between shapes of time- and frequency-plots of *RC* circuits.

Not only do the curves look vaguely similar but, to make things worse, details here seem tailor-made to deceive you:

Step response: In the *time RC* (time-constant), V_{cap} moves to about 0.6 of the applied step voltage (this fraction is $(1 - 1/e)$).

Frequency domain: At f_{3dB}, a frequency determined by RC, the filter's V_{out}/V_{in} is about 0.7 (this is $1/\sqrt{2}$).

Don't fall into that trap. Don't confuse these two quite-different plots.

A note concerning log plots: You may wonder why the curves we have have sketched in Fig. 2N.19 the curves in AoE introducing filters (§1.7.8), and those you see on the scope screen (when you "sweep" the input frequency) don't look like the tidier curves shown in most books that treat frequency response, or like the curves in Chapter 6 of AoE treating filters. Our curves trickle off toward zero, in the lowpass configuration, whereas these other curves seem to fall smoothly and promptly to zero.

AoE §1.7.8

RC Circuits

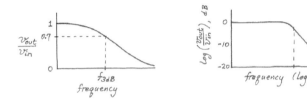

Figure 2N.21 Linear and log–log frequency-response plots contrasted.

This is an effect of the logarithmic compression of the axes on the usual graph. Our plots are linear; the usual plot ("Bode plot") is log–log: see Fig. 2N.21.

We will illustrate the contrast in appearance between linear and log–log plots in §2N.3.3 after a brief introduction to *decibels*.

2N.3.3 Decibels

AoE §1.3.2A

A "decibel"[10] is a unitless measure of a *ratio* between two values. Since it uses a logarithm of the ratio, it can help to describe values whose range is very large.

Since a decibel measures a *ratio*, it must always rely on a reference level. In the application most familiar to most of us, outside of electronics, *sound* level or loudness is measured relative to a level approximating the quietest audible sound. Sound measurement makes good use of the decibel because the range it needs to measure is enormous: the ratio of loudest sound to the quietest that is audible to a human ear can be as large as about three million to one (this is a ratio of sound pressure levels). This large quantity, awkward to express directly, shrinks to manageable form when expressed in decibels: 130 dB. For sound measurements, "0 dB" is the reference level for audibility.

In electronics usage, decibel describes a *ratio*, as always – but sometimes the magnitudes that are compared represent *power*, sometimes *voltage* or *amplitude*. Since power varies as the square of voltage, the two definitions of decibel look different:

$$dB = 10 \log_{10} P_2/P_1 = 20 \log_{10} A_2/A_1, \quad (2N.2)$$

where P represents power, A amplitude. It is the *amplitude* definition that will be useful to us.

Sometimes a label indicates the reference value to which a quantity is compared. For example, *dBm* uses 1 milliwatt as its reference (with some assumed load impedance). In this course, however, we use dB most often to describe the attenuations achieved by a *filter*, and in this context a few ratios continually recur.

Some dB values we often encounter:

- -3 dB = amplitude ratio of $1/\sqrt{2}$. This one comes up all the time, most often in the formula for the frequency where a filter attenuates enough so that we say it has "turned the corner" (but a "corner" that is so gentle that it hardly deserves the name, as you can see for example in Fig. 2N.27).
- -6 dB = amplitude ratio of $1/2$. This comes up in the description of the slope of *RC* filters. "-6 dB per octave" describes the rolloff of a lowpass filter, in jargon that could be translated to "halving amplitude for each doubling of frequency."
- $+3$ dB and $+6$ dB of course describe growth rather than attenuation: by amplitude factors of $\sqrt{2}$ or 2, respectively.

---

[10] A decibel is one tenth of a Bel – a unit defined by Bell Labs to measure signal attenuation in telephone cables, and named to honor Alexander Graham Bell. The Bel turned out (like Farad) to be impractically large, and is not used. You can find more lore on the decibel, if you like, in Watkinson's *The Art of Digital Audio* (3d ed., 2001), §2.18.

2N.3 Frequency domain view of RCs

- +20 dB = amplitude ratio of 10. This comes up in the alternative description of the slope of an *RC* filter: a lowpass, for example, falls at "−20 dB per decade," meaning "cutting amplitude by a factor of ten for each 10-fold multiplication of frequency."

Rise and fall on linear and log plots

Linear plots: In lab, when we watch filter attenuations on a scope, we see and perhaps draw *linear* relations of amplitude to frequency. (Some function generators can be set to *sweep* frequencies using a logarithmic sweep that compresses the horizontal axis. But we don't ordinarily use such generators, and such a *semi-log* plot does not look like the usual textbook and datasheet plot, which is log–log.) We see the lowpass roll off rather gently with its $1/f$ shape. On a log–log plot (standard in electronics references that describe filters) this lazy and curvy rolloff is displayed as a straight line, as you can confirm with a look at Fig. 2N.24.

First, in Fig. 2N.22, is a reminder of what a plot rising or falling at ±6 dB/octave looks like. We have labeled the axes "Vo/Vi" and "frequency," as for a filter.

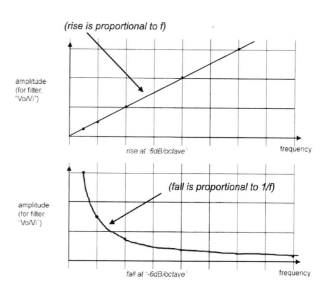

Figure 2N.22 Linear plot of rise and fall at ± "6 dB/octave".

This slope can also be described, as we said above, as ±20 dB/decade. We have not shown this in Fig. 2N.22 because we did not show a range sufficient to display a decade (a ten-fold) change in frequency.

Log–log plots: Log–log plots, like those in traditional "Bode plots," make the humble *RC* look more decisive. In particular, the lazy and curvy $1/f$ rolloff of a lowpass turns into a straight line that seems to dive resolutely and rapidly toward zero. The rise, too, looks tidier on a log–log plot, though not so drastically improved. In Fig. 2N.23 we have shown rises at "6 dB/octave" and "12 dB/octave." These are not measured filter results; they only show the behavior of curves on log–log paper.

Figure 2N.24 shows how a log–log plot can make an *RC* filter's response look more decisive, and can turn a curvy $1/f$ response into a tidy down ramp.

On these log–log plots, the behavior is easier to see if you describe it as "±20 dB/decade" and "40 dB/decade" rather than by *octave*.

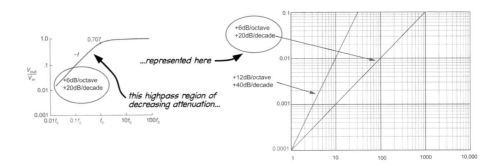

Figure 2N.23 Log–log plot of *RC* highpass filter's *rising* response.

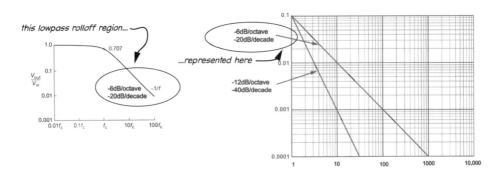

Figure 2N.24 Log–log plot of *RC* lowpass filter's *falling* response (often called "rolloff"). Lowpass rolloff figure on left side reproduced from AoE Fig. 1.104.

2N.3.4 Estimating a filter's attenuation

The shape of the frequency-response curves sketched in Fig. 2N.19 allows one to estimate the attenuation a filter will apply to a specified frequency range, given $f_{3\text{dB}}$. Here are some useful rules of thumb.

Attenuation in the **passband**

Attenuation will be under about 10%, if one keeps an octave (a factor of two in frequency) between *signal* and $f_{3\text{dB}}$. In a lowpass, for example, if you want to pass signals below 1 kHz, put $f_{3\text{dB}}$ at 2 kHz. In a highpass, if you want to pass signals above 10 kHz, put $f_{3\text{dB}}$ at 5 kHz.

In Fig. 2N.25 are highpass and lowpass filters, showing the passbands and the $f_{3\text{dB}}$ frequencies.

Details of passband: We rarely will need more detail about passband attenuation than the point mentioned above, that we lose 10% a factor of two away from $f_{3\text{dB}}$. (Specifically, we lose about 10% at $1/2\ f_{3\text{dB}}$ for a lowpass; at $2 \times f_{3\text{dB}}$ for a highpass.)

But, in case you are curious, Fig. 2N.26 shows passband attenuation in detail.[11]

Attenuation in the **stopband**

Where the filter attenuates substantially, amplitude falls or rises in proportion to frequency. This simple pattern is accurate well away from $f_{3\text{dB}}$ and is approximate close to $f_{3\text{dB}}$, as can be seen from Fig. 2N.27.

In a lowpass, for example, if $f_{3\text{dB}}$ is 2 kHz, then noise at 20 kHz ($10 \times f_{3\text{dB}}$) will be down to about 1/10 what its level would have been at $f_{3\text{dB}}$. By this reasoning, it will be down about $1/10 \times 0.7$, since the level is about 0.7 (by definition) at $f_{3\text{dB}}$. In fact, it will be attenuated somewhat less, because the $1/f$ shape is not achieved in the first octave or so above $f_{3\text{dB}}$.

---

[11] This is modified from Fig. 4.8 of Tektronix 'ABC's of Probes' (2011), p. 35.

2N.3 Frequency domain view of RCs

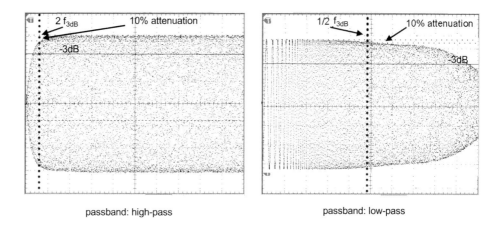

Figure 2N.25 Passbands for highpass and lowpass filters. Loss is about 10%, an octave from f_{3dB}. These are linear–linear plots; it is the (swept) waveforms that are shown, *not* (as in previous figures) a plot of response.

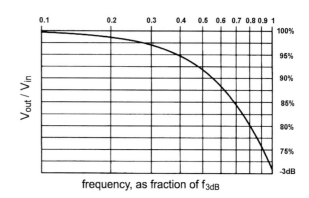

Figure 2N.26 Details of *RC* rolloff in the passband.

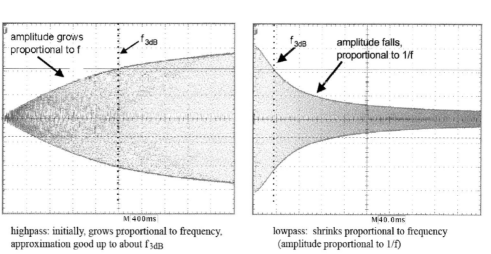

Figure 2N.27 Stopbands for highpass and lowpass filters. Growth or shrinkage with change of frequency.

RC Circuits

In Fig. 2N.27, one can see the deviation of the highpass response from a perfect straight line (that is, from strict proportionality to frequency) up to f_{3dB}. Far below f_{3dB}, the deviation from straight line would be slight.

It is harder to judge the lowpass deviation from a $1/f$ shape, but you may be able to see, if you look closely, that at $2 \times f_{3dB}$ amplitude is well above the 35% that one would predict if one expected a $1/f$ slope even in the first octave above f_{3dB}.

Still, the main point is not the imperfection of these approximations but their power. They let you give a quick and pretty-good estimate of what a filter will do to signals at a specified distance from f_{3dB}. That's valuable.

2N.3.5 Input and output impedance of an *RC* circuit

AoE §1.7.1D

If filter A is to drive filter B – or anything else, in fact – it must conform to our 10× rule of thumb, which we discussed earlier when we were considering only resistive circuits. The same reasons prevail, but here they are more urgent: if we don't follow this rule, not only will signals get attenuated; frequency response also is likely to be altered.

But to enforce our rule of thumb, we need to know Z_{in} and Z_{out} for the filters. At first glance, the problem looks nasty. What is Z_{out} for the lowpass filter, for example? A novice would give you an answer that's correct but much more complicated than necessary. They might say,

$$Z_{out} = Z_C \parallel R = \frac{R(-j/\omega C)}{R - j/\omega C}$$

Yow! And then this expression doesn't really give an answer: it tells us that the answer is frequency-dependent.

"Worst case" impedances

We cheerfully sidestep this hard work, by considering only *worst case* values. We ask, "How bad can things get?"

> We ask, "How bad can Z_{in} get?" And that means, "How *low* can it get?"

> We ask, "How bad can Z_{out} get?" And that means, "How *high* can it get?"

This trick delivers a stunningly easy answer: the answer is always just R! Here's the argument for a *lowpass* for example (see Fig. 2N.28):

Figure 2N.28 Worst-case Z_{in} and Z_{out} for *RC* filter reduces to just *R*.

worst Z_{in}: cap looks like a short: $Z_{in} = R$ (this happens at highest frequencies).

2N.3 Frequency domain view of RCs

worst Z_{out}: cap doesn't help at all; we look through to the source, and see only R: $Z_{out} = R$ (this happens at lowest frequencies).

Having an easy way to handle the filter's input and output impedances allows you to string together RC circuits just as you could string together voltage dividers, without worrying about interaction among them.

2N.3.6 Phase shift

<small>AoE §1.7.1</small>

You already know roughly what to expect: the differentiator and integrator showed you phase shifts of 90°, and did this when they were severely attenuating a sinewave. But you need to *beware* the misconception that because a circuit has a cap in it, you should expect to see a 90° shift (or even just noticeable shift). You should *not* expect that. You need an intuitive sense of when phase shifting occurs, and of roughly its magnitude. You rarely will need to calculate the amount of shift. In fact, as we work with filters we usually get away with ignoring the issue of phase shift.

How we get away with largely ignoring phase shift in RC filters

Before we get into some details, let's try to explain why we can come close to ignoring phase shift. Here is a start: a rough account of phase shift in RC circuits:

If the amplitude *out* is close to amplitude *in*, you will see little or no phase shift. If the output is much attenuated, you will see considerable shift (90° is maximum).

<small>AoE §1.7.9</small>

Figure 2N.29 shows curves saying the same thing for the case of a lowpass filter.

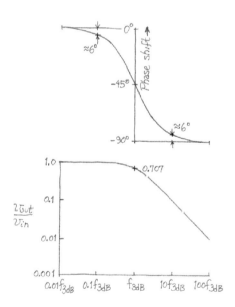

Figure 2N.29 Attenuation and phase shift (log–log plot).

As you can see, where the filter is *passing* a signal, the filter does not impose much phase shift. How much? If we assume that we can put f_{3dB} at twice the highest frequency that we want to pass (let's call that the highest "signal" frequency). Then the signal will be attenuated only slightly (about 10%) and will be shifted no more than about 25 degrees. Both attenuation and phase shift will be less at lower frequencies.

But, you may want to protest, more severe shifts occur over much of the possible input frequency range, shown in Fig. 2N.29. At many frequencies, phase shift is considerable: −45° at f_{3dB}, and more than that as frequency climbs, approaching −90°. Why not worry about those shifts?

We don't worry because those shifts are applied to what we consider not signal but *noise*, and we're just not much interested in what nasty things we are doing to noise. We care about *attenuating* noise, it's true; we like the attenuation, and are interested in its degree. But that's only because we want to be rid of it. We don't care about the details of the noise's mutilation; we don't care about its phase shift.

Why phase shift occurs

Why does this happen? Here's an attempt to rationalize the result:

- in an *RC* series circuit, the voltages across the *R* and the *C* are 90° out of phase, as you know;
- the voltages across *R* and *C* must sum to V_{in}, at every instant;
- the voltage out of the filter, V_{out}, is the voltage across either *R* or *C* alone (*R* for a highpass, *C* for a lowpass); and
- as frequency changes, *R* and *C* share the total V_{in} differently, and V_{out} thus can look much like V_{in} or can look very different. In other words, the phase difference between V_{in} and V_{out} varies with frequency.

Consider, for example, a lowpass filter. If much of the total voltage V_{in} appears across the cap, then the phase of the input voltage (which appears across the *RC* series combination) will be close to the phase of the output voltage, V_{cap}. In other words, *R* plays a small part, and V_{out} is about the same as V_{in} in both amplitude and phase. Have we merely restated our earlier proposition? It almost seems so.

But let's try a drawing.

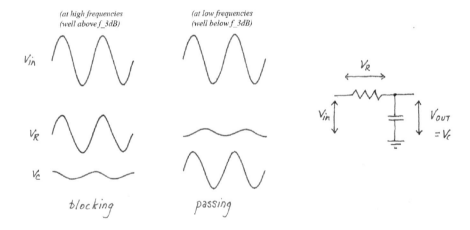

Figure 2N.30 *R* and *C* sharing input voltage differently at two different frequencies.

The sketch in Fig. 2N.30 shows what one would expect: little phase shift where the filter is *passing*, lots of phase shift where it is largely *attenuating*. Now let's try another aid to an intuitive understanding of phase shift: phasors.

2N.3.7 Phasor diagrams

AoE §1.7.12

These diagrams let you compare phase and amplitude of input and output of circuits that shift phases (circuits including *C*s and *L*s). They make the performance graphic, and allow you to get approximate results by use of geometry rather than by explicit manipulation of complex quantities.

The diagram uses axes that represent resistor-like ("real") impedances on the horizontal axis, and capacitor- or inductor-like impedances ("imaginary" – but don't let that strange name scare you; for our purposes it only means that voltages across such elements are 90° out of phase with voltages across the

2N.3 Frequency domain view of RCs

resistors). This plot is known by the extra-frightening name, "complex plane" (with nasty overtones, to the untrained ear, of "too-complicated-for-you plane"!). But don't lose heart. It's all very easy to understand and use. Even better, *you don't need to understand phasors* if you don't want to. We use them rarely in the course, and always could use direct manipulation of the complex quantities instead. Phasors are meant to make you feel better. If they don't, forget them.

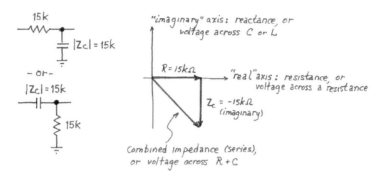

Figure 2N.31 Phasor diagram: "complex plane;" showing an *RC* at f_{3dB}.

Phasor diagram of an *RC* circuit

Fig. 2N.31 shows an *RC* filter at its 3 dB point, where, as you can see, the *magnitude* of the impedance of *C* is the same as that of *R*. The arrows, or vectors, show phase as well as amplitude (notice that this is the *amplitude* of the waveform: the peak value, not a voltage varying with time); they point at right angles to indicate that the voltages in *R* and *C* are 90° out of phase.

Phase: We are to think of these vectors as rotating counterclockwise, during the period of an input signal. So, thinking of the input as aligned with the hypotenuse, we see the voltage across the resistor *leading* the input (that is true of such a circuit: a *highpass*, which takes output across *R*). We see the voltage across the capacitor *lagging* the input (that is true of such a circuit: a *lowpass*, which takes output across *C*).

"Voltages?," you may be protesting, "but you said these arrows represent impedances." True. But in a series circuit, where the currents in the two elements are the same, the voltages are proportional to the impedances. So both interpretations of the figure are fair.

The total impedance that *R* and *C* present to the signal source is *not* 2*R*, but is a vector sum: it's the length of the hypotenuse, $\sqrt{2}R$. And from this diagram we now can read two familiar truths about how an *RC* filter behaves at its 3 dB point:

- The amplitude of the output relative to input is down 3 dB: down to $1/\sqrt{2}$: the length of either the *R* or the *C* vector, relative to the hypotenuse.
- The output is shifted 45° relative to the input: *R* or *C* vectors form an angle of 45° with the hypotenuse, which represents the phase of the input voltage. Whether output *leads* or *lags* input depends on where we take the circuit output, as we said just above.

So far, we're only restating what you know. But to get some additional information out of the diagram, try doubling the input frequency several times in succession, and watch what happens: each time, the length of the Z_C vector is cut to half what it was.

However, the first doubling also affects the length of the hypotenuse substantially; so the amplitude relative to input (let's assume a lowpass) is not cut quite so much as 50% (not quite so much as "6 dB"). You can see that the output is attenuated a good deal more than at f_{3dB} however, and also that phase shift between input and output has increased a good deal.

76 RC Circuits

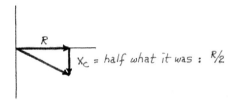

RC after a first doubling of frequency relative to Fig. 2N.31.

Again, on the second doubling of frequency, the length of the Z_C vector is halved, but this time the length of the hypotenuse is changed less than in the first doubling. So things are becoming simpler: the output shrinks nearly as much as the Z_C vector shrinks: nearly 50%. Here we are getting close to the ultimate slope of the filter's rolloff curve: −6dB/octave. Meanwhile, the phase shift between output and input is increasing too – approaching the limit of −90°.

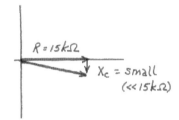

RC after a second doubling of frequency relative to previous diagram.

We've been assuming a *lowpass*. If you switch assumptions, and ask what these diagrams show happening to the output of a *highpass*, you find all the information is there for you to extract. No surprise, there; but perhaps satisfying to notice.

Phasor diagram of an *LC* circuit

Finally, let's look (Fig. 2N.32) at an *LC trap* circuit on a phasor diagram. (This is a sneak preview of a circuit we have not yet talked about, and a circuit element that you'll encounter in the next lab. We can't resist including it here, where we're in the thick of *phasors*.)

AoE §1.7.14

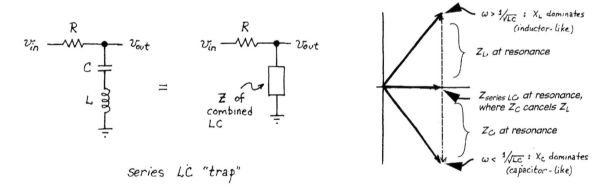

Figure 2N.32 *LC* trap circuit, and its phasor diagram.

This is less familiar, but pleasing because it reveals the curious fact – which you will see in Lab 3L when you watch a similar (parallel) *LC* circuit – that the *LC* combination looks sometimes like *L*, sometimes *C*, showing the phase shift characteristic of each – and at resonance, shows no phase shift at all. We'll talk about *LC*'s next time; but for the moment, see if you can enjoy how concisely this phasor

2N.3 Frequency domain view of RCs

diagram describes this behavior of the circuit (actually a *trio* of diagrams appears here, representing what happens at *three* sample frequencies).

To check that these *LC* diagrams make sense, you may want to take a look at what the old voltage-divider equation tells you ought to happen.

Here's the expression for the output voltage as a fraction of input:

$$\frac{V_{\text{out}}}{V_{\text{in}}} = \frac{Z_{\text{combination}}}{Z_{\text{combination}} + R}$$

But

$$Z_{\text{combination}} = \frac{-j}{\omega C} + j\omega L$$

Figure 2N.33 *LC* trap: just another voltage divider.

And at some frequency – where the magnitudes of the expressions on the right side of that last equation are equal – the sum is zero, because of the opposite signs. Away from this magic frequency (the "resonant frequency"), either cap or inductor dominates. Can you see all this on the phasor diagram?

Better filters

Having looked hard at *RC* filters, maybe we should admit that you can make a filter better than the simple *RC* when you need one. You can do it by combining an inductor with a capacitor, or by using operational amplifiers (coming soon) to obviate use of the inductor. We will try that method in a later lab. Figure 2N.34 shows a scope image showing three *RC* filters' frequency responses. Two of the three filters improve on the simple *RC* lowpass that we meet today.

AoE §6.2.2

By the way: "sweeping" frequencies for a scope display: The scope image in Fig. 2N.34 uses *frequency* for the horizontal axis – not *time*, as in the usual scope display. You may choose to use this technique in Lab 2L, and certainly will want to use it in Lab 3L (where it provides a pretty display of the dramatic response of the *LC resonant circuit*). The method is detailed in §2S.3 on p. 97. All we need say here is that at the left-hand edge of the scope display frequency is close to zero; the frequency climbs linearly to the right-hand extreme (where in this case the frequency reaches about 3.8 kHz). $f_{3\text{dB}}$ for all three filters is about 1 kHz.

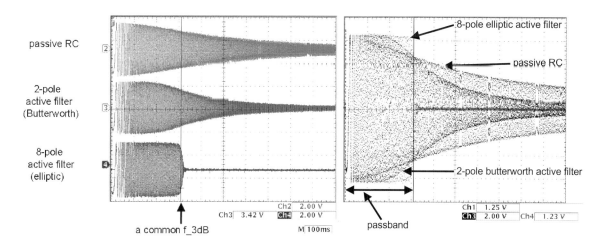

Figure 2N.34 Frequency response of three filters: improvements on the simple *RC*. Some things to look forward to. Left panel shows wider frequecy range; right panel shows more passband detail.

78   RC Circuits

1. The top trace in Fig. 2N.34 shows a passive *RC* filter. You can see that it's pretty droopy where it ought to be passing signals, and only very gradually "rolls off" (attenuates) signals at increasing frequencies.
2. The middle trace shows an "active" 2-pole[12] *RC* filter. Its frequency response improves on that of a passive filter (even a 2-pole passive filter) by some clever use of an amplifier and feedback (at the moment, the term "feedback" may sound pretty mysterious, to you. It will remain so until perhaps Lab 6L). This "active" filter improves on the passive *RC* in three senses: the "passband" is more nearly flat than for the passive filter, the rolloff (the way the output shrinks as frequency climbs beyond the "passband") is steeper, and the ultimate slope is steeper as well ($\propto 1/f^2$ rather than $\propto 1/f$). You will meet this filter in Lab 9L.
3. The bottom trace shows a fancier "active filter." This is an "8-pole" filter (like eight simple *RC* lowpass filters cascaded, but optimized so that it performs much better than a simple cascade of like filters would). You will meet this filter in Lab 12L. Its rolloff is, as you can see, spectacularly steep.

We don't want to make you a filter snob, though – and don't want to make you look down on the simple *RC*. It would not be in the spirit of this course to make you pine after beautiful transfer functions. Nearly always, a simple *RC* filter is good enough for our purposes. It is nice to know, however, that in the exceptional case where you need a better filter, such a circuit is available.

2N.4   Two unglamorous but important cap applications: "blocking" and "decoupling"

In part, we mention these to introduce some jargon that otherwise might bother you: the names "blocking capacitor" and "decoupling capacitor" might, at first hearing, suggest that these are peculiar, specialized types of capacitor. The names do not mean that. The names simply refer to the application to which the capacitor is put (this usage is reminiscent of the "load" resistor that appeared in Chapter 1N: not a special sort of resistor; just a resistor *considered as* nothing more than a load to the circuit under discussion).

2N.4.1   *Blocking* capacitor

A capacitor so used is included for the purpose of *blocking* a DC voltage, while passing a wiggle (an "AC" signal). In Fig. 2N.35, for example, the capacitor blocks the DC voltage of 10 V, permitting use of an input signal centered on ground.

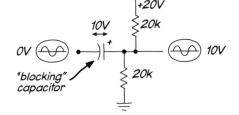

Figure 2N.35 A capacitor used to pass wiggles while blocking DC (permitting differing DC levels) is said to be a "blocking" cap.

You may be inclined to protest that the capacitor doing this job is also a *passing* capacitor. Yes, but we're accustomed to such usage (a *resistor* is also a *conductor*) – and the blocking capacitor differs from a piece of wire not in its passing but in its blocking powers.

---
[12] "2-pole" is jargon meaning, approximately, "behaves like two ordinary *RC*'s cascaded (that is, placed in series)."

You may further protest that you recognize this combination of cap and resistors as nothing new: it is a *highpass filter*, with the relevant R just R_{Thev} for the divider. Yes, you're right: it does behave like a highpass.

But it makes sense to call the cap *blocking* if the goal is to permit the peaceful coexistence of differing DC levels – that is, to block DC. This motive is not the same as the usual for a highpass, which is to attenuate low-frequency noise.

When blocking is the motive, the choice of $f_{3\text{dB}}$ becomes not critical: just make $f_{3\text{dB}}$ low enough so that all frequencies that you consider *signals* can pass. A huge C is OK, when the motive is only to block DC. No value is too large – though it may be foolish to choose a cap larger than needed, because large C values cost more and take more space than smaller ones, are often polarized, and are slower to settle.

A familiar case: scope input's "AC coupling"

When you switch the scope input to "AC" (as we suggest you ought to do only rarely), you put a blocking capacitor into the signal path. DC is blocked – and DC information is lost. That is why we advocate using DC coupling (which you can think of as "direct coupling") except in the exceptional case when you need to wash away a large DC offset in order to look at a small wiggle at high gain (a gain that would send a DC-coupled signal sailing off the screen).

2N.4.2 Decoupling or bypass capacitor

A very humble, but important application for a capacitor is just to minimize power-supply wiggles. Power supplies are not perfect, and the lines that link the supply to a circuit include some inductance (and some resistance, though the inductance is usually more troublesome).

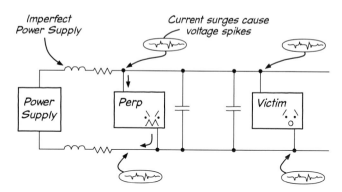

Figure 2N.36 A capacitor used to quiet a power supply is said to "decouple" one part of a circuit from another.

If one part of a circuit draws surges of current, those surges flowing through the stray inductance of the supply lines will cause voltage jumps. Those supply disturbances can "couple" between one part of the circuit (the "perp" or perpetrator, in Fig. 2N.36) and another (the "victim").

Decoupling capacitors tend to stabilize the power supply voltage. Such a capacitor provides a local source of charge for the *perp* – the circuit that needs a sudden surge of current – so not all of the current surge needs to flow in the longer path from the supply. A decoupling or "bypass" cap near the *victim* circuit simply tends to hold that supply voltage more nearly constant. In Fig. 2N.37, a digital oscillator performed the role of *perp*; about 6 inches distant, at the far end of a breadboard, the scope watched the power supply near the *victim*.

Small capacitors help a good deal ($0.1\mu\text{F}$). The cheapest, crummiest capacitors, the ceramics scorned for other purposes, work fine; in fact, they work a little better than the more expensive mylars that we prefer for filters, as you can see in Fig. 2N.37. Often a large capacitor (say, a $4.7\mu\text{F}$ tantalum) is paralleled with a small ceramic (0.01 to $0.1\mu\text{F}$).

Always the best general defense against such trouble is to provide heavy power supply lines or

80 RC Circuits

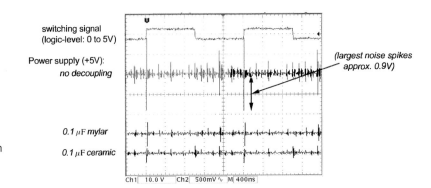

Figure 2N.37 Small capacitors between power supply and ground ("bypass") can stabilize the supply voltage.

"traces" on a printed-circuit board, with their low inductance. On high-quality printed circuits, entire layers are dedicated to ground and supply "planes." But even after doing this, a wise designer invariably adds decoupling caps (sometimes as many as a few dozen).

The use of a capacitor to "bypass" a circuit element is more general than the *decoupling* application we have just looked at. You will see a resistor "bypassed" by a capacitor, for example, in the high-gain amplifier of §5N.5.2. You will see a similar application in the "split feedback" of Chapter 9N, for example in the amplifier of Fig. 9L.8.

2N.5 A somewhat mathy view of RC filters

In this course, we do what we can to dodge mathematical explanations for circuit behavior. Often, math doesn't help intuition. But here we propose to see if a little math may help to make sense of the frequency response of an *RC* circuit. If this does not help, ignore it.

A highpass filter as voltage divider: Figure 2N.38 shows the highpass and its response.

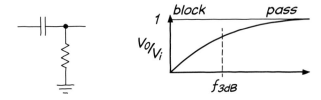

Figure 2N.38 Highpass filter.

And, treating it as a voltage divider, here is the fraction the output delivers:

$$\frac{V_{out}}{V_{in}} = \frac{R}{R + Z_C} = \frac{1}{1 + (Z_C/R)} = \frac{1}{1 - j(1/\omega RC)}$$

Let's try looking at what this expression is trying (perhaps not very clearly) to tell us.

The interesting term, here, is $-j(1/\omega RC)$. Depending on frequency, this term can take a magnitude that is...

- tiny,
- huge, or
- unity.

These are not the only possibilities, but these three define three distinct regions in the filter's behavior – the ones labeled "block," "f_{3dB}," and "pass" in Fig. 2N.38. Let's consider these regions, one at a time.

- If $-j(1/\omega RC)$ is *tiny*, which means $\omega \gg 1/RC$, then we're well above f_{3dB} and $V_{out}/V_{in} \approx 1$. This is the "pass region." The near absence of j from the expression (because its coefficient is tiny) says, "No phase shift." This fits your experience in Lab 2L, experience that says phase shift in an *RC* filter goes with attenuation.
- If $-j(1/\omega RC)$ has value unity, which means $\omega = 1/RC$, then we're at f_{3dB} and $V_{out}/V_{in} = \frac{1}{1-j}$. This has magnitude $\frac{1}{\sqrt{2}}$, more usually described as "−3 dB."[13] The presence of j in the expression promises phase shift. A phasor diagram like the one back in Fig. 2N.31 may be the best way to illustrate why that phase shift is 45°. The phase shift is the angle between the hypotenuse (V_{in}) and the voltage across R. The latter voltage "leads" V_{in} by 45°. Figure 2N.39 shows a short form of that figure again.

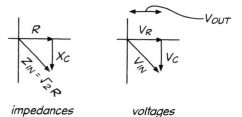

Figure 2N.39 Phasor diagram explains *geometrically* the *RC*'s 45° phase shift at f_{3dB}.

- If $-j(1/\omega RC)$ is *huge*, which means $\omega \ll 1/RC$, then we're well below f_{3dB} and $V_{out}/V_{in} \approx \frac{1}{-j(1/\omega RC)} = j\omega RC$. This expression says a lot: "j" says "+90° shift;" "ω" says that as frequency grows amplitude grows proportionately.

2N.6 AoE reading

Readings:

AoE: Chapter 1.4–1.7.12 but *omit* §§1.5–1.6.8, and §1.7.2 about inductors, transformers, and diodes (we'll reach those next time).

Treat as *optional* §§1.7.3–1.7.9, where complex impedances are presented. This patch is heavy on math, and out of character for this course. These sections have scared many students, who later learned that they can manage perfectly well in this course without such rigor. This mathematical treatment is not characteristic even of AoE, and *very* unlike the treatments we use in *Learning the Art of Electronics* . . . .

Appendix on Math Review. (World's shortest calculus course, if you feel you want such a refresher. But in any event, you'll survive this course without it).

Exercises:

Exercises in text.
Additional Exercises 1.39–1.42.

---

[13] You can get this result by multiplying $\frac{1}{1-j}$ by its "complex conjugate," $\frac{1}{1+j}$. Their product gives a magnitude squared that is real and equals 1/2. Taking its square root, we get $\frac{1}{\sqrt{2}}$, with the familiar value 0.7 or "−3 dB."

2L Lab: Capacitors

2L.1 Time-domain view

2L.1.1 *RC* circuit: time-constant

Here's another try in our continuing effort to make your labs more exciting – more *suspenseful*: we ask you to *do the first exercise* (measuring the *RC* "time-constant") with *unknown R, C values*. To make this game possible, we engaged the great Christo to *wrap* an *R*, *C* pair. The *R* is the skinny object; the *C* is the chubbier device. See Fig. 2N.6, which illustrates exponential charging and discharging.

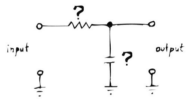

Figure 2L.1 *RC* circuit: step response.

Measure RC: Drive the circuit with a square wave at 500 Hz or less, and look at the output. Adjust frequency so as to get a useful image: too high, and you won't allow time enough to see the waveform move far; too low, and you'll see the full waveform, but using just a small portion of the scope screen, and thus your time measurements will be only approximate. In Fig. 2L.2 is a scope image suggesting both possible errors:

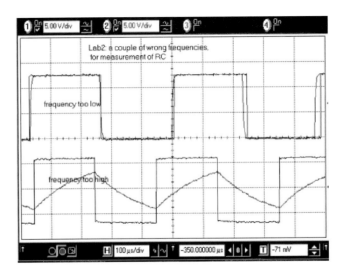

Figure 2L.2 A couple of wrong frequencies, for measurement of *RC*.

2L.1 Time-domain view

Be sure to use the scope's *DC* input setting, even though this is a time-varying waveform. (Remember the warning about the AC setting, last time?)

You will have no trouble determining *RC*. Measure the time constant by determining the time for the output to drop to 37% ($=\frac{1}{e}$).

Suggestion The *percent* markings over at the left edge of the scope screen are made-to-order for this task: put the foot of the square wave on 0%, the top on 100%. Then crank up the sweep rate so that you use most of the screen for the fall from 100% to around 37%.

Measure the time to climb from 0% to 63%. Is it the same as the time to fall to 37%? (If not, something is amiss in your way of taking these readings!)

Try varying the frequency of the square wave.

Deduce R and C values: You would have no trouble determining R if we allowed you to use an ohmmeter, but we don't allow that. See if, instead, you can use what you know of the limiting values of the *RC* circuit's *input impedance* to discover R, experimentally. Then you can solve for C.

In case this advice seems a little cryptic, here are some hints.

Hints:

- First, try to determine R – despite the fact that C is also present. Form a voltage divider with a known resistor ahead of the *RC* circuit. We suggest you start with R_{test}=1k. Apply a sinewave.

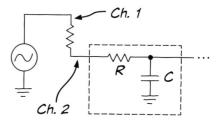

Figure 2L.3 Test setup to use R_{in} to reveal R.

Queries:
- How will you know that it is the effect of R that you are observing, rather than some combination of R and X_C? (*Hint:* do you see a phase shift between the waveforms on Channels 1 and 2?) *Note:* keep f_{in} under about 1 MHz, so as not to complicate your search with the effect of the BNC cable's capacitance to ground. That capacitance – about 30 pF/foot – becomes important at high frequencies; it forms a lowpass with R_{Thev} at the point marked "Ch. 2" in Fig. 2L.3. So, as you push the frequency high, looking for the disappearance of phase shift, you will be frustrated if you go too far!
- Once you have eliminated the pesky phase shift, what should you assume has happened to the value of X_C?
- Once you have a value for R, you're about done.

2L.1.2 Differentiator

Construct the *RC* differentiator shown in Fig. 2L.4. Drive it with a square wave at 100 kHz, using the function generator with its attenuator set to 20 dB. Does the output make sense? Try a 100 kHz triangle wave. Try a sine.

84 Lab: Capacitors

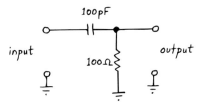

Figure 2L.4 *RC* differentiator.

Input impedance: Here's another chance to get used to quick *worst-case* impedance calculations, rather than exact and frequency-dependent calculations (which often are almost useless).

What is the impedance presented to the signal generator by the circuit (assume no load at the circuit's output)...

- ...at $f = 0$?
- ...at infinite frequency?

Questions like this become important when the signal source is less ideal than the function generators you are using.

2L.1.3   Integrator

Construct the integrator shown in Fig. 2L.5. Drive it with a 100 kHz square wave at maximum output level (attenuator set at 0 dB).

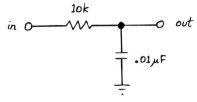

Figure 2L.5 *RC* integrator.

What is the input impedance at DC? At infinite frequency? Drive it with a triangle wave; what is the output waveform called? (Doesn't this circuit seem clever? Doesn't it remember its elementary calculus better than you do – or at least faster?)

To expose this as only an *approximate* or conditional integrator, try reducing the input frequency. Are we violating the stated condition (§2N.2.1):

$$V_{out} \ll V_{in}?$$

The differentiator is similarly approximate, and fails unless (§2N.2.1):

$$dV_{out}/dt \ll dV_{in}/dt?$$

Too large an *RC* tends to violate this restriction. If you are extra zealous you may want to look again at the differentiator of §2L.1.2 but this time increasing *RC* by a factor of, say, 1000. The "derivative" of the square wave gets ugly, and this will not surprise you; the derivative of the triangle looks odd in a less obvious way.

When we meet *operational amplifiers* in Chapter 3N, we will see how to make "perfect" differentiators and integrators – those that let us lift the restrictions we have imposed on these *RC* versions.

2L.2 Frequency-domain view

2L.2.1 Lowpass filter

Construct the lowpass filter[1] shown in Fig. 2L.6.

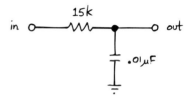

Figure 2L.6 *RC* lowpass filter.

What do you calculate to be the filter's −3 dB frequency? Drive the circuit with a sinewave, sweeping over a large frequency range, to observe its lowpass property; the 1 kHz and 10 kHz ranges should be most useful.

Find f_{3dB} *experimentally*: measure the frequency at which the filter attenuates by 3 dB (V_{out} down to 70.7% of full amplitude).

Note Henceforth we will refer to "the 3 dB point" and "f_{3dB}," not to the *minus* 3 dB point, nor f_{-3dB}. This usage is confusing but conventional; you might as well start getting used to it.

What is the limiting phase shift, both at very low frequencies and at very high frequencies?

Suggestion As you measure phase shift, use the function generator's SYNC or TTL output to drive the scope's External Trigger. That will define the input phase cleanly. Then if you are using an analog oscilloscope, use its *continuously-variable* sweep rate[2] so as to make a full period of the input waveform use exactly eight major divisions (or eight centimeters). The output signal, viewed at the same time, should reveal its phase shift readily; see Fig. 2L.7.

Check to see if the lowpass filter attenuates 6 dB/octave for frequencies well above the −3 dB point; in particular, measure the output at 10 and 20 times f_{3dB}. While you're at it, look at phase shift versus frequency; what is the phase shift for

$$f \ll f_{3dB},$$
$$f = f_{3dB},$$
$$f \gg f_{3dB}?$$

Finally, measure the attenuation at $f = 2f_{3dB}$ and write down the attenuation figures at $f = 2f_{3dB}$, $f = 4f_{3dB}$ and $f = 10f_{3dB}$.

---

[1] Aside: *Integrator* versus *Lowpass Filter*. "Wait a minute!," you may be protesting, "Didn't I just build this circuit?" Yes, you did. Then why do it again? We expect that you will gradually divine the answer to that question as you work your way through this experiment. One of the two experiments might be called a special case of the other. When you finish, try to determine which is which.

[2] On most scopes you'll invoke this by turning a little red knob on the larger *sweep rate* knob: when the red knob is turned counter-clockwise, it comes out of a clicked "detent" position, usually labeled "CAL." Once you've done that, the scope screen no longer is usable to read *time*. So, don't leave it that way when you finish your *phase shift* measurement!

86 Lab: Capacitors

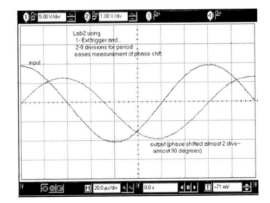

Figure 2L.7 It's easy to estimate phase shifts if you make a full *period* equal eight divisions.

> **Sweeping frequencies** This circuit is a good one to look at with the function generator's **sweep** feature. This will let your scope draw you a plot of amplitude versus *frequency* instead of amplitude versus *time* as usual. If you have a little extra time, we recommend this exercise. If you feel pressed for time, save this task for next time, when the *LC* resonant circuit offers you another good target for sweeping.
>
> You may want to look at §2S.3, our more detailed note on *sweeping*, but here is the strategy in brief.
>
> In order to generate such a display of V_{out} versus frequency, let the generator's *ramp* output drive the scope's horizontal deflection, with the scope in "X–Y" mode: in X–Y, the scope ignores its internal horizontal deflection ramp (or "timebase") and instead lets the input labeled "X" determine the spot's horizontal position.
>
> The function generator's **ramp** time control now will determine sweep rate. Keep the ramp *slow*: a slow ramp produces a scope image that is annoyingly intermittent, but gives the truest, prettiest picture, since the slow ramp allows more cycles in a given frequency range than are permitted by a faster ramp.

2L.2.2 Highpass filter

Construct a highpass filter with the components that you used for the lowpass. Where is this circuit's 3dB point? Check out how the circuit treats sinewaves: check to see if the output amplitude at low frequencies (well below the −3 dB point) is proportional to frequency. What is the limiting phase shift, both at very low frequencies and at very high frequencies?

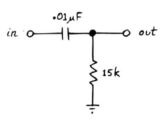

Figure 2L.8 *RC* highpass filter.

2L.2.3 Filter application I: garbage detector

The circuit in Fig. 2L.9 will let you see the "garbage" on the 110 V power line. First look at the output of the transformer, at **A**. It should look more or less like a classical sinewave. (The transformer, incidentally, serves two purposes – it reduces the 110 V AC to a more reasonable 6.3 V, and it "isolates" the circuit we're working on from the potentially lethal power line voltage.)

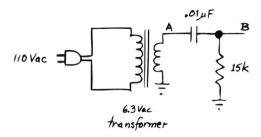

Figure 2L.9 Highpass filter applied to the 60 Hz AC power.

To see glitches and wiggles, look at **B**, the output of the highpass filter. All kinds of interesting stuff should appear, some of it curiously time-dependent. What is the filter's attenuation at 60 Hz? (No complex arithmetic necessary. *Hint:* count octaves, or use the fact – which you confirmed just above – that amplitude grows linearly with frequency, well below f_{3dB}.)

2L.2.4 Design: filter application II: selecting signal from signal-plus-noise

Now we will try using highpass and then lowpass filters to prefer one frequency range or the other in a composite signal, formed as shown in Fig. 2L.10. The transformer adds a large 60 Hz sinewave (peak value about 10 V) to the output of the function generator. Set the function generator frequency, initially, to around 10 kHz.[3]

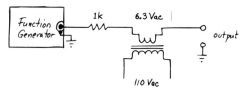

Figure 2L.10 Composite signal consisting of two sinewaves.

In order to choose the R value for your filter, you will need to determine the value of the output impedance[4] for the signal source you have constructed (function generator plus transformer). The function generator's R_{out} is 50 Ω; the series impedance of the transformer winding is negligible at the frequencies of interest to us; the 1k resistor is included, incidentally, to protect the function generator in case the composite output accidentally is shorted to ground.

First design – highpass: Design a highpass filter that will keep most of the "signal" and get rid

---

[3] If you have a two-channel function generator that supports adding the channels together, you can skip the transformer and just set your function generator add a 10 kHz signal to one at 60 Hz.

[4] We've been calling this Z_{out}, a name that certainly is not wrong – but this characteristic is not frequency-dependent, so it will be just as correct to call it R_{out}.

of most of the 60 Hz "noise." Assume that the frequency of what you consider "signals" may range between about 2 kHz and 20 kHz. As you design, consider:

- what is an appropriate f_{3dB}?
- what Z_{in} is appropriate for your filter?

Run the composite waveform ("signal" plus "noise") through your highpass filter.

Highpass filter (your design): Do you like the output of your filter? Is the attenuation of the 60 Hz waveform about what you would expect? (As you will gather gradually, the 60 Hz power lines are the most common and troublesome source of noise in the lab. This junk is often called "line noise.")

Second design: lowpass: Now let's change assumptions: let's suppose that we consider the 60 Hz "signal," and the function generator's 10 kHz "noise."

Design a lowpass filter that will keep most of the "signal" and get rid of most of the "noise."

Lowpass filter (your design): Now run the composite signal through your lowpass filter, and see if you like the result. If not, fix your design!

2S Supplementary Notes: *RC* Circuits

2S.1 Reading capacitors

2S.1.1 Why you may need this note

AoE §1.4.1

Most students learn pretty quickly to read resistor values. They tend to have more trouble finding, say, a 100 pF capacitor.

That's not their fault. They have trouble, as you will agree when you have finished reading this note, because the cap[1] manufacturers don't want them to be able to read cap values. The cap markings have been designed by an international committee to be nearly unintelligible. With a few hints, however, you can learn to read cap markings, despite the manufacturers' efforts. Here are our hints.

2S.1.2 Big caps: electrolytics

These are easy to read, because there is room to write the value on the cap, including units. "16 V" is the maximum voltage one can safely put across the capacitor without damaging the part. The *minus* sign and arrow indicate which is the negative terminal for this *polarized* capacitor.

Figure 2S.1 A big cap is labeled intelligibly.

Caps of 1μF and above usually are *polarized*

All of these big caps are *polarized* ("big," in our world, means 1 μF or more). That means the capacitor's innards are not symmetrical, and that you may destroy the cap if you apply the wrong *DC* polarity to the terminals:[2] the terminal marked + (or the one *not* marked "−," as in the capacitor of Fig. 2S.1) must be at least as positive as the other terminal.[3] (Sometimes, violating this rule will generate gas that makes the cap blow up; more often, you will find the cap internally shorted, after a while. Often, you could get away with violating this rule, at low voltages. But don't try.)

One of our students brought a trophy to class after inadvertently wiring a polarized cap backwards.[4]

If caps wired backwards always reacted so spectacularly, they would never have a chance to subvert our circuits. Unfortunately, a reversed cap usually turns quietly into a short circuit, making your circuit behave weirdly while offering no clue to what is wrong. (The goo that looks like spilled guts of the cap in Fig. 2S.2, incidentally, is not guts, but only glue that was used to fix the cap's remains to a piece of paper, for display.)

---

[1] "Cap" is shorthand for "capacitor," as you probably know.
[2] This is a rule for *DC*, sustained levels; a short-term swing, not long enough to develop a substantial reverse voltage across the capacitor, is harmless.
[3] They are not symmetrical because the dielectric is a thin oxide layer on the surface of one metal electrode; the other terminal of the capacitor is immersed in goo that forms the second electrode – in case that primitive explanation is of any help.
[4] Image thanks to Rick Montesanti, who told us proudly that the cap announced the error with a *bang* and fired junk twenty feet.

Figure 2S.2 Polarized cap that was hooked up with reversed DC voltage.

2S.1.3 Pretty big caps: tantalum

As the caps get smaller, the difficulty in reading their markings begins. Tantalums, which can provide quite large capacitance, often are physically large enough to be easy to read.

The tantalums often are silver colored cylinders. They are polarized, like the big electrolytics. A + marks the positive end, on the tantalum cap pictured.[5] The tantalum cap in Fig. 2S.3 uses the relatively ample surface area of its packages to state its value twice.

The "4R7" means pretty much what it says, if you know that the "R" (radix point) marks the decimal place: it's a 4.7 μF cap, and it can stand 50 V without damage.

The second marking, "475M", under "CS13B" (which is the capacitor type), states the value in exponential form, as if this were a resistor: 47×10^5 M. What's "M"? Microfarads? (Surely we're not to take the capital seriously: Megafarads?!)

Figure 2S.3 Tantalum cap.

But we must resist the plausible assumption that "M" is a unit. It is not. It indicates *tolerance*, instead: ($\pm 20\%$). (Wasn't that nasty of the labelers to choose "M?" Guess what's another favorite letter for tolerance. That's right: K, which looks like "kilo," but instead means $\pm 10\%$. Pretty mean!)

What units? If "M" is not a unit but a tolerance marking, what are the units? 10^5 *what*? 10^5 of *something small*. You will meet this question repeatedly, and you must resolve it by relying on a few observations. The only units commonly used in the US in capacitor labeling are

- microfarads: 10^{-6} Farad,
- picofarads: 10^{-12} Farad.

An intermediate unit, the nanofarad, nF: 10^{-9} Farad, is respectable in circuit diagrams, but fortunately not used to label capacitors. We say "fortunately" because we rely on the huge difference between "nF" and "pF" that allows us to guess which unit is intended, in a capacitor label.

The label "mF" (10^{-3} Farad) looks reasonable but is not used; instead, capacitors in this size range are measured in thousands of microfarads. For example, a cap will be described as "4,000 μF" rather than "4 mF." Strange, but dictated by tradition.

A Farad is a huge unit. The biggest cap you will use in this course is a few hundred μF. Such a cap is physically large. (We do keep a 1 *Farad* cap around, but only for our freak show.)[6]

So, if you find a little cap labeled "680," you know it's 680 **pF**.[7]

---

[5] For this particular series of caps, a metal nipple also marks the positive end. The nipple is much easier to spot than the small "+" mark, which is quite invisible if the cap is not rotated just right – but unfortunately there is only rarely any consistency linking the nipple and positive end.

[6] These giant caps are used as replacements for rechargeable batteries in calculators, portable computers, even flashing bike lights. Their advantage over batteries is long life: since their charge and discharge involve no chemical reactions, they should tolerate a great number of charge/discharge cycles.

[7] Except, of course, when it's *not* – that is, when the labeling scheme is exponential; see §2S.1.5.

2S.1 Reading capacitors

A picofarad is a tiny unit. You will not see a cap as small as 1 pF in this course.[8] So, if you find a cap claiming that it is a fraction of some unstated unit – say ".01" – the unit is μFs: here ".01" means 0.01 μF.

Beware the wrong assumption that a *picofarad* is only a bit smaller than a microfarad. A pF is *not* 10^{-9} F (i.e., 10^{-3} μF); instead, it is 10^{-12} F: a *million* times smaller than a microfarad.

So, we conclude, this cap labeled "475" must be 4.7×10^6 picofarads. That, you will recognize, is a roundabout way to say

$$4.7 \times 10^{-6} \text{ F}$$

We knew that was the answer, before we started this last decoding effort. This way of labeling is indeed roundabout, but at least it is unambiguous. It would be nice to see it used more widely. You will see another example of this *exponential* labeling in the case of the CK05 ceramics in §2S.1.5.

Figure 2S.4 Mylar: still big enough to be labeled clearly.

2S.1.4 A little smaller: polyester (mylar)

These are yellow cylinders, pretty clearly marked; see Fig. 2S.4.

Here " .01k" is just 0.01 μF, of course; by now you are sophisticated enough to know that "k" does not mean "kilo." Instead, it is another *tolerance* label, like "M" on the 4.7 μF cap above. The label "k" just means ±10%. These caps are not polarized; the black band marks the outer end of the foil winding. We don't worry about that fine point. Orient them at random in your circuits.

Figure 2S.5 Mylar ("polyester") cap, unrolled, reveals that it's a metal-foil coil, layers separated by mylar film.

2S.1.5 Small enough to be ambiguous: ceramic

Ceramics are the proletarians of the cap world: hard workers, not refined, not predictable, not precise, but good workers in difficult conditions.[9] They are used most often to "decouple" power supplies – that is, to stabilize the supplies with respect to ground. These ceramics are little rectangular boxes, or (the older style) little orange pancakes, and are used by the electronics industry in colossal numbers.[10] They act like capacitors even at high frequencies. The trick, in reading these, is to reject the markings that can't be units. As we'll see in a minute, we also need to be aware of the possibility of a non-standard labeling scheme.

CK05: These are little boxes, with their leads 0.2 inches apart. They are handy, therefore, for insertion into a printed circuit (or *were* handy, before the tiny *surface mount* parts took over). Figure 2S.6 shows the two sides of a CK05-packaged part.

Let's find the value, throwing out what doesn't interest us, along the way:

- 200 V: just what it sounds like: the maximum voltage one can safely put across the cap;
- K: this time not a tolerance but indicating the manufacturer (Kemet); and
- 9902: not the value, but the week of the part's *birthday*: week 2 of [19]99.[11]

So far, no value information. That must lie on the other side. There we see:

---

[8] There's good reason for that: stray capacitance in our breadboarded circuits is on this same scale. The stray capacitance between two adjacent columns on our plastic breadboards is about 2 pF. The input capacitance of a "10×" oscilloscope probe (a device you will meet in Lab 3L) is about 10 pF. The capacitance of three feet of BNC cable is about 90 pF.
[9] As AoE teaches, some ceramics perform well: the C0G (class-I) type.
[10] Wikipedia, https://LAoE.link//Ceramic_Capacitor.html, estimates annual production at around one trillion.
[11] Important for sentimental engineers who celebrate with all their components.

Figure 2S.6 CKO5 capacitor markings.

- CKO5, and on the next line...
- ...BX. These simply name the type of package, and
- 680k: here is the value, at last: 680 is in exponential notation; K is the tolerance, once more

So, this is a cap of 68pF (68×10^0 of something small, picofarads)

Possible ambiguity: old-style disc capacitors

Occasionally, you may run into a ceramic capacitor labeled acording to an older convention. Such *disk* capacitors come in relatively-large packages. In Fig. 2S.7 we show one of these, alongside a CK05 part.

Figure 2S.7 Some old disk caps are labeled using a non-exponential convention.

Having met and understood the CK05 labeling, one would assume that both these caps were 56 pF parts. The one on the right *is* a 56 pF cap. But the one on the left – in the older disc package – is a 560 pF part. Nasty, but true. The comforting thought, though, is that this disc package and its labeling scheme are nearly obsolete.

2S.1.6 Caps too small to label intelligibly: the coming thing!

We have shown examples of cap markings you are likely to run into as you do labs in this course. But when you start building printed circuits, you are likely to use the smaller *surface mount* packages. Only the largest of these even attempt labeling. The smaller ones are blank; it's up to you to recall which bin you took each from. If you get confused, all you can do is reach for another, from a properly-labeled package or drawer.

Surface mount cap: large enough for labeling

The cap in Fig. 2S.8 is of large enough value ($4.7\,\mu F$) that it needs a package that is large by surface-mount standards. So, though it is small, it does show labeling.

The enlarged view in Fig. 2S.9 shows the "475K" that we would expect for a $4.7\,\mu F$ cap.

2S.1 Reading capacitors

Figure 2S.8 Surface mount cap large enough to show labeling.

Figure 2S.9 Closer view of labeled SMT cap: "475K" states value.

Figure 2S.10 Smaller surface mount packages altogether give up the attempt at labeling.

Surface mount caps too small for labeling

As the surface-mount packages get smaller, however, they cease even to try to announce their values. Figure 2S.10 shows some caps that dare not speak their names.

2S.1.7 Tolerance codes

Just to be thorough – and because this information is hard to come by – we've listed below all the tolerance codes. These apply both to capacitors and resistors; the tight tolerances are relevant only to resistors; the strangely-asymmetric tolerance is used only for capacitors.

Supplementary Notes: *RC* Circuits

Tolerance Code	Meaning	
Z	+80%, −20%	(for big filter capacitors, where you are assumed to have asymmetric worries: too small a cap allows excessive "ripple;" more on this in Lab 3L and Chapter 3N)
M	±20%	
K	±10%	
J	±5%	
G	±2%	
F	±1%	
D	±0.5%	
C	±0.25%	
B	±0.1%	
A	±0.05%	
Z	±0.025%	(precision resistors; context will show the asymmetric cap tolerance "Z" makes no sense here)
N	±0.02%	

2S.2 *C* notes: trying for an intuitive grip on capacitors' behavior

2S.2.1 *C*s differ from *R*s, of course

This note tries some very simple arguments, in the hope that they may help you to strengthen your intuitive grip on the behavior of capacitors: not the sort of grip provided by $Z_C = -j/\omega C$, but the sort of grip that might be provided by a mental picture of the way voltages change on the two plates of the cap, as an input is applied.

Resistors live in the present: capacitors, in contrast, are fundamentally conservative: see Fig. 2S.11.

The voltage across a cap can't change abruptly, and *RC* circuits may be much influenced by what happened a while ago.[12]

Let's look for some intuition that will let us *expect* caps to behave as they do.

2S.2.2 Lowpass filter

A lowpass attenuates high frequencies If the input changes fast, the voltage on the capacitor cannot keep up with changes at the input, as we can see from Fig. 2S.12.

The output (V_{cap}) moves toward the input voltage – but then the fickle input quickly changes its level, and V_{cap} never quite catches up. So, only a diminished sketch of the input waveform reaches the filter's output.

A lowpass passes low frequencies If you are patient (or, more exactly, if the waveform is patient) and give the capacitor time to charge or discharge almost to the level of the applied voltage, then the input signal does reach the output: the *low* frequency is *passed*; see Fig. 2S.13.

---

[12] As you'll soon see (if you haven't noticed it on your own), the characterization as "conservative" applies most clearly to the *lowpass* configuration.

2S.2 *C* notes: trying for an intuitive grip on capacitors' behavior

Figure 2S.11 Top: Hippy Resistor: no sense of time. Bottom: Proustian Capacitor: ...believes that the past *does* matter.

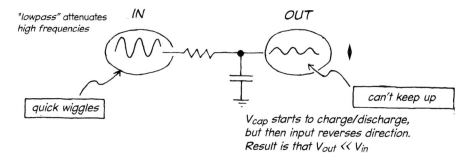

Figure 2S.12 A lowpass attenuates quick wiggles; output cannot keep up with input changes.

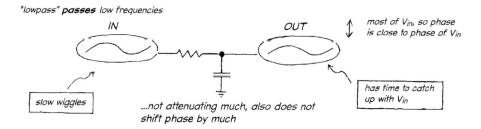

Figure 2S.13 A lowpass passes slow wiggles; output has time to catch up with input changes.

2S.2.3 Highpass

Highpass *does* transfer quick changes, from input to output

A quick edge passes: The highpass behavior is a little bit harder to get a grip on. I find that it helps to begin by thinking about the circuit in *time domain* (that is, by considering what you might see on an oscilloscope screen, at input and output) rather than in the more abstract *frequency domain* view.

Supplementary Notes: RC Circuits

Set up the *RC* in the so-called highpass configuration, and imagine applying a voltage *step* at the input, as in Fig. 2S.14.

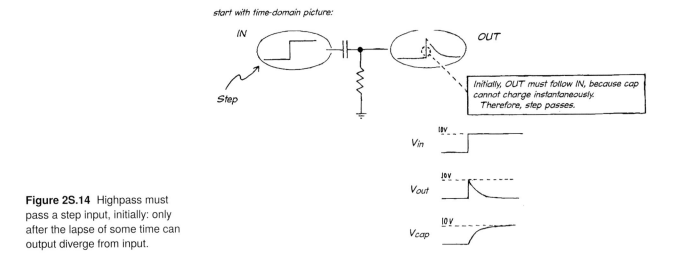

Figure 2S.14 Highpass must pass a step input, initially: only after the lapse of some time can output diverge from input.

A quick square wave passes: Repeat those quick edges, up, down, up . . . , and you are applying a *square wave*; see Fig. 2S.15. Such a waveform, like the quick edge, passes, too – so long as you don't allow it to sit still long. If you do allow it to sit high or low for a while, then you'll see the output look droopy.

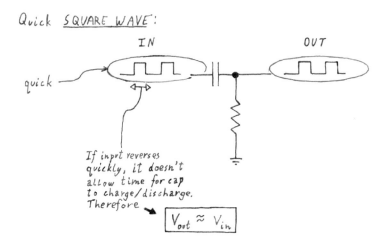

Figure 2S.15 Highpass can pass a quick square wave, too – if we don't let the input sit still for long.

A quick sinusoid also passes: A quick sinusoid[13] passes, too.[14], and for the same reasons: if we don't let the input sit anywhere for long, then the output voltage doesn't have time to move far from the input voltage. So, most of the input wiggle appears at the output (and, again, if the output *amplitude*

---

[13] Usually we don't speak so fancily – we call it a "sinewave."
[14] Fourier tells us that this isn't really a new point: what we called a "quick edge," says Fourier, can be described as a collection of sinusoids including some at *high-frequencies* So, to say, "the steep edge passes" is equivalent to saying, "the high-frequency sinusoids that form the edge pass." More on this in Lab 3L.

2S.3 Sweeping frequencies

looks like the input *amplitude*, then the output *phase* looks like the input *phase*). In other words, when you don't see attenuation, you'll not see phase shift (the same was true for the lowpass).

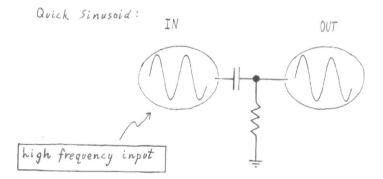

Figure 2S.16 Highpass will pass a "quick" (high-frequency) sinusoid.

...but the highpass does not pass low frequencies

No surprise, here: if we wiggle slowly, the output voltage has time to come to differ a great deal from the input voltage. So, the input can wiggle without being able to make the output wiggle much. In this case, the phase of whatever wiggle gets through will differ a lot from the phase of the input (this we haven't proven to you; but it seems plausible, does it not?).

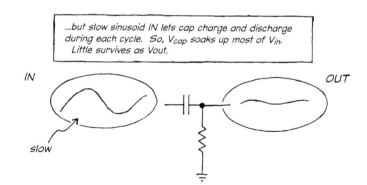

Figure 2S.17 Highpass does a poor job of passing slow wiggles; there's time for cap to charge and discharge.

2S.3 Sweeping frequencies

An oscilloscope can let one see at a glance what the frequency response of a circuit looks like. It can show amplitude versus *frequency* rather than amplitude versus *time*, as in the usual scope display.[15] All that's required is a function generator capable of "sweeping" frequencies – varying its output frequency in a regular manner. The function generators that we use in our labs can do this.

The best method for sweeping depends on whether the scope you are using is *analog* or *digital*. Our students begin with analog scopes, so our tips on sweeping frequencies also will begin by discussing these scopes.

---

[15] Strictly, even in X–Y mode, the scope shows amplitude versus time when the X-axis is driven by a repeating waveform like the sweep *ramp*. But when frequency changes with time, and the scope is properly synchronized, the horizontal axis represents *frequency* as well as *time*.

2S.3.1 Function generator's "continuous sweep" operation

We usually set the generator frequency by hand. But the generator can be set to sweep its main output frequency from a start frequency to a stop frequency.

Analog function generator

On our analog generators, this function is called "continuous sweep." When so used, the generator also puts out the *ramp* or *sawtooth* waveform that it uses to control the main generator's frequency. This waveform – from an output labeled "RAMP" on our generators – can be used to synchronize a scope display.

In Fig. 2S.18, one can see that the sinusoid frequency from the generator's output grows with the ramp voltage.

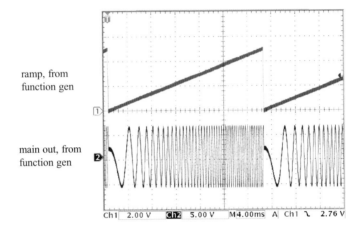

Figure 2S.18 Ramp and main outputs, with main frequency controlled by ramp voltage ("Continuous Sweep").

The *RAMP* amplitude is constant, but its repetition rate is adjustable. The frequency range at the generator's main output (a sinusoid, in Fig. 2S.18) is determined by START and STOP frequency settings. In our generators, those are set by two slider potentiometers (on the left edge of the generator front panel, in Fig. 2S.19), along with the usual *range* selection (three pushbuttons, on our generators).

2S.3.2 Swept display on analog scope, analog generator

The best way to sweep differs for analog and digital scopes, in our experience. The analog sweep is best done in so-called "X–Y" mode. The digital is best done in the usual timed-sweep mode.

X–Y display

The best method for an analog scope takes advantage of the scope's ability to substitute the Channel One input for its usual *horizontal* sweep timebase. Channel One is called the "X" input when this display mode is selected, because it controls the horizontal position of the scope's beam, and "X" is traditionally the horizontal axis of an X–Y plot used to visually compare two variables.

We illustrate the connections twice (perhaps overdoing it): first, Fig. 2S.19 showed what the setup looked like on our lab bench. Figure 2S.20 gives it as a sketch.

The signal that one applies to X is the *sawtooth* waveform available at the function-generator terminal labeled "RAMP." The left edge of the screen shows response at the START frequency, the right edge shows response at the END frequency, and the screen plots the variation between these extremes, as was evident in Fig. 2S.18.

2S.3 Sweeping frequencies

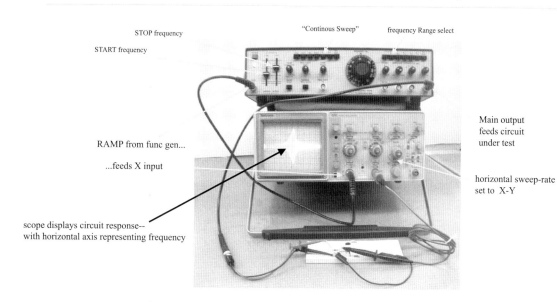

Figure 2S.19 Scope connections for a display of swept frequencies.

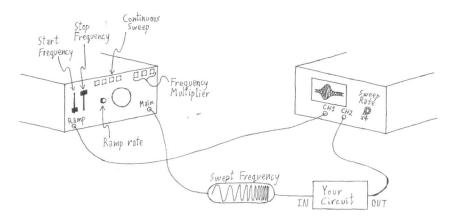

Figure 2S.20 Scope connections, for a display of swept frequencies.

Scope settings First, set up the scope:
- Set the horizontal sweep to X–Y. On the analog scopes, this is the extreme counterclockwise position for the sweep-rate knob.
- Set the Y channel (channel 2, on a 2-channel scope) to GND – temporarily.
- While feeding RAMP into the X channel, note the width of the trace: adjust X gain so that the trace just fills the screen. You need to be careful not to overfill the screen: too much gain will try to send the beam off the right-hand edge of the screen, where it will disappear or even double back. So reduce the X gain till the trace fills just half the screen (as on left, in Fig. 2S.22).
- Then double the gain, just filling the screen (as on the right in Fig. 2S.22).
- Now you can take the Y channel off GND, to DC. The scope now should show you amplitude out versus frequency.

Supplementary Notes: *RC* Circuits

Figure 2S.21 For analog scope X–Y display, set horizontal sweep to select that display.

X gain (Ch 1) too low: just half screen...

...X gain correct: fills screen

Figure 2S.22 Adjust X gain to use full screen width.

2S.3.3 Function generator settings

- Adjust the START and STOP frequencies until you see a response that looks reasonable:
  - For a lowpass or highpass filter, START should be zero frequency (or the lowest that is available). STOP you can adjust to get the display you want. At least, STOP should be well above f_{3dB}.
  - For the RLC resonant circuit of Lab 3L, you may want to START close to $f_{resonance}$ rather than at zero, to get a detailed view of the narrow passband. STOP must, of course, lie above $f_{resonance}$.
- Repetition rate:
  - in X–Y mode, you can freely vary the repetition rate (the ramp frequency) without disturbing the display's placement on the scope screen. That is why we recommend X–Y mode, rather than timed-sweep.
  - But the repetition rate does matter. If you set it too high, you will get a crude display, with too few cycles of stimulation reaching your circuit. In the case of the RLC, an excessive ramp frequency has even worse effects: it produces strange artifacts in the display. You will see bumps and depressions in the response, after $f_{resonance}$.
  - These occur because the resonant circuit continues to "ring" for a while, after it is stimulated. This ringing – persistent oscillation at $f_{resonance}$ – then interacts with the gradually changing input frequency.
    The two frequencies "beat:" sometimes the two add, reinforcing each other. In that case, a bump appears. At other times, the two frequencies subtract – oppose each other in phase. In that case, a depression appears in the response.
  - Therefore, you should use as low a ramp frequency as you can stand to watch: a low frequency repetition produces noticeable flicker on an analog scope.
    Low repetition rate does not produce flicker on a digital scope (which stores the waveform), but may try your patience by taking a long time to update the display. (At 1 s/div, for example, a screen update takes 10 seconds.) Find a comfortable compromise for the ramp rate.

2S.3 Sweeping frequencies

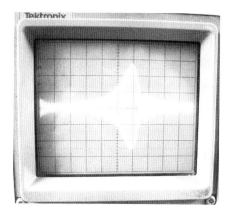

Figure 2S.23 If ramp repetition rate is too high, swept display goes bad.

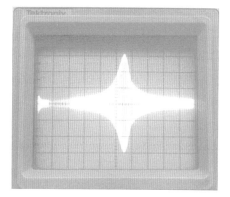

Figure 2S.24 Slow ramp repetition produces cleaner swept display.

Occasionally, timed scope sweep is better If you want to look at the response of *two* circuits, for comparison, on a two-channel scope, then X–Y is not available. You will need to use the scope's usual timed scope sweep (controlled by the "Horizontal" knob, as usual). This we ask you to do in Lab 9L, when you are to compare the frequency response of two filters: active versus passive. In the case of a digital scope, we recommend that you always use this familiar display method – as noted in §2S.3.4.

2S.3.4 Swept display on digital scope

A swept display is more straightforward, on a digital scope. There is no need to use the X–Y mode, though it is available. (We have found that it is somewhat buggy on our Tektronix TDS3014 scopes.) There is no need, because one can sweep very slowly without meeting the flicker problem that makes the adjustment of sweep on an analog scope a fussy operation, calling for continual readjustment.

So, use the usual *timed* scope sweep when using a digital scope. Use RAMP not to control the X-axis but to *trigger* the sweep. Use the falling edge, which is steep (therefore a well-defined trigger signal) and which just precedes the ramp that controls the main generator's output frequency.

The only issue that persists, from those mentioned in the notes on sweeping for analog scope (§2S.3.2), is the problem of artifacts that one can introduce by setting the ramp repetion rate too high.

Figure 2S.25 is a contrast between such a too-fast repetition rate (leftmost image) and a clean sweep, done more slowly (right-hand images). The rightmost image shows a nice wrinkle: the scope can outline the "envelope" of the frequency response for you if you choose the display option "envelope" rather than the usual "sample."

102 Supplementary Notes: *RC* Circuits

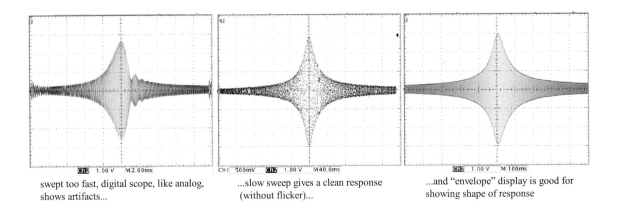

swept too fast, digital scope, like analog, shows artifacts...   ...slow sweep gives a clean response (without flicker)...   ...and "envelope" display is good for showing shape of response

Figure 2S.25 Repetition rate must be low, for clean digital display; "envelope" display suits sweep well.

2S.3.5 Swept display on digital scope, digital generator

No X–Y display, with digital function generator: *timed* sweep only

A *digital* function generator is in some ways better than analog: it allows great precision in its sweep. But it is in some ways worse: it offers no *ramp* output that can be used to drive an X–Y display. So, one is obliged to use a *timed* display. As a result, any change in function generator timing requires readjustment of the scope. You will want to use a function generator that includes a "trigger out" signal that indicates the beginning of the sweep.[16]

In Fig. 2S.26 we show the settings on an Agilent 33210A generator, and the resulting display, as we sweep the *RLC* circuit of Lab 3L. That circuit is resonant at about 16 kHz, and we have set up the generator to sweep a narrow range of frequencies, between 15.5 kHz and 16.5 kHz.

The generator includes a feature it calls "marker," generating a falling edge at a designated frequency. We have used that feature to find (visually) the peak of the *RLC*'s response, in Fig. 2S.26. Once that falling edge is lined up with the response peak, which occurs at the resonant frequency, then we can read that resonance value as the marker frequency. This frequency – about 16.09 kHz – is displayed in the lower-right corner of the figure.

This *marker* lets one find f_{resonant} approximately. But to get this value exactly one should use the method suggested in Lab 3L: watch input and output waveforms. Get close to resonance, then finely adjust f_{in} until *phases* of input and output match. When they do, you have found f_{resonant}.

Slow sweep works best

With a digital scope, since we don't need to worry about flicker, it is wise to use a slow sweep. The slow sweep prevents the ugly complication – visible in the right-hand image of Fig. 2S.27, a problem we saw back in §2S.3.3 and 2S.3.4.

The one-second sweep provides a pretty clean display (there are hints of artifacts, even here). The 30 ms sweep includes obvious beating between f_{in} and f_{resonant}, as we saw also in §2S.3.3 and Fig. 2S.23.

A stranger corruption appears as well: the seeming f_{resonant} is shifted when the circuit is swept fast. This is an illusion caused by the time required to drive the *RLC* into resonance. In the right-hand image of Fig. 2S.23 a delay of almost 1 ms appears between the time when the resonant frequency is

---

[16] When we went to equip our lab with new digital function generators, we discovered not all the ones we evaluated included this feature. It did not seem to be a matter of cost, so look for one that does.

2S.3 Sweeping frequencies 103

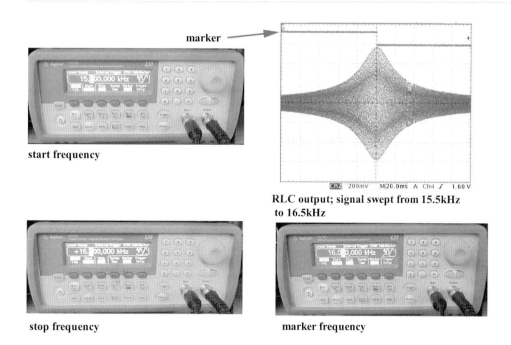

Figure 2S.26 Digital function generator permits precise sweep – but does not permit X–Y display.

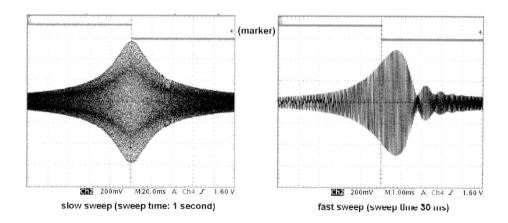

Figure 2S.27 Slow sweep provide an honest output; fast sweep corrupts the result with interactions between f_{in} and $f_{resonant}$.

applied (indicated by the falling edge of the timing signal) and the time when the RLC shows its peak V_{out}. So, to get an honest display, be patient: wait a second or so for your swept image.

2W Worked Examples: *RC* Circuits

2W.1 *RC* filters

2W.1.1 Filter to keep "signal" and reject "noise'

Problem: filter to remove fuzz

Suppose you are faced with a signal that looks like that in Fig. 2W.1: a signal of moderate frequency, polluted with some fuzz.

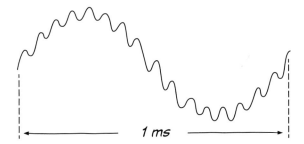

Figure 2W.1 Signal with fuzz added.

(1) Draw a *skeleton* circuit (no parts values, yet) that will keep most of the good signal, clearing away the fuzz.

(2) Now choose some values:

   (a) If the *load* has value ≥ 100k, choose R for your circuit.
   (b) Choose f_{3dB}, explaining your choice briefly.
   (c) Choose C to achieve the f_{3dB} that you chose.
   (d) By about how much does your filter attenuate the noise "fuzz"?

   What is the circuit's input impedance:

   (e) at very low frequencies?
   (f) at very high frequencies?
   (g) at f_{3dB}?

(3) What happens to the circuit output if the load has resistance 10k rather than 100k?

A solution

Skeleton circuit: You need to decide whether you want a lowpass or highpass, since the signal and noise are distinguishable by their frequencies (and are far enough apart so that you can hope to get one without the other, using the simple filters we have just learned about). Since we have called the lower frequency "good" or "signal," we need a *lowpass*: see Fig. 2W.2.

2W.1 RC filters

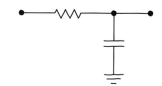

Figure 2W.2 Skeleton: just a lowpass filter.

Choose R, given the load: This dependence of R upon load follows from the observation that R of an *RC* filter defines the *worst-case* input and output impedance of the filter (see Chapter 2N). We want that output impedance low relative to the load's impedance; our rule of thumb says that 'low' means low by a factor of 10. So, we want $R \leq R_{\text{load}}/10$. In this case, that means R should be $\leq$10k. Let's use 10k.

Choose f_{3dB}***:*** This is the only part of the problem that is not entirely straightforward. We know we want to pass the low and attenuate the high, but does that mean put f_{3dB} halfway between good and bad? Does it mean put it close to good? ... close to bad? Should both good and bad be on a steeply-falling slope of the filter's response curve?

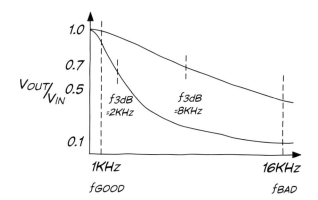

Figure 2W.3 Where should we put f_{3dB}? Some possibilities.

Assuming that our goal is to achieve a large ratio of good to bad signal, then we should not put f_{3dB} close to the noise: if we did we would not do a good job of attenuating the bad. Halfway between is only a little better. Close to signal is the best idea: we will then attenuate the bad as much as possible while keeping the good, almost untouched.

An alert person might notice that the greatest relative preference for good over bad comes when both are on the steepest part of the curve showing frequency response: in other words, put f_{3dB} so low that *both* good and bad are attenuated. This is a clever answer – but wrong, in most settings.

The trouble with that answer is that it assumes that the signal is a single frequency. Ordinarily the signal includes a range of frequencies, and it would be very bad to choose f_{3dB} somewhere *within* that range: the filter would distort signals.

So, let's put f_{3dB} at $2 \times f_{\text{signal (max)}}$: around 2 kHz. This gives us 89% of the original signal amplitude (this we claimed in Chapter 2N; you can confirm this result, if you like, with a phasor diagram or by direct calculation). Incidentally, the phase shift at that frequency is also moderate – around 25° lag. Again phasor diagram or calculation can confirm this value. At the same time we should be able to attenuate the 16 kHz noise a good deal (we'll see in a moment *how* much).

Choose C to achieve the f_{3dB} that you want: This calls for no more than plugging values into the formula for the 3 dB point:

$$f_{3dB} = \frac{1}{(2\pi RC)} \implies C = \frac{1}{(2\pi f_{3dB} R)} \approx \frac{1}{6(2 \times 10^3)(10 \times 10^3)} = \frac{1}{(120 \times 10^6)} \approx 0.008 \times 10^{-6}\,\text{F}$$

We might as well use a 0.01 μF cap. It will put our f_{3dB} about 25% low – 1.6 kHz; but our choice was a rough estimate anyway.[1]

By about how much does your filter attenuate the noise ("fuzz")? There are two quick ways to get this answer:

- Look at the ratio of f_{noise} to f_{3dB}. Attenuation will be roughly proportional, since amplitude falls off as $1/f$ (or, equivalently "–6 dB/octave").

  Here, $f_{noise}/f_{3dB} = 16\,\text{kHz}/2\,\text{kHz} = 8{:}1$. So, amplitude should be down to about 1/8 what it was at f_{3dB}. So the result is $\approx 0.125 \times 0.7 = 0.08$. Amplitude will be a bit more, though, since the slope does not reach the full $1/f$ shape till a few octaves above f_{3dB}.

- *Count octaves*: we could also say that the frequency is doubled three times ($= 2^3$) between f_{3dB} and the noise frequency. *Roughly*, that means that the fuzz amplitude is cut in half the same number of times: down to $(1/2)^3$: 1/8. Again, this is relative to the amplitude at f_{3dB}, where amplitude was already down to 70%.

What happens to the circuit output if the load has resistance 10k rather than 100k? Figure 2W.4 is a picture of such loading.

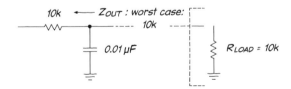

Figure 2W.4 Overloaded filter.

If you have gotten used to Thevenin models, then you can see in Fig. 2W.5 how to make this circuit look more familiar:

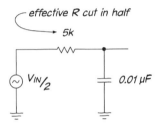

Figure 2W.5 Loaded circuit, redrawn.

The amplitude is down; but, worse, f_{3dB} has changed: it has doubled. Figure 3N.2 is a plot showing this effect.

---

[1] To get exactly our target f_{3dB} we can readjust R. We have fewer choices of capacitor value, but can choose any R (though it doesn't make sense, given a "10%" cap, to do better than specify R to 10%).

2W.1 RC filters

What is the circuit's input impedance?

(1) At very low frequencies? Answer: *very large*: the cap shows a high impedance; the signal source sees only the load – which is assumed very high impedance (high enough so we can neglect it as we think about the filter's performance).

(2) At very high frequencies? Answer: *R*: The cap impedance falls toward zero – but *R* puts a lower limit on the input impedance.

(3) At f_{3dB}?

This is easy if you are willing to use phasors, a nuisance to calculate, otherwise. If you recall[2] that the magnitude of $X_C = R$ at f_{3dB}, and if you accept the notion that the voltages across *R* and *C* are 90° out of phase so that they can be drawn at right angles to each other on a phasor diagram, then you get the phasor diagram of Fig. 2N.31, and can use a geometric argument to show that the hypotenuse – proportional to Z_{in} – is $R\sqrt{2}$.

2W.1.2 Bandpass

Problem: bandpass filter Design a bandpass filter to pass signals between about 1.5 kHz and 8 kHz. To do this, put the two f_{3dB} a factor of 2 from the nearest frequency. (Doing this will limit attenuation to about 10%, as you know from §2W.1.1.)

Assume that the *next* stage that your bandpass filter is to drive has an input impedance ≥ 1 MΩ.

A solution Roughly, Fig. 2W.6 shows the shape of the frequency response that we want.

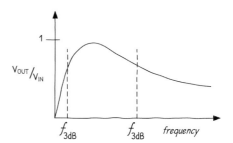

Figure 2W.6 Bandpass frequency response.

To get this frequency response from the filter we need to put *highpass* and *lowpass* in series. f_{3dB} for the highpass should be about half the lowest frequency of interest, so put it at about 750 Hz. f_{3dB} for the lowpass should be double the highest frequency of interest, so put it at about 16 kHz.

Viewed separately, the highpass and lowpass responses are shown in Fig. 2W.7. The horizontal axis is frequency, and runs from close to zero up to 30 kHz.

If the two filters are "cascaded" – placed in series, so that the poor old input has to pass through this double gauntlet – the result will be attenuation at the frequency extremes, and a bump in the middle: see Fig. 2W.8.

The linear frequency plot shows that the bandpass is pretty far from the ideal – a filter that blocks all of the bad and lets through all of the good. Instead of being shaped like a steep-walled mesa, it's more like a gentle foothill of a mountain range. The *log* frequency plot makes the response *look* a little more respectable – more as if there is a flat passband region; but of course the changed plot doesn't change the facts. Using simple *RC*s, we'll have to settle for this imperfect, slopey passband.

---

[2] See §2N.3.7: You can easily confirm that at f_{3dB}, $|X_C| = R$. For $|X_C| = \frac{1}{2\pi fC}$, $f_{3dB} = \frac{1}{2\pi RC}$, and if we plug f_{3dB} in as the frequency in the expression for $|X_C|$ – in other words, if we assume we are at f_{3dB} – we get $|X_C| = \frac{1}{2\pi(\frac{1}{2\pi RC})C} = R$.

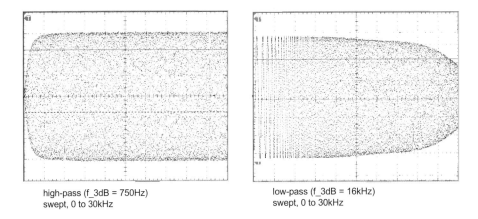

Figure 2W.7 Highpass and lowpass f_{3dB}'s at 750 Hz and 16 kHz, respectively.

high-pass (f_3dB = 750Hz) swept, 0 to 30kHz

low-pass (f_3dB = 16kHz) swept, 0 to 30kHz

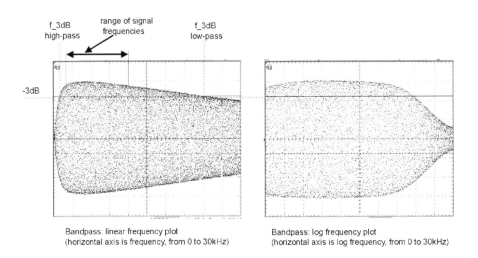

Figure 2W.8 Bandpass response that we want: linear and log frequency plots.

Bandpass: linear frequency plot (horizontal axis is frequency, from 0 to 30kHz)

Bandpass: log frequency plot (horizontal axis is log frequency, from 0 to 30kHz)

Choose Rs: Now we need to choose R values, because these will determine worst-case impedances for the two filter stages. The later filter must show Z_{out} low relative to the load, which is 1 MΩ; the earlier filter must show Z_{out} low relative to Z_{in} of the second filter stage.

So, let the second-stage R be 100k; the first-stage R is 10k:

...then calculate Cs: Here the only hard part is to get the filters right: it's hard to say to oneself, "The highpass filter has the lower f_{3dB};" but that is correct. Here are the calculations: notice that we try to keep things in *engineering notation* – writing "10×10^3" rather than "10^4." This form looks clumsy, but rewards you by delivering answers in standard units. It also helps you scan for nonsense in your formulation of the problem: it is easier to see that "10×10^3" is a good translation for "10k" than it is to see that, say, "10^5" is *not* a good translation.

Calculating the two C values is a purely mechanical process.

The lowpass:

$$f_{3dB-\text{low}} = 16 \text{ kHz} \Longrightarrow C = \frac{1}{2\pi \times 16 \times 10^3 \times 10 \times 10^3} \approx 1/10^9 = 0.001 \,\mu\text{F}$$

2W.2 RC step response

The highpass:

$$f_{3dB-high} = 0.75\,\text{kHz} \Longrightarrow C = \frac{1}{2\pi \times 0.75 \times 10^3 \times 100 \times 10^3} \approx 1/(0.45 \times 10^9) \approx 0.002\,\mu\text{F}$$

And the circuit looks as in Fig. 2W.9.

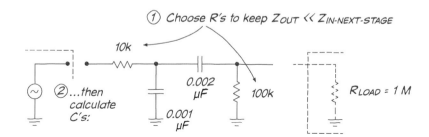

Figure 2W.9 The bandpass circuit.

2W.2 RC step response

We were struck recently by how difficult an "old" topic like RC behavior can be when it's a little different from the standard case. Several students convinced us we need more practice on such early learning. Here's a little workout on RCs.

2W.2.1 Problem: step response of RC circuit (time-domain)

Figure 2W.10 shows a capacitor feeding a pulse into several alternative circuit elements. Please draw the voltage waveform at the point where the capacitor meets "X" for each case. Show time and voltage scales, and label significant points on your waveform. As usual, we'll be happy with answers good to about 10%. Both R_{in_X} and C_{in_X} describe the effect of a C and an R in parallel; the far end of each is tied to a fixed voltage; let's assume that voltage is ground, just so all our drawings will look the same.

Note: We'd like you to draw V_{out}, not V_{cap}.

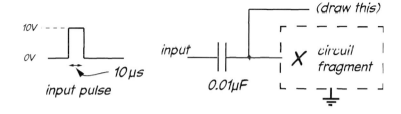

Figure 2W.10 Pulse, capacitively-coupled to various circuit fragments.

Here are the several alternatives for the circuit element labeled "X:"

- Idealized next stage: $R_{in_X} = \infty$, $C_{in_X} = 0$;
- Mixed next stage: $R_{in_X} = \infty$, but $C_{in_X} = 0.001\,\mu\text{F}$;
- Mixed next stage: $R_{in_X} = 100\text{k}$, $C_{in_X} = 0$;
- Mixed next stage: $R_{in_X} = 1\text{k}$, $C_{in_X} = 0$; or
- Non-ideal next stage: $R_{in_X} = 100\text{k}$, $C_{in_X} = 0.01\,\mu\text{F}$.

2W.2.2 Solution

Idealized next stage: $R_{in_x} = \infty$, $C_{in_x} = 0$: Since there is no path permitting current to pass to or from the second plate of the cap, $I_{cap} = 0$ and V_{cap} remains zero. Therefore $V_{out} = V_{in}$.[3]

Figure 2W.11 No R, no C.

Mixed next stage: $R_{in_x} = \infty$, but $C_{in_x} = 0.001\,\mu F$: This is a C–C divider (rare or impossible in life). V_{out} *shape* is same as V_{in}'s, but *amplitude* is attenuated according to the impedances of the two caps. $V_{out} = 0.9\,V_{in}$. (Again a DC path to ground is required on the output to make this circuit practical.)

Figure 2W.12 A capacitive divider.

Mixed next stage: $R_{in_x} = 100k$, $C_{in_x} = 0$: $RC = 1$ ms, $\gg$ pulse width. So V_{cap} will not change appreciably during pulse. Hence V_{out} looks like V_{in}. This case approximates the first case, of §2W.2.2. Again, $V_{out} = V_{in}$.

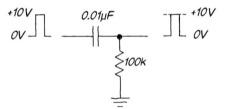

Figure 2W.13 Approximately the first case: all passes.

Mixed next stage: $R_{in_x} = 1k$, $C_{in_x} = 0$: $RC = 0.01$ ms, same as pulse width. So pulse output will decay to $V_{in} \times (1/e) \approx 37\%\,V_{in}$ during pulse. V_{out} steps to +10 V, decays to about 4 V, steps down to −6 V, then decays toward zero.

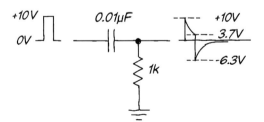

Figure 2W.14 Quick RC makes the exponential decays apparent.

---

[3] This is not a realistic case. A capacitor with no DC path to ground on the right side gradually would be charged by leakage currents. A practical circuit requires a resistive path to ground on the output to define the DC level at that terminal.

2W.2 RC step response

Non-ideal next stage: $R_{in_x} = 100k$, $C_{in_x} = 0.01\,\mu F$: Redraw as C–C divider driving R (this echoes our old Thevenin model, the R–R divider). As in the Thevenin model, the effective driving C is the two C values in parallel: $0.02\,\mu F$. So $RC = 2$ ms. Again pulse width $\ll RC$, so pulse survives – but at half amplitude.

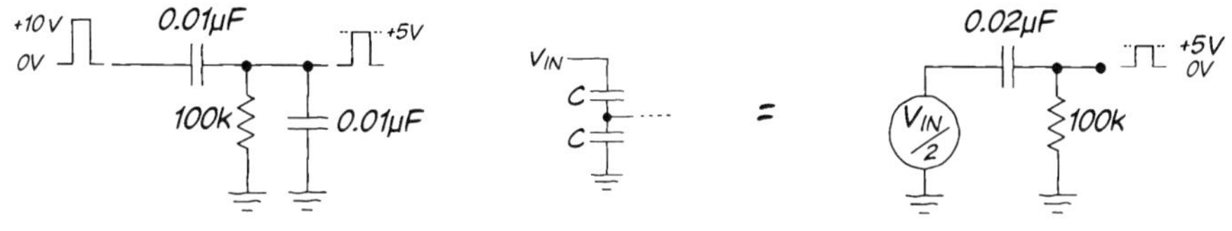

Figure 2W.15 A capacitive divider driving R: attenuates, but passes pulse.

3N Diode Circuits

Contents

3N.1	**Overloaded filter: another reason to follow our 10× loading rule**	**112**
3N.2	**Scope probe**	**113**
	3N.2.1 Defective 10× probe	114
3N.3	**Inductors**	**116**
3N.4	*LC* **resonant circuit**	**117**
	3N.4.1 Resonance	118
	3N.4.2 The meaning of Q	119
	3N.4.3 Testing whether Fourier was right	121
	3N.4.4 Response to a slow square wave: "ringing"	121
3N.5	**Diode circuits**	**122**
3N.6	**An important diode application: DC from AC**	**123**
	3N.6.1 Half-wave rectifier and clamp	123
	3N.6.2 Full-wave bridge rectifier	124
	3N.6.3 Zener diodes	126
3N.7	**Unregulated power supply**	**127**
	3N.7.1 Transformer voltage	127
	3N.7.2 Capacitor	128
	3N.7.3 Transformer current rating (*RMS* heating)	129
	3N.7.4 Fuse rating	129
3N.8	**Radio!**	**130**
	3N.8.1 Step 1: *LC* selects one "carrier" frequency	130
	3N.8.2 Step 2: detect "envelope" of AM waveform	131
	3N.8.3 What the result looks like	133
	3N.8.4 How to recover the AM information	133
3N.9	**AoE reading**	**135**

Why?

The sort of problem we mean to solve with the most important of today's circuits is the conversion of a sinusoidal power supply voltage – AC coming from the wall supply (often called "line" voltage) – to a constant DC level.

3N.1 Overloaded filter: another reason to follow our 10× loading rule

Remember our claim that our *10× rule of thumb* would let us design circuit fragments? Let's confirm it by watching what happens to a filter when we violate the rule.

3N.2 Scope probe

Suppose we have a lowpass filter, see Fig. 3N.1, designed to give us f_{3dB} a bit over 1 kHz (this is a filter you built last time, you'll recall).

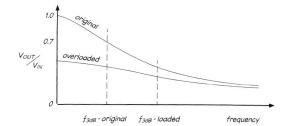

Figure 3N.1 *RC* lowpass: not loaded versus overloaded; redrawn to simplify.

If R_{load} is around 150k or more, the load attenuates the signal only slightly, and f_{3dB} stays put. But what happens if we put R_{load}=15k? Attenuation is the lesser of the two bad effects. Look at what happens to f_{3dB}:

Figure 3N.2 Overloaded filter: we get something worse than attenuation: excessive loading shifts f_{3dB}.

The shift of f_{3dB} from where it was designed to lie – in this case, a *doubling* of f_{3dB} – is much more serious than the excessive attenuation at DC. The DC attenuation can be fixed by boosting circuit "gain" in a later stage; but the shifted f_{3dB} cannot easily be fixed by doing something in a later stage. We hope this example will reinforce your faith in our "times-ten" rule – the rule that allows us to design circuit fragments as independent modules, confident that appending a new stage will not mess up performance of what we have designed.

3N.2 Scope probe

AoE Exercise 1.44

That mishap leads nicely into the problem of how to design a scope probe. We want to provide our scopes with higher input impedance.

You may be inclined to think we're awfully greedy, not to be content with the input impedance of a scope driven through a BNC cable: 1MΩ, in parallel with the capacitance of the cable. But the modest input impedance presented by cable and scope can be troublesome. It is particularly the stray capacitance that causes mischief.

This "1 MΩ" sounds nice enough – but consider what happens to the impedance of the cable as frequency climbs. Its capacitance to ground, along with the scope's input capacitance, adds up to about 100 pF, if the cable is about three feet long: this may still sound small – but at 1 MHz, it's a heavy load:

$$X_C = |Z_C| = \frac{1}{(2\pi f C)} \approx \frac{1}{(6 \times 10^6 \times 100 \times 10^{-12})} = \frac{1}{(0.6 \times 10^{-3})} \approx 1.6\,k\Omega$$

pretty low, and disastrously low when source impedance is high. We'll see this effect as we work our way to a good design for a probe. We'll first design the probe wrong.

The heavy capacitance of a bare BNC burdens the circuits you look at, and this burden can have effects even worse than just attenuation: it may make some of those circuits misbehave in strange ways. The most common misbehavior is a spontaneous oscillation.[1]

---
[1] The problem of unintended *oscillation* – so-called "parasitic oscillation" – will concern us much, in later labs (it is the main

Diode Circuits

So, we nearly always use "10×" probes with a scope: that's a probe that makes the scope's input impedance 10× that of the bare scope. If the bare scope looks like 1 MΩ in parallel with 100 pF – cable and scope – the "10×" probe should look about 10 times better: 10 MΩ in parallel with 10–12 pF.

3N.2.1 Defective 10× probe

Figure 3N.3 shows a defective design for a 10× probe Do you see what's wrong? It works fine at DC. But try redrawing it as a Thevenin model driving a cap to ground, as in the example we did at the start of these notes. The flaw should appear. What is f_{3dB}? Quite low, since the effective R driving the stray capacitance (R_{Thev}) is large, almost 1 M:

$$f_{3dB} = \frac{1}{2\pi RC} \approx \frac{1}{(2 \times 3 \times 10^6 \times 100 \times 10^{-12})}$$

So

$$f_{3dB} \approx \frac{1}{0.6 \times 10^{-3}} \approx 1.6 \, \text{kHz}$$

That's a disaster for a scope designed to work up to many tens of MHz.

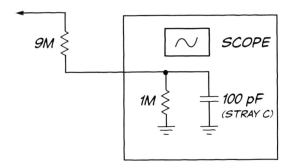

Figure 3N.3 Crummy 10× probe.

Remedy: We need to make sure our probe does *not* have this lowpass effect: scope and probe should treat alike all frequencies of interest (the upper limit is set by the scope's maximum frequency – for most in our lab that is up to 60 or even 100 MHz).

The trick is just to build two voltage dividers in parallel: one resistive, the other capacitive: see Fig. 3N.4. At the two frequency extremes one or the other dominates (that is, passes most of the current); in between, they share. But if each delivers $V_{in}/10$, nothing complicated happens in this "in-between" range.

What happens if we simply *join* the outputs of the two dividers? Do we have to analyze the resulting composite circuit as one, fairly messy thing? No. No current flows along the line that joins the two dividers, so things remain utterly simple.

So, a good probe is just these two dividers joined, as in Fig. 3N.5 (where the "stray" C is assumed to sum scope and cable effects). Practical probes make the probe's added C adjustable. This adjustment raises a question: how do you know if the probe is properly adjusted, so that it treats all frequencies alike?

topic of Lab 9L). It cannot occur until we add power-amplification to our circuitry. So, today, your circuits will not oscillate even when heavily loaded by a BNC cable. But in Lab 4L you may see exactly this misbehavior, since any of your transistor circuits can provide the amplification or "gain" necessary to sustain an oscillation: these parasitics are an interesting but troublesome novelty, coming soon.

3N.2 Scope probe

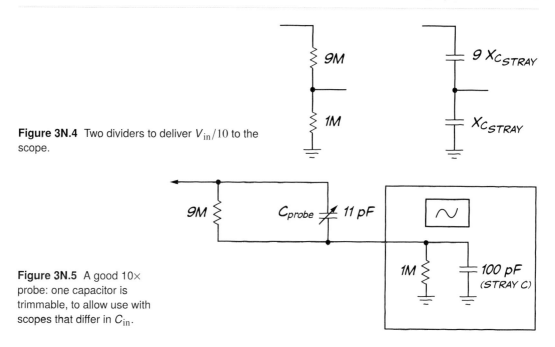

Figure 3N.4 Two dividers to deliver $V_{\rm in}/10$ to the scope.

Figure 3N.5 A good 10× probe: one capacitor is trimmable, to allow use with scopes that differ in $C_{\rm in}$.

Probe "compensation": one way to check: sweep frequencies...

One way to check the frequency response of probe and scope, together, would be to sweep frequencies from DC to the top of the scope's range, and watch the amplitude the scope showed. In Fig. 3N.6 we've sketched what the response would look like for adjustments of $C_{\rm compensation}$ that are wrong in each direction, and also correct.

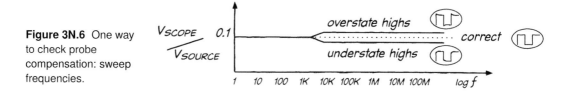

Figure 3N.6 One way to check probe compensation: sweep frequencies.

At low frequencies, the resistive divider dominates; at higher frequencies, above what we might call a "crossover frequency," the capacitive divider dominates. The two dividers ought each to deliver 1/10 amplitude. If the $C{:}C$ divider is wrongly adjusted, it delivers either more or less than that, as shown in Fig. 3N.6.[2]

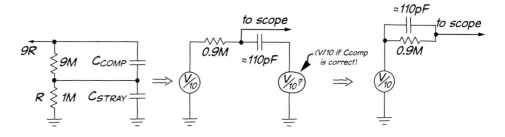

Figure 3N.7 Probe redrawn as equivalent RC network.

---

[2] The crossover frequency depends on the effective RC, where the circuit can be redrawn (modeled) as in Fig. 3N.7.

116 Diode Circuits

But this way of checking probe compensation is clumsy. It requires a good function generator, and would be a nuisance to set up each time you wanted to check a probe.

Probe "compensation": ...an easier way to check: Fourier, again

The easier way to do the same task is just to feed scope and probe a *square* wave, and then look to see whether the waveform looks square on the scope screen: see Fig. 3N.8. If it does, good: all frequencies are treated alike. If it does not look square, just adjust the trimmable C in the probe until the waveform *does* look square.

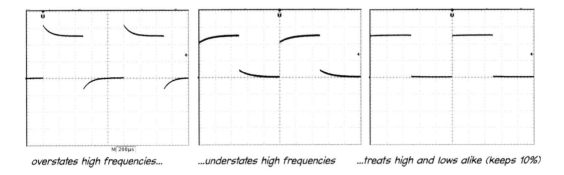

overstates high frequencies...   ...understates high frequencies   ...treats high and lows alike (keeps 10%)

Figure 3N.8 Using square wave to check frequency response of probe and scope.

Neat? This is so clearly the efficient way to check probe *compensation* (as the adjustment of the probe's C is called) that every respectable scope offers a square wave on its front panel. It's labeled something like *probe comp* or *probe adjust*. It's a small square wave (of fixed amplitude) at around 1 kHz.

3N.3 Inductors

In a physics course, inductors are treated with the same respect accorded capacitors, and the complementary behaviors of the two devices seem to support this view.

Here are the impedances of the two devices:

Impedance of capacitor $Z_C = -j/(2\pi f C) = -j/(\omega C)$.
Impedance of inductor $Z_L = j(2\pi f L) = j\omega L$.

The impedances of the two devices are complementary in two senses:

- the impedance of one device falls with frequency (true of Z_C), while the other rises with frequency (true of Z_L); and
- the phase shifts between current and voltage are opposite for the two devices, as is indicated by the opposed signs of $-j$ and j. Current *leads* voltage in the capacitor; it *lags* voltage in the inductor.

It looks as if you could use L or C equally well to make a filter. Below, for example, are two versions of a lowpass.

Both versions of the lowpass filter work. But only the RC is practical, at all but very high frequencies.[3] AoE §1.7.11
In this course – and in electronics generally – capacitors are used much more widely than inductors.

---

[3] "High" by the standards of this course: around 1 MHz and up.

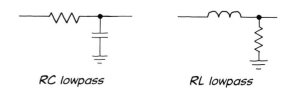

Figure 3N.9 In principle one can use *RL* or *RC* to form a filter....

RC lowpass   RL lowpass

The difference comes from the fact that inductors are relatively large and heavy (often including a core made of iron or another magnetically-permeable material), and that, owing to departures from ideal, they dissipate power.

In Fig. 3N.10 a self-righteous capacitor reminds us of this difference – and, like most of the self-righteous, it somewhat exaggerates its virtue.[4]

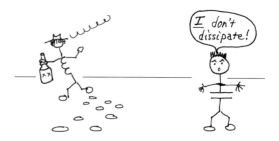

Figure 3N.10 Capacitors don't ordinarily dissipate much energy; inductors do.

That difference leads one to prefer capacitors and to avoid inductors altogether except at high frequencies (perhaps 1 MHz, or more, where a small value of inductance is sufficient to do the job), or in power conversion circuits (where they are ubiquitous).

The resonant RLC circuit that we use in Lab 3L to select a radio broadcast frequency illustrates a typical case where inductors work well. The relatively high frequency permits use of an inductor of small value and small size.

In circuits running below a few megahertz, where ordinary *operational amplifiers* perform well, clever circuitry even permits capacitors to emulate the behavior of inductors, without bringing along their nasty properties.

3N.4 *LC* resonant circuit

Figure 3N.11 shows Lab 3L's resonant RLC circuit. Before we acknowledge what's novel about this circuit – its *resonance* – let's take advantage of what we know of the impedances of capacitors and inductors to make a simple argument that this is, indeed, a *bandpass* filter: one that passes a range of intermediate frequencies, while attenuating both frequency extremes.

If we neglect the effect of the inductor – as in the left-hand circuit fragment in Fig. 3N.12 – we see a familiar RC lowpass. At high frequencies, it is *fair* to neglect the paralled inductor: its impedance is much larger than the impedance of the capacitor.

Toward the other end of the frequency range, the inductor's impedance is much less than that of the capacitor (recall that $Z_L = j\omega L = j2\pi f L$). In this frequency range, the RL forms a highpass.

---

[4] In some circuits, such as switching power supplies, the capacitor's "equivalent series resistance" (ESR) can dissipate power and degrade performance. See AoE §§6.2.1A and 9.6.3A. In Lab 11L you will find that we needed to specify a low-ESR capacitor for use with the switching regulator.

Diode Circuits

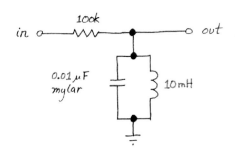

Figure 3N.11 Lab 3L's *RLC* circuit.

Over the full frequency range, then, the RLC forms a bandpass filter. So much, we can see without considering the LC's resonance.

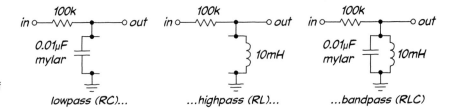

Figure 3N.12 A first approximation of the parallel-*RLC* circuit: highpass and lowpass.

3N.4.1 Resonance

AoE §1.7.14

But that rough analysis misses the interesting novelty of this circuit: the "resonance" to which we have referred. At some frequency – where the magnitude of the impedances of inductor and capacitor are equal – something startling occurs: the impedance of the parallel LC becomes very large.

It's not hard to persuade yourself that this happens, if you write out the formula for the impedance of the LC pair:

$$Z_{\text{LC-parallel}} = Z_C \parallel Z_L = \frac{Z_C \times Z_L}{Z_C + Z_L}$$

But, because $Z_C = -j/\omega C$ while $Z_L = j\omega L$, the opposite signs of the "j" indicate that at some frequency, where the magnitudes are equal, the two impedances should sum to zero. When this occurs, taking the denominator of the parallel impedance to zero, the parallel impedance "blows up" – becomes very large.[5]

The *resonant frequency*, where $|Z_C| = |Z_L|$ occurs when $|-j/\omega C| = |j\omega L|$ and $1/\omega C = \omega L$:

$$\omega^2 = 1/LC \Rightarrow \omega = 1/\sqrt{LC} \quad \text{or} \quad f_{\text{resonance}} = 1/(2\pi\sqrt{LC})$$

At this resonant frequency, where the impedance of the parallel LC becomes large, the circuit passes the largest-available fraction of its input. Ideally, that fraction would be 100%, but losses in the inductor mean that the maximum can be much less than 100% for large values of R. The RLC offers not just another way to make a bandpass; it permits making an extremely *narrow* passband, compared to what one can achieve with RCs.

The characteristic called "Q" describes just *how* narrow is the range of frequencies that are allowed to pass: $Q \equiv f_{\text{resonance}}/\Delta f$. Here Δf is the width at the amplitude that delivers half-power – the amplitude that is 3 dB below the peak, as sketched in Fig. 3N.13.

---

[5] "Infinite," you may want to say. But LC imperfections – largely caused by resistance and core losses in the inductor – spoil this result. But it is enough that the impedance becomes very large, and is extremely sensitive to small frequency changes.

The *RLC* does not work the way an *RC* filter does. It is more dynamic. It does more than form a frequency-selective voltage divider, although it does do that. It *stores energy*; it is more like a pendulum than like a coffee filter. The inductor–capacitor combination oscillates, once stimulated with a frequency close to its favorite – the frequency where it "resonates." In each cycle of oscillation, energy is transferred, back and forth, between inductor and capacitor.

When the *voltage* across the parallel pair is at a maximum, energy is stored in the electrostatic field between the plates of the capacitor; as the capacitor begins to discharge through the inductor, the current gradually grows, and when the *current* reaches a maximum, V_{cap} is zero. At that point in the cycle, all the circuit's energy is stored in the *magnetic* field around the inductor. The energy sloshes back and forth between capacitor and inductor.[6]

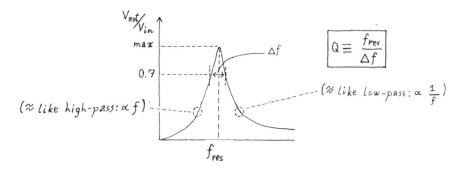

Figure 3N.13 Shape of *RLC* frequency response.

3N.4.2   The meaning of Q

Q measured using Δf: The lab asks you to estimate the circuit's Q ("quality factor," a term whose name apparently reflects the use of resonant circuits in radio tuners, where high selectivity is good).

Measured thus, Q is defined as follows, as we have said (Fig. 3N.13):

$$Q = \frac{f_{\text{resonance}}}{\Delta f}$$

where Δf" represents the width of the resonance peak between its "-3 dB" points: see Fig. 3N.14.

Q measured using decay time: Q can also be defined to describe how slowly energy leaks away from a resonant circuit, dissipated in its stray resistance:[7]

$$Q = \omega_0 \frac{\text{energy stored}}{\text{average power dissipated}}$$

Or, less abstractly, Q can be defined as

$$Q = 2\pi \times \{\text{number of cycles required for the energy to diminish to } 1/e\}$$

This latter definition isn't illuminating when you are looking at the frequency response of the *LC* circuit, as in the present exercise; this second definition will seem more helpful when you reach §3N.4.4, which looks at the *LC*'s response in the *time domain*. For the case of the *parallel LC*, one more rule is useful: $Q = \omega_0 RC$.

---

[6] See Purcell & Morin, §8.1, and especially Fig. 8.3, showing a damped oscillation.
[7] See Purcell & Morin, equation 8.12.

Diode Circuits

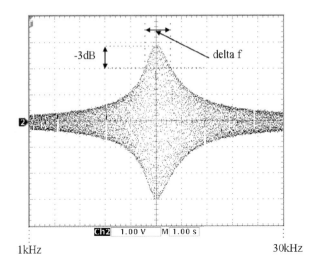

Figure 3N.14 Q measures the frequency-selectivity or *peakiness* of the resonant circuit.

1kHz   30kHz

Modifying Q: Q describes how efficient the RLC is, and depends mostly upon the characteristics of the inductor. But you can modify Q with your choice of R. Large R produces higher Q in a parallel LC, but lower amplitude out (the energy dissipated in the LC pair must be replaced; the larger the R feeding the LC, the larger the voltage drop across that R). So, one can trade off one virtue against another: high Q versus large amplitude.

If you're energetic and curious, you can look at the effect upon Q of substituting a 10k resistor for the 100k. Amplitude-out increases, while Q degrades. As usual, you're obliged to trade away one desirable trait to get another. But good Q is likely to be much more important than large amplitude; an amplifier can solve the problem of low amplitude.

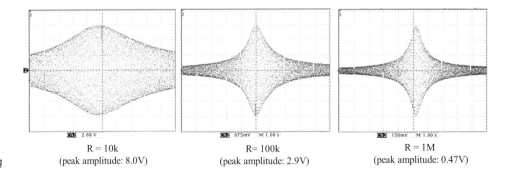

Figure 3N.15 Large R raises Q while diminishing output amplitude.

R = 10k (peak amplitude: 8.0V)   R= 100k (peak amplitude: 2.9V)   R = 1M (peak amplitude: 0.47V)

Figure 3N.15 shows the effect of three R values in the parallel RLC circuit: 10k, 100k and 1M. For the larger R values, Q does indeed improve – but at the expense of output amplitude. In Fig. 3N.15 we have adjusted scope gain so as to hold the *graphical* peak amplitude constant for the three images. We did this in order to make it easy to see the changes in the response *shape* (and Q). But note that the scope sensitivity increases from left to right (V/Div falls) – so that the amplitude out diminishes by more than a factor of 16. A smaller, secondary effect also appears, as R changes: the resonant frequency shifts slightly.[8]

In Lab 13L, where you are likely to use an RLC as part of an *FM demodulator*, you will take

---

[8] This is a "second order" for "light damping....," see Purcell, §8.1.

3N.4 LC resonant circuit

advantage of R's capacity to tailor the shape of the RLC response: and you will feel the tension between these two desirable characteristics, large amplitude versus high Q.

3N.4.3 Testing whether Fourier was right

The parallel RLC circuit lets you test one of Fourier's claims – in case you doubted it. Figure 3N.16 repeats a figure from today's lab, a sketch of how the Fourier series for a square wave begins.

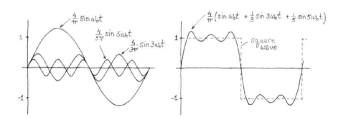

Figure 3N.16 Fourier series for square wave.

Less graphically, the Fourier series for a square wave of amplitude A is

$$\frac{4}{\pi}A(\sin(\omega t)) + \frac{1}{3}\sin(3\omega t)) + \frac{1}{5}\sin(5\omega t) + \cdots$$

Your LC can pick out these frequency components for you: and the result, in the lab circuit, is a very pretty perspective-like display (the resemblance to receding telephone poles or fenceposts beside a highway is striking).

3N.4.4 Response to a slow square wave: "ringing"

A slow square wave whose frequency is not at all critical (Lab 3L suggests you might try 20 Hz) stimulates the LC, even though 50 Hz is very far from the LC's resonant frequency of 16 kHz. How come? Because, as you know, Fourier teaches that a square wave includes in its edges high-frequency "harmonics" close to the resonant frequency. These harmonics are small (since the Fourier series for a square wave falls off like $1/f$).

In Fig. 3N.17 is a figure showing the resonant circuit of Lab 3L driven by a square wave of a frequency far below $f_{\text{resonance}}$ (square wave frequency $\approx$ 50 Hz).

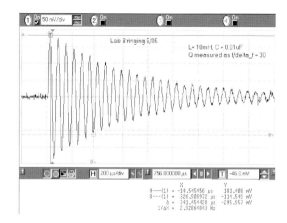

Figure 3N.17 Decay of RLC's ringing can reveal Q.

Notice that it is the *energy*, not *amplitude* that should decay to $1/e$ at the rate thus related to Q. Since the energy stored in the capacitor is proportional to amplitude *squared* (energy$_{\text{cap}} = \frac{1}{2}CV^2$), we want to see V^2 – not V – fall to 37%, and V to about 60%.

In Fig. 3N.17 such a point is marked by a cursor, and it seems to occur after about 5 cycles. This would indicate a Q of about 30.

Why does it decay? A student recently asked "Why does it decay exponentially?" – a good question. The short answer is "because the rate at which it loses energy is proportional to its amplitude." A slightly longer version of that answer would note:

- energy in the LC circuit is proportional to V^2 (if you consider the time in a cycle when all the energy is in the capacitor, for example, at that time energy$_{cap} = \frac{1}{2}CV^2$: here V is V_{RMS}: we are not interested in the sign of V or in its instantaneous value); and
- energy is dissipated in series resistance, R_{drive}, proportional to V^2/R_L (this is the power dissipated in the resistance).

Thus energy and rate of energy loss both are proportional to amplitude-squared. The *squared* factor only changes the time-constant, not the exponential shape of the decay.[9]

Poor grounding can evoke LC resonant waveforms

You will see such a response of an LC circuit to a step input whenever you happen to look at a square wave with an improperly grounded scope probe: when you fail to ground the probe close to the point you are probing, you force a ground current to flow through a long (inductive) path. Stray inductance and capacitance form a resonant circuit that produces ugly ringing. You might look for this effect now, if you are curious; or you might just wait for the day (almost sure to come) when you run into this effect inadvertently.

The scope images in Fig. 3N.18 show how shortening the ground return of a scope probe reduces ringing. The shorter ground return has less inductance, and pushes $f_{resonant}$ higher while storing less energy as well.

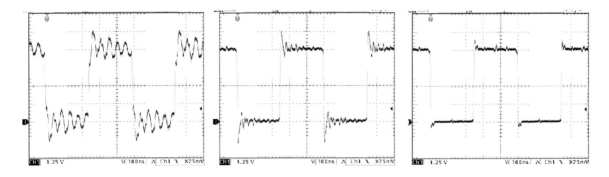

Figure 3N.18 Scope probe, poorly grounded, shows resonance of stray L and C. Left: no probe ground (except through scope power supply). Middle: probe grounded about 10″ from point probed. Right: probe grounded about 1/2″ from point probed.

3N.5 Diode circuits

AoE §1.6.1

Diodes do a new and useful trick for us: they allow current to flow in one direction only: see Fig. 3N.19.

The symbol looks like a one-way street sign, and that's handy: it's telling conventional current which

---
[9] See R.E. Simpson, *Introductory Electronics*, (2d ed., 1984), pp. 98–100.

Figure 3N.19 Diode as one-way current gate.

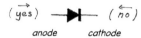

way to go. For many applications, it is enough to think of the diode as a one-way current valve, though you need to note also that when it conducts the diode doesn't behave quite like a wire. Instead, it shows a characteristic "diode drop" of about 0.6 V. This you saw in Lab 1L; Fig. 3N.20 may remind you of the curve you drew on that first day (you did *not* see the breakdown at ≈ -100 V, fortunately for your safety).

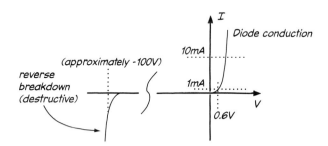

Figure 3N.20 Diode *I–V* curves: reverse current is in nanoamp range.

3N.6 An important diode application: DC from AC

Electrical power is distributed in *AC* form rather than DC, in order to minimize power losses in transmission lines. Using AC permits transmission at high voltages, and correspondingly low currents. AC is easily "stepped" up or down using transformers – which are simply two windings placed close together so that they are "coupled" by their magnetic fields. In Worked Example 1W.2 you can find an example of the efficiency advantage that goes with high transmission voltage.

Because power is transmitted in AC form whereas almost every electronic device needs a DC power supply, making this transformation from AC to DC is an important mission of diode circuits.

3N.6.1 Half-wave rectifier and clamp

AoE §1.6.2

We often use diodes within voltage dividers, much the way we have used resistors, and then capacitors. Figure 3N.21 shows a set of such dividers: what should the outputs look like?

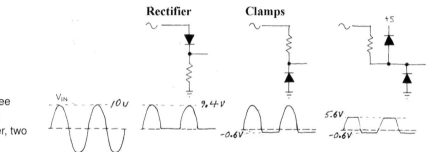

Figure 3N.21 Three dividers made with diodes: one rectifier, two clamps.

The outputs for the first two circuits look strikingly similar. Yet only the *rectifier* can be used to

124 Diode Circuits

generate the DC voltage needed in a power supply. Why? What important difference exists – invisible to the scope display – between the "rectifier" and the "clamp?"[10]

The bumpy output of the "rectifier" is still pretty far from what we need to make a power supply – a circuit that converts the AC voltage that comes from the wall or "line" to a DC level. A capacitor will smooth the rectifier's output for us. But, first, let's look at a better version of the rectifier.

3N.6.2 Full-wave bridge rectifier

AoE §1.6.2

This clever circuit in Fig. 3N.22 gives a second bump out, on the *negative* swing of the input voltage.

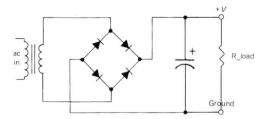

Figure 3N.22 Full-wave bridge.

Neat, isn't it? Once you have seen the circuit's output, you can see why the simpler rectifier is called *half-wave*.

Figure 3N.23 shows some details of the output of the full-wave rectifier, including the non-obvious fact that the bridge *input* drops *below* ground, on one half cycle.

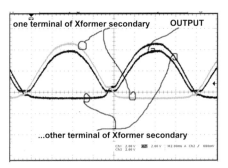

Figure 3N.23 Full-wave bridge rectifier.

Because of this wandering of the bridge input, the full-wave bridge can be used only with a "floating" source such as a transformer secondary. Neither of its inputs may be tied to ground, assuming that one of the two output terminals is to be defined as ground.

Rarely is there an excuse for using anything other than a full-wave bridge in a power supply, these days. Once upon a time, "diode" meant a $5 vacuum tube, and then designers tried to limit the number of these that they used. Now you just buy the bridge: a little epoxy package with four diodes and four legs; a big one may cost you a dollar.

Full-wave bridge in a "split supply"

A full-wave bridge can be used to generate a dual output: both positive and negative with respect to ground. Figure 3N.24 shows such a circuit.

---

[10] The answer is not the more obvious difference that one shows a 0.6 V offset when passing the signal, whereas the other does not; that difference would seem to favor the second circuit, the "clamp." No, what's important is the difference between the two *output impedances*. (Remember? We promised you that impedances would be a theme, in the analog part of this course.) The left-hand circuit – the rectifier – places only a conducting diode between source and load (the source usually is a power transformer). The clamp places a resistor between source and load. As a result, the rectifier provides the efficient way to build a power supply; the clamp does not.

3N.6 An important diode application: DC from AC

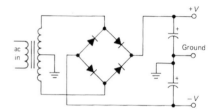

Figure 3N.24 Full-wave bridge used to generate both positive and negative ("split") outputs.

Evidently, the only difference from the single-supply of Fig. 3N.22 arises from use of the *center tap* on the transformer secondary, tying it to the point that we define as the *ground* of the split output. (We omit the load resistors from Fig. 3N.24; a load is, of course, assumed – if not, why build the supply?)

One might be tempted to think that the center tap is not necessary: that one could form a split supply simply by defining ground as the midpoint between the capacitors in Fig. 3N.24. That view would be wrong.

Absent use of the center-tap, the sharing of the total voltage, between positive and negative outputs would not be predictable; it would depend on the relative loading of the two outputs. The center-tap makes the two outputs independent, so that it is not just the total voltage, V_+ minus V_-, that the transformer determines. Instead, each of V_+ and V_- shows a voltage determined by the half-winding of the transformer.

If that proposition seems vague, and leaves you puzzled, you may want to trace the details of current flow on both half cycles of the line voltage, a flow sketched in Fig. 3N.25.

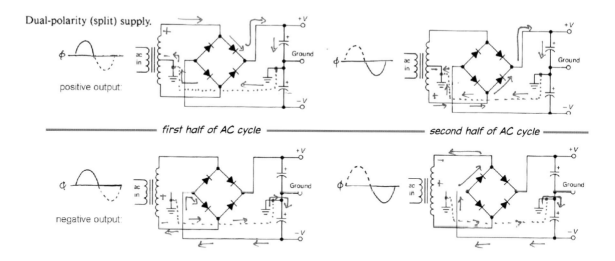

Figure 3N.25 Current flow in split-supply rectifier: positive and negative output flows shown separately.

Figure 3N.25 shows four cases: but there are, of course, only *two* half cycles to consider. We have done separate sketches for the positive and negative supplies, because we feared that thinking about both output polarities at once might be too much. You will notice, also, that we have drawn a dotted line between the two *ground* symbols, in order to make the return of current to the center-tap easier to envision.

In a minute, we'll proceed to looking at a full power supply circuit, which will include a full-wave bridge. But before we do, let's note one additional sort of diode: the zener.

126   Diode Circuits

3N.6.3   Zener diodes

You will not apply a simple zener in our labs (though you will meet a fancier *voltage reference* in Lab 11L). But we would like to mention the zener here alongside the other diodes that we'll meet in today's lab, 3L: the standard silicon diode (like the 1N914/1N4148 we use repeatedly today) and the *Schottky* diode, a low-forward-drop device that you are likely to choose as you build the AM radio detector late in 3L.

AoE §1.2.6A

The zener diode conducts at some low *reverse* voltage, and *likes to*! If you put it into a voltage divider "backwards" – that is, "back-biased," with the current running the wrong way down the one-way street – you can form a circuit whose output voltage is pretty nearly constant despite variation at the input, and despite variation in loading.

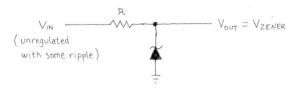

Figure 3N.26 Zener voltage source.

The plot of Fig. 3N.27 shows that the reverse conduction is not ideal: the curve is not vertical. The shape shows us why we need to follow AoE's rule of thumb that says "keep at least 10 mA flowing in the zener, even when the circuit is loaded."

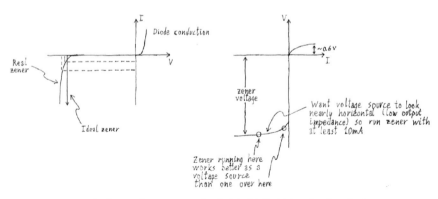

Figure 3N.27 Zener diode I–V curve: turned on its side, it reveals *impedances*.

Rotating the curve as in Fig. 3N.27 let's us see the curve's slope as a value in *Ohms*. The slope reveals $\Delta V/\Delta I$ – the device's *dynamic resistance*. That value describes how good a voltage reference the zener is: how much the output voltage will vary as current varies. (Here, current varies because of two effects: variation of *input* voltage, and variation in *output current*, or loading.) You can see how badly the diode would perform if you wandered up into the region of the curve where I_{zener} was very small.

A practical voltage source would never (well, hardly ever!) use a naked zener like the one we just showed you; it would always include a transistor or op-amp circuit after the zener, to limit the variation in *output current* (see AoE §2.2.4). The voltage regulators discussed Chapter 11 (and in AoE Chapter 9) use such a scheme, though integrated references are used in preference to zeners. At least you should

3N.7 Unregulated power supply

understand those circuits the better for having glimpsed a zener today. Now on to the diode application that you will see most often.

3N.7 Unregulated power supply

Figure 3N.28 shows a standard unregulated power supply circuit. We'll learn later just what "unregulated" means: we can't understand fully until we meet regulated supplies, supplies that use negative feedback to hold V_{out} constant despite variations in both V_{in} and output current.

AoE §9.5

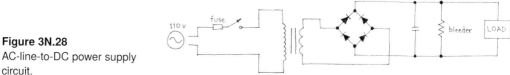

Figure 3N.28
AC-line-to-DC power supply circuit.

Now we'll look at a way to choose component values. In these notes, we will do the job incompletely, as if we were just sketching a supply. In Worked Example 3W you will find a similar case done more thoroughly. Assume we aim for the following specifications:

- V_{out}: about 12 volts;
- Ripple: about 1 volt;
- R_{load}: 120 Ω.

We must choose the following values:

- C size (μF);
- transformer voltage (V_{RMS});
- fuse rating.

We will postpone until Worked Example 3W two less fundamental tasks (because often a ball-park guess will do). Those elements that we will omit here are:

- transformer current rating; and
- "bleeder" resistor value.

Let's do this in stages. Surely you should begin by drawing the circuit without component values, as we did above.

3N.7.1 Transformer voltage

AoE §9.5.2

This is just V_{out} plus the voltage lost across the rectifier bridge. The bridge always puts two diodes in the path. Specify as *RMS* voltage: for a sinewave, that means $V_{peak}/\sqrt{2}$. Here, that gives V_{peak} at the transformer of about 14 V ≈10 V_{RMS}: see Fig. 3N.29.

Figure 3N.30 shows the surges of current that accompany the ripple. Current (labeled $I_{xformer}$ in the figure) is measured on the ground line. (Note the high scope sensitivity on that trace: 50 mV/div.)

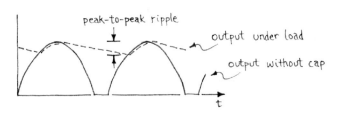

Figure 3N.29 Transformer voltage, given $V_{\text{out-peak}}$.

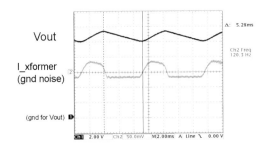

Figure 3N.30 Ripple on power supply: $I_{\text{load}} = 1\text{A}$, $C = 3300\,\mu\text{F}$.

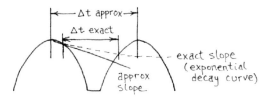

Figure 3N.31 Time details of ripple waveform.

3N.7.2 Capacitor

This is the interesting part. This task could be hard, but we'll make it pretty easy by using two simplifying assumptions. Fig. 3N.31 shows what the "ripple" waveform will look like.

We can estimate the ripple (or choose a C for a specified ripple, as in this problem) by using the general equation

$$I = C \frac{dV}{dt}$$

dV or ΔV is ripple; dt or ΔT is the time during which the cap discharges; I is the current taken out of the supply. To specify *exactly* the C that will allow a specified amount of ripple at the stated maximum load requires some thought. Specifically:

- What is I_{out}? If the load is resistive, the current out is not constant, but decays exponentially each half-cycle.
- What is ΔT? That is, for how long does the cap discharge (before the transformer voltage comes up again to charge it)?

We will, as usual, coolly sidestep these difficulties with some *worst-case* approximations:

- We assume I_{out} is constant at its maximum value; and
- We assume the cap is discharged for the full time between peaks.

Both these approximations tend to overstate the ripple we will get. Since ripple is not a good thing, we don't mind building a circuit that delivers a bit less ripple than called for.

AoE §1.6.3A

AoE §9.5.3A

Try those approximations here:

- I_{out} is just $I_{\text{out-max}}$, i.e., 100 mA;
- ΔT is half the period of the 60 Hz input waveform, i.e., $1/120\,\text{Hz} \approx 8$ ms.

What cap size does this imply?

$$C = I \times \frac{\Delta T}{\Delta V} \approx 0.1 \times 8 \times \frac{10^{-3}}{1\,V} \approx 0.8 \times 10^{-3}\,F = 800\,\mu F$$

That may sound big but it isn't for a power supply storage capacitor.

3N.7.3 Transformer current rating (*RMS* heating)

Supplement 3S

AoE §9.5.2

Because the current that recharges the capacitor comes in surges, rather than continuously, the transformer heats more than one would assume if its current flowed steadily, as the *load* current does. (See Worked Example 3W on power supplies for more on this topic.) The extra heating can be calculated using the root-mean-square ("RMS") value of the transformer current. Doing that is straightforward if you know the percentage of each cycle the transformer spends delivering current. That fraction is difficult to predict, but easy to observe.

Figure 3N.30 shows the ripple we saw on one of the lab's powered breadboards when we drew a steady 1A from the supply (here, we're digressing for a moment from the present design task, where the current is lower, at 100 mA).

During the time when the transformer feeds the storage capacitor, all the charge that is removed during a cycle by the load must be replaced. So, the magnitude of the current that flows during the recharge portion of the full cycle – roughly 3/8 in Fig. 3N.30 – must be proportionately larger than the steady *load* current. In this case, it must be 8/3 the steady 1A output current. Such a current spike heats more than a steady current that would deliver the same charge, so we must use this RMS current value to determine how large a transformer we need.

If, as in the scope image in Fig. 3N.30, the current flows during about 3/8 of the full cycle, the RMS value is

$$I_{\text{RMS}} = \sqrt{(\text{fraction of cycle during which current flows}) \times (\text{current})^2}$$
$$= \sqrt{3/8 \times (8I/3)^2} = \sqrt{3/8 \times 7.1\,I^2} = \sqrt{2.7\,I^2} \approx 1.6\,I$$

So, for a load current of 100 mA, you should specify a transformer that can handle at least 160 mA. You'd not want to put your specification right at the edge, so you might specify 200 mA.

3N.7.4 Fuse rating

The *current* in the primary is smaller than the current in the secondary, by about the same ratio as the primary *voltage* is larger. (This occurs because the transformer dissipates little power: $P_{\text{in}} \approx P_{\text{out}}$).

In the present case, where the *voltage* is stepped down about 11×, the primary current is lower than the secondary current by about the same factor. So, for 100 mA out, about 9 mA must flow in the primary.

In fact, the spikes of current that heated the transformer also heat the fuse, so the fuse feels its "9 mA" as somewhat more (see §3N.7.3); call it 20 mA. And we don't want the fuse to blow under the maximum current load. So, use a fuse that blows at perhaps four times the maximum I_{out}, adjusted for its heating effect: so 80 mA. A 100 mA, or 0.1 A fuse is a pretty standard value, and we'd be content with that. The value is not critical, since the fuse is for emergencies, an event in which very large currents can be expected.

130 Diode Circuits

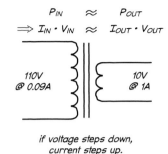

Figure 3N.32 Transformer preserves power, roughly; so, a 'step down' transformer draws less current than it puts out.

It's a good idea to use a *slow-blow* type, since on power-up a large initial current charges the filter capacitor, and we don't want the fuse to blow each time you turn on the supply. But note that it takes a pretty drastic overload to blow a fuse: a *fast-blow* fuse rated "1 A" does not blow at 1 A. The table below shows how long a small *fast-blow* glass fuse ("AGX"[11]) takes to blow, under several overloads:

% of Amp Rating	Time to Blow
110%	4 hours (min.)
135%	60 minutes (max)
200%	5 seconds (max)

3N.8 Radio!

In this section we'll meet the two elements essential to an AM ("amplitude modulation") radio receiver:

(1) a highly-selective bandpass filter that can pick out a single broadcast frequency; and
(2) a rectifier circuit that can *demodulate* the information encoded in the AM signal (with the help of an *RC* to knock out the high-frequency broadcast "carrier" signal).

3N.8.1 Step 1: *LC* selects one "carrier" frequency

Each broadcast station is alloted its peculiar frequency.[12] This is called a "carrier" frequency, chosen for its ability to propagate well, and conveying information by being varied or "modulated" in some way. The earliest and simplest of these modulation schemes, AM, does what the name suggests: varies the size of the carrier signal. It lets an audio signal in the range of a few kilohertz (talk or music) modulate a carrier of about 1 MHz. (In Lab 13L you will use FM, a method that is more robust than AM but not quite so simple. The FM of Lab 13L will use not radio frequencies but a much lower-frequency carrier, easier to demodulate.)

An *LC* is well-adapted to the task of selecting one station's carrier: it is the most sensitively frequency-selective circuit we now know in the course, and can easily be set to resonate at standard AM frequencies, say 1 MHz. One does not need to insert an *R* to feed the parallel *LC*; the antenna's impedance stands in for the *R* that we use in the first experiment of Lab 3L.

The *LC* not only rejects unwanted radio broadcasts, but also rejects the 60 Hz noise that is a good deal larger than any of the radio signals. One can see this effect in Fig. 3N.33. The left-hand image shows the 60 Hz sinusoid, strangely thick. The thickening is the ≈ 1 MHz "carrier." It is hard to notice

---

[11] Source: Datasheet for Cooper–Bussmann AGX Series Fast-Acting Glass Tube Fuse, current ratings 1/16 A to 2 A.
[12] Strictly, each station is assigned not one frequency but a *range* of permitted frequencies: 10 kHz wide, for AM broadcast, 200 kHz wide for FM ("frequency modulation").

the variation in thickness – but this variation carries all the information. In the right-hand image, the 60 Hz variation has disappeared, filtered out by the resonant circuit.

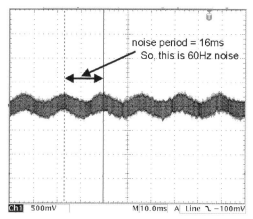

antenna signal, attached to nothing but scope probe

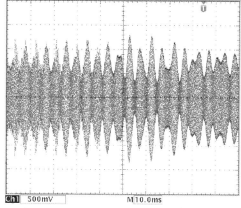

...antenna signal when attached to LC resonant at broadcast carrier frequency: 60Hz noise is gone, and signal (amplitude variation) is much larger

Figure 3N.33 Radio signal: raw, from antenna, and after *LC* frequency-selection.

A second effect, more interesting and surprising, is visible in the right-hand image of Fig. 3N.33. The resonant *LC* is doing something intriguingly new. The output amplitude of the *LC* is *larger than the input*, for the selected carrier frequency (notice that the scope sensitivity is the same for the two screen shots of Fig. 3N.33: the *LC* really *does* enlarge the modulating signal).

That's a trick no *RC* could perform.[13] This is possible only when the input is more like a *current source* than *voltage source*, as is the case for the antenna. The *LC* stores energy, and – like a pendulum, or a child's swing pushed repeatedly by an attentive parent – the *LC* amplitude can keep growing beyond the voltage-amplitude of the driving signal. Similarly, the child's swing can move far beyond the limited range of the parent's push if the pushes are properly timed, adding a little energy each time the child passes.

Here's another way to say why output can grow larger than input (this account is somewhat less metaphorical): if Q is large – as it is, here – then a good deal more energy is put in than is lost in a cycle when amplitude out equals amplitude in. So the amplitude grows until the rate at which energy is dissipated reaches the rate at which it's coming in. That happens when the amplitude is a good deal larger than the original input level. Hence the enlargement of the signal by this *passive* circuit ("passive" meaning 'no borrowing of power from a supply, as in an amplifier;' all the energy comes from the signal source).

3N.8.2   Step 2: detect "envelope" of AM waveform

If we sweep the scope fast enough to resolve the carrier frequency, we see its amplitude modulation, which looks like strange uncertainty about amplitude: see Fig. 3N.34.

The left-hand image in Fig. 3N.34 shows varying amplitudes superimposed; the right-hand image shows a single-shot trace. Now, the task is to recover the information that lies in the amplitude variation, while ignoring the high-frequency carrier.

---

[13] It's also a trick the *RLC* circuit of Lab 3L cannot perform. That circuit, driving the *LC* from a voltage source through a resistor, can never deliver amplitude greater than the input amplitude. This is because the "coupling" of source to *LC* through a resistor is "lossy" (in the jargon), whereas the coupling from antenna to *LC* in the radio circuit is not: it is capacitive, instead.

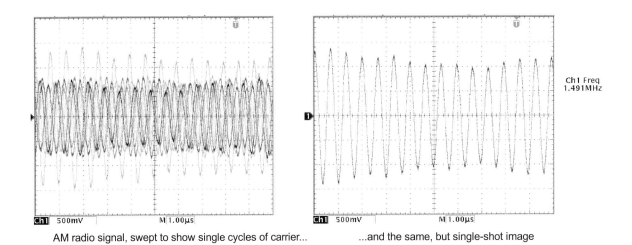

AM radio signal, swept to show single cycles of carrier...   ...and the same, but single-shot image

Figure 3N.34 Radio carrier resolved, showing its amplitude variation.

Remove the carrier: The detector should ignore or remove the "carrier" frequency (around 1 MHz). The carrier can be stripped away, now that it has done its job, which was to deliver the audio information. The carrier is placed at a high frequency (around 1.5 MHz in the images) because such signals travel much better than lower frequency signals.

Two concerns call for use of a radio *carrier*. One strong reason to use a carrier, rather than try to transmit the audio itself as a radio signal, is the antenna size that audio frequency signals would require. The 1 MHz AM signal is most efficiently received by an antenna a half-wavelength long: about 500 feet; we get by with a less-efficient antenna about 1/5 that long in our lab; if we were to transmit 1 kHz audio as a radio signal, an optimal antenna (half-wavelength) would be about 150 *kilometers* long! Cellphones, incidentally, can use very small antennas, thanks to their high carrier frequencies: most GSM phones, for example (TMobile and AT&T) use about 2 GHz, calling for an antenna about three inches long.[14]

Another powerful reason for using a carrier is the need to permit multiple broadcasts in a locality without mutual interference. That would not be possible if one simply transmitted audio frequencies without a carrier.

Removing the carrier sounds like a task for a lowpass filter. Well, it is and it isn't. A simple lowpass – meant to keep the "audio" signal (no more than 5 kHz, under the AM standard) while knocking out the carrier – would unfortunately get rid of everything: baby would exit with bathwater. So, before we apply a lowpass, we need another of Lab 3L's circuits . . . .

Detect the envelope: A lowpass alone would fail: if applied to the signal arriving from the *LC*, a lowpass would get rid of the carrier – and with it, all its amplitude variations – since the voltage swings of the carrier are *symmetric*. The integrating effect of the lowpass delivers a nicely-centered *zero* as output! What's needed, to prevent this sad result, is a way to *spoil the symmetry* of the *LC*'s output. If we rectify the *LC* output before applying the lowpass, we find that the shape of the envelope survives.

---

[14] A good article on antennas (and not very mathematical), is available at: `https://LAoE.link/Antennas_101.pdf`.

3N.8.3 What the result looks like

Conceptually, the *rectification* of the modulated carrier, and then the *lowpass* elimination of the carrier are successive steps. The envelope, embodying the audio information, appears at last in the bottom trace on the right in Fig. 3N.35. (Note that this *envelope* comes from the carrier shown above it, not from the left-hand scope image of Fig. 3N.35, which records a different moment in the radio broadcast.)

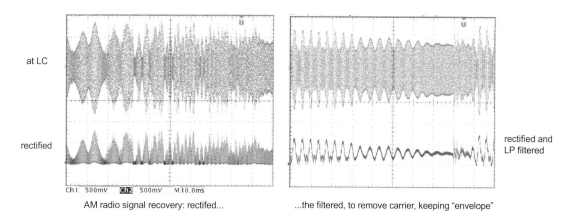

Figure 3N.35 AM signal recovery stages: rectification, then lowpass filtering.

3N.8.4 How to recover the AM information

Half-wave or full-wave?

- *Half-wave is enough...* Having met the *full-wave* rectifier as an improvement on the *half-wave*, for use in power supplies, you could quite reasonably suppose that the full-wave is needed for your radio. You might even fear that a half-wave could lose half the information conveyed by the AM signal – losing the excursions below ground, keeping only those above ground. But this turns out not to be the case. The symmetry of the carrier waveform means that we lose nothing by measuring its positive excursions only: we get the amplitude information. If you worry that the carrier may change amplitude between the positive and negative swings (indeed, *must* change amplitude, sometimes), don't worry: our detector will average the amplitudes of *hundreds* of cycles of the carrier, since one period of the highest audio signal (5 kHz) will encompass about 200 cycles of a 1 MHz carrier.
- *... half-wave is better than full-wave* In fact, there are good reasons to shun the full-wave rectifier, in this context:
  - *full-wave rectifier's input and output cannot share a single definition of ground.* If the output is referenced to ground (the usual scheme), then neither input terminal can be grounded. This truth, mentioned in §3N.6.2 and illustrated in Fig. 3N.22, was the basis for a warning that appears near the start of §3L.4.

In a power supply derived from the output of a transformer, this restriction is not troublesome; the transformer winding needs no reference to ground. And if one looks closely at input and output waveforms (while heeding the warning not to ground the input), one sees each input going a diode drop *negative* on the half cycle when it does not squirt positive charge into the output. We saw this negative excursion in Fig. 3N.23.

This property makes the bridge very inconvenient in the radio: either input or output must give up its connection to "ground." The antenna likes a connection to world ground; output doesn't absolutely need it, but the lack of it is very inconvenient. If you had to "float" your output ground, you could not use the scope to watch V_{out} in the usual way, because the scope ground is internally tied to world ground. After grounding one of the input terminals of the bridge, you could take the trouble to use the scope in *differential* mode to watch V_{out}, by attaching *two* probes to the output: one to the normal V_{out} terminal, the other to your floating "ground," then subtracting one from the other. But that's a nuisance.

- *The full-wave's two-diode drops can lose too much of the signal.* The signal coming from the antenna is only about a volt, so it's better not to lose *two* diode drops from its amplitude – even when using low-drop Schottky diodes.[15] The simple half-wave rectifier loses just one diode drop voltage.

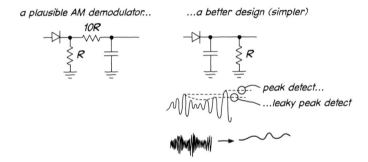

Figure 3N.36 Leaky peak detector is simpler than rectifier followed by lowpass.

Use a strange lowpass with the rectifier

The straightforward way to get rid of carrier and detect the envelope certainly seems to be to build a rectifier as usual (using a diode and an R to ground), then apply a lowpass (using a $10R$ resistor value, in order not to load the rectifier). That would work, but a simpler and sufficient circuit does the job with a single R. The diode feeds a cap, whose other end is tied to ground; diode and cap form a "peak detector," charging the cap to the highest input voltage (less a diode drop). Then, to allow the peak detector to droop when amplitude falls, we provide an R to ground (paralleling the cap), forming what we call a "leaky peak detector," as in Fig. 3N.36.

This circuit is hard to analyze in the frequency domain, but very easy in the time domain: we want the peak detector's decay-rate to be slow relative to the quick wiggles of the 1 MHz carrier, but quick relative to the wiggles of the audio waveform. Those two frequencies – carrier versus audio – are so far apart that it is very easy to find an RC time-constant somewhere between. In fact, the RC value is not very critical. In lab, you may want to experiment, listening to determine how sensitive your radio's sound output is to your detector's RC value.

---

[15] Forward voltage drop for the 1N5711 Schottky diode that we recommend for the lab radio is 0.4 V @ 1 mA.

3N.9 AoE reading

Reading:
> Finish Chapter 1, including §§1.5–1.6.8; and §1.7.2 on inductors, transformers and diodes – sections omitted last time.
> Appendix on drawing schematics.

> Appendix I: *Television: A Compact Tutorial*.

Problems:
> Problems in text and Additional Exercises 1.43, 1.44.

3L Lab: Resonant and Diode Circuits

3L.1 *LC* resonant circuit

3L.1.1 Response to sinusoid

Construct the parallel resonant circuit shown in Fig. 3L.1. Drive it with a sinewave, varying the frequency through a range that includes what you calculate to be the circuit's *resonant frequency*. Compare the resonant frequency that you observe with the one you calculated. (The circuit attenuates the signal considerably, even at its resonant frequency; the *L* is not perfectly efficient, but instead includes some series resistance.) As you watch for $f_{\text{resonance}}$ watch not for the frequency that delivers maximum amplitude out (this is difficult to determine), but instead watch for the other signature of resonance: the frequency where output is *in phase* with input.

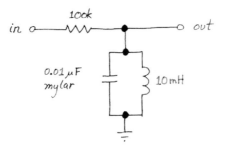

Figure 3L.1 *LC* parallel resonant circuit.

3L.1.2 *Q*: quality factor

Estimate the circuit's *Q* ("quality factor," a term whose name apparently reflects the use of resonant circuits in radio tuners, where high selectivity is good): *Q* is defined as

$$Q \equiv \frac{f_{\text{resonance}}}{\Delta f}$$

where "Δf" represents the width of the resonance peak between its "half-power" or -3 dB points. The skinnier the peak, the higher the *Q*; see Fig. 3L.2.

You can make a very *good* measurement of *Q* if you use a *frequency counter*[1] to reveal the small change in frequency between the points, below and above $f_{\text{resonance}}$, where output amplitude is down 3 dB. Note that this is "down 3 dB" relative not to amplitude *in* but relative to the maximum amplitude *out*: amplitude at resonance. Amplitude out never equals amplitude in, which it would if our components were perfect.

---

[1] Your DVM may include such a frequency counter. This will be necessary for those using analog oscilloscopes. Or, if you are using a digital scope, it will measure the frequency for you.

3L.1 LC resonant circuit

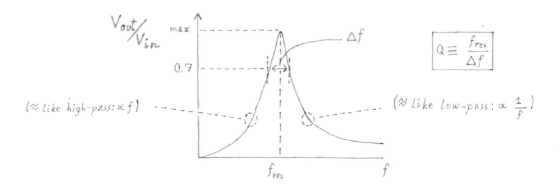

Figure 3L.2 You can measure Q, getting a precise Δf measure with a DVM.

If you're energetic and curious, you can look at the effect on Q of substituting a 10k resistor for the 100k. You'll notice that the amplitude out increases, with this reduced R. The fraction that survives, V_{out}/V_{in}, grows. But as amplitude out grows, Q degrades. As usual, you're obliged to trade away one desirable trait to get another. Good Q, however, is likely to be much more important than large amplitude; an amplifier can solve the problem of low amplitude.

3L.1.3   A pretty sweep

Use the function-generator's *sweep* feature to show you a scope display of amplitude-out versus frequency. (See §2S.3 if you need some advice on how to do this trick.)

When you succeed in getting such a display of frequency response, try to explain why the display grows funny wiggles on one side of resonance as you increase the sweep rate. *Clue:* the funny wiggles appear on the side after the circuit has already been driven into resonant oscillation; the function generator there is driving an oscillating circuit.

3L.1.4   Finding Fourier components of a square wave

This resonant circuit can serve as a "Fourier Analyzer:" the circuit's response measures the amount of 16 kHz (approx.) present in an input waveform.

Try driving the circuit with a *square* wave at the resonant frequency; note the amplitude of the (sinewave) response. Now gradually lower the driving frequency until you get another peak response (it should occur at 1/3 the resonant frequency) and check the amplitude (it should be 1/3 the amplitude of the fundamental response). With some care you can verify the amplitude and frequency of the first five or six terms of the Fourier series.

Figure 3L.3 shows the first few frequencies in the Fourier series for a square wave. You met this series earlier today as Fig. 3N.16.

3L.1.5   Classier: frequency spectrum display

If you *sweep* the *square wave* input to your 16 kHz detector, you get a sort of inverse frequency spectrum: you should see a big bump at $f_{resonance}$, a smaller bump at $\frac{1}{3} f_{resonance}$, and so on.

138 Lab: Resonant and Diode Circuits

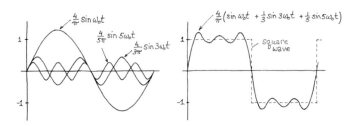

Figure 3L.3 Fourier series for square wave.

3L.1.6 Ringing

Now try driving the circuit with a low-frequency square wave: try 20 Hz (but note that any low frequency will do; what matters is the steep edge of the waveform with its high-frequency components. Or, if you prefer, think of the edge as putting a jolt of energy into the resonant circuit, energy that sloshes back and forth between L and C until it has been dissipated). You should see a brief output in response to each edge of the input square wave. If you look closely at this output, you can see that it is a decaying sinewave. (If you find the display dim, increase the square wave frequency to around 100 Hz.)

What is the frequency of this sinewave? (No surprise, here.)

Does it appear to decay exponentially? You may recall that we noted this behavior back in §3N.4.4.

Estimating Q from decay envelope: As we said in §3L.1.2, one can evaluate Q by noticing how fast the oscillation *envelope* decays away, after a stimulus to the RLC. Specifically (to quote ourselves),

$$Q = 2\pi \times (N_{\text{decay}})$$

where N_{decay} is the number of cycles required for energy to diminish to $1/e$ – where amplitude is down to about 60%, since energy is proportional to V^2 (see Chapter 3N).

When you see your own circuit's response to the slow square wave, count the cycles before decay to about 60%, and see if your Q, so calculated, matches the value you found by the method of §3L.1.2. See if you can confrm the claim we made in §3N.4.4, and illustrated with Fig. 3N.17, that poor probe grounding will produce ringing. While watching a square wave, remove the probe's ground clip. Satisfactorily ugly?

3L.2 Diode X–Y plots: a vivid way to display diode *I–V* behavior

We asked you to plot diode current and voltage, back in §1L.4. Part of the point was to get you used to the idea that not every device likes to follow Ohm's Law. Diodes don't.

A much less laborious way to show *I–V* behavior uses the oscilloscope to draw the plot for us.

Normally, the scope plots *voltage* (displayed as a vertical displacement, on what would be called the "Y" axis, in Cartesian coordinates) versus *time* (on the horizontal, or "X" axis). Occasionally, however, it's convenient to plot Y (vertical displacement) versus X (horizontal displacement), where horizontal displacement is determined not by time but by a second signal. This arrangement is described in §2S.3.2 and §2S.3.4 on "X–Y" displays.

Such a display lets you see at a glance the *I–V* behaviors of several sorts of diode. Here is the setup.

3L.2 Diode X–Y plots: a vivid way to display diode *I–V* behavior

3L.2.1 Setting up the diode X–Y plot

First a schematic. Fig. 3L.4 shows diode and resistor meeting at circuit *ground* – used as ground *only by the scope*. It is essential to *float* the generator's output, so that neither of its output lines is tied to ground. The Krohn-Hite function generators have a switch on the back to achieve this (see Fig. 5L.5). If your function generator can run on batteries or its AC plug or power adapter has only two prongs, its output is probably floating as well.

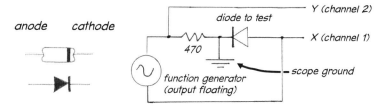

Figure 3L.4 XY diode plot wiring: floating function generator and scope channels tie to diode and resistor ends, junction is scope ground.

If you cannot float the output of your function generator, you can use an isolation transformer to drive the test circuit: see Fig. 3L.5.[2] Here, both the function generator and scope are connected to circuit ground.

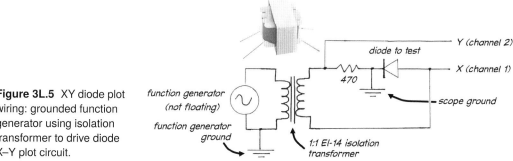

Figure 3L.5 XY diode plot wiring: grounded function generator using isolation transformer to drive diode X–Y plot circuit.

In order to generate a conventional plot output, with positive current vertical, positive diode-voltage to right, we must invert one of the two channels. We inverted Channel 2, the one measuring diode *current*. You can experiment with the inversion of the channels separately to get a display that you like.

Figure 3L.6 on the next page shows what the connections using a floated function generator output look like (we omitted the scope connections in this figure).

The oscilloscope's channels 1 and 2 monitor voltages across the diode and the resistor, respectively. The voltage across the resistor indicates diode current, of course. So the two channels show us *I* vs *V*. Since we wanted positive current to produce positive diode voltage, we were obliged to invert one or the other of the displayed channels. We chose to invert Channel 2, which shows current.

It is best to use a low frequency sinusoid to drive this circuit. We used 100 Hz, and set scope gain to 500 mV/div, except for the blue LED and the zener diode, which do not conduct until the applied voltage is above a couple of volts. Set the scope *display mode* to "X–Y," Ch1 vs Ch2, rather than use the normal X-versus-time display.

---

[2] Any 600 Ohm, 1:1 audio transformer should work. We bought a pack of 10 on Amazon for $7. See
https://LAoE.link/Isolation_Xformer.html. These are also available on eBay and AliExpress – search for "EI-14 1:1 transformer." These transformers have no polarity, so either side can be connected to the function generator.

Lab: Resonant and Diode Circuits

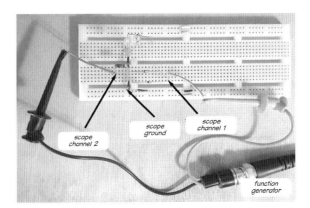

Figure 3L.6 What diode wiring may look like. Here, all the alternative diodes are shown in place. This helps one to compare responses.

3L.2.2 The several diodes

We suggest you try four diode types:

- Conventional silicon diode: 1N914/1N4148.[3]
- Schottky diode, low forward drop: 1N5817 or 1N5711.
- Light emitting diodes: LEDs in several colors:
  - red;
  - blue or white.
- Zener diode: this is the oddball in the set: its interesting behavior occurs when current passes through it *backwards*, relative to all the others. This is non-destructive reverse conduction, and it provides the zener's useful behavior: a way to hold an output voltage constant despite variation of input voltage. Zeners come in voltages from a few volts to several hundred volts.

3L.2.2.1 LED orientation

Whereas a band on a conventional diode indicates the cathode (the negative lead), LEDs use a different scheme. There are two and sometimes three ways to identify the LED cathode: see Fig. 3L.7.

- If the LED leads have not been cut, then the shorter leg is the cathode. That's easy.
- If leads have been cut, still you have two ways to spot the cathode:
  - One side of the plastic package *usually* is flattened; that is the cathode.
  - When you look through the side of the package, the *larger* structure (the "anvil") is the cathode. This method almost always is available to you. With a clear package the internal structure is very obvious.

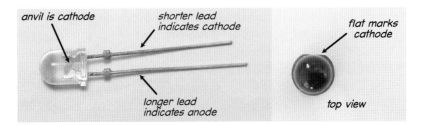

Figure 3L.7 Lead length or internal structure can identify LED leads.

---

[3] The 1N914 and 1N4148 are identical devices. Either may be used in the lab exercises.

3L.2.3 Sample X–Y plots

In case it helps to see some sample X–Y plots, we offer these examples – hoping, however, not to spoil your fun by revealing too much.[4]

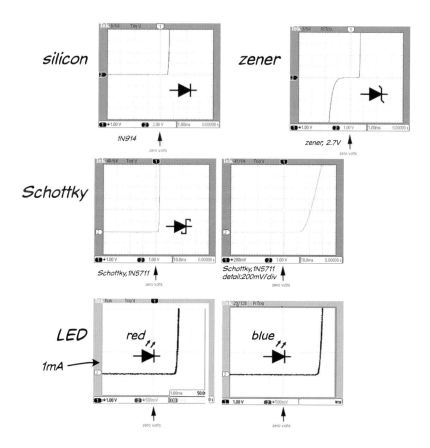

Figure 3L.8 *I* vs *V* plots for several different types of diodes.

A look at this set of plots may complicate our rule of thumb that "a diode drop" is about 0.6 V at 1 mA. While correct for the ordinary silicon diode, and a useful rule, you can see that it does not hold for all diodes.

3L.3 Half-wave rectifier

Construct a half-wave rectifier circuit with a 6.3 V AC (RMS) transformer[5] and a 1N914/1N4148 diode, as in Fig. 3L.9. Connect a 2.2k load, and look at the output on the scope. Is it what you expect? Polarity? Why is V_{peak} more than 6.3 V? (Don't be troubled if V_{peak} is even a bit more than 6.3 V×√2:

---

[4] Fig. 3L.8 are some plots we captured.
[5] You might well wonder where such a weird voltage came from: "6.3 V?" The answer comes from the history of vacuum tube radios. The tubes had filaments that needed to be heated in order to emit electrons. In early days they were heated by batteries: often by three lead–acid cells in series, each with open-circuit voltage of 2.1 V. When AC supplies were adopted to take the place of the battery, their *RMS* voltages were set to match the heating effect of the battery. Note that the 6.3 V RMS AC waveform has a peak value $\sqrt{2} \times V_{RMS}$ and provides the same power as a DC supply at the sinusoid's *RMS* value.

the transformer designers want to make sure your power supply delivers at least what's advertised, even under heavy load; you're here loading it very lightly.)

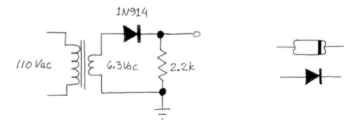

Figure 3L.9 Half-wave rectifier.

3L.4 Full-wave bridge rectifier

Now construct a full-wave bridge circuit, as in Fig. 3L.10. Be careful about polarities: the band on the diode indicates cathode, as in the figure. Look at the output waveform (but **don't** attempt to look at the input – the signal across the transformer's secondary – with the scope's other channel at the same time; this would require connecting the second "ground" lead of the scope to one side of the secondary. What disaster would that cause?[6]). Does it make sense? Why is the peak amplitude less than in the last circuit? How much should it be? What would happen if you were to reverse any one of the four diodes? (**Don't try it!**).

Don't be too gravely alarmed if you find yourself burning out diodes in this experiment. When a diode fails, does it usually fail *open* or *closed*? Do you see why diodes in this circuit usually fail in pairs – in a touching sort of suicide pact?[7]

Look at the region of the output waveform that is near zero volts. Why are there flat regions? Measure their duration, and explain.

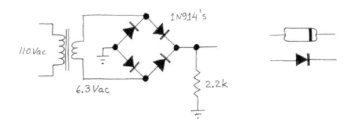

Figure 3L.10 Full-wave bridge.

3L.4.1 Ripple

Observe ripple, given C and load: Now connect a 15 μF filter capacitor across the output (**Important**: observe polarity). Does the output make sense? Calculate what the "ripple" amplitude should be, then measure it. Does it agree? (If not, have you assumed the wrong discharge time, by a factor of 2?)

---

[6] The disaster results from the consequent *shorting out* of one of the diodes. (Scrutinize Fig. 3L.10 to see why.) As a result, nothing limits the current in a second diode, which is guaranteed to burn out.

[7] This suicide pact makes troubleshooting the full-wave bridge interesting, in the case the output waveform looks wrong. Disconnect the transformer so that you can test the diodes (using a DVM's "diode test" function) and don't stop on detecting a single bad diode. If you find one dead diode, expect to find one more as well.

3L.4.2 Design exercise: choose C for acceptable ripple

Now suppose you want to let your power supply provide a current of up to 20 mA with ripple of about 1 V peak-to-peak. Your design task is to do the following:

- choose R_{load} so as to draw about 20 mA (peak); and
- choose C so as to allow ripple of about 1 V.

Draw your design and try your circuit. Is the ripple about right? (Explain, to your own satisfaction, any deviation from what you expected.) This circuit is now a respectable voltage source, for loads of low current. To make a "power supply" of higher current capability, you'd use heftier diodes (e.g., 1N4002) and a larger capacitor. (In practice you would always follow the power supply with an active *regulator*, a circuit you will meet in Lab 11L.)

3L.5 Design exercise: AM radio receiver (fun!)

To make this exercise fun, you'll need a strong source of radio signals: that requires a pretty good antenna (or a poor antenna whose signal has been amplified for you by someone who knows how to make a high-frequency amplifier).

We get a strong signal in our teaching lab by running about 30 feet of wire from the window of the lab to a fire escape on the next building. The antenna is nothing fancy: just an old piece of wire, insulated from the fire escape by a piece of string. It gives us almost a volt in amplitude.

If you looked (with a scope) at the signal coming from the antenna, it would look something like the image on the left in Fig. 3N.33 on page 131. After selection with a resonant circuit, the signal would look like the scope image on the right side of that figure. The 60 Hz noise is gone; that much is apparent. You cannot see a second benefit: the "carriers" of other radio stations also have been eliminated from the muddle of frequencies that came in on the antenna.

3L.5.1 A small but remarkable point

As we noticed in §3N.8.1, the resonant circuit not only selects a carrier frequency from among many; in addition, it makes the amplitude of the carrier that is selected much *larger* than in the original signal that came in the window.

3L.5.2 Detecting the AM signal

In order to detect an AM radio signal, you need to do two tasks – and you already know how to do them:

(1) rectify the signal (use a *Schottky* diode: 1N5711 or similar; its low forward-voltage will let it rectify a signal of just a few tenths of a volt); and
(2) then lowpass filter this rectified signal.

The output of your circuit will be a small audio signal (much less than a volt). It may be audible on old-fashioned high-impedance earphones; it can be made audible on an ordinary 8-ohm speaker if someone provides you with an audio amplifier with a gain of 20 or so (an LM386 audio amplifier works fine). You will recognize that we have offered you only a strategy, not part values. We said "select the carrier," but didn't say how. We said "rectify," but did not suggest a value for the R to ground. We said "lowpass filter," but did not suggest $f_{3\text{dB}}$. So, we have left to you some hard – and interesting – parts of the job. Here are some suggestions.

- To detect the carrier, use an *LC* circuit like the one you built at the start of this lab, but showing the following differences:
    - the resonant frequency should be around 1 MHz; and
    - you need no upper resistor in the "divider:" the antenna can drive The *LC* directly.
- The value of the resistor to ground is not critical; try 10k.
- The lowpass filter's job is to kill the carrier, keep the audio. Fortunately, these two frequencies are *very* far apart; so, you have a lot of freedom in placing f_{3dB}. The *form* of the lowpass you design may strike you as odd (though this depends on the way you chose to do the task: the "odd" configuration, described in §3N.8.4, Fig. 3N.36, uses the rectifier's resistor to ground as the *R* in the *RC* lowpass). Just make sure to put this in *time-domain* terms: that *RC* is very long relative to the period of the 1 MHz carrier," but short relative to the "signal" or "audio" period.

The reward we hope you will get is, of course, to *hear* the radio signal. (You will probably need to experiment with your *LC* circuit by tacking in small additional caps in parallel, in order to select a particular station (or you can cheat by doing what everyone else who ever built a radio does: use a variable capacitor!) If you lack both high-impedance earphones and an audio amplifier, then at least *look* at the fruits of your labor on the scope.

You should see something more or less like the AM waveforms we showed in Chapter 3N. We hope you will look, with scope, first at the "raw" unfiltered mixture of signals coming off the antenna (before connecting this to the parallel *LC*), then at the selected carrier (selected by the resonant circuit), then the rectified audio plus carrier, and finally the filtered and rectified output.

3L.6 Signal diodes

3L.6.1 Rectified differentiator

Use a diode to make a rectified differentiator, as in Fig. 3L.11. Drive it with a square wave at 10 kHz or so, at the function generator's maximum output amplitude. Look at input and output, using both scope channels. Does it make sense? What does the 2.2k load resistor do? Try removing it.

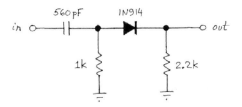

Figure 3L.11 Rectified differentiator.

Hint: You should see what appear to be *RC* discharge curves in both cases – with and without the 2.2k to ground. The challenge here is to figure out what determines the *R* and *C* that you are watching – and this problem is quite *subtle*![8]

3L.6.2 Diode clamps

AoE §1.6.6C

Construct the simple diode clamp circuit shown in Fig. 3L.12. Drive it with a sinewave from your function generator, at maximum output amplitude, and observe the output. If you can see that the clamped voltage is not quite flat, then you can see the effect of the diode's non-zero impedance. Perhaps you can estimate a value for this *dynamic resistance* (see §3N.6.3); try a triangle waveform if you attempt this estimate.

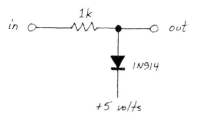

Figure 3L.12 Diode clamp.

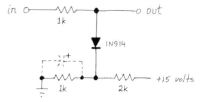

Figure 3L.13 Clamp with voltage divider reference.

AoE §1.6.6, Fig. 1.76

Now try using a voltage divider as the clamping voltage, as shown in Fig. 3L.13. Drive the circuit with a large sinewave, and examine the peak of the output waveform. Why is it rounded so much? (*Hint*: What is the impedance of the "voltage source" provided by the voltage divider? If you are puzzled, try drawing a Thevenin model for the whole circuit. Incidentally, this circuit is probably best analyzed in the *time* domain.) To check your explanation, drive the circuit with a triangle wave.

As a remedy, try adding a 15 μF capacitor, as shown with dotted lines (note polarity). Try it out. Explain to your satisfaction why it works. (Here, you might use either a time- or frequency-domain argument.) This case illustrates well the concept of a bypass capacitor. What is it bypassing, and why?

---

[8] When you remove the 2.2k to ground, the scope probe itself becomes important. The discharge path now is extremely slow: the probe's capacitance (perhaps 12 pF) discharges through the probe's 10M to ground. At a high input repetition rate, a square wave may produce what looks like a flat output.

3S Supplementary Notes and Jargon: Diode Circuits

3S.1 A puzzle: why *LC*'s ringing dies away despite Fourier

A student asked a good, hard question, recently. I was stumped, till the answer struck me – more or less the way the apple is said to have bonked Newton on the head – next morning as I cycled to work.

3S.1.1 The puzzle

A low-frequency square wave (say, 50 Hz) evokes a brief shiver from an *LC* circuit resonant at 16 kHz (the circuit we built in Lab 3L). We explain this response by noticing that the square wave has a harmonic at 16 kHz, though its amplitude is small. Our student put the puzzle this way: "Fourier says the 16 kHz component is present not just at the edges, but throughout the waveform. Why, then, is the resonant circuit not stimulated continuously? Why only at the edges?"

A good question.

3S.1.2 The solution

The key point is this: the resonant circuit is responsive to *a range of frequencies*, not just to 16 kHz. A first (and wrong) inference might be, 'Oh: then we should see still more activity between the edges, stimulated by several harmonics.' That inference is wrong because *it's only at the edges* that the phases of all the several harmonics coincide, reinforcing one another. Away from the edges, they tend to cancel: and that is why (of course? well, none of this seems obvious to me) the square wave is *flat* between its edges: the several harmonics conspire to cancel one another, once the edge-step has occurred.

The resonant circuit passes a *set* of harmonics, not just one; the members of this set cancel one another, away from the edges. This happens because the circuit's Q is wide enough to let through this set of frequencies.

The limited-Q explanation finds confirmation in the *time domain*, as well. The ringing after a square-wave's edge decays at a rate described by Q: high Q implies slow decay (Q, in this view, measures energy loss per cycle of oscillation).

And you saw something like this earlier in the Lab 3L exercise When the square wave frequency is *not* very far below $f_{resonance}$ (say, at 1/3 or 1/5 of $f_{resonance}$), the "ringing" – the 16 kHz harmonic – *does* persist all the way through, as the square wave sits high or low. In Fig. 3S.1 are images from Lab 3L showing the expected components of a square wave at 3, 7 and 9 times the frequency of the square wave. In all these cases, the fourier component persists across the "flat" section of the square wave. Why does it persist?

In the *time domain* the answer seems obvious: it hasn't had time to die away. But this persistence also is (of course?) consistent with the frequency-domain view: the nearest *other* harmonics are far enough from $f_{resonance}$ so that they are knocked out.

In other words, we can explain this case (where the ringing *does* persist), as we could explain the

3S.2 Jargon: passive devices

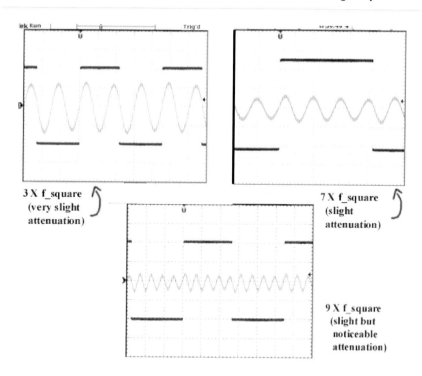

Figure 3S.1 Fourier components are sustained if *f*-square is not far from *f*-resonance.

case where the ringing died away, by referring to the effect of limited Q, in both time and frequency domains: in this case, the Q is high enough so that:

- in time domain: the ringing does not attenuate much, in the available time (the half-period of the square wave); and
- in frequency domain: the circuit is selective enough to keep the 16 kHz harmonic while killing any neighboring harmonics that would have cancelled the 16 kHz, away from the edges.

And in Fig. 3S.2 is the case that evoked the original question, where f_{square} is far below $f_{\text{resonance}}$, and the ringing dies away.

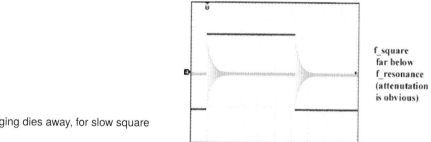

Figure 3S.2 Ringing dies away, for slow square wave, of course.

3S.2 Jargon: passive devices

choke Inductor (noun).

droop Fall of voltage as effect of loading (loading implies drawing of current); also, waveform distortion from effect of highpass filter, applied, say, to low-frequency square wave.

primary Input winding of transformer.

Q "quality factor" describes the sharpness of the peak of a frequency-selective RLC circuit:

$$Q = \frac{f_{\text{resonance}}}{\{3\text{ dB width}\}}$$

or, equivalently, Q can be defined as time for resonant oscillation to decay:

$$Q = \text{number of radians for energy to decay to } 1/e \text{ of its peak value}$$

ripple Variation of voltage resulting from partial discharge of power-supply filter capacitor between re-chargings by transformer.

risetime Time for waveform to rise from 10% of final value to 90%.

RMS Root mean square (more literally, "root mean of squares"). Used to describe power delivered by time-varying waveform. For sine, $V_{\text{RMS}} = V_{\text{peak}}/\sqrt{2}$. This is the DC voltage that would deliver the same *power* as the time-varying waveform.

secondary Output winding of transformer.

stiff When applied to a voltage source it means it "droops" little under load.

V$_{\text{peak}}$ "Amplitude." For example in $v(t) = A \sin \omega t$, the term A is peak voltage (see AoE §1.3.1).

V$_{\text{peak-to-peak}}$ or V$_{\text{p-p}}$ Another way of characterizing the size of a waveform; much less common than V_{peak}.

3W Worked Examples: Diode Circuits

3W.1 Power supply design

3W.1.1 Another power supply

Here we will do a problem much like the one we did more sketchily in Chapter 3N. If you are comfortable with the design process, skip to §§3W.1.6 and 3W.1.7, where we meet some new issues.

We are to design a standard *unregulated power supply* circuit. In this example we will look a little more closely than we did in class at the way to choose component values. Here's the particular problem.

Problem (Unregulated power supply) Design a power supply to convert 110 V AC 'line' voltage to DC. Aim for the following specifications:

- V_{out}: no less than 20 V;
- Ripple: about 2 V; and
- I_{load}: 1 A (maximum).

Choose:

- C size (μF);
- transformer voltage (V_{RMS});
- fuse rating (I);
- "bleeder" resistor value; and
- transformer current rating (current in secondary of transformer).

Questions: What difference would you see in the circuit output:

- If you took the circuit to Europe and plugged it into a wall outlet? There, the line voltage is 230 V, 50 Hz.
- If one diode in the bridge rectifier burned out (open, not shorted)?

3W.1.2 Skeleton circuit

First, as usual, we would draw the circuit without component values; see Fig. 3N.28. The fuse goes on primary side, to protect against as many mishaps as possible – including failures of the transformer and switch. Always use a *full-wave* rectifier (a bridge); most of the *half-wave rectifier* circuits you see in textbooks are relics of the days when "diode" meant an expensive vacuum tube. Now that diode means a little hunk of silicon, and they come as an integrated *bridge* package, there's rarely an excuse for anything but a bridge. A "bleeder" resistor is useful in a lab power supply, which might have no load: you want to be able to count on finding close to 0 V a few seconds after you shut power off. The bleeder achieves that. Many power supplies are always loaded, at least by a regulator and perhaps by the circuit they were built to power; these supplies are sure to discharge their filter capacitors promptly and need no bleeder.

3W.1.3 Transformer voltage

This is just the peak value of V_{out} plus the two *diode drops* imposed by the bridge rectifier, as usual. If V_{out} is to be 20 V *after* ripple, then $V_{out(peak)}$ should be two volts more: around 22 V. The transformer voltage then ought to be about 23 V.

When we specify the transformer we need to follow the convention that uses V_{RMS}, not V_{peak} (remember, V_{RMS} defines the *DC* voltage that would deliver the same power as the particular waveform). For a sinewave, V_{RMS} is $V_{peak}/\sqrt{2}$, as you know.

In this case,

$$\frac{V_{peak}}{\sqrt{2}} \approx \frac{23\,V}{1.4} = 16\,V_{RMS}$$

This happens to be a standard transformer voltage.

If it had not been standard, we would have needed to take the next higher standard value, or use a transformer with a 'tapped primary' that allows fine tuning of the step-down ratio.

3W.1.4 Capacitor

Figure 3W.1 is the "ripple" waveform, again. We have labeled the drawing with reminders that Δt depends on circumstances – in this particular, rather contrived, problem where we ask you to carry your supply to Europe. So Δt varies under the changed conditions suggested in the questions that conclude this problem.

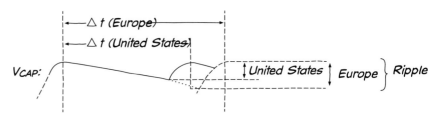

Figure 3W.1 Ripple.

Using

$$I = C\frac{dV}{dt}$$

we plug in what we know, and solve for C. We know that:

- dV or ΔV, the ripple, is 2 Vpp;
- dt or ΔT is the time between peaks of the input waveform: $\frac{1}{2\times 60\,Hz} \approx 8$ ms (in the United States);
- I is the peak output current: 1A.
  This specification of load *current* rather than load *resistance* may seem odd, at the moment. In fact, it is typical. The typical load for an unregulated supply is a *regulator* – a circuit that holds its output voltage constant; if the regulated supply drives a constant resistance, then it puts out a constant current despite the input ripple, and it thus draws a constant current from the filter capacitor at the same time).

Putting these numbers together, we get:

$$C = I \cdot \frac{\Delta T}{\Delta V} = 1\,A \cdot \frac{8 \times 10^{-3}\,s}{2\,V} = 4000\,\mu F.$$

Big, but not unreasonably so. You may want to call this "4 mF." That would not be wrong – but would be unconventional and would mark you as an outsider.

3W.1.5 Fuse rating

This supply steps the voltage down from 110 V to 16 V; the current steps up in proportion: about seven-fold. So the output (secondary) current of 1 A implies an input (primary) current of about $\frac{1}{7}$ A ≈ 140 mA.

But this calculation of *average* input current understates the heating effect of the primary current, and of the primary and secondary currents in the transformer. Because these currents come in surges, recharging the filter capacitor only during part of the full cycle, the currents during the surges are large. These surges heat a fuse more than a steady current delivering the same power, so we need to boost the fuse rating by perhaps a factor of 2, and then another factor of about 2 to prevent fuse from blowing at full load. (It's designed for emergencies.)

This set of rules of thumb carry us to something like:

fuse rating (current)=140 mA×2 (for current surges)×2 (not to blow under normal full load) ≈560 mA.

A 600 mA slow-blow would do. Why slow-blow? Because on power-up (when the supply first is turned on) the filter capacitor is charged rapidly in a few cycles; large currents then flow. A fuse designed to blow during a brief overload could blow every time the supply was turned on. The slow-blow has larger thermal mass: needs overcurrent for a longer time than the normal fuse, before it will blow.

3W.1.6 Bleeder resistor

Polite power supplies include such a resistor, or some other fixed load, so as not to surprise their users. Again the value is not critical. Use an R that discharges the filter cap in no more than a few seconds; don't use a tiny R that substantially loads the supply.

Here, let's let RC equal a few seconds: $\Longrightarrow R = \{$a few seconds$\}/C \approx $ 1k.

Before we go on to consider a couple of new issues, let's just draw the circuit with the values we have chosen, so far: see Fig. 3W.2.

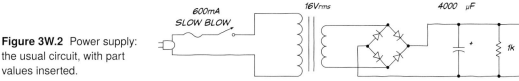

Figure 3W.2 Power supply: the usual circuit, with part values inserted.

3W.1.7 Transformer current rating

This is harder. The transformer provides brief surges of current into the cap. These heat the transformer more than a continuous flow of smaller current. We have made this point already today in Chapter 3N.

Figure 3W.3 is a sketch of current waveforms in relation to two possible ripple levels:

moderate ripple $I_{\rm RMS} = (\frac{1}{5}[5\,{\rm A}]^2)^{1/2} = \sqrt{5}\,{\rm A} = 2.2\,{\rm A}$;

tiny ripple $I_{\rm RMS} = (\frac{1}{20}[20\,{\rm A}]^2)^{1/2} = \sqrt{20}\,{\rm A} = 4.4\,{\rm A}$ *doubles* transformer heating.

The left-hand figure shows current flowing for about 1/5 period, in pulses of 5 A, to replace the charge drained at the steady 1 A output rate. The right-hand figure shows current flowing for 1/20 period, in pulses of 20 A.

Moral: a little ripple is a good thing. You will see that this is so when you meet voltage regulators, which can reduce a volt of ripple out of the unregulated supply to less than a *millivolt* at the point where it goes to work (where the output of the regulated supply drives some load).

152   Worked Examples: Diode Circuits

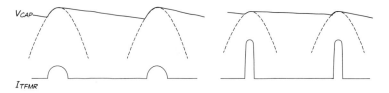

Figure 3W.3 Transformer Current versus Ripple: Small Ripple ⟹ Brief High-Current Pulses, and excessive heating.

Questions
What difference would you see in the circuit output:

(1) If you took it to Europe, where the line voltage is 220 V, 50 Hz?

(2) If one diode in the bridge rectifier burned out (open, not shorted)?

Solutions
(1) In Europe, the obvious effect would be a doubling of output voltage. That's likely to cook something driven by this supply (that's why American travelers often carry small 2:1 step-down transformers – though contemporary power supplies usually are of the *switching* type, and smart enough to accept happily the doubling of *line* voltage). It might also cook transformer and filter capacitor, unless you had been very conservative in your design (it would be foolish, in fact, to specify a filter cap that could take double the anticipated voltage; caps grow substantially with the voltage they can tolerate).

So you probably would not have a chance to get interested in less obvious changes in the power supply. But let's look at them anyway: the output *ripple* would change: ΔT would be 1/50 Hz = 10 ms, not 8 ms.

Ripple should grow proportionately. If load current remained constant, ripple should grow to about 2.5 V. If the load were resistive, then load current would double with the output voltage, and ripple would double relative to the value just estimated: to around 5 V.

(2) The burned out diode would make the bridge behave like a *half-wave* rectifier. ΔT would double, so ripple amplitude would double, roughly. Ripple frequency would fall from 120 Hz to 60 Hz.

This information might someday tell you what's wrong with an old radio: if it begins to buzz at 60 Hz, perhaps half of the bridge has failed; if it buzzes at 120 Hz, probably the filter cap has failed. If you like such electronic detective work, many pleasures lie ahead of you.

3W.2 Z_{in}

The problem: procedure to discover R_{in} and C_{in}
Describe a procedure for *measuring* R_{in} and C_{in} for the device in Fig. 3W.4. Note that your procedure should work even if the foot of the resistor shown to model R_{in} is *not* tied to ground. (This last requirement makes the problem harder than it otherwise might be).

Solution
The first step is to add a known R in series with the mystery box (let's call this "Box"). We will use two scope channels to watch the original signal and the point loaded by Box; see Fig. 3W.5.

3W.2 Z_{in}

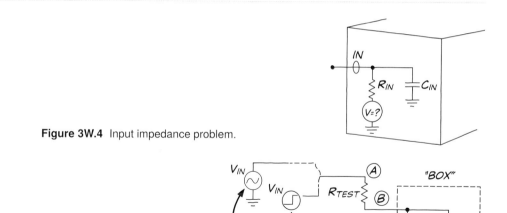

Figure 3W.4 Input impedance problem.

Figure 3W.5 Step 1: install known resistance in series with "Box".

Measure R_{in}

The challenge is essentially the same as the one presented by the exercise in §2L.1.1 that asks you to infer R and C once you have determined the RC product. You need to peel apart the effects of R versus C.

To measure R_{in}, our strategy will be to test the circuit at a frequency low enough so that the impedance of the capacitor is unimportant relative to that of the resistor: $X_C \gg R$. This will occur at a low frequency. But the point that calls for a little thought is *how* to determine when we have arrived at such a frequency. Here, we propose three alternative methods.

Measure R_{in} using DC

We can observe the open-circuit input voltage. That reveals the internal V. Use R_{test}, as in Fig. 3W.5, to form the usual voltage divider. If the internal V is zero, then we need an external DC source as input.

Measure R_{in} using a sinusoid

If we apply a sine to the input and watch points A and B, we need to make sure that whatever attenuation we see is caused by the R, not by X_C as well. An $R:R$ divider will show no phase shift; a divider with R on top and R parallel C will show phase shift, unless X_C is very large relative to the Box's R. So, we apply a low-frequency sinusoid, and watch for phase shift. If we see none, our measurement of R_{in} is not being corrupted by the effect of C_{in}. The scope images in Fig. 3W.6 show results with the scope inputs *AC* coupled.

Two possible hazards:

- At a very low frequency, if you AC-couple only the channel watching B, you may see a phase shift that is a pure scope artifact. AC-coupling the scope input is a good idea, if you use the method of centering the scope display on the 0% mark: since the lower part of the waveform goes off-screen, a *DC offset* in the signal can deceive you. But when you use AC-coupling you can see the blocking-capacitor's effect, at extreme low frequencies: below about 10 Hz.[1] This effect is visible in the leftmost image of Fig. 3W.6.

---

[1] The f_{3dB} for both our analog and digital scopes is about 7 Hz (highpass), when AC-coupled and fed by a BNC cable ("'1×"). When fed by a 10× probe, the scope shows an f_{3dB} so low it is not likely to trouble you: about 0.7 Hz. But in the present case, where one is likely to watch both channels AC-coupled, the phase shift is harmless: it would affect both channels equally. Fig. 3W.6, in contrast, shows channel 1 DC-coupled, channel 2 AC-coupled.

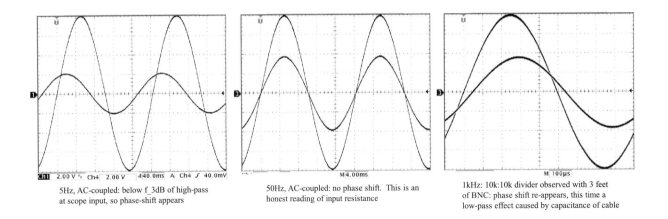

5Hz, AC-coupled: below f_3dB of high-pass at scope input, so phase-shift appears

50Hz, AC-coupled: no phase shift. This is an honest reading of input resistance

1kHz: 10k:10k divider observed with 3 feet of BNC: phase shift re-appears, this time a low-pass effect caused by capacitance of cable

Figure 3W.6 Using sinusoid, look for zero phase-shift to be sure you are reading R_{in}.

- At higher frequencies, you may see phase shift (lag, this time) caused by a lowpass filter formed by the combination of the test resistor and the stray capacitance of the BNC cable and scope channel with which you are watching point B. This effect is visible in the rightmost image of Fig. 3W.6.

In order to be sure that neither effect is fooling you, make sure you see no phase shift between the two channels, when you watch points A and B – as in the center image of Fig. 3W.6. The attenuation that you see there is caused by the circuit's R_{in} alone, not complicated by the effect of the capacitance to ground.

Measure R_{in} using a square wave

A square wave can reveal R_{in}, too. In Fig. 3W.7 the right-hand image shows the effect of a DC voltage at the foot of the unknown R of the box that we are measuring. We imposed this DC offset to remind ourselves of why we don't want to try to measure R_{in} by applying a DC voltage to the test setup. The DC voltage at the foot of R would throw us off, if we tried to infer R_{in} for this case. But the step input, like the sinusoid above, allows a measurement that is not thrown off by that DC voltage.

To infer R_{in}, compare input step amplitude to the amplitude that appears at the input of Box.

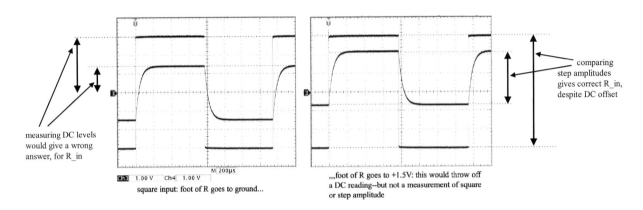

measuring DC levels would give a wrong answer, for R_in

square input: foot of R goes to ground...

,,,foot of R goes to +1.5V: this would throw off a DC reading--but not a measurement of square or step amplitude

comparing step amplitudes gives correct R_in, despite DC offset

Figure 3W.7 Square wave, like sinusoid, can allow one to measure R_{in}.

As Fig. 3W.7 illustrates, whereas a single DC reading at points A and B (points defined in Fig. 3W.5) would be thrown off by such a DC offset, in contrast, a measurement of step- or square-wave amplitude is not so distorted, as the right-hand image of Fig. 3W.7 indicates.

Discover C_{in}

Once one knows R_{in}, the next step is to measure RC. Given that, we will be able to calculate C_{in}. One can get RC in either of two ways, once more.

Use sinusoid: find RC, and then C, by measuring f_{3dB}

If one applies a sinusoid, as above, then one can gradually increase f_{in}, watching amplitude at B fall, until it is about 30% below its low-frequency level: now one has found f_{3dB}. Given f_{3dB}, and knowing R one can easily calculate C_{in}.

As you do this calculation, you need to recall that the effective R driving C_{in} is $R_{in} \parallel R_{test}$ (this is R_{Thev}).

Use square wave: find RC, and then C, by measuring RC directly

If you use a square wave, then you can measure RC directly, from the (time-domain) image on the scope screen. As usual, you need to choose an appropriate sweep rate: see Fig. 3W.8. This concern is not new to you, of course. You did this at the start of Lab 2L.

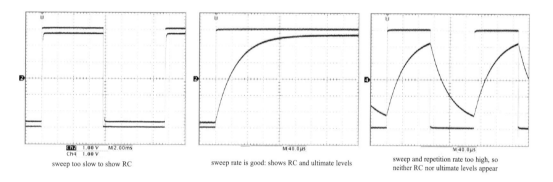

sweep too slow to show RC | sweep rate is good: shows RC and ultimate levels | sweep and repetition rate too high, so neither RC nor ultimate levels appear

Figure 3W.8 To measure RC, choose an appropriate sweep rate.

The middle image of Fig. 3W.8 would permit an RC reading – but one could do better, using still more of the scope screen to look at the time to rise from the lowest point to 63%.

Once you have RC, again you need to remember that the relevant R is R_{test} and R_{in} in parallel, as we said for the case of the sinusoid.

Part II

Analog: Discrete Transistors

4N Transistors I

Contents

4N.1	**Overview of Labs 4 and 5**	**159**
4N.2	**Preliminary: introductory sketch**	**162**
	4N.2.1 The name...	162
	4N.2.2 An intuitive model	162
	4N.2.3 The symbol	162
	4N.2.4 Assumptions	163
	4N.2.5 Starting simple: two views of transistor behavior	163
4N.3	**The simplest view: forgetting beta**	**163**
	4N.3.1 ...that a current source provides a constant output current	164
	4N.3.2 ...that a *common-emitter amplifier* shows voltage gain as advertised	165
	4N.3.3 ...that a follower follows	165
	4N.3.4 ...that a push–pull follower works, and also shows distortion	166
4N.4	**Add quantitative detail: use beta explicitly**	**166**
	4N.4.1 Digression on beta	166
	4N.4.2 Follower as impedance changer	167
	4N.4.3 Complication: *biasing*	169
	4N.4.4 Another complication: asymmetry can produce *clipping*	172
	4N.4.5 ...A remedy for clipping: the push–pull	173
4N.5	**A strikingly different transistor circuit: the switch**	**174**
4N.6	**Recapitulation: the important transistor circuits at a glance**	**175**
	4N.6.1 Impedance at the collector, by the way	175
	4N.6.2 ...and the switch stands apart	175
4N.7	**AoE reading**	**176**

Why?

What problems do today's circuits solve? They can modify signals in any of the following ways:

- improve the apparent impedance of a circuit element (boost R_{in}, reduce R_{out});
- hold current constant as voltage varies (make a "current source");
- amplify a voltage signal (a ΔV); and
- let a voltage signal switch current to a load On or Off.

4N.1 Overview of Labs 4 and 5

A novel and powerful new sort of circuit performance appears today and tomorrow: a circuit that can *amplify*. Sometimes the circuit will amplify voltage; that's what most people think of as an amplifier's job. Sometimes the circuit will amplify only current; in that event one can describe its amplification as a transformation of *impedances*. As you know from your work in the earlier labs, that can be a valuable trick.

The transistors introduced in AoE Chapter 2 are called **bipolar** (because the charge carrying mechanism uses carriers of both polarities – but that is a story for another course). Only at the very end of the analog section of this course will you meet the other sort of transistor, which is called **field effect** (FET) rather than "unipolar" (though they were called "unipolar" at first). The FET type was developed after bipolar, though, as noted later, the discoverers of the bipolar transistor stumbled upon it while trying to develop a FET. But the FET, in a form called MOSFET[1] has turned out to be more important than bipolar in *digital* devices. You will see much of these FET logic circuits later in this course. In *analog* circuits, bipolar transistors are widely used, but even there FETs are gaining.

You will sometimes apply transistors singly – as *discrete* parts.[2] More often, you will pair the brawn of a discrete transistor with the brains of an operational amplifier (op-amp)[3] in order to get high-current or high-power that is beyond the capacity of an ordinary op-amp. But an understanding of transistor circuits is important in this course not so much for this reason as because an introduction to discrete transistor design will help you to understand the innards of the *integrated* circuits that you are certain to rely on.

After toiling through this section, you will find that you can recognize in the schematic of an otherwise mysterious IC a collection of familiar transistor circuits. In an effort to de-mystify the op-amp, we conclude Lab 5L by asking you to build yourself an op-amp from an array of transistors. Recognizing familiar circuits in the schematic of an op-amp, you will consequently recognize the op-amp's shortcomings as shortcomings of those familiar transistor circuit elements.

This section on transistors is difficult. It requires that you get used to a new device and at the same time apply techniques you learned in the section on passive devices. You will find yourself worrying about impedances once again: arranging things so that circuit-fragment A can drive circuit-fragment B without undue loading; you will design and build lots of *RC* circuits. Often you will need a Thevenin model to help you determine an effective *R*. We hope, of course, that you will find this chance to apply new skills gratifying; but you are likely to find it taxing as well.

We say this not to discourage you, but, on the contrary, to let you know that if you have trouble digesting the rules for design with transistors, that's not a sign that there's something wrong with you. This is a *rich* section of the course – perhaps the most difficult.

With Lab 6L, op-amps arrive, and suddenly your life as a circuit designer becomes radically easier. Operational amplifiers, used with *feedback*, will make the design of very good circuits very easy. At that point you may wonder why you ever labored with those difficult discrete-transistor problems. But we won't reveal to you now that op-amps are easy – because we don't want to sap your present enthusiasm for transistors. In return for your close attention to their demands, transistors will perform some pretty impressive work for you.

Instead of yearning for op-amps, let's try to put ourselves into a state of mind approaching that of the transistor's three inventors, who found to their delight, two days before Christmas 1947, that they had constructed a tiny amplifier on a chunk of germanium. Probably they could not yet envision a time

---

[1] MOSFET is Metal-Oxide-Semiconductor Field-Effect transistor: a strange way to say that its input terminal is *insulated* – by the "oxide," SiO_2. More on MOSFETs appears in Chapter 12N.
[2] "Discrete" doesn't mean "polite, and able to keep a secret." The origins of the words *discrete* and *discreet* happen to be the same. But *discrete*, in this context, means "used singly," rather than as one of many elements in an integrated circuit.
[3] Operational amplifiers are high-gain amplifiers used to implement circuits that exploit negative feedback. Op-amps become our principal analog circuit elements, beginning in Chapter 6.

when there would be no more vacuum tubes;[4] no more high-voltage supplies; no more power wasted in heating filaments. The world could look forward to the microcomputer, the iPod, the smartphone, the – whatever comes next, made possible by amazingly tiny microcircuits.[5]

Here, in Fig. 4N.1 is a Nobel prize-winning device, looking wonderfully homemade. It's not *really* made from paper clips, Scotch tape and chewing gum.[6]

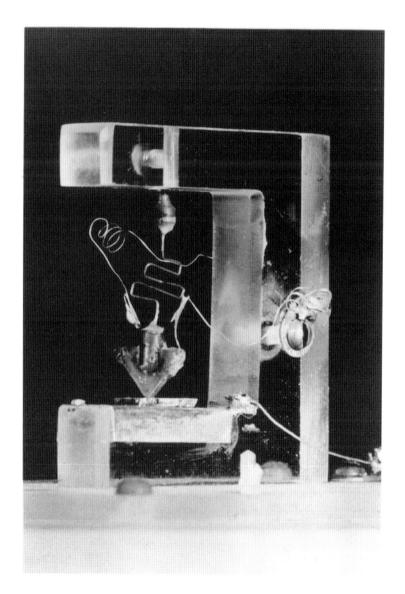

Figure 4N.1 The first transistor: point-contact type (1947).

---

[4] Well, almost no vacuum tubes. Rock'n roll guitarists and some mad audiophiles prefer the sound of tubes. The guitarists like the way the tubes sound when the amplifier is overdriven. The mad audiophiles like to spend money.
[5] For some excellent history of the discovery of the bipolar transistor, see Michael Riordan and Lillian Hoddeson, *Crystal Fire*, Norton (1997) p. 138ff., and a short article by authors that include the former head of Bell Labs: William Brinkman et al., "A History of the Invention of the Transistor and Where It Will Lead Us," *IEEE Journal of Solid-state Circuits* vol. 32, no. 12, Dec. 1997.
[6] Photo used with permission of AT&T Bell Labs.

4N.2 Preliminary: introductory sketch

4N.2.1 The name...

A transistor is a "**trans**[fer]" "[res]**istor**:" a thing vaguely like a resistor, but one that can be controlled – that can transfer a signal from input to output. In having *three* terminals, it differs from the passive devices we have met to this point (with the exception of the 4- or 5-terminal *transformer* – whose name similarly signals its behavior). The three terminals will make your first days with transistors challenging.

But take heart, if you feel this difficulty: soon you will pop out above the clouds into to the bright and cheery world of operational amplifiers, as we promised in §4W.1 We made the point there that we'd like you to know what's within an op-amp, so that its operation is not *simply* magical.

4N.2.2 An intuitive model

A transistor is a *valve*; see Fig. 4N.2. Notice, particularly, that the transistor is not a *pump*: it does not force current to flow; it permits it to flow, to a controllable degree, when the power supply tries to force current through the device.

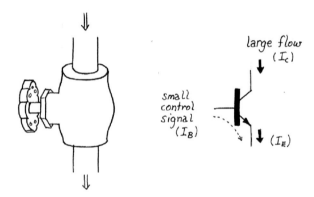

Figure 4N.2 A transistor is a valve (not a pump!).

4N.2.3 The symbol

The bipolar junction transistor (BJT) comes in two flavors: *npn* and *pnp*. The letters refer to the doping, positive (*p*) or negative (*n*), of the materials used to construct the device. The symbol apparently amounts to a sketch of the earliest prototype.[7] In Fig. 4N.3 we show the symbol for an *npn* transistor next to a fragment of the photo of the original Bell Labs transistor, rotated to put the emitter and collector above and below base.

The claim certainly looks plausible. The "base" was literally the base for the structure. Thomas Lee makes the intriguing observation that the bipolar transistor was *discovered* rather than invented by Shockley, Bardeen, and Brattain – because the three investigators stumbled upon the structure. They were not trying to form a bipolar transistor. Instead, they were troubleshooting their defective field-effect device, probing the surface of the "base" piece to investigate disturbances in current flow.

The symbol for a *pnp* transistor has the emitter arrow pointing toward the base, rather than away from it. In either type, current flows in the direction of the emitter arrow.

---

[7] Thomas H. Lee, *Planar Microwave Engineering*, Cambridge University Press (2004), p. 342.

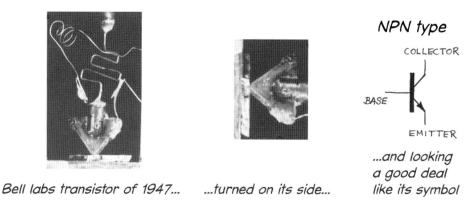

Figure 4N.3 Transistor symbol seems to sketch the original transistor.

4N.2.4 Assumptions

AoE §2.1.1

We initially look at the BJT operating in its *linear* region.[8] For the *npn* type, this means:

(1) $V_C > V_E$ (by at least a couple of tenths of a volt); and
(2) "things are arranged" so that $V_B - V_E$ is about 0.6 V (V_{BE} is a diode junction, and must be forward biased).

For a *pnp* transistor, reverse the polarities of these voltages and the directions of the resulting currents.

4N.2.5 Starting simple: two views of transistor behavior

We begin with *two views of the transistor*: one simple, the other *very* simple. (Later, we will complicate things.)

- Pretty simple: current amplifier: $I_C = \beta \times I_B$. See Fig. 4N.4.
- Very simple: say nothing of beta (though assume it's at work). Say only:

$$V_{BE} = 0.6 \text{ V}$$
$$I_C \approx I_E$$

Figure 4N.4 Transistor as *current*-controlled valve or amplifier.

4N.3 The simplest view: forgetting beta

We can understand – and even design – many circuits without thinking explicitly about beta (also called "h_{FE}"), the ratio of current out to current in: $\beta = I_C/I_B$. The simplest view will let you understand the following circuits. You can understand...

---

[8] The other regions are cutoff, where no current flows from collector to emitter, and saturation, where changes in the base-to-emitter voltage have little effect on collector-to-emitter current flow.

4N.3.1 ...that a current source provides a constant output current

See Fig. 4N.5: pretty charmingly simple, eh? It is nothing more than *Ohm's Law* that determines the magnitude of the output current, thanks to the predictability of the V_{BE} drop. Now turn to Fig. 4N.6 to see what happens if we vary the resistor on the collector – R_{load}. We'll also tack in a base resistor, which will make no difference except in an extreme case that we'll consider in a minute, a case when the current source fails.

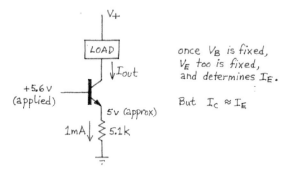

Figure 4N.5 Current source.

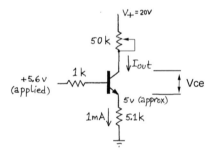

Figure 4N.6 Current source with variable load shown.

If we vary R_{load}, how does I_{load} behave? Suppose that R_{load} initially is 10k and we drop it to 1k. Does the current increase?

If you're inclined to say, "Yes," then you are showing nostalgia for the *passive* section with which this book began, and for the resistors of your acquaintance. You are asking the transistor to behave like those old friends – like an ohmic device: more voltage across it, so more current. But, no, it doesn't behave that way.

As you lower the value of R_{load}, it is true that V_C rises. But it does not follow that I_C grows. It does not, because $I_C \approx I_E$, and I_E is constrained – determined, in turn, by V_B. Unfamiliar, perhaps, but simple – and useful. If you take R_{load} all the way down to zero, does the current stay constant? Yes.

What if you take R_{load} to its maximum? Here, the answer is, "No. In this case, the current source fails." It fails because we have violated one of the assumptions that we mentioned in §4N.2.4.

In Fig. 4N.6, taking R_{load} beyond about 15k forces V_C too low: we can no longer satisfy the requirement that V_C exceed V_E, and the circuit cannot hold the current constant. The transistor can no longer do its magic – adjusting the voltage, V_C, as needed to hold I_C constant. When V_{CE} falls below a couple of tenths of a volt, the transistor is said to be *saturated*.

When that happens, none of the usual rules that we have been advertising will hold. The circuit does become ohmic, the transistor looking like a small-valued resistor; further increases in R_{load} simply reduce I_{load}. Setting R_{load} to 50k, for example, would take I_{load} down to about 0.4 mA (20 V across about 55k): plain old Ohm's law behavior.

This limitation on the range of the current source – called its *output voltage compliance* – reminds us that although the transistor circuits are clever, they still require us to use our heads. They don't work if we don't play by the rules.

4N.3.2 ...that a *common-emitter amplifier* shows voltage gain as advertised

AoE §2.2.7

Adding a collector resistor to the current source (as in Fig. 4N.7) is about all that's required to turn the source into a voltage amplifier.

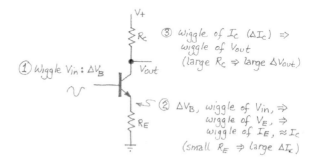

Figure 4N.7 Common-emitter amp.

One can regard the common-emitter amp as a two-stage circuit: see Fig. 4N.8. The first stage is a current source (with current controlled by V_{in}); this is a *V*-to-*I* block. The second is a resistor that converts that output current, and its variations, into a voltage: this is an *I*-to-*V* block.

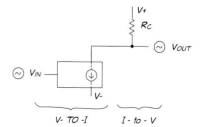

Figure 4N.8 Common-emitter amp, described as *V*-to-*I* block feeding an *I*-to-*V* block.

You may object that we're being a bit pompous in using the name *I*-to-*V* block for the humble resistor. Maybe so, but that is, indeed, the resistor's function in this circuit.[9]

4N.3.3 ...that a follower follows

AoE §2.2.3

The circuit in Fig. 4N.9 may look, at first glance, totally pointless: you start with a voltage wiggle; at the output you get the same voltage wiggle. Clearly it is not a voltage amplifier.

---

[9] And if you think we're giving a pretentious name to the resistor, consider a more spectacular instance: some bored engineers, apparently envious of their bosses' opportunities to go off and talk at conferences, got together to propose a paper. Their abstract described a new circuit element they had discovered: a very useful device that performs a bidirectional transformation between current and voltage, and one that – as an added benefit – kept the transformation *linear*! This novel device they named the *Linistor*. To their delight, they soon got the good news from the sponsors of the conference: they were invited to present the paper describing their invention. At that point, we are sad to report, they lost their nerve. The electronics world at large never got to hear about the linistor. The late Bob Pease tells this story in "What's All this Hoax Stuff?," *Electronic Design*, April 4, 1994 (available at https://LAoE.link/Hoax_Stuff.html).

But it turns out that it amplifies *current* or, equivalently *changes impedances*. We'll appreciate how this works when we add one more element to our description of transistor behavior, the current-multiplication factor, β (see §4N.4). With the present, very simple view we can only confirm that the follower follows.

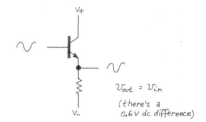

Figure 4N.9 Follower.

4N.3.4 ...that a push–pull follower works, and also shows distortion

AoE §2.4.1

Again, the circuit (Fig. 4N.10) is intelligible – but its virtues are not. We will see shortly why it is an extremely useful variation on the simpler *follower*.

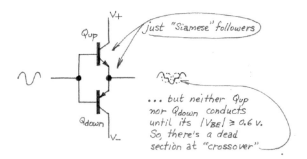

Figure 4N.10 Push–pull.

4N.4 Add quantitative detail: use beta explicitly

To understand why a *follower* is useful, we need to add a quantitative element that we omitted from the "simplest" view: we need to note that the *output current*, I_C (or the near-equivalent, I_E) is much larger than the *input current*, I_B. It is larger by the factor β. The notion of *current amplification* may be unfamiliar and uncongenial to you.

Though this is an accurate view of a follower, we more often describe the follower's behavior as *changing impedances*. That notion, too, takes getting used to. But the idea is simple, and builds on your familiarity with Ohm's law.[10] The point is simply this: voltage wiggles 'in' and 'out' are the same; currents are very different. The contrasting quotients ($\Delta V/\Delta I$) give *high* R_{in}, *low* R_{out}.

4N.4.1 Digression on beta

AoE §2.1.1

It can be hard to get used to our treatment of the transistor characteristic, β, a treatment that follows that in AoE.

---

[10] We're not renouncing our earlier point that the transistor's collector–emitter behavior is not ohmic (§4N.3.1, above). The impedances we speak of, here, are characteristics of input (at the base) and output (at the emitter).

On the one hand, we work hard to get across a piece of sound design advice: *don't ever design a circuit on the assumption that you know the value of β.* You should not, because the value is not predictable. It varies with individual transistors of a given type; it varies for an individual transistor as I_C varies; it varies with temperature.

On the other hand, we assume $\beta \approx 100$, for "small signal" transistors like the 2N3904. We don't know beta; beta is 100 (the king is dead, long live the king). This is just one more of Horowitz and Hill's permissions to "be wrong, but be wrong in the right direction."[11] We don't need to know beta as long as we *underestimate* its value, because beta is good.

4N.4.2 Follower as impedance changer

AoE §2.2.3B

The follower amplifies current. The same truth makes it an impedance changer: a circuit that offers high R_{in}, low R_{out}; see Fig. 4N.11.

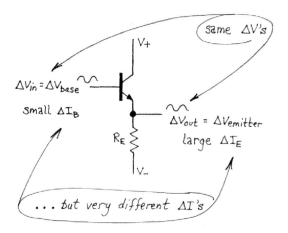

Figure 4N.11 How a follower changes impedances.

Figure 4N.12 shows a corny mnemonic device to describe this impedance-changing effect. Imagine an ill-matched couple gazing at each other in a dimly-lit cocktail lounge – and gazing through a rose-colored lens that happens to be a follower. Each sees what he or she wants to see!![12]

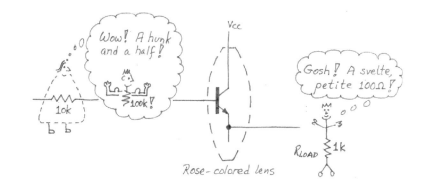

Figure 4N.12 Follower as rose-colored lens: it shows what one would *like* to see.

---

[11] See §4N.4.
[12] If Fig. 4N.12 seems very old-fashioned, feel free to interchange genders.

Estimating β by looking at R_{in}

In Lab 4L, we ask you to infer beta from the observed value of R_{in}. The experiment asks you to form a voltage divider using a known input resistor (33k, in the example just below) to feed the follower's base, and then to compare amplitudes on the two sides of this known resistor. Figure 4N.13 shows what we saw when we tried it.

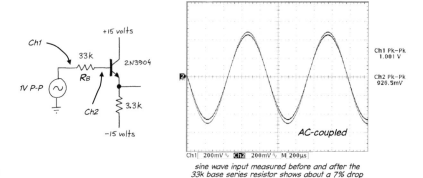

Figure 4N.13 Slight attenuation caused by follower's R_{in} can reveal β at work.

To get a clearer picture of R_{in} we have stripped away any DC difference due to $I_{B(quiescent)} \times R_B$ by *AC-coupling* the two signals in the scope image of Fig. 4N.13. (Usually, we urge you *not* to use AC coupling in the lab, because it knocks out interesting information; here, we're happy to hide the not-interesting DC difference.) AC-coupling helps one do a visual comparison – though the difference here is so slight that it would be hard to measure by eye. The oscilloscope helps us out by measuring the peak-to-peak amplitudes for us. We find we're keeping 93% to 94% of the original signal. Let's use that to estimate β.

If we take 93% as the truer reading (and because it gives a more conservative result), then we lose 7% and keep 93%. We keep about 13 parts while losing one. That means R_{in} is about 13 times as large as the known resistance of 33k.

We could now calculate R_{in}, and from this value infer β, in a two-step process. If you want to be lazier, though, try considering, as a sort of benchmark what you would see if β were 100, then compare what you see against that reference: see Fig. 4N.14.

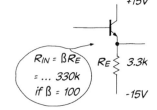

Figure 4N.14 A starting point or benchmark: estimate what R_{in} would be if β were 100.

If β were a hundred, the R_{in} would be 330k, just ten times R_{test} (we would see attenuation of just under 10%). But we see an R_{in} that looks about 13 times R_{test}. So, β must be about 130 rather than 100. This is plausible. Recall that we do not predict that β will *equal* 100. Instead, we say, in effect, "Call it 100, and you will safely underestimate β."[13]

---

[13] Here is yet another of Horowitz & Hill's licenses to be "right in the wrong direction." Compare our approximation, $Z_{in_{RC-filter}} = R$, and the use of 8 ms as Δt in the calculation of power-supply ripple as we mentioned in §4N.4.

4N.4.3 Complication: *biasing*

AoE §2.2.5

To this point – and early in Lab 4L – we have assumed the use of "split" power supplies (supplies of both polarities). Doing this allows our circuits to respond to signals symmetric about ground without *clipping* (that is, without the flattening that occurs when the waveform hits a limit).

The follower that you wire in the lab initially uses only a positive supply, and cannot follow an input voltage that swings below ground (or even below about one diode-drop positive). An input centered on ground evokes a half-wave-rectified output like the lower output trace in Fig. 4N.15.

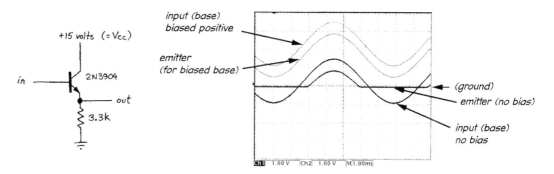

Figure 4N.15 A follower that uses only one supply must have its input biased positive.

This half-wave-rectified result reminds one how similar the *follower* may appear to be to a simple diode feeding a resistor. Wouldn't a diode and resistor (the first rectifier you met, back in Lab 3L), produce exactly the waveforms shown in Fig. 4N.15?

That is, wouldn't the two circuits of Fig. 4N.16 produce identical waveforms? Yes, to a very good approximation, they would.

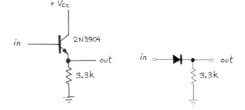

Figure 4N.16 Follower and half-wave rectifier: do they differ?

Figure 4N.17
Detail of "fraudulent" demo circuit.

A small fraud to make a point… In fact, we like to perpetrate a little fraud on our students during class, using the similarity of the two circuits. We talk about the follower, sketch the circuit, and then show a scope display of the follower's input and output. But – here's the fraud – we "forget" to turn on the power supply. The circuit doesn't seem to care.[14]

The very similar lower two traces of Fig. 4N.18 show how little difference the power supply makes to the voltage waveforms in this demonstration. How can this be? Is it possible that the power supply isn't necessary?

No. What this demonstration shows, instead, is that the oscilloscope, a voltage-sensing instrument, cannot directly show us what this circuit is doing.

---

[14] We should admit that the fraud requires us to insert a diode in series with the collector, so that the base–collector junction cannot conduct current into the positive supply, which would sink current even though switched off; see Fig. 4N.17.

Figure 4N.18 Emitter follower and a rectifier – implemented by failing to turn on the power supply: the voltage waveforms are hard to distinguish.

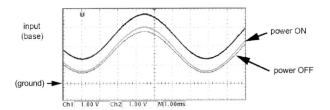

Making the follower's operation visible The follower is not a voltage amplifier. It is a *current* amplifier, instead – or, equivalently, it is an *impedance changer*. So, we need to insert a test *resistor* that will make the magnitude of *current* visible to the scope. At the input of each of the two competing circuits – follower versus rectifier, where the rectifier models the follower with power *off* – we'll put a 33k resistor. This will form a voltage divider that pairs it with the R_{in} of each circuit: see Fig. 4N.19.

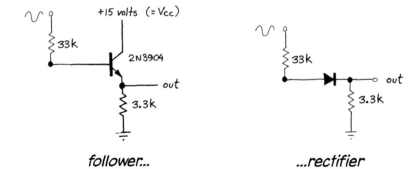

Figure 4N.19 Base resistor can make follower's *impedance change* or *current amplification* visible.

Now, what happens if we try the effect of power ON versus OFF as we watch signal source and emitter of the follower? It's a relief to see that power makes a giant difference to the follower's impedance: see Fig. 4N.20.

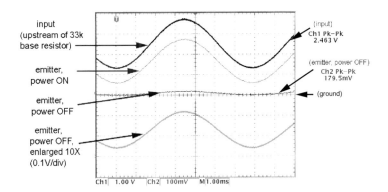

Figure 4N.20 Base resistor makes evident the follower's use of the power supply.

With power ON, the follower's input impedance is $\beta \times R_E$, as advertised: at least 330k. So the emitter waveform looks almost equal to the input swing (and shows the characteristic diode-drop of V_{BE}).

With power OFF, in contrast, the "follower" becomes just a diode rectifier, with V_{BE} serving as the diode. The input resistance is just 3.3k, and we see severe attenuation: less than 10% of the input swing survives.

This somewhat silly experiment reminds us that the transistor's output current – whether taken at emitter or collector – comes almost entirely from the power supply, not from the base. That's the key to the impedance-changing magic. The base current is just a small downpayment, needed to provide the big loan of power from the supply.[15] Incidentally, the experiment reminds us that the term "power supply" is named appropriately. It means what the phrase says.

A divider can provide needed bias In the lab exercises, we first provide this positive *bias* (pushing off-center) in the laziest way: simply by dialing up a positive offset from the function generator. But this is not a practical general solution.

The more general design scheme, as in Fig. 4N.21, allowing use of a *single* power supply, is to add a voltage divider that pulls the base positive. Thus the emitter can swing down as well as up, as the input swings negative and positive. A *blocking capacitor* allows base and signal source to adopt differing DC levels without a fight. The wiggles pass happily from source to base.

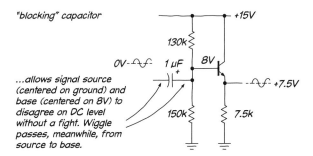

Figure 4N.21 Single-supply follower uses biasing.

The biasing divider must be stiff enough to hold the transistor where we want it (with V_{out} around the midpoint between V_{CC} and ground). It must not be too stiff: the *signal* source must be able to wiggle the transistor's base without much interference from the biasing divider.

The biasing problem is the familiar one: device A drives B; B drives C; see Fig. 4N.22. As usual, we want Z_{out} for each element to be low relative to Z_{in} for the next:

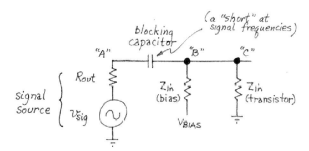

Figure 4N.22 Biasing arrangement.

You will notice that the biasing divider reduces the circuit's input impedance by a factor of ten, compared to what it would have been in a "split supply" design. This is regrettable; if you want to investigate complications, see the "bootstrap" circuit in AoE §2.4.3 for a way around this degradation.

You will have to get used to a funny convention: you us talk about impedances not only *at* points in a circuit, but also *looking* in a particular direction. For example: we will talk about the impedance "at the base" in two ways (illustrated in Fig. 4N.23):

- the impedance "looking into the base" (a characteristic of the transistor and its emitter load); and

---

[15] Well, this metaphor isn't quite right, since we never repay the loan. But maybe, on second thought, that's about right. Isn't that the way our present debt-ridden culture works?

- the impedance at the base, looking back toward the input (this characteristic is *not* determined by the transistor; it depends on the *biasing* network, and (at signal frequencies) on the *source impedance*.

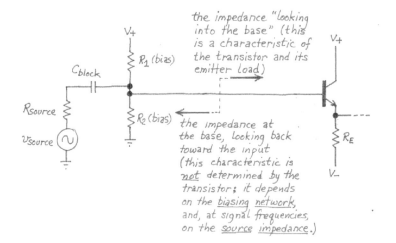

Figure 4N.23
Impedances "looking" in specified directions.

4N.4.4 Another complication: asymmetry can produce *clipping*

We ask you to admire the low R_{out} of the follower (and to measure this characteristic) in Lab 4L. Now, candor compels us to admit[16] that the follower can misbehave, despite its low R_{out}. Figure 4N.24 shows a hypothetical follower and a resistive "load."

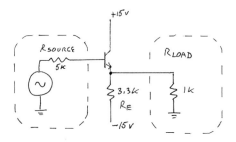

Figure 4N.24 Emitter follower, loaded. Can it drive this load?

Will this circuit work properly? Let's calculate the impedances 'in' and 'out'.

- R_{in} at the base: looks like $\beta \times (R_E \text{ parallel } R_{load}) \approx 75k$, taking β to be at least 100. Since this $R_{in} > 10 \times R_{source}$, we satisfy our "house rules" for impedances: OK.
- R_{out} looking back, at the emitter: R_E parallel $R_{source}/\beta \approx 50\,\Omega$ (the effect of R_E is negligible, as usual). This R_{out} is about 1/20 of R_{load}. So, again, no problem.

This is correct – but only for an input signal of modest amplitude. If the signal amplitude grows to more than about 4 V, then the circuit fails: the negative swings flatten – they "clip." You can see this occurring in Fig. 4N.25.

---

[16] People our age – is there any such person in our readership? – may recognize this phrase as the one Lyndon Johnson liked to offer as preamble to a big fib! This is not a fib. We are not crooks.

4N.4 Add quantitative detail: use beta explicitly

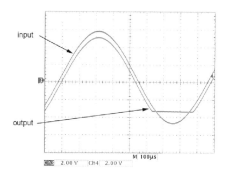

Figure 4N.25 Despite its low R_{out}, the follower can clip when loaded.

Why does this happen? And why only on the negative swings? The answer is simple, almost obvious, once you have heard it: the follower is radically asymmetric. It uses the *transistor* to drive current into the load; it uses a *resistor*, R_E, to pull current out of the load. The transistor is good at pushing current; the resistor is not so good at pulling current. At an extreme of negative swings, the transistor *shuts off*; see Fig. 4N.26.

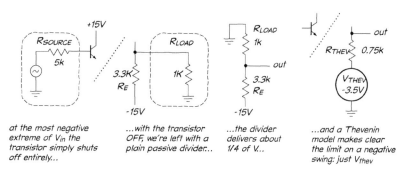

Figure 4N.26 Loaded follower clips when the transistor shuts off entirely, for an extreme negative V_{in}.

At that point, all the transistor magic that produced the nice low R_{out} – thanks to β and the "rose-colored lens" – dissipates like the magic that sustained Cinderella's beautiful carriage. The circuit turns back into a pumpkin: it becomes a simple resistive divider, and V_{out} gives up on its negative swings, sticking at a bit less than 1/4 of the negative supply ≈ −3.5 V.

4N.4.5 …A remedy for clipping: the push–pull

AoE §2.4.1

What is to be done? We can offer a dumb answer and a smart one. The dumb answer is to decrease the value of R_E so as to push the clipping voltage as close to the negative supply as needed. To go to −10 V, for example, we could cut R_E's value to 0.5k.

That doesn't sound bad. But it does sound bad if we change the numbers. Suppose our goal is to drive, as load, an 8 Ω speaker. To clip at −10 V calls for R_E of 4 Ω. That works, but consider the power dissipation when V_{out} is zero volts – in other words, the power dissipated in order to achieve *silence*. The quiescent current is 15 V/4 Ω, almost 4 amps; this current flowing between the 30 V power supply span dissipates almost 120 W – and this for silence!

So, we need a better answer, and that is the *push–pull* follower, one that replaces the pulldown *resistor* with a pull-down *transistor*, as in Fig. 4N.10, repeated in Fig. 4N.27 with added scope image showing crossover distortion.

174   Transistors I

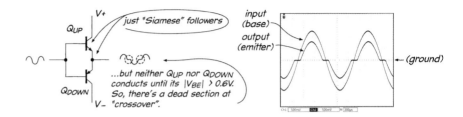

Figure 4N.27 Push–pull, again.

The push–pull is much more efficient than the simpler follower – dissipating little or no power with V_{out} at 0 V.[17] The push–pull, with its good efficiency, dominates the output stages of amplifiers. Your lab function generator uses one; your audio amplifier probably uses one;[18] so does the output of the operational amplifiers that you soon will meet.

A respectable push–pull must include a remedy for the cross-over distortion that troubles the simple design shown above. A pair of diodes or a pair of transistor V_{BE}'s can bias the bases of the output transistors apart; or an operational amplifier and feedback can be used, as you will demonstrate in Lab 6L.

AoE §2.4.1A

4N.5   A strikingly different transistor circuit: the switch

AoE §2.2.1

AoE presents this circuit first, as it introduces transistors, because the switch is easy to understand. We present it last, because it is so exceptional. Either way, it is important to keep it apart from all the other transistor applications in your thinking, because the *switch* deliberately violates a fundamental rule governing the design of all the other circuits: the *switch* operates in only two conditions, both pathological for an ordinary transistor circuit: it is totally ON ("saturated," in the transistor jargon) or totally OFF. Since it is saturated when ON, it does not show the high-input impedance we think of as one of the virtues of a transistor circuit.

Nor does the usual rule, $I_C = \beta \times I_B$ hold. A switch should be purposefully overdriven, with about 10 times the minimum base current to pass the required I_C. That saturates the switch strongly, keeping V_{CE} low to minimize power lost in the switch. The best switch would disappear electrically when ON, putting all power into the load, none into itself.

Skeptical students sometimes ask, "Why bother? You start with a switched input level (either High or Low), and that simply generates a switched output level (either High or Low). Pointless." (Sounds rather like the follower, which takes an input wiggle and delivers the same wiggle as output.) It's a fair question.

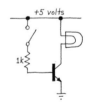

The answer is that the input level usually comes not from a manual switch but from a circuit fragment not strong enough to switch the "load." So, you might think of the switch circuit as (once again) a current amplifier. It resembles the follower, but it is simpler. It is also true that one can build a system of switches driving switches, and of these one can make useful "binary digital" circuits. Those digital circuits are the subject of the second half of this course. We concede that almost none of those digital circuits use *bipolar* transistors like those we are discussing now (they use MOSFETs instead – see Lab 12L); but they do rely on switching just like what we have been looking at.

Figure 4N.28 Switch: a useful circuit that may seem silly at first glance....

So, let's note that switches are important circuits – but let's also keep them in a compartment all

---

[17] "No power," you may insist, and you're right for the simple push–pull that we have pictured. The better version, however, which eliminates cross-over distortion by permitting a small current to flow through both transistors with V_{out} at 0 V does dissipate a little power. See AoE §2.4.1B.

[18] Not your digital music player, though. It is likely to use a power switch instead, for its still greater efficiency. See AoE §2.4.1C and the switching amplifier of Lab 12L, the LM4667.

4N.6 Recapitulation: the important transistor circuits at a glance

AoE §2.2.9B

To start you on the process of getting used to what bipolar transistor circuits look like, and to the crucial differences that come from what *terminal* you treat as output, Figure 4N.29 is a family portrait, stripped of all detail.

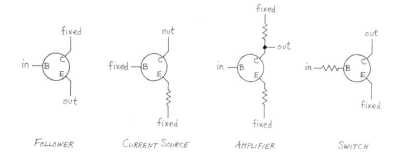

Figure 4N.29 The most important bipolar transistor circuits: sketch.

Note, among other differences in the circuits shown in Fig. 4N.29, that only the *follower* takes its output at the emitter. There, the impedance can be low.

4N.6.1 Impedance at the collector, by the way

An output at the collector, in contrast, provides *high* impedance, because at that terminal it is *current*, not *voltage* that is determined. So, the quotient $\Delta V_{CE}/\Delta I_C$ is very large (since ΔI_C is very small – ideally, zero). This is a case we foresaw back in §1N.2.1, when we promised you that the notion of *dynamic resistance* ($R_{dynamic}$) would become important to your understanding of transistor circuits. $R_{dynamic}$ at the collector is *very large*: ideally, infinite; practically, at least hundreds of kilohms. This fact is important not only for current sources, but also for the common-emitter amplifier, whose R_{out} is determined by R_C alone, since the effective R_{out} is R_C in parallel with the collector's huge $R_{dynamic}$.

4N.6.2 ...and the switch stands apart

The switch behaves entirely differently from all the other applications; it stands apart because it is run in the two conditions always *avoided* for the other circuits – as we said in §4N.5: OFF or fully ON ("saturated"). The uniqueness of the switch means you must use Fig. 4N.29 with care.

In Chapter 5 we will begin to use the more complicated *Ebers–Moll* model for the transistor. But the simplest model of the transistor, presented in this class, will remain important. We will always try to use the least complicated view that explains circuit performance, and often the very simplest will suffice.

4N.7 AoE reading

Reading:
    Chapter 2, §§2.1–2.2.8 and §2.4.1 (push–pull).
    Take a quick, preliminary look at §2.2.9 (transconductance). Next time we will look more closely at the Ebers–Moll view of transistors. In §4L.5.1 the high-gain amp presents for the first time a circuit that requires analysis with the Ebers–Moll model.

Problems:
    Problems in text and Additional Exercise 2.25.

4L Lab: Transistors I

4L.1 Transistor preliminaries: look at devices out of circuit

4L.1.1 Transistor junctions are diodes

Here is a method for spot-checking a suspected bad transistor: the transistor must look like a pair of diodes when you test each junction separately. But, *caution*: do not take this as a description of the transistor's mechanism when it is operating: it does *not* behave like two back-to-back diodes when operating (the circuits of Fig. 4L.1, if made with a pair of ordinary diodes, would be a flop, indeed).

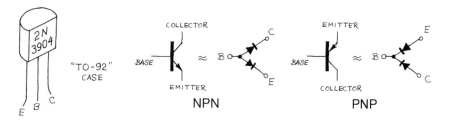

Figure 4L.1 Transistor junctions: (for testing, not to describe transistor operation).

4L.1.1.1 A game: discover transistor type and pinout, on a desert island

See if you can determine the type (*npn* or *pnp*) and *pinout* (base, emitter, collector leads) of a 2N3904 transistor by *experiment*. (On a desert island with DVM and a box of unknown transistors, you could sort and label the transistors; thus you could start the process of building a radio transmitter to summon your rescuers.) You can settle both *type* and *pinout* questions by flipping to the pinout section at the back of this book, of course.[1]

But try using a DVM, instead, employing its *diode test* function. (Most meters use a diode symbol to indicate this function.) The diode test applies a small current (a few milliamps: current flows from Red to Black lead), and the meter reads the junction voltage. Using the diode test function, you can determine the directions in which pairs of leads conduct. That will let you answer the *npn* versus *pnp* question. You can also distinguish *BC from BE* junctions this way: the BC junction is the larger of the two, and the lower current density across that *larger* junction is revealed by a slightly *lower* voltage drop.

4L.1.2 'Decouple' power supplies: fuzz warning

This is the first lab in which you may see high-frequency oscillations on your circuit outputs. At a modest sweep rate, this fuzz will look like a thickening of the trace; at a high sweep rate it may reveal

---

[1] We draw the *pnp* with the emitter at the top in Fig. 4L.1 because we like current to flow from top to bottom in our schematics. This often leads students to think that the *pinout* of the physical device is reversed between *npn*s and *pnp*s. It is not – the pinout of a 2N3906 is identical to that of the 2N3904 shown.

4L.2 Emitter follower

Wire up an *npn* transistor as an emitter follower, as shown in Fig. 4L.2. Drive the follower with a sinewave that is symmetric about zero volts (be sure the DC "offset" of the function generator is set to zero), and look with a scope at the poor replica that comes out. Why does this happen?[2]

If you turn up the waveform amplitude you will begin to see bumps *below* ground. How do you explain these?[3]

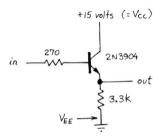

Figure 4L.2 Emitter follower. The small base resistor is often necessary to prevent oscillation.

Now try connecting the emitter return (the point marked V_{EE}[4] to -15 V instead of ground, and look at the output. Explain the improvement.

4L.2.1 Input and output impedance of follower

Measure Z_{in} and Z_{out} for the follower in Fig. 4L.3.[5]

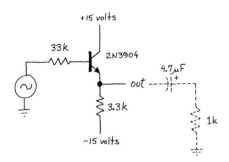

Figure 4L.3 Follower: circuit for measuring Z_{in} and Z_{out}. Disconnect the 1k *load* resistor while measuring Z_{in}.

---

[2] This is pretty simple: when V_{in} falls below about $+0.6$ V we are violating one of our transistor "ground rules:" we are failing to permit forward-biasing of the V_{BE} junction. I_C and I_E fall to zero.

[3] Powerful hint: the datasheet for the 2N3904 shows V_{BE} breakdown can occur – for a *reverse* bias across the BE diode – at a voltage as low as 6 V.

[4] This notation, repeating the subscript letter, is used to indicate power supply voltages: V_{EE} where the supply is negative (and applied to the emitter in an *npn* circuit), V_{CC} where the supply is positive (and applied to the collector.)

[5] We are calling this Z, but what interests us here is really R; we want to avoid effects that are frequency dependent. Particularly, we don't want to find ourselves studying attenuations caused by stray capacitance.

4L.2 Emitter follower

Measure Z_{in}

In the circuit of Fig. 4L.2 replace the small base resistor with progressively larger resistors, in order to simulate a signal source of moderately high impedance, i.e., low current capability.

Start with 33k, as shown in Fig. 4L.3; you are likely to see attenuation of only 5% to 10%. With such small attenuation, it is possible but tedious to infer β.

If you install larger resistors in place of this 33k, eventually you will see attenuation of about 50%, and then your observation (and your arithmetic) will become easy. You'll know, if you get to 50% attenuation, that your installed R is about equal to R_{in}. Use this data to make your inference of β.

Caution: Make sure that it is R_{in} that you are observing, and not an effect of C_{in} or C_{probe} that you are seeing. To make sure of this, adjust the input frequency until no appreciable phase shift appears between the original signal and what you see at the transistor's base.

With 1k "load" *detached*, measure Z_{in} for the circuit. In this case, that is the impedance looking into the transistor's base (the 33k series resistor, or the larger value you may have installed, is *not* a part of the follower. It is included to model a signal source with mediocre R_{out}). You can discover Z_{in} by using the scope's two channels to look at both sides of the base resistor. For this measurement the 3.3k emitter resistor is treated as the follower's "load". Use a small signal – less than a volt in amplitude.

The attenuation that you see may be *slight*, at least at first, and therefore hard to measure. If you are using a digital scope you can be lazy, using its amplitude-measurement functions. If you are using an analog scope, you need to work harder. The best way to measure a small attenuation using an analog scope is to take advantage of the *percent* markings on the scope screen.

Suggestions for measurement of Z_{in} (analog scope):

- Center the two waveforms on the 0% mark – these are the waveforms that appear on the two sides of the base resistor: let's call the waveforms "source" and "input".
- *AC couple* the signals to the scope, to ensure centering.
- Adjust the function-generator's amplitude to make the *source* peak just hit 100%.
- Now read the *input* waveform's amplitude in percent.

Does your result make sense? Is the follower transforming the impedance "seen through it," as promised? Once you have measured a value of Z_{in} you can infer the transistor's β (or "h_{FE}:" see AoE §2.2.3B). Make a note of this calculated β; in a few minutes you will want to compare it to the β you infer from your measurement of Z_{out}.

Measure Z_{out}

Replace any other base resistor you may have installed, *returning to the value of* 33k shown in Fig. 4L.3.[6]

Now measure Z_{out}, the output impedance of the follower, by connecting a 1k load to the output and observing the drop in output signal amplitude; again use a small input signal, less than a volt.

The procedure for measuring Z_{out} is slightly different from the one used to measure Z_{in}.

In order to measure Z_{out}, you need a two-step process:

- with *no load* attached, measure the amplitude of V_{out};
- then attach the 1k "load," and note the resulting attenuation;

---

[6] Larger R values make your work harder, not permitting you to ignore the effect of R_E, whose value strictly parallels the path through the transistor. R_{source}/β usually is so much smaller than R_E that can ignore the latter. A very large base resistor spoils that simplification. This exercise is hard enough without that complication.

- infer Z_{out}: you recall that you have built a voltage divider, where the upper "resistor" is the circuit's Z_{out} and the lower "resistor" in the divider is the load. The value of the *base resistor* is critical to your calculation, of course; this is the "source resistance" that the follower reduces by the factor β. So make note of what base resistor you are working with.

Infer β once again: in principle it should match the β you inferred from your measured Z_{in}.

The blocking capacitor used in measurement of Z_{out}: Why do we suggest that you use a blocking capacitor as you measure Z_{out}? The answer is rather subtle, in this case, because here you could get away with omitting the blocking cap; but in many other cases you could not get away with that.

We include the blocking capacitor to avoid disturbing *DC* levels in the circuit, while we watch what happens to time-varying (or *AC*) *signals*. In this case, the DC level at the emitter is approximately −1V; this is close enough to ground so that omitting the blocking cap would do no harm. Omitting the cap would alter that level only slightly. But in many other cases, where the emitter's DC voltage happened to be well away from ground, attaching a 1k *DC* "load" to ground would alter the DC or "bias" level appreciably. We do not want to do that; we mean only to see what the load does to *signal* amplitude (and we assume the signal is a voltage *wiggle*).

It's too bad that the cap in this case doesn't provide a more striking benefit. But we are trying to teach you a good habit: if you're interested in AC behavior, measure that behavior without disturbing DC conditions.

4L.3 Fixing a defect in the follower's R_{out}

4L.3.1 Simple emitter follower clips when heavily loaded…

With the emitter-follower of Fig. 4L.3 still set up, let's first embarrass it, to show its limitations. Then we will look at a remedy.

You established, in §4L.2.1, that the emitter follower can show admirably low Z_{out}, namely, an impedance lowered by the follower's "lens-like" effect. The follower's output impedance normally takes a value of about R_{source}/Beta.

Let's first make the follower look good, once more, even lowering its R_{out} somewhat: replace the 33k base resistor in Fig. 4L.3 with a 10k. Now, you would expect the follower's Z_{out} – which we can call "R_{out}," since we are not concerned with effects of frequency – to be no more than 100 ohms.[7] It should have no trouble driving a load of 1k ohms.

This is an argument made in §4N.4.4.

Let's see if the follower performs that way: drive the simple follower, as before, using a sinusoid at about 1 kHz and about 2 V peak-to-peak.

Attach and detach the 1k load. Confirm that the loading effect of the 1k resistor to ground is no more than about 10%. The actual attenuation under load is likely to be a good deal less.

Probably it goes without saying that the waveform at R_{load} should faithfully track the input waveform (both are likely to be centered on ground; but that is not important, here). We hope that your result confirms the expectation that R_{out} for the follower is low, as advertised, far below the 1k of R_{load}.

So far, so good: the emitter is showing off its low R_{out}. Now let's make trouble: increase the input amplitude, toward 10 V, and note what happens to the *loaded* output.

---

[7] By the next class, you will be a little more sophisticated, and will correct this estimate by adding the modeling element "little r_e," which has the small value of about 6 ohms, at the I_C of the circuit of Fig. 4L.3.

We hope you are somewhat surprised (although you won't be if you remember Fig. 4N.25 in this lab's class notes).

4L.3.2 ...Push–Pull follower can solve the clipping problem

A symmetrical "push–pull" pair of transistors solves the problem caused by the simple follower's asymmetry.

Modify your simple emitter follower into the "push–pull" version shown in Fig. 4L.4.[8] These small transistors have limited current- and power-handling capacity, so are not suitable for a practical push–pull driver. The circuit of Fig. 4L.4 would be pretty useless driving an 8-ohm speaker, for example.[9]

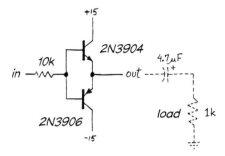

Figure 4L.4 Push–Pull follower.

We hope that you will see that the push–pull does solve the clipping problem that you saw in §4L.3.1 on the preceding page. The push–pull does, however, introduce a new problem. Watch the input and output as you gradually increase the amplitude of a 1 kHz sinusoid centered on ground, beginning at amplitude less than 0.5V. The push–pull should look very poor indeed. We will solve this problem in the first operational amplifier lab, Chapter 6L, when you meet an improved version of this push–pull circuit.

Don't let this new defect spoil your admiration for the push–pull. It is an important circuit, used in the output stage of many devices: audio amplifiers, motor drivers and also in the operational amplifiers that are central to the next stage of this course (for the latter examples, see Figs. 5L.15, 9S.1 and the detailed op-amp schematic in Fig. 9S.31).

4L.4 Current source

Construct the current source shown in Fig. 4L.5 (sometimes called, more exactly, a current "sink").[10]

Slowly vary the 10k variable load, and look for changes in current measured by the VOM. What

---

[8] We often use the name "push–pull" as a short form of "push–pull follower," though other push–pull configurations are possible, notably the push–pull switch used as the output stage of digital gates. See §14N.4.1 and Fig. 14N.25.

[9] The 2N3904 and 2N3906 can dissipate only about 0.6 W. With approximately 15 V across the transistor, this power limit would permit only a tiny voltage into an 8-ohm speaker (about 0.7 V). On a later day, in Lab 6, you will have a chance to try a practical push–pull, using power transistors. See §9L.4.

[10] The term "current source" unfortunately can be ambiguous. It describes the entire class of circuits that hold current constant despite voltage variation. But the term also can be used to describe a particular *sub-class*, the sort of circuit that feeds current into a load whose other other terminal is more negative. Such a *source* then is contrasted against a current *sink*, one like the one in Fig. 4L.5, which draws current *from* a load. Thus current sources come in two flavors, awkwardly called "sinks" and "sources." Context usually makes clear whether a reference to a "current source" indicates sense of current flow as well as constancy of current.

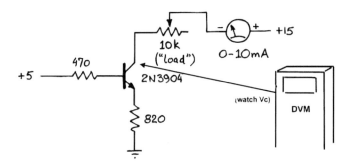

Figure 4L.5 Transistor current source.

happens at maximum resistance? Can you explain, in terms of *voltage compliance* of the current source ("compliance" is jargon for "range in which the circuit works properly")?[11]

Even within the compliance range, there are detectable variations in output current as the load is varied. What causes these variations?[12]

4L.5 Common-emitter amplifier

Wire up the common-emitter amplifier shown in Fig. 4L.6. What should its voltage gain be? Check it out. Is the signal's phase inverted?

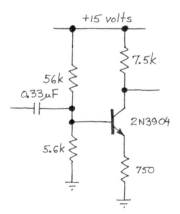

Figure 4L.6 Common-emitter amplifier.

Is the collector *quiescent* operating point correct (that is, its resting voltage)?[13] How about the amplifier's low frequency 3 dB point?[14] What should the output impedance be?[15] Check it by connecting a resistive load – let's say 7.5k – using a blocking capacitor. (The blocking cap, again, lets you test impedance at signal frequencies without messing up the biasing scheme.)

---

[11] The current source fails when the transistor saturates. The DVM will let you measure or infer V_{CE}. Thus you can look for the relation between saturation and the fall of I_{out}.
[12] Early effect is the principal cause: varying V_{CE} slightly varies the effective base width. See §5S.2.
[13] What level would be "correct?" The voltage that permits the widest swing without "clipping" – hitting either upper or lower limit. Roughly, that quiescent point would be half the supply voltage. A perfectionist would say, "Better to put it midway between the lower limit of about 1 V (V_E) and the supply. But with a 15 V supply we don't mind the lazier answer, "half the supply voltage." This will be our usual answer, for single-supply circuits.
[14] To calculate this you'll need to decide what is the effective *resistance* that should be paired with the blocking capacitor. C and that R form a highpass, as you know.
[15] Yes, just R_C. You've seen this point explained in Chapter 4N.

4L.5.1 Maximizing gain: sneak preview of Ebers–Moll at work

Now let's try a case that our first, simple, view of the common-emitter amplifier does not describe accurately: check out Fig. 4L.7 to see what happens if we parallel the emitter resistor with a big capacitor, so that at "signal frequencies" R_E is shorted out.

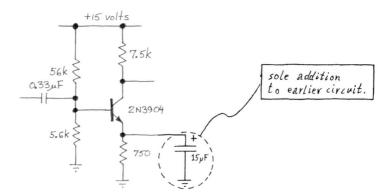

Figure 4L.7 Grounded emitter amplifier.

Modify your common-emitter amp to make the amplifier shown in Fig. 4L.7 (similar to Fig. 5N.17). What circuit properties does your addition of the *bypassing* capacitor affect? It affects gain, because R_E disappears from the gain equation. Does the C alter the biasing of the output (which was designed to be centered roughly at half the supply voltage)?[16]

Now drive this new circuit with a small *triangle* wave at 10 kHz, at an amplitude that almost produces clipping (you'll need to use plenty of attenuation – 40 dB or more – in the function generator). Does the output waveform look like Fig. 4L.8? (Compare to Fig. 5N.12.)

Installing the cap that bypasses R_E does *not* deliver infinite gain: sorry! In Chapter 5N you'll discover that we can predict the gain by adding to our transistor model a little resistor-like element in the emitter, a modelling device that we call (affectionately) "little r_e." In the present circuit, where we're running a quiescent current of about 1 mA, little r_e takes a value of about 25 Ω. See if your circuit's gain is consistent with that value for r_e.

Figure 4L.8
Large-swing output of grounded emitter amplifier when driven by a triangle wave.

This measurement will be difficult. First, you will need to reduce the function generator output to a level close to the minimum possible. Then you will find the input is immeasurably small, viewed with a 10× probe. This, then, is one of the rare occasions when you need to revert to a 1× display: use a BNC rather than a 10× probe as you watch the *input* signal.[17]

Typically, the gain you observe is around 250 – lower than you predicted. Low I_C can help explain that; so can any curvature that remains in the output waveform: the gain revealed by such a waveform is sometimes higher than the quiescent value, sometimes lower, but the *net* effect of the curvature is to *reduce* gain.

4L.6 Transistor switch

The circuit in Fig. 4L.9 differs from all those you have built so far: the transistor, when ON, is *saturated*. In this regime you should not expect to see $I_C = \beta \times I_B$. Why not?[18] In addition, β for this big power

---

[16] No: biasing is a simple *DC* effect.
[17] If your scope probe has a 10×/1× selector switch, just select 1×.
[18] One way to explain this is just to note that achieving a current as high as that would require either a smaller resistance on the collector or a larger power supply. When the transistor is saturated, the current is limited not by the transistor but by the *load*.

transistor is much lower than what you have seen for the *small signal* transistor, 2N3904 – at least at high currents. Minimum β for the MJE3055T is a mere *20*, at I_C=4A. But the transistor is big, and lives in a big package ("TO-220"), a package that can dissipate a lot of heat, especially if the metal package is thermally attached to a still bigger piece of metal (a "heat sink"). The large size of the transistor itself keeps current density down and saturation voltage low. You'll measure that $V_{CE(sat)}$ in a few minutes.

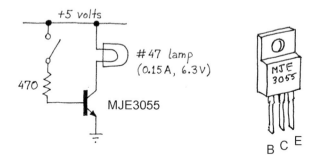

Figure 4L.9 Transistor switch: twist each transistor lead 90°, so that each fits easily into the breadboard's slots.

Turn the base current on and off by flipping the toggle switch – or, if you're too lazy to wire a switch, by pulling one end of the resistor out of the breadboard and touching it either to +5 or to ground. What is I_B, roughly? What is the minimum required β?[19]

4L.6.1 Saturation or "On" voltage: $V_{CE(sat)}$

Now let's make the transistor work harder. Replace the lamp with a 10 Ω *power* resistor (not an ordinary 1/4 W resistor; consider how much power the resistor must dissipate: $V^2/R = 25/10 = 2.5\,\text{W}$). A 1/4 W resistor would cook in this circuit.

Measure the saturation voltage, $V_{CE(sat)}$, with DVM or scope. Then parallel the base resistor with 150 Ω, and note the improved $V_{CE(sat)}$. Compare your results with the results promised by the data set out in Fig. 4L.10. Note that the curves assume heavy base drive: $I_B = I_C/10$, not I_C/β.

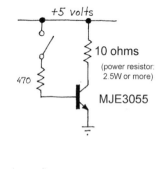

Figure 4L.10 Switch saturation: a heavier load current asks more of the switch here (data courtesy of ON Semiconductor).

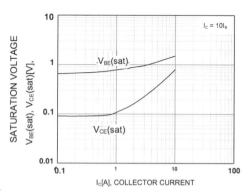

We will return to transistor switches later, in Lab 12L, when we'll meet the leading competitor for the tasks a '3055 can do: a big power field-effect transistor ("MOSFET"). At that time, we'll set up a

---

[19] You'll notice that we are *overdriving* the base, here, as is usual in a switch. Since base current here is about 10 mA, even at the specified minimum *beta* of 20, the transistor could pass more current than the load will permit. That's good: the switch will be well saturated, and its V_{CE} will therefore be low.

4L.7 A note on power-supply noise

competition between a '3055 and a MOSFET and see which does a better job of delivering power to the load.

4L.6.2 Switch an inductive load (more exciting)

Now replace the resistor at the collector with a 10mH *inductor* as in Fig. 4L.11. Replace the 5 V supply with 1.5 V from a single AA cell (we don't want to overheat the inductor – and the low cell voltage makes all the more impressive the voltage spike that soon will appear). Drive the 1 kΩ base resistor with a 1 kHz square wave from the function generator's "TTL" or "Sync" terminal (this is a "logic level" waveform: a square wave switching between ground and four or five volts).

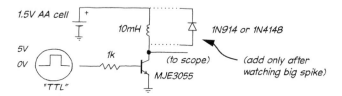

Figure 4L.11 Transistor switch with inductive load.

At the transistor's collector, you should find the inductor giving the transistor an alarming voltage spike (we saw more than 100V). When the transistor tries to turn off, the inductor tries to keep the current flowing. Its method is to drive the collector voltage up, until the transistor "breaks down," permitting the inductor to have its way – keeping the current flowing despite the transistor's attempt to turn it off abruptly.

A diode from collector to V_+ tames this voltage spike; such a diode clamp is a standard protection in circuits that switch inductive loads. Most transistors don't like to break down (this time we didn't mind since our goal was to show you this spike). Incidentally, we'll see a circuit that makes good use of this voltage spike when we look at switching power supplies later in Lab 12L.

4L.7 A note on power-supply noise

Note §9S.3 is devoted to noise problems. Here, we offer just a few pointers related to one form of noise.

In Figure 4L.12 is one of today's followers, swept rather slowly (at 20 μs/division). The follower has been fed a sine of about 15 kHz from the function generator. A strange *thickening* of the trace appears.

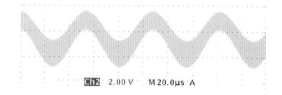

Figure 4L.12 Thickening of scope trace indicates a nasty "parasitic" oscillation.

Sweeping the scope faster (Fig. 4L.13, at 10 ns/div) resolves the "thickening" into a very fast sinusoid – up in the *FM* radio broadcast range. Decoupling the power supplies should eliminate this problem (see below).

Oscillations versus radio pickup

If you see this fuzz, try turning off your breadboard power supply. If the fuzz disappears, your circuit is guilty: it was causing the fuzz, by running as an unintended *oscillator* (we'll look closely at the

Figure 4L.13 Fast scope sweep can resolve the "thickened" scope trace.

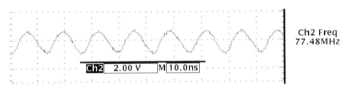

question how this can happen, in Lab 9L). If the fuzz persists, your circuit is innocent and you're probably seeing radio broadcast stuff picked up by your wiring. There's no quick way to eliminate that; you see it whenever your scope gain is very high and the point you look at is not of low impedance at those high frequencies.

Remedies If your circuit is oscillating – not merely picking up radiated noise – there are three remedies you might try, in sequence:

(1) Make sure that you are watching the circuit output with a scope *probe*, not with a BNC cable. The 10× heavier capacitive load presented by the BNC cable often brings on oscillation.
(2) Try shortening the power and ground leads that feed your circuit. Six inches of wire can show substantial inductive reactance at megahertz frequencies; shorter leads make the power supply and ground lines more nearly ideal – lower impedance – and harder for naughty circuits to wiggle.
(3) If the oscillation persists after those two attempted repairs, then add a "decoupling" or "bypass" capacitor between each supply and ground. Use a *ceramic* capacitor of value 0.01 or 0.1 μF; place it as close to your circuit as possible – again, inches can matter, at high frequencies.

Incidentally, we have listed this option last not because there's anything wrong with decoupling the supplies; soon you will be putting decoupling caps in routinely. We placed this remedy last just because it *is* so effective (and therefore mandatory): we wanted you to see first that the first two remedies also sometimes are sufficient.

4W Worked Examples: Transistors I

4W.1 Emitter follower

AoE §2.2.5A

AoE works a similar problem in detail: §2.2.5A. The example below differs in describing a follower for *AC* signals. That makes a difference, as you will see, but the problems are otherwise very similar.

4W.1.1 Problem: AC-coupled follower

Design a single-supply voltage follower that will allow this source to drive this load, without attenuating the signal more than 10%. Let V_{CC}=15 V, let the quiescent I_C be 0.5 mA. Put the 3 dB point around 100 Hz.

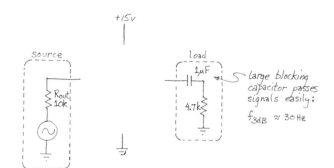

Figure 4W.1 Emitter follower (your design) to let given source drive given load.

4W.1.2 A solution

Before we begin, perhaps we should pause to recall why this circuit is useful. It does *not* amplify the signal voltage; in fact we concede in the design specification that we expect some *attenuation*, we want to limit that effect. But the circuit does something useful: before you met transistors, could you have let a 10k source drive a 4.7k load without having to settle for a good deal of attenuation? How much? We will try to explain our choices as we go along, in scrupulous – perhaps painful – detail.

Draw a skeleton circuit
Perhaps this is obvious; but start by drawing the circuit diagram without part values, as in Fig. 4W.2. Gradually we will fill those in.

Choose R_E to center V_{out}
To be a little more careful, we should say,

"we aim to center $V_{\text{out-quiescent}}$, given $I_{\text{C-quiescent}}$."

"Quiescent" means what it sounds like: it means conditions prevailing *with no input signal*. In effect, therefore, *quiescent* conditions mean *DC* conditions, in an *AC* amplifier like the present design.

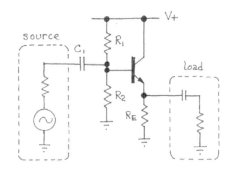

Figure 4W.2 Emitter follower skeleton circuit: load is AC coupled.

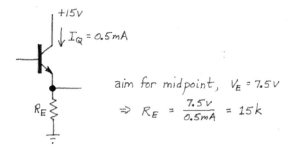

Figure 4W.3 Choose R_E to center V_{out}.

And anticipating some complications that we will meet in other transistor circuits, namely the *common-emitter* amplifier and the *differential amplifier*, we should acknowledge that, strictly, our goal is to center V_{out} *in the range available to it*. In the present case, that range does extend all the way from lowest voltage (ground) to most positive (V_+). It does not always behave so simply.

Center V_{base}

Here we'll be a little lazy: by centering the base voltage we will be sure to *miss* centering V_{out}. But we'll miss by only 0.6 V, and that error won't matter if V_{CC} is big enough. The error is about 4% if we use a 15V supply, for example.

Centering the *base* voltage makes the divider resistors equal; that, in turn, makes their R_{Thev} very easy to calculate.

Choose bias divider Rs so as to make bias stiff enough

Stiff enough means, by our rule of thumb, $\leq 1/10$ of the R that it drives. What it drives is the base, and we need to know the value of this $R_{in\ at\ base}$ at DC (that is, not considering *signal* frequencies). If we follow that rule, loading by the base will not upset the bias voltage that we aimed for (to about 10%).

The value of $R_{in\ at\ base}$ is just $\beta \times R_E$, as usual. That's straightforward. What is not so obvious is that we should *ignore* the AC-coupled *load*. That load is *invisible* to the bias divider, because the divider sets up *DC* conditions (steady state, quiescent conditions), whereas only *AC* signals pass through the blocking capacitor to the load. Figure 4W.4 spells out the process – but note that the process begins on the right, with the impedance that the divider is to drive. That concludes the setting of DC conditions. Now we can finish by choosing the coupling capacitor (also called "blocking capacitor;" evidently both names fit: this cap couples one thing, blocks another).

Choose blocking capacitor

We choose C_1 to form a highpass filter that passes any frequency of interest. Here we have been told to put f_{3dB} around 100 Hz.

4W.1 Emitter follower

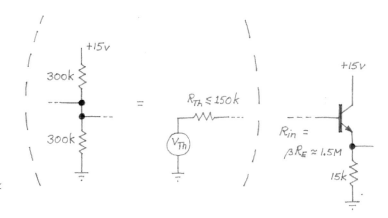

Figure 4W.4 Set $R_{\text{Thev-bias}} \ll R_{\text{in at base}}$.

The only difficulty appears when we try to decide what the relevant R is in our highpass filter: see Fig. 4W.5.

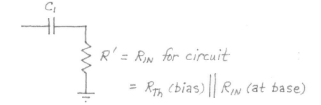

Figure 4W.5 What R for blocking cap as highpass?

We need to look at the input impedance of the follower, seen from this point. The bias divider and transistor appear in *parallel*.

Digression on series versus parallel Stare at the circuit till you can convince yourself of that last proposition. If you have trouble, think of yourself as a little charge carrier – an electron, if you like – and note each place where you have a *choice* of routes: there, the circuit offers *parallel* paths; where the routes are obligatory, they are in *series*. Don't make the mistake of concluding that the bias divider and transistor are in series because they appear to come one after the other as you travel from left to right.

So, $Z_{\text{in-follower}} = R_{\text{TH bias}}$ in parallel with $R_{\text{in at base}}$. The slightly subtle point appears as you try to decide what $R_{\text{in at base}}$ ought to be. Certainly it is $\beta\times$ something. But $\times$ what? Is it just R_E, which has been our usual answer? We did use R_E in choosing R_{TH} for the bias divider.

But this time the answer is, 'No, it's not just R_E,' because the *signal*, unlike the DC bias current, passes *through* the blocking capacitor that links the follower with its load. So we should put R_{load} in parallel with R_E, this time.

The impedance that gets magnified by the factor β, then, is not 15k but (15k $\parallel$ 4.7k), about 15k/4 or 3.75k. Even when increased by the factor β, this impedance cannot be neglected for a 10% answer:

$$R_{\text{in}} \approx 150\,\text{k} \parallel 375\text{k}$$

So, choose C_1 for f_{3dB} of 100 Hz. Then

$$C_1 = \frac{1}{2\pi\,100\,\text{Hz} \times 100\text{k}} \approx \frac{1}{6\times 10^2 \times 100 \times 10^3} = \frac{1}{60} \times 10^{-6} \approx 0.016\,\mu\text{F}$$

$C_1 = 0.02\,\mu\text{F}$ would be generous.

Worked Examples: Transistors I

Recapitulation

For people who hate to read through explanations in words, Fig. 4W.6 restates what we have just done.

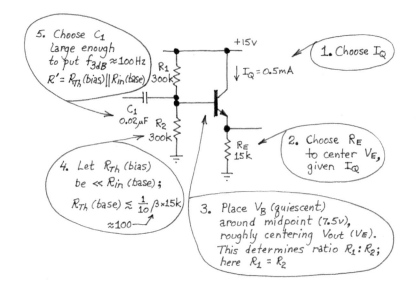

Figure 4W.6 Follower design: recapitulation.

4W.2 Phase splitter: input and output impedances of a transistor circuit

Problem Design a circuit that puts out waveforms inverted, with respect to each other

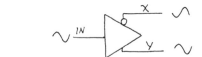

Figure 4W.7 Unity-gain phase splitter: generic.

Figure 4W.7 shows a circuit that you can make sense of with the tools from this, your first, day with transistors – our "simple" view of transistor operation. The circuit is called a "phase splitter," because it puts out two waveforms, 180° out of phase with each other. But it's useful to us, just now, because it gives a workout in calculating impedances – and that workout gives us a chance to review much of what we met in Lab 4L.

- Specifications:
  - Power supply: +20 V only.
  - Signal frequency range: 100 Hz and up.
  - $R_{\text{out-signal-source}} \leq 5\,\text{k}\Omega$.
  - Quiescent current (I_C).
- Calculate input and output impedances.

4W.2 Phase splitter: input and output impedances of a transistor circuit

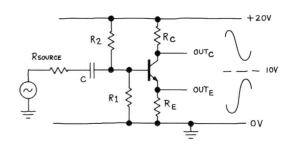

Figure 4W.8 Phase splitter, no component values.

Solution

Skeleton circuit: First, let's start with the skeleton circuit of Fig. 4W.8 and calculate what component values we ought to choose.

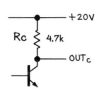

Figure 4W.9 Choose value for collector resistor.

Choose component values: We'll need the rules stated in the first transistor class:

- Simple: $I_C = I_B \times \beta$.
- Simpler: $I_C \approx I_E$; $V_{BE} = 0.6$ V.

The second rule is enough to let us calculate the voltage gain of this circuit; the first rule is necessary to let us calculate most input and output impedances. Let's do this in stages:

- Choose R_C, the collector resistor, see Fig. 4W.9

  In order to choose this value, we need to make an initial design choice – what the quiescent collector current should be. Let's be lazy and conventional and make it 1mA. Given that current, we can choose the R value by determining what *quiescent* voltage we would like to see at output Out_X.

  We don't want to reach, mechanically, for the answer, "Half the power supply." No. The basis for setting the quiescent voltage is to allow maximum swing without clipping – and this time the emitter voltage is not tamely sitting still (or nearly) as in the usual common-emitter amplifier. Instead, the emitter is swinging as widely as the collector swings – so, the range of voltage available to the collector is not the full power supply, but half of it.

  Therefore we should put $V_{\text{C\_quiescent}}$ at the midpoint of the range available to it. That range is 10–20 V, so the quiescent voltage should be 15 V. To get this result, we want to drop 5 V. So $R_C = 5\text{k}$. The nearest 10% value is 4.7k.

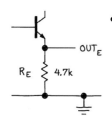

Figure 4W.10 Choose value for emitter resistor.

- Choose R_E, the emitter resistor, see Fig. 4W.10.

  Again the goal is to allow maximum swing, so we place $V_{\text{E\_quiescent}}$ at the midpoint of its range: we place it at 5 V. With 1 mA flowing, we need 4.7k again.

- The bias divider, see Fig. 4W.11.

  This is a little more involved.

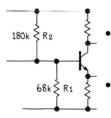

Figure 4W.11 Choose value for bias divider.

- Voltage: just a little above $V_{\text{E\_quiescent}}$. So, the exact target would be about 5.6 V. Our standards aren't as high as that, in a 20 V span. Let's be lazier and put the base voltage at 5 V rather than 5.6 V. That eases the arithmetic: we need 1/4 of the 20 V supply, so the ratio R2:R1 is 3:1.
- Values: we want the bias divider to be "stiff" relative to what it drives, the transistor base.
  - $R_{\text{in\_base}}$ This, we learned last time, is $\beta \times R_E \geq 500\text{k}$.
  - $R_{\text{Thev\_divider}}$ should be small relative to what it drives: $\leq 50\text{k}$.
  - R_1, R_2 values: $R_{\text{Thev\_divider}} = R_1 \parallel R_2$. Since $R_2 = 3R_1$, $R_{\text{Thev}} = \frac{3}{4}R$. So the smaller R should be about $\frac{4}{3}R_{\text{Thev}} \approx 66\text{k}$: we're fine with 68k. The upper resistor should be about 3× that. 180k is a little low – but we don't really mind since we're happy to put V_{base} a little above 5 V, recalling the 0.6 V V_{BE} drop.
- The blocking capacitor, see Fig 4W.12.[1] (We've inserted the C value in the figure; we'll explain how we chose that value.)

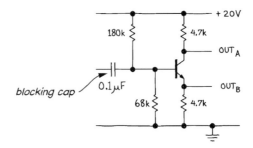

Figure 4W.12 Choose value for coupling/blocking capacitor.

We need an f_{3dB} as a design goal. Let's suppose we want to pass low-quality audio, so we can put f_{3dB} at 50 Hz.

After that choice, we need only to determine what effective R this C is paired with. That effective resistance is the bias divider in parallel with the transistor's base.[2]

You could calculate this – but, as usual, we prefer to be intelligently lazy. We know that $R_{\text{in\_base}}$ is much larger than $R_{\text{Thev\_divider}}$. Why? Because we designed it that way, a moment ago. So, we can neglect $R_{\text{in\_base}}$, and treat the effective R as only $R_{\text{Thev\_divider}}$. We needn't calculate that value, because we know it; it's the value we started with as goal when we designed the bias divider: 50k.

An exact C value would be about 0.06 µF: 0.1 µF will do nicely. There is no harm at all in pushing f_{3dB} a little lower than the target, here: the cap's function is only to block DC. If it passes signals at a little lower frequency than originally planned, good.

Now that the values are in place, we can calculate the impedances in and out.

R_{out} at collector: This is the straightforward one. Put the two paths in parallel, as usual (by analogy to R_{Thev} – a model that is useful for impedance measurements far beyond its original scope, which covered only *resistors*).

How should we treat the impedance seen "looking into" the collector? Step 1: don't fall for the error of thinking that the collector looks like the emitter, so that you'll see some *lens* effect using β. No. The collector behaves radically differently from the emitter.

Step 2: take advantage of the fact that, whereas the emitter is a *voltage* source, the collector is a *current* source: its *current* is determined. We have said this elsewhere, but let's say it again: when the current is determined, one can change voltage without changing current appreciably. So, the fraction $\Delta V/\Delta I$ is very large (ideally, infinite). That fraction is just our definition of effective *resistance* at a point (we sometimes call this "dynamic resistance").

Figure 4W.13 $R_{\text{out\_C}}$: turns out equal to collector resistor.

---

[1] As you know, this capacitor could just as well be called the "coupling capacitor." It *blocks* DC; it *couples* the AC signal.
[2] We know you wouldn't fall into the trap of thinking that these impedances are in *series*, just because you may see one drawn to the left of the other. A wire joins the two paths. We offered the same reminder earlier in §4W.1, the single-supply follower.

4W.2 Phase splitter: input and output impedances of a transistor circuit

So, this very-large impedance looking into the collector has no effect when paralleled with R_C: the R_C dominates and in this circuit defines R_{out} at the collector. As for the common-emitter amplifier, R_{out} at the collector is the collector resistor, R_C.

R_{out} at emitter: This is not so simple as the impedance at the collector. Here, we do see the transistor's "lens" effect, and to determine R_{out} at the emitter we "look through" the transistor toward the signal source. You worked through such a problem in Lab 4L when you measured R_{out} for the follower.

R_{out} is R_E in parallel with the other path, through the emitter. Let's write this out, first, with painful rigor. Then we'll throw out what doesn't matter much:

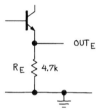

$$R_{\text{out\_E}} = R_E \parallel \left(r_e + \frac{R_{\text{Thev\_bias}} \parallel R_{\text{source}}}{(1+\beta)} \right)$$

Figure 4W.14
$R_{\text{out\_E}}$: rather complicated, but approximately R-signal-source/beta.

This looks pretty formidable. So, let's start throwing out what has little effect.

- R_E normally can go: it should be much larger than what parallels it.
- $R_{\text{Thev\_bias}}$ can go, since it defines Z_{in} for the circuit, as we saw above, and this Z_{in} must be large relative to $R_{\text{out-signal-source}}$.
- We cannot neglect r_e,[3] since its value at I_C=1 mA=25 Ω is not negligible compared with the value of $R_{\text{out-signal-source}}/\beta$. That value is 5k/100 = 50 Ω.

With the help of this intelligent laziness, then, we can reduce the formidable equation to the tame

$$R_{\text{out\_E}} \approx r_e + \frac{R_{\text{source}}}{\beta} = 25 + \frac{5k}{100}$$

The value for $R_{\text{out\_E}}$ calculated in this lazy way is 75 Ω; the exact value is 69 Ω. The error is about 8%, quite tolerable by our standards.

Z_{in}: Again, we will make it simple for ourselves by neglecting what doesn't matter much.

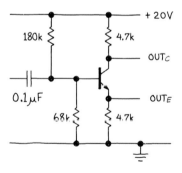

Figure 4W.15 Z_{in}: bias divider dominates.

- The capacitor: the key notion, here, is to recognize that what interests us is how this circuit behaves at *signal frequencies*. We don't care how it looks to what we consider *noise*: to frequencies below our input circuit's $f_{3\text{dB}}$. So, the C matters not at all: we sail right through it, as usual (if we didn't, we'd be using the wrong capacitor, and should go fetch a bigger one).
- The bias divider: we know this impedance is low relative to what the base looks like – because we designed it that way. So...

---

[3] We admit it's unfair to introduce this term before we reach its explanation in Chapter 5N. But we feel obliged to mention it here to make this solution as general as possible: r_e will not puzzle you after you've been through Chapter 5N.

- ...we can neglect the input impedance at the base, which has been designed to be at least 10× larger than $R_{\text{Thev\_bias}}$.

The short answer is that $Z_{\text{in}} \approx R_{\text{Thev\_bias}}$: i.e., 50k. This is within 10% of the exact answer, one that would take account of loading by $\beta \times (r_e + R_E)$; again, good enough, by our standards.

Perhaps we should admit that we've done something a little unrealistic here: we have neglected the *AC-coupled* load that may be attached to the emitter. That load should be included in the calculation, in a practical instance. In that case, we would amend the Z_{in} calculation:[4]

$$Z_{\text{in}} = R_{\text{Thev\_bias}} \parallel (\beta \times R_{\text{Load\_AC\_coupled}})$$

That load, unlike R_E (which forms a *DC* load for the bias divider), cannot be assumed to be large relative to $R_{\text{Thev\_bias}}$. And if such a load is to be attached, then we should require that $R_{\text{out\_signal\_source}}$ be lower than what is shown in this example: lower than 5k (make the limit perhaps 1k; at this value, it would permit $R_{\text{load}} \geq 400\,\Omega$). With such a load attached, Z_{in} would drop to 22k.

Figure 4W.16 shows the whole circuit.

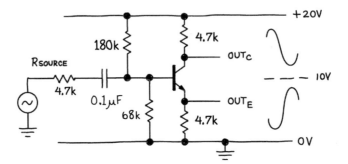

Figure 4W.16 Phase splitter circuit, with values.

Our main interest, in working through this example, was not in getting the refinements right. Instead, we hoped to remind you of a few basic notions, and to give you a chance to apply them:

- Output impedances are radically different at emitter and collector.
- AC and DC paths are to be distinguished – and "signal frequencies" to be distinguished from other frequencies (we do not design our circuit for the benefit of these other frequencies).

4W.3 Transistor switch

An ideal switch ought, when *ON*, to put maximum power into the load, while dissipating none itself. Fig. 4W.17 shows two possible *switch* circuits, and some questions about them.

Problem

1. **Collector current and switch power** What is approximate load current in the two cases? Assume that β is 100.
2. **Power in the switch** What power is dissipated in the transistor switch circuit in the two cases, assuming that saturation voltage, $V_{\text{CE(sat)}}$, is about 0.2 V? (Please include power in the base resistor.)
3. **Why is one better than the other?** Explain briefly why one is the preferred switch configuration.

---

[4] We could, below, take account of R_E that parallels the load, to be a little more accurate. But we prefer the simplification that allowed us to neglect the effect of $\beta \times R_E$, much larger than the effect of the bias divider.

4W.3 Transistor switch

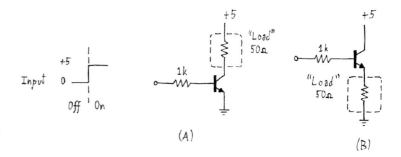

Figure 4W.17 Two possible "switch" circuits.

A solution

1. Collector current and switch power: Look at circuit **A** in Fig. 4W.17; it's a *switch* configuration:

$$I_B = 4.4\,\text{V}/1\text{k} = 4.4\,\text{mA}$$

This current is ample to saturate the transistor. The transistor saturates at perhaps 0.2 V, making $V_{\text{load}} \approx 4.8$ V.

So $I_{\text{load}} = 4.8\,\text{V}/50\,\Omega = 100\,\text{mA}$. (Note that it is the *load* that sets the current.)

Now look at circuit **B**, which is a *follower* rather than a switch. Redraw as a voltage divider with R_E, V_{BE}, and R_B. Note R_B is transformed by Q's beta into what looks like $1\text{k}/\beta = 10\,\Omega$, see Fig. 4W.18:

$$I_C \approx I_E = \frac{4.4\,\text{V}}{60\,\Omega} \approx 70\,\text{mA}$$

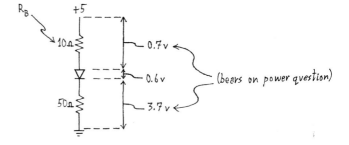

Figure 4W.18 Circuit B – a follower – redrawn as voltage divider to show where power is dissipated.

2. Power in the switch:

Circuit **A**, *switch*:

$$P = IV = P_{CE} + P_{BE} + P_{\text{base-resistor}}$$
$$= (0.1\,\text{A} \times 0.2\,\text{V}) + (4.4\,\text{mA} \times 0.6\,\text{V}) + [(4.4\,\text{mA})^2 \times 1\text{k}]$$
$$= 0.02\,\text{W} + 2.6\,\text{mW} + 20\,\text{mW} = 43\,\text{mW}$$

Circuit **B**, *follower*:

$$P = IV = P_{CE} + P_{BE} + P_{\text{base-resistor}}$$
$$= (70\,\text{mA} \times 1.3\,\text{V}) + (0.7\,\text{mA} \times 0.6\,\text{V}) + (0.7\,\text{mA} \times 0.7\,\text{V})$$
$$= 91\,\text{mW} + 0.4\,\text{mW} + 0.5\,\text{mW} \approx 92\,\text{mW}$$

Worked Examples: Transistors I

3. Why is one better than the other? The switch (**A**) puts more power into the *load*, less into the switch (less by a factor of about 1/2, as just above). The switch comes closer to putting the full power supply voltage across the load: 4.8 V versus 3.7 V for the follower.

If one puts the comparison into terms of *percentage* of total power that goes into the load, the contrast looks like this:

Switch (**A**):

$$P_{\text{load}} = 0.1, \text{A} \times 4.8, \text{V} = 0.48 \text{ W}$$

$$P_{\text{switch}} = 0.043, \text{W}$$

$$\Longrightarrow P_{\text{load}}/P_{\text{total}} = 0.48/0.52 = 92\%$$

Follower (**B**):

$$P_{\text{load}} = 0.07 \text{ A} \times 3.7 \text{ V} = 0.26 \text{ W}$$

$$P_{\text{follower}} = 0.09 \text{ W}$$

$$\Longrightarrow P_{\text{load}}/P_{\text{total}} = 0.26/0.35 = 74\%$$

In short – as you have gathered – when ON versus OFF is all we need, we prefer the *switch* configuration because of its *efficiency*. We also like the switch configuration for permitting output swing to any rail, independent of the base-drive range.

5N Transistors II

Contents

5N.1	**Some novelty, but the earlier view of transistors still holds**	**198**
	5N.1.1 "Why transistors are hard"	198
5N.2	**Reviewish: phase splitter**	**198**
	5N.2.1 Output impedances	199
	5N.2.2 Input impedances	199
5N.3	**Another view of transistor behavior: Ebers–Moll**	**200**
	5N.3.1 A case that calls for Ebers–Moll treatment	200
	5N.3.2 Ebers–Moll equation describes transistor's V_{BE} versus I_C...	201
	5N.3.3 Little r_e	202
	5N.3.4 r_e solves impedance problems, too	203
5N.4	**Complication: distortion in a high-gain amplifier**	**203**
	5N.4.1 Distortion remedy: emitter resistor – at the price of gain	205
5N.5	**Complication: temperature instability**	**205**
	5N.5.1 Temperature stabilizing: feedback, using emitter resistor	206
	5N.5.2 Temperature stability *and* high gain	207
	5N.5.3 Another temperature-stabilizer: Explicit DC feedback	209
	5N.5.4 One more way to stabilize the circuit: add a compensating transistor	209
5N.6	**Current mirror**	**210**
	5N.6.1 Simplest Mirror	210
5N.7	**Reconciling the two views: Ebers–Moll meets $I_C = \beta \times I_B$**	**210**
5N.8	**"Difference" or "differential" amplifier**	**211**
	5N.8.1 Why a differential amp?	211
	5N.8.2 A differential amp circuit	213
	5N.8.3 Differential amp evolves into "op-amp"	216
5N.9	**Postscript: deriving r_e**	**216**
5N.10	**AoE reading**	**217**

Why?

In this chapter we meet an amplifier sensitive to a *difference* between two inputs rather than to a difference from *ground*. This novelty permits implementation of the hugely important *operational amplifier*, which from the next class onward will be our principal analog building block.

5N.1 Some novelty, but the earlier view of transistors still holds

Today we encounter some familiar circuits that expose the limitations of our first view of transistors. High-gain amplifiers do this, with special clarity. But we will continue to use our first view whenever it suffices – because it is simpler.

The other important topic of this chapter is a circuit that does not require a new understanding of transistors, but builds a more complex circuit – and a very important one – out of circuit fragments that we met last time: the common-emitter amplifier and the follower.

The new circuit made up of these elements is a *difference amplifier* (often called a "differential amplifier;" we will use both terms), which when buffered with a *follower* output stage becomes an *operational amplifier*. We ask you to build such a circuit in Lab 5L, so that you can see that the op-amp is made up of modest and familiar elements. Armed with this comforting knowledge, you should be able to appreciate that the op-amp's wonderful performance – which you will witness next time – is not *just* the result of magic within the black box. The circuit's behavior is comprehensible.

The performance of op-amp circuits sometimes does seem magical – but we'll be pleased if you can enjoy that effect while also enjoying the comfortable sense that the device itself is just an aggregation of some old circuit friends of yours.

5N.1.1 "Why transistors are hard"

AoE §2.1

The basic difficulty in transistor circuits – or, at any rate, the basic *novelty* – arises from the fact that these devices have *three* terminals rather than the two that we are accustomed to in the passive devices we have met: resistors, capacitors, diodes, inductors.[1] We noted this novelty last time.

We have remarked before on the difficulty of discrete-transistor design, a difficulty that results perhaps from this three-terminal strangeness. On the other hand, two comforting thoughts: everyone else's treatment of transistors is much harder and more obscure than the one you will find in this book and in AoE. And here's a further comfort: soon you'll be using operational amplifiers, which will work very well while hiding from you the gritty difficulties of transistor circuit design.

5N.2 Reviewish: phase splitter

AoE §2.2.8

In Example §4W.2, we looked closely at the circuit in Fig. 5N.1, first putting together this design, and then calculating input and output impedances. Here, we'd like to use this circuit to bring out a few points that you saw in the first transistor lab, 4L.

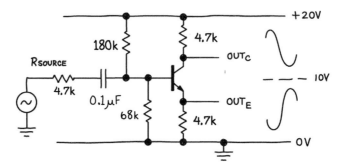

Figure 5N.1 Phase splitter.

---

[1] Perhaps you will object that your old friend the transformer has four terminals, so three terminals don't daunt you. If so, then you can skip this part of the class note.

5N.2 Reviewish: phase splitter

We will use our two simple views of transistor operation from last time, and these will be almost sufficient for the whole analysis. We will bump up against one inadequacy – which may help motivate moving on, after these preliminaries, to the second view of transistors. That second view will form the central part of this chapter.

We can make sense of nearly all features of the splitter of Fig. 5N.1 with two descriptions of transistor behavior from last time:

- Simple: $I_C = I_B \times \beta$.
- Simpler: $I_C \approx I_E$; $V_{BE} = 0.6$ V.

I_E...: The second of the two propositions (labeled "simpler") lets us calculate emitter current. The base voltage in Fig. 5N.1 is 1/4 of the 20 V supply, i.e., 5 V. The emitter voltage, V_E, is a diode drop lower – still approximately 5 V in a 20 V span (the 0.6 V error is fractionally small) – and the emitter current is just $V_E/R_E \approx 1$ mA (our favorite current, as you will gather: it makes the arithmetic so easy!).

...hence, I_C: The collector current, I_C, is just about the same as I_E.

...and quiescent output voltages: So we can calculate the quiescent (resting) voltages at emitter and collector: 5 V and 15 V (20 V less the 5 V drop across R_C). Are these reasonable? Yes, because they allow maximum swing without bumping ("clipping") on upswing or downswing.

5N.2.1 Output impedances

AoE §2.2.7A

At collector...: Here, R_{out} is just the value of the collector resistor, R_C: 4.7k (if you wonder *why*, take a look at the phase-splitter example, §4W.2.

AoE §2.2.7A

...at emitter: Here, we see the "rose-colored lens" effect that you saw in the follower of Lab 4L: R_{source} is reduced by the transistor's current-gain, β. (This is the first question for which we have needed the first of the two propositions above – the one called "simple".) We would therefore predict $R_{out} \leq 50\,\Omega$.

This is also the first point where our answer would be substantially wrong. The correct answer is about $75\,\Omega$, because of what we will soon call "little r_e," or "the intrinsic emitter resistance." (Intriguing new notion, coming later in §5N.3.3, and, we must admit, included in §4W.2, before you had heard the notion discussed.)

5N.2.2 Input impedance

This is dominated by the bias network – and thus is simpler than the case of the split-supply follower that you measured in Lab 4L. The R_{Thev} for the bias divider is much lower than the transistor's input impedance, so we do not meet that version of the "rose-colored lens" effect. (Of course, the *design* of the circuit, distinguished from the *analysis* that we are sketching here, did require paying close attention to the β transformation that determines R_{in} at the base. This process is spelled out in §4W.2 treating the splitter.)

5N.3 Another view of transistor behavior: Ebers–Moll

AoE §2.3

Some circuits that we worked with comfortably in the previous chapter will resist our first view, requiring use of the second, "Ebers–Moll" analysis. We just met such a circuit – one whose R_{out} is not as low as our simple view would predict. Our major example, however, will be another circuit, the common-emitter amplifier.

5N.3.1 A case that calls for Ebers–Moll treatment

The common-emitter amplifier of Lab 4L is easy to analyze: it's essentially a current source (the magnitude of the current varies with changes of V_{in}) and that current is fed to a resistor. That resistor converts the current back to a voltage – but scaled up to the extent that R_{C} exceeds R_{E}.

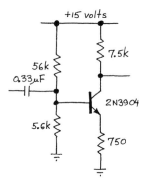

Figure 5N.2 Common emitter amplifier, gain of −10.

The amplifier's gain is $-R_{\text{C}}/R_{\text{E}}$, and we hoped, in Lab 4L, that you found this simplicity charming: the expression for gain says nothing about the transistor, relying purely on the attached resistors. So far, good.

But what happens if we get greedy for gain? Could we set the ratio of R_{C} to R_{E} as high as 100? 1000? What would happen if we made R_{E} zero? Maybe you would shy away from that last as mathematically offensive; but the defect in this design is not mathematical. You would run into the limitations of the transistor long before reaching the impossibility of an infinite-gain amplifier.

Let's go right to the hardest case: let's take R_{E} to zero. How do we handle this? We seem to be applying a wiggle to V_{BE}, but we have said that V_{BE} is fixed at 0.6 V.

AoE §2.3.4

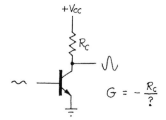

Figure 5N.3 Grounded-emitter amplifier. Infinite gain?

Well, we didn't mean *exactly* 0.6 V. This is one of those circuits that obliges us to recognize that V_{BE} varies slightly as I_{C} varies. In fact, the I_{C} versus V_{BE} curve looks just like the diode curve, already familiar to you. It differs only in slope: the transistor's curve is steeper: see Fig. 5N.4.

If we omit R_{E} from the amplifier, then we are applying the input signal directly to V_{BE}. The current, therefore, changes a lot – given the exponential shape of the curve: see Fig. 5N.5. The gain for the amplifier is large, therefore – but not, of course, infinite.

5N.3 Another view of transistor behavior: Ebers–Moll

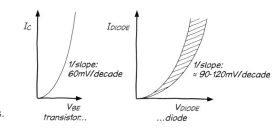

Figure 5N.4 Transistor *I* versus *V* looks like diode's.

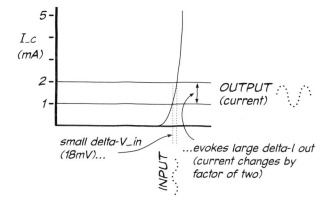

Figure 5N.5 An input applied to V_{BE} evokes large changes in I_C.

5N.3.2 Ebers–Moll equation describes transistor's V_{BE} versus I_C...

Ebers and Moll[2] describe the transistor as a voltage-to-current device – whereas our first view described it as a current-to-current converter (with β defining the multiplication factor). The two views are quite consistent, since the base current flows in a *diode*, the V_{BE} junction (see §5N.7).

We will write out the Ebers–Moll equation – but then we will whittle it down to make it more manageable. Then we rarely will refer to the full equation again. We prefer, instead, to model its consequences in forms that are more convenient. Here is the Ebers–Moll equation:

$$I_C = I_S \left(e^{\frac{V_{BE}}{kT/q}} - 1 \right) \tag{5N.1}$$

It turns out that kT/q is 25 mV at room temperature, and the "−1" is unimportant once the transistor is operating.[3]

So we can simplify the equation somewhat:

$$I_C \approx I_S \, e^{V_{BE}/25\,\text{mV}} \tag{5N.2}$$

Since I_S is a value that normally we don't know, Ebers–Moll usually boils down to the insight that the collector current is an exponential function of V_{BE}.

Only occasionally will we use the Ebers–Moll equation directly, most often to calculate relative values of I_C as V_{BE} changes. What happens, for example, if you increase V_{BE} by 18 mV? Let's call the old collector current I_{C_1}, and the new one I_{C_2}:

$$\frac{I_{C_2}}{I_{C_1}} = \frac{I_S \, e^{V_{BE2}/25\,\text{mV}}}{I_S \, e^{V_{BE1}/25\,\text{mV}}}$$
$$= e^{(V_{BE2}-V_{BE1})/25} = e^{\Delta V_{BE}/25}$$

---

[2] Their names make them sound like bearded German wise men living in Heidelberg. In fact, they're a pair of friendly-looking fellows from Ohio State.

[3] The "−1," delivering "-I_S", is important when the exponential term is negative: when the transistor is off. Then $I_C \approx -I_S$, which explains why I_S is called the "reverse leakage current."

where $V_{BE2} - V_{BE1} = \Delta V_{BE}$ is measured in mV. But that is just

$$e^{18\,mV/25\,mV} \approx 2$$

This is a number perhaps worth remembering: 18 mV ΔV_{BE} for a doubling of I_C; also sometimes handy: 60 mV ΔV_{BE} per *decade* (that is, 10× change in I_C).

Now, we will move still farther from the scary original equation, heading for an application of Ebers–Moll that is pleasantly straightforward.

5N.3.3 Little r_e

On our way to taming Ebers–Moll, we will take a strange preliminary step: we will turn the I_C versus V_{BE} plot on its side, as in Fig. 5N.6.

AoE §2.3.4

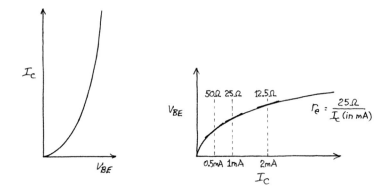

Figure 5N.6 Transistor transfer curve rotated: now slope is in *ohms*.

The reward for this odd redrawing comes immediately: we can model the transistor's *gain*[4] as a resistance. This "resistance" simply describes how much additional current you'd draw from the emitter if you tugged it down a bit (assuming the base voltage is fixed). (For a more elaborate exposition of how one might envision r_e, see §5N.9.) And, as you can see from the rotated plot in Fig. 5N.6, that effective resistance is wonderfully easy to calculate: simply divide 25 Ω by the collector current (in milliamps).

Grounded-emitter amplifier

If this notion still seems a little abstract, look at Fig. 5N.7 to see how wonderfully simple the problem of the grounded-emitter amplifier becomes when we draw in r_e as a little resistor.

Let's try it with some values inserted, as in Fig. 5N.8. The gain, R_C/r_e, can be evaluated if we

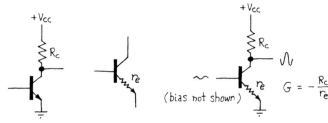

Figure 5N.7 Little r_e tames grounded emitter amplifier.

---

[4] This "gain" is peculiar – not the *change-in-voltage-out per change-in-voltage-in* that you used in Lab 4L. Instead, this is "transconductance" (g_m): *change-in-*current*-out per change-in-voltage-in*.

know I_C. Do we? Well, yes, if the circuit has been designed or set up properly: it should be set up so that $V_{\text{out-quiescent}}$ is at the midpoint of its available range, in this case 5 V. To get this result, I_C must be our favorite: 1 mA.

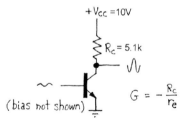

Figure 5N.8 Try r_e with some values inserted.

Given this I_C, we can evaluate r_e: it is just 25 Ω (*not* 25 kΩ, please note: we divide by collector current *in milliamps* (10^{-3} A), not by the collector current). So, the gain is $-5k/25 = -200$ (the minus sign indicates phase inversion, not attenuation, incidentally: when V_{in} moves up, V_{out} moves down). We will soon learn that this is the answer only *at the quiescent point*. Still, this is a good start.

5N.3.4 r_e solves impedance problems, too

AoE §2.3.4

Little r_e puts a floor under R_{in}: This model solves another pair of problems that we had not mentioned: what is R_{in} for the grounded-emitter amplifier? It would be zero if we did not include r_e. It becomes $\beta \times r_e$, which is ≥2.5k when we include r_e, and this is the correct answer.

AoE §2.3.3

r_e puts a floor under R_{out}: When our first view of the "lens effect" of a follower indicates an extremely low R_{out}, r_e brings things back down to earth.

For the circuit in Fig. 5N.9, our simple first view would predict R_{out} for this circuit might be under 0.5 Ω. But r_e shows us this is a little too good to be true. Increasing I_C can reduce R_{out}, but not to the level predicted by the simple view of transistors, unless one pushes on to very large I_C. Again r_e tames the extreme case.

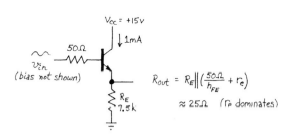

Figure 5N.9 r_e puts a realistic floor under R_{out}.

That concludes the good news about the grounded-emitter amplifier – a circuit that we soon will learn is not usable in the form we have presented. Now for some complications that oblige us to amend the circuit.

5N.4 Complication: distortion in a high-gain amplifier

AoE §2.3.4A

A gain of 200 is high. But evidently the gain is *not constant*, since I_C must vary as V_{out} moves (indeed, it is variation in I_C that *causes* V_{out} to move).

Figures 5N.10–5N.12 illustrate the funny "barn-roof" distortion[5] that you see if you feed this circuit a small triangle. First, in Fig. 5N.10, is the variation in gain that we would predict, given the variation in I_C as V_{out} swings.

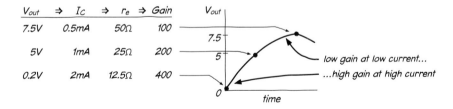

Figure 5N.10 Gain of grounded-emitter amp varies during output swing: gain evaluated at 3 points in output swing.

The plots in Fig. 5N.11 show how gain varies (continuously) during the output swing. This is bad distortion: -50% to $+100\%$.

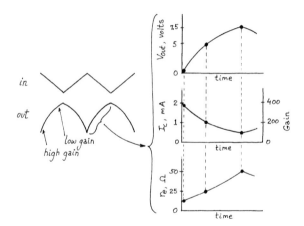

Figure 5N.11 During swing of V_{out}, I_C and thus r_e and gain *vary*.

And Fig. 5N.12 is a scope image confirming the predicted signal distortion. What is to be done?

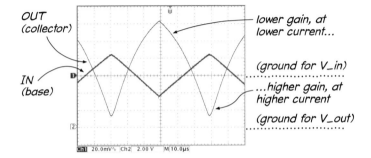

Figure 5N.12 Scope image of "barn roof" distortion. Note scope gain is 100× higher for input waveform.

---

[5] We used this term back in §4L.5.1 without explanation. The name is not standard, incidentally; we just happen to live near Vermont's dairy barns with their gambrel roofs.

5N.4.1 Distortion remedy: emitter resistor – at the price of gain

One cannot eliminate this variation in r_e, but one can make its effects negligible. Just add a constant resistance much larger than the varying r_e. That will hold the denominator of the gain equation nearly constant.

AoE §2.3.4B

With an emitter resistor added, the gain variation shrinks sharply: see Fig. 5N.13. Still r_e varies as widely as before; but its variation is buried by the big constant in the denominator. Circuit gain now varies only from a low of −9.1 to a high of −9.75: a −4%, +3% variation about the midpoint gain of −9.5.

Figure 5N.13 Emitter resistor cuts gain, but also cuts gain *variation*.

Punchline: an emitter resistor greatly reduces error (variation in gain, and consequent distortion). This we get at the price of giving up some gain. (This is one of many instances of Electronic Justice: here, *those greedy for gain will be punished: their output waveforms will be rendered grotesque*.)

We will see shortly that the emitter resistor helps solve other problems as well: the problem of temperature instability (see §5N.5), and even distortion caused by *Early effect*. How can a humble resistor do so much? It can because in the latter two cases the resistor is applying *negative feedback*, a design remedy of almost magical power. In the next segment of the course, in which we apply operational amplifiers, we will see negative feedback blossom from marginal remedy to central technique. Negative feedback is lovely to watch. Many such treats lie ahead.

5N.5 Complication: temperature instability

Semiconductor junctions respond so vigorously to temperature changes that they often are used as temperature sensors. If you hold V_{BE} fixed, for example, you can watch I_C vary exponentially with temperature.

But in any circuit not designed to measure temperature, the response of a transistor to temperature is a nuisance. Most of the time, the simple trick of adding an emitter resistor will let you forget about temperature effects. We will see below how this remedy works.

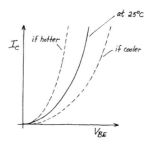

Figure 5N.14 Transconductance of bipolar transistor varies rapidly with temperature.

Preliminary warning. Do not look for a description of this temperature dependence in the Ebers–

Moll equation. That equation (mis-read) will point you in exactly the wrong direction: increasing T should shrink the exponent:

$$I_C = I_S \left(e^{V_{BE}/(kT/q)} - 1\right)$$

Don't be fooled: Ebers–Moll equation seems to say I_C falls with temperature. Not so.

But actual results are quite contrary to what this point seems to suggest: in fact, rising T increases I_C, and does that *fast*. The solution to the riddle is that I_S grows *very fast* with temperature, overwhelming the effect of the shrinking exponent.

Here are two formulations for the way a transistor responds to temperature:

- I_C grows at about 9%/°C, if you hold V_{BE} constant; and
- V_{BE} falls at 2 mV/°C, if you hold I_C constant.

The first of these formulations is the easier to grasp intuitively: heat the device and it gets more vigorous, passes more current. The second formulation often makes your calculation easier. But if you use the second, just make sure that you don't get the feeling that the way to calm your circuits is to build small fires under them!

5N.5.1 Temperature stabilizing: feedback, using emitter resistor

The remedy described here is simple, and widely used. It is also quite subtle. The left-hand circuit in Fig. 5N.15 is so unstable that it is useless. An 8°C rise in temperature saturates the transistor.

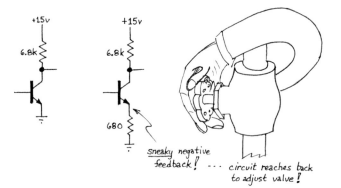

Figure 5N.15 An unstable circuit stabilized by emitter resistor. Note that we assume V_B is held constant.

Why does the right-hand circuit work better? How does the emitter resistor help, as I_C grows? Here is feedback at work. The circuit senses trouble as it begins:

- I_C begins to grow in response to increased temperature; and
- V_E rises, as a result of increased I_C (this is just Ohm's law at work).

But this rise of V_E diminishes V_{BE}, since V_B is fixed. Squeezing V_{BE} tends to close the transistor "valve." Thus the circuit slows itself down (as the somewhat-grotesque hand in Fig. 5N.15 is meant to suggest).

We don't claim that I_C will change not at all. Some ΔI_C with temperature is necessary in order to generate the error signal. But the emitter resistor prevents wide movement of the quiescent point.

The plot in Fig. 5N.16 shows how the emitter resistor's behavior – plotted as a straight "load line" intersecting with the transistor's shifting curve – limits the variation in I_C with temperature change.[6]

The larger the value of R_E, the stronger the feedback, and the less the variation in I_C with temperature.

---

[6] The "load line" plots V_{BE} as a function of current though R_E, assuming a fixed base voltage. Since the current through R_E must equal I_C under our simple view of transistor behavior, the quiescent I_C will be where the two curves meet.

5N.5 Complication: temperature instability   207

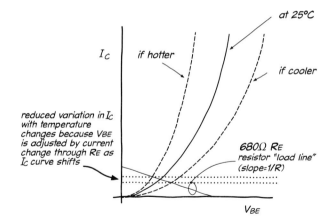

Figure 5N.16 Emitter resistor's effect plotted as "load line" intersecting transistor curves. The load line shows that as I_C increases, V_{BE} decreases because V_E (equal to $I_E \times R_E$) increases while V_B is held constant and $I_E \approx I_C$.

5N.5.2   Temperature stability *and* high gain

AoE §2.3.5A

You don't have to give up gain in order to get temperature stability – contrary to an impression we may have given back in §5N.5.1. You can get both stability *and* high gain, by arranging things so that R_E does its good stabilizing work while not reducing signal gain. The trick is to include R_E, as in a low-gain amplifier, but then to make it disappear at signal frequencies. R_E "disappears" at those frequencies because of the capacitor that parallels it, *bypassing* it: see Fig. 5N.17.

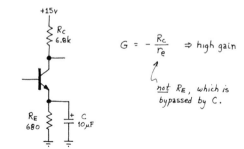

Figure 5N.17 Bypassed-emitter resistor: high gain plus temperature stability.

The circuit's gain, observed

The circuit in Fig. 5N.18 shows the same high gain as the unsatisfactory grounded-emitter amplifier. If we make the output swing very small, distortion is hardly noticeable:

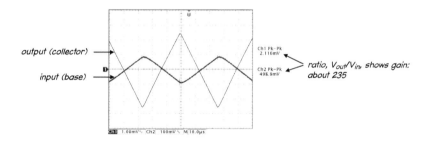

Figure 5N.18 Gain at quiescent point (small output swing) is pretty constant, and close to calculated value. (Note 100× difference in scope gains on the two channels.)

The gain at the quiescent point appears to be about -235 (the minus sign indicating only the signal's *inversion*). By calculation, we would predict gain of -272:

$$G \equiv \Delta V_{\text{out}}/\Delta V_{\text{in}} = -R_C/r_e = -7.5k/25 = -300$$

The difference between calculated gain and the gain observed in Fig. 5N.18 may be explained by the collector current being a little lower, in this circuit, than the target of 1 mA.

The circuit's temperature stability, observed

In Fig. 5N.19 the stability of this circuit is contrasted with the nasty behavior of the grounded-emitter amplifier of Fig. 5N.8. We fed a small triangle to both circuits, stable and unstable, and then warmed both with a heat gun.

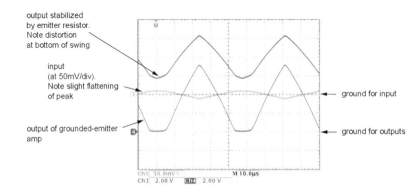

Figure 5N.19
Stable and unstable circuits contrasted: when heated, bad one clips.

The stable circuit's V_{out} stayed centered where it started; the unstable circuit's V_{out} drifted down close to ground, producing nasty clipping.

Curious detail: R_{in} degrades in saturation

A detail of Fig. 5N.19 is worth mentioning: the effect of saturation upon input impedance. The unstable output drifts down, and flattens ("clips") at ground. The transistor is *saturated* where the waveform clips: totally *on*. When that happens, the usual rules of transistor behavior do not apply (these are the rules that apply in what's called the "linear region" of transistor operation, where the device is neither totally *on* (saturated) nor totally *off* (cut off)).

When the transistor saturates its input resistance – usually $\beta \times R_E$ – becomes radically lower: no longer does any β multiplication occur. You can make out evidence of this effect in two details of the image in Fig. 5N.19:

- The input waveform itself is distorted – its peak flattened. That indicates that even the function generator, with its R_{out} of 50 Ω is getting overloaded by the transistor's R_{in}.
- The output of the stable circuit – a circuit driven by the function generator that drives the unstable circuit – also shows distortion, distortion produced by overloading of the signal source shared by the two circuits.

The moral seems to be *don't hang around with bad company*: even the virtuous temperature-stable circuit ends up corrupted by sharing an input with the naughty grounded-emitter amp.

…but note that distortion persists: The emitter resistor neatly solves the temperature-stability problem. The distortion, however, remains; it appears also, for both circuits, back in Fig. 5N.12. There is no way around this, if you want so much gain in one stage.

5N.5.3 Another temperature-stabilizer: Explicit DC feedback

AoE §2.3.5C

Figure 5N.20 shows feedback in a form more obvious than through R_E, but used to similar effect.

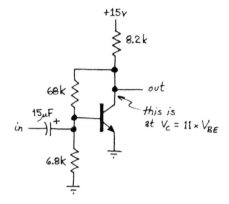

Figure 5N.20 DC feedback protects against temperature effects.

When you have played with some operational amplifier circuits in the coming labs, you will recognize this feedback as very similar to what you apply in those op-amp circuits: 1/11 of V_{out} is fed back to the input.[7] But here a subtlety is at work that is not usual in the op-amp circuits: the feedback affects DC levels, but *not* circuit gain. It does not affect gain because the function generator, whose R_{out} is low, is able to overwhelm the relatively feeble feedback signal at signal frequencies (those that get through the blocking capacitor). Yet at DC the feedback is strong enough to handle the problem of drift.

5N.5.4 One more way to stabilize the circuit: add a compensating transistor

AoE §2.3.5B

The circuit in Fig. 5N.21 *compensates* for any change in one direction by planting a circuit element that tends to change at the same rate in the opposite direction.

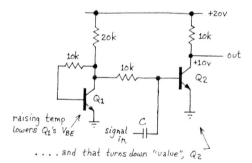

Figure 5N.21 Temperature stability through use of a second "compensating" transistor.

The circuit in Fig. 5N.21 works if both transistors live on the same piece of silicon, so that their curves drift equally, as temperature changes. Such shared drift is illustrated in Fig. 5N.22.

If the circuit is heated, Q_1's V_{BE} shrinks. But this shrinkage squeezes down Q_2 as both transistors get hotter. So Q_2's current does not grow with temperature. (The 10k resistor on the base of Q_2 makes the biasing circuit not too stiff: the signal source (presumed to be of impedance $\ll$10k) can have its way, as usual.) This circuit is a close kin to a "current mirror," a circuit introduced below and discussed in §5S.2. AoE also treats mirrors in §2.3.7.

---

[7] It is fed not quite to the *circuit* input but almost. The difference is only that it is fed to the base rather than to the far side of the blocking capacitor. As noted just below, this feedback affects DC levels, but not signals.

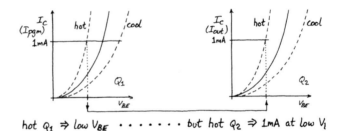

Figure 5N.22 Two matched transistors heated together can compensate to cancel drift.

5N.6 Current mirror

AoE §2.3.7

Here is another way to make a current source. Unlike the current source that you saw last time, this one requires the Ebers–Moll view of transistor operation: the view that I_C is determined by V_{BE} rather than by I_B.[8]

We offer you the mirror not because the current source of last time does not perform well, but because the mirror is so widely used in integrated circuits that it is worth our while to understand it.

5N.6.1 Simplest mirror

It is called a "mirror" because the output current matches an input current. The matching is achieved very simply: by applying equal values of V_{BE} to the two transistors – input and output. A mirror appears in LAoE Fig. 5L.10, and there's a fuller discussion of mirrors beginning at §5S.2. We've also added a lab exercise to familiarize you with mirrors, if you have the time: see §5L.3.

As often, the simplicity of the circuit is a little deceptive. The circuit uses an application of negative feedback that to me is not obvious at a first glance (see §5S.2.2).

In the lab exercise, we will ask you to build this simple mirror, using discrete transistors, so that you can see how it is imperfect. Then we'll let you improve it.

When you begin to use the LF411 operational amplifier in the first op-amp lab, you will be using a circuit that includes three mirrors (see §9S.4 – a formidably-clever circuit; don't expect yourself to understand it all, at this point).

5N.7 Reconciling the two views: Ebers–Moll meets $I_C = \beta \times I_B$

At first, we promoted the current-to-current view: $I_C = \beta \times I_B$. A little later we offered Ebers–Moll's account, $I_C = I_S \, e^{V_{BE}/V_T}$.

In case you're troubled by the thought that our two views may not be consistent, here's an argument, and then a picture, meant to reassure you:

- An argument: since the BE junction is a diode, feeding it a current (I_B) generates a corresponding V_{BE}. Ebers–Moll teaches that this V_{BE} is the cause of the resulting I_C. If we use this account to justify the first, simplest view of transistor operation – which treated base *current* as input – we are acknowledging and using Ebers–Moll.
- If Fig. 5N.23 does not make you feel better, forget it; you don't *need* to reconcile the two views if you don't want to.

---

[8] As you have gathered, these two "views" are consistent; they offer differing ways to describe one behavior.

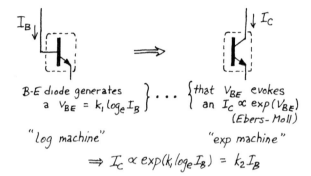

Figure 5N.23 Beta... and Ebers–Moll descriptions of transistor gain *reconciled*.

5N.8 "Difference" or "differential" amplifier

AoE §2.3.8

The differential amp is the last standard transistor circuit we will ask you to consider. It is especially important to us because it lets us understand the *operational amplifiers* that you soon will meet. These wonderful devices are in fact just very good differential amps, cleverly applied.[9]

5N.8.1 Why a differential amp?

A differential amplifier has an internal symmetry that allows it to cancel errors shared by its two sides, whatever the origin of those errors. Sometimes one takes advantage of that symmetry to cancel the effects of errors that arise within the amplifier itself: temperature effects, for example, which become harmless if they affect both sides of the amplifier equally. In other settings the shared error to be canceled is noise picked up by both of the amplifier's two inputs. Used this way, the amplifier picks out a signal that is mixed with noise of this particular sort: so-called "common-mode" noise.

You built a circuit back in Lab 2L that did something similar: passed a signal and attenuated noise. But that *RC* filter method works only if the noise and signal differ quite widely in *frequency*. The differential amp requires no such difference in frequency. It does require that the noise must be common to the two inputs, and that the *signal*, in contrast, must appear as a difference between the waveforms on the two lines. Such noise turns out to be rather common, and the differential amp faced with such noise can "reject" it (refuse to amplify it), while amplifying the signal: it can throw out the bad, keep the good.

An application: brain wave detector

Here is an example of a problem that might call for use of a differential amplifier. One can detect brain activity with skin contacts; the activity appears as small (microvolt range) voltage signals. The output impedance of these sources is relatively high, in the 1k to 10k range.

The feebleness of the signals makes their detection difficult. It is not hard to make a high-gain amplifier that can make the signals substantial. But the catch is that not only the signals but also *noise* will be amplified, if we are not careful. We can try to shield the circuit; that helps somewhat. But if the principal source of noise is something that affects both lines equally, we can use a differential amplifier instead – or as well; such a circuit ignores such "common mode" noise. The same difficulty applies to cardiac measurements – by the so-called EKG.

---

[9] At the risk of complicating the thought too much, we should acknowledge that the *clever* application is of fundamental importance. The cleverness lies in the use of *negative feedback*. It is this, and not just the good diff-amp, that provides the really impressive behavior of the op-amp circuits that you soon will meet.

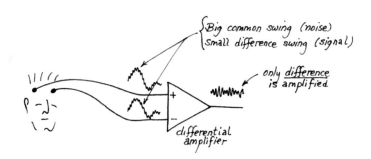

Figure 5N.24 An application for a differential amp: brain wave detection.

Line noise at 60 Hz will be coupled into both lines, and is likely to be much larger than the microvolt signal levels. A good differential amp can attenuate this noise by a factor of perhaps 1000 while amplifying the signal by, say, 100. An amp that could do that would show a "common-mode rejection ratio" – a preference for difference signals over common – of 10^5: 100 dB.

...and a similar demo: music pulled out of nasty 60 Hz noise

In a class demo, we made up a simple imitation of this sort of case. The differential signal source was an audio player (a CD player). Two unshielded wires ran from the CD player to a differential amplifier.[10]

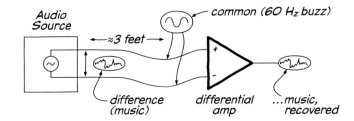

Figure 5N.25 Demonstration setup for pulling music signal from large common-mode noise.

A large 60 Hz voltage swing appears on each of the two wires running three feet or so in Fig. 5N.25. That 60 Hz noise is shown in Fig. 5N.26. A much smaller difference signal (music, in this case) looks like a barely noticeable tremor on the 60 Hz swing. One can just make out a small difference, here and there in the left-hand image of Fig. 5N.26: it looks like a slight darkening, as one of the nearly-identical-twin waveforms pokes out from behind its sibling.

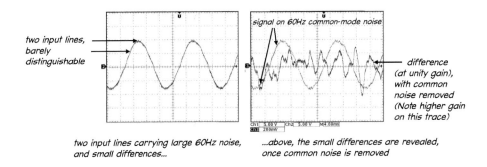

Figure 5N.26 Difference amplifier can pick out a small difference in presence of large common noise.

---

[10] This amp we built with an operational amplifier, not with discrete transistors, because doing so is much the easiest method. You will soon share our preference for op-amps over discrete designs, we are confident.

5N.8 "Difference" or "differential" amplifier

The difference between the two, however – shown in the right-hand scope image – is easy to make out once the large common-mode noise has been removed. Differential gain here is *unity*: so the "amplifier" here has done no more than remove the shared noise. (The "signal" looks as if it had been amplified in Fig. 5N.26; that's an illusion caused by the greater *scope* sensitivity for the output signal.) The effect, for a listener, is dramatic. The *original* signals sound like a loud 60 Hz buzz, no music audible; the *difference* signal sounds like music, unpolluted by any audible 60 Hz noise.

Differential signaling helps even digital circuits

When we reach the digital gates of Chapter 14N we will encounter *differential* transmitters and receivers, applied for the purpose of achieving good noise immunity in the presence of daunting common-mode noise.

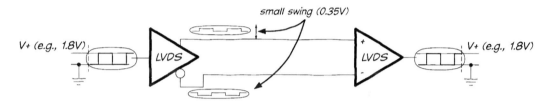

Figure 5N.27 Differential signaling applied in digital electronics: provides good noise immunity despite low supply voltage.

This method, used in the LVDS (Low Voltage Differential Signaling) circuits, can give good behavior even with the very low power supply voltages that are coming to dominate digital circuitry: supplies of 2.5 V, 1.8 V and lower. Figure 5N.27 does not show common-mode noise, but when we tried this circuit the circuit easily rejected common noise five or six times larger than the *signal*: see Fig. 14N.32. For a scope image showing rejection of drastic high-frequency noise by LVDS, see AoE Fig. 12.126.

5N.8.2   A differential amp circuit

> Compare AoE Fig. 2.63

Figure 5N.28 shows the lab's differential amp. This is the circuit you will build in Lab 5L. In order to achieve good matching of the two transistors, you will use an integrated array of transistors (a CA3096 array).

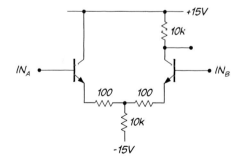

Figure 5N.28 Differential amp.

The circuit may look a bit daunting, but it is not hard to analyze. After we establish DC *quiescent* conditions, we will use a trick that AoE suggests: consider only pure cases – pure *common* signal, then pure *difference* signal.

Before we see what the circuit will do to input signals, let's first find the DC *quiescent* levels.

Quiescent points

V_{out}: Before you can predict V_{out} you need to determine currents.

Currents: If the bases are tied to ground, as we may assume for simplicity, the emitter voltages are close to ground, and it follows that point "A" is not far from ground (close to −1 V.). From this observation you can estimate I_{tail} (in the lower 10k resistor): it is about $14\,\text{V}/10\text{k} \approx 1.5\,\text{mA}$.

Since the circuit's inputs are at the same voltage, symmetry[11] requires that the 1.5 mA *tail* current must be shared equally by the two transistors. So I_C for each of them is about 0.75 mA.

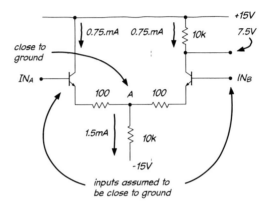

Figure 5N.29 Lab diff amp, showing quiescent currents and voltages.

From here $V_{\text{out\_quiescent}}$ is easy: centered as usual – but note that it is centered *not* between the supplies (that center would be 0 V). Instead, it is centered *in the range through which it can swing*. That is always the deeper goal. Since the floor on the output swing is defined not by the negative supply but by the emitter voltage, the floor is close to ground, and the proper $V_{\text{out\_quiescent}}$ is around 7.5 V.

Differential gain

Assume a pure *difference* signal: a wiggle up on one input, a wiggle down of the same size on the other input. It follows, you will be able to convince yourself after a few minutes' reflection, that the voltage at "A" of Fig. 5N.29 does not move.

That observation lets you treat the right-hand side of the amp as a familiar circuit: a common-emitter amp: see Fig. 5N.30.

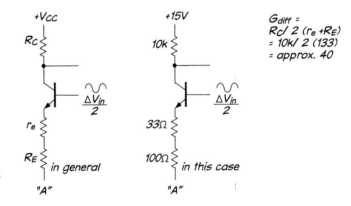

Figure 5N.30 Differential gain: just a common emitter amp again.

The gain might seem to be

$$G = -\frac{R_C}{r_e + R_E}$$

---

[11] "Symmetry?," you may want to protest. "There's a collector resistor on one side, and not on the other." True. But try to explain to yourself why that does not matter. The resistor does affect the *voltage* at each collector. But what influence does that voltage have on collector current? Not much. Ideally, none; Early effect says 'only a slight effect.'

That's almost correct. But you need to tack a factor of 2 into the denominator, just to reflect the way we stated the problem: the ΔV applied at the input to this "common emitter" amp is only *half* the difference signal we applied at the outset to the two inputs. You also might as well throw out the minus sign, since we have not defined what we might mean by positive or negative difference between the inputs.

So the expression for differential gain includes that factor of 2 in the denominator (*minus* sign discarded):

$$G_{\text{diff}} = \frac{R_C}{2\,(r_e + R_E)}$$

Common mode gain

Assume a pure *common* signal: tie the two inputs together and wiggle them. Now, point "A" is not fixed. Therefore, *this* common-emitter amp has much lower gain, because R_{tail} appears in the denominator of the gain equation. The denominator – which was just $2(r_e + R_E)$ for the differential case – now must include the much larger value, R_{tail}.

Again, that is *almost* the whole story. But another odd factor of 2 appears, to reflect the fact that another identical common-emitter amp – the other side of the differential amp – is squirting a current of the same size into R_{tail}. The result is that the voltage at A increases twice as far as one might otherwise expect from the increased current flowing in the right-side BJT alone; one can say this another way by calling the effective R_{tail} "$2R_{\text{tail}}$." Then we do not need to account for an additional current from the left-hand BJT due to the common mode input.

Let's redraw the circuit to show what's going on when we drive both inputs with the same signal ("common mode"). The redrawn circuit is Fig. 5N.31.

To calculate gain, we can consider just the right-hand common-emitter amplifier (peeled away, in

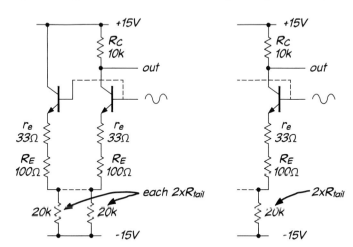

Figure 5N.31 Differential amp redrawn to show common-mode response.

Fig. 5N.31). And the common mode gain is

$$G_{\text{CM}} = -\frac{R_C}{r_e + R_E + 2\,R_{\text{tail}}}$$

The bigger we can make R_{tail}, evidently, the better. A current source in the tail, therefore, provides best common-mode rejection. Any respectable differential amplifier normally includes a current source in the tail – as your lab circuit does, once you have moved on after first trying a tail *resistor*. You used the resistor to measure a predictable common-mode gain, and for contrast with the improvement effected by use of the current source.

5N.8.3 Differential amp evolves into "op-amp"

In Lab 5L we invite you to carry on, once you have tried out the differential amp, adding two more stages so as to convert your differential amplifier into a not-bad operational amplifier.

Fundamentally, an operational amplifier is only a high-gain differential (or "difference") amplifier. But a good op-amp also includes a buffer to provide reasonably-low output impedance. Figure 5N.32 shows a block diagram of the lab 'op-amp,' with its gain of approximately 1000: pretty good.

> Compare AoE Fig. 2.91 (a circuit that includes internal feedback)

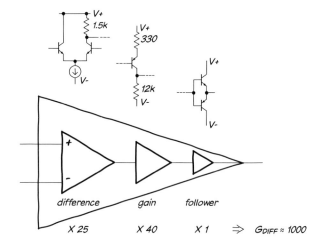

Figure 5N.32 3-stage op-amp of Lab 5L.

The IC op-amp you'll meet in Lab 6L provides gain a couple of hundred times better, along with better common-mode rejection. But this circuit offers a start – and one that we hope will demystify the triangles that you'll begin to invoke next time.

5N.9 Postscript: deriving r_e

Deriving the expression for r_e: The "intrinsic emitter resistance," r_e, is, as you know, the inverse of the slope of the transistor's gain curve: $\Delta I_{out}/\Delta V_{in}$ (see §5N.3.3). A little more formally, then, r_e is the derivative of the V as a function of I; that is, dV_{BE}/dI_C. Let's evaluate this, finding first the slope of the I–V curve.

If we write Ebers–Moll in the simplified form we use in Equation 5N.2

$$I_C \approx I_S\, e^{V_{BE}/25\text{mV}}$$

then

$$\frac{d(I_C)}{d(V_{BE})} = (1/25\text{mV})(I_S\, e^{V_{BE}/25\text{mV}}) \approx (1/25\text{mV})(I_C) = I_C/25\text{mV}$$

And r_e, the reciprocal of the slope of the gain curve, is

$$\frac{25\,\text{mV}}{I_C\ (\text{in amps})}$$

which we prefer to write as

$$\frac{25\,\Omega}{I_C\ (\text{in mA})}$$

Alternative 'derivation:' Tess o' Bipolarville: Here is an alternative argument to the same result: imagine a lovely milkmaid seated (in the summer twilight) on a stool, tugging dreamily at the emitter of a transistor whose base is fixed. She has pulled gently, until about 1 mA flows. What *delta*-current falls into her milkpail, for an additional tug?

Figure 5N.33 Dreamy milkmaid discovers the value of r_e, experimentally.

If the base is anchored, tugs on the emitter change V_{BE} a little; in response, I_C changes quite a lot. The squirts of ΔI_E (which we treat as equivalent to ΔI_C) reveal a relation between ΔV_{BE} and ΔI_C. The quotient

$$\Delta V_{BE}/\Delta I_C$$

is r_e, which behaves just like a resistor whose far end is fixed.

"Yes," muses the charming milkmaid, her milkpail now filled (with *charge*, in fact; but she hasn't noticed). "Just as I thought: r_e, though only a model, does behave for all the world like a little resistor. So that's why we draw it that way." And with that she rises, little suspecting what her discovery portends, and carries her milkpail off into the gathering dusk.

5N.10 AoE reading

AoE's Chapter 2 treats both bipolar transistors and elementary applications of operational amplifiers, so its material suits the topics of today's lab – though, as usual, it contains much that we do not require here. Here are some selected sections that are appropriate.

- Ebers–Moll: §2.3.
- §2.3 to end of chapter, but…
  - Omit §2.3.7 (current mirrors) (we *like* current mirrors, but want to lighten your load a bit).
  - Omit most details in introduction to negative feedback: §2.5; we will devote lots of attention to negative feedback soon – in the context of op-amp circuits. But highly relevant to us now is circuit B (non-inverting amp) of Fig. 2.85 in §2.5.
- The most important sections of reading are:
  - The Ebers–Moll model: §2.3;
  - differential amplifiers: §2.3.8;
  - push-pull output stages: §2.4.1; and
  - bootstrapping (a general notion of applications wider than just the emitter-follower that is described here, and shown in Fig. 2.80): §2.4.3.
- Problems: Exercise 2.28.

In this book: you might glance at §9S.4 for the scary full circuit for the '411 op-amp. This note shows a circuit like the one you build today – but with a great many refinements added. Worry about the refinements later.

5L Lab: Transistors II

Overview of the lab

To do *all* of today's lab is a challenge: the op-amp circuit is the most complex that you've built so far, and if some stage holds you up, you're likely to run out of time. But that shouldn't worry you. Only the *differential amp* (§5L.1) – not its conversion into an *op-amp* – is fundamental. We hope you'll get at least partway into the op-amp construction, because that experience will make you feel more at home with the op-amps that enter next time; but you need not finish the exercise.

In this Second Edition, we have added a set of lab exercises to familiarize you with the current mirror, a circuit we suggested you could skip in §5S.2. Nevertheless, it an interesting application of BJTs used in many analog integrated circuits to create stable, temperature-independent current sources.

5L.1 Difference or differential amplifier

Predict differential and common-mode gains for the amplifier in Fig. 5L.1 (don't neglect r_e). Note that you will build this using an IC array of transistors, not our usual 2N3904s.

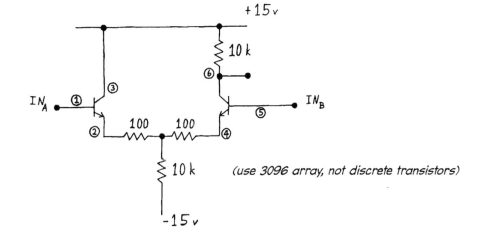

Figure 5L.1 First stage: differential amplifier, made from transistor array.

5L.1.1 Transistor array

We would like you to build the circuit on an array of bipolar transistors, the CA3096. These transistors are fairly well matched,[1] and will track one another's temperatures; such tracking helps assure temperature stability, as you know.

---

[1] CA3096 *npn* transistors show V_{BE} matching typically within 0.3 mV, max 5 mV (at I_C = 1 mA).

Figure 5L.2 shows the pinout which applies to both packages: the CA3096 in either DIP or SOIC. The "substrate," connected to pin 16, is the P-doped material on which the transistors are built. Since we don't ever want that implicit set of diodes to conduct, we should connect *substrate* to the most negative point in the circuit. Here, that is −15 V.

Figure 5L.2 CA3096 array of bipolar transistors.

If you have the CA3096 in a surface-mount package (SOIC), you (or *someone*) will need to solder it to an adapter, so that you can plug it into the breadboard.[2] Figure 5L.3 shows the DIP and surface mount versions of the '3096.

Figure 5L.3 '3096 in two versions: CA3096 DIP and SOIC mounted on DIP carrier.

Figure 5L.4 shows the way its pins are numbered, and the way it goes into the breadboard. It straddles the trench so that its 16 pins are independently accessible. (The same is true, of course, for the SOIC version in its carrier.)

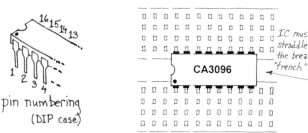

Figure 5L.4 DIP package pin numbering and insertion in breadboard.

5L.1.2 Setting up the test signals

Now you will use two function generators to generate a mixture of *common-mode* and *differential* signals.

5L.1.3 Setup I: using a function generator that can "float"

The signal setup is easy if you have a function generator that permits you to "float" its *common* lead – the one tied to the BNC cable's surrounding shield, a lead that until now we have always left tied

---

[2] We have used a company named Proto-Advantage (https://www.proto-advantage.com) to do this service for us.

5L.1 Difference or differential amplifier

to ground. Here, "float" simply means "disconnect from world ground." Not all function generators, however, permit such floating, and for these we suggest an alternative method in §5L.1.4.

Preliminaries

One generator will drive the other. In Fig. 5L.5 the scheme requires that you *"float" the driven generator*: find the switch or metal strap on the lab function generator that lets you *disconnect* the function generator's local ground from absolute or "world" ground. You will find such a switch or strap on the back of some generators.

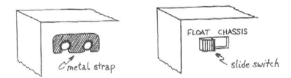

Figure 5L.5 Float the external function generator.

As you connect the two function generators to your amplifier, you will have to use care to avoid defeating the "floating" of the external function generator: recall that BNC cables and connectors can make *implicit* connections to absolute ground. You must avoid tying the external generator to ground through such inadvertent use of a cable and connector. So, you must avoid use of the BNC jacks on your breadboard; you must also avoid linking function generator to scope with a BNC for external trigger. You may find "BNC-to-mini-grabber" connectors useful to link function generator to circuit: these connectors do not oblige you to connect their *shield* lead to ground.

Composite signal to differential amplifier

Now let the breadboard's function generator (which cannot be "floated") drive the external function generator's *local ground* or "*common*" terminal. Use the output of the external function generator to feed your differential amplifier. That output can carry pure common-mode, pure differential, or a mixture of the two; see Fig. 5L.6. You will exploit this versatility later in §5L.1.5.

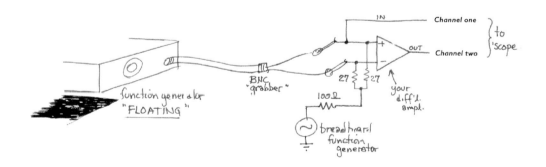

Figure 5L.6 Common-mode and differential signal summing circuit.

5L.1.4 Setup II: using a function generator that cannot "float"

If your main function generator lacks a switch that can float it, then you can use a transformer to provide the required separate control of *difference* and *shared* signals (usually called "differential" and "common-mode" signals).

Figure 5L.7 shows the arrangement, using a 6.3 V transformer – the same type that you used in Lab 3L. Note that this transformer is *not* to be plugged into the 120 V wall socket! We will, instead, use the external function generator to drive the transformer's power-input plug. Thus the external generator drives the *primary* of the transformer.

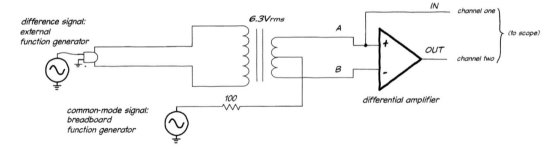

Figure 5L.7 Transformer allows two function generators to provide separate *differential* and *common-mode* signal sources.

The breadboard function generator driving the center-tap of the transformer *secondary*, provides equal signals to the two inputs of the amplifier.

Figure 5L.8 shows the signals that this arrangement can generate at the amplifier's inputs. These are the pure cases: only difference, or only common-mode.

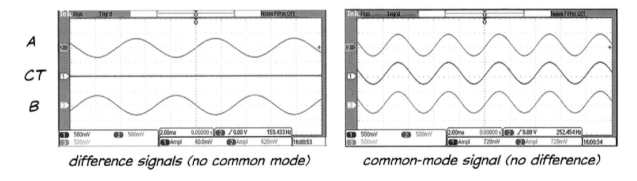

Figure 5L.8 Transformer use allows pure difference or pure common signals.

5L.1.5 A mediocre differential amp: resistor in "tail"

As a suggestion, try measuring common-mode and differential gain.

5L.1 Difference or differential amplifier

Common-mode gain (first try):

- **Measure common-mode gain:**
  - Shut off the differential signal (external function generator)[3] while driving the amplifier with a signal of a few volts amplitude. Does the common-mode gain match your prediction? If it is too high, the probable cause is the sneaking-in of a difference between the diff-amp's two inputs, so that the output you see includes some differential gain as well as common-mode. You can discover whether this is happening by simply shorting the diff-amp's two inputs together (parallel the series-pair of 27 Ω resistors with a length of wire). That piece of wire assures that the applied signal is true common-mode. If you do insert this wire, be sure to remove it after you measure G_{CM}. The two inputs must be permitted to diverge in the next step, where you measure differential gain.

Differential gain:

- **Measure differential gain:**
  - Turn on the external function generator while cutting common-mode amplitude to a minimum (there is no Off switch on the breadboard function generator).
  - Apply a small differential signal. Does the differential gain match your prediction? If the differential gain appears to be high by about a factor of two, recall that when you watch a wiggle at a single input, you are looking at about one-half the difference signal you are applying to the amplifier. If you doubt this, try watching both inputs, on the scope's two channels. You should find approximately equal and opposite (180°-shifted) waveforms on the two inputs.
  - Now turn on *both* generators and compare the amplifier's output with the *composite* input. To help yourself distinguish the two signals, you may want to use two frequencies rather far apart; but do not let this experimental convenience obscure the point that this differential amp *needs* no such difference. The method you used in Lab 2L to pick out a signal while rejecting noise did, of course, require such a difference.

This experiment should give you a sense of what "common-mode rejection ratio" means: the small amplification of the common signal, and relatively large amplification of the difference signal.

Nevertheless, this circuit still lets a large common-mode signal produce noticeable effects at the output. The improvement in the next step should make common-mode effects much smaller.

Common-mode gain (second try): Apply a current source to improve common-mode rejection. Replace the 10k tail resistor with a 1.5 mA current source. You may build this current source as you choose. The laziest way is to use a pair of current-limiting diodes made from JFET devices: see §12L.4. These are a part called 1N5294, rated at 0.75 mA ±10%. Two in parallel (see Fig. 5L.9) provide the desired 1.5 mA.

If this trick is too shabby and black-box for your taste, you know, of course, how to build current sources using bipolar transistors. Figure 5L.10 shows two possible circuits.

Replacing the tail resistor with a current source should reduce the common-mode gain a great deal. (What is common-mode gain if the output impedance of the current source is around 1M?[4]) You should see very good CMRR at low frequencies: say, 100 Hz. As frequency climbs, however, you'll see the output grow. This apparently results from capacitive coupling between input and output, as we argue in §5L.1.11.

---

[3] You may prefer to cut *amplitude* to a minimum, rather than shut off power to the generator.
[4] Common mode gain is approximately $R_C/(2 \times R_{tail}) \approx 10k/2M = 0.5 \times 10^{-2}$.

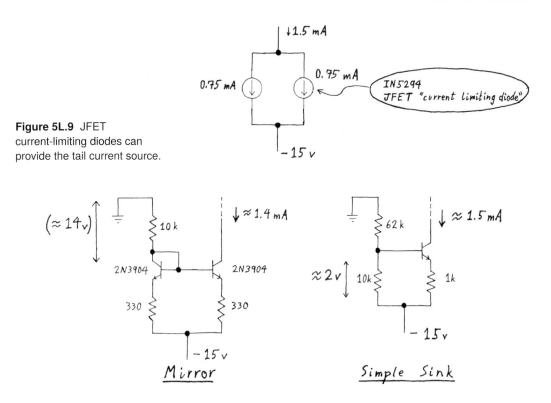

Figure 5L.9 JFET current-limiting diodes can provide the tail current source.

Figure 5L.10 Two alternative bipolar-transistor current sinks.

Common-mode and difference signals mixed: Note how this improved circuit treats a signal that combines common-mode and differential signals. Leave this circuit set up. You will be adding to it.

5L.1.6 A homemade operational amplifier

Here, we'll ask you to string together the three stages of the device that make up a standard operational amplifier. The op-amp is a just a good high-gain differential amplifier, so you can see that at this point in the lab you are partway to your destination.

An op-amp typically is a three-stage amplifier: a differential stage, a gain stage, and a push–pull output. Here, we ask you to add the two additional stages – a common emitter gain stage and a push–pull output – to the diff-amp you have built. These additions will convert this diff-amp into a modest op-amp. That device is, as you know, the building-block that you'll rely on in most of your analog designs, from Lab 6L onward. The op-amp you build today won't work as well as the IC version you meet in Lab 6L, but it should help you to gain some insight into what an op-amp is, and how it achieves its borderline-magical results.

Figure 5L.11 shows a block diagram that restates graphically the point we just made: an op-amp is a high-gain diff-amp with low output impedance.

We will ask you to modify your diff-amp somewhat to achieve higher gain, and to prepare it to drive the second stage conveniently. You'll test that; then the first two stages together; then the 3-stage amp. Finally, toward the end of this exercise, we'll ask you to apply overall feedback – a topic we have not yet discussed at any length. Perhaps you'll find the subject puzzling; we hope you won't mind this preview, even if the topic does come clear only later.

5L.1 Difference or differential amplifier

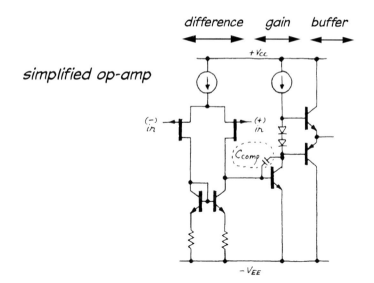

Figure 5L.11 Generic 3-stage operational amplifier.

5L.1.7 Stage 1: increase the gain of the bipolar differential amplifier

First circuit change: maximize gain: Remove the 100 Ω emitter resistors. Do you expect the circuit to lose *temperature stability*, with these gone? What happens to the *constancy of gain*?[5]

Test your views:

- to test temperature stability, watch V_{out} (with scope or DVM) as you try heating the CA3096 with your finger;
- to test constancy of gain, use a small *triangle* as input, and see whether you notice the "barn-roof" distortion that we saw in Lab 4L.

Second circuit change: move output quiescent point up: To get ready for addition of the next stage, change the collector resistor, R_C from 10k to 1.5k. This will violate our usual rule that calls for centering the output in the available range (here, 0 to −15 V). This change will also lower the circuit gain. But your circuit's modest gain is not so sad as it may seem. We hope CMRR will remain respectable. Calculate your circuit's new differential and common-mode gains – or, if you are energetic, *measure* these gains.

5L.1.8 Stage 2: gain stage: common emitter amplifier

When you reduced the R_C to 1.5k, you placed the diff-amp's output quiescent voltage close to the positive supply, because we want this output to drive a common-emitter amplifier made with a *pnp*. This second stage will provide most of the voltage gain in the circuit.

Figure 5L.13 shows the amplifier circuit that we propose. It's a conventional common-emitter amp (except that it probably looks annoyingly upside-down, to the *npn*-centric among us). The amplifier's input impedance is high enough not to load the preceding stage appreciably, as usual.

---

[5] The circuit is temperature-stable without emitter resistors, because the two transistors run at equal temperatures. Though their V_{BE}'s will change with temperature changes, their *sharing* of I_{tail} will not; so, the quiescent output voltage will not change as temperature changes. We noted this way of achieving stability back in §5N.5. See, especially, Fig. 5N.22. Constancy of gain, in contrast, will suffer: you will see distortion. Gain changes as V_{out} swings, for much the same reasons that cause "barn roof" distortion in a common-emitter amp, as noted in Chapter 5N. The diff-amp's distortion is shaped differently; §5S.1 explains this shape, in case you are curious.

LAB: Transistors II

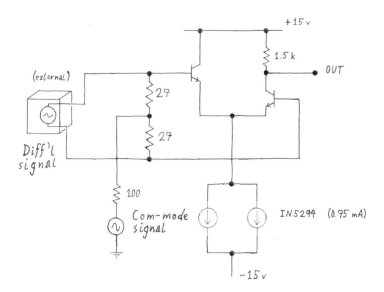

Figure 5L.12 Stage 1 diff-amp: preparing it to drive later stages.

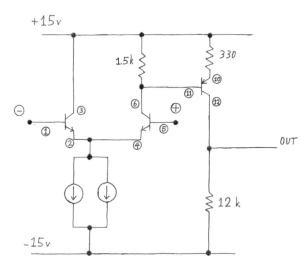

Figure 5L.13 Two-stage amplifier: differential and common-emitter.

Measure gain

Watch input and output of the CE amplifier stage, and measure this stage's gain. Then measure the overall differential gain, circuit input to circuit output (simply ground pin 1, applying a "mixed-mode" input at pin 5).[6]

You may need to tinker with the function generator's *DC offset* as you watch this high-gain amplifier, in order to make sure that neither first nor second stage clips.

5L.1.9 Stage 3: output stage: push–pull

In order to give the circuit low output impedance, we'll add a push–pull output stage. We won't bother to fix cross-over distortion, because we want to keep the circuit simple. In a minute, we'll let

---

[6] This sort of input mixes differential and common-mode signals, in fact – but harmlessly. The differential signal is the full amplitude of the applied signal; the common-mode is half that. As long as differential gain is much higher than common-mode, this sort of input is a very close equivalent to the true differential signal.

feedback try to undo this distortion. Fig. 5L.14 shows a push–pull voltage follower, made with two more transistors in the CA3096 array.

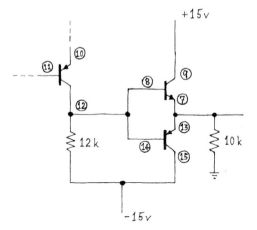

Figure 5L.14 Push–pull output stage (bipolar).

With this stage added, the baby op-amp is complete. The circuit – driven still with a mixed-mode input, and still running "open-loop" rather than with overall feedback – is in Fig. 5L.15.

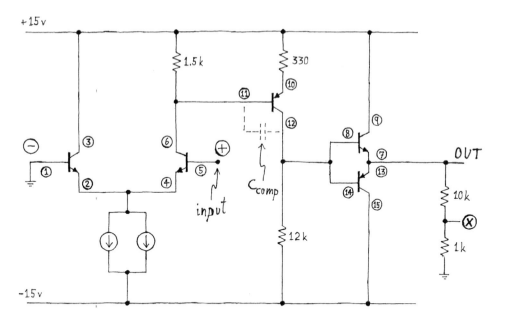

Figure 5L.15 Home-made op-amp: complete 3-stage circuit; still running open loop.

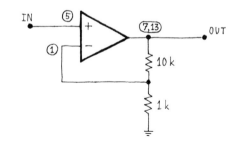

Figure 5L.16 A times 11 amp?

Feed a small sinewave differential signal to the input, at a low frequency: 1 kHz or under; watch input and output of this push-pull stage. You should notice cross-over distortion: dead sections in the output, while the push–pull's input is too close to zero to turn on either the *pnp* or the *npn* transistor. To show this *crossover distortion*, the circuit output *must cross zero*. You may need to adjust the DC-offset of the input signal, in order to center the output waveform.

5L.1.10 Trying feedback

Op-amps almost always use overall feedback. Let's try it with your circuit.

A X11 amplifier?

We'll first try the op-amp in the configuration that is normal for such devices: we feed back circuit output to circuit input (more precisely, we'll feed back a *fraction* of the circuit output). This arrangement is shown in Fig. 5L.16. We must keep the sense of feedback "negative:" output tending to diminish the input.

Connect the input at pin 1 to an attenuated version of the circuit output, marked X in Fig. 5L.15: 1/11 of the amplifier output that appears at pins 7 and 13. This connection, shown in Fig. 5L.16, will force the amplifier to try to drive this input, (1), to the voltage applied at the other input, pin 5. As a consequence, we will trick your circuit into amplifying by about 11× (this is the nominal gain, because we are feeding back one part in 11). Try it.

Gain is likely to be below the hoped-for X11, because our circuit gain is so modest.[7]

What's valuable and interesting about this *feedback* circuit is not, of course, that it delivers lower gain. As the British patent office reminded Mr. Black, *reduced* gain is not one's ultimate goal, in amplifier design.[8] Instead, one sacrifices gain for other desirable characteristics. In the present example, we hope you'll see two improvements in your amplifier's performance, now that you've applied negative feedback:

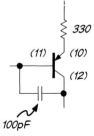

Figure 5L.17 Compensation capacitor can stabilize a feedback circuit by killing high-frequency gain.

- Perhaps the most interesting difference from the open-loop case that you tried one stage back is the disappearance of cross-over distortion from the circuit output – at least at modest frequencies.
  Feed a small sine, and continue to watch input and output of the last stage. We hope you'll find the output of the op-amp looking sinusoidal, while the input to the push-pull (pins 8, 14) looks strange – because feedback is forcing that point to do something to cancel the cross-over distortion. Pretty magical?
- A less striking virtue: the amplifier should not show the inconstant gain that causes "barn-roof" distortion in a triangular waveform. Such distortion somewhat troubled the pre-feedback design.

If your circuit begins to oscillate on its own, you'll need to reduce its high-frequency gain. The best way to do this is to place a small capacitor (try 100 pF) between collector and base of the common-emitter gain-stage transistor (pins 11, 12); see Fig 5L.17. This exploits Miller effect (see AoE §2.4.5), forming a lowpass filter whose apparent C is enlarged by the gain of this stage. Such reduction of high-frequency gain in order to achieve stability is called "compensation" and is routinely applied within op-amps.

A follower? (optional: risky business)

Oddly enough, the simpler circuit in Fig. 5L.18 – the voltage follower – is more difficult to stabilize than the 11× amplifier; that is, the follower is more likely to show those nasty parasitic oscillations.

---

[7] If you want to compare your circuit's gain against a theoretical estimate, see AoE §2.5.2. The circuit gain ought to be $A/(1+AB)$, where A is your circuit's "open-loop gain" (the differential gain you just measured), and B is the "fraction fed back" (here, 1/11).

[8] See AoE §2.5.1 and §6N.1. Of course, the point of that story is that Black had the last laugh!

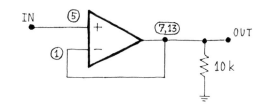

Figure 5L.18 Overall feedback imposed: a voltage follower.

If your circuit is very tidily built, you may be able to see a stable follower; some of these homemade op-amps, however, cannot be stabilized at unity gain, even with the compensation effort described above.

5L.1.11 Appendix: CMRR degradation as frequency climbs

Here's a fine point we referred to in §5L.1.5: an explanation for the observation that CMRR degrades as frequency of the common-mode input grows.

The waveform's phase and frequency-response provides a clue that what we're seeing does not result from any failure of the current source in the tail. Here are a couple of scope images, showing output for common-mode inputs. The first, in Fig. 5L.19, shows outputs for sinusoids at two input frequencies: 1 kHz and 10 kHz. The amplitude is much larger at the higher frequency.

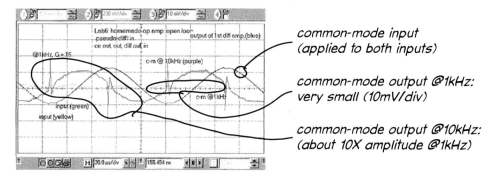

Figure 5L.19 Common-mode rejection diminishes at higher frequency: sinusoid inputs at 1 kHz versus 10 kHz. (Note that scope gain for output channels is 20× greater than for input.)

That sounds like *highpass* behavior, certainly. The scope image in Fig. 5L.20 says the same thing in another way, showing the output looking like a *differentiation* of the input.

Does your circuit behave the same way? If so, what you're seeing is capacitive feedthrough, from input (base of the input transistor) to output (collector resistor). A cascode could minimize this effect (see Miller Effect discussion in AoE §2.4.5). But let's not pause for such perfectionism now.

5L.2 AM modulator

Here's another application for a transistor differential amplifier. This exercise lets you work out the details of the design.

5L.2.1 Review: diff amp gain

If you maximize the differential amplifier's gain, remind us of the amplifier's gain? (Never mind numbers; tell us in general.)

230 Lab: Transistors II

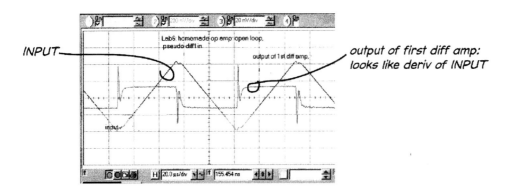

Figure 5L.20 Common-mode output acts like a differentiator: shows slope of triangle input.

```
Gain = _____
```

5L.2.2 Design: allow for varying the gain

If you replace the fixed current source that normally appears in the diff amp's "tail" – and that you applied in §5L.1.7 – with a *variable* current source, you can vary the value of "little r_e." Thus you can vary amplifier gain.

Design such a circuit and add it to the partially completed differential amplifier schematic in §5L.2.3.

We will apply a "carrier" frequency sinusoid of about 1 MHz to one input of the diff amp, while grounding the other input (this makes the carrier a "mixed-mode" input). We'll apply an audio-frequency signal to vary the value of the "tail" current about its quiescent value of 2 mA (it was a fixed value in Fig. 5L.12). So, the audio will vary ("modulate") the amplitude of the carrier frequency that is output.

Here we list some specifications for your circuit:

- Supply voltages: ±15 V.
- Quiescent current, each of the two transistors of diff amp: 1mA (this differs from the previous diff amp design; we hope the round number will ease your arithmetic slightly).
- "Carrier" frequency: approximately 1 MHz (this will come from your external generator when you build this circuit).
- "Signal" frequency (applied to vary the tail current): about 100 Hz to 3 kHz ("telephone quality" audio). This could come from breadboard function gen, but a second external generator or generator output may be more pleasant to use.
- Add a little bit of circuitry at the output of your amplifier to remove the ugly quiescent-level variation shown in the middle trace of Fig. 5L.22. Assume that the *load* for your circuit has very high input impedance (so high that you can neglect it).

We don't want to bore you: we understand that you've built a diff amp and a current source before. We remind you, in Fig. 5L.23 on page 232, of the high-gain diff amp design. The omission of the two R_Es provides the high gain – and the opportunity to *vary* this high gain.

If you decide to build your modulator on a standalone breadboard, the diff amp segment may look like Fig. 5L.21 before addition of the variable current source in the tail and the output conditioning components. You still may have this diff amp wired from the main exercise above, §5L.1.7.

So, your modest task is to design and build a circuit that will set the quiescent "tail" current at 2 mA, and then will let an audio signal vary the amplifier's gain. Doing this will "amplitude modulate" the

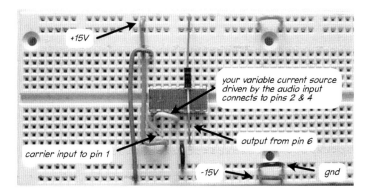

Figure 5L.21 The start of differential amp modulator wiring on a standalone breadboard.

"carrier," the 1 MHz input sinusoid. To make this current sink, you could use one of the extra *npn* transistors included in the CA3096 array, but we found it easier to do the job with a single discrete 2N3904.

Because modulating the diff amp's *gain* also shifts its quiescent level, the amplifier's immediate output is quite ugly (see middle trace in Fig. 5L.22). It is "ugly" in the sense that the waveform mixes amplitude variation (useful) with quiescent-point variation (useless artifact). We would like you to remove that latter variation, keeping only the AM signal, as in the third trace of the figure.

Although the variation in gain is slow (at low audio frequencies), note that the output will run at the higher *carrier* frequency. It is only its amplitude that varies slowly. So, don't be fooled into thinking that in order to pass the *slow* signal you need a lowpass. Though the signal frequency is low, the *AM signal* coming out of this stage necessarily includes the high-frequency carrier along with the slowly-varying amplitude. If this all seemed obvious to you, we apologize for belaboring a point.

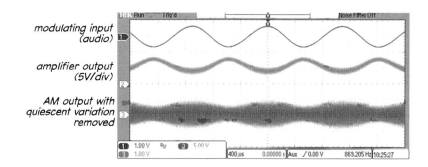

Figure 5L.22 AM modulator waveforms.

5L.2.3 Test your design

Complete the schematic of Fig. 5L.23 and build your circuit. To test your circuit, we suggest you use a small AM radio to let you listen to the fruits of your modulation. The transmitting antenna can be just a foot or so of hookup wire. The length is not at all critical: this makes a very inefficient antenna, but that's OK for present purposes.[9]

Just a test audio sinusoid will do. We find that a swept sine makes a good, distinctive test signal.[10]

---

[9] An efficient antenna, of the simplest design, using a wire strung well above ground, might be one half wavelength long: about 150 meters, at 1 MHz. Not convenient!

[10] In principle, the FCC isn't going to like what we're doing – but we expect that the power of our transmitters will be very low.

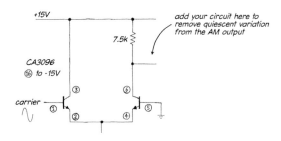

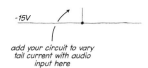

Figure 5L.23 AM modulator: Add your variable current source design to the diff amp tail and remove the varying low frequency DC from the output.

5L.3 Current mirror

We don't really expect you to continue through this section if you have just finished the differential amplifier. That is enough for one day. But if you are curious how non-linear, temperature dependent BJTs can be combined to create a very stable, linear current source, the following experiments are quite enlightening. The current mirror is also a fundamental circuit in most analog integrated circuits – it is what allows them to operate over a wide range of supply voltages and it is the high impedance current source used to reduce the common mode gain of the differential amplifier in operational amplifiers.

Current mirror: An overview

Here is one more way to make a current source. Unlike the first current source that you built in Lab §4L.4, this one requires the Ebers–Moll view of transistor operation: the view that I_C is determined by V_{BE} rather than by I_B.[11]

We offer you the mirror not because the current source of LAoE §4L.4 does not perform well, but because the mirror is so widely used in integrated circuits that it is worth our while to understand it.

It is called a "mirror" because the output current matches an input current. The matching is achieved very simply: by applying equal values of V_{BE} to the two transistors – input and output. A mirror appears in LAoE Fig. 5L.10, and there's a fuller discussion of mirrors beginning at §5S.2. As often, the simplicity of the circuit is a little deceptive. The circuit uses an application of negative feedback that to us is not obvious at a first glance: see §5S.2.2.

In the lab exercise, we will ask you to build this simple mirror, using discrete transistors, so that you can see how it is imperfect. Then we'll let you improve it.

When you begin to use the LF411 operational amplifier in Lab 6L, you will be using a circuit that includes three mirrors (see §9S.4 – a formidably-clever circuit; don't expect yourself to understand it all, at this point). The LF411 mirrors, incidentally, are not simple mirrors but the *Wilson* type (this clever design is described in §5L.3.2.4).

---

[11] As you have gathered, these two "views" are consistent; they offer differing ways to describe one behavior. See LAoE §5N.7.

5L.3 Current mirror

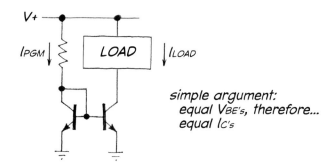

Figure 5L.24 Current mirror: conceptually very simple.

5L.3.1 Simple mirror

Build the simple mirror of Fig. 5L.25. Use two discrete transistors (2N3904, as usual). Ordinarily, one would not use discrete transistors. Instead, one would use a pair on an integrated circuit, to assure good matching of characteristics, and (even more important) to hold them at very nearly equal temperatures. We'll soon do that, but just now we are interested in the effect of temperature mismatch, so we need a design that is vulnerable to this problem.

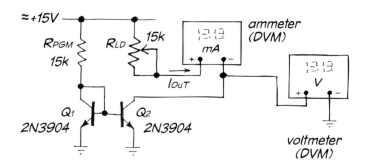

Figure 5L.25 Simple mirror.

You'll notice that the 15k value of the $R_{\rm PGM}$ is chosen to give us something close to our favorite current value, 1 mA. Adjust the variable $R_{\rm LD}$[12] to make $V_{\rm CE}$ of Q2 close to the same as that of Q1 (roughly 0.6 V). $I_{\rm out}$ should be close to 1 mA. To make your measurements easy, now adjust the breadboard's +15 V power supply so as to get an output current *very* close to 1.00 mA.

At its maximum value of 15k, the variable resistor standing in for $R_{\rm LD}$ will put Q2 close to its limit of operation: how?[13]

5L.3.1.1 Temperature effects

Unequal heating imposed by your fingers Now try warming either Q1 or Q2 with your fingers, and see whether you detect a change in $I_{\rm out}$. You may want to use your DVM's *microamp* range, if it's available, to measure with good resolution.

If you see $I_{\rm out}$ depart from 1 mA, both above and below the initial value, in response to your fingers, you are detecting a problem that you soon will solve. You will solve it, in §5L.3.2 on the next page, by using an integrated array of transistors.

Let the circuit stabilize, so that $I_{\rm out}$ returns to 1.00 mA.

---

[12] Use Bourns 3386F-1-153LF trimpot.
[13] That's right, with $V_{\rm CE}$ of just 0.6 V, Q2 is close to saturation. In saturation, Q2 would not perform as Q1 does; the mirroring would fail. But Q2 is *not* saturated at $V_{\rm CE}$ of 0.6 V, so the mirror should perform properly.

Unequal heating imposed by electrical power Now reduce the value of R_{LD} to increase Q2's V_{CE}. What happens to I_{out}? Does the value seem to drift upward, as if it had momentum? If so, why does it do that?[14] It receives more power, so its I_C grows; so, it receives more power, so.... You are watching a positive feedback loop. Eventually, I_{out} will level off. But we hope that what you have seen is enough to whet your interest in finding a solution to this temperature-instability problem.

5L.3.2 Improved mirrors

5L.3.2.1 IC array can make a better mirror

We can solve the temperature problem at a stroke, by building a mirror just like the one of §5L.3.1 on the preceding page, except that transistors Q1 and Q2 are neighbors on an array, the CA3096. This IC enforces close matching of temperature between Q1 and Q2, simply by placing the two near to each other on an IC. In addition (less important to us, in this exercise), the characteristics of the two array transistors match more closely than can be expected of a couple of discrete 2N3904s. So, the equal V_{BE} values should go with those of equal I_C.

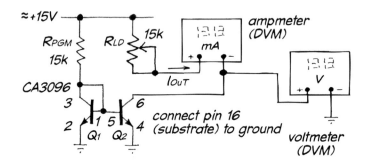

Figure 5L.26 Simple mirror, built with CA3096 array of matched transistors.

We cannot repeat the temperature-mismatch experiment of §5L.3.1.1, of course, because we cannot force the two transistors' temperatures apart. Instead, we'll have to assume the temperature problem due to differential *external* heating on the two transistors is gone, and then move on to observing a separate hazard, a result of *Early Effect*.[15]

5L.3.2.2 Early Effect I: simple mirror on IC array

Early Effect, in short, is a correction to the notion that the mirror holds current constant as voltage across the mirror varies. A good mirror comes close to doing that. But an accurate model of the mirror should add, in parallel with the ideal current source, a large-valued resistance. As voltage across the mirror grows, this modeling resistance delivers a slightly increased current. See Fig. 5S.6.

Adjust R_{LD}, as in §5L.3.1.1, so as to make the V_{CE} values for the two transistors very similar. Again adjust V+ so as to set I_{out} at 1 mA. Again measure the current using your DVM's microamp range.

Now reduce the value of R_{LD}, as you did before, and note the effect on I_{out}. Note the values of I_{out} with V_{CE} at three values: about 0.6 V, 7.5 V and 15 V.

Why does I_{out} vary as you adjust R_{LD}? Does I_{out} drift, for any particular setting of R_{LD}? What now prevents drift? What happened to the positive feedback loop?

If you pair measured changes in I_{out} with changes in V_{CE}, what R_{out} do you calculate for the mirror? For an argument explaining this result, see §5S.2.5. There, for a mirror using 2N3904 transistors and mirroring 1 mA, LAoE predicted R_{out} of about 100k ohms. You won't get exactly that value, but we

---

[14] Yes, Q2, with V_{CE} greater than that of Q1 gets hotter than Q1.
[15] Note, however, that unequal heating because of unequal power dissipation in the two transistors, due to unequal V_{CE}, is still a problem. We will see how to deal this with the Wilson mirror in §5L.3.2.4.

hope your result is not extremely different. The CA3096 transistors do differ from the 2N3904's, so an exact match is not possible. Beta for the '3096 is higher than for the '3904[16] – and with higher beta goes larger Early effect response.[17]

5L.3.2.3 Early Effect II: IC array mirror improved by emitter resistors

You are familiar with the stabilizing effect of an emitter resistor in a common-emitter amplifier. You saw in §5N.5.1 that it protected against thermal instability. Here, emitter resistors will offer a similar corrective, but this time applied against variation in I_C caused by Early Effect.

Install 470-ohm resistors on both emitters (pins 2 and 4).[18] Then run the same trial that you tried with the simple mirror (§5L.3.1 on page 233). The only setup changes you will need will be:

- instead of measuring V_C of Q2 relative to *ground*, in order to evaluate V_{CE} of Q2, use a DVM to measure the difference between V_C and V_E (which no longer sits at ground);
- readjust the positive supply, to get an initial I_{out} of 1 mA again, when the two V_{CE} values are equal, at about 0.6 V.

Again pair a few changes in I_{out} with changes in V_{CE}, to determine R_{out}. We hope it is better than it was for the simple mirror of §5L.3.1.[19]

How should one choose the value of the emitter resistors? The larger the value of the emitter resistor, the stronger the negative feedback. But, as usual, you don't want to overdo it. What do you lose if you use an R_E value larger than necessary?[20]

5L.3.2.4 Early Effect III: IC mirror improved by a third "cascode" transistor (forming a *Wilson* mirror)

We can reduce the error caused by Early effect without paying the price of a regrettable voltage drop across the R_E's inserted in §5L.3.2.3. We can do this by adding a third transistor that will soak up variation in voltage across the load, "hiding" these variations from Q2. This should eliminate almost all variation in I_{out} caused by Early Effect.

Here is the modified circuit, called a "Wilson mirror."[21] Its great virtue is that it operates close to the supplies, unlike the degenerated mirror.

As you try it out, recall that you will need to tie the voltmeter's lower lead not to ground, but to the emitter of Q3, in order to measure the relevant V_{CE}.

We found that this mirror provided much better R_{out} than what was available from the simpler mirrors of §5L.3.2.2 on the facing page or §5L.3.2.3.[22] A design that incorporates both remedies – Wilson's design and emitter resistors – can do even better. Texas Instruments' REF200 (noted in §11N.4.2, and diagrammed in Fig. 5S.14 on page 247), achieves a typical R_{out} of 100 Megohms using an improved Wilson current mirror.[23]

---

[16] 2N3904: 70 to 80 minimum at 1 mA, 300 max at 10 mA; CA3096: 150 to 500 at 1 mA.
[17] See AoE "X Chapters," §2x.5.1.
[18] Such an addition, as you probably recall, makes this circuit "degenerated," in the curious lingo of engineers.
[19] When we did this exercise, we were able to confirm this qualitative improvement, but the new R_{out} was not larger in proportion as $R_E + r_e$ is larger than the r_e of §5L.3.1. Our R_{out} was about 1.2M.
[20] Yes, you're right: you would waste some of the range through which your current source could operate properly. One of the special virtues of a mirror is that it can operate over nearly all of the supply range, unlike a conventional current source, which must set up an emitter voltage on the order of a volt, in order to operate reliably. That volt is lost to the range of the current source's "output compliance," the range of voltage over which it can hold current constant.
[21] Named after a Tektronix engineer named George R. Wilson. Wilson designed the circuit overnight, winning an informal competition with one of his colleagues, another eminent engineer: Barrie Gilbert, inventor of the analog mixer called the "Gilbert cell." `https://LAoE.link/Wilson_CM.html`.
[22] We measured about 3 megohms.
[23] See `https://LAoE.link/Improved_Wilson.html` for an explanation of this circuit.

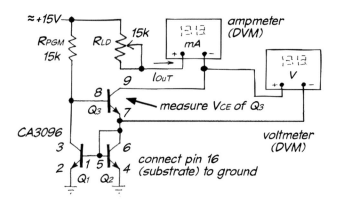

Figure 5L.27 Wilson mirror.

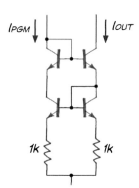

Figure 5L.28 TI REF200 current mirror achieves enormous R_{out} by using both Wilson design and emitter degeneration.

But we figure you will believe in Wilson and in emitter resistors without having to try the REF200. So, here we will call it a day.

5S   Supplementary Notes and Jargon: Transistors II

5S.1   Two surprises, perhaps, in behavior of differential amp

We have advertised the differential amp as just a pair of common-emitter amplifiers, and have promised you that there's not much new to understand here: you can use what you know from the two earlier labs where you build C–E amps. But students have noticed some effects that are new: not what one might expect from experience with a single C–E amp. One of these effects may not puzzle you for long, if at all; the other is quite subtle.

5S.1.1   Clipping of first-stage diff amp in Lab 5L op-amp

In the common-emitter amp, we are accustomed to see clipping at the positive supply, and close to ground, where the transistor saturates. So, the clipping shown by this differential amp, when its R_C has been reduced to 1.5k, is unfamiliar. The scope image in Fig. 5S.1 shows *two* outputs, because we have inserted a collector resistor above the left-hand Q, making the circuit perfectly symmetric, except in its *drive*. (This *drive* is what we called "mixed-mode" in §5L.1.8.)

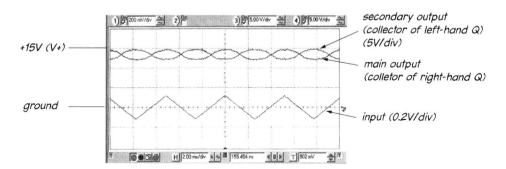

Figure 5S.1 Differential amp of Lab 5L, with R_C = 1.5k: two collector outputs shown; distortion is symmetric. (Note scope gain on input is 50× higher than on output channels.)

The flattening close the positive supply is nothing new. But two novelties appear:

- first, as the output swings *down*, the waveform flattens again, instead of growing steeper as in "barn-roof" distortion; and
- second, this flattening occurs not where the transistor *saturates* (which would be at a voltage close to ground) but at about 12.5 V *above* ground, where the transistor is very far from saturating.

This is unfamiliar, but we can make sense of this when we recall that the total current available to either transistor is limited – by the *tail* current source. So, the lowest collector voltage occurs not when the transistor *saturates*, but when it is hogging the entire *tail* current, leaving none for the other side. Strange, perhaps; but explicable.

5S.1.2 A closer view of the distortion: barn-roof reflected in a puddle?

We saw in Fig. 5S.1 that the diff amp's distortion is symmetric, unlike the "barn roof" distortion of a common-emitter amp. If we look closely at that distortion, when driving the amplifier less strongly than in that Fig. 5S.1, we find what looks like barn roof distortion *mirrored* – offset to the side and perhaps reflected in a barnyard puddle.[1]

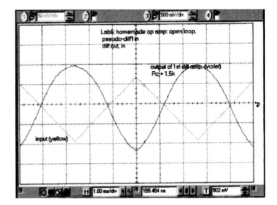

Figure 5S.2 Diff amp's distortion differs from C–E's: it's symmetric. (Note scope gain on input is 10× greater than on output channel.)

If you would like a mathematical argument – unusual in this course – to explain the symmetry in the distortion, we'll try one. The question we aim to answer is why gain should diminish on *both* both sides of the quiescent point. This is the barnyard–puddle symmetry. Here's the argument:

- Gain of the diff amp, with one input grounded:

$$G = -\frac{R_C}{r_{e_1} + r_{e_2}} \quad (5S.1)$$

- ... and at the quiescent point this is simple enough; the gain is just $G = -R_C/2r_e$.
- But, as you recall from your experience with a common-emitter amplifier, the gain varies as the output swings, because I_C has to vary in order to cause the output swing; and the sum of $r_{e_1} + r_{e_2}$ is not constant as the diff amp output swings – even though one r_e grows as the other shrinks.
- More specifically, if we call the current in the output transistor I_1 while calling the total "tail" current I_T (both assumed measured in milliamps), then:

$$G = -\frac{R_C}{(r_{e_1} + r_{e_2})} = -\frac{R_C}{\left(\frac{25}{I_1} + \frac{25}{I_T - I_1}\right)} = -R_C \left/ \frac{25(I_T - I_1 + I_1)}{I_1(I_T - I_1)} \right.$$
$$= -\frac{R_C}{25} \times \frac{I_1(I_T - I_1)}{I_T} = -\frac{R_C}{25} \times \left(I_1 - \frac{I_1^2}{I_T}\right) \quad (5S.2)$$

If I_1 gets all or none of the tail current, gain is zero. If I_1 is half the tail current, gain is at a maximum – about 22, for R_C=1.5k and I_T=1.5 mA, as in the lab circuit.

The gain curve looks like the plot of Fig. 5S.3. The figure makes evident that 50:50 sharing provides maximum gain.

---

[1] The output in Fig. 5S.2, appears to be centered on zero volts. This is an illusion that results from AC-coupling of the scope input.

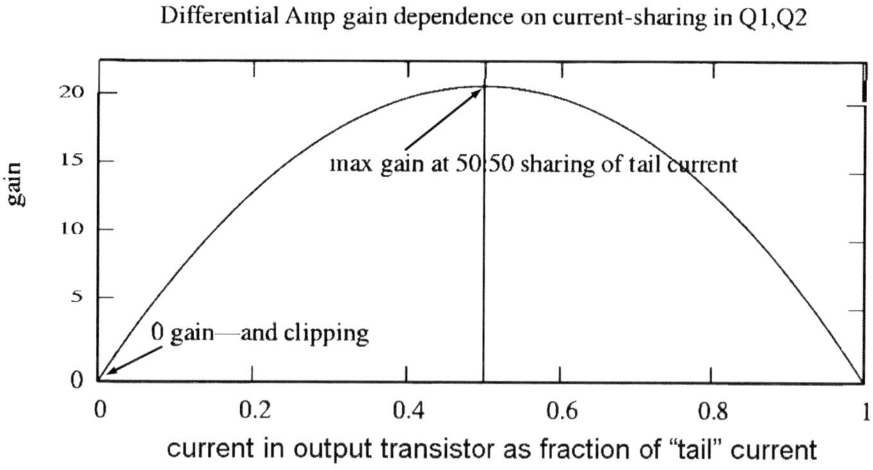

5S.2 Current mirrors: Early effect

Differential Amp gain dependence on current-sharing in Q1,Q2

max gain at 50:50 sharing of tail current

0 gain—and clipping

current in output transistor as fraction of "tail" current

Figure 5S.3 Plot of diff amp's calculated gain as sharing of tail current diverges from 50:50

An expression for the derivative of gain: $dG/dI_1 \propto 1 - (2I_1/I_T)$ says the same thing: the maximum gain occurs when I_1 is 1/2 the tail current.

And the flattening that we saw in Fig. 5S.1 occurs when the circuit gain falls to zero. This occurs when either transistor hogs all the current.

5S.2 Current mirrors: Early effect

5S.2.1 Mirrors, a topic you can skip

We concede that there's something funny about opening a section with the remark, "you can skip this." We treat mirrors this way because of our evident ambivalence about these neat circuits. We include them in the lab exercises but judge them less important to a designer than the transistor circuits that we have asked you to build, so if you don't get to them, that's ok. But mirrors come up fairly often for a *reader* of circuits. They abound in op-amp ICs. So, it is useful to consider how they work, even if you are unlikely to adopt a mirror as an element in your own designs.

Mirrors also are pedagogically useful, to demonstrate *Early effect*, a topic that we similarly removed from labs, but would like to make available to the curious. Early effect lets us make a quantitative estimate of the output impedance of a current source. To this point we have had to content ourselves with saying that current sources show output impedance that is *high*.

We will begin by introducing the current mirror. Then we will adopt it to watch Early effect in action.

Applying the Ebers–Moll view to circuits: current mirror

A current mirror makes no sense without the help of the Ebers–Moll view of transistors, illustrated in Fig. 5S.4.

AoE §2.3.7

Why is a mirror useful, given that we know other ways to make current sources? A mirror makes it easy to link currents in a circuit, matching one to another. Such linking is useful in integrated circuits. A mirror also can show the widest possible *output voltage compliance*. You can see a zoo of mirrors, adopted for both these motives, in the schematic of the operational amplifier that we use in many of our analog labs, the LF411 (see §9S.4).

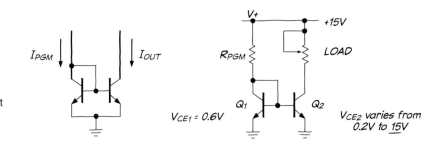

Figure 5S.4 Current mirror: Ebers–Moll view *required*.

It's easy to make $I_{program} = I_{out}$, and only a little harder to scale I_{out} relative to $I_{program}$.

Early effect and *temperature* effects both could spoil the neat equality between $I_{program}$ and I_{out}. We will learn later (see §5S.2.7) how to fight these problems. For now, let's leave the mirror in its simplest form, as shown above, and notice how the circuit works.

5S.2.2 Feedback sets mirror programming current...

Look at Fig. 5S.5. The *program* side of the current mirror looks simple, if a bit weird (base and collector joined?!). It's quite neat, though: an application of negative feedback.

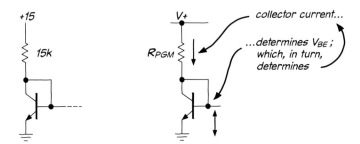

Figure 5S.5 Subtle negative feedback: programming side of the current mirror.

As you explain to yourself how this circuit fragment works – giving us our beloved 1mA – it's helpful to notice that nearly all of the current flows not in the base path, but from collector to emitter.[2]

5S.2.3 ...while equal V_{BE}'s assure equal I_C's: the "mirroring"

So, because of the equal V_{BE}'s of Q_1 and Q_2 (in Fig. 5S.4), $I_{out} = I_{program}$. So far, so neat.

The main virtue of this circuit is its ability to operate almost from "rail to rail" (jargon for "from one supply to the other"). The mirror will work until Q_2 saturates, so it shows wide "output voltage compliance." As R_{load} varies in the circuit of Fig. 5S.4, current stays pretty nearly constant at 1 mA. The voltage range from V_+ (15 V, in this case) down to about 0.2 V (just above Q_2's saturation voltage). This wide compliance helps explain the widespread use of mirrors in operational amplifiers, as we have argued above (another reason is the preference for transistors over resistors, in IC fabrication).

5S.2.4 Complications

A difficulty easily solved: temperature effects

If the temperatures of Q_1 and Q_2 diverge, the claim of 1:1 mirroring fails. You saw in Chapter 5N how strongly temperature affects I_C.

---

[2] Does it make sense? The base voltage – the same as collector voltage – is determined by the drop across $R_{program}$. But $I_C \times R_{program}$ sets a value for V_{BE}. Higher I_C drives down V_{BE}, so this is a negative feedback loop. It will stabilize, having found an I_C that generates a V_{BE} consistent with that I_C.

5S.2 Current mirrors: Early effect

But it is easy to hold Q_1 and Q_2 temperatures equal: just put the two transistors close together on an integrated circuit. Any other way of building a mirror would be downright perverse. So, temperature matching takes care of itself. (You saw this stabilizing technique applied to a common emitter amplifier, as well, in Chapter 5N, Fig. 5N.21: the "compensating" transistor shown in that mirror-like amplifier circuit keeps the amplifier temperature-stable.) It is true that V_{BE} changes with temperature, at a given I_C: as temperature rises, V_{BE} falls, for example. But that altered V_{BE}, produced by Q_1, is applied to a heated Q_2, whose 1 mA I_C requires just such a reduced V_{BE}.

5S.2.5 A harder problem, neatly illustrated by current mirrors: Early effect

AoE §2.3.7A

Another departure from the ideal simple *mirroring* is not so easily fixed: it is the slight degree to which collector current varies as voltage across the transistor varies.

Output impedance of a current source ideally is infinite (because the value R_{dynamic}, equivalent to $\Delta V_C/\Delta I_C$, is infinite if I_C does not change at all). But, in fact, I_C grows slightly as V_{CE} grows: so an actual current source behaves like a large resistor. In Fig. 5S.6 we model this imperfection.

Figure 5S.6 A non-ideal current source can be modeled as ideal parallel (large-) R.

Compare AoE Fig. 2.59

You may have met this model as a Norton Equivalent Circuit. Like the Thevenin model, it permits us to give a value to a circuit's degree of imperfection.

Figure 5S.7 shows another way to represent this behavior: as a non-zero slope in the I_C vs V_{CE} plot.

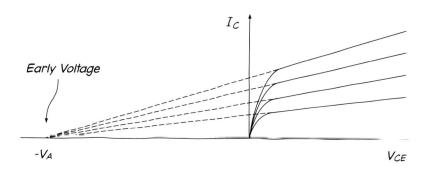

Figure 5S.7 Early voltage, where all the curves converge, provides a measure of departure from current constancy.

Strictly speaking, V_A is measured not relative to zero on the V_{CE} axis, but relative to the point at which the tangent is taken. The difference, however, between these two points is usually not substantial. V_{CE} ordinarily is much smaller than the value of V_A.

Figure 5S.8 gives the measured responses of three transistors as we increase V_{CE}.[3]

The leftmost transistor curves show the best current-source behavior, as you'll recognize. But in other respects (say, its impedance-changing virtues as a *follower*) that transistor, with its low β, might be considered the *worst* of the three transistors.

---

[3] Data taken from AoE Fig. 2.59. The middle plot is discussed in §2x.8 of *AoE – the x-Chapters*.

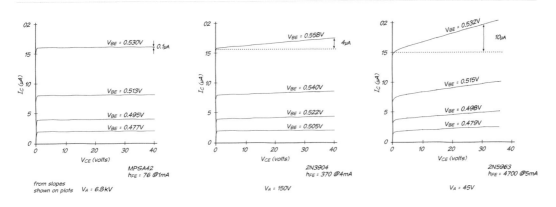

Figure 5S.8 Early effect: slopes of I_C versus V_{CE} reveal V_A. Taken from §2x.8 of *AoE – the X-Chapters*.

The mechanism of Early effect

The resistor-like response of the transistor to voltage across its C–E terminals is caused by *base width modulation*: an increase in V_{CE} widens a carrier *depletion region* between collector and base, shrinking the effective *base width* and thus increasing collector current. Early effect can be moderate or extreme, depending on the intrinsic base width: a thin base region responds more dramatically to a given narrowing because that narrowing constitutes a larger fractional change in the effective width.

So transistors with narrow base regions – which are those with *high β* – are especially sensitive to V_{CE} variation. This appears in the pairings of values for V_A and β shown below, for the three transistors plotted in Fig. 5S.8.

Transistor	β (typical)	Early voltage, V_A (from measured curves)
MPSA42	25	6.8 kV
2N3904	130	150 V
2N5963	1200 (min.)	45 V

V_A is useful – but also not ordinarily available as a transistor specification.[4] The curves of Fig. 5S.8 are difficult to measure, because temperature effects can muddle the attempt to see Early effect. If you have no V_A spec or curves to work with, you will have to settle for an estimate of V_A based on the transistors's β range. The estimate will be approximate, but helpful. For a small-signal transistor like the 2N3904, you can guess at a V_A of somewhere around 100 V.

5S.2.6 Calculating the results of Early effect

A V_A specification gives a quick impression of the quality of a transistor as current source. But it is rather an abstraction. It would be useful to be able to convert V_A – or some other Early effect specification – into a measure of the behavior of a particular circuit.

Two results of Early effect would be especially useful...

- a prediction of the *output impedance* of a current source; and
- a way to calculate a ceiling on the gain of a transistor amplifier, when its load (the impedance on the collector) is a current source.

We can, indeed, calculate both results.

---

[4] This omission of V_A from transistor datasheets seems odd. We asked the late analog wise man Robert Pease why, and he answered (in an email), "Hardly anyone asks for it. And those that need this info, know how to get it," meaning that they can measure ΔV_{BE} versus ΔV_{CE}. We remain puzzled by the omission, though.

5S.2 Current mirrors: Early effect

Calculating R_{out} for a current source

One way: use the slope of the I_C versus V_{CE} curves directly: If you are lucky enough to have a plot if I_C vs V_{CE}, like those in Fig. 5S.8, you can measure R_{out} from the plot.

For the 2N3904, for example, a simple current mirror (one with no emitter resistor) shows a slope that we have labeled for the particular curve that begins at about 17 μA:

$$\Delta I_C / \Delta V_{CE} = 4\,\mu A / 40\,V = 0.1 \times 10^{-6}\,\Omega^{-1}$$

The units are inverse ohms: Ω^{-1} or siemens. The inverse of this slope is the output resistance at the collector, R_{out}, a value in ohms: 10 MΩ. This is not a general answer, but an impedance at a particular output current (about 17 μA, in this case).

This is a start. We'd like to extend this to R_{out} at other currents. Let's recast the slope as a *percentage* change in current per volt. Fig. 5S.9 shows a hypothetical Early plot – similar to the 2N3904 curves, but showing a hypothetical Early voltage of 100 V rather than our measured V_A. We chose this V_A to keep the arithmetic as simple as possible.

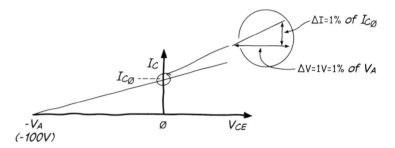

Figure 5S.9 Effect shown as fractional or percentage change per volt.

Assume a BJT with $V_A = 100$ V. A 1 V increase in V_{CE} – from an initial V_{CE} close to zero – is a 1% increase: see inset, Fig. 5S.9. The straightline ramp of the I_C plot, up from zero at V_A, enforces a resulting 1% increase in I_C.

This way to describe the Early effect slope is convenient, because it lets us calculate R_{out} at any initial current (I_{C_0}). At $I_{C_0} = 1$ mA, for example, the 1% $\Delta I_C = 0.01$ mA = 10 μA. So, R_{out} at 1mA is the inverse of this slope: 0.1M = 100 kΩ.

AoE §2.3.2D

Another way: Rather than use either V_A or the I_C versus V_{CE} slope directly, one can infer a useful relation, $\Delta V_{BE}/\Delta V_{CE}$. This relation – labeled with the symbolic representation η – will help when we estimate R_{out} for current sources:

$$\eta \equiv -\frac{\Delta V_{BE}}{\Delta V_{CE}} \quad \text{at a } \textit{fixed value of } I_C$$

This η is a little hard to get an intuitive grip on: it is the change in V_{BE} (a decrease) required to hold I_C constant as V_{CE} grows by a volt.

Using the hypothetical I_C versus V_{CE} curves again, from Fig. 5S.8, we can calculate the value of η. We saw a 1% increase in I_C for each 1 V increase in V_{CE}. (Incidentally, here comes one of the rare instances in which we use the Ebers–Moll equation explicitly.)

Ebers–Moll tells us that $I_C \approx I_S(e^{V_{BE}/25\,mV})$. (We ignore the insignificant "-1" from the full expression, as usual. See AoE §2.3.1. The value 25 mV is kT/q evaluated at room temperature.) The ratio of two collector currents is therefore

$$I_{C2}/I_{C1} = e^{(V_{BE2} - V_{BE1})/25\,mV}$$

If I_{C2} is 1% more than I_{C1}, as in the plot of Fig. 5S.8, Ebers–Moll lets us calculate the associated change in V_{BE}:

For this 1% change, then, $e^{\Delta V_{BE}/25\,\text{mV}} = 1.01$. We should admit the embarrassing fact that this describes a case *different* from the one that we mean to evaluate: it describes the case were V_{CE} is kept constant and an increase in V_{BE} causes the rise in I_C. Starting with this different case may strike you as a bit weird; it seems that way to us too. But hang on, the procedure does work.

$$\Delta V_{BE}/25\,\text{mV} = \log_e(1.01)$$
$$\Delta V_{BE} = \log_e(1.01) \times 25\,\text{mV} = 0.01 \times 25\,\text{mV} = 0.25\,\text{mV}$$

Now we can pop out of this thought experiment and admit that what we're after is the *reduction* in V_{BE} that we would need in order to hold I_C fixed as V_{CE} grows: this is the same value for ΔV_{BE} but is opposite in sign:

$$\eta \equiv -\Delta V_{BE}/\Delta V_{CE} = 0.25\,\text{mV}/1\,\text{V} = 0.25 \times 10^{-3}$$

For the moment, we will just note that we have found how to calculate this value. Soon, in §5S.2.7, we will put η to use.

5S.2.7 Improving R_{out} for a current source

Cascode or "Wilson mirror"

AoE §2.3.7B

One way to eliminate the effect of changes in V_{CE} is to add an extra transistor that prevents such V_{CE} changes. This trick – whose analogous design dates from vacuum-tube days – is called a "cascode" circuit.[5] The additional transistor simply passes the current fed it by the transistor that is to be protected from Early effect. This is the Wilson mirror discussed back in §5L.3.2.4. The pass-through transistor does feel wide variation in V_{CE}, but this transistor does not set the output current; it merely passes the current determined by the protected transistor. This design, illustrated in Fig. 5S.10, is easily implemented in an integrated circuit, and is the standard form for mirrors within operational amplifiers, including the one that we rely on in most of our op-amp labs, the LF411.

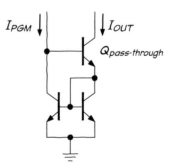

Figure 5S.10 Wilson mirror or "cascode" configuration: hides changes in V_{load}.

The circuit fragment shown at the left side of Fig. 5S.11, taken from the LF411 datasheet, looks at first like a bizarre new circuit component. Redrawn on the right, however, it reveals itself as the same circuit as that of Fig. 5S.10 (though done with *pnp*s rather than *npn*s, to *source* rather than *sink* current). Q_{13}, you will notice – which looks pretty strange in the original drawing – is redrawn more conventionally on the right as two distinct transistors, sharing common emitter and base connections.

---

[5] The word, though now applied to transistor circuits, reflects the date of its birth, 1939: it merges the word "cascade" with "cathode," a terminal name familiar to you from its use for diodes but one that did not carry over from vacuum tubes to transistors.

5S.2 Current mirrors: Early effect

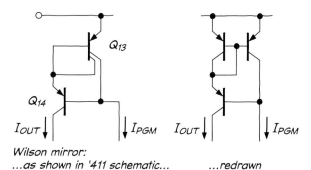

Figure 5S.11 Cascode or Wilson mirror in LF411 op-amp.

Wilson mirror:
...as shown in '411 schematic...   ...redrawn

Current source improvement by use of *emitter resistors*:

The characteristic that we called η, in §5S.2.6 will help us to calculate the improvement in R_{out} that emitter resistors can provide. Fig. 5S.12 shows a simple mirror with emitter resistors added.

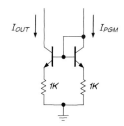

Figure 5S.12 Simple mirror improved by emitter resistors.

If you have accepted the argument (made in Chapter 5N) that installing an emitter resistor in a common-emitter amplifier stabilizes I_C, you will find it plausible that emitter resistors should tend to stabilize I_C in a mirror. In both cases, *negative feedback* is the key to the improvement.

We will try to do better than claim that the effect is plausible. We will use η to calculate the quantitative effect of these resistors.

Use η to calculate R_{out} for a mirror with R_E's: A rise of 1 V at the collector causes V_{BE} to shrink by $\eta \times 1$ V, thus causing a rise of emitter voltage by that amount: 0.25 mV (for the η value of 0.25×10^{-3} that we found in §5S.2.6).

Ohm's law says that this change in V_E will increase I_E and I_C by

$$\Delta V_E / R_E = 0.25 \times 10^{-3} \text{ V} / 10^3 \, \Omega = 0.25 \times 10^{-6} \text{ A}$$

At the collector, R_{out} will be the inverse of this result: 4 MΩ. Generalizing this result, we can see that a rise of 1 V at collector causes a rise η times smaller at the emitter. So the transistor, seen from its collector, behaves like a "lens": see Fig. 5S.13.

This lens metaphor is reminiscent of the follower that we likened to a rose-colored lens, but radically different in its result: this lens *increases* R_{out}, enlarging the emitter resistance by the factor $1/\eta$. In this example, that factor was 4000, converting the 1k R_E to an R_{out} of 4M.

Confirm η's utility, by calculating R_{out} for the original mirror: We earlier calculated R_{out} for the simple mirror using the slope of the I_C vs V_{CE} curves. Let's test our faith in η by recalculating that value using η instead, along with our usual Ebers–Moll sidestep, r_e. Let's see if we get the same result – 100k.[6]

---

[6] It would be strange if we did not – for we are simply running in reverse the calculation that led to our value for η.

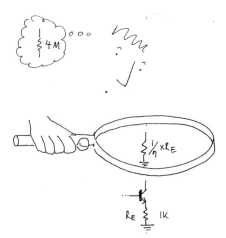

Figure 5S.13 Viewed from collector, transistor current source acts like a lens that magnifies R_E and r_e.

Using the result for the case where we installed R_E's, we note that we'll get a ΔV_{BE} of 0.25 mV for a 1 V ΔV_{CE}, and (having faith in our "little r_e" model) we view this as raising the voltage across the model r_e by that amount.

For the 1 mA current source, the value of r_e is 25 Ω at room temperature, so I_C grows...

$$\Delta I_C \approx \Delta I_E = \Delta V_{BE}/25\,\Omega$$

For an assumed 1V change in V_{CE}, this quotient is $[(\eta \times \Delta V_{CE}) = 0.25\,\text{mV}]/25\,\Omega$:

R_{out} is the reciprocal, $1/\Delta I_C$

$R_{\text{out}} = 25\,\Omega/0.25\,\text{mV} = 100\text{k}$, as we had hoped

We could put it more simply (and more metaphorically), as we did in Fig. 5S.13, saying that once again the Early effect "lens" enlarges what's at the emitter by the factor $1/\eta$, which we found to be 4000 in the present case. Here, where the circuit includes no R_E, we must use r_e evaluated to 25 Ω at 1 mA. So $R_{\text{out}} = 4000 \times 25\,\Omega = 100\,\text{k}\Omega$. This *lens* method seems the most straightforward.

...and a mirror can be improved by use of both *both* cascode and emitter resistors

Wilson mirrors are available as ICs, and one from Texas Instruments, REF200, which includes two current limiting elements plus one Wilson mirror,[7] uses *both* of the improvements described above: *cascode* and *emitter resistors*. These are shown in Fig. 5S.14.

This schematic makes the Wilson configuration much easier to recognize than in the LF411 datasheet – and adds the new element of emitter resistors.[8] With this double protection, the mirror in REF200 achieves spectacular R_{out}: a *typical* value of 100 MΩ.

---

[7] See the discussion of this part in Chapter 11N. A TI (formerly Burr-Brown) application note shows many current source circuits, including applications of the REF200 cascoded to achieve enormous R_{out} (>10 Gigohm). See https://LAoE.link/Current_Sources.pdf, "Implementation and Applications of Current Sources and Current Receivers," 2000.

[8] This circuit also adds a fourth transistor, a sort of *pass* element for the programming side, gilding the already impressive Wilson mirror's protections. This fourth transistor equalizes the two lowest transistors' V_{CE}'s at one V_{BE} drop, rather than one and two V_{BE}'s as in the standard Wilson mirror.

5S.2 Current mirrors: Early effect

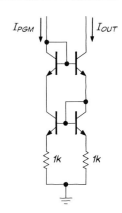

Figure 5S.14 REF200 IC mirror uses both cascode or Wilson configuration and additional stabilization by *emitter resistors*.

5S.2.8 Another consequence of Early effect: a ceiling on amplifier gain

We will see, in operational amplifiers, that transistor current sources routinely replace collector resistors,[9] in order to maximize gain of differential and common-emitter stages. When this is done, the impedance of that current source matters – but so does the output impedance of the amplifier transistor. A naive view might see a current source load as providing gain that was *infinite*. As usual, we can bring such hyperbolic impressions down to earth, using Early effect to quantify the ceiling on amplifier gain.

Gain of ordinary common-emitter amp: determined by R_C and emitter resistance

In the cases familiar to us from this course, a transistor amplifier converts changing collector current to changing collector voltage using a resistor on the collector. We have said that gain, in a common-emitter amplifier, is determined by the ratio of R_C to either R_E or to the resistor-like model of transistor gain that we call "little r_e".

Equivalently, the maximum gain of our usual amplifier can be described as $g_m \times R_C$, where $g_m \equiv 1/r_e$.[10] The "transconductance," g_m, is current-change *out* per voltage-change *in*: $\Delta I_C / \Delta V_{BE}$.

This formulation ignores Early effect, and until now ignoring Early has worked. In order to take account of Early effect, one would acknowledge that the resistance driven by the transistor's collector, strictly, is $R_C \parallel r_{\text{out-collector}}$ (let's call it r_o), where r_o quantifies Early effect. In the cases we have seen to this point, R_C has been so much smaller than r_o that there was no need to mention more than R_C.

Gain of a common-emitter amp using a good current source on the collector

Now, if we consider the case where we maximize gain by using a very-good current source in place of R_C, we need to take account of Early effect: we need to include the effect of the transistor's own r_o. To keep things manageable, let's assume an *ideal* current source as load (or if "ideal" bothers you, use a source like the REF200 mirror, with its r_o of around 100 MΩ).

In this case, it will be the transistor's r_o, not the familiar R_C, that will define the circuit's voltage gain. Gain will be $g_m \times r_o$, or (equivalently) r_o/r_e. At 1mA, for the transistor we have been assuming in our discussion, with its V_A of 100 V, this gain will be 100k/25 = 4000. High, but a long way from infinite.

---

[9] Collector resistor equivalents are "drain" resistors for similar amplifiers made with field-effect transistors, as in the LF411 op-amp.

[10] This definition – suggesting that g_m is derived from r_e – makes sense in this course, since we have used r_e repeatedly. In the bigger world, however, the definition is a little odd, since g_m is the concept much the more generally used.

5S.3 Transistor summary

Why this note?

We're guessing that you're feeling overloaded with information, as we approach the end of the discrete-transistor section of the course. You've heard important points and details, and there's some danger that details will elbow out more fundamental learning. So, here's a sort of 3-minute University version of what you've spent a week learning. Maybe it's a 6-minute University.

5S.3.1 Preliminary: why learn about discrete transistors?

Soon we'll be designing with op-amps, and the transistor designs we've studied (say, the common-emitter amp) will seem clumsy and defective and very complicated. Do you need to understand a C–E amp?

Maybe not, but the major argument *pro* is that you'll see such amplifiers within op-amps (second stage of '411, e.g.; and the differential amp that forms the first stage of every op-amp is essentially a pair of linked C–E amps). As we have said earlier, it's satisfying – though not necessary – to be able to understand how an op-amp works: to see that it's not *only* magic that lets an op-amp achieve its magical results.

Then there is a little room for discrete-transistor circuits: sometimes as high-frequency circuits, beyond the range of a garden-variety op-amp; more often, as higher-powered helpers that are controlled by an op-amp (say, a push–pull put within an op-amp's feedback loop, as in §6L.8).

5S.3.2 Bipolars

Main points, or truths worth recalling: bipolars

- Simple View (Chapter 4) (*note* that this simple view often is enough. Use it when it suffices.)
    - $V_{BE} \approx 0.6$ V: this truth (*not* contradicted by the Ebers–Moll view) often allows quick analysis and design.
    - $I_C \approx I_E$: follows from notion that β is large (that is, that $I_C \gg I_B$).
    - I_C is β times larger than I_B: we need this notion in order to understand input impedance of follower or of C–E amp, and to understand the follower's ability to change impedances ("rose-colored lens" effect): the follower boosts R_{in}, drops R_{out}.
- Fancier View: Ebers–Moll (Chapter 5)
    - I_C is determined by V_{BE} (rather than by I_B). More specifically, I_C is an exponential function of V_{BE}.[11]
    - Often we use this information – model it – with the curious device of "r_e," a model that shows a little resistor in the emitter lead of the transistor. This modelling device allows us to use our simpler (Lab 4L) view of transistors in order to calculate *gain* and *input impedance* of transistor circuits where R_E is very small or absent altogether.
- A few formulas.
    Input and output impedances of follower (rose-colored lens effect):[12]
    $$R_{in} = (1 + \beta) \times R_E; \qquad R_{out} = [r_e + R_{source}/\beta] \parallel R_E.$$

---

[11] We can reconcile this idea with Chapter 4's view (that I_C is determined by I_B) by recalling that the base-emitter junction is, after all, *a diode*, so that V_{BE} and I_B are linked. More specifically, I_B looks like an exponential function of V_{BE}. So, to say (as we do in Chapter 4) that I_C is a constant multiple of I_B is to say implicitly that I_C is an exponential function of V_{BE}, as Ebers–Moll teach us.

[12] Normally, we don't bother with the "$1 + \cdots$," since we don't begin to know β with that precision. We'll write β rather than the correct-but-silly "$1 + \beta$," in these formulas. Note, by the way, that "R_E" represents the whole impedance at the emitter (sometimes a *load* parallels the emitter resistor, for example).

- Usually we can simplify this by ignoring R_E, because that value usually is much larger than the resistance seen through the transistor "lens," R_{source}/β.
- If the circuit is *biased*, then $R_{\text{Thev\_bias}}$ is parallel to R_{source}: $[R_{\text{source}}||R_{\text{Thev\_bias}}]/\beta$. Again, R_{source} normally has been designed to be much lower than $R_{\text{Thev\_bias}}$, so the result usually boils down to R_{source}/β once more.
- R_{out} for C–E amp (taken at collector): $\approx R_C$, because $R_{\text{out}} = R_C \parallel Z_{\text{collector}}$, and $Z_{\text{collector}}$ is huge.[13]
- Gain of C–E amp: $-R_C/(r_e + R_E)$.
- Gain of differential amp:
  Differential gain is $G_{\text{diff}} = R_C/2(r_e + R_E)$.[14]
  Common-mode gain: $G_{\text{CM}} = -R_C/(r_e + R_E + 2R_{\text{tail}})$. This is approximately $-R_C/2R_{\text{tail}}$. A large value of R_{tail} permits low G_{CM}. A *current source* in the place of R_{tail} provides best – lowest – G_{CM}.
- ... and *switches* are weird: different from all the other circuits we have seen:
  - we want switches either to *saturate*, or to turn *off* – whereas all the other circuits call for avoiding both of these extremes
  - none of the usual linear-region transistor rules apply to switches:
    No, $I_C \neq \beta \times I_B$, in a switch: we want, instead, to overdrive the base, by about a factor of 10. Load current is determined by the *load*, not by $\beta \times I_B$.
    No, input resistance at the base does *not* equal $\beta \times R_E$ or $\beta \times r_e$. Since the switch is saturated, R_{in} is much lower than that. No helpful β limits the magnitude of base current.

5S.4 Important circuits

Most of us store circuits as graphic elements rather than as concepts. We invoke them like little linguistic idioms; we don't derive them each time from first principles or from formulas. Here, as a reminder, are some of your favorite transistor circuits.

- current sources:
  - single-transistor:

Current source (sink): single transistor

---

[13] That impedance is huge because that terminal behaves like a current source (or "sink"), and for any good current source, $\Delta V/\Delta I$ is very large. The transistor holds I_C nearly constant as V_C varies. *Early effect* describes the departure of the transistor from this ideal. But Early effect describes a small correction to the general truth that the transistor holds I_C constant.

[14] The factor of two in denominator reflects the way we set up the assumed input: a pure differential input looks like ΔV at one input, and $-\Delta V$ at the other input, so that the whole *difference* – or "differential" – signal is not ΔV but *two* ΔV.

- follower:
    - split-supply (the simpler circuit, but requires a second supply):

        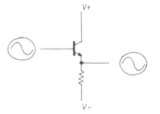

        Voltage follower ("emitter follower"): split-supply

    - single-supply (less elegant, but the cheaper form):

        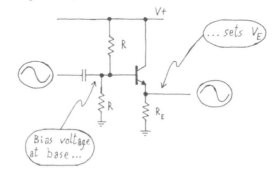

        Voltage follower "biasing"

    - ... and the input and output impedances:

        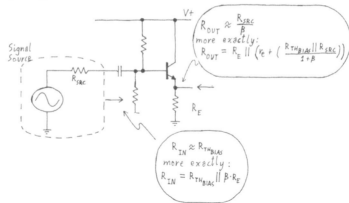

        Follower impedances

5S.4 Important circuits

- push–pull: solves the problem, in single-transistor follower, that the circuit's asymmetry makes it poor at *sinking* current from load (for the case of *npn*):

Push–pull follower

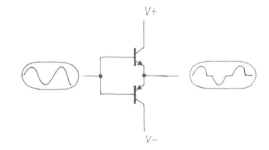

- amplifiers (voltage amps):
  - common-emitter amplifiers

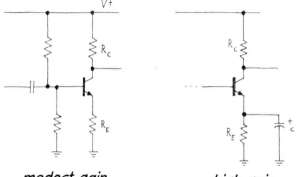

modest gain...   *...high gain*

$G = -R_C/r_e$, at "signal frequencies," where $X_C \ll R_E$

- common emitter impedances:

Impedances in and out, ce-amp

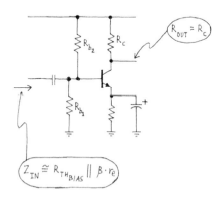

$R_{OUT} = R_C$

$Z_{IN} \cong R_{TH_{BIAS}} \parallel \beta \cdot r_e$

- differential amplifier:

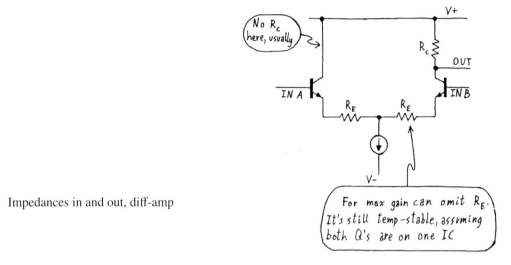

Impedances in and out, diff-amp

- Switch (something completely different!):

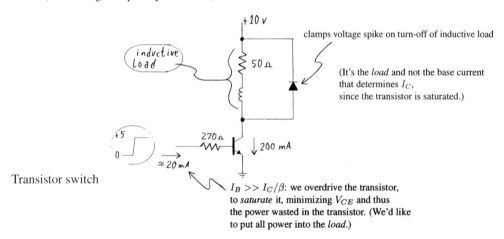

Transistor switch

5S.4.1 Fine points

Here are some refinements that we want you to be aware of, even though we don't expect you to be nimble with the details or with calculating the magnitude of these problems.

- three important ideas, but ideas you're not likely to treat quantitatively in this course:
  - temperature effects: the fact that transistors are very sensitive to temperature is a *fundamental* point; the formulas that predict response are not so important. Your designs should include protection against this sensitivity to temperature: most often, R_E does the job (providing *feedback*).
  - Early effect: describes the imperfection of a transistor current source. That is, describes the change in collector current with change in V_{CE}. Often negligible (usually, thanks to an R_E providing feedback); can cause substantial errors in a current mirror that lacks R_E.

- Miller Effect: a serious obstacle to design of high-frequency amplifiers. Miller effect results from capacitive feedback in any inverting amplifier, tending to oppose changes at the input. Because we don't build high-frequency amps in lab, we're not likely to feel the importance of Miller effect – but you will as soon as you try to design an amp that's to operate above, say, 1 MHz. That will happen *after* this course.
- Interesting ideas, but less important.
    - bootstrapping (though it's a nifty idea: make a circuit element disappear electrically, by making its two ends move the same way); and
    - Darlington connection.

5S.5   Jargon: bipolar transistors

biasing (see, for example, AoE §2.2.5). Setting *quiescent* conditions (see below) so that circuit elements work properly. To *bias* means, literally, to push off-center. We do that in transistor circuits to allow building with a single supply. The term is more general, as you know. (Compare AoE §1.6.6A, where a rectifying diode is *biased* into conduction.)

bootstrap (see, for example, AoE §2.2.5A). In general, any of several seemingly-impossible circuit tricks (source of term: "pull oneself up by the bootstraps:" impossible in life, possible in electronics!). In this chapter refers to the trick of making the impedance of a bias divider appear very large to improve the circuit's input impedance. Also *collector* bootstrap. See AoE §2.4.3A.

bypassed (-emitter resistor) (see, for example, AoE §2.2.5A). In common-emitter amp, a capacitor put in parallel with R_E is said to *bypass* the resistor because it allows *AC* current an easy path, bypassing the larger impedance of the resistor. Used to achieve high gain while keeping R_E large enough for good stability.

cascode (applied in Wilson Mirror: AoE §2.2.7). Circuit that uses one transistor to buffer or isolate another from voltage variation, so as to improve performance of the protected transistor. Used in cascode amplifier to beat Miller effect, in current source to beat Early effect.

clipping (illustrated in AoE §2.2.3D). Flattening of output waveform caused by hitting a limit on output swing. Example: single-supply follower will *clip* at ground and at the positive supply.

compliance (AoE §2.2.6D). Well defined in text: "The output voltage range over which a current source behaves well...."

complementary pair. A pair of *npn* and *pnp* transistors with similar characteristics. We use the 2N3904/2N3906 and the MJE3055/MJE2955 as complementary pairs in push-pull circuits.

Darlington pair (AoE §2.4.2). A connection between two BJTs such that the emitter of the first feeds the base of the second. The collectors are tied together to give a three terminal device that operates as a single transistor with a β that is the product of the two. However, the saturation voltage is increased by a diode drop and base-to-emitter voltage of the pair is the sum of the individual BJTs."

Early effect (See, for example, AoE §2.3.7A). Variation of I_C with V_{CE} at a given value of V_{BE} or I_C. Thus it describes transistors's departure from true current-source behavior.

emitter degeneration (AoE §2.3.4). Placing of resistor between emitter and ground (or other negative supply) in common-emitter amp. It is done so as to stabilize the circuit despite variation in temperature. (Source of term: gain is reduced or "degenerated." General circuit *performance* is much *improved,* however!)

impedance "looking" in a direction. Impedance at a point considering only the circuit elements lying in one direction or another. Example: at transistor's base impedance *looking back* one "sees" bias divider and R_{source}; *looking into base* one "sees" only $\beta \times R_E$.

Miller effect (AoE §2.4.5). Exaggeration of actual capacitance between output and input of an inverting amplifier, tending to make a small capacitance behave like a much larger capacitance to ground: (1+ Gain) times as large as actual C.

quiescent (-current, -voltage) (for example, see AoE §2.2.5). Condition prevailing when *no* input signal is applied. So, describes DC conditions in an amplifier designed to amplify AC signals. Example: $V_{\text{out-quiescent}}$ should be midway between V_{CC} and ground in a single-supply follower to allow maximum output amplitude (or "swing") without clipping.

split supplies (AoE §2.2.5B). Power supplies of both polarities, negative as well as positive. Used in contrast to "single supply."

transconductance (AoE §2.2.9). Well defined in AoE. Briefly, $g_m \equiv \Delta I_{\text{out}}/\Delta V_{\text{in}}$.

Wilson mirror (AoE §2.3.7). Improved form of current mirror in which a third transistor protects the sensitive output transistor against effects of variation in voltage across the load (third transistor in cascode connection, incidentally).

5W Worked Examples: Transistors II

5W.1 High-gain amplifiers

Problem How do high-gain amplifiers, see Fig. 5W.1, compare with respect to "linearity" or constancy of gain over the output swing? Explain your conclusion, briefly. *Assume* that each amplifier is fed by a properly-biased input.

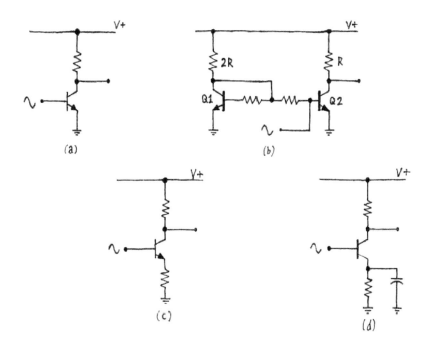

Figure 5W.1
High-gain amplifiers.

For case (b) we should provide some hints, because this circuit is very similar to a "current mirror," a circuit we didn't require you to learn. This circuit, mirror-like, will sink a current through Q1 of about $\frac{V+}{2R}$. That same current will flow in Q2, if we don't disturb things; Q2 is said to "mirror" the current passed by Q1, because the two V_{BE}'s are the same. This is the scheme that sets up the "biasing:" puts V_{out}, quiescent, at about 0.5 V+, as usual. Assume that the two transistors are well matched and live near each other on an integrated circuit. Enough said?

Solution All are the same except circuit (c). The others will show gain variation as the output voltage swings, because this output swing is caused by varying I_C, collector current, and the denominator of the gain equation for (a), (b), (d) is simply r_e, whose value varies with I_C. Circuit (c) differs because the constant value of R_E is added to the varying value of r_e. Circuit (c) gets its more constant gain at a price of course: the price is its diminished gain.

Another way to put the same point is to speak not – as we usually do – of r_e but instead of its inverse, g_m, the slope of the I_C versus V_{BE} curve. This slope, the transistor's *transconductance*, is proportional to I_C, and so is the circuit gain, when no R_E is present. Specifically, the circuit gain is $g_m \times R_C$. This formulation adds nothing new to our usual account which uses r_e, but perhaps makes the dependence on collector current more obvious.

Problem Which of the circuits in Fig 5W.1 are (is?) protected against temperature effects, and how? Explain the mechanism or mechanisms.

Solution Circuits (c) and (d) are temperature stable thanks to the emitter resistor, R_E, which provides the *negative feedback* described in Chapter 5N. Circuit (d) gives high gain along with stability (and also poor constancy of gain, as we have noted).

Circuit (a) is bias-unstable – and is, therefore a *bad* circuit.

Circuit (b) is a less familiar design, as the apologetic note in the problem statement suggests. This circuit is temperature-stable, but achieves it not by use of feedback but by taking advantage of the circuit's symmetry. This design is effective when the two transistors, Q1 and Q2 are well-matched and fabricated on one integrated circuit, as the problem statement declares they are.

Because the collector currents in the two transistors are equal, the two V_{BE}'s are equal. Imagine a change in temperature – let's imagine a rise of about 9°C. This drops the V_{BE} of Q1 by about 18 mV (see Chapter 5N: at constant I_C, $\Delta V_{BE}/\Delta \text{temp} = -2.1 \text{ mV}/°C$). If this were an unprotected circuit, like circuit (a), that change would reduce Q2's output to one half what it was: in other words the output quiescent voltage would be radically upset.

But because Q2 is heated just as Q1 is, Q2's current is unaffected by the heating of both transistors. Heating reduced the V_{BE} of Q1; but that reduced value evokes the *original* current from *heated* Q2. This stabilizing scheme might be called stabilization by *compensation* rather than by feedback: a change in a component, otherwise disturbing, is *compensated* for by a matching change in a second component. We will see this effect at work again in the next section.

5W.2 Differential amplifier

Problem Lab 5L begins with a diff-amp of modest gain (collector resistor is 10k, emitter resistors are 100 Ω, see Fig. 5W.2). Modify the circuit to maximize gain (without use of current sources). What is the modified circuit's differential gain? Common-mode gain?

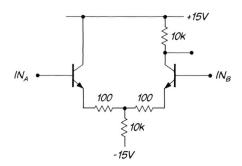

Figure 5W.2 Differential amplifier of Lab 5L.

Solution The change that maximizes gain (without using the fancy tricks of real op-amps – current mirror load, for example) is simply to remove the emitter resistors. Now the gain depends on r_e, whose quiescent value is about 33 Ω at the 0.75 mA collector current:

$$G_{\text{diff}} = R_C/2(R_E + r_e) = 10k/(2 \times 33\,\Omega) \approx 150$$

$$G_{\text{CM}} = -R_C/(R_E + r_e + 2R_{\text{tail}}) \approx -10k/20k = -1/2$$

Problem When you have maximized gain, does your circuit remain temperature stable? (Explain your answer.)

Solution Yes, as long as the two transistors run at the same temperatures, as they do when built on an integrated circuit, as in Lab 5L: the curves slide together, with temperature changes. If this were built with discrete transistors, it would not be temperature stable.

Problem (Tail of diff-amp) In Lab 5L, we suggest that you replace the 10k resistor in the tail with a current sink. Why? What is the argument that suggests a current sink will improve CMRR?

Solution CMRR is the ratio of good gain to bad gain, as you know: CMRR = $G_{\text{diff}}/G_{\text{CM}}$. The "tail" resistor influences common-mode gain but not differential, so boosting R_{tail} sounds like a good idea: the expression for common-mode gain (G_{CM}) shows $2R_{\text{tail}}$ in the denominator, whereas R_{tail} does not appear at all in the expression for G_{diff}. So a current source, with its large *dynamic resistance*, $\Delta V/\Delta I$, enlarges the effective R_{tail}, cutting G_{CM} while leaving differential gain untouched.

Problem Would increasing the value of R_{tail} by substituting a larger resistor – say, 100k – improve CMRR, relative to the circuit of Fig. 5W.2? Explain your answer.

Solution No, not at all. The larger *resistor*, unlike the current source, would cut the value of I_C by a factor of 10. To keep $V_{\text{out-quiescent}}$ properly placed (unchanged), we would need to boost the value of R_C by a factor of 10. Meanwhile, r_e would grow by the same factor, because of the reduction of I_C.

So, the sad result is that G_{CM} would be unchanged (still 1/2); G_{diff} would be unchanged: say 100k/(2×330)=150. CMRR, therefore, would be unchanged. We do indeed need a current source to improve CMRR. The answer, that CMRR is not improved, assumes emitter resistors removed, rather than included as in Fig. 5W.2.

5W.3 Op-amp innards: diff-amp within an IC operational amplifier

In Fig. 5W.3 is a much-simplified schematic of an op-amp, the '411 JFET-input device that you'll soon be using. We've included the gritty detail as well but we think you can answer by referring only to the left-hand, simplified figure. We will raise questions about each of the elements circled in Fig. 5W.3.

Observations

The *first stage* – differential input stage – uses JFETs rather than bipolars to achieve very low *bias current* (50 pA, typical, for the '411). The *second stage* – common-emitter amp, gain stage – uses bipolars because the bipolar's *exponential I–V* curve is steeper than the FET's *quadratic* curve (thus providing higher gain).[1]

Problem Why all those current sources? (The sources in this design are current mirrors, favorites of IC designers.) In particular, why are current sources, rather than resistors, used on *tail* and *output* of first stage, and why on the *collector* of the second stage?

[1] To be more precise, the bipolar shows higher "transconductance" – $\Delta I_{\text{out}}/\Delta V_{\text{in}}$.

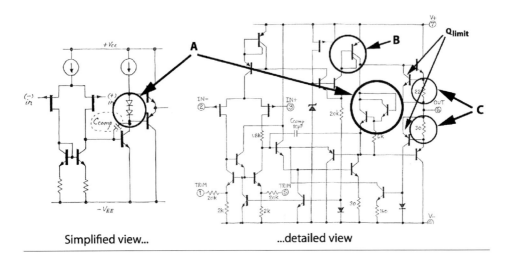

Figure 5W.3 Op-amp innards.

Solution

Why all those current sources? The current source on the tail of the input stage differential amplifier (the tail that's up in the air) is used to provide low common-mode gain, as usual (giving large CMRR: $G_{\text{diff}}/G_{\text{CM}}$). The common-mode gain when using bipolar transistors is, as you know, $G_{\text{CM}} = -R_C/(r_e + R_E + 2R_{\text{tail}})$: (for these FETs, one would change the labeling: $G_{\text{CM}} = -R_{\text{drain}}/(\frac{1}{g_m} + R_S + 2R_{\text{tail}})$). A large value for R_{tail} helps.

The current source makes R_{tail} huge because a good current source (as you well know) holds ΔI close to zero as voltage changes. In other words, a current source's $R_{\text{dynamic}} \equiv \Delta V/\Delta I = \Delta V/(\text{tiny})$ is huge. The huge effective R_{tail} makes common-mode gain tiny, and CMRR very large (100 dB, typical, for the LF411).

Current mirror on drains of diff-amp Qs:

Short answer: A short answer, based on the view of current mirror as just a current source (here, a "sink"), is just that the very large R_{dynamic} of the current source gives this differential stage high gain (the large R_{dynamic} replacing the usual "R_C" – here, an "R_{drain}," because these are field-effect transistors).

A longer answer: If you have looked into current mirrors a bit, then you know that their mission in life is to hold their two currents equal. (Strictly, to hold one current – I_{out} – equal to the other – I_{program}.)

The differential amplifier's two JFET transistors' mission, by contrast, is to make the two currents *unequal* in response to any difference between the two input voltages. When these two circuit fragments with their opposed goals meet, the result is very high voltage gain.

Suppose the quiescent current for each of the two transistor is 1mA (a number that is not realistic here: much larger than the '411's – but easy to work with). Now suppose that an input-voltage difference makes the currents *unequal*: say, the input or "programming" current becomes 0.9 mA (while the other Q's current grows to 1.1 mA since the total, set by I_{Qtail}, is held constant). The mirror will accept only that 0.9 mA, and the *difference* current, 0.2 mA, must squirt out to the next stage (to the common-emitter amplifier). The mirror has *doubled* the output current from the first stage, relative to the current that would have issued to the second stage, if the first stage had fed an ordinary current sink.

Or, to put this in terms we normally apply to voltage amplifiers, the impedance the mirror presents is not just large (as any current source's dynamic impedance is). It is *extra large* because just as the

5W.3 Op-amp innards: diff-amp within an IC operational amplifier

current applied to the R_{drain} grows (to 1.1 mA, in this case), the apparent impedance of this side of the mirror is boosted. The mirror not only tries to keep current constant (as any current source does); in this case it tries to shrink the current, further enhancing the ΔV response.

Probably the account in *current* terms is easier to follow than the one speaking of voltage gain.

AoE §2.3.7

Current source on collector of common-emitter stage ("B"): This current source, replacing the R_C that we are accustomed to putting on the collector of a common-emitter amplifier, again provides very high gain. If one did this with an ordinary amplifier (one that did not use overall feedback to make it well-behaved), this would be a "bad circuit." Used without feedback it would be useless because its output could only bang against one supply or the other. In an op-amp, using negative feedback as always, such extravagant gain is not a flaw but a virtue.

Problem What is the purpose of the elements labeled "A" in the two parts of Fig. 5W.3? (Note that the two parts are equivalent; if the detailed schematic is a bit much, consider the "simplified" diagram instead!)

Solution These two transistors drop two V_{BE}'s. You may want to think of them as two diodes in series, as shown in the simplified diagram on the left side of Fig. 5W.3. Since the top of this V_{BE}-stack drives the base of the upper transistor and the bottom of the stack drives the base of the lower transistor, the stack biases the bases apart, eliminating the dead zone that otherwise occurs close to 0 V in a push–pull. It is this dead zone that causes crossover distortion.

Soon you will see that feedback does a pretty good job of hiding the "dog" of crossover distortion (see Chapter 6N), you may wonder, "Why not let feedback solve this problem?" The answer is that feedback is not quick enough to do the job at high frequencies. In order to hide crossover distortion, the input to the push–pull must *slew* across about 1.2 V, up or down, and doing this takes time. During that slewing, a glitch will appear on the circuit output. Figure 5W.4 is a scope image of this effect in the home-made op-amp of Lab 5L. The circuit does hide most of the cross-over distortion, but a close look at the image – using the higher sweep rate of the right-hand image – shows that a glitch persists. Slewing past those two V_{BE}'s takes some time, and during that slewing, feedback, with its magic, fails.

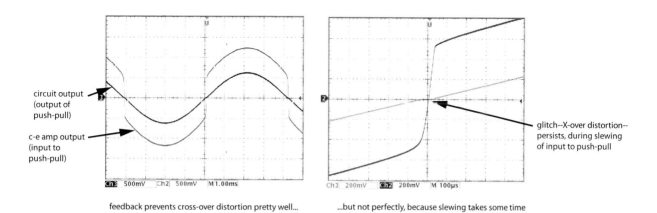

Figure 5W.4 Common-emitter stage slewing to hide crossover of output-stage push–pull: not perfect.

The time scale of Fig. 5W.4 is wrong for the '411, which slews much faster than our home-made op-amp of Lab 5L. But the problem for the '411 would be the same at some time scale, if the passive protections against crossover distortion had not been included.[2] The need for these passive protections reminds us of the limitations of feedback: it doesn't work well at high frequencies, when implemented with ordinary op-amps.

Problem Are the resistors that are labeled "C" essential? Explain.

Solution Yes they are. They prevent "thermal runaway" that otherwise would be very likely to destroy the push–pull transistors. The biasing-apart of the push–pull bases, described above, keeps both push–pull transistors just barely conducting, when input and output of the push–pull are at ground. This is the strategy that prevents a glitch when the circuit wakes up, in response to movement *away* from ground.

But that arrangement would be deadly without the emitter resistors. As a little current trickles through the push–pull transistors, it heats them slightly. But the heating increases their current, at a given V_{BE} (the V_{BE} imposed by the biasing network we met earlier). The increased current heats the transistors further. And so on. That's a positive-feedback loop, "thermal runaway." The emitter resistors provide the usual *negative* feedback to prevent runaway: if current grows, the emitter voltages move apart somewhat, reducing V_{BE}. You have seen this before, in the argument for including an emitter resistor in a high-gain common-emitter amplifier.

Those R_E's do another job in this circuit. They are part of a simple trick to limit the op-amp's output current to prevent damage by an overload. The scheme is so nifty that one can safely short a '411 output to its positive or negative supply.

Any current that flows in or out of the op-amp's output passes through one or the other of those resistors, developing a voltage drop. That drop is applied to one or other of the transistors labeled Q_{limit}, as its V_{BE}. When output current is sufficient to develop 0.5–0.6 V across that resistor, the Q_{limit} transistor begins to steal current from the base of the push–pull transistor. This effect – combined with the limited current available from the common-emitter stage – puts a safe ceiling on the total current available from the op-amp output. (This is a trick you will see repeated in the voltage regulators of Chapter 11N and Lab 11L.)

---

[2] Not all op-amps do include this passive protection against crossover distortion. The LM358, for example – the single-supply op-amp that we use in Labs 7L and 10L – does not include it.

Part III

Analog: Operational Amplifiers and their Applications

6N Op-Amps I

Contents

6N.1	Overview of feedback	263
6N.2	Preliminary: negative feedback as a general notion	266
6N.3	Feedback in electronics	267
6N.4	The op-amp Golden Rules	269
6N.5	Applications	270
	6N.5.1 A follower	270
	6N.5.2 The effect of feedback on the follower's R_{out}	270
6N.6	Two amplifiers	271
	6N.6.1 Non-inverting amplifier	271
	6N.6.2 A skeptic's challenge to the Golden Rules	271
	6N.6.3 Some characteristics of the non-inverting amp	272
6N.7	Inverting amplifier	272
	6N.7.1 What virtual ground implies	273
6N.8	When do the Golden Rules apply?	274
	6N.8.1 More applications: improved versions of earlier circuits	275
6N.9	Strange things can be put into feedback loop	277
	6N.9.1 In general...	277
6N.10	AoE reading	279

Why?

What problem do today's circuits address? The very general task of *improving* performance, through the application of *negative feedback*, of a great many of the circuits we have met to this point.

6N.1 Overview of feedback

Fig. 6N.1 shows a newspaper page on which Harold Black wrote one of his newly-conceived formalizations of feedback (this newspaper records not quite the basic notion, which he had sketched on a copy of the *New York Times* four days earlier, but rather an application of feedback in order to match impedances). He wrote this as he rode the ferry from Staten Island to work, one summer morning in 1927. (Was it chance, or caginess that led him to write it on a dated sheet?)[1]

We have been promising you the pleasures of feedback for some time. You probably know about the concept even if you haven't yet used it much in electronics. Now, at last, here it is.

---

[1] (Copyright 1977 IEEE. Reprinted, with permission, from Harold S. Black, "Inventing the Negative Feedback Amplifier," *IEEE Spectrum*, Dec. 1977). You may protest that the dated newspaper page could hardly prove the date of his invention. Couldn't he have dug out an old newspaper page to backdate his invention? No, because he was prudent enough to have each newspaper sketch witnessed and signed, as soon as he arrived at work.

Op-Amps I

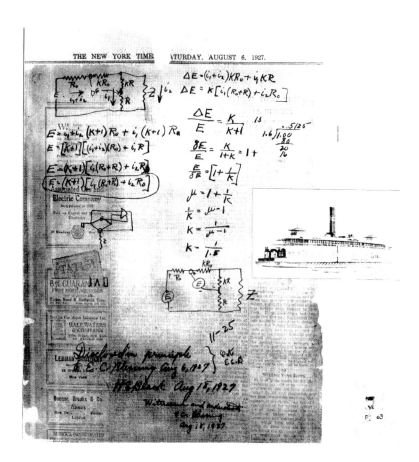

Figure 6N.1 Harold Black's notes on the feedback amplifier.

Feedback is going to become more than just an item in your bag of tricks; it will be a central concept that you find yourself applying repeatedly, and in a variety of contexts, some far from operational amplifiers. Already, you have seen feedback in odd corners of transistor circuits; you will see it constantly in the remaining analog labs; then you will see it again in a digital setting, when you build an analog-to-digital converter in Lab 20L, and then a phase-locked-loop in the same lab. It is a powerful idea.

Our treatment of negative feedback circuits begins, as did our treatment of bipolar transistors, with a simple, idealized view of the new circuits. At the conclusion of Lab 5L you built a simple "operational amplifier." From now on, we will rely on the integrated version of these little high-gain differential amplifiers. They make it easy to build good feedback circuits.

As this section of the course continues, we soon feel obliged once again to disillusion you – to tell you about the ways that op-amps are imperfect. But this is a minor theme. The major theme remains that op-amps and feedback are wonderful. We keep trying additional applications for feedback, and we never lose our affection for these circuits. They work magically well.

Lab 8L, introduces the novelty of *positive* feedback: feedback of the sort that makes a circuit unstable – or makes its output slam to one power-supply limit or the other. Sometimes that is useful, and sometimes it is a nuisance. In Lab 8L we look at the benign effects of positive feedback, including its use in *oscillators*. Lab 9L opens with one more fruitful use of such feedback, in an *active filter*.

The remainder of Lab 9L is devoted to the more alarming effects of positive feedback: *parasitic* oscillations that occur when unwanted positive feedback sneaks up on you, destabilizing a circuit. We

6N.1 Overview of feedback

will learn some protections against such mischief. Knowing how to apply such defenses may make you a hero when your group is plagued by mysterious fuzz on its scope traces.

Figure 6N.2 Righteous and deserving student, about to be rewarded for his travails with discrete transistors.

We then devote three labs to several particular applications of negative feedback:

- A motor-control loop using a technique called "PID" (describing the three modified error signals that are used in the feedback loop: "Proportional," "Derivative," and "Integral"). This is Lab 10L, the most complex circuit so far – and useful to let us see applications for many circuits that we have demonstrated as circuit fragments but not yet applied: the PID makes good use not only of differentiator and integrator but also depends upon a summing circuit, a diff-amp made with an op-amp, and a high-current motor-driver. We also look at the lock-in amplifier, a complex application used to detect a weak signal in the presence of overwhelming noise.
- Voltage regulators: these are specialized circuits designed for the narrow but important purpose of providing stable power supplies. We look at both linear and switching designs in Lab 11L.
- A group project, putting together a half dozen circuit fragments, some passive, most using feedback circuits. In this project, you students will design and combine the several circuit fragments to make a system that can transmit audio using frequency-modulated (FM) infrared light. This is Lab 13L (the number hops by one, to leave room for a final discrete-transistor lab, Lab 12L on MOSFETs, where feedback is not prominent).

With Lab 13L we conclude the analog half of the course, and in the very next lab you will find the rules of the game radically changed as you begin to build *digital* circuits. But we will save that story till later.

Op-Amps I

A piece of advice (unsolicited): How to get the greatest satisfaction out of the feedback circuits you are about to meet

Here are two thoughts that may help you to enjoy these circuits – first:

- As you work with an op-amp circuit, recall the equivalent circuit made without feedback, and the difficulties it presented: for example, the transistor follower, or the transistor current sources. The op-amp versions in general will work better, to an extent that should astonish you.

Having labored through work with discrete transistors you have paid your dues. You are entitled to enjoy the ease of working with op-amps and feedback.

Figure 6N.2 shows you climbing – as you are about to do – out of the dark valleys through which you have toiled, up into that sunny region above the clouds where circuit performance comes close to the ideal. You reach the sunny alpine meadows where feedback blooms. Pat yourself on the back, and have fun with op-amps.

A second thought:

- Recall that negative feedback in electronics was not always used; was not always obvious – as Harold Black was able to persuade the patent office (Black comes as close as anyone to being the inventor of electronic feedback).

The faded and scribbled-on newspaper that is shown at the start of this chapter is meant to remind you of this second point – meant to help us feel some of the surprise and pleasure that the inventor must have felt as he jotted sketches and a few equations on his morning newspaper. A facsimile of this newspaper, recording the second of Black's basic inventions in the field, appeared in an article Black wrote years later to describe the way he came to conceive his invention. Next time you invent something of comparable value, don't forget to jot notes on a newspaper, preferably in a picturesque setting – and then keep the paper till you get a chance to write your memoirs (and get a witness to confirm the date).

6N.2 Preliminary: negative feedback as a general notion

This is the deepest, most powerful notion in this course. It is so useful that the phrase, at least, has passed into ordinary usage – and there it has been blurred. Let's start with some examples of such general use – one genuine cartoon (in the sense that it was not cooked up to illustrate our point), and three cartoons that we did cook up. Ask yourself whether you see feedback at work in the sense relevant to electronics, and if you see feedback, is its sense positive or negative?

Now comes an example of feedback misunderstood. Outside electronics, the word "negative" has *negative* connotations: so, "negative feedback" sounds nasty, "positive feedback" sounds benign. A news story in the New York Times (October 8, 2008) concerning an economic slowdown began thus:

The technical term for it is "negative feedback loop." The rest of us just call it panic.

How else to explain yet another plunge in the stock market on Tuesday that sent the Standard & Poor's 500-stock index to its lowest level in five years – particularly in the absence of another nasty surprise.

What this reporter describes is, of course, *positive* feedback. But since the news is bad – "panic" – he can't resist his inclination to call it *negative*.

Sometimes a signal that usually would amount to negative feedback turns out to be positive (we will return to this topic, one that causes circuit instability, in Lab 9L). Is the observer applying positive or negative feedback to the two fellows on the left, in Fig. 6N.4? To decide, we need to judge whether his signal tends to increase or decrease the behavior that he senses.

6N.3 Feedback in electronics

Figure 6N.3 Feedback: same sense as in electronics? Copyright 1985 Mark Stivers, first published in *Suttertown News*, reproduced with permission.

Figure 6N.4 Freaks? Will he hurt their feelings?

The case in Fig. 6N.5 comes closer to fitting the electronic sense of *negative feedback*. In op-amp terms (not Hollywood's), who's playing what role?

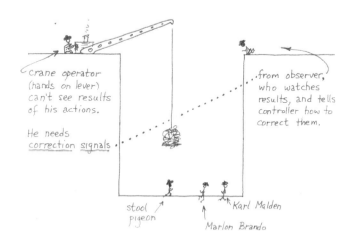

Figure 6N.5 Negative feedback: a case pretty much like op-amp feedback.

6N.3 Feedback in electronics

In conversation, people usually talk (as that reporter did above) as if "positive feedback" is nice, "negative feedback" is disagreeable. In electronics the truth is usually just the opposite. Generally speaking, negative feedback in electronics is wonderful stuff; positive feedback is nasty. Nevertheless the phrase means in electronics fundamentally what it should be used to mean in everyday speech.

AoE §2.5.1

Harold Black was the first to formalize the effects of negative feedback in electronic circuits. Here is the language of his patent explaining his idea:[2]

```
       Applicant has discovered how to use larger
       amounts of 'negative feedback than were contem-
35     plated by prior art workers with a new and im-
       portant kind of improvement in tube operation.
       One improvement is in lowered distortion arising
       in the amplifier. Another improvement is
       greater consistency of operation, in particular a
40     more nearly constant gain despite variable fac-
       tors such as ordinarily would influence the gain.
       Various other operating characteristics of the
       circuit are likewise rendered more nearly con-
       stant. Applicant has discovered that these im-
45     provements are attained in proportion to the
       sacrifice that is made in amplifier gain, and that
       by constructing a circuit with excess gain and re-
       ducing the gain by negative feedback, any de-
       sired degree of linearity between output and
60     input and any desired degree of constancy or
       stability of operating characteristics can be real-
       ized, the limiting factor being in the amount of
       gain that can be attained, rather than any limi-
       tation in the method of improvement provided
55     by the Invention.
```

He summarized his idea in an article that he published many years afterward:

... by building an amplifier whose gain is made deliberately, say 40 decibels higher than necessary (10,000-fold excess on energy basis) and then feeding the output back to the input in such a way as to throw away the excess gain, it has been found possible to effect extraordinary improvement in constancy of amplification and freedom from nonlinearity.[3]

One proof of the originality of Black's idea was the fact that the British Patent Office rejected his application, seeing Black as a dull fellow who had not understood that an amplifier should *amplify*, so that "throwing away... gain" showed the inventor's confusion. Black had the last laugh.

AoE §2.5.1

Lest we oversell Black's invention, let's acknowledge that Black did not invent or discover feedback. Our bodies are replete with homeostatic systems, like the one that holds our body temperature close to 99°F, whether we are shivering in the snow or sweating on a beach (with shivering and sweating part of the stabilizing mechanism). And Newcomen used a speed "governour" on his mine-pumping steam pumps: the spin rate raised or lowered the weights, closing down or opening the steam valve appropriately, so as to hold the speed nearly constant despite variations in load.

Open-loop *versus* feedback *circuits*: Nearly *all* our circuits, so far, have operated open-loop – with some exceptions noted below. You may have gotten used to designing amplifiers to run open-loop (we will cure you of that); you would not consider driving a car open loop (we hope), and you probably know that it is almost impossible even to *speak* intelligibly *open-loop*.

[2] Harold S. Black, "Wave Translation System," US Patent 2102671 (1937).
[3] *IEEE Spectrum*, Dec. 1977

6N.4 The op-amp Golden Rules

Examples of feedback without op-amps

We know that feedback is not new to you, not only because you may have a pretty good idea of the notion from ordinary usage, but also because you have seen feedback at work in parts of some transistor circuits: see Fig. 6N.6.

AoE §2.3.4B

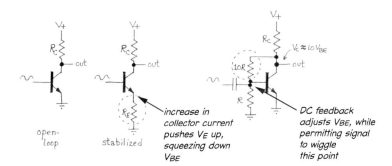

Figure 6N.6 Some examples of feedback in circuits we have built without op-amps.

Feedback with op-amps

AoE §4.1.1

Op-amp circuits make the feedback evident, and use a lot of it, so that they perform better than our improvised feedback fragments. The name "operational amplifier" derives from the assumption, early in their use, that they would be used primarily to do a sort of analog computing.[4]

Op-amps have enormous gain (that is, their *open-loop* gain is enormous: the chip itself, used without feedback, would show huge gain: $\approx 200,000$ at DC, for the LF411, the chip you will use in most of our labs). As Black proposed, op-amp *circuits* deliberately throw away most of that gain, in order to improve circuit performance.

6N.4 The op-amp Golden Rules

Just as we began our treatment of bipolar transistors with a simple model of device behavior, and that model remained sufficient to let us analyze and design many circuits, so in this chapter we start with a simple, idealized view of the op-amp, and usually we will *continue* to use this view even when we meet a more refined model.

AoE §4.1.3

The *golden rules* (below) are approximations, but good ones.

(GR#1) The output tries to do whatever is necessary to make the voltage difference between the two inputs zero.
(GR#2) The inputs draw no current.

Three observations, before we start applying these rules:

- Rule (2): we're confident that you understand that the "inputs" that draw no current are the signal inputs, labeled "+" and "−", not the op-amp's power supply terminals(!)[5]

---

[4] They seem to have been named in a paper of 1947, which envisioned their important uses as follows: "The term 'operational amplifier' is a generic term applied to amplifiers whose gain functions are such as to enable them to perform certain useful operations such as summation, integration, differentiation, or a combination of such operations." Ragazzini, Randall, and Russell, "Analysis of Problems in Dynamics by Electronic Circuits," *Proceedings of the IRE*, **35**, May 1947, pp. 444–452, quoted in *op-amp Applications Handbook*, Walt Jung, editor emeritus, (Analog Devices Series, 2006), p. 779.

[5] You may think this is obvious, but every now and then some students, impressed by all our praise of op-amps and their magic, fail to connect the op-amps power supply pins. The students wait expectantly for truly miraculous circuit performance – and are disappointed.

270 Op-Amps I

- Rule (1): the word "tries" is important. It remind us that it's up to us, the circuit designers, to make sure that the op-amp *can* hold its two inputs at equal voltages. If we blunder – say, by overdriving a circuit – we can make it impossible for the op-amp to do what it "tries" to do.
- More generally: these rules apply only to op-amp circuits that use *negative feedback*.

These simple rules will let you analyze a heap of clever circuits.

6N.5 Applications

6N.5.1 A follower

AoE §4.2.3

Figure 6N.7 shows about the simplest circuit one can make with an op-amp. It may not seem very exciting – but it is the *best* follower you have seen.

The Golden Rules let you estimate...

- ... R_{in} (try GR#2).
- ... V_{offset} (try GR#1); and compare this to the DC-offset we are accustomed to in a bipolar transistor follower.
- ... R_{out}? ....

Figure 6N.7
Follower.

6N.5.2 The effect of feedback on the follower's R_{out}

AoE §2.5.3C

R_{out} raises a much subtler question. The Golden Rules don't answer it, but you can work your way through an argument that will show that R_{out} is very low. The novel point, here, is that it is low not because the IC itself has low R_{out}; this value, rarely specified, is moderately low – perhaps around 100Ω. What's new is that *feedback* accounts for the very low R_{out} of this circuit – and of nearly all other op-amp circuits.[6]

Let's redraw the op-amp to resemble a Thevenin model: perfect voltage source in series with a series resistance that we might call "r_{out}." Our drawing will include the op-amp's high *open-loop* gain, A. Then we can try the thought experiment of tugging the output by a ΔV, and seeing how the output responds: see Fig 6N.8.

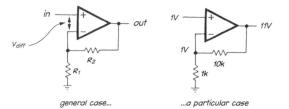

Figure 6N.8 Feedback lowers op-amp circuit's R_{out}.

If you try to change V_{out}, the circuit hears about it through the feedback wiring. It is *outraged*. Its high-gain amplifier squirts a large current at you, fighting your attempt to push the output away from where the op-amp wants it to stand. We'll leave the argument at this qualitative level, for now. We'll settle for calling the output impedance "very low." Later (mostly in Chapter 9N), we will be able to quantify the effects of feedback to arrive at a *value* for R_{out}.

---

[6] The one exception to this rule is an op-amp *current source*, whose output impedance should be *high* (the ideal current source, as you know, would show R_{out} infinite).

6N.6 Two amplifiers

6N.6.1 Non-inverting amplifier

AoE §4.2.2

This circuit is a bit more exciting than the follower: it gives us voltage gain.[7]

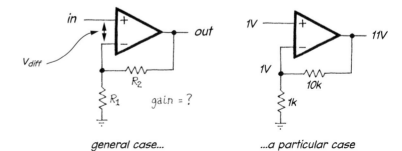

Figure 6N.9 Feedback lowers op-amp circuit's R_{out}.

Evidently, to satisfy the Golden Rules the op-amp has to take its output to 11 V, if we apply 1 V to the input. Generalizing, you may be able to persuade yourself that the gain is the inverse of the fraction fed back. So 1/11 is fed back, and gain is 11. The gain usually is recited more simply as

$$G = 1 + R_2/R_1$$

Students are inclined to forget this pesky "1 +." We hope you won't forget it. Sometimes, of course, it is sensible to "forget it" in the sense that you neglect it: when the R_2/R_1 fraction is very large. But throw out the "+1" only after you have determined that its contribution is negligible.

6N.6.2 A skeptic's challenge to the Golden Rules

Suppose that a skeptical student wants to take advantage of his awareness that the op-amp is a differential amplifier (and a circuit familiar from the second transistor lab); such a student might present a plausible challenge to the Golden Rules. This skeptic doesn't fall for the notion that the op-amp is a weird little triangle that relies on pure magic.

This person might object that our Golden Rules' analysis must be wrong, because the analysis is not consistent with our understanding of the differential amp that is the core of any op-amp. Our Golden Rules predict $V_{out} = 11$. But if GR#1 says that the op-amp has driven the input voltages to equality, then – Aha! We have caught a self-contradiction in the analysis! Equal voltages at the input of a diff amp cannot produce 11 V out; the Golden Rule must be wrong!

But, no; you won't fall for this argument. The way to resolve the seeming contradiction is only to acknowledge that the Golden Rules are approximate. The op-amp does not drive the inputs to equality, but to near-equality. How close? Just close enough so that your differential-amplifier view can work. In order to drive V_{out} to 11, the difference between the inputs – V_{diff}, in Fig. 6N.9 – must be not 0

---

[7] Incidentally, it was this sort of circuit that interested Black. He was concerned about making "repeater" amplifiers for the telephone company, in order to transmit voice signals over long distances. It was this need to send a signal through a great many stages – hundreds, in order to cross the country – that made the phone company more concerned than anyone else with the problem of distortion in amplifiers. Distortion that is minor in one or two stages can be disastrous when superadded hundreds of times.

It's a curious historical fact that the telephone company, which we may be inclined to view as a stodgy old utility, repeatedly stood at the forefront of innovation: not only Black's invention, but also bipolar transistors, field-effect transistors, and fiber-optic signal transmission were developed largely at "the phone company," if one includes within this company the famous Bell Telephone Laboratories ("Bell Labs").

but V_{out}/G_{OL}, where G_{OL} is the op-amp's open-loop gain (conventionally denoted "A"). For a '411 op-amp at DC, that indicates $V_{diff} \approx \frac{11\,V}{200,000} = 50\,\mu V$. Not zero, but pretty close.

Doing that calculation can help one to appreciate why it is useful to have all that excess gain to throw away (as Black put it): the Golden Rule approximations come closer to the truth as open-loop gain rises. $G_{OL} = 1,000,000$, for example (a value available on many premium op-amps) would reduce V_{diff}, in the example above, to just $10\,\mu V$.

6N.6.3 Some characteristics of the non-inverting amp

A couple of other characteristics are worth noting – perhaps they seem obvious to you:

- R_{in} is enormous, as for the follower.
- R_{out} is tiny, as for the follower.[8]

Incidentally, you might note that the *follower* can be described as a special case of the non-inverting amplifier. What is its peculiar R_1 value?[9]

6N.7 Inverting amplifier

AoE §4.2.1

The circuit in Fig. 6N.10 also amplifies well. It differs from the non-inverting amplifier in one respect that is obvious – it *inverts* the signal – and in some respects that are not so obvious.

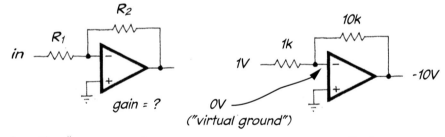

Figure 6N.10 Inverting amplifier.

inverting amp: general case...   *...a particular case*

GR#1 tells us that the inverting terminal voltage should be zero. From this observation, we can get at the circuit's properties

Gain: Let's look at the particular case: 1 V causes 1 mA to flow toward the inverting terminal. Where does this current go? Not into the op-amp (that is forbidden by GR#2). So, it must go around the corner, through the 10k feedback resistor. Doing that, it drops 10V. So, $V_{out} = -10V$.

Generalizing this result, you'll be pleased to see no nasty "$1 + \cdots$" in the gain expression: just $G = -R_2/R_1$. The *minus* sign means – as in the common-emitter amp – only that the circuit inverts: that output and input are 180° out of phase.

If you're not too proud to use a primitive way of looking at the circuit – a way that's based on its appearance, along with the fixed *virtual ground* at the inverting terminal – try thinking of the circuit

---

[8] We'll see later, when we look at the effects of feedback *quantitatively*, that the amplifier trades away a very little of its impedance virtues in exchange for gain. Still, this correction to the claim that amplifier and follower share good input and output impedances is only a minor change. The dependence of circuit impedances upon B, the fraction of output that is fed back, is treated in AoE §§2.5.3B and 2.5.3C. In this book we return to the topic in Chapter 9.

[9] For the follower, R_1 is infinite; so, gain goes to 1.

as a child's see-saw. It pivots on virtual ground. If the two arms are of equal length, it is balanced with equal voltages: its gain is −1. If one arm is 10 times longer, as in Fig. 6N.10, the long arm swings 10 times as far as the short arm. And so on.

Impedances, in and out: Now, let's consider the input impedance of the inverting amps of Fig. 6N.10. From the observation that the inverting input is fixed at 0 V also flow some of the circuit's peculiarities:

- R_{in}: let's see if we can get you to fall for either of two *wrong answers*:
  - Wrong answer #1: "R_{in} is huge, because input goes to an op-amp terminal, and GR#2 says the inputs draw no current."
  - Wrong answer #2: "Oh, no – what I forgot is that it's the output of the circuit that is at low impedance (we saw that back in §6N.6.1). So $R_{in} = R_1 + R_2$."

  What's wrong with these answers? To give you a chance to see through our falsehoods on your own, we'll hide our explanations in a footnote.[10]
- The inverting amp's inverting terminal (the one marked "−") often is called "virtual ground." (Why "virtual"?[11]) This point, often called by the suggestive name, "summing junction," turns out to be useful in several important circuits, and helps to characterize the circuit.

6N.7.1 What virtual ground implies

R_{in}: First, recognizing "virtual ground" – the point locked by feedback at 0 V – makes the circuit's R_{in} very simple: it is just R_1. So, compared to the non-inverting amp, the inverting amp's input resistance is only mediocre.

"Mediocre?," you may protest. "I can make it as large as I like, just by specifying R_1. I'll make mine 10M or 100M, if I'm in the mood." Well, no. Your defiant answer makes sense on this first day with op-amps, when we are pretending that they are ideal. But as soon as we come down to Earth (splashdown is scheduled for the next chapter), we're obliged to admit that things begin to go awry when the R values become very large. So, if we treat approximately 10M as a practical limit on R values, we must concede that the inverting amp cannot match the non-inverting amp's R_{in}.

What happens to the circuit's R_{in} if you omit R_1 entirely, replacing it with a wire? The resulting $R_{in} \approx 0$, alarming though it looks, is exactly what pleases one class of signal sources: it pleases a *current source*. In Lab 6L you will exploit this characteristic when you convert *photodiode* current to a voltage.

AoE §4.3.1D

Summing circuit

The fact that virtual ground is fixed allows one to inject several *currents* into that point without causing any interaction among the signals; so, the several signals sum cleanly. The left-hand circuit in Fig. 6N.11 shows a *passive* summing circuit. It is inferior, because the contribution of any one input slightly alters the effect of any other input.

In the lab you will build a variation on this op-amp circuit: a potentiometer lets you vary the *DC offset* of the op-amp output. The circuits shown in Fig. 6N.11 form a *binary-weighted* sum.[12] In principle, this is one way to build a digital-to-analog converter (DAC), though this method is used only rarely.

---

[10] The first wrong answer confounds two distinct questions: *circuit-* and *op-amp*-input impedances. These are not the same, here. It's true that a Golden Rule says the inverting input will not pass current; but no rule says that the *circuit* should not draw current, through R_1, a current that can pass *around* the op-amp, through R_2 to the op-amp output terminal. That is, of course, how the input current does flow. The second wrong answer neglects the fact that the voltage at the inverting terminal is locked at 0 V (by feedback), so that R_2 is quite invisible from the circuit input.

[11] Because it doesn't behave like ordinary (*real*) ground: current flowing to virtual ground bounces; it doesn't disappear, but runs off somewhere else (around the op-amp, using the op-amp's feedback path), where it remains available for measurement as it flows around the op-amp, using the op-amp's feedback path.

[12] Binary-weighted means that one can think of the three inputs as components ("bits") of a binary value, with each given double the weight of its "less significant" neighbor (you may find this notion easier to remember if you recall the weights of

274   Op-Amps I

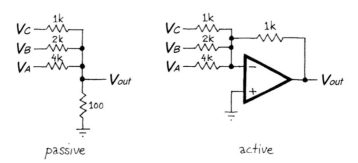

Figure 6N.11 Summing circuits.   *passive*   *active*

One more summing circuit: Does a summing circuit have to use the *inverting* configuration? No, but the non-inverting form is not so simple, and is difficult to extend beyond two inputs. Does the circuit in Fig. 6N.12 form the sum of V_A and V_B?

This is *almost* a summing circuit. The output is the *average*, rather than sum. An op-amp circuit could easily convert it to a sum (what gain is required?[13]). But unlike the inverting summer, this one doesn't easily extend to three, four and more inputs. So, the usual summing circuit uses the inverting configuration.

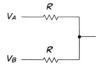

Figure 6N.12 Summing circuit?

Integrator, differentiator, and others

Virtual ground allows one to make a nearly-ideal integrator, as we'll see later. Again, the key is that the point where R meets C, a point that must be kept *close to ground* in the passive version, sits anchored *at ground*[14] in the op-amp version, see Fig. 6N.13.

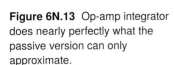

Figure 6N.13 Op-amp integrator does nearly perfectly what the passive version can only approximate.

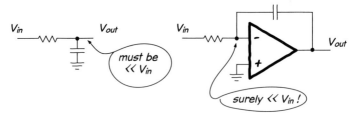

The same sort of improvement is available for the op-amp differentiator, though there it is less impressive, compromised by a subtlety that we are happy to postpone: concern for circuit *stability*. We will meet differentiator and integrator in Chapter 7.

6N.8   When do the Golden Rules apply?

AoE §4.2.7

Now that we have applied the Golden Rules a couple of times, we are ready to understand that they sometimes do not apply.

Try some cases: do the rules apply at all?

The question may seem to you silly: Fig. 6N.14(a) is pretty clearly not a golden-rule circuit; it's just a high-gain amplifier – either a mistake or an anticipation of the circuit we will meet as a "comparator"

Homer Simpson and his less significant neighbor, Ned Flanders). So, the output voltage in the right-hand circuit of Fig. 6N.11 is $V_{\text{out}} = -\left(\frac{V_C}{1} + \frac{V_B}{2} + \frac{V_A}{4}\right)$.

[13] That's right: 2.
[14] Well, OK: *very close* to ground; we'll see, next time, that some departure from the ideal is inevitable.

in Lab 8L. It uses *no feedback*. Figure 6N.14(b) works; Fig. 6N.14(c) does not, and the reason is that it uses the wrong flavor of feedback: positive.

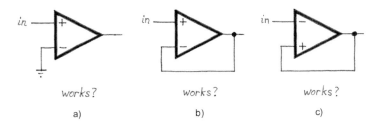

Figure 6N.14 Do the Golden Rules apply to these circuits?

So, generalizing a bit from these cases, one can see that the Golden Rules do not automatically apply to all op-amp circuits. Instead, they apply to op-amp circuits that...

- use feedback, and
- use feedback that is *negative*, and
- stay in the active region (don't let the op-amp output hit a limit, "saturating").

...Subtler cases: Golden Rules apply *for some inputs*

Will the output's *"attempt..."* to hold the voltages at its two inputs equal succeed in the two cases in Fig. 6N.15? No. But for the other polarity of input fed to each cicuit the answer would be "Yes." We will return to such conditional-feedback circuits next time.

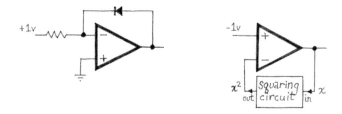

Figure 6N.15 Will the output's attempt succeed here?

6N.8.1 More applications: improved versions of earlier circuits

Nearly all the op-amp circuits that you meet will do what some earlier (open-loop) circuit did – but they will do it better. This is true of all the op-amp circuits you will see today in the lab. Let's consider a few of these: *current source, summing circuit, follower, and current-to-voltage converter*.

Current source: The left-hand circuit of Fig. 6N.16 is straightforward – but not usually satisfactory. For one thing, the *load* hangs in limbo, neither end tied to either ground or a power supply. For another, the entire load current must come from the op-amp itself – so, it is limited to about ±25 mA, for an ordinary op-amp.

The right-hand circuit is better. It gives you a chance to marvel at the op-amp's ability to make a device that's brought within the feedback loop behave as if it were *perfect*. Here, the op-amp will hide V_{BE} variations, with temperature and current, and also the slope of the I_C versus V_{CE} curve (a slope that reveals that the transistor is not a perfect current source; this imperfection is described by "Early effect," a topic discussed in §5S.2).

Do you begin to see *how* the op-amp can do this magic? It takes some time to get used to these wonders. At first it seems too good to be true.

276 Op-Amps I

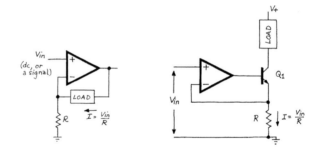

Figure 6N.16 Op-amp current sources.

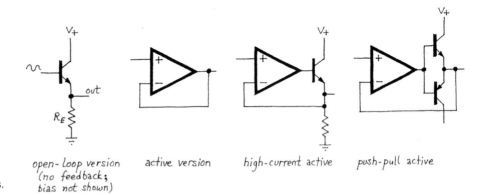

Figure 6N.17 Op-amp followers.

AoE §4.3.1E

Followers: Figure 6N.17 shows three op-amp-assisted followers, alongside a simple bipolar follower.
How are the op-amp versions better than the bare-transistor version? The obvious difference is that all the op-amp circuits hide the annoying 0.6 V diode drop. A subtler difference – not obvious, by any means, is the much better output impedance of the op-amp circuits. How about input impedance?[15]

Current-to-voltage converter: Figure 6N.18 shows two applications for an op-amp "transresistance amplifier." *A Puzzle*: if you and I can design an "ideal" current meter so easily, why do our lab multimeters **not** work that way? Are we that much smarter than everyone else?[16]

AoE §4.3.1C

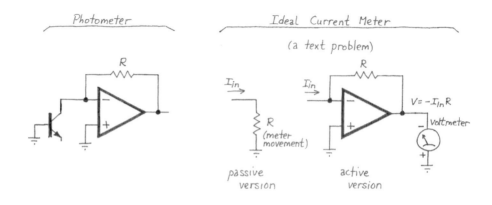

Figure 6N.18 Two applications for *I*-to-*V* converter: photometer; "ideal" current meter.

---

[15] Yes, the op-amp circuits win, hands down. Recall GR#2.
[16] Here's a thought or two on that question: as in the first current source (Fig. 6N.16), nearly the entire input current passes through the op-amp, so maximum current is modest, and the batteries powering the circuit will not last long.

6N.9 Strange things can be put into feedback loop

The push–pull follower within the feedback loop, shown in Figs. 6N.17 and 6N.20, begins to illustrate how neatly the op-amp can take care of and hide the eccentricities of circuit elements – like bipolar followers, or diodes.

6N.9.1 In general...

Figure 6N.19 illustrates the cheerful scheme, showing two distinct cases that fit all the op-amp circuits we will meet.

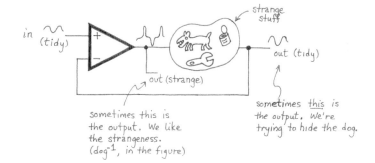

Figure 6N.19 Op-amps can tidy up after strange stuff within the loop.

Hiding the "dog"...

Sometimes, instead of treating the op-amp output as circuit output – and thus showing the inverse of the signal fed back, as we do in our op-amp voltage amplifiers for example – we use the op-amp to "hide the dog." The push–pull follower is such a case. The weirdness of the transfer function of the bare push–pull is not interesting or useful; it is a defect that we are happy to hide. Active rectifiers do a similar trick to hide diode drops.

To see the op-amp dutifully generating, at the input to the push–pull, whatever strange waveform is needed in order to hide the push–pull is quite dazzling the first time one sees it. See if you find it so. Figs. 6N.21 and 6N.22 show some scope images showing the process.

Crossover distortion in two forms...: We rigged a demonstration of the op-amp's cleverness (see Fig. 6N.20) by switching the feedback path between two points. The "silly" point, the op-amp output, produces a clean sinusoid in the wrong place while delivering nasty crossover distortion at the circuit output. The "smart" feedback point, the circuit's output, delivers a clean sinusoid where we want it.

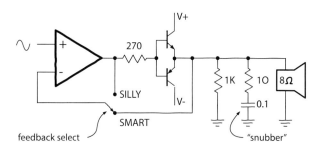

Figure 6N.20 Details of circuit set up to show op-amp cleaning up crossover distortion.

First, here is cross-over distortion, shown in two forms. The left-hand image of Fig. 6N.21 shows what this distortion looks like when the load is resistive. This might be called "classic" crossover distortion. The image on the right shows what the distortion looks like when the load is not a resistor but an 8 Ω speaker, whose inductance and self-resonances produce stranger shapes.

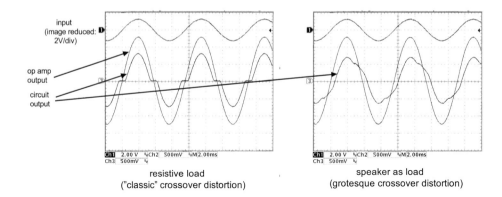

Figure 6N.21 Crossover distortion in two forms: resistive load, and speaker as load (stranger).

...Op-amp to the rescue: crossover fixed: Could the op-amp be clever enough to undo the strange distortion introduced by the push–pull – and rendered even stranger (asymmetric, too) by the speaker as load? Surely not!

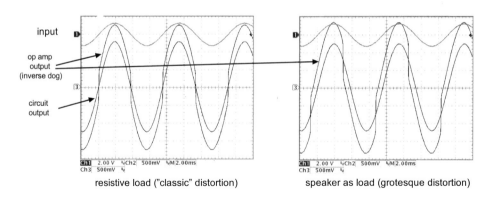

Figure 6N.22 Crossover distortion fixed by op-amp – which cleverly generates the inverse of the crossover transfer function.

Well, yes: the op-amp is that clever (at moderate frequencies): it generates just what's needed (dog^{-1}) to produce a clean output. Figure 6N.22 shows the cleanup that the op-amp provides. (Note that the input waveform is shown with scope gain reduced 4× relative to the output waveforms.)

AoE §4.3.1E

...Admiring the "dog": In other cases, as the comments in Fig. 6N.19 say, we want – and treat as circuit output – the *op-amp output*. In such cases, the "strange stuff" in the feedback loop is something we have planted in order to tease the op-amp into generating the inverse of that strange stuff ("inverse dog"[17]).

The "strange stuff" need not be strange, of course; if we insert a voltage divider that feeds back 1/10 of V_{out}, then V_{out} will shows us $10 \times V_{in}$.

But here are two more exotic cases: two examples where the "strange signal" (dog^{-1}) evoked by the "strange stuff" in the feedback is useful. The left-hand example in Fig. 6N.23, with the diode in the

[17] We would like to have included a visual representation, in the figure, of *inverse dog* – but we lost our nerve. What does *inverse dog* look like? Is it a dog with his paws in the air? Is it a cat? Is it a god? You can see why we abstained.

feedback loop, may look at first like a rectifier. But consider what it does for an always-negative input voltage.

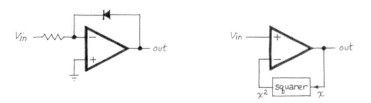

Figure 6N.23 Two cases where we plant strange stuff in loop to get "strange" and interesting op-amp output.

V_{out} is proportional to the *log* of V_{in}, for V_{in} negative; in the right-hand circuit, V_{out} is the square-root of V_{in}, for V_{in} positive. Both circuits fail for inputs of the wrong sign, as we meant to suggest when we first showed these two circuits, in Fig. 6N.15.

In Lab 6L you will be so bold as to put the oscilloscope itself inside one feedback loop. This is an odd idea, but one that produces entertaining results: see Fig. 6N.24.

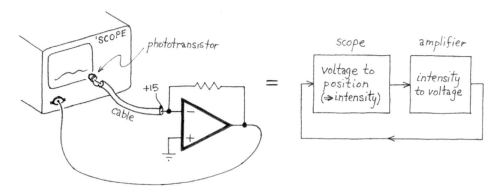

Figure 6N.24 Scope brought within feedback loop: adjusts location of CRT beam.

You'll see lots of nifty circuits in this chapter. Soon you may find yourself inventing nifty circuits. Op-amps give you wonderful powers. IC manufacturers show collections of application circuits for their devices. TI, for example, shows about 30 pages of circuits (reprinting National Semiconductor's application note 31: https://LAoE.link/TI_AN-31.pdf). Or for a sampling of application wisdom from the late analog wizard Jim Williams at Linear Technology, see https://LAoE.link/Williams_AN-6.pdf.

6N.10 AoE reading

- Chapter 2: Once again, introduction to negative feedback: §2.5 (this was assigned for Lab 5L, but remains a relevant introduction).
- Chapter 4: §§4.1–4.2.

6L Lab: Op-Amps I

6L.1 A few preliminaries

First a few reminders.

Mini-DIP package: You saw a DIP[1] in the previous lab. Fig. 6L.1 shows another, this time an 8-pin mini-DIP, housing the operational amplifiers that we will meet in this and later labs.

Figure 6L.1 Op-amp mini-DIP package.

The pinout (best represented by the rightmost image of Fig. 6L.1) was established by op-amps even earlier than the classic LM741[2] and is standard for single op-amps in this package. You will meet this pinout again in Lab 7L when you use the '741 and also the much more recent LT1150. Such standardization is kind to all of us users. Unfortunately, as parts get smaller, DIP parts are becoming scarce. Many new designs are issued in surface-mount packages only.

Power: Second, a point that may seem to go without saying, but sometimes needs a mention: the op-amp *always* needs power, applied at two pins; nearly always that means ±15 V in this course. We remind you of this because circuit diagrams ordinarily *omit* the power connections. On the other hand, many op-amp circuits make *no* direct connection between the chip and *ground*. Don't let that rattle you: the *circuit* always includes a ground in the important sense: a common reference treated as 0 V.

Decoupling: You should always "decouple" the power supplies with a small ceramic capacitor (0.01–0.1 μF), as suggested in Fig. 6L.2 and as we said in Lab 4L. If you begin to see fuzz on your circuit outputs, check whether you have forgotten to decouple. Most students don't believe in decoupling until they see that fuzz for the first time. Op-amp circuits, using feedback in all cases, are peculiarly vulnerable to such "parasitic oscillations." Op-amps are even more vulnerable to parasitic oscillations than the transistor circuits that evoked our warning in Chapter 4.

---

[1] "Dual in-line package."
[2] The pinout appears at least as early as the μA709, a Fairchild device introduced in 1965.

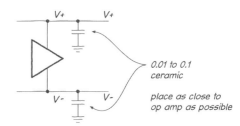

Figure 6L.2 Decouple the power supplies. Otherwise your circuits may show nasty fuzz (unwanted oscillations).

6L.2 Open-loop test circuit

Astound yourself by watching the output voltage as you slowly twiddle the pot in the circuit of Fig. 6L.3, trying to apply 0 V. Is the behavior consistent with the 411 specification that claims "Gain (typical) =200 V/mV?" Don't spend long "astounding yourself" however; this is a most *abnormal* way to use an op-amp. Hurry on to the useful circuits!

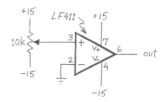

Figure 6L.3 Open-loop test circuit

6L.3 Close the loop: follower

Build the *follower* shown in Fig. 6L.4, using a 411. Check out its performance. In particular, measure (if possible) Z_{in} and Z_{out}, using processes to be described in later in this section.

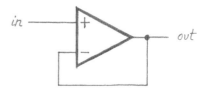

Figure 6L.4 Op-amp follower.

6L.3.1 Input impedance

Try to measure the circuit's input impedance at 1 kHz, by putting a 1 M resistor in series with the input. Here, watch out for two difficulties:

- *Beware* the finding "10 MΩ".[3]
- R_{in} is so huge that C_{in} dominates. You can calculate what C_{in} must be, from the observed value of f_{3dB}. Again make sure that your result is not corrupted by the scope probe's impedance (this time, it's the probe's capacitance that could throw you off).

[3] It's very easy to be so deceived, because this is the first circuit we have met that shows an input resistance much larger than R_{in} for the oscilloscope-with-probe. That R_{in} is 10 MΩ.

6L.3.2 Output impedance

The short answer to the value of the follower's output impedance is "low." It's so low that it is difficult to measure. Rather than ask you to measure it, we propose to let you show yourself that it is *feedback* that is producing the low output impedance.

Here's the scheme: add a 1k resistor in series with the output of the follower; treat this (perversely) as a part of the follower, and look at V_{out} with and without load attached. This is our usual procedure for testing output impedance. No surprises here: R_{out} had better be 1k.

Now move the feedback point from the op-amp's output to the point beyond the 1k series resistor – to apply "feedback #2," in Fig. 6L.5. What's the new R_{out}? How does this work? (Chapter 6N sketches an explanation of this result.)

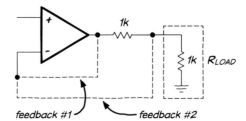

Figure 6L.5 Measuring R_{out} – and effect of feedback on this value.

If you're really determined, you can try to measure the output impedance of the bare follower (without series R). Note that no blocking capacitor is needed (why?[4]). You should expect to fail, here: you probably can do no more than confirm that Z_{out} is very low.

Do not mistake the effect of the op-amp's limited *current* output for high Z_{out}. You will have to keep the signal quite small here, to avoid running into this current limit. The curves in Fig. 6L.6 say this graphically. They say, concisely, that the current is limited to ±25 mA over an output voltage range of ±10 V, and you'll get less current if you push the output to swing close to either rail ("rail" is jargon for "the supply voltages"). This current limit is a useful self-protection feature designed into the op-amp. But we need to be aware of it.

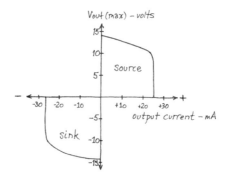

Figure 6L.6 Effects of limit on op-amp output current (LF411).

---

[4] No blocking cap because there's no DC voltage to block.

6L.4 Non-inverting amplifier

Wire up the non-inverting amplifier shown in Fig. 6L.7. You will recognize this as nearly the *follower* – except that we are tricking the op-amp into giving us an output bigger than the input. How much bigger?[5]

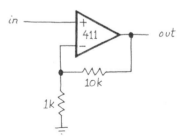

Figure 6L.7 Non-inverting amplifier.

What is the maximum output swing? How about linearity (try a triangle wave)? Try sinewaves of different frequencies. Note that at some fairly high frequency the amplifier ceases to work well: *sine in* does not produce *sine out*.

We are not asking you to quantify these effects today, only to get an impression of the fact that op-amp virtues fade at higher frequencies. In Lab 7L you will take time to *measure* the *slew rate* that imposes this limit on output swing at a given frequency. We are still on our honeymoon with the op-amp, after all; it is still *ideal*. "Yes, sweetheart, your slewing is flawless".

No need to measure input and output impedances again. Later, you will learn that you have traded away a little of some other virtues in exchange for the increased voltage gain; still, this amp's performance is pretty spectacular.

6L.5 Inverting amplifier

Construct the inverting amplifier drawn in Fig. 6L.8. If you are sly, you will notice that you *don't need to start fresh*: you can use the non-inverting amplifier, simply redefining which terminal is *input*, which is *grounded*. *Note*: Keep this circuit set up: you will use it again in §§6L.6 and 6L.7.

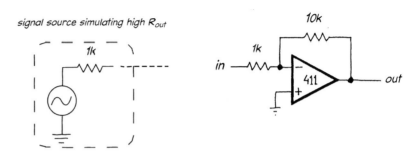

Figure 6L.8 Inverting amplifier.

Drive the amplifier with a 1 kHz sinewave. What is the gain? Is it the same as for the *non-inverting* amp you built a few minutes ago?

Now drive the circuit with a sinewave at 1 kHz again. Measure the input impedance of this amplifier

---

[5] Feeding back 1/11, we get $V_{out} = 11 \times V_{in}$. Looks familiar, from the final circuit of Lab 5L, does it not?

circuit by adding 1k in series with the signal source (simulating a source of crummy R_{out}). This time, you should have no trouble making the measurement. If you suppose that the 1k in series with your signal source represents R_{out} for your source, then what is the inverting amp's gain for such a source?[6]

Take advantage of the follower you built earlier, to solve the problem that we have created for you, the signal source's poor R_{out}. With the follower's help, your inverting amp's overall gain should jump back up to its original value (-10).

6L.6 Summing amplifier

Modify the inverting amplifier slightly, to form the circuit shown in Fig. 6L.9. This circuit sums a DC level with the input signal. Thus it lets you add a DC offset to a signal. (Could you devise other op-amp circuits to do the same task?[7])

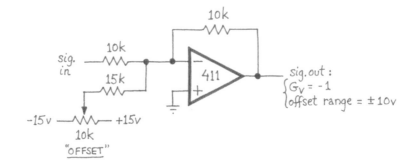

Figure 6L.9 Summing circuit: DC offset added to signal.

6L.7 Design exercise: unity-gain phase shifter

6L.7.1 Phase shifter I: using V_{in} and its complement

AoE §2.2.8A

The circuit in Fig. 6L.10 applies a signal and its inverse to an R and C in series. By varying R, one can make V_{out} more similar to V_{in} or similar to its complement, $V_{in\text{-inverted}}$. Thus phase can be adjusted over a range of 180°. So far, so simple.

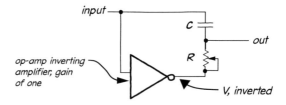

Figure 6L.10 Phase shifter, basic design.

The subtlety and cleverness of this circuit lies in the fact that the *amplitude* of the output always equals V_{in}, regardless of phase shift. (This is radically different from the behavior of, say, a lowpass filter, whose phase shift also can vary quite widely: between zero and almost $-90°$. But as the filter's phase shift varies, with changes of frequency, so does amplitude out.) Figure 6L.11 helps to explain this happy and surprising result.

---

[6] That's right: gain falls to half of what it was, because the effective "R_1" is doubled, in the gain equation $G=-R_2/R_1$.
[7] Probably; but we doubt they'd be as simple as this one.

6L.7 Design exercise: unity-gain phase shifter

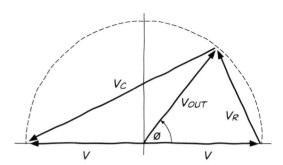

Figure 6L.11 Phasor diagram of phase shifter in Fig. 6L.10 (borrowed from AoE).

The input amplitude, fed to the series RC, is $2 \times V$, and this is shown on the horizontal ("real") axis of Fig. 6L.11. The voltages V_R and V_C are 90° out of phase; this behavior is familiar to us from our experience with RC filters. The voltages V_R and V_C sum to $2 \times V$. This relation is shown in the figure as the triangle whose hypotenuse – the sum – is the horizontal $2 \times V$.

The really nifty element in this result is the fact that V_{out}, represented as the distance from ground to the junction of R and C, is always equal to V_{in} or "V." (The point we describe as "ground" is simply the midpoint on the horizontal axis, midway between V_{in} and its complement.)

To put this another way, the R–C junction lives on a semicircle of radius V. That radius is shown, in Fig. 6L.11, as a phasor *arrow* joining origin to that R–C junction.

Now design your own op-amp adjustable phase shifter and try it out. As you choose your values for R and C, note that you want your maximum R (the pot value) to be much larger than X_C at the driving frequency. Don't feel disappointed when you notice that at a given pot setting, phase shift varies with input frequency. That's just the way this circuit works.

6L.7.2 Phase shifter II: phase shifter with voltage control

Figure 6L.12 is a less obvious route to the same goal. The advantage of this circuit over the one you designed in §6L.7.1 is the fact that the phase shift can be adjusted by varying a resistance *to ground*. This feature renders straightforward the use of an electronic signal (rather than your hand adjusting a pot) to control phase shift. You could, for example, use a JFET in its resistive range to serve as adjustable resistor;[8] then let a repeating waveform drive the FET to sweep the phase shift. If you were to mix that signal with the non-shifted original signal, the result might sound like a "flanger." You could try this today, if you have some time on your hands. (Not likely!)

Compare AoE §6.3.5

How does it work? This is a subtler version of the phase shifter that you designed in §6L.7.1. To see how it works, imagine taking the variable R value to extremes:

- when $R = 0$, this is just an inverting amp, unity gain; and
- when $R \gg X_C$, $V_{out} = V_{in}$.

These are the phase extremes, 180° apart.[9] Between these extremes, the phase at the non-inverting input is adjustable. The surprising virtue of *constant amplitude* results from essentially the argument that explains the simpler shifter (AoE §2.2.8A). We leave our explanation at that – a bit incomplete, "leaving the exercise to the reader" – because we don't want to stop you, at lab time, with a long digression on this topic.

---

[8] See §12L.4. You will also find such voltage-controlled resistance circuits discussed in AoE in §3.2.7.
[9] This circuit is similar to the "follower-to-inverter" circuit of AoE §4.3, Fig. 4.20.

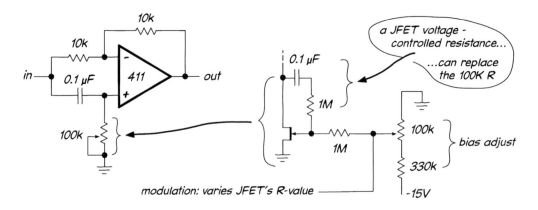

Figure 6L.12 Phase-shifter: second design.

6L.8 Push–pull buffer

Build the circuit shown in Fig. 6L.13. Drive it with a sinewave of 100–500 Hz. Look at the output of the op-amp, and then at the output of the push–pull stage (make sure you have at least a few volts of output, and that the function generator is set for no DC offset). You should see classic crossover distortion.

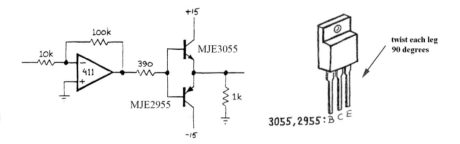

Figure 6L.13 Amplifier with push–pull buffer.

Listen to this waveform on the breadboard's 8 Ω speaker – or on a classier speaker, if you can find one. Your ears (and those of people near you) should protect you from overdriving the speaker. But it would be prudent, before driving the speaker, to determine the maximum safe amplitude given the speaker's modest power rating. The transistors are tough guys, but you can check whether you need to lower the power-supply levels on your breadboard, to keep the transistors cool, given the following power ratings:

- transistors: 75 W – if very well heat-sunk, so that case remains at 25°C. Much less (0.6 W) if no heat-sink is used, as is likely in your setup.
- speaker: 250 mW.

Now reconnect the right side of the feedback resistor to the push–pull output (as proposed near the end of Chapter 6N), and once again look at the push–pull output. The crossover distortion should be eliminated now. If that is so, what should the signal at the output of the op-amp look like? Take a look. (Doesn't the op-amp seem to be *clever*?)

Listen to this improved waveform: does it sound smoother (more flute-like) than the earlier waveform? Why did the crossover distortion sound harsh and metallic, more like a clarinet than like a flute – as if a higher frequency were mixed with the sine?[10]

If you increase signal frequency, you will discover the limitations of this remedy, as of all op-amp techniques: you will find a glitch beginning to reappear at the circuit output.[11]

6L.9 Current-to-voltage converter

In earlier exercises we met and solved a "problem" presented by the inverting amplifier: its R_{in} is relatively low. Once in a while, this defect of the inverting amp becomes a virtue. This happens when a signal source is a *current*-source rather than the much more common *voltage*-sources that we are accustomed to. A photodiode is such a signal source, and is delighted to find the surprising R_{in} at the inverting terminal of the op-amp: what is that R_{in} value?[12]

Photodiode: Use a BPV11 or LPT100 phototransistor as a photo*diode* in the circuit of Fig. 6L.14. These devices are most sensitive in near-infrared (around wavelength of 850 nm) but show about 80% of this sensitivity to visible *red* light. Look at the output signal. If the DC level is more than 10 V, reduce the feedback resistor to 4.7M or even to 1M.

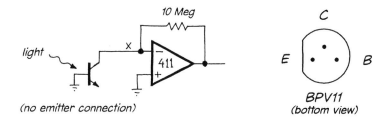

Figure 6L.14 Photodiode photometer circuit.

Figure 6L.15 A less good photodiode circuit.

If you see fuzz on the output – oscillations – put a small capacitor in parallel with the feedback resistor. The R is so big that a tiny cap should do: even 22 pF causes circuit gain to fall off at an f_{3dB} ($1/2\pi RC$) of less than a kilohertz. In the *phototransistor* circuit in Fig. 6L.16, with its smaller $R_{feedback}$, you would need a proportionately (100×) larger C. Why does this capacitor douse the oscillation? Because the oscillation can't persist if the circuit gain is gone at the (high) frequency where the circuit "wants" to oscillate. It wants to oscillate up there because there it finds large, troublesome phase shifts; in Lab 9L we'll see much more of this problem.

What is the average DC output level, and what is the percentage "modulation?" (The latter will be relatively large if the laboratory has old-fashioned fluorescent lights, which flicker at 120 Hz; much smaller with contemporary 40 kHz fluorescents, or with incandescent lamps.) What input photocurrent

---

[10] Well, because a higher frequency *is* mixed with the sine. The abrupt edges visible in the step from below to above the input waveform, as the latter crosses zero volts, include high-frequency components. Your ears recognize this, even if you have for a moment forgotten the teachings of Fourier.

[11] The circuit requires the op-amp to snap its output from a diode drop *below* V_{in} to a diode drop above, as the output crosses zero volts. That excursion of about a volt takes time. The op-amp can "slew" its output only so fast – and here that rate is much lower than the maximum "slew rate" that you will measure next time, because the op-amp input is not strongly overdriven during this brief transition. While this slewing is occurring, the op-amp is not able to make the output do what it ought to do. The glitch becomes visually noticeable in the scope display when the output waveform is steep and scope sweep rate is high.

[12] Yes, ideally this R_{in} is zero. In fact, it is very small (later, in Chapter 9N, you'll learn to calculate it as $R_{feedback}/A$, where A is the op-amp's "open-loop gain").

does the output level correspond to? Try covering the phototransistor with your hand. Look at the "summing junction" (point **X**) with the scope, as V_{out} varies. What should you see?[13]

Make sure you understand how this circuit is preferable to a simpler "current-to-voltage converter," a resistor, used as in Fig. 6L.15.[14]

Phototransistor: Now connect the BPV11 as a photo*transistor*, as shown in Fig. 6L.16 (the base is to be left *open*, as shown). Look again at the summing junction. Keep this circuit set up for the next stage.

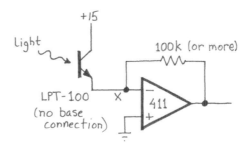

Figure 6L.16 Phototransistor photometer circuit.

Applying the photometer circuit: For some fun and frivolity, if you put the phototransistor at the end of a cable connected to your circuit, you can let the transistor look at an image of itself (so to speak) on the scope screen. (A BNC cable with grabbers on *both* ends is convenient; note that in this circuit *neither* terminal is to be grounded, so do not use one of the breadboard's fixed BNC connectors.) Such a setup is shown in Fig. 6L.17.

The image appears to be shy: it doesn't like to be looked at by the transistor. Notice that this scheme brings the scope itself within a feedback loop.

Figure 6L.17 Photosensor sees its own image.

You can make entertaining use of this curious behavior if you cut out a shadow mask, using heavy paper, and arrange things so that the CRT beam just peers over the edge of the mask. In this way you can generate arbitrary waveforms. If you try this, keep the "amplitude" of your cut-out waveform down to an inch or so. Have fun: this will be your last chance, for a while, to generate really silly waveforms:

---

[13] The "summing junction" is a *virtual ground* and should sit at 0 V. If it does not, then something has gone wrong. Most likely, the op-amp output has hit a limit ("saturated"), making it impossible for feedback to drive that terminal to match the grounded non-inverting input.

[14] The difference results from the photodiode's "disliking" large voltage variation. The passive circuit allows a changing voltage across the photodiode, and the diode varies its current somewhat in response. This variation distorts the relation between light intensity and diode current.

say, Diamond Head, or Volkswagen, or Matterhorn. You will be able to do such a trick again – at least in principle – once you have a working computer, which can store arbitrary patterns in memory in *digital* form. Practical arbitrary waveform generators use this digital method.

6L.10 Current source

Try the op-amp current source shown in Fig. 6L.18. What should the current be? Vary the *load* pot and watch the current, using a digital multimeter. This current source should be so good that it's boring. Now substitute a 10k pot for the 1k and use a second meter or a scope to watch the op-amp's output *voltage* as you vary R_{load} (in this case, just the 10k variable R). This second meter should reveal to you why the current source fails when it does fail.

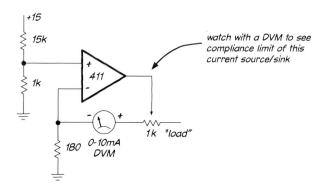

Figure 6L.18 Current source.

Note that this current source, although far more precise and stable than our simple transistor current source, has the disadvantage of requiring a "floating" load (neither side connected to ground); in addition, it has significant speed limitations, leading to problems in a situation where either the output current or load impedance varies at microsecond speeds.

The circuit of Fig. 6L.19 begins to solve the first of these two problems: this circuit sources a current into a load connected to ground.

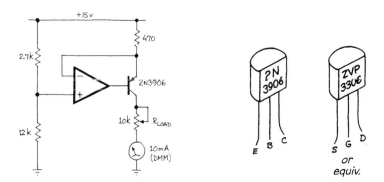

Figure 6L.19 Current source for load returned to ground.

Watch the variation in I_{out} as you vary R_{load}. Again, performance should be so good that it bores you. The 10K resistor will cause the current source to fail when you dial R up to about 2k. You can see why if you use a second meter to watch V_{CE}.

You may not see any current variation using the bipolar transistor: a 2N3906. If you manage to discover a tiny variation, then you should try replacing the '3906 with a ZVP3306, BS250P or VP0106

MOSFET (the pinouts are equivalent, so you can plug any of these in exactly where you removed the 2N3906).

Should the circuit perform better with FET or with a bipolar transistor?[15] Do you find a difference that confirms your prediction (that difference will be extremely small)? With either kind of transistor the current source is so good that you will have to strain to see a difference between FET and bipolar versions. Note that you have no hope of seeing this difference if you try to use a VOM to measure the current; use a DVM.

---

[15] The FET version performs a little better. The feedback circuit monitors I_E, whereas I_C is the output; the difference is I_B – a small error. The FET shows no equivalent to I_B: its input current is zero. So, in the FET version the *output* current is the same as the current *monitored*.

6W Worked Examples: Op-Amps I

The problem – just analysis this time: This is a rare departure from our practice of asking you to *design*, not to *analyze*. Inventing a difference amp[1] seemed a tall order, and, on the other hand, the difference amplifier's behavior seems far from obvious. So, here's a little workout in seeing how the circuit operates.

6W.1 Basic difference amp made with an op-amp

AoE §4.2.4

The difference amplifier made with an op-amp looks like Fig. 6W.1, which shows the general configuration and the particularly straightforward case of *unity* gain, the case that this note treats first.

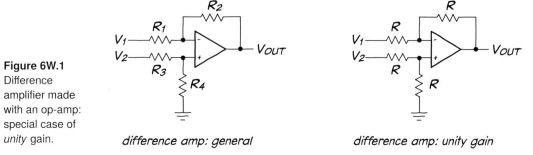

Figure 6W.1
Difference amplifier made with an op-amp: special case of *unity* gain.

difference amp: general   *difference amp: unity gain*

In general, as you know, the amplifier requires that the ratios of the resistors in the two paths, from V_1 and from V_2, must be well matched. Common-mode rejection depends on this matching.[2] The amplifier's differential gain is $R_2/R_1 = R_4/R_3$.

Seeing the amp's gain by trying some cases

It is possible to derive the circuit's gain using a little algebra (see §6W.1.5). But that sort of description often lacks persuasive power. So, first let's use another approach: we will try particular cases of input voltages, to satisfy ourselves that the circuit does what it ought to do. This technique is one more application of a common technique that we use when confronted by a novel circuit: try to redraw it so that it looks like a familiar circuit that we understand well. In §6W.1.1, for example, the diff-amp becomes a good old inverting amp.

---

[1] This circuit is more often called a "differential amplifier." But we prefer the less grandiloquent term here, used also in AoE §4.2.4. The term "differential" seems to suggest that this rather straightforward circuit is doing some challenging calculus operations for you. The word "difference" better describes what the circuit measures. Nevertheless, we should admit, we often revert to the more traditional term and find ourselves calling the circuit a differential amp. A simple sidestep is to call it a "diff-amp," as many of us often do.

[2] Because close matching is hard to achieve, the best approach may be to buy an integrated array of resistors tightly matched: the best, to 0.001% (see AoE §4.2.4) or to buy an integrated difference amp like TI/Burr–Brown's INA105 with gain within 0.05% of unity, or the INA149, described below in §6W.2, with 100 dB of common-mode rejection (CMR).

292 Worked Examples: Op-Amps I

Recall that in order to keep things simple we have made all R values equal, giving the difference amp *unity* gain.

6W.1.1 First case: Inverting amp

Ground V_2, apply a signal to V_1; see Fig. 6W.2. We're forcing the non-inverting input to 0 V, so the inverting input goes to *virtual* ground, and the circuit forms a traditional inverting amplifier, gain of -1.

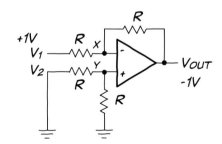

Figure 6W.2
Diff-amp: V_2 input grounded, gain is -1.

Case	Inputs		Internal Points		Output	Notes
	V_1	V_2	X	Y	V_{out}	
A	1	0	0	0	-1	inverting amp

6W.1.2 Second case: Non-inverting amp

Ground V_1, apply a signal to V_2; see Fig. 6W.3. Now the circuit forms a non-inverting amplifier of gain 2 for a signal applied at point **Y** (recall that $G = 1 + R_2/R_1$).

Figure 6W.3
Diff-amp:
V_1 input grounded, gain is +1.

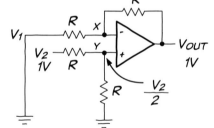

Case	Inputs		Internal Points		Output	Notes
	V_1	V_2	X	Y	V_{out}	
B	0	1	0.5	0.5	1	non-inverting amp gain of $2 \times (1/2)V_1$

At first glance gain of *two* may sound wrong – but not when you recall that the non-inverting signal is attenuated to one-half before it reaches **Y**. So, the non-inverting gain is +1, as one would hope.

6W.1.3 Third case: Equal inputs

Equal inputs should produce zero out. Try +1V to both inputs: see Fig. 6W.4. Equal inputs send equal voltages to the two op-amp inputs, forcing equal voltages at the end of the two divider-chains. So V_{out} must be ground. In other words *common-mode* gain is zero. (This will be true to the extent that the four Rs are well matched.)

6W.1.4 Fourth case: Non-zero inputs that differ by a volt

The output should be indifferent to any shared input, but should amplify the difference (1 V) by 1.

6W.1 Basic difference amp made with an op-amp

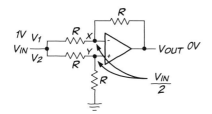

Case	Inputs		Internal Points		Output	Notes
	V_1	V_2	X	Y	V_{out}	
C	1	1	0.5	0.5	0	dividers match foot of each at gnd

Figure 6W.4 Diff-amp: equal inputs, output zero.

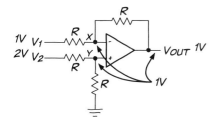

Case	Inputs		Internal Points		Output	Notes
	V_1	V_2	X	Y	V_{out}	
D	1	2	1	1	1	inv. amp but amp'ed with respect to V_Y and added to V_Y

Figure 6W.5 Diff-amp: two positive inputs, difference 1.

Yes. Here we might say there is a *common-mode* signal of 1.5 V, and a *difference* signal of 1 V. The amplifier ignores the common signal, and amplifies the difference by 1.

This case is exactly like the case that we called "non-inverting amp", §6W.1.2, except that both inputs are higher by 1 V. This common difference the amplifier ignores.

Alternatively, one might see this as an inverting amp that amplifies V_1 with respect to V_X and V_Y. In this case there is no difference to amplify, so the output is just the baseline voltage, V_Y.

6W.1.5 Gain, formally derived

If looking at particular cases and trying to liken them to familiar amplifier configurations seems laborious, maybe you'll prefer a more traditional analysis. First, let's redraw the circuit to add a few labels; see Fig 6W.6.

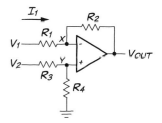

Figure 6W.6 Difference amplifier: a more general diagram.

Although we have generalized the diagram to permit gains other than unity, let us stick to unity gain, in order to make the arithmetic as easy as possible. So, again, let's assume all Rs are equal. They need not be for good differential performance; as we said at the outset, we only need $R_2/R_1 = R_4/R_3$.

The input current, I_1, is $(V_1 - V_X)/R$, and $V_X = V_Y = V_2/2$. So

$$I_1 = \frac{V_1 - (V_2/2)}{R}; \qquad V_{out} = V_X - (I_1 R) = V_2/2 - \left(\frac{V_1 - (V_2/2)}{R}\right) R$$

The Rs conveniently divide out, leaving $V_{out} = V_2 - V_1$.

294 Worked Examples: Op-Amps I

Though we looked, once more, at the unity gain case, the algebraic argument would work just as well for any matched set of resistor ratios: one where $R_2/R_1 = R_4/R_3$. If the resistor ratio is not unity, then of course V_{out} is not *equal to* the difference between the two circuit inputs but is a multiple of that difference.

Is this algebraic argument persuasive? Or do you prefer to look at the particular cases with which we began? Your answer will depend only on your tastes, of course.

6W.2 A more exotic difference amp: IC with wide common-mode input range: INA149

TI makes several integrated versions of the standard difference amp, ICs that include well-matched resistors delivering good common-mode rejection (100 dB for the INA149). In addition, one of these ICs, the INA149, adds a 5th resistor that makes sure that the op-amp is not overdriven even for an input as large as ±275 V: the fifth resistor divides such a large input voltage down to 13.5 V at the op-amp inputs. If you wonder why anyone would want such common-mode range, TI suggests, for example, that this range could be useful to monitoring current on a high-voltage line: the diff-amp could watch the voltage drop across a small resistor that carries that current.

Figure 6W.7 shows the circuit.[3] On the right side we have added labels in order to facilitate a discussion of the circuit.

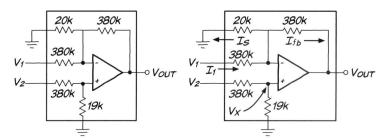

Figure 6W.7 Integrated difference amplifier, INA149.

This amplifier may look – if one recalls the general gain expression for a diff-amp (§6W.1) – as if it would deliver an attenuated output, with differential gain 1/20, reflecting the ratio of the 19k to the 380k input resistor on the V_2 side. But the circuit's differential gain turns out to be 1. We'll investigate how that result arises. We find the circuit's behavior far from obvious.

Again, we propose two ways to consider the circuit: first, using a couple of special cases that are easy to analyze; then chugging through some laborious algebra to get a general result.

6W.2.1 A couple of special input cases

As we found for the generic difference amp of §6W.1, some input cases are easy to analyze, and we'll try those to convince ourselves that, contrary to appearances, the circuit can indeed deliver a gain of 1.

Ground V_1: Grounding V_1 puts 380k and 20k to ground, in parallel: equivalent to 19k to ground. Thus we find a curious symmetry: V_2 is divided by 21 (=19k/(380k + 19k)). The inverting path divides its output voltage by the same fraction. So, $V_{out} = V_{in}$. You can put this argument differently, if you prefer: say V_2 is divided by 21, and the gain of the non-inverting amp driven by this value is 1 + 380k/(20k∥∥380k) = 1 + 380k/19k = 21. Gain thus is +1.

---

[3] The 20k and 19k resistors are shown grounded. They can be tied to other voltage references if an output with offset is wanted.

Ground V_2: This case is easier. Grounding V_2 puts both op-amp input terminals ("Vx" on diagram) at ground. So, the amplifier looks like a simple inverting amp (the 20k has no effect, conducting no current with 0 V across it). Gain is -1.

6W.2.2 A general solution

The algebra that generalizes these results is less easy and less fun. But here goes. We'll use the signals named on the right-hand side of Fig. 6W.7: V_1, V_2, V_X and the currents I_1, I_S and I_{fb}.

We'd like to find $I_{feedback}$. Knowing that value, and V_X, we will be able to calculate V_{out}:

$$V_X = V_2/21$$

Now let's calculate currents:

$$I_1 = (V_1 - V_X)/380k;$$

$$I_{feedback} = I_1 - I_S = \frac{V_1 - V_X}{380k} - \frac{V_x}{20k}$$

$$= \frac{V_1 - V_2/21}{380k} - \frac{V_2/21}{20k}$$

Now we can use that current to get V_{out}:

$$V_{out} = V_X - (I_{feedback} \cdot R_{feedback});$$

$$V_{out} = V_2/21 - \left(\frac{V_1 - V_2/21}{380k} - \frac{V_2/21}{20k}\right) \cdot 380k$$

$$= V_2/21 - (V_1 - V_2/21 - (V_2/21) \cdot 19)$$

$$= V_2/21(1 + 1 + 19) - V_1 = V_2 - V_1$$

This is no surprise, after the two special cases that we saw, where we grounded first one, then the other of the two inputs. But the result is reassuring.

6W.3 Problem: odd summing circuit

Suppose you would like to sum the signals provided by two transducers, giving the two transducers *equal* full-scale "weights." The summed output is to drive an analog-to-digital converter (ADC)[4], and the input range of that device defines the permitted output range for your *summer*. We will assume the transducers provide *DC* signals of just one polarity, as stated below.

Here are the relevant specifications:

ADC input range: 0 to +2.5 V
Signal source A: DC voltage range: 0 to 0.5 V
  Output impedance: $\leq 10\,M\Omega$
Signal source B: Amplitude range: 0 to -1 mA (current sink)
  Output impedance: $\geq 10\,M\Omega$
  Output voltage compliance: ≤ 0.1 V

---

[4] You needn't understand what this gadget does in order to answer this question.

6W.3.1 A solution

Getting our bearings: The list of specifications in §6W.3 may be hard to get a grip on, at first glance – but not once one notices that A is a voltage source, while B is a current source. That makes the large, important contrast clear; from there we can proceed to the details expressed in the numbers.

The *summing* circuit that we are to build will use an op-amp in its inverting configuration, as usual. Such a circuit sums *currents*, fundamentally; when we feed *voltages* to such a summer we place input resistors between signals source and the summing point, *virtual ground*. Those resistors convert input voltages to input *currents*. We will do that, as usual, for signal A.

But B, the signal that is a *current*, requires no such input resistor. Signal B, a current, goes straight to the point where currents are summed: virtual ground.

A skeleton circuit: A circuit with no component values will look like Fig. 6W.8.

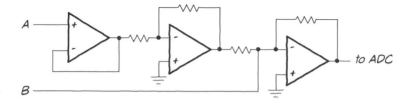

Figure 6W.8 A solution lacking component values.

Drawing the skeleton circuit is what requires some thought in this problem. Here are a couple of points that to us seemed not obvious.

- The tight "output voltage compliance" specification for B requires that B drive an op-amp summing junction. That's no surprise, or burden, since you probably were inclined to use this standard summing configuration, anyway. You must, of course, resist the tempting reflex (based on your experience with summers) to place a resistor between B and virtual ground.

    Why? (By the way: what goes wrong if you put, say, a 1k resistor between B and virtual ground?) This R does not interfere with the summing of currents at virtual ground. Rather, the R moves the voltage at B too far from ground: 1 V, at maximum current, far outside the 0.1 V compliance range.

- A needs a high-impedance buffer – the input *follower* shown in Fig. 6W.8, because the high output impedance of A otherwise interacts with the modest input impedance of the summing circuit.

    "Modest?," you protest? "Why should it be modest? I can make it huge, by using huge R values in the inverting stage that takes signal A." (The inverting stage is needed to cancel the inversion that is inherent in the summing circuit.)

    But you answer yourself: "Yes, modest. Even at 100 MΩ those Rs could introduce a 10% error, and we probably ought to limit ourselves to R values of about 10M, in order to avoid running into nasty side effects."

    Those side effects include large errors that result from I_{bias} flowing in unmatched resistive paths and, probably more troublesome, the introduction of an unintended *low-pass* effect within the feedback loop. Such a low-pass causes lagging phase shifts that can make the circuit unstable.[5]

    And you are too clever to fall for the following spurious argument: "I'll take advantage of A's R_{out} to form the inverting stage: I'll use a feedback resistor of 10M." That doesn't work because A's R_{out} is not specified to be 10M. Instead, as usual, the R_{out} value is given only as a *maximum*. You have seen this sort of specification many times before.

---

[5] We will devote most of Lab 9L to just such stability issues.

6W.3.2 …Calculating component values

Two component values matter:

- R_{FB} sets the full-scale V_{out}; and
- R feeding virtual ground sets the gain for signal A relative to signal B.

Figure 6W.9 finishes the design. This part of the design task doesn't require intelligence; just a little arithmetic.

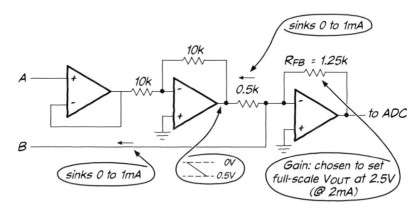

Figure 6W.9 Complete summing circuit, with component values.

Here are the arithmetic arguments:

R_{FB}: When the two signals, A and B, both are at their maximum we want $V_{out} = 2.5$ V. We want to give the two signals equal weight, so B's 1mA should produce $V_{out} = 1.25$ V. This sets the value of R_{FB} at 1.25k.

R feeding virtual ground: In order to give the two signals equal full-scale weights, this R should convert A's full-scale $V_{in} = 0.5$ V to 1 mA. An R of 0.5k will do this.

7N Op-Amps II: Departures from Ideal

Contents

7N.1	**Old: subtler cases, for analysis**	**299**
	7N.1.1 Do the Golden Rules apply?	299
	7N.1.2 Improving this active rectifier	301
7N.2	**Op-amp departures from ideal**	**302**
	7N.2.1 Should we care about these small imperfections?	303
	7N.2.2 Offset voltage: V_{OS}	304
	7N.2.3 Bias current: I_{bias}	306
	7N.2.4 Offset current: I_{offset}	308
	7N.2.5 Slew rate and rolloff of gain	309
	7N.2.6 Output current limit	310
	7N.2.7 Noise	310
	7N.2.8 Input and output voltage range	310
	7N.2.9 Some representative op-amp specifications: ordinary and premium	312
7N.3	**Five more applications: integrator, differentiator, difference amplifier, instrumentation amplifier, AC amplifier**	**313**
	7N.3.1 Integrator	313
	7N.3.2 Taming the op-amp integrator: prevent drift to saturation	314
	7N.3.3 A resistor T Network	316
	7N.3.4 An integrator can ask a lot from an op-amp, and can show the part's defects	317
	7N.3.5 An example of a case where periodic integrator reset is adequate	317
7N.4	**Differentiator**	**318**
7N.5	**Op-amp difference amplifier**	**318**
	7N.5.1 A curious side virtue: wide input range	320
7N.6	**Instrumentation amplifier**	**320**
	7N.6.1 A straightforward way to improve the op-amp difference amplifier	320
	7N.6.2 Three op-amp instrumentation amplifier ("IA")	320
	7N.6.3 Wheatstone bridge	321
7N.7	**AC amplifier: an elegant way to minimize effects of op-amp DC errors**	**323**
7N.8	**AoE reading**	**324**

Why?

We want to solve the problem of optimizing circuit performance by selecting from the great variety of available op-amps. We will try to make sense of the fact – not predictable from our first view of op-amps as essentially ideal – that there are not one or two op-amps available but approximately 30,000 listed (on the day of this writing) on one distributor's website (DigiKey).

Admitting that op-amps are not ideal marks the end our honeymoon with them. But we continue to

7N.1 Old: subtler cases, for analysis

admire them: we look at more applications, and as we do, we continue to rely on our first, simplest view of op-amp circuits, the view summarized in the *Golden Rules*.

After using the Golden Rules to make sense of these circuits, we begin to qualify those rules, recognizing, for example, that op-amp inputs draw *a little* current. Let's start with three important new applications; then we'll move to the gloomier topic of op-amp imperfections.

Big ideas from last time:

- Feedback and high gain allow elegantly simple design and analysis with the "Golden Rules."
- Some important applications: amplifiers, current source.
- Generalizing: op-amp circuits either exploit or hide the "dog" placed within the feedback loop.

7N.1 Old: subtler cases, for analysis

AoE §4.3.2D

Here is an issue we only glanced at last time: do the Golden Rules apply to a particular op-amp circuit? What about more general ones?

7N.1.1 Do the Golden Rules apply?

Before we begin to exploit the wonderfully useful Golden Rules, we should check that they apply. Do the rules apply to the circuit in Fig. 7N.1? The answer is a wishy-washy "Yes and No." Or, more precisely, they apply for positive inputs only.

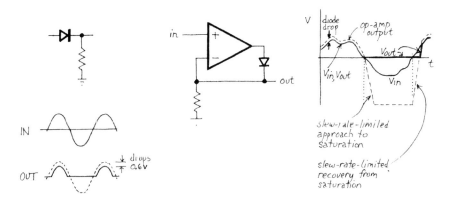

Figure 7N.1 A case that's a bit harder to analyze with the Golden Rules.

...Yes...: Positive inputs take the op-amp output positive (it helps to recall that the op-amp is a differential amplifier; take advantage of your understanding of the op-amp's innards). The positive output is able to feed current back to the inverting terminal, holding that terminal equal to V_{in}, as we would expect in a circuit obeying the Golden Rules. The circuit looks like a follower, with some extraneous elements added – a diode that's conducting, and an R to ground that has no effect except to load the output slightly. $V_{out} = V_{in}$.

300 Op-Amps II: Departures from Ideal

...And no: But for negative inputs, feedback fails: the diode blocks the signal from the op-amp output – which now is below ground. Lacking feedback, the op-amp simply snaps down to negative "saturation" (a volt or so above the negative rail, for an ordinary op-amp like the '411), and sits there.

That may sound like a disaster, but in this instance it is not. With the diode blocking current, the op-amp now is out of the circuit that links input to output. V_{out} is simply tied to ground through a resistor. The net effect of these two behaviors – Golden Rule follower in one case, disconnected op-amp in the other case – is to produce an output that is a rectified version of the input.

We included a passive rectifier in Fig. 7N.1 to remind you of the diode drop that the active version hides. The active version, having hidden the diode drop (the usual "dog") is capable of rectifying even signals whose amplitude is much smaller than 0.6V. That is its great virtue. Its weakness is that it works only at moderate frequencies: its output includes a brief glitch.

Trying it out: scope images The glitch is predictable. Fig. 7N.2 shows what it looks like: somewhat worse than one might predict, because it's not just *slewing* time that we must wait for, but also extra time to get out of the "saturated" state, in which the op-amp rests when feedback fails.

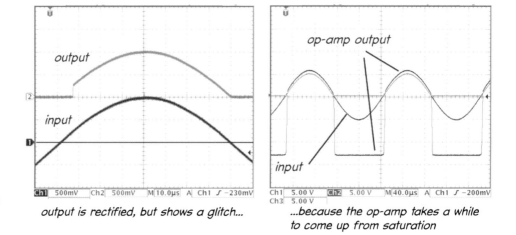

Figure 7N.2 Simple LF411 active rectifier produces an output glitch while recovering from saturation (scope gain: 500 mV/div.

output is rectified, but shows a glitch...   *...because the op-amp takes a while to come up from saturation*

The glitch duration is more than five times what a calculation of slewing-time alone would predict if one simply found the op-amp's specified *slew rate* (15 V/μs). The slewing that occurs here will be a good deal slower than that specified rate, which assumes extreme overdrive at the op-amp input.

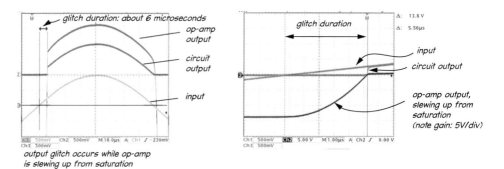

Figure 7N.3 Simple active rectifier: op-amp's recovery from saturation is slow (scope gain: 500 mV/div, except op-amp output in lower-right trace: 5 V/div).

output glitch occurs while op-amp is slewing up from saturation

You'll notice in §7N.2.5 that the maximum slewing occurs when the entire *tail* current flows in just one side of op-amp's input differential stage. This occurs only when a large *difference* voltage is applied between the two inputs. This does not occur during this brief glitch, where the voltage difference felt by the op-amp at first is very small, gradually growing to about 0.25 V.

It would be good if we could redesign the circuit to avoid saturating the op-amp; and we can.

7N.1.2 Improving this active rectifier

AoE Fig. 4.38

We can improve the circuit's performance – diminishing the glitch duration by at least 10 fold – by arranging things so that feedback *never* fails, and the op-amp therefore never saturates. The rectifier circuit in Fig. 7N.4 redesigns the rectifier for this purpose. One diode or the other always conducts.

For positive inputs, this is simply an inverting amplifier. For the other case – a case like the one that led the simpler rectifier of Fig. 7N.1 to saturate[1] – this circuit gets feedback through the second diode. While you're considering this circuit, use it as a chance to check your faith in op-amps. Specifically, does the diode between op-amp output and circuit output mess things up? (Have faith!)

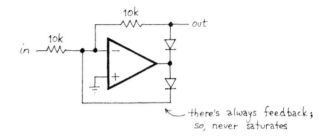

Figure 7N.4 One more diode improves the rectifier by preventing saturation.

This rectifier inverts. The left-hand image in Fig. 7N.5 shows the usual virtue of an active rectifier, a small input rectified without the subtraction of a diode drop, The right-hand image shows the changed op-amp output waveform that reduces the glitch: the op-amp output does not need to travel far (note the low scope gain: 0.5 V/div).

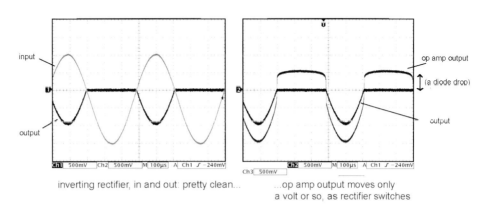

Figure 7N.5 Improved active rectifier.

Figure 7N.6, showing the transition at a higher sweep rate, confirms that the glitch is much reduced. Even at the high sweep rate of 400 ns/div in the right-hand image of Fig. 7N.6 – 25 times faster than in Fig. 7N.3 – the glitch is barely apparent.

[1] The polarities are reversed, between the two rectifiers, since the second one inverts.

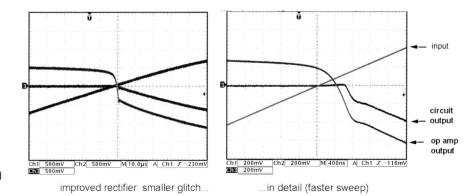

Figure 7N.6 Details of improved rectifier transition: nearly glitch-free.

improved rectifier: smaller glitch...   ...in detail (faster sweep)

R_{out} for the improved rectifier? This circuit shows a curious R_{out}. Although the op-amp always provides feedback (and thus diminishes the glitch of the simpler rectifier), the *circuit output* does not always show the low R_{out} that we expect from an op-amp circuit. When you have spotted the case for which this larger R_{out} applies, add another circuit fragment that solves that problem.[2]

7N.2 Op-amp departures from ideal

AoE §4.4.1

Let's admit it: op-amps aren't quite as good as we have been telling you. That's the sad news. But the happy news is that an amazing variety of op-amps exists – literally thousands of types – competing to optimize this or that characteristic. You can have low input current, low voltage-error (V_{offset}, coming soon in these notes), low noise – you just can't optimize all characteristics at once. You can find op-amps that run on high voltages (hundreds of volts), others that will run on a total supply of 0.9 V; some will deliver tens of amps as output current, others will subsist on nano-amps of supply current. Some are very fast.

Op-amp imperfections fall into two categories: DC and dynamic.

Dynamic errors: The dynamic errors affect performance at higher frequencies.

- Gain rolloff: the high gain that is essential to achieve the virtues of feedback dissipates as frequency rises. It is gone at the frequency f_T, where the open-loop gain is down to unity.[3] This characteristic often is specified as the "gain-bandwidth product," GBW.[4]
- The limited *slew rate* of the op-amp output imposes a limit on frequency and amplitude available *together* (small signals can be reproduced at frequencies higher than those available for larger signals).
- Noise injection (this is a subtlety that never becomes important in our lab circuits).

We will return to these errors later in these notes.

---

[2] Take a look at AoE if you're stumped, but if you don't have it to hand here's a powerful hint: the case that worries us is the one in which feedback flows through the lower diode. In that case, the upper diode is not conducting. So R_{out} is a disappointing 10k. A follower can remedy this defect.

[3] Actually, the benefits of feedback are gone when the open-loop gain of the op-amp decreases to the value of the closed-loop gain. So, an op-amp that is usable as a unity-gain follower to 4 MHz or so will only work to about 40 kHz when configured for a closed-loop gain of 100: see AoE §4.4.2A and Fig. 4.48.

[4] But see AoE §5.10.10A for a look at the subtle difference between f_T, and GBW.

7N.2 Op-amp departures from ideal

DC errors: The errors that concern us most are two DC errors which reveal that the Golden Rules exaggerate a bit.

- The inputs *do* draw (or source) a little current. This current, called I_{bias}, is the average of the currents flowing at the two inputs.
- The inputs are not held at precisely equal voltages. The principal cause of this inequality is V_{offset}, the op-amp's misconception concerning equality.[5]

Probably you are not surprised to hear this. Fig. 7N.7 gives three circuits that always deliver a saturated output after a short time. None would do this if op-amps were ideal.

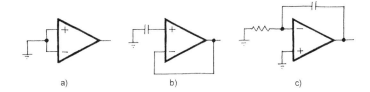

Figure 7N.7 Three circuits sure to saturate. Why?

No doubt you can figure out why they saturate. But the explanations are various enough so that you won't mind hearing them spelled out.

(a) This circuit would produce zero out only if V_{offset} were zero, as it cannot be.
(b) This circuit would produce zero out only if I_{bias} were zero, as it cannot be. In the case of a bipolar input stage, I_{bias} is the transistor's *base* current, and this current gradually will charge the capacitor. If the input stage of the op-amp is a FET, then I_{bias} is the transistor's *leakage* current. Either way, this current will charge the capacitor, so that the voltage at the non-inverting input must drift toward one supply or the other. (A '411 wired this way would drift toward the negative supply; see the discussion of a sample-and-hold made with a '411, in §12N.7.1.)
(c) If I_{bias} were zero, the voltage at the inverting terminal would be zero, and this case would be the same as Case (a). As we mentioned in discussing the integrator, the capacitor provides *no feedback* at DC (which you might also call "long-term feedback"). So the capacitor cannot save this output from saturation.

7N.2.1 Should we care about these small imperfections?

If one recites typical numbers for an op-amp's deviations from the ideal, those deviations sound pretty small. Here, for example, are some typical values for the '411: V_{offset}=0.8 mV, I_{bias}=50 pA, slew rate 15 V/μs; gain (open-loop) not infinite but 200,000 at DC. Listen to those units: "pA," "millivolts"? These sound so tiny! Do these departures from the ideal matter? Our answer is a wishy-washy "sometimes."

The number of op-amps available is enough to remind us that these small differences *can* matter. If they did not, one or a few op-amps would survive. Instead, there is room for an amazing variety of op-amps – and new ones are born every few months. For, optimizing one specification entails relaxing another. So it is worth learning to pay attention to the specifications – even while continuing to recognize that for a great many applications one can treat the op-amp as ideal, even if it is a mere "jellybean" device like the '411.[6]

---

[5] A second cause, usually smaller in magnitude, is the limited gain of the op-amp. We calculated the difference needed between the two inputs of the op-amp to generate an 11 V output, for example, back in §6N.6.2.
[6] "Jellybean" is a strange piece of jargon meaning cheap and common.

Op-Amps II: Departures from Ideal

Here are some more indications of the number available. Two leading manufacturers' websites show what each offers.

Texas Instruments The web site lists 1479 devices; their collection now includes all of National Semiconductor's couple of hundred op-amps.

Analog Devices The web site lists 980 devices; this includes about 350 of Linear Technology's op-amps as well as all of Maxim's designs.

And even a single manufacturer can make so many op-amps that it needs to sort them by type in order not to overwhelm its customers with one giant list. TI uses the following categories for its ordinary op-amps:

- Low Offset Voltage;
- Low Power;
- Low Noise;
- Low Input Bias Current;
- Wide Bandwidth;
- Wide Supply Voltage; and
- Single Supply Voltage.

And this set excludes more specialized op-amp types such as "Power Amplifiers and Buffers." The variety is daunting, despite this breakdown into sub-species.

Section 7N.2.9 includes a table from AoE giving the range of op-amp specifications available. A glance at this list reminds us – if we need reminding – that we can't have all virtues at once. Bipolars, for example, can beat MOSFET devices in *offset voltage* (V_{offset}) by a factor of perhaps 50 – but at the cost of input current (I_{bias}) several thousand times larger than for the CMOS part. That's to be expected, of course: an instance of electronic justice.

Another AoE table, in its §4.7, shows 34 "representative" op-amps, including the two that we rely on in most of our lab exercises – the LF411 and LM358. The set of specifications listed in that table is a bit wider than in the other – including input and output swing and power supply current. Does anything else in life come in so many tasty flavors?

Two circuits that make small imperfections matter: In two worked examples, we will look at two cases where the small DC errors *can* matter:

- a sensitive DC amplifier (§7W.1); and
- an integrator (§7W.2).

Now, let's look more closely at the most important of the op-amp's deviations from ideal performance.

7N.2.2   Offset voltage: V_{OS}

AoE §4.4.1A

The offset voltage is the difference in input voltage necessary to bring the output to zero.

This specification describes the amp's delusion that it is seeing a voltage difference between its inputs when it is not. One can represent this error as a DC voltage added to one of the inputs of an *ideal* zero-offset internal difference amp, as in the leftmost image of Fig. 7N.8.

This separation of idealized and realistic characteristics recalls the Thevenin model of §1N.4. Two consequences are suggested in Fig. 7N.8 – though the thought that one might hold the output at *zero* by applying just the right counter-error exaggerates what is possible.

The amplifier makes this mistake because of imperfect matching between the two sides of its

Figure 7N.8 V_{offset} represented as a voltage added to one input of an idealized op-amp; ...and consequences.

differential input stage. To achieve the impossible $V_{\text{offset}} = 0$, the op-amp would have to match perfectly not only the transistors of its differential amplifier but also the transistors of the current-mirror that serves as that first-stage's *load*. These elements are shown in a simplified op-amp circuit diagram, Fig. 7N.11. The full '411 schematic appears in Fig. 9S.31 on page 428. Figure 7N.9 shows the V_{OS} specifications for our principal op-amp.[7]

Symbol	Parameter	Conditions	LF411A			LF411			Units
			min	typ	max	min	typ	max	
V_{OS}	Input offset voltage	$R_S = 10\,\Omega, T_A = 25°C$		0.3	0.5		0.8	2.0	mV

Figure 7N.9 '411 spec: V_{offset}.

You can compensate for this mismatch by deliberately drawing more current out of one side of the input stage than out of the other, in order to balance things again. This may sound like a rash thing to do – after all, the manufacturer has done its best to achieve matching; is there any hope that your heavy hand can reach into the brain of the op-amp and improve things?!

Actually, the answer is "Yes," because you will make the correction while watching the op-amp output with a sensitive voltmeter. Thus you will know when you have optimized V_{offset}.[8] This correction is called *trimming offset*, and you will do it in Lab 7L. But this trimming is a nuisance, and the balancing does not last: time and temperature-change throw V_{offset} off again (see AoE §4.4.1B). It's a reasonable way to improve the performance of a "one-off"[9] lab instrument; it is not a reasonable approach for a circuit produced in large quantities (imagine the cost of adding a human adjustment step at the end of an assembly line; and even if the assembly line automates this adjustment, the operation adds to cost).

AoE §4.7

Self-trimming op-amps: "auto zero": Some op-amps are able to trim their own offsets by including – integrated with the main amplifier – a second "nulling" op-amp dedicated to "auto zeroing" the main amp.

The nulling amp works in two cycles – running at a few tens or hundreds of hertz – and thus is said to "chop" between the two conditions. This periodic nulling process, "chopper-stabilizing," is not the same as the early "chopper" op-amp design, which periodically disconnected the main amplifier's inputs, and thus provided a *sampling* effect that severely limited bandwidth. Chopper stabilizing, in contrast,

---

[7] The letter "A" after the op-amp part number, 411, indicates a "grade." The same device, but different specifications or operating temperature range. Suffix letters can also indicate the package type. Here, the suffix indicates better typical and worst case values for a number of parameters. In some devices, the part number itself changes to indicate different specifications: For example, the LM358A, LM258A, and LM158A are all the same part spec'ed for three different temperature ranges.

[8] You probably recognize that you – the relatively clumsy human twiddling a potentiometer – are able to perform this delicate surgical adjustment only because you are, while watching the output voltmeter, a part of a *feedback* loop.

[9] "One-off" is jargon for a circuit you make just *one* of, in contrast to something produced in quantity.

306　Op-Amps II: Departures from Ideal

leaves the main amplifier connected at all times, and therefore does not diminish high-frequency performance.[10]

The main benefit that results from this nulling is, of course, extremely low V_{offset}: as low as about 1 μV (typical) and 5 μV (max).[11] Secondary benefits include enormous open-loop gain and very good immunity to all DC and low-frequency errors: power-supply rejection, V_{offset} drift with time and temperature.

You will meet such a chopper-stabilized op-amp, the LTC1150, in Lab 7L. And we hope that the exercise of trimming down the '411's V_{offset} in that lab will not obscure the general truth that usually it will be wiser to:

- use a good op-amp, with low V_{offset}; and
- and design the circuit to work well with the V_{offset} of the amp you have chosen.

Figure 7N.10 offers a silly reminder that what we're trying to do when we trim offset is to undo an asymmetry built into the op-amp. And Fig. 7N.11 shows a more literal diagram of what trimming does: the trimming network reaches into the input stage of the op-amp, and draws more current out of one side than the other – just enough to offset the built-in imbalance of currents.

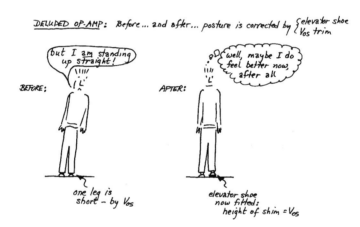

Figure 7N.10 Offset trim as shim in an elevator shoe.

7N.2.3　Bias current: I_{bias}

AoE §4.4.1C

The bias current, I_{bias}, is a DC current flowing in or out at the input terminals (it is defined as the average of the currents at the two terminals); see Fig. 7N.12.

For an amplifier with bipolar transistors at the input stage, I_{bias} is base current; for a FET-input op-amp like the '411, I_{bias} is a leakage current – tiny, but growing rapidly with temperature: see Fig. 7N.13.

See AoE Fig. 4.55

Balancing resistive paths to minimize the effects of I_{bias}: The bias current flows through the resistive path feeding each input; it can, therefore, generate an input error voltage, which may be

---

[10] In case you're curious to hear a quick account of how this nulling works, here goes.
  The nulling op-amp first nulls its own offset. This it does by tying its two inputs together and driving its output back to its own offset-null terminal (a single input that accomplishes what we do by hand on the '411, with a pot between two terminals). Then the nulling amp connects its inputs to the main amp's inputs and uses its output to drive the *main* amp's offset-null input. Capacitors store the two nulling voltages, which must be maintained during the periodic chopping. Negative feedback thus will drive the difference between the main amplifier's inputs very close to the ideal of equality.

[11] A very good explanation by an ADI engineer appears in an *Analog Dialog* note, "Demystifying Auto-Zero Amplifiers-Part 1", **34**, no.2, March 2000, https://LAoE.link/Auto-Zero_Amplifiers.html.

7N.2 Op-amp departures from ideal

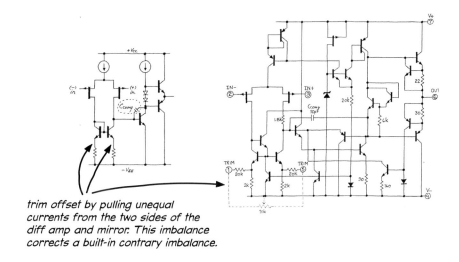

trim offset by pulling unequal currents from the two sides of the diff amp and mirror. This imbalance corrects a built-in contrary imbalance.

Figure 7N.11 Inside the '411. Schematics: simplified, and detailed views show how offset trim works.

Symbol	Parameter	Conditions		LF411A			LF411			Units
				min	typ	max	min	typ	max	
I_{os}	Input offset current	$V_S = \pm 15V$	$T_J = 25°C$		25	100		25	100	pA
			$T_J = 70°C$			2			2	nA
			$T_J = 125°C$			25			25	nA
I_B	Input bias current	$V_S = \pm 15V$	$T_J = 25°C$		50	200		50	200	pA
			$T_J = 70°C$			4			4	nA
			$T_J = 125°C$			50			50	nA

Figure 7N.12 '411 Specs: I_{bias} and I_{offset}.

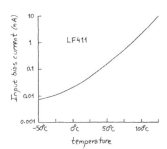

Figure 7N.13 '411 bias current: tiny, but grows fast with temperature.

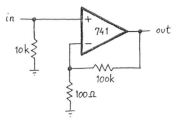

Figure 7N.14 Demonstration circuit: uses high-gain dc amp to make errors measurable.

amplified highly to generate an appreciable output error. The high-gain DC amplifier in Fig. 7N.14 shows a deliberate mismatch of resistive paths driving the two inputs. We sometimes used this circuit in order to demonstrate the effects of I_{bias} and V_{offset}:

308 Op-Amps II: Departures from Ideal

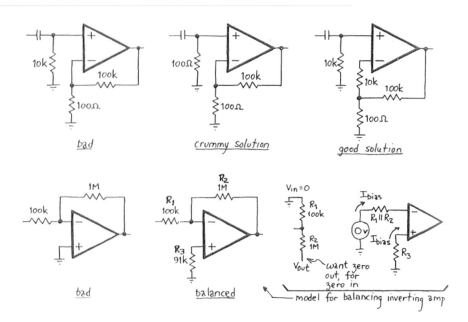

Figure 7N.15 Balanced resistive paths minimize output errors resulting from bias current.

The bias current of 80 nA (typical) flows through 10k in one path, through 100 Ω in the other,[12] developing a difference signal of about −0.8 mV at the non-inverting input.[13] This pseudo-signal would produce a spurious output of about −0.8 V, assuming no V_{offset} (the assumption that $V_{\text{offset}}=0$ is not realistic; but we like to think about one problem at a time).

... But you may not need to balance paths if I_{bias} is very low: But the circuit of Fig. 7N.14 doesn't show you I_{bias} effects unless you take the trouble to use an antique op-amp – a *'741* rather than a '411. A '411 gives no measurable effect from I_{bias}. The fact that this demonstration required a '741 suggests that often you will *not* need to worry about the effects of bias current; and for less than a dollar you can find an op-amp with a bias current 10^3 smaller than the '411's. The objection that balancing paths may not be necessary is fair; but you should know how to judge whether or not to worry about the effects of I_{bias}.

If I_{bias} effects are troublesome, then you should take the trouble to minimize this disturbance by matching the resistances of the paths to the two op-amp inputs. Figure 7N.15 gives some examples of circuits that do or do not balance paths.

Once you have balanced these resistive paths, I_{bias} no longer causes output errors. But a difference between currents at the inputs still does. That difference is called I_{offset}.

7N.2.4 Offset current: I_{offset}

AoE §4.4.1D

Offset current is the difference between the bias currents flowing at the two inputs. For the '411 the I_{OS} specification is about $\frac{1}{2} I_{\text{bias}}$; for the bipolar op-amps I_{OS} is smaller relative to I_{bias}. But recall how *tiny* I_{bias} is for the '411 and other FET-input devices.

I_{OS} predicts an error even when the resistances seen by the two inputs are balanced. Remedy? Use resistances of moderate value (under a few 10s of MΩ; see the argument for the clever *T* resistor trick noted later in §7N.3.2.)

---

[12] This is just R_{Thev} for the divider that drives the inverting input.
[13] The voltage is negative because the '741's I_{bias} is base current drawn by its *npn* input transistors.

7N.2.5 Slew rate and rolloff of gain

AoE §4.4.1K

The limited values of both slew-rate and gain turn out to be caused by a gain-killing capacitor designed into the op-amp. This is a component that we indicated, but did not explain, back in Fig. 7N.11. We will talk about this *compensation* device in Chapter 9N when we consider op-amp stability.

Gain rolloff: For the moment, we will only note that the result of this "compensation" capacitor (so-called probably because it "compensates" for the circuit's tendency to oscillate spontaneously) is to roll off the op-amp's gain at −6 dB/octave. The op-amp behaves, in other words, as if its output had been passed through a simple *RC* lowpass. In effect, this *has* happened, though in the internal second stage. So, the chip's very high gain, necessary to make feedback fruitful, evaporates steadily with increasing frequency – and is *gone* at a few MHz (typically 4 MHz for the '411; see Fig. 7N.16).

Symbol	Parameter	Conditions	LF411 min	LF411 typ	LF411 max	Units
A_{VOL}	Large-signal voltage gain	$V_S = \pm 15\,V$, $V_O = \pm 10\,V$, $R_L = 2k$, $T_A = 25°C$	25	200		V/mV
		over temperature	15	200		V/mV
GBW	Gain-bandwidth product	$V_S = \pm 15\,V$, $T_A = 25°C$	2.7	4		MHz

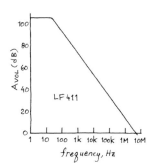

Figure 7N.16 '411 gain rolloff: spec and curves.

AoE §4.4.1J

The frequency, f_T, at which gain has fallen to unity, defines a frequency limit on the usefulness of *any* op-amp circuit.[14] This limit explains why not every circuit should be built with op-amps, wonderful though the effects of feedback are.

Slew rate: The compensation capacitor also explains the op-amp's slew rate. The slew rate, the maximum dV/dt of the op-amp output, is achieved when the maximum input-stage current – the entire *tail* current, issuing from a radically unbalanced differential stage – charges this compensation capacitor. For the '411, we can work backwards from the specified slew rate and the size of the compensation cap to infer the first-stage "tail" current, which is the maximum output from the differential stage:

$$I = C\,dV/dt = 10 \times 10^{-12} \times (15\,V/\mu s) = 150\,\mu A$$

Figure 7N.11 shows the compensation capacitor between base and collector of the high-gain second stage of the op-amp: the common-emitter amplifier.[15]

But we don't want to leave you with the impression that the '411's GBW of 4 MHz is a general ceiling on what you can expect. No, op-amps are available with almost any characteristic optimized (always at the expense of other traits). If you need an op-amp with GBW = 1 GHz, you can have one.[16]

---

[14] The parameter f_T is closely related, and usually identical, to "gain–bandwidth" product – "GBW." The two differ only in uncompensated or decompensated op-amps, which cannot stably run at unity gain. See Walter G. Jung, *IC Op-Amp Cookbook*, SAMS (3d ed., 1997), pp. 67–70.

[15] Well, that's almost a fair description; the circuit is complicated slightly by the presence of a follower driving the base of the common-emitter amp.

[16] For example, the TI/Burr–Brown's OPA640 promises f_T of 1.3 GHz. Current-feedback devices can run even faster.

7N.2.6 Output current limit

The current limit protects the small transistors of the output stage. The limit prevents overheating that otherwise would result when some clumsy user overloaded the amp. Most op-amps, including the '411, suffer no harm even from the blunder of shorting the output not just to ground but to either of the power supplies. We will postpone looking at how this current-limiting circuitry works until we reach Chapter 11N on *voltage regulators*.

You saw the effect of this current limit in a curve that appeared in Lab 6L. Figure 7N.17 shows this behavior again.

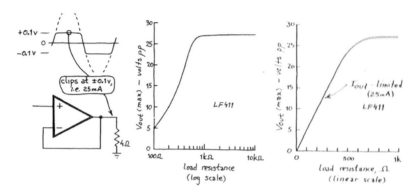

Figure 7N.17 Output current limit: output clips under load, despite very low R_{out}.

As you'd expect, high-current and high-voltage op-amps are available. But often you can get the high current you need by letting an op-amp serve as brains for a brawny current booster like the push–pull follower that you built in §6L.8. If however you insist on your right to be lazy, and want tens of amps at hundreds of volts, you can try Apex/Cirrus Logic: for example, MP165 offers 10 A (peak), 200 V, 100 W.

7N.2.7 Noise

Any amplifier adds some noise, as it works, and this effect – much too subtle to appear in our lab exercises – is specified for each op-amp. Two sorts of noise are specified: *voltage noise* (e_n) and *current noise* (i_n). The effects of the two add, but particular ranges of source resistance can make one or the other important. For high source resistance, current noise is the more important (the current noise generates a noise signal as it flows through R_{SOURCE}). The '411 datasheet boasts of its current-noise spec, among the "features" listed at the head of its datasheet: $i_n = 0.01$ pA/$\sqrt{Hz}$. Part of that head is reproduced in Fig. 7N.18.

7N.2.8 Input and output voltage range

Ordinary op-amps (like the '411)

Most op-amps cannot handle input voltages close to one or both power supplies, and cannot swing their outputs very close to one or both supplies. The '411 illustrates these limitations (power supplies assumed: '411A: ±20 V, '411: ±15 V). Note V_O and V_{CM} in Fig. 7N.19.[17]

---

[17] Complicating the specifications shown in Fig. 7N.19, the LF411 datasheet states, among its application hints, that the '411 will operate with a common-mode input voltage all the way to the positive supply but "...the gain bandwidth and slew rate may be reduced in this condition."

7N.2 Op-amp departures from ideal

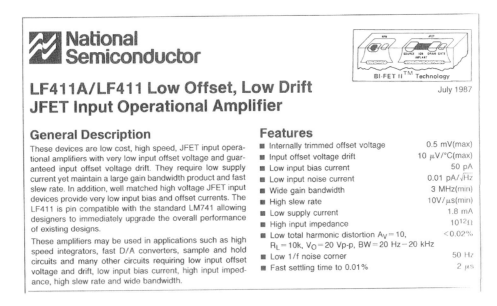

Figure 7N.18 LF411 datasheet boasts of its current–noise spec (courtesy of National Semiconductor).

Figure 7N.19 LF411 input and output voltage ranges do not go all the way to the supplies.

Figure 7N.20 LF411 datasheet confesses a weird hazard.

Application Hints

The LF411 series of internally trimmed JFET input op amps (BI-FET II™) provide very low input offset voltage and guaranteed input offset voltage drift. ...

Exceeding the negative common-mode limit on either input will force the output to a high state, potentially causing a reversal of phase to the output. ...

The consequence of exceeding "common-mode input range" can be startling. The '411 (like some other similar op-amps[18]) does a very weird thing if you take one or both inputs below the lower limit of that input range. This behavior is not advertised on page 1 of the datasheet, but instead is buried on page 8 among "Application Hints," shown in Fig. 7N.20.

We got stung by this misbehavior – which can turn negative feedback into positive – when we were breadboarding the "PID" motor-control circuit of Lab 10L. Eventually, we caught on, and substituted an op-amp (the LM358) that does not show this pathology.

The engineering community is aware of this strange behavior, and one company thought it might

---

[18] Our informant, the late Jim Williams (see below), lists several that do this same nasty trick; LF147, LF351, LF156. All are JFET-input devices, like the '411.

make some money by redesigning a popular op-amp to eliminate this phase reversal. The late Jim Williams, a longtime applications engineer and analog wise man at Linear Technology Co., told a striking story about this effort. His company made such an improved op-amp, and sent samples and simulation models to General Motors, which had been using a '411-like op-amp in the transmission control for its diesel locomotives.

GM reported back that when they simulated use of the "improved" op-amp, their locomotives now and then would jump into reverse. Everyone agreed to drop plans to substitute the "improved" op-amp. Apparently, GM had adapted its software to the nasty phase reversal, and when this reversal did not occur, trouble occurred. The case reminds one of the old folk tale (plausible enough) of a person who wakes up because a clock is silent – fails to chime the hour. GM's locomotives awoke (jumping into reverse) when their op-amps failed to flip phase.

Williams' story also reminds us why no one is in a hurry to plug an "improved" IC into a working design. In turn, this disinclination explains the curious fact that today's version of an old IC like the LM741 op-amp is much better than the original – but cannot boast of this on its datasheet. If it made such a boast – changing the datasheet – then it would no longer be a '741, and any designer proposing to buy this *improved* part for an existing design would need to persuade her bosses that it was worthwhile to run all the tests required before the new part could be adopted.

Rail-to-rail op-amps

Some op-amps show common-mode input range better than the '411s. "Single supply" types (like the '358 that you will meet today in Lab 7L) have a common-mode input range that includes the "negative supply" – usually just *ground* in this context. Another class of op-amps does better than that – the so-called "rail-to-rail" types.

Some accept *inputs* rail-to-rail; some can drive their *outputs* rail-to-rail; a few do both. Those are nice features – but like all other optimizations they come at the price of some other characteristic, such as offset voltage. So, you'll normally choose a rail-to-rail type only when you need that behavior. But we should admit that the need for this behavior is becoming more frequent as typical power supply voltages fall. With supplies of ±15 V, giving up a volt or so close to each rail isn't so bad. Giving up that much is unacceptable when the supply is, say, a single 3 V coin cell.

AoE §4.4.1F

A configuration permitting a *circuit* to show huge common-mode input range

A "difference" or "differential" amplifier made with an op-amp can permit a common-mode input range that extends far beyond its power supplies. That claim may seem to contradict the tight constraints mentioned in §7N.2.8, but it does not: it describes the wide swings permitted at *circuit* inputs rather than *op-amp* inputs. If that seems strange, recall that an ordinary *inverting* op-amp circuit could do the same trick, since V_{in} is not applied to the op-amp itself.

Using the difference-amp configuration, an op-amp IC from Texas Instruments[19] (INA149) promises an astonishing common-mode input range of ±275 V. The differential configuration, described in §7N.5 permits this result.

7N.2.9   Some representative op-amp specifications: ordinary and premium

AoE §4.4.1

Here is part of a table from AoE, illustrating the range of values for op-amp specifications. The "premium" values are the best available, and as AoE warns us, you can't combine the best in several specifications simultaneously; you must choose your op-amp according to what characteristics are most important to you.

---

[19] The design originated with Burr-Brown, which TI acquired.

7N.3 Five more applications

Parameter	bipolar		JFET-input		CMOS		Units
	jellybean	premium	jellybean	premium	jellybean	premium	
V_{OS} (max)	3	0.025	2	0.1	2	0.1	mV
I_B (typ)	50 nA	25 pA	50 pA	40 fA	1 pA	2 fA	@25°C
f_T (typ)	2	2000	5	400	2	10	MHz
SR (typ)	2	4000	15	300	5	10	V/μs

You may recognize the specifications for "JFET-input… jellybean" as almost exactly those of the '411 op-amp that you use in most of our labs' op-amp circuits.

7N.3 Five more applications: integrator, differentiator, difference amplifier, instrumentation amplifier, AC amplifier

7N.3.1 Integrator

To appreciate how very good an op-amp integrator can be, we should recall the defects of the simple RC "integrator" you met in Chapter 1N.

AoE §1.4.4

Passive RC integrator

To make the RC behave like an integrator, we had to make sure that

$$V_{\text{out}} \ll V_{\text{in}}$$

This kept us on the nearly straight section of the curving exponential-charging curve, when we put a square wave in. The circuit failed to the extent that V_{out} moved away from ground. But the output *had* to move away from ground, at least a little, in order to give an output signal: see Fig. 7N.21.

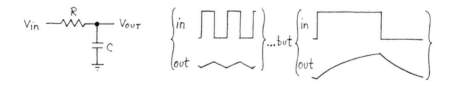

Figure 7N.21 RC "integrator:" integrates, sort-of, if you feed it the right frequency.

AoE §§4.2.6, 4.5.5

Op-amp version

The op-amp integrator solves the problem elegantly, by letting us tie the cap's charging point to 0 V, while allowing us to get a signal out: "virtual ground" lets us have it both ways: see Fig. 7N.22.

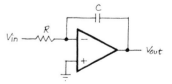

Figure 7N.22 Op-amp integrator: virtual ground is just what we needed.

As the scope image in Fig. 7N.23 shows, the circuit output can be larger than V_{in} without distorting the integration.

The op-amp integrator is so good that one needs to prevent its output from sailing off to saturation

314 Op-Amps II: Departures from Ideal

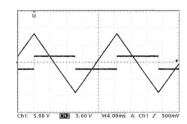

Figure 7N.23 Op-amp integrator eliminates requirement $V_{out} \ll V_{in}$.

(that is, to one of the supplies) as it integrates error signals: over time, a tiny lack of symmetry in the input waveform will accumulate; so will tiny op-amp errors. The scope image in Fig. 7N.24 shows the integrator output near negative saturation, then working for a while as the DC level drifts upward, till it bumps against the positive limit of the op-amp's range.

We should admit, therefore, that the integrator of Fig. 7N.22 was a defective circuit. We showed only a capacitor in the feedback path to keep the circuit conceptually simple. We will need to do more to make it usable.

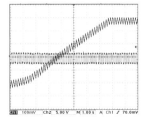

Figure 7N.24 Integrator without DC feedback drifts to saturation.

A simple way to describe what's wrong with the integrators in Figs. 7N.22 and 7N.7 is to point out that the circuit includes *no DC feedback at all* (this is a point we noted back in §7N.2). Such an arrangement guarantees an eventual drift to saturation.

So, practical op-amp integrators include some scheme to prevent the cap's charging to saturation.

7N.3.2 Taming the op-amp integrator: prevent drift to saturation

AoE Fig. 4.66

One remedy: a large resistor: A large resistor in parallel with the cap prevents saturation. This R this leaks off a small current, undoing the effect of a small input error current. The right-hand circuit of Fig. 7N.25 shows a so-called "T network," which can obviate the use of very large R values; this network is explained in §7N.3.3.

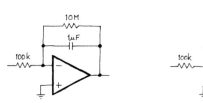

Figure 7N.25 Integrator saved from saturation by resistor parallel $C_{feedback}$.

When $R_{feedback} = 100 \times R_1$, as it is in Fig. 7N.25, then the leakage current through $R_{feedback}$ is only 1% of the signal current when V_{out} is comparable to V_{in}. Often, that imperfection in the integrator's performance is acceptable.

Effects of the resistor: Evidently, the resistor compromises performance of the integrator. We can figure out by *how much*. There are several alternative ways to describe its effects:

The resistor limits DC gain: In the circuit above, where R_{in} = 100k and $R_{feedback}$=10M, the DC gain is -100. So a DC input error of ± 1 mV $\Rightarrow$ output error of ± 100 mV. The integrator still works fine, apart from this error.

The resistor allows a predictable DC leakage: Suppose we apply a DC input of 1 V for a while; when the output reaches -1 V, the error current is 1/100 the input or "signal" current because, as we justed noted, $R_{feedback} = 100 \times R_{in}$. This error grows if V_{out} grows relative to V_{in}.

The resistor does no appreciable harm above some low frequency: At some low frequency, X_C becomes less than R, and soon R is utterly insignificant. As you know, $X_C = R$ at $f = 1/2\pi RC$. For the components shown in Fig. 7N.25 (where $R_{feedback}$=10M, C=1 μF) that frequency is about 0.02 Hz!

The resistor can become troublesome...: *At first, a bit of curvature shows the distortion...*

The effect of the feedback resistor becomes troublesome when V_{out} is much larger than V_{in}, as in Fig. 7N.26, where the curvature shows that we are not watching true integration, which should show a straight *ramp* waveform.

...at an extreme, the limited DC-gain becomes obvious: the circuit is then not *an integrator*

Pushing on in the same direction, if we let V_{out} exceed V_{in} by a great deal – an effect we get if we apply a very small input at very low frequency – then the feedback resistor destroys the integrating effect. The waveform in Fig. 7N.27 shows the compromised integrator revealing itself as no more than a $\times 100$ DC amplifier, after a while.

But don't let these distortions scare you away from using a feedback resistor in an integrator: to show them we had to go out of our way, applying extremes of low input amplitude and low frequency.

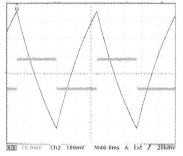

Figure 7N.26 Distortion of integral by feedback *R* becomes apparent if V_{out} is much larger than V_{in}. (Scope gains: input (small square wave), 10 mV/div; output (triangular), 100 mV/div).

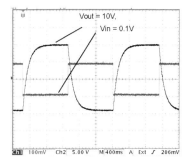

Figure 7N.27 Pushed to an extreme, the integrator is hopelessly compromised by the feedback resistor. (Scope gains: input (small square wave), 100 mV/div; output (exponential), 5 V/div).

7N.3.3 A resistor T Network

AoE Fig. 4.66

When a design calls for very large R values – as may occur for an integrator's feedback resistor, for example – a so-called "T network" permits use of modest resistor values, avoiding some of the harms that enormous Rs can introduce.

Figure 7N.28 offers a diagram intended to persuade you that the clever T resistor arrangement of Fig. 7N.25 can indeed make a 100k resistor look about 100× as large.

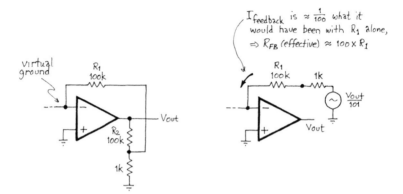

Figure 7N.28 How the T arrangement enlarges apparent R values.

Neat? The scheme is useful because the lower R_{Thevenin} of the T feedback network has two good effects:

- it generates smaller I_{bias} errors than the use of giant resistors would (see §7N.2.3); and
- it drives stray capacitance at the op-amp input better (avoiding unintentional lowpass effects in the feedback); this is not important here, but is in a circuit where no capacitor sits parallel to a big feedback resistor.

The T scheme does present one drawback: it amplifies input *voltage* errors – V_{offset}, §7N.2.2, even when the signal source is a current source. If this remark is cryptic, reconsider it after looking at §7W.1 where we contrast the case of two integrator drift rates, for *voltage* sources versus *current* sources.

Another remedy: a switch

AoE Fig. 4.18

A switch in parallel with the cap also can keep the integrator from drifting to saturation. This has to be closed briefly, from time to time, and this requirement can be a nuisance.

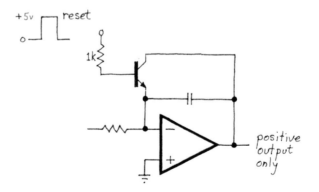

Figure 7N.29 Integrator saved from saturation by discharge switch.

The switch permits a more perfect integrator (no leakage current through a feedback R compromises its integration). Whether it is worthwhile to gain this advantage at the expense of rigging a resetting signal will depend on your application. Note the restriction that the transistor *reset* of Fig. 7N.29 fails if the integrator output goes negative. Only when we meet the *analog switches* made with MOSFETs, in Chapter 12N, will we be able to implement an transistor reset switch that works for both polarities of integrator output.

7N.3.4 An integrator can ask a lot from an op-amp, and can show the part's defects

A high gain amplifier like the one mentioned in §7N.2.3 can demonstrate effects of small input errors: V_{offset} and I_{bias}. So can an integrator, since it accumulates the effects of errors.

A demonstration circuit: integrator shows effects of op-amp imperfections

As a demonstration, we set up a contest between an old bipolar op-amp, the LM741, and an ordinary FET-input op-amp, the LF411. We built the circuit in Fig. 7N.30 twice:

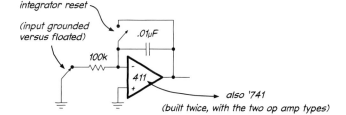

Figure 7N.30 Integrator circuit used to watch effects of V_{offset} and I_{bias}.

We first *grounded* the inputs to both of the twin integrators, and watched the drift. This appears in the leftmost image of Fig. 7N.31. The '411 wins – but not by a great margin: perhaps 15:1. Then we opened both inputs – mimicking what a *current* source would look like if connected to the integrator input. This produced an extreme victory for the FET-input (low I_{bias}) '411. In order to see *any* drift in the '411 integrator, we had to crank the scope sensitivity way up (to 50mV/div) – and we had to be patient.

7N.3.5 An example of a case where periodic integrator reset is adequate

A former student showed us this case, Fig. 7N.32, where manual resetting was quite practical (the student was collecting a measure of total response of a frog muscle to a stimulating pulse).

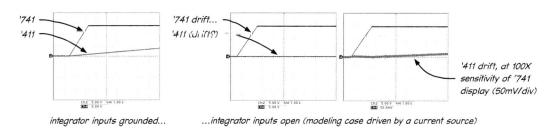

Figure 7N.31 Drift rates for competing integrators: FET-input device wins – but much more dramatically when driven by current source. (Scope gains: all 5 V/div except '411 drift in rightmost image, 50 mV/div)

Worked Example 7W.1 looks in greater detail at this contrast between the two op-amps and their errors,.

The duration of the integration was long, and the pace of the experiment slow enough, that it made

318 Op-Amps II: Departures from Ideal

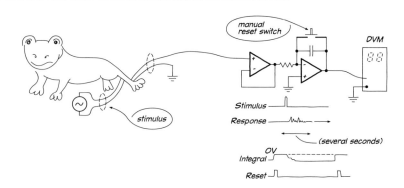

Figure 7N.32 A slow integration can make manual reset practical.

sense for this circuit to start with a manual reset (to clear the integrator), then apply the stimulus, then record the integrated value, all over the course of many seconds. An accurate – low-drift – integrator is required of course (and it was in pursuit of this that the former student returned to our lab).

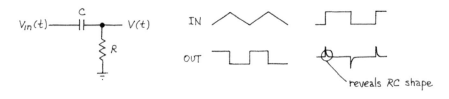

Figure 7N.33 *RC* "differentiator:" differentiates, sort-of, if *RC* kept very small.

7N.4 Differentiator

AoE §4.5.7

Again, the contrast with a passive differentiator helps one appreciate the op-amp version. Figure 7N.33 reminds us of the passive circuit, which you met in Lab 2L. To make the passive version work we had to make sure that

$$dV_{out}/dt \ll dV_{in}/dt$$

Again the op-amp version exploits *virtual ground* to remove that restriction.

But the differentiator is less impressive than the op-amp integrator. The op-amp version of the differentiator must be compromised in order to work at all: its gain must be ruined at high frequencies in order to keep it stable.

A practical differentiator, shown in Fig. 7N.34, turns into an integrator (of all things!) at some high frequency. This scheme is necessary to prevent unwanted oscillations (we will look more closely at this topic in Labs 9L and 10L). Figure 7N.35 is a preview of the stability problem that we look into later. The simple, impractical differentiator is marginally stable, and shivers in response to a steep waveform (the input signal is a square wave). That shiver is unacceptable, so we trade away high-frequency performance for decent operation in a restricted range.

7N.5 Op-amp difference amplifier

AoE §4.2.4

Figure 7N.36 shows a standard differential amp circuit made with an op-amp. The circuit's gain is R_2/R_1 (this is *differential* gain, of course; ideally, the common-mode gain is zero). Its common-mode

7N.5 Op-amp difference amplifier

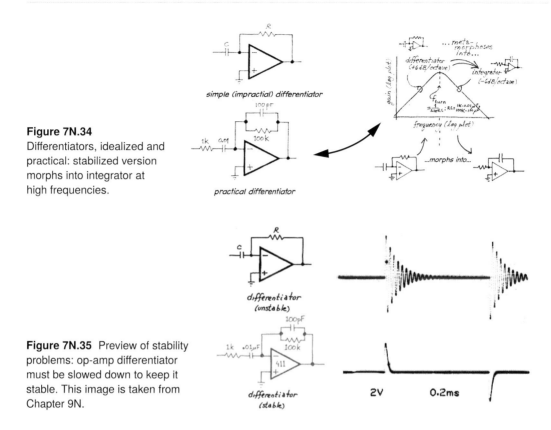

Figure 7N.34 Differentiators, idealized and practical: stabilized version morphs into integrator at high frequencies.

Figure 7N.35 Preview of stability problems: op-amp differentiator must be slowed down to keep it stable. This image is taken from Chapter 9N.

rejection can be very good if the ratios of the two dividers are well matched, as they can be when the resistor sets are integrated. In TI's INA194 diff-amp IC, for example, the resistors are integrated with the op-amp and laser-trimmed toward equality. CMR[20] for the INA194 is specified at 90 dB.

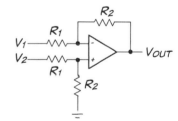

Figure 7N.36 Simple differential amp, made with an op amp.

In §6W.1 we offer some informal analysis of this standard difference amplifier, hoping to make its behavior comfortable to your intuition. We won't reproduce all that here, of course. But we can make a quick start in the direction of making sense of the circuit by noting that if the two input voltages are *equal*, then the two dividers must show equal voltages at their right-hand ends. One is grounded, so the op-amp output must also sit at zero volts. In other words, equal inputs produce $0\,V$ out. That make sense. You may want to try out some other particular cases: try grounding one input; try grounding the other. And so on.

---

[20] "CMR" in this case is the same as the more familiar "CMRR," the ratio of good gain to bad gain – since "good gain," differential, is 1.

7N.5.1 A curious side virtue: wide input range

A difference amp can tolerate a surprisingly-wide common-mode input range. Recall that common-mode input voltage implies *no* V_{out}; a small *difference* between inputs sharing a large common-mode signal implies only a *small* V_{out}. Example 6W.1, and Fig. 6W.7, discussed an integrated difference amplifier, INA149, with an amazing common-mode input range of ±275 V.

7N.6 Instrumentation amplifier

Now that you are acquainted with the difference amplifier, and have seen some specialized op-amp differential amplifiers described in §§6W.1 and 6W.2 (the latter amp designed for wide input voltage, well beyond the supplies), let's take a look at another particular flavor of diff amp. The *instrumentation amplifier* is a difference amplifier optimized for high input impedance and good CMRR.

7N.6.1 A straightforward way to improve the op-amp difference amplifier

AoE §5.15.1

If one wants to improve on the mediocre input impedance of a standard diff amp made with op-amps, surely the first thought would be to insert followers on the inputs: see Fig. 7N.37. This shows the design AoE describes as "naive."[21]

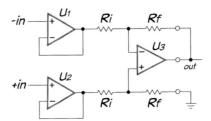

Figure 7N.37 Buffered Differential Amp: a first step

This design does improve input impedance but does not improve CMRR, as AoE points out, and still requires the usual excellent matching of resistances in the diff amp.[22] We can do better.

7N.6.2 Three op-amp instrumentation amplifier ("IA")

AoE §5.15.2

The ingenious circuit of Fig. 7N.38 is better because its first stage can provide differential gain, along with unity common-mode gain (thus it provides CMR). The diff amp stage then operates on an already-amplified signal, and can run at very modest differential gain – often just unity – and its CMRR (in dB) is added to the CMRR of the first stage.

Common-mode gain: If you drive the two inputs with a shared signal, no current flows in R_g. So, the two input op-amps operate just as followers. CM gain, then, is unity.

Differential gain: One way to calculate this gain recapitulates the way we analyzed the bipolar diff amp of Lab 5L. Assume a symmetric input: $+\Delta$ on one input, $-\Delta$ on the other. Then, if you chose to draw R_g as two series resistors each of $R_g/2$, with their junction labeled "A," we would have a case much like that described in §5N.7, Fig. 5N.29.

---
[21] See AoE, §5.15.1.
[22] AoE notes that the buffers even slightly degrade CMRR, putting two imperfect amplifiers in the signal path. AoE p. 357.

7N.6 Instrumentation amplifier

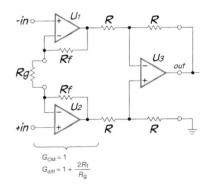

Figure 7N.38 Three op-amp instrumentation amplifier: a better design.

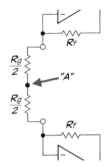

Figure 7N.39 A trick reminiscent of Lab 5L, to calculate differential gain.

Point "A" does not move in response to the symmetric input.

The gain of each op-amp, with respect to this fixed voltage, is $1 + \frac{R_f}{R_g/2} = 1 + 2\frac{R_f}{R_g}$. The total output voltage is double that, so you might be tempted to double this value. But total input voltage (as for the bipolar diff amp) is not ΔV but $2\Delta V$. So the differential gain is, indeed, $1 + 2\frac{R_f}{R_g}$.

If you don't like this trick of splitting R_g, you can get the same result by calculating currents that flow in response to a symmetric drive of $2\Delta V$.

7N.6.3 Wheatstone bridge

A common use of the instrumentation amplifier is to measure a small difference between two DC signals, as this requires a circuit that amplifies the difference between the signals while ignoring the common DC level. In this lesson's lab we create such as signal using a Wheatstone Bridge, to generate a small difference signal – of a few millivolts – for an IA to amplify. This bridge circuit is well adapted to generating a measurable *difference* voltage riding on a substantial common-mode voltage. The particular case that we ask you to try uses a bridge made up of four "strain gauge" sensors. The resistance of each changes very slightly when it is stretched or compressed. Strain gauges are used not only to create weigh scales, but also to monitor excessive deformation in structural members.

We do not ask you to work with a single strain gauge, first, because the work of attaching it to a piece of metal is a large task in itself. Second, a single gauge is vulnerable to temperature effects. Instead, we will use a manufactured "load cell" which uses four gauges installed to form a bridge. The bridge produces a signal larger than what single gauge could, and its symmetry protects against temperature effects.

Here is a schematic of the load cell:[23]

The strain gauges are mounted on a metal bar such that one of the gauges mounted on the upper

---

[23] Modified from: Daraceleste, https://commons.wikimedia.org/wiki/File:Wheatstone_bridge_with_gauges.jpg.

Op-Amps II: Departures from Ideal

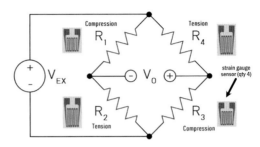

Figure 7N.40 Wheatstone bridge load cell.

surface and one on the lower surface of the cell are stretched when the cell end is pressed down. At the same time, two others also mounted on the upper and lower surfaces are compressed. If the load cell is clamped at one end to a rigid surface and not allowed to "droop," then the loading looks as in Fig. 7N.41.[24] The compression-tension pattern is proportional to the applied force by Hooke's law, as long as the force does not exceed the elastic limit of the bar.

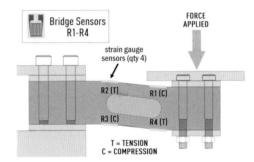

Figure 7N.41 Load cell strains proportional to applied force if one end of load cell clamped.

We need to measure the small difference voltage, V_O, and amplify it. The bridge behavior is simple, at a glance, but subtle if one looks closely.

7N.6.3.1 The simple, rough view

The excitation voltage, V_{EX}, may be DC or AC. Since the bridge's output voltage is proportional to the excitation voltage, higher is better but it is usually limited by the maximum voltage of the sensors in the bridge. If all R's are equal, the left-hand and right-hand dividers generate equal voltages at their midpoints and V_O, the differential signal fed to the instrument amplifier, is zero. If one R changes in value, the voltage division becomes unbalanced, and V_O becomes non-zero. V_O may be small, but we hope the IA will amplify this signal, ignoring the much larger common-mode signal (typically one-half V_{EX}) on which it rides.

7N.6.3.2 What's good about the Wheatstone bridge?

To appreciate what the Wheatstone bridge does for us, it helps to try doing the same task without the bridge. Suppose, as in today's lab, that our goal is to amplify a small signal that results from stretching or compressing a gauge. The strip's resistance is about 1k at rest.

If you hadn't met the bridge, how would you sense and amplify the effect of the bar's bending?[25]

We could run a current through the gauge to ground, and amplify the voltage across the part. Running

---

[24] Modified from: Daraceleste, https://commons.wikimedia.org/wiki/File:Strain_gauge_2b.jpg.

[25] You cannot capacitively couple the load cell's output to remove the DC component because the load on the strain gauges typically changes slowly or not at all over time.

7N.7 AC amplifier: an elegant way to minimize effects of op-amp DC errors

a constant 2.5 mA through it would give us a V_O of about 2.5 V. When we bend the bar, if we had just one strain gauge, we should see V_O vary a little.

But only *very* little. For example, trying this load cell, we found that strong finger pressure applied to bend the bar produced a change in voltage of about ±0.15 mV, at the junction of two opposed gauges (one stretched by the bend, the other compressed) connected to +5 V. Since the current is not changed by the flexing (one R growing as the other shrinks), what the change in voltage reflects is the small change in the resistance of the gauges.[26]

We would like to amplify this small signal perhaps by at least a few hundred. This small *difference* signal rides on a substantial DC voltage (≈ 2.5 V). How should we do this amplification?

We could compare this slightly-changed voltage to a reference voltage representing the un-stressed state: 2.5 V. So, we need a divider to generate the reference voltage.

Well, yes, we have re-invented the Wheatstone Bridge. The load cell that you will try today behaves better than the single strain gauge, as we have said in §7N.6.3 on page 321. In lab, we will use this load cell, with its built-in Wheatstone bridge, to drive an instrumentation amp.

7N.7 AC amplifier: an elegant way to minimize effects of op-amp DC errors

AoE §4.2.2A

If you need to amplify AC signals only, you can make the output errors caused by V_{os}, I_{bias} and I_{os} negligible in a clever way: just cut the DC gain to unity, illustrated in Fig. 7N.42.[27]

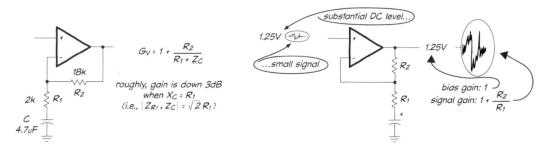

AC amp: general case...   ...and as applied to lab's single-supply microphone amplifier

Figure 7N.42 AC amplifier: neatly makes effects of small errors at *input* small at *output*.

This is the circuit used in Lab 7L to amplify a microphone's signal while using just a single power supply rather than the usual "split supply" that you are accustomed to in op-amp circuits. The use of one supply obliges us to bias the circuit input positive, presenting a case just like the one in Fig. 7N.42 on the right: a small wiggle rides a DC offset or bias. We want to amplify the wiggle, but not the DC bias.

What is f_{3dB} for the circuit shown in Fig. 7N.42? If one ignores the "1" in the gain expression, output amplitude is down 3 dB when the denominator in the gain expression is $\sqrt{2} \times R_1$. But that happens when $X_C \equiv |Z_C| = R_1$, and this occurs, as you well know, at $f_{3dB} = 1/(2\pi R_1 C)$. (Don't be fooled by

---

[26] We can infer the change in resistance that this small voltage difference implies. Current is 5 V/2k, 2.5 mA. The change in resistance, then, should be
$\Delta R = \frac{\Delta V}{I} = \frac{0.15\,\text{mV}}{2.5\,\text{mA}} = 0.06\,\Omega$
This fractional change of resistance is very small: $\frac{0.06\,\Omega}{1\,\text{k}\Omega} = \frac{6}{100{,}000}$, 60 parts in a million.

[27] The subscript v in the gain expression of Fig. 7N.42 indicates *voltage* gain in contrast to, say, *current* gain. Usually we omit this notation.

324 Op-Amps II: Departures from Ideal

the graphical detail that the capacitor is tied to ground, making the *RC* look something like a lowpass. You're not fooled. You recognize it as a *highpass* in sheep's clothing.)'

The *RC* analysis we need here is the same one that you used in choosing the emitter-bypassing capacitor for an amplifier back in Chapter 5N. Figure 7N.43 is a reminder of that circuit, a high-gain common emitter amp. The *single-supply* op-amp design, with its need for a *bias* voltage, is exceptional in our op-amp work, but you know it well from your work with discrete transistors.

AoE §2.3.5A

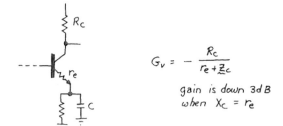

Figure 7N.43 Gain of AC amp is down 3dB when denominator (series impedance of *R* and *C*) is up to $R\sqrt{2}$.

$$G_v = -\frac{R_c}{r_e + Z_c}$$

gain is down 3dB when $X_c = r_e$

7N.8 AoE reading

- Reading
    - §4.4: a detailed look at op-amp behavior.
    - §4.5: a detailed look at selected op-amp circuits.
    - §4.6: op-amp operation with a single power supply.
    - §4.7: other amplifiers and op-amp types.
- Problems:
    - Additional Exercises 4.31–4.33.

7L   Lab: Op-Amps II

This lab introduces you to the sordid truth about op-amps: *they're not as good as we said they were last time!* Sorry. But after making you confront op-amp imperfections in the first exercise, §7L.1, we return to the cheerier task of looking at more op-amp applications. There, once again, we treat the devices as ideal.

On the principle that people should eat their spinach before the mashed potatoes (or is it the other way round?) let's start by looking at the way that op-amps depart from the ideal model. The integrator exercise does what it can to make the potentially dull topics of *offset-voltage* and *bias current* less dull by letting you undertake a demanding task of integration – one that you could not carry out without first overcoming the effects of these two op-amp imperfections.

We want you to get to know the integrator as a circuit important in its own right. But it also serves well to demonstrate the possibly troublesome effects of op-amp errors that can *sound* negligibly small: *bias* currents of picoamps, *offset* voltages below a millivolt. The integrator is unforgiving: it accumulates the effects of such errors, of course, holding a grudge for picocoulombs of charge delivered in the distant past, many milliseconds ago. We've also rigged a setup in which you get to integrate a signal from a piece of hardware (a disc drive or DC motor), for a refreshing change from feeding always from a function generator. One could almost imagine that this disc-drive motion integrator is a useful circuit.

7L.1   Integrator

Try the integrator: Construct the active integrator shown in Fig. 7L.1 using a '411 op-amp. The pushbutton is redundant at the moment; the 10M feedback resistor will keep the output from wandering far astray. Soon, we will remove the feedback resistor and rely on a manual pushbutton reset. But not yet. Keep this circuit set up after you finish this integrator section. You will use it again with the differentiator in §7L.2.

Try driving your integrator with a signal in the range 50 Hz to 1 kHz – all three waveforms listed below. This circuit is sensitive to small DC offsets of the input waveform (its gain at DC is about 500);

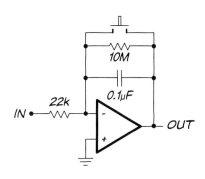

Figure 7L.1 Integrator.

the output is likely to go into saturation near the 15 V supplies. To prevent this, you will have to adjust the function generator's *DC-offset* control. That adjustment will be more manageable if you set the function generator's attenuation to 20 dB.

Try a square, sine and triangle waves. From the component values, predict the peak-to-peak triangle wave amplitude at the output that should result from a 2 V(pp), 50 Hz square wave input. Then try it.

Include feedback resistor: integrator tamed...: The 10M resistor tames this integrator – and somewhat corrupts it. The resistor can keep the output from wandering off to saturation, if you adjust the function generator carefully. But it makes the integration not quite honest – and not good enough for longer-term integration like what we mean to attempt soon. The feedback resistor permits a current to leak off the capacitor. For some applications this may not matter (the ratio of R_{feedback} to the input R is almost 500 to 1, so when V_{out} is comparable to V_{in} the error current caused by R_{feedback} is about 0.2% of the signal current). You'll probably be grateful for this feedback resistor while you try the integrator's response to various waveforms.

...Remove feedback resistor: integrator less tame but truer: When you remove the 10M resistor, the circuit will help you get a real gut feeling for the meaning of an integral. In fact, the attempt to adjust the generator's DC offset to keep V_{out} from saturating could drive you crazy. Don't suffer for long. Don't forget that the pushbutton can at least restore V_{out} to zero, for a fresh start.

7L.1.1 Integrator used to let one infer I_{bias} and V_{offset}

Drift; causation ambiguous: Ground the integrator's input, zero the output with the pushbutton, then release the button and watch the drift of V_{out}. You'll need a very slow scope sweep rate. Try 1 sec/div. (This is a good time to switch to using a digital scope, if one is available.) Note the drift rate. But note, also, that both I_{bias} and V_{offset} can contribute to this effect. To make things worse, the two effects may reinforce each other, or one may subtract from the other. Let's peel them apart.

Infer I_{bias} from drift rate with the effect of V_{offset} removed: Removing the effect of V_{offset} happens to be extremely easy: just float the input. Now V_{offset} does not force a current through the input resistor. So, V_{offset} does not contribute to the current that charges the capacitor. Given this simplification, see if your drift rate implies a value for I_{bias} that is in the '411's specified range: 50 pA typical, 200 pA max. You will need to be patient to measure this drift rate. Note it once you get a reading, and use the rate to calculate the value of I_{bias}.

Infer V_{offset} from drift rate: Now you should be in a position to estimate V_{offset} from the overall drift rate you saw as you started this lab. (You'll need to pay attention to the *sign* of each of these drift rates.) Compare your calculated value to the '411's specification: 0.8 mV typical, 2.0 mV max. What's your inferred V_{offset}?

7L.1.2 Making a low-drift integrator in two ways

Trim V_{offset} to a minimum ('411): We hope you found that V_{offset} accounts for most of your integrator's drift. Let's now solve that problem. Add the offset trim network of Fig. 7L.2. Ground the integrator's input, zero the output initially, with the reset pushbutton, then try to minimize the drift rate of the output.

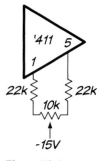

Figure 7L.2
Offset-trim network for '411 op-amp.

The trimming is a fussy task. Start with modest scope sensitivity – say, 1 V/div – and gradually increase sensitivity as you adjust V_{offset} down toward zero. Reset the integrator from time to time with the pushbutton. See if you can get the drift rate down to a few millivolts/second. When you've got it about as low as you can manage, see what residual V_{offset} your drift rate implies.

7L.1 Integrator

Now, lest you think that trimming is a really-satisfactory remedy for bad V_{offset} specs, heat or cool the '411 (with a heat gun, perhaps, or a can of component cooler), and see how good the drift looks. (In fairness to the '411, we should admit that even the good low V_{offset} chopper, coming soon, doesn't much like big changes of temperature. If you're energetic, you can repeat this test when you've wired up the chopper.)

Use a self-trimming op-amp: "chopper" amp, LTC1150: A "chopper" like the LTC1150 uses a second op-amp to drive a "nulling" input on the main amplifier – a single input that accomplishes what our manual offset-trim did, driving the difference between the two inputs of the main amplifier toward zero. Try this amplifier in place of the '411.

The pinout of this op-amp is the same as for the '411, so you can drop it into the circuit you wired for the '411 – except that *you must remove the trimming potentiometer, so as to leave pins 1 and 5 open.*

Ground the input – at the non-inverting input.[1] Watch the drift rate.

As for the '411, at this point the cause of drift is uncertain. Separate the effects of I_{bias} and V_{offset}, by eliminating the effect of V_{offset}, as you did for the '411. Infer the op-amp's V_{offset} from the drift rate that you saw with input grounded and keep a note of it. The LTC1150 promises V_{offset} of $\pm 0.5\,\mu\text{V}$ (typ), $10\,\mu\text{V}$ (max). We were chagrined to get values slightly *above* the specified maximum. We hope you'll do better![2]

7L.1.3 Apply the integrator: drive-motor position sensor

A person might reasonably ask, after watching an integrator in action, "But why would I want to integrate?" We have contrived one example of a signal that seems to want integrating to begin to answer this legitimate question. We'll admit that we cooked it up because it's kind of fun, rather than because it's useful. But it may set your imagination to conceiving other possible uses for an integrator.

Casting about for a signal whose integral might be interesting, we[3] thought of signals from coils moving in a changing magnetic field. The output voltage should be proportional to the rate at which the flux through the coil changes. We first tried pushing on a big audio speaker – and then thought of using the "voice coil" of a hard disc drive. The voice coil (so-called because the basic design was borrowed from audio speakers) is a simple motor that drives the drive head along the radius of the disc, so as to find or follow a disc track.

If opening up a dead disk drive is too much trouble, you can watch the same effect with the low-budget version shown on the left side of Fig. 7L.3. Those are two cheap little DC motors, to which we glued an arm cut from the plastic rotor at the far left. That rotor came from a hobby "servo motor" (a device that we will meet again in Lab 23L). The smaller motor is a 9 V device; the larger is a 3.3–5 V device.[4] Almost any DC motor will do, since you can adjust the integrator's sensitivity to your convenience (how?[5]).

We will use the DC motor not as a motor but as a *generator*. Instead of applying a voltage to move the disk-drive's "head" – or to spin the little DC motor – we will move the head by hand to generate

---

[1] We found that the location of our grounding connection on the breadboard caused drift-rate variations of more than a factor of two. The small currents flowing in the ground line seem to account for these microvolt variations at the input.
Connecting directly to the non-inverting input point on the breadboard provides the most honest shorted-input connection: zero in the sense that the op-amp sees zero.

[2] When we figure out what inflated our drift rate, we'll let you know.

[3] Well, "we" means the Royal Paul.

[4] If you want to buy a motor that we have tried, you can get the larger one (a Mabuchi FF-130RH) from many sources. We happened to buy ours from All Electronics, for a little more than $1.

[5] We know you know the answer: by adjusting the value of the integrator's input resistor.

Lab: Op-Amps II

Figure 7L.3 Alternative signal sources: DC motors; hard disc, with voice coil that positions drive head.

a voltage proportional to *velocity* of the head, or to the spin rate of the little motor. Integrating this signal, we should get a voltage proportional to the head's *position* (or the motor shaft's).

Query: Moving the head fast, or spinning the shaft fast, produces a large voltage input, moving either one slowly produces a small voltage input. If you move the head quickly from position A to position B – or rotate the motor shaft from angular position C to D – does the *rate* at which you make this movement affect the integrator's output?[6]

In this exercise our goal is to show this position on the scope screen. We don't want the display to wander appreciably, over the span of several seconds. So, we need a low-drift integrator like the two that we have just put together.

Let's apply your integrator – probably the most recent circuit, using the LTC1150, since that should still be wired up – to the signal put out by the disc-drive's voice coil. Watch the input signal (the signal from the two wires brought out from the disc's coil, or from the DC motor) and the integrator output. Try 2 V/division on the integrator output. Zero the integrator, and then move the disc head, inward and outward on the radius. We hope you like what you see coming from your integrator. We hope that you find you have built a circuit that displays the head's radial position – or the rotary position of the DC motor's shaft.

If you're not worn out, swap in the '411 for the chopper amplifier, and restore the trimming pot. (If you're lucky, your trim may still be good; if not, adjust it till you get an acceptably slow drift from your integrator, and then apply the disc-drive's signal.) See how the trimmed '411 does at displaying the drive's head position.

7L.2 Differentiator

The circuit in Fig. 7L.4 is an active differentiator. Try driving it with a 1 kHz triangle wave.

The differentiator is most impressive when it surprises you. It may surprise you if you apply it to a *sine* from the function generator: you might expect a clean cosine – unless you noticed the contrary result back in Lab 3L. In fact, some generators (notably the Krohn–Hite generators that we still use in

---

[6] Nope. Quick movement does drive the integrator harder – but for a shorter time than the slower movement (by definition of "quick" and "slower"!). So, the integrator output honestly measures the *position* of head or motor shaft. It does not care about rate of change of that position: it integrates the input voltage, but that voltage is the time-derivative of position – in case you like speaking of simple things in fancy mathematical terms, as we do not.

7L.2 Differentiator

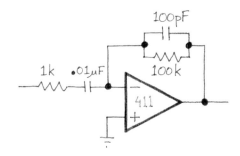

Figure 7L.4 Differentiator, including extra R and C for stability.

our lab) will show you, when their sine is differentiated, a waveform that reveals the purported *sine* to be a splicing of more-or-less straightline segments. This strange shape reflects the curious way the sine is generated: it is a triangle wave with its points whittled off by a ladder of four or five diodes. The diodes cut in at successively higher voltages, rounding it more and more as the triangle approaches its peak: see Figs. 7L.5 and 1N.12.

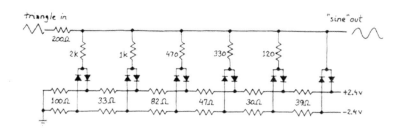

Figure 7L.5 Sketch of standard analog function-generator technique for generating sine from triangle.

You may even be able to count the diodes revealed by the output of the differentiator – though we've never been able to spot evidence of *all* the diodes.

AoE §4.5.7

A note on stability (preview of work to come): Here we are obliged to mention the difficult topic of stability, a matter treated more fully later in Chapter 9. Figure 7L.6 shows a simple differentiator – one with a single R and a single C. It necessarily lives at the edge of instability (sounds like a soap opera, doesn't it), because such a differentiator has a gain that rises at 6 dB/octave ($\propto f$). To build the differentiator as simply as in the figure would violate the stability criterion for feedback amplifiers (see AoE §4.9.3.)

In order to circumvent this problem, it is traditional to include both a series input resistor and a capacitor parallel to the feedback resistor, as shown in Fig. 7L.4. This additional R and C convert the differentiator to an *integrator* above some cut-off frequency. You saw a sketch of this frequency-response in Fig. 7N.34.

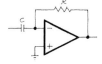

Figure 7L.6 Simple differentiator (unstable).

This compromising of the differentiator's performance is disappointing. You can watch the effect of this network by noting the phase shift between input and output as you gradually raise f_{in} from zero toward the breakover frequency and beyond. At breakover the phase shift should go to zero. At still higher frequencies you should see the phase shift characteristic of an integrator.

Incidentally, a faster op-amp (one with higher f_T) would perform better: the switchover to integrator must be made, but the faster op-amp allows one to set that switchover point at a higher frequency.

Integrate the derivative? A more intriguing way to see the imperfection of the differentiator is to feed its output to the integrator you built earlier, then compare *original* against *output* waveforms. Ideally, they would be identical – at least in phase (that is, apart from *gain* artifacts). Are they? Does the answer depend on input frequency?[7]

Sweeping frequencies (optional): You might also enjoy watching the frequency response of the op-amp differentiator, to confirm the pattern predicted in the sketch above: does the circuit show first gain that rises with frequency (differentiator behavior), then gain that falls (integrator behavior)?

7L.3   Slew rate

Begin by measuring slew rate and its effects with the circuit in Fig. 7L.7. We ask you to do this in two stages.

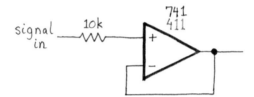

Figure 7L.7 Slew rate measuring circuit. (The series resistor prevents damage if the input is driven beyond the supply voltages).

7L.3.1   Square wave input

Drive the input with a square wave, in the neighborhood of 1 kHz, and look at the output with a scope. Measure the slew rate by observing the slope of the transitions. Note that full *slew rate* appears only when the op-amp's input stage is strongly unbalanced. So, if the slew rate seems low, make sure your input amplitude is sufficient.

7L.3.2   Sine input

Switch to a sinewave, amplitude a volt or so, and measure the frequency at which the output waveform begins to distort (this is roughly the frequency at which amplitude begins to drop as well). Is this result consistent with the slew rate that you just measured using a square wave? Watch out for a refinement, here: you will not see the op-amp's full *slew rate* until the op-amp inputs are radically unbalanced. A large square wave achieves this effect easily; a sinewave may not, unless it is both large and quick.

Now go back and make the same pair of measurements (slew rate, and sine at which its effect appears) with an older op-amp: a 741. The 741 claims a "typical" slew rate of 0.5 V/μs; the 411 claims 15 V/μs. How do these values compare with your measurements?

7L.4   Instrumentation amplifier

7L.4.1   Measuring a difference

There is a large class of sensors that respond to physical phenomena with a change of resistance. Examples include potentiometers (rotation or displacement), thermistors (temperature), and strain

---

[7] It should. The integrator is honest, but the differentiator differentiates only up to its breakover frequency, which you can calculate from its Rs and Cs. Use either pair.

gauges (bending). You might think that a voltage divider would be an easy way to convert a variable resistance to a variable voltage, however, in many cases higher accuracy can be obtained by using a Wheatstone bridge and making a differential measurement. Let's see why.[8]

7L.4.2 Measure the resolution of a simple voltage divider

Build the voltage divider circuit shown in Fig. 7L.8 below. You can use a slide switch or a pushbutton. Start with a resistance substitution box set for $330\,\Omega$ for R_{test}.[9]

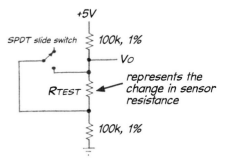

Figure 7L.8 Single-ended resistance change measurement using a voltage divider.

Use your multi-meter to measure the output voltage, V_O, with respect to ground. Open and close the switch to measure the voltage with and without the test resistor. This simulates measuring the sensor at two different physical states.

What is the smallest value of R_{test} for which you can reliably measure the difference between the two states? The output of the sensor we will be using in part 7L.4.5 on page 335 of this lab changes about $3.5\,\text{mV}$ (equal to $R_{\text{test}} \approx 280\,\Omega$) over its full range. Suppose you want to determine the output to one part in 20 (5%) of full scale? Will you be able to meet this specification with the single-ended circuit above?[10]

7L.4.2.1 Try a Wheatstone bridge instead

The problem with the simple divider is that you are trying to measure a very small change in the nominal $2.5V$ output of the voltage divider. This requires a measurement system with a high degree of precision. If, instead, you only measure the *difference* between two voltage dividers, you no longer have to worry about detecting a small change in a large value.

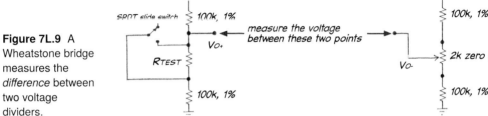

Figure 7L.9 A Wheatstone bridge measures the *difference* between two voltage dividers.

Add the second divider shown on the right in Fig. 7L.9 above.[11] Use a multi-turn 2k pot if you have

---

[8] This experiment is modeled on one developed by John Fox in his course at Stanford University.
[9] You may need to switch to individual resisters for the smallest values of R_{test}.
[10] Even if you think you could meet the specification with this circuit, you would need a way to subtract the 2.5 V DC from the divider output to isolate the voltage attributable to the sensor.
[11] This circuit is normally drawn with common excitation and ground to show its bridge configuration: see Fig. 7N.40.

one available.[12] Use your multimeter to measure the voltage *between* the two legs of the bridge as shown. Now $V_O = V_{O+} - V_{O-}$. With the switch closed, adjust the potentiometer so that the meter reads zero. Repeat the experiment in part 7L.4.2 above, trying smaller and smaller values for R_{test}. What is the smallest value of R_{test} you can repeatedly detect in the bridge configuration? Assuming a full scale change of 280 Ω in R_{test} corresponds to 500 g of weight, what is smallest weight you think you will be able to measure using the differential technique? Will you be able to determine the weight to within 5% of full scale?

7L.4.2.2 AC noise rejection

Now replace the 5 V supply in Fig. 7L.9 with a function generator set for a 200 mV, 5 Hz sine wave with a +5 V offset. This simulates a noisy excitation source. The noise is common to both sides of the bridge and is referred to as the "common mode signal."

Measure the *single-ended output voltage* between V_{O+} and ground (this is the same measurement you did in part 7L.4.2.) Can you measure a 330 Ω R_{test} value?

Move the black multimeter probe to the V_{O-} output of the bridge to make a *differential measurement* as in part 7L.4.2.1. Now can you measure the same smallest value for R_{test}? The differential measurement provides common mode rejection ("CMR"). This is especially important when the signal you are measuring is small compared to the noise in the measuring system.

7L.4.3 Build a home-made instrumentation amplifier

If we have persuaded you that a differential measurement can provide much better resolution than a single-ended one, then you recognize that you need the help of a differential amplifier. You want to amplify the difference between the voltages while ignoring the constant 2.5V DC common-mode voltage appearing at each terminal. This describes the function of the instrumentation amplifier. Its output is defined as

$$V_{\text{IA(out)}} = G_{\text{diff}} * (V_{O+} - V_{O-}) + G_{\text{CM}} * \frac{(V_{O+} + V_{O-})}{2}$$

Ideally, we want the common-mode gain, G_{CM}, to be zero and the differential gain, G_{diff}, to be whatever value is required to scale the sensor output to a reasonable range for the application.

Later in this lab we will use a commercial instrumentation amplifier, but first you will build one from individual op-amps to understand how the components come together to provide high input impedance, high differential gain and low common mode gain.

7L.4.3.1 The first stage

The three op-amp version of the instrumentation amplifier consists of two stages. The first stage shown in Fig. 7L.10 is called the *gain* stage.

You should see that the first stage of the instrumentation amplifier provides the same high input impedance as a unity gain follower, since the input signals are connected directly to the non-inverting inputs of the op-amps with no other path to ground. The harder question is how good is its common mode rejection?

Build the first stage with the LM358 op-amp: Build the first stage, shown in Fig. 7L.10 on the facing page, using the two op-amps in a LM358.

Use power supplies of ±15 V and two 1% 100k resistors for R_f. Choose R_g to give a differential gain of about 100 (see Chapter 7N for the gain formula).[13]

---

[12] If all you have is 1k pot – go ahead and try it. It may be adequate to adjust to zero volts differential output if you are lucky. Why? (Because it might not give you enough range to trim out the worst case error in the 100k, 1% R_{test} divider.)

[13] Does R_g need to be 1% as well? No, you don't really know the exact gain you need so a tolerance of 5% is fine.

7L.4 Instrumentation amplifier

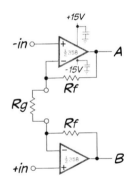

Figure 7L.10 First ("gain") stage of Instrumentation amplifier.

Common Input: Tie the two inputs together, and drive this single point from the breadboard's 1k potentiometer, configured as shown in Fig. 7L.11, which should deliver a voltage in the range ±5 V.

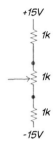

Figure 7L.11 Circuit to drive the IA common input between ±5 V.

Use two DVMs to watch the two output points, A and B in Fig. 7L.10, as you vary the input voltage. Note any differences between these. We would like to minimize this difference, of course, when the *input difference* is zero as these two outputs, A and B, will drive a difference amplifier in the next stage.

Let's think about what might account for any difference you see.[14]

Could R_f mismatch be the problem? How important is it to match the two R_f resistors? Consider how much current should flow in R_g when the two inputs are tied together. In an ideal op-amp, no current flows in R_g when the two inputs are tied together since −in, +in, A and B will all be at the same voltage. In addition, the resistors values used in this stage are not large enough for input bias current to cause an appreciable error with any modern op-amp. So R_f mismatch is unlikely to explain the error.

How about V_{offset} in the '358's? First, a thought experiment: suppose each op-amp in the LM358 has V_{offset} of 2 mV (the *typical* LM358 specification). Assume the inputs are tied together. Will the *gain stage* amplify this V_{offset} error?[15]

So, V_{offset} matters, but only in the unusual sense that a *difference* between the V_{offset} values of the two op-amps matters. In an integrated pair, the difference seems likely to be well under the V_{offset} value. The difference between V_{offset} values for the two op-amps in an LM358 package is not specified. But the V_{offset} value itself puts a ceiling on that difference.

---

[14] If you don't see any difference, try using a single multimeter to measure the *difference* between the two outputs.

[15] If your thought experiment ran as ours did, you noticed that, curiously, it is only a *difference* between the two V_{offset} values that should cause current to flow in R_g, and thus to be amplified. See AoE §5.15.3B. V_{offset} itself should appear as an output error: grounding both inputs, we would expect to see V_{offset} at each output, when the offset voltages of both op-amps are equal.

Try a low-V*_{offset} *op-amp: We may learn something more by trying a low-offset op-amp pair. Replace the LM358 in the gain stage circuit (Fig. 7L.10 on the previous page) with the two op-amps in an LT6020 (the pinout is the same). Its V_{offset} of 5 μV typical, 30 μV max, is about 200 times better than the '358's.

How close are the voltages at A and B for this circuit, when built with the LT6020? Is V_{offset} important in the gain stage? Yes, but only in the sense that a *difference* in V_{offset} is amplified by the differential gain of the first stage: $1 + \frac{2R_f}{R_g}$.

7L.4.3.2 Test a second stage difference amplifier

Before we build the full IA, let's try the second stage by itself: the difference amp (or "differential amplifier"). Wire the unity-gain diff amp using an LF411 op-amp as U3, again with power supplies of ±15 V: see Fig. 7L.12. Use four 10k 1% resistors.

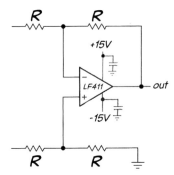

Figure 7L.12 Differential amplifier: stage two of the IA.

Two questions, before you take measurements: What effect do you expect V_{offset} for the op-amp to have on V_{out}? What do you expect that error to contribute to the diff amp output?[16]

What effect would you expect a mismatch in these R values to have on V_{out}?[17] With the two inputs of this stage still tied together, measure V_{out}, as you vary the common input voltage. What V_{out} would you hope for? What do suppose accounts for any departure from what you would like to see?[18]

7L.4.3.3 Complete the two-stage IA

Now let's link the two stages. Let the gain stage drive the diff amp: see Fig. 7L.13.

Again, tie the full-IA's two inputs together, and drive them from the potentiometer of Fig. 7L.11 on the preceding page. Vary the input voltage over its full range. What is your circuit's common-mode gain?[19]

7L.4.3.4 Differential Gain

Now tie one of the IA's two inputs to ground and drive the other with a small sinusoid at 100 Hz (you may need to use a voltage divider on the function generator output to avoid overdriving the IA). Does the IA V_{out} behave as you would expect? Are you applying a differential signal, common-mode, or a hybrid?[20]

---

[16] The V_{offset} of the LF411 is not outstanding (0.8 mV, typical) but you would not expect that error to contribute much to the output since the diff amp's gain is unity.

[17] Matching the two divider ratios is important as it determines the amount of common mode voltage from the first stage that appears at the output of the diff amp. With 1% resistors, the 2.5 V DC common-mode bridge voltage will be reduced by about 40 dB at the instrumentation amplifier output. See AoE 5.14.1.

[18] A perfect difference amplifier would output zero volts for any common mode input voltage. Here, we are constrained by resistor matching, so we would expect to see the output track a greatly attenuated version of the input common mode voltage.

[19] More commonly described by the term *common-mode rejection* ("CMR"), since the circuit aims to *reject* a common-mode input.

[20] We asked you this question when testing Lab 5L's differential amplifier.

7L.4 Instrumentation amplifier

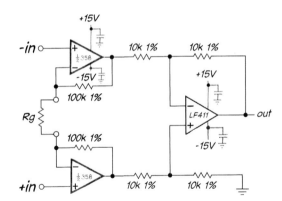

Figure 7L.13 Full IA: gain stage drives diff amp.

7L.4.4 Home-made IA's CMRR

What is your circuit's Common Mode Rejection Ratio in dB, where CMRR = $20\log_{10}(\frac{G_{\text{diff}}}{G_{\text{CM}}})$?

7L.4.5 Integrated IA

One doesn't need to work as hard as you did, to this point in this lab. You can buy an integrated IA that implements exactly the design you just built, and that includes resistors matched to better than the 1% tolerance that we used in our home-made IA. The integrated IA will perform better than the home-brew design. (As usual, you may be sore that we didn't just hand you the integrated IA at the outset; but we wanted you to appreciate the challenge of implementing such a good IA.)

The LT1167 Instrumentation Amp lets you set the gain with the choice of a single resistor, R_g, for a gain in the range 1 to 10,000. Gain is approximately $50k/R_g$.[21] Let's again start with a gain of about 100.

Use the potentiometer, as before, to drive the two inputs while they are tied together. See what V_{out} results.

Now, ground one input while driving the other input from the potentiometer, applying a difference signal.[22] What is the integrated IA's CMRR? How does what you measure compare to the LT1620 datasheet specification?

7L.4.6 Use a "load cell" to drive the instrumentation amp

We need to give the IA something to do with its powers – its high R_{in} and good CMRR. Let's feed it a small signal from a Wheatstone bridge. In this lab, we will use a 1kg (full-scale) *load cell*.[23] To make this device breadboard-able, you'll need to solder #22 solid wire to the four very fine wires (30 gauge?) attached to the load cell. The load cell is shown in Fig. 7L.14 on the next page.

The load cell uses four strain gauges, one in each of leg of the bridge, and is rated for 1000 grams full scale. The red and black wires are used to provide the excitation (i.e., power) to the Wheatstone bridge while the differential output signal is available on the green and white wires.

Such a load cell is described more fully in Chapter 7N.

In a scale, a typical application for a strain gauge, the cell would be mounted between a base and a load plate. Today, we will omit this structure and will do a stranger exercise, intended simply to show

---

[21] More exactly (if you care) gain = $49.4k/R_g + 1$.
[22] You will recognize that you are applying *both* difference and common-mode signals You may recall using the same sort of drive when testing differential gain in §5L.1.8. Differential gain is so much higher than common mode that, once again, this test is a good stand-in for applying a true differential input.
[23] This load cell is from Adafruit: `https://LAoE.link/Load_Cell.html`. You may substitute any similar part.

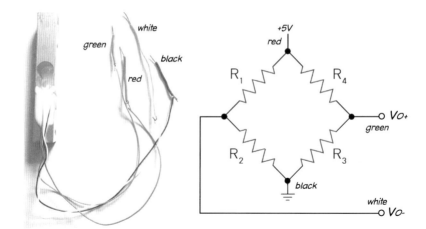

Figure 7L.14 Load cell with added wires, showing its included Wheatstone bridge.

off the gauge's sensitivity: we will let the load cell measure its deformation under its own weight, as we rotate it.

Start with the cell orientation (indicated by an arrow on one of its ends) horizontal. In principle, the IA output should be zero, but we saw several hundred millivolts out of the IA instead. We can live with that, because what interests us is not this baseline value but the contrast between UP orientation and DOWN. We can get that, despite a baseline far from zero.[24]

Note the values of V_{out} for orientations of $0°$, $+90°$, and $-90°$. Check that your values are more-or-less repeatable.

When a load cell is applied in the usual way – say, to make a scale – its deformation is much larger than in the case you just looked at (enlarged by mechanical leverage, or by application of a heavy load, as in a bathroom scale). We gave ourselves a tiny signal to work with, just to show off the sensitivity of the instrumentation amplifier.

7L.4.7 Let an integrated IA and analog-to-digital converter do the hard work

You can skip this part of the lab if you don't have an Arduino microcontroller handy. But if you do have an Arduino, you'll probably enjoy how easy this microcontroller and its associated analog-to-digital converter can make what was hard work for us, up to this point.

In §7L.4.5 on the preceding page you saw that an integrated instrumentation amp, the LT1167, made the task of sensing and amplifying the signal from a load cell pretty straightforward. Now we will try an IC that is even more impressive.

This is an amplifier with programmable gain (PGA) that feeds a high-resolution analog-to-digital converter. Figure 7L.15 provides a not-very-illuminating peek inside the part. The interesting parts – PGA and ADC – remain black boxes.

This part delivers its results in digital form. Its resolution is quite dazzling. It uses a *full scale* voltage of either ± 20 mV or 40 mV. It converts this to digital form using a 24-bit converter. Thus it slices the already-small range of its input into 2^{24} slices. Each slice, then, is at most $80 \text{mV}/2^{24} \approx 5$ nV.

The Sparkfun "breakout board" on which the part is wired makes the '711 almost embarrassingly easy to use.[25] We connect the '711 to an Arduino microcontroller using two lines that form a serial link called "SPI" (Serial Peripheral Interface). The '711 input amplifier takes its inputs from the load cell's Wheatstone bridge.

---

[24] Our *difference* readings were in the tens of millivolts.
[25] See https://LAoE.link/Load_Cell_Amp.html. The integrated circuit, HX711, is made by Avia Semiconductor (Xiamen), Ltd. Datasheet at https://LAoE.link/HX711_Datasheet.pdf. The SparkFun support library has been tested with ATmega328P, ESP8266, ESP32, and STM32 Arduino microcontrollers.

7L.4 Instrumentation amplifier 337

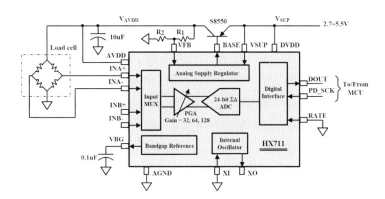

Figure 7L.15 HX711 24-bit ADC block diagram.

The wiring looks like Fig. 7L.16.

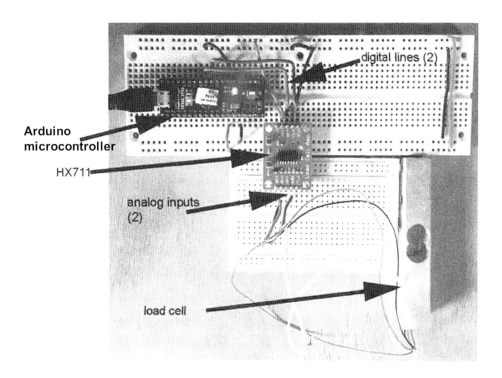

Figure 7L.16 HX711 wired to an Arduino microcontroller.

To complete our handoff of all the hard work to someone else, we then borrow a ready-made ADC-reading program and load it into the Arduino.

In a practical application, we would do something useful with the digital outputs of the '711. Today, we propose that you just have fun with the device: take a look at two ways to display the ADC's results. Doing this will at least let you appreciate the extreme sensitivity of this amplifier and converter. If you hold one end of the load cell while rotating the bar, the ADC readings will make the bending caused by gravity evident.

Open the Arduino IDE ("Integrated Development Environment"). Under Tools, choose the particular Arduino that you are using. From the Sparkfun website, get an HX711 test program called

"...calibrate." A tutorial explains what files you will need to load into your Arduino (HX711 library, and demonstration program): see https://LAoE.link/SparkFun_Calibration.html.

The calibration program assumes use of particular pins to form the serial link to the HX711. Near the head of the program pins are specified for Data and Clock:

```
#include "HX711.h"

#define DOUT  3
#define CLK   2
```

Make sure that your wiring and the listed use of DOUT and CLK match.

With library and program loaded, run the calibration program, and hold one end of the load cell, the other end unsupported.

Watch values as text One way to see the HX711's result is to watch the values output as text. To do this, under TOOLS choose SERIAL MONITOR. If you now rotate the load cell by hand, you should see the values vary slightly, reflecting the very small voltage changes delivered by the load cell's outputs. The changes will read as larger than ±1 V, obviously reflecting a very large assumed gain.

Watch values as plot A graphic display is more fun. You can get that by choosing, under TOOLS, SERIAL PLOT.

Initially, using the presets in the program (calibration factor -7050), the display will be insensitive to movement of the load cell. You can gradually decrease the large negative calibration factor, by holding down the A key on your keyboard. When this change has become sufficient, the plot will redefine the vertical axis so as to make changes that result from rotating the load cell quite evident, as in Fig. 7L.17.

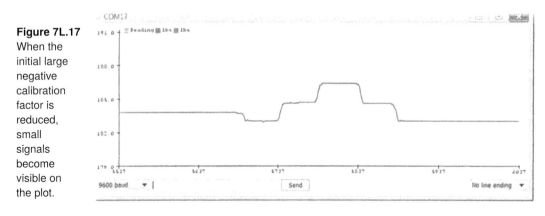

Figure 7L.17 When the initial large negative calibration factor is reduced, small signals become visible on the plot.

7L.4.7.1 Calibration code for Arduino and HX711 amplifier

As noted previously, you can get setup instructions and the calibration program from the Sparkfun website.[26]

But, in case you prefer to enter it from these notes, here it is. You will still need to visit the SparkFun site to get the HX711 library.

---

[26] https://LAoE.link/SparkFun_Instructions.html.

7L.4 Instrumentation amplifier

```
/*
 Example using the SparkFun HX711 breakout board with a scale
 By: Nathan Seidle
 SparkFun Electronics
 Date: November 19th, 2014
 License: This code is public domain but you buy me a beer if you use this and we meet someday (Beerware
license).

 This is the calibration sketch. Use it to determine the calibration_factor that the main example uses.
 It also outputs the zero_factor useful for projects that have a permanent mass on the scale in
between power cycles.

 Setup your scale and start the sketch WITHOUT a weight on the scale
 Once readings are displayed place the weight on the scale
 Press +/- or a/z to adjust the calibration_factor until the output readings match the known weight
 Use this calibration_factor on the example sketch

 This example assumes pounds (lbs). If you prefer kilograms, change the Serial.print(" lbs"); line to kg.
 The calibration factor will be significantly different but it will be linearly related to lbs
 (1 lbs = 0.453592 kg).

 Your calibration factor may be very positive or very negative. It all depends on the setup of your scale
 system and the direction the sensors deflect from zero state.
 This example code uses bogde's excellent library: https://github.com/bogde/HX711
 bogde's library is released under a GNU GENERAL PUBLIC LICENSE
 Arduino pin 2 -> HX711 CLK
 3 -> DOUT
 5V -> VCC
 GND -> GND

 Most any pin on the Arduino Uno will be compatible with DOUT/CLK.

 The HX711 board can be powered from 2.7V to 5V so the Arduino 5V power should be fine.

*/

HX711 scale;

float calibration_factor = -7050; //-7050 worked for my 440lb max scale setup

#include "HX711.h"

#define DOUT  3
#define CLK   2
void setup() {
  Serial.begin(9600);
  Serial.println("HX711 calibration sketch");
  Serial.println("Remove all weight from scale");
  Serial.println("After readings begin, place known weight on scale");
  Serial.println("Press + or a to increase calibration factor");
  Serial.println("Press - or z to decrease calibration factor");

  scale.begin(DOUT, CLK);
  scale.set_scale();
  scale.tare(); //Reset the scale to 0

  long zero_factor = scale.read_average(); //Get a baseline reading
  Serial.print("Zero factor: "); //This can be used to remove the need to tare the scale.
                                  //Useful in permanent scale projects.
  Serial.println(zero_factor);
}

void loop() {

  scale.set_scale(calibration_factor); //Adjust to this calibration factor

  Serial.print("Reading: ");
  Serial.print(scale.get_units(), 1);
  Serial.print(" lbs"); //Change this to kg and recalibrate the calibration factor if you follow SI units
                         //like a sane person
  Serial.print(" calibration_factor: ");
  Serial.print(calibration_factor);
  Serial.println();

  if(Serial.available())
  {
```

```
    char temp = Serial.read();
    if(temp == '+' || temp == 'a')
      calibration_factor += 10;
    else if(temp == '-' || temp == 'z')
      calibration_factor -= 10;
  }
}
```

7L.5 AC amplifier: microphone amplifier

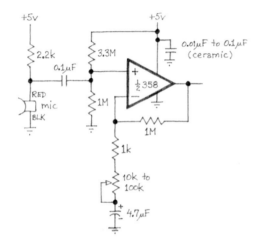
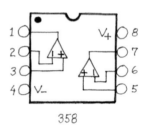

Figure 7L.18 Single-supply microphone amplifier.

7L.5.1 Single-supply op-amp

In this exercise you will meet a "single-supply" op-amp, used here to allow you to run it from the +5 V supply that later will power your computer. This op-amp, the '358 dual (also available as a "quad," the 324), can operate like any other op-amp, with $V_+=+15$ V, $V_-=-15$ V. But, unlike the '411, it can also be operated with V_-=GND, since the input operating common-mode range includes V_-, and the output can swing all the way to V_-.

Our application here does not take advantage of that hallmark of a single-supply op-amp, its ability to work right down to its negative supply. Often that is the primary reason to use a single-supply device. The op-amp millivoltmeter described in Worked Example 7W does exploit this capacity of the '358, unlike today's microphone amplifier.

Note Build this circuit on a private single breadboard strip of your own, so that you can save the circuit for later use if you decide to build the voice recorder of Worked Example §26W.3.

In Fig. 7L.18 the '358 is applied to amplify the output of a microphone – a signal of less than 20 mV – to generate output swings of a few volts. The "AC amplifier" configuration, you will notice, is convenient here: it passes the input bias voltage to the output *without amplification* because gain at DC is *unity*.

The microphone is an "electret" type (the sound sensor is capacitive: sound pressure varies the spacing between two plates, thus capacitance; charge is held nearly constant, so V changes with sound

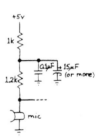

Figure 7L.19 Quieting power supply to microphone.

7L.5 AC amplifier: microphone amplifier

pressure according to $Q=CV$). The microphone includes a high-input-impedance field-effect transistor (FET) within the package, to buffer the electret.

The FET's varying output current is converted to an output voltage by the 2.2k pullup resistor. So the output impedance of the microphone is just the value of the pullup resistor: 2.2k.

Try isolating the power supply of the microphone as in Fig. 7L.19: You may find, after your best efforts, that your amplifier still picks up pulses of a few tens of millivolts at 120 Hz. Figure 7L.20 shows what the pulses look like.

Figure 7L.20
Ground noise on PB-503 breadboard: caused by current pulses recharging filter capacitor.

Probably you will have to live with these, unless you want to go get an external power supply (the adjustable supply you used in Lab 1L will do fine here). These pulses show the voltage developed in the ground lines when the power supply filter capacitor is recharged by the peaks of the rectifier output. They *shouldn't* be there, but they are hard to get rid of. They appear because of a poor job of defining ground in powered breadboards like the PB-503, and you can't remedy that defect without rewiring the innards of the PB-503.

7S Supplementary Notes: Op-Amp Jargon

Bias current (I_{bias}): average of input currents flowing at op-amp's two inputs (inverting, non-inverting).

Frequency compensation Deliberate rolling-off of op-amp gain as frequency rises: used to assure stability of feedback circuits despite dangerously-large phase shifts that occur at high frequencies.

Gain–bandwidth product (f_T): a constant describing gain and frequency-response of an op-amp; it equals the extrapolated frequency where op-amp open-loop gain has fallen to unity.

Hysteresis As applied to Schmitt trigger comparator circuits: the voltage difference between upper and lower thresholds.

Offset current Difference between input currents flowing at op-amp's two inputs (inverting, non-inverting).

Offset voltage (V_{offset} or V_{OS}): op-amp's input stage mismatch voltage: the voltage that one would need to apply, between the inputs, in order to bring the op-amp output to zero.

Open loop Circuit wired without feedback.

Rail-to-rail Said of inputs and outputs of an op-amp. "Rail" is jargon for power supply. Some devices allow rail-to-rail input range, some come close to this output range; some have both capabilities.

Saturation Condition in which op-amp output voltage has reached one of the two (±) output voltage limits, usually within about 1.5 V of the two supplies, but much nearer in rail-to-rail-output devices.

Schmitt trigger Comparator circuit that includes positive feedback.

Single-supply op-amp Device that accepts inputs close to (and sometimes below) its negative supply, which normally is ground. Device also can drive its output close to the negative supply (ground).

Slew rate Maximum rate (dV/dt) at which op-amp output voltage can change (assumes substantial voltage difference between op amp input terminals).

Summing junction Inverting terminal of op-amp when op-amp is wired in inverting configuration: inverting terminal then sums currents. That is, I_{feedback} is algebraic sum of all currents input to the summing junction.

Transresistance amplifier Current-to-voltage converter.

Virtual ground Inverting terminal (−) of op-amp when non-inverting terminal is grounded. Feedback tries to hold (−) at 0 V, and current sent to that terminal does not disappear into "ground," but instead flows through the feedback path (hence the ground is only "*virtual*").

7W Worked Examples: Op-Amps II

7W.1 The problem

Here's a statement of the design task, and some questions:

Design it: Design an integrator, using operational amplifiers, to the following specifications:
- it will ramp at +1 V/ms when a +1 V signal is fed to it;
- R_{out} for the signal source is unknown;
- you should include protection against saturation caused by long-term integration of small DC errors.

Questions:

Frequency response: Assume an integrator with a resistor paralleling the feedback capacitor. Approximately what is the *low-frequency* limit of your circuit's response, if we define that limit as the frequency where the response is reduced by approximately 3 dB?

Integrator errors: Assume an integrator with reset switch rather than one with a resistor paralleling the feedback capacitor. Given your integrator design, at what rate does its output move (V/second) when the input is in either of two conditions: *grounded* or *open*?
Assume you are using an LF411 op-amp, and use *typical* values.

Case 1: Input grounded (as if fed by a transducer that is a *voltage* source).

Case 2: Input open (as if fed by a transducer that is a *current* source). Note that for this case you should *remove* any input buffer you may have placed before the integrator.

Choose a better op-amp: Now find an op-amp from the small set shown in Fig. 7W.11 that will give around 5× better performance (integrator error $\leq 1/5$ the value you found for the '411 with input *grounded*).
Specify the op amp, and calculate its resulting dV/dt error (input grounded).

7W.1.1 A solution

A skeleton circuit: We need a standard op-amp integrator, with protection against saturation. Let's use a feedback resistor, 100× $R_{integrator}$, as usual. We also need two other op-amp circuits:

- an inverter to give the requested sense of output (positive out for positive in);
- an input buffer (a follower) because "R_{out} for the signal source is unknown."

Component values: For RC our plan to make the feedback resistor 100× $R_{integration}$ steers us toward an integration resistor (as we're calling the R that feeds the integrator) that is not extremely large. If, as usual, we put a 10M ceiling on our R values, then $R_{feedback}=10\text{M}$ and $R_{integration}=100\text{k}$. With an R value chosen, we know what current will flow for the specified 1 V input: $I_{\text{full-scale}} = 1\text{ V}/100\text{k} = 0.01$ mA. This full-scale current lets us choose C to be $I/\frac{dV}{dt} = \frac{0.01\text{ mA}}{1\text{ V/ms}} = 0.01 \times 10^{-6} = 0.01\ \mu\text{F}$.

344 Worked Examples: Op-Amps II

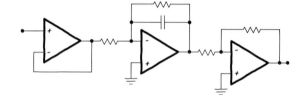

Figure 7W.1 Skeleton circuit for integrator (no component values).

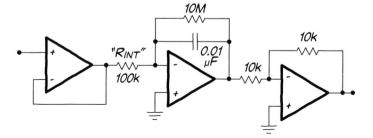

Figure 7W.2 Integrator with component values.

Frequency response: The feedback resistor will reduce V_{out} by 3 dB when $|Z_C| = R$.[1] That equality occurs at what we usually think of as "f_{3dB}": $1/(2\pi RC)$. The frequency takes the name "f_{3dB}," as you well know, because there the output of an RC filter (where R and C are in series) is down 3 dB.

But we should not think that we are looking at the same case, here. In this circuit, where R and C are in parallel, this frequency would better be called "$f_{crossover}$": this is the frequency where the magnitudes of currents flowing in R and C are equal. Below this frequency boundary, more of the current flows in the R; above this frequency, more of the current flows in the C. But this frequency is also the one where the integrator "will reduce V_{out} by 3 dB" *relative to what it would have been without the feedback resistor*. So, the reference to 3 dB remains appropriate.

After that long-winded digression, the calculation is straightforward. The feedback resistor will reduce V_{out} by 3 dB at $f = 1/(2\pi RC) \approx 1/(6 \times 10 \times 10^6 \, \Omega \times 0.01 \times 10^{-6} \, F) = 1/(6 \times 0.1) = 1.6$ Hz. This is very low, as we would hope. So, at most frequencies, the feedback resistor should not compromise the integration substantially.

Integrator errors (predict output drift rate, with no feedback resistor): Here are typical LF411 values: $V_{offset} = 0.8$ mV; $I_{bias} = 50$ pA.

Input grounded: The offset voltage V_{offset} produces an error in both the follower and integrator of Fig. 7W.2. Both errors contribute to the integrator's drift (error in the output inverter does not contribute to drift). Worst case, the two V_{offset} values may add.

The *current* that they cause to flow, charging the integrator capacitor – let's call it "I_{error}" – results from this double V_{offset} (worst case) across the integration resistor, R_{int}:

$$I_{error} = 2 \times V_{offset}/R_{int} = 1.6 \, \text{mV}/(100 \, \text{k}\Omega) = 16 \, \text{nA}$$

This current causes a drift:

$$dV/dt = I/C = 16 \, \text{nA}/(0.01 \, \mu\text{F}) = 1.6 \, \text{V/s} \qquad (7W.1)$$

The effect of I_{bias} (at 50 pA) is negligible compared with the 16 nA current caused by V_{offset}.

---

[1] You may want to take our word for this – or you may want to calculate the value of $Z_C \parallel R$ when $|Z_C| = R$.

Input open: For this case, we are modeling the behavior of an input transducer that is a *current source*. To simulate such a source we remove the follower that buffers the input stage, because a *current source* is happy to see the low input impedance that it finds at the inverting terminal (to put things anthropomorphically).

We also remove R_{int}, which serves no good purpose when the input signal is a current, and in fact can cause mischief. R_{int} simply moves the circuit input voltage away from ground, as currents flow. Such displacement can disturb a current source (one with small "output voltage compliance," a notion you met in Lab 4L).

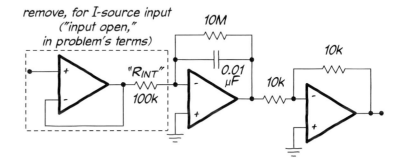

Figure 7W.3 Integrator input would be simplified if one assumed signal source was a *current source*.

Now V_{offset} has no effect on currents (because V_{offset}, which moves the inverting input away from ground, does not alter the magnitude of current arriving from the current-source that drives the integrator). Only I_{bias} causes the integrator output to drift:

$$dV/dt = I/C = 50\,\mathrm{pA}/(0.01\,\mu\mathrm{F} = 10\,\mathrm{nF}) = 5\,\mathrm{mV/s}$$

This is radically better (about 300 times better) than the result we got for a *voltage* source driving the same integrator. If you have a choice, you will choose a current source as your transducer.

Improve performance with a better op-amp: If your signal comes from a voltage source and you need better performance, you will need to find an op-amp better than the '411. To get a drift rate 5 times better than what the '411 provides – as the problem asks – we need to reduce the drift rate from the 1.6 V/s (see equation (7W.1)) to about 0.3 V/s.

Since we found that the '411's drift was caused by V_{offset} alone (the effect of I_{bias} was negligible), we need only shop for an op-amp with V_{offset} no more than 1/5 that for the '411, along with comparably low I_{bias}. We need $V_{offset} \le 0.8\,\mathrm{mV}/5 = 0.16\,\mathrm{mV} = 160\,\mu\mathrm{V}$, and low I_{bias}.

The drift rate for this op-amp (assuming, once again, that the integrator sees $2 \times V_{offset}$ because of the presence of the follower ahead of the integrator) should be

$$\frac{dV}{dt} = \frac{I}{C} = \frac{(2 \times V_{offset})/R_{integrator}}{C}$$
$$= \frac{0.1\,\mathrm{mV}/0.1\,\mathrm{M}\Omega}{0.01\,\mu\mathrm{F}}$$
$$= 1\,\mathrm{nA}/10\,\mathrm{nF} = 0.1\,\mathrm{V/s}$$

Yes (no surprise): the OPA111B does the job.

Op-amps with much lower V_{offset} are available if you need them. You will meet such a need in the next section. But there's no point choosing an op-amp much better than what we need, so we'll be content with the LF411 for the present task.

7W.2 Op-amp millivoltmeter

Problem Given

- a meter movement: 1 mA full-scale, 100 Ω,
- op-amp(s) of your choice,
- other parts as needed,

design a voltmeter to the following specs:

- use a single 15 V supply, if possible; if you can't manage that, then use split supplies, ±15 V;
- full-scale (input-) voltage: 10 mV;
- accuracy 1% of full scale;
- input resistance: ≥1 MΩ;
- reading for input grounded or open: 0 (to the usual 1% of full-scale range);

and if you're not exhausted after that – or perhaps along the way – add some desirable features:

- protection for the meter movement (from consequences of overdriving the input); or
- really fancy: valid reading for *either* input polarity, along with a polarity-indicating LED.

7W.2.1 Solution: millivoltmeter

A start: amplify...: Comparing input range to the output voltage range that we need to drive the meter movement will let us determine what gain we need. We'll also need to consider what form of amplifier is suitable.

Gain needed: Let's review the problem specifications that bear on gain:

- input voltage range: 0–10 mV;
- output voltage needed: 0–100 mV, since meter movement drops; 100 mV full-scale. It is described as 100 Ω and 1 mA full-scale

So a gain of 10 will do it.

What amplifier configuration: Should we use inverting or non-inverting? Either could satisfy the input-resistance requirement (1M). Let's see what the circuits would look like.

Non-inverting amp: This is the obvious choice, since this is a single-supply instrument and we want input impedance that is quite high. Figure 7W.4 shows a sketch of the circuit.

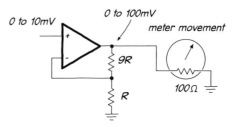

Figure 7W.4 Non-inverting amplifier, to drive meter movement.

The amplifier will need to be a "single-supply" type: its input must understand voltage levels down to zero, and its output must be able to go close to zero. A device like the LM358 that we use in §7L.5's microphone amplifier satisfies these requirements,

7W.2 Op-amp millivoltmeter

Current-source version of non-inverting design: It is also possible to apply a *current* rather than *voltage* to the meter movement; see Fig. 7W.5.

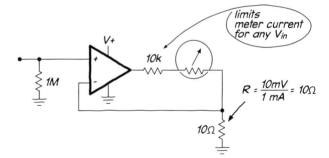

Figure 7W.5 Current, rather than voltage, applied to meter movement.

The circuit of Fig. 7W.5 includes a current-limiting resistor (the 10k), in case the input is overdriven. This protection anticipates that described in §7W.2.2.

Inverting amp: Could the job be done with an inverting amplifier? Yes, but this design is awkward if done with a single supply. It's straightforward if done with split supply; see Fig. 7W.6.

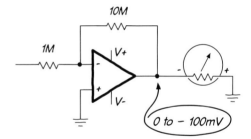

Figure 7W.6 Inverting design: straightforward if we use a split supply.

Use of a single-supply makes the task harder, because the circuit must operate close to its "positive" supply (well, the *more* positive supply – in this case, *ground*); see Fig. 7W.7.

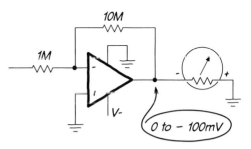

Figure 7W.7 Inverting amplifier, to drive meter movement.

The *inversion* means that the op-amp output will always be negative, so the op-amp will have to be powered in the curious way shown. Input of the op-amp now lives at the *positive* supply rail; the output runs from that rail to 100 mV below it. This circumstance rules out an ordinary single-supply op-amp like the '358. We would need a rail-to-rail op-amp, able to understand inputs all the way to the positive supply. This we can find: e.g., ADI's OP282 or AD8531 (the latter's power supply is limited to 6 V, a restriction common in CMOS parts; we could use a supply voltage lower than 15 V: say, 5 V).

The op-amp would also need to be able to take its output right up to the positive supply. Here, the configuration we have set up for ourselves is so odd that the characteristic does not seem to be specified; but probably the devices can sink current close to the positive supply (what is specified is the offset from positive supply while *sourcing* current).

So, it is possible to find an op-amp that fits the configuration – though we'll soon learn that this configuration is ruled out by the specifications of the amps that can handle what we require for this configuration: high-side, rail-to-rail input and output.

The op-amp specs the designs would require: So far, we have shown only how to get a gain of 10. We now need to look at the harder part of the task: getting accurate results.

We are asked to keep errors at ±1%. In addition, the meter should read *zero* (to within 1% of full-scale) when the input is *open*. We don't want to annoy the meter's users with a wandering output at times when nothing is connected to the millivoltmeter.

No wandering, with input open: The non-inverting design of Fig. 7W.4, with its astronomical R_{in}, would wander, if we did not add a resistor as shown in Fig. 7W.8. After a time, the input, and therefore output, would drift all the way to an extreme, pinning the meter movement at full-scale. To prevent this result, we must include a resistor to ground, as shown in Fig. 7W.8.

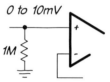

Figure 7W.8 A resistor to ground prevents drift, in the non-inverting design.

We chose the resistor value to satisfy the original problem specification, $R_{in} \geq 1\,\mathrm{M}\Omega$. The inverting design already includes this taming resistor; no change is needed in order to prevent that configuration from wandering.

What magnitude of input error is tolerable? We are to keep errors under 1%. Then we can tolerate $V_{\text{error,input}} = 0.01 \times V_{\text{full-scale}} = 0.1\,\mathrm{mV} = 100\,\mu\mathrm{V}$. This value we might call our "error budget."

What V_{offset} is tolerable? This total DC error will result from the sum of two effects: V_{offset} and I_{bias}. Since these two effects may have the same sign, we will allow each to use up only half the total error budget: we will allow $50\,\mu\mathrm{V}$ apiece.[2]

V_{offset}: We need to shop for an op-amp with V_{offset} around $50\,\mu\mathrm{V}$. That value is low – about 15 times smaller than what the '411 shows us. It is attainable. But we should not expect to find it in an op-amp that has been designed to optimize some other feature.

Hunting for a satisfactory op-amp, we find that this V_{offset} goal knocks out of contention the single-supply rail-to-rail op-amps required by the inverting design above. The two parts that could handle inputs up to the positive rail showed a V_{offset} of 3 mV and 25 mV, respectively. So, let's abandon the inverting design – which always seemed a bit perverse, anyway.

$V_{\text{offset}} \leq 50\,\mu\mathrm{V}$? Only a very few parts satisfy this specification, in the collection of "candidate op-amps" (see Fig. 7W.11):

- the LT1012A (at $25\,\mu\mathrm{V}$ max);

---

[2] Usually, one would assign budget fractions after seeing what dominates, rather than assume at the outset that both constraints are equally difficult.

- the MAX420E (5 μV, max) – a "chopper" – the sort that continually readjusts its offset to zero it, many times each second; and
- LMP2021MAX (5 μV) (another chopper) ; for this one we'd have to drop V_+ from 15 V to 5 V.

Two other parts comes close: OPA627B at 100 μV max, and OPA336, at 125 μV max – slightly exceeding our error budget. The OPA336 *typical* value of 60 μV would be acceptable. We will not here go into the hard question of when it might be all right to use something other than worst-case specifications.[3] We might keep these two in mind then.

If we are to do the task single-supply, then only the LMP2021 will do.

I_{bias}: The bias current flows in the 1M input resistor, generating an error voltage. Again, we'll assume that we can use up just half the error budget with this error: 50 μV. The bias current that we can tolerate, then, is

$$I_{bias} \leq 50\,\mu V / 1\,M\Omega = 50\,pA$$

This is a small value, but certainly available (even the humble LF411 comes close, showing 50 pA as its typical I_{bias}). The challenge will be to find a part that delivers this good specification while also giving low V_{offset}. If we mean to do the task single-supply, then we need also the peculiar virtues of an amp that can handle inputs and outputs close to ground.

Only the choppers keep the total error under 100 μV. The OPA627B at 100 μV is close enough, with its 5 pA I_{bias} (max), if we mean to use a split supply. The LMP2021 is the only part that does it all: keeps the input error under 100 μV and allows use of a single supply but is limited to a maximum supply voltage of 5.5 V.

7W.2.2 Refinements

Protect meter movement against overdrive: The designs of Figs. 7W.4 and 7W.6 could drive the meter movement all the way to the 15 V power supply. The op-amp's output current, typically limited to a few tens of milliamps, might not damage the movement. But it is easy to modify the circuit, just slightly, to protect the meter movement. Simply putting a series resistor ahead of the movement and then boosting the amplifier's gain makes sure that the movement could not be driven to more than 150% of full-scale; see Fig. 7W.9.

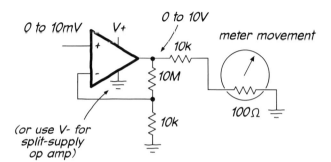

Figure 7W.9 Slight change to limit overdrive of meter movement.

---

[3] A knowledgeable friend of ours who designed digital circuits for a large computer maker told us that his company would have been unable to compete if they had not assumed that parts in their designs would work somewhat better than their worst-case, full-temperature-range specifications.

Allow both positive and negative inputs: This change requires use of a split supply. If the meter movement is of one polarity only,[4] then a four-diode *bridge* (familiar to you from power supplies) will put current through the movement in the forward direction regardless of the sign that reaches the bridge. The design of Fig. 7W.10 uses a *current-source* scheme (like that of Fig. 7W.5). Letting the circuit set a *current* rather than a voltage makes the design indifferent to voltage drops across the bridge diodes. The indicator LEDs are necessary to indicate input polarity.

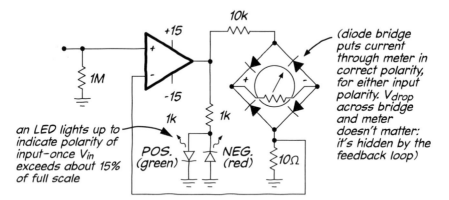

Figure 7W.10 A design permitting both positive and negative inputs. an LED lights up to indicate polarity of input–once V_{in} exceeds about 15% of full scale (diode bridge puts current through meter in correct polarity, for either input polarity. V_{drop} across bridge and meter doesn't matter: it's hidden by the feedback loop)

7W.2.3 The point of this exercise...

This exercise was fussy. What we wanted to show was that a very simple task becomes challenging if the design can tolerate only very small errors. We hope this example will help to give you a feel for why such an amazing variety of op-amps is available – with new ones continually arriving.

Candidate op-amps: All but one of those listed in Fig. 7W.11 are unsuitable for single-supply work. If you designed the meter with a split supply, all are eligible (though the single-supply part is limited to +5 V total voltage).

	V_{OS} (μV) typ	V_{OS} (μV) max	TCV_{OS} (μV/°C) typ	TCV_{OS} (μV/°C) max	I_b (pA) typ	I_b (pA) max	Comments
LM741C	2000	6000	–	15*	80,000	500,000	BJT, old, crummy
LF411	800	2000	7	20	50	200	JFET jellybean
OPA111B	50	250	0.5	1	0.5	1	precision FET
OPA627B	40	100	0.4	0.8	1	5	low-noise prec JFET
LT1012A	8	25	0.2	0.6	25	100	low-bias BJT, prec.
OPA336	60	125	1.5	–	1	10	CMOS, V_s=5.5 V max
MAX420E	1	5	0.02	0.05	10	30	chopper CMOS, V_s to ±15 V
Single Supply							
LMP2021	0.9	5	0.001	0.02	23	100	chopper, CMOS, V_s=2.2 to 5.5 V

* for "premium" grade only, otherwise not specified

Figure 7W.11 Candidate op-amps; most are eligible for split-supply only.

---

[4] Some movements are bipolar, starting with the needle at the center of its range of movement.

8N Op-Amps III: Nice Positive Feedback

Contents

8N.1	**Useful positive feedback**		**351**
8N.2	**Comparators**		**352**
	8N.2.1	What's a comparator?	352
	8N.2.2	Trouble with noise	353
	8N.2.3	Stabilizing with positive feedback	356
	8N.2.4	A good comparator circuit *always* uses positive feedback	356
	8N.2.5	An alternative way to provide hysteresis: AC feedback only	358
8N.3	***RC* relaxation oscillator**		**359**
	8N.3.1	Op-amp relaxation oscillator	359
	8N.3.2	The numbers…	359
	8N.3.3	A lazier way to build an *RC* oscillator	360
	8N.3.4	'555 *RC* oscillator/timer	361
	8N.3.5	Newer IC oscillators	363
8N.4	**Sinewave oscillator: Wien bridge**		**363**
	8N.4.1	Wien bridge's positive feedback network	364
	8N.4.2	Wien bridge's negative feedback network	364
8N.5	**AoE reading**		**367**

Why?

Some effects we would like to achieve with today's circuits:

- make a decisive comparison circuit; and
- make oscillators.

We will apply positive feedback for these purposes.

8N.1 Useful positive feedback

To this point we have treated positive feedback as a bad thing – as a result that you get when your pen slips, interchanging senses of feedback.

It turns out, however, that although negative feedback remains overwhelmingly the more powerful idea, with more applications, positive feedback can be useful. In this chapter we look at cases where it makes circuits decisive, and then at cases where it allows us to form *oscillators*. In Chapter 9N we will look again at positive feedback, but this time less cheerfully. We will meet some of the many instances in which positive feedback sneaks up on us and makes a well-behaved circuit begin to act crazy. Today, though, all our positive feedback is benign.

Op-Amps III: Nice Positive Feedback

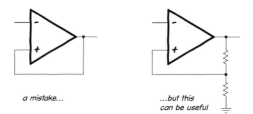

Figure 8N.1 Positive feedback isn't always a mistake.

a mistake...   *...but this can be useful*

8N.2 Comparators

AoE §4.3.2A

8N.2.1 What's a comparator?

A comparator is just a high-gain differential amplifier: such an amplifier "compares" its two inputs, though we have not described its performance that way till now. What's distinctive about the *comparator* is not so much the *device* as the use to which it is put. An op-amp can serve as a comparator, but – as you will confirm in the lab – an op-amp makes a second-rate comparator: it is slow. A chip designed for the narrow purpose of *comparing*, a special-purpose comparator like the '311, works almost 100× faster. And other comparators of more recent design are available with speeds nearly 100 times better than the '311s.

Comparator versus op-amp: Although an op-amp can sometimes do the job of a comparator, the contrary is not true: a comparator cannot serve in an op-amp's *negative-feedback* circuit. A comparator in such a circuit would be unstable. You would likely find its output oscillating at a high frequency. The op-amp has been tamed so that it's stable in a negative feedback loop. A comparator is not so restrained; it is designed for speed, and since it does not expect negative feedback, it is not designed to work with it.

Many comparators, like the LM311 that you will meet today, differ from an op-amp also in offering a versatile output stage.[1] This output stage is not a push–pull, as in an op-amp, but a *switch*.[2] In the '311, both emitter and collector are brought out for the user to connect; see Fig. 8N.2.

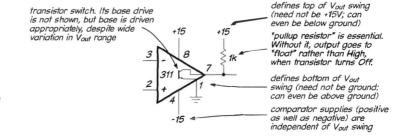

Figure 8N.2 '311 comparator, showing some differences from op-amp features.

A student may find the need to take care of emitter and collector a nuisance, at first (it's one more thing to think about – or two more). But comparators are designed this way for a good reason: to allow the user to determine both the top and bottom of the output swing. So, for example, the '311 makes

---

[1] Some comparators do use a push–pull switch in the output stage, for the greater speed permitted by its low output impedance. See, for example, Analog Devices' (previously Linear Technology's) 4 ns comparator, LT1715.

[2] The output transistor can be wired as a non-inverting switch if one places a pulldown resistor on the emitter (pin 1) and connects the collector (pin 7) to the high level voltage. This allows you to drive a ground-referenced load. Note that the sense of the inputs is reversed in this configuration, since the output on pin 1 is low when the transistor is off and high when it is on: see AoE Fig. 12.38.

it easy to provide a traditional "logic-level" swing of 0 V to 5 V, rather than an op-amp-like swing of ±15 V. Incidentally, the diagram shows the output transistor's *base* just hanging, but of course it doesn't just hang; the IC's designers took care to make it work properly with a wide range of output voltages – and independent of the power supplies for the differential amp that is the guts of the comparator.

Trying comparator versus op-amp: When a comparison task is not demanding, the op-amp and comparator perform (if you'll forgive the word) *comparably*. Figure 8N.3 shows the responses of a '411 and a '311,[3] when both are fed a large, low-frequency sinusoid. The comparator swing is quicker, but both comparator and op-amp do the job.

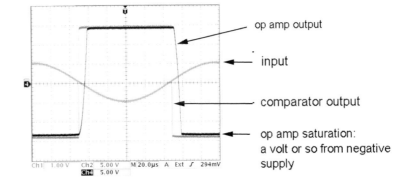

Figure 8N.3 Both op-amp and comparator can handle an easy task: large, low-frequency input. (Scope gains: input, 1 V/div; outputs, 5 V/div.)

Given higher-frequency input, the comparator keeps doing its job while the op-amp is left behind, see Fig. 8N.4.

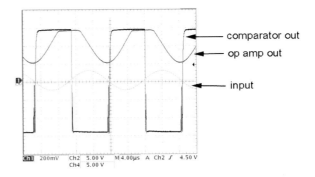

Figure 8N.4 Op-amp falls behind, with input at higher frequency. (Scope gains: input, 100 mV/div; outputs, 5 V/div.)

The output of the op-amp fed a high frequency sinusoid no longer looks like a square wave; it isn't quick enough even to swing its output as far as ground. The comparator output is square – but it is showing some signs of potential trouble, too. Both edges of the comparator output – rising and falling – show multiple transitions where there ought to be one. We'll see lots more of this misbehavior in a minute.

8N.2.2 Trouble with noise

We have sketched in Fig. 8N.5 what you might see if you fed these two simple comparators – implemented with '411 and '311 – a noisy input like the one shown. Assume that you want the

---

[3] The similarity of part numbers is not significant.

comparator to switch on the big, slow waveform's "zero-crossing." You do not want to switch on the little wiggles. (To make this hypothesis plausible, you might imagine that the slow waveform is 60 Hz, and our goal is to let a computer tell time by counting the 60 Hz zero-crossings.)

Compare AoE Figs. 4.31 and 4.33

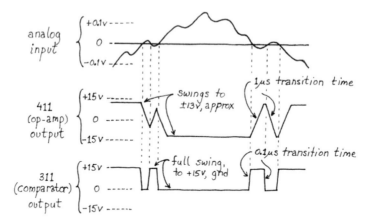

Figure 8N.5 A comparator without feedback will misbehave; the op-amp is *slow* as well.

Try it out: If you were to try making a 60 Hz clock reference this way – say, attenuating the output of a "line" transformer – you might get trouble. In Fig. 8N.6, we used a function generator rather than transformer to drive the comparator. Both images shows a comparator output – the bottom trace – that is pretty messy. Evidently, if this '311 output were our time reference, our clock would run fast!

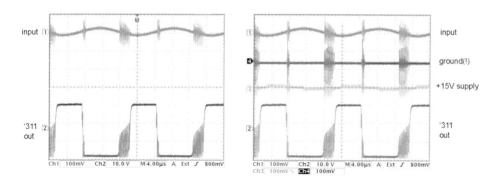

Figure 8N.6 Comparator without feedback has trouble when fed a small, gently sloping signal. (Scope gains: inputs, ground and supply, 100 mV/div; outputs, 10 V/div.)

Our hand-drawn figure, Fig. 8N.5, and the scope images in Fig. 8N.6, might lead you to think that the trouble comes from a defective input. After all, you can see wiggles on the input; maybe all we need is a better signal source. Or maybe we should lowpass filter the input to get rid of that fuzz? Plausible, but wrong.

The flaw in this view is that the causation runs backwards: in fact, it is the *output* activity that causes the noise on the *input*. The right-hand image in Fig. 8N.6 adds two significant traces: *ground* and the *positive supply*, +15 V. With ground itself shaking, even a glassy-smooth input waveform will seem, to the comparator circuit, to be shaking. Given the noisy ground and power supplies (we haven't shown the negative supply but it looks similar), it's not surprising that the comparator keeps changing its mind.

8N.2 Comparators

The case reminds me of my inglorious days as a grade-school outfielder. A fly ball would rise into the blue sky smoothly enough – but as it began to descend, and the earnest outfielder began to run to meet it, the baseball would begin to misbehave. It would start to dance up and down, making itself hard to catch: same problem as for the '311:

Figure 8N.7 Baseball dances the way comparator input dances – once the outfielder's reference frame starts jumping.

If we look closely at the *input* to a chattering comparator circuit, we see an effect a lot like the dancing baseball. The function generator's sinewave seems to get fuzzy at the worst possible time, just as the input crosses the comparator's threshold (in this case, ground). The two images in Fig. 8N.8 show that fuzz at two different time scales.

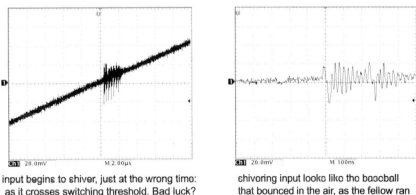

Figure 8N.8 Shiver of input signal, just at threshold-crossing.

input begins to shiver, just at the wrong time: as it crosses switching threshold. Bad luck?

shivering input looks like the baseball that bounced in the air, as the fellow ran

But this is not "bad luck," or the result of a perverse and defective function generator. The seeming-shiver is caused by the shuddering of *ground*. The function generator is not at fault (nor was the baseball actually hopping up and down as it descended).

The noise on the ground and supply lines of the '311 results from the intermittent surges of current that occur each time the output begins to switch. The output switches, thus sinking or sourcing a surge of current through supply and ground. This surge (in the inductance of the lines – more important than resistance) causes supply and ground to wiggle. That wiggle, felt by the high-gain differential input makes the comparator change its mind and decide not to switch. But shutting off the current disturbs the lines again, causing the switching to resume; and so on.

If, during these disturbances, the input signal remains close to the switching threshold, the circuit will switch repeatedly, and we will get the resulting craziness. You can solve this problem however. The standard solution is to provide some *positive* feedback.

8N.2.3 Stabilizing with positive feedback

Feedback variations: As we observed back in §8N.1, we need to broaden our view of positive feedback. Positive feedback is not just the feedback you get when you're confused. The three circuits in Fig. 8N.9, may begin to ease you toward a more nuanced view. One of the three case is a silly error – an utterly useless circuit; another is a second-rate, but not useless, circuit; and one is quite healthy.

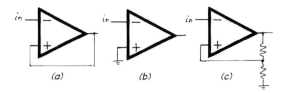

Figure 8N.9 Three ways to wire a comparator: only one is healthy.

Of the circuits in Fig. 8N.9, circuit (a) is the one that is doomed to live at saturation. If decisiveness is the mark of a good comparator circuit, then this *decider* carries the ideal of decisiveness too far; it is downright fanatical – and useless. It is almost impossible to get this circuit to change its mind.

Circuit (b), which uses no feedback, either positive or negative, is not necessarily useless, but is dangerous, as we have just seen. It can handle only the *easy* comparisons where the input moves rapidly across the threshold. The circuit is very likely to oscillate when fed a gently-sloping input.

Circuit (c) shows the usual, healthy comparator circuit. It applies *some* positive feedback to achieve decisiveness (our diagram of course hedges on the question *how much* positive feedback is used here). All our practical comparator circuits will use this tactic.

8N.2.4 A good comparator circuit *always* uses positive feedback

It turns out to be easy to make a comparator circuit ignore small wiggles like those shown causing mischief in Figs. 8N.5 and 8N.6. Just feed back a small fraction of the output swing, and – here is the novelty – feed it back in a "positive" sense, so that the output swing tends to confirm the comparator's tentative decision.

Such *positive feedback* makes the comparator decisive: the comparator uses positive feedback to pat itself on the back, saying (in the manner of many of us humans), "Whatever you've decided to do must be right." The act of switching tends to reinforce the decision to switch.

A comparator circuit that includes positive feedback is called a Schmitt trigger.[4]

Figure 8N.10 sketches the argument that a little hysteresis should let a comparator ignore small disturbances on the input.

Again compare AoE Fig. 4.33

Figure 8N.11 shows what happened when we tried applying this remedy to the gently sloping sinusoid that produced so much fuzz back in Fig. 8N.6. We have fed a small fraction of the output back to the non-inverting input (so, the feedback is, indeed, *positive*). Alongside the traces of the stabilized circuit we've included the earlier no-feedback scope image for comparison. A very little hysteresis does the job, in this case: about 20 mV is enough.

---

[4] Named after Otto H. Schmitt, an American graduate student who described such a circuit, made with vacuum tubes, in 1934. Does this humble circuit deserve to have one person's name attached to it?

8N.2 Comparators

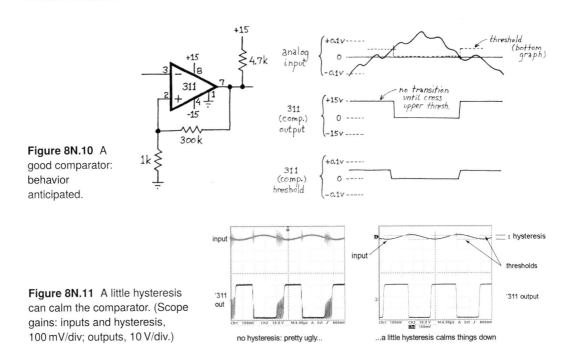

Figure 8N.10 A good comparator: behavior anticipated.

Figure 8N.11 A little hysteresis can calm the comparator. (Scope gains: inputs and hysteresis, 100 mV/div; outputs, 10 V/div.)

Hysteresis is familiar in non-electronic life: Hysteresis[5] and applications of positive feedback may be unfamiliar to you in electronics – but the notion is familiar to you in the rest of life. It is a form of dishonesty: changing the question after you decide. You agonize, for example, over whether to go to Harvard or Stanford; after you decide on Harvard, you tell yourself, "Of course. All that sunshine gets to be so tedious."

Strong hysteresis appears in college admissions from the college's point of view, too: when the college considers your application they ask themselves, "Is their oboe playing really exceptional? – or are they just another valedictorian with perfect SATs?" Once you're in, the question whether to expel you might be, "Was your drug bust a felony, or just another misdemeanor?" The standards change, because admissions committees like to tell themselves that whatever they decided must be correct – just as stable comparators do.

Getting elected to congress versus getting thrown out shows the same pattern: once you're in, you stay. Positive feedback makes a person feel good; makes them decisive and self-satisfied; negative feedback – the inclination to think the decision must be wrong because it's *my* decision – makes a person indecisive and, at an extreme, neurotic. Positive feedback can make you a contented dictator; lack of it can make you a Woody Allen.[6]

How much hysteresis? If hysteresis makes the comparator decisive, how much should you design in? As much as possible? What do you lose as you enlarge hysteresis? Hysteresis is effective medicine, but you should apply no more of it than you need because of its side effects; see Fig. 8N.12. It provides immunity to *noise* – blindness to disturbances on the input, up to some magnitude; but this same immunity is also blindness to *signal*. So a zero-crossing detector with symmetric hysteresis of 200 mV will not switch at all for an input whose peak-to-peak swing is less than 200 mV.

Often you will be obliged to guess at how much hysteresis you need. In the lab circuits we build,

---

[5] From the Greek for "falling short." Nothing to do with hysteria.
[6] See https://LAoE.link/Woody_Allen.html.

358 Op-Amps III: Nice Positive Feedback

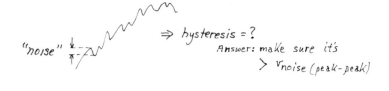

Figure 8N.12 Relation between noise and hysteresis.

100–200 mV usually is enough. If you are building a single, custom circuit, you can make hysteresis adjustable, as we do in lab, and find by experiment what level of hysteresis is required for stability.

A second side-effect of hysteresis sometimes matters: delay. So, even for an input signal large enough to switch the comparator, the switching will come *late*: the circuit will not be a true *zero*-crossing detector. The trick described next can get around that defect – at the expense of making the circuit somewhat sensitive to input frequency.

8N.2.5 An alternative way to provide hysteresis: AC feedback only

We can eliminate the *delay* that hysteresis normally imposes by using hysteresis that works only for a short time just after a transition: AC-coupled feedback achieves this, see Fig 8N.13. The two zero-crossing detectors shown below provide the same amount of hysteresis, but the AC-coupled one shifts the threshold only for a short time after the output switches. Thus it accurately detects the zero crossing, without delay.

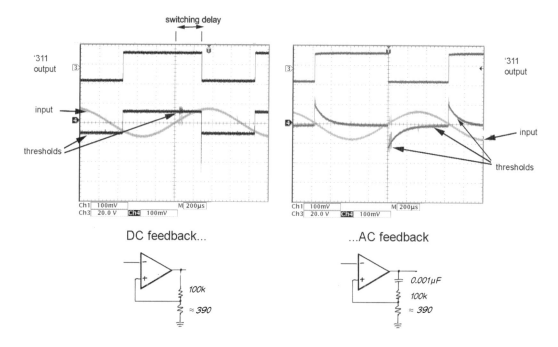

Figure 8N.13 AC-coupled feedback can eliminate the delay that DC hysteresis imposes. (Scope gains: inputs and thresholds, 100 mV/div; outputs, 20 V/div.)

This circuit would fail, chattering, if the input signal were very slow and gently sloped, so that it was still close to threshold at a time when the *RC* had decayed away, eliminating the threshold shift.

8N.3 RC relaxation oscillator

Oscillators: Curiously, we seem to be backing into a discussion of oscillators, having met some nasty, accidental repetitive switching before we ever tried to build an oscillator. The multiple transitions or fuzz produced by a comparator circuit that lacks hysteresis (for example, Fig. 8N.6) are, indeed, oscillations. Now, we want to learn how to produce such waveforms, but in a form that is controlled and predictable.

It's pretty obvious that a circuit that produces *purposeful, controlled oscillations* can be useful. Humankind has been using such things ever since starting to measure time with the help of a pendulum rather than a sundial or water clock.

Conceptually, an oscillator is something that keeps talking to itself – and contradicting itself. But the contradiction must include a delay, as well. The circuit on the left in Fig. 8N.14 contains insufficient delay; it talks to itself, but it is so quick to contradict itself that it barely quivers around the switching threshold (we concede that the quiver is an oscillation; but we'd like something larger). The circuit on the right does oscillate with a full swing, having been given extra delay, a delay provided by a couple of inverters placed in series.

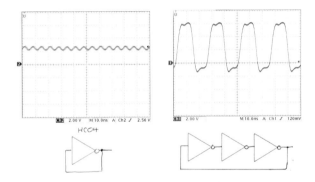

Figure 8N.14 Self-contradiction, with sufficient delay, brings oscillation.

The inverting amplifiers shown are *logic* gates, essentially single-supply comparators, with their switching thresholds placed at one half the supply voltage.

8N.3 RC relaxation oscillator

8N.3.1 Op-amp relaxation oscillator

Figure 8N.15 gives the lab circuit, which you recognize as just a Schmitt trigger like the one in Fig. 8N.10 (here, given unusually large hysteresis) – but this Schmitt trigger is wired in a new way: it is feeding *itself*:

AoE §7.1.2A

8N.3.2 The numbers...

What are the thresholds? How long does V_{cap} take to move between the thresholds?

If you're inclined to answer this question by using $e^{t/RC}$ to calculate the time for the RC formed by 100k and 0.1 μF to charge between one threshold and the other (about $1/11 \times |V_{\text{Positive or Negative: 15 V}}| \approx \pm 1.4$ V), you're not wrong. But you are working too hard.

The excursion of approximately 2.8 V in a span of about 16 V is a small enough fraction so that the current is nearly constant.[7] That makes the problem amenable to the easier formula $I = C dV/dt$, with

[7] The 16 V is the maximum voltage difference between the capacitor's starting voltage – about 1.4 V positive or negative –

360 Op-Amps III: Nice Positive Feedback

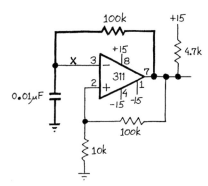

Figure 8N.15 *RC* relaxation oscillator.

I the midpoint value, 15 V/100k. As we say in Lab 8L, the answer obtained with this shortcut is within about 1% of the more exact answer that emerges from the laborious calculation of the exponential charging time. We further simplify the problem by ignoring the effect of the pullup resistor, present on charge-up, but absent from discharge-down. We can afford that 5% error on one swing.

8N.3.3 A lazier way to build an *RC* oscillator

When you want an oscillator in a hurry, you can plug an integrated Schmitt trigger inverter in place of the comparator circuit of Fig. 8N.3. This logic gate (called a 74HC14) has hysteresis built in, and thresholds at approximately half the supply voltage, as for the earlier simple oscillators of Fig. 8N.14. Figure 8N.16 shows such an oscillator in two forms: one, on the left, uses a resistor as in the comparator relaxation oscillator of Fig. 8N.15. The circuit on the right uses a current-limiting diode in place of the resistor to produce a *sawtooth* waveform. Note that the output of that circuit is taken at the capacitor, *not* at the output of the gate, as for a square-wave oscillator.

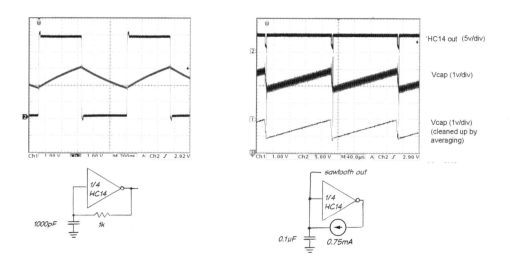

Figure 8N.16 *RC* relaxation oscillator made with integrated Schmitt inverter rather than op-amp.

and the voltage at the other end of the 100k feedback resistor – about 15 V positive or negative. We are not describing the *output* voltage swing of the comparator.

8N.3 RC relaxation oscillator

The weakness of this simple oscillator is that its frequency is not very predictable. Here are the specifications, taken from Lab 8L:

- threshold voltages:
    - negative-going: 1.8 V (typical), 0.9 V (min), 2.2 V (max)
    - positive-going: 2.7 V (typical), 2.0 V (min), 3.15 V (max)
- hysteresis: 0.9 V (typical), 0.4 V (min), 1.4 V (max)

Which of these matters to determine $f_{\rm osc}$?[8] An oscillator built with the HC14 is not highly predictable, but it sure is simple!

The op-amp "relaxation oscillator" of Fig. 8N.15 is a classic, and sometimes provides the best source of a square wave at moderate frequencies.[9] And the relaxation oscillator made with a logic gate provides an easier way to get an approximately equivalent result. But often the easiest method of all is to use an IC dedicated to the task of implementing oscillators.

8N.3.4 '555 RC oscillator/timer

[AoE §7.1.3]

The '555 is a peculiar, not-particularly-elegant square-wave oscillator that is even older than the '741 (the '555 was born in 1971) – but it may be the most popular IC ever produced.[10] Its dominance is fading, but it is widely sourced, and one still can find small books of '555 application notes. In the lab you will use an improved version: a part labeled either "7555" or "...C555," or "...555C," made of CMOS MOSFET's[11] for low power, and capable of running faster than the original, bipolar device.

Because the '555[12] contains a *flip-flop*, a device you have not met, it's a little hard to explain. We'll try, though: Fig. 8N.17 shows a diagram of its insides, and two ways to wire the part.

The left-hand figure shows the output driving the timing capacitor, exactly as in the op-amp relaxation oscillator of Fig. 8N.15 (and as in the logic-gate version of Fig. 8N.16). We hope that this scheme is beginning to look familiar.

The right-hand circuit of Fig. 8N.17 shows the traditional '555 wiring – a bit strange. This scheme does not connect the timing RC to the output at all, relying instead on a dedicated *discharge* transistor to discharge the capacitor as needed. This arrangement makes an already ornate circuit still more complex; but it has the virtue of isolating the timing circuitry from any effect of a *load*. The left-hand circuit of Fig. 8N.17, in contrast, would change frequency if a load attenuated its $V_{\rm out}$.

[Compare AoE Fig. 7.8 with Fig. 7.9]

As in the op-amp relaxation oscillator, the capacitor voltage here moves between two thresholds, always frustrated. The relaxation oscillators of §§8N.3.1 and 8N.3.3 achieved two thresholds by the use of hysteresis. The '555 uses a simpler scheme: two comparators, each with a fixed threshold. The flip-flop is a memory element that remembers which comparator was tripped most recently. Its output, "Q," is set to ground when the capacitor voltage reaches the threshold of the "too high" (upper) comparator and stays there until it is set to $V_{\rm CC}$ when the capacitor voltage reaches the threshold of the "too low" (lower) comparator. Similarly, the output then remains at $V_{\rm CC}$ until the high limit is reached

---

[8] The magnitude of hysteresis matters. Just where this span is placed – whether it goes from 0.9 V to 1.8 V or from 1.8 V to 2.7 V – does not matter.

[9] One of our heroes, the late Bob Pease, formerly of National Semiconductor, used this design, for example, in order to make a super-low-power 1 Hz oscillator. His circuit is a single-supply version of the relaxation oscillator of §§8N.3.1 and 8N.3.3, using a low-power comparator and drawing 1.4 µA at 6 V. https:/LAoE//Oscillator_Stuff.html.

[10] According to its designer, it was the top selling IC for around 30 years, and by 2004 had sold more than a billion units. See https://LAoE.link/555_Timer.html.

[11] Oof! That's a mouthful of acronyms, is it not? Spelling it out, though, is even worse: "CMOS" is Complementary Metal-Oxide Semiconductor; "MOSFET" is Metal Oxide Semiconductor Field Effect Transistor.

[12] We'll call the chip '555 even though the version you will use is CMOS and may like to be called *7555*; for the purpose of this explanation of its operation, the version of the '555 does not matter at all.

Op-Amps III: Nice Positive Feedback

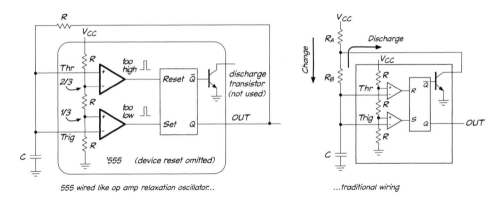

Figure 8N.17 '555 oscillator.

again. The "$\overline{Q}$" output of the flip-flop is just the opposite of the "Q" output and is used to control the discharge transistor.

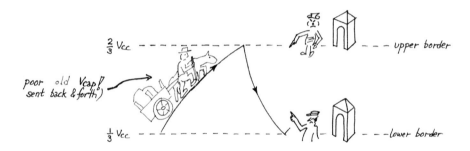

Figure 8N.18 A very informal explanation of the '555's operation.

The two comparators act like unfriendly border guards at the $\frac{1}{3}V_{CC}$ and $\frac{2}{3}V_{CC}$ boundaries. When the capacitor voltage crosses either frontier,[13] it finds itself sent back toward the other – like the stateless refugee in Fig. 8N.18. The oscillator configuration on the left side of Fig. 8N.17 charges and discharges the capacitor through a single resistor, R, resulting in a square wave output with a 50% duty cycle.

In the traditional wiring, shown on the right side of Fig. 8N.17, the border guards send the capacitor voltage up and down by turning *on* or *off* the discharge transistor. When the '555 output is low, this transistor is on, discharging timing capacitor C through resistor R_B. When the '555 output is high, the transistor is off, allowing the capacitor to charge toward the positive supply through the series connection of R_A and R_B. Since the capacitor charges through *both* resistors but discharges through *only one* of them, the output is always high longer than it is low, meaning it always has a duty cycle >50%.[14]

You can vary the '555's waveform *at the capacitor* by replacing the resistors R_A and R_B with other elements: a current source or two, or – for R_B, but not R_A – even a piece of wire. Can you picture the waveforms that result – at the capacitor, not at the '555 "output" terminal – for the several possible configurations?[15]

---

[13] Would it be too silly to think of the upper border voltage as the frontier of Upper Volta?
[14] A caution here. The '555 datasheet formula for duty cycle calculates the percentage of time the output is *low*, not high; a non-standard way of describing duty cycle.
[15] Of course you can. So can we: using a *wire* rather than a resistor in the discharge path produces a quick sharp discharge,

The '555 has other applications, too: it can form a *voltage-controlled* oscillator. We invite you to try this in Chapter 8L, and you will use this technique in the final analog project. The '555 also can be wired to put out a single pulse when triggered (behaving like a so-called "one-shot"). And with some cleverness you can tease it into doing a great many other tricks.

8N.3.5 Newer IC oscillators

AoE §7.1.4A

You will not be surprised to hear that by now, more than a half century after the birth of the '555, other IC oscillators have appeared. Some are just easier to wire: the LTC6906, for example, integrates the capacitor, so that a single resistor sets f_{osc} (frequency range 10 kHz to 1 MHz). Unfortunately for our purposes, this part is not offered in a "through-hole" DIP package. Many newer parts run at voltages lower than the '555's minimum of 4.5 V. Some can be programmed from a microcontroller (an extreme case is one from Dallas/Maxim that lives in a 3-pin TO-92 package: DS1065 (30 kHz to 100 MHz)). Lots of alternatives are available; but the '555 remains a useful workhorse.

8N.4 Sinewave oscillator: Wien bridge

AoE §7.1.5B

This oscillator – the only sine generator you will build in this course – is clever, and fun to analyze. How does it put out a sine, rather than the usual square or triangle waves (those are the waveforms that are easy to generate)? Its clever gain control prevents the *clipping* that would produce the usual square output. Its positive feedback network favors just one frequency above all others. Only this "fittest" frequency survives – and if only one frequency survives, this single surviving frequency will, by definition, be a pure sinusoid. This circuit, like many oscillators, uses both positive and negative feedback. For the purpose of analysis, we have separated these two feedback senses below, so that we can consider them separately.

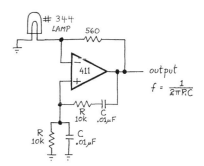

Figure 8N.19 Wien bridge sine oscillator.

while the slow *RC*-curvy charge persists. One disgruntled student complained that they didn't see anything useful about such a "shark-fin" waveform. They had a point.

The same circuit with *current source* replacing the charging resistor, while keeping the wire in the discharge path produces the quite-respectable waveform called *sawtooth*. These are widely useful: analog pulse width modulation (PWM), demonstrated in §8L.4, applies such a sawtooth. The digital equivalent is demonstrated in §25L.1.4.

If you put current sources in both charge and discharge paths, the waveform on the capacitor becomes a *triangle*. See AoE §7.1.3E.

8N.4.1 Wien bridge's positive feedback network

The positive feedback network is frequency selective, and at the most favored frequency the network passes back to the + input the maximum, 1/3 of the output swing. It treats all other frequencies – above and below its favored frequency – less kindly. Figure 8N.20 sketches the behavior.

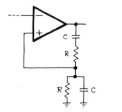

Figure 8N.20 Wien bridge positive feedback: fraction of output fed back.

At that favored frequency, the phase shift also goes to zero. The preferred frequency – the one at which the oscillator will run – turns out to be $1/(2\pi RC)$.[16]

8N.4.2 Wien bridge's negative feedback network

The negative feedback – redrawn in Fig. 8N.21 to look more familiar – adjusts the gain, exploiting the lamp's current-dependent resistance (the lamp is rated at 14 mA at 10 V; this suggests an approximate value for its *resistance* – but the key to the circuit's cleverness is that the R value is not a constant). The use of the lamp was the key contribution of Hewlett to the design of a low-distortion sine generator. His patent application of 1942 describes the elegantly simple scheme.[17]

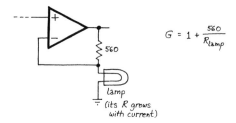

Figure 8N.21 Self-adjusting gain: negative feedback of Wien bridge, redrawn.

He sounds almost embarrassed by the simplicity of his solution ("a small incandescent lamp, or similar device..."). But simple is neat – with this invention he and his friend David Packard launched a company.

The lamp's peculiar *I* versus *V* curve: You may, conceivably, remember doing an experiment back in the first lab of this course, in which we asked you to plot *I* versus *V* for two "mystery boxes." One plot showed a straight line: that was a resistor. The other showed a bend: that was a small incandescent lamp bulb.

The lamp (#344) used in today's lab circuit is not the one used then (#47), but shows a similar bend. This bend reflects the effect of filament heating, since the resistance of a metal grows with

---

[16] Yes, this seems almost too good to be true. We won't prove the point to you. To prove that this frequency gives the maximum feedback signal is a messy exercise in phase-sensitive calculus. But you can pretty readily confirm, at least, that at this frequency the impedance of the series RC is $\sqrt{2}R$, and the impedance of the parallel RC is half that. Thus 1/3 is fed back.

[17] US patent 2,268,872 (1942). His use of a lamp to control gain was anticipated, by about one year, by Meacham. *Bell System Technical Journal* **17**, cited in Wikipedia article on the Wien Bridge.

8N.4 Sinewave oscillator: Wien bridge

Figure 8N.22 Excerpt from Hewlett's patent on use of a lamp to regulate output amplitude.

> Amplitude control to prevent the oscillations from building up to such a large value that distortion occurs, is obtained according to my invention by non-linear action in the amplifier circuit. In order to produce this non-linear variation, I provide for resistance R₃ a small incandescent lamp, or similar device in which the resistance increases rapidly with increased current flow, the lamp being heated by the plate current of the tubes 10 and 11 or by an auxiliary means, so such a temperature that its resistance will vary rapidly with a small change in current. Thus, when the oscillation amplitudes tend to increase, the temperature of the lamp R₃ increases with a resulting increase in resistance thereby causing a greater negative feedback, thus reducing the amplification. Similarly, as the oscillations decrease in amplitude, the current through the lamp is reduced permitting the lamp to cool with an accompanying decrease in resistance and reduction of the negative feedback, thus increasing the amplitude of the generated oscillations. As a result the system operates at substantially a constant amplitude which is preselected to be below the value at which grid current flows. As a result no distortion of the wave form takes place.

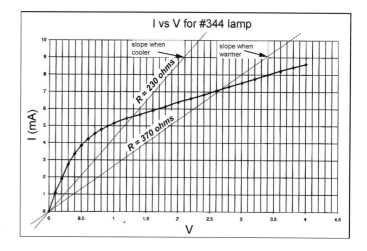

Figure 8N.23 *I* versus *V* for #344.

temperature. We have inserted two straight lines in Fig. 8N.23, to show the lamp's effective resistance at a two arbitrary *average* *I*-versus-*V* points. The lamp's resistance in your lab circuit lies between the two lines that we have drawn.[18]

Note that the curve that is plotted shows not the response of the lamp to quick changes of *I* or *V* but rather the *I*, *V* combinations that the lamp settles into when allowed many seconds at a given *average* – long enough for its temperature to stabilize.[19] The temperature results from the heating effect of that particular *I*, *V* product. This notion is rather subtle, and may make your head spin briefly, as you first consider what the lamp's plot shows you.

Convince yourself that the sense of variation in lamp resistance does tend to stabilize gain at the

---

[18] We saw amplitude out of our Wien bridge at about 9 V. This is an *rms* value of about 6.4 V, and rms is what matters for the lamp plot, which describes essentially the lamp's response to temperature.
To sustain a 6.4 V oscillation at the output, the Wien bridge input – what the lamp "feels" – is 1/3 as large: about 2.1 V. At this rms value, the plot shown in Fig. 8N.23 would predict a lamp resistance of about 330 Ω. Such an *R* value would then call for a feedback *R* of 660 Ω in order to give the required gain of three. We use 560 Ω in the lab, not 660 Ω, so the lamp for the oscillator that ran at 9 V amplitude must have followed a curve slightly different from that in Fig. 8N.23 – but only slightly different.

[19] Strictly, the *I* and *V* values plotted show the *rms* average, since what concerns us is its heating effect. This is precisely the significance of rms values: the rms value of a sinusoid, for example, is $1/\sqrt{2} \times V_{\text{peak}}$, and this is the *DC* voltage that would deliver the same power as the rms value of the sinusoid.

necessary value. Don't forget that Fig. 8N.23 plots I on the vertical axis, as usual; a plot of *resistance* would interchange the axes, and would show slope *rising* with I and V (and temperature).

What value of gain *is* necessary to sustain oscillations without clipping? It may take you several seconds to answer; it takes the op-amp much less time.[20]

Rubbery behavior: The detail of performance that is hardest to explain is the so-called "rubbery behavior" that results if one touches the non-inverting terminal while the circuit is running. Figure 8N.24 is a sketch, while Fig. 8N.25 is a scope image showing the response to a finger's touch.

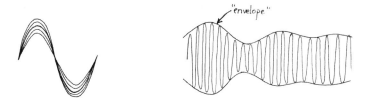

Figure 8N.24 Rubbery behavior" of Wien bridge oscillator when one touches the non-inverting terminal: at two sweep rates.

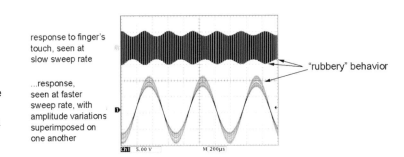

Figure 8N.25 Wien bridge lives close enough to the edge of instability so that a finger's touch makes it shiver.

The finger's touch disturbs the circuit probably because the finger's capacitance upsets the carefully adjusted equilibrium of the *gain*-setting circuit.[21] To explain why the circuit goes rubbery when poked, recall that although the negative feedback network is stable, it is only marginally stable, because the lamp (with its long time-constant of response) looks like an extreme lowpass filter within the loop. This situation is reminiscent of the jumpiness of the differentiator, which you saw back in Chapter 7. This circuit, with its automatic gain adjustment, would not be satisfactory as a general purpose amplifier; it is too jumpy, overshooting in its gain adjustments, and then taking a long time to stabilize again. It turns out that the jumpier the circuit – that is, the longer it takes to stabilize after a disturbance, because "damping" is slight – the more perfectly it holds a single frequency when not disturbed (electronic justice once again?). The oscillator is satisfactory in our application, because normally we do not disturb it: we do not poke it with a finger.

The circuit is also sensitive to mechanical vibration, and this vulnerability explains why electronic analogues often are used in place of the lamp.[22]

How good is our Wien bridge? The answer is "pretty good." We compared the lab circuit's frequency spectrum against that of the function generator on the powered breadboards: see Fig. 8N.26. The lab sinusoid looked much better than the breadboard's.

---

[20] Now that several seconds have elapsed, let's announce the result: yes, we need a gain of 3 to sustain an oscillation.
[21] In lab, you might try a 10 pF capacitor to ground, as a stand-in for your finger. We found they had very similar effects.
[22] See, for example, AoE §7.1.5B, Fig. 7.20B, or Jim Williams' Wien bridge oscillator in his application note "High Speed Amplifier Techniques," *Linear Technology* AN47, August 1991, p. 49, Fig. 114 at
https://LAoE.link/Williams_AN47.pdf, each using a JFET as the voltage-controlled resistance in place of the lamp.

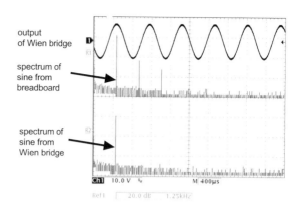

Figure 8N.26 Frequency spectra of lab's Wien bridge versus breadboard's sine generator.

We should admit that we put up our champion – the lab circuit – against a pretty weak competitor: an IC called an ICL8038, an inexpensive part that produces a sinusoid by whittling the points off a triangle. This is the same strategy used by our good lab function generators; the latter do their whittling more discreetly, as is not surprising, given that they cost well over 100 times as much as the '8038.

But the good quality of the Wien's sine has not let it take over the world. The whittled-triangle-wave method remains dominant in analog generators, because this method makes it easy to provide a wide range of frequencies (0.3 Hz to 3 MHz on our lab generators, for example). And such methods now compete with a more recent method: digital synthesis, the playing out of a table of sine values to a digital-to-analog converter (a "DAC"). We'll meet DAC's later in this course.

AoE §7.1.5A

Sine from square – but not now: Also in the digital part of the course you will meet an analog–digital hybrid method that produces a clean sinusoid: feed a square wave to a steep analog filter, while adjusting the filter's f_{3dB} to preserve only the square wave's fundamental – a pure sine.

For now, the Wien bridge gives us our cleanest sinusoid.

8N.5 AoE reading

Reading
- Chapter 4 (op-amps):
  §4.3.2A: comparators.
  §4.3.2B: Schmitt trigger (the usual comparator circuit).
  §4.3.3: triangle-wave oscillator.
  §4.6.4: voltage-controlled oscillator.
- Chapter 7 (oscillators):
  From start through §7.1.5B, Wien bridge oscillator.
  For lab, the '555 is probably the most important; the Wien bridge is the most interesting.
  §7.2.1E: long pulse using '555 again.
  §7.1.7A, Table 7.2 listing a wide variety of oscillators.
- Chapter 12 (logic interfacing):
  §12.1.7A: comparators (more detail than in Chapter 4).

Problems:
- Additional Exercise 4.34 (comparator, using slow op-amp).

8L Lab: Op-Amps III

Positive feedback: good and bad

Until now, as we have said in Chapter 8N, we have treated *positive* feedback as evil or as a mistake: it's what you get when you get confused about which op-amp terminal you're feeding. Today you will qualify this view: you will find that positive feedback can be useful: it can improve the performance of a comparator; it can be combined with negative feedback to make an oscillator ("relaxation oscillator": positive feedback dominates there); or to make a negative impedance converter (this we will not build, but see AoE §4.107, Fig. 4.104: there, negative feedback dominates). And another clever circuit combines positive with negative feedback to produce a *sinewave* out (this is the "Wien bridge oscillator").

Later, in Chapter 9, we will see what a pain in the neck positive feedback can be when it sneaks up on you.

8L.1 Two comparators

Comparators work best with positive feedback. But before we show you these good circuits, let's look at two poor comparator circuits: one using an op-amp, the other using a special-purpose comparator chip. These circuits will perform poorly; the faster of the two, the restless circuit, will help you to see what's good about the improved comparator that *does* use positive feedback.

8L.1.1 Op-amp as comparator

You will recognize the "comparator" circuit in Fig. 8L.1 as the very first op-amp circuit you wired, where the point was just to show you the "astounding" high gain of the device. In that first glimpse of the op-amp, that excessive gain probably looked useless. Here, when we view the circuit as a comparator, the very high gain and the "pinned" output are what we want.

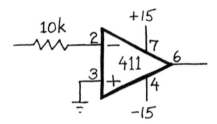

Figure 8L.1 Op-amp as simple comparator.

Drive the circuit with a sinewave at around 100 kHz, and notice that the output "square wave" output is not as square as one would hope. Why not?[1]

---

[1] The rise and fall are slow, moving at a rate defined by the op-amp's *slew rate*, a limitation that you recall from Lab 7L:

8L.1.2 Special purpose *comparator* IC

Now substitute a LM311 comparator for the LF411. The pinouts are *not* the same. You will notice from Fig. 8L.2 that the '311 output stage looks funny: it is not like an op-amp's, which is always a push–pull; instead, two pins are brought out, and these are connected to the *collector* and *emitter* of the output NPN transistor, respectively. These pins let the user determine both the top and bottom of the output swing. If one uses +5 (with an external "pullup" resistor) and ground to pin 1, for example, the output provides a "logic level" swing, 0 to +5 V, compatible with traditional digital devices. In the later circuits, you will keep the top of the swing at +15 V and the bottom of the swing to −15 V, as in Fig. 8L.2.[2] So arranged, the '311 output will remind you of op-amp behavior.

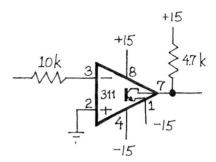

Figure 8L.2 '311 comparator: no feedback.

Does the '311 perform better than the '411?[3]

Parasitic oscillations: A side-effect of the '311's fast response is its readiness to oscillate when given a "close question" – a small voltage difference between its inputs. Try to tease your '311 into oscillating, by feeding a sinewave with a *gentle slope*. With some tinkering you can evoke strange and lovely waveforms that remind some of the Taj Mahal in moonlight – but remind others of the gas storage tanks on the Boston's Southeast Expressway. Judge Figure 8L.3 for yourself.[4]

Minimizing oscillations: remedies other than positive feedback: You can sometimes stabilize a comparator *without* the remedy of *hysteresis*, which we are about to promote. Let's see how far these efforts can carry us. First, to get some insight into *why* the '311 is confusing itself, keep one scope probe on the oscillating output and put the other on the ±15V power supply line, close to the '311. *AC-couple* this second probe, and see if the junk that's on the output appears on the supply as well. Chances are, it does.

Keep those ugly '311 oscillations on the scope screen, and try the following remedies:

- Put ceramic *decoupling* capacitors on positive and negative power supplies. If the oscillations stop, tickle them into action again, by reducing the slope of the input sine.

---

[1] 15 V/μs for the '411. Thus a rail-to-rail transition takes a couple of microseconds – and a bit more because the op-amp needs a little time to recover from *saturation*, the state in which this no-feedback circuit obliges it to spend most of its time.

[2] Most datasheets label pin 1 in the DIP package as "ground." This is misleading since pin 1 can be connected to any voltage from the negative supply up to within a few volts of the positive supply. As long as you arrange for the BJT collector (pin 7) to be pulled up to a higher voltage than the emitter (pin 1), the output will still switch correctly.

[3] This question reminds us of a question posed to us by a student a few years ago: 'I can't find a '411. Is a '311 close enough?' This person no doubt was recalling the many times we had said, "Put away that calculator! We'll settle for 10% answers; 2π is 6," and so on.

[4] Two undergraduates produced this handsome waveform: Elena Krieger and Belle Koven, October 2004.

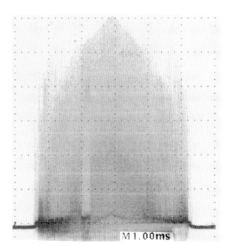

Figure 8L.3 '311 indecision: gas storage tanks? Taj?

- Short pins 5 and 6 together: these adjustment pins often pick up noise, making things worse. Again tease the oscillations back on, if this remedy temporarily stops them.

8L.1.3 Schmitt trigger: using positive feedback

The *positive feedback* used in the Fig. 8L.4 circuit will eliminate those pretty but harmful oscillations. See how little hysteresis you can get away with. As input, use a small sinewave (<200 mV) around 1 kHz, and adjust the pot that sets hysteresis until you find the border between stability and instability. Then watch the waveform at the *non-inverting input*. Here, you should see a small square wave (probably fuzzy, too, with meaningless very high frequency fuzz *not* generated by your circuit; perhaps radio, visible now that you have the gain cranked way up). This small square wave indicates the *two* thresholds the '311 is using, and thus (by definition) just how much hysteresis you are using. Finally, crank the hysteresis up to about three times that borderline value (for a safety margin).

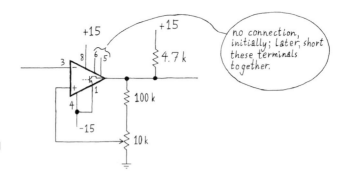

Figure 8L.4 Schmitt trigger: comparator with positive feedback (and *hysteresis*).

If you describe your circuit as a "zero-crossing detector," how late does it detect the crossings? Could you invent a way to diminish that lateness?[5]

Leave this circuit set up for the next experiment.

---

[5] You could – except that the world has already invented the method. See §8N.2.5, especially Fig. 8N.13, describing *AC* positive feedback.

8L.2 Op-amp RC relaxation oscillator

Fix hysteresis at its maximum value, in the circuit of Fig. 8L.4, by replacing the potentiometer of the preceding circuit with a 10k resistor. Then connect an RC network from output to the comparator's inverting input, as shown in Fig. 8L.5. *This feedback signal replaces any external signal source; the circuit has no input.* Here, incidentally, you are for the first time providing *both* negative and positive feedback.

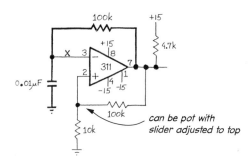

Figure 8L.5 *RC* relaxation oscillator.

Predict the frequency of oscillation, and then compare your prediction with what you observe. You can save yourself time by assuming that the capacitor *ramps*, as if fed a constant current equal to 15 V/R_{feedback} (as pointed out in §8N.3.2).

8L.3 Easiest *RC* oscillator, using IC Schmitt trigger

Here's another way to get a square wave output, timed by an *RC*. It is very similar to the *RC* relaxation oscillator of the previous section but uses an IC that has hysteresis built in. The output is a "logic-level" swing, 0 to 5 V, *not* ±15 V. The circuit's weakness is the imperfect predictability of the part's hysteresis, an uncertainty that makes the frequency of oscillation uncertain; but logic output and simplicity make it a circuit worth meeting.

8L.3.1 *RC* feedback

Figure 8L.6 shows all there is to it. Use a capacitor of 1000 pF,[6] and choose a feedback *R* to get $f_{\text{osc}} \approx 1$ MHz. As you choose *R*, note the following specifications:

- threshold voltages:
  - negative-going: 1.8 V (typical), 0.9 (min), 2.2 (max);
  - positive-going: 2.7 V (typical), 2.0 (min), 3.15 (max).
- hysteresis: 0.9 V (typical), 0.4 (min), 1.4 (max).

The range of possible values makes prediction of f_{osc} look discouraging – but use *typical* values to choose your *R*, and see how close your result comes to the target of 1 MHz (or "period of 1 µs," to speak in terms more appropriate to your scope use, where you will see period rather than frequency).

---

[6] This value was chosen to be large relative to the scope probe's roughly 10 pF load, so that you can probe the capacitor without changing f_{osc} appreciably.

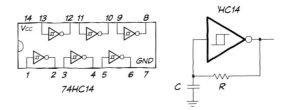

Figure 8L.6 *RC* relaxation oscillator.

8L.3.2 Sawtooth?

One student complained that the waveform of §8L.3.1 looks silly – like a sharkfin. Like that fellow, you may prefer an oscillator that produces a *sawtooth*. That you can easily achieve by replacing the feedback *R* with a *current-limiting diode*: the 1N5294 that you met as you were building a differential amplifier, in Lab 5L It is rated at 0.75 mA. A larger *C* will produce a prettier sawtooth: try 0.1 μF. Note the range of the sawtooth's swing, because we are about to use this waveform.

8L.4 Apply the sawtooth: PWM motor drive

You now have met the two elements of a "pulse-width modulation" (PWM) circuit: a ramping oscillator output, and a comparator. Let's put them together to make a PWM driver. We'll use it to spin a DC motor. Fig. 8L.7 has the circuit, with some details omitted.

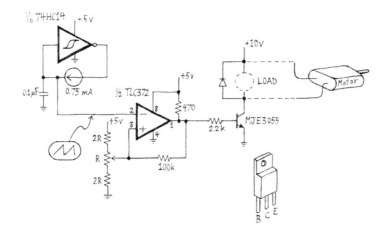

Figure 8L.7 PWM circuit; some details left to you.

Give your comparator circuit a little hysteresis: 50–100 mV should suffice. Note that the transistor loads the circuit somewhat, so that output swing rises not to 5 V but to about 4 V.

While you are testing the circuit, and perhaps adjusting values, you may want to install a *resistor* in place of the motor – 100 Ω or more. Look at the *sawtooth* and *comparator output* waveforms as you vary the threshold using the potentiometer.

Are you able to adjust the comparator-output's "duty cycle" over a wide range – from perhaps 10% to 90%? ("Duty cycle" is jargon for "percentage of period for which the output is high.") If the range is not so wide as you'd like it to be, tinker with the values of the resistors above and below the threshold potentiometer. Then check that the transistor switch is doing its job.

When you are satisfied, replace the resistor that has been serving as test *load* with a small DC motor.

You can use the 3–12 V motor you used to drive the integrator in Lab 7L.[7] We have shown a 10 V supply to be sure we don't overheat the 12 V motor. But you'll get away with +15 V if you don't have a 10 V supply handy.

You may want to stick a bright pointer on the motor shaft – or perhaps the vane is still in place from Lab 7L – to make it easier to judge how fast it spins as you adjust the duty cycle. Try loading the motor – applying some braking with your fingers. You should find torque that's not bad, at low duty cycles, whereas simply dropping the voltage to the motor – as your grandmother's sewing machine *rheostat* did – gives poor torque at low speed.[8]

8L.5 IC *RC* relaxation oscillator: '555

The '555 and its derivatives have made the design of moderate frequency oscillators easy. There is seldom any reason to design an oscillator from scratch, using an op-amp as we did above. The ICM7555 or LMC7555 or TLC555 is an improved '555, made with CMOS. (We will refer to the CMOS part as '555 for brevity, though you may be using the LMC part.) It runs up to 1 MHz (7555) or 3 MHz (LMC7555), versus 100 kHz for the original, bipolar '555, and its very high input impedances and rail-to-rail output swings can simplify designs.

8L.5.1 Square wave

Connect a 7555 in the '555's classic relaxation oscillator configuration, as shown in Fig. 8L.8. Look at the output. Is the frequency correctly predicted by

$$f_{\text{osc}} = \frac{1}{(0.7[R_A + 2R_B]C)}?$$

Now look at the waveform on the capacitor. What voltage levels does it run between? Does this make sense?

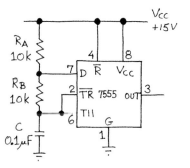

Figure 8L.8 7555 relaxation oscillator: traditional '555 configuration.

Sharkfin (?) oscillator, again: Now replace R_B with a short circuit. What do you expect to see at the capacitor? At the output? (A detail: the cap voltage now will fall a good way *below* the lower threshold value of $V_{CC}/3$; this occurs because the 7555 takes a while to respond to the crossing of that threshold, and the voltage now is slewing down *fast*.)

---

[7] We use the Mabuchi FF–130SH–11340. But any motor that can withstand 10 V will do.
[8] See *The Art of Electronics: the X Chapters*, Horowitz and Hill, §9x.5 PWM for DC Motors.

50% duty cycle: The 7555 can produce a true 50% duty-cycle square wave, if you invent a scheme that lets it charge and discharge the capacitor through a *single* resistor.[9] See if you can draw such a design, and then try it. *Hint:* the old '555 could not do this trick; the 7555 can because of its clean rail-to-rail output swing. When you get your design working, consider the following issues:

- In what way does the output waveform of this circuit differ from the output of the traditional '555 "astable" (as an oscillator sometimes is called)?
- Is the oscillator's period sensitive to loading? See what a 10k resistive load does, for example.

If your design is the same as ours, then the frequency of oscillation should be

$$f_{\text{osc}} = \frac{1}{(1.4RC)}$$

a result that is the same as for the "classic" configuration, except that it eliminates the complication of the differing charge and discharge paths. The "classic" design yields the same equation, to a very good approximation, as long as R_A is much smaller than R_B. Does the value of f_{osc} that you *measure* for your design match what you would predict?

Finally, try $V_{\text{CC}} = +5$ V with either of your circuits to see to what extent the output frequency depends on the supply voltage.

8L.5.2 Triangle oscillator

AoE §7.1.3E

Two 1N5294 current regulator diodes can give you a triangle waveform at the capacitor. A full-wave bridge and one 1N5294 could also do the trick. This scheme works best with higher supply voltages (e.g., 15 V) so the current source compliance and diode drops are less than $\frac{1}{3}V_{\text{CC}}$. If you're in the mood, try out your design for the *triangle* generator.

8L.6 '555 for low-frequency frequency modulation ("FM")

Here is a quick preview of a technique we will use in the analog project lab, 13L, to send audio across the room as flashes of light. The rate at which a light-emitting diode (LED) flashes will convey the audio information. A '555 can do the *encoding*, converting time-varying voltages (the music waveform) to variations in frequency. A simple parallel RLC circuit like the one you built in Lab 3L can decode the FM, converting it back to a time-varying voltage (the music waveform, recovered).

The '555 can do this modulating because it brings out (to pin 5) a point on the 3-resistor divider that defines the upper comparator threshold. If a signal pushes that pin above its normal resting voltage of $2/3 \times V_{\text{CC}}$, the '555 oscillation slows; if it pushes pin 5 down, the oscillation speeds up.

---

[9] Well, "invent" may be a bit exaggerated if you have read Chapter 8N.

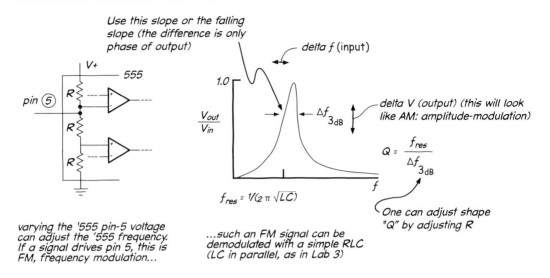

Figure 8L.9 A '555 can implement frequency-modulation; a parallel RLC can demodulate such a signal.

If you're short of time, you may want to limit yourself to the easy task of confirming that this FM scheme works. You can drive pin 5 with a low-frequency sinusoid (try 100 Hz at amplitude of 0.5 V) while watching the '555 output on a scope. Don't forget that you need a *blocking capacitor* between the function generator and pin 5, since pin 5 normally rests at 10V when the '555 is powered by +15 V. You will see what looks like jitter on the '555 output. If it is hard to interpret, try reducing the driving frequency well below 100 Hz.

8L.7 Sinewave oscillator: Wien bridge

Curiously enough, the sinewave is one of the most difficult waveforms to synthesize. (Your function generators make a sine by chipping the corners off a triangle, as you may recall.) The Wien bridge oscillator does it by cleverly adjusting its own gain to prevent clipping (which would occur if gain were too *high*) while keeping the oscillation from dying away (which would occur if gain were too *low*).

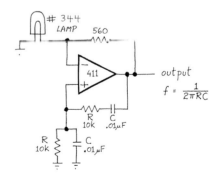

Figure 8L.10 Wien bridge oscillator.

The frequency favored by the positive feedback network should be

$$\frac{1}{2\pi RC} \tag{8L.1}$$

See whether your oscillator runs at this predicted frequency.

At this frequency, the signal fed back should be *in phase* with the output, and 1/3 the amplitude of V_{out}.[10] The negative feedback provided by the other path (the one that includes the lamp) adjusts the gain, exploiting the lamp's current-dependent resistance. Convince yourself that the sense in which the lamp's resistance varies tends to stabilize gain at the necessary value. What gain is necessary to sustain oscillations without clipping?[11]

You can reduce the output amplitude by substituting a smaller resistor for the 560 Ω in the negative feedback path.[12] Try poking the non-inverting input with your finger and note the funny rubbery behavior at the output. Try sweeping the scope slowly as you poke: now you can watch the slow dying away of this oscillation of the sine's *envelope*.

If you're energetic, you can confirm that this sine is much cleaner than the function generator's, by putting it through an op-amp differentiator. The differentiator fed by the Wien oscillator should show an output that looks like a very convincing sinusoid, not like the *zigguratoid*[13] that you saw when you differentiated the function generator's "sine" back in Labs 7L and 2L. Or, if you happen to be using a digital scope that includes an FFT, instead of using a differentiator you can simply compare the frequency spectra of the two sinusoids – function generator versus your little Wien bridge, as we did in Chapter 8N.

---

[10] At this frequency, where {the magnitude of X_C} = R, the impedance of R and C in series (the upper branch of the divider) is $\sqrt{2}R$ whereas the impedance of the R and C in parallel (the lower branch of the divider) is $\frac{\sqrt{2}}{2}R$. Hence the result that 1/3 of V_{out} is fed back, at this favored frequency. More surprising perhaps is the fact that this divider delivers a waveform in phase with the input.
[11] Yes, you're right: exactly *three*.
[12] Pretty smart circuit, eh? Adjusts lamp's R value to suit your new feedback resistor!
[13] We hope you will not spend your afternoon seeking this waveform in an electronics reference work.

8W Worked Examples: Op-Amp III

8W.1 Schmitt trigger design tips

8W.1.1 First easy case: setting thresholds

Setting particular thresholds exactly can be a pain in the neck. The process may put you through tedious algebra. But there are two easy cases.

Thresholds symmetric about zero: Suppose, for example, that you want thresholds at ±1 V, and output swing is ±15 V. The feedback divider pulls threshold equally far above and below ground: the foot of the feedback divider is tied to ground.

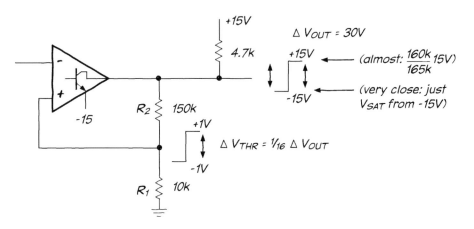

Figure 8W.1 First easy case: thresholds symmetric about zero.

The divider is to deliver 1 V out of 15 $\Rightarrow R_2 \approx 15 \times R_1$ (or 14×, if you want to be more precise). The key notion is that the foot of the divider is put at the midpoint of the output swing. You'd get the same tidy result if the thresholds were to be symmetric about 2.5 V while the output swung between 0 and +5 V: say, thresholds at 2.4 V and 2.6 V.

Second easy case: thresholds very far from symmetric: In this case you may, for example, want thresholds at 0 V and +0.1 V, and output swings 0 to +5. Here the feedback divider pulls *up* but not *down*. So just design a divider that pulls the threshold up to 0.1 V:

$$R_1/(R_1 + R_2) \times 5V = 0.1 \text{ V}$$

That means V_{Thresh} is about one part in 50, and R_2 is about 50×R_1.

378 Worked Examples: Op-Amp III

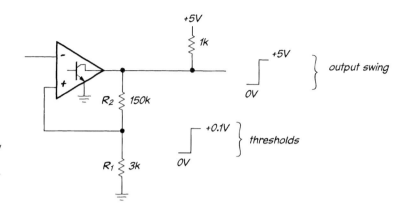

Figure 8W.2 Another easy case: thresholds pulled in only one direction ('very far from symmetric').

8W.1.2 An easier task: determining hysteresis

Suppose the task given is not "*put thresholds at* 1 V *and* 1.1 V," but instead "*set hysteresis at* 0.1 V; *put thresholds close to* 1 V."

The two formulations look just about equivalent, but the second task turns out to be much easier than the first. Let's try it. Suppose output swing is 0 to +5 V: see Fig. 8W.3.

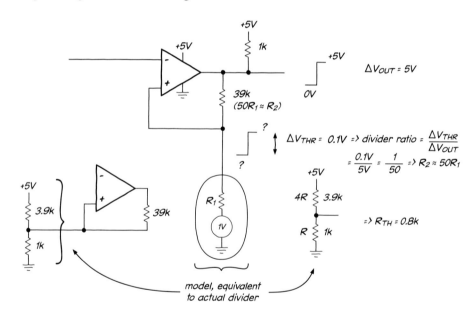

Figure 8W.3 Aiming for particular hysteresis, near a target voltage: threshold and hysteresis considered *separately*.

Here's the process:

Hysteresis: This is determined entirely by the divider ratio. Again we want 0.1 V/5 V: one part in 50; again $R_2 \approx 50 \times R_1$.

Thresholds "close to 1 V:" This is determined by the voltage to which the foot of the feedback divider is tied (at the moment, let's assume we have a handy source of this 1 V, an ideal voltage source; in fact, we are going to use a voltage divider, and then we will treat V_{Thev} as that 'voltage at the foot...' If this issue isn't yet worrying you, forget this comment until later).

You'll provide this 1 V with a voltage divider – but it has some R_{Thev}. Does that mess things up? No: let R_{Thev} be the value you want for R_1.

8W.1 Schmitt trigger design tips

This settling for approximate thresholds may seem like a cheap trick. Often it is not: often what you want is not named threshold voltages, but an approximate threshold, and appropriate hysteresis. Then this shortcut fits the formulated goals nicely.

A wrinkle: adjustable threshold: Can you see a way to make threshold adjustable while holding hysteresis constant? (Possible help: recall that op-amps are cheap.)

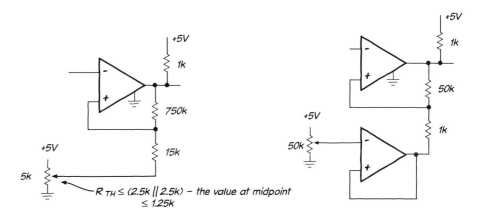

Figure 8W.4 Two ways to hold hysteresis constant despite changes of threshold.

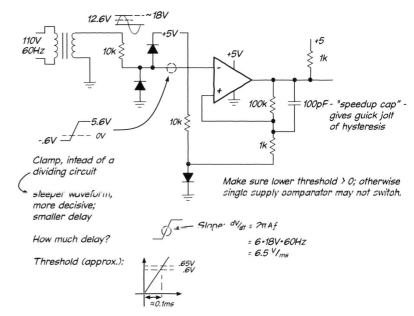

Figure 8W.5
Zero-crossing detector.

A potentiometer whose $R_{\text{out}} \ll R_1$ works, as in the left-hand figure of Fig. 8W.4. If you are too lazy to think about R_{Thev}, the op-amp buffer of the right-hand figure also solves the problem.

Problem (Design a zero-crossing detector) We want to know when the AC line voltage crosses zero (within a few $100\,\mu$s). We are to use a single-supply comparator, powered from +5 V, and we have a 12.6 V transformer output available; the transformer is powered from the "line" (60 Hz).

Compare AoE Fig. 4.85

Solution See Fig. 8W.5.

380 Worked Examples: Op-Amp III

Problem (Schmitt trigger, thresholds specified) Design a Schmitt trigger, using a '311 powered from ±5 V, to the following specifications:

- output swing: 0 to +5 V
- thresholds: ±0.1 V (approximate)

Solution See Fig. 8W.6. This problem could be painful if we were not willing to approximate; so let's feel free to use approximations. It helps to consider the two cases separately: what happens when the '311 output is low (0 V); and what happens when the output is high (5 V).

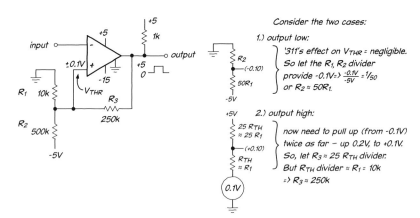

Figure 8W.6 Solution: aiming for particular thresholds, but willing to miss by a bit.

8W.2 Problem: heater controller

The goal is to design a comparator circuit – using a single-supply device (TLC372) rather than the usual '311 – that will keep your coffee at a temperature that you like. We'll put a temperature sensor on the hotplate, and will hold the plate's temperature at a level that is adjustable, and always somewhat below the boiling point.

8W.2.1 Temperature sensor and comparator

The temperature sensor, an LM50, puts out a voltage proportional to temperature (celsius) – with a half-volt constant tacked in. We will use this sensor to monitor the heater's temperature.

Our single-supply comparator has the bottom end of its output switch internally tied to ground, unlike the '311, which lets us define that emitter voltage.

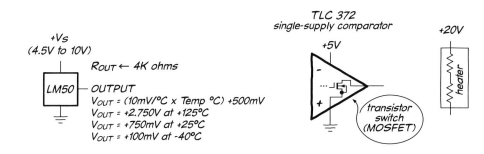

Figure 8W.7 Temperature sensor and comparator.

8W.2.2 Comparator circuit (your design)

Show how to use this sensor, and a potentiometer controlled manually, to turn the heater coil ON or OFF as the temperature moves below or above the pot's setpoint. Here are the details:

- temperature adjustment range: 50–100°C;
- hysteresis: 5°C; and
- heater is a resistive and somewhat-inductive immersion heater. It likes to be driven with 20 V DC.

8W.2.3 A Darlington transistor pair

In case you need such a device,[1] here is some data that you may use in your design (*typical* specifications, rather than the usual min and max), if you decide to use the part.

A DigiKey search came up with a TI part, TIP110, promising lots of β – the betas of the two transistors multiply, since the first transistor's I_E provides base current for the output transistor. Figure 8W.8 shows some *typical* curves for this part. Let's use these – rather than the usual worst-case specs, just for the novelty of taking data from curves rather than from a list of numbers.

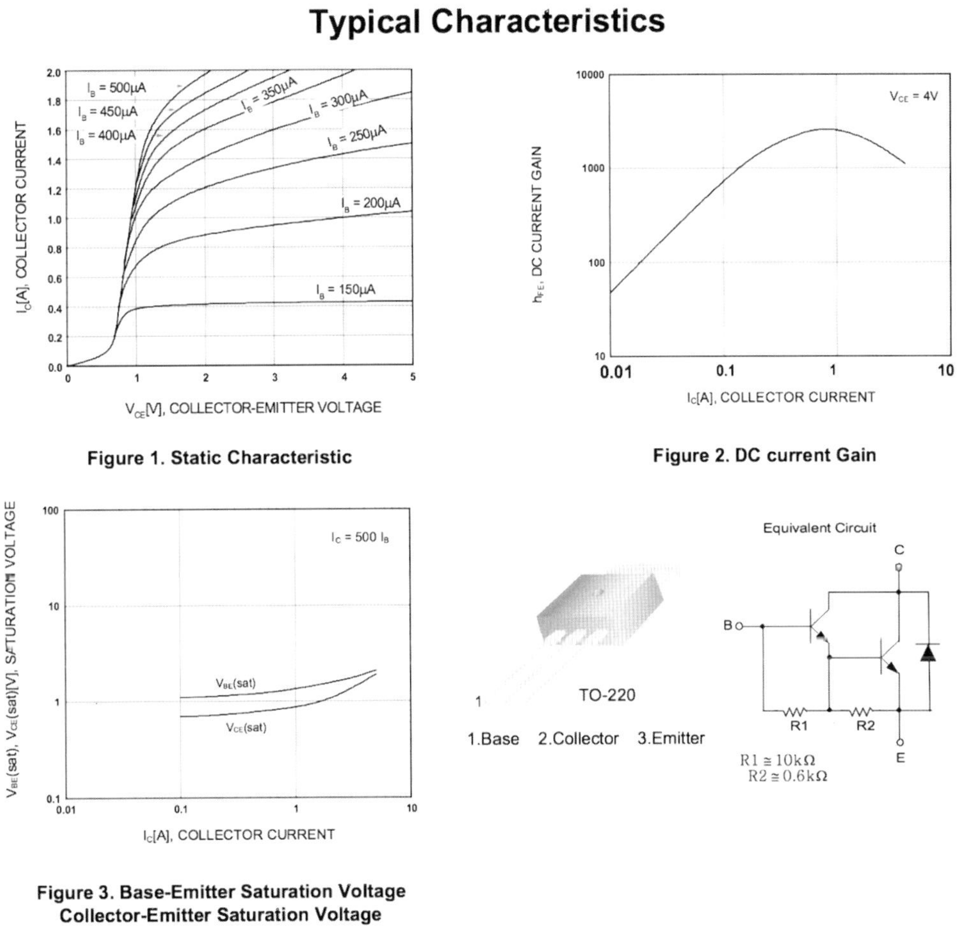
Typical Characteristics
I_B = 500μA
I_B = 450μA
I_B = 400μA
I_B = 350μA
I_B = 300μA
I_B = 250μA
I_B = 200μA
I_B = 150μA
I_C[A], COLLECTOR CURRENT
V_CE[V], COLLECTOR-EMITTER VOLTAGE
Figure 1. Static Characteristic
V_CE = 4V
h_FE, DC CURRENT GAIN
I_C[A], COLLECTOR CURRENT
Figure 2. DC current Gain
I_C = 500 I_B
V_BE(sat)
V_CE(sat)
V_BE(sat), V_CE(sat)[V], SATURATION VOLTAGE
I_C[A], COLLECTOR CURRENT
Figure 3. Base-Emitter Saturation Voltage
Collector-Emitter Saturation Voltage
TO-220
1
1.Base 2.Collector 3.Emitter
Equivalent Circuit
C
B
E
R1
R2
R1 ≅ 10kΩ
R2 ≅ 0.6kΩ

Figure 8W.8 Some typical spec curves for a Darlington transistor pair, TIP110.

[1] Does this sound like a hint?!

The resistors integrated with the transistor pair speed *turn-off* (not an issue in this case). The diode that parallels the Darlington pair normally will not conduct.

Though we propose to use *typical* specifications, let's try to give our design a factor-of-two safety margin. In your solution, tell us approximately what portion of the 20 V supply you expect to see dropped across the load.

8W.2.4 Solution: heater controller

Temperature range to voltage range: The sensor adds a 500 mV offset to its reading of 10 mV/°C. So the temperature range, 50–100°C, maps to this voltage range:

$$\text{Lowest} = \text{offset} + \text{minimum reading} = 500\,\text{mV} + (50°\text{C} \times 10\,\text{mV}/°\text{C}) = 1\,\text{V}$$

$$\text{Highest} = \text{offset} + \text{maximum reading} = 500\,\text{mV} + (100°\text{C} \times 10\,\text{mV}/°\text{C}) = 1.5\,\text{V}$$

So the manual control – a potentiometer – needs to be wired to span this range. We have plugged in some arbitrary values in Fig. 8W.9 (chosen to make our arithmetic easy).

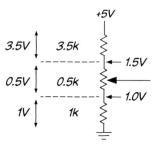

Figure 8W.9 Potentiometer sets voltage, 1–1.5 V, temperature, 50–100°C.

Comparator circuit: We'll need some positive feedback, to get the required hysteresis, so the skeleton circuit will look like Fig. 8W.10.

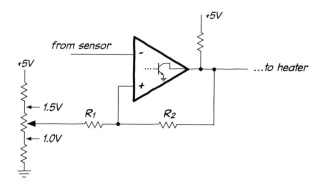

Figure 8W.10 Skeleton comparator circuit (no values).

Component values: Now let's choose some values.

- R_{pullup}: oddly enough, we'll need to postpone this choice until we've designed the switching circuit. For now, it's enough to say that this resistor will be so much smaller than R_2, in Fig. 8W.10, that we will be able to neglect it as we calculate hysteresis.
- R_1, R_2. These will set the amount of hysteresis – and once again we need to postpone their choice until we have calculated R_{Thev} for the manual-adjustment potentiometer.

- So we'll need to do that R_{Thev} calculation:
  - As usual, we need to consider *worst case*, and then make sure that the variation in R_{Thev} as one adjusts threshold doesn't appreciably alter the amount of hysteresis.[2]
  - Worst case is the *maximum* R_{Thev}. That will come when the two "paths" seen from the slider – up to the +5 supply, and down to ground – are most nearly equal. That will occur with the slider at the top.
    At that point the two paths look like 3.5k ∥ 1.5k ≈ 1k.
- Choose R_1: knowing $R_{Thev} \approx$ 1k lets us choose R_1. As usual, we want it much larger than what's driving it. Make $R_1 = 10$k.
- Now R_2: we want 5°C hysteresis, which translates to 50 mV. So we want to feed back a small part of the 5 V output swing: 50 mV/5 V = 50/5000 = 1/100. So $R_2 \approx 100 \times R_1$. Let $R_2 = 1$M.

8W.2.5 Heater driver

Can the '372 drive the load? We want to be able to put 1 A through the heater. The comparator itself can't do that, so we'll use a transistor switch. Our rule of thumb for driving a switch calls for $I_B \approx I_C/10$ – *not* I_C/β, because we want to drive the switch strongly into saturation.

That rule would call for 100 mA from the comparator – and that is more than the small output transistor of the TLC372 can handle (it can sink 20 mA max). So, we'll need to use either a MOSFET switch (a device that it's not fair to invoke till we've met them, in Lab 12L), or a Darlington transistor pair.[3] We've already tipped our hand, in the *problem* statement, so let's assume we'll use the TIP110 Darlington pair.

8W.2.6 Using the Darlington switch

Curve 2 in Fig. 8W.8 shows plenty of gain – near the part's maximum – at 1 A: current gain (β) better than 2000. So, we'll assume a β of 1000. By our normal rule-of-thumb, we'd drive the transistor 10× harder, as if β were 100. But that rule does not quite translate to the Darlington, since the main (output) transistor cannot saturate; only the first can. Further, we don't need to rely on a rule of thumb because the "static characteristic" plot in Fig. 8W.8 shows a plot of input current (I_B) versus output current (I_C). From these curves we can see just what base drive is needed (and then we'll double that drive, to be cautious).

It appears that 300 μA in gets 1 A out (that's β of about 3000). We can double the input drive to 600 μA. We probably should be even more generous, since doubling base current does not, in general, produce a doubling of the Darlington's I_C. Let's go to 1 mA (a nice, round number – and our favorite current, in this course).

That value permits us, at last, to calculate what value we need for R_{pullup}. The input voltage for the Darlington will be about 1.5 V.[4] So, we'll put about 3.5 V across the comparator's R_{pullup} when it is driving the switch ON. To get the 1mA that we want, therefore, we can use $R_{pullup} = 470\,\Omega$, and the base resistor $\approx$ 3.5 V/1 mA = 3.5k. We'll use 3.3k.

---

[2] We will not be so careful as to calculate the fractional *variation* in hysteresis that can result with varying pot settings. We'll settle for the usual, simpler "times-ten" rule: $R_{Thev} \ll R_1$.

[3] Or we could use the similar Sziklai transistor pair, in principle – but you won't find these for sale on DigiKey as ICs. So, we'll stick with Darlington.

[4] A quick view of the schematic would indicate that the input voltage would be "two $V_{BE} \approx 1.2$ V;" a close look at the V_{BE} saturation curves, however, in Fig. 8W.8 suggests that the input voltage is somewhat higher. This is getting very fussy, indeed; but we might as well use the detailed information that we happen to have.

Finally, we need a *protection diode* across the load, in order to protect the Darlington from voltage spikes each time the switch turns OFF. The inductive load otherwise may damage the Darlington.

8W.2.7 The complete circuit

And here it is, all put together in Fig. 8W.11. Values shown in parentheses are 10% or 1% resistor values – "3.5k" happens to be a strange value to specify.

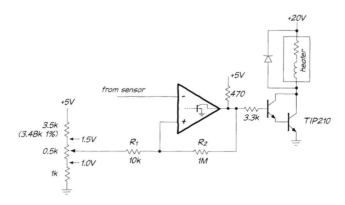

Figure 8W.11 Full heater-control circuit.

…What part of the +20 V will be dropped across the load?

This is one way to ask "what is the ON voltage of the switch." We learn, from two sources, that this ON voltage should be under 1 V at 1 A:

- The third plot in Fig. 8W.8 shows about 0.9 V "collector–emitter saturation voltage."
- The first plot shows the same thing for strong base drive and $I_C = 1$ A.

So about 19 V of the 20 V supply will be applied to the load: pretty good efficiency.

9N Op-Amps IV: Parasitic Oscillations; Active Filter

Contents

9N.1	Introduction	385
9N.2	Active filters	386
9N.3	Nasty "parasitic" oscillations: the problem, generally	388
9N.4	Parasitic oscillations in op-amp circuits	388
	9N.4.1 Occasionally, an odd object in the loop flips phase, at all frequencies...	389
	9N.4.2 ...but much more often, feedback changes flavor only at high frequencies	390
	9N.4.3 Why the op-amp includes a 90° lag	392
	9N.4.4 ...Given the op-amp's phase lag, what can you do for stability?	392
9N.5	Op-amp remedies for keeping loops stable	393
	9N.5.1 Rolloff high frequency gain	393
	9N.5.2 The photodiode circuit	393
	9N.5.3 Driving a long coaxial cable	394
9N.6	A general criterion for stability: loop gain where phase shift approaches 180°	398
9N.7	Parasitic oscillation without op-amps	400
9N.8	Remedies for parasitic oscillation	402
	9N.8.1 First remedy: quiet the supply (shrink the disturbance fed back):	402
	9N.8.2 Another remedy: kill high-frequency gain	402
9N.9	Recapitulation: to keep circuits quiet...	403
9N.10	AoE reading	404

Why?

What are we trying to do today? We want to try two tasks, of which the second is the larger and more fundamental:

- Problem: make an improved filter, using op-amps.
- Problem: fend off unintended ("parasitic") oscillations.

9N.1 Introduction

Today we glance at *active filters* to remind you that such filters are available when you need a filter better than an ordinary *RC* such filters are available. Then we give most of our attention to *nasty* oscillations.

In the previous chapter we saw that positive feedback can be useful – though it is the underdog in feedback circuitry, less important than the great strategy of negative feedback. It can make a switching circuit decisive, and it allows construction of *oscillators*. Now we turn to the dark side of positive feedback: those cases where it sneaks up on us, causing oscillations that we didn't want (and usually did not expect). A bitter formulation of Murphy's Law in this special context says – describing the perverse ways that "parasitic" oscillations are inclined to pop up – "Oscillators won't, amplifiers will."

We will look at some cases of parasitic oscillations in order to get familiar with the problem. Then we will look at remedies, circuit changes that can prevent such behavior. It is the remedies, of course, that we are after. If nasty fuzz pollutes the output of a circuit in your lab, and you can make a few changes that make it go away, you will be a hero. This is a black art – sometimes involving literal hand-waving,[1] as well as some hand-waving explanations – and it is a skill that is right at home in a course based on *The Art of Electronics*.

9N.2 Active filters

AoE §6.2.4

As we point out in §9S.2, the motive for using these filters rather than a simple RC is to approach nearer to the ideal in filter behavior: flat *passband*, abrupt *rolloff*, then much attenuation in the *stop band*. An op-amp-assisted two-stage RC can improve both characteristics – flattening the passband and steepening the rolloff – as the frequency sweep in Fig. 9N.1 illustrates.

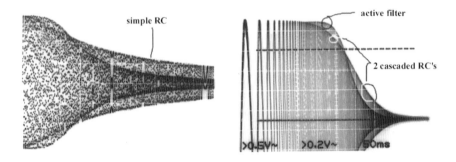

Figure 9N.1 Active versus passive lowpass: frequency responses contrasted.

The circuit in Fig. 9N.2 uses aptly placed positive feedback to achieve both improvements: it flattens the passband by giving the signal a boost just below f_{3dB}, where the passive RC is droopy; it provides a $1/f^2$ ultimate slope, as even a simple cascading of two RCs would; more magically, it provides a steeper transition between passband and stopband than the passive cascade could deliver.

AoE §6.3.1

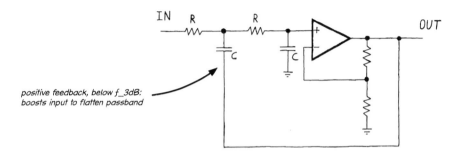

Figure 9N.2 VCVS second order active filter ("two-pole").

AoE §6.2.5

Furthermore, simply by altering the gain of the amplifier one can choose to optimize one or another

---

[1] "Literal" hand-waving? Yes. Waving your hand near a circuit that shows parasitic oscillations while observing the effect of your hand's proximity (important for its stray capacitance) sometimes can provide a clue to where in the circuit an oscillation originates.

filter characteristic: flatness of passband (best in the so-called "Butterworth" filter) or steepness of rolloff (best in "Chebyshev" design) or waveform shape preservation (best in "Bessel").[2] The filter of Fig. 9N.2, which is the design you will try in lab, is a "two-pole" design. When you need an even better filter, you can cascade additional stages.

Unfortunately, the method we have promoted elsewhere in this course, in which the several stages of our circuits are independent modules to be strung together, works poorly for filters. You would not design a four-pole active filter by cascading two identical two-pole filters. Instead, you would be obliged to do the harder work that ordinarily we eschew: analyze the entire four-pole problem. In practice, this may mean consulting a table like those in AoE for the gains of the several stages: see AoE §6.3.2.

Or you can make the process even more mechanical than that. You can download a program from any of several IC vendors: Analog Devices, Microchip, LTC, or Texas Instruments, then let that program do the design work for you.[3] In Fig. 9N.3 we show some screen prints from TI's Filter design Tool. You describe the filter performance that you'd like, as on the left-hand image.

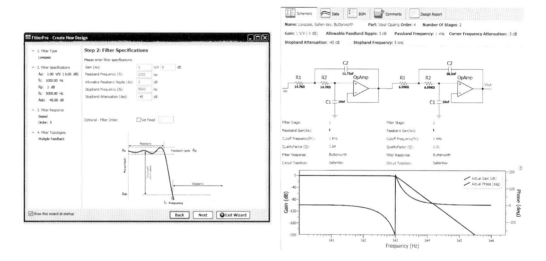

Figure 9N.3 TI's Filter Design Tool will do the design work for you. We offer the figure for a glimpse of response curve and circuit diagram; we concede that details of specifications may not readable in this figure.

TI's design tool draws the circuit schematic, with component values, as on the right side of Fig. 9N.3. The specified filter behavior was demanding enough to call for a 4-pole filter. So the tool cascaded two stages resembling the VCVS that we showed in Fig. 9N.2, except using unity gain (the filters of Fig. 9N.3 are called "Sallen-and-Key" configurations[4]). TI's software holds your hand to an embarrassing degree: not only does it draw the circuit, specifying part values that you'd otherwise have had to look up in a table, it also politely asks whether you'd like to see component values shown with *exact*, 1%, or 5% tolerances, and then produces a bill of materials showing all the parts you need to buy.

Why is TI so nice to us? Well, they may hope that we will buy their op-amps, once we're using their tool, but probably their highest hope is that we'll buy their IC dedicated to implementing active filters,

---

[2] The Bessel filter preserves waveform shape by providing phase shift that varies linearly with frequency, for "constant time delay:" AoE 6.2.6B.
[3] Analog Devices: https://LAoE.link/ADI_Filter_Wizard.html, Microchip: https://LAoE.link/Microchip_FilterLab.html, or search for "FilterLab;" TI: https://LAoE.link/TI_Filter_Design.html.
[4] For Sallen-and-Key filters see AoE §6.2.4E.

the UAF42.[5] The UAF42 is a neat part, with on-chip capacitors trimmed to 0.5% tolerance – but it is expensive ($16 for one, as we write). So, you may want to use the TI software, then shop for your own op-amps.

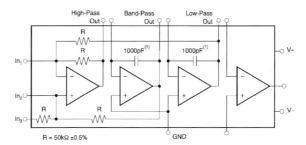

Figure 9N.4 TI's active-filter IC, UAF42.

9N.3 Nasty "parasitic" oscillations: the problem, generally

The conditions that lead to unwanted oscillations of course resemble those that lead to wanted oscillations – those that we produce with purposeful *oscillators*. The requirements are just two, for sustained oscillation. The circuit must show...

Gain:[6] otherwise the disturbance has to die away (as it does, say, in the *LC* resonant circuit hit with a square wave). More specifically, as the disturbance appears at the output and is fed back to the input, then appears again at the output, it must not get smaller: net gain around the loop must be at least 1.

Positive feedback: the circuit must talk to itself, patting itself on the back, saying, "Good. Do more of what you're doing." (In speaking of *gain*, just above, we were indeed assuming positive feedback; but the two requirements are distinct.)

Sometimes the presence of some kind of feedback is obvious, as it has been in all our op-amp circuits; the subtlety of the problem then lies in figuring out why the flavor has changed from the benign *negative* to the nasty *positive*. Occasionally – as in the case of the discrete follower discussed in §9N.7 – it is far from obvious that any feedback at all is present. If the circuit nevertheless oscillates, we will be obliged to do some experimenting, and perhaps even *thinking*, to determine how the feedback occurs. Our ultimate goal will always be to find remedies that prevent the oscillation.

9N.4 Parasitic oscillations in op-amp circuits

Op-amp circuits, which ordinarily rely on negative feedback, are evidently ripe for trouble. If ever something comes along to change the "flavor" of feedback from negative to positive, a placid amplifier can become an oscillator.

---

[5] Inherited from Burr–Brown, a company that TI bought.

[6] It is *voltage* gain that is required. Strictly, it would be better to say "power gain," since a transformer can provide voltage gain, but it cannot sustain an oscillation. The inventors of the transistor were keenly aware of the voltage versus power distinction. Walter Brattain, one of the transistor's three inventors, recorded in his lab notebook, Dec. 15, 1947, "got voltage amp about 2 but not power amp." A day later, he noted, "... power gain 1.3 voltage gain 15." An important step forward. From Bell Laboratory archives, cited in Wikipedia article on history of the transistor.

We refer in this discussion to "voltage gain" in the hope that it is a clearer and more familiar concept. Unless output current is reduced proportionately to rise of voltage, as in a transformer, voltage gain will be accompanied by the necessary *power gain*.

9N.4 Parasitic oscillations in op-amp circuits

Any old op-amp circuit looks like the one sketched in Fig. 9N.5. We're familiar with the notion that we can tease the op-amp into doing what we like by putting this or that blob within the loop – but today we will find that the op-amp can rebel, going crazy when asked to accept blobs of certain types.

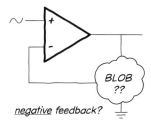

Figure 9N.5 Is it stable? Depends on the blob.

9N.4.1 Occasionally, an odd object in the loop flips phase, at all frequencies…

Does the circuit in Fig. 9N.6 apply negative feedback? "Sure," you may be inclined to answer. "The feedback goes to the inverting terminal." But what if inside the blob lives an *inverter*? "Oh," you answer. "That's silly – it would never happen. But if it did, we'd just swap inputs."

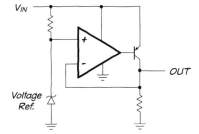

Figure 9N.6 This probably looks right; but it isn't. The feedback, here, is *positive*.

You're pretty much right: such a case is at least extremely rare – but the circuit shown *is* just such a case (this is a "low-dropout" voltage regulator that you will meet in Lab 11L). And you're right that the remedy is to swap inputs. We've done that in Fig. 9N.7 – and the circuit *looks* shocking, but in fact does make sense. This feedback makes sense because within the loop lies an inverting circuit: a PNP common-emitter amplifier.

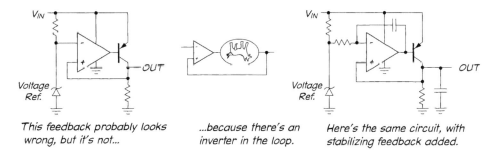

This feedback probably looks wrong, but it's not…

…because there's an inverter in the loop.

Here's the same circuit, with stabilizing feedback added.

Figure 9N.7 This probably looks wrong. It is a very rare circuit – where a full-time inverter sits within the feedback loop.

To try to persuade you that this circuit is useful, and not just a stunt that we cooked up, let's spell out its virtue: in contrast to the usual scheme, in which a discrete follower is included within the feedback loop (see Fig. 6N.17), this one avoids inserting a V_{BE} drop between V_{in} and V_{out}. As a result, the difference between V_{in} and V_{out} can be as low as 100 mV. Hence the circuit's name, low-dropout.

390 Op-Amps IV: Parasitic Oscillations; Active Filter

On the right side of Fig. 9N.7 we have shown some extra stabilizing that normally must be added. But the strange-looking feedback of both circuits is correct, in any case; not crazy.

An example of inverted feedback often is recited in courses on controls: two people share an electric blanket with dual controls, one for each side of the bed. Accidentally, the controls for the two sides get swapped. As one person begins to feel chilly, he turns up the heat. You can imagine the uncomfortable night that ensues.

9N.4.2 ...but much more often, feedback changes flavor only at high frequencies

Figure 9N.8 shows a much more common case: the dog is not upside-down, but only *slow*. The signal fed back arrives a bit late. Figure 9N.8 shows the sleepy dog lying within the loop. Figure 9N.9 shows the electronic equivalent: a lowpass inside the loop.

Figure 9N.8 Something with slow response, inside loop, can cause trouble.

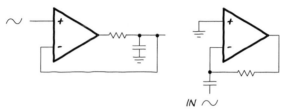

Figure 9N.9 Lowpass in loop can cause trouble.   Lowpass in loop?   This may look implausible.   Is this more plausible? Yes, it's a differentiator (and a jumpy one)

At low frequencies, the lowpass does no harm. Since the input is changing slowly, the signal fed back, though a little out of date, is still quite similar to an up-to-date description. So feedback works as usual.

But, if the input and output are changing fast, then an out-of-date description can be entirely misleading: it can be as much as 90° wrong as it goes into the op-amp, and as much as 180° wrong when one includes the additional 90° lag imposed by the op-amp itself. (This lag in the op-amp is surprising, we admit; see §9N.4.3.) So, the simple differentiator shown in Fig. 9N.9 shivers when fed a square wave. The square wave is especially disturbing because of the high-frequency components in its edges, see Fig. 9N.10.

> AoE §4.5.7, and Fig. 4.69

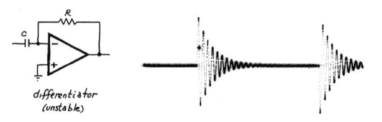

Figure 9N.10 Differentiator response to square wave: unstable when extra *R* and *C* are omitted: lowpass in feedback brings trouble.

That jumpiness is not acceptable, so the differentiator normally is tamed by the addition of the additional *RC* pair shown in Fig. 9N.11. These kill the circuit's gain at high frequencies. An alternative way to say this is that they halt the phase lag that grows dangerously with frequency – and then reverse that phase shift.

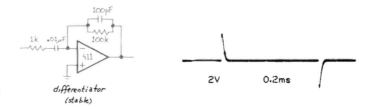

Figure 9N.11 Differentiator response to square wave: stable when extra R and C are added.

The differentiator that you built in §7L.2 included the stabilizing second RC pair shown in Fig. 9N.11. So in that lab exercise you did not see the shivering instability.

You will notice that we were not able to solve the differentiator problem the way we handled the odd case of a simple inverter within the feedback loop, the rare instance of Fig. 9N.4.1. In the differentiator circuit we cannot we cannot swap inputs as we could for the low-dropout regulator, because swapping inputs on the differentiator would give positive feedback at low frequencies and for a DC input.[7] At DC this positive feedback would lock up the circuit, with the output stuck at one rail[8] or the other.

It's not hard to think of homely analogies to this problem, where information concerning the results of actions is slow getting back to the person in control.

Imagine, for example, that you've decided to play a perverse game of seeing whether you can drive a car using someone else's vision. You wear a blindfold, and your passenger tells you whether to steer left or right to stay in your lane. If you don't drive too fast, and if the passenger gives continual quick advice, you may manage. You won't manage if a car suddenly swerves in front of you. Your response won't be quick enough for that. But for a slow drive on a country lane, the process might work.[9]

But if your passenger is a slow-talking and perversely calm cowboy, who won't be hurried and says something like, "Well, young feller, I think you need to just mosey a shade, just a shade I'm saying, not a lot, over westward, that is to say leftward.....," you won't be able to keep up, unless you're driving along at a walking pace. The incoming advice is likely to be out of date. Soon a drastic correction in the opposite direction will be called for – and the car will be weaving right and left all over the road: oscillating.

This is a time-domain way of describing a problem that may be clearer in frequency domain. Looking at the lowpass in the feedback loop in Fig. 9N.9, one can see that at low frequencies the feedback signal will be pretty much in phase with the op-amp output, and the capacitor will be having almost no effect: feedback is negative and healthy. At high frequencies, however (what we mean by "high" depends, of course, on the value of RC), the phase shift between op-amp output and the signal that reaches the op-amp's inverting input can be considerable, approaching $-90°$.

But, you may protest, $90°$ does not flip a signal; we need a $180°$ change before we're in trouble. Yes, you're right. But the catch, the circumstance that makes $90°$ deadly, is the fact – usually hidden from us as we watch op-amps in circuits – that the op-amp includes a $90°$ lag *within itself*. The op-amp, strange to tell, behaves like an *integrator*, beginning at a very low frequency. For the LF411, integrator behavior begins at about 20 Hz (not 20 kHz, but 20 Hz).

---

[7] You may fairly object, "But I'm not going to look for the derivative of a DC level!" True, but we cite DC as the extreme of "low frequency," just providing a case that is easy to analyze.

[8] You've probably picked up this piece of jargon: "rail" to mean either "limiting voltage, close to power supply," as here, or "power supply," as in *rail-to-rail-output op-amp*, the type that can drive its outputs close to both supplies.

[9] Please do not try this; this is a *thought* experiment!

Frequency compensation to stabilize an op-amp: Figure 9N.12 is a plot showing gain rolloff of a "compensated" op-amp, and the associated phase shift between input and output. We have inserted the numbers that fit the LF411 op-amp that you have been using in the labs. The plot shows how the '411's gain rolls off with frequency, and shows the resulting phase shift between non-inverting input and output (note that this description assumes we are operating the op-amp *open-loop*, as we almost never do; but this open-loop behavior of the amplifier is crucial to its proclivities in the usual, *closed-loop* circuit). We explore in §9S.1 how op-amp *frequency compensation* is achieved.

AoE §§4.9, 2.5.4B

AoE §4.9.1

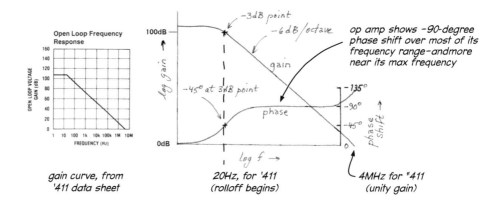

Figure 9N.12 A compensated op-amp like the '411 shows 1/f rolloff and −90° phase shift.

9N.4.3 Why the op-amp includes a 90° lag

AoE §4.9.2A

Even stranger, at first hearing, is the fact that this 90° lag within the op-amp, and the sad $1/f$ rolloff of gain, is deliberately *imposed* by the op-amp's designers. Why? A short answer is that without *compensation* things would be worse: the amplifier's several stages form a cascade of lowpass filters, and these filters eventually would provide a deadly 180° lag that would make the amplifier unusable. So a 90° lag may be surprising, but it's much better than what we would get without compensation.

The effect of the op-amp's built-in 90° lag (and normally it is even a shade more[10]) is that in order to produce accidental positive feedback we need add only about another 90°, *not* 180°. And any lowpass filter can provide that 90° phase lag.

9N.4.4 . . . Given the op-amp's phase lag, what can you do for stability?

If the feedback network is purely resistive, there is no problem: the network causes no phase shift, so the circuit is unconditionally stable. But when a lowpass characteristic sneaks in, we have to take care.

A non-remedy: "I won't apply high frequencies": A naive first thought surely might be to try to exploit the truth that feedback trouble comes only at relatively high frequencies. So, why not just *apply only low frequencies*?

Because you can't do that. It may be true that a particular op-amp circuit goes unstable only above 50 kHz, and that you plan to use the amplifier for signals below 10 kHz. That plan will not save you.

The problem is that the circuit responds not just to the signals that we choose to apply (perhaps rather low frequencies). It responds also to *noise*, which one must assume is present *at all frequencies*.

---

[10] As much as about 45° more, at f_T. See Fig. 9S.8.

This is one of the sad truths that underlies the gloomy dictum, "amplifiers will...." The consequence is that we must design all our amplifier circuits so that they will not oscillate even when driven with high-frequencies.

Some valid remedies: The compensated op-amp was designed to be stable with resistive feedback, not to be stable with a lowpass in the loop. So, when a lowpass filter is present, we need to forestall oscillations. Here are some ways to do that:

- Impose a rolloff of high-frequency gain, in addition to the rolloff built into every compensated op-amp. This method is illustrated in the photodiode circuit of Fig. 9N.14.
- Bypass the troublesome blob at the higher frequencies where it would introduce dangerous phase lag. We will see this method, which we call "split feedback," later on page 397.
- Adopt an extreme remedy, altering the op-amp itself. Design a peculiar custom op-amp that can undo harmful phase lags. This is the strategy used in the "PID" loop,[11] a design that you will apply in Lab 10L.

We will illustrate the first two of these remedies below. Altering the op-amp we will postpone until you meet Lab 10L's PID loop.

9N.5 Op-amp remedies for keeping loops stable

9N.5.1 Rolloff high frequency gain

The circuits in Fig. 9N.13 are vulnerable to parasitic oscillations, in various degrees, because of lagging phase shifts (lowpass effects) within the feedback loop.

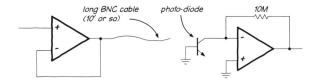

Figure 9N.13 Two circuits vulnerable to parasitic oscillations.

9N.5.2 The photodiode circuit

In the photodiode circuit, the implicit or accidental lowpass is far from obvious. There, it is formed by the large feedback R that drives the stray capacitance at virtual ground (a few pF). The 10M R is not reduced by the paralleled photodiode, because the photodiode is a current source. Its dynamic resistance, $\Delta V/\Delta I$, is very large.

Suppose that stray capacitance at the inverting input is, say, 3 pF. In that event, the large feedback resistor forms a lowpass in the feedback loop with f_{3dB} of around 5 kHz. At that frequency, phase shift in the feedback loop would be a lagging 45° – not enough to bring on oscillation. But at higher frequencies, the phase lag grows, approaching the deadly 90°.

For the photodiode circuit (one you built in Lab 6L), the best remedy is to parallel the large feedback resistor with a small capacitor – say, 10 pF, or even smaller: see Fig. 9N.14.[12] This capacitor will

---

[11] PID stands for Proportional, Integral, Derivative, describing the three functions of feedback *error* that are applied in such a control loop. We devote the first half of Lab 10L to this technique.
[12] Even 1 pF might be enough. Since caps smaller than 10 pF are hard to find, you might try making a "gimmick capacitor" by twisting two pieces of insulated solid wire together. You get about 1pF per inch and the capacitance varies depending on how tightly you twist the wires together. Robert Pease, *Troubleshooting Analog Circuits*, Newnes, (1991, 1993, p. 20). See also https://LAoE.link/Gimmick.html

compromise the circuit's response to quick optical signals. Circuit gain is down 3 dB at a little above 1 kHz (that is where $X_C \approx 10M$). To put the same point differently, above that frequency the feedback circuit no longer forms a lowpass, with its lagging phase shift. Instead, it forms a C–C divider, with its phase shift moving back toward *zero* as frequency climbs beyond the crossover frequency.

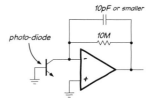

Figure 9N.14 Photodiode circuit, stabilized.

9N.5.3 Driving a long coaxial cable

The lowpass in the case of the long cable is formed from the considerable stray capacitance of a BNC cable (about 30 pF/foot) and the op-amp's small but non-zero output impedance – on the order of 50–100 Ω,[13] but crucially altered by feedback. This output *resistance* in fact is transformed by feedback into a seeming output *inductance*, forming a lowpass with a resonant frequency much lower than the f_{3dB} that a simple RC would provide. It is this seeming *inductance* that explains the frequency of the oscillation brought on by the capacitive load. A resistive simple R_{out} sounds like a plausible explanation of the oscillation; but the numbers don't work.

AoE §4.6.2

See AoE Fig. 4.53

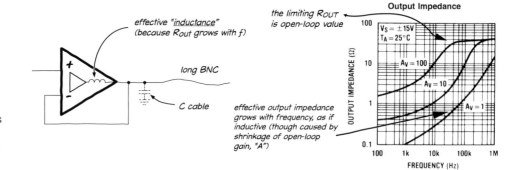

Figure 9N.15 Op-amp's Z_{out} behaves as if inductive, pushing down frequency of dangerous phase lag.

If one assumed that this R_{out} paired with the cable's capacitance formed the lowpass effect that brings on oscillation, one would predict oscillations at a frequency far above the amplifier's f_T. At such a frequency, the amplifier's gain is long gone, and oscillation is not possible. Plugging in the numbers for the present case – $R_{out} \approx 40\,\Omega$, $C_{load}=450$ pF (assuming a 15-foot cable) – one would calculate the accidental-filter's $f_{3dB} \approx 8$ MHz. At this frequency, the '411 op-amp has no gain at all. No oscillation could occur.

But a view of the amplifier's output impedance as *inductive* does fit the observed oscillation. The op-amp output acts *inductive* not because of actual inductance in the IC but because the impedance of the closed-loop circuit grows linearly with frequency – as the impedance of an inductor does.[14]

---

[13] Op-amp datasheets normally do not specify a hardware R_{out} (that is, a value for the IC itself, apart from feedback). But the '411 datasheet suggests a value of about 40 Ω. This we infer from the plot of R_{out} (Fig. 9N.15), which levels off at around 40 Ω at high frequencies if B, the fraction fed back is small (0.1 or 0.01). At these frequencies, the AB product is small, so the effective R_{out} – as improved by feedback – hits the limit of the R_{out} apart from feedback.

[14] This occurs – as you will see in §9S.5 – because the circuit's effective R_{out} is the barebones "hardware" R_{out} divided by the

9N.5 Op-amp remedies for keeping loops stable

This inductance driving C_{load} forms a resonant circuit, and resonant at a frequency consonant with the observed oscillation.

The oscillation in Fig. 9N.16 occurs at around 3.4 MHz. This is close to the op-amp's maximum frequency – where its "phase margin" – the difference between its internal phase shift and the deadly 180° – is down to perhaps 50°. If resonance is close to this frequency, the resonant circuit can provide the necessary extra phase lag (as we've said, a little less than 90° will suffice, close to the op-amp's f_T). Figure 9N.16 shows an LF411 follower driving a square wave into 15 feet of coaxial cable.

Figure 9N.16 illustrates how spooky and whimsical such oscillations can be: the square wave only *sometimes* sets the circuit into oscillation. The square wave is the most disturbing signal we can provide, since its edges include stimulation over a wide range of frequencies. But the square wave in Fig. 9N.16 brings on oscillation only for signals that are below ground: apparently, the output impedance of the circuit differs above and below ground (where a different set of transistors drives the output, in the push–pull stage). And the right-hand image in Fig. 9N.16 shows that not even the negative edges always bring on oscillation.

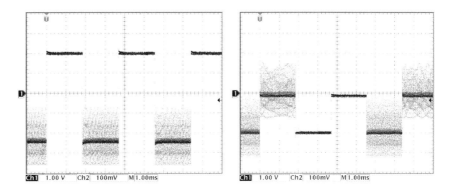

Figure 9N.16 A long coaxial cable can push an op-amp follower into oscillation.

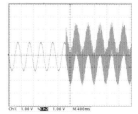

Figure 9N.17 Sinusoid may not bring on oscillation – but then it may.

A sinusoid is less disturbing, and may get through without provoking the oscillation. It seems to be getting safely through, in the left-hand half of the trace in Fig. 9N.17. But halfway through the scope's sweep we managed to bring on oscillation simply by pressing the function-generator's range button: the momentary interruption of the signal was enough to send the circuit off the rails. The lesson seems to be that capacitive loading of op-amp circuits is hazardous. We need remedies.

---

quantity $1 + AB$, and A falls with frequency, at the rate $1/f$. AoE – The x Chapters explores this point thoroughly, showing scope images to demonstrate that the op-amp's output impedance truly is inductive, even including the expected phase I-to-V phase difference. In *Troubleshooting Analog Circuits*, (p. 100), Bob Pease warns, similarly, against assuming that an op-amp output is simply resistive.

Remedies: 1. Feed back less: One solution is oddly counterintuitive: instead of using a follower, use an amplifier to drive the cable. Give it, say, a gain of five and the circuit may calm down.

At first, you're likely to protest, "More gain? But gain is just what an oscillator requires." Yes, but it's *loop gain* – the gain going round the loop – that matters. This is a notion that the Wien bridge illustrates nicely: the Wien circuit sustains an oscillation at the frequency where it is attenuated to 1/3 in the feedback loop – if the circuit has a gain of 3. Attenuation greater than 1/3 in the feedback loop, at that gain, would kill the oscillation.

Giving the cable driver "gain" necessarily *feeds back less*, since the gain is $1/B$, where B is the fraction fed back.[15] Figure 9N.18 shows the effect of gradually feeding back less than one. By the time we've fed back just 1/5 (for a gain of 5), the circuit is utterly placid.[16]

See AoE Fig. 4.102

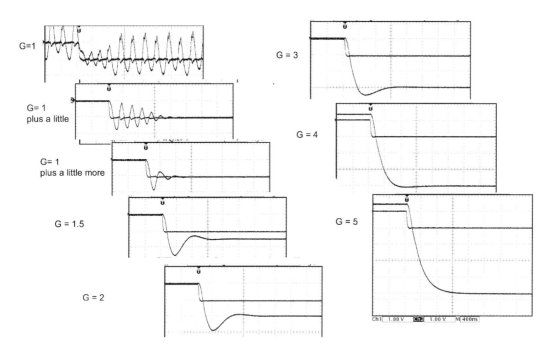

Figure 9N.18 Op-amp driving a capacitive load can be quieted by feeding back less – giving it gain.

These scope images give us our first chance to see that there are *degrees* of stability; it is not just a question of Yes/No: it does/does not oscillate. The second trace of Fig. 9N.18, for example (gain of one "... plus a little"), does not sustain an oscillation. But this ringing probably would be troublesome enough to call for a remedy.

We will meet this notion that there are degrees of stability in two other settings: in op-amp compensation (treated in §9S.1), where *phase margin* describes how closely the op-amp approaches the brink

---

[15] The closed-loop gain is actually $\frac{A}{1+AB}$ where A is the open-loop gain of the op-amp. But for large values of A, this approaches $1/B$.

[16] In case you are intrigued by the point that the op-amp output acts inductive, you may enjoy describing differently the effect of "feeding back less" – that is, of changing from follower to amplifier. Feeding back less – reducing B – raises the closed-loop output impedance, increasing the value of the apparent "inductance." This change tends to push $f_{resonance}$ down, by a factor $\sqrt{2} \times$ ratio-of-apparent-L-values. By itself, this change would be destabilizing. But, at the same time, a contrary, stabilizing effect outweighs this change. Feeding back less pushes the unity-gain frequency down by a factor proportional to the shrinkage of the fraction fed back – outweighing, or "outrunning" – the rate at which the resonant frequency has fallen. All this is well explained, with helpful plots, in *AoE – The x Chapters*, on advanced topics in operational amplifiers.

of disaster, the $-180°$ shift; and in the PID loop of Lab 10L, where we will be able to adjust the gain of our home-made "op-amp" to flirt with such instability.

Remedies: 2. Push the phase-shifted point outside the feedback loop: A series resistor of $100\,\Omega$ or so can provide a very simple remedy for the long cable problem.

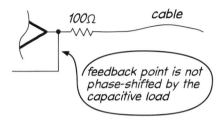

Figure 9N.19 Simple remedy for long-cable problem: series resistor.

The phase shift caused by the cable's capacitance still occurs – and, in fact, is aggravated. But the point in the circuit where most of this dangerous phase-shift occurs now lies *outside* the feedback loop. We have paid a price for stability: the circuit's R_{out} is degraded by the series resistor.

Remedies: 3. A refinement of long-cable remedy: "split" feedback: With a few more parts, we can have stability along with the original follower's good low R_{out} at lower frequencies. Figure 9N.20 shows the modified long-cable circuit, "splitting" the feedback path.

AoE Fig. 4.76

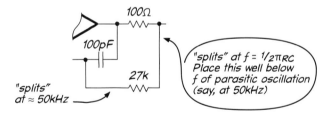

Figure 9N.20 Refined remedy for long-cable problem: "split" feedback.

Place the "crossover" frequency – at which most of the signal passes through the capacitor, bypassing the phase-shifting network – safely below the frequency at which the circuit oscillates.

Such a frequency might sound excessively high for audio, for example. But note that you don't want to push the crossover frequency lower than necessary. The benefits of feedback are vitiated at that crossover frequency. The correction of push–pull distortion, for example, would be spoiled by too low an $f_{\text{crossover}}$. This low crossover would slow the op-amp's drive that is needed so as to compensate for the push–pull's effects. You may recall the snap above or below $|0.6\,\text{V}|$ that the op-amp needed to provide in order to cure crossover from §6N.9.

The "split feedback" remedy is a powerful one. We use it in the push–pull driver circuit that concludes Lab 9L, and again in the motor-driver circuit of the PID in Lab 10L.[17]

See AoE §4.6.2 once again

One more long-cable remedy: use a specialized driver IC: Some followers are designed specifically to tolerate heavy capacitive loads (for example, Analog Devices' AD817, or TI/National's LM8271/2). This type simply slows down when connected to a heavy capacitive load: the value of the

---

[17] Other, subtler remedies also are available. Some are described in Bob Pease's Application Note "Linear Brief," National Semiconductor LB-42 (https://LAoE.link/Pease_LB-42.pdf), and more briefly in his book *Troubleshooting Analog Circuits*.

"compensation" capacitor that reduces high-frequency gain is effectively increased by C_{load}.[18] So, the circuit is self-stabilizing.

This solution may look appealing – but when you want to stabilize a circuit in a hurry, adding a series resistor may beat waiting a day or so for delivery of a specialized IC.

9N.6 A general criterion for stability: loop gain where phase shift approaches 180°

AoE §4.9.1A

"Loop gain" describes what it sounds like: the gain on a trip around the feedback loop: through the amplifier IC (whose gain is called A), then through the feedback network (whose "gain" – often an *attenuation* – is called B). The gain all the way round the loop, then, is AB. In general, we like high loop gain. It is what gives feedback circuits their several virtues – constancy of gain, low R_{out}, and so on. But it is also what makes op-amp circuits vulnerable to oscillation.

In order to keep a circuit stable, one must be sure that the loop gain, AB is less than one at the frequency where the phase shift reaches the deadly 180°.

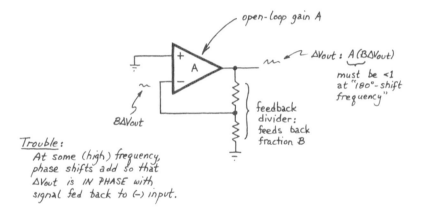

Figure 9N.21
Stability conditions summarized.

Assume there is noise around at all frequencies, disturbing the circuit input. That disturbance – call it ΔV_{in} – is amplified to produce a ΔV_{out} that is A times as big; then some or all of that ΔV_{out} is fed back to the input. A fraction B is fed back, so $AB\Delta V_{in}$ is fed back. If, at a frequency where 180° of phase shift occurs, this fed-back signal is as big as the original, or bigger, the circuit oscillates. In other words, $AB \geq 1$ at $-180°$ brings oscillation.

This view suggests remedies:

- limit the size of A:
- limit the size of B;
- or limit both.

In any case, we need to limit the product, AB. Let's look separately at the ways to limit A, then B.

AoE §4.9

Shrinking A: "frequency compensation": The gain curve in Fig. 9N.22 of the "uncompensated" amplifier shows too much gain at the frequency where phase-shift reaches 180°. An uncompensated op-amp wired as a follower ($B=1$) would oscillate.

[18] If this notion is puzzling, take a look at §9S.1.

9N.6 A general criterion for stability

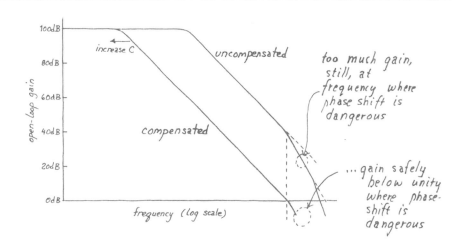

Figure 9N.22 Compensated versus uncompensated op-amps.

The compensated amplifier is made stable through the deliberate *early* rolloff of its gain with frequency. Its gain has been reduced, so as to fall to unity before the frequency that would impose that deadly phase shift. Op-amp compensation is discussed more thoroughly in §9S.1. In Lab 10L we will find that one can tinker with more than just the gain-rolloff of A; one can even alter its phase-shifting. Exciting complications to come!

Shrinking B: We thrashed this point pretty thoroughly on page 396: give the circuit gain, and it is less inclined to oscillate.

In summary: limit loop gain, AB: For stability, you need to hold the *loop gain* below unity at frequencies where dangerous phase shifts appear. Figure 9N.23 shows an op-amp's gain plot reminding us of the meaning of *loop* gain in contrast to *circuit* gain.

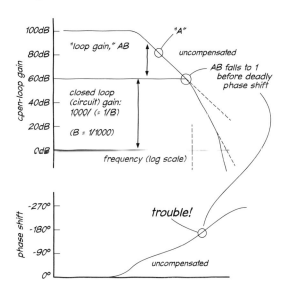

Figure 9N.23 AB determines feedback stability, as well as feedback virtues (illustrated with uncompensated op-amp, barely stable at gain of 1000).

Figure 9N.23 reiterates a point we tried to illustrate with Fig. 9N.21, summarizing conditions for stability. It is the value of AB at the dangerous phase-shift frequency that determines whether a feedback circuit is stable or not. High AB can be worrisome.

But AB also measures the potency of the *good* effects of feedback; it measures the capacity of the op-amp to improve circuit performance. In §9S.5 we look more closely at the way in which loop gain depends on both circuit configuration and chip properties: the way it depends, in other words, on B, and on A.

You have felt this sort of tension before, where reaching far for one thing you want (here, performance improvements through feedback), you may give up something else that you need (here, circuit stability). In discrete common-emitter amplifiers, for example, you were able to trade off *gain* versus *linearity*. Of the two concerns attracting us here, stability is the more urgent. With that paramount goal in mind we must hold AB safely low. We will settle for the circuit performance that results.

9N.7 Parasitic oscillation without op-amps

Surely this circuit cannot oscillate...: Feedback need not be explicit, as it is in op-amp circuits, in order to cause trouble. Figure 9N.24 is a circuit that oscillates because of very sneaky effects. Wouldn't you nominate this circuit as *least likely to oscillate*? Isn't it a sure-fire dud?[19] No voltage gain, and no feedback: surely, such a limp thing cannot sustain an oscillation.

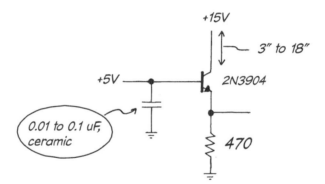

Figure 9N.24 Follower as *oscillator*.

...but it does oscillate: It oscillates, and fast enough so that we like to use it to jam FM broadcasts in the lab[20] – just to convince everyone that these parasitic oscillations can, indeed, be troublesome. This big oscillation completely silences the FM broadcast in the lab (we hope we're not jamming more than our lab room).

Symptoms of a parasitic: One of your circuits might oscillate at a time when you were not watching it with a scope. What clues might you get to the fact that a parasitic buzz was occurring? Two possible clues come to mind.

First clue: if you're very lucky, you may be listening to FM radio, and may notice reception going bad. This is not likely.

Another clue, in the present case, appears if you happen to use a DVM to check the circuit output. You may notice "impossible" readings, as we did, recently: see Fig. 9N.26.

Emitter more positive than base? It looks that way. But this is a fake reading: the oscillation has fooled the DVM, which accidentally has rectified a part of the fast sinusoid.

---

[19] Oxymoron, anyone?
[20] There happens to be an aggressive local station that calls itself "Jammin' 94.5," and with that name it just seems ripe for jamming.

9N.7 Parasitic oscillation without op-amps

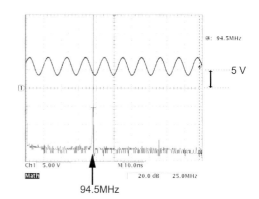

Figure 9N.25 Jammin' Jammin' 94.5 with the discrete follower. (Scope settings: 5 V/div, 10 ns/div.)

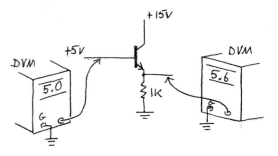

Figure 9N.26 "Impossible" DC readings may reveal a parasitic oscillation.

If you're inclined to protest that you'd not be likely to pick up either of these clues, you may be right. Good evidence lies in a story that Pease tells, in which this very circuit caused a parasitic oscillation that went undetected until the computers that included the circuit failed their FCC radio-frequency emissions tests (see footnote on page 402).

How can it oscillate? To answer that riddle one must answer two subsidiary riddles:

- How does the circuit achieve voltage gain (to sustain oscillations)?
- How does the circuit provide positive feedback?

The answers emerge if one takes into account imperfections of the circuit. The idealized follower cannot oscillate; a real one can.

- The gain appears if one draws explicitly the inductance always implicit in the power supply,[21] along with stray capacitance, see Fig. 9N.27.

    When you build this circuit in §9L.2, we encourage you to use an outrageously long power-supply lead, to exaggerate the stray inductance present in even well-built circuits. You can confirm that the collector supply line looks like a resonant circuit (a parallel LC like the one you built in Lab 3L): put your hand close to the wire, or even grab it. The parasitic frequency will change, because of the changed stray *capacitance*, now enhanced by your hand.

- The feedback appears if one notices that a disturbance on the power supply, which here is the collector, can disturb the emitter (through CE capacitance) in a sense that increases the collector disturbance: in other words, here is positive feedback.[22]

Cf. AoE §7.1.5D esp. Fig. 7.31

[21] Ballpark value? Perhaps 500 nH/meter, though the value depends on wire diameter and distance to other conductors.
[22] Here we mean that one "notices" this path conceptually. You will not be able to observe these details with your scope. Attaching a scope lead to the collector is almost certain to stop the oscillation, as noted below in §9N.8.1.

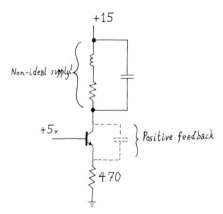

Figure 9N.27 Follower redrawn to include feedback, and inductance that can provide gain.

This circuit, as redrawn, is nearly identical to the current source whose oscillations are explained in AoE. There the circuit is likened to a purposeful oscillator called a Hartley *LC* oscillator.

Most of us settle for learning some rules of thumb that stop oscillations when they appear; it is not easy to model a circuit in detail, including stray inductance and capacitance. Another of those electronic Murphy's laws holds, however: if the circuit can find a frequency at which it could sustain an oscillation, it will find that frequency and oscillate – irritating its designer.[23]

9N.8 Remedies for parasitic oscillation

9N.8.1 First remedy: quiet the supply (shrink the disturbance fed back):

You may be able to stop the oscillation by quieting the collector: this "shrinks the disturbance," in the terms we used just above. Two rather obvious ways to do this come to mind:

- keep the power supply lead (+15 V) short (this reduces the inductance of the line, putting the resonant frequency so high that the transistor lacks gain at that frequency); and
- install a small ceramic *bypass* capacitor on the collector.

In fact, the oscillation is likely to be strangely fragile – but this characteristic can make it the more troublesome. You may suspect that the collector is oscillating, and decide to probe it to find out. But no, you find it's not oscillating. But the act of probing (that is, of attaching the probe's 10 pF of capacitance to ground) may have stopped the oscillation – though only while you probe it, of course. You may need to investigate by attaching a short length of wire to the probe tip and then "flying over" the circuit, at a distance of a centimeter or two to sense oscillations without killing them. Spooky, indeed.

9N.8.2 Another remedy: kill high-frequency gain

Even when your circuit is stuck with some disturbance fed back, of a phase that could bring oscillation, the oscillation will not occur if the circuit gain is insufficient. You may see some transient junk, but not a sustained oscillation. The base resistor in a follower circuit has this effect: spoiling the circuit's gain at the high frequencies where oscillation otherwise could begin.

---

[23] This circuit is not just a pedagogue's concoction. It seems to have caused real mischief to some engineers who used essentially this circuit to drive a computer's "Reset" pin. The computers then failed FCC interference-radiation tests because each computer included this unintended little radio broadcasting machine!

One explanation of the** base resistor **remedy: redraw as common base amplifier: Figure 9N.28 is a sketch of our transistor, viewing the circuit as a *common base* amplifier, an amplifier driven from its emitter. You probably have not seen an amplifier driven this way before (except in passing, when you may have seen in AoE, a Miller-killer made with a common-base amp, base firmly fixed. See §5S.5 and AoE Fig. 2.84(C)). Here, as suggested in Fig. 9N.28, we are doing the opposite of what that high-frequency configuration achieved. We *want* to spoil the high-frequency gain.

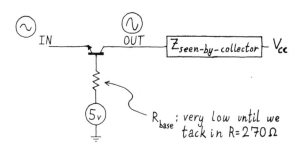

Figure 9N.28 Follower redrawn as common-base amplifier.

The gain of this *common base* circuit is

$$\text{gain} = \frac{R_C [\text{or } Z_C]}{r_e + [(\text{resistance driving base})/(1+\beta)]}$$

Here, with the base driven (or "held") by a stiff voltage source, the denominator of the gain equation is just "little r_e;" but we can diminish the gain in a way equivalent to the way we cut the gain of the common-emitter amp: by tacking in a resistor that enlarges the denominator of the gain equation. Here, we insert a resistor in series with the *base*; in the common-emitter amplifier we achieved an equivalent reduction of gain by inserting a resistor in series with the *emitter*.

AoE §2.4.5B

Another explanation of the** base resistor **remedy: Miller effect: There is another way to describe what the base resistor achieves; perhaps you'll prefer this other description, though we have barely mentioned *Miller effect* in this book. Figure 9N.29 shows our follower with two competing *feedback* paths drawn in – C_{CB} and C_{CE}. The latter provides *positive* feedback, as you know; C_{CB} provides *negative* feedback by moving the base voltage. It is this feedback, tending to spoil the gain of a high-frequency inverting amplifier, that is called Miller effect.

How much negative feedback? The fraction of a collector-wiggle fed back is set by the voltage divider formed by C_{CB} and R_{Thev} that is driving the base. By increasing the latter from a value close to zero to $270\,\Omega$, we much increase the strength of the negative feedback. Thus we tip the feedback competition in favor of the good guys, *negative* – and the oscillation should stop.

Remedies: The remedies, then, are fundamentally the same as those we apply when we need to stop or prevent op-amp oscillations:

- shrink the disturbance that is fed back; or
- diminish the circuit's (high-frequency-) gain.

9N.9 Recapitulation: to keep circuits quiet...

Some of the methods that tend to prevent parasitic oscillations are obvious; some are not. Among the obvious we might classify these:

404 Op-Amps IV: Parasitic Oscillations; Active Filter

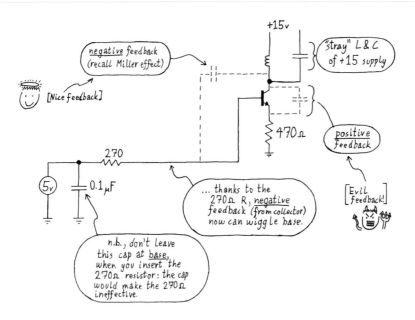

Figure 9N.29 Competing feedback paths: stabilizing versus destabilizing.

- keep power supply leads short and wide (minimize inductance);
- decouple power supplies (as usual).

Among the less obvious protections we might classify these:

- include a base resistor on any discrete transistor follower; and
- provide a "split" feedback path in an op-amp circuit that is to drive a troublesome load – a load that can impose substantial phase lags (see page 397).

In Lab 10L we will return to these stability questions, and there will try yet another remedy: we will design a sort of custom op-amp that can *undo* some troublesome phase lags. Parasitic oscillations always will lurk, threatening your circuits. You'll stand a better chance of fending them off if, boyscout-like, you are *prepared*.

9N.10 AoE reading

Reading:
- AoE:
    Chapter 7: §7.1.5E: parasitic oscillations.
    Chapter 4:
        §4.9: op-amp frequency compensation.
        §4.5.7: differentiators.
        §4.6.2: capacitive loads.

9L Lab: Op-Amps IV

Today we look first at one more *benign* use of positive feedback, an *active filter*, and then spend most of our time with circuits that oscillate when they should not. In this lab, of course, they "should," in the sense that we want you to see and believe in the problem of *unwanted* oscillations. On an ordinary day, the oscillations that these circuits can produce would be undesirable, and would call for a remedy. Some of you have met these so-called "parasitic oscillations" in earlier labs. Consider yourselves precocious. In the exercises below, you will try first to bring on oscillations, then to stop them.

If any topic in electronics deserves the title "Art of...," taming oscillations must be that topic. With some trepidation, we invite you to try to make these nasty events occur. We quoted, in Chapter 9N, the standard variation on Murphy's Law that says,"Oscillators won't; amplifier circuits will." Since today's circuits purport to be amplifiers rather than oscillators, probably they "will...."

9L.1   VCVS active filter

We ask you to build two versions of a two-pole filter: a simple cascade of two passive RCs, and an active lowpass. The two filters have equal f_{3dB}s.[1]

If you have a *resistor substitution box*, use that to set the value of the active filter's R_{gain}. The substitution box makes changing values easy and, unlike a *variable resistor*, lets you know what value you are applying.

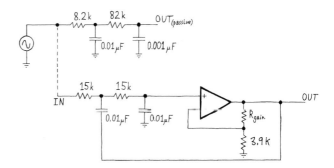

Figure 9L.1 VCVS: a particular implementation, with cascaded RCs added for comparison.

Set R_{gain} to 2.2k, providing the response with the flattest passband ("Butterworth"). Confirm that both circuits behave like a lowpass filters; note f_{3dB}, and note approximate attenuation at $2f_{3dB}$ and $4f_{3dB}$. We hope you will find the -12dB/octave slope that is characteristic of a 2-pole filter, though you won't see that full steepness in the first octave above f_{3dB}.

---

[1] They may look as if they will not show the same f_{3dB}: indeed, the RC values of active and passive filters are *not* the same: 2 kHz for the passive, 1 kHz for the active. Yes, the RCs are unequal. But only the active filter has $f_{3dB} = 1/(2\pi RC)$. The *passive* filter's two elements share an RC value. But the entire passive filter shows an f_{3dB} that is about half the f_{3dB} of each stage, because where each filter stage drops *3 dB*, the entire filter drops *6 dB*.

Filter Type	R_{Gain}	Gain
best time delay (Bessel)	1k	1.3
flattest (Butterworth)	2.2k (2.3k)	1.6
steep (Chebyshev)	4.7k (4.3k)	2.1
nasty peak (no one claims this one!)	6.8k	almost 3
OSCILLATOR!	10k	3.5

We hope, also, that the simple cascaded *RC* looks wishy-washy next to the active filter. The simple *RC* and the active filter should show the same f_{3dB}.

9L.1.1 Sweep both filters

Presumably you have been watching the outputs of both filters on the scope, as you drive the two filters with a common input. For a vivid display of the two filters' frequency responses you will want to *sweep* frequencies automatically. You have done this before, using the function-generator's *sweep* function, but this time you may have to do the task a little differently from the way you did it earlier if you used an analog scope.

This time, you *cannot use* the *X–Y* display mode. Instead, use a conventional sweep (this allows you to watch two output signals, not just one), while *triggering* the scope on the function generator's RAMP output (use the steep falling edge of the ramp). If you are using a digital function generator, use the trigger out signal: see §2S.3.5.

9L.1.2 Effects of varying the amount of positive feedback: other filter shapes

While *sweeping* frequencies, try altering the shape of the active filter's response by changing R_{gain}. Since you are changing not only shape but also output amplitude, you will need to adjust scope gain each time you change R_{gain}. It is useful to keep the trace sizes of active and passive filters comparable, because the passive response provides a convenient reference against which to judge the active responses. Your substitution box may not offer exactly the values shown in the table (we chose the shown values using AoE's Table 6.2, §6.3.2). Do the best you can. In parentheses we have shown the more-exact value that is desirable, in case you're a perfectionist.

The last two cases, in which we deliberately overdo the positive feedback, are pointless in a filter – but for today's lab they may be useful. They remind us of the boundary we are moving toward in this lab, where positive feedback becomes harmful, and which we'll soon reach.

9L.1.3 Step response; waveform distortion

Watch the circuit's response to a 200 Hz square wave, and note particularly the *overshoot* that grows with circuit gain. If you are feeling energetic, you might test also the claim that the tamest of the filters (with $R=1$k), which shows the best step response, also shows the least waveform distortion. The $R=4.7$k filter should show most distortion. Try a triangle as test waveform. The contrast will not be very striking: we saw only a little distortion, from the worst of the filters.

9L.2 Discrete transistor follower

How can the familiar and mild-mannered circuit in Fig. 9L.2 oscillate? To answer that riddle one must answer two subsidiary riddles (as you have seen in §9N.7).

9L.2 Discrete transistor follower

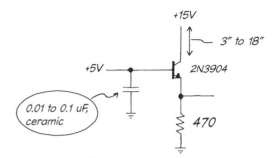

Figure 9L.2 Follower as *oscillator*.

- How does the circuit achieve voltage gain (to sustain oscillations)?
- How does the circuit provide positive feedback?

The answers depend on the truism that power supplies are not perfect, and that stray capacitance can let the transistor talk to itself.

Build the circuit, allowing three to six inches of wire on the +15 V lead, and watch the *emitter* (if you try to watch the collector instead, you are likely to kill the oscillation with the scope probe's capacitance). See what "ground" is doing at the foot of the emitter resistor, as well. (To measure this voltage you should ground the scope lead some distance away.)

If the circuit fails to show oscillations, try worsening your supply by making the path for +15 V to your circuit more circuitous: pass it through a wire a couple of feet long. When the path is crummy enough (that is, inductive enough), the oscillation should begin.

Remedies

You saw, we hope, a high-frequency oscillation (when we tried the circuit, we saw a sinewave of a couple of volts at about 60 MHz; with care, and shortening the supply lead, we were able to coax it up toward 100 MHz). Try the remedies that are in your bag of tricks. We propose these below.

Quiet the supply (shrinking the disturbance fed back): You can stop the oscillation by quieting the collector: this "shrinks the disturbance," in the terms we used in Chapter 9N. There are two ways of doing this.

Shorten supply leads: This isn't much fun, and you may believe that it works even before you try it. But a short +15 V supply lead probably will stop the oscillation. If you do this, revert to the long, ugly leads, so as to keep the trouble coming.

Decouple the noisy supply: You can quiet the collector with a capacitor to ground. This cap "isolates" the power lines from the transistor circuit. Or, to say the same thing in other terms, provides a local source of charge when the transistor begins to conduct more heavily: the power supply and ground lines need not provide this surge of current, and therefore will not jump in response; instead, the local bucket of charge will provide the needed current.

Try a *ceramic* capacitor, about 0.01–0.1 μF. If this fails, take a look at our notes on stability in §9S.3.

But, again, remove that decoupling cap after trying it. We want oscillations, to test the more interesting remedies that follow.

9L.2.1 Diminish high-frequency gain

With the parasitic oscillation screaming away, now let's see if we can stop it by spoiling the high-frequency gain of this follower (which, after all, was not intended as an amplifier!).

Remedy 1: base resistor: See if the remedy works: add a resistor of few hundred ohms between the +5V source and the base. Make sure you *remove* the decoupling cap that was making the base voltage "stiff" at high frequencies. We hope the base resistor stops your oscillation. If not, try larger values of R.

Remedy 2: ferrite bead: Another way to tame this oscillation – arguably more elegant, because it kills the high-frequency oscillation without degrading low-frequency R_{out} – is to place a *ferrite bead* on the base lead (or in series with the lead, if the transistor is surface-mount, as most now are). This increases the base lead's inductance, and also makes it "lossy," so that at high frequencies it looks resistive rather than inductive (inductive is risky, because it can resonate with stray capacitance, bringing on oscillations again). A typical bead's frequency response is shown in Fig. 9L.3.[2]

Ferrite Bead Simplified Model and Simulation
A ferrite bead can be modeled as a simplified circuit consisting of resistors, an inductor, and a capacitor, as shown in Figure a. R_{DC} corresponds to the dc resistance of the bead. C_{PAR}, L_{BEAD}, and R_{AC} are (respectively) the parasitic capacitance, the bead inductance, and the ac resistance (ac core losses) associated with the bead.

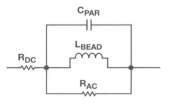

(a) Simplified circuit model

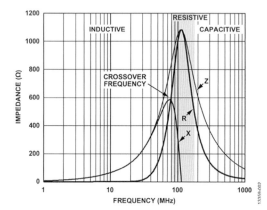

(b) Tyco Electronics BMB2A1000LN2 measured ZRX plot.

Figure 9L.3 Frequency response of typical ferrite bead: "lossy" resistive behavior helps stability.

The frequency-response plot in Fig.9L.3(b) shows, separately, the responses of the several elements of the bead model, and also its overall response, labeled "Z." The trace labeled "X" shows the response of the inductor (rising with frequency, before the resistive response dominates); the high-frequency rolloff reflects the response of the capacitance in the bead.

Our usual hookup wire, 22 gauge, is too heavy to fit through the bead. Use thinner wire, for example 30-gauge polyvinylidene fluoride (trade name Kynar). If you form two loops, you will quadruple the effective impedance.[3] Figure 9L.4 shows what the bead looks like on a wire.

9L.3 Op-amp instability: phase shift can make an op-amp oscillate

9L.3.1 Simple op-amp amplifier, with various loads

Lowpass filter in the feedback loop: The circuit in Fig. 9L.5 is a familiar non-inverting amplifier, with a heavy capacitive load that makes it marginally stable. As load, use about ten feet or so of BNC

---

[2] The curves are typical, presented in an "Analog Dialogue" note by Analog Devices:
https://LAoE.link/Ferrites_Demystified.html.

[3] A second loop doubles the magnetic field, for a given current, and doubles the response to the magnetic field.

9L.3 Op-amp instability: phase shift can make an op-amp oscillate

Figure 9L.4 Ferrite bead with fine wire looped through it.

cable (30pF/foot). If you don't find a long BNC, join several shorter sections with links or with a BNC "Tee".[4]

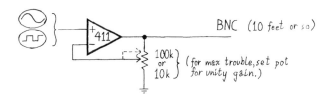

Figure 9L.5 Load capacitance can make a feedback circuit unstable.

Your first task is to coax this circuit into oscillation. To do this, *minimize* the gain: turn the pot to feed back *all* of the output signal.

The circuit may not oscillate, at first. If it does not, drive it with a square wave. That should start it, and once started it is likely to continue even when you turn off the function generator, or change to a sinewave.

Remedy 1: Shrink the disturbance fed back: Now – with the circuit oscillating – gradually turn the pot so as to shrink the fraction of V_{out} that is fed back. (What are you doing to *gain*, incidentally?) When the oscillation stops, see how the circuit responds to a small square wave (make sure the output does not saturate). If you see what looks like ringing (or like the decaying oscillation of a resonant circuit), continue to shrink the fraction of V_{out} that is fed back until that ringing stops.

We hope that you now have confirmed to your satisfaction that stability can depend on the gain of a feedback circuit. This fact explains why so-called "uncompensated" op-amps are available: if you know you will use your op-amp for substantial gain, you are wasting bandwidth when you use an amp compensated for stability at unity gain.

Other remedies: But suppose you want unity gain, and need to drive a long coaxial cable? You could try any of several solutions. Laziest would be to shop for an op-amp or buffer designed to be stable when loaded with a lot of capacitance ("a lot" is something above, say, 100 pF).

Try another pair of remedies, here. Try them separately. See if either is sufficient. If neither is, you might try them together.

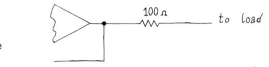

Figure 9L.6 Series *R* can take shifted point outside feedback loop.

---

[4] At high frequencies, this might be a poor idea, but the T works fine at our op-amp frequencies.

- Series *R*: take the load outside the feedback loop. Put a 100 Ω to 220 Ω resistor in *series* with the BNC cable, and *outside* the feedback loop, as in Fig. 9L.6. See if this stabilizes the amp.
- Snubber: connect a small-valued R AC-coupled to ground, as in Fig. 9L.7.

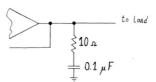

Figure 9L.7 Snubber.

The snubber is a little mysterious. It appears to kill high frequencies in an unusually crude way: by loading the output heavily at high frequencies – a range where Z_{out} is rising, as the impedance of the snubber shrinks toward its minimum, the value of *R* alone. (The impedance of the snubber shown in Fig. 9L.7 falls to $\sqrt{2}R \approx 14\,\Omega$ at around 160 kHz).

9L.4   Op-amp with buffer in feedback loop

The circuit in Fig. 9L.8 resembles others that you have built before. The difference this time is that we'll be looking for trouble.

Use big power transistors: MJE3055, MJE2955. You used these back in the push–pull exercise of §6L.8: terminals are BCE seen from the front (the side with the part number labeling).

The key change that makes trouble very likely is our adopting an op-amp that has gain higher than '411's. We ask you to use an LT1215, a device whose gain starts higher (10^6 at DC, versus the '411's 200,000) and extends out to 23 MHz (versus the '411's 4 MHz). So, the op-amp has not only more virtue but also more potential vice, at frequencies where phase-shifts inside the feedback loop become dangerously close to converting negative feedback into positive. We hope it will buzz for you, so that you can try a new remedy.

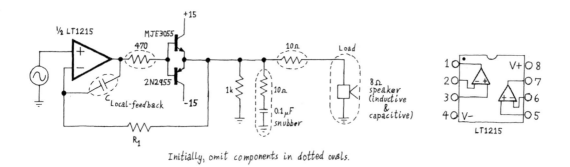

Figure 9L.8 Push–pull buffer in feedback loop.

(Note that the pinout of this dual op-amp matches that of the LM358, standard for duals, but does not match that of the more familiar LF411.) Try the circuit without speaker attached: feed the circuit a sinewave of a volt or so, at around 1 kHz, and confirm that the circuit follows, without showing crossover distortion.[5]

---

[5] If you're "lucky," you may be blessed with a parasitic oscillation even in this preliminary test.

9L.4 Op-amp with buffer in feedback loop

Now, as load, add an 8 Ω speaker. (If the sound is obnoxious, lower the amplitude.) We hope you will see oscillations at the output of your amplifier (they may not affect the *sound*, however).

Persuade your circuit to show parasitic oscillations, brought on by the low-frequency sinewave. Determine the approximate *frequency* of the parasitic fuzz, so as to decide who's to blame, op-amp or transistors. If the frequency is in a range where the op-amp still has substantial open-loop gain, you can safely assume the op-amp is to blame. Oscillations caused by the transistors are likely to be much faster. (We got the handsome oscillations shown in Fig. 9L.9.)

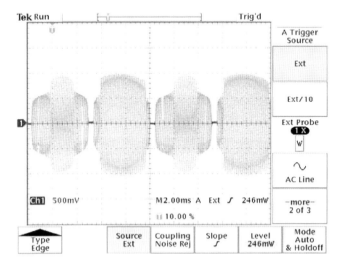

Figure 9L.9 Push–pull fuzz: no split feedback, speaker attached.

Cheat, if you must, to get an oscillation: If you cannot get your circuit to misbehave with speaker as load, try an installed capacitive load: say, 0.001 μF to ground. Experiment with the C_{load} that brings on oscillation. But be warned that too large a C, curiously enough, will prevent oscillation, simply by overburdening the op-amp, in the manner of the *snubber* circuit.

Remedies: Then try (separately) the following remedies; see if you can stop the oscillations.

- Decouple the power supplies (if you haven't already done so).
- Add a base resistor of a few hundred ohms, if you suspect the transistors.
- Add local (or "split") feedback: 100pF direct from op-amp output to the inverting input: see Fig. 9N.20. You will need an R in the *long* feedback path from circuit output, to mate with this new C. The idea is to let the R provide feedback at moderate frequencies, C to provide feedback at the higher frequencies where phase-shifts make the circuit unstable. Choose R to equal X_C at some frequency below the frequency at which you have observed the oscillation.
- Add a snubber RC (10 Ω, 0.1 μF, again). See if the snubber *alone* can stop the oscillation (remove "split feedback" to do this test).
- Once the op-amp is stabilized – with split feedback or with snubber – see if removing the base resistor disturbs the circuit.
- Try a small resistor (start with a value under 10Ω) *outside* the feedback loop, in series with the speaker or C_{load} (nothing new to you: see Other remedies on page 409).

We hope you can quiet the amplifier. If you can't, ask for help!

9S Supplementary Notes: Op-Amps IV

9S.1 Op-amp frequency compensation

AoE §§4.9, 2.5.4B

We have noted elsewhere that all the op-amps we meet in this course use internal "frequency compensation" that makes them stable – at least, if we refrain from putting strange things within their feedback loops. Frequency compensation, surprisingly enough, means deliberate rolling-off of the amplifier's gain. This may seem perverse – especially given Black's fundamental insight that "excess gain" is exactly what is necessary to achieve the benefits of negative feedback.[1] But the need for stability is so fundamental that every other concern must give way in its pursuit.

The '741 op-amp, a hugely successful part, seems to owe its success to the fact that it was the first internally compensated op-amp – and therefore delightfully easy to use,[2] since it is 2 caps and a resistor on three pins. About 45 years after its introduction by Fairchild (the company that developed the first fully integrated circuit) the '741 still occasionally appears in new designs – at least, in designs from developing countries, where its wide availability and low price may be especially appealing. The '411 op-amp that we use in most of our labs uses a design fundamentally the same as the '741's, and includes internal compensation, now standard. The major difference between the two parts is the '411's use of field-effect transistors at the input stage.

A look inside the '411 circuit shows a capacitor planted between output and input of the high-gain stage; this capacitor performs just as the cap does when placed between output and inverting terminal of an op-amp: it forms an *integrator*.

This rolloff and phase shift (or "integrator behavior") is imposed because it is better than the alternative: excessive phase shifts that would bring instability.

9S.1.1 Waveforms to persuade you the op-amp really *does* integrate

A skeptical reader might well reject what we have just said: haven't you seen, over and over in the lab, that an op-amp does *not* show phase shift? You have watched input and output of followers and amplifiers on the scope. A follower *follows*, faithfully reproducing the input waveform, does it not?

You are entitled to be skeptical. But the way to reconcile what you have observed with what we have just claimed is to recall that every op-amp circuit you have seen – with the unimportant exception of a couple of comparator circuits – *included negative feedback*. That feedback hides the op-amp's internal integration, just as it hides any other annoying transfer functions that are put within the loop. It hid the push-pull's crossover distortion; in general, it hides the "dog" that we sketched in §6N.9.

Only if one runs the op-amp *open loop* does the op-amp's integration become visible. Doing this

---

[1] By now, the language of Harold S. Black's patent probably is becoming familiar to you (patent 2,102,671 [1937]): "...by constructing a circuit with excess gain and reducing the gain by negative feedback, any desired degree of linearity between output and input...can be realized, the limiting factor being in the amount of gain that can be attained...."

[2] The μA709 required the user to optimize two RC networks for compensation. Walter Jung, Chapter H, "Op-amp History," in Analog Devices' Op Amp Applications Handbook, 2005, at https://LAoE.link/Op_Amp_Handbook.html.

9S.1 Op-amp frequency compensation

experiment is a bit of a chore, so we have not asked you to build this circuit. One must carefully adjust the DC-offset of the input to keep the op-amp from saturating; then apply a small wiggle. When we did this, we got the waveforms of Fig. 9S.2, which expose the op-amp's integration. At the lowest

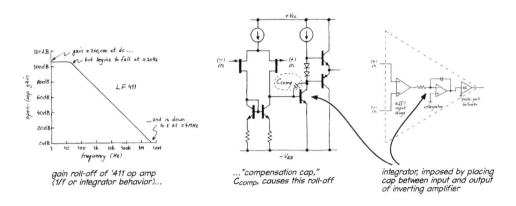

Figure 9S.1 Gain rolloff for compensated LF411 op-amp – and its cause.

frequency – 4 Hz in scope capture image *(a)* – only a little phase shift appears because we are below the frequency at which the internal compensation capacitor has a significant effect. At 100 Hz in image *(b)*, the full −90° shift of the integration is evident. And, just in case we haven't convinced you yet, image *(c)* shows a square wave integrated into a triangle wave at 100 Hz.

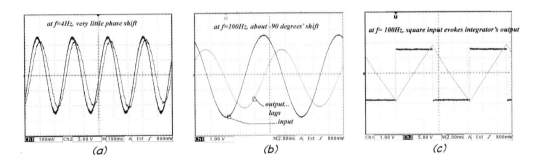

Figure 9S.2 Op-amp's internal integration becomes visible if one runs it *open loop*.

9S.1.2 Why compensation? Considering the hazards to an uncompensated op-amp

AoE §4.9

The uncompensated op-amp would need help: This rolling-off of gain must occur in order to keep the circuit stable – whether the rolloff is done for us ("internal compensation") or by our attaching an external "compensation" capacitor. To see why it is necessary, we need to note that the op-amp, a multi-stage circuit (see Fig. 9S.3), behaves like a series of lowpass filters.

AoE §4.9.1

It behaves like this because its several stages form a series of non-zero source resistances – two of them actually *current sources*[3] – driving stray capacitances. If we didn't intervene, we would find that the cumulative phase shift grows alarmingly: see Fig. 9S.4.

We cannot tolerate a phase shift as large as 180° under conditions where *loop gain* – the signal amplified and then fed back – is as much as unity. Loop gain, you may recall usually is written as the

---

[3] Of course, the current sources here do fit the description, "non-zero source resistances."

414   **Supplementary Notes: Op-Amps IV**

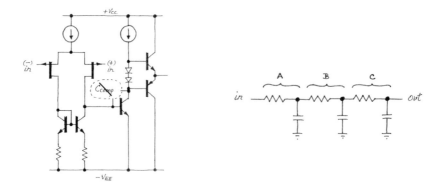

Figure 9S.3 An uncompensated op-amp behaves like two or three lowpass filters, cascaded.

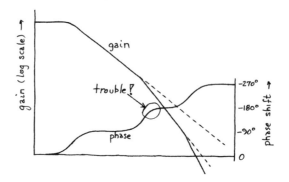

Figure 9S.4 An uncompensated op-amp would show deadly phase shift.

product AB, where A is the op-amp's open-loop gain (the characteristic plotted in Fig. 9S.4) and B is the fraction fed back.

Two possible solutions...: Two solutions are available, in principle (this we have said in Chapter 9N), for stabilizing such an op-amp: reduce B, by feeding back less than all of V_{out}; or reduce A by deliberately degrading the amplifier's gain as frequency climbs.

AoE §4.9.1A

Limit B: even an uncompensated op-amp can be stable provided very little is fed back....: The first of these remedies would look like Fig. 9S.5: reduce the fraction fed back by putting a divider in the feedback loop (this is a gain-of-1000 amplifier).

...but uncompensated op-amp as follower goes crazy: An uncompensated version of the classic 741 op-amp has too much gain where its phase-shift hits $-180°$, and wired as a follower it does, indeed, go berserk: see Fig. 9S.6.

...Limit A: but the normal method is to limit high-frequency gain: As you can imagine, designers prefer to work with an op-amp that is stable even when not asked for very high gain. Best of all is an op-amp (like all that we use in lab) that is stable even when the *entire* output voltage is fed back (in other words, when it is wired as a follower). To tolerate 100% feedback ($B=1$), the op-amp must have its gain ruined before phase shift reaches $-180°$. Figure 9S.7 shows a follower and the rolloff of gain that it requires.

AoE §4.9.1

9S.1 Op-amp frequency compensation

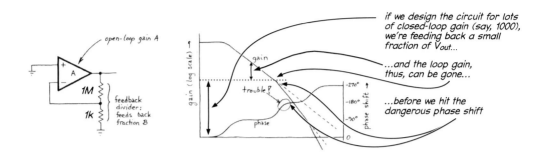

Figure 9S.5 An abnormal solution: even an uncompensated op-amp could work provided little enough is fed back.

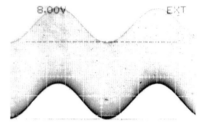

Figure 9S.6 Uncompensated op-amp, wired as follower, oscillates.

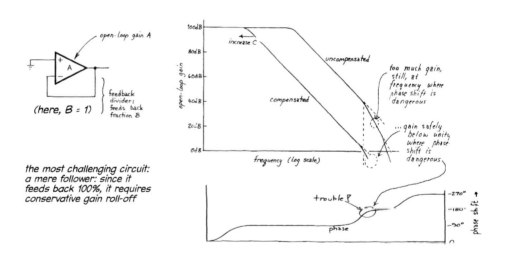

Figure 9S.7 The usual solution: rolloff open-loop gain (A) for stability even at B = 1.

Phase margin: As the compensated op-amp approaches its maximum frequency – and its gain approaches unity – its phase shift flirts with trouble: moves beyond $-90°$, but stopping short of disaster. The distance from "disaster" – from $-180°$ – is called the op-amp's "phase margin." Most op-amps put that at between 45° and 60°. The plot in Fig. 9S.8, from the '411's datasheet, shows about 50° margin (wrongly labeled "phase" rather than "phase margin").

This detail reminds one why even somewhat less than a $-90°$ shift imposed by something put into the feedback loop can send a circuit off the rails.

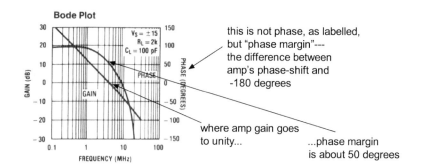

Figure 9S.8 Op-amp phase shift approaches 180°, but stops safely short.

9S.1.3 Decompensated and uncompensated op-amps

Uncompensated op-amps: One can buy an op-amp that leaves it to the user to install that gain-killing capacitor (external to the IC, of course). Early op-amps worked this way; after the '741 they became exceptional. Such an "uncompensated" amplifier is offered because internally-compensated op-amps give up gain in order to make the devices stable in the hardest case, the follower circuit. Fig. 9S.7 contrasted the gain rolloff of a generic compensated versus uncompensated op-amp. At unity gain, the uncompensated op-amp would be unstable; its compensation could be adjusted to fit the closed-loop gain in a particular application.

Decompensated op-amps: An intermediate possibility is a "decompensated" op-amp:[4] one that would be unstable at unity gain (where you feed back 100%) but stable at some closed-loop gain such as 10 (where you feed back only 1/10). If you know that your circuit will not push circuit gain below 10, then you can afford higher open-loop gain at a given frequency. The Linear Technology LT6230, for example, a 215 MHz part, is offered in two versions: one compensated the other decompensated – rigged to be stable for gains of 10 or more. The curves in Fig. 9S.9 (pretty hard to read, because they are so dense with information) contrast the two versions.[5]

Since the circuit with gain of 10 runs with reduced B, the circuit can be stable with open-loop gain, A, boosted, as well. You'll notice that the decompensated part shows higher gain at every frequency. It can afford that, because it is the *loop gain* – the product AB – that matters, and a gain-of-ten circuit holds this product to a level that is safe for this op-amp.

AoE §4.9.2B

9S.2 Active filters: how to improve a simple *RC* filter

Figure 9S.10 shows a linear amplitude-versus-frequency plot for an ordinary *RC* lowpass filter. This response disappoints us in two ways:

- Its pass-band is not as flat as we would like. That's evident in the plot shown: not much like the flat pass-band of a textbook drawing!
- Its fall-off is not as steep as we would like. We have gotten used to "−6 dB/octave," characteristic of our simple *RC* lowpass, but sometimes we'd like more.

---

[4] AoE points out that a better name would be "under-compensated."
[5] Linear Technology: LT6230/LT6230-10 datasheet at `https://LAoE.link/LT6230_Datasheet.pdf`.

9S.2 Active filters: how to improve a simple RC filter

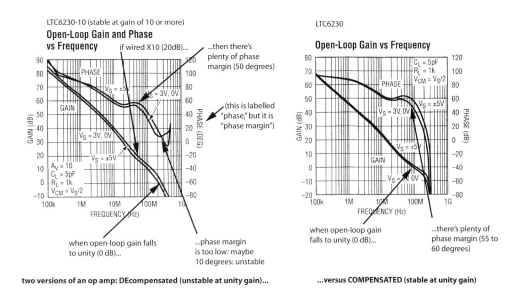

Figure 9S.9 This op-amp comes in both compensated and decompensated versions. The latter offers higher gain, but is unstable as a follower.

Figure 9S.10 *RC* lowpass response (linear plot).

Simply cascading *RC*s does not solve both problems: it makes the ultimate fall-off twice as steep (−12 dB/octave) – but it does only a very little to boost the droopy passband.

At first glance, the cascading seems to make the *pass-band* even worse: droopier. This is illustrated and confirmed with a scope photo in Fig. 9S.11[6] comparing a single *RC* with f_{3dB}=1 kHz, against a cascade of two *RC* lowpass filters, each with f_{3dB}=1 kHz.

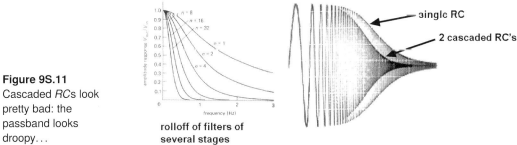

Figure 9S.11 Cascaded *RC*s look pretty bad: the passband looks droopy...

The frequency that was f_{3dB} now is f_{6dB}: the output is down 6dB. But to compare passbands fairly, one ought to compare filters with the same f_{3dB}'s. A comparative plot so "normalized," makes the cascade look better; see Fig. 9S.12. In fact, the cascade's passband is *less* droopy than the single *RC*'s – a result that may contradict your intuition.

---

[6] The left picture is taken from AoE Fig. 6.2; the two parts of Fig. 9S.11 can't be compared directly, because the scope image is *log–linear* (log frequency, but linear amplitude), whereas AoE plots here are linear on both axes.

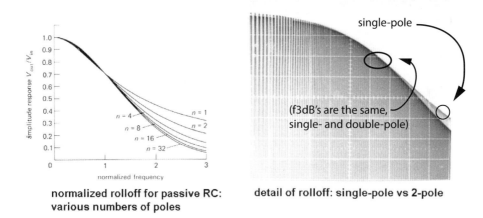

normalized rolloff for passive RC: various numbers of poles

detail of rolloff: single-pole vs 2-pole

Figure 9S.12 Normalized plot of frequency response of single versus 2-cascaded *RC* lowpass filters: cascade looks better. The left-hand image reproduces AoE's Fig. 6.2B.

If nothing better were available, we would live with these droopy passbands, in order to get the steep ultimate rolloff. However, *LC* filters can give better performance – and, as you will see when you do Lab 9L, so can op-amp *active filter* circuits, which are able to mimic the performance of *LC* filters without the use of inductors.

To get a better filter *shape* – that is, frequency response – we need a design trick that lets us flatten the passband while getting a good steep rolloff.

In order to flatten the pass-band of a simple cascaded two-*RC* lowpass, we need to give the filter output a boost (a kick in the pants) just where it is most wishy-washy: just below f_{3dB} (*below*, in the case of a lowpass, anyway). On the other hand, in order to steepen the ultimate fall-off, we need only duplicate the effect of 2 similar *RC*s in series. Soon, you may find yourself getting fussier, and caring about finer points, like the steepness of the *initial* rolloff (close to f_{3dB}). But for a start, let's try whether we can see intuitively how a particular active filter manages to flatten the passband of an *RC* lowpass while achieving the −12 dB/octave rolloff of a "2-pole" filter (the ultimate slope you would see from two *RC* lowpass filters in series).

9S.2.1 An intuitively-intelligible active filter: VCVS

AoE §6.3.1

AoE calls the filter in Fig. 9S.13 "even partly intuitive," implicitly acknowledging that most active filters are depressingly non-intuitive: their designs reflect messy math, and therefore force one toward cookbook design methods that are not instructive. The VCVS[7] "Sallen–Key" filter shown in the figure makes both of the improvements that we called for.

Active filter flattens the passband...: Positive feedback boosts the passband, feeding back what amounts to a *bandpass* kick: see Fig. 9S.14. Why "bandpass?": the feedback is *highpass*, but the source of the feedback is the output of a *lowpass* (the basic circuit we are building); the product of the two is bandpass, hence the well-placed and well-shaped kick in the pants.

What we have described as a bandpass kick could also be described as akin to bootstrapping C_1 – and then some: not only does C_1 not burden the signal fed by R_1 at frequencies below f_{3dB}, it actually boosts the signal that is fed to the passive lowpass formed by $R_2 C_2$. This is because, in this passband frequency range, the far side or "foot" of C_1 is driven *in-phase* with the input signal and at *larger amplitude*.

---

[7] VCVS: "voltage controlled voltage source:" a name that does nothing to demystify this circuit!

9S.2 Active filters: how to improve a simple *RC* filter

In the "stop band", the first capacitor, C_1, *loads* R_1 instead of boosting it, since the circuit output – applied to the foot of C_1 – is small on account of the attenuation imposed by the R_2C_2 lowpass. Deep into the stop band, the filter behaves like the cascade of two simple *RC* lowpass sections. (It's not quite as clean as that, since the two sections do not show the 1:10 impedances that would prevent their interaction.) Hence the steep ultimate rolloff (−12 dB/octave).

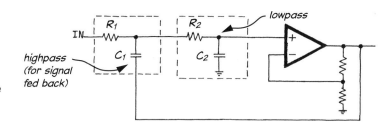

Figure 9S.13 2-pole active filter: *VCVS* (lowpass), one variant of Sallen–Key filter.

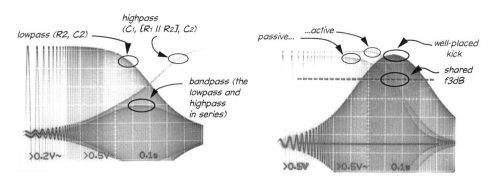

Figure 9S.14 Frequency response of active bandpass filter from Fig. 9S.13 in passband.

...Active filter can achieve very-steep rolloff: A mystery not explained by this 'kick in the pants' argument is the very steep rolloff achievable close to the cutoff (or f_{3dB}) frequency. Phase shifts, ganging up fortuitously explain it, we suppose: not a very satisfactory answer, but about all that's available in this non-mathematical view.

Figure 9S.15 shows the filter's response plotted against the response of a simple *RC* and two cascaded *RC*s all sharing a common f_{3dB}. The 2 pole active filter flattens the passband while steepening fall off.

See AoE Fig. 6.30, comparing responses of three filter types

The active filter can be tuned to optimize what you choose: Adjusting the gain lets one optimize one filter characteristic or another, always at the expense of some other filter behaviors. One gain gives flattest passband (called a "Butterworth" filter); another gain (higher) provides steepest rolloff (called a "Chebyshev" filter); yet another gain (lower) preserves waveshape best ("Bessel"). In Fig. 9S.16 we show the effect of three gain settings.

Since it is *positive feedback* that boosts the passband, we need to use this medicine judiciously: too much, and we'll first make a bump in the passband, in place of droop – and then, if we really overdo it, we'll kick the circuit into uncontrolled, continuous oscillation. Figure 9S.16's oscilloscope images show a range of possibilities. These images result from various amounts of positive feedback (circuit gain varies, too; these plots have not been normalized for constant amplitude in the passband).

Higher-order filters: If you want a 3- or 4-pole or higher-order filter, you can cascade active 2-pole sections. Here our intuitive grip slips almost entirely: the several sections are *not* alike, and the explanations for the strange coefficients (gains, in the VCVS form, and f_{3dB} values) become mercilessly mathematical.

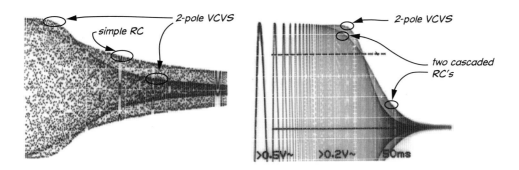

Figure 9S.15 Frequency response of 2-pole VCVS filter compared to simple RC, and to a pair of RCs cascaded.

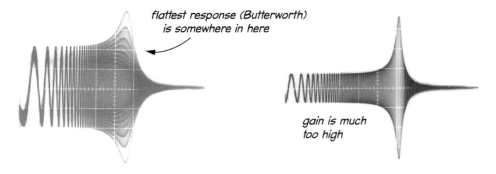

Figure 9S.16 Filter's frequency response depends on strength of positive feedback: multiple exposure shows several response shapes.

We find some comfort in the thought that the big, ugly equation that describes the desired response of, say, a 4-pole filter (which shows powers up to four in frequency) can be factored into the somewhat cozier quadratic form that you have already built – a modest 2-pole filter. But apart from that, these fancy filters are distressingly mysterious. If the mystery doesn't offend you, you'll find you can build good active filters by using the tables provided in AoE (table in §6.3.2: lowpass, four alternative shapes).[8]

For the ultimate in cookbookery, as pointed out in Chapter 9N, you can invoke TI/Burr–Brown's filter design program, FilterPro. You may feel like a dope, but the program will deliver a good design. However you proceed, we hope that today's glimpse of an active filter in lab will embolden you to use these circuits.

AoE §6.3.3

Compare AoE Fig. 6.27

---

[8] And here's a friendly and clear book, if you feel like learning more about active filters: Don Lancaster's *Active Filter Cookbook* (Sams, 1975).

Transient response; shape preservation: Fig. 9S.17 shows responses in the time domain of the VCVS 2-pole lowpass, at three gain settings: tamest (best time delay and shape preservation); flattest; and steepest. (These are the gains you will try for yourself, in the lab.) You can see that overshoot grows with gain, and that the *steepness* of the response also grows with gain. As usual we pay a price in once characteristic when we improve some other.

The bottom image in Fig. 9S.17 shows a less striking contrast among the three filters' treatment of a triangle. The bottom image shows the Bessel doing a somewhat better job of preserving the triangle's shape, as advertised.

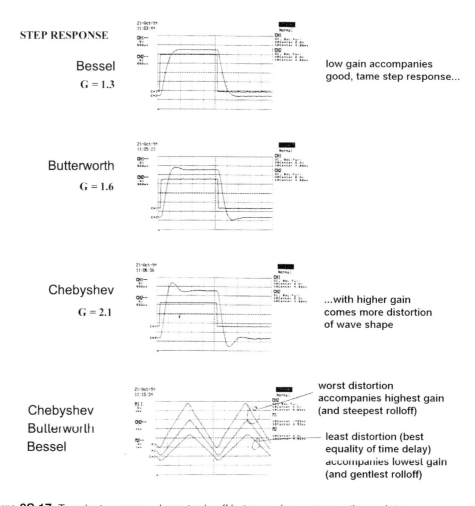

Figure 9S.17 Transient response shows trade-off between shape preservation and steepness.

9S.3 Noise: diagnosing fuzz

Who's to blame if your circuit output (and sometimes *input*, as well!) looks fuzzy? 'An oscillation,' probably – though not certainly. But what's oscillating? Usually, you can find out pretty easily. Sometimes, it's *not* easy!

9S.3.1 Preliminary question: is it a parasitic oscillation?

Is the oscillation "parasitic?" Not necessarily. *Something* is oscillating, but the fuzz may not be your circuit's fault. Consider a few other likely causes. It will help to know the frequency of the fuzz but, oddly enough, it may not be easy to judge at once whether the offending noise is at a frequency much *lower* or much *higher* than your signal's. You may, at a glance, know only that there's fuzz. Adjusting the scope's sweep rate should settle the question.

Then, if the frequency is much lower, usually the problem is 60 Hz pickup; if the frequency is very much higher, probably you're seeing pollution by radio or TV broadcast. In-between frequencies may indicate parasitic oscillations.

9S.3.2 One pollution: line noise

An example: The lower trace in the scope image of Fig. 9S.18, shows a sinewave at around 1 kHz, with a displeasing fuzziness or thickening that the focus knob would not fix. What's wrong?

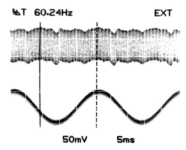

Figure 9S.18 Stable low-frequency bumps on your higher-frequency signal waveform indicate *line* noise. (Sweep rates: top trace, 5 ms/div, bottom trace, 0.2 ms/div.)

It could be either high-frequency or low-frequency noise. "Line noise" at 50 or 60 Hz is a troublemaker whose work is easy to spot: try triggering the scope on *line*, and drop the scope's sweep rate down to a few ms/div.[9] That's what we did to get the top trace of Fig. 9S.18. The wobbles stabilized for us. When this occurs, the effect will be evident at a glance. In addition, the waveform in the image repeats at 60 Hz, as the cursors indicate (along with the $1/\Delta T$ calculation displayed above the upper trace). That's not surprising, but does tell us a little more than we can learn from the fact that *line* triggering stabilizes the display. Sometimes *line* noise repeats at 120 Hz rather than 60 Hz, as we'll see in a moment.

Figure 9S.19 shows the circuit that produced these waveforms: we drove the high-frequency input signal through a 10M series resistor, then watched the op-amp input (as if we were trying to measure R_{in}). The measurement point is driven very flimsily, and thus is especially vulnerable to noise.

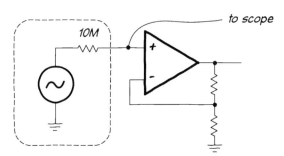

Figure 9S.19 High-impedance drive at op-amp input makes that point vulnerable to noise pickup.

---

[9] We expect that by now you have picked up this piece of jargon meaning "noise resulting from the 60 Hz oscillation of the power line voltage" (50 Hz in much of the rest of the world).

9S.3 Noise: diagnosing fuzz

The troubleshooting point, reiterated: If you see stable bumps when triggering on *line*, then your bumps are *synchronized* with the 60 Hz line power supply, and almost surely they are *caused* by that source.

- If the bumps repeat at 60 Hz (or 120 Hz, the repetition rate of the full-wave-rectified waveform put out by the power supply), still there is more than more one likely cause:
  - your circuit may be picking up the 60 Hz, at some *high-impedance* input (we just saw this in the circuit of Fig. 9S.19, and it's the more common cause);
  - alternatively, you may be looking at the effect of ground noise – a voltage difference between the grounds on your several instruments, caused by currents flowing in long ground "returns." Make sure that all your instruments are plugged into a single power strip or, anyway, a single power outlet. This "ground loop" problem is explained a bit further in the next subsection.
  - If the bumps are of *one polarity and repeating at 120 Hz*, you're probably looking at the effect of ground- and supply-current surges that occur when your power supply filter capacitors recharge, each half-cycle of the *line* input. This you cannot fix – except with the sidestep of finding a quieter supply, as we suggest in §9S.3.3.

9S.3.3 Another pollutant: ground noise from ground loop

Figure 9S.20 illustrates the second cause, not pickup but ground noise resulting from plugging instruments into differing outlets. The top trace in the figure shows a sinewave as it comes into a lab breadboard from a function generator. The trace looks sinusoidal, but fuzzy. Slowing the sweep and triggering on *line*, as usual, we see what the second trace shows. The bumps stabilize, so that this seems to resemble the *pickup* case.

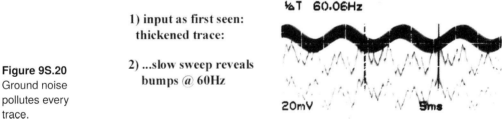

Figure 9S.20 Ground noise pollutes every trace.

1) input as first seen: thickened trace:

2) ...slow sweep reveals bumps @ 60Hz

But look at the circuit: Fig. 9S.21. The fuzz is not the fault of the generator, and it is not pickup. This time the impedance at the input of the circuit is *low*: not the 10Mz,Ω of the earlier case, but the 50 Ω of the function generator. So "pickup" is not likely.

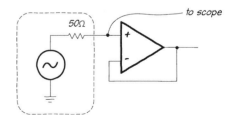

Figure 9S.21 Circuit showing bad 60 Hz noise at input: this one is not vulnerable to pickup.

Moving the oscilloscope power plug from a power strip a couple of feet down the bench into a power strip shared with function-generator and power supply solved the problem. The input now looks like trace 3 in Fig. 9S.22. (The sweep rates in Figs. 9S.20 and 9S.22 happen to differ.) The line noise is barely apparent in the latter.

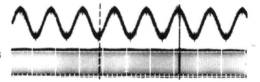

Figure 9S.22 Ground loop fixed: traces lose their 60 Hz noise.

The ugliness of Fig. 9S.20 resulted from the fact that two instruments, scope and function generator, by chance were plugged into separate room outlets. These separate outlets may not merge their third "world-ground" lines for a considerable distance. So, this is a ground loop at work, one formed by the long ground lines. The line-frequency "noise" signal is caused by a voltage drop as currents flow in the non-zero impedance of the ground lines. There are two likely causes of such currents. An AC line filter (of the sort present in many instruments) can cause as much as a milliamp or so to flow in the third line, the "world ground line" (this can occur when hot and neutral lines are linked by a capacitor to world ground, as they are in some filters; look up, for example, Corcom filters). With several such devices attached to the building's power lines, the flow of 60 Hz currents in ground lines could develop as much as volt or so of noise. A second possible cause could be noise currents that are induced by a 60 Hz magnetic field cutting the (building-size) ground loops. Such fields could arise from currents flowing in other power lines in the building, or from inductive devices such as motors and transformers; these are often present in industrial and commercial buildings.

Another ground loop instance: The scope image in Fig. 9S.23 shows the same ground loop behavior, but to more spectacular effect. "What is this mess?," you may find yourself exclaiming if your carefully built circuit evokes such a display.

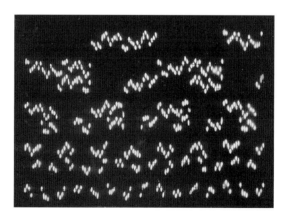

Figure 9S.23 Ground noise that can occur when instruments do not share a common ground definition: it can make a display unintelligible!

The image in Fig. 9S.23 is trying to show us a 4-trace "scope multiplexer" display (something you may build in a digital lab later in this course). The image in Fig. 9S.24 shows what it ought to look like. The display in this latter figure is correct: the traces are *chopped*, and this is what a 4-bit counter should look like. The hideous image that we showed in Fig. 9S.23 comes, once again, from a scope that is plugged into an outlet strip different from the one used for breadboard and function generator: pretty awful!

9S.3 Noise: diagnosing fuzz

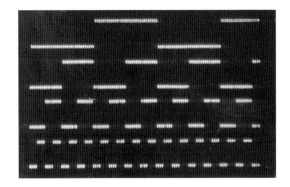

Figure 9S.24 4-channel scope display as it should look.

More commonplace ground noise: poorly implemented ground: Here's the *third* case – where the bumps are of *one polarity and repeating at 120 Hz*. The PB-503 breadboard has a flimsy ground that jumps during the supply-current surges that occur when its filter capacitors recharge, each half-cycle of the *line* input. For this problem, there's not much you can do, short of substituting a quieter power supply. Our PB-503 powered breadboards show a lot of this ground noise; a good lab power supply, like our Agilent units, shows no such defect. Lab 7L suggests such a change of supply in the discussion of the microphone amplifier circuit.

Figure 9S.25 illustrates this effect. The top trace shows what you're likely to see first: again a thickening or fuzziness of the output trace. (This time, the thickening is slight.) Trigger on line, slow the sweep rate, and the wiggles stabilize; then the 120 Hz repetition rate reveals the cause is not 60 Hz pickup, but the power-supply problem. The most revealing signal is the bottom trace: the ground line itself is bumpy! Time to reach for a better supply.

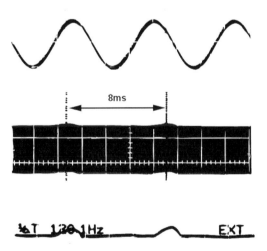

Here's the initial observation: a slightly thickened signal trace.

Now if you suspect the noise is line or power supply, trigger on "line" and slow the sweep rule. If the bumps stabilize, the supply is the cause.

A close look at the ground line itself shows the supply noise (caused by the current surges as the filter cap is charged on each half cycle of 60 Hz).

Figure 9S.25 Ground noise pollutes every waveform observed in the polluted circuit (sweep rate much higher in top trace than in lower two).

9S.3.4 Yet another pollutant: RF pickup

At the other frequency extreme, you may notice fuzz – usually low-amplitude – that you can resolve by sweeping the scope very fast. The lower trace in Fig. 9S.26 shows such indeterminate fuzz. If you can trigger on this fuzz and sweep fast, you can confirm that it's in the multi-MHz range. This we have done in the upper trace of the figure. You then can complete your confirmation of *radio* as the culprit by *turning off* your circuit and watching its input. If your circuit is not to blame, the input noise will persist. This test works for the low-frequency circuits that dominate in this course; it will not work if your circuit is capable of amplifying received RF noise. In that case, turning off power may turn off the observed noise, leaving you uncertain whether your circuit is receiving or causing the high-frequency oscillation.

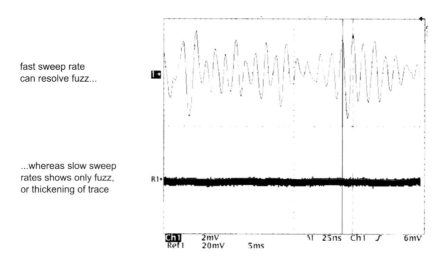

Figure 9S.26 Vague fuzz sometimes will resolve into high-frequency sine, if you sweep fast (top trace shows fast sweep, bottom trace slow sweep). Note differing scope gains on the two traces.

There's no easy remedy for this problem. In principle, what you want to do is shield your circuit – or at least minimize wire lengths at high-impedance points. Low impedance, too, makes your circuit less vulnerable to such pickup. It is useful, in any case, to know that the fuzz is not your circuit's fault: you won't now spend time chasing through your circuit to find causes for its misbehavior.

9S.3.5   True parasitic oscillations

If it is *an oscillation, what's oscillating?* Often the *frequency* of oscillation can settle this question.[10] Let's look at two examples.

Example 1: Discrete-transistor follower: Look back at Fig. 9L.2, which shows a case where there's only one suspect.[11] If the output trace is fuzzy, the fault is the transistor's; no detective work required, here, but this easier case may help us to solve the second case, in Fig. 9S.28. Figure 9S.27 shows what the follower's output looks like at two sweep rates.

Example 2: Op-amp and power push–pull: Here's a harder case: who's to blame for the fuzz at the output of the circuit in Fig. 9S.28? At a glance, looks as if either op-amp or transistors could be to blame. Let's look at the output of this circuit.

First, in Fig. 9S.29, there's the rather spectacular craziness that appears when we sweep fairly slowly – as we would if we were looking only for the audio output waveform.

---

[10] For this observation and for a good discussion of stability problems, we are indebted to T. Frederiksen, *Intuitive IC Op Amps*, p. 154ff. National Semiconductor Corporation (1984).

[11] Could the fuzz be radio pickup? Not likely, because the point of observation, the transistor's emitter, should present a very low impedance.

9S.3 Noise: diagnosing fuzz

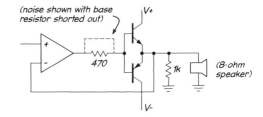

Figure 9S.27 Follower's fuzz resolves into RF frequency – but this time it is *not* pickup!

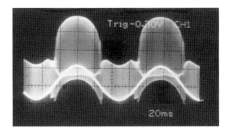

Figure 9S.28 A circuit whose oscillations could be blamed on either op-amp or transistors.

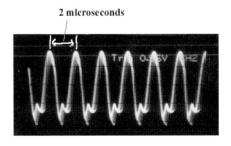

Figure 9S.29 Wild parasitic fuzz on push–pull output, driving speaker.

A faster sweep reveals the frequency of oscillation: see Fig. 9S.30. This oscillation frequency – a few hundred kilohertz (500 kHz) – is implausibly low for a transistor's parasitic oscillation, as in the follower that we saw a little earlier (in §9L.2). So, we blame the op-amp, and apply op-amp remedies (a shorter path that bypasses the push-pull at high frequencies: see Fig. 9L.8). Apparently it's the op-amp that's buzzing. We're in a frequency range where accidental lowpass elements in the feedback loop can cause phase shifts that approach their maximum values, and the op-amp still has gain. So, negative feedback has turned positive.

Figure 9S.30 Fast sweep reveals frequency of parasitic oscillation once more.

Why the standard remedy, decoupling, sometimes fails: Sometimes an oscillation gets *worse* when you add the usual small ceramic decoupling capacitor to the power supply. Probably, by bad luck, the capacitance that you just added formed a *resonant circuit* with the supply-line's stray inductance. Try a ceramic of 10× different value; try an electrolytic in parallel with the ceramic; or replace the ceramic with a tantalum cap. If all else fails, try a small series resistor between supply and cap –

428   Supplementary Notes: Op-Amps IV

though here you of course give up some of the nice low impedance you were after when you put the cap in. If the cap does form part of a resonant *LC*, this *R* – usually considered a vice in a capacitor – lowers the *Q* of the unwanted resonant circuit, quieting it.

9S.4   Annotated LF411 op-amp schematic

You very seldom *need* to know in detail what's going on within an op-amp, but it's fun and satisfying to look at a scary schematic, for example Fig. 9S.31, and realize that you can recognize familiar circuit elements.

If the sheer quantity of circuitry here didn't scare you off, you might recognize at least the following elements without our help: a differential amp at the input stage; a common emitter amp, next; a push–pull output – and a hall of mirrors (if you've dared to look at §5S.2). The current limit at the output mimics a trick you will see in voltage regulators (Chapter 11N).

Some of the details are subtle, though, so we've done what we can to explain through our annotations. We offer this schematic both as a reviewing device and as a reward for the hard work you put into discrete transistors. We hope that this exercise makes you feel knowledgeable – lets you feel that you are beginning to be able to read schematics that would have meant nothing to you just a few weeks ago.

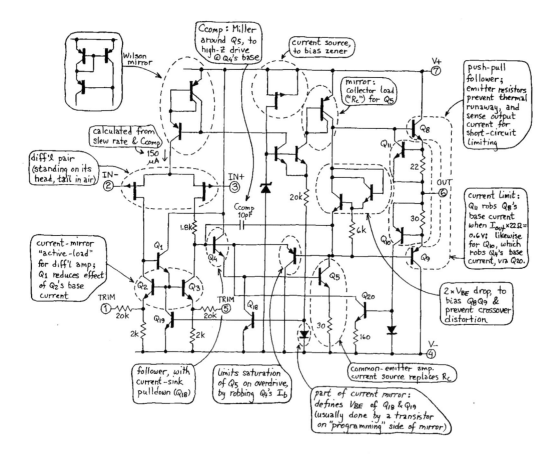

Figure 9S.31 Annotated schematic of LF411 op-amp.

9S.5 Quantitative effects of feedback

AoE §2.5.3

AoE §2.5.2

We will leave to Example 9W.1 detailed calculations of the way that *circuit* gain varies with *open-loop* gain. Here, we will settle for an intuitively appealing sketch of the way feedback does its magic. Figure 9S.32 is a generalized model of feedback.

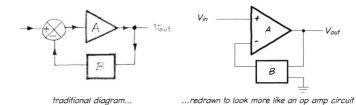

Figure 9S.32 Generalized feedback block diagram.

It is easier to understand if we redraw the *B* block to show a fraction of V_{out} fed back, as in the usual case. Figure 9S.33 probably looks pretty familiar.

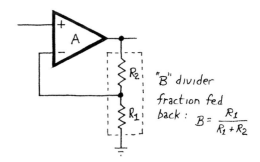

Figure 9S.33 Block digram redrawn to show *B* as voltage divider.

- B = *fraction fed back* (a characteristic of the circuit, not the chip); and
- A = *open-loop gain* (a characteristic of the chip).

And let's take note of the product AB, called *loop gain*. Since high values of both A and B enlarge the benefits of feedback, you'll not be surprised to find (in §9S.5.1 on the next page and later) that when we measure feedback's effects we will use that AB product.

These notions are enough to let us speak quantitatively of circuit characteristics that until now we could only describe with adjectives: "big," "small," "very good." These adjectives might describe R_{in}, R_{out}, and constancy of gain.

We will derive the expression for circuit *gain*, and try to show only qualitatively how feedback improves input and output impedances. Try R_{in} or R_{out} for the non-inverting amp: see Fig. 9S.34. (You saw such an argument for R_{out} in §6N.5.2.)

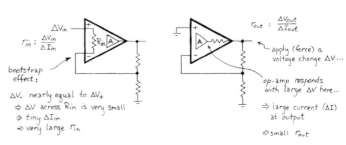

Figure 9S.34 Qualitative argument: how feedback improves R_{in} and R_{out} (non-inverting amp).

9S.5.1 R_{in}

R_{in} is *bootstrapped*: wiggle V_{in} and the inverting terminal wiggles about the same way; so, ΔV across the input is very small, and so, therefore, is ΔI. Differential R_{in} is improved by the factor $(1 + AB)$, it turns out. This is bootstrapping, an effect we mentioned back in §9S.2.1 when discussing the VCVS filter. The argument is that if the far side of some component moves as the near side does (being pulled up "by its bootstraps"), then the component disappears electrically. But this argument becomes silly when R_{in} is 10^{12}, as it is for the '411: then C_{in} dominates the input impedance. Unfortunately, most of C_{in} is *not* bootstrapped.

AoE §§2.5.3B, 2.4.3

9S.5.2 R_{out}

The argument for R_{out} is similar: wiggle the output (apply a little ΔV); the amp responds by moving point X a lot, putting a large voltage across its $R_{out\text{-}chip}$. But this sources or sinks a large current, I_{out}. No doubt you can see the nice result that follows – magically-low effective R_{out}. Using A and B we can now make this argument quantitative, and we will do that in §9S.5.4.

AoE §2.5.3C

9S.5.3 Gain

We have said that op-amp circuit gain turns out to be $1/B$, where B is the fraction fed back. Feeding back 1/10 in a non-inverting amp evokes a gain of 10, for example.[12] Let's justify that result, using just our definitions of A and B.

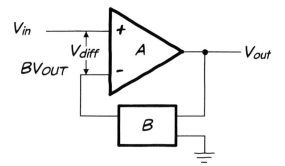

Figure 9S.35 Getting a general expression for circuit gain.

By "gain" we mean v_{out}/v_{in}.[13] We know that the op-amp is a differential amplifier that amplifies v_{diff} by its open-loop gain, A. Let's just look at how it amplifies that difference, and what feedback does to its gain:

$$v_{out} = A \times (v_{in} - Bv_{out}) = Av_{in} - ABv_{out}$$
$$v_{out}(1 + AB) = Av_{in}$$
$$\text{Gain} = \frac{v_{out}}{v_{in}} = \frac{A}{(1 + AB)} = \frac{1}{(1/A + B)}$$

When A is large – and $AB \gg 1$ – then Gain $\approx 1/B$. This is the usual case to which we are accustomed

---

[12] You may or may not recall, from a time when you were younger – back in Chapter 6 – that this evoked behavior is what we called the op-amp's ability to generate "inverse dog." The "dog" was the transfer characteristic of whatever was put inside the feedback loop.

[13] The lower-case v, describing the amplifier's "small-signal" response, is equivalent to what we more usually write: $G = \Delta V_{out}/\Delta V_{in}$. We need not make much of the difference between V and v or ΔV here, because this is a DC amplifier in any case. Thus it is unlike the discrete-transistor amplifiers that have seen, which were AC-coupled. Only ΔV (properly also called v) interested us when we worked with those.

– and it was this constancy of gain that interested Black most of all as he began his thinking about using feedback to stabilize telephone "repeater" amplifiers. Of course, A shrinks with frequency, and as it does the wonderful effects of feedback shrink too.

9S.5.4   A quantitative example

Try R_{in} or R_{out} for the non-inverting amp, assuming the following characteristics.

Assumed op-amp specifications:

- op-amp's open-loop R_{in} (differential, between its two inputs) = 1 MΩ;
- op-amp's open-loop R_{out} = 100 Ω;
- the amplifier circuit has nominal closed-loop gain of 10; and
- the chip's open-loop gain is 1000 at 4 kHz.[14]

R_{in} and R_{out} are each *improved* by the factor $(1 + AB)$. (This argument is strongly reminiscent of the argument that a bipolar transistor improves what's on the far side of it by the factor $(1 + \beta)$ this is what we called the "rose-colored lens" effect.)

How large is this factor $(1 + AB)$? The term A is given as 1000; and B is the fraction of the output swing that is fed back: here, it is divided by 10 (that is, B=0.1; this is, of course, how the amp achieves its gain of 10). So, the product AB is 100. Therefore the hardware characteristics of the chip are improved by that factor:

R_{in} (differential) is boosted from 1 MΩ to 100 MΩ

R_{out} is *reduced* from 100 Ω to 1 Ω

Notice that these effects are not quite the same as the effect of the transistor's β, which operated on whatever impedance was on the far side of the transistor; the op-amp's $(1+AB)$ operates on what's inside the feedback loop: one does not look *through* the op-amp.[15]

Important points: The virtues of a feedback circuit depend on both A and B: a circuit works best where AB, the loop gain, is large. The quantity A falls with frequency; B is greatest in a follower, least when you look for very high gain (electronic justice, once again: if you want lots of circuit gain, you'll give up some benefits of feedback).

A refinement that carries good news: The gain rolloff imposed on "compensated" op-amps makes them behave open loop like integrators, above a few tens of hertz. This notion is familiar to you. That phase shift, $-90°$, looks at first as if it would do nasty things to our gain calculations. We recited a rule saying that gain for a finite-gain op-amp is $A/(1 + AB)$, and we used this formula to calculate effective circuit gain in §9S.5.4. What happens to this calculation if we complicate it by admitting the truth that for almost all frequencies the open-loop gain, A, lags the input by 90°? In complex or phasor notation that means describing the open-loop gain as $-j|A|$.

At first glance, this certainly seems like a nasty complication – and a disappointment in this course, where in general we have been assuring you that phase shifts usually don't concern us. Already, we have carved a big exception to that reassurance: phase shifts must concern us when we consider the

---

[14] This is true for the LF411.
[15] A partial exception, perhaps, to this rule occurs for the inverting configuration, which makes R_{source} relevant to the calculation of B, the fraction fed back.

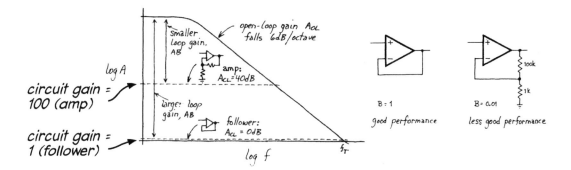

Figure 9S.36 *A* falls with frequency; *B* is highest for follower.

stability of feedback circuits. So far, that has given us challenges. Now, for a change, the phase shift is going to pay us a surprising dividend, *improving* the gain of a finite-gain op-amp.

AoE makes this point by proposing a hypothetical case, which we will restate here, spelling out the steps in AoE's calculation.

A hypothetical case: gain-of-10 circuit, limited *A*: *A* and *B* are similar to those stated in the example of §9S.5.4 on the previous page. But this time, *A* is less: $A = 100$, $B = 0.1$.

AoE §2.5.4B

Circuit gain, ignoring phase shift: Circuit gain: $G = \frac{A}{1+AB} = \frac{100}{1+10} = 9.1$.

This is a 9% error: disappointing; given the Golden Rules, we hoped for something better. It turns out that we are entitled to such hopes.

Circuit gain, acknowledging phase shift: Now let's accept the ugliness of complex notation, in hopes of finding better gain – something closer to the idealized view of op-amp behavior expressed by the Golden Rules. Let's write the open-loop gain not as "100" but as "$-100j$," and then let's see what results.

Circuit gain: $G = \frac{A}{1+AB} = \frac{-100j}{(1+(-100j)\times(0.1))} = \frac{-100j}{1-10j}$.

So far, this looks a lot like the no-shift calculation. But now let's allow the $-j$ to do its work.

To find the magnitude of this gain, G, we need two steps:

1. Clean up the denominator, getting rid of the j, by multiplying numerator and denominator by the complex conjugate of the denominator, $1 + 10j$:

$$G = \frac{-100j}{1-10j} \times \frac{1+10j}{1+10j} = \frac{-100j + 1000}{1+100} = \frac{-100j + 1000}{101} = -0.99j + 9.9$$

2. Given the complex result, $a + bj$, find its magnitude, $|G| = \sqrt{a^2 + b^2}$:

$$|G| = \sqrt{9.9^2 + 0.99^2} = \sqrt{98.01 + 0.98} = \sqrt{98.99} = 9.95$$

This error is just 0.5% – considerably better than the way things looked when we ignored *phase*.

9W Worked Examples: Op-Amps IV

9W.1 What all that op-amp gain does for us

We use the Golden Rules to calculate *gain* if, say, we feed back one part in 100. The Golden Rules rely on an assumption that op-amp gain is very high (because, in Black's words, "...improvements are attained in proportion to the sacrifice that is made in amplifier gain..."). Assuming enormous gain, as we do when we rely on the Golden Rules, we say the gain of the circuit in Fig. 9W.1 is 100.

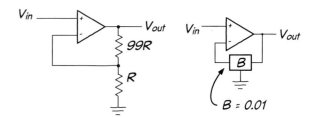

Figure 9W.1 Feed back 1%.

Let's see what happens as we assume several less-than-ideal values for op-amp *open-loop* gain, A, in the circuit from Fig. 9W.1.

Problems: circuit gain at several frequencies

9W.1.1 Assume no phase shift in A

First, let's assume that the op-amp's open-loop gain A shows *no phase shift*.

Problem What is the circuit's gain for:

- $A = 100$?
- $A = 1000$?
- $A = 100{,}000$?

Solution In general, as you know, $G = A/(1 + AB)$. Here, $B = 0.01$, so, for the several values of A the values of the product AB are 1, 10 and 1000.

At the highest gain of 100,000, we get $G = 100\text{k}/(1 + 1000) = 100$ (to within 0.1%).

At $A = 1000$, we get $G = 1000/(1 + 10) = 91$.

At $A = 100$, we get $G = 100/(1 + 1) = 50$. This sounds very sad. In a minute, we'll see that this overstates the degradation of gain.

9W.1.2 Two cases contrasted: no phase shift versus $-90°$ phase shift in A

Now we will ask similar questions, but this time trying two contrasting assumptions about the op-amp's open-loop gain. First we will assume no phase shift, as in the preceding question, 9W.1.1. Then we will assume, more accurately, that A includes a 90° lag, as it does above a few tens of hertz. At first glance this phase lag sounds like bad news. But it actually improves results for circuit gain. It turns out that the results of §9W.1.1 are gloomier than real life.

Problem Design a non-inverting amp with gain of 20 (using the Golden Rules). Assume an ideal op-amp as usual. Calculate gains for finite A.

Solution The circuit feeds back one part in 20:

$B = 0.05$.

Problem Calculate gains assuming no phase shift.
What is circuit gain if op-amp $A=100$?

Solution *Assuming no phase shift.*
We assume this time that $A=100$.

> This hypothetical gain is lower than what we usually have seen in our lab circuits. For the '411 op-amp this would be its open-loop gain at 40 kHz. We set A low in this example in order to emphasize the helpful effect of the op-amp's phase shift.

The product $AB = 5$.

$$G = \frac{A}{(1+AB)} = \frac{100}{(1+5)} = 16$$

This is 20% below the hoped-for "Golden Rule" gain.

Problem Calculate gains assuming $-90°$ phase shift.
What is circuit gain if we acknowledge the phase lag, and put the open-loop gain $A = -100j$?

Solution *Assuming $-90°$ phase shift.* Now we write the open-loop gain in the complex notation that indicates the lagging phase shift: $A = -100j$.

$$G = \frac{A}{(1+AB)} = \frac{-100j}{(1-5j)}$$

We did a similar calculation in §9S.3. Here, we can find the magnitude of this complex quantity using a shortcut: the gain is the ratio of *magnitudes* $V_{\text{out}}/V_{\text{in}}$.[1] Then $G = 100/\sqrt{(1+25)} = 19.6$. Much better: just 2% below the Golden Rule gain. Taking account of the phase shift tends to restore a person's faith in the Golden Rules, does it not?

9W.2 Stability questions

Problem (Three marginally-stable circuits) Figure 9W.2 shows three circuits that may buzz, either on transitions or continuously, because of parasitic oscillations. For each, we'd like to hear a brief explanation of what circuit elements are likely to cause trouble, and we'd like you to show the most important remedies.

For the third of the circuits – the push–pull follower within the feedback loop – assume an LF411 op-amp, and consider two hypothetical frequencies for the parasitic oscillations: these frequencies may suggest alternative explanations for the trouble. Let us suppose two cases showing two widely differing frequencies: $f_{\text{parasitic}}$ at 100 kHz as one case, 20 MHz as the other.

---

[1] This is because $\left|\frac{X}{Y}\right| = \frac{|X|}{|Y|}$.

9W.2 Stability questions

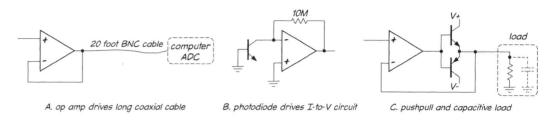

A. op amp drives long coaxial cable   B. photodiode drives I-to-V circuit   C. pushpull and capacitive load

Figure 9W.2 Three marginally stable circuits.

Solution

Why it's likely to buzz: Incidentally, as we acknowledge in Chapter 9N, such a long cable would need to be treated as a "transmission line" in order to work properly at high frequencies (above a few MHz) or with high edge-rates. Here, we will assume that our frequencies are low enough so that we can neglect such effects – as the frequencies in our lab circuits nearly always are.

Figure 9W.3 Op-amp drives long coaxial cable.

The capacitance of the coax cable, about 30 pF/foot for 50 Ω cable, forms a lowpass filter when this stray capacitance is driven by the non-zero R_{out} of the op-amp – an impedance that in fact is even more troublesome than it seems when we call it "R_{out}." The lowpass can add a lagging phase shift approaching 90° (see Chapter 9N). Added to the op-amp's built-in 90°+ lag, that shift can turn negative feedback positive.

The op-amp output impedance actually looks *inductive* because the value of R_{out} rises with frequency, as the op-amp's open-loop gain falls, diminishing the benign effects of feedback.[2] This inductive behavior brings dangerous 90°-lagging phase shifts into the frequency range where the op-amp still has substantial gain.

Remedies: long cable:

Simplest Remedy: A small series resistor (say, 100 Ω) calms this circuit by putting the phase-shifted point *outside* the feedback loop: see Fig. 9W.4. The price of this improvement, however, is a much-enlarged output impedance for the circuit (100 Ω or so, in this case), whereas one of the great virtues of a follower should be its low Z_{out}.

Figure 9W.4 Simple stabilizing remedy for op-amp driving a long coaxial cable: series resistor.

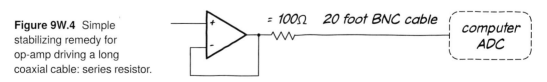

---

[2] See §9S.1.

Split-feedback Remedy: A "split" feedback network can give stability without degrading R_out: see Fig. 9W.5. The frequency at which the signal "crosses over" from the resistive to the capacitive path is determined by the RC value of R_split and C_split.

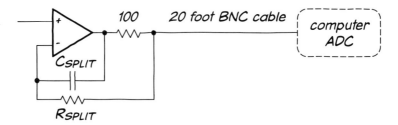

Figure 9W.5 Improved stabilizing remedy for op-amp driving a long coaxial cable: two feedback paths, for high versus low frequencies.

If for example, we saw a parasitic oscillation at 1 MHz, we might put $f_\text{crossover}$ around 100 kHz.

$$\Rightarrow RC_\text{crossover} = 1/(2\pi \cdot f_\text{crossover})$$
$$\approx 1/(6 \cdot 0.1 \times 10^6) = 1.6 \times 10^{-6}$$

Many RC combinations would do, such as $C=100$ pF, $R=15$k.

At high frequencies, the feedback network reduces to that in Fig. 9W.4. At low frequencies, feedback hides the 100 Ω resistor, giving the low output impedance that one looks for in a follower.

Higher gain (a cheap trick remedy): giving the circuit some gain mean reducing the fraction of the output signal that is fed back. That may be enough to stabilize the circuit. Figure 9N.18 illustrates the increasing *degree* of stability that results from increasing gain in this circuit.

Special op-amp (an even cheaper trick – though it may cost more): A heavy capacitive load can be made manageable also by simply buying the unusual op-amp that is designed for such a load. These devices permit the output loading simply to slow down their output changes. See §9N.5 and AoE §4.6.2.

Op-amp I-to-V converter driven by photodiode: Sketching in the stray capacitance at the "virtual ground" (see Fig. 9W.6) reminds us that the large feedback resistor meeting a current source (the photodiode) forms a lowpass within the feedback loop. The effective R is 10M, driving this small stray capacitance (perhaps 10 pF: the sum of the effects of the op-amp input, the photodiode, and the circuit layout). There is no voltage source at the input – as there would be in, say a ×100 inverting amplifier. Such an amplifier, in contrast to this circuit, would drop the value of the effective R that drives the stray C. The amplifier would use an input resistor of 10k. This low R then would drive C_stray. In the photodiode circuit, the 10M driving even just 5 pF puts f_3dB for this accidental lowpass at 3 kHz: low enough so that oscillations are likely.

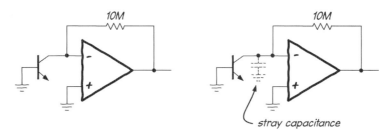

Figure 9W.6 Stray C makes photodiode circuit vulnerable to oscillation.

9W.2 Stability questions

Remedy, photodiode circuit: feedback capacitor: A small C to parallel the large feedback R tames the circuit: see Fig. 9W.7. Even a tiny C will be effective. $C_{stab}=1$ pF would dominate the feedback network (with effects noted just below) above about 16 kHz.

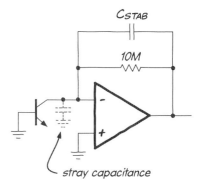

Figure 9W.7 A feedback C can quiet the photodiode amplifier.

The effect of the feedback capacitor could be described in a couple of ways.

- For the first, how this additional C alters the signal that is fed back. C_{stab} forms a C–C divider, at frequencies where $X_{C(stab)}$ is much less than the 10M of $R_{feedback}$. This C–C divider imposes *no* phase shift, so leads to no stability problem.
- A second way is to consider circuit gain: without the feedback C, the circuit shows gain rising with frequency until this rise collides with the falling-gain curve of the op-amp itself. This condition brings instability, just as it did in the simplest op-amp differentiator (see Chapter 9N, and AoE §4.5.7).

    The feedback capacitor limits the circuit gain to a constant value, above frequencies where $X_{C(stab)}$ dominates the feedback network. This condition is stable.

Op-amp drives push–pull and capacitive load: Capacitive loading is the principal problem, once again, but the presence of the push–pull makes the *follower* another suspect – as we saw with the single-transistor follower of §9L.2.

The problem statement proposes two hypothetical cases, with parasitic oscillations at differing frequencies: $f_{parasitic}$ is 100 kHz or 20 MHz.

The significance of the frequency of parasitic oscillation: The frequency of the parasitic oscillation can give a strong – and even conclusive – clue, if we're wondering whether to blame op-amp or discrete transistors.

A 20 MHz parasitic cannot be blamed on a '411 op-amp. The '411's open-loop gain falls to unity at 4 MHz.

A 100 kHz parasitic is not at all likely to arise from the discrete transistor circuit, since we rely on the impedance of stray inductance to account for voltage gain in this configuration (see §9N.7). At this low frequency, the impedance of such inductance is about 1000 times lower than at the frequencies we observed in the discrete-follower exercise from §9L.2. The op-amp is much more likely the cause of the 100 kHz oscillation.

We will tailor the remedies to the likely cause of the trouble.

Remedy, push–pull: at 100 kHz ***apply op-amp remedies:*** The most effective op-amp remedy – once we have decoupled the power supplies, because we'd be embarrassed to be caught forgetting that standard precaution – probably is the "split" feedback strategy that works also on the problem of the long coaxial cable.

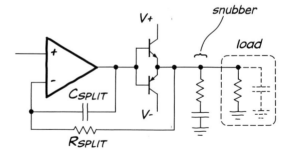

Figure 9W.8 Push–pull stabilizers, if op-amp is to blame.

The R and C values in the split-feedback network define the crossover frequency, as you know (we applied this in the case of a long cable on page 435). We could put the crossover at about 50 kHz with the values shown in Fig. 9N.20: R=27k, C=100 pF.

The "snubber" values are quite standard and not critical, reflecting a conventional guess at the values that will so load the circuit as to calm it. We'll use 10 Ω and 0.1 μF as usual. Not very interesting, but sometimes helpful.

Remedy, push–pull: at 20 MHz **apply discrete transistor remedies:** Here, the strategy is to spoil the high-frequency gain, and for this purpose an R_{base} of a few hundred ohms is effective. We hope you demonstrated this remedy to yourself in the Lab, see §9L.4.

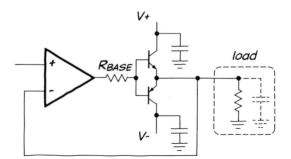

Figure 9W.9 Push–pull stabilizers if discrete-transistors are to blame.

There is no harm, of course, in applying both sets of remedies, in advance. Certainly we ought always to decouple power supplies and keep supply and ground paths short and of low impedance. Including a base resistor on a follower also should now be habitual, too. We separated the two sorts of remedies in this exercise: those appropriate to op-amp oscillations and those appropriate to discrete-transistor parasitics. There is no reason to be so rigid in applying protections to forestall parasitics. We might as well be generous with our protections.

10N Op-Amps V: PID Motor Control Loop and Lock-in Amplifier

Contents

10N.1	**PID controller**	**439**
	10N.1.1 Examples of real problems that call for this remedy	440
	10N.1.2 The PID motor control loop	440
	10N.1.3 Designing the controller (custom op-amp)	442
	10N.1.4 P: Proportional-only circuit: predicting how much gain the loop can tolerate	444
	10N.1.5 Derivative, D	446
	10N.1.6 Integral	452
10N.2	**Lock-in amplifier**	**453**
	10N.2.1 The lock-in amplifier: the basic strategy	453
	10N.2.2 Modulation	453
	10N.2.3 Effects of multiplying	456
	10N.2.4 A non-mathematical explanation for noise cancellation	457
10N.3	**AoE reading**	**457**

This chapter looks at two complex applications of operational amplifiers, the PID (Proportional, Integral, Derivative) control loop and the Lock-in Amplifier. Each application could a fill chapter in itself and we don't expect you to get through the notes and labs in a single session. Nevertheless, both applications provide an opportunity to apply all that you have learned in the book up to now to solve difficult real world problems: the first to tame feedback loops with built-in 90 degree phase shift making compensation difficult and the second to pull tiny signals out of overwhelming noise. Let's start by taking a look at the PID control loop...

10N.1 PID controller

Why? The problem PID sets out to solve

The problem is a familiar one: how to keep a feedback loop stable, despite lagging phase shifts within the loop. We saw a collection of troublesome circuits in Lab 9L. We found that even the lag introduced by a simple lowpass filter could upset a feedback loop. To stabilize such circuits, we learned several techniques:

- cut the amplifier's high-frequency gain – say, by paralleling the feedback path with a small capacitor;
- "split" the feedback paths, so that at the troublesome high frequencies feedback bypasses the thing that causes a lag in the loop; and
- enhance negative feedback relative to positive, as in the discrete follower.

Op-Amps V: PID Motor Control Loop and Lock-in Amplifier

The novelties in today's challenge are two-fold:

- this time, it is not just a lowpass filter but an *integrator* that is placed within the feedback loop;
- and, given this difficulty – for which none of our earlier remedies was sufficient – we need to do more than tinker with what's in the loop: *we need to modify the control amplifier itself.*

This sounds pretty radical, and it is. We don't plan to crack open an integrated IC, of course. Instead, we will put together a circuit that mimics an op-amp, but with characteristics that we can control.

10N.1.1 Examples of real problems that call for this remedy

The circuit that you will put together in Lab 10L is called a "PID" loop, where PID stands for Proportional, Integral, Derivative. These names describe the three functions of loop *error* – difference between input voltage and the signal fed back – that are provided by the PID's homebrew "op-amp." If this is puzzling, hang on; we'll be spelling out all three elements in the pages that follow.

PID applications PID loops are used in a variety of settings where something rather sluggish is to be controlled. An automatic elevator, for example, is designed to stop with elevator floor lined up with floor level – and without much overshoot; a car's "cruise control" keeps speed nearly constant, without much oscillation about the target speed; industrial processes such as chemical mixing or heating use PIDs;[1] fly-by-wire airplanes use electronics rather than cables to link cockpit controls with tail and wing surfaces, and responses must be fast, but stable.

In the first half of Lab 10L we mimic what might be a fly-by-wire control: we try to make a DC motor's shaft position match the shaft position of a potentiometer turned by hand.

The Wien bridge recalls criteria for oscillation, and for stability: As you know well, having grappled in Lab 9L with circuits that wanted to buzz, lagging phase shifts within the loop cause trouble. Rather than drum again at the points made in that lab, let's get at the criteria for stability through the back door: by recalling how we had to design the Wien Bridge oscillator in order to get a sustained sinusoid from the circuit.

The Wien bridge sustained an oscillation if net gain on a trip around the loop was just *unity*. If gain was less than one, the oscillation died away. If gain was more than one, the circuit oscillated – but also clipped its output, spoiling the sinusoid.

Similarly, in any feedback circuit, an oscillation will die away if loop gain is less than one at the frequency where phase shift hits 180 degrees.[2] In our encounters with op-amp *compensation*, we met one way to ensure stability: roll off the open-loop gain as frequency rises. Let's start with a primitive version of this remedy, since gain-cutting is the most familiar, among our standard remedies. This we will describe as limiting the "P," or Proportional, gain. Later, we will move on to fancier solutions.[3]

10N.1.2 The PID motor control loop

Potentiometer-to-motor loop: it looks very simple The task we undertake here looks easier than it is. Certainly the goal is modest. Fig. 10N.1 shows the scheme.

---

[1] The history of PID use in industry accounts for some of the strange language in this field. As you will see, for example, the thing controlled often is called "plant."

[2] Always −180 is the hazard, *not* +180; but inversion of the sense of feedback is the main point, so let's stress that by speaking of the deadly 180, rather than emphasize the sign, here.

[3] In this discussion, as in the lab, we will add these elements not quite in the order mentioned in the name PID: we will use P first, then D, finally I.

10N.1 PID controller

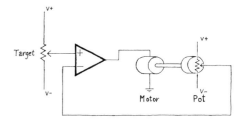

Figure 10N.1 Basic motor-position control loop: very simple!

What could be simpler? Not much, on paper. But the challenge turns out to lie in keeping the circuit *stable*. The issue is fundamentally similar to the one you met in Lab 9L when you noticed that a humble lowpass in an op-amp's feedback loop could turn negative feedback into *positive*. The problem there arose from the $-90°$ phase shift imposed by the op-amp itself (and even greater phase shift, close to its unity-gain frequency). That $-90°$ meant that we couldn't afford to insert much lag into the loop.

Today's circuit is harder. We are stuck with an *extra* $-90°$ shift, or integration – quite apart from the one ordinarily imposed by the op-amp itself. This additional integration forces us to alter our methods.

This time, the integration comes from the nature of the stuff we are putting inside the loop: a motor whose shaft position we are sensing. Since we cannot afford the phase shift of an ordinary op-amp, we must build ourselves a sort of *custom* op-amp: one that provides modest gain and no phase shift.

The integration within the loop The "integration" or *lag* inside the loop results from a phase-shift that is inherent in the design of this particular *load*. Specifically, the lag results from the fact that we aim to control the motor's *position*, whereas we drive the motor with an error signal that is a *voltage*.[4] An implicit integration results.

That follows because a DC motor spins at a rate proportional to the applied voltage;[5] this spin rate, persisting for a while, causes a change of position. Hence the unavoidable integration within the loop.

To make this last point graphically vivid, Fig. 10N.2 shows how the position pot responds to a *square-wave* input to the motor. The triangular output looks a lot like what you saw in the integrator of Lab 7L, doesn't it? – apart from the inversion inserted by the lab integrator.

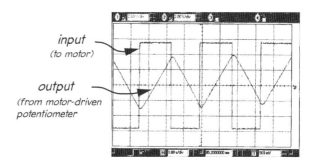

Figure 10N.2 Motor-drive to position-sensing potentiometer forms an *integrator*.

---

[4] This voltage is an amplified "error signal:" a measure of the deviation of circuit output from target. In the motor control circuit, both target and measured outputs are DC voltages.

[5] An easy way to convince yourself of this, if it's not already plausible enough, is to recall that if you spin the shaft of a DC *motor* with your fingers, the motor acts like a DC *generator*. It generates a voltage proportional to its spin rate. So, when a voltage drives a DC motor, the motor responds with a voltage – so-called "back EMF" ("electromotive force," or voltage). That back EMF almost equals the applied voltage; the spin rate levels off where the difference between applied voltage and back-EMF permits a current just sufficient to replace the energy the motor loses to friction, its mechanical loading, and to the motor's heating.

10N.1.3 Designing the controller (custom op-amp)

We need to go back toward first principles, as we design this control loop. We know that we need gain to get the benefits of negative feedback – but we suspect that we cannot get away with the usual large gain of a standard op-amp (we may need to "roll off" gain; or we may want the gain always low). And we know we cannot afford the phase lag of an ordinary op-amp – a lag of at least 90°. Can we design such an amplifier?

Yes. As we learned when we built an op-amp out of transistors, the device is essentially a differential amplifier. Everything else is, relatively, a detail. So let's start by building a differential amplifier that does *not* show an op-amp's −90° lag.

Figure 10N.3 shows a standard differential amplifier – the one you will use in the PID lab. It provides just *unity* gain, so we'll need to follow it with a gain stage.

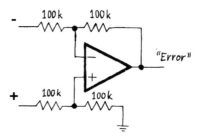

Figure 10N.3 Differential amp, unity gain.

No phase shift...: Perhaps you don't believe our claim that this differential amplifier can show no phase shift. How can it? – given that it was built with an ordinary op-amp, and we have been insisting that an op-amp includes a 90° lag. The lack of phase shift in the diff amp is rather amazing – though perhaps not to you who have built followers and other op-amp circuits that did not betray the integration hidden within the op-amp. But the claim is correct: the phase shift of the op-amp is just one more regrettable "dog" that feedback dutifully hides from us. Our PID loop will not see the op-amp's internal phase-shift; feedback has made that shift harmless.

...and modest gain: Even after we have eliminated our home-made op-amp's phase shift, the feedback loop will be vulnerable to instability, because of delays or lags that may arise in addition to delay caused by the motor-pot's *integration*. So we will give our amplifier much less gain than is provided by the usual device (like the '411, with its gain of 200,000 at DC). In addition, it will be convenient to make our amplifier's gain adjustable. See Fig. 10N.4 for a sketch of how the circuit would then look.

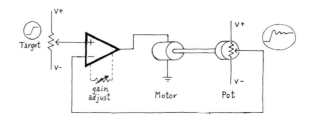

Figure 10N.4 Proportional-only drive: now we're ready to see how much gain our loop can tolerate.

Now we have a set up that should allows us to make a stable loop. But how are we to determine how much gain to apply? In the lab, our answer is entirely experimental: we increase the gain gradually until we see instability – and then we back off. In this section and the next we will try to describe what is happening as we approach that borderline of instability.

10N.1 PID controller

Aside for a conceptual diagram of the loop Here is a way to describe the loop that is more formal than what we usually like to impose on you. Still, you may be slightly interested to see the terms that control-theory people like to use. You'll notice, here, the curious use of the word "plant" to describe the thing that is controlled. (In our case, 'plant' is motor and potentiometer.) The word reflects the industrial beginnings of control theory, as we have said earlier.

The "controller" is our homemade control amplifier. If Fig. 10N.5 puts you off, don't lose heart; in a moment we will draw it to look more like the op-amp loops we are accustomed to. Figure 10N.6 is the same diagram redrawn in a more familiar form: the motor-pot "plant" is the usual "dog" within the loop.

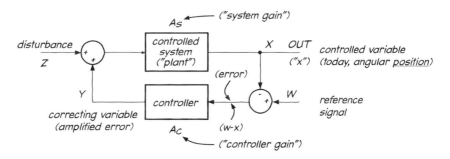

Figure 10N.5 Loop to control motor shaft position: block diagram in form used in class discussion.

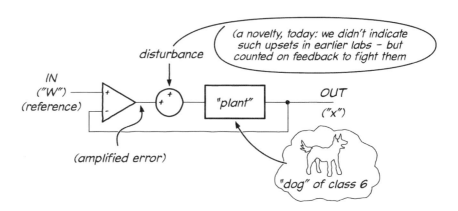

Figure 10N.6 Loop to control motor shaft position: block diagram in form familiar from op-amp discussions.

These two diagrams do not tell you anything you didn't know from a statement of the original problem. But the first diagram, particularly (Fig. 10N.5), may help you to see that the signal that travels around the loop passes through the "plant" and the "controller," and that stability will depend upon what a trip around the loop does to a signal – or to a disturbance.

Giving explicit attention to a "disturbance" is an incidental novelty, here. So is the naming of the transfer function or *gain* for each of the two blocks: *plant* or "system" gain is labeled A_S while *controller* gain is named A_C. This assignment of labels allows us to write a simple expression for the gain applied to anything that makes it around the loop ("signal" or "noise" or "disturbance"). This product, defined in equation 10N.1, is what we called "loop gain" in the usual op-amp setting. The loop gain here is:

$$A_{\text{loop}} = A_C \cdot A_S : \quad \text{controller gain times system gain} \tag{10N.1}$$

444 Op-Amps V: PID Motor Control Loop and Lock-in Amplifier

In this loop, we are stuck with A_S – the behavior of the "plant." The challenge, as we set up and trim the PID loop, will be to adjust judiciously the other gain, A_C – the *controller* gain.

We will first adjust the *magnitude* of A_C to maintain stability. Then, when we add the derivative D, we will be adjusting also the *phase* of A_C. It will turn out that phase and magnitude interact: proper phase behavior permits higher gain.

10N.1.4 P: proportional-only circuit: predicting how much gain the loop can tolerate

"Plant" frequency response (system gain: A_S) Figure 10N.7 plots the frequency response of a hypothetical motor-to-position element (this is A_S only; we're not yet looking at the "loop"). The plot is hypothetical, but quite close to what you're likely to see in the behavior of your PID lab.

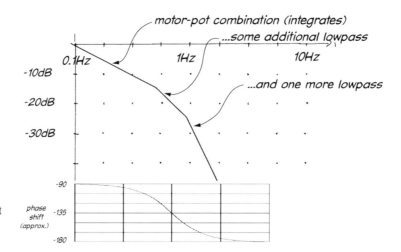

Figure 10N.7 Gain and phase response of motor-pot combination: –90° lag, from the lowest frequencies.

If this plot seems rather abstract, here's an attempt to say in words what the downward-sloping *gain* plot says about the motor-pot combination. Our words will try an informal time-domain explanation for the frequency-domain rolloff.

The downward-sloping plot indicates that for a given input amplitude (a sinusoid), at low frequencies the motor shaft will have time to turn quite a long way before the input signal reverses sign. At higher frequencies, it will turn only a little way in the available time (roughly a half period). If this sounds like what we say about the voltage on a capacitor in an *RC* circuit, when we consider it in time domain, that's appropriate. The processes are very similar.

You'll notice that the plot differs from the usual op-amp frequency response plot in that its rolloff begins, at the lowest frequencies, with a 90° degree lag. This is simply a consequence of the *voltage-drive/position-sense* mechanism we noted above; there is no phase-shift free region, as there is for an op-amp. But more important than this difference is the similarity to the op-amp plot: both plots allow us to anticipate where trouble could begin.

We can expect that we *may* get trouble where *plant* shift approaches 180°. This happens at the bend in the gain plot of Fig. 10N.7, where a *second* 90° lag is added to the integrator's constant –90°. When this second 90° kicks in, *negative* feedback may have been transmogrified into *positive*.

But the fact that the *plant* can provide this scary phase shift does not settle whether the circuit will oscillate. Whether this phase shift causes oscillation depends on whether we provide enough *controller* gain. Recall the Wien bridge design: in order to sustain an oscillation we needed gain sufficient to make up for attenuation in the positive feedback path.

In designing our PID loop, we will not go right to that edge of instability; but we need to know where that edge is. We want to keep our eye on that "corner" frequency, the boundary of danger.

Controller gain: A_C In Fig. 10N.8 we have added to the A_S plot that you just saw, two hypothetical values of A_C. The term A_C gain, which multiplies A_S, slides the A_S curve up, forming the loop gain $A_C \cdot A_S$ in the plot of Fig. 10N.8. In both cases, A_C is flat with frequency.

In one case, A_C is low enough so that loop gain falls to less than unity before the dangerous *corner* frequency. In the other case, where A_C is set higher (labeled $A_{C'}$), the loop gain exceeds unity at the corner frequency. This higher A_C would make the circuit oscillate.

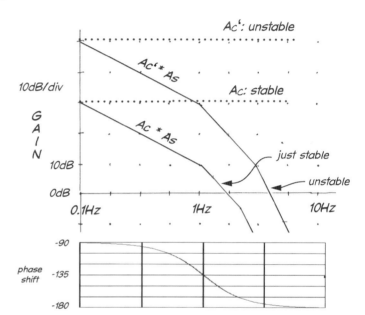

Figure 10N.8 Proportional loop: stable at modest P gain, unstable at a little more P gain.

In the lab circuit, you will discover the corner frequency by finding the frequency of "natural oscillation." This is PID jargon for the frequency where oscillation sets in, when the gain is just a little too high. You discover this frequency by gradually increasing A_C (here, just P gain), until oscillation begins. In our lab circuit it is just a few Hz: perhaps 3 Hz. In lab you will be able to restore stability by reducing the controller gain, A_C.

Degrees of stability: phase margin You will also find that stability is a matter of *degree*: as you adjust the gain upward, approaching the oscillation condition, the loop will become more and more jumpy. This jumpiness is easiest to see in response to a disturbance deliberately imposed.

In the lab, you can disturb the circuit in either of two ways. You can change the "target" input voltage abruptly (applying a *step* change either by giving a sharp twist to the pot or by applying a slow square wave from a function generator). An alternative (perhaps more fun) is to apply a step "disturbance:" use your fingers to turn the motor shaft away from its resting position and watch the circuit try to recover. When the loop is close to oscillation, the output will overshoot repeatedly, before settling to its final position.

Given a plot of A_S ("system" or "plant" frequency response, including phase shift) one can see just how far the loop phase shift lies from the deadly $-180°$, at a given A_C (controller gain). That distance, in degrees, is called the "phase margin." This is a notion you have seen before, in discussion of op-amp *frequency compensation*. Op-amps are "compensated" to provide between 45 and 60 degrees of margin.

Figure 10N.9 illustrates the response of a loop to a square wave, at several phase margins. In life, it may be more likely that you would run the process in reverse: observe the response and infer the phase margin.

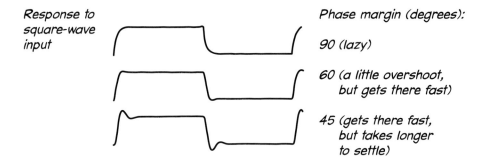

Figure 10N.9 Step response shows effect of various phase margins on stability. After Tietze, U. & Schenk, C., *Electronic Circuits: Handbook for Design*, Second edition, Springer (2008), p. 1105.

In the lab PID circuit, you will find that at lower A_C the circuit is, indeed, tamer, less disposed to overshoot. But low A_C also has some harmful effects: the loop response is quite slow, and the circuit tolerates a good deal of residual error.

Figure 10N.10 shows response for low P gain. The response looks like an *RC*: makes sense, since the motor drive diminishes as the *error* diminishes (while the position draws nearer to the target).

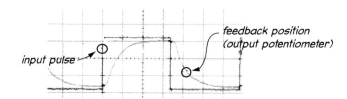

Figure 10N.10 Low P gain (four) keeps stability, but at a price.

We can increase P gain somewhat, and get better performance in some respects (smaller residual error, and quicker response). But at this higher P gain the circuit shows some overshoot: see Fig. 10N.11.

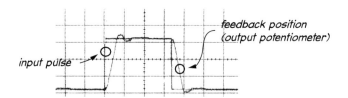

Figure 10N.11 Middling P gain (sixty) produces some overshoot.

And if we push P gain still higher, we go close to the edge of instability. In Fig. 10N.12, the downward step produces several cycles of ringing, and the upward step produces continuous ringing. With P gain that high, the circuit is not usable. But it can be made stable, even at this higher gain, if we add in some D, the derivative of the error signal.

10N.1.5 Derivative, D

Since it is the *lagging* phase shifts that cause the mischief, can we improve the loop's response by injecting some phase shift of the other flavor, *leading*? Yes, we can, if we do it carefully. Let's start by

10N.1 PID controller

Figure 10N.12 High P gain (eighty) produces some overshoot and ringing: barely stable.

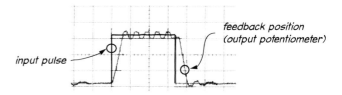

looking at what we would *like* to see as an effect of injecting such *derivative*, and then we'll look at how we might determine how much D to add.

Figure 10N.13 shows the original plot of loop gain, $A_C \cdot A_S$. This time we have added a dotted line to show how we'd like the corner to get straightened out by the A_C response.

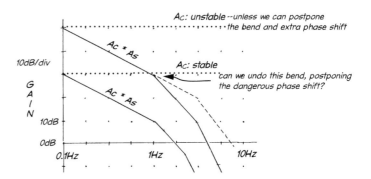

Figure 10N.13 What we'd like: to undo another phase lag.

In order to get that result – in the $A_C \cdot A_S$ product – we need to put an *upward* tilt into the controller gain, A_C. That is, we need to make the controller response look like the *derivative* beginning at the frequency where trouble otherwise would begin.

How do we get that result? We generate the derivative, D, of the error term, and add that to the P error, amplifying both to get a bent A_C curve rather than the flat ones of Fig. 10N.8. We will use the sum of these two error functions[6] to provide the correction signal, in our negative feedback loop.

To put the upward bend at the right frequency – the frequency where the system's corner occurs – we arrange things so that at this frequency the magnitude of D equals P. In Fig. 10N.14, D grows steadily with frequency, and when summed with P, it takes over at the corner frequency. Thus it jumps in just in time to save the day: in time to undo that second 90° lag. Adding D to this circuit has permitted us to use higher gain than we could manage with P alone, and the higher gain strengthens feedback, enhancing its usual benefits.

An intuitive explanation for what the derivative does for the loop If you find the frequency-domain account we have just given a bit abstract, consider the rough arguments made in this subsection, trying to make sense of the benefits that the D term provides.

The problem is easiest to get a grip on if one considers a loop that is quiet and settled – and suddenly driven by a step change at the input. That abrupt change of input makes the *error* function, P suddenly step from zero to gain × input-step. What does the *derivative* add? It adds a strong additional signal

---

[6] The curve in Fig. 10N.14 will look strange to you if you have not been looking at log plots recently. The derivative, D, seems to *take over* from the P rather than to *sum* with it. At lower frequencies, A_C looks like P alone, ignoring D; at higher frequencies, it looks like D alone.

Yes. But that makes sense if we recall that, on the log plot, as soon as D climbs noticeably above P its value is so much the larger of the two that their sum is virtually the same as D alone. So the lower frequencies are totally dominated by the P term, the higher frequencies totally dominated by D.

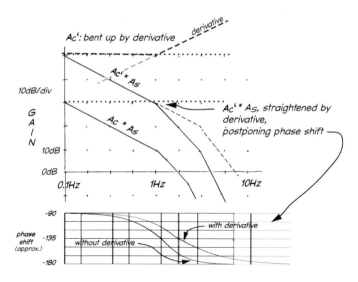

Figure 10N.14 D term added to P term makes stable a loop that otherwise would oscillate.

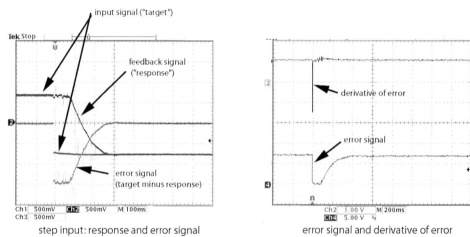

Figure 10N.15 Loop response to a step change of input: error and derivative of error. (Scope gains: 500 mV/div in left-hand image; 1 V/div for derivative trace, 5 V/div for error signal in right-hand image.)

in the same direction, because not only has the error appeared, but it has appeared rapidly, climbing abruptly from zero to its new value. Figure 10N.15 shows such waveforms.

The loop does not seem, at first, to respond at all; then it gradually drives the response to match the input, and the error is driven down toward zero once more. The right-hand image in Fig. 10N.15 shows the strong kick that the derivative gives, as the error steps. This kick tends to speed the response of the loop.

If we sweep the scope faster to see the step response in greater detail, as in Fig. 10N.16, we see the derivative in this case giving a kick several times stronger than the correction signal provided by the P error alone (note the different scope gains for the *error* versus *derivative* displays). The left-hand image in Fig. 10N.16 squeezes the large derivative signal on-screen by reducing the scope gain for that channel by a factor of five. The right-hand image uses a 1V/division scale for all channels, showing how strong the D signal is relative to P.

And a homely interpretation of what D does for the loop: The D term – the derivative of the error – is a little hard to grasp intuitively. Here's a try: suppose the problem that the loop means to

10N.1 PID controller

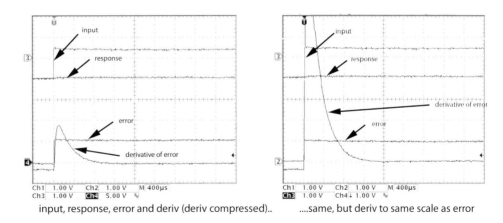

Figure 10N.16 Faster scope sweep, and D correction signal stronger than the bare P. (Scope gains: 1 V/div except derivative in left-hand image.)

solve is to move a car into a parking place that has a concrete wall at the front. Our controllers appear, in this drama, as two valet-parking attendants. The first is a P-only controller – a cautious little old man, well past retirement age. His strategy for getting the car parked and not hitting the wall, is simply to go very gently and slowly. He gets there, eventually – but as he sees the wall growing nearer (the *error* term shrinking), he drives more and more slowly, finally just inching along.

The other valet-parker is a cooler P-plus-D controller. He's a teenage punk, wearing a baseball hat backwards. He hops into the driver's seat, guns the car, squirts into the parking place and hits the brakes. He gets the car parked fast.

That may seem metaphoric nonsense. Figure 10N.17 shows waveforms from Fig. 10N.15 that we claim describe the same behavior, perhaps making the valet-parkers more plausible. The main effect is a strong boost at the start – hitting the gas hard when a new error appears. The lesser effect is a gentle braking near the end of the process. Perhaps the high D gain puzzles you, given that we have designed the loop to make D and P gains equal at the corner frequency. The D gain is high where the error suddenly steps because the steep edge of this error waveform contains high frequencies – as Fourier taught.

How to calculate the needed derivative gain How do we calculate the necessary differentiator gain? To avoid complications, let's assume that the gain of the P path is unity. Then our goal is to arrange things so that the derivative contribution, D, is equal to the P contribution at the frequency where trouble otherwise would occur. The D should keep the loop stable, until yet another lowpass cuts in; at that point, we should have arranged to make the loop gain safely low: less than unity, so that a disturbance must die away.

Let's start by reminding ourselves what we mean by differentiator "gain;" then we'll calculate what RC we need for stability. Gain for a differentiator, by definition is

$$\frac{V_{\text{out}}}{dV_{\text{in}}/dt}$$

We know that, for the op-amp differentiator,

$$V_{\text{out}} = I \times R_{\text{feedback}}$$

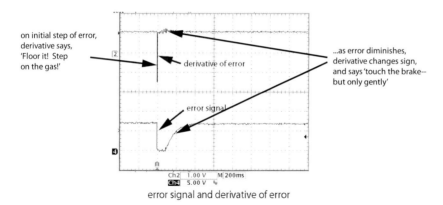

Figure 10N.17 Derivative effects likened to valet-parker's stepping on gas, then on brake. (Scope gains: error 5 V/div; derivative 1 V/div.)

and this I is just

$$C \times \frac{dV_{in}}{dt}$$

So (neglecting the sign of the gain; the op-amp version inverts)

$$\text{gain}_{(deriv)} = \frac{V_{out}}{dV_{in}/dt} = \frac{R \cdot C \cdot dV_{in}/dt}{dV_{in}/dt} = RC$$

So RC defines the differentiator's *gain*.[7]

If $V_{in} = A\sin(\omega t)$, then $V_{out_{deriv}} = \omega RCA \cos(\omega t)$. We want to find the value of RC, the differentiator's gain, that would set $V_{out_{deriv}}$ equal to $V_{out_{proportional}}$. Let's treat P gain as unity; then we want both P and D to equal V_{in}.

If we set the D gain equal to unity, then

$$V_{out_{deriv}} = V_{out_{proportional}} = V_{in}$$

and, equivalently,

$$\omega RCA \cos(\omega t) = A\sin(\omega t)$$

Consider just the maximum amplitudes of V_{in} and dV_{in}/dt (where sin and cos terms equal 1). We want to set these amplitudes equal to each other, and both are equal to the input amplitude, A. Then

$$\omega RCA = A \quad \text{hence} \quad \omega RC = 1, \quad \text{or} \quad RC = \frac{1}{\omega}$$

In other words,

$$RC = \frac{1}{2\pi f}$$

To paraphrase this equation in words: RC should be about 1/6 of the period of natural oscillation.

Figure 10N.18 is a scope image to show the beneficial effect of adding some D. Here the P gain is high enough so that if P were the only feedback the circuit would oscillate continuously. Now, with D added, the circuit gets where it's going fast, and does not overshoot.

---

[7] Perhaps you find this a rehearsal of the obvious! Just a look at units make it very plausible that RC should be the differentiator's gain: input is volts/second; output is volts; the conversion factor needs units of seconds.

10N.1 PID controller

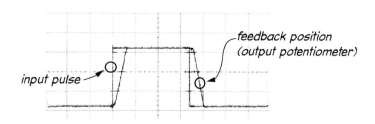

Figure 10N.18 Derivative can stabilize loop, and get you there faster (overall gain: 120; D resistor: 220k).

10N.1.5.1 But you don't want too much D

Too much D makes the circuit timid...: As you might expect, and can see in Fig. 10N.19, too much D brings trouble again: first, the circuit becomes timid. As it approaches its destination it loses its nerve, puts on the brakes too early.

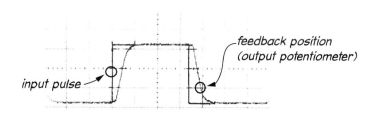

Figure 10N.19 Too much D makes the circuit timid, slow to get where it's headed (overall gain: 120; D resistor: 680k.

...still more D makes the circuit oscillate: More surprising, if you inject still more D, the circuit becomes unstable once again. Fig. 10N.20 tries to make an argument to show why this is so (phase shift plot shows exaggerated abrupt change of phase, misleading if you take this image too literally).

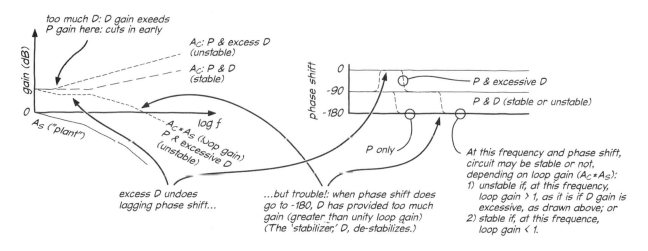

Figure 10N.20 Excessive D can bring on oscillations, again, by pushing total $A_C \times A_S$ gain too high.

We introduced the *derivative* of error as a stabilizing device. In Fig. 10N.14, D permitted overall gain that otherwise would have caused instability by undoing a lagging phase shift. That worked because controller gain, A_C, was low enough so that when multiplied by the "plant" gain, A_S, the loop gain had fallen below unity before we reached the deadly phase shift.

Where D gain is excessive, in contrast, as in Fig. 10N.20, D cuts in too early (phase shifts in this figure, incidentally, are shown as unrealistically abrupt).[8] The overall controller gain, A_C, thus is excessive, and as a result the loop gain is excessive as well. The result is an oscillation, as it was in the earlier case when we used P alone and high P gain made A_C excessive. Excessive loop gain brings trouble at the frequency where phase shift hits $-180°$, regardless of the cause of the excessive gain.

10N.1.6 Integral

Adding the last of the three error functions, an *integral*, I, of the error, confers benefits usually less important than those provided by adding D. It does not help stabilize the circuit. Instead, it diminishes long-term errors, driving these toward zero.

The effect of I is evident if we look at what it can do for a very-low P gain circuit – one that we saw before in Fig. 10N.10. On the left of Fig. 10N.21 is that low P-gain response. On the right is the same circuit with an integral of error added in.

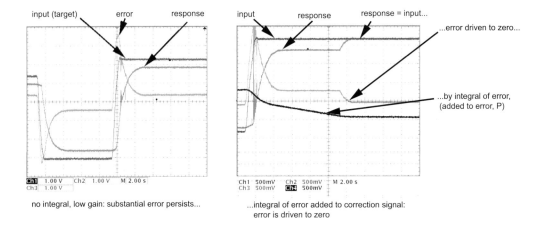

Figure 10N.21
Integral can drive long-term errors down, even at low gain. (Scope gains: 1 V/div in left-hand figure; 500 mV/div in right-hand figure.)

The integral of error drives the loop tight in against the target – but only after a delay of almost five seconds in the present example, shown in the right-hand image of Fig. 10N.21.[9]

Good question: how can the PID loop be stable, with I added? We have made much of the point that this loop includes an inherent 90° lag, so that adding another $-90°$ would make the circuit unstable. Yet, the *integral* term does insert another $-90°$ (lagging) signal. It does this without bringing instability because the integral's effect is long gone before the crucial point, where gain goes to unity.

Why it is OK to go to dangerous phase shift, as long as one returns before the unity-gain frequency, we leave "as an exercise for the reader."[10] Let us know, when you figure out a good intuitive explanation for this marvel![11]

... now on to the lock-in amplifier.

---

[8] "Cuts in early" is just another way to say "D gain is excessive," since D "cuts in" on the graph where D gain exceeds P gain.
[9] This delay is caused by "static friction" or "stiction," as noted in Lab 10L. The motor does not begin to move until its drive voltage reaches some minimum level.
[10] We do this, of course, because we're not sure of the right answer!
[11] Wise people tell us that it *is*, indeed, OK to go to 180° shift and beyond, as long as one returns to less than 180° at the unity-gain point. You will see this truth demonstrated again when you meet phase-locked loops in the digital part of this course. So far, intuition fails us here: we can't explain it because we can't really understand it.

10N.2 Lock-in amplifier

Today's lab asks you to design a lock-in amplifier, a circuit well adapted to pick out a signal from a noisy environment.[12]

10N.2.1 The lock-in amplifier: the basic strategy

The basic strategy is a two-step process:

- Multiply a signal of interest by a reference frequency; this process one can describe as "modulating" the signal of interest.[13] The motive for doing this multiplication can be described as moving the signal away from DC and very low frequencies, where noise is likely to be the most troublesome. The very simplest example of such noise might be the optical noise presented by ambient light when the signal is a weak light source, as it is in today's lab circuit. Subtler forms of low-frequency noise include $1/f$ noise.[14]
- This multiplication produces a waveform with signal information now residing in "sidebands" at $f_{\text{reference}} \pm f_{\text{signal}}$.[15]

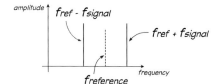

Figure 10N.22 Double sidebands, produced by multiplying reference by signal ("modulating" the signal).

- Stage two: to recover the signal, multiply the modulated reference by a synchronized reference. This will extract the original signal. Ambient (DC) and low-frequency noise, and any noise not very close to the reference frequency, can be cancelled out.

This may sound rather hocus-pocus. We'll try to spell out how the process works.

10N.2.2 Modulation

In today's lab we will try to detect an illuminated LED, despite the presence of other optical signals: lots of ambient light, and some time-varying light that comes from room illumination. We will use a photodetector at a distance where the weakness of the LED's output, relative to other light sources, should make it difficult to sense.

---

[12] This lab is modelled on one developed by John Fox in his course at Stanford University. He kindly let us look at his course materials, and offered much advice.

[13] A potentially confusing usage: in this context this reference signal sometimes is called the *carrier* (in the Analog Devices datasheet for its AD630 "Balanced Modulator/Demodulator," for example). To say that the carrier "modulates" the signal is to invert the usage of "carrier" relative to its usage in traditional AM radio (using *amplitude modulation*). There, we say that the *signal* modulates the *carrier* We will live with this inconsistency of language.
Most sources say that the reference modulates the signal of interest (AoE §8.14.1); others say that the signal modulates the reference signal (Analog Devices datasheet for AD630: "...balanced modulator topologies accept two inputs, a signal (or modulation) input applied to the amplifying channels and a reference (or carrier) input applied to the comparator.")
Perhaps the two usages do not disagree in any important way, since signals A and B are multiplied to produce the modulated output, and in the product neither multiplier is privileged.

[14] See AoE §8.1.3.

[15] In traditional AM radio, the carrier is preserved, and used to ease detection; in the lock-in amplifier we will be discussing, the carrier or "reference" is not preserved; the output contains only the sidebands.

Is the LED on? If we want to detect whether a distant LED is on or off, our detection process might be defeated by ambient light, as well as by other light sources – perhaps even flicker from fluorescent room lights. We would like to find a method that protects against such noise sources. Our goal will be to determine whether the signal source is present.

If we *modulate* the optical signal, converting it from a DC level to an AC signal, then we can cancel noise that is not synchronized with the modulated signal. This we noted at the outset, in §10N.2.1 on the preceding page.

Demodulation To *de*-modulate the received signal, we multiply it periodically by +1, then −1. We will time the multiplications using a reference signal whose frequency and phase match the reference used in modulation. The signal, as demodulated, will be full-wave rectified, and the time average of this demodulated signal input will be positive and proportional to the DC optical signal.

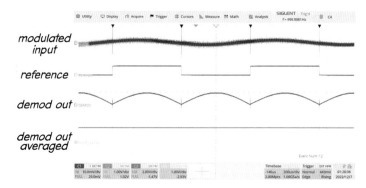

Figure 10N.23 The lock-in demodulator full-wave rectifies an input multiplied by a reference that matches the modulating reference, in frequency and phase.

The power of this process appears when we consider what this +1, −1 alternation does to all non-synchronized signals – that is, everything that we consider "noise." Those noise frequencies will be multiplied by +1, then −1, over and over, and after many cycles they will average to zero. We will carry out the averaging with a simple lowpass filter.

Yes, phase matters Probably it is evident that phase of the demod reference relative to modulating reference matters, but a look at scope images will make the point more vivid. Zero phase difference delivers the strongest output, as shown in Fig. 10N.23. Incidentally, an extreme mismatch, of 180°, also would produce an equally strong output – but an output whose average is negative: see Fig 10N.24.

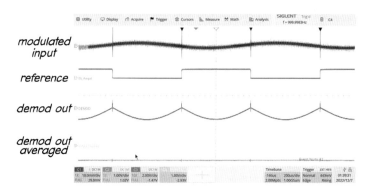

Figure 10N.24 An extreme phase shift of 180 degrees produces a strong negative output.

If, by bad luck, delays in the signal path should cause a 90-degree difference between the two phases, the output would go to *zero* – as it is intended to do for *non-signals*. Figure 10N.25 on the next page shows this result.

10N.2 Lock-in amplifier 455

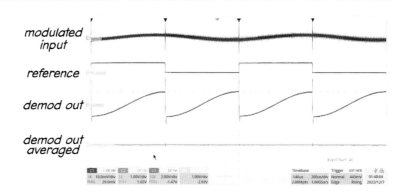

Figure 10N.25 Ninety-degree phase mismatch, between input and reference, kills demodulator output entirely.

The fourth signal trace, at bottom, shows the zero-voltage of the averaged demodulator output.

Your lab lock-in will include an adjustable phase shifter, an element of a normal lock-in amplifier, even though you may not find that your circuit introduces phase differences that need correction.[16]

The lock-in rejects non-synchronized signals ("noise") Let's watch what happens to noise not synchronized with the time-varying input signal.

Fig. 10N.26 shows the result: the top trace shows a large low-frequency sinusoid, which functions as the *reference* in this case, along with a tiny square wave riding that sinusoid. The square wave is what we consider "signal." The small square wave and its reference run at 1 kHz. The injected noise runs at a few hundred Hertz. The noise gets multiplied by +1 and −1 at times not synchronized to its waveform, and averages to zero, just like a DC input.

The *signal*, in contrast, gets enhanced: multiplied by +1 when positive, −1 when negative, as in Fig. 10N.23 on the preceding page. It is full-wave rectified, and averages to a positive value.

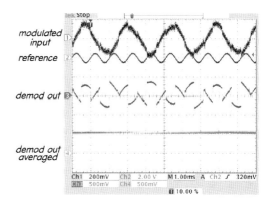

Figure 10N.26 The lock-in rejects non-synchronized time-varying signal, as it rejects DC.

The ornate Demod Out waveform in Fig. 10N.26 is hard to interpret. It includes a large component not synchronized with the reference, and this averages to zero. In other words, the injected noise has no effect on the output. The synchronized square wave, in contrast, is rectified and generates the positive output voltage.

---

[16] Commercial lock-in amplifiers also include a phase-locked loop (see Chapter 20N) to recover the reference from the modulated input input signal when it is not otherwise available.

10N.2.3 Effects of multiplying

A mathematical argument for the demodulation After modulation, in which the input signal was multiplied by a reference, $f_{\text{reference}}$, the information resides in the sidebands above and below that reference frequency, as we claimed in §10N.22 on page 453.

Let us assume that all signals are sinusoids – because this makes the argument a little simpler – and that the reference and modulated input signal frequencies are the same. In that case, the result of this multiplication is two outputs, a sum and a difference, because of a trigonometric identity:

$$\cos X \times \cos Y = \frac{1}{2}(\cos(X - Y) + \cos(X + Y)) \tag{10N.2}$$

This is, of course, just what produced the sidebands mentioned in §10N.22.

When the *demodulator* and reference frequencies ("X" and "Y" in equation 10N.2) are the same, the term $\cos(X - Y)$ goes to one. So, the output is a DC level, one-half the amplitude of the input signal. This is the result that interests us. We would also get a term at double the input frequency. That term we will filter out.

We assumed that there was no phase difference between reference and input signals when we claimed that "$\cos(X - Y)$ goes to one." If, instead, there is a phase difference, then the first term is not *cos(zero)* but *cos(phase-difference)*. To maximize the output amplitude we want to take the phase difference to zero, and to achieve this result your circuit will include an adjustable phase shifter.

An easy way to multiply: square wave modulation and de-modulation Though we could use a sinusoid as reference, as in the discussion above, in today's lab we will use, instead, a *square wave* as demodulator reference. We do this because our simple lab demodulator needs a square wave to switch it. This square reference works in almost the same way as the sinusoidal reference, except that it introduces the effects of the square wave's harmonics (illustrated in Fig. 10N.27). These will not trouble us, because our lowpass filter will remove them.

To illustrate these harmonics, we modulated a sine wave input of 1 Hz with a reference of 10 Hz.[17] These values are oddly near to each other, and the reference frequency strangely low; but the low values make the waveforms easy to interpret. We can see the effect of the square wave components, at $f_{\text{reference}} \pm f_{\text{input}}$, for the first six harmonics of the square wave.

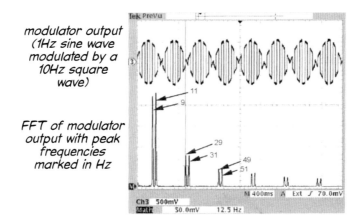

Figure 10N.27 Simple ±1 multiplication behaves like a reference at each harmonic of the applied reference frequency.

The effect shown in Fig. 10N.27 is more complex than what we will see in today's lab, where

---

[17] The illustration shown in Fig. 10N.27 shows results of modulation using an integrated modulator, AD630, rather than the DG403 implementation that we propose in today's lab. The AD630 uses a square-wave reference, as does our lab modulator.

reference and signal will be matched in both frequency and phase. But the multiplication effect is evident. As expected, we get the input sinusoid multiplied by not a single modulation frequency (10 Hz, in this case), but by each of the harmonics of the 10 Hz square wave: frequencies of 10 Hz, 30 Hz, 50 Hz. . .

10N.2.4 A non-mathematical explanation for noise cancellation

Given our demodulator that alternates between multiplying by +1 and −1 it is plausible that if we timed the alternation just right we might be able to enhance what we like – our modulated "signal." We could multiply by +1 when it was positive, for example, and −1 when it was negative. The result would be full-wave rectification, achieved just by judicious timing: This case is, of course, just what we get when the demodulator reference frequency matches the modulated input frequency, with no phase difference. This produces the maximum DC output signal when lowpass filtered. This is what our demodulator will try to achieve in today's lab. Any noise, on the other hand, is unlikely to be in phase with the modulation reference and will average to zero when filtered.

A slightly more quantitative view... The lock-in's discrimination between signal and noise can be described more exactly. Rather than say, as if the lock-in would detect only a single frequency, "what we get when the demodulator reference frequency matches the modulated input frequency" we can note that inputs may lie close to the target signal frequency but not just at that frequency. So, discrimination is a matter of degree.

The lowpass filter at the lock-in's output determines how finely the circuits discriminates between signal and noise. For a lowpass filter of width Δf, inputs that lie within Δf of the signal that interests us will be detected. Those outside that range will be rejected.

10N.3 AoE reading

- §15.6.2 (analog PID).
- §8.1 (noise).
- §8.1.3 (1/f noise).
- §8.14 (bandwidth narrowing).

10L Lab: Op-Amps V

The PID control loop and the lock-in amplifier are each significant builds and we don't expect you will be able to complete both of them in a single lab session. Feel free to do one, the other, or both as your time and interest permit. We will note that these are two of the more interesting labs, not only because they are complete systems that solve difficult problems, but also because they demonstrate sophisticated analog circuit design using many of the components and techniques from previous labs.

10L.1 PID controller

Introduction: why bother with the PID loop?

Today's feedback loop...: Today's circuit looks straightforward: a potentiometer sets a *target position*; a DC motor tries to achieve that position, which is measured by a second potentiometer. *Lags* cause the difficulty: the correction signal is likely to arrive too late to solve a problem that the circuit senses. If that happens, the *remedy* can make things worse.

This motor control circuit is a classic feedback network called a PID circuit. Its response ultimately will include three functions of the circuit *error* signal: P (proportional) I (integral); and D (derivative). Stability is the central issue.

...And why it interests us: This feedback problem holds two sorts of interest for us.

- Pedagogical appeal:
    - It gives us a chance to apply – and to apply in concert – a collection of circuits that you have seen either only as fragments, or only on paper:
        ○ differential amplifier, made of op-amps (this we have met only on paper);
        ○ differentiator;
        ○ integrator;
        ○ summing circuit; and
        ○ high-current driver (with motor as load).
- PID confronts a classic control problem; it provides a scheme with many practical applications.

Putting the pieces together: Students often tell us that they like to build circuits that *do* something – in contrast to circuits that produce just images on a scope screen. Today's circuit qualifies: it guarantees to make a little DC motor squirm ("squirm" when it's unstable; tamely spin, then stop, when it's stable).

This PID circuit is by far the most complex in this course to date. That makes it a good setup for improving your debugging skills (this is a "glass-half-full" way to say that you're very likely to make some wiring errors today). We often boast that in this course, *bugs are our most important product*. Today's circuit is likely support this boast. (The debugging will be especially challenging if your sloppy

10L.1 PID controller

lab partner fails to keep leads short and color-coded, and forgets to bypass power supplies. We know *you* wouldn't make such errors.)

Finally, we're happy to let this lab reinforce a concern first discussed in Lab 9L: stability. Although today's circuit problem is singular – integrator within the loop – and the remedy more subtle than usual, the general stability problem is one that confronts us in almost every circuit that has gain.

10L.1.1 PID motor control

The task we undertake here is one we have described in Chapter 10N. We will repeat some of what appears there to save you the trouble of refering back when you do the lab. But 10N is more thorough than what we write here.

Our goal, most simply stated, is just to use a feedback loop to get one DC voltage to match another; and since both voltages come from potentiometers, the goal can also be described (maybe sounding more exciting) as making the position of one potentiometer shaft mimic that of another. We want to be able to use our fingers to turn a potentiometer by hand and see a motor-driven pot mimic our action. Such controls are sometimes offered on fancy audio equipment, so that the equipment can be controlled either by twisting a knob, or by using a remote that controls the knob from across the room. A more impressive PID application is remote surgery.

10L.1.1.1 The motor-pot assembly

The motor-pot gadget comes in two forms. The scope images and details of frequency response in most of Lab 10L and Chapter 10N use the rotary-potentiometer version (on left side in Fig. 10L.1). The other version, shown to the right in that figure uses a *slider* pot rather than rotary, and responds much faster. Use whichever is available.[1]

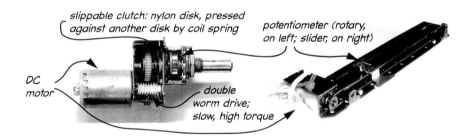

Figure 10L.1 Two motor-pot assemblies: rotary and slider potentiometers.

The clutch in the rotary version serves two purposes: it permits a human hand to control the potentiometer directly when the motor is stopped. It also protects the motor against stalling and overheating if the motor drives the pot to one of its limits. The slider version lacks this feature; a stalled motor will draw around 800 mA – but apparently without damaging the motor. Because of the lack of a slip-clutch in the slider version, we recommend the rotary.

10L.1.1.2 The motor control loop

Figure 10N.1 on page 441 was a minimal sketch of today's circuit. The special difficulty we're facing now comes from the fact that the motor-to-potentiometer block, our circuit's load, *integrates* voltages.

---

[1] The rotary type can be, for example, ALPS RK16812MG099; the slider version shown is COM-10734 from Sparkfun Electronics. The potentiometer "taper" is not critical. Linear is good; audio taper is OK (the ALPS part uses a B3 audio taper).

So, we can't use an ordinary op-amp as the *triangle* in the feedback loop shown the figure. The extra −90° shift, or integration, in this circuit forces us to alter our methods. Since we cannot afford the phase shift of an ordinary op-amp, we build ourselves a *custom* amplifier: one that provides modest gain and no phase shift. We will apply the radical remedy: altering not the load in the loop but the op-amp itself. Figure 10L.2 is a reminder of Fig. 10N.4 – redrawn to suggest that we now will be able to tinker with the amplifier's gain. Figure 10L.2 only shows the simplest configuration – "P". Later, we will alter also the phase behavior.

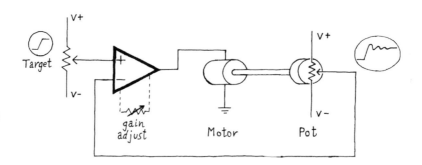

Figure 10L.2
Proportional-only drive will cause some overshoot; gain will affect this.

We will first try this circuit with its gain adjusted *low*, and we expect to find the circuit fairly stable. Then, as we increase gain, we should begin to see overshoot and ringing; if we push on to still higher gains, we should see the circuit oscillate continuously. At the end of this chapter we append some scope images illustrating just such responses to variations in simple "proportional" gain.

10L.1.1.3 Motor driver

Let's start with a subcircuit that is familiar: a high-current driver, capable of driving a substantial current (up to a couple of hundred milliamps). We'll use the power transistors you've met before: MJE3055 (*npn*) and MJE2955 (*pnp*). The motor presents the kind of troublesome load likely to induce parasitic oscillations, as in the final exercise of Lab 9L. We need therefore the protections that we invoked there: not only decoupling of supplies, but also both a *snubber* and *split feedback* that bypasses the troublesome phase-shifting elements.

We are trying hard here to *decouple* one part of the circuit from the others: the 15 μF caps should prevent supply disturbances from upsetting the *target* signal. Similar caps at the ends of the motor-driven potentiometer aim to stabilize the feedback signal.

You may also want to use an *external power supply* to provide the *motor's* ±15 V supplies if you have such an extra supply handy. We suggest this not for decoupling but because the motor's maximum current exceeds the breadboard's 100 mA rated output and might disturb those supplies even if one inserted plenty of decoupling caps. The external supply, unlike the breadboard supply, can provide the necessary current. (But we have also built this lab successfully *without* this separate power supply.)

Note: Your motor pot's resistance value may differ from the 10k shown in Fig. 10L.3. If so, scale the resistors appropriately. If your motor's potentiometer has value 100k, for example, just scale the 4.7k Rs up by a factor of ten.

Note also that you *must not use '411 op-amps*. The '411 has the nasty property – common to "bi-fet" devices – that it can flip its output phase if the input voltage goes below its specified common-mode input range.[2] The result is that if an input to any of the amplifiers in this loop momentarily swings to

---

[2] The op-amp output is forced *high* in this event. If the input going too far negative is the non-inverting, this changes the flavor of feedback from negative to positive.

10L.1 PID controller

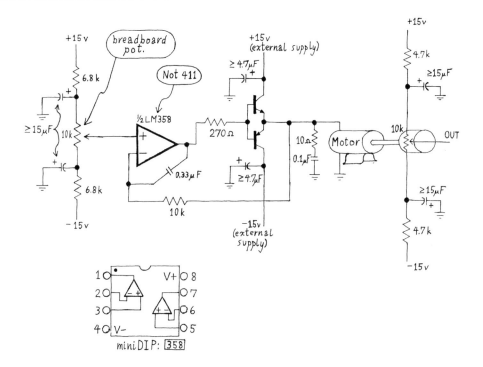

Figure 10L.3 Motor-driver.

within about a volt of the negative supply, the loop is very likely to get hung up by this nasty positive feedback.[3]

Wire up the two potentiometers as well as the motor-driver itself. The resistors at the ends of the two potentiometers – 6.8k resistors on input, 4.7k resistors on the motor pot – restrict input and output range to a range of about ±7 V to keep all signals well within a range that keeps the op-amps happy. The difference in R values makes sure that the input range cannot exceed the achievable output range.

You can test this motor driver by varying the input voltage and watching the voltage out of the motor-driven pot. Don't be dismayed if you see a good deal of hash on the scope screen. This hash may look very much like a parasitic oscillation, familiar to you from Lab 9L. Figure 10L.4 shows what we saw when watching the motor drive with the motor moving.

But if we look at this hash more closely, as in Fig. 10L.5 we find some clues that it is not the usual parasitic oscillation at work. The spikes seem to be the effects of the DC motor's brushes breaking contact periodically with the motor's commutator. One clue is the fact that noise is not continuous, but seems to be a set of narrow spikes at a low repetition rate. The other clue – pretty conclusive – is the fact that the spike voltages exceed the power supply: this effect looks a lot like the behavior of an inductor (the motor winding), angry each time the commutator switches the current *off*. So, don't let this hash worry you. It's ugly, but we'll live with it.

Any V_{in} more than a few tenths of a volt should evoke a change of output voltage. You will hear the motor whirring, and will see the shaft slowly turning (the motor drive is geared down through a two-stage worm- and conventional-gearing scheme).[4] After perhaps 20 seconds, the pot will reach its limit and will cease turning. But that's all right: a clever clutch scheme, mentioned back in §10L.1.1.1,

---

[3] This hazard is not hypothetical; we first breadboarded this circuit with '411s – and were forcibly reminded of the part's nasty phase-inversion by the occasional lock-up failure of the loop.

[4] See photo of motor-pot innards in Fig. 10L.1.

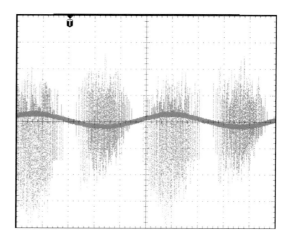

Figure 10L.4 Motor drive hash looks like a parasitic oscillation.

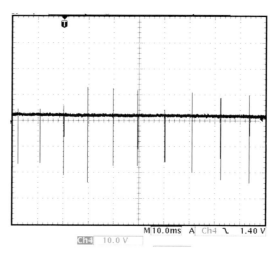

Figure 10L.5 Motor drive hash seen in greater detail: not parasitic oscillation after all.

allows the motor to slip harmlessly when the pot reaches either end of its range. If the *signs* of V_{in} and the *change* in V_{out} *do not match*, then be sure to interchange leads of the output pot, to make them match. We don't want a hidden inversion, here. It would upset our scheme when we later close the loop.

10L.1.1.4 Pseudo op-amp

Now we do a strange thing: we use three op-amps to make a rather crummy op-amp-like circuit, as in Fig. 10L.6. This is the circuit we have sometimes referred to as our "control amplifier."

The first stage you recognize as a standard differential amp: it shows unity gain. The second stage simply inverts;[5] the third stage seems to be doing no more than *undoing* the inversion of the preceding circuit, along with permitting adjustment of gain. That is true at this stage; but we include this circuit because soon we will use it, fed by two more inputs, as a *summing* circuit. So used, it will put together the three elements of the *PID* controller: Proportional, Integral, and Derivative. In the present P-only circuit[6] we also use the third amplifier to vary the overall gain of our home-made op-amp.

The entire circuit then is simply a differential amplifier with adjustable gain. And this gain is always

---

[5] This inversion is included to let this signal share a polarity with the "Derivative" and "Integral" signals to be generated shortly; these signals will come from circuits that necessarily invert.

[6] We call this "P," as we call the other signals, soon to be added, "D" and "I," although all are inverted. Strictly, then, this is "–P." We omit the "minus" in these labels, thinking it easier to refer to P than to minus P.

10L.1 PID controller

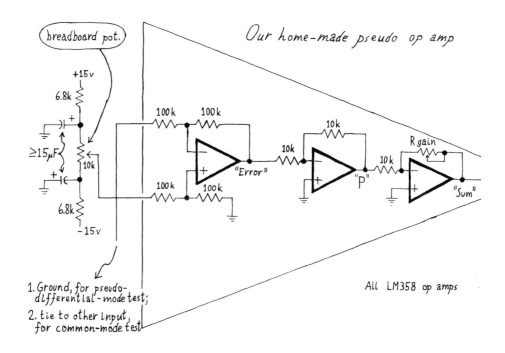

Figure 10L.6 Differential amp followed by gain stage and an inversion.

low relative to the very high values we are accustomed to in op-amps. We need the modesty of this gain, and we need its lack of phase shift between input and output. Both characteristics contrast with those of an ordinary op-amp as you know. The fixed, high gain of an ordinary op-amp, along with its *integrator* behavior beginning at 10 or 20 Hz, would get us into trouble today, turning negative feedback into positive.

Check common-mode and differential gains

Common-mode gain...: We suggest that you use a resistor substitution box to set the *summing* circuit's gain. Set the gain at ten, and see whether a *common-mode* signal – a volt or so applied from the input pot applied to both inputs – evokes the output you would expect. (Do you expect *zero* output?)

...(Pseudo-)differential gain: Then ground one input (the 100k that feeds the first op-amp's *inverting* input) using the level from the potentiometer as input. Watch that input, and the circuit output, with the *R* substitution box value set to 100k: see if you get the expected gain of +10.

A couple of features of this test may bear explanation.

- Yes, the gain is *positive* when the input pot drives the *non-inverting* input to this home-made op-amp, since *two* inverting stages follow the diff-amp.
- We are applying a "mixed-mode" signal by grounding one input of the diff-amp and driving the other. (You did this also in Lab 5L, as you drove the home-made "op-amp.") Since the *differential* gain is so much higher than the *common-mode*, this *mixed-mode* signal works almost as a true differential signal would. An applied signal of v appears as a *differential* signal of magnitude v, combined with a *common-mode* signal of magnitude $v/2$. Given even a mediocre CMRR, this modest *common-mode* signal mixed with the *differential* is harmless.

A DVM may be handier than a scope, at this point, to confirm that the output of this chain of three op-amp circuits shows a mixed-mode gain of +10, while you drive the input with the *input* potentiometer voltage. When you finish this test, leave the output voltage close to zero volts.

10L.1.1.5 Drive the motor

You have already tested the motor driver. Let's now check the three new stages – those that form the pseudo op-amp – by letting their output feed the motor-driver. *Ground* the inverting input for this test, as shown by a short dotted line in Fig. 10L.7. You do *not* yet need to make the connection to that terminal from the output potentiometer – a connection shown as an alternative, longer dotted line in that figure.

Confirm that you can make the motor spin one way, then the other, as you adjust the input pot slightly above and then below zero volts. (The motor-driven pot, as we have said, fortunately can take the pot to its limit without damaging pot or motor.)

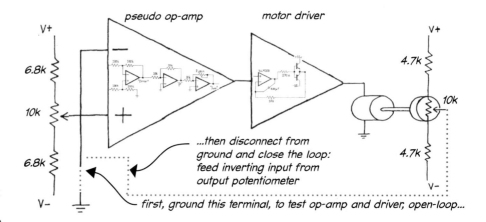

Figure 10L.7 Op-amp plus driver: first try open loop to test diff-amp, gain stage, sum and motor drive; then close the loop.

10L.1.1.6 Close the loop

Now let's close the loop. Reduce the *gain*, using the *R substitution box*: set it to about 1.5 (R_{gain}=15k). Replace the *ground* connection to the inverting input of our "pseudo op-amp" with *the voltage from the output potentiometer*. This connection is shown as the longer dotted path in Fig. 10L.7. Make sure to disconnect the inverting input from the *ground* that you tied it to in §10L.1.1.5.

Watch V_{in} on one channel of the scope, $V_{output-pot}$ on the other. *If a digital scope is available, this is a good time to use it* because a very slow sweep rate is desirable: as low as 0.5 second – or even 1 second – per division.

Several ways for testing the loop: Two or three methods are available to you for testing the new setup.

- **Two ways to drive the input**
  - **Square wave from function generator**. A function generator can provide a small square wave (±0.5 V, say) at the lowest available frequency (about 0.2 Hz on our generators). This input can temporarily replace the *manual* input potentiometer. This is probably the best choice since it provides a consistency you cannot achieve by hand.
  - **Manual step input**. You may, however, prefer the simplicity of *manually* applying a "step input" from the *input* pot: a step of perhaps a volt.

  The *output* pot should follow – showing a few cycles of overshoot and damped oscillation.

10L.1 PID controller

- **An alternative test: disturb the output, and watch recovery**. A second way of testing the circuit's response is available if you prefer (and you may want to try this in any case, after looking at the response to a step input): leave the input voltage constant, then manually force the pot away from its resting position simply by turning the knob of the *output* pot. Let go, and watch the knob return to its initial position – showing some overshoot and oscillation as when the change was applied at the input pot.

You start with a very low gain (1.5), which should make the circuit stable, even in this P-only form. Now use the substitution box to dial up increasing gain. At R_{gain} = 220k (|gain| = 22) we saw some overshoot and a cycle or two of oscillation, evident in the motion of the motor and pot shaft. If this shaft were controlling, say, the rudder of an airplane, this effect would be pretty unsettling. The circuit works – but it would be nice if we could get it to settle faster and to overshoot less.

Increasing the gain, at R_{gain}=680k (|gain| = 68), we were able to make out several cycles of oscillation (the bigger, uglier trace in Fig. 10L.8 shows the motor drive voltage; there the oscillation is more obvious).

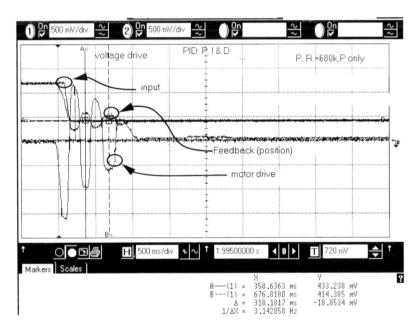

Figure 10L.8 P only: gain is high enough to take us to the edge of oscillation.

With a little more gain (R_{gain}=1M, in our case; |gain| = 100) and the application of either a step change at the input, or a displacement of the output pot by hand, we see a continuous oscillation. Find the gain that sets your circuit oscillating, and then *note the period of oscillation* at the lowest gain that will give sustained oscillation. We will call this the period of "natural oscillation," and soon we will use it to scale the remedies that we'll apply against oscillation.

10L.1.2 Add derivative of the error

Well, of course we *can* get it to settle faster; we *can* improve performance. (If we couldn't, would the name of this sort of controller include the I and D in PID?) We can speed up the settling markedly, and even crank up the P gain a good deal once we have added this derivative. Thinking of the stability difficulty as a problem of taming the phase shifts of sinusoids – as we did for op-amps generally – we can see that inserting a derivative into the feedback loop will tend to *undo an integration*.

The integrations are the hazard, here: one is built in – the translation from motor speed to motor position. Additional integrations resulting from lagging phase shifts can carry us to the deadly −180° shift that converts *nice* feedback into *nasty* – the sort that brings on the oscillation you have just seen.

10L.1.2.1 Derivative circuit

The standard op-amp differentiator in Fig. 10L.9 can contribute its output to the summing circuit. Here we show the entire prior circuit (Fig. 10L.7), with the differentiator added. The differentiator's gain is rolled off at about 1.5 kHz. Again we recommend that you use a *resistor substitution box* to set the D gain if such a box is available.

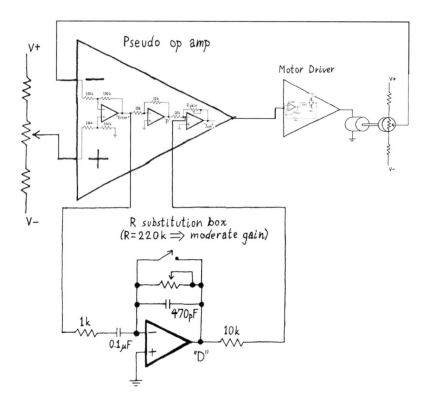

Figure 10L.9 Derivative added to loop.

Figure 10L.9 also shows a switch in the feedback path that permits you to kill the derivative when you choose to.

How much derivative? Our goal in adding derivative is to cancel the extra phase shift otherwise caused by a lowpass effect that brings on instability. How do we know at what frequency this trouble occurs, and therefore how to set the frequency-response or (equivalently) *gain* of the differentiator?

It turns out that you already have this information: you got it by measuring the frequency (or period) of "natural" oscillation, back in §10L.1.1.6. There, as you know, you gradually increased the P-only gain till you saw that an input disturbance would evoke either an output that took a long time to settle or else a continuous oscillation. (When *we* ran that experiment, for example, we got a "natural oscillation" period of roughly 0.6 second using the rotary motor-pot. With the slider pot, that period was about 40ms. If you are using a slider version, adjust your D gain accordingly – do not use the values we note below, which apply to the rotary version.)

We aim to make the derivative contribution, D, equal to the P contribution at the "corner" frequency where sustained oscillation would occur. *RC* defines the differentiator's *gain*. You'll find an argument for this proposition in Chapter 10N in case you need to be persuaded. In order to make the D gain equal to the P gain at the frequency of "natural oscillation," we want $RC = 1/(2\pi f)$, where f is that oscillation frequency.

A scaling rule of thumb: frequency of natural oscillation dictates D gain: If, as this formula suggests, RC should be about 1/6 of the period of natural oscillation, then for our $T_{\text{oscillation}}$=0.6 s we'd set RC to about 0.1 s, or a bit less.[7]

Let's make this value adjustable, though, because we want to be able to try the effect of more or less than the usual derivative weight: if you have a second *resistor substitution box*, use it to set the differentiator's gain (RC). Otherwise, use a 1M variable resistor. Watching the position of the rotator will let you estimate R to perhaps 20%; the midpoint value certainly is 500k, and 750k is close to the 3/4-rotation position. The differentiator's output goes into the summing circuit installed earlier, through a resistor chosen to give this D term weight equal to the P's.

We hope you will find this D to be strong and effective medicine. Once it has tamed your circuit's response – eliminating the overshoot and ringing – crank up the P gain, to about twenty ($R_{\text{sum\_gain}}$=220k) or more. Is the circuit still stable? Try more D. Does an excess of D cause trouble? The scope image of the circuit's response will let you judge whether you have too much or too little D: too little, and you'll see remnants of the overshoot you saw with P-only; too much D, and you'll see an RC-ish curve in the output voltage as it approaches the target: it chickens out as it gets close. And if you keep increasing the D gain still further, as we said in the Chapter 10N, especially §10N.1.5, the circuit goes unstable once again: it oscillates.

Switch: The toggle switch across the feedback resistor will let us cut D in and out; the switch seems preferable to relying, say, on a very-large variable R to feed the summing circuit. We find it can be hard to keep track of multiple pot settings to know whether we're contributing D or not. A switch makes the ON/OFF condition easier to note.

10L.1.3 Add integral

Adding the third term – the I of PID – can drive residual error (a difference between the input pot voltage and the output pot voltage) to zero. Figure 10L.10 is a diagram of the full PID circuit with the integrator added.

Two details of the integrator may be worth noting:

Two polarized caps placed end-to-end: This odd trick works to permit use of *polarized* capacitors in a setting that can put either polarity across the capacitance. The effective capacitance is, of course, only one half the value of each capacitor. We use polarized caps only because large-value caps like these 15 μF parts are hard to find in non-polarized form.

Seeming absence of DC feedback: at first glance, this integrator seems doomed to drift to saturation, since the integrator includes neither of our usual protections against such drift – feedback

---

[7] See, e.g., Tietze and Schenk *Electronic Circuits: Handbook for Design*, Second edition, Springer (2008). A less formal approach appears in David St. Clair's *Controller Tuning and Control Loop Performance*, Straight-Line Controls, Inc. at `https://LAoE.link/PID_Control.html`, whereyoucanpurchaseasimulatorordownloadademothatallowsyoutotrytheauthor'srules. Aneasytousesimulator,alongwithagoodtutorial,appearsin\url{https://LAoE.link/OnlinePID_Sim.html}.The\emph{simulation}letsyoutry(asyouwouldexpect!)theeffectofvaryingPgainandofaddinginDandI--justaswedotoday. If we use a convenient C value of 0.1 μF, the R we need is about 1M. Finally, an Excel spreadsheet PID simulator we use in lecture is available at `https://LAoE.link/ExcelPID_Sim.html`.

resistor or momentary discharge switch. But neither is necessary here because overall feedback – all the way around the large loop, from input pot to output pot – makes such unwanted drift impossible. In short, there *is* DC feedback, despite appearances to the contrary.

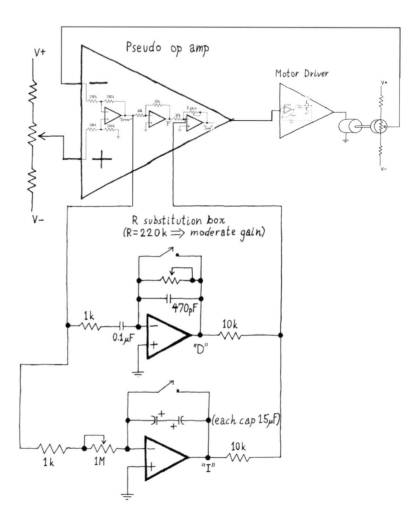

Figure 10L.10 Integral added to complete the PID loop.

Watching the effect of I: In today's circuit, the residual error is hard to see on the scope, so adding I will not reward you as adding D did. Your best hope will come if you cut the P gain very low: try R_{gain}=100k, so that the circuit feedback ought to tolerate a noticeable residual error, when not fed an I of the error. If you have been using a function generator to provide step inputs to your circuit, now replace that signal source with the manually-adjusted pot input. Slow the scope sweep rate, to a rate that permits you to see the multi-second effect of the integration.

If you are using a digital scope, you will be able to watch input ("Target"), output ("Motor pot"), and *Integrator* signals responding to a step input applied from your input potentiometer. If you are patient, you can even make out the effects of the motor and pot's "sticktion" (a cute term for "static friction"): the motor and pot do not move smoothly in response to a slowly-changing input (here, the

I term). Instead, the motor fails to move till the I voltage reaches some minimal level; then output voltage jumps to a new level, and waits for another shove. You can see these effects in some of the scope images in §10L.1.4. at the end of this chapter.

Too much I again brings instability: It sounds dangerous, doesn't it? – tacking in an integral term when integration, plus other lagging phase shifts, are just what threatens the circuit's stability. It *is* dangerous, as you can confirm by overdoing the I. You should be able to evoke continuous oscillation, as in the dark days before you knew about the stabilizing effect of D. Yet, remarkable though this fact is, some I does improve loop performance – driving long-term error toward zero – and need not bring on instability.

10L.1.4 Scope images: effect of increasing gain, in P-only loop

See Fig. 10L.11.

10L.2 Lock-in amplifier

Lock-in demonstration using a weak optical signal

In this lab we attempt to detect a feeble optical signal, which we create with a 10 or 20 mV sine wave driving an LED, in a room that includes substantial ambient light, and some time-varying optical signals, perhaps from fluorescent lights. Our method will be to multiply the optical signal source by a "reference" frequency of about 1 kHz. Then we will detect the original signal ("demodulate" it) by multiplying the received signal by the same reference. It is important that the demodulation reference be the same as the modulating signal not only in frequency, but also in *phase*.[8]

10L.2.1 The overall scheme

A function generator will drive an LED at about 1 kHz – this *reference* signal modulates the DC level driving the LED. The reference will also be passed, as a synchronizing signal, to the demodulator, which receives a weak version of the LED's signal. The demodulator's output will be a DC level indicating the presence (and amplitude) of the received signal.

10L.2.2 Attenuator and phase shifter

In this initial testing phase, rather than use the actual optical signal, we will use a small (and adjustable) signal from our function generator as the lock-in amplifier input. We will feed this signal to the demodulator while observing the effect of various phase differences between the test input and the reference waveform.

Since the optical receiver will have quite a bit of gain, we need to attenuate our test signal. Figure 10L.13 shows the attenuator we will use while testing the demodulator.[9]

You have seen in this chapter's notes, the relative phase between the modulated input signal and the demodulator reference is critical. We will get the strongest response from our demodulator when the two phases match. So, we propose two alternative methods to adjust the phase of the reference signal.

---

[8] We will include the ability to adjust the phase of the demodulator reference slightly in order to compensate for any phase shifts in the transmission environment and/or the signal processing circuitry.

[9] Except for the optical transmitter (Fig. 10L.15), which runs on a single 15 V supply, all op-amps and comparators in the lock-in run on ±15 V.

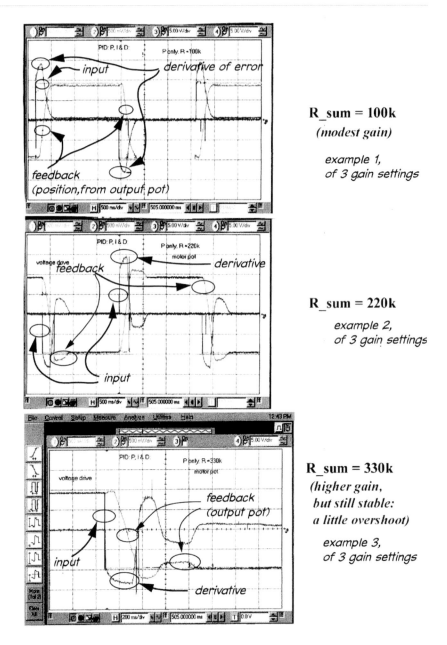

Figure 10L.11 Increasing P-only gain brings increasing overshoot.

- If your function generator offers a pair of outputs and the option of adjusting the phase between these outputs, no need to build anything. You can exploit this feature of the generator, using channel 1 to provide the input sine wave modulating reference and channel 2 to provide the demodulator reference, a 0 V to 5 V square wave at the same frequency.

- If you have only a single channel function generator or you cannot control the phase between the two channels, you can build the shifter circuit shown in Fig. 10L.14 on the facing page to provide the same functionality.

10L.2 Lock-in amplifier

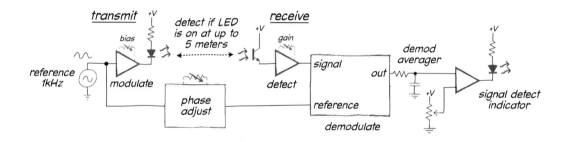

Figure 10L.12 Transmit-receive task, using lock-in amplifier.

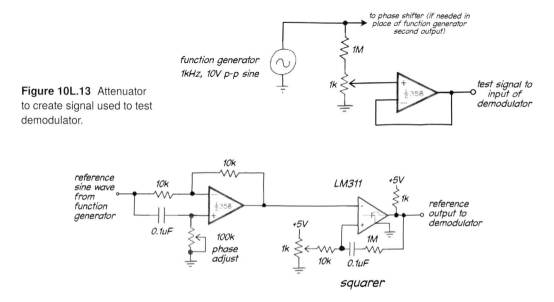

Figure 10L.13 Attenuator to create signal used to test demodulator.

Figure 10L.14 Phase shifter to adjust demodulator reference phase relative to the modulator reference.

10L.2.3 Transmission hardware

Once we have the receiver/demodulator working, we will replace the test signal with the actual optical input. The reference will "multiply" (i.e., modulate) the optical signal by varying the current passed through an infrared LED. Figure 10L.15 on the next page shows the circuit. The 10k pot sets the DC bias through the IR LED while the reference AC signal varies the LED current around that bias. Build this circuit on a separate small breadboard so you can move it away from the receiver. You can use any N-Channel power MOSFET you have available such as the IRLZ34 or IRL2505PBF.

The op-amp can run single-supply on 15 V. You could use an LED power supply voltage as low as 5 V; we got better results with the higher supply, although a side effect is extra power dissipated in the MOSFET.

Once you have built and tested the modulator, you can set it aside.[10]

---

[10] Most cell phone cameras have some sensitivity to IR light. You should be able to tell that your transmitter is working by looking at the IR LED through your camera and seeing if the brightness increases as you increase the bias.

Lab: Op-Amps V

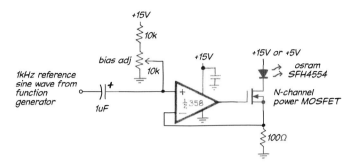

Figure 10L.15 Optical transmitter: reference modulates brightness of LED transmission.

10L.2.4 Detector hardware

Since we expect a weak optical signal, we need to provide lots of gain in the optical receiver. We will first drive the receiver and modulator with the attenuated sine wave test signal directly to make sure everything works and see how the lock-in functions. Once everything is working correctly, we will replace the test signal source with an IR photo-transistor and see if we can receive the modulated signals from the transmitter.

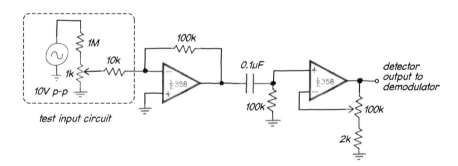

Figure 10L.16 Detector amplifier, fed by test signal.

The input divider creates a test signal from near zero to 10 mV peak-to-peak. Use your scope to watch the input signal and the amplifier output. If you have a switchable gain oscilloscope probe, set the input signal probe to *times-one*. (You will need to set the scope to reflect this change to *times-one* so the units display correctly.[11])

Turn the test circuit pot to its maximum value and look at the signals. The two op-amps provide a gain of up to 500. If you have a digital scope you may want to try reducing its bandwidth and averaging several scans to reduce the noise visible on both input and output. Use the test circuit pot to try reducing the input amplitude. What is the smallest input signal amplitude you can measure?

10L.2.5 Demodulator

One can buy a demodulator from Analog Devices, the AD630, a complex IC that does all the hard work for you. But we think you'll find it more fun (and illuminating) to build your demodulator from op-amps and an analog switch. The modulated signal is a tiny 1 kHz sine wave with lots of added noise. As discussed in the notes (see §10N.2.3), an easy way to multiply is to use a square wave rather

---

[11] Don't forget to set the probe back to X10 when you are done.

10L.2 Lock-in amplifier

than a sine wave to synchronously multiply the received signal by +1 or −1. This is easily implemented by using a DG403 SPDT analog switch to select between the received signal or an inverted version of it, depending on the sign of the reference.[12]

We need only feed this demodulator an INPUT and a REFERENCE, and then lowpass filter its output to get a measure of average signal amplitude. Since it is possible that there could be some phase delay in the transmission and reception of the IR signal, we have included the ability to adjust the phase between the transmitter reference and the demodulator reference. As noted in §10L.2.2, you can use either a two-output function generator or a phase-adjust circuit to adjust the phase between these two reference operations in order to maximize the demodulator's output.

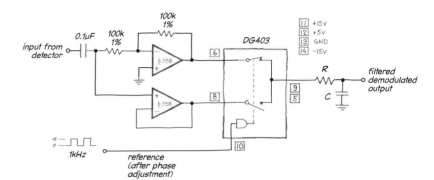

Figure 10L.17 Demodulator using DG403 analog switch.

The lowpass filter on the DG403 output should be fast enough to respond to changes in the signal you are looking at but slow enough to minimize ripple on the filtered output. A lower cutoff also improves the signal-to-noise ratio of the lock-in. For determining if the LED is on or off, 1 Hz or so should be adequate. A low cutoff frequency means you need a large capacitor which must be *non-polarized*, because, as you know, demodulator output can go negative for some phase differences between input and reference. Two ways one might handle this requirement are shown in Fig. 10L.18.

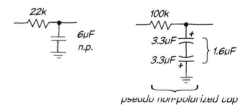

Figure 10L.18 The averager requires a large, non polarized capacitor.

You can always add a buffer after the averager if your application requires lower output impedance.

10L.2.6 Testing it out

Use scope and a DVM to measure the filtered DC output. Adjust the test circuit pot for a 10 mV sine wave.

Try adjusting the phase difference at the demodulator in order to see how phase affects the demodulated signal. At a phase difference of 90 degrees, as you know, the output voltage should go to zero, and at a phase shift of 180 degrees the output should go negative.

With phase adjusted for a maximum output, use the test circuit pot to reduce the amplitude of the

---

[12] There are two identical SPDT switches in the DG403. If you find it easier to wire up your demodulator using the other one, feel free to do so. Our choice of one of the two pairs of switches, reflected in the pin numbering shown in Fig. 10L.17, was arbitrary.

test signal as much as you can and still see a difference in the DVM reading between the cases of the signal generator test signal ON and OFF. You will still need to supply the signal generator output to the demodulator.

10L.2.6.1 Add a yes/no detection indicator

Let's add an LED to indicate the demodulator's result: signal detected or not. Since the demodulator output can go negative, we need a split-supply comparator, our old friend the LM311: see Fig 10L.19.

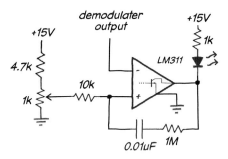

Figure 10L.19 Comparator will indicate signal Detected/Not-detected.

10L.2.7 Try optical detection

Once you have verified that your receiver and demodulator work, it is time to replace the test circuit with the IR photo-transistor and see how feeble a signal you can detect.

Remove the test circuit shown in the dotted lines in Fig. 10L.16 and replace it with the photo detector circuit shown in Fig. 10L.20.

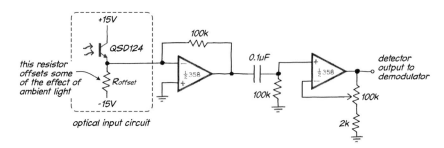

Figure 10L.20 Detector amplifier, fed by optical receiver.

The detector output voltage from the I-to-V circuit goes negative for any optical input, including ambient light, so you may want to include the pull-down resistor, R_{offset}. You can choose a value that allows a resting (no-signal) voltage far from op-amp saturation. The 100k feedback resistor value is only a suggestion. We want gain, but must choose a value that avoids saturation in the first op-amp of this receiver.

10L.2.7.1 Share reference signals

You will need a long coax BNC to carry the sine wave *reference* signal to the transmit breadboard. Probably you already have the square *reference* wave connected to the receive-and-demodulate breadboard.[13] If your circuit shows power supply ugliness corrupting the receiver, you might power that section of the circuit with a pair of 9-volt batteries.

---

[13] Both references run at 1 kHz, but your circuitry of §9W.1.2 on page 434 allows you to adjust the relative phases of the two references.

10L.2.7.2 Try detecting the IR signal

Watch the modulating sine wave at the input to the transmitter (there called "Reference"). You will want to reduce the amplitude significantly – probably to about a volt peak-to-peak, initially.

Start with transmitter fairly close to the receiver until you have everything working. Adjust the transmitter LED bias to get a signal at the output of the transconductance amplifier (the first stage of Fig. 10L.20). Adjust the R_{offset} resistor if necessary to keep the first stage amplifier from saturating.

10L.2.7.3 Adjust the phase

Adjust the phase as necessary to get the maximum output signal from the demodulator. Once the system is working correctly, move the transmitter away from the receiver and start reducing the sine wave amplitude. You can test if you are receiving the modulated signal by blocking the IR LED with your hand. The DVM reading should slowly go to zero with the signal blocked and return with it unblocked.

Comparator: adjust the threshold level of the LM311 comparator circuit, and confirm that the LED indication matches what the DVM readings suggest.

10L.2.8 Add optical noise – if you dare

If you are ambitious, you can complicate things for your demodulator by feeding optical noise. An easy way is to drive an IR LED like the principal emitter directly from a function generator (whose 50-ohm output resistance will limit the current to a safe value). Figure 10L.21 shows a result we got using the Analog Devices AD630 demodulator (rather than the DG403 home-brew demodulator), when we tried that. (You saw a similar noise-rejection scope image, Fig. 10N.26, in today's class notes.)

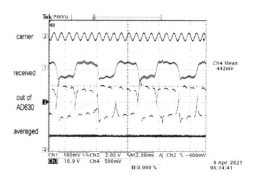

Figure 10L.21 Optical noise, not synchronized, does not disturb the lock-in.

You must make sure that this optical noise does not cause the receiver to *saturate*, so it is wise to watch the output of the detector's two stages, as you do this experiment.

The noise rejection we are watching, here, is not impressive compared to the best that the integrated AD630 offers. The datasheet promises the AD630 can pick out a signal 100 dB below the noise level. Here, our signal is about 20 dB below the noise. But the test does demonstrate the method.

11N Voltage Regulators

Contents

11N.1 Evolving a regulated power supply — **477**
    11N.1.1 Unregulated supply — 477
    11N.1.2 Zener sets V_{out} — 478
    11N.1.3 Zener plus discrete-transistor follower — 478
    11N.1.4 Zener or reference plus op-amp follower — 478
    11N.1.5 Reference plus op-amp follower plus "pass" transistor — 479
    11N.1.6 Stabilized circuit — 479
    11N.1.7 Current limit: a complete regulator circuit — 480
    11N.1.8 Dropout voltage — 481

11N.2 Easier: 3-terminal IC regulators — **482**
    11N.2.1 Fixed output: 78xx — 482
    11N.2.2 Variable output: 317 — 483

11N.3 Thermal design — **484**
    11N.3.1 Thermal transfer, in general — 485

11N.4 Current sources — **486**
    11N.4.1 An IC for intermediate currents — 486
    11N.4.2 A bipolar IC for low currents — 486
    11N.4.3 JFET as current source — 487

11N.5 Crowbar overvoltage protection — **487**

11N.6 A different scheme: switching regulators — **488**
    11N.6.1 Preliminary: getting used to inductors' peculiar proclivities — 488
    11N.6.2 Three switching configurations: "Boost," "Buck" and Invert — 490
    11N.6.3 Efficiency — 490
    11N.6.4 Implementing the feedback — 491
    11N.6.5 Switchers aren't always what you want — 492

11N.7 AoE reading — **493**

Why?

What problem do we meet today? We try to design a circuit that provides an output *supply* voltage that is constant despite fluctuations that may arise in both input voltage and output current loading.

Voltage regulators: One could argue quite plausibly that this is not a topic in its own right; it is only one more application of negative feedback, and could fit quite well into the op-amp chapter. Isn't a regulator just a follower driven by a reference? Yes, it is – though we'll refine the *follower* soon. But this function is needed so often that specialized ICs have evolved to do just this job, so that one almost never does use an op-amp. And power supplies and their regulators are so universal in instruments of

11N.1 Evolving a regulated power supply

all kinds that AoE assigns them a chapter of their own. We follow this scheme, giving them a day in the lab.

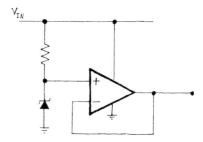

Figure 11N.1 Voltage "regulator:" just another use of feedback.

11N.1 Evolving a regulated power supply

AoE §9.5

11N.1.1 Unregulated supply

In the beginning – well, in Chapter 3N and Lab 3L, anyway – there was the unregulated power supply. Look back to Fig. 3N.28: it provided a DC level from the AC "line" voltage. Why isn't this good enough? How does it fall short? Let us count the ways:

- You can see from Fig. 11N.2 that output shows some ripple – and it is not a good idea to try to solve that problem by boosting capacitor size. Doing that would reduce ripple, but at the cost of increasing transformer heating, calling for a larger and heavier transformer.

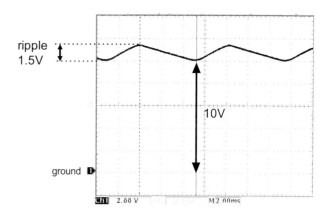

Figure 11N.2 Unregulated power supply shows substantial ripple. (Scope gain: 2 V/div.)

- V_{out} varies somewhat, with loading ...
    - The transformer's output impedance is mediocre: in Lab 3L, for example, we saw about 20% droop between light and full loading from our nominal 6.3 V transformers;
    - Ripple amplitude increases with increased loading for a given filter capacitor size, reducing average V_{out}.
- V_{out} can vary widely because of large variation in the *line* voltage (nominally 120 V, but varying by about ±10%, and occasionally more at times of heavy loading).[1]

---

[1] For example, one state's (Illinois) regulations require that "voltage variations as measured at any customer's point of delivery shall not exceed a maximum of 127 volts nor fall below a minimum of 113 volts for periods longer than two minutes in each instance" on "120 V" lines. See https://LAoE.link/Line_Voltage.html.

11N.1.2 Zener sets V_{out}

A zener diode can stabilize V_{out}: see Fig. 11N.3. What's wrong with this scheme?

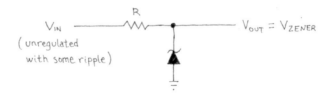

Figure 11N.3 Zener voltage source.

- It's inefficient: to provide 100 mA of output current at full load, for example, it must idle with 110 mA flowing through the zener (110 because we want at least 10 mA through the zener even under full load).
- The zener's voltage varies somewhat with variation in its current, a variation that necessarily occurs as loading varies.
- Tolerances for even the *nominal* zener voltage are only mediocre.

11N.1.3 Zener plus discrete-transistor follower

AoE §2.2.4

A follower helps: it permits large I_{load} and large variation in this load current without calling for large currents in the zener: see Fig. 11N.4. This is better, but still disappointing.

- V_{BE}, somewhat variable, causes variation in V_{out} even for constant V_{zener}.

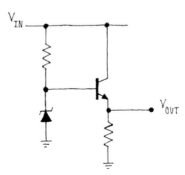

Figure 11N.4 Zener plus discrete follower: regulator?

11N.1.4 Zener or reference plus op-amp follower

An op-amp can hide the V_{BE} drop and its variations, and can give us much lower R_{out}: see Fig. 11N.5.

On the right of Fig. 11N.5 we have made a small improvement: we replaced the zener with an IC voltage reference (2.5 V: same as Chapter 11L) providing a more precise initial voltage (±3% in the grade you'll meet in lab), and good constancy over wide variations in current (20 μA to 20 mA).

This circuit lacks some refinements we will want to add – and it has one great flaw:

- its output current is small: limited to the op-amp's 25 mA (typical for the '411; current limits for all ordinary op-amps are similar).

11N.1 Evolving a regulated power supply

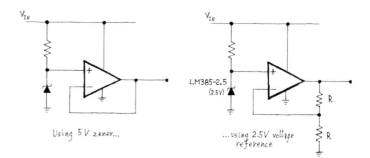

Figure 11N.5 Zener plus op-amp follower: regulator?

11N.1.5 Reference plus op-amp follower plus "pass" transistor

Figure 11N.6 is almost what we need. It can provide a large current. One could boost that further by using a *Darlington* pass transistor configuration (with its beta-squaring), or a power MOSFET (a field effect transistor that needs essentially zero input current). The op-amp – with its feedback that encompasses the pass transistor – gives the circuit very low R_{out}.

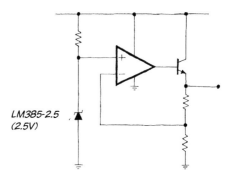

Figure 11N.6 Zener plus op-amp plus pass transistor: regulator?

However, it has a flaw that makes it unpredictable, and perhaps unusable. The output drives a power supply line which will be studded with decoupling capacitors. These are very likely to cause phase lags that push the op-amp into oscillation – as you know from your experience in Lab 9L.[2]

11N.1.6 Stabilized circuit

To keep the circuit stable despite capacitive loading, we use the methods of Lab 9L – primarily the *splitting* of the feedback path: see Fig. 11N.7.

The effective R that pairs with the feedback C to set $f_{crossover}$ is R_{Thev} for the feedback divider. This is a useful circuit. We'll add one more feature and call it done: we'll add a *current limit*.

---

[2] You might think we could say oscillations are certain, with these capacitive loads. Oddly though, extreme capacitive loading can stop oscillations. So we probably ought to say only that this circuit is vulnerable to parasitic oscillations, and must be stabilized.

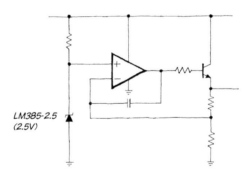

Figure 11N.7 Stabilized circuit.

11N.1.7 Current limit: a complete regulator circuit

To make the circuit foolproof, you should design it to survive abuse by fools: people who short the output to ground, from time to time.[3] All reasonably designed power supplies include this protection, implemented through the addition of one more transistor: see Fig. 11N.8.

AoE Fig. 9.2

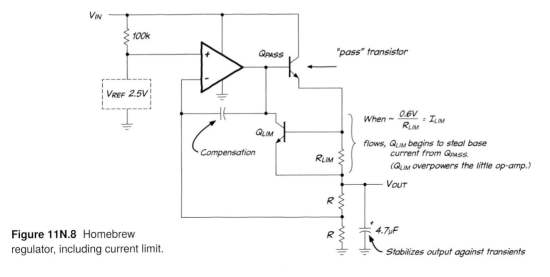

Figure 11N.8 Homebrew regulator, including current limit.

The *current-limiting* scheme: Q_{limit} begins to direct current away from the pass transistor if the output current grows too large. Note that this scheme depends on the truth that the op-amp's output current is limited; if not for that limit, the new transistor would actually *add* to the output current, making things worse.

Op-amps use the same current-limit scheme: The output stage of op-amps uses the same sort of current limit. You will recognize in Fig. 11N.9 the limit circuit in the output stage of the LF411, for example.

Full '411 circuit appears in §9S.4.

---

[3] This collection of "fools" encompasses all of us of course.

11N.1.8 Dropout voltage

Any regulator of this type ("linear," rather than the "switching" type that you will see in §11N.6) needs some minimum difference between input and output voltage. This is called the "dropout voltage," because the output drops out of regulation if you don't fulfill this requirement.

Most regulators need 2–3 V. You will find this minimum plausible when you recall the V_{BE} drop in the follower, the need of most op-amps for a volt or so between V_{out} and the positive supply, and then the current-limit circuitry.

Specialized low-dropout regulators can get by on a few tenths of a volt drop between V_{in} and V_{out}. This feature can be important in battery-powered designs, where the battery voltage may gradually droop. In such a circuit a low-dropout regulator could function long after a conventional linear regulator would have dropped out. But note that *switching* regulators may be better still for battery-powered devices. See §11N.6.

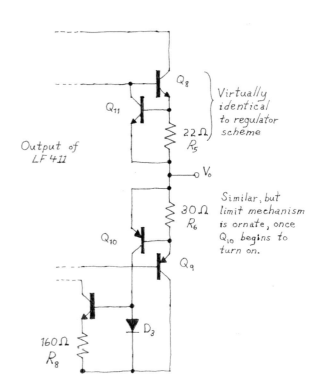

Figure 11N.9 Current limit in op-amp ('411) looks like regulator's limit circuit.

AoE §9.3.7

Low-dropout regulator: The strategy used to reduce the dropout voltage is to reconfigure the *pass* transistor. In a conventional regulator design like that of Fig. 11N.1.7, where the pass transistors is wired as a *follower*, a V_{BE} drop lies between input and output and the op-amp itself must lose some voltage between its positive supply and its output.

The low-dropout regulator wires the pass transistor not as follower but as what might be described as an inverting amplifier: see Fig. 11N.10.

This is a circuit we mentioned back in Chapter 9N, as a rare case with an inversion within the feedback loop. The feedback capacitor is required – providing *split* feedback. The resistor between voltage reference and inverting input permits feedback to that terminal despite the "stiffness" of the good voltage reference.

AoE §9.3.12F

Voltage Regulators

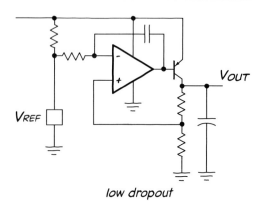

Figure 11N.10 Low-dropout regulator: pass transistor is reconfigured. Note funny feedback!

A large bypass capacitor on the output also tends to tame the circuit's tendency to be jumpy.[4] When you don't need the low-dropout feature you ought to use the ordinary configuration, which is more stable.

11N.2 Easier: 3-terminal IC regulators

Now that you have paid your dues by re-inventing the regulator, we'll let you consider some regulators that are much easier to use.

11N.2.1 Fixed output: 78xx

The whole circuit of Fig. 11N.8 – reference, op-amp and pass transistor – plus somewhat more, is available on one chip. The simplest of these regulators is embarrassingly easy to use: this is the *three-terminal* fixed-output type in Fig. 11N.11.

AoE §9.3.2

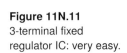

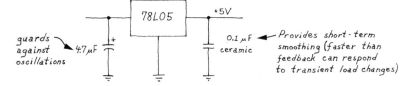

Figure 11N.11 3-terminal fixed regulator IC: very easy.

You may find it gratifying to notice – in the schematic of Fig. 11N.12 – that the designers of the 78xx regulators seem to have been looking over our shoulder as we evolved the linear regulator.

The circuit is just what we designed, apart from use of a Darlington pass transistor and the application of a peculiar boost (labeled ΔV_{CL}) to the base voltage of the current-limiting transistor. The diagram also omits the necessary stabilizing elements.

This device, like other IC regulators, limits not only its output current but also its own *temperature*. Such a thermal shutdown protects the regulator when what is excessive is not *current* alone but *power*. (You will demonstrate this protection in the lab, by putting a large voltage drop across the regulator: $V_{in} \gg V_{out}$.) The thermal limit also protects you, the designer, against the possibly destructive effects of inadequate heat-sinking (see §11N.3).

---

[4] Compare AoE §9.3.2 noting role of a large capacitor to tame a negative voltage regulator with a similar pass-transistor configuration, the LM7905.

11N.2 Easier: 3-terminal IC regulators

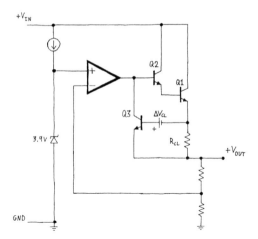

Figure 11N.12 Simplified circuit of the classic 3-terminal regulator, LM7805, looks a lot like our recent design.

11N.2.2 Variable output: 317

The *317* is almost as easy to use as the simpler 78L05, and is more versatile: it allows you to adjust V_{out}. So you need not stock an IC for each voltage that you may need. It can also be used to rig up an easy current source for values above about 10 mA.

The left-hand sketch in Fig. 11N.13 shows a simplified version of the '317's circuitry.[5] The right-hand sketch shows the device wired to source 10 mA into a load returned to ground (or to a negative supply).

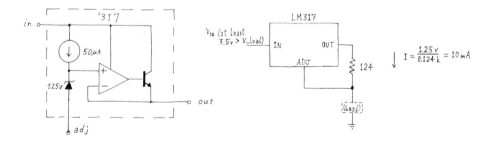

Figure 11N.13 LM317 variable regulator: simplified circuit, and use as current source.

The '317 likes to hold a *constant* voltage (1.25 V) between its two output terminals, as the diagram above indicates; it uses no "ground" terminal.

We have shown the '317 as current source not because this is the most common use for the part; it isn't. Instead, we find this the easiest way to understand the '317's *voltage* regulation: just let it source this fixed current through a resistor to ground. The fixed current times this resistance defines the voltage at ADJ. Then $V_{\text{out}} = V_{\text{ADJ}} + 1.25$. So, the circuit in Fig. 11N.14, for example, permits adjustment of V_{out} through the range 5–10 V.

The notion that the device feeds a fixed current to R_1 is only one way to see its operation. You may prefer the formula

$$V_{\text{out}} = 1.25V(1 + R_2/R_1)$$

where R_1 is the current-setting resistor, R_2 the resistor to ground. But tidy formulas always threaten to obscure how the circuit achieves what it does.

---

[5] A more detailed version showing current limit and Darlington pass transistor appears in AoE Fig. 9.9.

A circuit example: adjust V_{out}: Figure 11N.14 shows a sample circuit to bring these abstractions down to earth. The two resistors below ADJ let the 10 mA current drop 3.75 V to 8.75 V, to give the advertised V_{out} range.

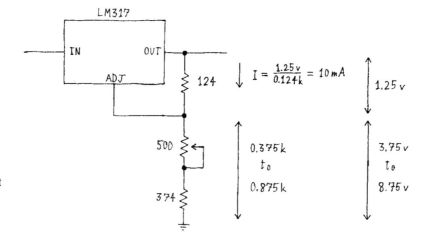

Figure 11N.14
Example of '317 circuit with V_{out} adjustable through the range 5 V to 10 V.

But don't get carried away: V_{out} range is limited: One might be tempted to think the '317 all-powerful: make a 1000 V supply by installing a 100k R to ground? No. V_{out} is limited to about 37 V because of a 40 V limit on difference between V_{in} and V_{out}: if the output is shorted to ground with V_{in} very large (substantially greater than 40 V) the part will be destroyed.

'317 peculiarities: maintain minimum I_{out}: Part of the '317's cleverness lies in the fact that it lacks a ground terminal.[6] But the side effect of the lack of a ground terminal is the fact that the '317 is *powered* by the current that passes through the device – from IN terminal to OUT. So, the regulator fails if the user sets that current too low. The *minimum* output current is specified as sometimes 5 mA, sometimes 10 mA.[7] Use 10 mA, to be safe; it follows that R_1 should be about 120 Ω: or 124 Ω if you're using 1% values.

11N.3 Thermal design

Overheating is an issue we have not considered until now (except perhaps in specifying large transistors for the push–pull drivers of several op-amp labs). We're obligated to consider it when designing power supplies because the currents are much larger than those we have been using to this point. To keep a part from overheating, we need to let it dissipate heat at the same rate that its electrical power generates heat.

So much is obvious. But you might not anticipate that the rate at which heat can be dissipated is calculable with methods exactly analogous to the use of Ohm's law. All we need to do is accustom ourselves to the changed units.

---

[6] "Why is this clever?," you may protest. Well, this peculiarity allows it to act as a floating current source and thus to allow adjustment of V_{out} simply through adjustment of a single resistor.

[7] The datasheet from National Semiconductor (now absorbed into Texas Instruments) is ambivalent on this question: as it specifies device properties, in its tables of "... Electrical Characteristics," the datasheet recites Iout $\geq$ 10 mA. But it uses half that value in all the suggested circuits that appear later within the same datasheet.

11N.3 Thermal design

11N.3.1 Thermal transfer, in general

A look at the *units* will reveal the analogy between thermal and electrical flows and resistances. In place of voltage (by which we always mean voltage difference) we use temperature difference. Temperature difference drives the flow of heat, just as voltage difference drives the flow of current. The rate of thermal flow, a rate of energy or heat transfer, we measure as power: in watts, the W in the equation below.

The resistance to heat transfer is analogous to electrical resistance. Here, its units are °C/Watt: temperature difference per rate-of-heat-transfer. In the familiar electrical case, resistance in ohms is the equivalent: voltage difference per current (which is rate-of-charge-transfer).

Electrical $I = V/R$

Thermal   Rate of heat transfer = (Temp Difference/Thermal Resistance)
$W = $ (Temp difference$/\Omega_{\text{thermal}}$)

Figure 11N.15 represents the analogy graphically, while, to make things less abstract, Fig. 11N.16 is an exploded sketch of an IC and its heat sink, showing where these thermal resistances appear. The subscripts in the figure bear explaining. For example, $R_{\theta JC}$ is the thermal resistance between *junction* (the guts of the IC) and the *case*.

Figure 11N.15 Thermal resistance calculations are much like what you're used to with Ohm's law.

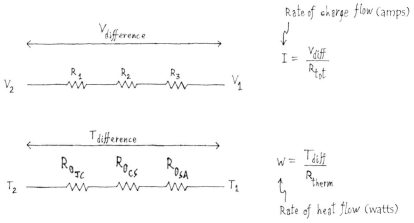

Figure 11N.16 Illustrating where the thermal resistances occur.

486 Voltage Regulators

If you know how much power your part needs to dissipate, then you use the following:

$$R_{Thermal} = \frac{T_{Junction} - T_{Ambient}}{Power}$$

to determine what total thermal resistance your circuit can tolerate. You may be stuck with some of the resistances – such as *junction-to-case*, a value determined by IC design. But you have a choice of heat-sinks, and you can choose one effective enough to keep the total $R_{Thermal}$ acceptable. We go through such an exercise in Chapter 11W.

11N.4 Current sources

We have noted that the '317 can serve as a current source – but only for currents of about 10 mA and up. Let's look at some current-source ICs that are more versatile.

11N.4.1 An IC for intermediate currents

The LT3092 (see Fig. 11N.17) can *sink or source* 0.5–200 mA. It can do both because it is a *two terminal* device. Placement of the *load* determines which job the part performs. This versatility contrasts with the behavior of bipolar transistor current sources that we have seen, where *npn* can only *sink*, and *pnp* can only *source*. (On the other hand, you *have* met a two-terminal current-limiting device: the JFET described in §11N.4.3, the part that you used to define the "tail" current of Lab 5L's differential amplifier.)

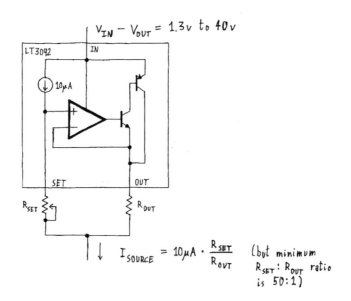

Figure 11N.17 LT3092 two-terminal current limiter.

The dynamic impedance (R_{out}) of the LT3092 is spectacular at DC (100MΩ) and good at low frequencies: about 3 MΩ @ 1 kHz.

11N.4.2 A bipolar IC for low currents

For lower currents (50–400 μA) the REF200 (see Fig. 11N.18) is very neat, and provides enormous R_{out} at DC, as does the LT3092. In §11W.2 we have posed some puzzles asking how to apply the three elements of this part to sink or source a variety of currents: 50, 100, 200, 300, 400 μA. The mirror, as indicated in the figure, holds its right-hand current equal to the current fed into its left-hand side.[8]

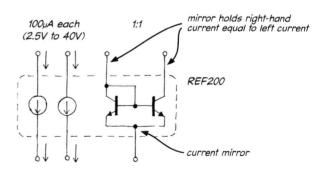

Figure 11N.18 REF200 low-value current-source IC.

11N.4.3 JFET as current source

You met this two-terminal current source in Lab 5L, now shown in Fig. 11N.19. It was pleasantly easy to use: it comes in a glass package that looks like a diode. Within the diode-like package is a JFET – a junction field-effect transistor, a transistor type that you will meet again in the next chapter to implement an automatic gain control circuit. (In passing, we also note that JFETs give the '411 op-amp its high input impedance.) The JFET's two control terminals are shorted together within the package, and in this configuration the JFET runs at a fixed current.

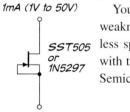

Figure 11N.19 JFET (Junction Field Effect Transistor) two-terminal current source.

AoE §9.1.1C

You have experienced what is appealing about this device: it is very easy to use. It does have weaknesses. Compared with the LT3092 and the REF200 its dynamic impedance (R_{out}) is good but less spectacular: about 1 MΩ @ 1 mA. Its serious drawbacks are its price, (several dollars, varying with the value of *current*[9]) and its poor tolerance (± about 20% for the SST-505 and for the Central Semiconductor parts like 1N5294, as well).

11N.5 Crowbar overvoltage protection

The failure of a regulator could feed an excessive voltage to a lot of precious electronics. The most familiar hazard is the possibility of overvoltage (or "surge") applied to an entire computer.[10] Figure 11N.20 shows a circuit that shuts down the supply (clamps it to approximately 1 V) when the voltage climbs too high. (ICs are available to do the voltage sensing, too.)

The word "crowbar" apparently refers to the image of someone (courageous? foolhardy?) shutting down a huge power supply by shorting it to ground with a massive piece of steel.

---

[8] The mirror is drawn as a simple mirror, but in fact is a Wilson or cascode mirror, as noted in Chapter 5S.

[9] We were charmed to discover, in the strange pricing structure of these parts, the first objective evidence ever adduced for the proposition that most practicing engineers are as lazy as we. When 1 mA versions of this part were offered, they carried a dramatic premium over parts like the 0.75 mA value that we use. Why? Because everyone likes to do Ohm's law arithmetic with the value *1*, and we lazy people drive the price of that part way up.

[10] In the concluding labs for this course, in which they had constructed a computer from many ICs, one pair of students accidentally demonstrated this vulnerability in horrible form: they used a variable supply for their home-made computer, adjusting it always to 5 V: well, *almost* always. One day when they failed to check this 5 V value, they cooked many ICs in their computer. Finding and replacing the destroyed parts was a painful lesson in the hazards of overvoltage. The present version of the microcomputer lab breadboard includes a *crowbar* clamp like the one described in this section. So such disasters are now very much less likely.

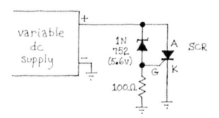

Figure 11N.20 Crowbar overvoltage protection (the circuit Fig. 11L.8).

Since the SCR – the diode-like element on the right in Fig. 11N.20 – turns ON when its control terminal ("gate," G) reaches about 0.6 V, much as a transistor does, the circuit of Fig. 11N.20 fires and clamps its output if the input reaches about 6.2 V. If this circuit is to succeed in protecting downstream circuitry, the clamp itself must be stronger than the failing upstream regulator: a fuse should be placed upstream, to blow when the clamp fires.

A peculiarity of the SCR distinguishes it from the somewhat similar power transistor: once turned on, it remains *on* after the gate drive has been removed. Thus the crowbar circuit operates like a circuit breaker, locking the supply in shutdown state. Only turning off the SCR current, by shutting down the faulty power supply (or by blowing an upstream fuse), can release the SCR.

11N.6 A different scheme: switching regulators

AoE §9.6.1

A linear regulator is doomed to inefficiency, since its strategy for holding V_{out} constant as V_{in} varies is simply to soak up the difference. A linear regulator fed a ripply 10 V can put out a very constant 5 V – but the power wasted in the regulator, in this case, would equal the power delivered to the load. Efficiency in such a case is 50%. One cannot do much better, because uncertainty about the *line* voltage, and the need to keep ripple from dipping below the regulator's dropout voltage obliges one to put V_{in} 4 or 5 V above V_{out}. What is to be done?

The answer is to adopt a method that is quite different. Don't soak up the difference between V_{in} and V_{out} in order to maintain V_{out} at a given level. Instead, use a *switch* to send intermittent gushes of current into a storage element as needed. When a gush is not needed, leave the switch turned *off*.

11N.6.1 Preliminary: getting used to inductors' peculiar proclivities

Because we have not used inductors much in this course, we may owe you a reminder of the inclination of inductors to keep a current going, once it is flowing. All the switching power supply configurations (shown below in §11N.6.2) exploit this behavior.[11]

We hope you demonstrated this property for yourself in a switch exercise back in Lab 4L. In class we repeat this exercise as a demonstration, provoking an inductor to show off its flywheel-like behavior. The circuit we use is just a switch to ground (it happens to be a MOSFET – the field-effect transistor that you will meet next time), with an inductor linking the switch to a positive supply. The positive supply is a single AA cell, at about 1.6 V: see Fig. 11N.21.

We wire it first as in "stage one." The square wave input turns the FET switch ON; current ramps up in the inductor; the switch turns OFF – and the inductor wants to keep current flowing. Figure 11N.22 shows the inductor's response.

As the switch opens, the inductor (indignant!) insists on keeping the current flowing, by driving the

---

[11] But note that many popular topologies do not. See, for example, the "forward converter" of AoE Fig. 9.74, which uses a transformer to transfer energy rather than store it.

11N.6 A different scheme: switching regulators

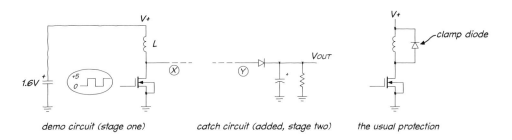

Figure 11N.21 Inductor switching demonstration: voltage spikes can be caught to generate large V_out.

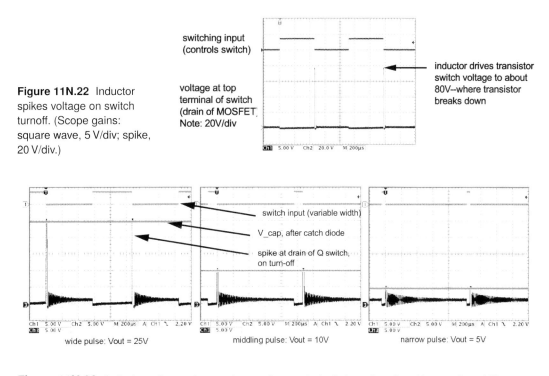

Figure 11N.22 Inductor spikes voltage on switch turnoff. (Scope gains: square wave, 5 V/div; spike, 20 V/div.)

Figure 11N.23 A diode and capacitor can be used to catch the inductor's spikes (three pulse widths shown). (Scope gain: 5 V/div.)

switch voltage (at **X** in Fig. 11N.21) higher and higher – until, at about 76 V, the switch "breaks down" and conducts.

This breakdown is not good for a transistor switch. Normally, one would protect the switch with a diode as shown on the far right of Fig. 11N.21. For purposes of the demo, however, we want to show these large spikes – and in a moment we will put them to use. Note that the *supply* voltage is a modest 1.6 V.

This perverse behavior of the inductor can be put to use. The voltage spikes can be used to source current into a capacitor in repeated surges. The capacitor thus can charge to a voltage much higher than the original supply voltage – as in the demonstration.[12]

---

[12] An old fashioned camera's flash circuit often works this way: it starts with a battery voltage of perhaps 3 V, and when you turn on the flash you may hear the circuit singing. If so, you are hearing the switching of an inductor, used to charge a storage capacitor up to 300 to 400 V to fire the xenon flash tube when you press the shutter release.

490 Voltage Regulators

Figure 11N.23 shows such an application of the seemingly mischievous voltage spikes. The capacitor voltage rests close to the voltage of the repeated spikes, which are shown for three input pulse widths.

By varying the width of the input pulse, we varied V_{out}, as shown in Fig. 11N.23, from 5 to 25 V. The ON time of the switch, determined by the pulse width, determines how close the inductor comes to its maximum current. At the pulse repetition rate shown, we found that further widening of the pulse did not raise V_{out}.

11N.6.2 Three switching configurations: "Boost," "Buck" and Invert

AoE §9.6.4

The demonstration – which took 1.6 V up to 20 V or so – used one of three basic circuit configurations available to switching regulators. Appropriately, that configuration is called "boost," labeled "B" in Fig. 11N.24. (All three circuits are drawn showing Schottky diodes for their low forward voltage drops; for even greater efficiency, a MOSFET switch can be used in place of the diode.)

The first configuration is a step-down, called "buck." As help in getting a grip on inductor behavior we offer a sketch of voltage and current waveforms in the buck circuit. Note that the transistor switch is turned ON by taking its control input to ground (the transistor is a *p*-channel MOSFET, analogous to a *pnp* bipolar transistor).

Figure 11N.24 Three switcher configurations.

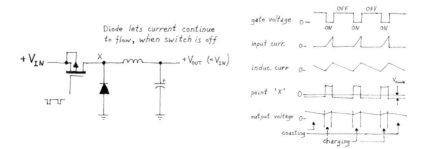

Figure 11N.25 Current and voltage waveforms in a buck (step-down) switcher.

The *ripple* on the output voltage of Fig. 11N.25 may look familiar – it may look like the 120 Hz ripple of an unregulated supply. Yes and No. Note the radical difference in frequencies: the switching frequency usually is in the range 100 kHz to 1 MHz, to allow use of small inductors and capacitors. The "droop" time here is on the order of a microsecond rather than the 8 ms of a powerline-driven full-wave rectifier.

11N.6.3 Efficiency

AoE §9.6.1

You can convince yourself with a few seconds' thought that a *switch* that is either fully ON or fully OFF ideally dissipates no power at all. Consider the two cases in Fig. 11N.26.

In life, things aren't quite so good. Some power does get dissipated: in the inductor; in the switch, with its non-zero R_{on} and capacitance; in the diode, which conducts while the switch is OFF (up from *ground* – startling those of us unaccustomed to inductors' strange habits). Nevertheless, the results can come close to the ideal; much closer than a linear regulator can. Efficiency of 80–95% is feasible.

11N.6 A different scheme: switching regulators

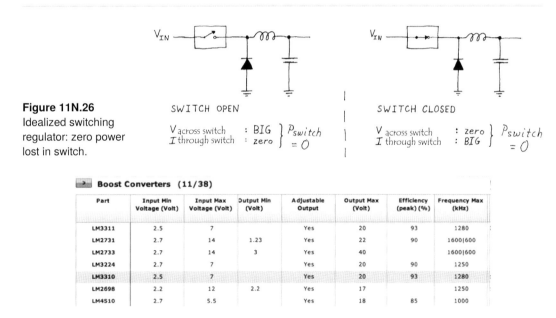

Figure 11N.26 Idealized switching regulator: zero power lost in switch.

Figure 11N.27 Excerpt from National Semiconductor table of switching regulators.

Figure 11N.27 contains part of a table describing National Semiconductor's[13] *boost* switching regulators (those that generate $V_{out} > V_{in}$). Note the impressive efficiencies, all at 85% or more.

Efficiency is the great strength of switchers, but they have other virtues: they can generate $V_{out} > V_{in}$, as in the examples we just looked at. They can change the *sign* of a voltage input – or can do both transformations at once. Those tricks allow powering circuits at a variety of voltages (such as 3.3 V, 2.5 V, 1.8 V) from a single source.

For battery-powered devices, like a cell phone, the importance of efficiency is obvious: it permits long battery life. For line-powered circuits the importance is less obvious. The efficiency is valuable there rather because it permits supplies that are lighter and smaller and less hot than a linear supply.

11N.6.4 Implementing the feedback

We have not mentioned, to this point, how the switch is controlled in a switching regulator. Negative feedback compares V_{out} against a reference voltage. In this, the switcher behaves like the linear regulator. But the feedback circuit does not simply hold the switch steadily ON when V_{out} is low, OFF when V_{out} is higher than the reference. Such a scheme could produce very large ripple and could require huge inductors and capacitors.

Instead, the switching occurs continuously – or nearly so – and what is varied is the *duty cycle* of the switch: the percentage of time it spends ON in each cycle. (It is also possible to vary frequency but varying duty cycle is the more usual scheme.)

The switching regulator that you meet in the power-supply lab provides a hybrid scheme, varying not duty cycle or frequency but the duration of a burst of high-frequency switching cycles (it describes itself as a "gated oscillator switcher;" others call this design "Pulse Burst Modulation" or "hysteretic conversion".[14] This note points out the good feedback stability, i.e., freedom from loop oscillation, of PBM.).

When V_{out} falls a little below the reference voltage (by an amount set by a built-in hysteresis circuit),

---

[13] National was swallowed up by Texas Instruments in 2011.
[14] See a good note by Analog Devices: https://LAoE.link/Switching_Regulators.pdf.

the regulator fires a burst of 20 kHz cycles, driving V_{out} up. Once V_{out} rises high enough, all switching terminates and the regulator's power consumption falls close to zero.

A result of this scheme is ripple at frequencies well below the 20 kHz switch rate: as low as 2 kHz in the cases shown in Fig. 11N.28. This is in a frequency range that can be very annoying to human ears,[15] so it seems a strange choice by the designers of the part.

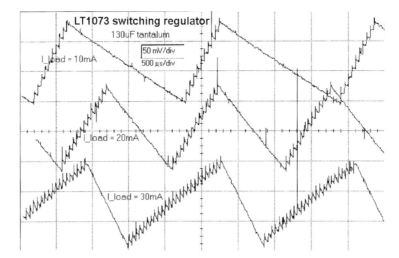

Figure 11N.28 Ripple on output of 1073 switch-burst regulator.

The motives for this scheme are two. One is to minimize the regulator's power dissipation: when the switching shuts off, current draw falls from hundreds of milliamps to about 100 μA, and under very light loading the circuit can be idle for "seconds at a time," the datasheet promises. A second virtue of this sort of design – in which the feedback loops relies on hysteresis with bang–bang behavior – is that it is not vulnerable to the instabilities that can trouble continuous-feedback designs. You know, from your experience in the *nasty oscillators* and PID labs, that ordinary feedback loops can be upset, especially by capacitive loading.

11N.6.5 Switchers aren't always what you want

They are noisy...: For digital devices, which shrug at low-level noise, switchers are the right choice. But the switching noise voltage that always appears on their outputs can rule them out for sensitive analog circuits.

AoE §9.5.5, especially Fig. 9.53

...and can be difficult...: And we should mention another reason why many people reach for a good old linear regulator: a switching supply can be subtle and difficult to design. It can be hard to keep them stable. And it is not only students in an introductory course like this one who might shy away from switchers. IC manufacturers know that switching power supplies have a reputation that tends to scare engineers.

...So manufacturers offer help: National Semiconductor/TI offer a series that they call "Simple Switchers," and since the name may not be enough to persuade its customers it also provides a web design program that steers you to particular ICs offered by them[16] given your design goals, and finishes

---

[15] "Ears,", you protest again? Who is listening to this switching frequency? Well, ideally, no one. We were unable to hear our lab circuit. But inductors can sing at the frequency which they are driven, and so can capacitors. So it's a good idea to put switching noise above the audible range.

[16] https://LAoE.link/TI_Power_Designer.html. For some reason their web service doesn't seem aware of their competitors' parts.

AoE §9.6.5

the design for you (much as TI's FilterPro does for active filters): see Fig. 11N.29. This web application goes so far as to make up a "Bill of Materials" (BOM) for you: see Fig. 11N.30.

Figure 11N.29 National/TI's Webench leads the timid by the hand, in switcher design.

Figure 11N.30 Webench makes it really easy.

You do feel a bit dimwitted when you use this service. But you're not required to admit that this is how you "designed" your switching supply!

And, yes, switchers can be hard even for smart engineers: Linear Technology, also feeling the designer's pain, offers a long application note by their late wizard, Jim Williams, entitled *Switching Regulators for Poets: A Gentle Guide for the Trepidatious*.[17] Williams, who was not a beginner at electronics, wrote:

Before this effort, my enthusiasm level for switchers resided somewhere between trepidation and terror. This position has changed to one of cautiously respectful optimism.

If Jim Williams could feel optimistic, so can you.

11N.7 AoE reading

Chapter 9:

Unregulated supply: §9.5;
Evolving a linear regulator: §9.1;
IC regulators: §9.3;
Switching regulators: §9.6.1ff.

---

[17] This note was written thirty-odd years ago, but the fundamental difficulties of switching regulators have not changed much since then – except for the arrival of delightful crutches like National's Webench. The note is AN25 (1987), available at `https://LAoE.link/JW_AN25.pdf`.

11L Lab: Voltage Regulators

This lab begins with a trial of your design for a home-made voltage regulator. There is nothing very new here: the only elements you have not seen in a previous lab are:

(1) the voltage reference – a super-duper zener, in effect; and
(2) the current-limit included in the bipolar version described below.

The MOSFET version – an alternative regulator design – allows you to try the new transistor type as you try the regulator. Both home-made regulators raise stability issues that you will recognize from your experience with Lab 9L's "nasty oscillators." This exercise is not realistic: you are not at all likely to design a regulator from parts; we hope, though, that designing one once will give you insight into how a linear regulator works.

In the remainder of the lab, you will try first IC *linear regulators* – very straightforward; then an IC *switching regulator*. The switcher uses feedback to stabilize V_{out}, as the linear regulator does, but it regulates the output voltage in a way that is unfamiliar to us in this course: by switching an inductor, and exploiting the inductor's efforts to keep current flowing when the switch opens. The switching regulator can achieve effects that at first glance seem magical; V_{out} greater than V_{in}; V_{out} negative, for V_{in} positive. We hope you'll be impressed by this little IC.

11L.1 Linear voltage regulators

11L.1.1 Voltage reference

The homemade voltage regulator that we ask you to build in §11L.1.2 is required to put out a voltage that remains constant despite variations in loading. This behavior is familiar to us from all the low R_{out} devices we have used and built, including all our voltage *followers*. But today we ask our regulator to do something we have not seen before: hold its output constant despite variation in its *input* voltage.

A zener can provide this constancy, as you know. We will use a zener later in §11L.1.6. But for better constancy of V_{out}, which we want from our homemade regulators, we will rely on a *voltage reference* IC, the LM385–2.5.

Its behavior resembles that of a zener – but is much better. It shows very good constancy of V_{out} as its current varies over a wide range. This you can see in the left-hand plot of Fig. 11L.1.

As you read the plot, notice that the scales marked on left and right vertical axes, indicating voltage variation for the two devices, differ by a factor of almost 200.

11L.1.2 Voltage regulator: your design

Please design, then try out, a voltage regulator built to satisfy either of the descriptions set out below. In addition to performance specifications, each design problem includes a specification of available

11L.1 Linear voltage regulators

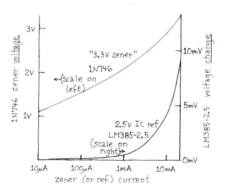
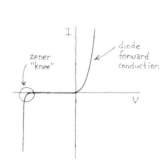

Figure 11L.1 *I–V* curves for zener and IC voltage reference compared.

parts. After specifying the task, just below, we will state more specifically what we mean when we ask you to *try out* the circuit.

Design task

Option A: Bipolar 5 V regulator, with current limit
- V_{out}: 5 V (±5%);
- V_{in}: 8–25 V;
- Current limit: 100 mA, more or less;
- Kit of parts:
    - op-amp: 1/2 LM358 (dual op-amp, single-supply; see Lab 9L);
    - "pass" transistor: 2N3055 (pinout: BCE, seen from front);
    - current-limit transistor: 2N3904;
    - voltage reference: LM385–2.5, a 2.5 V super-zener ;
    - frequency compensation: start with *none*; add it when you've seen the problem.

Figure 11L.2 Pinout of LM385 voltage reference.

Hints
- Look at the '723 IC's schematic, AoE §§9.2 and 9.2.1: in effect, you are building a homebrew '723;
- when you calculate the resistor you need to get a 100 mA limit, note two complicating facts:
    - the '358 provides a rather high maximum output current: 40 mA, typical; most of this will be *added* to your circuit's output current, under current-limit conditions;[1]
    - when this is occurring, the current-limit transistor will be passing enough current so that the usual rule V_{BE}=0.6 V will not hold. Note that V_{BE}≈0.6 V at 1 mA, for a '3904; V_{BE} rises about 60 mV per decade of current, so at, say, 35 mA you may see V_{BE} of about 0.7 V, and in your particular circuit you may see as much as 0.75 V.

Or, an alternative design (a little more ticklish, because of its tendency to *oscillate*).

Option B: Low-dropout 5 V regulator This design differs in omitting the current-limit (normally very desirable!), so that you can minimize the *dropout voltage* – the minimum difference between regulator input and output. That difference defines the minimum power wasted in a

---

[1] You could limit the magnitude of the current that the current-limit transistor could add to I_{out}: insert a resistor on the base lead of the *pass transistor*. Put this resistor upstream of the I_{limit} transistor tap. 300 Ω, for example, would limit current to about 25 mA for a V_{in} of 8 to 10 V.

Lab: Voltage Regulators

linear regulator, and – just as important, in a battery-powered instrument – determines just how low V_{in} can fall before the output fails.

- V_{out}: 5 V (±5%).
- V_{in}: from slightly more than 5 V up to 25 V.
- Kit of parts:
    - op-amp: 1/2 LMC6482 (dual op-amp, single-supply, rail-to-rail; see LM358 in Lab 7L) for pinout;
    - "pass" transistor: *Bipolar:* 2N3906 (same pinout as 2N3904: EBC) or *MOSFET*: ZVP3306 ($R_{on} = 14\,\Omega$; pinout: SGD, seen from front), BS250P ($R_{on} = 14\,\Omega$) or VP0106 ($R_{on} = 11\,\Omega$);
    - voltage reference: LM385–2.5, a 2.5 V super-zener; and
    - frequency compensation: start with *none*; add it when you've seen the problem.

Hint
- Beware the inversion by the transistor circuit, which is *not a follower*. Instead, this is an inverting amplifier. We will need to take feedback from the transistor's *drain* (FET drain is equivalent to bipolar collector): I_D grows as gate voltage *falls*. That fact needn't cause you any difficulty, as long as you are aware of it; if you know about this inversion, you know how to keep the sense of feedback negative *in fact* (though the feedback may look shocking *on paper!*).

11L.1.3 Trying your circuit

Oscillations: Your circuit probably does *not* work well if you have followed our instructions: we have asked you to omit "frequency compensation," and your circuit output is very likely to oscillate, at least when presented with a capacitive load. Add a ceramic capacitor in the range 0.01–0.1 μF, between V_{out} and ground. Such a cap normally stabilizes a power supply, but it is likely to de-stabilize the bipolar regulator, at least, looking like a capacitive load. Incidentally, make sure that you watch the output with a *scope*: if you make the mistake of thinking of this as a DC circuit which one can therefore understand with DC voltmeter alone, you will find that oscillations can cause DC errors that are puzzling indeed.

Once you have seen the oscillation problem, try to solve it. In both regulators, a feedback capacitor between op-amp output and the inverting input[2] should do the job (killing the high-frequency gain[3]). This is the remedy we have called "split feedback." What value is necessary? Note the relevant R: the R of the lower-frequency feedback path (here, R_{Thev} for the feedback divider). Choose RC to bypass the nasty stuff well below the frequency of oscillation that you have observed.

In addition, you should place a couple of capacitors to ground, at the output: a small ceramic (for high-frequencies; say, 0.01–0.1 μF) and also a big tantalum (a little less good at high frequencies: try 4.7 μF or bigger).

Vary V_{in}: A regulator should provide constant output despite variations in input – including the 120 Hz "ripple" that rides an unregulated input. To save time, instead of applying *ripple*, just vary V_{in} (from a variable lab power supply) by hand, and watch V_{out}. You're not likely to see any variation in V_{out}, if your circuit is working right – until you take V_{in} so low that the circuit "drops out" of regulation.

---

[2] Note that even for the low-dropout design – where the feedback from circuit output goes to the *non-inverting* input because of an inversion effected by the PMOS circuit – the *high-frequency* feedback provided by the feedback capacitor must go to the *inverting* input in order to give feedback the proper sense.

[3] You may prefer to say, as we did in Lab 9L, that the cap provides a separate feedback path, bypassing the network that produces dangerous phase shifts at high frequencies.

11L.1 Linear voltage regulators

Measure the dropout voltage: A regulator's "dropout voltage" is just the minimum difference between V_{in} and V_{out} required in order to let the regulator hold V_{out} where it ought to be. Measure $V_{dropout}$ for your homemade regulator. What characteristic of what device(s) in your circuit imposes this minimum on V_{in}?[4]

The answers differ for the two design options, as you would expect. $V_{dropout}$, the minimum difference between V_{in} and V_{out}, is the sum of...

Option A: Using an *npn* pass transistor
- minimum difference between the op-amp's positive supply and its output: about 1.5 V...
- ...plus V_{BE} for the pass transistor (roughly 0.6 V)...
- ...plus the drop across R_{lim} in the current-limiting circuit (up to 0.6 V).

Option B: The low-dropout version
- Assuming that the you use a PMOS FET, which needs several volts between its input terminal ("Source") and control terminal driven by the op-amp ("Gate"), we don't need to worry about the specification we called "minimum difference between the op-amp's positive supply and its output" above; op-amp output will be well below the positive supply voltage.
- Instead of V_{BE} for the pass transistor, we need consider only its minimum voltage between in and out terminals (its "Source" and "Drain"). For low values of V_{DS} the MOSFET behaves like a small-valued *resistance*, so this minimum depends on output current and the transistor's R_{on}.
  In this circuit you can see that this R_{on} value is important. The small MOSFET would drop almost 1.5 V at 100 mA; the big one would drop a mere 4 mV.
- Since the low-dropout version lacks circuitry to limit its output current, there is nothing further to consider in accounting for the dropout voltage of this version.

Loading: Now see to what extent V_{out} remains constant independent of *loading*. As "load," use 10 Ω resistors, placed in parallel (why not smaller ones?). (A current meter placed in series between V_{out} and this load will let you measure the load current.) Don't go overboard in loading the low-dropout version, by the way, since it has no built-in current limit. Feel free to go overboard with the current-limited version – once you are convinced that the current limit works: you can even *short* the output to ground.

You should find – if you care to look – that the low-dropout regulator's dropout voltage varies with loading. Once you have decided what's imposing the minimum dropout voltage, you'll recognize why this relation between dropout and loading is necessary.

Figure 11L.3
78L05 3-terminal 5V regulator.

11L.1.4 Three-terminal fixed regulator

This device, shown in Fig. 11L.3, is embarrassingly easy to use. It is so handy, though, that it's worth your while to meet it here. It protects itself not only with current limiting, but also with a thermal sensor that prevents damage from excessive power dissipation ($I_{out} \times [V_{in} - V_{out}]$), an overload that could occur even though the current alone remained below the limiting value. You will watch this thermal protection at work, and incidentally will see the effect of *heat sinking* upon the regulator.

Watch this thermal protection by providing a load that draws less than the chip's maximum current of 100 mA. Use two 120 Ω resistors in parallel to ground; see Fig. 11L.4. (Check, with a quick calculation, that you are not overloading these 1/4-watt resistors.)

---

[4] This is a rhetorical question! See below.

Lab: Voltage Regulators

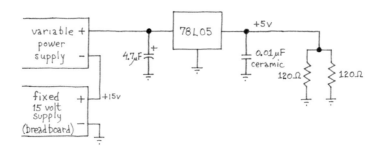

Figure 11L.4 78L05: demonstration of thermal protection.

Note: We suggest you *stack* the variable supply on top of the breadboard's fixed +15 V supply. This scheme *requires* that you *float* the negative terminal of the variable supply; that terminal must *not* be tied to world ground.

A calculation, for the zealous: This demonstration is fun even if you don't predict what V_{in} will bring on current limiting, but you can make the experiment especially satisfying if you try to predict what voltage across the regulator will bring on thermal limiting. That is not hard, though it might take you a while to dig it out of the datasheet. The thermal specifications state a *maximum permissible junction temperature* of 125°C and give the *thermal resistance* between junction and ambient, which in this case means both the ambient air and the circuit board to which the regulator's leads are attached. You will be using the TO92 package – plastic – which dissipates substantial heat through its leads (even the length of these leads matters, you'll notice, from the curves shown in Fig. 11L.6).

To calculate the maximum power the package can dissipate before overheating the junction, plug in the values here.

78L05: Thermal Specifications

Package	Thermal Resistance (typical, max)	
	$R_{\theta JC}$	$R_{\theta JA}$
TO-39	20°C/W, 40°C/W	140°C/W, 190°C/W
TO-92	180°C/W	190°C/W

Figure 11L.5 Push-on heat sink for T092 package.

Definitions:

$R_{\theta JC}$ = thermal resistance, junction to case

$R_{\theta JA}$ = ... junction to ambient: includes JC and CA (case to ambient) in series

Assume that the device begins to limit when its junction temperature reaches about 150°C – but don't be shocked if your calculation indicates the device may be tolerating a higher temperature, perhaps as high as 200°C. Note units of *thermal resistance*, R_θ: °C/watt.

Experiment: dropout voltage and thermal self-protection: Gradually increase V_{in} from close to 0 V. As you proceed, note:

- the dropout voltage; and
- the V_{in} that actually evokes the chip's thermal self-protection.

11L.1 Linear voltage regulators

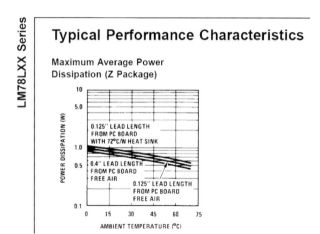

Figure 11L.6 78L05 max power dissipation depends on lead length (T092 package).

By the way, *what* does the chip do to limit its power dissipation? How will you know when the chip has begun to take this protective action?[5]

When you notice the chip limiting its own power dissipation, you will be able to call off this self-protection by cooling the chip. Try putting a bit of wet tissue paper to the 78L05's fevered brow. It should recover at once. Then try a push-on heat sink instead. You may have to crank up V_{in} to see the self-protection begin again. Now fan the regulator, or blow on it. Does the output recover, once more?

11L.1.5 Adjustable three-terminal regulator: 317

This regulator allows you to select an output voltage by use of two resistors. You can make a variable output supply by replacing one of them with a variable resistor. The 317 lets you stock one chip to get all the positive supplies you need (at least, up to 1 A output current); it also lets you trim to exactly 5 V, for example, if the 78L05's 5% tolerance is too loose. In addition, the 317 is easy to wire as a current source. In other respects this regulator is much like the 78L05: it includes both current and temperature sensing to protect itself from overloads.

Wire up the circuit in Fig. 11L.7. Try $R = 750\,\Omega$; what should V_{out} then be? Measure it.

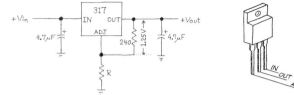

Figure 11L.7 317 voltage regulator circuit.

You should be aware of a *quirk* in the '317's specifications (a quirk that stung Tom, recently): the device does not regulate properly unless the current from its output is at least 10 mA. So you must avoid large resistor values in the feedback path, unless you can be sure that a substantial load current will be drawn at all times.

Replace R with a 1k pot, and check out the 317's performance as an adjustable regulator. What is the minimum output voltage ($R=0$)?

---

[5] Yes, the IC limits its output current to limit its $I \times V$ power dissipation. The evidence will be the fall of V_{out} from its normal level of 5 V.

11L.1.6 Crowbar overvoltage protection

Here's a little circuit that can protect against the potentially horrible effects of a power supply failure, by clamping the supply voltage close to ground in case that voltage exceeds some threshold. Here, we have set the threshold around 6 V: about right, as the start of danger in a 5 V supply (we chose 5 V because it is a standard computer supply voltage). The "softness" of the zener's *knee*[6] gives this circuit only approximate control of the threshold. You might prefer, in practice, to use an integrated overvoltage sensor that includes a precise voltage reference[7]; some of these ICs include the SCR, as well.

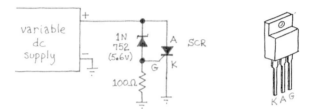

Figure 11L.8 Crowbar circuit, and pinout of SCR: G = gate; K = cathode; A = anode.

The *silicon controlled rectifier (SCR)* shown in Fig. 11N.20, is a device you have not seen before in these labs. It behaves more or less like a power transistor: to turn it on you need to provide some current at its gate which will accept current if you bring its voltage up to around 0.6 V above the cathode.

The SCR differs from a transistor in *latching itself ON* once it begins to conduct. To turn the device off, you must stop the flow of current by some external means: in this circuit by shutting off the power supply. Evidently the SCR is well-suited to this application: trouble turns it *ON*; only someone's intervention then can revive the power supply. Note that the crowbar does *not* shut off if the supply voltage simply attempts to revert to a safe level, such as 5 V in the present case.

In a practical circuit, you would add a capacitor to ground at the SCR's gate, to protect against triggering on brief transients. Today, your power supply should be quiet – since it powers nothing other than this protection circuit! So, we have omitted that protection capacitor.

Try the circuit by gradually cranking up the supply voltage, using either an external variable supply or the breadboard's 15 V supply, which (on the PB-503) is adjustable with a built-in potentiometer. When you finish this experiment, incidentally, you'll probably want to set the breadboard's adjustable supply back to +15 V, since you're likely to expect that voltage next time you use this supply.

11L.2 A switching voltage regulator

The switching regulator's strategy is wonderfully simple: instead of soaking up the difference in voltage between V_{in} (unregulated) and V_{out} (regulated), it alternately connects and disconnects V_{in} and V_{out} (with a suitable filter to smooth V_{out}). In principle, the regulator – now reduced to a clever *switch* – need not dissipate power. At all times one or the other of the switch's V or I is zero. (That's the *ideal*; *real* switching regulators can give 80% efficiency routinely, 95% in some designs.)

---

[6] Some lonely Dilbert must have invented this piece of jargon.
[7] For example, ON Semiconductor MC3423. See AoE §9.13.1D.

11L.2.1 Switching regulator IC: LT1073

Figure 11L.9 shows a block diagram for the LT1073 switching regulator, a low-current device that is remarkably easy to use (the datasheet boasts "no design required" as its first selling point: a bit embarrassing for us aspiring designers – but also evidence that switching regulator design has scared away lots of engineers!).

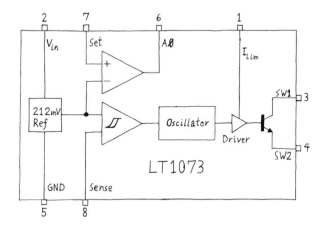

Figure 11L.9 LT1073 switching regulator: block diagram.

11L.2.2 Step-up switching regulator

Figure 11L.10 shows a stunt you can't do with a linear regulator: get +5 V from a single AA cell. It's feedback as usual – but what's not usual is that this time feedback controls an oscillator that gives a kick to the inductor (at about 20 kHz) when V_{out} falls below the target.

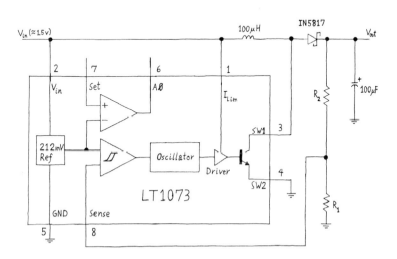

Figure 11L.10 Step-up ("boost") regulator, using LT1073.

A design choice: feedback divider: Choose R_1 and R_2 to give an output voltage of about 5 V, given the value of the internal voltage reference (212 mV). You'll get clean output waveforms if you keep R values under 100k, though much larger values are permitted.

502 Lab: Voltage Regulators

Output capacitor: Use a tantalum 100 μF capacitor on the output, or a Sanyo OS-CON type if you can find one: this capacitor, with its lower series-resistance (and somewhat lower inductance), should give your circuit less of the spiky noise that appears when a jolt of current is injected into the capacitor, on switch turn-off. To check your circuit's response to variable loading, use the circuit in Fig. 11L.11, including a meter that will let you measure I_{load}. *Note* that the variable resistor must have a higher power rating than our usual pots (which are 1/2 W parts). This circuit calls for a 1 W pot.[8]

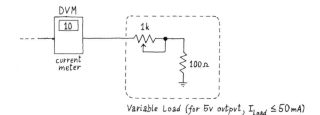

Figure 11L.11 Variable LOAD circuit. (Potentiometer should be rated at 1 W or more.)

Use a scope to watch V_{out} and the switch terminal, SW1 (pin 3). The ripple of about 150 mV on V_{out} (embroidered with somewhat spiky steps) reflects the *hysteresis* of the comparator on the '1073: the regulator lets V_{out} droop a bit, then kicks it upstairs.

Query: if the comparator hysteresis is 5 mV (typical), why is the apparent hysteresis here so much larger?[9] Figure 11L.12 shows what we saw for three load currents.

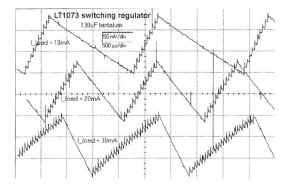

Figure 11L.12 Ripple of switching regulator, and the stairstep rise in response to regulator's 40 kHz switching. (Scope gain: 50 mV/div, sweep rate: 500 μs/div.)

Reducing ripple (optional): The 1073 includes a ×1000 amplifier, and this can be used to reduce the output hysteresis that the circuit tolerates. All you need do is insert the amplifier into the feedback path (the 470k resistor in Fig. 11L.13 is simply a *pullup*: the amplifier's output is not the usual push-pull stage).

---

[8] The calculation of the power rating for a variable resistor is simple, but also subtle. One might think that if we adjust the variable resistor in Fig. 11L.11 to, say, 10 Ω, this 10 Ω value will dissipate little power (about 20 mW, at 50 mA). But the flaw in this reasoning is that the fraction of the pot where this power is to be dissipated is so small that it cannot unload the heat.
So, one should calculate power that would need to be dissipated with a full 50 mA load current through the *entire* pot, even though this will not occur in this circuit (the variable resistor at 500Ω would pass much less than 50 mA). It is the *heat per unit length* of the resistive element that must not exceed specification.
Hence our result of about 1 W: we calculate I^2R power when 50 mA flows through the 0.5k pot (about 1.25 W). Power is lower at all but the minimum R value.

[9] That's right: because we're feeding back only a fraction of V_{out}, where we are watching the ripple that "reflects" the effects of the comparator's hysteresis.

11L.2 A switching voltage regulator

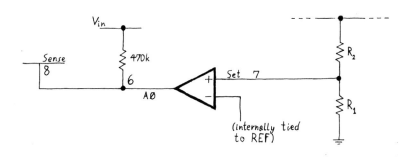

Figure 11L.13
Op-amp in feedback path reduces output ripple.

Query: How does this amendment reduce ripple?[10]

11L.2.3 Step-down switching regulator

AoE Fig. 9.62: typical waveforms for such a step-down switcher

The circuit in Fig. 11L.14 is similar to the step-up, but this time the transistor switch lies between V_{in} and the inductor, rather than between inductor and ground. When the switch in this step-*down* configuration turns off, the inductor draws current through the diode, from ground. To do this, it takes the voltage at the diode slightly *negative*: pretty weird, if you're not accustomed to seeing inductors at work.

Feed this circuit from a variable external power supply. It should work with an input that varies from a few volts positive to 30 V (the max permitted). The capacitor on the input helps stabilize the supply when the switch connects to the inductor. Again, let R_1 and R_2 provide a +5 V output.

Can you infer from the behavior of the output waveform what the regulator is doing to hold V_{out} constant, as you vary V_{in}?[11]

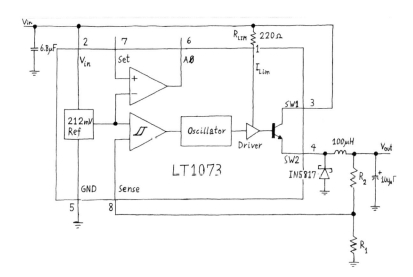

Figure 11L.14
Step-down ("buck") switching regulator.

---

[10] Nothing fancy, here: the internal amplifier's gain of 1000 makes the circuit responsive to changes at the SET input 1000 times smaller than those at the SENSE input. This would seem to reduce our observed output ripple of about 120 mV to well under a mV. The datasheet's promise, however, is more modest: "a few millivolts." This modesty is puzzling. See Switching Regulator datasheet: see especially the first 10 pages: https://LAoE.link/LT1073_Datasheet.pdf.

[11] Probably you'll see a variation in the switch's *duty cycle*.

11L.2.4 Negative from positive

The step-down can be tricked into providing a *negative* output from a positive input. All you need do is redefine "ground" in the circuit of §11L.2.3 and Fig. 11L.14 as *OUT*, while redefining the signal labeled V_{out} in Fig. 11L.14 as *ground*. Strangely simple, eh?

Try it – remembering that *all* your instruments that define ground (power supply, scope) must agree on this redefinition of ground.

11W Worked Examples: Voltage Regulators

11W.1   Choosing a heat sink

Problem   Choose a heatsink adequate to unload the heat developed by a '317 regulator that must handle a 5 V drop at 1 A (for example, we may assume $V_\text{in} = 10$ V, $V_\text{out} = 5$ V). Along the way, choose the best '317 package.

This problem is very similar to one done with slightly different numbers in AoE §9.4.1A.

Solution

Find thermal resistance specifications for our parts: Our goal will be to unload the 5 W (5 V at 1 A). We need to establish some values:

- maximum permitted junction temperature;
- temperature difference between junction and ambient; and
- total permitted thermal resistance in that path given those two values.

When we have done that, we can calculate how big a heat sink will be required to balance the books: that is, to take out heat at the rate it is being generated.

- Junction temperature: the '317 can stand a junction temperature of 125°C. We'll not design right to the limit. Let's treat the design maximum as 100°C.
- Ambient: you might expect 25°C, but recall that this regulator is likely to live in a box with other electronics. So, let's be conservative and expect ambient of 50°C.

What thermal resistance can we tolerate?   Given the temperature difference (50°C) and the power to be dissipated (5 W), we can calculate what total thermal resistance is acceptable.

$$R_\text{Thermal} = \frac{T_\text{Junction} - T_\text{Ambient}}{\text{Power}}$$

$$R_\text{Thermal} = \frac{50°\text{C}}{5\,\text{W}} = \frac{10°\text{C}}{\text{W}}$$

This is the total thermal resistance that we can tolerate between the hot junction and the outside world.

Without a heat sink?   The datasheet in Fig. 11W.1 tells us thermal resistance of the part itself, junction to ambient, assuming no heat sink.[1]

The TO-220 case gives a thermal resistance of 80°C/W, junction to ambient. That won't do. To make the no-heatsink case still worse, the specification assumes ambient at 25°C, unrealistically low. We need a heat sink.

---

[1] Data is from the Fairchild Semiconductor datasheet for the LM317 in a TO-220 package.

Figure 11W.1 Thermal resistances from 317 datasheet.

Thermal Characteristics

Values are at $T_A = 25°C$ unless otherwise noted.

Symbol	Parameter	Value	Units
P_D	Power Dissipation	Internally Limited	W
$R_{\theta JA}$	Thermal Resistance, Junction to Ambient	80	°C/W
$R_{\theta JC}$	Thermal Resistance, Junction to Case	5	°C/W

Heat sink is required: Using a heatsink places several thermal resistances in the path from junction to ambient. The steps will be *junction-to-case*, *case-to-sink* and, finally, *sink-to-ambient*. We will use a thermally-conductive gasket (with its R_{CS}) for insulation, because the case is at the V_{out} voltage, then a heat sink (with its R_{SA}):

$$R_{\text{Thermal\_total}} = R_{JC} + R_{CS} + R_{SA}$$

The total must be kept under 10 (let's drop the units for the moment). We know that R_{JC} uses 5, leaving us 5. Now it is time to turn to a catalog of available parts. We can use the DigiKey website searching their tables that describe available gaskets and heat sinks.

There we find a gasket at 0.07. The heat sink must be a shade under 5 to keep our total under 10. We used the "Natural" column – meaning no fan assumed. We find a couple at 4.4. Figure 11W.2 shows fragments of the part search screen. We could use either of these two heatsinks. Now our total thermal resistance is about 9.5, so safely under 10.

Compare Parts	Image	Digi-Key Part Number	Manufacturer Part Number	Manufacturer	Description	Thermal Resistance @ Forced Air Flow	Thermal Resistance @ Natural	Material	Material Finish
☐		HS410-ND	7023B-MTG	Aavid Thermalloy	BOARD LEVEL HEATSINK 1.95" TO220	2.5°C/W @ 400 LFM	4.4°C/W	Aluminum	Black Anodized
☐		6398BG-ND	6398BG	Aavid Thermalloy	HEATSINK TO-220 PIN BLACK	2°C/W @ 400 LFM	4.4°C/W	Aluminum	Black Anodized

Figure 11W.2 Searching for a heat sink.

This was a somewhat fussy design task – but in your real design life, doing these calculations is worth the trouble. It's nice to avoid the embarrassment of seeing your circuits melt, or go up in flames.[2]

11W.2 Applying a current-source IC

Problem Use the REF200 current-source IC to generate several output currents.

We introduced the REF200 IC in Chapter 11N: see Fig. 11N.18. The manufacturer promises that

---

[2] A former student sent us a note asking that we add the topic of thermal management to the course. He went through our class without ever hearing about the issue – and he reported that the first circuit he built at his new job embarrassed him (and *us*, we should say) by overheating.

its two 100 μA sources and the *current mirror* can be used to sink or source a variety of currents: 50, 100, 200, 300, 400 μA.

Some details:

Current mirror: If you skipped the current mirror exercise in §5L.3, the short description of a mirror's behavior is as simple as what is stated in Fig. 11N.18: it holds equal the currents sunk by its top two terminals. The left-hand current serves as the "programming" current; the right-hand current matches that current. For an explanation of mirrors see §5S.2.

Sink versus source: The 100 μA elements are *two-terminal*, equally adept at sourcing or sinking. In the proposed problem, the 50 μA circuit-solutions will not be so versatile, by the way: different designs are needed for *sink* versus *source*. The configurations for the other current values can be true 2-terminal designs.

Some of these designs are difficult.

Solution *A variety of output currents from the REF200.* Aren't some of these solutions quite subtle? – the 50 μA source, and the 300 μA and 400 μA, especially.[3]

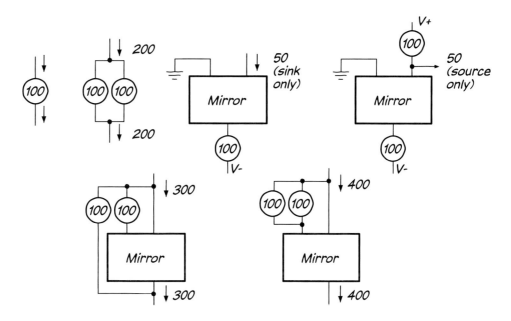

Figure 11W.3 Applications for the REF200 dual current source: 50 μA through 400 μA.

Was this exercise fun, or frustrating? We don't think we came up with the 50 μA source solution without peeking at the TI datasheet.

---

[3] These applications appear on the datasheet for the REF200 on the Texas Instruments site.

12N MOSFET Switches and an Introduction to JFETs

Contents

12N.1 Why we treat FETs as we do — **509**
    12N.1.1 How we'll get away with treating FETs in a day: some ideas carry over from bipolars — 509
    12N.1.2 ...Just another three-terminal "valve"... — 509
    12N.1.3 Some competing symbols for the MOSFET — 510
    12N.1.4 Why bother with FETs? — 511

12N.2 Power switching: turning something ON or OFF — **512**
    12N.2.1 How on is ON? — 512
    12N.2.2 How hard is the thing to drive? — 513
    12N.2.3 Effects of FET capacitances on switching — 513

12N.3 A power switch application: audio amplifier — **514**
    12N.3.1 Waveforms — 515

12N.4 Logic gates — **516**
    12N.4.1 A primitive logic inverter... — 516
    12N.4.2 ...An elegant logic inverter: CMOS — 517

12N.5 Analog switches — **517**
    12N.5.1 A first-try: single-MOSFET analog switch — 517
    12N.5.2 An improved analog switch: CMOS — 518
    12N.5.3 Imperfections — 518

12N.6 Applications — **519**
    12N.6.1 Integrator reset — 519
    12N.6.2 Lots of other applications... — 520
    12N.6.3 A particularly intriguing application: switched-capacitor filter — 520
    12N.6.4 Sample-and-hold — 521
    12N.6.5 An overview of considerations in choosing components — 522
    12N.6.6 Standard design issues — 522
    12N.6.7 Summary of sample-and-hold errors — 523

12N.7 Testing a sample-and-hold circuit — **524**
    12N.7.1 Its behavior (showing defects) — 524
    12N.7.2 Charge injection — 525
    12N.7.3 Bigger C can make charge injection harmless — 526
    12N.7.4 Speed limits during sampling — 526

12N.8 AoE reading — **528**

12N.1 Why we treat FETs as we do

AoE §3.1.5

Are we pushing the breathless pace of this course too far in proposing to dispose of Field Effect Transistors (FETs) in a day? We gave more time to bipolar transistors, and much more to operational amplifiers.

We think we can get away with this hasty treatment, because the FETs that are most important operate as switches, and we think you can pick up those *switching* applications quickly. Linear transistor applications, like those we struggled with a few weeks back, are much harder – as we suspect you noticed.

The FETs that you will try in Chapter 12L are primarily the insulated-gate type or MOSFET.[1] This is the type that wholly dominates digital electronics, and it is within digital devices that the overwhelming preponderance of transistors live. From Chapter 14N on, we will begin seeing MOSFETs in *digital gates*. But for today we will see MOSFETs used in two other settings: first, as *power switches* (where they compete with the bipolar switch, an application you met back in Lab 4L) then as *analog switches* (also called "transmission gates").

12N.1.1 How we'll get away with treating FETs in a day: some ideas carry over from bipolars

Some fundamental similarities between bipolar transistors and FETs give you a big head-start, as you meet these new devices. We will say just a little in this chapter about one large class of FETs: the *junction* type, JFETs.[2] Chapter 12S gives a cursory introduction to these devices and we have added a lab exercise to explore their use as a voltage controlled resistor. For a more complete discussion of the JFET, see AoE Chapter 3. We concentrate today on the other class of FETs, the sort with an insulated gate. The symbols for the two types indicate the differences: in one case, a diode junction at the input, in the other case an insulator (the "oxide" in the MOS sandwich).

12N.1.2 ...Just another three-terminal "valve"...

AoE §3.1.1A

A FET, like a bipolar transistor, is a three-terminal *valve* controlled by a voltage between two of the terminals. And if we concentrate on the type most often used – the so-called "*n*-channel" type, analogous to *npn* bipolars – we find the polarities of the control voltages and current flow are familiar. Figure 12N.1 shows symbols for the new part, and your old friend the *npn*.

The symbol for the MOSFET, which includes a gap, makes visible the FET's great virtue: astronomically high input resistance. When you met the bipolar *follower* we trumpeted its ability to boost input impedances by the factor β. The MOSFET does dramatically better. The gap shown in the symbol represents an insulating SiO_2 layer. In class we sometimes do a silly demonstration, illustrated in Fig. 12N.2, of the MOSFET's high input impedance by using a student as a "wire" to convey a little charge from teacher to MOSFET.

The teacher touches one hand to +5 V or ground to control the lamp that the MOSFET drives. Not a very useful circuit, but a fun demo. If the student lets go of the MOSFET input while the lamp is lit, the lamp stays lit. (Why?[3])

---

[1] This type is named strangely, as if to provide a recipe for the geek stranded on the proverbial desert island and wanting to building a transistor: "Metal Oxide Semiconductor" FET: MOSFET. The British acronym, IGFET ("Insulated Gate..."), makes more sense.

[2] You met these in passing as current-limiting diodes in 5L, and 11N.

[3] This effect only underlines the fact of the MOSFET's enormous input resistance. The lamp stays lit because stray capacitance at its input (gate) holds whatever charge is placed on it, since the MOSFET's input current is almost zero. Many

510 MOSFET Switches and an Introduction to JFETs

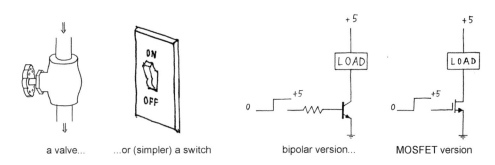

Figure 12N.1 MOSFET switches resemble bipolar switches.

a valve... ...or (simpler) a switch bipolar version... MOSFET version

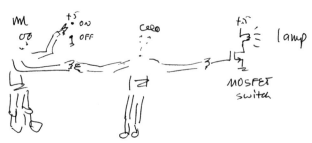

Figure 12N.2 Class demonstration meant to persuade people of MOSFET's high R_{in}.

charged teacher can light lamp, using student as conductor

The FET works more like a vacuum tube than like a bipolar transistor (not that "vacuum tube" is likely to mean much to you[4]): the presence of an insulator at the MOSFET's input means that only the "field effect" can reach and control the *channel*, the conducting region of the transistor.

12N.1.3 Some competing symbols for the MOSFET

Unfortunately, a variety of MOSFET symbols coexist, some more helpful than others. Usually we will use the rightmost of the three symbols in Fig. 12N.3.

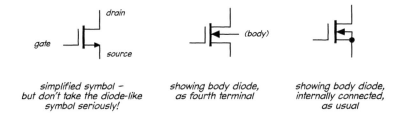

Figure 12N.3 MOSFET schematic symbols.

simplified symbol – but don't take the diode-like symbol seriously!

showing body diode, as fourth terminal

showing body diode, internally connected, as usual

All of these symbols – unlike some which we'll refrain from showing – help the reader by indicating which terminal is source (the one nearer the gate input), and which the drain. The leftmost symbol is appealingly simple, but we worry that the diode-like symbol on the source terminal might mislead

---

semiconductor memories exploit this behavior: "dynamic" and "flash" memories, EPROM's and EEPROM's, for example. Some of these devices remember by holding a small charge on the gate of a MOSFET for many *years*. We will meet these FET-memories again in Chapter 18N.

[4] No; you're too young. But the man who taught one of the authors in his first electronics course had learned with tubes and despised bipolar transistors because of their addiction to input (base) current. When FETs came along he breathed a great sigh of relief and recognition: a proper electronic valve had returned! It's probably not by chance that FETs were the devices the researchers at Bell Labs were trying to implement when they stumbled onto the bipolar transistor. The FET's behavior was more familiar – though the mechanism, one should admit, is not much like the tube's.

12N.1 Why we treat FETs as we do

you: no such diode exists in the device (see Fig. 12N.5 for a literal picture of what's in the device). The right-hand two symbols show the "bulk" or "body diode," the implicit junction between substrate and channel.

The middle symbol, showing the body as a fourth terminal, is appropriate to some integrated-circuit schematics. But power MOSFETs always have body tied to source, as in the rightmost symbol. This internal connection implements an implicit *diode* between source and drain, as shown in Fig. 12N.4.

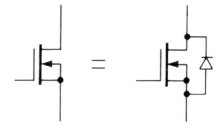

Figure 12N.4 "Body" or substrate forms a diode linking source to drain.

Because of this diode connection, a power MOSFET cannot be used as a bidirectional switch: drain voltage must not be allowed to go a diode drop below source.

A glimpse of the MOSFET's innards: A hasty sketch of the MOSFET's structure may help to make less mysterious the odd "body diode" sometimes included in the MOSFET schematic symbol. This diode is always present and is an implicit result of the MOSFET's construction, not a diode added to the device. Figure 12N.5 is a sketch of the structure of an *n*-channel MOSFET.

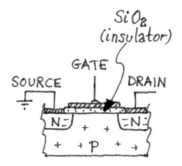

Figure 12N.5 MOSFET construction. "Body" diode is implicit substrate-to-channel junction (no channel shown, here).

The *p*-type body forms a diode with the channel – and must never be forward biased. So the body of an *n*-channel FET must be at least as negative as the source, which is more negative than the drain.[5]

Having braced you with a sense that you already have a partial familiarity with FETs, let's push on into some of their peculiarities.

12N.1.4 Why bother with FETs?

Exhausted by your struggles with bipolar transistors, but triumphant, perhaps you wonder why we should bother you with another sort of device. We consider them because they can do some jobs better than bipolar transistors. Consider MOSFETs when you want:

---

[5] If the "drain" terminal happens to be carried more negative than the "source," this more negative terminal *becomes* the effective *source*. See AoE §3.1.3. The near-symmetry of the FET permits this result, drain and source differing only in capacitance. But note that an implicit diode prevents taking drain more than a diode drop below source (*n*-channel), in the case of a power MOSFET. See §12N.1.3 and AoE §3.5.4E.

- very high input impedance;
- a bidirectional "analog switch"; and
- a power switch.

The first of these FET virtues is the most important: enormous input impedance. Well, strictly, it is the input *resistance* that is enormous. The considerable input *capacitance* of a large MOSFET, like the power switches you will meet in today's lab, can make input *reactance* troublesome at high frequencies.

12N.2 Power switching: turning something ON or OFF

AoE §3.5

You built a bipolar switch long ago in Lab 4L; today you'll build the equivalent circuit with a "power MOSFET" (= just "big, brawny" MOSFET). Figure 12N.6 shows both; what's to choose between them?

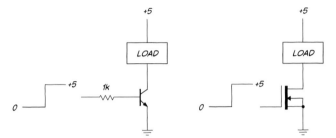

Figure 12N.6 Two power switching circuits: bipolar and MOSFET, doing the same job.

The most important issues are:

- how much power the switch wastes when ON; and
- how easy or hard the thing is to drive.

12N.2.1 How on is ON?

An ideal power switch puts all power into the load, dissipating none itself. So the two transistors compete in how low they can hold their outputs when ON. The contrasting specifications are the bipolar's *saturation voltage* $V_{CE(sat)}$ versus the FET's $R_{DS(on)}$ (watch out for the nasty fact that "saturation" means something quite different when applied to FETs! See AoE 3.1.1A).

The bipolar wastes some power also in its V_{BE} junction, whereas the FET wastes none in its gate where no current flows.

AoE §3.6.3A

MOSFET switches are easily paralleled: The FET also shows a nice behavior that makes it easy to parallel MOSFETs for high currents (and low $R_{DS(on)}$): at high currents and high V_{GS}, MOSFETs have the good sense to *reduce* their currents as they heat, whereas bipolars do the opposite. This difference leads to two contrasts: first, several MOSFET *switches* can be paralleled without "ballast resistors" – emitter resistors that provide negative feedback to a bipolar switch, see Fig. 12N.7. The bipolars, if paralleled without these resistors, would give more and more of the shared current to the hottest transistor. Soon it would be getting little or no help from its partners, and would burn out.

The bipolar's perverse response to temperature – the hotter it gets, the more current it wants to pass – leads to "current hogging" not only among several paralleled parts but also to hogging on the

12N.2 Power switching: turning something ON or OFF

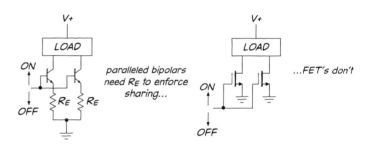

Figure 12N.7 MOSFET *switches*, unlike bipolars, can be paralleled without "ballast" resistors

microscopic scale: if a particular location in a bipolar junction gets hotter than its neighboring regions, it passes more than its share of current, and gets hotter still.

Local current hogging on the bipolar transistor makes the device behave like a transistor with a lower power rating once the voltage across the device is appreciable (AoE's Fig. 3.95 shows the effect beginning at just 10 V).

Bipolars are, in effect, workaholics – whereas MOSFETs behave more like the rest of us: they are inclined to slough off work instead of hogging it. This doesn't mean a bipolar cannot do a particular job; it just means that you will need a bigger transistor than you might expect, because of this effect. Lest you get too excited about the MOSFET's good sense, however, be warned that the rule permitting easy paralleling of MOSFET *switches* does not apply to MOSFET *linear* circuits. These usually run at lower currents where they show the nasty positive temperature response of bipolars: they pass more current as they get hotter.

12N.2.2 How hard is the thing to drive?

We have made much of the MOSFETs giant input impedance *at DC*. You'll see evidence of this in the lab, when you find that your finger can turn the switch on or off (if you touch your other hand to +5, then ground). But away from DC, the FET's higher *capacitances* begin to tip things against it.

Strange to tell, the large capacitances of big MOSFETs have given rise to specialized MOSFET-driver ICs (the sort of strange and highly specialized thing that Darwin used to come upon in the Galapagos). These can deliver a very large current for a short time, in order to charge the FET's input capacitance.

12N.2.3 Effects of FET capacitances on switching

The values of input capacitance – C_{GS} and C_{GD} – are big (100s of pF for a MOSFET that can handle a few amps at low $R_{DS(on)}$). But feedback makes things even worse. The C_{GD} gets exaggerated by the quick swing of the drain as the device switches: that big dV/dt causes a large flow of current in C_{GD}. You'll see that effect in the lab. It appears as a strange kink or hesitation in the movement of gate voltage, as the MOSFET switches. You see it only when you provide weak gate drive, as in Fig. 12N.8.

Now the 10k resistor slows the charging of stray capacitances at the gate, and the FET's switching is slow; see Fig. 12N.9.

The exaggeration of C_{GD} by the slewing of drain voltage in a direction opposite to the movement of V_G is an instance of *Miller effect*. That is a problem that bedevils anyone trying to make a high-frequency inverting amplifier. (And it is a topic that we have not pursued in this book, though you have glimpsed Miller effect as a way to understand how a base resistor stabilizes a discrete emitter follower back in §9N.8.2.)

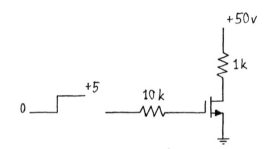

Figure 12N.8 Weak gate drive can expose slowing effect of MOSFET capacitances.

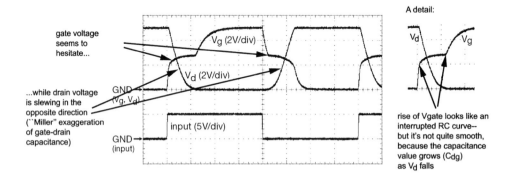

Figure 12N.9 MOSFET capacitances can slow switching.

IGBTs: We will note only in passing an integrated hybrid of MOSFET and bipolar transistors called "Insulate Gate Bipolar Transistors" (IGBT). These show the high input impedance of a MOSFET, but better ON voltage than a MOSFET, for high-voltage loads. Since we'll not meet such loads in this course, we'll not be trying these devices, and refer you to AoE if you need to know more.

12N.3 A power switch application: audio amplifier

You have seen the efficiency that a *switching* voltage regulator can achieve in the LT1073 of Lab 11L. Ideally, the switch dissipates no power whereas a linear regulator is obliged to soak up the difference between V_{in} and V_{out}. A linear regulator is not likely to exceed about 50% efficiency; a switching regulator can approach 90%.

An audio amplifier can achieve a similar gain in efficiency by replacing the valve-like behavior of a push–pull amplifier with ON/OFF switching. (Such an amplifier is labeled "Class D," in contrast to the "Class AB" biased push–pull.) The difficulty that results is the generation of switching noise, as you will see in Lab 12L when you try such an amp.

This electrical noise need not produce *audible* noise, however. The switching frequency is placed far above audio range (into at least hundreds of kilohertz). Nevertheless, a look at the waveforms in Figs. 12N.10 and 12N.12 may make you uneasy. The noise that appears on the sinusoid at the top of each scope image is noise on the circuit *input*. The abrupt drive pulses disturb the ground line, and radiate high-frequency electrical noise. This disturbance occurs despite the usual efforts to quiet the supplies: decoupling capacitors of $1\,\mu\text{F}$ and several of $0.1\,\mu\text{F}$. The application notes for the part confess the hazards posed by the narrow and steep output pulses:

> From an EMI standpoint, this is an aggressive waveform that can radiate or conduct to other components in the system and cause interference.

12N.3 A power switch application: audio amplifier

If your application cannot stand such noise, you will prefer a less efficient (and less "aggressive") push–pull output stage.

The switching amplifier puts out very short *pulses* at the full supply voltage (5 V in today's application, §12L.3).

These pulses are effectively *filtered* by the speaker's inductance – applying brief ramps of current to the speaker. This exploitation of the speaker's inductance allows a "filterless" design (thus small and cheap), with high efficiency: 70% to almost 90% best case. Small battery-powered devices like cellphones or portable music sources need that sort of efficiency.

12N.3.1 Waveforms

Sinusoid: Figure 12N.10 shows an input sinusoid at about 1 kHz and the output pulse streams that the LM4667 Class D amplifier applies to the two ends of the speaker. The switching noise is alarming to look at – but not to listen to. On a properly designed printed circuit board, in contrast to the breadboarded circuit here, the switching noise would be much less extreme. See, for example, the clean waveforms in AoE Fig. 2.73 The two output waveforms (the traces labeled 2 and 3 in the figure) show first one then the other dominating, to drive the speaker coil first in one direction and then in the other, as the voltage of the input waveform crosses its midpoint. The waveforms somewhat resemble the *pulse-width modulated* waveforms that you saw in Lab 7L, though "pulse-density modulation" would be a more accurate description of these pulse streams.

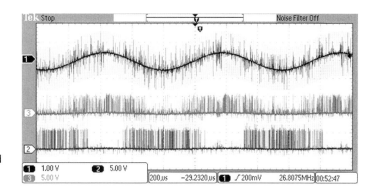

Figure 12N.10 Switching audio amplifier showing input polluted by switching noise (top trace) and two output pulse streams.

Detail of the pulse stream, and noise: Figure 12N.11 shows the pulses in more detail, along with the disturbance that they cause to the entire circuit, including the displayed *input* sinusoid. The somewhat irregular rate at which the pulses occur reveals the quite subtle circuitry within the amplifier, which uses a technique called "delta–sigma" modulation to produce the output pulse streams, using these pulses to achieve a correct *average* speaker current, over a history of many pulses. Note that the pulse rate lies far above the frequency of the audio signal that is to be reproduced.

A music waveform: Finally, Fig. 12N.12 shows a music waveform passed by this amplifier. The switching hash is very obvious – but we hope that when you try this in lab you can confirm that the audible effect of this hash is slight.

Figure 12N.11 Detail showing output pulses and noise coupled to input sinusoid.

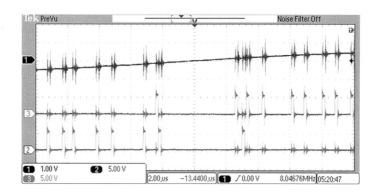

Figure 12N.12 Music waveform as input; pulse streams that result.

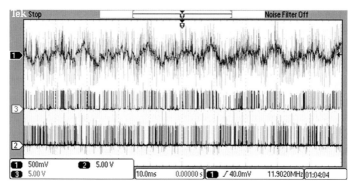

12N.4 Logic gates

A power switch like the one shown in Fig. 12N.6 is essentially the same circuit as the simple *logic inverter* on the left-hand side of Fig. 12N.13. The inverter differs from the power switch, however, in that the purpose of the *logic* device is not to put power into a load; quite the contrary.

Instead, the purpose of the circuit is only to generate an output voltage level (high or low) that is determined by a high or low input, while minimizing power dissipated everywhere in the circuit. The input of a MOSFET clearly is well suited to saving power, since the gate draws almost no current. But some cleverness is required in order to make sure that the remainder of the circuit also operates without dissipating power. Such cleverness appears in the better logic gate shown on the right in Fig. 12N.13.

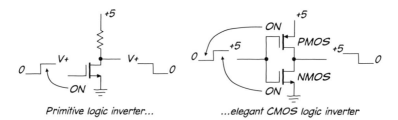

Figure 12N.13 Two logic inverters.

Primitive logic inverter...   *...elegant CMOS logic inverter*

12N.4.1 A primitive logic inverter...

The very simple "logic gate" on the left in Fig. 12N.13 works – but not well. As you will confirm in Lab 14L, its output cannot rise fast because stray capacitance at its output must be charged through the pullup resistor. This makes the switching slow. In addition, the circuit is inefficient because when

V_{out} is *low*, the resistor dissipates power. If that's not bad enough, the gate also has such high output impedance in the high state that its noise immunity is poor. So, such an inverter rarely is used.

12N.4.2 . . . An elegant logic inverter: CMOS

The right-hand inverter in Fig. 12N.13 is an elegant circuit – and the simplest example of the sort of MOSFET logic that has taken over the world of digital electronics. The name for this pairing of *n*-channel and *p*-channel MOSFETs is CMOS, with "C" standing for Complementary. You will demonstrate the superiority of this inverter in Lab 14L. But even without trying the circuit one can see from its symmetry that it shows none of the three weaknesses of the single-transistor inverter. We see in the next section another good application of CMOS within *analog switches*.

12N.5 Analog switches

AoE §3.4.1

MOSFETs compete with bipolars as power switches; as "analog switches" – devices that are to pass or block a *signal* rather than just the flow of current to a load – the MOSFETs have the application to themselves.

12N.5.1 A first-try: single-MOSFET analog switch

One can appreciate the CMOS analog switch by comparing it – as we did for the CMOS logic inverter – to a less clever design. Figure 12N.14 shows such a simple gate: a single MOSFET used to pass or block a signal. In order make sense of a detail shown in the rightmost circuit of the figure, we must acknowledge the MOSFETs "body" diode. This is an element of the FET that we've not used before.

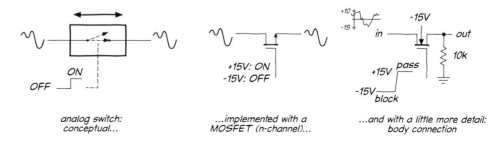

Figure 12N.14 An analog switch done with one MOSFET.

The "body" diode is shown connected to -15 V. Since that is the most negative point in the circuit, this connection makes sure that this diode never conducts.

In the case of the analog switch, source and drain swing as widely as the input and output signals, and typically both can swing positive and negative. So one must take care to tie the body to a voltage that can never take the wrong sign relative to the in and out signals. If body were instead tied to *source* as in a power MOSFET, then a negative input signal, while the switch was OFF, could cause the body diode to conduct.

The analog switch made with one MOSFET works only as long as the input signal remains safely below the positive control voltage that is applied to the *gate*. Otherwise, when the input signal swings close to the positive power supply, the MOSFET will feel its V_{GS} shrinking and eventually will turn itself off.

MOSFET Switches and an Introduction to JFETs

12N.5.2 An improved analog switch: CMOS

AoE §3.4.1A

The neat remedy for the limitation of the single-MOSFET design is to add a second MOSFET, of the complementary *p* type, in parallel. This second transistor is driven with the complement of the control signal that drives the gate of the *n*-MOSFET so the two transistors are either both OFF or both ON.

It is a nifty Jack Sprat circuit, as you can confirm if you imagine applying an input signal that swings all the way from negative to positive supply. At one extreme of input swing – near the positive supply, say – *n*-MOS fades, but *p*-MOS is fully on; at the other extreme of input swing, Jack and Mrs. Sprat exchange roles. And in the middle, the two transistors both work fairly well. Good CMOS switches – like the one you will meet in the lab – handle such wide signal swings happily, with nearly constant R_{on}).

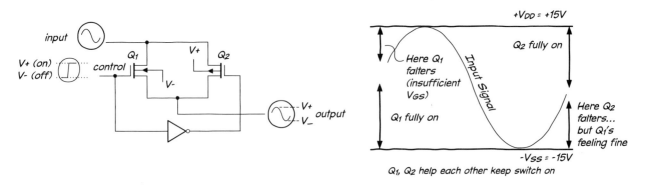

Figure 12N.15 CMOS analog switch: paralleled *n*-MOS and *p*-MOS help each other out.

12N.5.3 Imperfections

AoE §3.4.2

A perfect transistor analog switch would be fully ON or fully OFF: it would behave like the mechanical switch. A real analog switch, like a real power switch, falls short of course.

When ON, it looks like a small resistor (the DG403, which you'll use in the lab looks like 30 Ω or less); when OFF, it looks like a small capacitance linking input to output (DG403: about 0.5 pF); it also leaks a little: a steady current may flow (DG403: ≤0.5 nA). Whether or not these imperfections matter to you depends on the circuit.

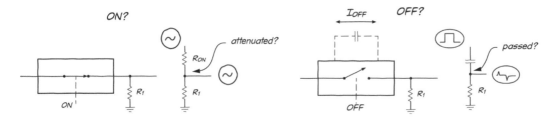

Figure 12N.16 Analog switch departures from simple ideal: ON or OFF.

To get a sense of when such imperfections may be troublesome, we will look at a particular case – one that we do as an in-class demonstration. We will see the effects of some of the DG403's characteristics on a particularly-demanding circuit, a *sample-and-hold*, in §12N.6.4. But before we do that hard work, let's have some fun, just noting some applications for these devices.

12N.6 Applications

Analog switches will solve many problems that otherwise would be difficult. Here is small set of applications, to start your imagination stirring.

12N.6.1 Integrator reset

Here's a task that seems utterly straightforward: reset an integrator.

Manual reset: It is easily done with a mechanical pushbutton. The "No" on the right-hand sketch in Fig. 12N.17 warns against a common error. It is a *plausible* error: just force the integrator output to zero. What's wrong with that?[6]

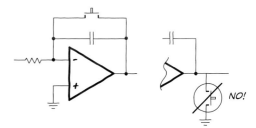

Figure 12N.17 Manual integrator reset is straightforward – if you do it right.

Transistor reset: Letting a control signal effect a reset turns out not to be so straightforward. Would those in Fig. 12N.18 work? Yes, *half* the time: when the integrator output was positive. When, instead, it went negative, reset would lose control: the more negative terminal would become the effective reference voltage, and both sorts of transistor would begin to turn ON even though the control signal were held at 0 V – a voltage intended to turn the switch off.

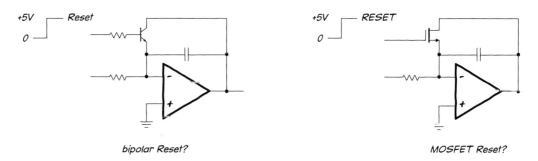

Figure 12N.18 Two attempts at transistor reset for integrator.

An easy reset that works: Replacing the single-transistor switch with an analog switch solves the problem. The device in Fig. 12N.19 can handle an integrator output that goes to either supply extreme, assuming the switch is powered from the same supply voltages as the op-amp.

---

[6] Well, doing that does not discharge the capacitor. Instead, it forces the far end of the capacitor away from virtual ground. The capacitor may or may not eventually discharge depending on what is driving the integrator. A bad reset scheme, in any case.

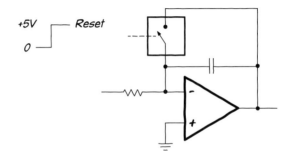

Figure 12N.19 Analog switch easily solves the reset problem, handling both output polarities.

12N.6.2 Lots of other applications...

Multiplexers and their uses: An analog switch can steer a signal here or there. Doing that for the purpose of sharing a resource is said to "multiplex" that resource. Examples abound – such as a telephone line shared among several subscribers who take turns in a scheme mediated by a switching office. In today's lab you will build a humbler "mux" circuit: a "scope multiplexer" that lets two input signals share a single scope channel. In a computer a device like an analog-to-digital converter (ADC) can be shared among many sources, since ordinarily it is used for just one process at a time. Simply route all those signal sources through an analog multiplexer that feeds the ADC; convert one source, then move on to the next. The resource is time-shared.

AoE §3.4.3

Once you have met the analog switch you will think up lots of other applications. It is a good addition to your bag of tricks, solving problems that, like the integrator reset, would be difficult or impossible without this device.

12N.6.3 A particularly intriguing application: switched-capacitor filter

AoE §6.3.6

Late in Lab 12L you will have a chance to try a novel sort of lowpass filter – one that foreshadows the replacement of analog functions with digital, a change coming soon in the schedule of this course. This filter uses analog switches to meter out charge in discrete packets to simulate the behavior of a resistor. This scheme allows a binary digital signal (ON versus OFF) to control the effective resistance by varying the *frequency* with which the "doses" are delivered.[7] This probably sounds very abstract. A diagram will show how simple the notion is. Figure 12N.20 shows the lab circuit.

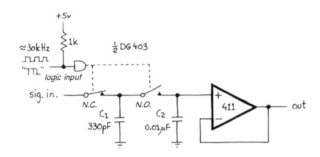

Figure 12N.20 Switched-capacitor filter (lab circuit).

The two analog switches are controlled by the same logic input signal and operate out of phase –

---

[7] The delivery of charge in discrete packets has affinities with digital methods, but we don't mean to suggest that this switched-cap filter is a digital device. It is not, since the controlling input signal is a *continuously*-varying frequency. And the doses of charge that are delivered, though discrete, can vary continuously in their magnitude. The circuit output, more obviously, is an analog quantity, a voltage.

12N.6 Applications

the normally open switch (N.O.) is open when the normally closed switch (N.C.) is closed and vice versa. The square wave labeled "TTL" input transfers a small dose of charge between the smaller cap and the larger, on each cycle. These transfers gradually carry the output voltage toward the level of the input voltage.

Figure 12N.21 illustrates a homely analogy in which voltages are represented (as usual) by water levels. An energetic person with a soup spoon tries to keep the bucket's water level equal to the level of water in the tidal basin (well, given the speed of tides, he doesn't need to be very energetic; but maybe you see what we're trying to show).

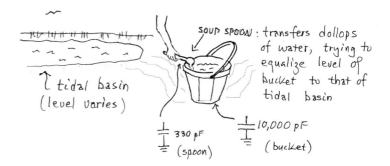

Figure 12N.21
Switched-cap circuit likened to spoon and bucket.

If you're thinking that this seems a cumbersome way to mimic the behavior of a resistor, consider how easy it is for a computer to vary the frequency of a square wave. That is all that's required to control f_{3dB} of this filter.

If you have time at the end of today's lab, you can play with an integrated version of this sort of filter. It is an extremely steep active filter,[8] but it uses the switched-cap scheme rather than resistors, so that you can control f_{3dB} simply by twiddling the frequency knob on a function generator. This easy control of f_{3dB} will be useful in later labs, as well, where you will use the filter to remove artifacts introduced by analog-to-digital conversion.

12N.6.4 Sample-and-hold

AoE §3.4.3C, Fig. 3.85, §4.5.2

This is an important circuit and provides a good test bed for consideration of the effects of switch imperfections.

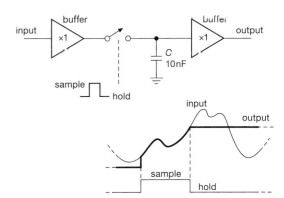

Figure 12N.22 Sample-and-hold: simple scheme.

The sample-and-hold is useful whenever an analog value must be held constant for a while. The

---

[8] This is an 8-pole *active* "elliptic" filter (MAX294). Its radical fall-off with frequency is shown in Fig. 12L.14.

most frequent need for this function is in feeding an analog-to-digital converter, a device that (in most of its forms) takes a substantial time to carry out the conversion. While converting, the ADC likes to look at an unchanging input voltage.

Normally a buffer is included on the input side, as shown in Fig. 12N.22, in order to drive the capacitor promptly to the *sampled* voltage. In the lab circuit we omit the left-hand buffer for simplicity, relying instead upon the low source resistance of the function generator. The output buffer is always included to prevent leakage of the charge that is *held* on the storage capacitor.

12N.6.5 An overview of considerations in choosing components

Before we look at a particular case, with part values, let's first scan the concerns that will guide the choice of the storage capacitor.

Choosing C: What is at stake in the choice of C? What good and bad effects result from use of a large C? – from use of a small C?

A large C makes it easy to *hold* a voltage, but hard to *sample* it (hard to *acquire* the voltage quickly).

Sampling: During *sampling* we want V_{cap} to come close to V_{in}, and to do this fast. The storage cap is charged through an R value that is the sum of all resistances driving C, usually dominated by $R_{DS(on)}$ of the analog switch (much larger than the R_{out} for the op-amp circuit that buffers the signal source).

Simply evaluating RC allows a first approximation of *acquisition* time: count RCs to see how long the *sample* pulse must persist in order to drive error below a particular level – say 5 RCs to get under 1% for example. Subtler effects may also apply: a large storage cap may call for a substantial *current*, perhaps beyond what an ordinary op-amp can provide. And you may need to specify an op-amp capable of driving a capacitive load – the sort of load that, as you know, can bring instability; see AoE §4.5.2.

Hold: During the *hold* interval, *droop* is the concern, and this droop will result from the sum of all error currents flowing into or out of the capacitor. Those currents are the switch's leakage current, I_{off}, the op-amp's I_{bias}, and the capacitor's self-discharge (probably negligible compared to those two other currents).

AoE §3.4.2E

Charge injection: This is a nasty complication – one that may push the necessary capacitor size far above what would be necessary considering only *droop*. Charge injection describes the jolt of charge the switch delivers as it is switched OFF. This occurs because of imperfect symmetry in the switch: internal control voltages, driving the gates of the two paralleled MOSFETs, swing in opposite directions in order to turn off both FETs. Those signals are capacitively coupled to the output. If their couplings were just equal, the two effects would cancel. To the extent the couplings do not match, a net charge is delivered to the storage capacitor – and at the worst possible time: just when it is disconnected from the signal source. So the injected charge adds a voltage step to the saved value. The smaller the storage capacitor, the larger the voltage step caused by charge injection.

Later, in §12N.7.2, we will look at some errors observed in a demonstration sample-and-hold.

12N.6.6 Standard design issues

Here is a recapitulation of the design questions likely to arise in design of a sample-and-hold.

- What size C? (We know now that we lose something at either extreme.) Big enough to keep droop tolerable: use $I = C\,dV/dt$.

- I is the sum of all leakage currents (analog switch, capacitor's self-discharge, op-amp buffer's bias current).
- dt or Δt is the hold time – usually just long enough to allow an ADC to complete its conversion.
- dV or ΔV is the tolerable change of voltage during conversion: the judgment of what is tolerable must be somewhat arbitrary; let's suppose we will let the sample-and-hold contribute a total error of 1/2 the resolution of the converter. We have already used up about 1/2 this total "error budget" in the sampling stage: V_{cap} did not quite reach V_{in}, remember? So we can allow droop to soak up the remainder of the budget: 1/4 resolution of the converter.

This result puts a lower limit on C. The *sampling* concerns put an upper limit, for a given sampling time. As usual, we are caught between two competing concerns.

AoE §3.4.2E

Effects of "charge injection" in the lab analog switch (DG403): For the DG403, the charge injected is 60pC. That doesn't sound big until you look at the ΔV it causes when dumped into a small cap. If we use $C = 100\,\text{pF}$, 60 pC is troublesome:

$$Q = CV; \quad \Delta V = \Delta Q/C = 60\,\text{pC}/100\,\text{pF} = 0.6\,\text{V}$$

The hazard of charge injection, therefore, pushes us toward larger cap size; so does the problem of *feedthrough*: the transmission of high-frequency signal when the switch is turned off.

In short, we need a cap a good deal bigger than what we might choose if we considered only the problem of droop.

12N.6.7 Summary of sample-and-hold errors

Figure 12N.23 summarizes sample-and-hold errors with switching delay added. The effect of charge injection, called "hold-step," here is a negative-going error; in the example later in this chapter (see Fig. 12N.26), the sign of the effect is the opposite. The switching *delays* provide a further reason why fairly long sample or acquisition times are necessary. The DG403 takes around 100ns to switch; pretty clearly we would be kidding ourselves if we did a droop calculation and an RC acquisition calculation and concluded that we could acquire the data in 10 ns.

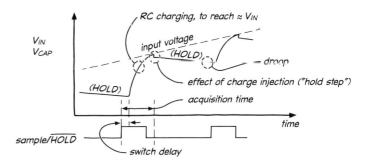

Figure 12N.23 Sample-and-hold errors.

Finally, to give you some sense of what values to expect, here are some specifications for a better analog switch and a good integrated sample-and-hold. The ADG1211 analog switch specifies charge injection at 1 pC – 60 times better than the DG403 that we use in our lab circuit (see §12N.7.2). The LF198 integrated sample-and-hold promises a maximum *hold-step* of 1 mV – about 100 times better than what we achieve with our first try at a homemade sample-and-hold in §12N.7.2.

12N.7 Testing a sample-and-hold circuit

Figure 12N.24 shows a sample-and-hold, made with '411s and the analog-switch that you'll meet in today's lab, the DG403. We'll look at some waveforms that describe how well it performs.

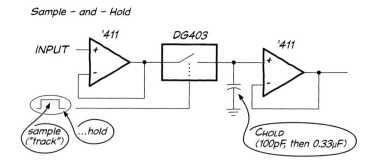

Figure 12N.24 Sample and hold as built for demo.

12N.7.1 Its behavior (showing defects)

Droop: A small C charges up quickly, but also loses its charge quickly. Figure 12N.25 shows the circuit's *droop*. From the *droop rate* one can infer the discharge current.

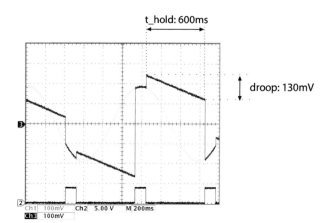

Figure 12N.25 Droop rate reveals leakage current (C = 100 pF). (Scope gains: 100 mV/div except Sample pulse on Channel 2: 5 V/div.)

$$I_{\text{discharge}} = C dV/dt = 100\,\text{pF} \times 130\,\text{mV}/600\,\text{ms} \approx 100\,\text{pf} \times 215\,\text{mV/s}$$
$$= 100 \times 10^{-12}\,\text{F} \times 0.215\,\text{V/s} \approx 22\,\text{pA}$$

Is this current caused by switch leakage, op-amp bias current, or both? The consistent downward slope of V_{cap} reveals the answer: apparently op-amp bias current dominates.

Switch leakage could not account for the droop because V_{cap} *falls* in the region low on the image in Fig. 12N.25 between first and second *sample* pulses; it falls here even though V_{in} lies *above* V_{cap} for most of this *hold* period. So the cap is not discharging toward V_{in}, across the switch.[9]

---

[9] It is possible, but unlikely, that the switch is discharging the cap toward the switch's negative supply voltage. The DG403 datasheet is not so specific as to indicate the sign or path of the OFF-state leakage current, but I'm told that leakage through the channel is likely to be much larger than leakage across oxide to gates, or to substrates. The datasheet supports this view too by implication, as it specifies "Switch OFF Leakage Current." It lists two input conditions, with switch input and output

12N.7 Testing a sample-and-hold circuit

The consistent downward droop also fits the *sign* of the '411s bias current.[10]

This argument is convincing – but let's check by trying an op-amp with extremely low bias current, a CMOS part, then look at the droop again. We replaced the '411 (JFET input: I_{bias}=50 pA, typical (200 pA, max)) with an LMC662CN (CMOS: I_{bias} = 2 fA, typical (2 pA, max)). Figure 12N.26 gives the result contrasted with the '411s. The droop occurs over about 0.6 seconds.

The droop rate for the circuit using the CMOS buffer is very low. One can barely make it out, even if one waits 20 seconds, as in Fig. 12N.27.

The current this droop-rate implies is extremely low. Using two horizontal cursors we estimated the droop at about 12 mV in 18 seconds. We can calculate the implied discharge current:

$$I_{discharge} = C \times dV/dt = 100 \times 10^{-12}\,\text{F} \cdot 0.66 \times 10^{-3}\,\text{V/s} = 66 \times 10^{-15}\,\text{A} = 66\,\text{fA}$$

This is in a plausible range for I_{bias} of the CMOS op-amp.[11] It seems to follow that the analog switch is leaking hardly at all.

12N.7.2 Charge injection

A good op-amp buffer can minimize droop, even with a small capacitor. But, as we noted back in §12N.6.5, a small C makes the circuit vulnerable to another error that is harder to handle: the effects of *charge injection*. When the switch opens, it delivers a jolt of charge (60 pC, for the DG403, typical).

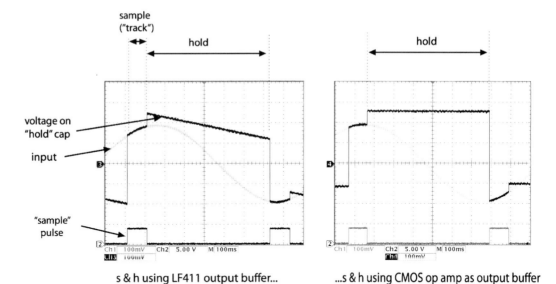

Figure 12N.26 CMOS op-amp buffer reduces droop rate of sample-and-hold. (Scope gains: 100 mV/div except sample pulse, Ch. 2: 5 V/div.)

Delivered to a small cap – 100 pF in this case – this produces a big, spurious step (0.6 V we'd predict; the step visible in our scope images is smaller, but still very troublesome). The "hold-step" is apparent in Figs. 12N.26 and 12N.28.

The voltage hop just at the moment when the circuit ought to be recording the input level is bad.

---

at opposite extremes, ±15 V, and in each of two orientations; the leakage currents are shown to go to extremes of *both signs*. Apparently, the signs of the leakage currents correspond to the orientation of the voltage across the switch; if not, specifying these extremes of V-across-switch would not seem relevant.

[10] The '411s p-channel JFETs offer a junction that is always back-biased, and that junction, if back-biased, would leak toward the two other JFET terminals, both more negative than gate: *source* and *drain*. Leakages to substrate or package could contribute, but the fact that I_{bias} grows exponentially with temperature along with a note in the '411 datasheet support the

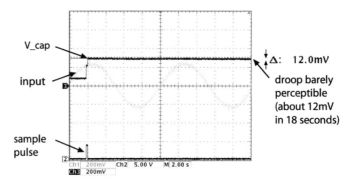

Figure 12N.27 Sample-and-hold droop with CMOS buffer: barely measurable, at almost 20 s hold. (Scope gains: 200 mV/div except sample pulse, Ch. 2: 5 V/div; note time scale change, relative to Fig. 12N.26: here, 2 s/div.)

To make things worse, the magnitude of the *charge-step* is not constant with input voltage. The step is large at the top of the input waveform's swing, much smaller at the bottom of the sinusoid. So it cannot be subtracted away (say, to correct the digital value read by an ADC fed by a sample-and-hold). This variation is apparent in the scope image, Fig. 12N.28.

12N.7.3 Bigger C can make charge injection harmless

To improve the sample-and-hold, it looks as if we need a much larger C. That's true: it will diminish the effect of charge injection, and will slow the droop. Figure 12N.29 shows that the *hold-step* effect of charge injection disappears when we replace the 100 pF storage C with one much larger, 0.33 μF: this works but it exacts a price as we'll see next.

12N.7.4 Speed limits during sampling

The much larger *hold* capacitor solves the charge injection problem; but it is bound to cost us something. Yes: it slows acquisition of the sample. In both images of Fig. 12N.30 the storage C is now 0.33 μF

---

view that *junction* current dominates. See the '411 I_{bias} versus temperature datasheet curve under DC Electrical Characteristics, and Note 7 in that datasheet. National's datasheet can be seen at /https://LAoE.link/LF411_Datasheet.pdf.

[11] 66 fA is surprisingly far above *typical* I_{bias}, but it is just a few percent of the *maximum* specified value. Probably the cause is leakage in our solderless breadboard.

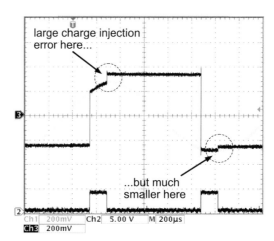

Figure 12N.28 Charge injection produces large error with small C (100 pF). (Scope gains: 200 mV/div except sample pulse, Ch. 2: 5 V/div.)

rather than the original 100 pF. The voltage V_{cap} never quite catches up with V_{in} during the brief acquisition time provided by the sampling pulse.

We can make out two causes for delay in *acquisition* of V_{in}.

The ramp: The straight ramp looks, at first glance, like an op-amp's *slew-rate* limit. But the value of its slope (0.04 V/μs for the left-hand image of Fig. 12N.30) is much too low for this op-amp (LMC662: slew rate about 1 V/μs). The cause must be something else: yes, it's the op-amp's limited output current, which must charge or discharge the cap. One can infer $I_{\text{out-max}}$ from the slope of this curve:

$$I_{\text{out-max}} = C\, dV/dt = 0.33\, \mu\text{F} \times 0.04\, \text{V}/\mu\text{s} \approx 0.013\, \text{A} = 13\, \text{mA}$$

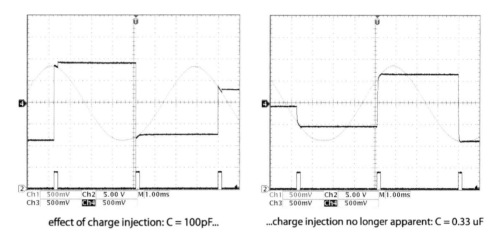

Figure 12N.29 Big *hold* capacitor slows droop and shrinks the effect of charge injection. (Scope gains: 500 mV/div except sample pulse, Ch. 2: 5 V/div.)

That's about right for the LMC662.[12]

The curvy final section: The curvy section at the end of the ramp you recognize as an *RC*: here the FET switch is in its resistive region and its R_{on} drives the big cap. If *RC* is about 10 μs (a very rough

---

[12] LMC662C I_{out} limit spec'd at 13 mA for 5 V supply. Our circuit uses a split supply, ±5 V.

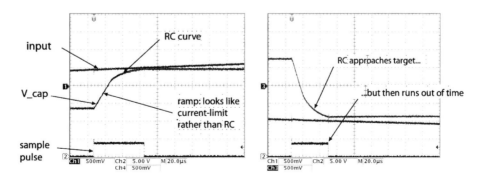

Figure 12N.30 A big *C* slows acquisition: a brief sampling pulse doesn't let V_{cap} reach V_{in}. (Scope gains: 500 mV/div except sample pulse, Ch. 2: 5 V/div.)

estimate – for this piece of an *RC* curve is hard to judge), then

$$R = RC/0.33\,\mu F = 10, mus/0.33\,\mu F \approx 30\,\Omega$$

This value fits quite well too.[13]

A sampling pulse needs to be brief, since everyone's always in a hurry to get a digital conversion done. So the slowing effect of the large *C* is troublesome. We are reminded once again of a general truth: design usually is constrained so that optimizing one characteristic (here, suppressing the charge-step effect) exacts a price in some other characteristic (here, acquisition time). Using a large *hold* capacitor is not a good solution.

You will have gathered that it is difficult to build a good sample-and-hold. We really want a switch with low charge injection, rather than a mediocre switch and then a big *C* to suppress the effect of the injection.

In fact, we nearly always do laziest, easiest thing to get a good sample-and-hold: we buy one from people who make these for a living. For example, the Analog Devices AD7569 we used to use in the microcontroller labs includes an integrated sample-and-hold, an 8-bit ADC, and an 8-bit DAC, all designed to work together. The designers of this part took care to keep the hold-step effect below the level that matters, and to keep the storage cap small enough so that acquisition was pretty fast: adequate, anyway, to meet the requirements of the specified conversion time.[14]

12N.8 AoE reading

Chapter 3: An introductory overview: §3.1 (introduction, FET characteristics) through §3.1.4 (linear and saturated regions).
Gate current, especially "dynamic gate current": §3.2.8C.
Most important is a very long section on FET switches, §3.4. Ignore JFET switches, here.
A short discussion of the lab's switched-capacitor filter design appears in §6.3.6.

---

[13] R_{on} for the DG403 is spec'd at 20 Ω, typical, 45 Ω, max.
[14] The AD7569 part does an 8-bit conversion in 2 µs, using something like 200 ns – one clock cycle – to acquire the analog level in its on-chip sample-and-hold.

12L Lab: MOSFET Switches and a JFET AGC

This lab opens with a quick look at a "power MOSFET" (≈ "big MOSFET") as power switch. The MOSFET offers an alternative to the bipolar power transistor that you met back in Lab 4L. The power MOSFET exercise should go quickly: we want you to see the MOSFET's enormous DC input resistance, but also the large input capacitance that makes it hard to switch *fast*. Then the remainder of the lab is given to trying applications for the so-called *analog switch* or *transmission gate*: a switch that can pass a signal in either direction, doing a good job of approximating a mechanical switch – or, more precisely, the electromechanical switch called a *relay*.

One of the more important circuits that uses this analog switch uses an op-amp as well, and that fact is one reason why we chose to postpone this FET lab till now (MOSFETs might seem to belong just after the two bipolar transistor labs). This combined application is the *sample-and-hold*, often used in a circuit that converts a signal from analog into digital form. The sample-and-hold holds an input value constant during the conversion process. The sample-and-hold also serves us, here, as a good test bed for the analog switch: it reveals the imperfections of the device.

We conclude by creating an automatic gain control circuit with a JFET used as a voltage-controlled variable resistor.

12L.1 Power MOSFET

This exercise repeats a task you carried out back in §4L.6 using a bipolar transistor. Here you will use a MOSFET to do the job. Both of the MOSFETs listed below are designed to turn ON with only a "logic-level" *high* input voltage applied to V_{GS}. In some respects the MOSFET is much better than the bipolar equivalent. Its great strength, as you are about to confirm, is its huge input impedance, at least at DC.

Part	$R_{DS(on)}$ (millohms, max)	$V_{GS(on)}$ max	Mfr
IRL2505PBF	13	4.0 V †	Infineon
IRLZ34	70	4.0 V †	Vishay
† $V_{GS(th)}$ max is specified at 2.0 V but that threshold voltage specification is not very useful, because it's specified at a low current (0.25 mA for both devices). You're not likely to be running these power MOSFETs at such a tiny current. A better indication of necessary V_{GS} drive is the fact that $R_{DS(on)}$ is specified at V_{GS} of 4 V and 5 V. We have here adopted 4 V as $V_{GS(on)}$.			

530 Lab: MOSFET Switches and a JFET AGC

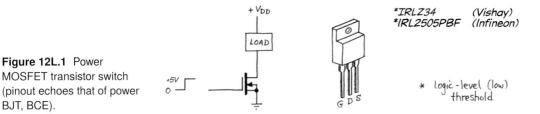

Figure 12L.1 Power MOSFET transistor switch (pinout echoes that of power BJT, BCE).

12L.1.1 Power

Both transistors are high-current devices (104 A for the Infineon part, 30 A for the Vishay) – but this big number may be misleading. These specifications assume that the *case* temperature somehow has been held at 25°C: more or less impossible at high current.[1] The *power* limits are more significant.

Again, these can be confusing. The power limits for the devices are high: 200 W for the Infineon, 88 W for the Vishay. But, again, these specifications assume the case is held at 25°C. With the case at 75°C, for example – more likely, at high current – the Infineon part can dissipate about 140 W, the Vishay 59 W; with the case at 100°C Infineon can dissipate about 100 W, the Vishay 44 W.

AoE §§3.5.1B and 9.4.1A

So we are obliged to worry about the temperature rise caused by electrical power in the transistor. The datasheets specify the *thermal resistances* for the transistor, allowing us to calculate what power it can dissipate with and without heat-sinking.

Let's look at the Infineon part in detail: without a heat sink, junction-to-ambient ($R_{\theta JA}$), when standing vertical, is $R_{\theta JA} = 62°C/W$ (max).

Adding a heat sink helps a lot, of course. A biggish heat sink for the TO-220 package offers 6.5°C/W (Aavid 7022BG). The thermal gasket's R_Θ adds about 0.5°C/W, giving a total thermal resistance, from junction to ambient, of

$$R_{\Theta JA} = R_{\Theta JC} + R_{\Theta CA} = (0.75 + 7.0) \approx 7.8°C/W$$

If we run the transistor junction temperature to 150°C (25°C short of the limit) and assume (conservatively) that the transistor may live in a box at 50°C, the temperature difference is 100°C. Let's see what power the transistor can unload, with and without heat sink.

- Without heat sink (or fan), transistor standing vertical:

$$P = \text{rate-of-heat-transfer} \cdot \text{temp-difference} = (1/R_\Theta) \cdot \Delta T$$
$$= (1/R_{\Theta JA}) \cdot \Delta T = (1/62) \cdot 100 = 0.016 \cdot 100 \approx 1.6\,W$$

- With heat sink: $P = (1/7.8) \cdot 100 \approx 13\,W$

These values are a long way from the impressive number, 140 W, specified with the case at 25°C.

12L.1.2 Input impedance

Build the circuit shown in Fig. 12L.2. Use a #47 lamp as load and confirm that the FET will switch when driven through the 10k resistor at low frequencies (switch the input between 0 and +5 V, moving the input wire *by hand*). High input impedance is the MOSFET's great strength, as you know.

To get a more vivid sense of what "high input impedance" means, let the input side of the 10k resistor float, then touch it with one hand, and touch your other hand alternately to ground and the

---

[1] The Infineon datasheet cheats even more. The specified max current is only a *calculated* value since the TO-220 *package* can only withstand a continuous current of 74 A.

12L.1 Power MOSFET

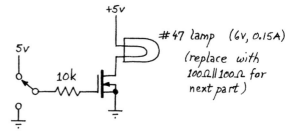

Figure 12L.2 Setup for "measuring" input impedance. (Rotate each leg of the TO-220 package 90° to ease insertion into breadboard.)

+5 supply. Impressive? Now try letting go of the 10k resistor after switching the FET on or off. Why does the FET seem to remember what you last told it to do?[2] This exercise, frivolous though it seems, foreshadows some of the strange results you will get if you ever forget to tie the input of a MOS logic device either high or low: misbehavior that is spookily intermittent!

12L.1.3 Switching at higher frequencies

Effect of input capacitance. Now use a signal generator, with a 0 to +5 V square wave, to drive the MOSFET switch. Keep the 10k resistor in place, for the present. Watch V_{gate} and V_{out} as you increase the driving frequency from about 10 kHz. What goes wrong?[3] The waveform at the gate should look pretty strange.

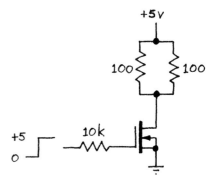

Figure 12L.3 FET switch, again: speed limit.

Solve the problem by replacing the 10k resistor with a value that works better at high frequencies. Now see how the switch looks at a few hundred kHz.

12L.1.4 Power dissipation: bipolar (BJT) versus MOSFET

Quien Es Mas Macho? Here, to make an otherwise humdrum experiment a little more exciting, we propose that you treat this exercise as a contest between brawny champions of two competing types, MOSFET and bipolar. In some respects, the transistors are similar: they use the same big package (called TO–220), and carry some ratings that are at least in the same ballpark. The bipolar is specified to carry as much as 10A continuous (again, apart from limits on heating, as we noted of the MOSFETs in §12L.1.1).

---

[2] Your particular MOSFET may or may not "remember." The maximum leakage rate is rather high. See datasheet.
[3] Yes, the 10k has trouble driving the gate capacitance – especially C_{gd}, which is exaggerated (doubled) by the circumstance that the output (drain) swings in a direction opposing the input (gate) swing. §12N.2.3 makes this point.

You can't tell from the specifications we have noticed so far – maximum current and power – which transistor will win. The answer will depend mostly on the voltage drop across the transistor, a measure that is specified in different ways for FETs and bipolars. But the MOSFET should look pretty good in a contest where the goal is to put available power into the load rather than into the transistor switch.

Set up the contest by putting the same heavy load current through both transistors (about 0.5 A), and measure the voltage across each. *Use a resistor rated at 5 W or more*: if you forget this advice and use a .25 W resistor, you may be reminded by smoke or even by burned fingers.

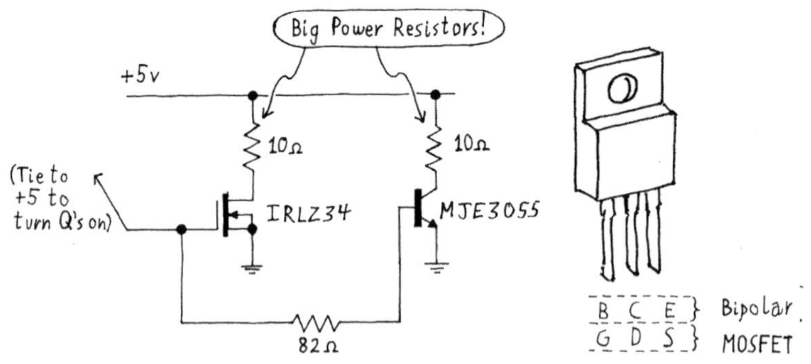

Figure 12L.4 MOSFET versus bipolar switch contest. (Rotate each leg of the transistor 90°, to ease insertion into breadboard.)

How ON is the ON switch? MOSFET versus bipolar: From this number – V_{DS} or V_{CE} – you can calculate $R_{DS(on)}$ for the FET – and an equivalent value for the bipolar if you like, though the term "$R_{DS(on)}$" is not used for bipolars. Then (making the FET look still better) you can calculate power dissipation in each transistor, remembering to include the considerable power dissipated as a result of *base* current.

For a vivid confirmation of the fact that most of the power in these switch circuits is going into the load and not the transistor switch, you can put a finger on load resistor, then transistor. You should find the resistor much warmer than the transistor

12L.2 Analog switches

The CMOS analog switch is likely to suggest solutions to problems that would be difficult without it. This lab aims to introduce you to this useful device. You'll see from Fig. 12L.5 that schematically it is extremely simple: it passes a signal or does not.

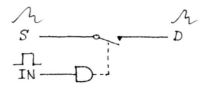

Figure 12L.5 Analog switch: *generic*.

The switch we are using has especially nice properties: it is switched by a standard logic signal, 0 to +5 (High, +5 = ON). But it can handle an analog signal anywhere in the range between its supplies,

12L.2 Analog switches

which we will put at ±15 volts. It also happens to be a *double-throw* type, nicely suited to selecting between two sources or destinations.

Figure 12L.6 shows the switch and its pinout. As signal source use an external function generator; as source of the "digital" signal that turns the switch on or off use one of the slide switches on the breadboard. The easiest to use is the 8-position switch: put one of its slides in the ON position; now that point will be high or low, following the position of the slide switch just to the right of the "DIP" ("dual in-line package") switch.

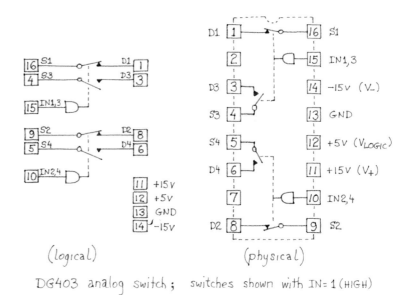

Figure 12L.6
DG403(Maxim) analog switch: block diagram and pinout.

Caution: each DG403 package contains two switches. Tie the unused "IN" terminal to ground or to +5 (this makes sure the logic input to the switch does not hang up halfway, a condition that can cause excessive heating and damage.)

And a *reminder*: this IC uses *three power supplies*: connect all of them!

12L.2.1 Switch imperfections

We'd like you to see that the switch isn't perfect. Specifically, we ask you to measure its *resistance* when ON and its *feedthrough capacitance* when OFF. But the fun and the main interest of the remainder of today's lab lie in the *applications* of analog switches. So, hurry through these preliminary measurements.

R_{ON} (lab): Ideally, the switch should be a short when it is ON. In fact, it shows a small resistance, called R_{on}. Measure R_{on}, using the setup in Fig. 12L.7. Use a 1 kHz sine of several volts amplitude. Confirm that the switch does turn ON and OFF, and measure R_{on}.

Feedthrough: The circuit of Fig. 12L.8 makes the switch look better: its R_{on} is negligible relative to the 100k resistor. Confirm this.

When the switch is OFF, does the signal pass through the switch? Try a high-frequency sine (≥ 100 kHz). Try a square wave. If any signal passes through the OFF switch, why does it pass?[4]

---
[4] Capacitance between input and output couples the two terminals though the switch is OFF.

534 Lab: MOSFET Switches and a JFET AGC

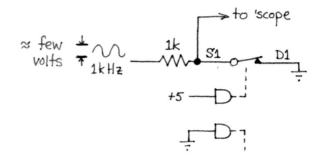

Figure 12L.7 R_{on} measurement.

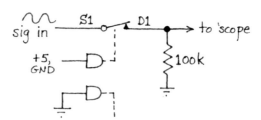

Figure 12L.8 More typical application circuit (R_{on} made negligible).

Note: As you do this calculation, don't forget that you are looking at the output with a scope probe whose capacitance (to ground) may be more important than its large R_{in}: you're really watching a *capacitive divider* at work. As long as you don't forget this probe capacitance, you should have no trouble calculating the switch's C_{DS}.

12L.2.2 Applications for the analog switch

These applications offer the fun in today's lab. You may not have time for all of these circuits. We hope you'll try the sample-and-hold application anyway.

Chopper circuit: Figure 12L.9 gives a cheap way to turn a one-channel scope into two channel (and on up to more channels if you like). (Query: what are the limitations on this trick?[5])

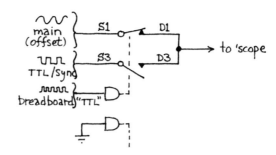

Figure 12L.9 Chopper circuit: displays two signals on one scope channel.

For a stable display, trigger on one of the input signals (preferably the square TTL[6] signal) not on

---

[5] The main limitation is speed. The scheme will entirely miss a brief transient that occurs while the circuit has chopped to the other input. In addition, $R_{DS(on)}$ combined with the scope's capacitance imposes a minimum rise and fall time.

[6] TTL stands for Transistor–Transistor Logic, a now-obsolete logic family, fabricated from bipolar transistors. You will encounter TTL in Lab 14L.

the chopper's output, where the transients will confuse your scope. If you are using a digital scope that offers a "dots only" display option, use that: it draws no line linking the two waveform traces. The images in Fig. 12L.10 show the contrasting cases.

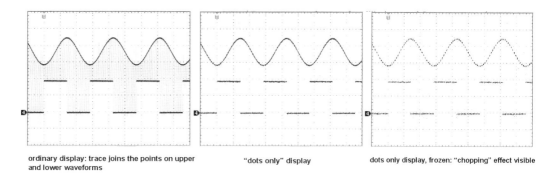

ordinary display: trace joins the points on upper and lower waveforms   "dots only" display   dots only display, frozen: "chopping" effect visible

Figure 12L.10 Digital scope can hide traces between the two displayed waveforms.

The rightmost image in the figure shows the chopper with scope *stopped*. It makes the distinct points of the chopped display visible: a dot on the sinusoid, then on the square wave, then back to the sinusoid....

Sample-and-hold: This application is much more important. It is used to *sample* a changing waveform, *holding* the sampled value while some process occurs (typically, a conversion from analog into digital form).

Try the circuit of Fig. 12L.11. Can you infer from the droop of the signal when the switch is in *hold* position, what leakage paths dominate? (This will be hard, even after some minutes of squinting at the scope screen; don't give your afternoon to this task.)

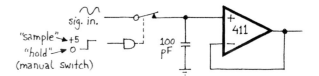

Figure 12L.11 Sample-and-hold.

Query: How does one choose a value for *C*? What good effects, and what bad, would arise from choice of a cap that was: (a) very large; (b) very small?[7]

Can you spot the effect of *charge injection* immediately after a transition on the control input? Compare the specified injection effect and the voltage effect you would predict, given the specification (≤ 60 pC, typical, for the DG403) and the value of your storage capacitor.

Optional: a dynamic view of charge injection: If you inject charge periodically, by turning the switch on and off with a square wave, you can see the voltage error caused by charge injection in vivid form.[8]

---

[7] A very large cap droops only slowly and doesn't show large jumps in response to charge injection; but it is slow to charge to the input voltage, requiring a long *sample* pulse. That long pulse slows the sample-and-hold process.

[8] Thanks to two undergraduates for showing us this technique: Wolf Baum and Tom Killian (1988).

Lab: MOSFET Switches and a JFET AGC

The *held* voltages, when we did the exercise, rode above the input by a considerable margin; you will notice the margin varies with the waveform voltage. Why?[9]

Good sample-and-hold circuits evidently must do better, and they do. See, e.g., the AD582 with charge injection of ≤ 2 pC; and see AoE §3.4.2E.

Negative supply from positive (flying capacitor): You know that a switching regulator can generate a negative output from a positive input, with the help of an inductor (you may have built such a circuit in Lab 11L). For low-current applications the circuit in Fig. 12L.12, which requires no inductors, is sometimes preferable. The trick is to do a little levitation by shoving "ground" about in a sly way. A similar use of "flying capacitors" can generate a voltage larger than the input voltage. This circuit provides only a small output current, as you can confirm by loading it.

Such circuits often are put onto an integrated circuit that would otherwise require either a negative supply or a second positive supply, higher than the main supply.[10]

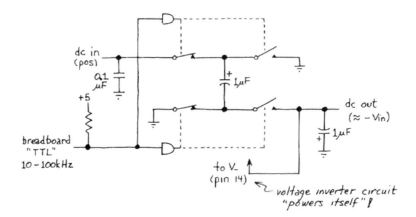

Figure 12L.12 Flying capacitor voltage inverter.

Exercise. Switched-capacitor filter I: built up from parts: Use the function generator to feed a sinusoid to the filter of Fig. 12L.13 while clocking the circuit with a *TTL* square wave from the powered breadboard. This filter's f_{3dB} is set by the clock rate. This makes it a type convenient for control by computer.

The effective RC simulated by this spoonful-at-a-time charge circuit depends on the relative sizes of the two capacitors (how big the spoon is, relative to the bucket) and on how fast the transfers are made ($f_{switching}$):

$$f_{3dB} = \frac{C_1/C_2}{2\pi} \times f_{clock} \quad (12L.1)$$

Given the values used here, this formula predicts

$$f_{3dB} = (0.03/2\pi) \times f_{clock} \quad (12L.2)$$

Try the circuit and compare its f_{3dB} with the predicted value. Does the filter behave generally like an RC filter: does it show the same phase shift at f_{3dB} that you would see in an RC lowpass? Does f_{3dB}

---

[9] Apparently the capacitances between the IC's two internal switching signals and its output – capacitances that ideally would be equal so that their effects would cancel – vary somewhat with voltage across the switch.

[10] Semiconductor memories, for example, sometimes require such a supply. Once upon a time, some memories demanded that the user supply +12 V as well as +5 V. The flying-capacitor trick ended this demand placed on the user; the chip solved the problem internally, and the ICs that lacked the internal step-up were driven out of the market.

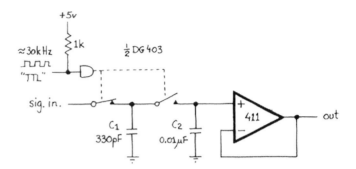

Figure 12L.13
Switched-capacitor lowpass filter.

vary as you would expect with clock frequency? Because this is a "sampling" filter, you can expect it to get radically confused when the input signal changes quickly relative to the rate at which the square wave transfers samples (the "sampling rate"). This radical confusion is called "aliasing."

We'll look in depth at sampling-rate issues, including aliasing, when we reach analog-to-digital conversion in Lab 20L. For now, be warned that you should expect trouble when the frequency of a sinusoidal input approaches one-half the "sampling rate" applied to the analog switch control inputs.

Do you see feedthrough of the clock signal? Sometimes a passive lowpass filter is placed at the output of the switched-cap filter to attenuate that noise.

Exercise. Switched-capacitor filter II: integrated version: This integrated filter is more complicated than the one you just built up from parts, as you might expect. Integrated switched-capacitor filters are available in several forms. Essentially, they are op-amp *active filters* like the one you glimpsed in Lab 9L, except that they use switched capacitors to simulate the performance of a resistor. This design permits easy control of the effective *RC*, as you saw in the preceding exercise. Some integrated filters allow the user to determine the filter type.

The one you will meet here is committed as a *lowpass*; that makes it easy to wire. It is an *eight-pole elliptic* filter – one that sacrifices other virtues in order to achieve spectacularly steep rolloff. You'll see this filter again in the chapter on analog to digital interfacing. Here we just want you to discover how easy to use it is and how sharp is its rolloff.

Like your simple switched-cap filter, this one shows an f_{3dB} proportional to the filter's *clock* rate. That is the feature that later will be especially useful to us when we will want to vary f_{3dB} widely without rebuilding circuits.

Build the circuit shown in Fig. 12L.14, clocking the device with the breadboard TTL signal. Let the external function generator provide an analog input to the filter. Make sure to add a *DC offset* to the input signal, so that the signal stays between about 1 V and 4 V. This filter, you'll notice, uses a single supply, not the split supply that we used with your home-made switched-cap filter on page 536. Hence the need for a DC offset.

To test the filter, try setting the breadboard TTL-oscillator frequency to about 100 kHz, placing f_{3dB} around 1 kHz ($f_{3dB}=f_{\text{clock-filter}}/100$, as you will have gathered). Feed the filter a sinewave centered on 2.5 V; keep the amplitude below about 1 V; vary f_{in} from far below f_{3dB} to far above. Again, the filter should fail when the input frequency climbs above about half the TTL-oscillator frequency (which sets the filter's "sampling rate"): aliasing, once again.

You should be able to see the filter's abrupt rolloff. You will also notice the discrete character of the filter's output: the filter output shows "steps" rather than a smooth curve.

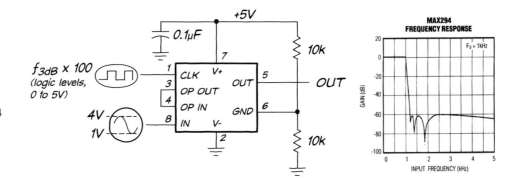

Figure 12L.14 MAX294 switched-capacitor lowpass filter (MAX294 frequency response is shown relative to f_{3dB}, which is adjustable).

Two optional filter tests, for fun:

Peel away frequency components of a square wave: As you know from your experience in *compensating* scope probes, a square wave can quickly reveal a circuit's frequency response. Try this: feed your filter a square wave at around 100 Hz, and take f_{clock} gradually down from its maximum. Can you make out the frequency components $5 \times f_{square}$, and then $3 \times f_{square}$, shortly before you strip the square wave down to its fundamental sine?

Fancier demonstration of controllable f_{3dB} (for the energetic): The prettiest picture of the filter's performance appears if you *sweep* the input frequency, and watch the effect of varying f_{clock} to the filter. f_{3dB} should be variable over a wide range: up to about 1 kHz if you use the *breadboard* oscillator as its clock; up to around 25 kHz if you use a higher clock rate from an external function generator's TTL output.

Is *feedthrough* of the chip's clock noticeable at the output? Can you confirm the steep rolloff that is claimed: 48dB/octave, close above f_{3dB}? Don't be shocked if you find some bumps in the "stopband:" Fig. 12L.14 shows that the MAX294 rolls off like a rock – but then bounces a bit (like a rock hitting a marble floor?).

12L.3 Switching audio amplifier

A traditional audio amplifier, like the push–pull followers that we are accustomed to, is obliged to soak up power as it delivers power to the load. For example, if a push–pull is powered from ±2.5 V and delivers a sinusoid of 2.5 V peak-to-peak, then at best half the supply voltage is dropped across the load (±1.25 V), the rest across the push–pull's transistors. Efficiency, therefore, is below 50%.[11] A higher supply voltage would impose still lower efficiency at this low amplitude.

A switching amplifier gets around that problem, using the strategy that you met in the switching regulator of Lab 11L. The switch, which is ON or OFF and never in-between, ideally dissipates no power. All power goes to the load.

We invite you to try an integrated amplifier that uses this method, a National/TI LM4667. It provides *two* output pins to drive both ends of the speaker. This technique doubles the maximum voltage available to drive the speaker, relative to what would be available with a single-ended output, which would tie one end of the speaker to ground.

---

[11] Efficiency works out to be $(\pi/4) \cdot (V_{pk})/(V_{supply})$. In this case that comes to about 40%. You can work this out for yourself if you enjoy solving an integral.

You have seen scope images in Chapter 12N that provide a foretaste of the electrical noise you can expect to see from this amplifier.

12L.3.1 The circuit

The circuit of Fig. 12L.15 is one of the first *surface mount* ICs you have used (in contrast to "through-hole" parts),[12] mounted on an adapter that allows plugging it into a breadboard.

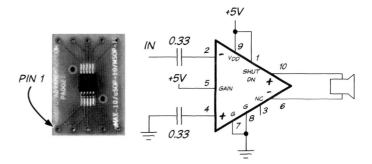

Figure 12L.15 LM4667 switching amplifier, wired for singled-ended input and 6 dB gain.

Wire the circuit of Fig. 12L.15 on a single breadboard strip that you can put aside and keep.[13] That way you can use it later in the course to amplify audio signals in order to drive a speaker.

Add *decoupling* capacitors as close to the part as you can place them (to minimize the stray inductance of the supply lines): use 1 μF and several 0.1 μF ceramics.

Drive the circuit with a small sinusoid: perhaps half a volt amplitude at a few hundred hertz,[14] and listen as you install the decoupling caps. Experiment with their number and placement. You are likely to be persuaded that decoupling is worthwhile.

Figure 12N.10 showed scope images of this circuit's waveforms, and as we said in Chapter 12N, the switching noise is extreme. But we hope that with the help of judicious supply decoupling you can make the switching noise barely *audible*.

12L.4 A Glimpse of JFETs

We barely touch Junction Field Effect Transistors ("JFETs") in this course. Today, we will say only enough about JFETs to be able to understand that a JFET can serve as a voltage-controlled resistance.[15] This is new to us. So far, we have seen transistors used as voltage-controlled *current sources* – not at all the same thing.

The JFET, like the bipolar transistors that you are familiar with, is a 3-terminal device in which the current is controlled by the voltage difference between two of its terminals ("gate," G, and "source," S, analogous to base and emitter in the bipolar transistors that you met in Chapters 4 and 5). The third terminal "drain," D, can be thought of as analogous to the BJT's collector.

Unlike BJTs, however, in JFETs the drain current is controlled by the gate's electrostatic field, and

---

[12] You may have used such a version of the CA3096 transistor array in Lab 5L.

[13] The input capacitors may bother an astute observer: isn't this just like the *bad* op-amp circuits we have warned you about? The amplifier input seems to lack the required *DC path to ground*. It looks that way, but internal to the amplifier there is such a path to ground. This does not mean that the input is tied to zero volts; instead, in this single-supply circuit, it is *biased*, as usual. The inputs are biased to about 1.2 V.

[14] Values chosen so as not to torture people who may be working near you.

[15] We are ignoring, then, their *major* real-world application: low-noise amplifiers with high input impedance (i.e., low input current). See AoE §§3.2, 3.3, 8.6, Tables 8.2 and 8.3b, and AoE-x §3x.1–3x.4.

no input current is required (or desired!). In that way they are similar to MOSFETs: they come close to the ideal of an input that draws no current. In a JFET (which historically preceded the MOSFET), the gate overlies the drain–source channel, but, instead of being insulated by an oxide layer (the "O" of MOSFET), it forms a semiconductor junction (the "J" of JFET) akin to a diode. As long as the gate is negative with respect to the source terminal, this junction is reverse-biased, and only a tiny leakage current flows in the gate (−1 nA max in the device we are using). However, unlike with a MOSFET, you must not forward bias the gate-to-source (or gate-to-drain) junction, or exponentially increasing diode-like gate current will flow.

A consequence is that JFETs run at maximum current[16] when the gate is at the source voltage ("zero gate bias"), and bringing the gate negative reduces the drain current.[17] The name for this is *depletion mode*, as contrasted with the *enhancement mode* MOSFETs we have seen, where a positive gate–source voltage is required to make them conduct.[18]

12L.4.1 Plot the JFET's VITC when $V_{GS} = 0$

Use the circuit of Fig. 12L.16 to measure I_D ("drain" current, equivalent to collector current in a bipolar transistor) of a 2N5485 N-Channel JFET as a function of Drain-to-Source voltage (V_{DS}). This relationship changes as the Gate-to-Source voltage becomes more negative. Here we just want you to plot the Voltage-Current Transfer Characteristic (VITC) curve for a single V_{GS} of zero volts.

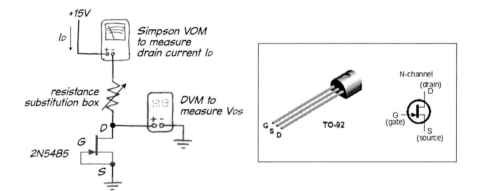

Figure 12L.16 Circuit to measure drain current of a JFET as a function of drain-to-source voltage when $V_{GS} = 0$.

Measure the Drain current on the Simpson and Drain-to-Source voltage on the DVM as you increase the load resistance (which will cause the Drain-to-Source voltage to decrease). For most of the range, from 15 V down to around 3 to 4 V you will not see much change in current, so you can take measurements every volt or so. As you approach $V_{DS} = 0$, take measurements more often to get enough data points to show the voltage/current relationship in that region. Plot the result with V_{DS} on the horizontal axis and I_D on the vertical axis. Figure 12L.17 shows what we got when we tried this experiment.[19]

You should see two distinct regions in the *IV* curve. Above some value of V_{DS}, the plot looks horizontal just like a current source. Indeed, you used a JFET in this region as a current-limiting

---

[16] Officially called I_{DSS}.
[17] Reaching zero at the "gate-source cutoff voltage" $V_{GS(off)}$ (sometimes called the "pinch-off voltage" V_P).
[18] The polarities described in this section assume use of N-channel FETs.
[19] We used the X–Y plot feature in Excel to create this graph.

12L.4 A Glimpse of JFETs

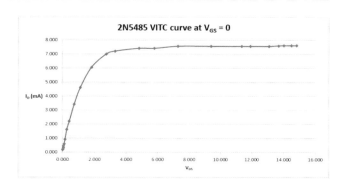

Figure 12L.17 I_{DS} vs V_{DS} graph for our 2N5485.

"diode," applied to provide constant "tail" current in the differential amplifier of Lab 5L. This part was the 1N5294 (see §12S.1.3). This behavior is available because the JFET passes its maximum current with *zero* back bias on V_{GS}: see Fig. 12S.10.

If, instead of shorting the gate to source, one varies V_{GS}, one can vary the output current, I_D. Figure 12L.18 (reproducing LAoE Fig. 12S.5) shows a family of curves plotting drain current versus voltage across the transistor, for several values of applied V_{GS}. It reveals a region where the I–V curves come close to showing the straight-line behavior of a resistor.

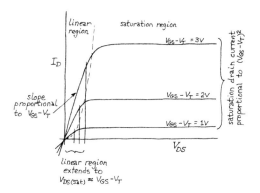

Figure 12L.18 At low *textitV*$_{DS}$, JFET *textitI* – –V behavior approximates that of a resistor.

That straight-line region is the one that we will exploit today. Figure 12L.19 shows that region in greater detail (reproducing LAoE Fig. 12S.7). Today, our goal will be to implement a voltage-controlled resistance (let's call this a "R_V"), as a necessary element in a circuit whose *gain* can be controlled by a voltage.

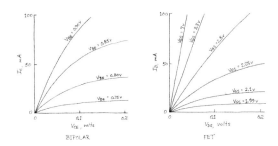

Figure 12L.19 Comparison of bipolar and FET low-voltage regions.

The FET curves look pretty good – pretty straight. But Fig. 12L.19 does not make clear a useful property of the JFET so applied: the straight line behavior works also in the third quadrant of the plot – where V_{DS} is negative (as long as we do not forward bias the JFET PN junction).

It turns out we can make these *IV* curves even straighter by exploiting a trick mentioned in LAoE §12S.1.1, footnote 3: we add, to the V_{GS} signal, one half of V_{DS}. Sounds like hocus-pocus, but it works and is easily done: see Fig. 12L.20.[20]

12L.4.2 Try the Voltage-Controlled Resistance

Fig. 12L.20 shows a circuit applying this trick, using a voltage divider feeding back 1/2 of V_{DS}, adding this to the applied V_{GS}.

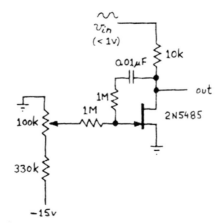

Figure 12L.20 JFET can function as voltage-controlled resistor. Circuit adds $\frac{1}{2}V_{DS}$, to straighten *I–V* curve in linear region.

Wire the circuit of Fig. 12L.20, and watch the input (at about 1 kHz) and output as you adjust the *gate* voltage. If you keep the sinusoidal V_{in} small, the output should remain sinusoidal. As you increase the amplitude of V_{in} you will begin to see distortion. Why? (Hint: see Fig. 12L.19 on the preceding page.)

What is the maximum amplitude for which you get a clean output? Max V_{in} for clean output: _____

What do you estimate (from your observations) to be the approximate range of *R* values for this voltage-controlled resistance? What you will observe will be a fraction, V_{out}/V_{in}. From this fraction, you can estimate *R* for the R_V.

- Minimum *R*: _____
- Maximum *R*: _____

Record rough values relating your R_V's resistance to the reverse bias applied to its gate. The exact values do not matter, but we'd like you to get an impression of what voltages your circuit may need to apply to the R_V's gate.

Figure 12L.21 shows what our numbers looked like.

To save yourself time, you might enter your numbers alongside ours, to work out your R_V's effective *R* values.

---

[20] A look at the equation that describes the resistance Drain-to-Source (r_{DS}) explains why adding $\frac{1}{2}V_{DS}$ works. See AoE, §3.2.7, equation 3.9: $1/r_{DS} = 2\kappa[(V_{GS} - V_{Thev}) - \frac{V_{DS}}{2}]$. The quantity $V_{GS} - V_{Thev}$ defines how *ON* the transistor is driven. The messy little $\frac{V_{DS}}{2}$ needs to be cancelled, to make the *I–V* behavior linear, or truly *resistive*. You may complain that we have just moved the hocus-pocus topic to "where did that equation come from?" Fair enough. But we will not try to justify the equation.

12L.5 Applying the voltage-controlled resistance

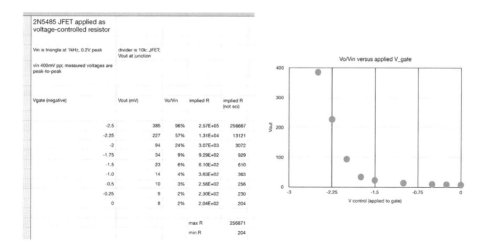

Figure 12L.21 JFET Voltage-controlled resistance divider response plot.

12L.5 Applying the voltage-controlled resistance

We will use a R_V like that of Fig. 12L.20 on the preceding page to make an amplifier whose gain is controlled by a voltage that measures output amplitude. Thus, we can implement an automatic gain control circuit ("AGC").

An AGC holds its output amplitude roughly constant, as input amplitude varies. AGCs for radios became common in the 1930s, to handle the problem of time-varying signal strength. An AGC was especially important in a car radio, because, there, signal strength can vary especially widely (drive into a hollow, say, and then up onto a ridge). By around 1930, the early Motorola brand car radios included AGC's.[21] An audio AGC is also widely used in simple voice recorders. Your cellphone uses a similar principal in its RF power controller to regulate its radio output signal, so as to be strong enough to reach a distant cell tower, but quiet enough, when near a tower, not to drown out other users.

12L.5.1 Voltage-controlled gain (again)

Before we add feedback to control gain, let's confirm that varying the JFET's gate voltage can vary the gain of your amplifier. We'll just use the R_V as R_1 in a standard non-inverting amplifier. To give us plenty of maximum gain, we'll use a large feedback resistor, 150k. We also include an input attenuator that will allow an input as large as 10 V in amplitude, without exceeding the low limit on amplitude across the JFET (see your finding in §12L.4.2 on the facing page).

Here's our test setup for the manual control of gain using the R_V circuit of Fig. 12L.20:

We suggest you use an LT6020 dual op-amp (a low-offset part you may have met in the instrumentation-amp lab). We found its output waveforms cleaner than the LM358's, in this circuit.

Here, you will have another chance to check the maximum input amplitude that your circuit can handle before distortion appears.

12L.5.2 Sensing the average output amplitude

Here's another circuit you have built before. The last time you met it, it was a key element in the AM demodulator circuit of Lab 3L.

---

[21] The name "Motorola," as you may have recognized, merges "Victrola," the home sound source of the day, and "motor," to advertise the car-friendliness of this radio.

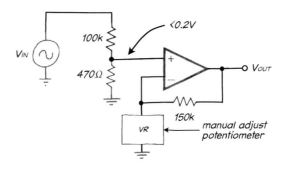

Figure 12L.22 Manual gain adjustment to test use of R_V in amplifier.

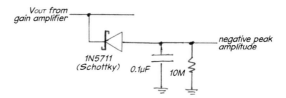

Figure 12L.23 Leaky peak detector with diode oriented to generate a negative voltage.

We suggest using a Schottky diode, for its low voltage drop (0.4 V @ 1 mA, under 0.3 V @ 0.1 mA, typical). We could, instead, have used another op-amp to make an active rectifier that hides the diode drop. But the small drop is not troublesome in this circuit. We can simply set the target average amplitude (defined by the "reference" divider of Fig. 12L.25, below) at the smaller value, reduced by the amount of the expected diode drop.

The circuit of Fig. 12L.23 is so simple that you may not want to bother to test it.

Figure 12L.24 shows a block diagram of the AGC that we have in mind.

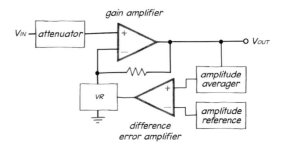

Figure 12L.24 Block diagram of AGC using the R_V and averager.

To complete the AGC, we need only add the differential error amplifier circuit, and the divider that provides the amplitude reference voltage. You can use the second op-amp in the LT6020 or LM358 package to build the diff amp. Since the averager discharge path now will be through the input to the diff amp, remove the 10M used in the test circuit sketched in Fig. 12L.23.

Figure 12L.25 on the facing page shows a difference amplifier and voltage reference.[22] The voltage reference is set low in order to allow input amplitude range to go quite low (down to perhaps 0.5 V, in our tests).

The gain shown for the diff amp is high (at 200), but this gain is, of course, very much less than the gain of the bare op-amp (600,000, for the LT6020 at DC).

Could we have simplified the circuit, wiring the op-amp as a simple comparator, thus using its full open-loop gain? That sounds simple – but recall our concern, tested in Lab 9 on parasitic oscillations, to roll off gain in an op-amp circuit. In general, as you know, we must make sure to take gain down

---

[22] The 220 pF capacitors attenuate feedthrough of the input frequency, which we saw disturbing V_{gate} of the 2N5485.

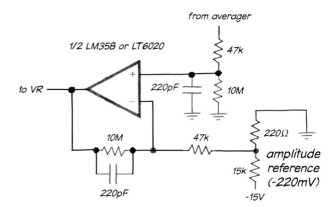

Figure 12L.25 Difference error amplifier, comparing average amplitude against reference.

to unity below a frequency where phase shifts can change negative feedback to positive. Injecting the full open-loop gain of a second op-amp within the main amplifier's feedback loop sounds scary. It is scary, and does not work.[23]

12L.6 Put the pieces together: AGC

Now put your circuit fragments together to make an AGC: a circuit that holds its output amplitude at about 3/4 volt (peak) as you vary the input amplitude over a range of approximately 0.5 V to 10 V. At least, that's an attainable goal.

Start with a sinusoid at 1 kHz. You may find that the AGC's performance degrades as frequency climbs. Explore the performance of your AGC. Perhaps you'll find ways to improve on the designs we've suggested.

---

[23] We could not resist trying it though. We did, indeed, get a self-sustaining parasitic oscillation, even with no input applied. For a detailed look at the stability of a composite amplifier, see §4x.4.1 of *AoE – the x-Chapters*.

12S Supplementary Notes: MOSFET Switches

In Chapter 12N we treat MOSFET switches and some of their applications. This note says a little more about both MOSFETs and the junction type, JFETs. You can safely ignore this note if you're only concerned with MOSFET applications, as we are in our lab exercises.

12S.1 A physical picture

The operation of a FET is easier to describe than the operation of a bipolar transistor. You will recall that we did not even try to describe *why* a bipolar transistor behaves as it does.

AoE Fig. 3.5

A FET, in contrast, just cries out for diagramming: a glance at Fig. 12S.1 will remind you of the FET's greatest virtue, its very high input impedance. The input terminal looks like either an *insulator* (so-called MOSFET type) or a *p–n* junction (in the so-called JFET) – one that is *never forward-biased*, unlike the bipolar transistor's base-emitter junction.

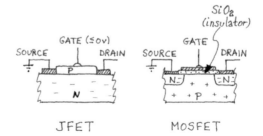

Figure 12S.1 A first view of JFET and MOSFET: a slab of semiconductor.

So, no *current* flows at the control terminal: you just apply a *voltage*; the electric *field* modifies the channel. Hence the name, of course.

The MOSFET diagram in Fig. 12S.1 makes the transistor looks as if it would not conduct, drain to source – since a positive voltage at drain would reverse-bias the right-hand $n{:}p$ junction. This is correct – until we give it some help.

The MOSFET begins to conduct if one applies a positive voltage to the gate to induce a layer of n-type (electron-rich) region that can link the two n regions at drain and source. The "enhanced" MOSFET then resembles the JFET that has not been depleted: a conducting "channel" links source to drain. The MOSFET plotted in Fig. 12S.2 (like most, but not all) is OFF until you apply a gate voltage to turn it ON. Such a device is said to be of *enhancement mode*, conducting only when you give it some help (when you "enhance" its conduction).

The JFET or MOSFET – with a slab of doped semiconductor between its end terminals, source and drain[1] can be expected to conduct if we apply a voltage difference between its drain and source

---

[1] Perhaps the plumbing analogy suggested by the names bothers you: current goes from drain to source? Isn't that what

12S.1 A physical picture

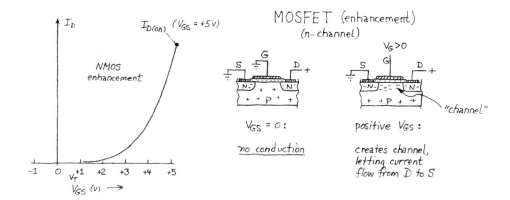

Figure 12S.2 MOSFET conducts if positive V_{gate} induces region to link drain and source n-type regions: it's OFF till you turn it ON.

terminals. At low voltages[2] across the device (V_{DS}), JFET or MOSFET behavior is *Ohmic*: current proportional to V_{DS}.

So far the behavior makes some intuitive sense. We'll soon (§12S.1.1) tackle the harder question of why at larger V_{DS}, *current source* behavior replaces Ohmic.

For comparison with the MOSFETs curve, Fig. 12S.3 shows the I_D-versus-V_{GS} curve for a JFET: ON until you do something to turn it OFF (perhaps only in degree). A negative voltage applied to the gate narrows the conducting channel.

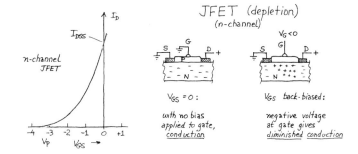

Figure 12S.3 JFET I–V curve looks like MOSFETs – but slid to quadrant left of $V_{\text{GS}}=0$: it's ON till you turn it OFF.

Such devices are said to operate in "depletion mode." All JFETs and a few MOSFETs operate this way.

The need of a *depletion-mode* device for a negative control voltage dooms it as a general-purpose switch. The *enhancement-mode* MOSFET, in contrast, can work with a single supply to generate an output capable of switching another switch of the same type; see Fig. 12S.4.

Thus one can build a large system of single-supply MOSFET switches, using enhancement-mode devices (we're hoping this suggestion brings computers to mind). The switches we have drawn are not

---

happens when the sewer gets blocked? The explanation for these strange names must be that the fellows developing the FET – the same people who named the bipolar's "collector" and "emitter," though here the case is muddied by the pnp configuration where a flow of "holes" dominates – were thinking like physicists, not like engineers: they seem to have considered electron flow rather than conventional current. This seems to be the case even though this transistor type is named "bipolar" because current flow depends in part on flow of both "holes" and electrons.

[2] Low voltages here means less than $V_{\text{GS}} - V_T$. That *difference* voltage measures the degree to which the gate is driven ON. V_T is the threshold voltage at which I_D rises from zero to a small but measurable value – usually specified as a few nano-amps. See AoE §3.2.7.

548 Supplementary Notes: MOSFET Switches

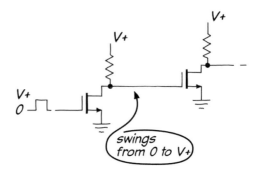

Figure 12S.4 A MOSFET switch (enhancement) can drive another with just one supply.

of the type used in large arrays of logic devices. We saw in Chapter 12N that a *CMOS* configuration is much more efficient (AoE §3.4.4A).

One cannot do this with depletion-mode transistors.

12S.1.1 A go at explaining the FET's constant-current behavior

The style of this course surely is to ignore the details of physics that might explain what's going on within a transistor, but FETs do tempt one to hope for intuition concerning their behavior. We'll give a try to this effort, while not expecting serious success.

Linear region versus current-source regions: resistive behavior at low V_{DS}... We claimed in §12S.1 that it is plausible that a continuous slab of doped semiconductor should conduct. It is also plausible that the current it passes should be proportional to the voltage applied (between drain and source). This is the behavior one sees for low V_{DS}, as one can see in the leftmost, the "linear," region of Fig. 12S.5.

AoE §3.2.7

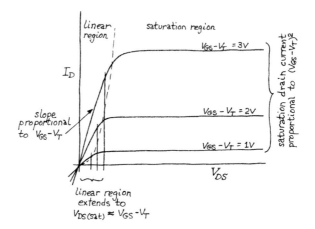

Figure 12S.5 FET is resistive at low V_{DS}, but behaves like a current-source at higher V_{DS}.

This family of curves looks much like the set of curves that describe I_C versus V_{CE} for a bipolar transistor, where each curve would be determined not by V_{GS} but by either V_{BE} or I_B. In Fig. 12S.6 are bipolar and FET curves, placed side by side for comparison (the contrast in current-region slopes is a bit unfair to the bipolars: note the much more sensitive current scale for the bipolar). The horizontal axis shows the voltage across the transistor. Each curve is defined by a particular input control level: V_{GS} or V_{BE}.

12S.1 A physical picture

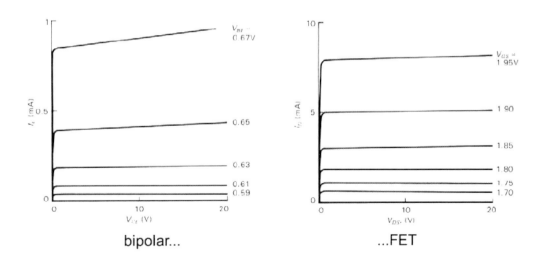

Figure 12S.6 Families of *I* and *V* curves, bipolar compared with FET (AoE Fig. 3.2).

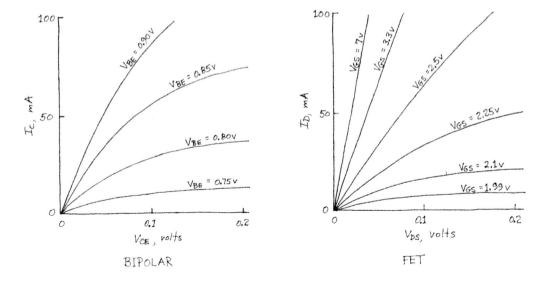

Figure 12S.7 At low voltage across transistor, FET is more linear than the bipolar transistor (AoE Fig. 3.2).

You may wonder whether a bipolar transistor might not be capable of the same *resistive* behavior. The answer is that its *I–V* is less linear than the FETs, at low voltages: see Fig. 12S.7.

A simple circuit trick can further straighten the FETs curves, it turns out – making the FET a useful voltage-controlled resistor.[3]

So far, so simple: the resistive behavior at low voltage sounds reasonable. A harder challenge is to get some understanding of the constant-current behavior that is much more typical of the FET.

---

[3] The "trick," discussed in §12L.4.1, and its footnote 19, is to use an $R{:}R$ voltage divider to feed half of V_{DS} to the gate. See AoE §3.2.7A, for more on this topic.

...Saturation region: **current-source** *behavior at higher* V_{DS} Resistive behavior, intuitively comfortable for us resistor-philes, is exceptional for an FET. Ordinarily the FET, like a bipolar transistor, passes a *current* determined by its control signal – V_{GS} for the FET. This current-source behavior appears in Fig. 12S.5 over all except the lowest range of V_{DS} values. The nearly horizontal current plot resembles what one would see for a bipolar transistor. Why? We'll attempt an informal explanation.

As one gradually increases V_{DS}, the rising voltage near the drain diminishes the width of the conducting region there, and eventually that conducting channel "pinches off." The leftmost image of Fig. 12S.8 sketches the non-conducting case with $V_{GS}=0$. The two to the right show conduction, first in the linear region, then in the pinched off region. It is the rightmost image that attempts to picture the *constant current* or "saturation" region of FET operation.[4]

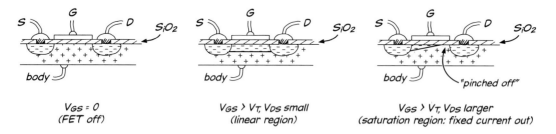

Figure 12S.8 MOSFET conduction: three regimes.

A first (and quite reasonable) guess would be that current I_D would fall to zero when V_{DS} rose enough to pinch off the channel. But I_D does not do that. Instead, if V_{DS} continues to rise, I_D levels off. The region between pinch-off point and drain is not as rich in carriers as is the induced "channel". But the field there is strong enough to carry electrons that have drifted through the channel on to the drain.

Here's our attempt at an intuitive understanding of this effect. If this is too silly for your tastes, take a look at the texts cited below.

The pinched-off region forms a bottleneck, a narrow region where a kind of traffic jam occurs, see Fig. 12S.9. As V_{DS} rises, much of this drop occurs over the short length of that pinched section, pushing up the current density in this bottleneck; then, if V_{DS} grows further, increasing the field (drivers at the drain end lean on their car horns) the bottleneck region grows longer, and traffic flow or I_D levels off at its *saturation* value. In short, further increases in V_{DS} cause two opposing effects that nearly cancel: stronger field, but reduced carrier mobility.

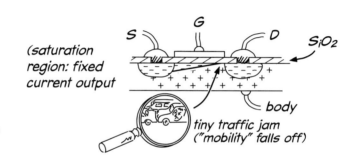

Figure 12S.9 Tiny traffic jam to explain FET "saturation:" mobility falls off as field strengthens.

---

[4] You will recognize what a bad choice of language this is: to use the word "saturation" to describe normal FET operation, while the word means something radically different when applied to bipolar transistors or operational amplifiers. But we are stuck with this usage.

12S.1 A physical picture

So current stays roughly constant.[5]

12S.1.2 Saturation not quite constant: channel-length modulation resembles Early effect

AoE §3.1.1A

You may have noticed that the *I–V* curves across FETs resemble those for bipolar transistors not only in rough outline, as we suggested on page 548 and Fig. 12S.6, but also in their upward tilt. That tilt, described by Early effect for bipolars, results from *channel-length modulation* in an FET. The mechanism is similar to that for Early effect. As V_{DS} grows, the length of the conducting channel is diminished by the growth of the pinched-off region. So current slightly increases (*very* slightly: the dynamic impedance of the 1N5294, a JFET current-limiting diode that we use in several labs, for example, is high: above 1MΩ [specified at $V_{DS} = 25$ V]).

AoE §3.2.2 and Fig. 3.20

12S.1.3 An application for constant-current behavior

A JFET, with its ability to hold current constant even at $V_{GS} = 0$, makes a handy two-terminal current-limiting device – as you may recall from Lab 5L, when you used such a thing in the "tail" of the differential amplifier. The curves you saw back in Fig. 12S.5 remind us that we cannot expect *constant-current* behavior however, unless we take care to keep V_{DS} above the level where *resistive* behavior ceases. For the 1N5294 that minimum voltage is 1.2 V.

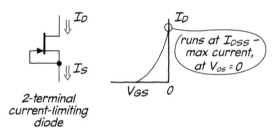

Figure 12S.10 JFET forms simple current-limiting diode.

---

[5] For a fuller explanation and handsome diagrams see Burns, S. and Bond, P., *Principles of Electronic Circuits*. West (1987), §5.2; see also Jaeger, R. and Blalock, T., *Microelectronic Circuit Design*. McGraw-Hill (4th ed., 2011), §4.24.

13N Group Audio Project

Contents

13N.1	**Overview: a day of group effort**		**552**
	13N.1.1	Circuits that do something, again…	552
	13N.1.2	…and circuits of your own design	552
	13N.1.3	A description of the technical task	553
	13N.1.4	The elements of the audio-transmission chain	553
13N.2	**One concern for everyone: stability**		**555**
13N.3	**Sketchy datasheets for LED and phototransistor**		**556**

Why?

Part of what we aim for today is a review, since the circuit includes an unusually broad variety of elements. What you will achieve is the wireless transmission of an audio signal, using optical encoding.

13N.1 Overview: a day of group effort

Today's lab and class differ from all the others in the course. In class, we want to hear you do the teaching: tell your classmates about the piece of the project that you designed. We want your audience to tell you (politely) how you might improve your design. The lab differs from all the others in that it is a collaboration. Your fragment won't do anything very entertaining until it is joined with the work of your classmates.

13N.1.1 Circuits that do something, again…

Our motive for trying this class and lab format was to respond to students' pleas that they be given a chance to build circuits that "do something." Apparently, people get tired of seeing only scope traces. The earlier build-an-op-amp and then the PID lab also tried to respond to this yearning. Today's lab offers a similar chance to put many circuit fragments together. This lab differs from PID and the others, however, in asking you to do all the *design* work.

13N.1.2 …and circuits of your own design

Generally, we have not dared to write labs this way. We feared that the process would be too slow: you would draw a design, find its defects, correct them, then try the amended design. It doesn't sound like work that could fit within an afternoon. Trying that process every day would have made this quite a different sort of course – a good one, but not one that could gallop through the large amount of material that we attempt. We do, however, dare to try this method on this one day, because we are confident that you will come up with working designs: each of the several circuit elements is well within the range of your skills.

We hope that the exercise will serve several purposes. It will be fun, and should provide satisfying confirmation that you have learned something in the past weeks.

13N.1.3 A description of the technical task

Figure 13N.1 is a sketch of what we want to build. The notes below will spell out what the several boxes ought to do.

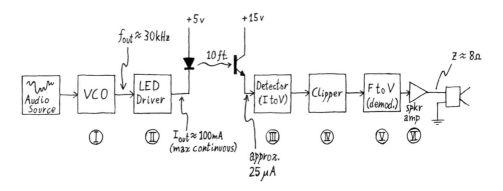

Figure 13N.1 A block-diagram of the group audio project.

You will send music across the room – or, at very least, across a gap of a few feet – "wirelessly." In Marconi's day, and again more recently, wireless meant "using radio." Here it does not. We will transmit using infrared "light."[1] More particularly, we will encode the audio *information* onto a *carrier* that is a frequency above human hearing – but far short of radio. We put our carrier at about 30 kHz. We will encode the information – the time-varying voltages – as *frequencies*. This scheme is called, as you know, Frequency Modulation, FM.

FM is appealing here because of its good noise immunity and also because it allows us to drive the LED at full brightness (when ON), permitting maximum range.[2]

The variation in frequency need not be large. In order to convey an audio signal up to about 3 $mrkHz$, a frequency variation of well under 3 kHz can be sufficient.

13N.1.4 The elements of the audio-transmission chain

Here is a short catalog of the elements of today's circuit. In some cases, our descriptions will be purposely vague because we don't want to short-circuit the design process that we hope you will go through.

On the transmission side:

Modulator: This is a voltage-controlled oscillator (VCO). We use a '555 oscillator to convert an audio signal of time-varying voltage into a square wave of varying frequency. This circuit's voltage-to-frequency function won't be linear, but works quite well enough for today's purposes. You could use a more perfect VCO (such as, say, the VCO included in the 74HC4046 that you will meet in Lab 20L). But for this extra effort you would get no audible improvement over today's rougher design.

---
[1] Is it "light" if humans can't see it? A nice question, which we'll not stop to worry about.
[2] Perhaps you will protest that "noise immunity" and "range" are two ways of saying the same thing. You have a point.

Details: (Questions of interest to designers of this stage; perhaps too detailed for the rest of you).

The needed frequency variation is slight and we can tolerate a somewhat nonlinear response – because our standards for this task are not especially high, and because we anticipate nonlinearities in the demodulator, in any case.

There are easy and hard ways to vary the frequency of a '555 oscillator. The hard ways probably would use a voltage-controlled current source to vary a sawtooth frequency. The easy way takes advantage of the *control* terminal (pin 5 for the DIP package), letting the audio signal vary the threshold voltages.

Consider duty cycle: does your design achieve close to 50% duty cycle (that is, percentage of period during which the waveform is *high*)? Does duty cycle matter?

Make the center frequency adjustable by manipulation of a potentiometer wired as variable resistor. Don't forget to provide a constant R along with the adjustable so that the circuit doesn't go crazy if you happen to twiddle the pot to an extreme.

The carrier frequency: We chose a frequency of around 30 kHz, intermediate between extremes that could have caused difficulty:
- we wanted it not too low: not in a range audible to humans, because some of the carrier is likely to persist after demodulation – noise mixed with the audio signal; and
- we wanted it not too high: not in a range where our op-amps and comparators begin to falter.

LED driver: This takes the relatively delicate square wave that is output by the VCO (a '555 oscillator) and makes it strong enough to drive a high-current infrared LED. The LED is capable of running 100 mA continuous current, and drops about 2 V when conducting.

Details, LED driver:

LED current is specified (maximum continuous current; for current specification, see sketchy datasheets in §13N.3.) Does it follow that your driver should be a *current source* (or "sink")?

Consider what the signal from the '555 output looks like electrically, and what the load looks like (LED dropping about 2 V).

Try to minimize power dissipation. The current is large; you don't want a voltage higher than necessary. If you use a resistor, make sure it won't get hot enough to burn your fingers.[3]

On the receiving side:

Photodetector: This converts the current passed by an infrared-sensitive phototransistor to a voltage. The voltage swing should be about as large as you can manage, taking care always to avoid clipping the waveform.

Details:

Note that ambient light will include some infrared, so that the phototransistor current never will fall to zero (despite the transistor's dark case designed to block visible light). Note also that the phototransistor current flows in one direction only.

You're likely to take the current from the phototransistor's *emitter* (compare the photo circuits of Lab 6L). But don't infer that the circuit behavior you see will resemble that of an emitter follower (Lab 4L) where you last took an output from emitter.

Clipper: This circuit takes the detector output, perhaps small, and transforms it into a square wave swinging ±15 V. (The large amplitude is helpful to the later stages.) The uniform amplitude

---

[3] Do this not by wearing thick gloves but by considering power dissipated versus the resistor's power rating.

imposed by this circuit is important: variations in amplitude that appear in the *detector's* output would corrupt the ultimate demodulated audio output if allowed to persist.

Details:

Note that the output of the detector will include a DC component (unless the detector designers have removed that for you). That DC level is not information; only the frequency of the varying voltage carries information.

To keep the output clean, you will need hysteresis. Make this adjustable right down to zero since you'd like your circuit to be as sensitive as possible – permitting maximum range for the entire circuit.

Demodulator: This takes the square wave of varying frequency and converts it, in two stages, into a time-varying amplitude: an audio signal like the one that drove the VCO on the input side.

- Stage one: uses a filter with steep skirt to convert changes in frequency to changes in amplitude.
- Stage two: uses a conventional AM demodulator to produce the audio waveform.

Details:

Any filter with a steep response will do, but the easiest circuit probably is an *RLC* resonant circuit. You will recall from Lab 3L, especially §3L.1.2, that this circuit's $\Delta_{amplitude}/\Delta_{frequency}$ response can be made very steep.

For best conversion of $\Delta_{frequency}$ to $\Delta_{amplitude}$ you'd like a response curve that is close to a straight-line ramp. We suggest that you not strain to calculate the best shape for this response curve. Instead, in the rough-and-ready style of this course, make the *shape* of the circuit's frequency response adjustable.

Recall what circuit elements in Chapter 3L's resonant circuit affected the shape of the response: large R provided large Q values (steep curve); but large R also provided a lower output amplitude (entire curve became small, though peaky). In short, large R produces a small pointy Matterhorn; smaller R produces a large gently-sloping Blue Ridge Mountain.

Consider phase: does the demodulator's output voltage rise or fall as frequency rises? Does the answer to this question matter?

Amplifier: Boost demodulator output to drive an 8 Ω speaker.

Details:

This is a circuit you have built before. Make its gain adjustable (since a loud output can be maddening during the time when the group is trying out and adjusting the whole circuit). You probably need not set minimum gain below one (and this choice will permit use of a high-impedance form of the amplifier).

Note that you do not want to put any *DC* current through the speaker. Only wiggles produce sound; a DC level produces only heat.

The major challenge will be to keep this circuit *stable*.

13N.2 One concern for everyone: stability

The several elements of the circuit will be wired each on its own breadboard strip. Then these will be brought together, linked and powered. This is a setup peculiarly vulnerable to parasitic oscillations. Against this hazard:

- *Decouple* all power supplies, on every individual strip, with ceramic caps of 0.01–0.1 μF.

- Do all you can to keep power supply leads *short*. (In our course, we provide power through bus connectors pulled off old breadboard strips; the low inductance and substantial capacitance of power buses helps: see Fig. 13N.2.)

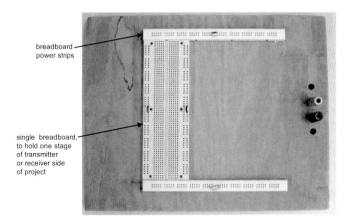

Figure 13N.2 Power bus strips help to minimize power supply impedance, against parasitic oscillations.

13N.3 Sketchy datasheets for LED and phototransistor

The LED and phototransistor work in the near-infrared, nearly matching peak sensitivity to peak emission wavelength. Both are narrow-angle devices, chosen to maximize range – but having the drawback that the emitter must be well aimed. Here are some of their specifications.

Device	central wavelength (nm)	emission/reception Angle	current	rise & fall time (μs)
LED: TSTS7100	950	±5°	250 mA (max cont.)*	0.8
PhotoQ: QSD124	880	±12° (half power angle)	6 mA (ON, min.)	7
* But 100 mA is more usual for this part				

And here are pinouts.

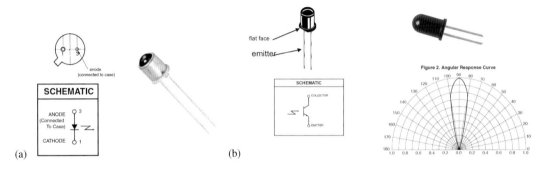

Figure 13N.3 (a): Infrared LED pinout: TSTS7100. (b): Infrared phototransistor pinout: QSD124.

13L Lab: Group Audio Project

13L.1 Typical waveforms

Figure 13L.1 gives a preview of what's coming in the form of waveforms at several points. Before you have built the circuits, these waveforms may be a bit cryptic; but trying to understand these plots may help you to get a grip on the whole project. Then we'll finish these lab notes with some suggestions on how to test your circuits.

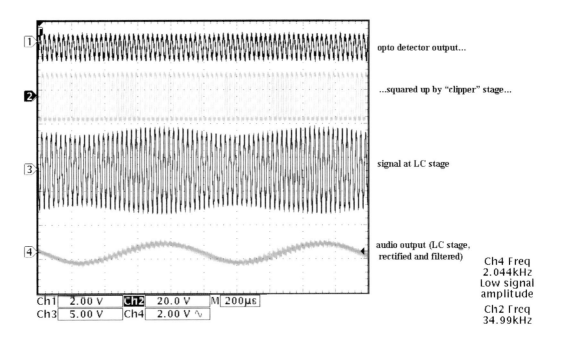

Figure 13L.1 Waveforms at several stages of the IR transmission chain.

13L.1.1 How much do you need to know about stages that abut yours?

Not much, we expect. If your abutters surprise you, we expect you'll be nimble enough to adjust on project day. It's certainly fair to ask them, beforehand – but probably not necessary.

13L.1.2 How much advice do we want to give you beforehand?

Not much because it's helpful to the whole group to have design issues aired before the group rather than settled backstage. We hope that the design you present will *not* be perfect. Perfect designs are not very instructive. And please don't test your design beforehand by building it; a real discussion stopper is a line like "I don't know why the value is 10k – but I built it and it worked." For reviewing purposes, it's much better to discuss rationales.

13L.2 Debugging strategies

Here are some debugging tips that can simplify our task: we'll pair modules that talk to each other; when these pairs work, we'll string together all the elements of the chain. You'll find a block diagram of that entire chain in Fig. 13N.1.

13L.2.1 Debug VCO (*V*-to-*f*) and demod (*f*-to-*V*)

These adjustments are particularly ticklish. VCO people can adjust the center frequency. Use a DVM as frequency-meter, to watch this frequency. Note what frequency works best (that is, produces the cleanest sine from the demod block) in case you mess up an adjustment and need to return to that center frequency. We find a 1k resistor in series with the pin 5 input helps to protect the CMOS '555 from damage. It does not affect circuit performance (the internal divider's resistors present an equivalent resistance of 33k).

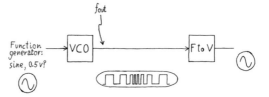

Figure 13L.2 Testing VCO and demodulator.

Before testing with the VCO, demods will have done the best they could on their own: demods can do their preliminary test by feeding their circuit a square wave whose frequency is swept. The demod circuit should output a fairly good replica of the *ramp* shape that evokes the *swept* square wave.

In testing the linked VCO-to-demod pair, VCO people will try to find their comfortable place on the slope of the "mountain" that is the shape of demod's frequency response. Demod people meanwhile, can adjust shape of that mountain: they can adjust the Q of their circuit, looking for a region that is pretty linear.

Don't drive yourselves crazy by letting both VCO and demod people twiddle at the same time: take turns! When demod puts out a pretty good reproduction of the source sine, leave the adjustments as they are. It will be harder to optimize response at the later stage when you will be working with the entire chain of circuit elements.

13L.2.2 Optical link only

Optical detector people probably have at least a gain-control to adjust; perhaps also a DC-offset. If you're not sure whether the LED is alive, use an infrared detector card (if it has just come from storage in a dark drawer, it will need a few minutes in room light for activation).[1]

---

[1] Laboratory quality IR detector cards are expensive, but you're likely to find inexpensive versions on eBay. Your cellphone camera may or may not show IR, depending on its camera's filtering.

13L.2 Debugging strategies

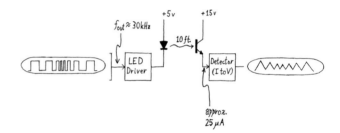

Figure 13L.3 Testing optical link only.

13L.2.3 Optical recovery, squared-up

Occasionally, attaching the comparator as load will destabilize the detector. If this happens (indicated by fuzz on the detector output), make sure you have decoupled all your power supplies, and that you have minimized power-supply lengths.

Clipper people will have a hysteresis-adjust to play with. They want minimum hysteresis (for max sensitivity) that is consistent with stability.

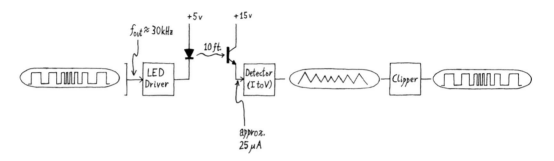

Figure 13L.4 Adding clipper to optical-link test.

13L.2.4 Sine-to-sine, with audio amp

Again, stability is the hardest issue here: the amp is straightforward (make sure the transistors aren't tiny 2N3904s and 2N3906s). Testing the amp by feeding it from the demod block gives a better test than just taking a function generator as source for the amp.

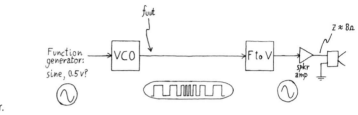

Figure 13L.5 Trying to recover audio, testing speaker driver.

Part IV

Digital: Gates, Flip-Flops, Counters, PLD, Memory

14N Logic Gates

Contents

14N.1	**Analog versus digital**	**564**
	14N.1.1 What does this distinction mean?	564
	14N.1.2 But why bother with digital?	565
	14N.1.3 Alternatives to binary	566
	14N.1.4 Special cases for which digital processing obviously makes sense	568
14N.2	**Number codes: Two's-complement**	**568**
	14N.2.1 Negative numbers	568
	14N.2.2 Hexadecimal notation	569
	14N.2.3 BCD encoding	570
14N.3	**Combinational logic**	**570**
	14N.3.1 Digression: a little history	570
	14N.3.2 DeMorgan's theorem	572
	14N.3.3 Active-high versus active-low	573
	14N.3.4 Missile-launch logic	575
	14N.3.5 Implementing combinational functions	575
	14N.3.6 Considering gates as "Do this/do that" functions	577
14N.4	**Gate types: TTL and CMOS**	**578**
	14N.4.1 Gate innards: TTL versus CMOS	579
	14N.4.2 Thresholds and noise margin	579
14N.5	**Noise immunity**	**580**
	14N.5.1 DC noise immunity: CMOS versus TTL	580
	14N.5.2 Differential transmission: another way to get good noise immunity	582
14N.6	**More on gate types**	**584**
	14N.6.1 Output configurations	584
	14N.6.2 Logic with TTL and CMOS	585
	14N.6.3 Speed versus power consumption	585
14N.7	**AoE reading**	**586**

Why?

We aim to apply MOSFETs to form gates capable of Boolean logic and to look at some such binary operations.

564 Logic Gates

14N.1 Analog versus digital

14N.1.1 What does this distinction mean?

The major and familiar distinction is between *digital* and *analog*. Along the way, let's also distinguish "binary" from the more general and more interesting notion, "digital."

- First, the analog versus digital distinction.
  - An *analog* system represents information as a continuous function of the information (as a voltage may be proportional to temperature, or to sound pressure).[1]
  - A *digital* system, by contrast, represents information with *discrete* codes. The codes to represent increasing temperature readings could be increasing digital numbers (0001, 0010, 0011, for example) – but they could use any other code that you found convenient. The digital representation also need not be *binary* – a narrower notion mentioned just below.
- Now the less important binary versus digital distinction: binary is a special case of digital representation in which only two values are possible, often called True or False. When more than two values are to be encoded using binary representations, multiple binary digits are required. Binary digits, as you know, are called *bit*s (*b*inary dig*it*).

A *binary* logic gate classifies its inputs into one of two categories. You can think of the gate as a comparator – one that is simple and fast: see Fig. 14N.1.

Figure 14N.1 Two comparator circuits: digital inverter and explicit comparator, roughly equivalent.

The digital gate resembles an ordinary comparator (hereafter we won't worry about the digital/binary distinction; we will assume our digital devices are binary).

How does the digital gate differ from a comparator like, say, the LM311?

Input and output circuitry: the '311 is more flexible (you choose its threshold and hysteresis; you choose its output range). Most digital gates include *no* hysteresis. The exceptional gates that include hysteresis usually are those intended to listen to a computer bus (assumed to be extra noisy) and the inputs to some larger ICs such as the FPGA you will meet in our labs, which includes 200 mV of hysteresis on all of its inputs.

Speed: the logic gate makes its decision (and makes its output show that 'decision') at least 20 times as fast as the '311 does.

Simplicity: the logic gate requires no external parts, and needs only power, ground, and input and output pins. It can work without hysteresis because the logic signals presented to its inputs transition decisively and fast.

---

[1] We don't quite want to say that the representation – say, the voltage – is *proportional* to the quantity represented since a log converter, for example, can make the representation stranger than that.

14N.1.2 But why bother with digital?

Is it not *perverse* to force an analog signal – which can carry a rich store of information on a single wire – into a crude binary high/low form? Let's consider the cons and pros. A caricaturing of "digital audio" appears in Fig. 14N.2, where the sound of the Stradivarius is converted to arcade-game quality. We doubt this represents your understanding of *digital audio*.

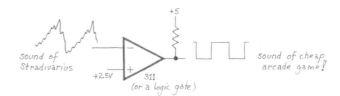

Figure 14N.2 Naive version of "digital audio": looks foolish!

Disadvantages of digital:

Complexity – more lines are required to carry the same information.[2]

Speed – processing the numbers that encode the information sometimes is slower than handling the analog signal.

AoE §10.1.1

Advantages of digital:

Noise immunity – the signal is born again at each gate; from this virtue flow the important applications of digital:
- allows stored-program computers (which may look like a wrinkle at this stage, but turns out, of course, to be hugely important);
- allows transmission and also unlimited processing without error – except for round-off/quantization; that is, deciding in which discrete binary bin to put the continuously variable quantity; and
- can be processed out of "real time:" at one's leisure. (This, too, is a consequence of the noise immunity already noted which allows storing and recovering digital information relatively easy.)

Can we get the advantages of digital without loss? Not without *any* loss, but one can carry sufficient detail of the Stradivarius' sound using the digital form – as we try to suggest in Fig. 14N.3 – by using *many* lines.

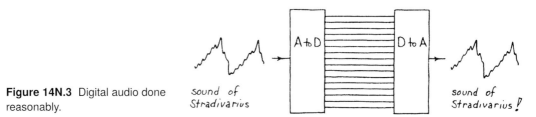

Figure 14N.3 Digital audio done reasonably.

---

[2] "More lines" often does not apply to *transmission* of signals because serial transmission is widely used. But at least in its inner workings, digital processing does encode values as multiple parallel bits.

A single bit allowed only two categories: it could say, of the music signal, only "in high half of range" or "in low half." Two bits allow finer discrimination: four categories (top quadrant, third quadrant, second quadrant, bottom quadrant). Each additional line or bit doubles the number of slices we apply to the full-scale range.

Our lab microcomputer uses 10 bits for converted analog signals.[3] Commercial CDs and most contemporary digital audio formats use 16 bits, permitting $2^{16} \approx 65,000$ slices. To put this another way, 16-bit audio takes each of our finest slices (about as small as we can handle in lab, with our noisy breadboarded circuits) and slices that slice into 2^6 sub-slices. This is slicing the voltage very fine indeed.

In Fig. 14N.4 we demonstrate the effect of 1, 2, 3, 4, 5, and finally 8 bits in the analog–digital conversions (both into digital form and back out of it, to form the recovered analog signal that is shown).

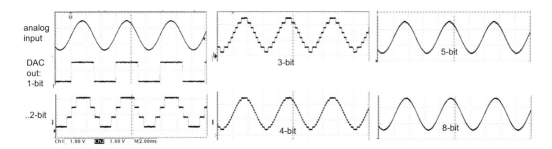

Figure 14N.4 Increasing the number of bits improves detail in the digital representation of an analog quantity.

14N.1.3 Alternatives to binary

AoE §§7.1.9G and 14.4.5C

Binary is almost all of digital electronics, because binary coding is so simple and robust. But a handful of integrated circuits do use more than two levels. A few use three or four voltage levels, internally, in order to increase data density: NAND flash memory for example, uses four levels or even eight. Data storage density is increased proportionately.[4] This density comes at the cost of diminished simplicity and noise immunity. The better noise immunity of the older 2-level parts sometimes is promoted as one of their selling points, but multi-level now dominates NAND flash.

More important instances of non-binary digital encoding occur in data *transmission* protocols. Telephone modems ("modulator–demodulators") confronted the tight limitation of the system's 3.1 kHz frequency limit (300 Hz to 3.4 kHz), and managed to squeeze more digital information into that bandwidth by using both multiple-levels (more than binary) and *phase* encoding. One combination of the two called "16QAM" squeezes 4 bits worth of information into each "symbol," pushing the data rate to 9600 bits/second. Does that not sound quite *impossible* on a 3.4 kHz line? It is, of course, possible. The process of increasing coding complexity has continued, pushing data rates still higher.

A plot like a phasor diagram – with "real" and "imaginary" axes – can display the several phases as well as amplitudes of single-bit encodings. The plot in Fig. 14N.5 shows "4QAM" single-amplitude,

---

[3] The analog-to-digital converter in our microcontroller can be configured to digitize an analog input into either 2^8, 2^{10}, or 2^{12} slices but the built-in digital-to-analog converter can only convert a 10-bit result back to an analog signal, so that is what we will use.

[4] Fun fact (according to Wikipedia): a quarter byte (2 bits) sometimes is called a "crumb," joining its foody companions, bit, byte and nibble (a 4-bit binary value).

four-phase-angle encoding that permits four unique "symbols" (or, "two bits per symbol") by summing two waveforms that are 90° out of phase (the "quadrature" of the QAM acronym).[5]

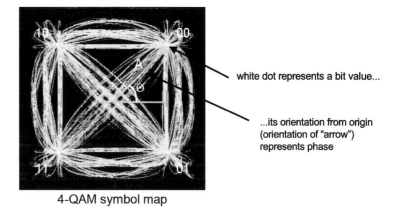

Figure 14N.5 "Constellation" diagram, drawn by vector display – phasor-like: showing QAM (phase- and amplitude-encoded digital information).

Denser encodings are possible using the same scheme. The constellation diagrams in Fig. 14N.6 show 16QAM which carries 4 bits of information in each "symbol," one symbol being represented by one dot on the diagram.[6]

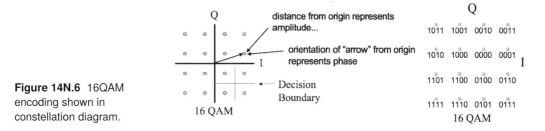

Figure 14N.6 16QAM encoding shown in constellation diagram.

The QAM waveforms (seen in time-domain) are exceedingly weird: see Fig. 14N.7.[7]

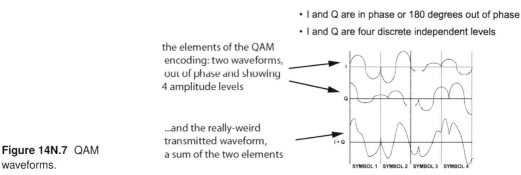

Figure 14N.7 QAM waveforms.

Complex encoding schemes like QAM are appropriate to data transmission, but – luckily for us – the storage and manipulation of digital data remains overwhelmingly *binary*, and therefore much more easily understood.

---

[5] Source: National Instruments, "Quadrature Amplitude Modulation" https://LAoE.link/NI_QAM.html. This tutorial shows the relation between the time-domain waveform and this phase-and-amplitude constellation plot.

[6] Source: Acterna Digital QAM Signals Overview and Basics Testing White Paper (no longer available but see https://LAoE.link/QAM_tutorial.html for a similar discussion).

[7] Source: Acterna, tutorial, again.

568   Logic Gates

14N.1.4   Special cases for which digital processing obviously makes sense

In some cases the information is *born* digital so no conversion is needed. The information may be numbers (as in a pocket calculator) or words (as in a word processor). Since the information never exists in analog form (except perhaps in the mind of the human), it makes good sense to manipulate the information digitally: as sets of codes. To do otherwise would be perverse (store your term paper as a collection of voltages on capacitors?).

14N.2   Number codes: Two's-complement

AoE §10.1.3

Binary numbers may be familiar to you already: each bit carries a weight double its neighbor's. That's just analogous to decimal numbers as you know: it's the way we would count if we had just one finger. The number represented is just the sum of all the bit values: $1001 = 2^3 + 2^0 = 9_{10}$, where "$9_{10}$" means "nine in base-ten notation."

	DECIMAL			BINARY		
	10^2	10^1	10^0	2^2	2^1	2^0
	2	1	3	1	0	1
	$200 + 10 + 3 = 213$			$4 + 0 + 1 = 5_{10}$		

Figure 14N.8 Decimal and binary number codes compared.

14N.2.1   Negative numbers

AoE §10.1.3C

But 1001 *need not* represent 9_{10}. Whether it does or not in a particular setting is up to us, the humans. *We* decide what a sequence of binary digits is to mean. Now, that observation may strike you as the emptiest of truisms, but it is not quite Humpty Dumpty's point. We are not saying 1001 means whatever I want it to mean; only that sometimes it is useful to let it mean something other than 9_{10}. In a different context it may make more sense to let the bit pattern mean something like "turn on stove, turn off fridge and hot plate, turn on lamp." And, more immediately to the point, it often turns out to be useful to let 1001 represent not 9_{10}, but a *negative* number.

The scheme most widely used to represent negative numbers is called "two's-complement." The relation of a 2's-comp number to positive or "unsigned" binary is extremely simple:

> the 2's-comp number uses the most-significant-bit (MSB) – the leftmost – to represent a *negative* number of the same weight as for the unsigned number.

So 1000 is +8 in unsigned binary; it is −8 in 2's-comp. And 1001 is −7. Two more examples appear in Fig. 14N.9: the 4-bit 1011 can represent a negative five, or an unsigned decimal eleven; 0101, in contrast, is (positive-)5 whether interpreted as "signed" (that is, 2's-comp) or "unsigned."

	Unsigned	2's-Comp
1011	$8 + 2 + 1 = 11_{10}$	$-8 + 2 + 1 = -5_{10}$
0101	$4 + 1 = 5_{10}$	$\ldots = 5_{10}$

Figure 14N.9 Examples of 4-bit numbers interpreted as *unsigned* versus *signed*.

This formulation is not the standard one. More often you are told a rule for forming the 2's-comp representation of a negative number, and for converting back. This AoE does for you:

AoE §10.1.3C

14N.2 Number codes: Two's-complement

To form a negative number, first complement each of the bits of the positive number (i.e., write 1 for 0, and vice versa; this is called the "1's complement"), then add 1 (that's the "2's complement").

You may find it easier to form and read 2's-comp if, as we've suggested, you simply read the MSB as a large negative number, and add it to the rest of the number, which represents a (smaller) positive number interpreted exactly as in ordinary unsigned binary.

AoE §10.1.3D

Two's-complement may seem rather odd and abstract to you just now. If you get a chance to *use* 2's-comp it should come down to earth. When you program a microcomputer, for example, it is used to tell the machine exactly how far to "branch" or "jump" forward or back as it executes a program. In the example of Fig. 14N.11 below we will say this by providing a 2's-comp value that is either positive or negative. But first a few words on the *hexadecimal* number format.

14N.2.2 Hexadecimal notation

Let's digress to mention a notation that conveniently can represent binary values of many bits. This is *hexadecimal* notation – a scheme in which 16 rather than 10 values are permitted. Beyond 9 we tack in 5 more values, A, B, C, D, E and F, representing their decimal equivalents 10 through 15.

The partial table in Fig. 14N.10 illustrates how handily hexadecimal notation – familiarly called "hex" – makes multi-bit values readable, and "discussable."

	Binary	Hexadecimal	Decimal
	0111	7	7
	1001	9	9
	1100	C	12
	1111	F	15
	10011100	9C	156

Figure 14N.10 Hexadecimal notation puts binary values into a compact representation.

You don't want to baffle another human by saying something like, "My counter is putting out the value One, Zero, Zero, One, Zero, One, One, Zero." "What?," asks your puzzled listener. Better to say "My counter is putting out the value 96h." One can indicate the *hex* format, as in this example, by appending "h." Or one can show it, instead by writing "0x96."[8]

Now, back to the use of 2's-comp to tell a computer to jump backward or forward. The machine doesn't need to use different commands for "jump forward" versus "jump back." Instead, it it simply adds the number you feed it to a present value that tells it where to pick up the next instruction. If the number you provide is a negative number (in 2's-comp), program execution hops back. If the number you provide is a positive number, it jumps forward.

Perhaps the example in Fig. 14N.11 will begin to persuade you that there can be an interesting, substantial difference between "subtracting A from B" and "adding negative A to B." Subtraction – if taken to mean what "subtraction" says – requires special hardware; addition of a negative number, in contrast, uses the same *hardware* as an ordinary addition. In the example, we are *subtracting* 4 by *adding* -4 (F...FFFC in 2's-comp) to the present value. That's what makes the use of 2's-comp a tidy scheme for the microcomputer's *branch* or *jump* operations. (The overflow bit – which always occurs on a backwards jump – is discarded.)

---

[8] If you use the "h" convention, you must start a hex value with a digit when programming. A human might understand "A0h" to be 160_{10}, but most programming languages will interpret it as the name of a variable or label. Using "0A0h" indicates to the assembler or compiler you are specifying a numeric value.

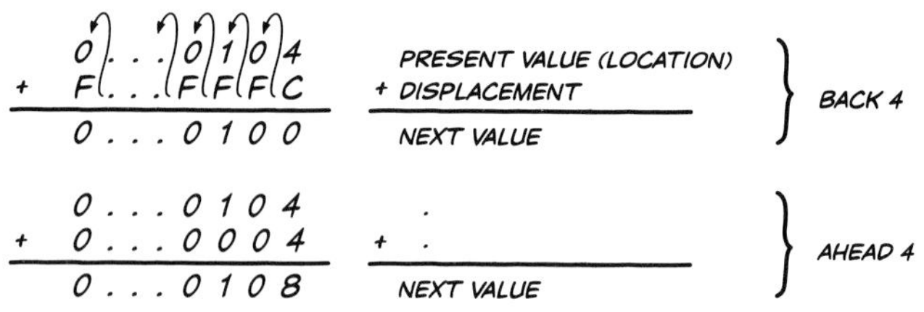

Figure 14N.11 Example of 2's-comp use: plain *adder* can add or subtract, depending on sign of addend: computer jump displacement.

```
    0...0 1 0 4        PRESENT VALUE (LOCATION)
+   F...F F F C      + DISPLACEMENT              ⎫
    ─────────────      ────────────────────      ⎬ BACK 4
    0...0 1 0 0        NEXT VALUE                ⎭

    0...0 1 0 4        .
+   0...0 0 0 4      + .                         ⎫
    ─────────────      ────────────────────      ⎬ AHEAD 4
    0...0 1 0 8        NEXT VALUE                ⎭
```

14N.2.3 BCD encoding

Humans are so used to having ten fingers that they just don't like displays showing binary values. BCD (Binary Coded Decimal) encoding groups 4-bit binary values to represent the digits 0_{10} to 9_{10}. Binary values 1010 and above are not used. The advantage of BCD encoding is it makes it easy to display numeric values that normal people can read. The disadvantage is that it requires more bits for a given range. For example, an 8-bit (two hex digits) value can encode 256 values in binary but only 100 in BCD.

14N.3 Combinational logic

Explaining why one might want to put information into digital form is harder than explaining how to manipulate digital signals. In fact, digital logic is pleasantly easy, after the struggles of analog design. We will begin with a discussion of *combinational logic*, that is logic functions whose output is purely dependent on the *current* value of the inputs. Later, we will look at *sequential* circuits, logic with memory of past inputs.

14N.3.1 Digression: a little history

Skip this subsection if you take Henry Ford's view that "History is more or less bunk."

Boole and DeMorgan: It's strange – almost weird – that the rules for computer logic were worked out pretty thoroughly by English mathematicians in the middle of the 19th century, about 100 years before the hardware appeared that could put these rules to hard work. George Boole worked out most of the rules in an effort to apply the rigor of mathematics to the sort of reasoning expressed in ordinary language. Others had tried this project – notably, Aristotle and Leibniz – but had not gotten far. Boole could not afford a university education and instead taught high school while writing papers as an amateur, years before an outstanding submission won him a university position.

Boole saw assertions in ordinary language as propositions concerning memberships in *classes*. For example, Boole wrote, of a year when much of Europe was swept by revolutions,

...during this present year 1848...aspects of political change [are events that]...timid men view with dread, and the ardent with hope.

And he explained that this remark could be described as a statement about *classes*:

...by the words timid, ardent we mark out of the class...those who possess the particular attributes...[9]

Finally, he offered a notation to describe compactly claims about membership in classes:

---

[9] "The Nature of Logic" (1848), in I. Grattan-Guinness and G. Bornet, eds., *George Boole: Selected Manuscripts on Logic and its Philosophy*, Birkhäuser, (1997), p. 5.

...if x represent "Men", y "Rational beings" and z "Animals", the equation

$$x = yz$$

will express the proposition

"Men and rational animals are identical"

So claimed Mr. Boole, the most rational of animals! And he then stated a claim that suggests his high hopes for this system of logic:

> It is possible to express by the above notation any categorical proposition in the form of an equation. For to do this it is only necessary to represent the terms of the proposition... by symbols... and then to connect the expressions ...by the sign of equality.[10]

Given Boole's ambitions, it is not surprising that the sort of table that now is used to show the relation between inputs and outputs – a table that you and I may use to describe, say, a humble *AND* gate – is called by the highfalutin name, "truth table." No engineer would call this little operation list by such a name, and it was not Boole, apparently, who named it so but the philosopher Ludwig Wittgenstein, 70-odd years later.[11] Since Boole's goal was to systematize the analysis of *thinking*, the name seems quite appropriate, odd though it may sound to an engineer.

Before we return to our modest little electronics networks, which we will describe in Boole's notation, let's take a last look at the sort of sentence that Boole liked to take on. This example should make you grateful that we're only turning LEDs ON or OFF, in response to a toggle switch or two. He proposes to render in mathematical symbols a passage from Cicero, treating conditional propositions.

Boole presents an equation, $y(1-x) + x(1-y) + (1-x)(1-y) = 1$, and explains,

> The interpretation of which is: – *Either Fabius was born at the rising of the dogstar, and will not perish in the sea; or he was not born at the rising of the dogstar, and will perish in the sea; or he was not born at the rising of the dogstar, and will not perish in the sea.*

One begins to appreciate the compactness of the notation, on reading the "interpretation" stated in words. You'll recognize, if you care to parse the equation, that Boole writes $(1 - x)$ where we would write $\bar{x}$ or x^* (that is, x false).[12]

So, if we call one of Boole's propositions B (re Fabius' birth), the other P (re his perishing), in our contemporary notation we would write the function, f as

$$f = (B \cdot \overline{P} + \overline{B} \cdot P) + (\overline{B} \cdot \overline{P})$$

This rather indigestible form can be expressed more compactly as a function that is false only under the condition that B and P are true: $\overline{f} = B \cdot P$.

The truth table would look like:

B	P	f
0	0	1
0	1	1
1	0	1
1	1	0

And probably you recognize this function as NAND (NOT AND).

---

[10] "On the Foundations of the Mathematical Theory of Logic..." (1856), ibid., p. 89.
[11] Wittgenstein's sole book published in his lifetime is the apparent source of the name. If "truth table" sounds a bit intimidating, how about the title to Wittgenstein's book : *Tractatus Logico-Philosophicus* (1921)!
[12] S. Hawking, ed., *God Created the Integers*. Running Press (2005), p. 808.

Comforting truth #1: To build *any* digital device (including the most complex computer) we need only the three logic functions in Fig. 14N.12. (It is worth noting that the NOT is implied by the circle or "bubble" alone. The triangle is a buffer providing high input impedance and low output impedance.[13] When ever you see a bubble, think NOT even if the triangle is missing.) Note that there are multiple ways to indicate signal inversion. We will use the bar ($\overline{A}$) and asterisk (A*) interchangeably, but you may see "!" or "/" placed before Boolean variables to indicate inversion as well.

All logic circuits are combinations of these three functions, and only these.[14] The AND and OR function may have more than two inputs. The output of the AND is true if *all* its inputs are true. The OR output is true if *one or more* of its inputs are true.[15]

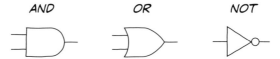

Figure 14N.12 Just three fundamental logic functions are necessary.

(You will also encounter the exclusive-OR gate in Fig. 14N.13. While useful, it is not a fundamental logic function and can be constructed from the gates above. The XOR gate always has exactly two inputs.)

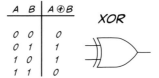

Figure 14N.13 Exclusive-OR (XOR) gate and truth table.

Comforting and remarkable truth #2: Perhaps more surprising, it turns out that just *one* gate type (not *three*) will suffice to let us build any digital device. The gate type must be NAND or NOR; these two are called "universal gates."

Figure 14N.14 Universal gates: NAND and NOR.

Augustus De Morgan – a penpal of Boole – showed that what looks like an AND function can be transformed into OR (and vice versa) with the help of some inverters. This is the powerful trick that allows one gate type to build the world.

14N.3.2 DeMorgan's theorem

This is the only important rule of Boolean algebra you are likely to have to memorize (the others that we use are pretty obvious: propositions like $A + A^* = 1$).

Theorem (DeMorgan's theorem)   You can swap AND and OR shapes if at the same time you invert all inputs and outputs. For the graphical form see Fig. 14N.15.

---

[13] Indeed, we will occasionally place the bubble *before* the triangle to emphasize we are inverting an active-low input signal.
[14] But as we shall see shortly, as long as you have a NOT function, you only need either an AND or an OR (but not both) to implement any Boolean expression.
[15] The exclusive-OR gate in Fig. 14N.13 might be closer to what you imagine when you think of "or," but it is not a fundamental logic function.

14N.3 Combinational logic

Figure 14N.15 DeMorgan's theorem in graphical form.

When you do this you are changing only the *symbol* used to draw the gate; you are not changing the *logic* – the hardware. (That last little observation is easy to say and *hard* to get used to. Don't be embarrassed if it takes you some time to get this idea straight.)

So any gate that can *invert* can carry out this transformation for you. Therefore some people actually used to design and build with NANDs alone back when design was done with discrete gates. These gates came a few to a package, and that condition made the game of minimizing package-count worthwhile; it was nice to find that any leftover gates were "universal." Nowadays, when designs usually are done on large arrays of gates, this concern has faded away. But DeMorgan's insight remains important as we think about functions and draw them in the presence of the *active-low* signals of a type described in §14N.3.3 below.

This notion of DeMorgan's, and the "active-low" and "assertion-level" notions will drive you crazy for a while if you have not seen them before. You will be rewarded when you meet signals that are "active-low." Such signals spend most of their lives close to 5 V (or whatever voltage defines a logic high), and go low (close to 0 V) only when they want to say "listen to me!"

14N.3.3 Active-high versus active-low

AoE §10.1.2A

Signals come in both flavors. Figure 14N.16 shows two forms of a signal that says "Ready".

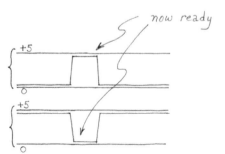

Figure 14N.16 Active-high versus active-low signals.

A signal like the one shown in the top trace of Fig. 14N.16 would be called "Ready;" one like the lower trace would be called "$\overline{\text{Ready}}$," the "bar" indicating that it is active-low: "true-when-low." Signals are made active-low not in order to annoy you, but for good (but now somewhat antiquated) hardware reasons. We will look at those reasons when we have seen how gates are made. But let's now try manipulating gates, watching the effect of our *assumption about which level is true*.

AoE §10.1.7

Effect on logic of active level: active-high versus active-low: When in §14N.2.1 we considered what the bit pattern 1001 *means* when treated as a *number*, we met the curious fact – perhaps pleasing – that we can establish any convenient convention to define what the bit pattern means. Sometimes we want to call it 9_{10} ([*positive*] nine); sometimes we will want to call it -7_{10} (*negative* 7). It's up to us.

The same truth appears as we ask ourselves what a particular *gate* is doing in a circuit. The gate's operation is fixed in the hardware (the little thing doesn't know or care how we're using it); but what its operation *means* is for us to interpret.

That sounds vague, and perhaps confusing; let's look at an example (just DeMorgan revisited, you will recognize). Most of the time – at least when we first meet gates – we assume that high is true. So, for example, when we describe what an AND gate does, we usually say something like

"The output is true if both inputs are true."

The usual AND truth table of course says the same thing, symbolically.

A	B	A · B
0	0	0
0	1	0
1	0	0
1	1	1

or:

A	B	A · B
F	F	F
F	T	F
T	F	F
T	T	T

AND gate using *active-high* signals – and truth table in more abstract form, levels not indicated.

The right-hand table – showing Truth and Falsehood – is the more general. The left table is written with 1s and 0s which tend to suggest *high* voltages and *low*. So far, so familiar.[16]

But – as DeMorgan promised – if we declare that 0s interest us rather than 1s, at both input and output, the tables look different, and the gate whose behavior the table describes evidently does something different.

A	B	A · B
0	0	0
0	1	0
1	0	0
1	1	1

or:

A	B	A · B
T	T	T
T	F	T
F	T	T
F	F	F

Example of the effect of *active-low* signals: AND gate (so-called!) doing the job of OR'ing lows.

We get a 0 out if *A* **or** *B* is zero. In other words we have an *OR* function if we are willing to stand signals on their heads. We have then a gate that *ORs lows*; Fig. 14N.17 shows its behavior. This is the correct way to draw an AND gate when it is doing this job, handling active-low signals.

Figure 14N.17 AND gate drawn to show that it is OR'ing lows.

It turns out that often we need to work with signals "stood on their heads:" active-low. Note, however, that we call this piece of hardware an *AND gate* regardless of what logic it is performing in a particular circuit. We call it *AND* even if we draw it as in Fig. 14N.17. To call one piece of hardware by two names would be too hard on everyone.

We will try to keep things straight by calling this an AND *gate*, but saying that it performs an OR *function* (*OR*'ing lows, here). Sometimes it's clearest just to refer to the gate by its part number: "It's an '08." We all should agree on that point, at least!

---

[16] AoE declares its intention to distinguish between 1 and high in logic representations (§10.1.2A). We are not so pure. We will usually write 1 to mean high, because the 1 is compact and matches its meaning in the representation of binary numbers.

14N.3.4 Missile-launch logic

Here's an example – a trifle melodramatic – of signals that are active-low: you are to finish the design of the circuit in Fig. 14N.18 that requires two people to go crazy at once in order to bring on the third world war:

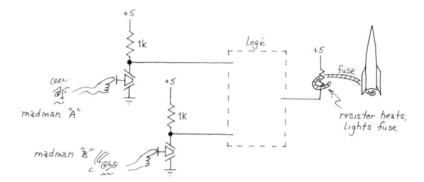

Figure 14N.18 Logic that lights fuse if both operators push "Fire" at the same time.

How should you *draw* the gate that does the job?[17] What is its conventional *name*?[18]

Just now, these ideas probably seem an unnecessary complication. By the end of the course – to reiterate a point we made earlier – when you will have met many devices whose control signals are active-low, you will be grateful for the notions "active-low" and "assertion-level symbol."[19]

14N.3.5 Implementing combinational functions

14N.3.5.1 Implementation by inspection

Some of the time the combinational logic you need to design is so simple that you require nothing more than a little skill in *drawing* to work out the best implementation. For such simple gating, the main challenge lies in getting used to the widespread use of *active-low* signals.

Figure 14N.19 shows an example: "Assert Enable if RD* or WR* is asserted and NOW* is asserted."

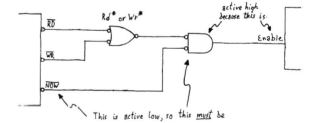

Figure 14N.19 Easy combinational logic; some signals active-low: don't think too hard; just *draw* it.

You can make this problem difficult for yourself if you think too hard: if you say something like, "Let's see, I have a low in and if the other input is low I want the output high – that's a NOR gate…" – then you're in trouble at the outset. Do it the easy way, instead.

---

[17] You need a gate that responds to a coincidence (AND) of two *lows*, putting out, for that combination, a *low* output. This is a gate that ANDs *lows*, and you should draw it that way: AND shape, with bubbles on both inputs and on the output.

[18] The name of this gate is OR even though in this setting it is performing an AND function upon active-low signals.

[19] We must admit that this is much less of an issue in this new edition. We previously used a legacy microcontroller with half a dozen active-low control signals and external peripheral devices with dozens more. Our new microcontroller has a single active-low control signal, "$\overline{\text{RESET}}$," and most of the peripherals we use are incorporated within the microcontroller. Nevertheless, the discrete 7400 series logic we use was designed with active-low control signals, and getting your head around the concept is still important.

Here is the easy process for drawing clear, correct circuits that include active-low signals:

1. draw the *shapes* that fit the description: AND shape for the word AND, regardless of active levels;
2. add inversion bubbles to gate inputs and outputs wherever needed to match active levels;
3. figure out what gate types you need (notice that this step comes *last* – and comes *never* if you use a programmable logic device and a logic compiler).

In general, bubbles meet bubbles in a circuit diagram properly drawn: a gate output with a bubble feeds a gate input with a bubble, and your eye sees the cancellation that this double-negative implies: bubbles pop bubbles. This usual rule does *not* apply in the exceptional case when your logic is looking for *dis*-assertion of a signal. We'll see such cases in §15N.3.9. It also often does not apply to edge-triggering, where it is *timing* that determines one's selection of a clock edge ("leading" versus "trailing").

14N.3.5.2 Implementation using sum-of-products

Sum-of-products is a way of creating a binary expression for *any* active-high truth table as the OR of a series of AND terms. The advantage of sum-of-products is that the truth table can be implemented with only two levels of logic (AND gates followed by an OR gate) as long as the inputs and their complement (inverse) are available. Say we have the following two-input truth table with output Z:

A	B	Z
0	0	0
0	1	1
1	0	0
1	1	1

The first 1 output, when A is low and B is high can be represented by the Boolean equation $Z = \overline{A} \cdot B$. Note that this equation is 0 for *every other combination of A and B*. The second 1 output, when A and B are both 1, is $Z = A \cdot B$. Since this equation is also 0 for every other combination of the inputs, we can OR the two equations together to get the Boolean equation for the full truth table: $Z = \overline{A} \cdot B + A \cdot B$.[20] Implementation is straightforward: AND the inputs to get the product terms then OR them to get the sum.

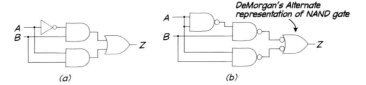

Figure 14N.20 Pure sum-of-products implementation of expression (a) and universal gate (NAND) implementation (b).

While the first implementation (a) looks cleaner, the second (b) using the DeMorgan's alternative representation of the NAND gate as an OR function of active-low inputs is more common. This is because inverting gates typically respond faster than non-inverting gates.[21]

Occasionally, an application will not care what the output is for a particular combination of inputs. In that case the truth table will use an "X" for the output to indicate you *don't care* what that output is. You normally choose the output value that gives the simplest sum-of-products equation. For example, if you need to implement this truth table:

---

[20] Sum-of-products equations can be simplified using standard arithmetic techniques by considering AND as multiplication and OR as addition. Unlike arithmetic equations, it is also permissable to duplicate product terms if helpful. In this example, using the distributive property, $Z = B \cdot (\overline{A} + A)$. But $\overline{A} + A = 1$ so $Z = B$.

[21] In addition, if this were the only circuit you were building, it would take three DIP packages to build the implementation in (a), while only one 74HC00 quad NAND to build the one in (b).

14N.3 Combinational logic

A	B	Z
0	0	0
0	1	X
1	0	0
1	1	1

You can choose either $X = 0$, in which case $Z = A \cdot B$, or you can choose $X = 1$, in which case $Z = B$. The latter is certainly easier to wire up.

14N.3.6 Considering gates as "Do this/do that" functions

A truth table describes a logic function fully, so you may be inclined to think that you can leave your understanding of gates at that level: just write out (or memorize) the truth tables for some functions: AND, OR, and XOR. But sometimes it helps to think of a two-input gate as using one of its inputs to operate upon the other. That probably sounds cryptic. Some examples will make the point.

14N.3.6.1 AND as Pass/Block* function:

Treat input A as *control*. Then the AND gate will *pass* B if A is high, *block* B if A is low: see Fig. 14N.21.

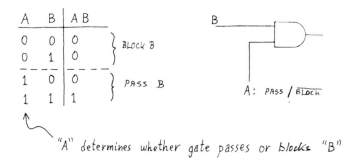

Figure 14N.21 AND gate viewed as Pass/Block*.

You will see AND used this way to implement a *synchronous clear** in a counter described in Chapter 16N.

Note that when the gate does what we describe as "block," it is driving its output *low*, not simply passing "nothing" and leaving the output an open circuit as a 3-state buffer would (see §14N.6.1). So such a blocked signal must be fed to an OR gate, if combined with a "passed" signal, as in a 2:1 multiplexer: see Fig. 14N.22.

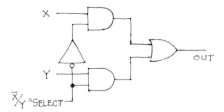

Figure 14N.22 AND gates, when "blocking" must be fed to an OR, which ignores lows.

The OR gate has the good sense to ignore a *low*. Simply joining the two AND outputs would fail of course, setting up a fight between the two gates.[22]

---

[22] This point is noted in §14W.1.

Logic Gates

AND as "IF" Here's another way of describing how the AND function occasionally can help: think of the AND function as passing or permitting a signal *IF* a condition is fulfilled (pass B if A is asserted).

Ordinary language occasionally uses the word "and" to mean "if" – well, it did in Shakespeare's time, anyway. No doubt you recall these lines from Romeo and Juliet, Act III, Scene 5:[23]

> An you be mine, I'll give you to my friend;
> An you be not, hang, beg, starve, die in the streets, . . .

For "an" and "and" in this passage we would, of course, write "if."

14N.3.6.2 XOR as Invert/Pass* function

An XOR gate can do the useful work of acting as a *controllable inverter*. The XOR will *invert B* if A is high, *pass B* if A is low: see Fig. 14N.23. You will see the XOR put to this use on page 688.

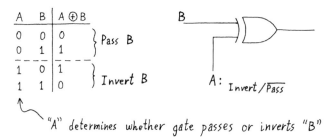

Figure 14N.23 XOR gate viewed as controllable inverter.

14N.3.6.3 OR as Set/Pass* function

The OR gate will *set* the output (force output high) if A is high, *pass B* if A is low: see Fig. 14N.24.

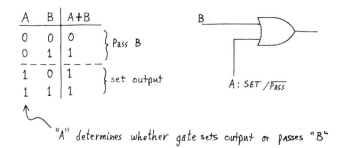

Figure 14N.24 OR gate viewed as Force/Pass* function.

14N.4 Gate types: TTL and CMOS

AoE §12.1.1

Older logic gates were constructed from bipolar transistors. These devices are known as *TTL* (transistor-transistor logic). The newer CMOS (complementary metal-oxide semiconductor) logic we use is built from N and P type MOSFETs.[24] Both logic families are numbered 74*XXNN* where "XX" is the technology and "NN" identifies the part. Different technologies with the same NN number are identical in pinout and function, e.g., a 74LS00 and a 74HC00 are both quad two-input NAND gates, although the former is **L**ow **P**ower **S**chottky TTL, while the latter is **H**igh **S**peed **C**MOS.

---

[23] Just kidding. We had to look them up, too.
[24] Some older microcomputers use NMOS technology, gates made with only N type MOSFETs, and are designed with voltage levels compatible with TTL.

14N.4 Gate types: TTL and CMOS

TTL (and the oddball HCT CMOS technology described in §14N.4.2) run on 5 V (±0.5 V) while HC CMOS runs on any supply voltage between 2 V and 6 V.

14N.4.1 Gate innards: TTL versus CMOS

AoE §§10.2.2, 10.2.3

A glance at Fig. 14N.25 should reveal some characteristics of the gates.

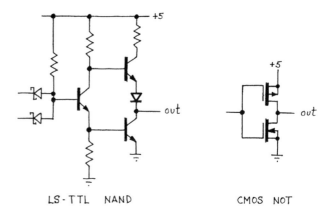

Figure 14N.25 TTL and CMOS gates: NAND, NOT.

Inputs: You can see why TTL inputs *float high*, and CMOS do not.
Threshold: You might guess that TTL's threshold is off-center – low, whereas CMOS is approximately centered.
Output: You can see why TTL's high is *not* a clean 5 V, but CMOS' is.
Power consumption: You can see that CMOS passes *no* current from +5 V to ground, when the output sits either high or low; you can see that TTL, in contrast, cannot sit in either output state without passing current in (a) its input base pullup (if an input is pulled low) or (b) in its first transistor (which is ON if the inputs are high).

14N.4.2 Thresholds and noise margin

All digital devices show some noise immunity. The guaranteed levels for TTL and for 5 V CMOS show that CMOS has the better noise immunity. Even at running on 3 V, CMOS is significantly better than 5 V TTL: see Fig. 14N.26.

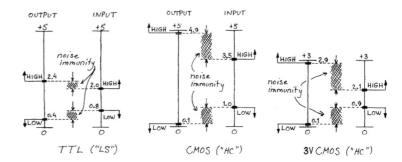

Figure 14N.26 Thresholds and noise margin: TTL versus CMOS at 5V and 3V.

Curious footnote: TTL and NMOS devices are so widely used that some families of CMOS, labeled *74xCTxx*, have been taught TTL's bad habits on purpose: their thresholds are put at TTL's nasty levels ("CT" means "**C**MOS with **T**TL thresholds"). We may occasionally need to use such gates (74HCTxx)

580   Logic Gates

where we are obliged to interface between devices running on different supply voltages. When we have a choice, however, we will stick to straight HC CMOS due to its better immunity to noise and its ability to run on the same 3.3 V supply as our FPGA and microcontroller.

Answer to "Why is the typical control signal active low?" We promised that a look inside the gate package would settle this question, and it does. TTL's asymmetry explains this preference for active-low. If you have several control lines, each of which is inactive most of the time, it's better to let the inactive signals *rest high*, and let only the active signal be *asserted low*. This explanation applies only to TTL, not to CMOS. But the conventions were established while TTL was supreme, and they persist long after their rationale ceases to apply.

Here's the argument: two characteristics of TTL push in favor of making signals active-low.

A TTL *high* input is less vulnerable to noise than a TTL *low* input Although the *guaranteed* noise margins differ by a few tenths of a volt, the *typical* margins differ by more. So, it's safer to leave your control lines high most of the time; now and then let them dive into danger.

A TTL input is easy to drive high In fact, since a TTL input *floats high*, you can drive it essentially for free: at a cost of no current at all. So, if you're designing a microprocessor to drive TTL devices, make it easy on your chip by letting most the lines rest at the lazy, low-current level, most of the time.

Both these arguments push in the same direction; if a control line is inactive for most of the time, making the inactive state high provides better overall immunity to noise than making the inactive state low.[25]

14N.5   Noise immunity

All logic gates are able to ignore noise on their inputs, to some degree. The test setup below shows that some logic families do better than others in this respect. We'll look first at the simplest sort of noise rejection, *DC noise immunity*; we'll find that CMOS, with its nearly-symmetric high and low definitions does better than TTL (the older, bipolar family), with its off-center thresholds. Then we'll see another strategy for resisting noise, *differential* transmission.

14N.5.1   DC noise immunity: CMOS versus TTL

Figure 14N.27 shows the test setup we used to mix some noise into a logic-level input. We fed this noisy signal to four sorts of inverting logic gates, one TTL, three CMOS.

Progressively increase noise level ...: First, moderate noise: all gate types succeed, in the left-hand image of Fig. 14N.28, all gates ignore the triangular noise. These gates, in other words, are showing off the essential strength of digital devices.

In the right-hand image of Fig. 14N.28, the TTL and *HCT* parts fail. They fail when the noise is added to a *low* input. HCT fails when TTL fails and this makes sense since its thresholds have been adjusted to match those of TTL.[26] CMOS did better than TTL because of its larger noise margin.

---

[25] The phrase "control line" may puzzle you. Yes, we are saying a little less than that every *signal* is active-low. Data and address lines are not. But every line that *has* an active state, to be distinguished from inactive, makes that active state *low*. Some lines have no *active* versus *inactive* states: a data line, for example, is as active when low as when high; same for an address line. So, for those lines designers leave the active-high convention alone. That's lucky for us: we are allowed to read a value like 1001 on the data lines as 9; we don't need to flip every bit to interpret it. So, instead of letting the *active-low* convention get you down, count your blessings: it could be worse.

[26] You may be wondering why anyone would offer HCT whose noise-immunity is inferior to garden-variety HC CMOS. HCT

14N.5 Noise immunity

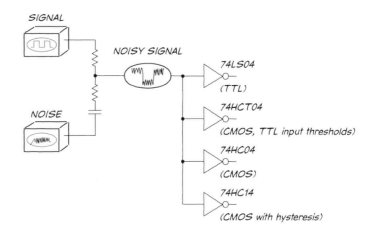

Figure 14N.27 DC noise immunity test setup: noise added to signal, fed to 4 gate types.

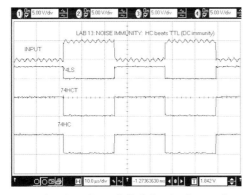

moderate noise: all digital gates ignore it

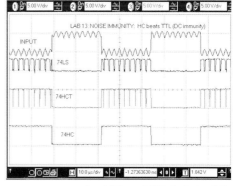

more severe noise: TTL and TTL-imitator fail

Figure 14N.28 Moderate noise: all gate types ignore the noise; more noise fools TTL gates. (Scope settings: 5 V/div.)

Much noise: all but Schmitt trigger fail: When we increase the noise level further, even CMOS fails. But we included one more gate type, in the test shown in Fig. 14N.29, a gate that does even better than the CMOS inverter. That one gate not fooled by the large noise amplitude, shown on the bottom trace, is one with built-in hysteresis (a 74HC14).

Seeing this gate succeed where the others fail might lead you to expect that all logic gates would evolve to include hysteresis. They don't though, because hysteresis slightly slows the switching and the concern for speed trumps the need for best noise immunity. This rule holds except in gates designed for especially noisy settings. Gates designed to receive ("buffer") inputs from long bus lines, for example, often do include hysteresis.

Below, in §14N.5.2, we'll meet a gate type that takes quite a different approach – and, in fact, is proud of its small swing. Fig. 14N.30 shows a chart in which this gate type, called LVDS (Low Voltage Differential Signaling) boasts of its tiny signal swing:[27]

---

exists to allow upgrading existing TTL designs by simply popping in HCT, and also to form a logic-family bridge between TTL-level outputs and CMOS inputs when both devices are powered from 5V. We use HCT in a third way to interface the outputs of 3.3V CMOS logic to devices powered by 5V. See §21N.1.1.

[27] Texas Instruments LVDS Owner's Manual: https://LAoE.link/LVDS_Guide.pdf. Figure 14N.30 is derived from a figure in an earlier version of the LVDS manual published by National Semiconductor.

582 Logic Gates

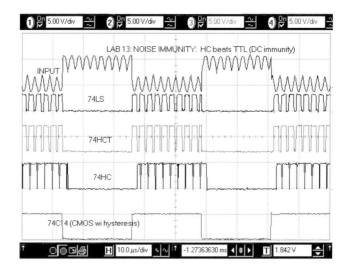

Figure 14N.29 Much noise: all gate types fail except a gate with hysteresis (HC14). (Scope settings: 5 V/div.)

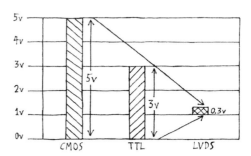

Figure 14N.30 Small signal swing can be a virtue, but requires differential signaling.

The small swings of LVDS gates afford two side benefits: low EMI (emissions: "Electro Magnetic Interference") and reduced disturbance of the power supply. We call these "side" benefits because the fundamental great strength of the low swings is that the logic can give good noise immunity at low supply voltages, as we'll argue in §14N.5.2.

14N.5.2 Differential transmission: another way to get good noise immunity

Recent logic devices have been designed for ever-lower power supplies to reduce power consumption – 3.3 V, 2.5 V and 1.8 V – rather than for the traditional +5 V. The trend is likely to continue. Such supplies make it difficult to protect gates from errors caused by noise riding a logic level. The 0.4 V DC-noise margin available in TTL and HCT is possible only because the supply voltage is relatively large. Compressing all specifications proportionately would, for example, give 50% less DC noise margin in a 2.5 V system – and things would get worse at lower supply voltages. The problem is most severe on lines that are long, like those running on a computer backplane.

A solution to the problem has been provided by *differential* drivers and receivers. These send and receive not a single signal, but a signal and its logical complement, on two wires; see Fig. 14N.31 (we first met this method in §5N.8, Fig. 5N.27). Typically, noise will affect the two signals similarly, so subtracting one from the other will strip away most of the impinging noise. This is a process you saw in §5L.1, the analog differential amplifier lab; here we find nothing new, except that the technique is applied to digital signals. The differential drivers called LVDS transmit *currents* rather than voltages; these currents are converted to a voltage difference by a resistor placed between the differential lines

14N.5 Noise immunity

at the far end of the signal lines, the receiving end. (This is nice, for the resistor "terminates" the lines, forestalling "reflections." See Appendix C.)

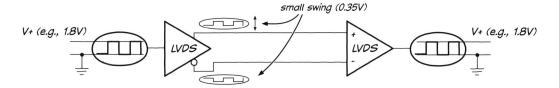

Figure 14N.31 Differential signals can give good noise immunity despite low power supply voltages.

In Fig. 14N.32 we put a standard 0/5 V logic (TTL) signal into a differential driver and injected about a volt of noise (a triangular waveform).[28]

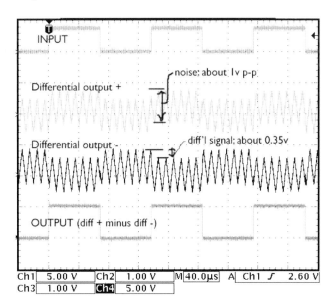

Figure 14N.32 LVDS signals: differential signals can survive noise greater than the signal swing.

The two middle traces in Fig. 14N.32 show the output of the driver IC: a differential pair of signals, one in-phase with the TTL input, one 180° out of phase. The driver converts the *voltage* TTL input to *currents* flowing as diff$_+$ and diff$_-$; at the receiver, a terminating resistor (100 Ω) converts the currents back into voltages.[29] The differential swing is small: about 0.35 V – dwarfed by the triangular noise in Fig. 14N.32. But since the differential receiver looks at the *difference* between diff+ and diff-, it reconstructs the original TTL cleanly, rejecting the noise.

This success, with the small differential swing, illustrates how differential signaling can provide good noise immunity to logic that uses extremely low supply voltages. The LVDS specification requires a voltage swing of only 100 mV at the receiver.

These differential gates show two other strengths: (1) they are fast (propagation delays of transmitter and receiver each under 3ns); and (2) they emit less radiated noise than a traditional voltage gate (as noted above). They do this first because of the small voltage swings; and further, because of the

---

[28] We did this in a rather odd way: by driving the *ground* terminal on the transmitter IC with this 1V (peak-to-peak) triangular waveform.

[29] The terminating resistor serves another good purpose: matching the "characteristic impedance" of the transmitting lines, it prevents ugly spikes caused by "reflection" of a waveform that might otherwise occur as it hit the high-impedance input of the receiving gate. Again we refer you to Appendix C for more on this topic.

symmetric *current* signals that travel in the signal lines (currents of opposite signs, flowing side-by-side), signals whose magnetic fields tend to cancel.[30]

14N.6 More on gate types

14N.6.1 Output configurations

Active pullup: All respectable gates use *active pullup* on their outputs (often referred to as push–pull or totem pole), to provide firm highs as well as lows. You will confirm in the lab that the *passive-pullup* version (labeled NMOS in Fig. 14N.33) not only wastes power but also is *slow*. Why slow?[31]

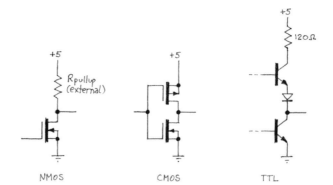

Figure 14N.33 Passive versus active pullup output stages.

Open-collector/open-drain: Once in a great while "open-drain" or "open-collector" is useful.[32] You have seen this on the '311 comparator. It permits an easy way to OR multiple signals: if signals A, B and C use active-low open-drain outputs, for example, they can share a single pullup resistor. Then any one of A, B or C can pull that shared line low. This arrangement often is called "wired OR."

Another use of the open collector/drain output is to drive loads from higher supply voltage than +5 V. For example, the 74LS06 is a hex inverter, identical in pinout to the 74xx04, but with 30 V open collector outputs capable of sinking 40 mA. You could use this device to turn on a string of LEDs arranged in series with a single current limiting resistor: see Fig. 14N.34.

AoE §10.2.4C

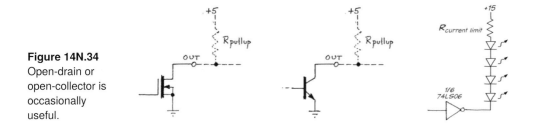

Figure 14N.34 Open-drain or open-collector is occasionally useful.

AoE §10.2.4A

---

[30] See Texas Instruments LVDS guide: https://LAoE.link/LVDS_Guide.pdf.
[31] Slow because the inevitable stray capacitance must be driven by a mere pullup resistor, rather than by a transistor switch. You can make the pullup resistor smaller, but then you increase the power dissipation when the MOSFET is on.
[32] "Open-collector" for bipolar gates; "open-drain" for MOSFET gates.

14N.6 More on gate types

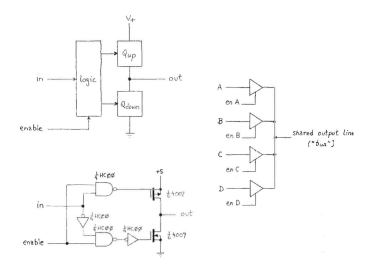

Figure 14N.35 Three-state output: conceptual; the way we build it in the lab; driving a shared bus.

Three-state: *Very* often *three-state* outputs are useful:[33] these allow multiple drivers to share a common output line (then called a "bus") as long as only one device is enabled to drive the bus at a time. These are widely used in computers.

Beware the misconception that the "third state" is a third output voltage level. It's not that; it is the *off* or disconnected condition (often referred to as "Hi-Z"). Fig. 14N.35 shows circuitry to implement this output stage, first conceptually, then the way we'll build it in Lab 14L.

14N.6.2 Logic with TTL and CMOS

The basic TTL gate that we looked at a few pages back in §14N.4.1 was a NAND; it did its logic with diodes. CMOS gates do their logic differently: by putting two or more transistors in series or parallel as needed. Fig. 14N.36 shows the CMOS NAND gate you'll build in the lab, along with a simplified sketch, showing it to be just such a set of series and parallel transistor switches. The logic is simple enough for you not to need a truth table. The output will be pulled low only if both inputs are high – turning on both of the series transistors to ground. Thus it implements the NAND function.

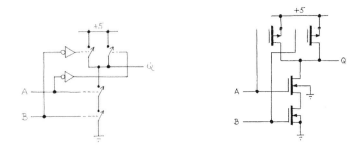

Figure 14N.36 NAND gate built with CMOS.

14N.6.3 Speed versus power consumption

The plot in Fig. 14N.37 shows tradeoffs available between speed and power-saving for some "standard logic" families. As you can see from this figure, everyone is trying to snuggle down into the lower

---

[33] You will often hear these called "Tri-State." That is a trademark belonging to National Semiconductor (now absorbed by Texas Instruments), so "three-state" is the correct generic term.

Logic Gates

left corner, where you get fast results for almost nothing. Low voltage differential signaling and faster CMOS (TI's "LVC," and the still faster AUC, for example) seem the most promising, at the present date.

And arrays of gates – PALs and FPGAs – can show speeds better than those indicated in Fig. 14N.37 because stray capacitances can be kept smaller on-chip than off.

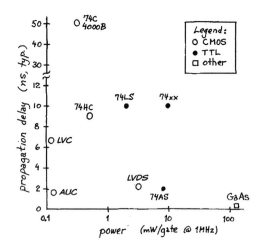

Figure 14N.37 Speed versus power consumption: some present and obsolete logic families.

14N.7 AoE reading

Chapter 10 (Digital Logic):
- §10.1 Basic Logic Concepts
- §10.2 Digital ICs: CMOS and bipolar (TTL)
  - for big picture of logic family competition see Fig. 10.22.
- §10.3 Combinational Logic.
  - §10.3.1 Logic identities are useful; DeMorgan's theorem is important.
  - we don't worry about Karnaugh mapping: §10.3.2.
  - ...and don't study the long catalog of available combinational functions: §10.3.3.
  - take a quick look at §10.6: Some typical digital circuits.

14L Lab: Logic Gates

After configuring your breadboard for digital circuits, this lab invites you to look at integrated circuit logic gates; first examining their characteristics and foibles, then using them to carry out some Boolean logic operations.

We then look within the black box, in effect, by putting together a logic gate from transistors. The point here is to appreciate why the IC gates are designed as they are, and to notice some of the properties of the input and output stages of CMOS gates. We will concentrate in this lab, as we will throughout the course, on CMOS. To overstate the point slightly, we might say that we will treat ordinary TTL as a venerable antique.

Finally, we look at the three-state output, which allows you to violate our sacrosanct rule that outputs should not be connected together.

But all of the work we do in this lab is rather antique – because logic now is seldom done with little packages of a few gates. Normally, logic networks are built from large arrays of gates that are programmed to carry out a particular function. Soon you will be taking advantage of a modest version of such devices containing about 5000 logic cells, each cell consisting of several dozen gates. But today we'll just use one or a few gates at a time because that's a good way to start getting a grip on what a gate does.

14L.1 Set up

Some *ground rules* in using logic:

1. Never apply a signal beyond the power supplies of any chip. That means...
2. ...for the logic gates that we use, keep signals between 0 and V_{CC} (+3.3 V or +5 V).
3. ...for the field programmable logic array (FPGA) and microcontroller, keep signals between 0 and +3.3 V.

    This rule, in its general form – "stay between the supplies" – applies to analog circuits as well; what may be new to you is the nearly-universal use of single supply in digital circuits. Where you can get into trouble is when you turn off the power but not the input signal(s). Then *any* input voltage can cause excess current and damage the device.[1] If you cannot be sure all input signals will be off when power is off, you can add a series resistor ($\approx 470\,\Omega$) to your inputs to limit the current to ≤ 10 mA. This slows down the response a bit but should not be a problem with most of the circuits we build.

4. Never connect gate outputs together (unless they have open collector/drain or three-state outputs – and even then the latter must be configured so only a single gate is not in the high impedance state at a time).

---

[1] This occurs because the inputs to logic gates are protected against electrostatic discharge (ESD) by back-biased diodes to V_{CC} and ground: see Fig. 14L.10. However, when the power is off $V_{CC} = 0$ V and any input voltage that forward biases the protection diodes can cause currents in excess of the maximum allowed.

5. Power all your circuits from +5 V and ground only – until we reach the FPGA and microcontroller which are powered with 3.3 V instead. This applies equally to CMOS (in its traditional 5 V form) and to TTL.[2] Power and ground pins on ordinary 7400 series digital parts (not complex ICs like FPGA and microcontrollers) are the diagonally-opposite corner pins, as in Fig. 14L.1.[3]

To simplify things, we will use external inputs (from your external function generator and the breadboard function generator, switches and buttons) of ground for a low and 3.3 V for a high, both for circuits running on +5 V and those running on +3.3 V. While the worst case minimum voltage for a 74HC part to recognize an input as a high is $0.7 \times V_{CC}$ or 3.5 V, the typical value is 2.4 V.[4] It is extremely unlikely you will have a problem with 5 V logic not recognizing a 3.3 V input as a high, but if necessary you can adjust the variable positive supply up to 3.5 V. If you do, just remember to set it back to 3.3 V when you connect those inputs to the FPGA and microcontroller.

14L.1.1 Logic probe

The logic probe is a gizmo about the size of a thin hot dog, with a cord on one end and a sharp point on the other. It tells you what logic level it sees at its point; in return, it wants to be given power (+5 V and ground) at the end of its cord (n.b., the logic probe does *not* feed a signal to the oscilloscope even if the cable ends in a BNC connector!).

However, if you do find a BNC connector on the probe cord, connect +5 V to the center conductor, using one of the breadboard jacks. If, instead, you find that the cord ends with alligator clips or a pair of strange-looking grabbers, use these to take hold of ground and +5 V.

If your probe has a switch for the type of logic, you should set the probe to "TTL" if the device you are testing is a 74LS or 74HCT part. Use the "CMOS" setting for a 74HC part when powered from +5 V. For 3.3 V logic, you should always use the "TTL" setting. In general, you can leave the probe on the "TTL" setting all the time, although be aware that the probe could show a valid high TTL value (≥ 2 V) when the voltage is less than the actual worst case high minimum voltage for CMOS logic running on 5 V (≥ 3.15 V).

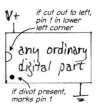

Figure 14L.1 Most ordinary digital parts take power and ground at their corner pins.

How to use the probe: Most logic probes use different colored LEDs to distinguish high from low – and to distinguish both from "float" (simply "not driven at all; not connected"). This ability of the probe is extremely useful. (Could a voltmeter or oscilloscope make this distinction for you?[5])

Use the probe to look at the output of the breadboard function generator when it is set to *TTL*. Crank the frequency up to a few kilohertz. Does the probe blink at the frequency of the signal it is watching? Why not?[6]

14L.1.2 Adjusting the breadboard power supply

All our digital logic will run on either +5 V or +3.3 V. The breadboard already has a fixed +5 V power supply (on the red binding post) but you will need to set the adjustable positive supply (yellow binding

---

[2] Lower supply voltages are widely displacing 5 V. At the time of writing, 3.3 V, 2.5 V and 1.8 V are common supplies. For the most part we will use +5 V when building standalone digital circuits and +3.3 V when using CMOS parts with the FPGA and microcontroller.
[3] But *be careful!* There are some odd exceptions like the 74HC4050 level shifter and the 74xx90 counter. It never hurts to check the datasheet when using a new (to you) part.
[4] Philips Semiconductor HCMOS family characteristics, March 1988. https://LAoE.link/HCMOS_Specs.pdf.
[5] No! The special virtue of the logic probe is the ability to distinguish a logic low from the plain old "zero volts" that a DVM might show you. The DVM cannot distinguish a *float* – no connection to anything – from a good solid connection to ground, which the logic probe understands is what we mean by low.
[6] Why not? Because your eyes are not quick enough to see a 1 kHz blink rate (you probably can't notice a blink rate beyond about 25 Hz). The logic probe slows fast switching to the lazy rate of a few Hz, so that we humans can see it.

14L.1 Set up

post) to +3.3 V. The adjustment knob is very sensitive at lower outputs and the circuits elements we use will be damaged by voltages above +3.6 V so you should adjust it carefully. Also, you will want to be very careful not to move or even touch the adjustment knob once set. It is a good idea to use hot melt or silicon glue (both easily removable when you go back to the analog labs) to keep the knob from moving once you have the voltage set. It could save you from accidentally damaging the expensive FPGA or microcontroller. We suggest always checking the variable positive supply voltage with a multimeter to make sure it is less than 3.6 V before connecting components to it. In our lab we take this one step step further and add a dedicated fixed 3.3 V supply: see online Chapter 14O at `https://LAoE.link/LAoE_Chapter_14O.pdf`.

14L.1.3 LED indicators

The "LOGIC INDICATOR" LEDs on the breadboard are buffered by voltage comparators and present a high input impedance (100k to ground). You should set the upper switch to correspond to the supply voltage of the logic you are connecting the indicators to and the lower switch to the logic type. Like the logic probe, the breadboard does not light either LED if the voltage is not a valid value for the logic and V_{CC} voltage selected.[7] Nevertheless, the +5/TTL setting (high lit if ≥ 2.2 V; low lit if ≤ 0.8 V) will work most of the time for both +5 V and +3.3 V logic.

To appreciate what the logic probe did for you earlier, try looking at a fast square wave, using one of these LEDs rather than the logic probe: use the breadboard oscillator (TTL) at a kilohertz or so. Does what you see make sense? You may now recognize that the logic probe stretches short pulses to make them visible to our sluggish eyes: it turns even a 30 ns pulse into a flash of about one-tenth of a second (the faster probes can do this trick with even narrower pulses).

14L.1.4 Switches

Switches available on the powered breadboards: The PB-503 includes three sorts of switch on its front panel:

- Two debounced pushbuttons (at the lower left corner of the breadboard, marked PB1, PB2). These deliver an *open collector* output, and that means that they are capable of pulling to ground *only*, never to the positive supply. To let such an output go to a logic *high*, you will need to add a *pullup* resistor to 3.3 V (to be compatible with both 5 V and 3.3 V logic). We also ask you to add a small series resistor to protect the output transistor in case of a short circuit directly to power: see Fig. 14L.2.

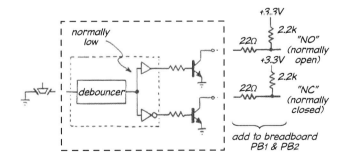

Figure 14L.2 Pullup and current limiting required on open-collector output (debounced pushbuttons on PB-503).

Note that what looks like a discrete transistor in the figure is included within the breadboard (the

---

[7] Some early versions of the PB-503 breadboard did not include switches to set the voltage and logic type and only included a single LED to indicate a high level.

larger dashed box); you need add *only the series and pullup resistors*. Be sure to take your output from the intersection of the two resistors and *not* the white terminal block when you connect a switch to your circuit.

- Eight slide switches (not debounced), fed from an additional +5/+V slide switch (marked "LOGIC SWITCHES," Fig. 14L.3). Each switch outputs either 0 V (ground) or the voltage selected by the ninth slide switch. You should leave that switch in the "+V" position, connected to the adjustable supply set to +3.3 V. This should work with both 5 V and 3.3 V logic.[8]

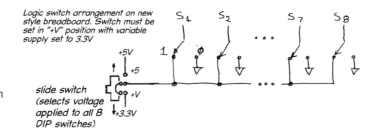

Figure 14L.3 Logic switches with high output voltage set by common slide switch.

Older PB-503s use a different arrangement where each switch in an eight-position DIP switch is either open circuit (floating) or connected to a common voltage, ground or +5 V, selected by a slide switch: see Fig. 14L.4. To get eight independent levels you need eight *pullup* resistors to 3.3 V, while setting the slide switch always *low* ("GND").[9]

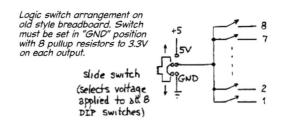

Figure 14L.4 DIP switch in line with output of one common slide switch on older breadboards.

- Two uncommitted slide switches ("SPDT SWITCHES"). These are on the lower right, and are *bouncy* (not debounced, anyway). To make them useful, tie one end to ground, the other to +3.3 V, and use the *common* terminal as output. You might as well wire these now, use them today, and then leave them so wired for use in later labs.

Another way to generate digital inputs: An easy way to provide levels where you want them is to wire a DIP switch so that one side ties directly to the ground bus. The DIP switch should be inserted *upside-down*, so that the slider *down* position (which physically looks like a low) closes the switch and generates a logic low. The switch shown in Fig. 14L.5 is providing a 00 combination.

If the switch is wired directly to an IC's inputs, the pullups may have to run sideways. Again, the switch in Fig. 14L.6 is shown providing a 00 combination, this time going to the two inputs of a NAND gate, in a 74HC00.

---

[8] However, if the adjustable supply is set higher than 3.6 V or the slide switch moved to the 5 V position you could damage the FPGA and/or microcontroller. We placed a piece of electrical tape over the top half of the voltage select slide switch to remind us never to switch to +5 V.

[9] Again, moving the slide switch to the "5 V" position will damage 3.3 V logic.

14L.1 Set up

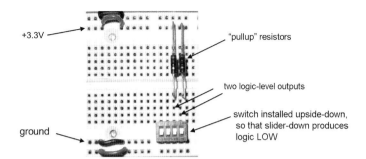

Figure 14L.5 DIP switch installed on breadboard can provide logic levels.

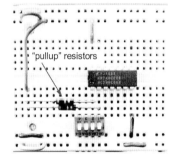

Figure 14L.6 DIP switch can provide logic levels just where you want them.

14L.1.5 Function generator level shifter

The breadboard function generator is adjustable in frequency and voltage and includes a slide switch to select between a sine, triangle or square wave output. It also includes a dedicated "TTL" output on the terminal block which provides a fixed zero to 5 V square wave when the square wave shape is selected. To work with both 5 V and 3.3 V logic, you need to add a "level shifter" to this output to convert the TTL signal out of the breadboard function generator to a zero to 3.3 V signal. The device we use is a "74HC4050," which contains six independent level shifters. We will only need to use one. Each level shifter is a buffer designed to withstand a digital input "high" signal of up to 15 V. However, its output high voltage is limited to whatever its V_{CC} pin it is connected to: see Fig. 14L.7.[10]

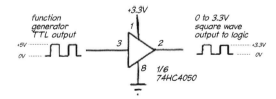

Figure 14L.7 3.3 V level shifter for function generator TTL output.

---

[10] Warning! This is one of these oddball digital circuits we warned you about that does *not* use the upper left and lower right corner pins for power and ground.

14L.2 Input and output characteristics of integrated gates: TTL and CMOS

You are going to compare NAND gates from two logic families: TTL and CMOS. Wire up and test a 74LS00 and a 74HC00 as follows:[11]

14L.2.1 Measure output voltage levels

Power and ground: As we suggested back in §14L.1, a 14-pin DIP like the ones you meet in this lab usually takes power and ground at its corner pins: Northwest and Southeast when the part is oriented horizontally: pins 14 (V_{CC}: +5 V) and 7 (ground).

Input signals: Use the breadboard Logic Switches or other source to allow you to select either 0 or +3.3V to each of the two inputs of one of the NAND gates in each DIP package, driving both the TTL and CMOS simultaneously. If you want to be really fancy and lazy, show the two inputs on two of the powered-breadboards LEDs, and the output on a third LED. Doing that helps you to see inputs and outputs simultaneously, as if you were looking at one line of a truth table.

74XX00 pinout: In Fig. 14L.8 the TTL part is 74LS00; the CMOS part is 74HC00.

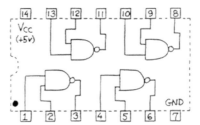

Figure 14L.8 NAND gates: TTL and CMOS.

Note: for the CMOS part (but not TTL), tie all the six unused input lines to a common line, and temporarily ground that line. Take the trouble to tie the unused inputs together, rather than ground each separately, because soon we will want to drive them with a common signal (in §14L.3.1).

Now note both *logic* and *voltage* levels out as you apply the four input combinations. (Only one logic-out column is provided below because here TTL and CMOS should agree.) As a *load*, use 10k resistors from output to ground.

INPUT		OUTPUT Logic Levels	Volts: TTL	Volts: CMOS
0	0			
0	1			
1	0			
1	1			

---

[11] The "74" indicates that these parts follow the part-numbering and pinout scheme established by the dominant logic family, Texas Instruments' 74xx TTL series. "LS" stands for Low Power Schottky, a process that speeds up switching over the original TTL design. At the time when the *LS* prefix was chosen (1976), TTL was thought to go without saying; thus there's no *T* in the designation in contrast to CMOS, the late-bloomer, which always announces itself with a *C* somewhere in its prefix: HC, HCT, AC, ACT, etc. In the 74HC00, "C" indicates CMOS; "H" stands for "high speed": i.e., speed equal to that of the then-dominant TTL family, 74LS.

14L.3 Pathologies

14L.3.1 Floating inputs

If you took our advice earlier and connected the NAND inputs to two of the powered-breadboards LEDs, you will need to disconnect them for the following experiments.

TTL: Disconnect both inputs to the NAND, and note the output *logic* level (henceforth we will not worry about output voltages; just logic levels will do). What *input* does the TTL "think" it sees, therefore, when its input floats?[12]

AoE §10.8.3B

CMOS: Here the story is more complicated, so we will run the experiment in two stages.

1. Floating input: *effective logic level in*: Tie HIGH one input to the NAND, tie the other to 6 inches or so of wire; leave the end of that wire floating, and watch the gate's output with a logic probe or scope as you hold your hand near the floating-input wire, or take hold of it at an insulated section. (Here you are repeating an experiment you did with the power MOSFET in §12L.1.2.) Try touching your other hand to +5 V, to ground, to the TTL oscillator output, or hold your free hand near the transformer of the breadboard's internal power supply. We hope that what you see will convince you that floating CMOS inputs are less predictable than floating TTL inputs, although we urge you to leave *no* logic inputs floating.
2. Floating input: *Effect on CMOS power consumption*:

    You may have heard that you should not leave unused CMOS inputs floating. Now we would like you to *see* why this rule is sound (though, like most rules, it deserves to be broken now and then).[13]

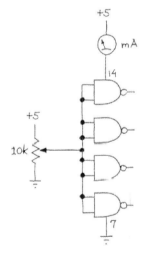

Figure 14L.9 Test setup: applying intermediate *input* level raises power consumption.

Tie the two NAND inputs to the other six, earlier grounded; disconnect the whole set from ground,

---

[12] It thinks it sees a high – though a high that is vulnerable to noise. A look at the innards of the gate shown in Fig. 14N.25, reveals the internal resistive pullup on the base of the first transistor. That transistor therefore is held on when inputs are open, as if driven by a genuine high input.

[13] In particular, we don't want you to feel obliged, when breadboarding circuits in this lab, to drive all unused inputs. That's required in a circuit that you build in permanent form, but your lab *time* is more valuable than the extra *power* that your circuit may consume when inputs are left floating – and your time is more valuable even than the IC that conceivably might be damaged by your failure to drive all its unused inputs.

and instead connect it to a potentiometer that can deliver a voltage between 0 and +5 V. Rotate the pot to one of its limits to apply a valid logic level input to all four of the NAND gates in the package.

Now (with power off) insert a current meter (VOM or DVM) between the +5 V supply and the V+ pin (14) on the CMOS chip: see Fig. 14L.9. Restore power and watch the chip's supply current on the meter's most sensitive scale. The chip should show you that it is using very little current: low power consumption is, of course, one of CMOS' great virtues.

Now switch the current meter to its 150 mA scale (or similar range) and gradually turn the pot so that the inputs to all four NAND gates move toward the threshold region where the gate output is not firmly switched high or low. Here, you are frustrating CMOS' neat scheme that assures that one and only one of the transistors in the output stage is on. When input voltage is close to $V_{DD}/2$, the typical switching threshold, both transistors can be partially on. You can see on the current meter dial the price for this inelegance: power consumption thousands of times higher than normal.

Floating inputs thus are likely to cause a CMOS device to waste considerable power. Manufacturers warn that this power use can also overheat and damage the device. Floating inputs are also prone to oscillation. In this course we will sometimes allow CMOS inputs to float while breadboarding, as we have said. But you now know that you should never do this in any circuit that you build to keep.

14L.3.2 Effect of failing to connect power or ground to CMOS logic gate

Now we ask you to do, purposely, what students often do inadvertently, as they breadboard logic circuits: omit power and ground connections.[14] The effect is not what one might expect. Perhaps you would expect an un-powered IC to deliver a constant low output, or a floating output.

But the protection clamps on input and output of each gate complicate the behavior. Figure 14L.10 shows what the clamps look like for 74HC parts.[15]

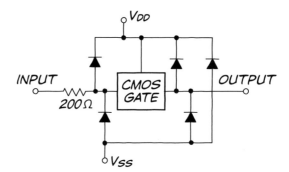

Figure 14L.10 74HC logic gate protection circuitry.

The clamps allow inputs (and sometimes outputs) to provide current that can power the IC. So, a *high* input can feed the +5 V supply line through the upper clamp diodes (though these don't *look* connected); a low can feed the ground line through the lower clamp diode – though both +5 V and ground so driven would reach compromised levels. So if you apply both a high and a low to inputs somewhere on the chip, you have powered the chip! As you can imagine, the result is a very strange pattern of misbehavior: everything seems to be working – until an unlucky combination of input levels occurs.

---

[14] You may be sure that *you* wouldn't make this mistake – but many others do. One term, the error became so common in our class that the teaching assistants began to distribute candy rewards to the students who did not make this error: who applied power and ground first, rather than postpone the chore till after they had wired the more interesting connections.

[15] Fairchild/National App. Note AN248, Input Protection Circuits and Handling Guide for CMOS Devices. Available at https://LAoE.link/AN-248.pdf.

If you let the output of an unpowered 74HC00 NAND gate drive a 10k load resistor (with the other end tied to ground) and put that output on one of the breadboard's buffered LEDs (to make it easy for you to see the logic level out) you may see the HC00 NAND generating an XOR function. Why?[16] We should confess we can't promise the XOR result: variations in protection circuitry among manufacturers make the outcome unpredictable.

14L.4 Applying IC gates to generate particular logic functions

Now let's have some fun with these gates. First we'll do a couple of tasks with NANDs to get used to the remarkable fact that with NANDs you can build *any* logic function. Then we'll invite you to apply any of the standard logic functions, including XOR, to make a digital *adder* then a *comparator*.

Most people prefer these brain teasers to a session of wiring. Don't let these problems bring your lab work to a dead *stop*! Please don't give any of these problems more than ten minutes of your precious lab time. You can always finish these brain teasers at home.

14L.4.1 NAND applications

BOTH: Use NANDs (CMOS or TTL) to light one of the breadboard's buffered HIGH LEDs when both inputs are high.

EITHER: Use NANDs (CMOS or TTL) to light one of the HIGH LEDs when either of the inputs is low (here we mean a plain *OR* operation, not exclusive-or, by the way). (Trick question! Don't work too hard.)

14L.4.2 Adders using any gates you like

Half adder: Use any gates to make a half adder. A half adder adds two bits together and outputs the sum and a carry according to the following truth table:

INPUT		OUTPUT	
A	B	C_{OUT}	SUM
0	0	0	0
0	1	0	1
1	0	0	1
1	1	1	0

Truth table for a half adder: A and B are the inputs, SUM is $A + B$ truncated to one bit and C_{out} is the carry out.

Use your breadboard's logic switches for the inputs and display the outputs on the Logic Indicator LEDs.

Hint: the XOR function is a big help here. (XOR is 74HC86; pinout is same as for '00: in fact, *all*

---

[16] The argument for XOR is that only this input combination (either 0 1 or 1 0 on the gate's two inputs) provides both +5 and ground and therefore makes the gate put out the HIGH that a NAND ought to put out in response to a 01 or 10 input. 00 input provides no +5, so the gate cannot put out a high (as the NAND should). More problematic is 11: the gate is not powered, but some 74HC00's (Motorola's, for example) put out a HIGH for this input, while others (National's) put out a low. In neither case is the gate functioning. The HIGH from the Motorola chip seems to be simply passed through, with the entire IC presumably hanging close to +5 V.

2-input 74XX gates use the same pinout – except the oddball '02, which is laid out backwards. See Appendix B.) You can think of XOR as a 1-bit *equality*/inequality* detector.[17]

Full adder: Use any gates to make a full adder. A full adder adds two bits plus a carry together and outputs the sum and a carry according to the following truth table:

INPUT			OUTPUT	
C_{IN}	A	B	C_{OUT}	SUM
0	0	0	0	0
0	0	1	0	1
0	1	0	0	1
0	1	1	1	0
1	0	0	0	1
1	0	1	1	0
1	1	0	1	0
1	1	1	1	1

Truth table for a full adder: A and B are the inputs, C_{IN} is the carry in from the previous place, SUM is A + B truncated to one bit and C_{OUT} is the carry out.

Another hint: Do not take apart your half adder to build the full adder – you are going to need it in the next problem. And you should use new logic switches and indicators for the inputs and outputs of the full adder as well.

Two-bit binary adder Build a two-bit binary adder out of your half adder and your full adder:

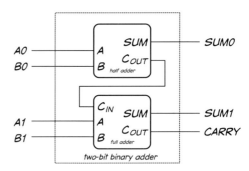

Figure 14L.11 Two-bit binary adder.

14L.4.3 Digital comparators using any gates you like

Two-bit equality detector: Use any gates to make a comparator that detects equality between two 2-bit numbers. (This circuit, widened, is used a lot in computers, where a device often needs to watch the public "address bus," to respond upon seeing its own distinctive "address.")

---

[17] Remember that an asterisk ("equality*") indicates a signal that is active-low. The same low/high meanings could be written with an overbar: "equality/inequality."

14L.5 Gate innards; looking within the black box of CMOS logic

Two-bit A>B detector: Use any gates to make a comparator that detects when one of a pair of two bit number (call it A) is larger than the other (call it B). You need not make the circuit symmetrical: it need not detect B>A, only A>B versus A ≤ B.

Warning: this circuit can get pretty complicated. We'd like to urge you not to use a lot of lab time on this problem.

14L.5 Gate innards; looking within the black box of CMOS logic

In the following experiments we will use a CD4007 (or CA3600) to investigate the gritty details of what lies within a logic IC. This part is an array of complementary MOS transistors, as shown in Fig. 14L.12.

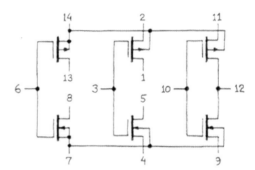

Figure 14L.12 '4007 (or "CA3600") MOS transistor array.

14L.5.1 Compare two inverters

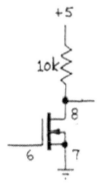

Figure 14L.13 Simplest inverter: passive pullup.

Passive pullup: Build the circuit in Fig. 14L.13 using one of the MOSFETs in a '4007 package. Be sure to tie the two "body" connections appropriately: pin 14 to +5 V, pin 7 to ground. This will look familiar to you: it is one more instance of the convention we mentioned earlier, in §14L.1: corner pins carry power and ground. (In fact, you will find that this is automatic for the particular FETs in the package that we show below; but you should be alert to this issue as you use MOSFETs.)

Confirm that this familiar circuit does invert, as you drive it with a TTL level from an *external function generator* (the breadboard oscillator is too slow to make this circuit look really bad, as we soon will want it to). The function generator's TTL output provides a V_{OH} that is not high enough to satisfy CMOS's preference for a high of close to 5 V (though TTL usually *will* make the CMOS gate switch).[18] The usual remedy is to "pull up" the TTL output, with a resistor of a few kΩ to +5 V.[19] Watch the output on a *scope* (voltmeter or logic probe will not do, from this point on).

Now crank up the frequency as high as you can. Do you see what goes wrong, and why?[20] Draw what the waveform looks like.

---

[18] If you don't recall TTL's V_{OH} value, you might look again at the V_{OH} numbers you got in §14L.2.
[19] This pullup may work on your function generator; it did not work on ours (Krohn–Hite) because their outputs labeled "TTL" are not true TTL: their TTL output cannot be pulled above 4 V. A real TTL gate rises easily to +5 V when pulled up. But 4 V is sufficient to drive the CMOS gates.
[20] The rise is slow, because the usual stray capacitance is driven by the 10k pullup resistor, forming a lazy *RC*.

Lab: Logic Gates

Active pullup: CMOS Now replace the 10k resistor with a *p*-channel MOSFET, to build the inverter in Fig. 14L.14. Look at the output as you try high and low input frequencies. The high-frequency waveform should reveal to you why all respectable logic gates use *active pullup* circuits in their output stages. (The passive-pullup type, called *open-collector* for TTL or *open-drain* for MOS, appears now and then in special applications such as driving a load returned to a voltage other than the +5 V supply, or letting several devices drive a single line. Usually, that last case is better handled by the *3-state* output described in §14L.5.2 below.)

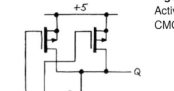

Figure 14L.14 Active pullup: CMOS inverter.

14L.5.2 Logic functions from CMOS

CMOS NAND: Build the circuit in Fig. 14L.15 and confirm that it performs the NAND function.

Figure 14L.15 CMOS NAND.

CMOS three-state: The three-state output stage can go into a *third* condition besides high and low: *off*. This ability is extremely useful in computers: it allows multiple *drivers* to share a single driven wire, or *bus* line. Here, you will build a buffer (a gate that does nothing except give a fresh start to a signal), and you will be able to switch its output to the OFF (high impedance) state. It is a "three-state buffer."

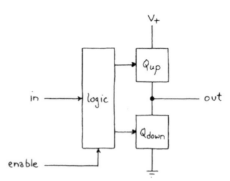

Figure 14L.16 Three-state buffer: block diagram.

The trick, you will recognize, is just to add some logic that can turn off *both* the *pullup* and *pull-down* transistors. When that happens, the output is disconnected from both +5 and ground; the output then is off, or "floating." One usually says such a gate has been "three-stated" or put into its "high-impedance" (often abbreviated as "Hi-Z") state.

If you're in the mood to design some logic, try to design the gating that will do the job using NANDs along with the '4007 MOSFETs. Here's the way we want it to behave:

14L.5 Gate innards; looking within the black box of CMOS logic

- If a line called `enable` is low, turn *off* both the *pullup* and *pull-down* transistors. That means:
  - drive the gate of the upper transistor *high*;
  - drive the gate of the lower transistor *low*.
- If `enable` is high, let the input signal drive one *or* the other of the upper and lower transistors on: that means
  - drive the gates of the upper and lower transistors the same way – *high* or *low* – turning one on, the other off (because one is *p*-channel, the other *n*-channel, of course).

That probably sounds complicated, but the circuit is straightforward. If you're eager to get on with building, borrow our solution, shown in Fig. 14L.17.

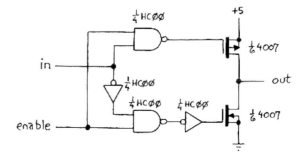

Figure 14L.17 Three-state circuit: qualifies gate signals to Q_{up} and Q_{down} using 74HC00 NAND gates.

Use a slide switch to control the 3-state's enable input, as shown in Fig. 14L.18. A second slide

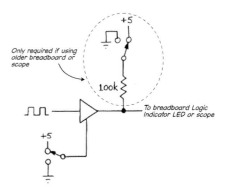

Figure 14L.18 Circuit to demonstrate operation of 3-state buffer.

switch drives a 100k resistor tied to the 3-state's output. The slide switch can be set to ground or +5 V: this slide switch, feebly driving the common output line through the 100k, is included so that we can *see* whether the 3-state is enabled or not. A scope or voltmeter by itself is not able to detect the "off" condition, as you know.[21]

Test your circuit by driving the input with the breadboard oscillator, and watching the effect of `enable`. When the 3-state is *disabled*, even the feeble 100k should be able to determine the output level. When *enabled*, the 3-state should drive its square wave firmly onto the common output line. This "common output line" provides our first glimpse of a *data bus*.

At the moment, before you have seen applications, this trick – the 3-state's ability to disappear, electrically – may not seem exciting. Later, when you look at connecting peripheral devices to a computer, you will find 3-states invaluable if not exciting.

---

[21] You do not need to add the 100k resistor and switch if you have the newer version of the breadboard with both green and red LED logic indicators. Since neither LED lights if the input is floating (high impedance), you can just connect the output of your circuit to one of the indicators to see which of the three states it is in.

We include Fig. 14L.19 to suggest why 3-states might be useful. It illustrates a single line "bus" or shared line, similar to the 3-state output data line we will use to connect SPI protocol peripheral devices to our microcontroller at the end of the course. Parallel computer buses would use multiples of this bus line; at least eight and often many more.[22]

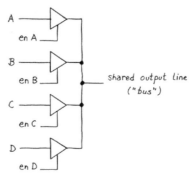

Figure 14L.19 Three-states driving a shared "bus" line (*do not bother to build this*).

---

[22] However, modern computer designs tend to exploit fast *single* line buses sending data *serially*. Personal computers have replaced parallel hard drive connections with SATA (*Serial* AT Attachment) hard drives.

14S Supplementary Notes: Digital Jargon and Logic Interfacing

14S.1 Digital glossary

Active-high /-low Defines the level (high or low) in which a signal is "True," or – better – "Asserted" (see next term). We avoid the former because many people associate "True" with "High," and that is an association we must break.

Assert Said of a signal. This is a strange word, chosen for its strangeness, to give it a neutrality that the phrase "Make true" would lack (because many people seem predisposed to think of True as High).

We can say, with equal propriety, "Assert WR* (or $\overline{\text{WR}}$)," and "Assert RESET51." In the first case it means "take it low;" in the second case it means "take it high" (because the signal names reveal that the first signal is *active low*, the second *active high*).

Assertion-level symbol This is a very strange phrase, meant to express a notion rather hard to express: it is a logic symbol, of the *two* that deMorgan teaches us always are equivalent, *chosen to show active levels*. So, a gate that ANDs lows and drives a pin called EN*, should be drawn as an AND shape with bubbles at its inputs and output, even though its conventional name may be OR. (The conventional name, note, assumes that signals are *active-high*.)

Asynchronous Contrasted, of course, with synchronous: asynchronous devices do not share a common clock, and may have *no* clock at all. Most flip-flops use an *asynchronous* clear function, also called "jam-type." Usage is complicated by the possibility that an *asynchronous function*, such as RESET*, can be an element of a *synchronous* circuit such as a counter. Such a combination occurs, for example, in the 74HC161 counter.

Clear Force output(s) to zero. Applied only to sequential devices. Same as *Reset*.

Combinational Said of logic. Same as "combinatorial." Output is a function of present inputs, not past (except for some brief propagation delay); contrasted with *sequential* (see below).

Decoder A combinational circuit that takes in a binary number and asserts one of its outputs defined by that input number. (Other decoders are possible; you will build in a later lab a binary-to-seven-segment decoder.) For example, a '139 is a dual 2-to-4 decoder that takes in a 2-bit binary number on its two select lines. It responds by asserting one of its four outputs.

Demultiplexer ("demux") Combinational circuit that steers a single time-shared input to one or another destination (this input may be a bus rather than a single line, as on the 8051 microcontroller which uses a single time-multiplexed 8-bit bus for both the low eight bits of external addresses and for data).

Don't care states Used in the output of a truth table to indicate that we do not care if the output is a one or a zero with that particular combination of inputs. Normally indicated by "X" as the output. Often allows simplification of the Boolean expression. You will also see *don't cares* on inputs in datasheet truth tables to indicate the input is ignored when some other input or condition is asserted. For example, if a device has a clear input, other inputs which control the device outputs will usually have no effect when clear is asserted.

Edge-triggered Describes flip-flop or counter *clock* behavior: the clock responds to a transition, not to a level. By far the most widely used clocking scheme.

Enable The meaning varies with context, but generally it means "Bring a chip to life": let a counter count; let a decoder assert one of its outputs; turn on a 3-state buffer; and so on.

Float Said of input or of a 3-stated output when driven neither high nor low ("Hi-Z").

Flop Lazy baby talk for "flip-flop," which is baby talk for "bistable multivibrator." But babies in this case express themselves better than old-fashioned physicists or engineers, who were inclined to say all those ugly syllables. (Flop is not to be confused, incidentally, with the acronym FLOP, a piece of computer jargon that stands for "Floating Point Operation.")

Hold time Time *after* a clock edge (or other timing signal) during which data inputs must be held stable. On new designs hold time normally is *zero*.

Jam clear, jam load Asynchronous clear, load: the sort that does not wait for a clock.

Latch Strictly, a *transparent latch* (see below), but often loosely used to mean *register* (edge-triggered) as well.

Latency Delay between stimulation and response. Often arises in response to an *interrupt*, where prompt response can be important.

Load A counter function by which the counter flops are loaded as if they were elements of a simple *D-flop* register.

Multiplexer ("mux") Combinational circuit that passes one out of *n* inputs to the single output; uses a binary number code input on *select* lines to determine which input is so routed. For example, a 4:1 mux uses 2 select lines to route one of its four inputs to the output. A "demultiplexer" does the complementary operation (see above).

One-shot (or monostable) Circuit that delivers one pulse in response to an input signal (usually called *trigger*), which may be a level or an edge. Most *one-shots* are timed by an *RC* circuit; some are timed by a clock signal, instead.

Preset Force output(s) of a sequential circuit high. Same as *Set*.

Propagation delay Time for signal to pass through a device. Timed from crossing of logic threshold at input to crossing of logic threshold at output.

Register Set of D flip-flops, always edge-triggered (see *latch*).

Reset Same as *Clear*: means force output(s) to zero.

Ripple counter Simple but annoying counter type; asynchronous – the Q of one flop drives the clock of the next. Slow to settle, and obliged to show many false transient states.

Sequential Of logic: circuit whose output depends on past as well as (in some cases) on present inputs. Example: a T (toggle) flip-flop: its next state depends on both its history and the level applied to its *T* input before clocking. Contrasted with *combinational* (above).

Set Same as *Preset*: means set output high (said only of flops and other sequential devices).

Setup time Time *before* clock during which data inputs must be held stable. Always *non-zero*. Worst-case number for a 74HC: about 20 ns.

Shift register A set of D flops connected Q-of-one to D-of-the-next. Often used for conversion between parallel and serial data forms.

State Condition of a sequential circuit, defined by the levels on its flip-flop outputs (Qs). A counter's state, for example, is simply the combination of values on its Qs.

State machine Short for finite state machine or FSM. A generalized description for any sequential circuit, since any such circuit steps through a determined sequence of states. Usually reserved for the machines that run through non-standard sequences. A toggle flip-flop and a decade counter are FSMs, but never are so-called. A microprocessor executing its microcode is a state machine. So is an entire computer.

14S.2 Interfacing among logic families

Sum-of-Products A standardized way of converting an active-high truth table into a Boolean equation. Each high output is resented as an AND of the inputs and their complements to create the product terms. The product terms are OR'ed together (SUM'ed) to give the result.

Synchronous Sharing a common clock signal (syn-chron means same time). Synchronous functions "wait for the clock." The advantage of synchronous operation is the circuit will always be in a known state after the clock edge once the longest propagation delay has elapsed.

Three-state Describes a logic *output* capable of turning *off* to a high impedance state ("high-Z") as well as driving a high or low logic level. Same as *tri-state*.

Tri-state A trade name for *three-state*: Was a registered trademark of National Semiconductor, now part of Texas Instruments.

14S.2 Interfacing among logic families

This topic is dull, but sometimes necessary. There is no question that a 74HCxx gate can drive a 74HCxx gate – and if you find yourself using antique TTL gates (say 74LSxx to 74LSxx), they will understand each other. But sometimes you will be obliged to mix logic families, usually in order to get a function not available in the main logic family of your design. Then you need to know whether the two devices can drive each other – and if they cannot, how to solve the problem.

14S.2.1 Five-volt parts: trouble driving CMOS

AoE §12.1.3

Most of the time you are likely to use CMOS logic. In Lab 14L you explored the nice properties of CMOS inputs and outputs: output highs and lows are clean – within a few tenths of the positive supply or ground; input thresholds are comfortably far from those good output levels. The inputs draw no DC current to speak of (a microamp max over full temperature range; typically very much less than that); they present a load that is a modest capacitance (10 pF, max) They can drive a substantial load, sinking or sourcing 4 mA (74HCxx).[1] They're a pleasure to work with.

But if you were to ask an old TTL part to drive a 74HCxx part it might not work. The problem is that the TTL output *high* voltage (V_{OH}) is inadequate: less than the required CMOS input *high* voltage (V_{IH}). TTL's V_{OH} is only 2.4 V; HC's V_{IH} is a little more than 2/3 of the supply voltage: 3.15 V when powered at 4.5 V. Figure 14S.1 illustrates the problem.

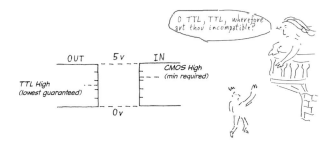

Figure 14S.1 TTL V_{OH} inadequate to drive CMOS V_{IH}.

What is to be done? TTL clearly needs a boost, or CMOS needs to reach lower. Both remedies are possible. CMOS can "reach lower" if we substitute 74HCTxx for 74HCxx. In order to solve exactly the problem we are discussing, HCT uses TTL input thresholds at the expense of giving up CMOS' symmetry and its superior noise rejection. Alternatively, TTL can "get a boost" if we use a resistor to

---
[1] These are specifications for ON Semiconductor's 74HC00.

pull up its output to 5 V. To understand the effectiveness of the simple pullup one must notice that the inadequate TTL *high* voltage (2.5 V worst case, about 3 V to 3.5 V typical) is not locked at that level; it is not a low-impedance source like the TTL *low*. Rather, it is a source that runs out of gas at this TTL *high* voltage – but does not object if some kindly pullup resistor wants to come along and finish the job. In the case of TTL parts, the "running out of gas" results from the use of an NPN pullup transistor (in contrast to the PMOS pullup of a CMOS output): see Fig. 14N.25.

As you choose that pullup resistor, just avoid extremes: a very large R would make the rise slow, because the R must charge stray capacitance; using a very small R would overload the driving gate, asking the TTL output to sink more current than it can handle (this maximum current is specified as I_{OL}). A few kΩ is fine for LS TTL; with the weaker signals available from some microprocessors you need to beware overloading: you must pay close attention to that I_{OL} limit.

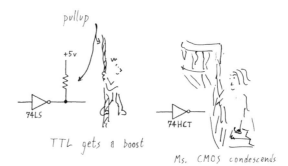

Figure 14S.2 TTL can meet CMOS – with a little help.

Much that isn't TTL behaves like TTL: You may be inclined to protest that you're not so old-fashioned to use TTL (which is indeed nearly obsolete, even in its late "LS" form – low-power Schottky). Maybe so – or at least the occasions when you meet TTL will be rare. But a strange fact makes this "TTL" problem more pervasive than it seems: many devices that are not made of the bipolar transistors used in true TTL parts use TTL input and output levels nevertheless. NMOS devices – those made with *n*-channel transistors only, omitting the *p*-channel that lets CMOS do such a good job of pulling its output high – understandably cannot provide good V_{OH} levels. But the startling bad news is that some devices made of CMOS use the lower TTL levels, presumably because that choice slightly eases the designer's and fabricator's job, letting them use NMOS pullup transistors rather than PMOS. Always read datasheets careful when using a new device to make sure it is compatible with what it is driving and what is driving it.

AoE §12.1.2A

14S.2.2 Parts powered by lower voltages

AoE §12.1.3

The old TTL levels live on in some low-voltage parts: As supply voltages come down from 5 V, the lukewarm logic highs are disappearing – the V_{OH} that runs out of gas far below V_{supply}. In a 3.3 V design, no one dares to waste 1.5 V as TTL logic did. But TTL logic levels remain important quite far from their origin in bipolar logic.

As you design a circuit powered from a voltage lower than 5 V – say, 3.3 V – you may run into cases where you need a function that is available only in a 5 V version. Sometimes such interfacing is easy; in other cases, where voltage disparities are larger, a special *translator* gate will be required.

You may encounter the problem in either direction. Suppose the output of a 5 V device needs to be connected to 3.3 V logic whose inputs are limited to 3.3 V max. In the lab, we used an inexpensive 74HC4050 to provide one-way high-voltage-to-low-voltage level translation for the on-board function.

14S.2 Interfacing among logic families

This device is just a hex buffer whose output level is set by the V_{CC} you connect it to but whose inputs can tolerate up to 15 V as a high level. This reduced the breadboard's TTL function generator output to a level safe in anticipation of when we connect it to our expensive 3.3 V FPGA and microcontroller: see §14L.1.5.

For connecting 3.3 V logic outputs to 5 V devices or for bidirectional interfacing, Maxim makes a "universal" bidirectional translator for linking parts supplied from voltages as disparate as 1.2 V and 5.5 V (MAX3370 is the single-line part). NXP (formerly Philips Electronics) developed a simpler bidirectional translator using a single MOSFET. This ingenious circuit is described in §14S.2.3 below.

14S.2.3 A bidirectional level translator

Here we provide a chance to marvel at an ingeniously simple design to the problem of interfacing logic running on different voltages.

The present problem is to design a *bidirectional* translator that will allow a single line linking a lower-voltage part to a higher be driven from *either end*. The bidirectional quality is what makes this task not standard.

Such bidirectional lines are not common, but they do come up at least in the I-squared-C ("Inter-Integrated Circuit") interface – which you will find briefly described in Chapter 26N. Even before you feel a need for such an interface, you may enjoy seeing the ingenuity of this one.

This solution, described by Philips in an application note AN97055, uses a single MOSFET and a pair of resistors to link digital parts powered by differing supplies.[2] Figure 14S.3 is a sketch of how the interface works.

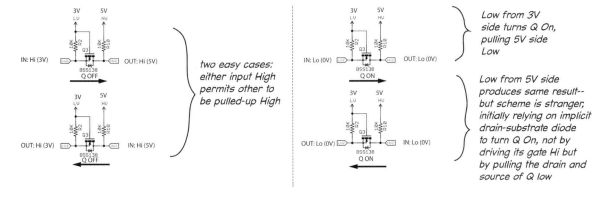

Figure 14S.3 Philips bidirectional logic-level translator. (Source: Philips/NXP application note AN97055.)

---

[2] Available at http://LAoE.link/Phillips_AN97055.pdf. Adafruit and Sparkfun among others have wired up implementations of this interface idea.

14W Worked Examples: Logic Gates

14W.1 Multiplexing: generic

AoE §10.3.3C

In this section we look at three ways of designing a small *multiplexer*:

- using ordinary gates; and
- using 3-states;
- using analog switches.

The notion of multiplexing, or *time-sharing*, is more general and more important than the piece of hardware called a multiplexer (or "mux"). You won't often use a mux, but you use multiplexing continually in any computer, and in many data-acquisition schemes.

Figure 14W.1 illustrates multiplexing in its simplest, mechanical form: a mechanical switch might select among several sources. The sources might be, for example, four microphones feeding one remote listening station of the secret police. Once a month (or once each hour, if s/he's ambitious) an agent flips the switch to listen to another of the four mikes. The motive for multiplexing, here, as always, is to limit the number of lines needed to carry information.

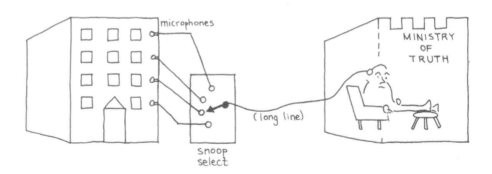

Figure 14W.1 Mux in electro-mechanical form: to let four sources share one output line.

This case is a little far-fetched. More typically, the "scarce resource" would be an analog-to-digital converter (ADC).[1] And the most familiar example of all must be the *telephone* system. Without the time-sharing of phone lines, phone systems would look like the most monstrous rats' nests: picture a city strung with a pair of wires (or even one wire) dedicated to joining each telephone to every other telephone! Figure 14W.2 illustrates what even a network of 8 phones would look like.

A similar argument makes sharing lines efficient in a computer, where the shared lines are called "buses." Here the claim is complicated by the recent emergence of fast *serial* links, in preference to

---

[1] Indeed, our microcontroller, which you will see shortly, has only a single ADC but includes a multiplexer to allow you to select one of 20 analog inputs to connect to it.

14W.1 Multiplexing: generic

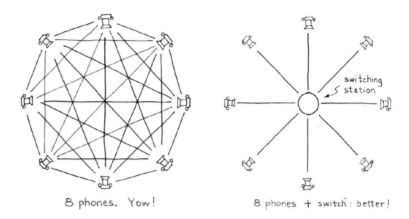

Figure 14W.2 Argument for multiplexing: limit the number of wires running here to there in a telephone system.

parallel lines, to link a computer's processor to memory and peripherals. These serial links now predominate. Such point-to-point communication is much less vulnerable to transmission line reflections and timing skew than is the "multi-drop" scheme of a parallel bus. Thus the serial links permit higher data rates.

AoE §14.5

But even where the external links use just a few wires (a differential pair in each direction, for example, in PCI express) still parallel buses persist *within* the IC, out of sight, and serializer/deserializer (*serdes*) blocks provide the translation, as the right-hand image of Fig. 14W.3 tries to suggest.

Figure 14W.3 Computers traditionally used external parallel buses; now more often serial, point-to-point.

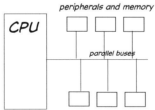 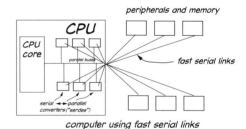

14W.1.1 Multiplexing: hardware

A mechanical multi-position switch can do the job; so can the transistor equivalent, a set of analog switches. Logic gates can also do the job – though these allow flow in only one direction, unlike the other two devices. The digital implementation requires a little more thought if you haven't seen it before. And even the analog switch implementation requires a *decoder* as soon as more than two signals are to be multiplexed.

We need two elements:

1. *Pass/Block* circuitry, analogous to the closed/open switch; and
2. *Decoder* circuitry, that will close just one of the pass/block elements at a time.

Let's work up each of these elements, first for the implementation that uses ordinary gates.

Pass/Block An AND gate will do this job more or less: see Fig. 14W.4. To *pass* a signal, hold one input high; the output then follows or equals the other input; to *block* a signal, hold an input low. This case is a little strange because this forces a *low* at the output – and this forced *low* is not the same as the result of opening a mechanical switch (but is essential to the design as you will see).

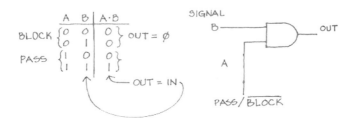

Figure 14W.4 AND gate can do the Pass/Block* operation for us.

Joining outputs The outputs of the AND gates may *not* be tied together. Instead, we need a gate that "ignores" lows, passes any highs (as the "blocking" ANDs will be putting out lows). An OR behaves that way and is just what we need. Figure 14W.6 shows a 2:1 mux made with gates:

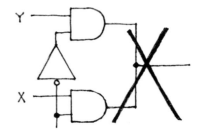

Figure 14W.5 Don't do this! The outputs of ordinary gates may *not* be tied together.

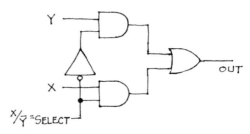

Figure 14W.6 2:1 mux.

Decoder If the mux has more than two inputs we need a fancier scheme to tell *one and only one* gate at a time to *pass* its signal. The circuit that does this job, pointing at one of a set of objects, is called a *decoder*. It takes a binary number (in its *encoded* form, it is compact: n lines encode 2^n combinations, as you know) and translates it into 1-of-n form (*decoded*: it's not compact, so not a good way to transmit information, but often the form needed in order to make something happen in hardware).

The decoder's job is to detect each of the possible input combinations. Figure 14W.7, for example, is the beginning of a 2-to-4-line decoder.

Decoders are useful in their own right (later you will you will create a BCD-to-seven segment decoder in programmable logic; the equivalent of the discrete 74LS47). The decoder is *included* on every multiplexer and in every multiplexing scheme that goes beyond two signals.

14W.1 Multiplexing: generic

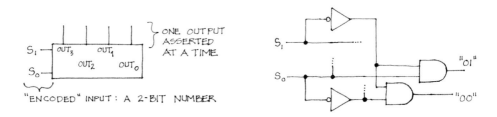

Figure 14W.7 2-to-4-line decoder: block diagram; partial implementation.

14W.1.2 Worked example: Mux

AoE §10.3.3A

Problem (4:1 Mux, designed 3 ways)   Show how to make a 4-input multiplexer using: (a) ordinary gates, (b) transmission gates, and (c) gates with 3-state outputs.

Solution

Decoder: All three implementations require a decoder, so let's start with this.

We need to detect all four possibilities and might as well start, as in the earlier example, by generating the complement of each *select* input. (Does it go without saying that we need *two select* lines to define four possibilities? If not, let's say it. If this isn't yet obvious to you, it will be soon.)

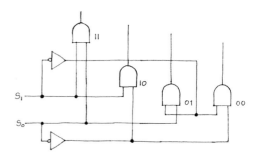

Figure 14W.8 1-of-4 decoder.

Pass/Block gating: ordinary gates: You know how to do this job with AND gates. The circuit in Fig. 14W.9 generates all the four input combinations, and applies them to ANDs as Pass/Block* as in Fig. 14W.4.

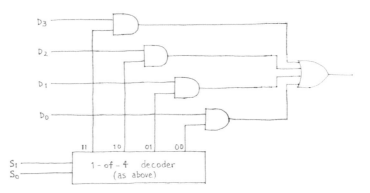

Figure 14W.9 4-1 mux, gate implementation.

Pass/Block gating: transmission gates: This is easier. (You *do*, of course, need to remember what these "transmission gates" or "analog switches" are. If you don't recall them, from Lab 12L, go back and take a second look.)

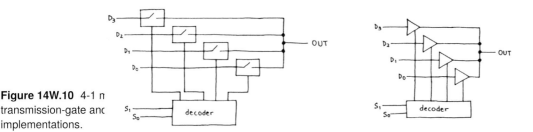

Figure 14W.10 4-1 n transmission-gate and implementations.

Again we use the decoder. This time we *can* simply join the gate outputs, since any transmission gate that is *blocking* its signal source *floats* its output, unlike an ordinary logic gate, which drives a *low* at its output when blocking. This scheme is sketched in Fig. 14W.10.

Three-states: This is a snap after you have done the preceding case. Again use the same decoder, and again you *can* join the outputs. A 3-state that is off, or blocking, does not fight any other gate. That's the beauty of a 3-state, of course.

Problem Under what circumstances would (b) [the transmission-gate implementation] be preferable?

Solution In general it is preferable *only* for handling *analog* signals. (For this purpose it is not just *preferable*; it is *required*.) For digital signals it is usually inferior, since it lacks the all-important virtue of digital devices: their noise immunity, which can also be described as their ability to clean up a signal: they ignore noise (up to some tolerated amplitude), and they put out a signal stripped of that noise and at a good low impedance. The analog mux, like any analog circuit, cannot do that trick; to the contrary, the analog mux slightly degrades the signal. At very least, it makes the output impedance of the signal source worse, by adding in series the mux's R_{on} (around 100 Ω).

We should not overdo this point: it is not wrong to use an analog mux, or a single analog switch, to route digital signals. In fact, analog switches with low R_{on} can provide the fastest way to connect or disconnect to a bus – outrunning a conventional three-state at the expense of losing signal regeneration. These devices are available, marketed as "Zero Delay Logic" (the QuickSwitch[2]). And transmission gates are used within fully-custom ASICs (application specific ICs). But analog switches normally are used – as the name surely suggests – for analog signals.

The analog version, incidentally, can pass a signal in either direction, so it works as a *demultiplexer* as well as mux. But that is not a good reason to use it for digital signals. If you want to demux digital signals, use a *digital* demux.

14W.2 Binary arithmetic

This worked example looks at five topics in binary arithmetic:

Two's-complement: versus *unsigned*, and the problem of *overflow*.
Addition: a hardware design task, meant to make you think about how the adder uses *carries*.

---

[2] This is a product of the company, Integrated Device Technology. It promises very low propagation delay (exaggerated as "Zero Delay"), because of its extreme simplicity. In fact, delay will occur, but caused mostly by the RC product of the switch's R_{on} and stray capacitance of the line that is driven.

14W.2 Binary arithmetic

Magnitude comparators: done laboriously by hand.

Multiplication: an orderly way to do it contrasted with a foolish way.

ALU: foreshadowing the microprocessor, this exercise means to give you the sense that the processing guts of the CPU are simple, made of familiar elements.

14W.2.1 Two's-complement

Here's a chance to get used to 2's-complement notation. We want to underline two points in these examples:

- A given set of bits has no inherent meaning; it means what we choose to let it mean, under our conventions (this is a point made in Chapter 14N as well).
- A sum (or product or other result), properly arrived at, may nevertheless be wrong if we *overflow* the available range.

Both of these points are rather obvious; nevertheless, most people need to see a number of examples in order to get a feel for either proposition. Overflow, particularly, can surprise one: it is not the same as "a result that generates a carry off the end." Such a result may be valid. Conversely, a result can be wrong (in 2's-complement) even though no carry out of the MSB occurs. This is pretty baffling when simply *described*. So let's hurry on to examples.

Let's suppose we feed an *adder* with a pair of four-bit numbers. We'll note the result, and then decide whether this result is correct, under two contrasting assumptions: that we are thinking of the 4-bit values as *unsigned* versus *two's-complement* numbers.

Problem (Two's-complement) Suppose that you feed a 4-bit adder the 4-bit values A and B, listed below. Please write–

1. The maximum value one can represent in this four-bit result –
   - unsigned; and
   - in 2's-complement.
2. The sum.
3. Then what the inputs and outputs mean in decimal, under two contrasting assumptions:
   - first, the numbers are (4-bit-) *unsigned*; and
   - second, the numbers are (4-bit-) *two's-complement*.
4. Finally, note whether the result is valid under each assumption.

We have done one case for you as a model.

Worked Examples: Logic Gates

IN:	A	B	A plus B	Valid?
Binary:	0111	1000	1111	
Decimal:				
Unsigned:	7	8	15	yes
2's-comp.:	+7	−8	−1	yes
Binary:	0111	0111		
Decimal:				
Unsigned:				
2's-comp:				
Binary:	0111	1010		
Decimal:				
Unsigned:				
2's-comp:				
Binary:	0111	0100		
Decimal:				
Unsigned:				
2's-comp.:				
Binary:	1001	1000		
Decimal:				
Unsigned:				
2's-c:				

Try these. Then see if you agree with our conclusions, set out below.

Solution (Two's-complement)

The maximum values that one can represent with four bits

- Unsigned: 15_{10}.
- In 2's-complement: $+7_{10}, -8_{10}$.

IN:	A	B	A plus B	Valid?
Binary:	0111	0111	1110	
Decimal:				
Unsigned:	7	7	14	yes
2's-c:	+7	+7	−2	no
Binary:	0111	1010	0001	
Decimal:				
unsigned:	7	10	1	no
2's-c:	+7	−6	+1	yes
Binary:	0111	0100	1011	
Decimal:				
unsigned:	7	4	11	yes
2's-c:	+7	+4	−5	no
Binary:	1001	1000	0001	
Decimal:				
unsigned:	9	8	1	no
2's-c:	−7	−8	+1	no

Problem What is the rule that determines whether the result is valid?

Solution For unsigned, isn't it simply whether a carry-out is generated? It is.

For *two's-complement* the rule is odder: *if the sign is altered by carries*, the result is bad. A carry into the sign bit *can* indicate trouble – but only if there is no carry out of that bit; and vice versa.

In other words, an XOR between carries in and out of MSB indicates a 2's-complement overflow. The microcontroller you will meet in Lab 22L uses similar logic to detect just that overflow (indicated by its "V" condition flag, in case you care at this point). You can test this *XOR* rule on the cases above if you like.

14W.2.2 Addition

You may find it entertaining to reinvent the *adder*. As a reminder, the table below shows the way a single-bit *full-adder* should behave (the name "full" indicates that the adder can take in a *carry-in* signal so that such full-adder modules can be cascaded).

INPUT				OUTPUT	
Carry-in:	*A*	*B*		**Carry-out**	**Sum**
0	0	0		0	0
0	0	1		0	1
0	1	0		0	1
0	1	1		1	0
1	0	0		0	1
1	0	1		1	0
1	1	0		1	0
1	1	1		1	1

You will notice that you can, if you like, look at the *output* as a two-bit quantity rather than look at it as distinct *sum* and *carry*. The advantage of the latter view is just that it invites extension to larger word widths: when a 4-bit sum overflows, it too generates a *carry*.

Problem (Design a one-bit full adder.) Use the truth table above, if you need it, and design a full adder. Figure 14W.11 illustrates its block diagram.

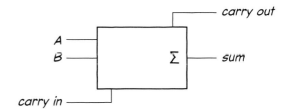

Figure 14W.11 Full adder.

Solution The *sum* is *A* XOR *B* when carry-in is low; it's the complement of XOR when carry-in is high. Does that suggest what the sum function is, as a function of the *three* input variables that include carry-in? Yes, it's XOR of all three (output 1 if an *odd* number of inputs are high): see Fig. 14W.12.[3]

Carry-out? You may be able to see the pattern after staring a while at the truth table. Or you might reason your way to the function: carry-out should be asserted when two or more of the input bits

---

[3] Since XOR is a two-input logic function, you must do this in two steps. However, the order in which you XOR the pairs is unimportant.

614 Worked Examples: Logic Gates

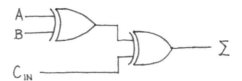

Figure 14W.12 Sum is C_{in} XOR A XOR B.

are asserted. You might conceivably want to learn the antique method of Karnaugh mapping (see Fig. 14W.13 and let the map reveal patterns: patterns that show chances to ignore a variable or two).

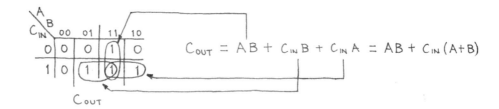

Figure 14W.13 Carry-out Karnaugh map.

One way or another, you can arrive at a sketch of the one-bit full adder (Fig. 14W.14).

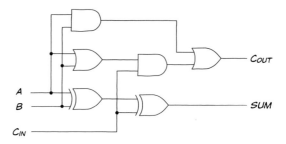

Figure 14W.14 Full adder.

14W.2.3 Digital comparators

AoE §10.3.3E

Detecting equality: Many computer peripherals need to know when a value arriving on the address bus is equal to the peripheral's reference address. A set of XOR functions can detect such equality, bit by bit (see Fig. 14W.15). We invited you to design an equality detector in Lab 14L, and no doubt you figured out that all you needed was to detect equality, bit-by-bit, with XOR gates fed by corresponding bits of a multi-bit word: A_1 XOR B_1, A_0 XOR B_0. Then AND lows from the several XORs (the name of the gate that AND lows is "NOR," though the name, of course, does not describe the gate's function in this case).

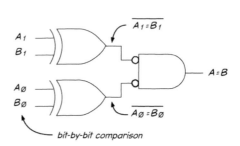

Figure 14W.15 XORs can detect equality, bit-by bit.

14W.2 Binary arithmetic

Detecting inequality: A>B ("magnitude comparator"):

Problem (2-bit greater-than circuit.) Use any gates you like (XORs are handy but not necessary) to make a circuit that sends its output high when the 2-bit number A is greater than the 2-bit number, B. Figure 14W.16 shows a black box diagram of the circuit you are to design.

Figure 14W.16 2-bit greater-than circuit.

Note that this problem can be solved either systematically, using Karnaugh maps, or with some cleverness: consider how you could determine equality of two one-bit numbers, then how to determine that one of the one-bit numbers is greater than the other; then how the more- and less-significant bits relate; and so on. We'll do it both ways.

Solution (1. Systematic.) Figure 14W.17 works, though we don't suppose anyone wants to learn Karnaugh mapping – a technique briefly described in AoE §10.3.2 – for the pleasure of solving such problems.

Solution (2. Brainstorm the problem.) Figure 14W.18 shows a more fun way to do it. The brainstormed circuit is no better than the other. In gate count it's roughly a tie. Use whichever of the two approaches appeals to you. Some people are allergic to Karnaugh maps; a few like them.

14W.2.4 Multiplication

Multipliers are much less important than adders; you should not feel obliged to think about how multiplication works. (Multipliers are important in digital signal processors, but not in the work we will do in this book.) Think about it if the question intrigues you.

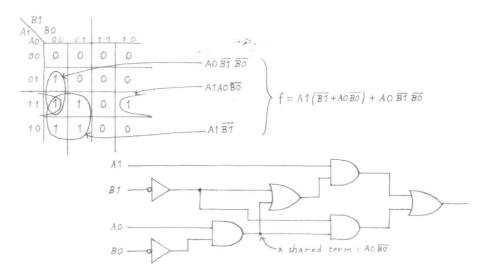

Figure 14W.17 2-bit greater-than circuit: systematic solution.

616 Worked Examples: Logic Gates

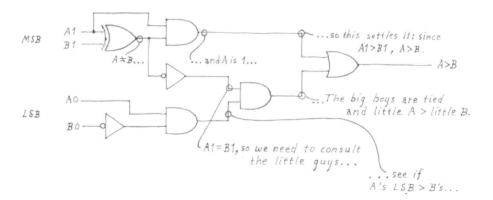

Figure 14W.18 2-bit greater-than circuit: brainstormed solution.

One way to discover the logic needed to *multiply* would be to use Karnaugh maps or some other method to find the function needed for each bit of the product.

But even this small example may be enough to convince you that there must be a better way. A 3×3 multiplier (with its *six* input variables) would bog you down in painful Karnaugh mapping; but larger multipliers are commonplace. There is a better way.

This turns out to be one of those design tasks one ought *not* to do with Karnaugh maps, nor with any plodding simplification method. Instead, one ought to take advantage of an orderliness in the product function that allows us to use essentially a single scheme, iterated, in order to multiply a large number of bits. The pattern is almost as simple as for the adder.

Binary multiplication can be laid out just like decimal if we want to do it by hand – and the binary version is much the easier of the two: see Fig. 14W.19.

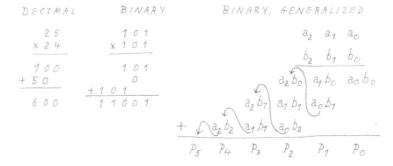

Figure 14W.19 Decimal and binary multiplication examples – and a generalization of the pattern; binary is easier!

Once you have seen the tidy pattern, you will recognize that you need nothing more than 2-input AND gates, and a few *adders*. With this method, you don't have to work very hard to multiply two 4-bit numbers – a task that would be daunting indeed if done with Karnaugh maps. (You need only *half* adders, not *full* adders; but you might choose to use an IC adder like the 74HC83, rather than build everything up from the gate level.) Let's try that problem.

Problem (3×3 multiplier.) Design a 3 × 3 multiplier along the same lines, this time using two 4-bit full adders (74HC83) and as many 2-input gates as you need.

Solution Figure 14W.20 is a circuit you'll want to buy, not build! But you can see that you could easily extend this to make a 4 × 4 or 16 × 16 multiplier.

14W.2 Binary arithmetic

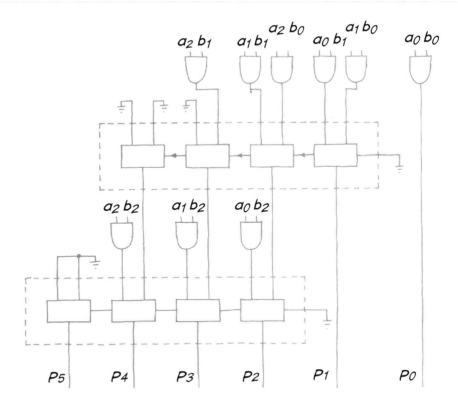

Figure 14W.20 3×3 multiplier using IC adders.

14W.2.5 Arithmetic logic unit (ALU)

As a wrap-up for this discussion of arithmetic here is a device central to any computer: a combinational logic circuit that does any of several logical or arithmetic operations. The ALU takes two operands and n control signals (which select which of up to 2^n operations to perform) and outputs the result. To design this does not require devising anything new; you need only assemble some familiar components.

Problem (Problem: design a 1-bit ALU.) Figure 14W.21 is a block diagram of a simplified ALU. Let's give it a carry-in and a carry-out, and specify that the carry out bit be low in any case where a carry is not generated (for example, A OR B).

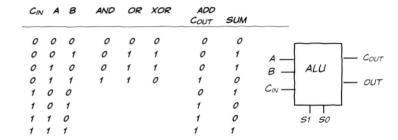

C_{IN}	A	B	AND	OR	XOR	ADD C_{OUT}	SUM
0	0	0	0	0	0	0	0
0	0	1	0	1	1	0	1
0	1	0	0	1	1	0	1
0	1	1	1	1	0	1	0
1	0	0				0	1
1	0	1				1	0
1	1	0				1	0
1	1	1				1	1

Figure 14W.21 1-bit ALU: block diagram.

The two select lines, S_0 and S_1 in Fig. 14W.21 should determine which operation the ALU performs (use any select code you like).

Worked Examples: Logic Gates

Solution (Design it with gates and a mux.)   All we need is to combine standard gates and an adder with a multiplexer: see Fig. 14W.22.

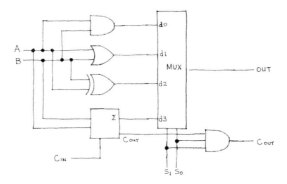

Figure 14W.22 A simple ALU.

14O Online Content: Logic Gates

14O.1 Adding a fixed 3.3 V supply to the PB-503 breadboard

Available at https://LAoE.link/LAoE_Chapter_14O.pdf.

15N Introduction to Programmable Logic

Contents

15N.1	**The hardware**		**621**
	15N.1.1	A bit of history	621
	15N.1.2	Back to the present: The FPGA	621
15N.2	**Creating the bitstream**		**623**
15N.3	**Introduction to Verilog**		**624**
	15N.3.1	Writing Verilog	624
	15N.3.2	The form of a Verilog file	624
	15N.3.3	Verilog literals	625
	15N.3.4	Scalar vs. vector signals	626
	15N.3.5	Some useful Verilog operators	626
	15N.3.6	Structural vs. behavioral design	628
	15N.3.7	Modules are hierarchical	629
	15N.3.8	But what about the inputs and outputs?	630
	15N.3.9	Verilog can tame active-lows	631
15N.4	**Digital logic recap**		**635**
15N.5	**AoE reading**		**635**

Why?

You have probably noticed by now that digital logic uses a lot more devices and signal lines than analog circuits. Instead of a single analog signal carrying all the information about a value, say 0 to 10 V representing 0 to 100°C to a tenth of a degree, a digital system would need ten binary data lines to represent temperature with the same precision. While you might have enjoyed wiring up a few digital gates to create the two-bit adder in the last lab, building a 16-bit adder out of gates would be a bit tedious.

In addition, packaged logic gates are low density, typically containing only a few gates.[1] That means any reasonably complex digital systems might need tens or hundreds of DIP packages. Because signals have to travel between packages, systems built with discrete logic are limited in speed as well.

If you are building thousands or millions of an electronic gadget, you can have a custom integrated circuit built which contains the digital logic for your product.[2] While the price per part is relatively low for a custom IC, the non-recurring engineering costs (NRE) can be quite high, in the hundreds of thousands to millions of dollars. As long as you make enough of them to amortize these costs, full custom is the way to go.

For designs unable to recover the NRE costs, there is "programmable logic." Seeing "programmable,"

---

[1] This is known as small scale integration or SSI. More complex devices containing a few hundred gates, such as counters, are known as medium scale integration (MSI).
[2] Known as an *Application Specific Integrated Circuit* or *ASIC*.

you may think of memories, which are indeed "programmable logic devices." But *Programmable Logic Devices* or PLDs is a term reserved for devices organized differently from memories, and designed, broadly speaking, to replace discrete gates rather than to store data and code as memory ordinarily does. Programmable devices have superseded the small packages of gates like the 74HC00 that provide only four 2-input NANDs per packages for example. A PLD is much more versatile than a cabinet full of 7400 series gates, and very much denser, and therefore makes a design cheaper to implement and much easier to amend. The appeal of PLDs is irresistible.

As a result, this method has displaced the use of discrete packages of standard logic gates for moderate to very complex designs in smaller volume. You purchase an integrated circuit containing thousands to millions of gates that can be configured electronically. Then, using a "logic compiler," you tell the chip how you want these gates connected to implement your design. (We will discover the happy fact that a logic compiler can also make the pesky problem of active-low signals very easy to handle while simultaneously simplifying our Boolean equations for us – goodbye Karnaugh maps.)

15N.1 The hardware

15N.1.1 A bit of history

AoE §11.2.1

PLDs (the generic term) have been around since the 1970's. Early devices were "mask programmable," that is only the final metal interconnect layer of the integrated circuit was customized for each application. This was a relatively low NRE step compared to a full custom design but still too expensive for one-off designs and turnaround time was measured in weeks.

After some experimentation with the form of internal logic, one dominant scheme appeared: the so-called PAL (a trade name for Programmable Array Logic – a clever rearrangement of letters to form a user-friendly acronym). Fuse programmable PALs became commonly available in the 1980s. The internal gates in the PAL were arranged to support sum-of-products combinational logic using a programmable AND array followed by a fixed OR array: see Fig. 15N.1(a). Fuse programmable devices contain narrow sections of conductive material in the integrated circuit that can be overheated ("blown") in a controlled manner to create open circuits; hence use of the term "burn" for programming the device. Typically, an external programming tool was used to selectively open fuses to obtain the desired behavior. These devices were one time programmable (OTP), once a fuse was blown there was no way to stitch it back together. In Fig. 15N.1(b), the fuses have been blown to implement the unsimplified Boolean equation from §14N.3.5.2.

The simplicity of this structure made the parts fast – somewhat faster than little packages of gates, because internal capacitances were lower than printed-circuit capacitances. Reduced capacitances not only let logic levels change faster but also permitted the gates to omit driver stages. The result was greater speed.

With PALs, a circuit designer could create custom combinational logic in a few minutes at relatively low cost. A variety of PALs appeared: some with active-low outputs, others active-high; some with flip-flops included; some with a few wide ANDs, others with narrower but more numerous ANDs feeding the OR. This variety was good in a way, and also a nuisance, requiring users to stock lots of parts.

15N.1.2 Back to the present: The FPGA

Modern programmable devices, Field Programmable Gate Arrays (FPGAs) are much more flexible. As its name implies, not only can you reconfigure them multiple times, but they can and must be

Introduction to Programmable Logic

Figure 15N.1
Simplified fuse programmable logic device (a) and device after programming (b).

configured in-circuit. FPGAs use volatile internal memory to store their configuration information (called a "bitstream") and it must be reloaded each time power is applied.[3]

Unlike PALs, all or most of the external I/O pins in the FPGA can be configured as inputs, outputs or bidirectional depending on an application's requirement. In addition, the programmable logic blocks of the FPGA are much more flexible than the fixed sum-of-products organization of the PAL. You can configure the FPGA to replicate almost any digital circuit that can be built in custom silicon, with the added benefit that the design can be changed by sending the end user a new bitstream file (but with the disadvantage of higher cost per unit).

The basic unit of an FPGA is the logic cell, consisting of a look up table (LUT), a flip-flop, some combinational logic with a multiplexer (i.e., a digital switch) to speed arithmetic operations between cells and another multiplexer to select between combinational operation without the flip-flop and sequential operation with: see Fig. 15N.2.[4]

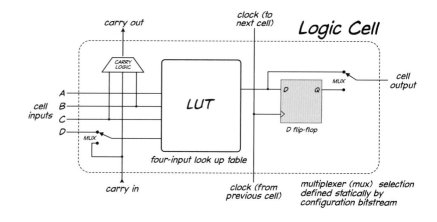

Figure 15N.2 FPGA logic cell.

The LUT is a small memory that implements (in this example) a 16 entry truth table. In the previous chapter we saw that any truth table can be implemented in gates using sum-of-products. In the FPGA, a truth table is duplicated directly in memory with the inputs used as an address to select the value to output. Unlike the fixed OR gate structure in the PAL, the LUT allows any four-bit combinational logic function to be implemented by loading the memory with the appropriate data.

In order to enable fast multibit arithmetic, the FPGA handles the calculation of the carry in a binary addition operation with special hardware (the "carry logic" block). The carry could be calculated just

---

[3] Most applications store the bitstream in nonvolatile flash memory, similar to what is used in USB drives. An embedded microprocessor, or in some cases the FPGA itself, loads the bitstream at power up or after a reset to configure the device.

[4] For now, we will concentrate on combinational operations, so you can ignore the flip-flop.

by routing the output of one block to the next, but then the design would have to wait for the carry to ripple through each cell and the answer would not be available until n (the number of bits in the addends being being summed) propagation delays later. The extra carry logic hardware allows the carries to be handled in parallel with the look up operation.[5] The multiplexer ("mux") shown on cell input D can be used to route the previous carry bit to the LUT when the cell is being used for arithmetic operations, since it must be included in the sum. It is set to either the input D position or the carry position when the bitstream is loaded and cannot change unless the configuration is updated with a new bitstream.

The output of the LUT goes to a one bit memory element (the "D flip-flop") and to a cell output multiplexer which selects between the output of the LUT for combinational operations and the output of the flip-flop for sequential logic. Like the input mux, it is set at configuration and does not change. For the moment, assume the output is connected to the LUT. We will discuss flip-flops and sequential logic in a later chapter.

In addition to logic cells, FPGA typically include fixed blocks of memory and digital signal processing (DSP) multiply-and-accumulate slices. Larger FPGAs may include full embedded microcontrollers, serial interfaces, clock generators and computer bus interface blocks. The smallest FPGAs contain hundreds of logic cells and cost a few dollars; the largest contain several million cells and cost considerably more.[6]

FPGAs are somewhat more difficult to work with than PALs, whose delays are quite predictable because of the rigidity of their structure. Timing in the FPGA depends on how the elements are laid out – and even delays in signal paths matter, now that gate delays have been cut so low. So the fine points of FPGA design are difficult, although we will keep our designs simple so as to not push any limits.[7] In the lab we will use a smallish version of such an IC containing a mere 5280 LUT cells (and we will only need a small fraction of these to create some impressive applications).[8]

15N.2 Creating the bitstream

Even the most powerful FPGA is quite useless until a bitstream is loaded specifying the interconnects between the logic cells and the connections between cells and external I/O pins. We use a human readable Hardware Description Language (HDL) to define how the programmable logic is connected. In our case we will use Verilog, an HDL originally created in the 1980s to allow simulation of electronic system. Verilog competes with another HDL named VHDL.[9]

While much of Verilog syntax resembles the C programming language, it is important to understand that a HDL is *not* procedural. Unlike software code which is a series of instructions executed sequentially in the order written, Verilog is a set of written instructions describing circuit connections. When you wire your breadboard you do it sequentially because you only have two hands, but when you are

---

[5] In addition, since each LUT has only one output, you would need almost twice as many logic cells to implement an adder. One set of n cells to calculate each sum bit and a second set of $n - 1$ (since the first bit only requires a half-adder) to calculate the carry.

[6] As of fall 2023, the highest performance Xilinx VU57P FPGA containing 2.85M logic cells, gigabytes of RAM, almost 10,000 DSP slices and ten 100G Ethernet interfaces was listed at DigiKey for $169,659.60.

[7] For an introduction to FPGAs that is unusual for being entertaining as well as authoritative, see Clive Maxfield's *The Design Warrior's Guide to FPGAs: Devices, Tools and Flows*, Newnes (2004).

[8] Our Lattice Semiconductor CE40UP5K also includes a megabyte of static RAM, eight DSP blocks, several built-in oscillators, two types of serial interfaces and three high current LED outputs, none of which we will be using. Its speed is comparable to discrete HC logic gates, with propagation delays of 16 to 20 nanoseconds.

[9] VHDL, a bit hard to pronounce, is better than what it stands for: "Very [High Speed Integrated Circuit] Hardware Description Language." It was developed at the request of the US Department of Defense, whereas Verilog arose in private industry.

done the circuit runs with all the wires connected no matter in what order you built it. The same is true of a HDL – be careful about using signals thinking they will be changed because their logic was defined earlier in the Verilog code.

The completed Verilog code is put through a logic compiler program called a synthesizer to create a netlist, a process called synthesis. The netlist itself is generic, a floor planner followed by a place-and-route (PNR) program figure out how to implement the netlist using the resources on a particular FPGA. In addition to the netlist, these programs need to know what internal signals are connected to which external pins of the device. There may be other constraints as well, for example certain signals may be timing sensitive and need to be kept close together or routed through internal buffers to minimize skew. Finally, the bitstream must be generated from the output of the place-and-route step and downloaded into the nonvolatile storage of the target system.

15N.3 Introduction to Verilog

AoE §11.3.4B

Let's get started. As you know from your experience with other programming languages, the refinements are many, the options can be dazzling. But our goal in this course is modest: only to let you try out some of the capabilities of Verilog and FPGAs. Later, when you undertake a more ambitious design, no doubt you will want to get yourself one or more Verilog reference texts and then explore the rich possibilities.[10]

15N.3.1 Writing Verilog

Structurally, Verilog is similar to C (or Arduino which is a subset of C). With some [annoying] exceptions, lines are terminated in a semicolon (";") meaning you can continue a long statement on multiple lines. Anything following a double forward slash ("//") is a comment and ignored by the synthesizer up to the end of the line.[11] Multiline comments are allowed using the C convention of including them between "/*" and "*/". Verilog files typically use a ".v" extension.

15N.3.1.1 Best practices
Always include a header at the top of your file with (at minimum):

- the name of the file;
- the name of the creator;
- the date of creation; and
- a description of what the code does.

Use comments, indents, and white space to make your code clearer. It is also good idea to include the name of the programmer, the date, and the purpose of the update when modifying your code.

15N.3.2 The form of a Verilog file

A Verilog file consists of one or more modules. You can think of a module as analogous to an IC with input pins, outputs pins and logic inside. The example in Fig. 15N.3 is a file containing one module that implements a two-input AND gate and a two-input OR gate on the same inputs:

---

[10] Here are two that we like: J. Bhasker, *Verilog HDL Synthesis* (1998): extremely concise, offering little discussion but at least one example of each topic treated; S. Palnitkar, *Verilog HDL* (2003): more discursive, and including fewer examples, but more detailed discussion of each. Online Chapter 15O describes some useful resources for learning and simulating Verilog without actual hardware.

[11] But note that our FPGA development environment uses a few special comments which are ignored by any other development system to set the name of the top level module and assign I/O pins to signals. See Chapter 15S.

15N.3 Introduction to Verilog

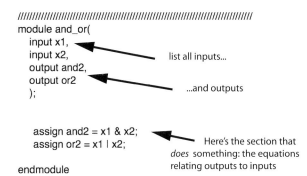

Figure 15N.3 Verilog source file for AND OR functions.

To declare a module, the "module" keyword is followed by the name of the module, a port (signal) list enclosed in parentheses and a concluding semicolon.[12] The port list is a declaration of the inputs and outputs to the module separated by commas. Ports must be declared as "input," "output," or "inout" (for bidirectional signals).[13] Verilog is case sensitive so "X1" would not be the same signal as "x1".

Ports can be of two types, wire or reg. The default type is wire, which is what we are using for this combinational logic example; we will use reg for the sequential logic discussed in the next chapter. We did not have to include the "wire" keyword for these signals (but we could have: e.g., "input wire x1,"). However, you must include the "reg" keyword if you are not using the default wire type.

With the preliminary definitions of interface signals done, we can get down to business, which is done in the two "assign" lines. *Assign* is the keyword that defines the logic linking the output (for example, in the first assign "and2") with the relevant inputs (here, x1 and x2). Assign outputs may only be used with wire signals. The & means AND; | means OR. Assign is only appropriate for combinational logic; sequential circuits are designed with a different form, discussed in a later chapter.

Finally, we end the module with an "endmodule" statement (with no semicolon!).

15N.3.3 Verilog literals

Verilog numeric literals are number constants. They are specified with a width followed by an apostrophe, then the radix followed by the value. The literal:

```
1'b1
```

is the single bit value 1. The number before the apostrophe is the number of bits in the numeric constant. Here, 1' tells the synthesizer this is a one-bit scalar value. The "b" says the value that follows is in binary. Other options are "d" for decimal, "h" for hex and "o" for octal (base 8). Radix specifiers are case insensitive so we could have written this constant as "1'B1" instead. You can include an underscore in a literal numeric value for readability, it is ignored by the compiler:

```
8'b1011_1110    // 190 decimal specified as an 8 bit binary value
8'd190          // same 8-bit literal specified in decimal
8'hBE           // (or 8'hbe) same 8-bit literal specified in hex
```

Individual wires can be set to 0, 1, X and Z where X represents an unknown or *don't care* and Z is high-impedance. X and Z are case insensitive. The following is a valid literal value:

```
4'b10XZ    // LSB is high-Z, next higher bit is unknown value
```

Literals lacking a radix specifier are assumed to be decimal values. If the width specifier is omitted, the default is 32 bits.

---

[12] We use the terms "port," "signal" and "variable" interchangeably in discussing Verilog.
[13] This is the form for the Verilog-2001 standard. There is an older, alternative form where the type and direction of a port is not included in the module definition. We will discuss this version later in the one place where we are obliged to use it.

15N.3.4 Scalar vs. vector signals

Verilog signals can be either scalar (single bit) or vector (multi-bit). To create a new one-bit wire signal, use the "wire" keyword followed by the signal name. Here we create four separate one-bit signals.

```
wire A0;      // LSB of four-bit value
wire A1;
wire A2;
wire A3;      // MSB of four-bit value
```

Alternatively, we could create these signals as a four-bit vector by specifying a *range* before the signal name:

```
wire [3:0] a;   // four signals grouped together
```

The range ("[3:0]") allows you to address an individual bit (a *bit-select*) in a vector. The most significant bit of the vector is the left hand value in the range, while the value on the right is the least significant bit. For example:

```
assign a[3] = 1'b1;   // set the MSB in the vector high
assign A0 = a[0];     // set a scalar wire to the vector's LSB value
```

You can also specify groups of contiguous bits (a *part-select*) with fixed starting and ending bit values:

```
assign a[1:0] = 2'b00;   // set the two LSBs in vector low
```

An *indexed part-select* is specified by position and size. This form is useful because the position (but not the size) can be a variable. The left value is the starting bit and the right is number of bits. If the separator is +: you count up from the starting bit and if -: you count down.

```
assign a[0 +: 2] = 2'b00;   // selects a[1:0]
assign a[3 -: 2] = 2'b11;   // selects a[3:2]
```

Concatenation: allows you to build vectors from scalars and smaller vectors:

```
assign a = {A3, A2, A1, A0};   // set vector to value of scalar signals
assign a[1:0] = {A1, A0};      // set two LSBs of vector to scalar values
```

You can use concatenation to rotate a vector (a common ALU operation):

```
wire [7:0] vsrc;
wire [7:0] vdest;

assign vdest = {vsrc[0], vsrc[7:1]};   // rotate right by 1
assign vdest = {vsrc[6:0], vsrc[7]};   // rotate left by 1
```

The *replication* operator repeats a value multiple times:

```
assign a = {2{1'b0}, 2{1'b1}};   // set vector to 0011
assign a = {4{A0}};              // set all bits of vector to value of wire A0
```

15N.3.5 Some useful Verilog operators

Bitwise operators: In the example of §15N.3, we used the binary AND and OR Boolean operators & and |. These, along with the binary ^ for XOR and the unary ~ for NOT are bitwise operators. You can combine the NOT with the other three operators to get binary NAND, NOR and EXNOR. The bitwise operators evaluate to an n-bit result when given n-bit operands:

```
       ~1        // NOT result is 0
     0 ~& 1      // NAND result is 1
     ~0111       // NOT result is 1000
 0111 & 1000     // AND result is 0000
```

15N.3 Introduction to Verilog

Reduction operators: If you need typing practice you might enjoy calculating parity (a scalar value that indicates if the number of 1's in a vector is even or odd) by XORing each bit in the vector individually:

```
wire p;
assign p = a[0] ^ a[1] ^ a[2] ^ a[3];   // set p to parity of vector a
```

You might enjoy it less if the vector is 32 bits wide. In that case you can use the reduction operators which look just like the six binary bitwise operators but operate on vectors. Just put a bitwise operator in front of a vector and Verilog applies the bitwise operator between each bit in the vector to generate a scalar result.

```
wire p, allones;
assign p = ^a;           // calculate parity of vector a
assign allones = &a;     // allones = 1 only if a = 1111
```

Logical operators: As in C, Boolean operations on vector (multi-bit) operands can either be bitwise or logical. The logical operators are && for AND, || for OR and ! for NOT. For scalar signals, there is no difference between the logical and Boolean operators. This is not true for vector operands. The logical operators always return a scalar result. Any vector not equal to zero is considered true. You will want to be careful to use the correct operator for what you are trying to accomplish (compare these examples with the bitwise operators above).[14]

```
        !1         // result is 0, same as bitwise ~1
      !0111        // NOT true is 0
 0111 && 1000      // true AND true is 1
```

Logical shift operators: The operators, << and >> shift signals left and right respectively by an integer number of bits. The shift operation is of the form:

```
<value to shift> <shift operator> <number of bits to shift>
```
For example

```
wire [7:0] a;
assign a = 8'b10100010;
assign a = a << 2;       // a now 10001000
assign a = a >> 3;       // a now 00010001
```

In a left shift, a zero is shifted into the low order positions while overflow bits shifted past the MSB are discarded. If no overflow occurs, left shift is identical to multiplying by $2^{bits\text{-}to\text{-}shift}$. In a right shift, zeros are shifted into the MSB position and bits shifted out of the LSB position are discarded. This provides an integer divide-by-power-of-two, always rounding down.

Ternary conditional operator: ? works the same as it does in C. The general form of the operator is:

```
condition ? value_if_true : value_if_false
```
If the condition in the first expression before the ? is true, the result is the second expression, the one between the ? and the :. If it is false, the result is the third expression after the :.

---

[14] We have a bad habit of using ! to invert scalars. Please forgive us, we are looking into rehab.

Equality operators: You have already seen that Verilog uses a single equals sign to assign the value of an expression to a wire. As in C, equality is tested by a separate equality operator == which returns a scalar true/false result.[15] Use != to test for inequality. These return a scalar result.

Relational operators: The relational operators are:

```
>    // greater than
<    // less than
>=   // greater than or equals
<=   // less than or equals
```

These are useful in conditional statements and return a scalar result.

Operator precedence: You could memorize the precedence of the various Verilog operators or just do what we do, use a lot of parentheses.[16] However, the unary operators have highest precedence so (~A &~B) is unambiguous.

15N.3.6 Structural vs. behavioral design

So far, we have used Verilog to duplicate what we would have built with discrete gates. This is called a *structural model* of the system we are trying to synthesize. The *behavioral model* describes the function of the system algorithmically. Neither is inherently better than the other, but one may make the code more understandable than the other. However, unless you are starting from a gate representation you may find the behaviorial model easier to implement since it does not require a creating a truth table and deriving the Boolean expression for the logic.

Consider the circuit in Fig. 15N.4 below.

Figure 15N.4 A circuit to allow a signal to pass or not.

This circuit connects the input signal to the output when the `pass` control signal is asserted. When `pass` is not asserted, the output is blocked (low). The Verilog code for the structural model is straightforward but does not give much indication of the purpose of the logic:

```
assign out = in & pass;
```

A behaviorial model, on the other hand, makes the purpose of the circuit much clearer:

```
assign out = pass ? out : 1'b0;
```

This assign uses the Verilog ternary conditional operator to make it clear that we are not just ANDing two signals to see when both are high, but rather that `pass` is controlling whether the input is connected to the output or not.

---

[15] Verilog goes one better than C by adding a second equality operator, ===. The normal == operator returns X if any operand includes a Z or X. The triple equals compares all four possible bit values and always returns true or false. The !== operator does the same thing but returns the not equals result. However, since there are no gates in the FPGA that can compare an unknown to an unknown (or even know that a bit is unknown since every bit is high or low), === and !== cannot be synthesized into logic.

[16] For a complete list of Verilog operators and their precedence see https://LAoE.link/Verilog_Operators.html.

15N.3.7 Modules are hierarchical

Luckily, we do not have to build our entire system in one module. We can build our FPGA system the same way we would build our design with discrete logic, a functional block at a time. Let's assume we want to implement the two-bit binary adder from the last chapter using programmable logic. Rather than define all the logic in the top level module, we will define a one-bit full adder module and use that to build the larger device.[17]

Figure 15N.5 shows the truth table for a full adder and an implementation with logic gates:

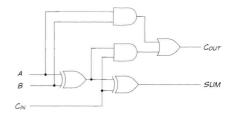

Figure 15N.5 Full adder truth table and implementation.

Here is a Verilog structural module that implements the full adder:

```
// full adder built with gates
module full_adder(
  input   a,
  input   b,
  input   cin,
  output  sum,
  output  cout
);

  assign sum = (a ^ b) ^ cin;
  assign cout = ((a ^ b) & cin) | (a & b);

endmodule
```

We need to use two of these adder modules in the top level module, connecting them as shown in Fig. 15N.6 to create our 2-bit adder.[18]

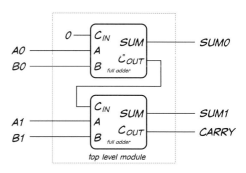

Figure 15N.6 Block diagram of 2-bit adder using full adder modules.

---

[17] As we saw in the previous chapter, you only need a half adder in the least significant bit position of a multi-bit addition because there is no carry in. Here we will use the full adder in both bit positions to keep the design to only two modules; the full adder and a top level module to implement the two-bit adder.

[18] Now you can see why the FPGA logic cell in Fig. 15N.2 includes special circuitry to calculate the arithmetic carry in parallel. In the architecture of Fig. 15N.6, the result is not available until the [cell propagation time] * n (the number of bits).

```
// two-bit adder using two full adder modules
module FPGA_TOP(
    input   A0,         // LSB of addend A
    input   A1,         // MSB of addend A
    input   B0,
    input   B1,
    output  SUM0,       // LSB of result
    output  SUM1,       // MSB of result
    output  CARRY
);


wire carry0;            // define an internal signal

//instantiate adders using port connection by name method
full_adder LSBSum(.a (A0), .b (B0), .cin (1'b0), .sum (SUM0), .cout (carry0));
full_adder MSBSum(.a (A1), .b (B1), .cin (carry0), .sum (SUM1), .cout (CARRY));

endmodule
```

We prefer a convention of all uppercase names for signals that are connected to external (to the FPGA) pins in the module definition. After the definition, we define a wire, `carry0`, to connect the carry out of the least significant sum to the carry in of the most significant sum. We then *instantiate* two instances of the full adder module to handle the two bits in the input addends. Instantiation consists of stating the name of the module we want to create an *instance* of, then giving a unique name to the new instance followed by a port connection list.

There are two ways to create the connection list; we have used the "port connection by name" method which links ports by name. The form is ".[port name](signal name)" where [port name] is the port name in the declaration of the module we are instantiating's port list. Signal name is the signal we are connecting this port to in the current module. Port connections are separated by commas. The white spaces in the connection list are optional but improve readability. Since we are explicitly linking ports by name, the order of ports in the connection list does not have to match the module definition order. If there are module outputs that you are not using, you do not have to include them in the port connection list.

The alternative method of specifying the port connection list is by using an ordered list. Here you do not have to include the port name from the module definition, but you do have to include *all ports in the same order as the module declaration*. For example, had we chosen to use the "port connection by order" method, the adder instantiations would have looked like this:

```
full_adder LSBSum(A0, B0, 1'b0, SUM0, carry0);
full_adder MSBSum(A1, B1, carry0, SUM1, CARRY);
```

All the ports in the instances of the full adder are connected to external pins with two exceptions. There is never a carry into the least significant digit of an integer addition, so we set the carry in signal to the first adder to false with a constant value of zero.

The other connection is the internal connection of the carry out of the LSB to the carry in of the MSB. The `carry0` wire we defined is used to make this connection between the output of one adder instance to the input of the other. Note how similar this is to wiring devices by hand.

15N.3.8 But what about the inputs and outputs?

Astute readers (as you surely are) have probably noticed by now that we have been cagey about how the top level module inputs and outputs are connected to the outside world. That is because – it

15N.3.9 Verilog can tame active-lows

15N.3.9.1 But first, let's agree on some terminology

Suppose we want to draw gates that will:

Assert $\overline{\text{DOIT}}$ if:
- $\overline{\text{WRITE}}$, VALID and $\overline{\text{TIMED}}$ all are asserted, as long as neither $\overline{\text{PROTECTED}}$ nor $\overline{\text{BLOCK}}$ is asserted;
- or if either $\overline{\text{WILLY}}$ or NILLY is asserted.

The only hard part: agreeing on what we mean: Curiously enough, this exercise presents the only point in the course when students sometimes get *angry*, disgusted with us teachers for talking nonsense, or worse. It's as if we're suddenly reciting blasphemous and inconsistent remarks about the students' religions.

Why? Because the phrase, "$\overline{\text{WRITE}}$ is asserted" is ambiguous. We don't want to insist that our understanding of this phrase is the only reasonable one; we only want to ask you to join us in what we admit is only a *convention* – a way of understanding this phrase. We want you to understand it as we do; that way, we'll not misunderstand each other.

We take "$\overline{\text{WRITE}}$ is asserted" to mean that the signal named "$\overline{\text{WRITE}}$" is *low*. The *bar* over the signal's name here means that the signal is *active-low*.

[AoE §10.1.2A]

Indignant students sometimes protest that "$\overline{\text{WRITE}}$ is asserted" is to say that WRITE is false. If WRITE is false, it is low, and its complement, "$\overline{\text{WRITE}}$" is *high*.

Such a protest is not wrong; it just does not follow the convention that we are imposing – and which is widely recognized: many signals (in fact most on older devices) are *active-low*, and one indicates this active-low quality with a "bar" or asterisk. Such a signal normally is *born* active-low, rather than being the complement of an active-high signal of the same name.

We did write the word "WRITE" (without a bar over its name) to the left of the boundary and circle (often called a "bubble"), in Figure 15N.7. But we did not thereby mean to imply that any such signal *exists*. Probably it does not exist, and in those drawings there is no point indicated where one could apply a probe to test an active-high WRITE signal. Only the active-low "$\overline{\text{WRITE}}$" is available for probing.

In the case of "$\overline{\text{DOIT}}$," we used this active-low labeling because the label is placed to the left of the bubble, out in the world where the signals run and are available for probing. On the leftmost part of the diagram, in contrast, we used labels without bars (showing "WRITE," for instance), because a bubble intervenes between this fictitious point and the outside world where "$\overline{\text{WRITE}}$" travels.

Well, we have expended a lot of words to make a small point – but it is one on which we must agree, if we are to be able to talk to one another about signals that are active-low.

Figure 15N.7 shows an implementation of the logic that we described in words.

It's not *quite* as simple as "Let bubbles meet bubbles." For most of the signals shown in Fig. 15N.7, "bubbles meet bubbles," so that one can see inversions cancelling each other. But you shouldn't apply such a rule mechanically.

---

[19] For a sneak preview, see the full code for two-bit adder at `https://LAoE.link/FPGA/15N_TwoBitAdder.v`.

632 Introduction to Programmable Logic

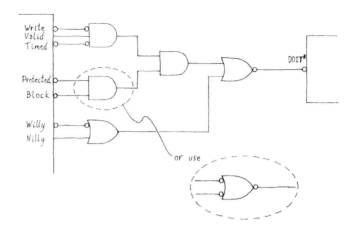

Figure 15N.7 Gates drawn to implement the logic described earlier.

The simple-minded rule does work when you are using gates to detect *assertion* of signals. It does *not* work – as we noted back in §14N.3.5 – when a gate is to detect *dis*-assertion. The leftmost two-input AND gate in the figure is used to detect such a *disassertion* of the signals $\overline{\text{PROTECTED}}$ and $\overline{\text{BLOCK}}$. So, in this case, bubbles do *not* meet bubbles. The mismatch, obvious to the eye, reveals at a glance that the job of this AND gate is to sense falsehood or disassertion.[20]

DeMorgan sometimes allows two equally good symbols: DeMorgan teaches that two alternative symbols always are available for drawing a logical function, and we have shown these alternatives in Fig. 15N.7 for the logic used for sensing disassertion or falsehood. Usually, one of the two deMorgan-equivalent symbols shows more clearly than the other what is going on in the circuit. You should make the choice that seems to you to represent best what your circuit is doing.

Once in a while, however – as in the case of the logic applied to $\overline{\text{PROTECTED}}$ and $\overline{\text{BLOCK}}$ – the two alternative symbols are equally good. The one we drew first senses that *both signals are false*, i.e., disasserted. The alternative form, shown lower, senses that *either signal is asserted* and uses that event to *disassert* the input to the next gate. The results are logically identical of course. Use whichever you prefer. The OR shape does the better job of reflecting the language of the design specification, "neither ...nor"... But the AND shape, sensing "both signals false," shows logic that may be easier to think about.

---

[20] Beware the Jabberwock of inconsistency. In the National Instruments datasheet for the 74HC16X counter, the clear and load signals are shown as active-high inside the device with a bubble connecting them to external pins. Fine, except that the pins are labeled "CLEAR" and "LOAD." This is **wrong**. Both functions are active-low and the names of the signals at the pins

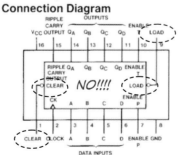

should be reflective of this to avoid errors. You will also find datasheets with both a bar over the internal signal name *and* a bubble leading to the outside world – again, this is wrong. Only one indication makes sense. Two implies a "double negative" leaving you unsure if the signal is active-high or active-low.

15N.3.9.2 Now, let Verilog simplify our lives

In addition to saving you hours pushing tiny wires into the breadboard (which you may instead spend typing HDL code into the computer) the logic compiler can diminish the nastiness of "active-low" signals like the ones above.

The technique that can make active-lows not so annoying is to create an active-high equivalent for each active-low signal: we define a signal that is the complement of each active-low signal.[21] Then one can write logic equations in a pure active-high world and stop fretting about the active levels of particular signals. Doing this not only eases the writing of equations; it also produces code that is more intelligible to a reader.

Figure 15N.8 is a simple example: a gate that ANDs together two signals, one active-high, the other active-low. The gate's output is active-low. It's easy enough to draw this logic.[22]

Figure 15N.8 Schematic of simple logic mixing active-high and active-low signals.

In Verilog, the equation for this logic can be written in either of two ways.

An ugly version of this logic: It is simpler, though uglier, to take each signal as it comes – active-high or active-low. But the result is a funny looking equation. Here is such a Verilog file:

```
module actlow_ugly_oct11(
    input a_bar,
    input b,
    output out_bar
    );

  // see how ugly things look when we use the mixture of active-high and
  // active-low signals:
  assign out_bar = !(!a_bar & b);
endmodule
```

The file begins with a listing of all signals, input and output. We have appended "_bar" to the names of the signals that are active-low simply to help us humans remember which ones are active-low. The symbol "!" (named "bang") indicates logical inversion (see also §15N.3.5).[23]

The equation to implement the AND function is studded with "bangs," and these make it hard to interpret. None of these bangs indicates that a signal is false or disasserted; all reflect no more than the active level of signals.

A prettier version of this logic: If we are willing to go to the trouble of setting up an active-high equivalent for each active-low signal (admittedly, a chore), our reward will be an equation that is easy to interpret.

---

[21] This newly defined signal exists only "on paper" (i.e., inside the synthesizer's memory) for the convenience of the human programmer; no extra hardware is required for the implementation.

[22] We allowed a logic compiler to draw this for us; a human would not have drawn that final inverter; a human would have placed a bubble at the output of the AND shape. (This was drawn by the Xylogic's development environment; unfortunately the synthesizer we use in the lab does not output schematics.)

[23] We are cheating a bit here. Logical inversion does the same thing as the bitwise inversion operator with scalar operands like these, but is *definitely not* the same with vector operands. Nevertheless, we use it here because it prints clearer on paper in this book. We suggest you do not repeat our bad example and instead get in the habit of using the bitwise operator (tilde) when you intend a bit operation, scalar or otherwise.

Here, we define a pair of active-high equivalents after listing the actual signals as in the earlier file; we do this for the one active-low *input* and for the active-low *output* (rhe odd line "wire..." tells the compiler what sort of signal our homemade "out" is):[24]

```
module actlow_pretty_oct11(
   input a_bar,
   input b,
   output out_bar
   );

   wire out;

   // Now flip all active-low signals, so we can treat
   // ALL signals as active-high as we write equations
   assign a = !a_bar;     // makes a temporary internal signal "a"
                          // never assigned to a pin

   assign out_bar = !out; // this output logic looks reversed but it's correct
                          // because the active-high "out" will be an input
                          // to this logic, used to generate the OUTPUT

   // then see how pretty the equation looks using the all-active-high
   // signal names

   assign out = a & b;
endmodule
```

The equation says what we understand it to mean – active levels apart: "make something happen if both of two input conditions are fulfilled."[25]

In short, the logic compiler allows us to keep separate two issues that are well kept separate:

1. What are the active levels – high or low? This is disposed of near the start of the file in the "pretty" version, above, by creating internal active-high versions of external active-low signals; and
2. What logical operations do we want to perform on the inputs? (This is the interesting part of the task).

If all this Verilog code is hard to digest on first reading, don't worry. We will continue to look closely at Verilog and its handling of active-low signals as we go along. It takes some getting used to.

---

[24] We are taking a shortcut here. The internal connection "a" is not defined, but the logic compiler assumes the undefined destination of an *assign* is a wire. This can be dangerous if you later mistype the variable name; the WebFPGA IDE will not give a warning, instead it will create a new signal (and your circuit will not work). You can avoid this problem by adding `default_nettype none` to your Verilog files and explicitly declaring the type of every signal.

[25] You programmers out there are shaking your heads, thinking "That can't work – you assigned out_bar *before* you set the value of out to a & b." Ah, but you are thinking *procedurally*! Here we are forced to write our Verilog sequentially but we are *defining hardware* so it does not matter in what order we make the connections. Imagine you are wiring this circuit using 74HCXX gates. It does not matter whether you first connect a and b to an AND gate and *then* connect the AND gate to an inverter, or if you connect the inverter first then the a and b inputs. Be sure to think of Verilog as a recipe for connecting logic elements before you test your circuit, like building a breadboard, not as a series of instructions to execute sequentially.

15N.4 Digital logic recap

- Digital provides good noise immunity.
- Noise immunity allows most of what's wonderful about digital:
  - *binary* coding of information permits electronics of extreme simplicity, thus small size and low cost;
  - noise immunity is fundamental to data storage, and transmission through adverse conditions; and
  - the fact that information is encoded permits processing to detect and correct errors (e.g., in digital music playback).
- Since noise immunity is fundamental, better noise immunity is important: 5 V *and* 3.3 V CMOS beats old TTL. For signal transmission, differential signaling is best, and works at the low voltages now common.
- Arrays of gates – PALs, PLDs and FPGAs – are taking over from small packages of just a few gates.
- A logic compiler, used to program such parts, normally takes care of minimization (optimization).
- Many digital control signals are *active-low* (for historical reasons). A logic compiler can hide the confusion that can result when active-high and active-low signals are mingled.
- Binary numbers: 2's-complement represents negative numbers. Simple scheme: MSB is negative, in 2's-comp. But signed/unsigned is a matter for humans to interpret; the binary representation does not reveal which is intended.

15N.5 AoE reading

Chapter 10 (Digital Logic):
- §10.1.4G Basic logic: hardware description language.
- §10.5.4 Programmable logic devices.

Chapter 11 (Programmable Logic Devices):
- §11.1 A brief history.
- §11.2 The hardware:
  - see end of 11.2.6 for the reason we use an online FPGA development tool.
- §11.4 Advice.

WebFPGA Docs at `https://LAoE.link/WebFPGA_Docs.html`
- Appendix D. Verilog Logic Operators.

15L Lab: Programmable Logic

15L.1 Introduction to FPGA combinational logic

Please check the errata page at https://LAoE.link/Installing_the_FPGA.html before you begin to see if there have been any updates or changes to the procedure below.

15L.1.1 Setting up the WebFPGA

The WebFPGA breakout board pairs the Lattice iCE40 FPGA with a 1 Mbit flash memory, a 16 MHz precision oscillator, an ARM Cortex-M4 microcontroller, and a 3.3 V voltage regulator along with several buttons, LEDs and a micro-USB connector, providing all the hardware necessary to use the FPGA.

You will need to supply 5 V power to the WebFPGA, either through the USB or directly to the breakout board. The USB 5 V input is connected to the onboard voltage regulator through a Schottky diode. We will connect the breadboard 5 V supply to the "5 V" pin on the WebFPGA through a Schottky diode as well: see Fig. 15L.3. That way, the higher voltage supply will power the board when both are connected without potentially damaging the other. The 5 V input is regulated down to 3.3 V by the voltage regulator to drive all the components except the Neopixel RGB LED (which runs off 5 V). The output of the voltage regulator is available at the pin labeled "3V3." It may be used to supply up to 100 mA to external devices.

Start by mounting the WebFPGA on the bottom row of the breadboard. You will leave it there even as you work on other labs that do not require the FPGA. We suggest mounting it on the bottom-left side of the breadboard so it is easy to wire to the logic slide switches, but make sure your microUSB cable clears the pushbutton resistors so you can program the WebFPGA.

Add wires to connect the three WebFPGA ground pins to breadboard ground and connect the breadboard +5 V supply rail to the "5 V" input pin on the WebFPGA through a 1N5817 Schottky diode. Be sure to bypass the +5 V supply to ground near the connection to the WebFPGA. Figure 15L.1 shows the WebFPGA installed with power and ground connected.

We are going to connect the breadboard LOGIC SWITCHES to eight I/O pins on the WebFPGA. Since voltages higher than 3.6 V can damage 3.3 V logic, check the breadboard "+V" variable supply and make sure it measures no more than 3.6 V before proceeding. We suggest anchoring the +V voltage adjustment knob in place with silicon adhesive or hot melt glue (both of which are easily removed later): see Fig. 15L.2.[1] As noted in the previous chapter, we also add a fixed 3.3 V supply to the breadboard: see online Chapter 14O at https://LAoE.link/LAoE_Chapter_14O.pdf. Also, make sure the voltage select slide switch for the LOGIC SWITCHES is set to "+V" (3.3 V). We put a small piece of tape over the upper half of this switch to keep it from being moved to the "+5" position.

---

[1] We are being cautious because very small movements of the adjustment knob causes large changes in the V+ output voltage. The WebFPGA is relatively expensive and voltages greater than 3.6 V may damage it.

15L.1 Introduction to FPGA combinational logic

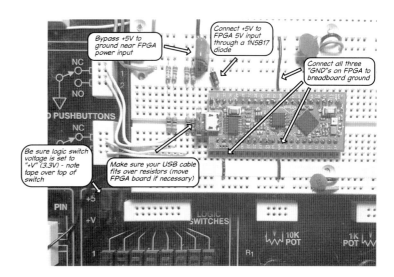

Figure 15L.1 Installing the WebFPGA on the breadboard.

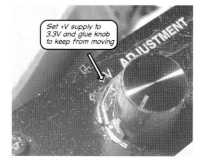

Figure 15L.2 V+ supply adjusted to 3.3 V with knob glued in position.

Next, connect the eight LOGIC SWITCHES to the WebFPGA. Switch S_1 connects to I/O pin 04, while switch S_8 connects to I/O pin 11: see Fig. 15L.3. This completes the initial WebFPGA installation.

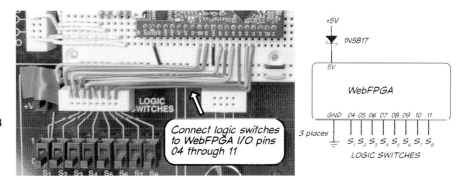

Figure 15L.3 Initial WebFPGA setup.

15L.1.2 Test your WebFPGA installation

With the breadboard power off, connect a microUSB cable between your computer and the WebFPGA. The green power LED should light. If it does not, check your wiring.

Set up the development environment as described in Chapter 15S.[2] Now open up the Web IDE at

---

[2] Unless you are using Linux, this consists of connecting an external LED to the FPGA through a current limiting resistor and opening a web page with a browser that supports WebUSB. Linux users have one additional step.

https://LAoE.link/WebFPGA_dashboard.html and select "Connect Device (F9)." Choose "Cascadia WebUSB FPGA Programmer" and press "Connect." Load the `WF\_neopixelx1\_fading_demo.v` example and click "Run Synthesis." It may take up to two minutes for synthesis to complete.[3] When you see "Bitstream loaded in browser. Ready to flash!," click "Flash Device." Once loaded, the Neopixel on the board should cycle through its colors. Disconnect the microUSB cable and power on the breadboard. The Neopixel should continue to display the RGB pattern. You are now ready to write Verilog code and program the FPGA.

15L.1.3 Get familiar with WebFPGA I/O

Follow the directions in §15S.1.3 to control first the onboard LED, then an external LED from the white user button. Make sure you understand how variables in your Verilog code are mapped to external pins on the WebFPGA through the `@MAP_IO` directive in the code comments.

15L.1.4 Create some simple combinational logic

Wire five unused WebFPGA I/O pins to breadboard LOGIC INDICATORS LEDs 1 through 5. It should not matter how the two level selection switches are set, but the best choices are "+V" and "CMOS."

Write Verilog code, run synthesis and flash the WebFPGA to implement the following combinational functions of logic switches S_7 and S_8 on the indicator LEDs:

LOGIC INDICATOR 1 = NOT S_7
LOGIC INDICATOR 2 = S_7 AND S_8
LOGIC INDICATOR 3 = S_7 NOR S_8
LOGIC INDICATOR 4 = S_7 XOR S_8
LOGIC INDICATOR 5 = S_7 AND (NOT S_8)

Test all [four] combinations of the input switches to check your logic functions.

15L.1.5 Build a multiplexer two ways

A multiplexer takes two inputs, A and B, and a control signal, `select`, and outputs A when `select` is low, and B when `select` is high. Here is the truth table for a one-bit mux:

select	A	B	out
0	0	0	0
0	0	1	0
0	1	0	1
0	1	1	1
1	0	0	0
1	0	1	1
1	1	0	0
1	1	1	1

Structural model: Build and test a one-bit multiplexer structurally in Verilog using gates.

---

[3] Although, if a bitstream already exists for a source file it will be immediately available. That is likely to be the case for the example files.

Behaviorial model: Build and test the multiplexer behaviorally using the Verilog conditional operator.

15L.1.6 Build some digital comparators

2-bit equality detector: Make a comparator that detects equality between two 2-bit numbers.

2-bit A > B detector: Make a comparator that detects when one of a pair of two bit number (call it A) is larger than the other (call it B). You need not make the circuit symmetrical: it need not detect $B > A$, only $A > B$ versus $A \leq B$.

15L.1.7 Build a more efficient adder

The two-bit full adder we used to explain Verilog hierarchy in §15N.3.7 consisted of two identical full adders for simplicity. The more efficient design you built out of gates in Lab §14L.4.2 used a half adder in the least significant position. Build a hierarchical two bit adder in the WebFPGA using both a full adder and a half adder as shown in Fig. 15L.4.

Use the following logic indicators for output.

LOGIC INDICATOR 6 = CARRY
LOGIC INDICATOR 7 = SUM1
LOGIC INDICATOR 8 = SUM0

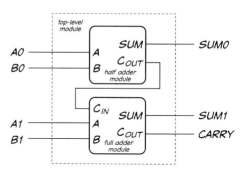

Figure 15L.4 Two-bit binary adder.

15L.2 Active-low design

You have been asked to design a voting system that allows three people (A, B, and C) to vote – each by pressing a *normally open* pushbutton if in favor of a proposal – and then lights one of four LEDs to indicate if none, one, two or three people voted *for* the proposition. You may use the N.O. outputs of the two pushbuttons (PB_1 and PB_2) already available on the breadboard but you will need to add an additional pushbutton for the third vote. (Does the added pushbutton need to be debounced as well?[4]) The output LEDs should be driven through MOSFETs as shown: see Fig. 15L.5.[5]

---

[4] No. Any switch bounce lasts, at most, a few milliseconds. No human will see such a short flicker in the output.
[5] If you are short on time you can connect the outputs to logic indicators instead and skip the paragraph asking you to build two of the circuits with 74HC devices.

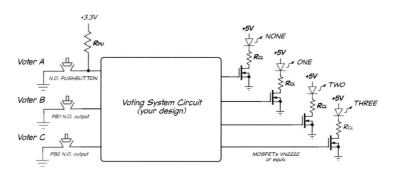

Figure 15L.5 Block diagram of voting system.

Design the circuit: Initially, you may use any type of gate with as many inputs (with or without inverting bubbles) as you like, as well as inverters. Please use SUM-of-PRODUCTS design techniques and be sure to use the form of the gate that best expresses what is happening (i.e., apply deMorgan's Theorem as necessary to make the function of the circuit clear).[6]

Build and test your NONE and THREE designs with 74HC logic devices: Add the third pushbutton, build the four output indicator LED circuits and use discrete gates to build **only** the two circuits to light the NONE LED if *no buttons* are pressed and the THREE LED if *all three pushbuttons* are pressed. Note that you will have to convert your "any gates allowed design" to only use actual devices available in 74HC logic. Be sure to document what you *actually* build.

Build the complete circuit using the WebFPGA: Now build the full voting circuit using the WebFPGA. Why is it ok to run the LEDs from +5 V when the WebFPGA can only tolerate 3.3 V max?[7]

15L.3 Solutions

Our solutions to the lab problems... but please try to do them on your own first.

   https://LAoE.link/FPGA/15L_CombLogic.v.
   https://LAoE.link/FPGA/15L_MuxTwoWays.v.
   https://LAoE.link/FPGA/15L_TwoBitComp.v.
   https://LAoE.link/FPGA/15L_TwoBitAGB.v.
   https://LAoE.link/FPGA/15L_TwoBitAdder.v.
   https://LAoE.link/FPGA/15L_Vote3.v.

---

[6] You may find it easier to draw the circuit for each output on a separate sheet. Otherwise, it gets very messy.
[7] The 3.3 V FPGA logic is isolated from the 5 V by the MOSFETs.

15S Supplementary Notes: The WebFPGA

15S.1 Introducing the WebFPGA

AoE §11.2.6

There are a number of [relatively] low-cost, small FPGA breakout boards that will work with the exercises in this book.[1] What sets the WebFPGA apart from the other choices is its development environment. To use an FPGA, you need a way to edit the Verilog code, synthesize it, preferably simulate its operation, define I/O pin connections, floorplan and place-and-route the internal cells and cell connections for optimal cell use and proper timing, then finally create the bitstream and download it to the device you are using. Each FPGA manufacturer offers its own tool set to accomplish these tasks.[2] One thing in common with all of these is that they are huge and difficult to learn. As *The Art of Electronics* points out:

> These software tools are not terribly easy to learn, and they have a distressing tendency to change with time. They are often afflicted with bugs, requiring skillful workarounds. It can be a full-time job keeping up.

In addition, they usually do not support all three major operating systems: Windows, macOS, and Linux. For example, the Lattice development tool for the iCE40 device we use is a 679 megabyte download and only available for Linux and Windows. It is not uncommon to spend more time learning the development environment than learning how to use the FPGA.

15S.1.1 A simplified development environment

To avoid these limitations, we have chosen the WebFPGA to create programmable logic circuits in the lab. This board is supported by a simple to use, no-installation, browser-based integrated development environment (IDE). This allows you to compile and download code to the FPGA from any browser that supports WebUSB (an API standard to allow control of USB devices from a browser). As of late 2022, current versions of Chrome, Opera, and Microsoft Edge are supported on Android, Chrome OS, Linux, macOS, and Windows version 10 and up.[3] With the WebFPGA, you can begin developing programmable logic systems in a few minutes without installing and learning a massive tool kit.[4]

While you can create and edit files in the WebFPGA IDE, it does not allow you to save files to disk so you may prefer to use an external editor and copy/paste your code to the IDE. We are fond of the free Notepad++ editor (`https://LAoe.link/notepad-plus-plus.html`) but there are a number of other excellent program editors available on the web.

---

[1] See online Chapter 15O at `https://LAoE.link/LAoE_Chapter_15O.pdf` for a list.
[2] There are also several open source tools for some FPGAs created by enterprising individuals who have reverse engineered the bitstream for certain devices. See, e.g., `https://LAoE.link/OS_Tools.html`.
[3] See `https://LAoE.link/WebUSB.html` for an up-to-date list of browser and versions that support WebUSB. Although not listed, we have successfully used the Brave browser (which is based on Chrome) with the WebFPGA IDE.
[4] If you prefer to use the open source tools or Lattice's iCEcube2, go right ahead. The only thing that changes from our directions is how you assign the top-level module ports to I/O pins.

Supplementary Notes: The WebFPGA

Linux users will need to install a udev rule in order for the user space to have read/write permissions to the board:[5]

```
# Store the file here:
#   /etc/udev/rules.d/60-webfpga.rules

SUBSYSTEM=="usb", ATTR{idVendor}=="16d0", ATTR{idProduct}=="0e6c", TAG+="uaccess"
```

Be sure to reload the rules with

`sudo udevadm control --reload-rules && sudo udevadm trigger`

15S.1.2 The hardware

The WebFPGA is built around the Lattice UltraPlus ICE40UP5K.[6] This FPGA is a 3.3 V device; all inputs and outputs must be kept between 0 and 3.3 V. However, the WebFPGA breakout board operates on USB power or 5 V as it includes a voltage regulator to power the components (and supplies 3.3 V out at up to 100 mA for external circuitry). The WebFPGA has a user pushbutton, a yellow user-controlled LED and a single Neopixel RGB LED for on-board I/O: see Fig. 15S.1. The board uses a standard micro-USB cable to download bitstreams to the onboard flash memory from your computer (or Android phone if you are truly hard core).

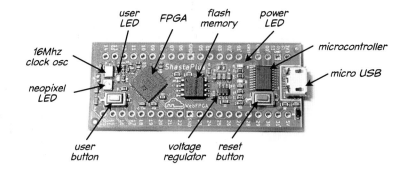

Figure 15S.1 WebFPGA with major components labeled.

Information on the WebFPGA is available at https://LAoE.link/WebFPGA.html. The schematic is at https://LAoE.link/WebFPGA_Schematic.html.

15S.1.3 Connecting Verilog signals to external devices

In any FPGA design environment, you need a way to tell the logic compiler which ports (i.e., the input and output signal names) in the top-level Verilog module should be connected to which external pins on the FPGA. For example, suppose we want to turn on the built-in yellow LED on the WebFPGA board when the on-board user pushbutton is pressed. If you look at the schematic for the board, you can see that the white (user) pushbutton is connected to signal IOT_51a which is pin 42 on the FPGA, and the cathode of the yellow LED is connected to signal IOT_42b, which is pin 31. The anode of the yellow LED is connected to +3.3 V through a current limiting resistor so the LED is on when that FPGA pin is low (active low).

In most development environments you would specify the FPGA pin names or numbers to connect to port names. In the WebFPGA environment, the on-board resources are assigned special port names which map the names to the correct pins. The available resources and their port names are:

---

[5] Download from https://LAoE.link/FPGA/60-webfpga.rules.
[6] Datasheet at https://LAoE.link/ICE40_Datasheet.pdf.

15S.1 Introducing the WebFPGA

Name	FPGA pin	notes
WF_LED	31	Yellow user LED (output - active low)
WF_CLK	35	16 MHz clock (input)
WF_BUTTON	42	User pushbutton (input - low when pressed)
WF_NEO	32	NEO Pixel LED (output)
WF_CPU1	11	These are four connections to the
WF_CPU2	12	on-board microcontroller used
WF_CPU3	13	for programming the FPGA. We will not be
WF_CPU4	14	using these in the lab.

We can use these predefined signal names to write a Verilog module to turn the yellow LED on and off as we press the pushbutton without having to reference any physical pin numbers at all:[7]

```
////////////////////////////////////////////////////////////////////
// Company: Learning the Art of Electronics
// Engineer: David Abrams
//
// Create Date:    2022-12-17
// Module Name:    fpga_top
// Project Name:   ButtonLED.v
// Target Device:  WebFPGA
// Description: Turns WebFPGA user (yellow) led on when user (white)
//              button pressed to demonstrate control of on-board
//              resources.
////////////////////////////////////////////////////////////////////
module fpga_top (
  input WF_BUTTON,
  output WF_LED
);
  assign WF_LED = WF_BUTTON;
endmodule
```

Note that only the top level module can connect ports to pins in the WebFPGA environment. The default top level module is the one named "fpga_top."

This works for the on-board resources, but the WebFPGA board is not very useful if you cannot connect signals to external devices. The board has 32 header pins labeled "00" through "31" connected to FPGA I/O pins, most of which can be used for general purpose external input or output: see Fig. 15S.2. Note, however, that a few of the pins have multiple uses. Unless you are short of I/O pins, it is best to use the 24 pins labeled "FPGA user I/O" and avoid using the dual-use pins (except G1, G3 and G6).[8]

Suppose, instead of the on-board LED, you decide to control an external LED connected between the WebFPGA +3.3 V output and the pin labeled "14" through a current limiting 470 Ω resistor as shown in Fig. 15S.3.

The pin labeled 14 on the WebFPGA board is connected to signal IOT_48b which is pin 36 on the FPGA. In most development environments, a separate Physical Constraints File (.pcf) or a pin configuration form would be used to describe the mapping between the FPGA IC's pin numbers and ports of the Verilog module. However, since the WebFPGA environment only accepts a single source

---

[7] Code available at https://LAoE.link/FPGA/15S_ButtonLED.v.
[8] Pins 00, 01, 02, and 31 are used for SPI communications with the on-board microcontroller and flash memory; pin 13 is connected to the 16 MHz clock; pins 16, 28, and 29 have open-drain outputs – see §15S.1.3.1 below to use them. While the pins labeled "XX/GX" have input clock buffers connected internally, they can be used for normal I/O (except for pin 13/G0, which is already in use for the 16 MHz clock input). Pin 3 does not work in an edge-triggered always@ sensitivity list but is a normal input or output otherwise.

644 Supplementary Notes: The WebFPGA

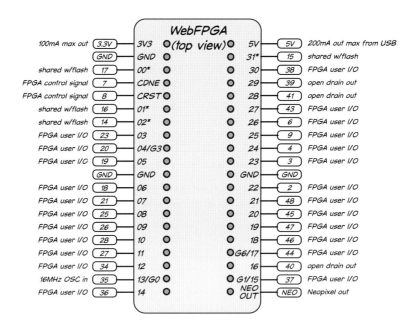

Figure 15S.2 WebFPGA header pin connections.

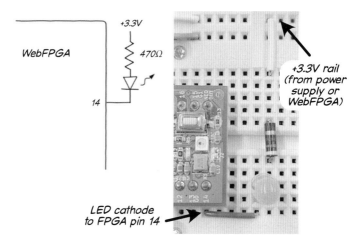

Figure 15S.3 External LED and resistor connected between +3.3 V and FPGA header pin marked "14."

file, it provides an alternative way to assign ports to pins. We will call this port "EXT_LED" in the Verilog module.[9] In a comment of your Verilog code, include the line:

```
// @MAP_IO EXT_LED 14
```

Note that the space between the double slash and the @ is required for the IDE to recognize the pin assignment.[10]

This tells the place-and-route step to assign the Verilog identifier "EXT_LED" to the FPGA pin connected *to the header pin labeled 14 on the breakout board*. You never have to worry about what FPGA IC pin the header pin is connected to. You can @MAP_IO to any pin number between 0 and 31.

---

[9] The WebFPGA documentation recommends assigning ports connected to external pins a name in ALL_CAPS. While not required, it is a good idea to distinguish externally connected ports.

[10] FPGA synthesis tools typically ignore comments entirely. WebFPGA has added a preprocessor to scan the input file comments for their special directives and generate a constraint file that the synthesis and place & route software used in their backend server understand.

However, only signals in the top level module can be mapped to external WebFPGA pins. Since the mapping command is a comment, your Verilog code will still synthesize with no changes in other development environments (although in that case you will have to use that environment's pin mapping mechanism to route the EXT_LED signal to the *correct FPGA pin number*, not the board pin label).

There is one other WebFPGA synthesizer directive:

```
// @FPGA_TOP ButtonExtLED
```

Normally, the compiler expects the top level module to be named "fpga_top." This directive allows you to substitute a more descriptive module name (a good idea). For example, here is the button controlled external LED module ready to synthesize using a descriptive module name:[11]

```
//////////////////////////////////////////////////////////////////////
// Company: Learning the Art of Electronics
// Engineer: David Abrams
//
// Create Date:    2021-12-17
// Module Name:    ButtonExtLED
// Project Name:   ButtonExtLED.v
// Target Device:  WebFPGA
// Description: Turns on an external led connected to pin 14 when user
//              button pressed to demonstrate use of @MAP_IO and
//              top level module name directives.
// Additional Comments: Uses LED w/ cathode connected to pin 14 of board
//                      and anode connected to +3.3V through 470 ohms.
//////////////////////////////////////////////////////////////////////

//*****************************************************************
// The comments in this block are directives to
// the WebFPGA environment.
//
// Directive to set name of top level module (default is fpga_top)
// @FPGA_TOP ButtonExtLED
//
// Directive to connect the named signal below
// to a specific WebFPGA output pin.
// @MAP_IO EXT_LED 14
//*****************************************************************

module ButtonExtLED (
  input WF_BUTTON,
  output EXT_LED
);

  assign EXT_LED = WF_BUTTON;  // both button and LED are active low

endmodule
```

15S.1.3.1 Using the open-drain I/O pins

Three of the header pins in Fig. 15S.2 are marked as "open drain out." These pins are open-drain N-Channel MOSFET outputs designed to sink high current to drive external RGB LEDs. If you run out of normal I/O pins, you can use these for input and output, although you will need to add a pullup resistor on undriven inputs (such as an SPST pushbutton) and outputs will only be able to sink current, not source it. You will also need to instantiate a special SB_IO_OD primitive as shown in the following example:[12]

---

[11] Code available at https://LAoE.link/FPGA/15S_ButtonExtLED.v.
[12] Code available at https://LAoE.link/FPGA/15S_OpenDrain.v.

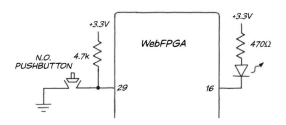

Figure 15S.4 External LED and button connected to open-drain I/O pins.

```
//////////////////////////////////////////////////////////////////
// Company: Learning the Art of Electronics
// Engineer: David Abrams
//
// Create Date:    2021-12-17
// Module Name:    fpga_top
// Project Name:   OpenDrain.v
// Target Device:  WebFPGA
// Description: Turns on external led connected when external button
//              pressed. Both LED and button are connected to open
//              drain I/O pins to demonstrate how to configure them
//              as normal I/O.
//////////////////////////////////////////////////////////////////


//****************************************************************
// @MAP_IO PIN_OUT 16   // open-drain connected to LED cathode
// @MAP_IO PIN_IN  29   // connected to SPST pushbutton switch
//****************************************************************

module fpga_top (
  input PIN_IN,
  output PIN_OUT
);

wire mywire;        // wire to connect pins together

SB_IO_OD #(
  .PIN_TYPE(6'b000001),   // configure as simple input, no output capabilities
  .NEG_TRIGGER(1'b0)
) pin_in_driver (
  .PACKAGEPIN(PIN_IN),    // the physical pin to use as input
  .DIN0(mywire)           // where to put data from that pin
);

SB_IO_OD #(                // open-drain IP instance
  .PIN_TYPE(6'b011001),   // configure as output
  .NEG_TRIGGER(1'b0)
) pin_out_driver (
  .PACKAGEPIN(PIN_OUT),   // connect to this physical pin
  .DOUT0(mywire)          // and output the state of this signal
);

endmodule
```

15S.2 Verilog resources

While the WebFPGA online development environment allows you to get started immediately synthesizing and testing Verilog code, it does have its limitations. Unlike most resident environments, it does not allow you to view a hierarchical expandable schematic/block diagram of your synthesized design. In fact, it does not provide any graphic output at all. In addition, there is no option to simulate your code, and error reporting is minimal.

We offer a solution to the latter two issues, some recommended resources for learning Verilog, as well as possible alternatives to the WebFPGA in the online Chapter 15O at `https://LAoE.link/LAoE_Chapter_15O.pdf`.

15W Worked Examples: Programmable Logic

15W.1 Digital comparators

In the last chapter's Worked Examples, we looked at several digital comparators constructed out of gates. We certainly could translate those to structural models in Verilog, but that misses the point. The advantage of an HDL is it frees us from truth tables, Boolean equations, and the need to implement the result with logic gates. Instead, we can describe the desired result behaviorally.

Detecting equality: A=B

Problem (Design a 2-bit comparator.)   Design a module to return true when the two 2-bit inputs are identical.

Solution (Let Verilog do the work.)   In Verilog this is merely a matter of using a conditional operator:

```
/////////////////////////////////////
module equals_comparator2(
    input [1:0] a,
    input [1:0] b,
    output aequalsb
    );

    assign aequalsb = (a == b) ? true : false;

endmodule
```

Detecting inequality: A>B

Problem (Design a 2-bit comparator.)   Design a module to return true when the 2-bit number A is greater than the 2-bit number, B.

Solution (Verilog makes it easy again.)   Just change the conditional operator you are using from "==" to ">".

```
/////////////////////////////////////
module gt_comparator2(
    input [1:0] a,
    input [1:0] b,
    output agtb
    );

    assign agtb = (a > b) ? true : false;

endmodule
```

Verilog easily expands the design

Designing a magnitude comparator for more than two bits would be a chore if done by hand. It's embarrassingly easy with Verilog. Just change the vector size in the definitions of *a* and *b* to make them 8-bit rather than 2-bit quantities by replacing "[1:0]" in the module definition to "[7:0]" in two places.[1]

```
//////////////////////////////////////////
module gt_comparator8(
    input [7:0] a,
    input [7:0] b,
    output agtb
    );

  assign agtb = (a > b) ? true : false;

endmodule
```

Expanding in other senses would be straightforward, too: adding other outputs such as *A=B* and *A<B* and *A≥B*. We won't do that. But you may be getting a sense that you quite easily could get Verilog to do this for you.

---

[1] Later, we will see how to semi-automate this process with the "parameter" construct.

15O Online Content: Programmable Logic

15O.1   Online FPGA resources

15O.2   Alternatives to the WebFPGA

Available at `https://LAoE.link/LAoE_Chapter_15O.pdf`.

16N Flip-Flops

Contents

16N.1	**Sequential circuits generally, and flip-flops**		**651**
	16N.1.1	The simplest flip-flop: the "latch," made from a few transistors	652
	16N.1.2	The simplest flip-flop again: the latch made with two gates	652
	16N.1.3	Switch debouncers	653
	16N.1.4	But SR usually is not good enough	654
	16N.1.5	Edge-triggering	655
	16N.1.6	Flip-flop characteristics	658
16N.2	**Applications: more debouncers**		**661**
	16N.2.1	D-flop and slow clock	661
	16N.2.2	Positive-feedback debouncer	662
16N.3	**Another flop application: shift-register**		**662**
16N.4	**Flip-flops in Verilog**		**663**
	16N.4.1	The simplest: an SR flip-flop	663
	16N.4.2	A Verilog D flip-flop	665
	16N.4.3	A warning about multiple always blocks	667
16N.5	**AoE reading**		**667**

Why?

In this chapter we meet circuits that remember states of binary signals. Such circuits can be more versatile than the merely *combinational* circuits that we met earlier. Computers and most of the other digital devices we rely on would not be possible without a way to remember past events. For example, it is hard to increment a count if you cannot remember what the current count value is.

16N.1 Sequential circuits generally, and flip-flops

AoE §10.1.4A

The combinational circuits that we looked at in the past few chapters "live in the present," like the resistive circuits we met discussing analog circuits on the first day. *Sequential circuits* offer a contrasting behavior that neatly parallels the contrast we saw between resistive and RC circuits: *RC* circuits care about the *past*, and thus offer behaviors that are more complex and more interesting than those of purely resistive circuits (§2S.2, Fig. 2S.11). The same is true of sequential digital circuits in contrast to combinational.

If combinational logic were the whole of digital electronics, the digital part of this course would now be just about finished. You could close the book and congratulate yourself.

But, no: the interesting part of digital lies in sequential circuits. These permit building counters, memories, processors – and computers.

Flip-Flops

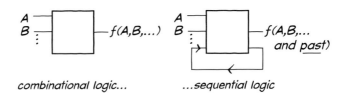

Figure 16N.1 Combinational versus sequential logic circuits: block diagrams contrasted.

Flip-flops are easy to understand. A harder question arises, however, when we look at flops in detail: why are *clocked* circuits useful? We will work our way from primitive bistable circuits toward a good *clocked* flip-flop, and along the way we will try to see why such a device is preferable to the simpler design.

16N.1.1 The simplest flip-flop: the "latch," made from a few transistors

The feedback shown in the block diagram of the sequential logic in Fig. 16N.1 appears in Fig. 16N.2 on the right as the link between the output of each MOSFET transistor switch and the gate controlling the other switch. This is circuit is *bistable*; that is it will will lock itself into one of two "states:" one or the other switch will be held ON while the other is held OFF.

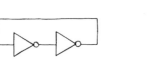
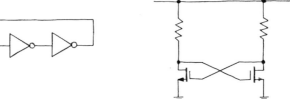

AoE §14.4.3; compare Fig. 14.20

Figure 16N.2 Simplest flip-flop: cross-coupled inverters.

At first glance, this circuit seems to be quite useless: a circuit that is stuck forever – and, even worse, unpredictable. But the circuit is not really stuck, because it is easy to force it to a desired state. If you turn OFF the transistor that is ON for a moment (by connecting the gate of the ON transistor to ground), then the ON switch turns off and, as a result, the switch formerly OFF turns ON. Thus the momentary grounding of an FET's gate can *flip* the circuit; doing the same to the other transistor can *flop* it back to its other state.

The name flip-flop provides a charmingly simple description of the circuit's behavior. (Engineers know how to speak simply; more hifalutin' people like to obscure the circuit's behavior by calling it a "bistable multivibrator" (!)).

This transistor latch can store one bit of information; in an integrated version, the pullup resistors would be implemented with transistors, as shown in Fig. 16N.3. Two more transistors would permit both writing and reading (that is, *writing* by forcing the cross-coupled transistors to the desired state, *reading* by simply connecting the Qs to the outside world).

16N.1.2 The simplest flip-flop again: the latch made with two gates

Figure 16N.4 shows the cross-coupled latch done with gates rather than plain transistors. This is the version you will build in lab. This circuit is commonly called a *set-reset flip-flop* or *SR flip-flop*. Such a circuit is also known as a *latch* because it saves (or latches) a momentary signal.

AoE §10.4.1

The bubbles on the inputs of the NAND gates remind us that these gates will respond to *lows*: it is

16N.1 Sequential circuits generally, and flip-flops

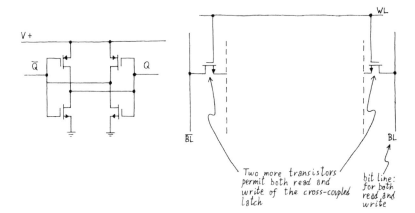

Figure 16N.3
Simplest flip-flop as applied to store one bit in a static memory (RAM).

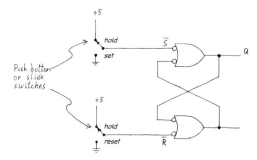

Figure 16N.4 Cross-coupled NANDs form a simple SR latch.

a *low* at the $\overline{S}$ input that will "set" the flop (that is, will send Q high); a low at the $\overline{R}$ input will "reset" the flop (that is, will send Q low).[1]

As was true for the equivalent latches built with transistors, the useful property of the circuit is its ability to *hold* the state into which it was forced once the forcing signal is removed. What input levels, then, let it perform this *holding* task?[2] (Those input levels might be said to define the circuit's resting state.) Circuits like this, where the outputs can change anytime the inputs change, are known as *asynchronous* circuits.

This device works, but is hard to design with. It is useful to *debounce* a switch; but nearly useless for most other sequential circuits, in this simplest, barebones form.

16N.1.3 Switch debouncers

AoE §10.4.1A

A common property of mechanical switches is *switch bounce*. When a switch moves from one position to another, the metal contacts hit with enough force to cause them to move apart and break contact, then come back together to make contact again. This can occur several times over a few hundred microseconds or so. If the switch is controlling an LED viewed by human, bounce is a non-issue. However, if the circuit connected to the switch responds to *edges*, i.e., transitions from low to high

---

[1] The name we provide to *signals*, in this case $\overline{S}$ and $\overline{R}$, reflect their active state, hence the overbars. The names we apply to controls like the switches reflect their function. In this case moving the upper switch to the ground position *sets* the output Q high so we label that switch position "set." We will continue to apply this distinction between signal names and functional labels to future circuit schematics as well.

[2] The input combination for "holding" or "remembering" the last state into which it was pushed, is 11: high on both inputs, disasserting both $\overline{S}$ and $\overline{R}$.

or visa-versa as many sequential circuits do, multiple transitions can be a problem. Thus the need to debounce mechanical switches.

There are several ways to debounce a switch. The SR provides one way – for *double-throw* switches only (not for the simpler sort of switch that simply makes or breaks contact between two points). Figure 16N.5 shows an SR latch wired as debouncer, and waveforms at S∗ and R∗, and at the debounced output. It shows the SR flop's ability to *remember* the last state into which it was forced. When S∗ rises to +5, and bounces low a few times, Q does not change in response to this bounce; the bounce simply drives the flop alternately into SET and REMEMBER modes. When R∗ first goes low, it forces Q low; when R∗ bounces high, Q does not change in response, because the flop simply remembers the low state into which it was forced. Again, it alternates between RESET and REMEMBER modes.

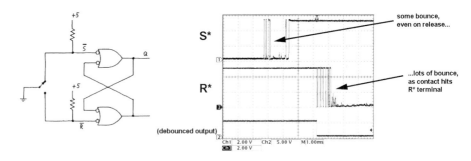

Figure 16N.5 Switch bounce: first hit sets or resets flop: bounce does not appear at output.

16N.1.4 But SR usually is not good enough

But in other settings this flop would be a pain in the designer's neck. To appreciate the difficulty, imagine the circuit in Fig. 16N.6 made of such primitive latches and including some feedback. Imagine trying to predict the circuit's behavior for all possible input combinations and histories. Painful.

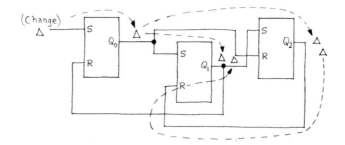

Figure 16N.6 Example meant to suggest that asynchronous circuits are hard to analyze and/or design.

Designers wanted a device that would let them worry less about what went on at all times in their circuits. They wanted to be able to get away with letting several signals change in uncertain sequence, for part of the time. Figure 16N.7 shows the basic scheme they had in mind. This permits less perfect circuit behavior without alarming results.[3]

The *gated SR latch*, or *transparent latch* takes a step in the right direction. This circuit is just the NAND latch plus an input stage that can make the latch indifferent to signals at S and R: the input stage works like a camera's *shutter*.

AoE §10.4.2

---

[3] Does this waveform remind you of the average undergraduate's week, except that a student's "do right" duty cycle may be lower: perhaps a narrow pulse on Sunday evening? If so, the student's week is more like the timing of the *edge-triggered* flop of §16N.1.5.

16N.1 Sequential circuits generally, and flip-flops

Figure 16N.7 Relaxing circuit requirements somewhat: a designer's goal.

The two-input SR version shown on the left-hand side of Fig. 16N.8 normally is simplified to its D form shown on the right. This single input is more convenient and eliminates the problematic input case for the SR, in which *both* inputs are asserted.

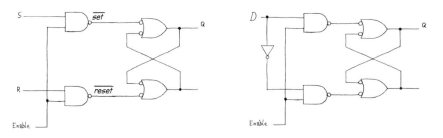

Figure 16N.8 Clocked or gated SR latch (left), and the more useful transparent D latch (right).

The transparent latch's behavior, shown in Fig. 16N.9, achieves more or less what we wanted; but not quite. The trouble is that it's still hard to design with this circuit. Consider what sort of clock signal you would want, to avoid problems with feedback. You would need an ENABLE pulse of just the correct width: wide enough to update all the flops, but not so wide that an output change would have time to wrap around and call for a second state change. Ticklish.

Figure 16N.9 Transparent latch moves in the correct direction: now we can stand screwy input levels part of the time.

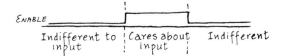

Because the simple NAND latch is so hard to work with, the practical flip-flops that people actually use nearly always are more complex devices that are *edge-triggered*.

16N.1.5 Edge-triggering

AoE §10.4.2A

Edge-triggered flip-flops care about the level of their inputs only during a short time just *before* (and in some rare cases after) the transition ("edge") of the timing signal called *clock*: see Fig. 16N.10.

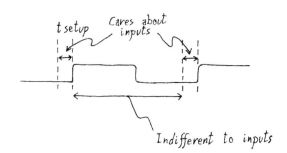

Figure 16N.10 Edge-triggering relaxes design requirements.

An older design, the JK flip-flop, called by the nasty name, *master–slave*, sometimes behaved nastily as well and was generally replaced by the edge-trigger circuit in Fig. 16N.11.[4] The master–slave offers the single virtue that it is easy to understand, but we will not examine it here. (If you are curious to meet this circuit, see AoE §10.4.2.)

The behavior called *edge-triggering* may sound simple, but it usually takes people a longish time to take it seriously. Apparently the idea violates intuition: the flop acts on what happened *before* it was clocked, not after. How is this possible? (*Hint:* you have seen something a lot like it on your scopes, which can show you the waveform as it was a short time *before* the *trigger* event. How is *that* magic done?[5])

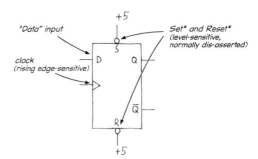

Figure 16N.11 74HC74 edge-triggered D flip-flop.

D (for data) flip-flop: In this device, the D input value just before the rising edge of the clock (during the *setup time*) is sampled and the Q output is set to that value shortly after the clock edge (the *propagation delay*). Figure 16N.12 shows its crucial timing characteristics. A D-flop only *saves* information: it does not transform it. But that simple service is enormously useful.

AoE §10.4.2C

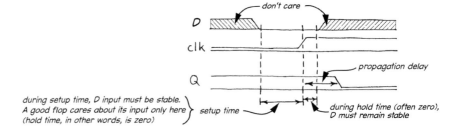

Figure 16N.12 D-flop timing: case where D is low, showing setup time, hold time, and propagation delay; hold time usually zero.

Digression on the signal name "clock": The name "clock" for the updating signal introduced in §16N.1.5 is potentially misleading. It suggests, to a thoughtful reader, several notions that do not apply:

- the term suggests something that keeps track of, measures, or accumulates time;
- or, even short of that, it suggests at least something that *ticks* regularly: a sort of strobe or heartbeat.

Neither notion applies to the *clock* of a flip-flop. (A clock may be – and fairly often *is* – a steady square wave; but it need not be.[6]) A clock is no more than a trigger signal that determines *when* the flip-flop updates its memory.

---

[4] Some master–slave JK flops showed a pathology called "ones-catching": they were sensitive to events that occurred while the clock was at a single level, high, rather than being responsive only to a *change* in clock level.
[5] An answer to this riddle appears in §16N.1.6, especially Fig. 16N.23.
[6] It is not a steady wave, for example, when the clock signal comes from a pushbutton. In a commonly used circuit, a user presses a button which applies a rising edge to a D flip-flop whose D input is tied high, capturing the press and starting some process. When the process is complete, the system resets the flip-flop.

16N.1 Sequential circuits generally, and flip-flops

A better word for the signal might be "update." But we are stuck with the term "clock." Just get used to disbelieving its metaphor. The engineers who named it were not competent poets.

Triggering on rising- versus falling-edge: The D-flop shown on the left in Fig. 16N.13 responds to a *rising* edge. Some flops respond to a *falling* edge instead. A flop never responds to *both* edges.[7] As with gates, the default assumption is that the clock is *active high*. An inversion bubble indicates active low: a *falling edge* clock.

Figure 16N.13 Clock edges: rising or falling edge.

Edge- versus Level-sensitive flops and inputs: Some flip-flops or flop functions respond to a clock-like signal, but not to the *edge* itself; that is, the device does not lock out further changes that occur after the level begins. The **transparent** latch is such a device. We met this circuit back in Fig. 16N.8.

Examples of level-sensitive inputs:

- A "transparent latch," like those used in some displays, such as the nifty but largely obsolete Hewlett Packard 5082-7300 series LED display with integrated latch and decoder driver shown in Fig. 16N.14.[8]

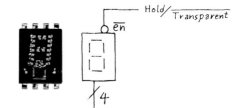

Figure 16N.14 Level-sensitive transparent latch: display follows 4-bit input while $\overline{en}$ is low; latched when high.

- The latches included on the AD558 Digital-to-Analog Converter, which you will meet in §20L.1.1, work as shown in Fig. 16N.15. The transparent behavior allows one to ignore the input register, getting a "live" image of the digital input, simply by tying the active-low enable input low.[9] This makes it possible to use the device with a microcomputer, which requires the input data to be latched, or used stand-alone, which typically does not.
- Reset/set inputs on a flip-flop. The $\overline{\text{RESET}}$ (sometimes labeled $\overline{\text{CLEAR}}$) and the $\overline{\text{SET}}$ (sometimes labeled $\overline{\text{PRESET}}$) signals reach into the D-flop's *output latch*, and prevail over the fancy edge-trigger circuit that precedes the out latch; see Fig. 16N.16. These so-called "jam type" inputs take effect at once (asynchronously).[10]

AoE §10.4.2A

---

[7] Well, *hardly* ever. Flip-flops using both edges seem to be confined to programmable logic and microcomputers.
[8] See datasheet at https://LAoE.link/HP_Display.html.
[9] Well, "inputs" plural. Both the $\overline{CE}$ and $\overline{CS}$ inputs must be tied low to put the latch in transparent mode.
[10] You might be wondering what happens if you assert both set and reset simultaneously? That tends to be implementation dependent and may vary between manufacturers, even for the same device part number. It is worth following the guidance of the doctor whose patient complained "Doc, it hurts when I do this." The response: "Well, don't do it."

Flip-Flops

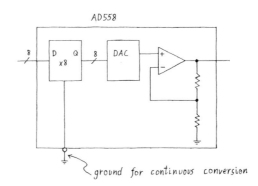

Figure 16N.15 Transparent latch on DAC allows one to use the DAC as continuous converter, if necessary.

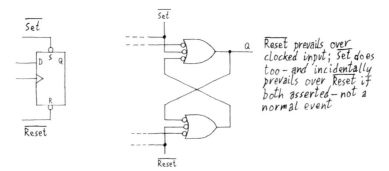

Figure 16N.16 Reset* and Set* on 74xx74 D-flop.

This set-or-reset scheme is available on many flip-flops, but not all – the quad 74HC175 includes a common clear but not a set – and is not quite universal on counters (although they almost always have a way to set the counter to a known value). But a reset is never treated as a clock itself – even in the cases (called "synchronous clear") where the action of resetting "waits for the clock." The edge-triggered behavior is reserved for *clocks*, with only the rarest exceptions (like a microprocessor's interrupt-request lines).[11]

Edge-triggering usually works better than level-action (that is, makes a designer's tasks easier); edge-triggering therefore is much the more common configuration. In addition, it makes fully *synchronous* systems possible, ensuring that all signals are in a stable state a known delay after the clock edge.

16N.1.6 Flip-flop characteristics

It is possible to design sequential logic either with clocks ("synchronous") or without ("asynchronous"). Yet almost every sequential circuit you will see is synchronous. Why? Because "breaking the feedback path" eases design and analysis of sequential circuits: see Fig. 16N.17 for example. Both these circuits are intended to toggle the output state under an input signal's control (EN for the circuit on the left and clk for the one on the right). The feedback circuit on the left that we built using a transparent latch does not work; it oscillates at a bit more than 60 MHz: see Fig. 16N.18. The edge-triggered self-contradictor, on the right in Fig. 16N.17, is well behaved, as you will see in the lab.

AoE §10.4.2C

The beauty of edge-triggered, synchronous circuits: The *clock* edge is a knife cleanly slicing between *causes* of changes – all to the left of the clock edge, in the timing diagram – and the *results* of changes, all to the right of the clock edge. The idea is illustrated in Fig. 16N.19.

---

[11] These INT* lines are not quite true edge-triggered inputs, but "pseudo edge-sensitive;" they respond to a transition, detected as a difference between levels before and after a processor clock. Their behavior is, however, very close to edge-sensitive.

16N.1 Sequential circuits generally, and flip-flops

In saying this, we assume zero *hold* time, though this specification is not universal. Different manufacturers of one part, the 74HC74 for example, made different choices: some provide zero hold time (STMicrosystems, and Texas Instruments); others specify non-zero hold time (3 ns) (ON Semiconductor and NXP/Philips) – see Fig. 16N.20.

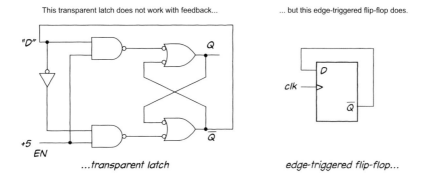

Figure 16N.17 Transparent latch can be unstable. Clocked device (edge-triggered) makes feedback harmless: instability is impossible.

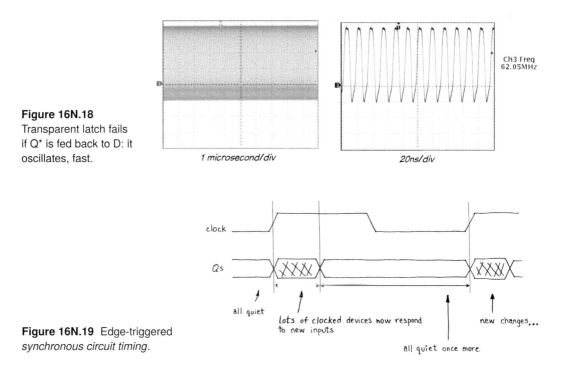

Figure 16N.18 Transparent latch fails if Q* is fed back to D: it oscillates, fast.

Figure 16N.19 Edge-triggered *synchronous circuit timing.*

There is nothing magical about achieving $t_h = 0$: the IC designers simply adjust the relative internal delay paths on the *data* and *clock* lines to shift the "window" of time during which data levels matter (called "aperture" in some other settings, such as analog–digital conversion). Other things equal, sliding the window to the right to introduce hold time, shortens set-up time. But zero hold-time is desirable; keeps timing simple and worry-free, whereas a non-zero hold-time obliges one to specify and worry about such exotica as *minimum* propagation delay, a characteristic not normally shown on datasheets.

660 Flip-Flops

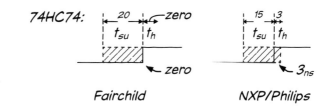

Figure 16N.20 An IC designer can trade hold-time against set-up time: two 74HC74 designs.

A particular example: The circuit in Fig. 16N.21 works fine, as we noted back in §16N.1.6 and as you will demonstrate in Lab 16L.2.2 – unless you try to push the clock speed very high. In that case, trouble reappears: see Fig. 16N.22.

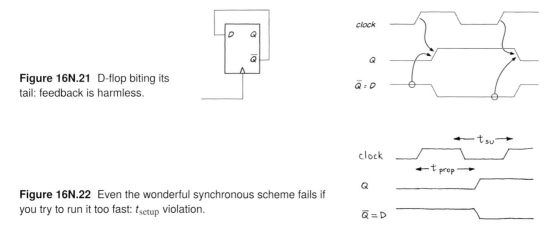

Figure 16N.21 D-flop biting its tail: feedback is harmless.

Figure 16N.22 Even the wonderful synchronous scheme fails if you try to run it too fast: t_{setup} violation.

When it's clocked too fast this fails, because D changes during t_{setup}. This produces an unpredictable output. The circuit may even hang up, refusing to make up its mind for a strangely long time (a device so hung up is called "metastable;" see §17N.1.5 and AoE waveforms at §10.4.2D).

Set-up time: "Set-up time" sounds like a technical detail, but it's a concept you need in order to get timing problems right.

AoE §10.4.2A

Detailed view (in the special case of the 74HC74 D-flop): **Set-up time** is the time required for a change of D level to work its way into the flop – to the point labeled with a star symbol in Fig. 16N.23. Incidentally, while you're looking at the guts of the '74 flop you can gather from this figure how the "edge-trigger" effect works.

- Edge-triggered (positive or *rising* edge, in the present case – and in most flip-flops) means that information *enters the flop when the clock rises*. The diagram shows that only when CLOCK rises to a logic high can the output SR latch be changed – Set or Reset, depending on which of its two inputs, C_1 or C_2, is forced low.
- Edge-triggered also means that information *does not enter the flop after the clock has risen*. This lockout occurs because the C_1 and C_2 signals are fed back; one or the other of those two inputs to the SR latch will be low, and that signal wrapped back locks this low in place (by preventing a low at the other C), regardless of later changes in the level of D.

The new information needs to work its way through a "pipeline" that is two gates deep in order to set up the flop to act properly on a clock. You might guess that this process could take as long as two gate delays, around 20–25 ns, and that is in fact about the value specified for the 74HC74 flop's setup

16N.2 Applications: more debouncers

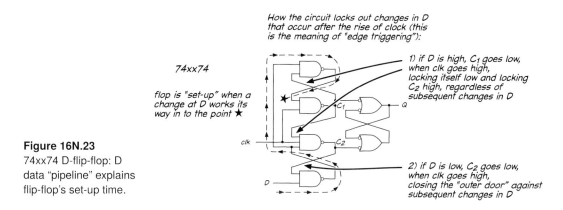

Figure 16N.23 74xx74 D-flip-flop: D data "pipeline" explains flip-flop's set-up time.

time. (In fact, internal gate delays are less than the delay of a packaged gate; but the scale remains about right.)

16N.2 Applications: more debouncers

16N.2.1 D-flop and slow clock

A D-flop can debounce a switch if you have a relatively slow clock signal available. The strategy is to make sure that the flop cannot be clocked twice during bounce. If it cannot, then the output – Q – cannot double back.

Unlike the SR debouncer, this method introduces a delay between switch-closing and output. Occasionally that matters (for example, in the reaction timer described in §21L.2.2); usually it doesn't. This debouncing method has the virtue of working with the simpler, cheaper switch: the single-pole type (SPST) typically available in pushbuttons. Fig. 16N.24 shows the dead-simple debounce circuit.

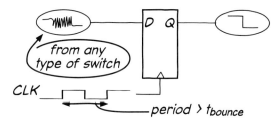

Figure 16N.24 Another debouncer: D-flop and slow clock.

In Figs. 16N.25 and 16N.26 are some waveforms making the point that the clock period must exceed bounce time. When the clock runs too fast, as in Fig. 16N.25, the flop *passes* the bounce to the output: not good.

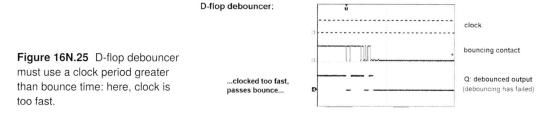

Figure 16N.25 D-flop debouncer must use a clock period greater than bounce time: here, clock is too fast.

A slower clock, as in Fig. 16N.26 does the job. The main drawback of this debounce method – in

662 Flip-Flops

addition to the uncertain *delay* – is the need to generate the slow-clock signal. If you need to debounce a single switch, it is not an appealing method. If you need to debounce several lines, the slow-clocked flop method makes sense. In any case, you will want to ensure a greater margin between the expected bounce time and the clock period than we show here.

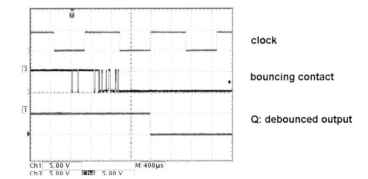

Figure 16N.26 D-flop debouncer must use a clock period greater than bounce time: here, period *is* (barely) long enough.

16N.2.2 Positive-feedback debouncer

The simplest debouncer, for a double-throw switch ("SPDT") is just a non-inverting CMOS gate with feedback to create an asynchronous bi-stable circuit: see Fig. 16N.27. At first glance, this may look too simple to work – but work it does because when the switch contact bounces, it does not bounce back far enough to touch the other contact but merely *opens the circuit*. The key notion is that the gate input *never floats*; the gate input is always driven by the gate output, even when the switch is in transition, not touching either +5 or ground.

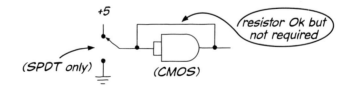

Figure 16N.27 Another debouncer: positive-feedback with CMOS gate.

16N.3 Another flop application: shift-register

AoE §10.5.3

The circuit in Fig. 16N.28 is about as simple a flop circuit as one could imagine: just a sort of daisy chain.

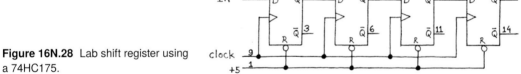

Figure 16N.28 Lab shift register using a 74HC175.

A shift-register generates predictable, orderly delay; it shifts a signal *in time*; it can convert a *serial* stream of data into *parallel* form, or vice versa. Serial to parallel is the easy direction: see Fig. 16N.29.

Such a circuit is useful because it permits using a small number of lines to transmit many bits of information (the minimum number of lines for a simple shift register would be two: data and clock; some schemes – more complex – can omit the clock).[12] Converting from parallel to serial is not quite so simple. It requires *multiplexer* logic feeding each D (except the first): one must be able first to steer parallel data into each D "loading" the shift register, then to revert to the normal daisy-chained shift-register wiring in order to shift out the serial stream.

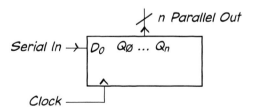

Figure 16N.29 Shift-register can convert data from serial form into parallel.

The shift register is *synchronous*. A synchronous circuit sends a common clock to all of its flip-flops; so they all change at the same time.[13] This is the sort of design one usually strives for. (In the next chapter we will compare this to an *asynchronous* sequential circuit – the ripple counter.)

16N.4 Flip-flops in Verilog

Code available at https://LAoE.link/FPGA/16N_FlipFlops.v.

16N.4.1 The simplest: a SR flip-flop

A set-reset flip-flop can be implemented using discrete gate packages either with cross-coupled gates or using a D flip-flop with asynchronous set and clear inputs: see Fig. 16N.30.

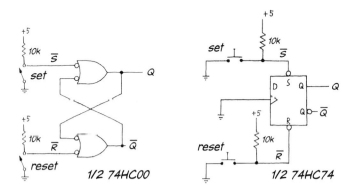

Figure 16N.30 SR flip-flops made with NAND gates and using a D flip-flop IC.

This device can be designed in Verilog in either of two ways as well, more or less mimicking the two circuits sketched in Fig. 16N.30:

- one design literally implements the cross-coupled NAND logic, as a combinational circuit; and
- the other uses the Verilog `always` block construct.

---

[12] We will glimpse these clock-less serial protocols later in the course: we will see it in the traditional RS232 serial scheme (§26N.1.1), and in the more recent USB (which, however, uses two data lines because it uses differential signaling for good noise immunity at modest voltages).

[13] Except to the small extent that propagation delays of the several flops may differ.

Cross-coupled NAND implementation: This design is direct translation of the NAND SR latch design. It makes Verilog nervous, because it involves a "loop," or a self-contradiction. Some Verilog compilers may give a warning similar to "the following signal(s) form a combinational loop: q_bar, q." But it works.[14]

```
// simple SR flip-flop using combinational logic
module sr_ff(
  input s_bar,
  input r_bar,
  output q
);

  wire q_bar;
  assign q = ~s_bar | ~q_bar;
  assign q_bar = ~r_bar | ~q;
endmodule
```

Sequential logic implementation: Verilog uses the *always* procedural block to define sequential logic.[15] The form of an always block is:[16]

```
always @ (<sensitivity list>)    // spaces are optional
begin
  <statements>                   // implement your logic here
end
```

Unlike the assign statement, which continuously updates the destination signal as the righthand expression changes (i.e., it creates a wire connecting the expressions on either side of the "=" sign), the always block only causes the statements between "begin" and "end" to evaluate when a sensitivity is triggered, otherwise the output does not change.

The sensitivity list can be either a signal, in which case the *level* of the signal triggers the sensitivity, or a condition (which is triggered by either a rising or falling signal *edge*). A signal will cause the always block statements to be continuously evaluated as long as the signal is in the triggered state. A condition only evaluates the always block once after the edge occurs. You can combine sensitivities of the same type (signal or condition) with "or" (case sensitive) to create more complex systems.[17]

The SR flip-flop is asynchronous, so we will use the levels of the set and reset inputs to cause the output to change.[18]

```
// SR flip-flop using sequential always block
module sr_ff(
  input s_bar,
  input r_bar,
  output reg q
);

  always @(!s_bar or !r_bar)
```

---

[14] A simulation of this device is available at https://LAoE.link/SRFF_NAND.html.

[15] The always block may also be used for combinational logic. It is particularly useful when you need to use a conditional operator other than "?".

[16] You should consider the begin and end statements as the equivalent of curly braces in C. Like the curly braces, they are optional if you only have a single statement following the conditional. We take advantage of this in print to make the examples clearer and shorter, but you may want to get in the habit of always including begin and end, even with a single statement, because sooner or later you will add a second statement to a conditional but forget to add the grouping and then spend hours trying to figure out why your design does not work. Ask us how we know.

[17] WebFPGA pin 3 does not work in an edge-triggered sensitivity list but is a normal input or output otherwise. We use pin 30 for our external clock input.

[18] A simulation of this version is available at https://LAoE.link/SRFF_always.html.

```
    begin
      if (!s_bar)
        q <= 1'b1;
      else if (!r_bar)
        q <= 1'b0;
    end
endmodule
```

- We have to declare the output, q, as a `reg` because it is assigned its value in the `always` block. "Reg" does not mean hardware register or flip-flop, it means the signal holds its value until it is overwritten. It is a requirement that all signals or variables assigned in an always block be declared as `reg`.
- The sensitivity list monitors the listed signals. In a level-sensitive list, as in the example above, whenever any listed signal changes level the `always` block is evaluated. When that occurs here, the output is set high if `s_bar` is asserted or low if it is not and `r_bar` is asserted (the conditional structure prioritizes set over reset).
- The other new construct is the non-blocking assignment operator "<=". This functions similarly to `assign`, which cannot be used within always blocks. This seemingly superficial change is important and quite subtle. In general, you should use non-blocking assignment for sequential logic and the blocking "=" assignment operator for combinational logic.[19]
- `if (<test>) .. else if (<test>)` is one of the two new conditional statements allowed in an always block. (We will discuss the other, the `case` statement, in the chapter on memory.) The line that follows the conditional test is executed when the "if" condition is fulfilled; it falls through to the "else" if the condition fails. ("if... else" sets may be nested.) You can use `begin` and `end` to group multiple statements in a conditional branch. While you can use the ? conditional operator in an always block, you cannot use `if` or `case` with an `assign`.
- When `always@` is used to create combinational logic (usually because you want to use `if` or `case`), the sensitivity expression "`always@(*)`" is often used. This causes the block to evaluated when *any* signal in the block changes state.

16N.4.2 A Verilog D flip-flop

AoE Fig. 10.78

To create a D flip-flop, we need a way to only evaluate the always block on the rising edge of the clock. We do this by using the posedge condition in the sensitivity list. The always block is evaluated only when the specified signal edge occurs. This is known as a *clocked* always statement.

```
module d_flip_flop(
  input d,
  input clk,
  output reg q
  );

  always @(posedge clk)
  begin
    q <= d;
  end
endmodule
```

---

[19] Non-blocking means all the statements in the always block are evaluated in parallel, all at the time specified. The blocking forms will be executed in order (as in an ordinary computer program). Use of a blocking assignment after "always@(posedge...)" in sequential logic is permitted but not recommended because of potential timing problems. See Bhasker, §2.18. Chapter 16S provides an example of what can go wrong using blocking assignments in sequential logic.

Again we must declare any signals set within the always block as `reg` and set their value using non-blocking assignments.[20]

Building a better flip-flop: Suppose we want to upgrade our simple design to the equivalent of the 74HC74 flip-flop in Fig. 16N.30, with a $\overline{Q}$ output and asynchronous reset ($\overline{R}$) and set ($\overline{S}$) inputs.

It might seem we first need to add another `reg` variable for $\overline{Q}$ and set it in the always block, but it is simpler (and uses fewer WebFPGA resources) to just invert the Q output with an `assign` statement before (or after – remember order does not matter) the clocked `always` block. In addition, you might be tempted to change the sensitivity list to something like:

```
always @(posedge clk or !s_bar or !r_bar)
```

However, if you do, the WebFPGA synthesizer will complain "No support for synthesis of mixed edge and level triggers." So how do we combine the synchronous edge-triggered clock with the asynchronous jam inputs? Easy, we cheat. We make all inputs edge-sensitive and check the level of jam inputs first in the always block, *before* we set the output to the input:

```
// Design for 74HC74 equivalent
module d_flip_flop_set_clear_q_bar(
  input d,
  input clk,
  input s_bar,
  input r_bar,
  output reg q,
  output q_bar
);

  assign q_bar = !q;   // use combinational logic for inverted output

  always @(posedge clk or negedge s_bar or negedge r_bar)
  begin
    if (!r_bar)
      q <= 1'b0;
    else
     if (!s_bar)
        q <= 1'b1;
    else
      q <= d;
  end
endmodule
```

This seems wrong. How can we use edge triggering to get asynchronous, level-sensitive behavior? The answer is that when either set or reset is asserted, the corresponding signal goes from high (not asserted) to low (asserted) and this falling edge triggers the always block *immediately*, which sets the output appropriately. This looks like an asynchronous response because it does not depend on the clock. When the clock edge *does* happen, the always block logic checks the reset and set inputs first; if either is still asserted it keeps the output reset or set and ignores the D input. Only if neither jam input is asserted on the clock edge does the conditional `if` statement set the output to D.[21] We have arbitrarily chosen to make reset override set if both are asserted simultaneously.[22]

---

[20] Simulation at https://LAoE.link/D_flipflop.html.
[21] Simulation at https://LAoE.link/74HC74_flipflop.html.
[22] Remember what the doctor said: "Don't do it."

16N.4.3 A warning about multiple always blocks

A Verilog design may have multiple *always* blocks, each triggered by a different sensitivity list. What you cannot do is have different blocks try to modify the same signal as an output. This may work in the simulator but if you try to synthesize this design you are likely to get an error something like "net is constantly driven from multiple places." The synthesizer is complaining that you are trying connect multiple outputs to the same signal. You can resolve this by either combining the blocks that modify the signal or set a flag in one block to indicate to another that a signal it controls needs to be modified. (You can always use a signal in multiple always blocks as an *input*, as long as you only modify it in no more than one block.)

16N.5 AoE reading

Chapter 10 (Digital Logic I):
- §10.1.4A Combinational versus sequential logic.
- §10.4 Sequential logic.
- §10.5 Sequential functions available as ICs, see especially:
    - §10.5.1 Latches and registers;
    - §10.5.3 Shift registers.
- §10.8 Logic pathology, especially:
    - §10.8.2 Switching problems;
    - §10.8.3 Weaknesses of TTL and CMOS.

Chapter 12 (Logic Interfacing):
- §12.1.4 Driving digital logic inputs.

Specific advice:
> You need not worry too much about Karnaugh mapping. It was a useful device in the era before computer logic compilers. These have now made Karnaugh mapping obsolete (unless you are converting truth tables to logic by hand).
> Concentrate on flip-flops in §10.4.

WebFPGA Docs at `https://LAoE.link/WebFPGA_Docs.html`.
- Appendix C. DOs, DON'Ts and MUSTs.

16L Lab: Flip-Flops

16L.1 A primitive flip-flop: *SR* latch

This circuit, Fig. 16L.1, the most fundamental of flip-flop or memory circuits, can be built with either NANDs or NORs. We will build the NAND form. It is called an *SR* flip-flop or latch because it can be "Set" or "Reset." In the NAND form it also is called a "cross-coupled NAND latch." Build this latch using the pushbutton switches from your lab kit, *not* the debounced pushbuttons on the breadboard. Record its operation noting which input combination defines the "memory state;" and make sure you understand why the state is so called.

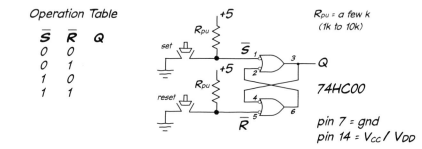

Figure 16L.1 A simple flip-flop: cross-coupled NAND latch.

Leave this circuit set up. We will use it shortly.

16L.2 D type

Practical flip-flops It turns out that the simple *latch* is very rarely used in circuit design. A more complicated version, the *clocked* flip-flop, is much easier to work with. The simplest of the clocked flip-flop types, the D, simply saves at its output (Q) what it saw at its input (D) just before the last clocking edge. The particular D flop used below, the 74HC74, responds to a *rising* edge.

The D flop is the workhorse of the flop stable. You will use it hundreds of times for each time you use the fancier JK flip-flop (a device you may have read about, but which we are keeping out of the labs because it is almost obsolete). Most likely you will never use a JK.[1]

16L.2.1 Basic operations: saving a level; reset

The D's performance is not flashy, and at first will be hard to admire. But try.

Feed the D input from a breadboard slide switch. Clock the flop with a "debounced" pushbutton. The pushbutton switches on the left side of the breadboard will do. Note that these switch terminals need

---

[1] As far as we can tell, none of the 74LS JK flip-flops have been brought over to 74HC technology, a sure sign of obsolescence.

pullup resistors, since they have *open-collector* outputs. (You should have set this up in §14L.1.4.) Use the "NC" pushbutton output so that *pressing* the button creates a rising edge.

Warning: a surprising hazard. The value of the resistor used as *pullup* on the clock input turns out to matter. A large pullup value (say, $\geq 10k$) is likely to cause mischief: that is explored on page 670.

Disassert $\overline{\text{RESET}}$ and $\overline{\text{SET}}$ (sometimes called $\overline{\text{CLEAR}}$ and $\overline{\text{PRESET}}$), by tying them high.

Note that the '74 package includes *two* D-flops. You needn't bother to tie the inputs of the unused flop high or low: that is good practice when you build a permanent circuit (averts possible intermediate logic state that can waste power, as you saw in Lab 14L) – but would slow you down unnecessarily, as you breadboard circuits.

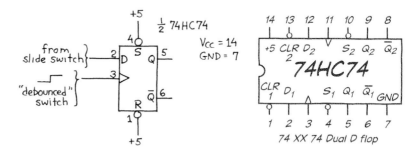

Figure 16L.2 D-flop checkout.

- Confirm that the D-flop ignores information presented to its input until the flop is clocked.
- Try asserting $\overline{\text{RESET}}$. You can do this with a wire; bounce is harmless here. (Why?[2]) What happens if you try to clock in a high at D while asserting $\overline{\text{RESET}}$?
- Try asserting $\overline{\text{SET}}$ and $\overline{\text{RESET}}$ at the same time (something you would never purposely do in a useful circuit). What happens? (Look at *both* outputs.) What determines what state the flop rests in after you release both?[3] (Does the answer to that question provide a clue to why you would not want to assert both $\overline{\text{SET}}$ and $\overline{\text{RESET}}$ in a circuit?)

16L.2.2 Toggle connection: version I: always change or "divide-by-two"

The feedback in the circuit in Fig. 16L.3 may trouble you at first glance.[4] (Will the circuit oscillate?) The *clock*, however, makes this circuit easy to analyze. In effect, the clock breaks the feedback path. Build this circuit and try it.

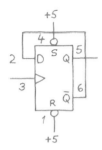

Figure 16L.3 D-flop biting its own tail.

- First, clock the circuit manually using a debounced pushbutton (but see page 670 concerning the value of the pullup resistor).
- Then clock it with a zero to 3.3V square wave from the function generator (the breadboard generator is less good than an external generator, with its higher $f_{\max}$). Be sure to confirm the generator output does not go below zero volts or exceed five volts *before connecting the output to your circuit*. Watch the clock and Q on the scope (you will want to trigger on Q). What is the relation between f_{clock} and f_Q? Now you know why this humble circuit is sometimes called by the fancy name "divide-by-two." What is the duty cycle of the Q output?

---

[2] The first low achieves the reset. Bounce – causing release and then reassertion of Reset* causes no further change in Q.

[3] The result is unpredictable, because it depends on which of the two inputs – S* or R* – has the last word: that is, which is the last to cross its threshold, rising from low to high.

[4] See Chapter 16N for difficulties that feedback introduces into non-clocked sequential circuits.

Looking for Trouble?

Here's a side excursion for the adventurous.

We said "clock the circuit manually," and assume that you would carry out this step by using one of the PB-503's *debounced pushbuttons* – probably the NC, so that you get a clock upon pressing the pushbutton:

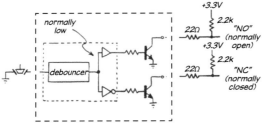

Breadboard's debounced switches: these require a pullup resistor.

You need to choose a value for the pullup resistor. At first glance, the value doesn't seem to matter, as long as it's not so tiny that it overloads the transistor switch. But in fact the pullup value *does* matter. To help you see how the flop is behaving, display its Q on one of the breadboard's LEDs. Now, try R_{pullup}=2.2k. This should work well: the flop should toggle each time you press the button.

Now try R_{pullup}=1M. This should work badly: you should see the Q *sometimes* toggle – and at other times appear not to respond at all.

This occurs because of the slow rising edge provided by the over-large R_{pullup}. This big R, driving stray capacitance, produces a slowly-rising edge that an edge-triggered device finds troublesome. The problem is exactly the one that you saw causing weird instability in the LM311 comparator, back in §8L.1.2 (remember the "Taj Mahal by moonlight"?).

You will find a scope image detailing the effect of such a slow edge in §17N.1.1. Use a scope to watch clock and Q, triggering on Q, and see if you can make out what goes wrong when the value of R_{pullup} is too large.

Then restore the small R_{pullup} of 2.2k to clock your later circuits properly.

- Crank up the clock rate to the function generator's maximum, and measure the flop's *propagation delay*. In order to do this, you will have to consider what In and Out voltages to use as you measure the time elapsed. You can settle that by asking yourself just what it is that is "propagating." If the answer to this question is that it is "a change of logic level" that propagates, then what is the appropriate voltage at which to measure propagation delay?[5]

16L.2.3 Toggle connection. Version II: change when told to or T-flop

A more useful *toggle* circuit uses an input to determine whether the flop should change state on the next clock. This behavior is properly called "T" or "Toggle." (The preceding circuit – which toggles always – is not called a T-type; the best name for it is probably "divide-by-two.")

Show how to use combinational logic and a D-flop to make such a T-flop. Figure 16L.4 shows how the circuit should behave.[6]

To exercise your T-flop, clock it from the function generator, control it with a manual switch, and watch clock and Q on a scope.

Keep this T-flop set up; we will use it again.

---

[5] Since it is logic-level *change* that is propagating, we should use the voltage that typically defines a level change. For 5 V CMOS that is 2.5 V.

[6] Hint: Remember that an XOR can serve as a controllable inverter if you use one input to the XOR as Signal, the other as Control as you saw in Chapter 14N.

16L.3 Switch bounce, and three debouncers

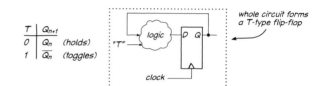

Figure 16L.4 T-flop behavior.

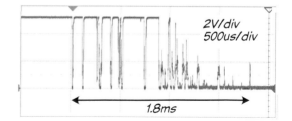

Figure 16L.5 A microswitch normally open contact, pulled up to 5 V through 33k, bouncing from high to low.

16L.3 Switch bounce, and three debouncers

Figure 16L.5 is a scope trace showing a microswitch pushbutton bouncing its way from a high level until settling to a low.

To see the harmful effect of switch bounce, clock the divide-by-two flip-flop from Fig. 16L.3 with a (bouncy-) ordinary switch such as a microswitch pushbutton as in Fig. 16L.6. When you press the button, a spring forces the grounded common terminal to connect to the normally open (n.o.) switch terminal, pulling the clock input to ground. However, the energy of the contacts colliding causes them to bounce apart, meaning the n.o. terminal is temporarily floating until the spring forces the terminals together again. While it is floating, the pullup resistor pulls the flip-flop clock input high. This can occur several times until the closing energy is dissipated and the contacts remain connected as seen in Fig. 16L.5.[7]

Watch the flip-flop's outputs on an LED. The bouncing of the switch is hard to see on an analog scope (easy on a digital), but its effects should be obvious in the erratic behavior of the flip-flop when you *close* the switch (which should not affect the *rising-edge* clock input).

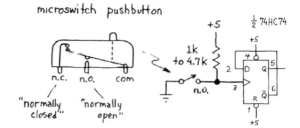

Figure 16L.6 Switch bounce demonstration.

16L.3.1 Watching bounce (optional for analog scope enthusiasts)

Use a digital scope to capture the switch bounce on the flip-flop clock input. Use "Single" mode and trigger on the falling edge of the signal.

Switch bounce is hard to see on an analog scope, because it does not happen periodically and because the bounces in any event do not occur at exactly repeatable points after the switch is pushed.

---

[7] See https://LAoE.link/Switch_Bounce.html. You may have seen this effect when a marble or ball bearing is dropped on a hard surface.

672   Lab: Flip-Flops

You can see the bounce, at least dimly, however, if you trigger the analog scope in normal mode with a sweep rate of about 0.1ms/cm. You will need some patience, and some fine adjustments of trigger level. Some switches bounce only feebly. We suggest a nice snap-action switch like the microswitch type.

16L.3.2   Eliminating switch bounce: three methods

AoE §10.4.1A

Cross-coupled NANDs as debouncer   Return to the first and simplest flip-flop (which we hope you saved), the cross-coupled NAND latch, also called an SR flop. As input use the bouncy pushbutton. Ground the switch's *common* terminal, and make sure to include pullup resistors on both flop inputs (they should be there, still); see Fig. 16L.7.

Why does the *latch* – a circuit designed to "remember" – work as a debouncer?[8]

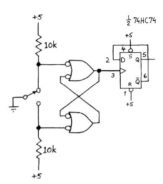

Figure 16L.7   NAND latch as switch debouncer.

CMOS buffer as debouncer (positive feedback)   Figure 16L.8 shows a much simpler way to debounce when you use a double-throw switch (SPDT) like the one used in the preceding exercise, debouncing with NANDs, as in Fig. 16L.7: just use a non-inverting CMOS gate (or two cascaded inverting gates) with its output talking to its input.[9]

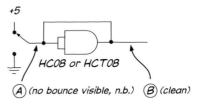

Figure 16L.8   CMOS AND gate as switch debouncer.

This one is no fun to *watch* – because *no bounce* will be visible, even at the input. Once you see why this is so, you will have understood why this circuit works as a debouncer. In particular, what happens during the time when the switch input to the AND is connected to neither +5 nor ground?

D-flop as debouncer; discovering duration of switch bounce   A plain D-flop can debounce, if clocked appropriately. Wire up this circuit to test the notion. Figure 16L.9 suggests using one flop of a 4-flop '175 because you are about to use this part in another circuit.[10]

---

[8] It does because during a *bounce*, the inputs revert to their *memory* state (both high), in which they simply hold the state into which they were pushed when the particular input earlier went *low*.

[9] This circuit does not work reliably with TTL: it's a CMOS special.

[10] The '175 is more efficient than using the 74HC74 when the circuit you are building requires more than two flip-flops and connects all their clock and clear inputs together. It does not have a set input, but much of the time that control goes unused anyway.

16L.4 Shift register

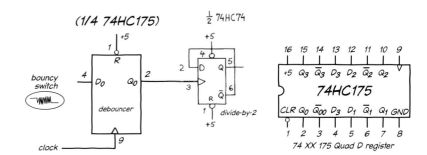

Figure 16L.9 D-flop as debouncer.

Start with a high-frequency square wave driving the D-flop clock (say, 100 kHz or more). You should see evidence of bounce in the divide-by-2's misbehavior. Now lower the clock rate to 100 Hz. Bounce should cease to trouble the divider. Raise the clock frequency till evidence of bounce reappears. Note the clock period (you'll get the best information by watching the clock on the scope of course; you'll get a ball-park *frequency* from the dial on the function generator). That clock period reveals how long your switch is bouncing.

16L.4 Shift register

Build the circuit in Fig. 16L.10 (a shift-register that we will evolve into a *digitally-timed one-shot*). We use a convenient structure in this circuit that includes four flops with a common clock. Such a configuration is called a "register," and here it is applied to the particular use as *shift-register*. The circuit delays the signal called "in," and *synchronizes* it to the clock. Both effects can be useful. You will use this circuit in a few minutes as a *one-shot* – a circuit that generates a single pulse in response (maintaining the metaphor[11]) to a "trigger" input (here, the signal called *in*).

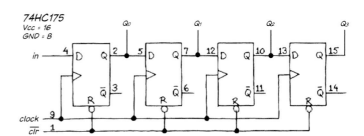

Figure 16L.10 4-bit shift-register.

Clock the circuit with a logic signal from an *external* function generator; use the breadboard's oscillator (or the second channel of the external function generator if available) to provide in. Let f_{clock} be at least $10 \times f_{\text{in}}$.

16L.4.1 One flop: synchronizer

- Use the scope to watch *in*, and Q_0; trigger the scope on *in*.

---
[11] As you may have gathered already, engineers have at least some of the characteristics of good poets: they keep the images simple and vivid. They call the pedant's "monostable multivibrator" a one-shot; they call the pedant's "bistable multivibrator" a "flip-flop," as you know. They call the active-pullup output stage of a TTL gate "totem pole," because that's what it looks like. The synchronizing color-burst in a TV signal sits on the waveform's "back porch." There's lots of this vivid imagery in engineering. Can the language of the social sciences offer any comparable pleasures?

Lab: Flip-Flops

- What accounts for the *jitter* in signal Q_0?
- Now trigger on clock, instead. Who's jittery now?
- Which signal is it more reasonable to call jittery or unstable? (Assume that the flops are clocked with a system clock: a signal that times *many* devices, not just these 4 flip-flops.[12])

16L.4.2 Several flops: delay

- Now watch a later output – Q_1, Q_2, or Q_3, along with *in*. (We'll leave the triggering to you, this time.)
- Note the effect of altering f_{clock}.

16L.4.3 Several flops plus some gates: double-barreled one-shot

A little logic – used to detect particular *states* of the shift-register – can produce pulses of fixed duration in response to an input pulse of arbitrary length (except that the pulse must last long enough to be sure of getting "seen" on a clock edge: in other words, the input pulse must last longer than a clock period). The one-shot's input signal is called the *trigger*.

In Fig. 16L.11 we show a pair of output pulses we would like you to produce. If you fill in the shift-register waveforms, you will discover the logic you need in order to produce those pulses.[13]

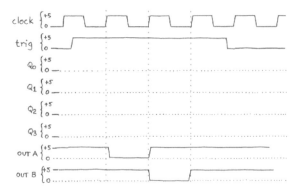

Figure 16L.11 Timing diagram for digitally-timed one-shot.

Incidentally, this all-digital circuit – whose output pulse is timed by the clock – is not what people usually mean when they say "one-shot:" the most common form uses an *RC* to time the pulse width. Such a one-shot is sometimes more convenient, but lacks the great virtue of *synchrony* with the rest of the digital circuit.

Draw your own design and checkout the following.

- Slow-motion: first use a manual switch to drive *Trigger*, and set the clock rate to a few hertz. Watch the one-shot outputs on two of the breadboard's buffered LEDs. Take *trigger* low for a second or so, then high. You should see first one LED then the other flash low, in response to this low-to-high transition.
- Full-speed: when you are satisfied that the circuit works, drive *trigger* with a square wave from one function generator (the breadboard's) while clocking the device with an external function generator (at a higher rate).

---

[12] Since we treat the clock as the reference timebase, when we trigger on clock we notice the uncertainty of the input waveform's timing. It makes sense to say that the input is "jittery."

[13] Hint: A pair of NAND gates added to the shift-register does the job.

How would you summarize the strengths and weaknesses of this one-shot relative to the more usual *RC* one-shot?[14]

16L.5 That was fun, let's do it again in Verilog

16L.5.1 Basic operations

Create a module to emulate a 74HC74 type flip-flop in Verilog. Then create a top-level module to connect the breadboard switches and LEDs to the WebFPGA and instantiate your D flip-flop in it. (We ask you to use a separate top level module both to get practice in hierarchical design and because it will set you up for the exercises following.) Test it with the WebFPGA following the instructions in §16L.2.1. Does your PLD flip-flop match its discrete sibling in every way?

16L.5.2 Toggle connection: divide-by-two

Now connect your flip-flop as the divide-by-two described in §16L.2.2. Follow the directions in that section to test the circuit. How does it compare to the discrete part?[15]

16L.5.3 Toggle connection: T flip-flop

Add whatever you need to turn your D flip-flop into the toggle flip-flop described in §16L.2.3. How does it compare to the one you built out of discrete parts?

Wait - that's silly: We just asked you to do some structural Verilog design mimicking your discrete gate and D flip-flop circuit to create a T flip-flop using programmable logic. It makes much more sense (and is simpler) to just design a toggle flip-flop module directly. Go ahead and create a toggle flip-flop module and compare its performance to the structural version. Is it faster than the version built out of gates and a D flip-flop?[16]

16L.5.4 Shift register

Build the 4-bit shift register in §16L.4 in the WebFPGA two ways. First design the shift-register using D flip-flops (structural design), then with vectors using the concatenation or shift operator (behavioral design).

Structural version: Here is the module definition for the D flip-flop version:

```
module SR_4BITWCLR_DFF (
  input clock,
  input clr_bar,
  input in,
  output [3:0] q
);
```

[14] Well, in case you're interested, here are our views: Strength: the digitally-timed one-shot's great virtue is that its output is *synchronous* with the system clock, so its pulse begins and ends at a predictable time, shortly after the clock edge, and safely away from t_{setup}. The only weaknesses are the greater complexity of the circuit relative to a traditional *RC*-timed one-shot, and the circuit's *latency*: the output can be delayed as much as one full clock cycle from the rise of trigger. The digital version also will not respond at all to a trigger signal that is too brief.

[15] It certainly should operate identically, but the propagation delay may differ.

[16] This depends on how smart the synthesizer optimizer is. Since our development environment does not show us what the actual design it comes up with is, all we can do is try both and compare them experimentally.

Lab: Flip-Flops

Your Verilog should implement the circuit shown in Fig. 16L.10 using your 74HC74 D flip-flop module.

You will need to add the comment block used by WebFPGA to assign the module name and I/O pins. It is easiest to wait to assign the actual WebFPGA pin numbers until you build the breadboard test setup. That way you can wire devices the in the most direct way, and then assign the pin connections to match what you built.

Connect the clock input to a debounced pushbutton and connect the $\overline{\text{CLR}}$ and IN inputs to two of the eight LOGIC SWITCHES. The four shift-register outputs should go to LOGIC INDICATOR LEDs. The pushbutton should shift when pressed (not when it is released).

Test your design by clearing the shift-register, then shifting in a constant one and watching it move across the outputs. Clear the shift register again and shift in a single one. Watch it propagate over to $Q3$. Make sure you have no stuck bits and that clear works properly.

Behavioral version: Now modify your shift register Verilog code to use the concatenation or shift operator (no explicit flip-flops) and test it again.

```
module CAT_4BITWCLR (
  input clock,
  input clr_bar,
  input in,
  output reg [3:0] q
);

<your code goes here>
```

Build the one-shot: Modify one of your shift register designs to implement the double-barreled one-shot in §16L.4.3. Test it with a high-speed clock (1 kHz should be fine) and a debounced pushbutton for the trigger to make sure it works. You should see the output pulses OUT A and OUT B shown in Fig. 16L.11 on your scope in response to a trigger input. (Hint: use single trigger mode on the scope and use the trigger signal as the source.)

16L.5.5 Some other flip-flop circuits

If you have time, feel free to try out the flip-flop circuits in Chapter 16S. You can build the structural version using a D flip-flop and gates or just build the functional circuit in the accompanying Verilog code.

16L.6 Solutions

Our solutions to the lab problems ... but please try to do them on your own first.
  https://LAoE.link/FPGA/16L_FlipFlops.v.
  https://LAoE.link/FPGA/16L_ShiftRegisters.v.

16S Supplementary Notes: Flip-Flops

16S.1 Flip-flop tricks

We've grouped here some flip-flop configurations that you have not yet seen. Despite this section's title, we don't mean to disparage these circuits. They are extremely useful. We've included Verilog code to implement each of these circuit fragments.[1]

Set-Reset flip-flop: We've already looked at using a SR flip-flop as a latch to capture a transitory set or reset signal. See §16N.1.1.

What's good and bad about this simple SR latch?

Virtues: Simplicity; useful as a debouncer – but only for a double-throw switch (SPDT).

Vices: Reset fails if Set persists during attempted Reset. So edge-triggered variation is usually preferable.

Edge-triggered "flag": This circuit has an async rising edge set and an async active-low level reset (Fig. 16S.1). The Q of this flop often serves as a "flag" – a signal that persists until deliberately knocked down.

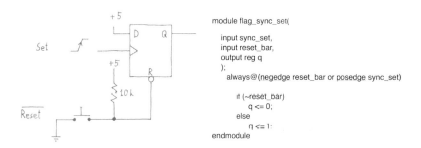

Figure 16S.1 Edge-triggered flop can be reset while the signal that set Q high persists.

Virtue: Flop can be reset despite persistence of the signal that set Q high.

Sticky flag: Synchronous rising clock edge set, active-low async reset. If the asynchronous clock of Fig. 16S.1 isn't satisfactory – for example, because the SET line in that figure may carry glitches – then a *synchronous* flag is required, as in Fig. 16S.2.

Virtue: Synchronous clocking protects against vulnerability to glitches on SET line of Fig. 16S.1 (assuming that glitches occur *after* clocking of the flop in Fig. 16S.2, as is usual).

---
[1] Code available at https://LAoE.link/FPGA/16S_FFTricks.v

Supplementary Notes: Flip-Flops

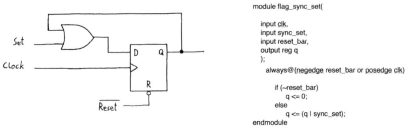

Figure 16S.2 Sticky synchronous set.

Virtue: As with async clocked circuit of Fig. 16S.1, flop can be reset despite persistence of the signal that set Q high.

Another sticky flag: Synchronous set, synchronous active-high reset. One more gate (see Fig. 16S.3) makes the sticky flag of Fig. 16S.2 fully synchronous: the now active-high Reset also waits for the clock.

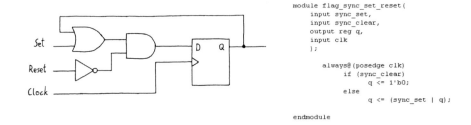

Figure 16S.3 Sticky flag II, synchronous set, synchronous reset.

Virtue: Fully synchronous.

A generalized circuit: synchronous Do_This, Do_That flop circuit Redefining the logic that feeds the flop's D input as a 2:1 multiplexer, one can draw – and think of – the circuit in Fig. 16S.4 as something more general than the preceding ones, whose only aim was to set or reset the flop. This circuit could be used to distinguish, say, parallel-load versus serial operation for a shift register.

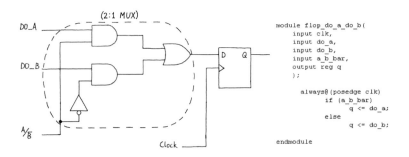

Figure 16S.4 Generalized flop circuit: mux provides synchronous Do_This, Do_That circuit.

Virtue: Fully synchronous again.

16S.2 Blocking versus non-blocking assignments

Here's a point you should feel free to ignore. This example is for those who are troubled by the odd distinction between these two sorts of "assignment," a distinction mentioned back in §16N.4.1.

An example can make this abstract-seeming difference clear. Let's use Verilog to design a 3-bit shift-register.[2] Figure 16S.5 shows what we intend to build. We will design this twice: once correctly – and once not, as if we had not digested the difference between *blocking* and *non-blocking* assignments.

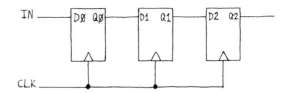

Figure 16S.5 3-bit shift-register: the design we want to implement.

Let's do it wrong: blocking version: Blocking assignments are executed in the order they appear within a sequential block. It is called blocking because it completes the assignment of left-hand side from right-hand side without permitting interruption by any other Verilog operation. As Cummings puts it, "A blocking assignment 'blocks' trailing assignments in the same *always* block from occurring until after the current assignment has been completed."[3] Blocking assignments behave as you probably are accustomed to expect in computer code: one line executes, then the next.

Let's design the shift-register with blocking assignments, letting IN generate Q_0 on the clock, Q_0 generate Q_1... and so on. Here's the code:

```
module shift_reg_blocking(
    input clk,
    input in,
    output reg q0,
    output reg q1,
    output reg q2
    );
  always@(posedge clk)
    begin
        q0 = in;
        q1 = q0;
        q2 = q1;
    end
endmodule
```

This code will compile properly. But it won't build what we intended. The `always` block takes action upon the rising edge of clk as we expect. The several assignments all are made in response to a single clock edge. So far, so good.

But in the blocking implementation, on the clock edge Q_0 is set to IN *before* Q_1 is set to Q_0 and this occurs *before* Q_2 is set to Q_1. The result is all three outputs get set to IN on a single clock edge. Not what we wanted! The schematic in Fig 16S.6 shows what Verilog built from this code. It's not a shift register at all but a flop with three labels assigned to its single output.

---

[2] This example appears in the notes for MIT's 6.111, Spring 2004, (https://LAoE.link/MIT_6.111.pdf) and also in a paper by Clifford Cummings, "Nonblocking Assignments in Verilog Synthesis, Coding Styles that Kill," (https://LAoE.link/Styles_that_Kill.pdf). Bhasker treats the topic in his §2.18.
[3] Cummings, §3.

680 Supplementary Notes: Flip-Flops

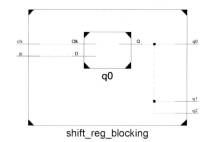

Figure 16S.6 Schematic of Verilog's implementation of blocking "3-bit shift-register" design. It does not work.

Let's do it right: non-blocking version: When we use *non-blocking* assignments, on each clock edge the three flop outputs take the *old* levels that were present at their inputs just before the clock rose. As the MIT notes put this, "... all assignments [are] deferred until all right-hand sides have been evaluated (end of the simulation timestep)."[4]

Because each assignment takes in an *old*, pre-clock level, this design does, indeed build a shift-register, as the Verilog schematic in Fig. 16S.7 confirms.

```
module shift_reg_non_blocking(
  input clk,
  input in,
  output reg q0,
  output reg q1,
  output reg q2
);

  always@(posedge clk)
    begin
      q0 <= in;
      q1 <= q0;
      q2 <= q1;
    end

endmodule
```

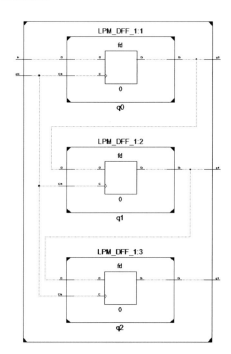

Figure 16S.7 Code and schematic of Verilog's implementation of non-blocking shift register. This works.

The example supports the point, which you probably believed anyway, that you should use *non-blocking* assignments for sequential designs.

---

[4] MIT notes, p. 7. The use of the word "simulation" in what is a discussion not of simulation but of *synthesis* is odd – and reminds us of the discomfiting fact that Verilog was born as a simulation program. Its use for synthesis came later.

17N Counters

Contents

17N.1	**But first some circuit dangers and anomalies**	**681**
17N.1.1	A flip-flop vulnerability: slow clock edge	681
17N.1.2	A flip-flop vulnerability: a fast clock edge	682
17N.1.3	A glitchy circuit: timing diagram can help	682
17N.1.4	Avoid gating clocks	684
17N.1.5	Metastability	685
17N.2	**Counters: ripple vs. synchronous**	**685**
17N.2.1	Ripple counters	685
17N.2.2	Synchronous counters	687
17N.3	**Designing a larger, more versatile synchronous counter**	**689**
17N.3.1	We need four T-flops...	689
17N.3.2	...we need to tell the several Ts when to change, and we add Carry$_{in}$ and Carry$_{out}$...	690
17N.4	**Some useful counter functions**	**691**
17N.5	**Creating divide-by-N counters**	**692**
17N.6	**Counters in Verilog**	**695**
17N.7	**AoE reading**	**698**

Why?

The problem we'd like to solve today is how to measure and record a number of digital events (that is, we would like our circuit to "count").

17N.1 But first some circuit dangers and anomalies

Now that we have seen that sequential circuits are almost always (the SR and transparent latches being the exceptions) designed with edge-triggered logic, we need to look at what can go wrong with edge triggering if we are not careful.

17N.1.1 A flip-flop vulnerability: slow clock edge

The fact that an edge-triggered *clock* input responds to each edge (either rising or falling – not both, for a particular part) makes such a terminal vulnerable to *switch bounce*; indeed, one rarely debounces a signal unless it goes to a clock. Luckily, we know know several ways to debounce a switch.

But edge-sensitive inputs are also vulnerable to a subtler hazard: a clock edge that is not steep enough – an edge that is *slow*.

682 Counters

Figure 17N.1 shows what a divide-by-two counter did when the clock came from a switch pulled up by a resistor that was too large (100k). The large pullup R conspired with stray capacitance to produce an edge whose slope was too gentle. The result was that initial switching of the flip-flop output from high to low caused a shiver in the power supply, which caused the clock edge to shiver, which triggered the flip-flop a second time back to the high state. The scope images show the Q0 output switching twice when we expected only a single transition.

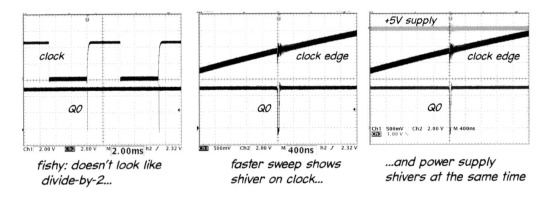

Figure 17N.1 Slow edge can cause trouble for an edge-sensitive input.

This is very similar to the trouble we saw back in the comparator circuit of Lab 8L when we fed a gently-sloping waveform into a circuit that lacked hysteresis. As we noted back in Chapter 14N, a few logic devices *do* include hysteresis. The 74HC14 hex inverter is identical the 74HC04 except that the inputs of the former include about one volt of hysteresis. The Lattice FPGA we use includes 200 mV (typically) of hysteresis on every input.

17N.1.2 A flip-flop vulnerability: a fast clock edge

A fast clock can cause problems as well. A very fast high-to-low transition can create enough ground bounce to create a second rising edge as the signal bounces back from the falling edge. Ground bounce occurs when when a fast signal transition, discharging a capacitive load, causes the device's ground to jump due to parasitic inductance between the chip ground and system ground. We saw a similar effect with the comparator circuit in Fig. 8N.8.

You will especially see this on circuits with insufficient grounding (e.g., solderless breadboards), inadequate supply decoupling capacitors, long signal lines, and signals driven by fast logic, like 74AC drivers. This is why you will sometimes see small-value resistors (10 to 50 ohms, typically) in series with a clock signal, preferably near the source.[1]

17N.1.3 A glitchy circuit: timing diagram can help

Sometimes, the problem is not how fast the clock edge transitions, but the circuit creating the edge. Figure 17N.2 is a commonplace and useful combinational circuit, a 2:1 multiplexer, designed with conventional gates, and built using a PAL.[2]

In order to illustrate a potential problem, we have wired the circuit for the dullest possible operation.

AoE §10.4.4B

See §14W.1

---
[1] Thanks to Jim MacArthur for this insight.
[2] But the problem illustrated occurs no matter how the circuit is built as long as the logic used has a non-zero propagation time.

17N.1 But first some circuit dangers and anomalies

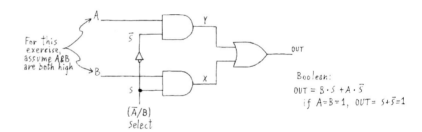

Figure 17N.2
Two-to-one mux; glitchy case discussed here assumes both data inputs high.

Common sense – or a Boolean analysis, if you want to be so academic about it – says that the output must be a high always (the mux's job is to select A or B: high or high). Since a piece of wire to V_+ would give the same output, we should think of the wiring of Fig. 17N.2 as illustrating a particular input combination that might be present some of the time.

A timing diagram, drawn carefully enough as in Fig. 17N.3 to show gate delays, predicts that the circuit will not be so well-behaved.

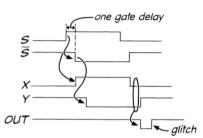

Figure 17N.3 A timing diagram for the mux predicts a glitch.

A scope looking at S (select) and OUT might hide the glitch from you – if you weren't looking for it. The slow sweep rate in the leftmost image of Fig. 17N.4 makes the brief low pulse very hard to see. It appears in the middle image, swept faster. And it is measurable – though evidently very brief (this mux was made with a PAL, and the internal gate delays that produce the glitch are small, smaller than if it were built with discrete gates). Brief though it is, it probably is sufficient to be detected by another device made with the same technology.

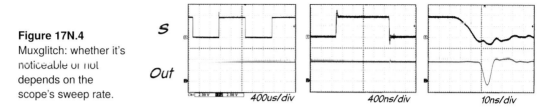

Figure 17N.4
Muxglitch: whether it's noticeable or not depends on the scope's sweep rate.

Remedies? Integrated muxes do not produce this glitch – even though their published circuit schematic looks just like ours. Why? We're not sure. It may be because the designers of ICs can match delays where they need to. The IC mux may include a delay in the non-inverted-S path, a delay that matches the delay imposed by the inverter in the $\overline{S}$ path.[3]

But the designers have another option available, to prevent the glitch – and if you are building the mux

---

[3] When asked to come up with a solution for this problem, students often try to add an extra gate to delay the non-inverted S signal. This is not a great solution; non-inverting gates have longer propagation delays compared to inverting gates so the chance of matching the paths exactly is small, unless you happen to be designing the IC chip. Read on to see a much better solution guaranteed to work.

yourself this is your only option. One can add what sometimes is called a "redundant" gate to prevent this particular glitch (see Fig. 17N.5).[4] This gate – not truly redundant, as preventing a glitch is a necessary operation, hardly "redundant" – forces *OUT* high when both inputs are high, regardless of the level of *S*.

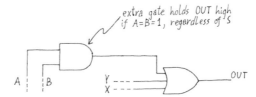

Figure 17N.5 Extra gate can eliminate an anticipated glitch.

The lesson? We offer this example not to make you a nervous glitchophobe. Glitches won't jump out of your circuits at random. We hope this example will illustrate several points.

1. If transients on the outputs of a circuit could cause mischief, one needs to consider the circuit dynamically, not just in Boolean, static terms (there are plenty of circuits, of course, where you know output transients would be harmless).
2. A timing diagram can help to disclose the possibility of such glitches.
3. An output glitch like the one illustrated in Fig. 17N.4 results from what is called a "hazard."[5] The cause of the problem is the need for two internal signals to change at the same time. The logic compiler that produces code for the PAL will eliminate any gate that is *logically* "redundant," like the extra gate installed to prevent this glitch – unless you tell the compiler's minimizer to back off and permit this non-minimal implementation.

The hard part of the task, evidently enough, is to spot the possibility of the glitch before it bites you. If you see it coming, you can prevent it without much effort.

17N.1.4 Avoid gating clocks

Try to avoid controlling a synchronous circuit using a Pass/Block* on the clock input: see Fig. 17N.6. If the control signal is asynchronous with respect to the clock, this circuit can generate spurious clock edges, leading to unexpected behavior. Typically, you can find another way to enable or disable the circuit (many digital devices include an enable input that inhibits their operation).

If you absolutely have to gate the clock, you can first use a flip-flop to synchronize the control signal with the clock and then use the flip-flop output as the gating signal. That way you will always inhibit the clock when it is in a known state.

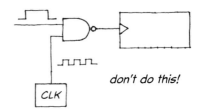

Figure 17N.6 Danger, Will Robinson![6]

---

[4] Old-timers who recall Karnaugh maps, which use shapes or "covers" to group or cover logic *Ones*, may call this remedy a "redundant cover."
[5] And this hazard, producing a transient false state, is called a "static hazard," in case anyone cares. The difference between the condition technically called a "hazard" and one called a "race" doesn't interest us much. We *are* interested in the resulting glitch.
[6] See https://LAoE.link/Danger.html.

17N.1.5 Metastability

AoE §10.4.2D

If you violate the setup time rules for a bistable device like a flip-flop, the output may end up in a metastable state, neither a valid 1 or 0. This state can persist for a very long time (relative to the normal propagation delay time) before it finally resolves to the correct value. The goal of synchronous design is to ensure the inputs and outputs of a circuit are stable before applying the next clock edge, but in systems with inputs asynchronous to the clock it is always possible for an input to change just as the clock occurs. Jim MacArthur offers this practical advice:

> Whenever a digital designer designs a circuit where a clocked input could change at the same moment as the clock, a little light should go on in their head, labeled "Warning: Metastability Hazard." Naturally, this situation will happen whenever an asynchronous event is introduced into a synchronous system. Nevertheless, there are ways to reduce the possibility of the metastability causing mayhem. The accepted practice for FPGA design is to immediately synchronize asynchronous inputs with the system clock. This tends to significantly reduce the possible harm metastability can cause. (Nothing can eliminate it entirely, but enough synchronization can increase the MTBF from ms to millennia.)
>
> This isn't just theoretical. Last year I encountered a bug with my go-to state machine, basically a triggerable event generator running at 100 MHz on a Lattice FPGA. I had neglected to synchronize a trigger input that was tested by a conditional jump, the result of which was that every few million loops, the state machine would jump to a non-existent location and crash. A simple input flop fixed the problem.

As you can imagine, this kind of infrequent, seemingly random problem could be very difficult to track down. Prevention (in the form of a flip-flop clocked by the system clock on asynchronous inputs) makes much more sense.

17N.2 Counters: ripple vs. synchronous

We are going to start by building binary counters from flip-flops. Since each flop can remember one-bit, we need N flip-flops for an N-bit counter. We also need to choose how we let each higher order bit know when to toggle. We have two choices; one results in an *asynchronous (i.e., ripple) counter*, the other a *synchronous counter* (which is what we generally prefer).

17N.2.1 Ripple counters

AoE §10.4.2E

Ripple counters are simple, but their weakness is that their Q outputs do not change all at the same time. The progressing waveform visible in Fig. 17N.9 gives the counter its name.

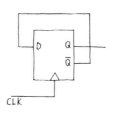
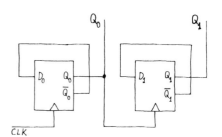

Figure 17N.7 Simple counters: "divide-by-two"; "divide-by-four" (ripple type).

Because the several flops change state in succession rather than all at once, not only are they slow to reach a stable state but also they necessarily produce *transient* states that ought not to be present.

The divide-by-two counter to the left in Fig. 17N.7 is a convenient way to create a 50% duty cycle square wave from a fixed frequency input clock of any duty cycle. Of course, this input clock must be at twice the desired output frequency. You built this in the last lab.

The two-bit counter to the right in Fig. 17N.7 is a *down* counter, because the second flop changes only when the first is *low* before the clock edge: see Fig. 17N.8. It is easily transformed into an *up* counter.[7]

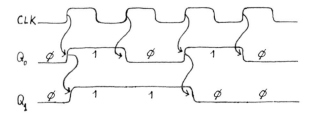

Figure 17N.8 "divide-by-four" ripple counter operation.

An integrated ripple counter, like the 74HC393, forms an *up* counter by using a falling-edge clock. Its ripple delays are less than those shown by the counter built from 74HC74 D-flops shown in Fig. 17N.9. The '393 flop-to-flop delays are about 4 ns (Fig. 17N.10) driving a low capacitance load. While this sounds small, these delays can add up. In the 74HC4040 12-stage ripple counter, the delay between the clock edge and the MSB changing is typically 55 ns but increases to almost 190 ns driving a 50 pF load. That means you could have to wait 55 to 190 ns before you are sure what the count is. If, say, you were to use this type of counter to access the next instruction in a computer's program memory, you would severely limit the speed at which you could run the computer. In modern digital systems, counters of 16 to 32 bits are not uncommon, so the delay using a ripple counter would be much worse.

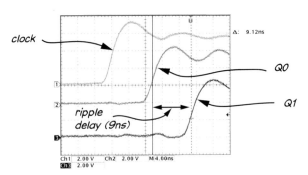

Figure 17N.9 2-bit ripple counter from flip-flops shows delay between Qs – thus false transient states. (Scope settings: 2 V/div, 4 ns/div.)

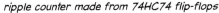

ripple counter made from 74HC74 flip-flops

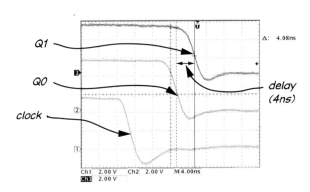

Figure 17N.10 Integrated ripple counter shows less delay between Q's – but still ripples. (Scope settings: 2 V/div, 4 ns/div.)

74HC393 ripple counter

---

[7] How? Just treat the $\overline{Q}$ as outputs; or, equivalently, treat Q as an output but use $\overline{Q}$ to do the ripple *clocking*.

Virtues of ripple counter

- Simple. Therefore, an IC counter can hold more ripple stages than synchronous (but note that this difference is unimportant for a counter implemented in a PLD with its rich stock of gates).
- Fast. Since there is no logic between stages, a ripple counter's maximum clock frequency is limited only by the propagation delay and setup time of a single flip-flop.
- In some applications, its weaknesses are harmless. For example,
  - frequency dividers (where we don't care about relative timing of *in* and *out*, and don't need to look at the several Q outputs in parallel, but care only about the relative frequency between input and output), and
  - slow counters (driving a display for human eyes, for example: since we humans can't see the false states).

17N.2.2 Synchronous counters

AoE §10.4.3F

17N.2.2.1 What synchronous means

A synchronous circuit sends a common clock to all of its flip-flops; so they all change at the same time.[8] This is the sort of design one usually strives for.

The defining signature of the synchronous counter (or any synchronous circuit), on a circuit diagram, is just the connection of a common clock line to all flops (hence syn-chron: same-time). The synchronous counter – like synchronous circuits generally – is preferable to the asynchronous ripple type. The latter exist only because their internal simplicity lets one string a lot of flops on one chip (e.g., the "divide by 4k" counter we have in the lab – the '4040, or the more spectacular MC14536: a 24-bit, programmable divide-by-up-to-16M).

Figure 17N.11 shows a generic synchronous circuit; there's no clue to what it may be doing, but its synchronous quality is obvious. The advantage of synchronous design is we only have to wait one flip-flop propagation delay for all the outputs to be valid no matter how many bits we are looking at.

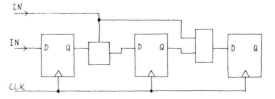

Figure 17N.11 Generic synchronous circuit: common clock is its signature feature.

Virtues of synchronous counter (and synchronous circuits in general)

- Settles to valid state faster, after clock (wait for just *one* flop delay).
- Shows no false intermediate states.[9]

17N.2.2.2 Synchronous counting requires a smarter flip-flop

Making a counter that works this way requires a flip-flop smarter than the one used in the ripple counter. Whereas the ripple counter's flops change state whenever clocked, the sync counter's flops are clocked continually, and should change only sometimes. More specifically, in a two-bit synchronous counter the second flop in the timing diagram Fig. 17N.13 and circuit Fig. 17N.14, Q_1, should change on every second clock; if there were a third flop, Q_2, it should change on every fourth clock, and so on.

---

[8] Except to the small extent that propagation delays of the several flops may differ.
[9] Even a synchronous circuit can show false transients states if *skew* is substantial among the propagation delays of its several flip-flops. Unequal capacitive loading can have this effect. See AoE 10.4.3F.

In order to get this behavior from flops that are clocked continually we need a flip-flop (to implement Q_1 here) that will, under external control, either *hold* its present state or *toggle* (flip to the other state). That is called "Toggle Flip-Flop" or "T-flop" behavior.

And XOR can make a D-flop into a T: It turns out that an XOR gate is just what we need in order to convert a humble D-flop into the smarter T-flop. We noted, back in §14N.3.6.2, that the XOR function can be described as one that lets one input determine whether the other input is *passed* or *inverted*. With the help of the XOR, then, we can make a D-flop into a T – one that changes on the next clock only if its T input is driven high: see Fig. 17N.12. You built this circuit in §16L.2.3.

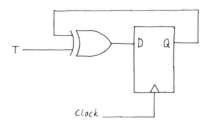

Figure 17N.12 An XOR as pass*/invert can convert a D flip-flop into a T.

17N.2.2.3 The counter behavior we want

Figure 17N.13 shows a timing diagram for a synchronous divide-by-four, noting the cases where we want Q_1 to change and those where it should *hold*. Compare to the ripple counter in Fig. 17N.8. We will connect D- and T-flops as required, to get this desired behavior.

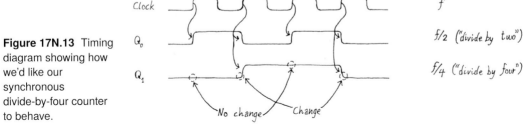

Figure 17N.13 Timing diagram showing how we'd like our synchronous divide-by-four counter to behave.

Since the low-order flop, Q_0, changes on every clock, it doesn't need to be smart: just tie Q_0* back to D_0, as in the ripple counter. It's Q_1 that needs to be a little smarter.

In two cases shown in Fig. 17N.13 Q_1 holds its present state after clocking (these cases are labeled "No change," in the figure); in two other cases Q_1 changes state. What we need to arrange then, is to have the second flop's T input driven high each time we want change on the next clock, and driven low when we want Q_1 to hold on the next clock.

Setup-time: an important subtlety: In making a synchronous design, it is essential to drive all the inputs properly *before* the clock edge (in our case, the rising edge). It won't do therefore, to look at Fig. 17N.13 and say, "Oh, I see: Q_1 should change whenever Q_0 falls." That information – "Q_0 has fallen" – comes too late. Our circuit needs to drive T_1 appropriately *before* the next clock edge. More precisely, it needs to drive T_1 early enough so that D_1 is in the correct state by *setup-time* before the clock edge. (For the 74HC74, t_{setup} is 16 ns, max; the *typical* value – versus *max* – is just a few nanoseconds.)

So Q_1 will behave as we want if we tell it, "Change on the next clock if Q_0 is high." The simple wiring in Fig. 17N.14 implements that design.

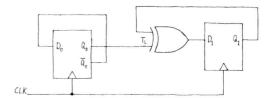

Figure 17N.14 2-bit synchronous counter.

17N.3 Designing a larger, more versatile synchronous counter

Let's carry on, moving toward a counter that is more versatile and thus closer to the sort of integrated counter one might buy (if you chose to buy such a smallish IC rather than build the part from a PLD). Here are some features we'd like in our improved counter, which we propose to build at the modest size of four bits (thus it will be a "divide-by-16" if we don't mess with its "natural" 2^n count length).

- It should be easily "cascadable" – that is, easily linked with other similar counters to form a counter of larger capacity. That means it needs a *carry in* as well as a *carry out*.
- It should include a *clear* function, and this function should be *synchronous*.[10]
- It should include a *load* function – that is, one that allows the user to load the four flops in parallel to permit counting from an arbitrary initial value.

We'll work our way to this result in stages.[11]

17N.3.1 We need four T-flops...

Figure 17N.15 are four T-flops, made from Ds. The shared clock line makes the flops' timing *synchronous*. We have made even the LSB flop a T type, as we did not in the simpler two-bit sync counter of §17N.2.2.3. In that simpler counter we made the LSB flop toggle on every clock, willy-nilly. The design we are working toward this time is not so simple: our Carry$_{in}$ function will require a controllable T on the LSB.

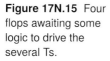

Figure 17N.15 Four flops awaiting some logic to drive the several Ts.

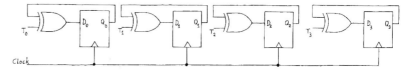

And Fig. 17N.16 shows a timing diagram to illustrate how we want the counter to behave. To save ourselves effort in drawing, we have started the diagram not at zero, but at count 12_{10}.

---

[10] Well, in this example at least. Sometimes we would prefer an asynchronous reset. We will see examples of each a bit later.
[11] We will make the clear and load active low because most 7400 series counters use active low control signals. There is no reason not to make these active high instead if you are creating a counter from scratch in a CMOS PLD.

17N.3.2 …we need to tell the several Ts when to change, and we add Carry$_{in}$ and Carry$_{out}$…

As in the case of the little 2-bit counter you designed last time, the rule is like the rule that governs a car's odometer: let the higher digit (in this case, a *binary* digit: a *bit*) change only when all the lesser bits roll over to zero. So, in order to get synchronous behavior, we need to tell a flop, "change on next clock" whenever a rollover is imminent. This rollover is imminent when all the lesser flops are high.

So, we just AND lesser Qs to drive any particular T. While we're at it, we'll include in our circuit sketch in Fig. 17N.17 the Carry$_{in}$ and Carry$_{out}$ that will allow cascading this counter with other similar parts.

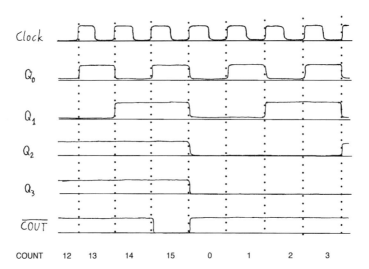

Figure 17N.16 Timing diagram describes the way we want our 4-bit UP counter to behave.

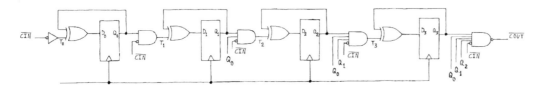

Figure 17N.17 Four T-flops driven to achieve synchronous UP count; Carry$_{in}$ and Carry$_{out}$ added to allow cascading.

One detail of the Carry$_{out}$ logic may strike you as odd: the Carry$_{out*}$ AND (more or less a NAND gate) includes not only the four Qs but also the Carry$_{in}$ for this counter stage. The logic includes Carry$_{in*}$ because that will be driven by a Carry$_{out*}$ from a lesser stage if the counter is part of a cascade.

Thus the Carry$_{out*}$ that includes Carry$_{in*}$ says not only "I'm full" (speaking for this single stage) but also "We're all full" (speaking for itself and for all lesser stages similarly designed). This simple and neat scheme permits cascading multiple counter stages simply by tying Carry$_{out*}$ of each stage to Carry$_{in}$ of the next stage. Without this inclusion, we could only cascade two counters before we would have to add external gates to AND all proceeding carry outs to the carry in of the next stage.

Cascading counters: Any respectable counter allows "cascading" several of the devices to form a larger counter. The carries (in and out) implemented in §17N.3.2 would permit us to form a 12-bit counter, for example, with no additional logic by stringing together three 4-bit blocks of counters like the one in Fig. 17N.17. From the perspective of a multi-device counter:

Carry$_{in}$ This is an enable: when asserted, it tells the counter to pay attention to its clock. Notice that all chips are tied to a common clock line in Fig. 17N.18. Do *not* drive one counter's *clock* with a carry out. If you do, you're reverting to a ripple scheme, not a fully-synchronous counter. On an up/down counter, this enable is sometimes labeled as "carry/borrow in."

Carry$_{out}$ This signal warns that the counter is *about to roll over* or overflow. In the case of a natural-binary up counter, it detects the condition *all ones* on the flop outputs. In a *down* counter, Carry$_{out}$ would detect the condition *all zeros*. On an up/down counter, the carry is sometimes labeled as "carry/borrow out."

Notice that this signal must come *before* the roll-over, not after, because it has to tell the next (more-significant) counter what to do on its next clock.

A refinement in C_{out*}, worth noting: The need to include C_{in*} in the gating for C_{out*} appears only in a case where more than two counters are cascaded. We want the *third* counter – the high-order – to increment only if *both* the lesser counters are full. We don't want it to increment just because its immediate neighbor is full.

Figure 17N.18 illustrates how easy it is to cascade three 4-bit counters to create a synchronous 12-bit counter with properly designed Carry$_{out*}$ and Carry$_{in*}$. Such cascading is easy – but if carried on for many stages would limit the counter's speed because of "ripple delay" of the carries. Adding external *carry lookahead* logic to the design allows faster clocking at the cost of more complexity.[12]

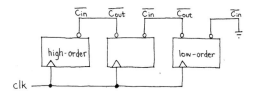

Figure 17N.18 Cascading three integrated counters: *easy!*.

17N.4 Some useful counter functions

Integrated counters are available with a variety of features; and you can implement any of these features yourself, using a PLD.

At a minimum, an N-bit counter needs a clock input and N data outputs.

A selection of counters showing sync/async functions appears in AoE §10.5.2

Clearing: Clearing (or "resetting") could be called a special case of *loading* (see below). However, it only requires one additional pin and is so often useful that nearly all counters include it. Clear is available in two styles:

Asynchronous or "jam" clear. The clearing happens a propagation delay or so after Clear* or Reset* is asserted (say, 5–10 ns); the clearing does *not* wait for the clock. This type of clear is familiar to you from your experience with discrete flip-flops, whose clears always are *jam* type.[13]

---

[12] Perhaps you remember back in §15N.2 we mentioned that our FPGA's logic cell included special "carry logic" to avoid the delays caused by requiring the carry to ripple through each cell when building a multi-bit adder.

[13] Be careful with terminology. A synchronous counter with an asynchronous clear is *still* a synchronous counter; never call such a *counter* asynchronous – only the clear function ignores the clock.

692 Counters

Synchronous clear. The clearing is *timed* by the clock: on assertion of Clear* or Reset*, *nothing happens* in response until the next clock edge. Implementing such a Clear just requires including the clear control signal in the input to the counting flip-flops: see Fig. 17N.19.

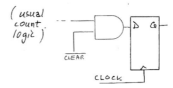

Figure 17N.19 Synchronous clear: requires only an AND gate.

Loading: Many counters allow you to load a value "broadside" (i.e., *parallel load*) into the flops. This requires N data inputs plus a load control input:

Load When you assert LD*, the counter is transformed into a simple register of D-flops: on the next clock edge those flops simply take in the values presented on the data inputs. (This description fits the so-called "synchronous" load. An "asynchronous" or "jam" load also is available – it works like the jam clear described above.) Async load is convenient for loading initiated by a manual pushbutton; the sync version is better when the load is initiated by a circuit signal.

When you release LD*, the counter becomes a counter once more. In §17N.5 we'll look at an example of the use of load to create a "Divide-by-N" counter.

Cascading counters to form a larger device: We like to be able to string counters together, much as we string flops together, to form a device with larger capacity. Synchronous counters always permit this: one can simply feed Carry$_\text{out}$ from one stage to Carry$_\text{in}$ of the next. And the properly designed Carry$_\text{out}$ logic will be asserted only when all prior stages are full, as required.[14]

Count direction: Only needed for up/down counters. A single control input selects whether the counter increments or decrements on the clock edge.[15] This also controls when the Carry$_\text{out}$ signal is asserted: on the maximum (terminal) count for up counting or all zeros for down.

BCD encoding: A 4-bit BCD counter counts from 0000 to 1001 then increments back to 0000. Its carry out is asserted on an up count of 1001. By cascading several of these together, you can display count values in base 10 at the expense of needing more bits to handle a given range. For example, two 74HC161 binary counters can resolve 256 values but two 74HC160 BCD counters can only count 100 values.

17N.5 Creating divide-by-N counters

AoE §10.5.2D

Dividing by powers of two with binary counters is easy; you just select the nth output to get a 2^n divider. So if you need a divide-by-four you use the second output (usually labeled Q_1). Dividing

---

[14] The 74HC16X series counters commonly call Carry$_\text{out}$ either "TCO" (Terminal Count Out) or "RCO" (Ripple Carry Out). They do not have a pin named "Carry$_\text{in}$" but instead have two inputs *enables*, ENT and ENP. Both enables must be high for the counter to increment on a clock edge but only ENT is AND'ed with the four counter outputs to form the carry out signal, so make sure you connect that one to the carry out of the lower order counter if you cascade more than two counters.

[15] Except for the oddball 74HC192/3 4-bit counter which has separate clocks, one for up and one for down.

17N.5 Creating divide-by-N counters

by other integers is a bit more complicated. We have to reset or reload the counter (depending on the technique we are using) to get the divider modulus desired. We particularly want you to see the difference between using *synchronous* and *asynchronous* functions to accomplish this task.

Integrated counters are available with both sorts of resets: the 74HC161 is a 4-bit binary counter with an asynchronous clear; the 74HC163 is the same counter with synchronous clear. All four counters in the 74HC16X family have identical pinouts.[16]

The 74HC16X series of counter are made by a number of manufacturers. All work identically but for some reason each decided to come up with its own names for the control signals:

NI (and new TI)	old TI	Nexperia	Fairchild/Renesas/Thompson
clk	cp	cp	clock
clr	mr	mr	clear
enp	pe	cep	enable p
ent	te	cet	enable t
load (ld)	spe	pe	load
rco	tc	tc	ripple carry [output]

We will use the signal names in the first column.

The *only* difference between the '161 and '163 is asynchronous versus synchronous clear (load is always synchronous).[17] It's not hard to state the difference: a synchronous input "waits for the clock," before it is recognized; asynchronous (or *jam*) inputs take effect at once (after a propagation delay of course); asynchronous functions do not wait for the clock. Either synchronous or asynchronous load or reset overrides the normal counting action of the counter.

It's difficult to see why the difference matters without looking at examples. Here are some.

Problem (Divide-by-13 counter) Given a divide-by-16 counter (one that counts in natural binary, from 0 through 15), make a divide-by-13 counter (one that counts from 0 through 12, or through some other set of 13 states). Decide whether you want to use clear or load, and whether you want these functions to be synchronous or asynchronous.

Solution (A poor design: use an asynchronous clear) Figure 17N.20 gives a plausible but bad solution: detect the unwanted state 1101 (13_{10}) and clear the counter on that event.

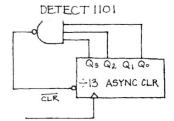

Figure 17N.20 A poor way to convert natural binary counter to divide-by-13.

Why is this poor? The short answer is simply that the design obliges the counter to go into an *unwanted* or false state. There is a glitch: a brief invalid output. You don't need a timing diagram to tell you there is such a glitch; but Fig. 17N.21 shows the false state lasts just long enough to clear the counter.

In some applications you might get away with such a glitch (after all, ripple counters go through

---

[16] The 74HC160 and 74HC162 are BCD counters with async and sync clear respectively.

[17] Indeed, if CLR* is tied high in your application, they are identical and you can use either.

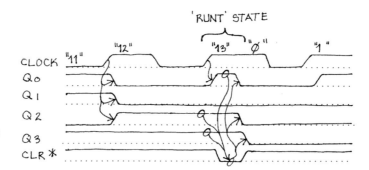

Figure 17N.21 Poor divide-by-13 design: false 14th state between 12 and 0.

similar false transient states, and ripple counters still are on the market). You could get into still worse trouble though: the CLR* signal goes away as soon as state "13" is gone; the quickest flop to clear will terminate the CLR* signal; this may occur before the slower flops have had time to respond to the CLR* signal; the counter may then go not to the zero state but instead to some unwanted state (12, 9, 8, 5, 4, or 1). That error would be serious; not just a transient. (A very similar example is treated in AoE §10.5.2D.) Even if the counter works properly, it becomes difficult to debug circuits with very short signals, particularly if the counter's clock is orders of magnitude slower than the CLR* signal.

Solution (A proper design: synchronous) A counter with a synchronous clear or load function – like the 74HC163 – allows one to modify count length cleanly without putting the counter into false transient states and with no risk of landing in a wrong state. It is also ensures that all signals last for a minimum of one full clock period, making debugging easier.

To cut short the natural binary count, restricting the machine to 13 states, we again need to detect a final state and drive CLR*. This time we need to detect the **12** state, not the 13, as before. On detecting 12, the logic tells the counter *before* the clock to clear on the next clock.

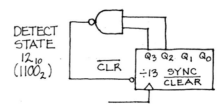

Figure 17N.22 Synchronous divide-by-13 from divide-by-16 counter.

In case you need convincing, Fig. 17N.23 is a timing diagram showing the clean behavior of this circuit. The CLR* signal sits asserted for a full clock cycle before it is acted upon. The synchronous-clear divide-by-13 circuit works nicely. It's too bad though that it requires a NAND.[18] Can we do any better?

AoE §10.6.1

Use of the load function instead of the clear can achieve nearly the same result with a single inverter instead of the NAND *for any divider modulus*. Figure 17N.24 shows the divide-by-13 created with a 74HC161 counter (a 74HC163 would work as well) using its synchronous load.

This use of Load rather than Clear has some funny side effects.

- Using Load in this manner can oblige one to use a strange set of states (starting from three, say, and counting up to 15, then loading three again in order to define 13 states); this would be all right

---

[18] We admit this "one extra gate" becomes ridiculous when one uses a PLD. But let's assume we really want to do the job with this IC counter.

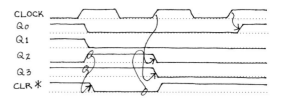

Figure 17N.23 Proper count modification: using synchronous clear.

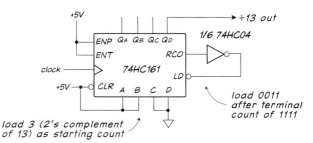

Figure 17N.24 Divide-by-13 counter using synchronous Load function.

if the frequency alone interested you, but it would a problem if you wanted to see the counts 0 through 12.

- You might be tempted to use RCO or $\overline{\text{LD}}$ as the divide-by-n output but RCO is the combinational AND of the four internal counter 'flops. If, say, Q_B goes to 1 before Q_A goes to 0 in transitioning from a count of 1101 to 1110 you could get a short glitch that could be disastrous if you are using that output to clock another counter or a flip-flop. Use Q_D for your divide-by output. See the worked examples for a more detailed look at this issue.

- Or, if we want to use states 0,...,12, then use of Load requires a *down* counter. This is the technique we use in Lab 17L to make a counter of variable modulus: load the initial value; count to zero, then use the Carry/Borrow signal to load once more. ("Borrow," by the way, is just a *down*-counter's "Carry" signal).

We should not make too much of these odd effects: the number of states a counter steps through always bears a slightly funny relation to the value loaded or detected: if you load and count down, *states* = (*count*+1); if you load and count up, *states* = (2's-complement of count loaded); if you detect and clear, *states* = (*count-detected*+1). So, things are tough all over, and it doesn't matter much which scheme you choose.[19]

Synchronous load and clear functions are nice: they support the ideal of fully-synchronous design. But synchronous functions do not entirely *replace* asynchronous because sometimes the synchronous type is a decided nuisance. For example, if you need to reset a counter on an event that is asserted for less than a clock cycle, only an asynchronous reset is guaranteed to work.

17N.6 Counters in Verilog

Counters the hard way: You certainly could build Verilog counters structurally, using individual flip-flops, but it is much easier to build behavioral counters, particularly if you need a lot of bits.

---

[19] If you want to make it easy on yourself, set the load inputs to 0001 and count up to n to create a divide-by-n counter. An advantage of this scheme is load is *always* synchronous in the '161 and '163 so you can use whatever you have on hand.

Counters the easy way: The simplest behavioral Verilog binary counter just increments a vector in an always block. Here is how we create a 4-bit, synchronous up counter:[20]

```verilog
module simple_counter (
  input clk,              // "wire" is implicit
  output reg [3:0] count  // declare a reg signal to hold the count
);

  always @(posedge clk)
  begin
    count <= count + 1;
  end
endmodule
```

You can use a *part-select* to access each bit of the counter. A part-select of the MSB, count[3], will give you a divide-by-16 output at a 50% duty cycle. If you need a larger counter, just change the size of the count vector.

Unlike a counter implemented in software, we do not have to test for the terminal count in a Verilog counter because this code creates *hardware*. We defined count as four binary bits, so the synthesizer assigns four flip-flops to the counter. When you increment a four-bit counter, 1111 increments to 0000 since four bits can only distinguish 16 values.

Non-binary counters: A 4-bit BCD counter is only slightly more complex.

```verilog
module bcd_counter (
  input wire clk,
  output reg [3:0] count
);

  always @(posedge clk)
  begin
    if (count == 4'b1001)   // a 4-bit BCD counter counts from 0000 to 1001
      count <= 4'b0000;
    else
      count <= count + 1;
  end
endmodule
```

Here, a part-select of the MSB will give you a divide-by-10 output (but not at a 50% duty cycle).

A caveat: A common mistake is to not declare enough bits to get the count you need. If you don't have enough bits . . .

```verilog
// 1HZ, 50% counter - but does not work b/c count vector too small
module counter_1hz (     // create 1 Hz, 50% DC output from WebFPGA 16MHz clock
  input WF_CLK,
  output ctr_out
);

  reg [21:0] count;        // declare a [too small] vector to hold the count
  reg tgl_1hz;
  assign ctr_out = tgl_1hz;

  always @(posedge WF_CLK)   // 16MHz clock
  begin
    if (count == 8000000)    // but 22 bits only counts 0 to 4,194,303 decimal
```

---

[20] Counter examples available at https://LAoE.link/FPGA/17N_Counters.v.

17N.6 Counters in Verilog

```verilog
    begin
      tgl_1hz <= ~tgl_1hz;   // does not work
      count <= 1;
    end
    else
      count <= count + 1;
  end
endmodule
```

...the synthesizer will not warn you about it and your conditional will never evaluate as true. Change count to a 23-bit vector ("[22 : 0]") to see it work properly.

An up/down counter: This 8-bit counter counts up or down depending on a control input.

```verilog
// 8-bit up/down counter
module updn_counter (
  input clk,
  input up_dn_bar,
  output reg [7:0] count
  );

  always @(posedge clk)
  begin
    if (up_dn_bar == 1)    // counting up
      count <= count + 1;
    else                   // count down
      count <= count - 1;
  end
```

Adding an asynchronous counter reset works the same way as the flip-flop clear:

```verilog
module areset_counter (
  input clk,
  input reset_bar,
  output reg [3:0] count
  );

  always @(posedge clk or negedge reset_bar)
  begin
    if (!reset_bar)
      count <= 4'b0000;
    else
      count <= count + 1;
  end
endmodule
```

To turn this into a synchronous reset, just remove "`or negedge reset_bar`" from the sensitivity list.

To add a load function, test for $\overline{\text{LOAD}}$ asserted and set count to the input data. To include a carry in, just leave the count unchanged unless it is asserted. Here is a 4-bit binary up counter with synchronous, active low reset and load (reset prioritized over load) with a carry in (enable) and a combinational carry out signal (tco):

```verilog
module fancy_counter (
  input clk,
  input reset_bar,
  input load_bar,
  input enable,         // count enable or carry in
  input [3:0] data,
  output reg [3:0] count,
  output tco            // carry out
```

```
);

always@(posedge clk)    // all functions are synchronous
begin
  if (!reset_bar)
    count <= 4'b0000;
  else if (!load_bar)
    count <= data;
  else if (!enable)     // if not enabled, do nothing
    count <= count;
  else
    count <= count + 1;
end

assign tco = &count & enable;
endmodule
```

17N.7   AoE reading

Chapter 7 (Oscillators and Timers):
- §7.2.4 Timing with digital counters.

Chapter 10 (Digital Logic):
- §10.4.2D Metastability.
- §10.4.3B Example: divide-by-3.
- §10.4.3F Synchronous binary counters.
- §10.5.2 Counters.
- §10.6.1 Modulo-n counter: a timing example.
- §10.8 Logic pathology.

17L Lab: Counters

17L.1 Counter lab

In this lab you start by building both synchronous and ripple counters out of discrete flip-flops. You then move up from the modest "divide-by-four" to an 8-bit "fully synchronous" counter. We will let it show off some of its agreeable features, notably its *synchronous load*, and then we will put it to use in a programmable *divide-by-n* machine. You will finish up by building an exact emulation of the 74HC16X counter in Verilog.

17L.2 Counters: ripple and synchronous

17L.2.1 Ripple counter

If you wire a flop to toggle on *every* clock (the easiest way to do this is with a D-flop[1]) and cascade two such flops, so that the Q of one drives the clock of the next, you form a "divide by four" circuit. One flop changes on *every* clock; the other changes on *every other* clock. Evidently, you could extend this scheme to form a divide-by-a-lot. Today, we won't go further than divide-by-four.

Wire a pair of D-biting-its-tail flip-flops; cascade the two flops to form just such a divide-by-four *ripple* counter.

- Watch the counter's outputs on two LEDs while clocking the circuit at a few hertz. Does it divide by four? If not, either your circuit or your understanding of this phrase is faulty. Fix whichever one needs fixing.

    This counter behaves oddly, in one respect: it counts *down*. An easy way to make it count *up* (or are we only making it *appear* to count up?) is just to watch the $\overline{Q}$ outputs rather than the Q's. (Ripple counters usually are built with *falling-edge* clocks, instead; then the Qs do count up.)

- Now clock the counter as fast as you can, and watch Clock and first Q_0 then Q_1 on the scope. Trigger on Q_1. (Why?[2])

- Watch the two Qs together and see if you can spot the rippling effect that gives the circuit its name: a lag between changes at Q_0 and Q_1. (If you are using an analog scope, you will have to sweep the scope about as fast as it will go, while clocking the counter fast to make the display acceptably bright.)

---

[1] You don't need the fancier "T" behavior to build a ripple counter. You *will* need it to make a synchronous counter.
[2] You want to trigger on the signal the happens *less often* so the two signal stay synchronized. If you triggered on the LSB, it would look stable but the horizontal location of the next most significant bit could wander depending on if the trace starts at the `00 -> 01` transition or the `10 -> 11` transition.

17L.2.2 Synchronous counter

The timing diagram in Fig. 17L.1 shows how a *synchronous* divide-by-four counter ought to behave.

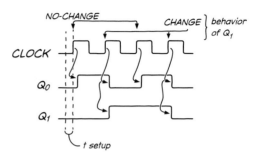

Figure 17L.1 Timing diagram showing behavior wanted from synchronous divide-by-four counter.

As you know, the diagram is the same as for a ripple counter – except for one important point: in the synchronous design, *all flops change at the same time* (at least, to a tolerance of a few nanoseconds). Achieving this small change in behavior requires a thorough redesign of the counter.

You will need a T-flop to make a synchronous counter. Then the key is to find, on the timing diagram, the pattern of pre-clock conditions that determine whether Q_1 ought to change (toggle). To permit synchronous behavior, we must look at conditions *before* the clock edge (during set-up time); no fair to say something like "let Q_1 change if Q_0 falls;" that scheme would carry us back to the ugly ripple behavior.

Once you have discovered the pattern, drive T_1 appropriately, and try your design..

See if you can use the scope to confirm that the *ripple* delay now is gone: watch Q_0 and Q_1 again. Synchronous counters are standard, ripple counters rare. In many, if not most, counter applications, the ripple counter's slowness in getting to a valid state and its production of transient false states are unacceptable.

Suppose you wanted to add another T flip-flop to get a divide-by-eight counter. What else would you have to add to get the desired behavior? Can you think of a general rule of what is needed to increase an N-bit synchronous counter to $N + 1$ bits?[3]

17L.3 Test a divide-by-3 built two ways

Build the divide-by-3 circuit in Fig. 17L.2 using a 74HC161 synchronous 4-bit counter with asynchronous reset. Be sure to hook up power and ground (not shown).

- Drive the counter with a 10 kHz clock and look at the Q_B output. Is this a divide-by-3?[4]
- Now look at the Q_A output. Is it also a divide-by-3? (Hint: you might have to look carefully.)[5]
- Finally, look at the $\overline{\text{CLR}}$ input. At first, it looks like it is always high, but if you look closely you should see the short low-going pulse that is resetting the counter.

Reconfigure this circuit to create a divide-by-three with a synchronous reset. (You can either

---

[3] To add the third stage, you need a T flip-flop and a two-input AND gate. To add an $N + 1$ stage, you need a T flip-flop and an N-input AND gate. (You could use a two-input gate and just AND the T input and Q output of the Nth flip-flop to get the T input of the $N + 1$ stage, but then you would have to wait for $N - 2$ AND gate delays before you could clock the counter again, eliminating much of the advantage of the synchronous design.)

[4] It should be if you built it correctly.

[5] Q_A might look like a divide-by-3, but if you expand the trace you should see a short (10 to 20 ns) high-going pulse before the counter resets. If you tried to use this to clock another device, it might act like a divide-by-two.

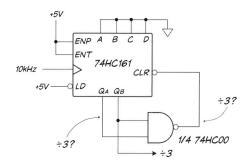

Figure 17L.2 Divide-by-3 using async reset.

substitute a 74HC163 or just use the $\overline{\text{LD}}$ input to load 0x0000 into the counter. You will also have to change the count which causes the reset.) Look at the same signals (Q_A, Q_B and the signal that clears the counter). While the asynchronous reset divide-by-three counter worked, you had to deal with pulse widths orders of magnitude shorter than the clock. None of the signals in the synchronous reset divider are shorter than one clock period. That is why we prefer divide-by-N counters using synchronous reset.

17L.4 8-bit counter

The 74LS469 is a versatile 8-bit up/down counter.[6] Here are its features.

- It counts up or down.
- It includes three-state outputs: this feature allows you tie the counter outputs directly to a computer bus if you need to. The outputs are high impedance (floating) when $\overline{\text{OE}}$ is not asserted.
- It can be loaded with an initial value (although only *synchronous* load is supported). Today, we will use $\overline{\text{LD}}$ to allow us to make a *divide-by-strange-number*.
- It includes carry in ($\overline{\text{CBI}}$) and carry out ($\overline{\text{CBO}}$) pins that normally would be used to allow cascading several of these units, to form a bigger counter.[7] The carry in tells the counter to either count or to hold its present value despite clocking. (You may recognize carry in as just equivalent – on the counter-wide scale – to the T input that controlled the single T-type flip-flop that you built in Lab 16L from a D-flop and an XOR gate.) We will take advantage of this carry in function – which can also be described as an *enable* – in forming the stopwatch of §17L.6 The carry out is useful to cascade counters or to capture counter overflow.
- The counter does not include a separate reset. If you want to set it to zero, you have to use the load function with the load inputs set to all zeros.

17L.4.1 Wiring the counter

Since the '469 is an LS technology part it runs on +5 V. Any of its outputs connected to digital logic should go to LS or HCT parts since the input of HC parts are not guaranteed compatible with TTL output voltage levels. However, if you do want to use it with HC parts, connecting the LS output to a 10k to 33k pullup resistor ensures the CMOS input sees a valid logic 1 when the TTL output is high. If you are using the WebFPGA to simulate a 74LS469, then all inputs must be limited to ≤ 3.3 V.

---

[6] The 74LS469 is considered an obsolete device and may be hard to find. For alternatives, see online chapter https://LAoE.link/LAoE_Chapter_170.pdf.

[7] $\overline{\text{CBI}}$ and $\overline{\text{CBO}}$ stand for Carry/Borrow In and Carry/Borrow Out respectively. In an UP/DOWN counter, this carry sometimes is called "borrow" on the down count, by analogy to subtraction. We'll call it just "carry" for both counting directions.

Start by setting up the counter so that you can see it count:

1. Use the eight breadboard logic switches to set the counter load value. If you make S1 the MSB and S8 the LSB, you can enter values as they would be written on paper (LSB on the right). You do not have to disconnect the switches from the FPGA, they can drive both devices at once.
2. Show the counter outputs on the breadboard LOGIC INDICATORS. The indicator LED switches should be set to "TTL" and "+5 V."
3. Manually clock the counter from one of the debounced pushbutton switches. (The counter should change when you press the button, not when you release it.)
4. Connect the $\overline{LD}$ input to the N.O. output of the other debounced pushbutton (or you can use a slide/toggle switch to apply ground or +3.3 V).
5. Connect the $\overline{U}/D$ input to a slide switch (either one of the uncommitted ones on the breadboard or add one from your lab kit). Set up the switch so it can apply either +3.3 V or ground to the counter $\overline{U}/D$ input.[8]
6. Either ground the $\overline{CBI}$ input or connect it to a slide switch to select 0 V or +3.3V.
7. Connect the $\overline{CBO}$ output to a LED connected through a resistor to ground. The counter output is spec'ed at a bit over 3 mA so use a high efficiency LED and choose the current limiting resistor accordingly.
8. Make sure to enable the 3-state outputs. If you connect the $\overline{OE}$ input to a slide switch to choose between 0 or 5 V, you can test the three-state outputs.

Optional: DIP switch controls inputs. You can use a 4-position DIP switch to control some or all of the $\overline{LD}$, $\overline{CBI}$, output three-state enable ($\overline{OE}$) and $\overline{U}/D$ inputs if you prefer. Figure 17L.3 shows what the DIP switch controlling its four signals might look like.

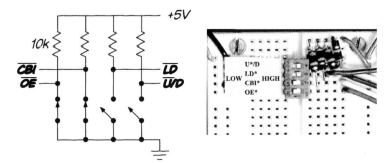

Figure 17L.3 DIP switch used to control four counter signals.

17L.4.2 Test the counter

Now test the counter to see if it works correctly.

Normal counting: Pressing the clock pushbutton should advance the counter when the $\overline{U}/D$ input is low and decrement it when it is high whenever $\overline{CBI}$ is asserted. The counter should not count if $\overline{CBI}$ is not asserted. The $\overline{CBO}$ output should be asserted on a count of 0xFF when $\overline{U}/D$ is low and a count of 00H when high but only if $\overline{CBI}$ is asserted.

Try loading: You should be able to load any value from the logic switches into the counter by clocking the counter when $\overline{LD}$ is asserted no matter what the state of $\overline{CBI}$ is.

---
[8] The 74LS469 datasheet calls this control "$\overline{UD}$", but that makes no sense.

17L.4.3 Try the counter at higher speed

Now replace the pushbutton clock signal, which was clocking the counter, with a logic-level signal from the function generator. Try f_{clock} of a few kHz.

Watch clock and Q_0, then clock and Q_7 (triggering in both cases on the Q – and the *slowest* Q when watching more than one).

In the unlikely event that your counter's behavior is erratic you may want to clean up the clock, by passing the function generator's TTL signal through a local logic gate that has hysteresis: see Fig. 17L.4.

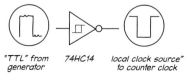

Figure 17L.4 HC14 can clean up a messy clock.

Then watch Q_0 and Q_7: do you see any delay of the higher-order Q relative to the lower-order, as you did in even the small ripple counter you built in §17L.2.1?[9] Now take a look at $\overline{\text{CBO}}$. You will need to trigger on the carry – and note that it is only one clock wide, so may look very narrow indeed if your scope screen's sweep covers many clocks.

If you use a couple of spare scope channels to watch Q_0 and Q_1 you can see the difference between carry on an UP count and carry on a DOWN count. If you feel like playing with the logic LED outputs again, drop the clock rate way down so that you can see individual states. Check that counting up and counting down behave as expected.

17L.5 Build a selectable divide-by-N counter

Configure the 74LS469 as a selectable divide-by-N counter. If you let the counter's $\overline{\text{CBO}}$ line drive its $\overline{\text{LD}}$ pin, it will reload itself each time it overflows. The operation of your circuit should be as follows:

- Use the logic switches on your breadboard to set a divide value. Setting 00000001 divides by 2 (the counter loads 00000001, decrements to zero on the next clock then reloads 00000001 on the clock edge after that). Setting higher binary values divides the clock by the logic switch value plus 1.
- Use LS or HCT logic or pullup resistors on any '469 outputs connected to HC devices. (You do not need pullups to display the outputs on the breadboard indicator LEDs if set to "TTL" and "5 V.") Be sure to remove any sources (like a switch) driving counter input signals that you need to connect to a different output. Remember, you cannot connect two outputs together!
- Continue to display the eight counter outputs on the breadboard indicator LEDs and display the $\overline{\text{CBO}}$ on the stand-alone LED. This LED will display a once per divide cycle signal (i.e., the led should light up once per divide-by-N count cycle, perhaps only for a short time.)

You can initially manually clock the counter from one of the debounced pushbutton to debug it, then connect it to the function generator or breadboard oscillator TTL output to run it at about 10 kHz. Your scope can display the frequency of the input and output to see if you are dividing correctly.

---

[9] We hope not.

17L.5.1 Make horrible music

To make more vivid the power of this loadable counter to vary its *modulus* – the number of states it steps through – let's listen to the counter's output frequency. Let $\overline{\text{CBO}}$ drive a transistor switch (a MOSFET is easiest), which in turn drives the breadboard's speaker, as shown in Fig. 17L.5.[10] If you want to annoy your neighbors with a louder tone, let $\overline{\text{CBO}}$ drive a toggling flip-flop: the 50%-duty-cycle signal that comes out of the flop makes more noise than the narrow $\overline{\text{CBO}}$ pulse does. Without this flop, the low duty-cycle of the $\overline{\text{CBO}}$ pulse can make it barely audible for some count-lengths.

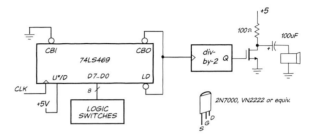

Figure 17L.5 Music maker: divide-by-two makes output volume greater, and independent of frequency.

Fixing an interesting flaw in this circuit:

The problem: glitches on $\overline{\text{CBO}}$: When we tried this, we were surprised to find that the $\overline{\text{CBO}}$ signal – which detects the counter's state *zero* during *down* counting – carried nasty glitches, glitches substantial enough to clock the *divide-by-two* flip-flop that we intended to provide a 50% duty cycle. The glitches made nonsense of our music machine: see Fig. 17L.6.[11]

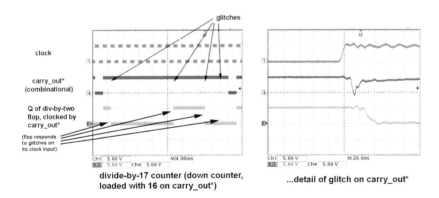

Figure 17L.6 Glitches on carry can cause disaster when driving a clock.

The remedy: synchronize the $\overline{\text{CBO}}$ signal with a flop: You'll notice that the glitches in Fig. 17L.6 occur *after* the clock edge. This is the usual case, since it is the changing signals, brought about by

---

[10] If you have not seen the notation of a signal line with a small slash and a number *n* next to it before, it is a way of indicating a set of *n* parallel related signal lines. If you find yourself drawing schematics with 32-bit buses, you will be very grateful for this shortcut.

[11] To be fair, this happened with a different counter – one we made out of a PAL – but this problem is not uncommon in synchronous systems. Since such circuits only care that the signals are stable just before the clock edge, designers don't worry about unexpected transitions on combinational logic outputs at other times. We describe a fix for this problem in Worked Example §17W.1.

clocking, that trigger the glitches. Such glitches are entirely harmless to a flip-flop clocked by the signal that evoked the glitches – in this case, the clock of the main counter.

Figure 17L.7 illustrates a remedy for these bad carries: a synchronizing circuit that blocks the glitches. The glitches will not pass through the flop, because they occur *after* the edge that updates this synchronizing flip-flop.

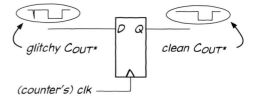

Figure 17L.7 A synchronizer flip-flop can clean up C_{out*}.

If your music machine is troubled by carry glitches, you could include such a synchronizer upstream of the divide-by-two flip-flop that is to drive the speaker but you can save the extra flip-flop by combining the synchronizer with the divide-by-two. How? Well we already have seen such a synchronous device, the toggle flip-flop. Figure 17L.8 shows a synchronous divide-by-two to connect to the glitchy 74LS469 $\overline{CBO}$ signal. It is a slightly modified version of our T flip-flop from §17N.2.2.2. Here the flop-flop output toggles when the input is *low* rather than high (saving us an inverter on $\overline{CBO}$).[12]

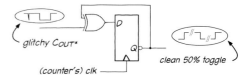

Figure 17L.8 A modified toggle flip-flop both synchronizes the glitchy C_{out*} signal line and divides by two.

Sketch your clean music maker circuit (so that if you look back in a month or so, you'll rediscover how clever you were).

Once things are working right, the logic switches should act like the keyboard of a crude musical instrument. Do you hear the pitch fall by an *octave* when you change from divide-by-X to divide-by-$2X$?

Out of tune because of the zero state: For very low divide-by values, your instrument goes out of tune because of the extra state *zero* included in each down-count cycle.

For example, if you load 16_{10} from the logic switches, it counts down from 16 to zero, then reloads 16. So, it goes through 17 states, not 16. When you load a 32 from the keypad you might expect a doubling of period – but this evokes 33 states – not quite an octave different from 17. In fact, that small difference amounts to almost exactly a semi-tone, as your ear may confirm for you.

17L.5.2 Create a 1 Hz clock source

Another use of a divide-by-n counter is to create a timing source from the power line frequency.[13] Build the line frequency clock source in Fig. 17L.9 and connect it to your divide-by-n counter. Set your counter to divide by 60 (or 50) and confirm that the output is a 1 Hz signal. Older line-operated digital clocks use this method to create their timing signal.

---

[12] Yet another thanks to Jim MacArthur for this idea.
[13] You could use a crystal oscillator but, surprisingly, the AC line is more accurate over the long run because it is adjusted by the power company to *average* to exactly 60 Hz (or 50 Hz) over time, a process known as Time Error Correction (TEC). See https://LAoE.link/TEC.html.

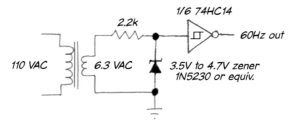

Figure 17L.9 A line frequency clock source.

17L.6 Counter applications: stopwatch

We know this is a long lab. We offer these exercises because we think you'll find them fun, and instructive. Do what you can. Don't feel guilty if you don't build the stopwatch. You may get a chance to build it using the FPGA in the digital project lab, 21L.

A very slight alteration of the 8-bit counter will allow you start and stop the counter: you need only add a manual switch to control the level of the $\overline{\text{CBI}}$ and a pushbutton to $\overline{\text{LD}}$ (used as a reset with the D inputs all set low).

The circuit sketched in Fig. 17L.10 can measure the length of time a signal spends low by replacing the HOLD/$\overline{\text{COUNT}}$ switch with the signal to measure (and, of course, you could add an inverter to measure the time a pulse is high). This period-measuring circuit then could be put to any of a number of uses. In Lab 21L for example, we propose that you use a similar FPGA-based counter to measure the period (or half-period, to be a little more accurate) of the waveform coming from a 555 RC oscillator. If you then hold the R constant and plug in various Cs, you will find that you have built a *capacitance meter*.

Connect the $\overline{\text{LD}}$ input to a NO pushbutton with pullup resistor. You might want to connect the $\overline{\text{CBI}}$ input to the other pushbutton's NO output as well, if you don't mind holding it down while timing. Otherwise, leave it connected to a slide switch. Drive the counter clock with a 10 Hz signal from the function generator or breadboard generator; this gives you a stopwatch able to time up to 25.5 seconds with 0.1 second resolution.[14]

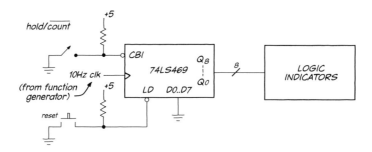

Figure 17L.10 Stopwatch block diagram.

First confirm that you can start and stop the counter by taking $\overline{\text{CBI}}$ low, then high, using the pushbutton or switch. You can clear the counter, as you confirmed earlier in this lab, by pressing the RESET pushbutton. You now have a clumsy stopwatch (and you have to read the output in binary).

---

[14] You could use 100 Hz instead to get 0.01 sec resolution, but only for timings under 2.56 sec.

17L.7 Counters in Verilog

Be sure you only apply 3.3 V signals to the inputs of the WebFPGA in the following experiments. When you connect to external HC logic devices, run them on 3.3 V as well.

17L.7.1 Design and test a 74HC161 counter

Design a Verilog version of the 74HC161 counter. Use the 74HC161 datasheet to make sure you match the actual part exactly. Here is the module definition:[15]

```
module CTR_74HC161 (
  input clk,
  input clear_bar,
  input enable_p,
  input enable_t,
  input load_bar,
  input A, B, C, D,
  output Qa, Qb, Qc, Qd,
  output RCO
);
```

Connect your WebFPGA 74HC161 to breadboard resources as follows:

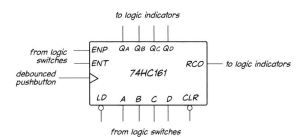

Figure 17L.11 WebFPGA 74HC161 test connections.

Test your counter: Connect the clock input to a debounced pushbutton and the $\overline{\text{CLR}}$, $\overline{\text{LD}}$, ENT, ENP and the A–D load inputs to the eight LOGIC SWITCHES. The four counter outputs and the TCO output should go to LOGIC INDICATOR LEDs. Make sure the pushbutton increments the counter when pressed (not when it is released). Test all the 74HC161 functions for proper operation. Refer to the datasheet if necessary.

Create a 74HC163: Modify your design to be a 74HC163 counter and test it. This should be a fairly simple Verilog modification.

Test a divide-by-N: Connect an (actual) 74HC04 inverter between TCO output and $\overline{\text{LD}}$ input of your WebFPGA 74HC16X counter (remember to disconnect the Logic Switch from the latter and run the inverter on 3.3 V).[16] Connect the clock input to a function generator 0 to 3.3 V output (again disconnect the pushbutton switch) and set it for an approximately 10 kHz square wave. Use your oscilloscope to

---

[15] There is a test bench at https://LAoE.link/74HC161_counter.html to test your code. You should make a copy before adding your counter code so you can save your design. See Appendix E, online at https://laoe.link/Testbenches.pdf.

[16] Ok, making you wire up a real inverter when you could just add one line of code to your Verilog design to do the same thing is silly, but we want to pretend the WebFPGA is an actual 74HC16X counter.

measure the input and output frequencies and try various (or all) load values to see if your settable divide-by-N circuit works properly.

Optional: *try the divide-by-3 again:* If you are bored, you can repeat the experiment in §17L.2 using your WebFPGA 74HC16X counter.

Leave the WebFPGA and inputs and outputs connected as is. You will be using your 74HC16X counter again in the next lab.[17]

---

[17] Solutions available at https://LAoE.link/FPGA/17L_Counters.v.

17S Supplementary Notes: Digital Debugging

17S.1 Digital debugging tips

Here are a few thoughts on debugging digital circuits.

Use a logic probe, not DVM or – worse – your eyes This should go without saying, but we're not sure it yet does. We find it *boring* to stare at a wire, trying to see if it *goes* where it should. It's more fun, faster, and more instructive to probe that same wire seeing whether it *behaves* as it should, particularly since the logic probe captures short pulses and indicates when voltages are not legal logic levels.

Probe chip pins, not the breadboard: Usually, but not always, breadboard and chip match: an IC's pin may be folded under, or it may be failing to make contact with the breadboard because the breadboard socket contact has spread with rough use. So, be skeptical: poke the pins.

Figure 17S.1 A case where probing the breadboard would not be good enough to reveal the IC's behavior (bottom view).

Disconnect loads from a misbehaving signal: Suppose one bit of the '541 3-state buffer won't follow its data input; it shows neither high nor low when the buffer is enabled. You can detect this most easily with a logic probe but a scope would show a voltage intermediate between legal logic levels. Whose fault is this fishy output level? The '541's? Or is it something "downstream" on that data line? The downstream device may be driving the line, producing "bus contention" – a fight between a high and a low that produces the intermediate level.

It's easy to settle that question if you pull out the line that the '541 pin drives. Now enable the '541 and watch its output pin: OK? look at the breadboard at that position. OK? Look at each of wires that earlier was plugged into that slot. Often you will find '541 OK, but one of the wires asserting a high or low when it should be only listening. If you do, you know to continue your search *downstream*.

Corollary: Build and *test each functional block* of your circuit before moving to the next. It is unlikely every circuit you build will work the first time, either because of a build error or a design flaw. If you build your entire circuit and it does not work, how do you know where do you start looking for a problem? It is much more productive to build and test each functional block (defined as the minimum build that is testable) before moving on. When you build the next block, test it before connecting it to the previous build. If you connect a working block to a working previous circuit and it no longer works, you know the problem is in the interface between the two. (P.S. This advice applies to analog circuit construction and debugging as well.)

Check power supplies: Here's a dull rule – dull, but useful: before looking for subtler causes, make sure your misbehaving chip has *power*.

That may sound silly to you. You may be inclined to reject this advice, thinking, "If the chip didn't have power it would act dead, and my circuit is misbehaving in stranger ways. It doesn't just act dead." Such a thought rests on an error: it is hard to remember – but necessary – that a CMOS circuit can work *without power applied to its V_{DD} and ground pins*. This is a point you may recall from the Lab 14L.

CMOS can do this perverse trick because protection diodes on the input lines allow *inputs* (or outputs – but these are much less likely to drive the chip) to provide power and ground (at slightly degraded levels). Figure 17S.2 shows the typical input- and output-protection scheme.

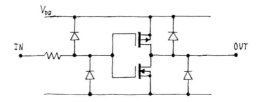

Figure 17S.2 CMOS input-protection diodes can power a chip through *inputs*.

Missing power and ground connections can produce strange intermittent failures: If your circuit *sometimes* fails, perhaps it lacks power, ground, or both. It may be taking the missing connection through an input. So, to take a simple example, imagine that you are using one gate of an AND package, as in Fig. 17S.3.

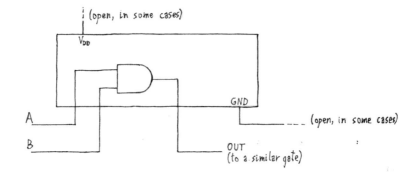

Figure 17S.3 Hypothetical use of a gate in an AND package.

If you omit one or more power and ground connections, you may find the following puzzling behavior patterns, as you may have seen in Lab 14L:[1]

---

[1] As we said there, we found that National Semiconductor (now TI/National Semiconductor) 74HC00's behave this way; Motorola's give a high for the High–High input combination.

17S.1 Digital debugging tips

Input	Will this gate work?		
A B	Both V_{DD} & GND open	Only V_{DD} open	Only GND open
0 0	NO	NO	YES
0 1	YES	YES	YES
1 0	YES	YES	YES
1 1	NO	YES	NO

If you fail to provide power through the power and ground pins then the gate will work *sometimes*. It will work by stealing power from a data line – as long as it finds both highs and lows available at either input or output. The cases that *don't work*, above, are the ones where *ins and outs are unanimous*.

So when a gate is showing weird misbehavior you may save a good deal of time if you can humble yourself sufficiently to poke power and ground pins early on to see if they're connected to the power buses.

Strange floats: three common causes: A logic probe sometimes will show you a *float* that you don't expect. (We are using the word "float" as shorthand for *an intermediate logic level*, as in §17S.1: neither high nor low. Often such a level is not the result of a truly *floated* output, which can be achieved only by an OFF 3-state.) Such a float is a sign of a specific kind of trouble. Sometimes it looks even stranger: you see not just a float, but a fast *oscillation*. To detect the float you should use a logic probe, and if it has a switch you need to set it to the appropriate technology (CMOS versus TTL): we have gotten fooled by ignoring that switch. The probe will detect the fast oscillation too if you set its switch (if any) to Pulse rather than Memory.

If you do detect an *oscillation*, then it's time to get a scope: see §17S.1.1.

A strange float usually results from one of the following three causes.

1. The source has been disconnected by mistake (a bad connection; perhaps the stripped length of a wire stuffed into a hole is too short).
2. The source logic chip may lack power or ground. Don't forget (as noted above) that a CMOS part lacking power and ground can appear to work fine, part of the time, by stealing power through its inputs.
3. Two outputs may be fighting. To check this, just disconnect all but one source, as suggested above.

Probe the fishy bit once you have discovered which one is fishy. Often the stubborn "1" will turn out to be not a true "1" but a *float* (the data displays may treat floating inputs as logic highs, in the manner of TTL, or they may simply show a value earlier driven and now held on stray capacitance).

The most common cause for the surprising float is the dullest: #1 above: simply a bad (open) connection in the signal path.

Strange values on data or address displays: treat these as valuable clues:
One likely pattern: 0 goes to 2; 4 goes to 6; 8 goes to A... Step through the sequence slowly while writing the sequence down in binary. Usually the misbehavior jumps out at you once you see the sequence in as individual bits (for example two signals are connected by a short circuit). If the pattern does not declare itself then try writing an additional column of values: the *expected bit pattern*. Table 17S.1 below gives some sample misbehaviors made intelligible by the binary listing:

Hypothetical hexadecimal-character sequences, where plain increment was expected:
Another likely pattern: 0 goes to 0; 1 goes to 8; 8 goes to 1; 3 goes to C...; or the count sequence is strange... Again, write out the pattern or sequence. Usually you will find bits interchanged. In the

Supplementary Notes: Digital Debugging

Table 17S.1 Misbehavior

Sequence:		
Hex	Binary	⇒ Diagnosis
3	0011	
2	0010	
3	0011	
6	0110	Bit 1 is always high;
7	0111	(possibly floating)
6	0110	
7	0111	
A	1010	
Hex	Binary	⇒ Diagnosis
0	0000	
1	0001	Look at how fast each bit toggles:
8	1000	Bit 0 changes every time: Good!
9	1001	
4	0100	There is a bit that changes every second clock –
5	0101	it behaves, in other words, like bit 1; but the bit that behaves
C	1100	like bit 1 is in the wrong position: it's the *leftmost* bit.
D	1101	Meanwhile, the slowest bit is in the bit 1 position – whereas it ought to be bit 3. Apparently, bit 1 and bit 3 are swapped; to put it another way, the top 3 bits have been rotated (This mistake is easy to make on an HP hex display which has screwy pinouts.)

example we have just described, data lines D3 and D0 are interchanged; so are D2 and D1. In other words, the ribbon cable feeding the four data lines needs to be rotated 180°.

17S.1.1 A harder puzzle: oscillations

A mysterious *oscillation* is probably the second-hardest problem to debug. (The hardest problem is a malfunction that is *intermittent*.) In digital circuits, surprise oscillations do not result from the usual "parasitics" that you learned to know and love [hate?] in the analog section: since digital circuits normally are immune to small noise signals, they won't amplify and propagate these as an analog circuit can.

But it's not too hard to make an unintended oscillator by wiring an unintended feedback loop – one that contradicts itself. That sounds like negative feedback, which in our analog experience would be self-stabilizing. But in a digital circuit, with its fast transitions, negative feedback with the inevitable gate-delay often produces an oscillation. You can call it "output contradicting input." That name makes the oscillation seem plausible. The *period* (or frequency) of oscillation often will provide a clue to what's the likely feedback path. Below are two cases to illustrate the point.

Case 1: A fast oscillation indicates a short feedback path: If we see an oscillation as fast as the one shown in Fig. 17S.4 – period just a couple of gate delays, we can guess that the feedback path runs through *a single gate*. Recall that the period shows *two* passes through the gate, so *two* gate delays. The HC14 waveform reproduces a case we saw recently in the lab: someone tied output to input, on this gate, an inverting Schmitt trigger. The HC14's typical gate delay (from its spec sheet) is about 11 ns at 5 V. This oscillation runs faster than that; but we're in the right range.

Case 2: A slower oscillation indicates a feedback path through a slower device: About the only point that's easy about Case 2 is the conclusion that this loop passes through far more than a

17S.1 Digital debugging tips

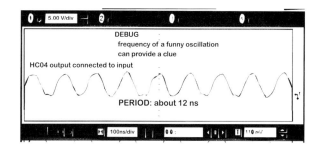

Figure 17S.4 Case 1. A nasty oscillation: path is through a single gate.

single gate. The period again should reflect *two* contradictions, so we'd expect the path delay to be half the period, or around 60–65 ns. This is too slow for a gate, and too fast for a processor cycle (the CPU clock period we were using is about 90 ns; the quickest processor action takes several clocks). This delay is about right for *memory*.

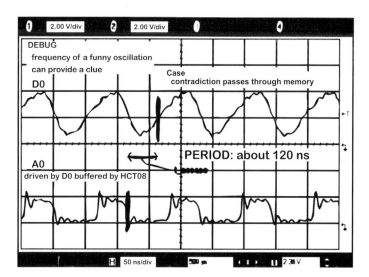

Figure 17S.5 Case 2. A second nasty oscillation, this one too slow to blame on one or two gate delays.

In fact, we generated the waveform in Fig. 17S.5 by passing a contradiction through *memory*: we tied A0 to D0, then wrote in data that made D0 "contradict" A0: at address 0 we wrote *1*; at address 1 we wrote *0*. The contradiction should take t_{access} to propagate. A simple short, we're sad to admit, did not produce an oscillation for this case: the address drive (CPU) is stronger than the data drive (CMOS RAM). So we cheated a little, buffering D0 with a 74HCT08. That makes it a fair fight and produces the oscillation (the 'HCT08 also contributes perhaps an additional 6–9 ns of delay, as well).[2]

In general, don't suffer too long (>5 minutes) with a pesky problem. Ask for help! A second pair of eyes can be invaluable.

---

[2] The oscillation is much faster than the specified t_{access} would let one predict for this 100 ns RAM; but such devices usually run faster than this *worst-case* spec. So again we're close enough to the expected value so that the period of oscillation helps our diagnosis.

17W Worked Examples: Applications of Counters

17W.1 Using carry out as a clock signal

In a synchronous counter, Carry$_{out^*}$ is generated by ANDing the Carry$_{in^*}$ input with all the counter's Q outputs. It usually works well because it is applied to a synchronous input, Carry$_{in^*}$, of the next higher order counter: see Fig. 17N.17. It may not work well if it is put to an *asynchronous* use clocking an edge-triggered device. The problem is that the combinational Carry$_{out^*}$ can show glitches. This can occur if the several Qs on which it depends show sufficient *skew* – that is, change at different times, despite their common clock. A combinational Carry$_{out^*}$ for a PLD counter we built showed such glitches: The detailed image, on the right in Fig. 17W.1, makes clear that the timing of the glitch – *after* the clock edge – makes it harmless to a *synchronous* circuit, which samples its inputs only during *setup time* just before the active clock edge.[1] This immunity is the great virtue of synchronous circuitry.

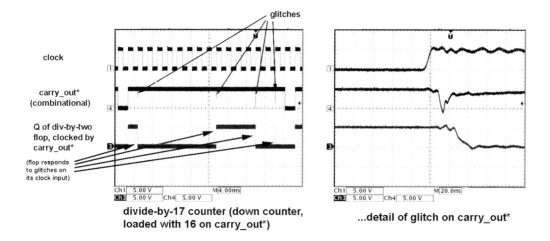

Figure 17W.1 Combinational, asynchronous Carry$_{out^*}$ can show glitches – substantial enough to clock a flop.

But when one wants to use Carry$_{out^*}$ to drive a circuit that responds to an edge of Carry$_{out^*}$ itself – like the toggling flop shown as the bottom trace of Fig. 17W.1, or like an overflow-detecting flip-flop, clocked by the Carry$_{out^*}$ – then the glitches can cause trouble, and we need a remedy.[2]

---

[1] Some devices show non-zero hold time, as you may recall. But even these parts rarely show hold times that would reach so far as the glitch in Fig. 17W.1.

[2] We were surprised to find these glitches severe in *down* counting by in our PAL Up/Down counter, and barely apparent in *up* counting. Apparently, some mismatch between rise-times and fall-times accounts for this difference – since the Carry$_{out}$ depends on *lows* for the Down direction, *highs* for Up.

17W.1 Using carry out as a clock signal

To clean up the Carry, one must make it *synchronous* before using it. Instead of detecting the state *maximum-count* (which warns of a rollover on the next clock), a *synchronous* Carry uses a combinational circuit to detect the state one *before* the maximum state (for an up counter; minimum state, for a down counter), and feeds that to the D of a flop clocked by the counter clock: see Fig. 17W.2.

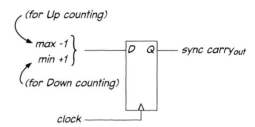

Figure 17W.2 A D-flop can form a synchronous carry.

This synchronizing technique is simple enough; it is not so easy to make a satisfactory synchronous Carry$_{out}$ for an up/down counter.[3]

Such a Carry$_{out*}$, being synchronous, is clean: it does not pass glitches that occur in the disorderly time soon after the system clock edge. Here, in Fig. 17W.3, is such a synchronous carry (this is the design we adopted for our PLD counter).

Noise still appears on the Carry$_{out*}$ line – but the noise pulses are not substantial enough to matter, as Fig. 17W.3 attests: the flip-flop ignores those noise pulses, responding only to the legitimate Carry$_{out*}$ signal (the flop is a divide-by-two, clocked by Carry$_{out*}$). Contrast the tidy behavior of the divide-by-two here with the nasty effects of carry glitches in Fig. 17W.1.

The right-hand image in Fig. 17W.3 shows the prettifying effect of adding a decoupling capacitor on the supplies of the PLD counter.[4] This decoupling does not affect the behavior of the circuit, since the noise was harmless – but it does make the waveform prettier. (And we hope you are getting into the habit of always bypassing your power supplies in any case.)

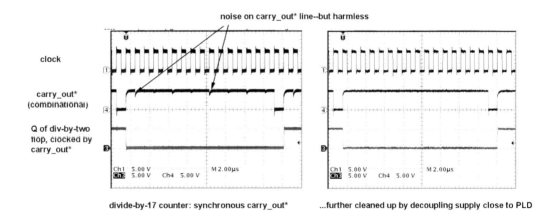

Figure 17W.3 Synchronous Carry$_{out*}$ is clean, unlike the asynchronous combinational carry.

---

[3] Nearly always, the scheme sketched in Fig. 17W.2 works. It can fail if one changes the count direction on either Max or Min count, since the circuit detects only the state preceding Max or Min.

[4] The decoupling capacitor is 0.1 μF ceramic, applied at the PLD socket. Jim MacArthur suggests this easy-to-remember rule: any IC with a clock input needs a decoupling capacitor as close as possible to its supply pin(s).

17W.2 Counting as a digital design strategy

Because it is easy to make digital circuits that count with accurate time measurement, it often turns out that a good way to make a digital device designed to measure some quantity is to build what in effect is a *stopwatch* measuring the duration of a cleverly generated pulse. More specifically, here's the idea:

1. build a counter that counts clock edges (this is a sort of "watch");
2. add gating that lets you start and stop the watch (making it a "stopwatch");
3. build some circuitry that provides a waveform whose period is proportional to a quantity that interests you (call this quantity "input"); and
4. use the stopwatch to measure period, and thus to measure "input."

Here is an example to illustrate the technique.

Digital voltmeter (or ADC; voltage input): The circuit of Fig. 17W.4 measures the time a ramping voltage takes to reach the voltage labeled "INPUT." The higher the input, the longer the time and the larger the resulting readout. This is a "single-slope" ADC. A slightly more complex design, called "dual slope," is preferable (see §20N.3.2); but its strategy is similar to the single-slope design, shown in the figure.

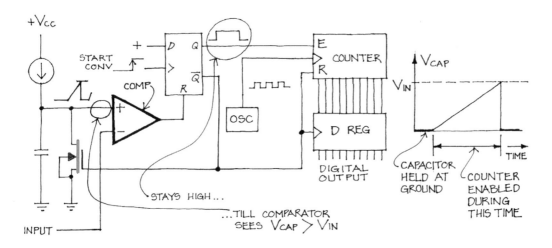

Figure 17W.4 Example: measure period to measure *voltage*.

17W.3 Using a counter to measure period, thus many possible input quantities

Counting as a digital design strategy: As noted in the previous example, we can build a variety of instruments using the following generic two-stage form:

- an application-specific "front end" generates a pulse whose duration is proportional to some quantity that interests us; and
- a "stopwatch" measures the duration of that front-end pulse.

17W.3 Using a counter to measure period, thus many possible input quantities

Here we will first look examples of circuits that fail in an attempt to use this arrangement, and then we will go through a longer design exercise where we try to do the job right.

17W.3.1 Three failed attempts

First failure: digital piano-key speed-sensor: Figure 17W.5 is a flawed scheme: part of a gadget intended to let the force with which one strikes the key of an electronic instrument determine the loudness of the sound that is put out. The relation between digital count and loudness shows a nasty *inverse* relation.

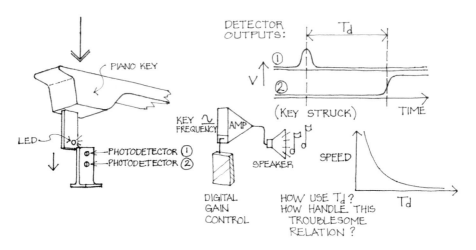

Figure 17W.5 First failure: measuring period to measure how hard a key was hit.

We offer this as a cautionary example: probably this piano design is a scheme worth abandoning. The design could be made to work with some clever post-processing of the signal; but to call for that suggests there must be a better arrangement.

Second failure: latching a final count: In Fig. 17W.6 a small error makes this big circuit useless. The clearing and latching are driven by a shared pulse. So the cleared value is saved: zeros forever.

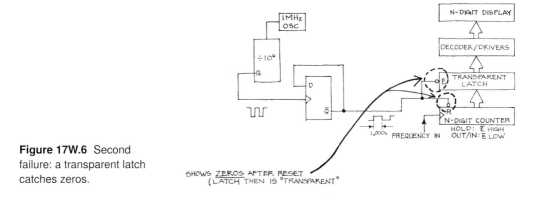

Figure 17W.6 Second failure: a transparent latch catches zeros.

Third failure: again missing a final count: As the note in Fig. 17W.7 says, this circuit is not perfect. It is much better than the transparent-latch design of Fig. 17W.6, but once in a while it may show a spurious count. The output register is clocked while the counter is clocking, so the count value may change during the register's setup time. Not much is at stake: an occasional odd display lasting one second is the danger.

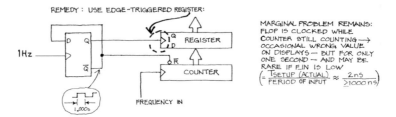

Figure 17W.7 Third failure: an edge-sensitive register violates setup time.

17W.3.2 Trying to get it right: sonar ranger

Now that you're getting good at designing these circuits (or at least good at finding fault with the designs of others), let's do an example more thoroughly.

An ultrasonic sonar sensor generates a high pulse between the time when the sensor transmits a burst of high frequency "ultrasound" and the time when the echo or reflection of this waveform hits the sensor.[5] The timing diagram in Fig. 17W.8 shows this graphically.

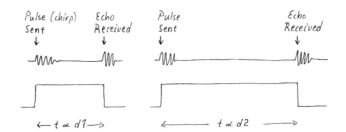

Figure 17W.8 Timing of sonar ranging device.

So the duration of the high pulse is proportional to the distance between the sensor and the surface from which the sound bounces. Only a couple of bursts are sent per second.

Problem Design hardware that will generate a count that measures the pulse duration, and thus distance. Let your hardware cycle continually, taking a new reading as often as it conveniently can. Assume that the duration of the pulse can vary between $100\,\mu$s and $100\,$ms. You are given a 1 MHz logic-level oscillator.

In particular:

- Make sure the counter output is saved, then counter is cleared to allow a new cycle.
- Choose an appropriate clock rate, so that the counter will not overflow, and will not waste resolution.

Solution We do this in several steps. First, just draw the general scheme; the block diagram in Fig. 17W.9. It looks almost exactly like all our "stopwatch" examples; a counter is gated by the input signal, the count is latched and displayed, then the counter is reset for the next cycle. The difference lies only in the device that generates the *period* to be measured by our stopwatch.

---

[5] It was designed originally for use in Polaroid cameras; now used often in robotics. See, for example, `https://LAoE.link/Sonar_Sensor.pdf`.

17W.3 Using a counter to measure period, thus many possible input quantities

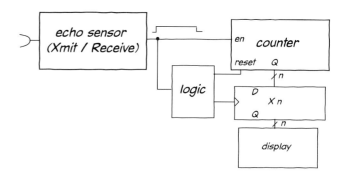

Figure 17W.9 Sonar device: digital readout. A block diagram.

The rest of our work is only to take care of the timing details. As usual, it is these that present the only real challenge.

We need to save the final count, then clear the counter, at the end of each cycle. Your job is complicated by the circumstance that the register on the display board we are using has a nasty *transparent latch*. One way to handle that difficulty is by adding the complexity of the double-barreled one-shot we built with the shift register. The first pulse would make the latch transparent; when that goes low, the second pulse would clear the counters. Here we will choose *edge-triggered* flip-flops instead. Nearly always, edge-triggered devices ease a design task.

Let's suppose we decide to use a custom 16-bit PAL counter we designed for another task. This device offers a synchronous clear and an asynchronous load; we will have to decide which to use for the clearing operation. The counter also offers a Carry$_{in}$, that we can use as an enable to start and stop the counter. How's Fig. 17W.10 look, using the counter's synchronous RESET*? Is the timing scary? Will the count get saved, or could the *cleared* value (just zeros) get saved? The timing diagram says it's OK: the D-flops certainly save their data before the counter is cleared on the next clock edge.

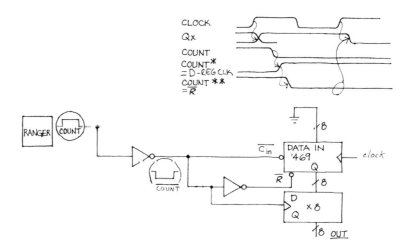

Figure 17W.10 Trying to make sure the count is saved before it is cleared.

To get some practice in thinking through this problem, look at another case, Fig. 17W.11, that's a little different: this hypothetical counter uses an *asynchronous* ("jam") clear. The problem is that we assert ($\overline{\text{CLEAR}}$) to the counter first (when count goes low), *then* invert this signal to clock the D register.

Depending on which is faster, the counter's asynchronous reset or the inverter's propagation delay, this could work like an elaborate equivalent to the cheaper circuit of Fig. 17W.12.

We should stick with the earlier circuit then: the one that clocks the D-flops slightly before clearing

720 Worked Examples: Applications of Counters

the counter and that uses a *synchronous* clear. A ghost of a problem remains however. Whether we should worry about it depends on whether we can stand an occasional error.

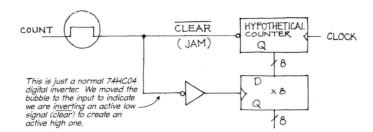

Figure 17W.11 Another example: failing at the same task.

Sometimes the counter Qs will be changing during the *setup time* of the D register. That can lead to trouble. This is the defect that worried us in the circuit of Fig. 17W.7. The most bothersome trouble would be to have some of the register's D-flops get *old* data (from count *N*) while others get *new* data (from count *N*+1). That's worse than it may at first sound: it implies not an error of *one* count, but a possibly huge error: imagine that it happens between a count of 7Fh ("h" indicating "hexadecimal") and 80h: we could (if we were very unlucky) catch a count of 8Fh. That's off by nearly a factor of two.

This will happen *very* rarely. (How rarely will the Qs change during setup time? *Typical* time during which flop actually cares about the level at its data input ("aperture," by likeness to a camera shutter, apparently) is 1–2 ns (versus 20 ns for *worst case* t_{setup}); if we clock at 1 MHz, that dangerous time makes up a very small part of the clock period: 1 or 2 parts in a thousand. So, we may get a false count every thousand samplings. If we are simply looking at displays that does not matter at all. If, on the other hand, we have made a machine that cannot tolerate a single oddball sample, we need to eliminate these errors.

Figure 17W.12 Sad equivalent to the bad save and clear circuit.

A very careful solution: This problem exists because the ranger's COUNT control signal is asynchronous to the counter's clock. Here is one way to solve the problem: *synchronize* the signal that stops the counter (with one flop); delay the *register* clock by one full clock period, to make sure the final count has settled. Perhaps you can invent a simpler scheme than that shown in Fig. 17W.13.

Problem (A nice addition: overflow flag.) Can you invent a circuit that will record the fact that the sonar ranger has *overflowed* – i.e., warn us that its latest reading is not to be trusted?

Hints: flip-flops remember. A good circuit would clear its warning as soon as it ceased to apply: when a valid reading had come in. But be careful – the warning should only become visible at the end of the measurement when the new distance value has been latched.

Solution Here's one way.

- Let the *falling edge* of the C_{out} signal clock a flop that is fed a constant high at its D input.[6] Call the Q of this flop *overflow*.
- When the count period finally does end, let the *overflow* Q get recorded in a *warning* flop that holds the warning until the end of the next measurement.
- Meanwhile, to set things up properly for the next try, let the end of the count period clear the *overflow* flop so that it will keep an open mind as it looks at the duration of the *next* period.

---

[6] A reminder: an up counter's carry out is asserted when the counter reaches its maximum count and dissasserted on the next clock edge, when it overflows to zero.

17W.3 Using a counter to measure period, thus many possible input quantities

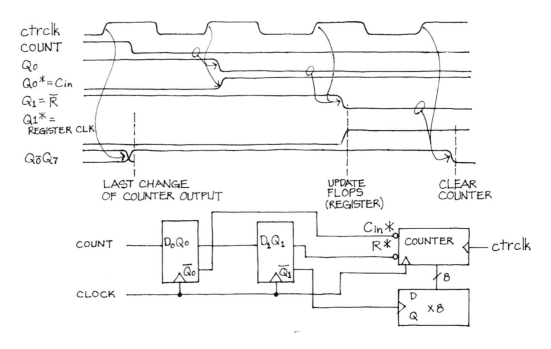

Figure 17W.13 Delaying flop clock to make sure we don't violate setup time.

However, this scheme has a problem. Since the counter's carry is simply an ANDing of several Qs, it could glitch on a transient state of the counter. We saw this possibility in §17W.1, where a scope image (Fig. 17W.1) shows some of these events. Such a glitch would trigger a spurious overflow indication. In Fig. 17W.14, we've added a flip-flop to synchronize C_{out} with the counter's clock to eliminate this possibility. Since the sync flop delays C_{out} by one clock cycle, we now trigger the overflow flop on the *rising* edge of C_{out}, which occurs when the count reaches its maximum value, but does not clock the *overflow* flop until one clock cycle later. (We don't want to wear you out; we only want to alert you to the sort of timing issues that make digital design not so obvious as it appeared in Chapter 14N, when all was combinational, and timing problems did not arise.)

The details of this design are fussy, but we will see one of the methods used here several more times: feeding a constant to a D input to exploit the nice behavior of an *edge-triggered* input.

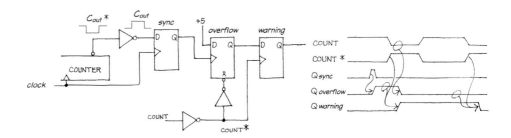

Figure 17W.14 One flop synchronizes C_{out} to eliminate glitches, one records fact of overflow, and another tells the world when the final count is ready.

17W.3.3 Recapitulation: full circuit of sonar ranger

Figure 17W.15 is, with few words, a diagram of a no-frills solution to the sonar ranger problem. The diagram omits the overflow warning logic drawn above, and the "very careful..." logic. It uses small 4-bit decoders and displays: not a likely choice now that LCD and even pretty OLED ("organic LED") displays are available.

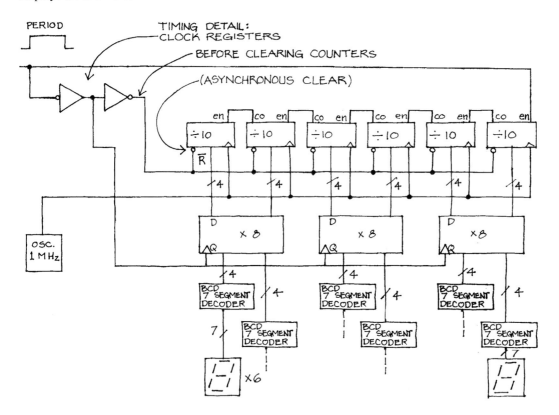

Figure 17W.15 Sonar period-measuring circuit: full circuit.

17W.4 Bullet timer

Here's an application of flip-flops and counters: a circuit that will measure muzzle velocity of a bullet. The bullet is fired at two wires that are 10 cm apart. Breaking the first wire starts a counter; breaking the second stops the counter. A competition rifle has a muzzle velocity of about 3500 feet per second. Your instrument should compare rifle performances to that speed. (This problem resembles the *reaction timer*, a problem we will see in a future lab, but is a bit simpler.)

Here are some specifications for the circuit.

- To sense the bullet's passage, use a thin wire with one end grounded, the other pulled up to +5 V through a resistor. When the bullet breaks the wire, the voltage rises to +5 V. Assuming that stray capacitance is 50 pF, choose a pullup value that will not introduce an appreciable error.
- Decide whether you need to worry about the equivalent of bounce – intermittent connection to ground as the bullet passes, perhaps scraping along the severed wire.

17W.4 Bullet timer

- You want 0.1 μs resolution. Decide how many bits your counter needs. (You'll need to estimate the time for the bullet of the "competition rifle" to move 10 cm.)
- Cascade 4-bit counters (74HC162s: synchronous BCD counters – 0 to 9 count – with synchronous $\overline{\text{CLEAR}}$ and $\overline{\text{LOAD}}$) to do the counting for you. Use enough '162s to measure perhaps 30% variation in bullet speed. Make your cascaded counter synchronous.
- Your time-keeping clock is a 30 MHz crystal oscillator. Divide it down to get the needed resolution.
- Include a manual pushbutton that generates a signal named $\overline{\text{ARM}}$. Pushing this button clears a flip-flop; this flop will go high when the bullet hits the first wire.
- The output of this flop should start the counter when the bullet hits the first wire.
- Let the bullet's breaking the second wire clear the flop and stop the counter.
- Display the result using the integrated 7-segment LED display w/ latch and decoder shown in Fig. 16N.14. These take a four bit binary input and display the hex value on an LED matrix. If we don't need the latch, we can tie the $\overline{\text{EN}}$ input low to make it transparent.
- Include a circuit to detect *overflow* of your multi-digit counter. Light a red LED on overflow; let assertion of the $\overline{\text{ARM}}$ signal clear the overflow signal. Assume you are allowed to use an *asynchronous* Carry$_{\text{out}}$ signal.

17W.4.1 Bullet timer solution

We'll present possible designs for several fragments of a circuit that would perform as demanded – and then we'll sketch the whole circuit. You may find the explanations offered with each fragment excessive. If you do, skip to the one-page solution at the end, which includes short explanations of most design choices.

Starting and stopping the counter: The bullet will break one wire to start the counter, another to stop it. To keep the counter going between these two events we need an element that will hold this "counting" state: a flip-flop.

The bullet needs to switch the level of a signal, twice. Let it break a wire that ties a signal to one level, while a resistor pulls to the other level.

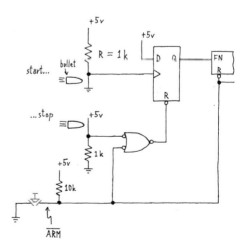

Figure 17W.16 A flop will start counter, then stop it.

Worked Examples: Applications of Counters

Generate the clock signal: Given our 30 MHz oscillator, and our need to resolve to $0.1\,\mu s$, which implies need for a 10 MHz clock rate, we need a divide-by-three counter. Figure 17W.17 is a circuit that uses a counter with *synchronous* clear in order not to generate nasty glitches.[7]

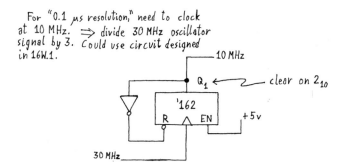

Figure 17W.17 10 MHz clock from 30 MHz oscillator.

Details: timing questions: The time we are trying to measure is quite brief, and we wish to resolve it to $0.1\,\mu s$, so we should make sure the delays in our circuitry don't compromise its performance.

RC delay when wire breaks The pullup/pulldown resistor forms an RC with stray capacitance which we have estimated at 50 pF. How small should we make the R? Let's keep RC well under the $0.1\,\mu s$ resolution. Keep it at, say, a half or quarter of that – 50 ns or 25 ns: see Fig. 17W.18. A pullup of 1k probably is OK; $470\,\Omega$ is a little better.

Figure 17W.18 Keep delay small from R_{pullup} and C_{stray}.

Chatter or bounce, as bullet passes? Is this harmless? It is. The bullet's length is not negligible relative to the 10cm distance between Start and Stop wires, so if one imagines the bullet scraping contact during that time, perhaps we have a problem to solve: one that looks like switch bounce. But in fact there is no problem: the flip-flop, as drawn, responds only to the first transition (on clock) then first low (on the reset); the flop acts like a debouncer. This is a sufficient argument; but in addition it seems very probable that the wire breaks contact cleanly on the initial break so that the electrical signal does *not* show chatter (bounce-like noise) in any event.

How many 4-bit counters needed? How long may the bullet take to travel the 10 cm gap between Start and Stop wires? Average speed is given as 3000 feet/s; average time then (see below) is about

---

[7] A 74HC162 counter is shown. This is a BCD or "divide-by-ten." A '163 natural binary would work as well. Both offer *synchronous* Clear*. You could even use a '160 or '161 if you connected all the data inputs ground and cleared the counter using the load input.

110 μs. If we want to resolve to 0.1 μs, we'll need four decimal digits (though the most significant counter will not show more than the value "1"; on a DVM this would be called "3 1/2 digits"), so we'll use four 4-bit counters, each a "divide-by-ten."

Once we know the range of times we need to measure (said to be 30% more or less than the average), and resolution wanted, the counter use is pretty pedestrian: see Fig. 17W.19.

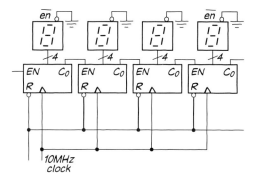

Figure 17W.19 Four divide-by-ten counters will measure time for us.

Detect overflow: Detecting overflow is more interesting. We need to use the highest-order counter's *carry-out* signal to sense overflow – because an overflow warns us that we cannot trust the recorded count. (We admit that overflow is wildly unlikely in this design. But it's a technique worth learning; so, here goes, though you may object that you have seen this at least twice before.)

We need to get straight whether the carry indicates an overflow or *impending* overflow. In fact, for a properly-designed synchronous counter, the carry indicates the latter: it means that overflow will occur on the *next* clock. So, it is the termination of *carry-out* that will indicate overflow. And, since we want the lighting of the overflow indicator to persist, we'll need a flip-flop to keep the LED lit till the user turns it off.

Simplest overflow detector: async use of $Carry_{out}$***:*** Figure 17W.20 shows a straightforward way to watch for overflow. The termination of a $Carry_{out}$ signal indicates the counter has overflowed ($Carry_{out}$ is asserted on maximum count – one short of overflow).

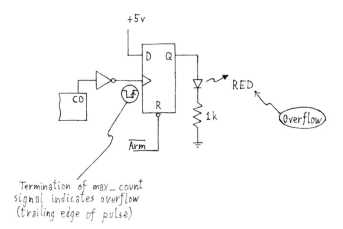

Figure 17W.20 Overflow sensor keeps LED lit after overflow.

This plan *may* work but we have noted in §17W.1 that glitches on the $Carry_{out}$ line could throw off the simple overflow detector that we have sketched here. The detector can malfunction if the counter's Qs, though synchronous, show enough *skew* so as to change not-quite-simultaneously. Such a timing mismatch could produce a spurious Carry$_{out}$ signal.

One can solve this problem by using a *synchronous* carry to indicate overflow.

If one creates the counter from scratch in programmable logic, it is easy to design a synchronous carry into the counter. Here is the Verilog `always` block to implement such a carry in a 16-bit up counter: it senses terminal count *minus one* and latches that as the carry out. (Minus one because the latch flip-flop delays the output signal by one clock cycle.) This synchronized (and glitch free) co signal can safely be used in the circuit of Fig. 17W.20.[8]

```
always@ (posedge clock)
  count <= count + 1;
  if (count == 16'b 1111111111111110) // maximum count minus one
    cout_sync <= 1; // this will assert CO on next clock, state max
  else
    cout_sync <= 0;
end
```

Using commercial IC counters, we don't have that option, but can make a near-equivalent by adding a *synchronizing* flip-flop that will not pass glitches. Such a circuit is sketched in Fig. 17W.21.

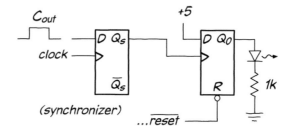

Figure 17W.21 Synchronizer added to counter's carry-out can protect overflow flop from glitches.

Glitches on the asynchronous Carry$_{out}$ line will occur soon *after* the clock edge; they will not pass through the synchronizer. Figure 17W.22 makes this point graphically: the signal *sync'd_Carry_out* does not pass the glitches that appear on Carry$_{out}$.

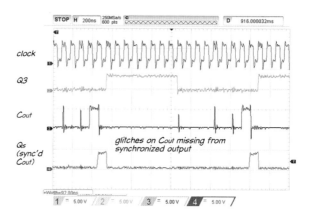

Figure 17W.22 Glitches in the PAL counter on Carry$_{out}$ counting down can be cleaned up by a synchronizing flop.

This remedy may well not be necessary. To produce the glitches shown in Fig. 17W.22 we had to

---

[8] Of course, this raises the question if you are going to synchronize the carry out so it can be used with an overflow flip-flop, why not just implement that flip-flop in the PLD and output an overflow signal directly? In the counter we built, it was because we did not have a spare I/O pin.

try hard.[9] But the synchronizing strategy is one worth remembering. Asynchronous circuits – ready to respond at any time – cannot tolerate noise on their clock lines; synchronous circuits can – so long as the noise does not occur during setup time, as normally it does not.

A full solution (circuit diagram): Figure 17W.23 pulls together the several pieces of the design – with most of the explanations included.

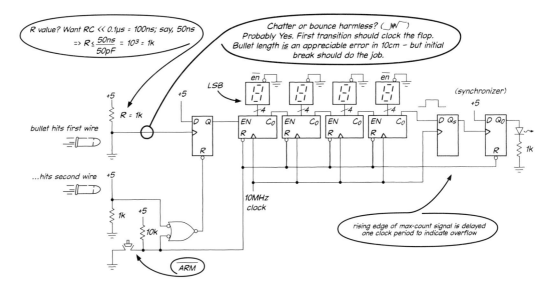

Figure 17W.23 Full circuit for a solution.

---

[9] We had to tease the circuit by slowing the clock edge. This slow edge apparently clocked the several flops at slightly different times. Capacitively loading some of the outputs (Q0,...,Q2) was ineffective, probably because the flop outputs are buffered before being brought out to the pins. And inducing the glitches by slowing the clock edge is a bit meretricious: the slow edge becomes unreliable as a synchronizing signal. Just as it can clock the several flops within a synchronous counter at different times, it can also clock the synchronizer at a time different from the time when it clocks the counter. Such a mismatch can undo the cleanup – passing the glitch to the output of the synchronizer, if the synchronizer's clock is a little late.

17O Online Content: Counters

17O.1 74LS469 alternatives

Available at https://LAoE.link/LAoE_Chapter_170.pdf.

18N Memory

Contents

18N.1	**Buses**		**729**
	18N.1.1	When are three-states needed?	729
18N.2	**Memory**		**730**
	18N.2.1	Memory describes a way of organizing storage	730
	18N.2.2	Memory types; memory jargon	731
	18N.2.3	How memories are specified	733
18N.3	**Memory in Verilog**		**736**
	18N.3.1	Verilog ROM	736
	18N.3.2	Verilog RAM	738
	18N.3.3	Larger memories in Verilog	740
18N.4	**AoE reading**		**741**

18N.1 Buses

These are just lines that make lots of stops, picking up and letting off anyone who needs a ride. The origin of the word is the same as the origin of the word for the thing that rolls along city streets.[1]

Power buses carry V+ and ground to each chip; data and address buses go to many ICs in a computer. Control signals may run to many parts, although we usually do not describe these as bus signals.

While these concepts are germane to parallel buses as well, serial buses have, for the most part, supplanted them in the embedded controller world. (At least externally; parallel buses are still used extensively *inside* computers and microcontrollers.)

18N.1.1 When are three-states needed?

AoE §14.1.6

Answer: whenever more than one *driver* (or "talker," to put it informally) is tied to one wire. This is one of the two cases where outputs can be connected together (the other is open collector/drain outputs).

Figure 18N.1 depicts a typical configuration for an embedded microcomputer system. Several external peripheral devices, here a write-only liquid crystal alphanumeric display (LCD), a read/write memory (RAM), and an analog to digital converter, are connected to a microcomputer using a serial peripheral interface (SPI) bus. The bus is controlled by the microcomputer which has an internal serial communications processor (SERCOM) to handle the communications protocol. The micrcomputer asserts a chip select (CS) to a select a device (only one CS can be asserted at a time), and the SERCOM begins sending clock pulses to all the devices on the CLK line. It sets its data out (DO) line high or low

---

[1] If Latin isn't too Harvardy for you: "bus" is just short for "omnibus," "to or for all." The city bus, an innovation in early 19th century England, was given this name to contrast it against a carriage that one would hire for a single point-to-point ride.

before each rising clock edge until all eight bits have been sent. The enabled peripheral listens on its data in (DI) line, saves each byte sent and sends data back to the microcomputer on its DO line. The microcomputer stores the returned data before sending the next rising clock edge.

Where in Fig. 18N.1 are three-states needed? (In case you need reminding, a three-state is an output stage capable of disconnecting entirely, rather than only switching the output to ground or positive supply, as in a conventional gate.) The presence of more than one *receiver* (or "listener") does *not* call for 3-states.

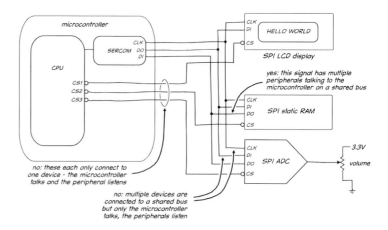

Figure 18N.1 Where are 3-states needed?

The chip select lines only connect two devices point-to-point and the microcomputer always drives the CS line and the peripheral only listens – so these are normal signal lines. The line connecting the DO of the microcomputer to the DI of all peripherals is a bus, but the peripherals only listen while the microcomputer only talks, so no 3-state is needed there. Only the line that allows each peripheral's DO to send data to the microcomputer's DI pin has multiple talkers on it. Each peripheral which is not selected *must* use a 3-state output driver to keep its DO line in a high impedance state while the selected device talks.

18N.2 Memory

AoE §14.1.2

We have spoken rather glibly of a "memory" without explaining what it is. Let's remedy that omission. A *memory* is an array of flip-flops (or other devices each of which can hold a single bit of information), organized in a particular way to limit the number of lines needed to access this stored information. A memory is used most often, as you know, to hold data and program for a computer; it can also be used to generate a combinational logic function – as you will see in Lab 18L.

18N.2.1 Memory describes a way of organizing storage

The leftmost image in Fig. 18N.2 shows a *register* capable of storing eight bits. A register is just a collection of flip-flops, with a common clock input, and all Ds and Qs brought out in parallel. If those 8 bits are stored, alternatively, in *memory* form, then inputs and outputs are multiplexed each onto one line; and a further improvement allows abolishing even the IN versus OUT distinction, so that IN and OUT become the bidirectional DATA line.

At first glance the memory looks silly: a lot of extra internal hardware just to save a few pins? Let's count pins:

18N.2 Memory

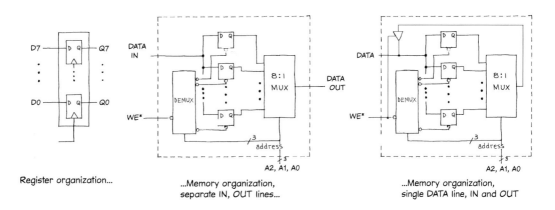

Register organization... ...Memory organization, separate IN, OUT lines... ...Memory organization, single DATA line, IN and OUT

Figure 18N.2 Memory as a way of organizing data storage, minimizing external connections. When WE* (write enable) is asserted, data is written to the memory; otherwise data is read from it. You will often also see a chip enable or chip select input which keeps the output data line in the high impedance state when not asserted.

Function	Number of Lines if using Regiser	Memory
data in	8	1
data out	8	1 or 1 shared between IN and OUT
clock/write	1	1
power, ground	2	2
address	0	3
total	19	7 or 8

Pin count: register versus memory: The memory advantage doesn't look exciting – until you crank up the number of stored bits to a reasonable level. Then the memory reveals that it is the *only* feasible way to store the information (try counting pins on a 1Mbit memory reorganized as a *register*, for example).

Memories normally squeeze away one more pin, as shown on the right in Fig. 18N.2, combining input and output into one pin. WE* is used to turn off the 3-state that normally would be showing the memory's content as output so data can be written *to* the device using the same data line for input.

Memory organization allows access to only one stored value at a time, whereas the register delivers all its bits in parallel. But the memory scheme matches nicely with the sequential nature of computer processing. Note however – speaking of parallel versus sequential – that the RAM we have sketched in Fig. 18N.2 is just *one bit wide*. While there are exceptions, stand-alone static RAM often uses eight such structures in *parallel*, storing an 8-bit value at each location.[2] The one-bit-wide memory is a scheme we offered only for its simplicity.

That's all there is to the *concept* of a memory. Now on to the *jargon*.

18N.2.2 Memory types; memory jargon

AoE §14.4

One can slice the universe of memory types in several ways:

RAM versus ROM: can one write to the memory?

---

[2] Even the serial access RAM devices used with modern microcontrollers are typically organized 8-bit wide internally, reading a byte at a time out through a serial bus: see §26W.2.

- If **No** it's called a **ROM** ("read-only memory"). Only the manufacturer can write to a ROM. The name obviously overstates the part's character: a memory that no-one could write to would be thoroughly useless.[3]

- If **No, not unless you work pretty hard**, it's called a **PROM** ("programmable read-only memory"). A user can write to a PROM, though not so easily as one can write to a RAM (see below).
    - **PROMs come in several flavors** (none of them nowadays called PROM).[4]
        - **EPROM** (E for Erasable): the oldest. To erase these you'd have to remove them from the circuit and put them in a UV box for 30 minutes or so.[5] That was a nuisance, so they have been supplanted by EEPROMs (electrically-erasable PROMs) of various sorts. These put a small charge on the floating gate of a MOSFET. In this they resemble dynamic RAMs – but the charge persists for 10 or 15 years, rather than for a handful of milliseconds. AoE §§14.4.5B, C

        - **EEPROM**: these can be erased a byte-at-a time, and can be erased in-circuit. That is obviously more convenient than the EPROM scheme. Our old PALs used this technology.

        - **Flash memory** (so-called because it is erased quickly): good for mass storage, but not byte-erasable – must be erased in larger blocks. USB memory sticks, SD cards and our microcontroller program memory all use flash.

- If **Yes, easily** the memory is called "RAM," a misnomer, since *all* these semiconductor memories are "random access memories."[6] But the name has stuck.
    - **RAMs come in two main flavors.**
        Storage technology: flip-flops, called "static" RAM; versus capacitors, called "dynamic" RAM. One can tell that these names must have been invented by the manufacturers of the dynamic memories: the ones that use capacitors. Capacitor based memories aptly could be named "forgettories" (to steal a word from AoE), since they forget after a couple of *milliseconds*! These memories need continual reading or pseudo-reading, to "refresh" their capacitors. Despite their seeming clumsiness, *dynamic* RAMs enjoy the single advantage that their remembering unit is very small and uses little power; so most large memories are dynamic. We used external static RAM in our old microcontroller labs because it required no refreshing. In fact, this type of memory is so simple and power-thrifty that you can leave it powered continuously for a year or more using a 3V coin cell. Our new microcontroller doesn't offer a choice; all the RAM is static and internal to the controller. AoE §§14.4.3, 14.4.4
        Blurring the static/dynamic distinction: "Pseudo-static" RAM blurs this contrast, providing the ease of use of a static RAM with almost the full density of dynamic: the refresh process is handled internally and is not apparent to the user. These have

---

[3] Signetics (now part of Phillips Semiconductor) famously created a datasheet for a "Write-Only-Memory" (https://LAoE.link/WOM.pdf) released as an April Fool's joke in 1973.

[4] Going forward, we will refer to memory that we write once or infrequently just as "ROM."

[5] If you think the acronyms are excessive, in this field of RAMs and ROMs, consider this one, perhaps the mother of all excessive acronyms: the sort of EPROM that was offered in a cheaper package omitting the quartz-erasing window. This packaging reduced the part price more than tenfold. Lacking a window, it could not be erased. The idea behind this packaging was that a user could work out a design using an EPROM, then when satisfied with the programmed content would specify the cheaper package. These parts were called OTP EPROMS: "one-time-programmable erasable programmable read-only memories." Luckily for the world, these are obsolete.

[6] The name may seem puzzling. It does not mean that what the memory delivers is a randomly-chosen entry of course. It means that entries may be recalled or stored, in any order, and access times for all stored values are substantially equal. This equality contrasts most radically against a medium like tape and also, in lesser degree, against all types of disk storage.

largely displaced traditional static RAM except where its extreme low power consumption is important or where extreme speed is needed (as in fast *cache* memory).

- **Synchronous RAMs**. Both static and dynamic RAMs come in a more complex form with a clocked state machine governing the asynchronous innards. We will not look into these but instead refer you to AoE §§14.4.3C and 14.4.4B.

AoE §14.4.1

- **Volatility:** Does it remember, without power? In general ROM and PROMs of all sorts do remember (that is, they are "non-volatile"); RAMs do not (they are "volatile"). Again, some memories straddle this boundary:
    - **Chubby RAM** (more official name: Battery Backup SRAM, BBSRAM). Figure 18N.3 shows DIP and surface-mount versions of this device. You can guess why the package is fat.[7]

Figure 18N.3 Chubby RAM acts like RAM, but remembers for 10 years. Figure used with permission of Dallas/Maxim.

These are as easy to use as ordinary RAM, but are expensive ($10 to above $30 for the STM M48T35 and Dallas DS1244Y – which also include a real-time clock).
- A variation on this scheme is a battery *socket* with battery and power-down/up circuitry.
- Other backup schemes are used, too, such as "shadow-RAM": it's a RAM with a non-volatile memory attached that can be updated from the RAM before power-down (for example, Cypress CY14B101KA.) These memories offer packages smaller than those for the battery-backed devices, unlimited number of RAM store cycles, and automatic non-volatile storage on power-down. A capacitor sustains the supply voltage long enough to accomplish the back-up storage.

- **New Technology:** AoE describes several new technology non-volatile memories, including Ferroelectric RAM (FRAM), Magnetoresistive (MRAM), and Phase-change RAM (PRAM or PCM), as having "read and write access at full speed, unlimited endurance, and long retention." While promising, none of these have yet replaced static or flash memory devices in common usage.

AoE §14.4.5D

18N.2.3 How memories are specified

Organization: "number-of-words × word-length." For example, the RAM we use in the worked example of Chapter 24W is 32K × 8 or 32KiB × 8: 32,768 words (or locations), each 8 bits wide.[8]

Units: "K" and "KiB" versus "k": The use of the term "kilo-" in "kilobyte" offers potential for ambiguity. How many bytes in a kilobyte? This may sound like a silly question – but is a kilobyte 2^{10} bytes (1024) or is it 10^3 bytes?

One answer – not very satisfactory – might be, "Oh, who cares; the two values differ by only 2.4%." One response to that might be, "Well some lawyers care." At least two lawsuits complained about the use of 10^3 to define kilobyte, and recovered some damages.[9] Even within the electronics industry,

---

[7] Well, in case you don't like guessing, we'll reveal that the fat package houses a lithium coin cell.
[8] If this is your first encounter with the unit "KiB," see an explanation just below.
[9] See https://LAoE.link/Safier_v_WDC.html.

manufacturers don't agree. Some makers of hard disk drives say a "kilobyte" means 10^3 bytes. Doing this made their drives seem a little bigger than if they had adopted the 2^{10} convention (but not much bigger) – hence one of the lawsuits. Memory manufacturers generally use "kilobyte" to mean 2^{10}. Flash drive sizes normally are stated in Gigabytes: 2^{30} bytes, not a power of ten.[10]

But it would be nice to be rid of that ambiguity. An upper-case K is one convention that can distinguish the computer "Kilobyte," 2^{10}, from the scientific meaning of "kilo" as 10^3. Another convention, established in 1998 but not yet widely used, uses the unit "KiBi," abbreviated "KiB." "KiBi" is easier to distinguish from an ordinary kilo. With it arrived the corresponding "MiBi" and "GiBi," and so on. In this book you will find us using upper-case K to denote 2^{10} bytes.

Memory part numbers tend to mention the number of *bits*, perhaps because this is a more impressive number. But it is a fair description of total memory size. The 32K × 8 SPI static RAM of the sort that we use in Chapter 24W calls itself 23K256 (256K bits).

Synchronous vs. asynchronous: Memory can be synchronous, in which case data is written into the memory or read out in response to a clock edge, or asynchronous. If the latter, the memory acts like a latch on write, the data is saved continuously while the write strobe is asserted. The output data for asynchronous read changes whenever the address changes.

Multi-port: The type of memory shown in Fig. 18N.2 is called single-port memory. There is a single set of address inputs and data cannot be read and written at the same time. A dual-port memory has two separate sets of both address and data lines so it can be accessed by two processes simultaneously. This type of memory is common in video displays where one set is write-only (from a computer or video processor) and one set is read-only to a display: see Fig. 18N.4(b). A dual-port memory where both interfaces allow full read/write access is useful in multi-processor systems to allow the separate processors to communicate with each other and pass data back and forth. (Although multiprocessor systems can be designed to time-share a single port memory as well.)

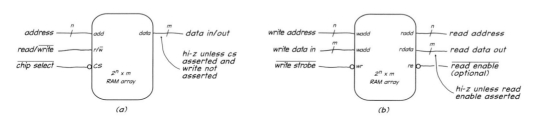

Figure 18N.4 Asynchronous single port (a) and dual port (b) RAM.

A FIFO (First-In-First-Out) is a type of dual-port memory with a write data interface and a read data interface but no address selection. As its name implies, the read interface returns the oldest value written. Typically, this type of memory includes FIFO empty and FIFO full flags and is used to interface asynchronous processes with different timing requirements. For example, in a communications interface data may be written into the FIFO as it arrives, then processed by a computer in the same order at a later time.

Conventional (parallel) or serial interface: Older memory exposed the memory organization depicted in Fig. 18N.2 directly as its user interface. In a $2^n \times m$ memory, the user selected the word to

---

[10] The exceptional flash drive company that chose 10^9 – paid some damages for overstating drive size by 7%.

read or write using a set of n parallel address lines and the data was written or returned as a set of m parallel signals. Additional pins enabled the device (CE or CS) to allow multiple devices to be used, a 3-state enable (OE) to share the bus, and control lines to select read or write mode. A 32K × 8 RAM requires 15 address lines and 8 data lines plus control inputs, all in a 28 pin package: see Fig. 18N.5(a). This worked fine with older microcomputer designs that included a parallel bus to connect to external memory and peripheral. It is what we used with the 8051 microcontroller in the previous edition of this book.

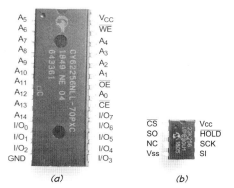

Figure 18N.5 256 Kbit (32K × 8) static RAM with conventional (a) and SPI (b) interfaces.

Modern microcontrollers include a multitude of *internal* peripherals, all of which vie for access to a limited set of external I/O pins. In order to allow for the maximum utilization of package pins, most newer designs, including the ARM Cortex-M0+ you will soon meet, have forsaken parallel expansion buses and switched to serial buses. Figure 18N.5(b) shows the same 256 Kbit capacity static RAM with an SPI interface in an 8-pin DIP package (one pin of which is unused). Internally, the SPI RAM has the same 32K × 8 memory array, however, it is accessed by transmitting commands and data over the clock (SCK), serial data in (SI) and serial data out (SO) signal lines. While the serial device is a bit more complicated to use, it is much easier to wire up and, if you need more storage, you can plug in a larger capacity (but still 8-pin) device and you merely have to update your software to handle the increased address range. A parallel interface would likely require a larger package and a circuit board redesign.

Memory timing: The most important conventional memory characteristic, once you have settled on a type and size, is *access time*: the delay between presentation of a valid address ("your" job) and delivery of valid data (the memory's job).[11]

AoE §14.4.3A

Static RAM timing: Figure 18N.6 shows a timing diagram for a conventional (i.e., parallel interface) 32K × 8 static RAM, assuming it is 70 ns speed grade (middling, among the speeds available for this part). Don't let all the details overwhelm you. The number that matters most is the time between presenting an *address* and getting stable *data*: this is *access time*, in its usual sense. Another time, a little less important, is the time between asserting *Chip Select** (or Enable*) and getting stable *data*. For this RAM, those times are the same. Since CS* usually is generated as a function of address, it's the CS*-to-data time that will tell you how soon data can be ready.

[11] Here we refer to conventional memory, where the address to read from is presented to the device in parallel. In a serial memory, the speed of the interface protocol could also be a limiting factor.

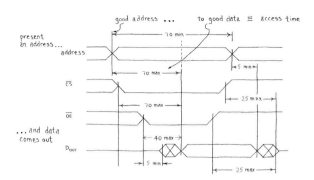

Figure 18N.6 Read timing for 70 ns parallel static RAM.

18N.3 Memory in Verilog

While not very efficient, small memories can be built using FPGA logic cells. The advantage of using logic cells is that the Verilog code is portable from FPGA to FPGA. Larger memories usually take advantage of dedicated memory arrays in the FPGA. These vary from device to device and the code to utilize them varies as well.[12]

18N.3.1 Verilog ROM

Verilog ROM is useful for decoders and combinational logic (as we have seen there is a one-to-one correspondence between a truth table and a Boolean expression).

You create ROM with the Verilog `case` statement. This construct is similar to the switch...case control structure in C and Arduino. Unlike C, there is no fall-through, only the matching item's statements are evaluated, so no `break` is required between case items. Case provides a cleaner structure than a series of nested `if ... else if` conditionals. The form of the `case` statement is:

```
case (<case_expression>)
  case_item1: <single statement>
  case_item2: begin <multiple statements> end // use for multiple statements
  case_item3, case_item4: <statement>         // multiple items do same thing
  ...
  default: <single statement>
endcase
```

The `<case_expression>` is evaluated, then compared to each `case_item` sequentially until a match occurs. You may concatenate signals to create the expression. When a match occurs, the statement[s] after the ":" are executed. If there is more than one statement, they must be bracketed with `begin ... end`. If no match occurs and there is a `default` item, its statement[s] are executed, otherwise no action occurs. Case can only be used in an always block.

Example: A keypad decoder. The keypad in Fig. 18N.7 consists of a matrix of 16 switches. The scanning hardware (which may also be implemented in the FPGA) provides the row and column of the pressed key and asserts the `pressed` flag when any of the 16 keys are pressed. We want to create a module that translates the 4-bit row/col switch position into a BCD value from 0 to 9 and sets a "valid" flag while the `pressed` flag is asserted. If no key is pressed or the pressed key is non-numeric, the

---

[12] In some development environments, the synthesizer may automatically infer and use dedicated memory for larger Verilog memory arrays. The WebFPGA IDE will infer memory if the code format conforms to the instructions in the iCEcube2 User Guide, Section *About Inferring Memory* available at https://LAoE.link/iCEcube2_Guide.pdf.

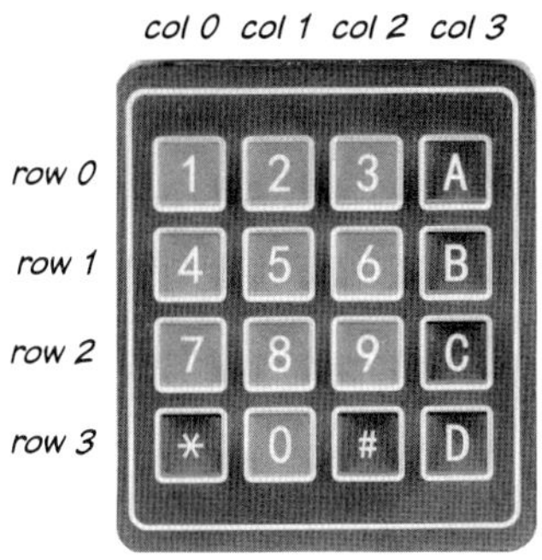

18N.3 Memory in Verilog

Figure 18N.7 A 4 × 4 matrix keypad.

valid flag should be cleared (in which case we don't care what the output is). We can use a lookup table (i.e., a ROM) to make the translation:[13]

```verilog
module num_decode (
  input [1:0] row,         // row of pressed key (0 to 3)
  input [1:0] col,         // column of pressed key (0 to 3)
  input pressed,           // asserted while any key is pressed
  output reg [3:0] out,    // four bits required for 0 to 9
  output reg valid         // flag asserted when valid key pressed
);

  always @(*)
  begin
    valid = 1;
    case ({pressed, row, col})
      5'b10000: out = 1;
      5'b10001: out = 2;
      5'b10010: out = 3;
      5'b10100: out = 4;
      5'b10101: out = 5;
      5'b10110: out = 6;
      5'b11000: out = 7;
      5'b11001: out = 8;
      5'b11010: out = 9;
      5'b11101: out = 0;
      default: begin out = 0; valid = 0; end
    endcase
  end
endmodule
```

Since this is a combinational function, we use (*) for the sensitivity list and = as the assignment operator: see §16N.4.1. The valid flag is set to 1 before the case statement and only changed in the default case. You could instead set it in each matching item statements, but this is easier saving us a begin and end on each item match. With either construct, the synthesizer should resolve this to valid = !<default case>.

The case expression is formed from the pressed flag, the row and the column concatenated into a single 5-bit vector. We then compare this new vector with each case item. If there is a match, we set the output to the digit on the keypad for that key. If not, the default case sets the output [arbitrarily] to 0 and the valid flag to 0.

A shortcut to consolidate case items: If one or more bits in the case expression have special meaning that overrides the others, it would be nice not to clutter the case statement with all possible

---
[13] Available at https://LAoE.link/FPGA/18N_ROM_KeyDcdr.v.

cases. For example, in what is known as "one hot" encoding, each bit in a vector indicates the current state of a sequential circuit. While you only expect one bit at a time to be set, you would like to design the code to be robust enough that it works even if more than one bit is set (but should that happen you decide the higher order bit has priority).

Rather than compare all possible numeric values from 8′b10000000 to 8′b11111111 to test for the MSB operation, you can use the casex variant which allows *don't cares* (x, z or ?) in case items. Here is a module that converts an 8-bit one hot vector to a 3-bit binary value. While it uses x for don't care, you could use z or ? as well. In this design, a bit should always be set but we include a default to go to a known state if it is not.[14]

```
module onehot_encoder (
  input [7:0] op_bit,      // each bit signifies a different operation
  output reg [2:0] encoded_op,
  output reg no_op
);

  always @(*)
  begin
    casex (op_bit)
      8'b1xxxxxxx: encoded_op = 7;  // if MSB set, don't care about others
      8'b01xxxxxx: encoded_op = 6;  // can use z, x, or ? for don't care
      8'b001xxxxx: encoded_op = 5;
      8'b0001xxxx: encoded_op = 4;
      8'b00001xxx: encoded_op = 3;
      8'b000001xx: encoded_op = 2;
      8'b0000001x: encoded_op = 1;
      8'b00000001: encoded_op = 0;
      default: encoded_op = 7;     // a bit should always be set but just in case...
    endcase
  end
endmodule
```

18N.3.2 Verilog RAM

RAM is created with Verilog arrays. An array is a group of wires or vectors (usually the latter) that can be accessed individually by an index. Vectors and arrays look similar but operate differently:

```
wire [7:0] vector1;   // declaration for a group of 8 signals
wire array1 [7:0];    // declaration for array of 8 wires
```

Here vector1 consists of 8 signals that can be accesses simultaneously or in subsets (part-select) of 1 to 7 signals and vector1[3:0] selects the low nibble of 8-bits; it returns a 4-bit vector. An array element is selected with a single value, the index of the element you want. So array[0] selects the first element in the array, a single bit value, while array[7] selects the last element, also a single bit. You cannot select multiple array elements at a time.

Arrays can be multi-dimensional:

```
wire out;
reg array2 [15:0];        // declaration of a one-dimensional array of 16 wires accessible 1 at a time
reg array3 [15:0][15:0];  // declaration of a two-dimensional array of 256 wires accessible 1 at a time

assign out = array2[12];   // out set to 13th element in array2
assign out = array3[1][8]; // out set to 23rd element in array3
```

[14] Available at https://LAoE.link/FPGA/18N_ROM_OneHot.v.

18N.3 Memory in Verilog

In creating RAM, we typically define arrays of vectors. Here is a small 32×8 RAM with separate input and output lines, synchronous write and asynchronous read:[15]

```
module ram_32x8 (
    input wire  [4:0] address,    // need 5 bits to specify 32 locations
    input wire  [7:0] data_in,
    input wire  wr_clk,           // data_in stored on rising clock edge
    output wire [7:0] data_out    // data_out changes when address changes
);

    reg [7:0] ram_array [31:0];   // declare an array of 32 8-bit vectors

    always @(posedge wr_clk)      // synchronous write
    begin
     ram_array[address] <= data_in;
    end

    assign data_out = ram_array[address]; //asynchronous read
endmodule
```

You could make the read synchronous by declaring `data_out` as `reg` and moving its assignment (using "`<=`") into the `always` block. You would also need to add a control line to select whether you are reading or writing on the clock edge.

Using parameters: Parameters are a way of specifying a constant as a text identifier in a Verilog file. They are akin to `#define` in C. In addition to clarifying the purpose of constants in Verilog code, they are useful when creating memory to allow you to change the memory organization in a module merely by changing the parameter values for width and depth. Unfortunately, this is the one place where we have to revert to the old method of module definition.[16] Here is an asynchronous RAM module whose size is configurable by two parameters and which uses the same data bus for input and output. This means bus must be set to high impedance on a write operation. It implements the device shown in Fig. 18N.4(a). With the default parameters shown, it implements a 16×8 RAM.[17]

```
module ram_p (address, rw_bar, cs_bar, dataio);

    // change following two parameters to define memory size
    parameter ADDR_W = 4;       // number of bits in memory address
    parameter DATA_W = 8;       // number of bits in each memory location

    // the following calculates memory size as 2 ^ <bits in address>
    parameter MEM_SIZE = 1 << ADDR_W;

    // now we can declare the module signals based on the parameter values
    input wire [ADDR_W-1:0] address;
    input wire rw_bar;                    // data latched on low level
    input wire cs_bar;
    inout wire [DATA_W-1:0] dataio;       // used for both input and output

    // the memory array is declared as an internal reg
    reg [DATA_W-1:0] ram_array [MEM_SIZE-1:0];   // declare RAM array

    // active low is confusing; create active high versions of control signals
    wire cs = !cs_bar;
```

---
[15] Available at `https://LAoE.link/FPGA/18N_RAM_32x8.v`.

[16] The current Verilog standard allows you to define parameters in the module definition; however, the WebFPGA IDE does not appear to support this method.

[17] Available at `https://LAoE.link/FPGA/18N_RAM_Param.v`.

```
  wire write = !rw_bar;

  always @(*)
  begin
    if (write && cs)
      ram_array[address] = dataio;     // asynchronous write to latch
  end

  // the output data lines are set to high impedance ('bz) if the chip
  //   is not selected or write is asserted (write mode)
  // if the chip is selected and write is not asserted (read mode)
  //   the data in the ram is placed on the data lines
  assign dataio = (write || !cs)? 'bz : ram_array[address];
endmodule
```

Data is written while `rw_bar` input is low and read while it is high as long as `cs_bar` is asserted. The data bus must be high impedance when writing to the RAM or when the RAM is not selected.

Another shortcut using loops: Often, particularly in repeating structures like memory, you need to duplicate or initialize hardware. Verilog provides four non-synthesizable looping statements that can be used in an `always` block to simplify repetitive construction:

```
// These two can be used to duplicate hardware structures ...
   for (<initial value>, <end test>, <increment>)
   repeat (<repeat count>)
// ... the next two are normally used in test benches.
   forever
   while (<!end condition>)
```

A `for` loop can be used to initialize memory:

```
for (i = 0; i < MEM_SIZE; i = i + 1) // ++ is not supported in Verilog
   memory[i] = 0;
```

A more complex example at https://LAoE.link/FPGA/18N_Life.v creates an implementation of Conway's game of life.[18] This version consists of an 8 × 8 matrix of 256 "cells" (flip-flops) each of which is set to a value on the next clock edge that depends on the current state of the eight cells surrounding it. (Cells wrap around left-right and top-bottom.) The calculation of each cell's next value is identical. If you look at the Verilog code, a Verilog `for` statement is used to replicate the complicated next cell state calculation 256 times. This is much easier than writing a case statement duplicating this combinational logic equation 256 times for each cell. The `for` loop index is used to select the cell for each hardware construction. It is important to remember that the `for` loop is *only evaluated once*, when the synthesizer creates the hardware to implement the 256 cells from the FPGA's resources. Once the bitstream is loaded, all 256 cells calculate and update in parallel each clock cycle.

18N.3.3 Larger memories in Verilog

The Verilog code in §18N.3.2 uses one FPGA logic cell for each bit of RAM memory. ROM arrays require multiple cells to implement the combinational `case` logic. Since the LUTs available are often limited compared to memory requirements, most FPGAs include dedicated memory blocks of various sizes for use with larger memory arrays, either RAM or ROM. These are supported with special "macro" modules unique to the device you are using.[19] See Chapter 18S for examples of how to use the memory blocks in the iCE40 on the WebFPGA.

---

[18] See https://LAoE.link/Conways_Life.html.
[19] In some cases, the development tool for the FPGA you are using may infer a memory block automatically from a normal Verilog array declaration: see, e.g., 18S.2.1.

18N.4 AoE reading

Chapter 14 (Computers, Controllers and Data Links):
- §14.1.2 Memory.
- §14.3 Bus signals and interfacing.
- §14.3.4A Bus signals: bidirectional versus one-way.
- §14.4 Memory types.
- §14.4.6 Memory wrapup.
- §14.7 Serial buses and data links, especially:
    - §14.7.1 SPI.

18L Lab: Memory

Today you will first build and test a simple RAM in the FPGA. You will then build a seven-segment decoder ROM and connect it to the output of the FPGA 74HC161 counter you designed in the last lab to show the count on an LED display.

Note: This is a long but [we hope] fun lab. We normally ask students to design the parallel-to-serial converter (§18L.2.3 Item 3) as homework before coming to lab.

18L.1 Build a Verilog RAM

18L.1.1 Start with a single-port version

Build and test the single-port RAM shown in Fig. 18L.1 using the WebFPGA.

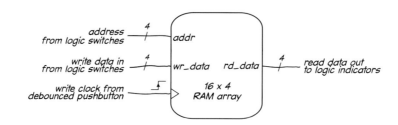

Figure 18L.1 RAM to build using the WebFPGA.

- The memory size is 16 locations, each holding a 4-bit value (16×4).
- The memory has separate data in and data out buses. (This is to allow you to use the LOGIC SWITCHES for data input. If you used a single `inout` data bus, you would have to wire 3-state drivers to the data switch outputs.)
- Use four of the logic switches to select the address and the other four to set the write data. (Feel free to build a larger RAM if you like by using the SPDT switches and/or DIP switches.)
- Display the read data out (`rd_data`) on four of the logic indicators.
- Use one of the debounced pushbuttons for the write clock.
- The memory has asynchronous read:
  - The read data bus (`rd_data`) should always display the current value of the memory location selected by the address input (`addr`).
- The memory has synchronous write:
  - On the rising edge of the write clock, the value on the write data bus (`wr_data`) should be written into the memory location selected by the address input.

Try to write your Verilog code using parameters so the module can be reused in other applications. Test your design by loading each of the 16 locations then reading them back. If you load the location's address into each location, it is easy to tell if the data read back is correct.[1]

18L.1.2 Make your RAM dual-port

Convert your design from §18L.1.1 to a two-port ram by adding four more address lines for the read data: see Fig. 18L.2.

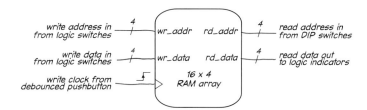

Figure 18L.2 Block diagram of 16 × 4 dual-port RAM.

The dual-port RAM should work identically to the single port version with the following changes:[2]
- The memory has asynchronous read:
  - The read data bus (`rd_data`) should always display the current value of the memory location selected by the read address input (`rd_addr`).
- The memory has synchronous write:
  - On the rising edge of the write clock, the value on the write data bus (`wr_data`) should be written into the memory location by the write address input (`wr_addr`).

Since you only have eight logic switches on your breadboard, you will have to use a DIP switch or four slide switches to set the read address. If you use DIP switches they will require pullup resistors to +3.3 V.

18L.2 Build a display decoder ROM

18L.2.1 Introduction

AoE §12.5.3A

In this part of the lab, you will add a seven-segment LED display to the output of the binary counter you built in the last lab to show the counter value. A seven-segment LED display consists of seven LEDs arranged in a figure eight so that the digits 0–9 can be displayed by selecting which LEDs are turned on. You will implement six additional characters for the counts of A–F so that you can display all the outputs of the counter: see Fig. 18L.3.

Figure 18L.3 Seven-segment display characters for binary data.

---
[1] Our solution available at https://LAoE.link/FPGA/18L_RAM16x4.v.
[2] Our solution available at https://LAoE.link/FPGA/18L_RAM2P_16X4.v.

In a seven-segment display, the seven LEDs (and optionally a decimal point LED) are arranged with all the anodes (*common anode*) or all the cathodes (*common cathode*) connected together. We will be using a common anode display in which the anode is connected to +5 V and segments are lit up by connecting the corresponding LED cathode with an electronic switch to ground through a current limiting resistor or using a switchable current source: see Fig. 18L.4.

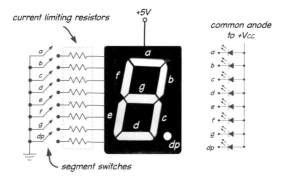

Figure 18L.4 Schematic and typical application of a common-anode 7-segment display with decimal point. Segments are labeled a–g.

The display we are using is a four-digit multiplexed display. This means that there are four seven-segment displays with separate common anodes for each digit but with the segment cathodes for all four digits connected together (note that this display also includes a decimal point LED for each digit for a total of eight cathode connections).[3] A schematic of this display is shown in Fig. 18L.5.

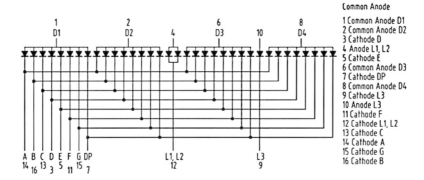

Figure 18L.5 Schematic of 4-digit, 7-segment, common anode 7-segment display.

The advantage of a multiplexed display is that it allows multiple digits to be displayed with only one extra anode connection per added digit. The disadvantage is that you can only display one digit at a time, so you need switches to connect each anode to +5 V sequentially, powering a single digit at a time. In practice, you close one anode switch, leaving the others open, and then close the cathode switches to display the character desired in that position. You then move to the next digit and do the same thing. You continue round-robin through all the digits at a rate fast enough that the user sees all the digits on simultaneously. This is known as a "time multiplexed" display.

Even though the display we are using has four digits, in today's lab you will connect only one anode to +5 V, leaving the other three unconnected, to use it as a single digit display. In the digital project lab (Chapter 21L), you will add anode switches and create FPGA code to multiplex the display to show all four digits.

The FPGA outputs cannot sink the current required to light the LEDs, so we will connect it to a

---

[3] The display also includes two stand-alone LEDs (L1 and L2) forming a colon between the center digits so it can be used in a clock design. There is another stand-alone LED (L3), but we have no idea what it is for.

18L.2 Build a display decoder ROM

TLC5916 8-Channel Constant Current LED Sink Driver: see Fig. 18L.6. This device has eight outputs that are either off (open circuit) or on (sinking a constant-current to ground). The constant current value for all eight outputs is set by a single programming resistor, R_{ext}. The individual outputs are set by shifting eight bits into an internal shift register (1 = on; 0 = off). An output latch keeps the LED segments from changing while a new digit is being shifted in (data is latched when LE is low). In this lab, you will replace the output pins that connected the counter outputs to the logic indicators with three output pins connected to the TLC5916 to send data to it from a parallel-to-serial converter. The counter outputs will be rerouted to a binary-to-seven segment decoder ROM to provide the parallel inputs to the converter: see Fig. 18L.7.

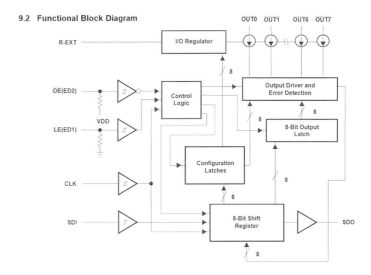

Figure 18L.6 Block diagram of the TLC5916 constant current sink driver.

The TLC5916 must be powered from the 3.3 V rail to ensure its input logic levels are compatible with the WebFPGA outputs. The display, however, must run on +5 V to provide adequate voltage for the LED voltage drop and driver current source compliance.

The WebFPGA will need to provide the TLC_CLK, TLC_SDI and TLC_LE output signals to the CLK, SDI and LE inputs of the TLC5916 respectively. The TLC5916 includes an output enable ($\overline{OE}$) to blank the display without altering the shifted data and a serial data out (SDO) to allow you to cascade several drivers together. We will not be using either of these functions in this lab. The $\overline{OE}$ should be grounded to enable the display at all times.[4]

The hardware: To convert the four-bit 74HC16X counter output from the last lab to digits on the seven segment display, you will need to construct a 16 × 7 ROM in the WebFPGA programmed as a Binary-to-Seven Segment decoder. The decimal point on the display will show the counter RCO/TC output value. That means you will want to route the RCO counter output through the decoder ROM as well, since it turns all segments off to handle blanking the display. The inputs to the WebFPGA remain the same as the last lab, with the exception of an additional switch input to blank the display. The new hardware consists of the TLC5916 and the LED display. The FPGA output pins that previously connected the counter outputs to the logic indicators now drive the TLC5916. Figure 18L.7 shows a schematic of the hardware to build for this lab, with the final internal WebFPGA functions included in block diagram form (we will add and test each block one at a time).

The 74HC16X counts 0x0 to 0xf. Its outputs go to the seven-segment decoder to produce eight

---
[4] To save an FPGA I/O pin, we will blank the display by turning all the segments off instead.

Lab: Memory

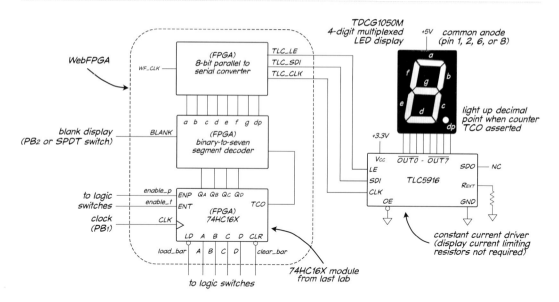

Figure 18L.7 PLD binary counter with seven-segment LED display.

(including the decimal point) parallel segment signals. These segment signals go into the parallel-to-serial converter to create the three signals that drive the TLC5916.

You will want to build this addition neatly, since we will be using this hardware in several future labs. You should implement the simplest, most direct connections between the WebFPGA and the TLC5916 inputs and between the TLC5916 outputs and the LED display segment cathodes without worrying about the segment order. You can always modify either the pin mapping or the ROM table in the WebFPGA to get the correct output display. Figure 18L.8 shows our build.

18L.2.2 Verilog code development philosophy

Structurally, you will want to integrate the counter, decoder ROM, and parallel-to-serial converter together to build the counter/seven segment display system by creating a top-level parallel-to-serial converter module into which you instance separate, self-contained counter and decoder modules. All the inputs to the top-level module inputs control the counter, except for the BLANK input. The BLANK input and the outputs of the counter are routed to the decoder ROM. The decimal point should light up when RCO/TC is asserted. The eight outputs of the decoder ROM are connected to the parallel-to-serial converter which generates the outputs of this new top-level module, the pins going to the TLC5916 display driver. By using modules, you simplify the Verilog code in the top-level module to mainly wiring. This may also make the design easier to understand and, if you have tested each module for proper operation, there should be no surprises when you wire them together. Once you get a module working, you shouldn't need to make any modifications to it to add the next block.

18L.2.3 Procedure

1. The first step is to choose a value for the TLC5916 current programming resistor R_{ext}. Use the datasheet for the TDCG1050M Clock Display to determine a reasonable value for the segment

18L.2 Build a display decoder ROM

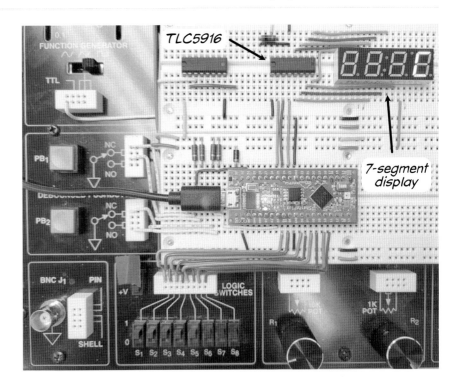

Figure 18L.8
Decoder ROM hardware.

current.[5] You will want to choose a current on the high side of the reasonable range (but do not exceed the maximum DC forward current), since in a later lab you will multiplex the display and each digit will be on only 25% of the time. Next look at the TLC5916 datasheet to determine a value of R_{ext} which will supply the selected current.[6] Think about why it is acceptable to drive the LED display from +5 V when the WebFPGA is limited to 3.3 V on its I/O pins.[7]

2. Next, add the display and the TLC5916 display driver to your breadboard. Wire the circuit as shown in Fig. 18L.7. As mentioned earlier, don't worry about which TLC5916 pin connects to which display pin, rather wire in the most straightforward manner possible. However, *be sure to annotate the schematic* with what you actually build. The eighth driver output lights the decimal point used to display the counter carry output signal (TCO).

Anytime you build a circuit, you should test it with the minimum extra hardware possible before trying to use it in a more complicated application. With the FPGA, it is a good idea to start with the simplest test to verify the hardware works, and then add functionality incrementally. Trying to debug a complex FPGA design without testing as you go along is quite difficult, since you won't know where to begin. In addition, trying to debug a large system with multiple errors can be extremely frustrating, as fixing a problem may not provide feedback in the form of a working system. So, rather than jump right in and try to construct the counter, decoder and converter all at once, first test your external circuitry with a minimal FPGA configuration consisting of a clock, a switch and pushbutton:[8]

---

[5] Available at https://LAoE.link/TDCG1050.pdf, although any multiplexed, common-anode four-digit LED display will work.
[6] Hint: See Fig. 19 at https://LAoE.link/TLC5916.pdf.
[7] The FPGA outputs are *only* connected to the TLC5916, which is powered from 3.3 V. The constant current TLC5916 outputs are compliant up to 20 V even though the part itself is limited to a supply voltage between 3 V and 5.5 V. (See Table 7.3 *Recommended Operating Conditions* in the TLC5916 datasheet.)
[8] Complete code available at: https://LAoE.link/FPGA/18L_Hardware_Test.v.

```
module TLC5916_Test(
    input WF_CLK,           // 16MHz system clock for timing
    input SDATA,            // NC pushbutton for data to shift into TLC5916
    input LATCH,            // high transparent, low latched
    output TLC_SDI,         // outputs to TLC5916
    output TLC_CLK,
    output TLC_LE
);

  reg [22:0] clkdiv;        // clock divider counter to get 2Hz TLC clock

  always@(posedge WF_CLK) begin
    if (clkdiv == 23'd8000000)      // MSB of counter is clock
       clkdiv <= 23'b0;
    else
       clkdiv <= clkdiv + 23'b1;
    end

  assign TLC_LE = LATCH;            // TLC latch controlled by logic switch 1
  assign TLC_SDI = SDATA;           // NC pushbutton for data shifted in
  assign TLC_CLK = clkdiv[22];      // clock LED segments at 2Hz
endmodule
```

With the LE logic switch set high (transparent), pressing and holding the pushbutton should light up the segments, one every half second until all segments are on (but not necessarily in "a" through "g" order). Releasing the button should cause the segments to go out sequentially in the same order. With the latch enabled (low), pressing the pushbutton should have no effect on the display; however, if you hold the button down for more than four seconds then switch the latch to transparent mode, all segments should light up at once.

3. Once you have your hardware working correctly, begin adding FPGA components incrementally by building and testing the parallel-to-serial converter: see Fig. 18L.9.

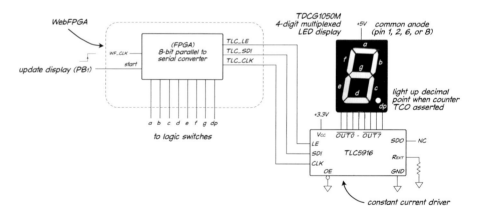

Figure 18L.9 Initially, test only the FPGA parallel-to-serial converter module.

Design the Verilog 8-bit parallel-to-serial converter. This module (Fig. 18L.10) has three inputs and three outputs. The inputs are the parallel 8-bit data to serialize (DATA_IN), a rising edge start signal (START), and a system clock (WF_CLK) which is used to clock the sequential converter elements. The three single-bit outputs are the output data stream (TLC_SDI), a clock signal (TLC_CLK) whose rising edge is used by the TLC5916 receiver to store the serial data, and an active low transparent latch enable (TLC_LE) that keeps the display driver outputs from changing while data is being shifted in.

18L.2 Build a display decoder ROM

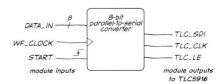

Figure 18L.10 Parallel-to-serial converter to drive the TLC5916.

To use the converter, you first set the 8-bit DATA_IN to the value to send, then supply a rising edge on the START line. The START is specified as rising edge triggered so that the START signal may to go low at any time after initiating a send. The converter should always transmit a full eight bits of output data once started, even if start goes low during the output transmission.

Once the START signal is asserted, the TLC_LE output should go low at or before the first rising edge of TLC_CLK and remain low until the last bit is sent.

To send a bit, the TLC_SDI output is set to the value of the bit to send, then the TLC_CLK goes from low to high to generate a rising edge that the TLC5916 uses to shift in the new bit. This is repeated eight times to send all eight DATA_IN bits. Bits may be sent either LSB first or MSB first (since you will be able to adjust the output of your seven segment decoder to either format).

You may assume DATA_IN remains stable until the conversion is complete (i.e., while TLC_LE is low).

The input clock (WF_CLK) is continuous and is used for the internal logic of the converter. However, the output clock (TLC_CLK) is *not* continuous and *should only be enabled for the eight output data bits*.[9]

The output clock does not have to be at the same frequency as the input clock (and it is likely that it will not be in most implementations). Note that it is poor practice to gate a clock (i.e., use a gate to turn a clock on and off) unless you do so synchronously with the clock signal.

Figure 18L.11 is a timing simulation of one possible solution to this problem shown sending 0xD2 then 0x5A (LSB first).[10] While this implementation shows TLC_SDI changing to the next bit on the falling edge of TLC_CLK, that is not necessary. It is acceptable to set TLC_SDI to the next bit on the rising edge of TLC_CLK, as the TLC5916 will have already captured the previous output bit by the time the new output propagates to TLC_SDI.

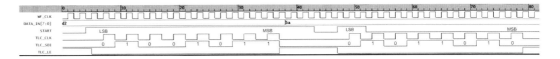

Figure 18L.11 Timing of one possible parallel-to-serial converter implementation.

Hints:

- The TLC5916 clock (TLC_CLK) *does not* have to be the same as the converter clock (WF_CLK). It is easier if it is not. We used an output clock at one-half the frequency of the input clock.
- Thinking about the specification, you need a flip-flop to capture the rising edge of START, a counter to count the eight output bits (and create the output clock), and a device to serialize the parallel data. The later can be either a shift register or (since the specification states that the input data stays stable during the conversion) a multiplexer. We used the latter.[11]

---

[9] There is an alternative design that uses a continuous clock and pulses the TLC_LE signal high then low *after* the eighth bit has been clocked into the TLC5916 but *before* the next rising edge of the output clock. Feel free to implement this scheme instead if you prefer.

[10] A higher resolution version of this image is available at https://LAoE.link/FPGA/18L.DEA009.png.

[11] If you get stuck, a block diagram similar to what we implemented is available at https://LAoE.link/FPGA/18L.DEA012.png.

Lab: Memory

- Use the LSB of the counter that counts eight bits to create the shift clock. Then the next three higher order bits count the number of bits shifted. Stop the counter when all eight bits have been shifted out.
- There are two overarching design possibilities for this module. In the first, you send the 8-bit data *once* after the rising edge of START. This requires a flag (i.e., a flip-flop) to latch the START signal and then remember that you are done after sending the data by clearing the flag until the next START rising edge. If you use a fast enough clock, you do not have to use TLC_LE (the output will be shifted out to the display so quickly you will not see it). You can leave the latch transparent (high). The other design sends the 8-bit data continuously as long as start is high (however you must always shift out 8-bits before stopping). This design is simpler; however, here you must use the display latch or all segments will flash as data is shifted into the TL5916 repeatedly.
- Don't worry about the bit order the first time you write the code. You can always correct it once you have your system working by changing the order that the input switches are connected to your shift register or multiplexer.
- There is an EDAPlayground testbench available at https://LAoE.link/18L_P2S_Test.html. if you want to test your design. This is what we used to create Fig. 18L.11.

Here is a header for the module. You will need to change the @MAP_IO pin assignments to match the connections to your WebFPGA.[12]

```
//////////////////////////////////////////////////////////
// Company: Learning the Art of Electronics
// Engineer: David Abrams
//
// Create Date:    11/3/2020
// Design Name:    18L_P2S_test.v
// Module Name:    P2S_Test
// Description:    Module to send logic switch values as
//                 serial data to TLC5916.
//////////////////////////////////////////////////////////

//*********************************************
// The comments in this block are directives to
// the WebFPGA environment.
//
// Set name of top level module (default is fpga_top)
// @FPGA_TOP P2S_Test
//
// @MAP_IO start      30 // PB1 NC output: rising edge starts data output
// @MAP_IO TLC_SDI    25 // output to TLC5916: serial data in
// @MAP_IO TLC_CLK    24 // output to TLC5916: rising edge clock
// @MAP_IO TLC_LE     23 // output to TLC5916: transparent when high
//
// @MAP_IO segdata[0]  4 // inputs from logic switches
// @MAP_IO segdata[1]  5
// @MAP_IO segdata[2]  6
// @MAP_IO segdata[3]  7
// @MAP_IO segdata[4]  8
// @MAP_IO segdata[5]  9
// @MAP_IO segdata[6] 10
// @MAP_IO segdata[7] 11
//*********************************************

module P2S_Test(
```

[12] Header available at: https://LAoE.link/FPGA/18L_P2S_Test_Hdr.v

18L.2 Build a display decoder ROM

```
    input WF_CLK,           // 16MHz system clock runs converter
    input [7:0] segdata,    // the 8 parallel data bits to sent to TLC5916 serially
    input start,            // rising edge triggers new output cycle if idle
    output TLC_SDI,         // outputs to TLC5916 - add "reg" if needed
    output TLC_CLK,
    output TLC_LE
);
```

4. Add your serial-to-parallel converter to the FPGA. If everything is working, you should be able to turn on each segment (and the decimal point) individually by selecting one of the eight logic switches and starting the conversion with the pushbutton. This will also allow you to determine the mapping between the segment data vector (`seg_data`) and the actual display segments.

5. Once you are sure your wiring is correct and your converter works, design a binary-to-seven segment decoder ROM module. This module should translate the four-bit binary input value along with the decimal point and blanking inputs into the eight parallel outputs that light the display segments. Fig. 18L.3 shows the segments to light for counter values 4'b0000 through 4'b1111. Your decoder should take a vector input, `ctr_value`, for the counter value. The blank input should turn off all segment LEDs and the decimal point when asserted.

   Here is the module definition:

```
module ROM_Test(
    input WF_CLK;           // 16MHz system clock
    input [3:0] ctr_val,    // character (4-bit) to display
    input start,            // rising edge triggers new output cycle if idle
    input BLANK,            // active high blanking
    input dec_point,        // decimal point on or off
    output TLC_SDI,         // TLC5916 inputs (FPGA outputs)
    output TLC_CLK,
    output TLC_LE
);
```

6. Add your seven-segment decoder ROM to the parallel-to-serial converter module. Use logic switches to provide the four-bit binary and decimal point decoder inputs. Use the second pushbutton or a SPDT switch for the blanking input (your choice if it is active high or low). Figure 18L.12 shows the system at this intermediate step.

   Hint: Create your decoder as a separate module, then instantiate it in the parallel-to-serial converter using it as the top level module. You should not have to make any changes to the function of your converter other than using the decoder outputs as the ROM inputs rather than the logic switches.

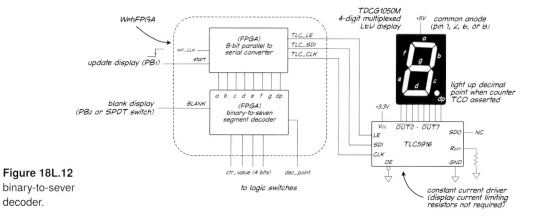

Figure 18L.12 binary-to-seven decoder.

7. Once you have this working, modify your Verilog code to eliminate the manual start input and automatically trigger the parallel-to-serial converter whenever any of the eight segment signals does [or could] change. One way to do this is by saving the segment data in a register every time you output to the TLC5916, then comparing the saved data to the decoder output for changes. When you detect a change, set the start signal. You will need to reset start after it has been recognized by the converter; however, you can only modify a particular signal in *a single* `always` block. It is not legal to set a signal in one block and clear it in another, so you will need to set start and clear it in the same `always` block.

8. Finally, add your 74HC16X counter module from the last lab to supply the ctr_value and dec_point (i.e., TC) input signals to the ROM decoder.[13]

```
module Counter_Display(
    input WF_CLK,              // 16MHz system clock - outputs TLC bits at 8MHz
    input CLK,                 // clock to 74HC16X ctr
    input BLANK,               // active high blanking
    input clear_bar,
    input enable_p,
    input enable_t,
    input load_bar,
    input A, B, C, D,
    output TLC_SDI,            // TLC5916 inputs
    output TLC_CLK,
    output TLC_LE
);
```

This module corresponds to the final block diagram of Fig. 18L.7. The display decimal point should light up when the RCO/TC counter output is asserted (i.e., on a count of 4'b1111 if ENT is high).

Test your working system using the counter inputs the same way you tested the 74HC16X counter in the last lab. You will probably want to manually clock the counter with a debounced pushbutton. The big difference will be that the count will now be displayed in hex on the seven-segment display rather than in binary on the logic indicators. Make sure the decimal point lights up on a count of "F" (and extinguishes if ENT is dissasserted) and that the BLANK pushbutton turns off all seven segment LEDs and the decimal point on the display.

9. (Optional) If you have time and are interested, try replacing *just one* LED segment output of the ROM with a combinational logic equation. (Or just take a look at page 4 of this datasheet: https://LAoE.link/74LS247.pdf.)[14] That should convince you that the ROM approach is much easier.

Leave the TLC5916 and LED display build on your breadboard. We will use it again in the digital project lab.

18L.2.4 Solutions

Here are our solutions to the parts of §18L.2.3 in case you get stuck **but please try to get your own version working *before* looking at our answers.** You will only learn FPGA design by doing it (struggles and all).

- §18L.2.3, Item 2: https://LAoE.link/FPGA/18L_Hardware_Test.v.

---

[13] In case you did not get the counter working, here is one you can use: https://LAoE.link/FPGA/17L_Counters.v.
[14] This device simplifies the problem by not decoding inputs A–F.

- §18L.2.3, Item 3: https://LAoE.link/FPGA/18L_P2S_Test_BD.v is a direct structural implementation of the block diagram in Footnote 11; https://LAoE.link/FPGA/18L_P2S_Test.v is a behaviorial implementation that contains both single output and repeating output versions; https://LAoE.link/FPGA/18L_P2S_Test_Continuous_Clk.v is a version that demonstrates using a continuous clock.
- §18L.2.3, Item 6: https://LAoE.link/FPGA/18L_Decoder_ROM_Test.v.
- §18L.2.3, Item 7: https://LAoE.link/FPGA/18L_Decoder_Auto_Start.v.
- §18L.2.3, Item 8: https://LAoE.link/FPGA/18L_Counter_Display.v.

18S Supplementary Notes: Memory Blocks

So far, we have been using (wasting?) one FPGA logic cell flip-flop to create each bit of RAM memory and one or more logic cells for each ROM bit. This is inefficient and, for anything more than a small memory array, could leave us without enough logic cells to implement the rest of our design or require a larger, more expensive device. Most FPGA manufacturers include blocks of stand-alone memory that are higher density and take up less space than using logic cells. The Lattice iCE40UP5K in the WebFPGA, is no exception, it includes two types of block memory:[1]

- Thirty (30) Embedded Block RAM (EBR) memory blocks, each 4 Kbits in size. These can be configured as 2-, 4-, 8- or 16-bit wide memory and have separate read and write ports (dual-port memory). They can be used as lookup tables, general purpose RAM memory, or FIFOs, and can be initialized by the bitstream. By disabling write access, an EBR becomes a fixed ROM. Multiple EBRs can be cascaded to create larger blocks of memory.

- Four (4) Single Port RAM (SPRAM) memory blocks, each containing 256 Kbits. While these cannot be initialized, they be cascaded to provide up to 1 Mbit of general purpose RAM. As their name implies, they are single-port memory, although they have separate data in and data out buses. The natural size of each block is 64K × 16, but with some complexity other data widths can be used.

18S.1 Accessing FPGA resource blocks

Normally, FPGA memory blocks are not used by the `case` and `array` constructs we discussed in 18N to implement ROM and RAM.[2] Lattice provides a library of "macro" primitives to access a number of common constructs and special features in their iCE40 family devices. These are essentially predefined modules in the synthesizer and they are instantiated in the same way as any module you have written.[3] We will take a look only at the EBR memory in this example, but using other macros for SPRAM, DSP blocks, or serial I/O involves similar steps. (However, the Lattice documentation is fairly terse – you may find it helpful to search the web for examples.)

---

[1] See iCE40 UltraPlus Family datasheet at `https://LAoE.link/Ice40UP_Datasheet.pdf`. Note that this is the only Verilog code in the book that is device dependent. It will not work on other FPGAs outside the iCE40 UltraPlus family.
[2] There are exceptions. The WebFPGA IDE will infer memory if the code format conforms to the instructions in the iCEcube2 User Guide, Section *About Inferring Memory* available at `https://LAoE.link/iCEcube2_Guide.pdf`.
[3] See the Lattice ICE Technology Library guide at `https://LAoE.link/Ice40_Library.pdf`.

18S.2 Using WebFPGA EBR memory blocks

The EBR memory blocks are described in detail in the Technical Note *Memory Usage Guide for iCE40 Devices*.[4] Each block consists of a write interface, with write address lines, a write input data bus, a synchronous write clock, and several control lines. The read interface has read address lines, a read output data bus, a synchronous read clock, and two enable control lines. When used as a ROM, with preloaded data, the write interface is not used.

The technology library includes primitives to instantiate an EBR in either 2-, 4-, 8- or 16-bit wide memory with a either positive or negative edge clocking. We will be using the SB_RAM256x16 configuration (positive clocking), although we will only be using the lower eight bits of the 16-bit output word.[5]

This example initializes the memory with a bit pattern, disables the write interface to create a ROM, and reads data out of the memory sequentially at a 10 Hz rate, displaying the result on the logic indicators. Here is the code:[6]

```
//////////////////////////////////////////////////////////////////////
// Company: Learning the Art of Electronics
// Engineer: David Abrams
//
// Create Date:    2023-01-27
// Project Name:   18W_EBR_ROM.v
// Module Name:    EBR_ROM
// Target Device:  WebFPGA
// Description: Example of using Lattice iCE40 EBR memory block as ROM
//              Loads the ROM with data to make the logic indicator LEDS
//              do a little dance.
//////////////////////////////////////////////////////////////////////

//*********************************************
// The comments in this block are directives to
// the WebFPGA environment.
//
// set name of top level module (default is fpga_top)
// @FPGA_TOP   EBR_ROM
//
// Commands to connect the named signals below
// to specific WebFPGA output pins.
// @MAP_IO Q[7]       25  // logic indicator 1
// @MAP_IO Q[6]       24  // logic indicator 2
// @MAP_IO Q[5]       23  // logic indicator 3
// @MAP_IO Q[4]       22  // logic indicator 4
// @MAP_IO Q[3]       21  // logic indicator 5
// @MAP_IO Q[2]       20  // logic indicator 6
// @MAP_IO Q[1]       18  // logic indicator 7
// @MAP_IO Q[0]       17  // logic indicator 8
//*********************************************

module EBR_ROM (
  input WF_CLK,
```

---

[4] Available at https://LAoE.link/Ice40_EBR_Guide.pdf.

[5] We are not using the 512 × 8 configuration because it interleaves the output data between two sets of data lines, complicating the code. If you needed more than 256 memory locations, you could instantiate a SB_RAM512 × 8 EBR module, then multiplex the 16-bit output into an 8-bit result using the LSB of the memory to select between the even output lines (when LSB = 0) and the odd ones (when LSB = 1). See Ibid, Table 4.4, Mode 1. We leave this as an exercise for the reader – or possibly an exam problem.

[6] Available at https://LAoE.link/FPGA/18S_EBR_ROM.v.

```verilog
  output [7:0] Q
);

parameter MEMSZ = 11 * 16;   // number of words of data in ROM (rows * words/row)

// Create a 10Hz memory address clock from the 16MHz WF_CLK
  reg [19:0] divctr;     // divide by 800,000 takes 20 bits
  reg mclk;              // 10Hz memory address clock

  always@(posedge WF_CLK) begin
    if (divctr == 20'd800000) begin
       divctr <= 20'd0;
       mclk <= ~mclk;
    end else
       divctr <= divctr + 20'd1;
  end

// Create 8-bit memory address counter to increment at 10Hz
  reg [7:0] raddr;   // address counter for 256 locations (8 bits)

  always @(posedge mclk) begin
    if (raddr == (MEMSZ - 1))
       raddr <= 8'b0;
    else
       raddr <= raddr + 8'b1;
  end

 wire [7:0] dummy;        // place to put unused byte from ROM

// Instantiate a 4Kbit (256 X 16) EBR ROM initialized with LED data
SB_RAM256x16 ram256x16_inst (
   .RDATA({dummy,Q}),   // read data goes to logic indicators
   .RADDR(raddr),       // read address increments every 100ms
   .RCLK(mclk),         // same clock that increments address reads next data word
   .RCLKE(1'b1),        // both read enables are always asserted
   .RE(1'b1),
   .WADDR(8'b0),        // write ports (waddr, wdata and mask) not used for ROM
   .WCLK(1'b1),         // write clock not used for ROM
   .WCLKE(1'b0),        // write enables not asserted for ROM
   .WDATA(16'b0),
   .WE(1'b0),           // write enables not asserted for ROM
   .MASK(16'b0)
);

// Define data to load into memory from bitstream
defparam ram256x16_inst.INIT_0 =
256'h0000_0001_0002_0004_0008_0010_0020_0040_0080_0040_0020_0010_0008_0004_0002_0001;
defparam ram256x16_inst.INIT_1 =
256'h0000_0001_0002_0004_0008_0010_0020_0040_0080_0040_0020_0010_0008_0004_0002_0001;
defparam ram256x16_inst.INIT_2 =
256'h0000_0001_0003_0007_000F_001F_003F_007F_00FF_007F_003F_001F_000F_0007_0003_0001;
defparam ram256x16_inst.INIT_3 =
256'h00FF_007F_003F_001F_000F_0007_0003_0001_0000_0001_0003_0007_000F_003F_007F_00FF;
defparam ram256x16_inst.INIT_4 =
256'h0080_0040_0020_0010_0008_0004_0002_0001_0080_0040_0020_0010_0008_0004_0002_0001;
defparam ram256x16_inst.INIT_5 =
256'h0080_0040_0020_0010_0008_0004_0002_0001_0080_0040_0020_0010_0008_0004_0002_0001;
defparam ram256x16_inst.INIT_6 =
256'h00FF_007F_003F_001F_000F_0007_0003_0001_00FF_007F_003F_001F_000F_0007_0003_0001;
defparam ram256x16_inst.INIT_7 =
256'h0000_0080_00C0_00E0_00F0_00F8_00FC_00FE_00FF_007F_003F_001F_000F_0007_0003_0001;
```

18S.2 Using WebFPGA EBR memory blocks

```
defparam ram256x16_inst.INIT_8 =
256'h00AA_0055_00AA_0055_00AA_0055_00AA_0055_00AA_0055_00AA_0055_00AA_0055_00AA_0055;
defparam ram256x16_inst.INIT_9 =
256'h0018_0000_0018_0024_0042_0081_0000_0081_0042_0024_0018_0000_0018_0024_0042_0081;
defparam ram256x16_inst.INIT_A =
256'h0000_0000_0000_0081_0042_0024_0018_0000_0018_0024_0042_0081_0000_0081_0042_0024;
endmodule
```

There are three hardware components to this example. The first is a counter and flip-flop used to create the 10 Hz clock from the WebFPGA's 16 MHz input clock. This requires a 20-bit counter to divide 16 MHz by 800,000 ($2^{20} = 1,048,576$) which is then used to clock a divide-by-2 flip-flop. The result, mclk, is the 10 Hz, 50% duty cycle memory clock.[7]

Next we have an 8-bit counter capable of addressing all of the EBR memory locations. However, we may not need all 256 words, so we check and reset the counter when it counts up to the last valid data in ROM. In this case we declare a parameter, MEMSZ, for the number of locations actually used. We have let the synthesizer calculate the value of MEMSZ based on the number of 256 bit initialization statements we are using.

Finally, we instantiate the EBR memory block with the SB_RAM256 × 16 macro, disabling the write interface and using the clock and address counter we created to read sequential words out of the memory at a 10 Hz rate and display them on the breadboard's logic indicators. We permanently assert both read enables as we want to read data on every rising clock edge.

The DEFPARAM following the EBR instantiation initializes the memory data. DEFPARAM works just like PARAMETER, however it is capable of overriding a parameter statement in the module itself. While the Lattice documentation is unclear if an EBR block is set to an initial value, we tested a block without explicitly initializing it with DEFPARAM and it contained all zeros.

When we synthesized the code above, the bitstream used 29 flip-flops (20 for the divide-by-800,000 counter, one for the MCLK flip-flop and 8 for the address counter, but none for the ROM memory). It also confirms that an EBR block is being clocked by MCLK as we expect: see Fig. 18S.1.

```
Placement of Design:       LUT4s:41( 0.78%)
                           FLOPs:29( 0.55%)
                           IOs:  17(43.59%)
                           CLOCK:EBR_ROM|mclk    FREQ: 168.82 MHz
                                                 target: 1.00 MHz
Packer: DRC check was successful. Now doing packing.
Packer: Packing was successful. Logic cells used: 41
Routing of Design: Successful. Nets:87 Iterations:2.
Writing of Design Netlist: Successful.
```

Figure 18S.1 Resource usage of EBR_ROM module.

18S.2.1 Inferred ROM blocks

The disadvantage of using the Lattice EBR macros is that the design becomes less portable. While [hopefully] the code should work with other Lattice devices, another manufacturer is likely to have a different method or macro to access its internal memory blocks.[8]

---

[7] There is no real need for a 50% duty cycle clock here but we thought it would be nice to show you how it is done. If you are connecting the clock to an external pin, a 50% duty cycle is much easer to see on the scope than just setting the output high for one cycle every 1,600,000 clock cycles.

[8] It turns out this code did not work on the other Lattice-based (but different device family) FPGA board we tried – it uses a different format macro to access its EBR memory. It is good practice to encapsulate any macro primitives in a generic module so that you only have to change that one module if you switch manufacturers or device families.

Supplementary Notes: Memory Blocks

One way to avoid this problem is to use the normal Verilog structures for creating memory. We rewrote the EBR ROM code above to use the `case` statement in accordance with Lattice's guidelines for the code to be inferred as ROM. The resulting Verilog is available at https://LAoE.link/FPGA/18S_CASE_EBR_ROM.v. While the bitstream still uses an EBR memory block and only 29 flip-flops, the number of logic cells used has ballooned more than threefold, from 41 to 129: see Fig. 18S.2. It appears that inferred memory is much less efficient than incorporating memory blocks using the Lattice primitives, although we are not sure why.[9]

Figure 18S.2 Resource usage of CASE_EBR_ROM module using case operator to infer ROM.

```
Placement of Design:    LUT4s:129(  2.44%)
                        FLOPs:29(  0.55%)
                        IOs:  17(43.59%)
                        CLOCK:CASE_EBR_ROM|mclk    FREQ: 168.82 MHz
                                    target: 1.00 MHz
Packer: DRC check was successful. Now doing packing.
Packer: Packing was successful. Logic cells used: 129
Routing of Design: Successful. Nets:167 Iterations:4.
Writing of Design Netlist: Successful.
```

---

[9] This is a limitation of the WebFPGA IDE. If we could see a schematic of what the synthesizer created for each design, we would be able to figure out what happened.

19N Finite State Machines

Contents

19N.1	State machine: new name for old notion	760
19N.2	Designing any sequential circuit: beyond counters	761
19N.3	Designing an odd two-bit counter	761
	19N.3.1 FSM Design	762
	19N.3.2 FSM Implementation	765
	19N.3.3 Assigning state names	770
19N.4	Selecting the FSM clock frequency	770
19N.5	Recapitulation: several possible sequential machines	771
19N.6	AoE reading	771

Why? Finite State Machine design methodology provides a rigorous way to design synchronous systems. It applies not only to hardware but to software design as well (as we shall see when we study embedded microcontrollers). If your abstract state machine design meets the systems specification, implementation in discrete logic, PLDs or software is primarily a mechanical process of translation. In addition, since each processing cycle is confined to a single state, debugging is localized as well. Thus FSM design offers an alternative to ad hoc solutions that often fail to catch corner cases or introduce subtle bugs in a system's implementation.

19N.1 State machine: new name for old notion

By now you have seen a variety of counters: ripple and synchronous built from flip-flops, IC counters like the 74HC16X series, and counters implemented in programmable logic. A counter is a special case – approximately the simplest – of the more general device, *state machine*. This term is short for the more formal "Finite State Machine" (FSM).[1]

An FSM is a sequential device that walks through predictable sequences of "states." (A state, you will recall, is defined as the set of outputs on the device's flip-flops.) That sounds oddly abstract; but if you apply the notion to a simple counter you will see that the idea is simple and, indeed, familiar. Figure 19N.1 shows a directed graph called a "state transition diagram" or "flow diagram" of a 2-bit up-counter along with a "state table" that describes – somewhat abstractly, we admit – how the counter behaves.

---

[1] Such a machine is distinguished from a "Turing machine" – a theoretical model of a general-purpose computer. A Turing machine is assumed to have infinite number of states (and memory) as an FSM does not.

760 Finite State Machines

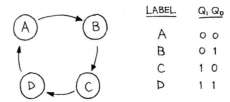

Figure 19N.1 2-bit counter: flow diagram and state table.

19N.2 Designing any sequential circuit: beyond counters

Figure 19N.2 is a highly-general diagram showing any clocked sequential circuit.[2]

AoE §10.4.3

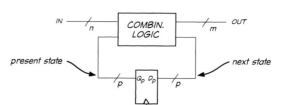

Figure 19N.2 Sequential circuit: the general model of a FSM.

As you can see, the D flip-flops will take the levels at the D inputs and transfer them to their Q outputs after the next clock. The Q of circuit will go from its *present state* to the values on the Ds on the next clock; thus the values on the Ds show the circuit's *next state*. With p flip-flops, this FSM can have up to 2^p states. AoE shows how to use this notion to design a divide-by-three counter. We will do nearly the same task – but using different state assignments.

Figure 19N.2 shows *outputs*; in counters – the only state machines we have worked with, so far – the outputs have been simply the flip-flop Qs, apart from sometimes a Carry$_{out}$. The outputs may depend on the machine's *state* alone, or may depend upon *state and inputs*. The two types are given the names "Moore" (state alone) versus "Mealy" (state and inputs).[3]

A counter with a Carry$_{out}$ that sensed only the levels of the counter's Qs would be a Moore machine. A counter whose Carry$_{out}$ depended also on Carry$_{in}$ (as it usually does) would be a Mealy machine. We concede that learning such names is not edifying. But you may want to be able to recognize the terms. We will concentrate on designing Moore FSMs.

19N.3 Designing an odd two-bit counter

Let's design a two-bit counter with an unusual count sequence using FSM design techniques. The first three steps are common to all implementations – discrete flip-flops, PLDs or software.

Specification: Here what we want for our counter: see Fig. 19N.3.

---

[2] An FSM need not be clocked. The more general model would define the block we have shown as a set of D flops as just a "delay" block. Design of non-clocked FSMs is more difficult than what we undertake in this book, and only rarely necessary.

[3] Mealy and Moore wrote important papers that defined these ideas in the mid-1950s. Moore was a professor of computer science; Mealy worked at Bell Labs.

19N.3 Designing an odd two-bit counter

- The counter counts 00, 11, 10 then back to 00.
- The counter has three outputs – two bits for the count and an active-low LED that lights when the count is 00.
- The counter has one input, an active-low enable ($\overline{\text{ENAB}}$) that inhibits counting when not asserted.[4]
- The counter has an *asynchronous* active-low reset.[5]

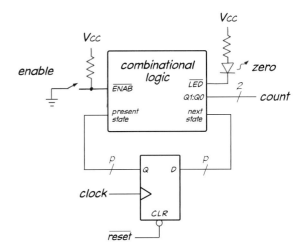

Figure 19N.3 Block diagram of our odd counting FSM.

19N.3.1 FSM design

Step 1: Define the states. The first two design steps, defining the states and drawing the state transition diagram are the most critical. If you do not define the states correctly or your state diagram does not match the specification, the FSM will not work. On the other hand, if you correctly complete these steps implementation in any form is straightforward, with a high probability of success. In the first step, you need to adequately capture all the state information necessary to implement the specification. If you have too few states, your machine will not work; however, if you have too many, some may be redundant but your system may still operate correctly.

In this problem, we need three states to remember the current count. Try to give your states functional descriptions rather than refer to the inputs and outputs as 1s and 0s. Instead of "The $\overline{\text{LED}}$ output is low…" say "The LED indicating a count of 00 is lit." We give states letters beginning with "A," which we usually make the reset state if one is specified. State descriptions should indicate what the state is waiting for or expects but should not include state transition information (we will add that in the next step). For example, you would not include the enable input in a state description for this FSM because enable only affects transitions between states. However, since this is a Moore machine you should include output information as part of the state description.

- State A – Count is 00. LED indicating a zero count is on. This is the (asynchronous) reset state.
- State B – Count is 11. LED is off.
- State C – Count is 10. LED is off.

---

[4] We don't count the clock as an input to the FSM. Its purpose is to tell the FSM when to transition from the present state to the next state.

[5] The reset is not an FSM input because it is asynchronous. Had we specified a *synchronous* reset, it would have been included as just another ordinary input to the FSM.

Step 2: Draw the state transition diagram. The directed graph in Fig. 19N.4 describes the transitions between states in our counter. Each state is indicated by a circle containing the state letter. The outputs for each state are shown next to the state circle. Each arrow exiting a state specifies a transition that occurs on the rising edge of the FSM clock. Next to the arrow are the input values that cause that particular transition. We prefer to describe the outputs functionally and make all inputs active-high, and then convert to actual signals values later in the implementation (but you can use actual signal levels as long as you make clear what are using). If your FSM has an asynchronous reset state, indicate it with an arrow. If the reset is synchronous, no explicit annotation is required as it will be included in the state transition arrows.

Your diagram *must* have 2^n transition arrows coming out of *each* state, where n is the number of inputs. If not, the diagram is missing information and incomplete. It is acceptable, however, for one arrow to represent multiple transitions. For example, if our oddball 2-bit counter had a synchronous reset, states C and D could have arrows transitioning to state A marked with "1X" where the 1 indicates that reset is asserted (again we prefer to use active-high signals until implementation) and the X means we do not care whether the enable input is asserted or not. We would count this arrow as two transitions in verifying the required 2^n transitions out of a state.

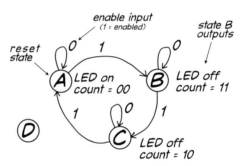

Figure 19N.4 State transition diagram of odd sequence divide-by-three counter.

The diagram shows a fourth state, "D" floating outside the sequence. This reminds us that there will be an unused fourth state if we implement this FSM with flip-flops. In the next step, we will use "XX" as the next state for state D to indicate that we don't care what comes next. This is a little dishonest, since we do care to a limited extent: we'll admit shortly that we do not want the counter to get stuck in this unused state. But it is traditional to specify *don't care* as the next state for any unused state. The motive is to constrain the circuit as little as possible, to minimize the logic that determines the next state.

The circuit will, of course, *care* about that next state: when the design is finished that next state will be as fully determined as any other. In fact in a few minutes we will see what that next state is in our design.

Step 3: Create the Present State/Next State table. The state transition diagram fully captures the operation of the abstract finite state machine described in the specification (with the possible exception of the active state of the signals). Nevertheless, we always translate it into a Present State/Next State table to simplify implementation: see Fig. 19N.5. The PS/NS table contains one row per state and three sets of columns. The first column is the current (i.e., the present) state. To the right of the current state are 2^n columns, one for each possible input combination, containing the next state to transition to on the clock edge for that present state and input combination. Finally, in a Moore FSM, there are m columns showing the outputs for each present state.

Creating the PS/NS table is a matter of translating the graphical state transition diagram information

19N.3 Designing an odd two-bit counter

to tabular form. You may decide to switch from functional descriptions ("LED on") to signal values in anticipation of implementing the FSM. However, we prefer to continue using active high signal names.

	enable (1 = enabled)		COUNT	
PS	0	1	$Q_1\ Q_0$	LED
A (rst)	A	B	0 0	1 (on)
B	B	C	1 1	0 (off)
C	C	A	1 0	0 (off)
D	X	X	X X	X
	NS			

Figure 19N.5 Present State/Next State table for odd 2-bit counter.

Once we have the PS/NS table, our paths diverge depending on how we intend to implement the FSM.

Step 4: Assign binary values to states. All implementations require us to assign binary values to states but only in FSMs built with flip-flops and logic does it matter how we make the assignments. If you plan to implement the FSM using behavioral Verilog or software, just assign the states sequential numeric values.[6]

However, when implementing using flip-flops the choice of state assignments influences the complexity of the combinational logic. In general, choosing an assignment that gives the fewest number of bit transitions between state transitions results in the simplest logic. In the oddball counter, however, assigning the output values to the states results in the simplest output logic (since the counter output value is then just the flip-flop Qs).

For a flip-flop implementation we must redraw the PS/NS table with state letters replaced by their binary assignments. We have also replaced the functional output description with a logic level (the active-high inverse of the $\overline{\text{LED}}$ output).[7]

PS Q_1Q_0	NS (ENAB = 0) D_1D_0	NS (ENAB = 1) D_1D_0	LED 1 = ON
A=00	00	11	1
B=11	11	10	0
C=10	10	00	0
D−01	XX	XX	X

Step 5: Create the Boolean equations for the output and the next state inputs. The PS/NS table shows the binary state values before and after each clock edge. If we plan to use D flip-flops to design this circuit, the next state also defines the pair of values we must feed to the D inputs of the two flip-flops, and we have so labeled the Next State (NS) columns.

We need to determine the Boolean equations for the D inputs to the flip-flops in order to design the combinational logic for the FSM. These days, with large numbers of gates available on a logic array, minimizing the logic equations is usually not worthwhile and we might as well just use the

---

[6] The exception is if you decide to use "one-hot" state encoding, which assigns one flip-flop per state. One-hot encoding reduces the complexity of the combinational logic at the cost of additional flip-flops. In an FPGA, where the lookup table in each logic cell is small and can only realize a simple Boolean equation but each logic cell includes a flip-flop, one-hot encoding may result in fewer total logic cells used.

[7] You can skip this step and the next one and go right to implementation if you are planning on a behavioral Verilog or software implementation. All you need for those is the abstract PS/NS table.

unsimplified sum-of-products equations. We won't do that now, because we would like to show the danger that can result from such use of don't cares.

Recall that the XX means we can assign high or low, as we please, to make the circuit easy to build. If we assign the *don't cares* for the next state in the D row as follows, we can write the Boolean equations by inspection:

PS Q_1Q_0	NS (ENAB = 0) D_1D_0	NS (ENAB = 1) D_1D_0	LED 1 = ON
A=00	00	11	1
B=11	11	10	0
C=10	10	00	0
D=01	**01**	**01**	1

LED = $\overline{Q_1}$ or $\overline{\text{LED}} = Q_1$ which is what we want since the LED is active-low;
$D_0 = (Q_0 \cdot \overline{\text{ENAB}}) + (\overline{Q_1} \cdot \text{ENAB})$;
$D_1 = (Q_1 \cdot \overline{\text{ENAB}}) + ((Q_1 \oplus Q_0) \cdot \text{ENAB})$.

We got these functions by assuming convenient levels for the Xs. Now let's see what Next State we have provided for the unused state. D_1 is low, D_0 is high. Oops! That means that state D goes to D.

Why should we suddenly care? We said at the outset that we *don't care*. But we didn't quite mean it, as we admitted earlier. In fact, there is one outcome that we do care about avoiding: the machine must not get stuck in one of its unused states, like state D.

And if you're inclined to protest, "But it will never go to state D," then you're being too optimistic. On power-up the machine could go to D. And even if you build a power-on reset circuit to force the counter to a legitimate state (say, state A) when you turn on power, a glitch can carry the circuit into D. So, it's never safe to design a machine that would get stuck in an unused state.

The remedy is simple: enter some other valid next state in place of XX, and redesign the circuit's logic. Since we don't expect the FSM to ever be in state D, we don't really care what state it goes to. As long as it goes to a valid state it will resume normal counting.

Suppose instead we make state B (coded as 11) the next state from D (coded as 01)?

PS Q_1Q_0	NS (ENAB = 0) D_1D_0	NS (ENAB = 1) D_1D_0	LED 1 = ON
A=00	00	11	1
B=11	11	10	0
C=10	10	00	0
D=01	**11**	**11**	1

Now there are no stuck states, the equation for D_0 does not change and the equation for D_1 becomes a bit less complicated:

$\overline{\text{LED}} = Q_1$;
$D_0 = (Q_0 \cdot \overline{\text{ENAB}}) + (\overline{Q_1} \cdot \text{ENAB})$;
$D_1 = Q_0 + (Q_1 \oplus \text{ENAB})$.

19N.3.2   FSM implementation

Implementation using flip-flops and logic gates.   Once you have the logic equations for the output and next state D inputs, implementation is just a matter of wiring up logic to realize these equations in the "combinational logic" block of Fig. 19N.3. Figure 19N.6 shows the divide-by-three circuit designed as a state machine.[8]

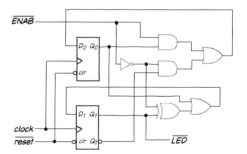

Figure 19N.6 Oddball counter, from manual design.

This – or the similar design problem included in Lab 19L presents probably your last occasion to do such a task by hand. Logic compilers like Verilog will normally handle the detail work for you.

Implementation using flip-flops and a memory. *Memory* offers another way create an FSM. We have seen the duality between Boolean logic and truth tables several times already. We can program a ROM with the PS/NS table of our design to implement the combinational logic block of Fig. 19N.3.

The *address* lines of the memory can carry the *input* variables and present-state bits; the *data* lines of the memory can deliver the *output functions* while also generating the *next-state* information: see Fig. 19N.7. Memory makes it easy to provide many outputs: just use a memory with a wider data word. To provide for longer sequences of operations, just increase the number of flip-flops. To build a machine that does many alternative tricks, increase the number of inputs – which means only adding address lines: that is, use a memory that stores more data words.

Not only is a ROM-based FSM much easier to wire up than discrete logic, if you discover an error in your Boolean equations while debugging, all you need to do to correct it is to burn a new ROM, rather than spend hours rewiring. Also, since it is difficult to find EEPROMS with fewer than 16K memory locations, you have an almost infinite supply of states to add enhancements when the person who provided the specifications comes in and says, "You know, it would great if we could just add. . . "[9]

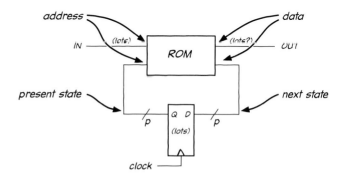

Figure 19N.7 FSM using memory in place of combinational logic.

To implement our oddball counter, we need an 8×3 or 8×5 memory. The latter allows us to divorce the state assignment from the output count. Here is what that ROM would look like with the states assigned sequentially, beginning with A = 00. We use the actual active-low values of the input and

---

[8] A simulation of this FSM is available at https://LAoE.link/FPGA/FSM_Counter_Sim.html.
[9] Although you would likely create FSMs of any complexity in programmable logic, so these advantages are probably moot.

output to avoid the need for external inverters. State D (shown in bold) is still a *don't care* so we need to program its next state to any state *but* D to avoid a stuck state.[10]

State	ROM Address	ROM Output			ROM Data
	$\overline{\text{ENAB}}\ Q_1 Q_0$	$D_1 D_0$	$O_1 O_0$	$\overline{\text{LED}}$	
A	000	01	00	0	0x08
B	001	10	11	1	0x17
C	010	00	10	1	0x05
D	**011**	**00**	**00**	**0**	0x00
A	100	00	00	0	0x00
B	101	01	11	1	0x0f
C	110	10	10	1	0x15
D	**111**	**00**	**00**	**0**	0x00

A circuit like the one shown in Fig. 19N.7, if given eight inputs, would behave like an 8-bit microcontroller's CPU. Its eight input lines permit a repertoire of roughly 256 different tricks.[11] This state machine is the processor's *instruction decoder*. The "instructions" are eight-bit values fed to the decoder's inputs. Each evokes a particular sequence of operations.

The outputs of the instruction decoder implement the responses to the instructions. The response is a sequence of fine-grained operations necessary, say, for copying the contents of one on-chip register to another. If the instruction means "copy the contents of register 0 to register 1," then the FSM outputs might carry out three steps, something like this:

- turn on 3-state for Reg 0, putting Reg 0 contents onto an internal data bus;
- clock Reg 1, capturing the value from the internal data bus;
- turn off 3-state for Reg 0.

This hypothetical operation is a very simple one. Complex operations, such as a processor's arithmetic *divide*, involve a great many successive actions.

Implementation in structural Verilog. You can always implement your FSM structurally, duplicating what you would build with discrete logic by instantiating flip-flops and connecting the inputs and outputs of the FSM and the flip-flops through combinational logic. The only time this really makes sense is if you already have done the hard work of figuring out the Boolean equations for the outputs and next state. It also results in Verilog code that is totally obscure – without the state transition diagram is is near impossible to figure out what it does. Take a look at the structural code that implements our oddball counter.[12]

```
module Odd_Counter_FSM (
  input   CLOCK,
  input   RESET_BAR,
```

---

[10] Be careful, this is *not* a PS/NS table, although it is derived from one. It is the actual data to program into the ROM. The "State" column is informational. The five bits in the three columns to the right of the double vertical line are the FSM output and next state for each possible address (present state and enable) combination and are programmed as the ROM Data at the address location just to the left of the double line. D_1 and D_0 go to the flip-flop D inputs while O_1, O_0 and $\overline{\text{LED}}$ are the FSM outputs. A simulation of this FSM is available at https://LAoE.link/FPGA/FSM_ROM_Counter_Sim.html.

[11] "Roughly?" you protest. "Why not *exactly*?" It would be exactly 256 if all codes were implemented, and no code were used to allow adding a second byte that can expand the instruction set.

[12] We have only printed the top level module here. Full code including the flip-flop definitions is available at https://LAoE.link/FPGA/19N_Odd_Counter_Struc.v.

19N.3 Designing an odd two-bit counter

```
    input  ENAB_BAR,
    output LED_BAR,
    output Q1,
    output Q0
);

    wire d0, d1, q0_bar, q1_bar, enab;

    // create active-high enable
    assign enab = !ENAB_BAR;

    // instantiate flip-flops, connect to Q outputs and clock and reset inputs
    dff DFF0(.clk(CLOCK), .d(d0), .rst_bar(RESET_BAR), .q(Q0), .q_bar(q0_bar));
    dff DFF1(.clk(CLOCK), .d(d1), .rst_bar(RESET_BAR), .q(Q1), .q_bar(q1_bar));

    // implement boolean equations for next state and LED output
    assign LED_BAR = Q1;
    assign d0 = (Q0 & ENAB_BAR) | (q1_bar & enab);
    assign d1 = Q0 | (Q1 ^ enab);

endmodule
```

You could also create a structural version using flip-flops and a ROM. There is an example of a ROM-based Verilog FSM included as a bonus implementation in Chapter 19W.

Implementation in behaviorial Verilog. Verilog can do the fussy work for you in the design of a state machine. We use the `case` statement to implement the PS/NS table directly; you don't have convert the PS/NS table to binary or create Boolean equations. This provides several advantages. The FSM becomes self-documenting, you can easily recreate the PS/NS table or even the state transition diagram from the Verilog code. In addition, converting the PS/NS table to Verilog is largely mechanical; if your table properly represents your specification you should be able to create a working FSM in the FPGA with little additional design effort: see Fig. 19N.8.

```
always@(*) begin
   casex ({enab, cur_state})
      {1'b1, A}   : next_state = B;
      {1'b1, B}   : next_state = C;
      {1'b1, C}   : next_state = A;
      {1'bX, D}   : next_state = A;
      default     : next_state = cur_state;
   endcase
end                // end always
```

	enable (1 = enabled)	
PS	0	1
A (rst)	A	B
B	B	C
C	C	A
D	A	A
	NS	

Figure 19N.8 Correspondence between PS/NS table and behaviorial case statement.

Here is the behaviorial version of the oddball counter in "triple always" format.[13] This means three `always@` blocks are used – one triggered on the clock edge to copy the next state values to the present state register, a combinational block to calculate the next state from the current state and the input, and a third combinational block to create the output.

```
/////////////////////////////////////////////////////
// Company: Learning the Art of Electronics
// Engineer: David Abrams
//
```

[13] Code available at https://LAoE.link/FPGA/19N_Odd_Counter_Behav.v.

```verilog
// Create Date:    02/20/2023
// Design Name:    19N_OddCtrFSM_Behav.v
// Module Name:    Odd_Counter_FSM
// Description:    Behavorial Verilog FSM for odd count
//                 divide-by-three counter
//////////////////////////////////////////////////////////

//********************************************************
// The comments in this block are directives to
// the WebFPGA environment.
//
// Set name of top level module (default is fpga_top)
// @FPGA_TOP   Odd_Counter_FSM
// @MAP_IO CLOCK        30    // NC debounced pushbutton
// @MAP_IO RESET_BAR     3    // NO debounced pushbutton
// @MAP_IO ENAB_BAR      4    // logic switch 1
// @MAP_IO LED_BAR      25    // logic indicator 1
// @MAP_IO Q1           18    // logic indicator 7
// @MAP_IO Q0           17    // logic indicator 8
//********************************************************
module Odd_Counter_FSM (
  input   CLOCK,
  input   RESET_BAR,
  input   ENAB_BAR,
  output reg LED_BAR,
  output reg Q1,
  output reg Q0
);
  // create active high-signals
  wire enab, reset;
  assign enab = !ENAB_BAR;
  assign reset = !RESET_BAR;

  // assign binary values to states
  parameter A=2'b00, B=2'b11, C=2'b10, D=2'b01;

  // current and next state variables
  reg [1:0] cur_state;
  reg [1:0] next_state;

  // First @always block: State Transition Logic
  // Creates FSM cycle and implements asynchronous reset

  always @(posedge CLOCK or posedge reset) begin
    if (reset)
      cur_state <= A;
    else
      cur_state <= next_state;
  end

  // Second always@ block: Next State Logic
  // Creates combinational logic to determine next state
  // this implements the transitions in the directed graph
  always@(*) begin
    casex ({enab, cur_state})
      {1'b1, A}  : next_state = B;
      {1'b1, B}  : next_state = C;
      {1'b1, C}  : next_state = A;
      {1'bX, D}  : next_state = A;            // eliminate stuck states
      default    : next_state = cur_state;    // otherwise no change in state
```

```
        endcase
    end                  // end always

  // Third always@ block: Output Logic
  // (This is sometimes just done in an "assign" statement)
  always@(*) begin
    {Q1, Q0} = cur_state;
    case (cur_state)
      A:       LED_BAR = 1'b0;
      B:       LED_BAR = 1'b1;
      C:       LED_BAR = 1'b1;
      default: LED_BAR = 1'b1;
    endcase
  end
// The third always@ block could be replaced by assign
// but you must remove "reg" from output declarations
// assign LED_BAR = (cur_state == A) ? 1'b0 : 1'b1;
// assign {Q1, Q0} = cur_state;

endmodule
```

This format is overkill for such a simple circuit but useful in a more complicated design, a CPU for example. Often, you can just `assign` the outputs and possibly the next state calculation as well, as noted in the comments above. However, variables assigned in an always block must be declared as `reg` while those using `assign` must be wires. Do pay attention to the difference between sequential and combinational `always@` blocks. The former uses <= for assignment, while combinational blocks just use =.

19N.3.3 Assigning state names

In our code here and in the worked example, we have assigned state names to match the alphabetic characters in the state transition table. We have done this to make the correspondence between the PS/NS table and the Verilog or microcomputer code clear, so that you can see how straightforward it is to translate from the former to the latter.

In our actual designs, we use much more descriptive state identifiers. In the worked example, we would call state A "IDLE," state B "BLINK_START," state C "BLINK_OFF," state D "BLINK_ON" and state E "BLINK_END." We urge you to use descriptive state names as well. The easier you make it to understand the code, the less you have to document externally.

19N.4 Selecting the FSM clock frequency

In our odd count FSM, the clock frequency was controlled by whatever we wanted to count. However, in general the frequency of an FSM is chosen to be fast enough to be responsive to the system requirements. The response of an FSM is limited by the clock frequency since state only changes on the clock edge. Often there is no upper limit on the FSM clock other than the speed of the hardware or controller used for implementation. However, in some cases you may want the clock to be a specific frequency so it can be used for timing: see Chapter 19W for an example where the clock is used to set the blink frequency of an output LED. If implemented behaviorally in programmable logic or a microcontroller, you can select a higher frequency clock to get a quick response but use a counter to stay in any given state for a longer time. This is difficult to do with a discrete version because the additional flip-flops required to implement the delay increase the complexity of the design.

You may also want to limit the clock frequency in order to debounce input switches. Since the FSM only cares about the FSM inputs on the rising edge of the clock, bouncing that lasts for less than the FSM clock period will be ignored.

19N.5 Recapitulation: several possible sequential machines

If there are *no* external inputs, the circuit knows only one trick; it might be a counter, for example (it could be binary or decade; it could even count in some strange sequence; but it can perform just one operation). Its next state will depend upon its present state alone.

If there is *one* external input: then the FSM can do either of two operations. This could be an up/down counter for example.

If there are *eight* external inputs, as for an 8-bit microcontroller, then this sequential machine would be capable of running many alternative sequences: 256 of them. The eight input lines select one out of its large repertoire of tricks; these eight input lines are said to carry "instructions" to the processor. A sequence of instructions constitutes a program.

19N.6 AoE reading

Chapter 10 (Digital Logic):
- §10.4.3 Combining memory and gates: sequential logic.

19L Lab: Finite State Machines

In this lab, we would like you to design the control logic for a reaction timer as a Finite State Machine and then implement it in several ways. You will first build the control logic FSM using flip-flops and combinational logic. Then you will replace the combinational logic with a ROM or RAM built from the WebFPGA. After that, you will build the same FSM fully within the WebFPGA, first using flip-flops, then using behaviorial design.

19L.1 Design the control logic for a reaction timer

19L.1.1 The specification

The goal is to create a signal named DURATION which goes high and lights a green LED when one user presses a START button and goes low, turning off the LED, when a second user presses the STOP button. (The plan is to later add circuitry to time the length of the DURATION pulse.) However, if the STOP button is pressed before the START button, a red CHEAT LED illuminates. Once either the START or STOP button has been pressed, a RESET button must be pressed before another run can be performed. The reset is asynchronous. Fig. 19L.1 shows an overview of the control logic.[1]

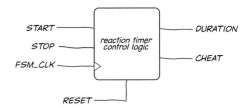

Figure 19L.1 Reaction timer control logic inputs and outputs.

We'd like the circuit to work regardless of how long a finger stays on a button and we want to implement the cheat function so a simple AND gate is not adequate. This is a perfect application for a FSM design.

19L.1.2 The hardware for the flip-flop version of the FSM

The block diagram for a flip-flop version of the FSM is shown in Figure 19L.2. Your job is to design the block labeled "combinational logic."

The two input buttons "START" and "STOP" are active high and not debounced. You should use the pushbuttons that fit into the breadboard for these buttons, not the debounced buttons built into the breadboard. The "RESET" button is active low. You can use one of the debounced pushbuttons on your

---

[1] We ask our students to first try to design this logic before they learn about FSM design. Most find it difficult to create a correctly working design or they use more flip-flops than necessary. If you like a challenge, stop now and give it a try. Then compare the experience with using the FSM design strategy that follows.

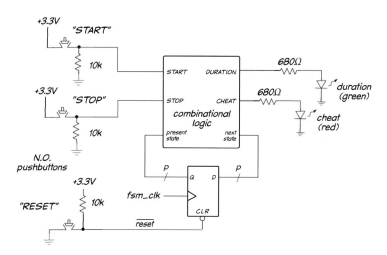

Figure 19L.2 Block Diagram for FSM reaction timer control logic.

lab kit (although this signal does not need to be debounced) or you can use the microswitch we used to look at switch bounce or a slide switch.

There are two active high outputs, one to drive the green "DURATION" LED and one to light the red "CHEAT" LED. The flip-flops are clocked from the 3.3 V level-shifted output of the function generator or from an external function generator set for a zero to 3.3 V clock. You can use either the 74HC74 or the 74HC175 flip-flops (but it should be obvious which one will be easier to use).

All logic should run from the +3.3 V supply, since we will replace the combinational logic block with the WebFPGA in the second part of the lab.

Operation of the reaction timer:

- One of the players must press the RESET button to prepare the reaction timer. Both LEDs should then be off.
- The first player surreptitiously presses the START pushbutton to light the DURATION LED and start the timer. (You will have the opportunity to display the time for the duration signal in a later lab.)
- The second player presses the STOP pushbutton as *quickly as possible* after the DURATION LED lights to stop the timer and extinguish the DURATION LED.
- If the second player presses the STOP pushbutton before the DURATION LED lights, the red CHEAT LED should light.
- Once the STOP pushbutton has been pressed, any changes in the signals from the START and STOP pushbuttons should be ignored. Pressing either button multiple times should have no effect on which LED (if any) is lit.
- The RESET button must be pressed before another run can be performed.

19L.1.3 Design the abstract FSM

Step 1: Determine the states for this FSM. List them ("A", "B", "C", etc.) and describe what each state means functionally including what the outputs should be.[2] Mark the reset state. Hint: Remember that the FSM can only change states on the rising edge of the clock so *each state lasts for at least one clock cycle*. Also, each output can only have *one* value for each state.

---

[2] Functionally means describing the states in terms of how they function *as a reaction timer* not logic levels. For example, the state of the LEDs should be described as either "ON" or "OFF" not "0" or "1."

19L.1 Design the control logic for a reaction timer

Step 2: Draw the state transition diagram for this finite state machine. States should be represented by vertices (circles) labeled "A", "B", "C", etc., and draw arcs with arrows between them to show the state transitions. The vertices should indicate the value of the outputs for each state and the transitions should be labeled with the value of the input signals. Indicate the starting (reset) state.

Decide whether the STOP and/or START pushbuttons need to be debounced. Also, choose a clock frequency for your FSM and be able to justify your choice.

Step 3: Create a PS/NS table from your state graph. Each row should show the present state followed by the next state as a function of the present state and the input values along with the *functional value* for each of the two output LEDs ("ON" or "OFF"). Put the present states in a column on the left side of your table and the input values in a row at the top of your table. There should be only one row for each current state. Include columns showing the value of each output as well.

19L.1.4 Implement the FSM using discrete flip-flops and logic gates

Assign binary values to states. Redraw your PS/NS table with binary values ($Q_0, Q_1, \ldots, Q_P$; $D_0, D_1, \ldots, D_P$) instead of functional states. Your table should also include binary states for the outputs. Some assignments of states give simpler logic equations than others, particularly if only one bit changes between states.

Create Boolean equations for the next state. Figure out the simplified Sum-of-Products Boolean algebraic combinational equation for each "next" bit ($D_0, D_1, \ldots$, etc.). Make sure there are no unwanted stuck states.

Create Boolean equations for the outputs. Show the simplified Sum-of-Products algebraic combinational equation for your state machine outputs (DURATION and CHEAT). Pay attention to whether the asserted state is high or low based on the schematic of the external connections to your design shown in Fig. 19L.2.

Draw the schematic. Draw a schematic of your completed reaction timer control logic FSM (the rectangular block labeled "Combinational Logic" in Figure 19L.2). Show the part numbers of the flip flops and logic you use.[3] Your schematic should label inputs and outputs to match those in the FSM block of Figure 19L.2.

19L.1.4.1 Build and test your FSM control logic design

Build on the upper breadboard so you do not have to disturb the WebFPGA. You do not have to disconnect any switches connected to the WebFPGA even if they are also used as inputs to your discrete design. Make sure your reaction timer control system works properly. **Remember to use only +3.3 V power in this build.**

---

[3] Hint: Use DeMorgan's Theorem to minimize the number of different gate packages you will need.

19L.1.5 Replace the discrete logic with a memory

We would like you to replace the discrete combinational logic generating the output and next state computation with either a RAM or ROM built from the FPGA.[4] You can use the two-port RAM you designed in §18L.2 or modify the keypad decoder ROM from §18N.3.1.[5]

The number of address lines required for the memory used in this part of the lab is the number of flip-flops in your FSM plus the number of inputs. The data width is equal to the number of flip-flops plus the number of output signals. So the size of the memory you need is $2^{(n+p)} \times (m + p)$, where n is the number of inputs, m is the number of outputs and p is the number of flip-flops.

You will want to write out the contents of the ROM/RAM in tabular form. The rows of your table are the memory addresses formed by the concatenation of the flip-flop Q present state outputs and the FSM's inputs. These select a value in the memory for the next state and set the outputs for the current state. The memory outputs are the D inputs to the flip-flops concatenated with the output signal lines.

If using a ROM, create it and test it with logic switches for the address inputs and logic indicators for the outputs. Then disconnect the FSM flip-flop D inputs from the combinational logic and connect them to the ROM outputs. Connect the output LEDs to the ROM as well. Leave the logic indicators connected to the ROM outputs for debugging in case your FSM does not work properly. Disconnect the ROM address inputs from the logic switches and connect them (in the right order!) to the FSM inputs and Q outputs.

If you are using RAM, load it with your table data using logic switches for the `wr_addr` and the `wr_data`. You can test it by connecting the `rd_data` outputs to the logic indicators and using DIP switches or slide-switches for the `rd_addr`.

Once you are sure the RAM is working, replace the `rd_addr` switches with the FSM inputs and the Q outputs of the state flip-flops. Then disconnect the flip-flop D inputs and the FSM output LEDs from the discrete logic and connect them to the RAM read data outputs `rd_data`. Again, you can leave the `rd_data` outputs connected to the logic indicators for debugging.

Test the FSM to see that it is working properly with memory in place of the logic.

19L.1.6 Implement the reaction timer control logic in the FPGA

19L.1.6.1 Implement it structurally using flip-flops

Write Verilog code to implement the reaction timer control logic using D flip-flops and combinational logic within the WebFPGA. This is essentially identical to what you did in §19L.1.4 above but using programmable logic rather than ICs and wires. You should define the flip-flops in a separate module and instantiate them in the top level module. Here is the module definition:

```
module FSM_REAC_TMR_LOGIC(
  input START,
  input STOP,
  input CLOCK,
  input RESET_BAR,
  output DURATION,
  output CHEAT
);
```

---

[4] If you are short of time, you may prefer the ROM or just skip this part. However, the RAM is more interesting since you can change the operation of the FSM by reprogramming the RAM with the logic switches. This means you can quickly create any FSM with four states or fewer that uses up to two input buttons and two output LEDs with the hardware in Fig. 19L.2. With the ROM you must modify your Verilog code, re-synthesize and download the new bitstream to change the FSM's operation.

[5] If you skipped the WebFPGA RAM in the last lab you can use ours at: https://LAoE.link/FPGA/18L_RAM2P_16X4.v.

Depending on your code, you may have to change some of the module ports to type "reg." You can clock your FSM from the function generator or divide down the WebFPGA 16 MHz clock to the desired FSM frequency. To use the 16 MHz clock:

```
// @MAP_IO CLOCK 13
```

Use external buttons for RESET, START, and STOP, and external LEDs for DURATION and CHEAT (not the LED and button on the WebFPGA).[6]

Optional: If you are feeling ambitious, you can replace the Boolean logic in this implementation with a ROM created as a Verilog module. It should be the same ROM module used in §19L.1.5.

19L.1.6.2 Implement the control logic using behavioral coding

Write Verilog code to implement the PS/NS table from §19L.1.3 using behavioral code (i.e., the *case* statement). You can use the same top level module definition as in the structural implementation.

19L.2 Solutions

Here are our solutions to the parts of this lab in case you get stuck. **As always, we suggest you try to get your own version working *before* looking at our answers.**

- Design of the abstract FSM §19L.1.3
  - Define States: `https://LAoE.link/FPGA/19L.1.3_Define_States.html`
  - State Transition Diagram: `https://LAoE.link/FPGA/19L.1.3_State_Diagram.html`
  - PS/NS Table: `https://LAoE.link/FPGA/19L.1.3_PSNS_Table.html`
- Implement with flip-flops and logic §19L.1.4
  - Assign Binary Values: `https://LAoE.link/FPGA/19L.1.4_Binary_Table.html`
  - Boolean Equations for NS: `https://LAoE.link/FPGA/19L.1.4_NS_Equations.html`
  - Boolean Equations for Outputs: `https://LAoE.link/FPGA/19L.1.4_Output_Equations.html`
  - Draw the Schematic: `https://LAoE.link/FPGA/19L.1.4_Schematic.html`
  - Our Hardware Build: `https://LAoE.link/FPGA/19L.1.4.1_Build.html`
- Implement with flip-flops and memory §19L.1.5
  - ROM and RAM Implementations: `https://LAoE.link/FPGA/19L.1.5_FF_Memory.html`
- Implement the reaction timer FSM in the FPGA §19L.1.6
  - Structural Implementation: `https://LAoE.link/FPGA/19L.1.6.1_Reac_Tmr_Struc.v`
  - Behavioral Implementation: `https://LAoE.link/FPGA/19L.1.6.2_Reac_Tmr_Behav.v`
  - Here is simplified behaviorial implementation that only uses a single `always@` block. It works but is (in our opinion) less clear and does not lend itself to more complex designs: `https://LAoE.link/FPGA/19L.1.6.2_Reac_Tmr_Behav_Shrt.v`

---

[6] There is a testbench you can use to test your design at `https://LAoE.link/FPGA/RT_FSM_Testbench.html`. This testbench assumes you are using the FSM clock as input to your module. If you are using 16 MHz divided down you will need to modify your code to not divide the simulation clock to get useful results.

19W Worked Examples: Finite State Machines

19W.1 Complex blink five ways

Here, we'd like you to show you how to do one task many ways. This is a favorite device of exam-writers; this kind of question lets the teachers feel that there's some coherence in the digital material. Students may not feel the same way about these questions.

We are going to implement the same finite state machine in five different hardware configurations:

1. with flip-flops using discrete logic gates for the combinational logic;
2. with flip-flops using a ROM for the combinational logic;
3. in the FPGA structurally duplicating method 1 above;
4. in the FPGA with a behavioral design; and
5. in a microcontroller.

We present this problem early – at a time when the last method, using a microcontroller, is premature. Ignore that section of this chapter if you feel uncomfortable with controllers and revisit it when we come to them later in the course.[1]

Here's the specification for the system. The user interface consists of one pushbutton and one LED (both active low). The block diagram in Fig. 19W.1 should help you see what we're getting at.

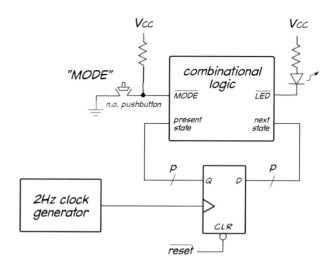

Figure 19W.1 Block diagram of the complex blink FSM.

- Start with the button not pressed and the LED off. This is the [asynchronous] reset state.

---

[1] However, in this problem we use an Arduino, so you may already be familiar with how it works. We will ask you to implement the complex blink FSM using a Cortex-M0 microcontroller in a C development environment in a later lab.

- Press and hold the button. The LED comes on.
- Release the button. The LED blinks at a 1 Hz rate.
- Press and hold the button. The LED goes off.
- Release the button. The LED stays off.

No matter how we intend to implement this FSM, the first three steps are identical: 1) define the states; 2) draw the state transition diagram; and 3) create the present state/next state (PS/NS) table.

19W.1.1 Step 1. Define the states

This system requires five states: see Fig. 19W.2. Because it is a Moore FSM, and must have fixed outputs during each state, it requires two states to implement blinking: one with the LED on and one with it off. An interesting design issue is whether to start the blink with the LED on or off. It might seem like it does not matter (and in terms of the specification it does not), but from a usability standpoint, having State C turn the LED off provides feedback to the user sooner (within 500 mS) after releasing the button in State B. If State C turned the LED on, the user would press the button, the LED would come on (State B), then when they released it, it could take up to 500 ms to enter state C then another 500 ms before it went out (until it reached State D). It is better to provide feedback to the user sooner rather than later that something is happening, which is why we turn the LED off in State C.

Students often add a sixth state – "F: button released, LED off" to this FSM. There is nothing wrong with this but State F is redundant, it is the same as State A. The FSM works fine with the extra state, it is just a bit harder to design. This is in contrast to not having *enough* states, which results in an FSM that does not work properly.

A: LED off and not blinking; waiting for button press - reset state
B: button pressed to start blinking; LED on while button down
C: button released; LED off and blinking; waiting for button press
D: button released; LED on and blinking; waiting for button press
E: button pressed to stop blinking; waiting for button release; LED off

Figure 19W.2 The complex blink FSM requires five states.

19W.1.2 Step 2. Draw the state transition diagram

Figure 19W.3 shows the operation of the complex blink FSM. We prefer to use active-high inputs on our diagram, but you can use the actual input values (active-low in this case) as long as you are clear you are doing so and the signal name has an active-low indicator (e.g., an overbar). Similarly, we prefer to describe the output, in this case an LED, functionally as either "on" or "off," but a logic level is acceptable as long as it is clear what it means.

The state transition diagram is a complete description of the abstract system. It is also the last real design step; if the diagram is correct, the steps that follow are almost purely mechanical, no further consideration of the specification should be necessary. Therefore, it is a good idea to test the state transition diagram against the specification, since this is where most students mess up (often by not having enough states). The first test is easy, make sure there are 2^n arrows coming out of each state, where n is the number of inputs. Next walk through the specification moving from state to state to make sure the diagram actually implements what is specified. A typical error is not waiting for an input to become disasserted before moving back to the initial state. This results in a system the repeats itself when an input is asserted even though the specification is edge-triggered (i.e., the system should respond *once* to an input transition, not to the input level).

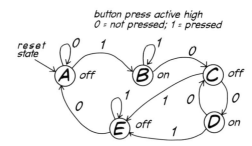

Figure 19W.3 State transition diagram for complex blink FSM using active-high inputs signals.

One limitation of this system is that it requires a 2 Hz clock to get a 1 Hz LED blink (500 ms in each of States C and D). This results in a sluggish response to user input as the system only checks the input button every 500 mS. This is fairly easy to fix in the FPGA and microcontroller, you increase the clock rate and add a counter to stay in States C and D multiple clock cycles to get the correct blink rate: see §19W.1.9. It is harder to do in the discrete design where you have to add additional states and flip-flops to use a faster clock rate.

19W.1.3 Step 3. Create the present state/next state table

From this point forward, the procedure is pretty much mechanical. If the state transition diagram is correct and you do not make any mistakes, it is possible to implement the FSM by a straightforward translation of the present state/next state table into a circuit design or HDL/software code.[2] The present state/next state table is just a conversion of the transition diagram to table form, making it easier to implement. The PS/NS table should have one row for each state with each possible input combination in a separate column: see Fig. 19W.4. We are still using active high signals, since there is no cost to using active-high signals in the FPGA and microcomputer.

	MODE (button pressed = 1)		
PS	0	1	LED
A	A	B	0 (off)
B	C	B	1 (on)
C	D	E	0 (off)
D	C	E	1 (on)
E	A	E	0 (off)
	NS		

Figure 19W.4 Present State/Next State table for complex blink FSM.

At this point, the design paths diverge depending on whether we are implementing the FSM with discrete flip-flops (or a structural FPGA design) or building a behaviorial FPGA or microcontroller software implementation.

19W.1.4 Step 4. Assign binary values to states

All implementations require us to assign numeric values to the states. For hardware implementations using p flip-flops, where 2^p is greater than or equal to the number of states, binary values between 0 and 2^{p-1} are used. The exception is *one-hot encoding* which uses one flip-flop for each state. One-hot

---
[2] One exception is in the discrete design where the selection of binary values for state assignments can influence the complexity (but not the correct operation) of the logic.

encoding uses more flip-flops (equal to the number of states) but can result in simpler combinational logic since only a single flip-flop is asserted at a time. For example, the outputs in a Moore FSM using one-hot are just the OR of the bits in the present state variable representing states with the output asserted (assuming active-high outputs): see https://LAoE.link/One_Hot.html.

In a microcomputer, the numeric assignments to states do not affect the implementation.[3] However, in hardware implementations the assignments can have an impact on the complexity of the Boolean equations for next state and output. The general rule is to try to keep the number of bit transitions between state transitions as low as possible (a single transition is best). This usually results in the simplest logic. If the number of states is less than 2^P, *don't cares* can further simplify the logic, but you must watch out for stuck states. In the complex blink FSM, three flip-flops are required to encode five states. We initially just assigned states in binary order:

PS	NS (MODE = 0)	NS (MODE = 1)
A=000	000	001
B=001	010	001
C=010	011	100
D=011	010	100
E=100	000	100
101	XXX	XXX
110	XXX	XXX
111	XXX	XXX

but then we noticed that if we chose E=101, the least significant bit of the next state was equal to 1 when MODE was 1, so we changed it and set the *don't cares* to match the most common used state value in each column, guaranteeing no stuck states:

PS	NS (MODE = 0)	NS (MODE = 1)	LED
A=000	000	001	0
B=001	010	001	1
C=010	011	101	0
D=011	010	101	1
100	000	101	X
E=101	000	101	0
110	000	101	X
111	000	101	X

19W.1.5 Step 4. Create Boolean equations for next state and output

Applying sum-of-products results in the following equations for next state and output:

$D_0 = \overline{Q_2} \cdot Q_1 \cdot \overline{Q_0} + \text{MODE}$
$D_1 = \overline{Q_2} \cdot (Q_1 + Q_0) \cdot \overline{\text{MODE}}$
$D_2 = (Q_2 + Q_1) \cdot \text{MODE}$
$\text{LED} = \overline{Q_2} \cdot Q_0$, but since the output is active low: $\overline{\text{LED}} = \overline{\overline{Q_2} \cdot Q_0}$

---

[3] In fact in C it is often easiest to just use an `enum` (enumeration) data type for the state assignments and let the compiler choose the values.

19W.1.6 Implementation 1. Discrete flip-flops and logic gates

Figure 19W.5 is the schematic of the version of the complex blink FSM we built using flip-flops and discrete gates. With judicious use of Boolean identities and DeMorgan's theorem, we were able to keep the build to only five DIP packages, a LMC555 timer for the 2 Hz clock, a 74HC175 quad D flip-flop, one 74HC00 quad two-input NAND and two 74HC10 triple three-input NANDs. Why did we decide not to debounce the two switches?[4] When building synchronous systems, the 74HC175 provides more flip-flops per package (with the clocks and reset already connected) than using the individual flip-flops in the 74HC74. The only thing missing is a set input and that is rarely needed as long as we assign the reset state the binary value of zero. We sketched out our design but assigned pin numbers as we went along, wiring the shortest path from one device to another and annotating the schematic with what we built.

We included LEDs on the outputs of the state flip-flops and clock for debugging and demonstration. Debugging finite state machines can be difficult because they are feedback systems. Typically, if our FSM does not work, we will replace the Q outputs from the flip-flops with logic switches and test the logic by setting the switches to each possible state and seeing if the next state logic outputs are correct. That usually catches a logic wiring error. If the logic is correct but the FSM is still not working, we replace the clock with a debounced pushbutton and step through to see that the system properly implements the state transition diagram. (If it does, it means our FSM design does not work and we have to go back to the review the first three design steps again.)

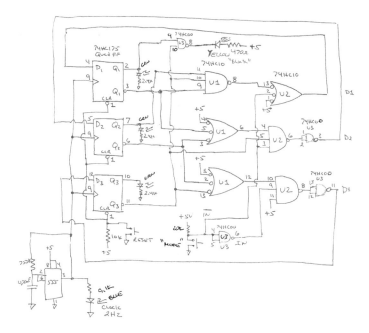

Figure 19W.5 The complex blink FSM we built with discrete flip-flops and logic gates.

19W.1.7 Implementation 2. Discrete flip-flops and a ROM

The disadvantage of using discrete gates in an FSM is not only the work required to wire it up, but it is quite difficult to change if you made a mistake in the original design or if you want to change

---

[4] We certainly don't have to debounce the reset pushbutton, since any switch bounce just resets the state flip-flops several times over the bounce period. We don't have to debounce the MODE pushbutton because we only look at it every 500 mS, much longer than any bounce could last.

the operation of the systems. Since the logic just implements the PS/NS truth table, we decided to implement the truth table directly with a ROM. Not only is the wiring significantly easier, changing the design is merely a matter of of reprogramming the ROM. Figure 19W.6 shows the complex blink FSM we built still using discrete flip-flops, but the next state and output are computed using a 16×4 read only memory.[5] Here there is no need to use anything but a straight binary state assignment, since there is no logic to try to simplify. You can see the state assignments in the PS/NS table in the upper right of the figure and the resulting ROM programming below it. The unused state assignments all transition to State A to avoid stuck states.

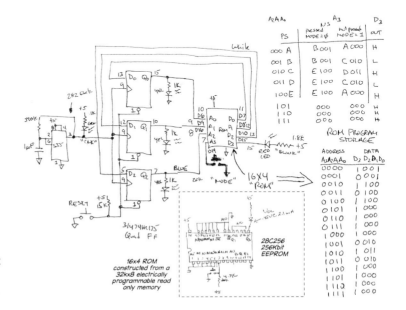

Figure 19W.6 Our complex blink FSM build using discrete flip-flops and ROM.

19W.1.8 Implementation 3. Structural Verilog in the WebFPGA

All the Verilog implementations use the hardware build shown in Fig. 19W.7. The clock and state LEDs are optional but are useful for debugging and/or demonstration. The structural implementation just reproduces the discrete version from §19W.1.6 in the FPGA using the Boolean equations in §19W.1.5.[6] Because this is just a reproduction of the discrete hardware in the FPGA, state 3'b100 is still skipped and the FSM goes from state 3'b010 or 3'b011 to state 3'b101 when the MODE button is pressed while blinking.

```
//////////////////////////////////////////////////////////////
// Company: Learning the Art of Electronics
// Engineer: David Abrams
//
// Create Date:  2/12/2023
// Design Name:  19W_Complex_Blink_Struc.v
// Module Name:  ComplexBlink
// Description:  Structural implementation of the complex blink FSM
//               in Verilog running on the WebFPGA.
//////////////////////////////////////////////////////////////
```

---

[5] Actually, we used a $32k \times 8$ electrically programmable read only memory but only used the lower four address bits and lower four output data bits as shown in the figure inset (wasting 99.98% of the available storage space).

[6] Code available at https://LAoE.link/FPGA/19W_Complex_Blink_Struc.v.

Worked Examples: Finite State Machines

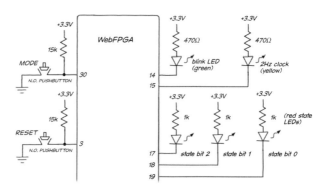

Figure 19W.7 Complex blink FPGA build for both structural and behavioral implementations.

```
//***********************************************
// The comments in this block are directives to
// the WebFPGA environment.
//
// Set name of top level module (default is fpga_top)
// @FPGA_TOP   ComplexBlink
//
// @MAP_IO MODE_BAR    30
// @MAP_IO RST_BAR      3
// @MAP_IO LED_BAR     14
// @MAP_IO CLK2HZ      15
// @MAP_IO STATE_LED0  19
// @MAP_IO STATE_LED1  18
// @MAP_IO STATE_LED2  17
//***********************************************

module ComplexBlink(
  input  WF_CLK,
  input  MODE_BAR,
  input  RST_BAR,
  output LED_BAR,
  output CLK2HZ,
  output STATE_LED0,
  output STATE_LED1,
  output STATE_LED2
);

// create a 2Hz clock
  reg [22:0] clk_div;             // counter to slow down 16MHz clock to 2Hz

  // create a 2Hz clock by dividing the 16MHz clock by 8,000,000
  always @(posedge WE_CLK) begin
    clk_div <= clk_div + 1;
    if (clk_div >= ('d8000000))
      clk_div <= 'd0;
  end

  assign CLK2HZ = clk_div[22];

  // now build the FSM

  wire q0, q1, q2, q0_bar, q1_bar, q2_bar; // need some wires to connect things
  wire d0, d1, d2;
  wire mode = ~MODE_BAR;
```

```verilog
  // create next state and output from the equations of step four of the FSM design
  assign LED_BAR = ~(q2_bar &  q0);
  assign d0 = (q2_bar & q1 & q0_bar) | mode;
  assign d1 = q2_bar & (q1 | q0) & MODE_BAR;
  assign d2 = (q2 | q1) & mode;

  // wire up the flip-flops
  FF_74HC175   FF_Q0 (.d(d0), .clk(CLK2HZ), .r_bar(RST_BAR), .q(q0), .q_bar(q0_bar));
  FF_74HC175   FF_Q1 (.d(d1), .clk(CLK2HZ), .r_bar(RST_BAR), .q(q1), .q_bar(q1_bar));
  FF_74HC175   FF_Q2 (.d(d2), .clk(CLK2HZ), .r_bar(RST_BAR), .q(q2), .q_bar(q2_bar));

  // connect the state output LEDs

  assign STATE_LED0 = q0;
  assign STATE_LED1 = q1;
  assign STATE_LED2 = q2;

endmodule

////////////////////////////////////////////////////////////
// Design for 74HC175 equivalent flip-flop
module FF_74HC175(
  input d,
  input clk,
  input r_bar,
  output reg q,
  output q_bar
);

  assign q_bar = !q;   // use combinational logic for inverted output

  always @(posedge clk or negedge r_bar)
  begin
    if (!r_bar)
      q <= 1'b0;
    else
      q <= d;
  end
endmodule
```

Bonus implementation: See https://LAoE.link/FPGA/19W_Complex_Blink_Struc_ROM.v for a structural version of the complex blink FSM that uses a ROM rather than Boolean equations to determine the next state and output.

19W.1.9 Implementation 4. Behavioral Verilog in the WebFPGA

The structural implementation only makes sense if you already have the Boolean equations or ROM data for the next state and output. Even then, it is not self-documenting – you need to record the binary state assignments, truth tables, and sum-of-products calculations somewhere to understand where the design came from. A behavioral model makes more sense in most cases as it is not only self-documenting, but it also saves you the work of figuring out the Boolean equations for the output and the flip-flop inputs. Since we let the synthesizer simplify the logic, we just assign binary values to states sequentially.[7]

---

[7] Code available at https://LAoE.link/FPGA/19W_Complex_Blink_Behav.v.

Worked Examples: Finite State Machines

```verilog
//////////////////////////////////////////////////////////////
// Company: Learning the Art of Electronics
// Engineer: David Abrams
//
// Create Date:  2/12/2023
// Design Name:  19W_Complex_Blink_Behav.v
// Module Name:  ComplexBlink
// Description:  Behavioral implementation of the complex blink FSM
//               in Verilog running on the WebFPGA.
//////////////////////////////////////////////////////////////

//*********************************************
// The comments in this block are directives to
// the WebFPGA environment.
//
// Set name of top level module (default is fpga_top)
// @FPGA_TOP   ComplexBlink
//
// @MAP_IO MODE_BAR     30
// @MAP_IO RST_BAR      3
// @MAP_IO LED_BAR      14
// @MAP_IO CLK2HZ       15
// @MAP_IO STATE_LED0   19
// @MAP_IO STATE_LED1   18
// @MAP_IO STATE_LED2   17
//*********************************************

module ComplexBlink(
  input   WF_CLK,
  input   MODE_BAR,
  input   RST_BAR,
  output  LED_BAR,
  output  CLK2HZ,
  output  STATE_LED0,
  output  STATE_LED1,
  output  STATE_LED2
);

  // define states
  parameter    BLINK_OFF = 3'd0,
               BLINK_START = 3'd1,
               BLINK_LEDOFF = 3'd2,
               BLINK_LEDON = 3'd3,
               BLINK_STOP = 3'd4;

  // create a 2Hz clock by dividing the 16MHz clock by 8,000,000
  reg [22:0] clk_div;

  always @(posedge WF_CLK) begin
    clk_div <= clk_div + 1;
    if (clk_div >= ('d8000000))
      clk_div <= 'd0;
  end

  assign CLK2HZ = clk_div[22];

  // create active high versions of mode pushbutton and output LED
  wire button;
  reg outled;
  assign button = ~MODE_BAR;
  assign LED_BAR = ~outled;
```

```verilog
// variables to hold present state and next state
reg [2:0] curState = 0;          // current state
reg [2:0] nxtState;              // next state

// setup LEDs to show current state
assign STATE_LED0 = curState[0];
assign STATE_LED1 = curState[1];
assign STATE_LED2 = curState[2];

// first always@ block creates the FSM cycle.  On positive edge
// of clock go to next state unless reset button pressed
always @(posedge clk_div[22]  or negedge RST_BAR) begin
  if (!RST_BAR)
    curState <= BLINK_OFF;       // asynchronous reset
  else
    curState <= nxtState;
end


// next always@ block defines combinational logic to determine next state
always @(*) begin
nxtState = curState;  // default if valid state is current state (no change)
  case (curState)
    BLINK_OFF    : if (button) nxtState = BLINK_START;
    BLINK_START  : if (!button) nxtState = BLINK_LEDOFF;
    BLINK_LEDOFF : if (button) nxtState = BLINK_STOP; else nxtState = BLINK_LEDON;
    BLINK_LEDON  : if (button) nxtState = BLINK_STOP; else nxtState = BLINK_LEDOFF;
    BLINK_STOP   : if (!button) nxtState = BLINK_OFF;
    default      : nxtState = BLINK_OFF;  // unused states go to reset state
  endcase
end                    // end of always @(*)

// third always@ block creates FSM output
always @(*) begin
  case (curState)
    BLINK_START: outled = 1'b1;
    BLINK_LEDON: outled = 1'b1;
    default:     outled = 1'b0;
  endcase
end

endmodule
```

Yet another bonus implementation: As we mentioned earlier in §19W.1.2, the requirement for a 500 ms clock period to get a 1 Hz blink rate makes the system sluggish to respond to user input. In the FPGA implementation it is fairly easy to increase the FSM clock frequency and then add a counter to get the correct blink rate.

Here is a version of the behaviorial design that runs the FSM clock at 8 Hz, ensuring recognition of user input in 125 ms or less: https://LAoE.link/FPGA/19W_Complex_Blink_Behav_8Hz.v. It instantiates a counter module, `delay_ctr`, that counts the 8 Hz clock and sets a scalar output, `endtime`, that goes true after four clock periods. The main FSM loop resets this counter before transitioning to either State C or D, and then stays in those states until either the MODE pushbutton is pressed (which means go to State E) or until `endtime` goes high, meaning it is time to transition to the alternative LED blink state. From the state transition diagram (Fig. 19W.3), States C and D are entered from States B, C and D, so each of these states must reset the counter whenever they set `nxtState` to either State C

786   Worked Examples: Finite State Machines

or State D. States C and D must then dissassert the delay counter reset to allow the counter to begin counting the blink state cycles.[8]

19W.1.10 Implementation 5. A microcontroller version

An FPGA is overkill both in cost and speed for such a simple and slow system. An ATtiny85 Arduino microcomputer costs only a few dollars, cheaper than using flip-flops and gates, and is easier to wire up.

Hardware The hardware is pretty obvious, but we've drawn it in Fig. 19W.8. The hardware should help to make the code intelligible. The labels beginning with a "P" are Arduino pin numbers used in the code. This device comes in an 8-pin DIP with five available I/O pins. That's not enough to bring out the both the 2 Hz clock and the state LEDs, so we have eliminated the clock LED. Arduino devices have optional internal pullup resistors when pins are configured as inputs, meaning that we do not have to include them externally on the pushbuttons.

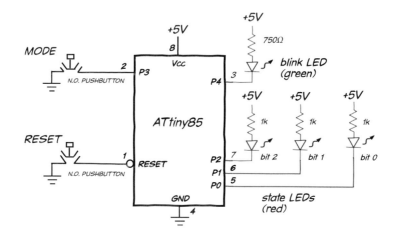

Figure 19W.8 Arduino based complex blink FSM. Input pins have internal pullups.

```
/////////////////////////////////////////////////////////////
// Company: Learning the Art of Electronics
// Engineer: David Abrams
//
// Create Date:   2/12/2023
// Design Name:   19W_Complex_Blink_Arduino.ino
// Description:   Implementation of the complex blink FSM
//                on an Arduino ATtiny85.
//
//           States:
#define A    1 // Blinking mode off; pushbutton released; LED off
#define B    2 // Pushbutton pressed to start blinking mode; LED on
#define C    3 // Blinking mode on; pushbutton released; LED on
#define D    4 // Blinking mode on; pushbutton released; LED off
#define E    5 // Pushbutton pressed to stop blinking mode; LED off
/////////////////////////////////////////////////////////////

const int greenLED = 4;      // assign Arduino pin numbers to constants
const int mode_bar = 3;
```

---

[8] You could run the clock much faster, as long as the clock period is longer than the MODE pushbutton bounce. We chose 8 Hz so you could more or less still see the clock frequency on the diagnostic clock LED.

19W.1 Complex blink five ways

```
const int StateBit0 = 0;      // LEDs to display state
const int StateBit1 = 1;
const int StateBit2 = 2;

const int FSM_Freq = 2;       // FSM frequency in Hertz

int cycTime = 1000/FSM_Freq;  // calculate state delay in milliseconds
int curState = A;             // current/next state

//functions to abstract active low button and LED to active high
boolean isBtnPrsd(void) {
    return !digitalRead(mode_bar);
}

void LEDOut(boolean onoff) {
    digitalWrite(greenLED, !onoff);
}

// setup code runs once - initializes I/O
void setup() {
 pinMode(greenLED, OUTPUT);
 pinMode(mode_bar, INPUT_PULLUP);

 pinMode(StateBit0, OUTPUT);
 pinMode(StateBit1, OUTPUT);
 pinMode(StateBit2, OUTPUT);
}

void loop() {
    // display Current State on three external LEDs (active low)
    digitalWrite(StateBit0, !(curState & 0x01));
    digitalWrite(StateBit1, !(curState & 0x02));
    digitalWrite(StateBit2, !(curState & 0x04));

    switch (curState) {
       case A:                        // State A: blink is off - waiting for button press
          LEDOut(false);              // LED is off in state A
          delay(cycTime);
          if (isBtnPrsd())            // see what state we go to next
             curState = B;            // if button pressed, turn blink on so next state
          else
             curState = A;            // not really necessary, already is the current state
          break;

       case B:                        // State B: blink mode is starting
          LEDOut(true);               // LED is on in state B
          delay(cycTime);
          if (isBtnPrsd())            // see what state we go to next
             curState = B;            // wait until button is released to begin blink
          else
             curState = C;            // and go to the first of the two blink states
          break;

       case C:                        // State C: Blink Mode - LED Off
          LEDOut(false);              // LED is off in state C
          delay(cycTime);
          if (isBtnPrsd())            // see what state we go to next
             curState = E;            // if button pressed - end blink
          else
             curState = D;            // and go to blink LED on state
```

```
          break;

     case D:                        // State D: Blink Mode - LED On
        LEDOut(true);               // LED is on in state D
        delay(cycTime);
        if (isBtnPrsd())            // see what state we go to next
           curState = E;            // if button pressed - end blink
        else
           curState = C;            // go to blink LED off state
        break;

     case E:                        // State E: Blink mode is ending
        LEDOut(false);              // LED is off in state E
        delay(cycTime);
        if (isBtnPrsd())            // see what state we go to next
           curState = E;            // wait for button released to go to starting state
        else
           curState = A;            // go to off state
        break;

     default:
        curState = A;               // default is off state
        break;
  } // end of switch statement
} // end of loop()
```

An Arduino program consists of a `setup()` function that is called once, and a `loop()` function that runs forever.[9] Here the setup function only needs to configure the I/O pins used as either inputs or outputs. The MODE pushbutton input is configured as an input with internal pullup. The reset pin does not need configuration as it has a permanent internal pullup resistor.[10]

Two helper functions convert the active-low pushbutton and blink LED to use active high logic. It is good practice to encapsulate hardware in interface functions. By isolating hardware from the application you can change hardware or the microcontroller itself without having to modify the main loop. The means that once your application runs correctly, you need only debug the interface function if a hardware change causes it to malfunction.

The main loop runs once every FSM period. It first updates the output state indicator LEDs then implements the PS/NS table of Fig. 19W.4 using a `switch` statement. This is similar to the `case` statement in Verilog, however, once a case item matches the Arduino/C version, it continues executing statements to the end of the switch statement (i.e, it "falls through" to each of the following case items). Normal practice is to limit each case item to its own set of statements by adding a `break` before the next case item, as we have done here.

In each state cycle, we update the FSM outputs (here just the blink LED) at the beginning of the state cycle, then set the next state based on the input MODE button at the end of the cycle, mimicking a hardware implementation. Translation from the present state/next state table to the Arduino `switch` statement is purely mechanical as shown in Fig. 19W.9.

As with the FPGA in §19W.1.9, you could improve the response time in a software implementation by increasing the FSM clock rate, then stay in each blink state for multiple cycles.

---

[9] Code available at https://LAoE.link/FPGA/19W_Complex_Blink_Arduino.ino.
[10] See Fig. 8-1 in the ATtiny datasheet at https://LAoE.link/FPGA/19W_ATtiny_Datasheet.pdf.

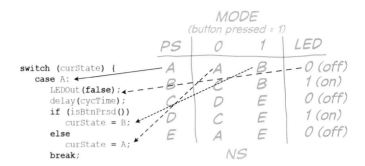

Figure 19W.9 Equivalence of PS/NS table to Arduino case item.

19W.2 Takeaway

No matter how you are planning to implement your design, the finite state machine technique provides a rigorous method to create a system that is robust, easy to implement, easy to debug, and easy to maintain. If your state transition diagram correctly meets your specification, implementation should require little additional design.

We started insisting on FSM design of microcontroller systems in our courses several years ago after spending one too many nights up late trying to untangle students' spaghetti code of deeply nested `if then else` statements containing hidden bugs. While there are still bugs in student code, we no longer have to stay up so late to find them.

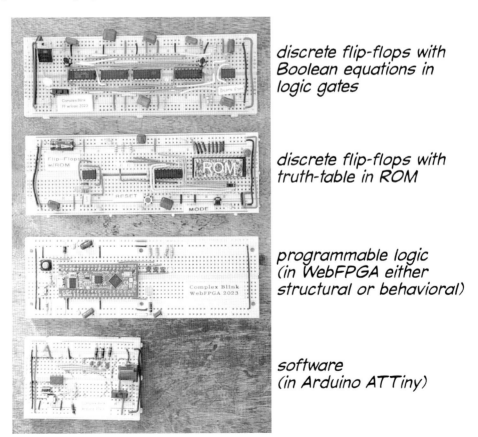

Figure 19W.10 Complex Blink FSM implemented many ways.

Part V

Digital: Analog–Digital, PLL, Digital Project Lab

20N Analog ↔ Digital; PLL

Contents

20N.1	**Digital ⇔ analog conversion, generally**	**794**
	20N.1.1 Slicing across the vertical axis: *amplitude or voltage resolution*	794
	20N.1.2 Slices across the horizontal axis: sampling rate	795
	20N.1.3 Effect of an inadequate sampling rate (aliasing)	796
20N.2	**Digital to analog (DAC) methods**	**798**
	20N.2.1 Thermometer DAC	798
	20N.2.2 An op-amp current-summing circuit	798
	20N.2.3 $R-2R$ ladder	798
	20N.2.4 Switched-capacitor DAC	800
	20N.2.5 1-bit DAC; pulse-width modulation (PWM)	801
20N.3	**Analog-to-digital conversion**	**802**
	20N.3.1 Open-loop	803
	20N.3.2 Dual-slope	804
	20N.3.3 Closed-loop	805
	20N.3.4 A binary search: example	807
	20N.3.5 Delta–sigma ADC	808
20N.4	**Sampling artifacts**	**814**
	20N.4.1 Predicting sampling artifacts in the frequency domain	814
	20N.4.2 Aliasing is predictable	816
20N.5	**Dither**	**816**
20N.6	**Phase-locked loop**	**817**
	20N.6.1 Phase detectors: simplest, XOR	817
	20N.6.2 Phase detectors: edge-sensitive	819
	20N.6.3 Applications	820
	20N.6.4 Stability issues much like PID's	822
20N.7	**AoE reading**	**824**

Why?

Problem: convert a range of voltages to a range of digital codes. The complementary process – digital to analog – is intellectually less challenging and less various, but useful.

20N.1 Digital ⇔ analog conversion, generally

AoE §13.2

Analog versus digital: The task of converting from either form to the other can include challenging *analog* tasks. In order to convert from analog to digital (ADC, A/D) a circuit must classify an input (usually a voltage), putting it into the correct digital category; often the circuit must do that fast.

Running in the other direction (DAC, D/A) is conceptually easier. But it is hard to do this precisely (making the conversion good to many bits), and the digital processing will have added extraneous *non*-information, like "steppiness" in the sampled and recovered waveform; this junk must be cleaned away.

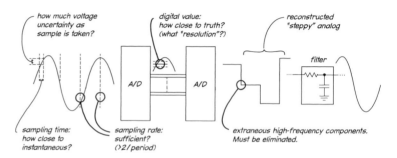

Figure 20N.1 Analog in, to ADC; then digital out, to DAC: analog reconstructed.

That cleaning job can present a substantial analog task; but it is one that is accomplished by the lowpass filter that cleans up the output of every digital-audio player. The filtering can be accomplished in part within the digital domain. Some of this processing, called "oversampling," involves interpolating pseudo-samples between actual samples.[1] Digital filtering, including the "oversampling" that boosts the apparent sampling rate, eases the task of the final analog stage, a smoothing lowpass filter.

What sampling does not mean: We will return to the question of how one must sample a waveform, in order to convert it to digital form. But let's dispose of one possible misconception at this early point: if, as in Fig. 20N.1, the sinusoid is sampled three times during one period, the *sampled* value is not an *average* of the waveform value during that time interval. Instead, it is the value present during the very short sampling pulse. The ideal sampling would be instantaneous; an actual sampling pulse may last a few tens of nanoseconds.

ADC: how fine to slice, in voltage (vertical) and time (horizontal): If you mean to convert a sinewave to digital form then back again, you run at once into the problem of *how much information* you need to carry into the digital domain. You must be content with a limited number of slices both in the vertical direction – amplitude resolution – and in the horizontal direction, where the fineness of "slicing" is determined by the sampling rate.

20N.1.1 Slicing across the vertical axis: *amplitude or voltage resolution*

This is a pretty straightforward issue. You decide, as usual, how big an error is tolerable. In this course, many of the converters you build or use in the lab will be *8-bit* devices; thus we'll settle for 256 voltage slices, each worth between 10 and 16 mV (how large the slice is depends on the full-scale voltage, since each slice is $V_{\text{full-scale}}/2^{(\text{number-of-bits})}$).

---

[1] Interpolation is one way of describing the process of digital filtering.

Commercial digital audio, at 16 bits, would cut each of our 8-bit slices into another 256 sub-slices (*16*-bit resolution ⇒1 part in 64K).[2] Our breadboarded circuits would bury any such pretended resolution in noise, as you know if you recall how big the usual fuzz is on your scope screen when you turn the gain all the way up. In general, the level of noise present in the analog signal sets a limit on what resolution is useful in a converter. In audio conversion there is an additional boundary: one gets no advantage by resolving an audio waveform to a level below what a human can hear.

Quantization error: Since going from analog to digital entails forcing a continuous input into bins – a sort of procrustean fitting of a smoothly-varying original into a limited set of alternatives – the process necessarily introduces errors. Figure 20N.2 sketches an analog input, and the levels (the nearest available) to which the input was assigned, in an imagined conversion into digital form.[3]

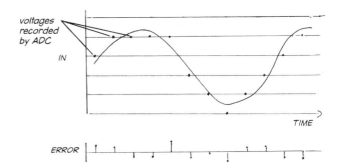

Figure 20N.2 Quantization error: a maximum of 1/2 the smallest voltage slice (after Pohlmann *Principles of Digital Audio*, McGraw-Hill, 2011).

This figure is offered in order to persuade you that the maximum error for the conversion process will be 1/2 the size of the smallest "slice." Such errors are an inevitable result of the classification of continuously varying *analog* voltages into discrete *digital* bins. Thus, quantization necessarily adds this noise at the level of 1/2 LSB.

20N.1.2 Slices across the horizontal axis: sampling rate

Here the result is surprising to anyone who has not considered the question before: Nyquist pointed out the curious fact that a shade more than *two* samples per period of a sinewave will carry a *full* description of that sinewave.[4] Thus one can reconstruct the original with just that pair of samples (this works only if one has many cycles of those two samples). Shannon recited the rule thus:[5]

Theorem *If a function $f(t)$ contains no frequencies higher than W cps, it is completely determined by giving its ordinates at a series of points spaced $1/2W$ seconds apart.*
   This is a fact which is common knowledge in the communication art.

What Nyquist does not say: We would like to inoculate the reader against a possible misunderstanding of Nyquist's rule: we would like to underline the truth that exactly two cycles per period is *not*

---

[2] The size of the smallest "slice" (least significant bit, LSB) is fantastically tiny on, say, a portable player, whose full-scale voltage may be under 2 V. For example, 16-bit resolution and 2 V full-scale implies an LSB value of 30 μV.

[3] This figure is modeled on one offered by Pohlmann, K., *Principles of Digital Audio*, 6th ed. McGraw Hill (2011), Fig. 2.7, p. 31. The darkened points represent samples taken – and the representation may confuse you since each dot shows in *analog* form the supposed *digital* representation of the original input. Nevertheless, we think the information this figure offers does serve to make less abstract the notion of "error <1/2 LSB".

[4] The rule, suggested but not clearly stated by Nyquist in 1928, was proven by Claude Shannon in 1949, and sometimes is called the Nyquist–Shannon Sampling Theorem.

[5] Claude Shannon, *Proceedings of the Institution of Radio Engineers*, **37**,(1), 10–21, (1949), quoted by Nika Aldrich in his superlative book, *Digital Audio Explained for the Audio Engineer*, 2nd ed. Sweetwater (2005), p. 111.

sufficient. Aldrich makes one sufficient argument for this truth: given just two points, it is possible to draw a sinusoid at that frequency and *of any amplitude*. So the information is not sufficient to satisfy the essential requirement: that just one waveform can be drawn through the sampled points.[6]

If reproducing a sinusoid of a single frequency strikes you as a rather sterile academic exercise, recall Fourier's wisdom, noting that any waveform can be expressed as a sum of sinusoids – and note that if we can sample adequately the *highest* frequency component of a signal we necessarily get plenty of information about lower frequencies.[7]

Here's the idea, illustrated in Fig. 20N.3 with an eight-bit AD7569 combined ADC and DAC. We are not pushing for extreme stinginess in sampling here; instead, we pick up about *four* samples per period. The scope image shows a round trip: analog input converted to digital and back again.

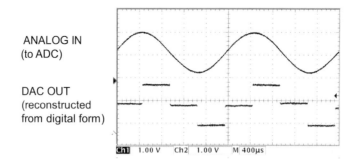

Figure 20N.3 Analog in, to ADC; analog out reconstructed from DAC.

It is not hard to believe that those steppy edges could be smoothed away. Indeed they can, by a good lowpass filter – a filter much better than a simple *RC* lowpass. Figure 20N.4 shows the steppy output smoothed by such a lowpass. The filtered output does, indeed, look like the original analog input.

We will return to the question of what sampling rate is required, and how good a lowpass filter is needed later in this chapter and again but more thoroughly in Chapter 20S.

Evidently, the sampling rate required depends upon the frequency of the analog waveform: if the waveform includes many frequency components, then it is the *highest* that concerns us; the lower frequencies are easier to catch. Commercial digital audio again provides a useful case in point. That system aims to store and recover signals up to 20 kHz. To do this, it samples at 44.1 kHz. This can be described as "10% oversampling:" a sampling rate just 10% above Nyquist's theoretical minimum.

20N.1.3 Effect of an inadequate sampling rate (aliasing)

If you don't do what Nyquist told you to do, you get into trouble. You don't just fail to get what you meant to get; you get nonsense. Sampling at too low a rate, you get an artifact, a fake signal that will be irreversibly commingled with the true signal. The fake is said to be an "aliased" signal because it is translated from its true frequency. Strange to tell, the true signal also is captured, but the presence of the alias usually corrupts the signal hopelessly.[8]

---

[6] Aldrich, op.cit. p. 123. Don't forget the important assumption, which Aldrich emphasizes, that the only permitted waveform is a *sinusoid*. No fair improvising other odd shapes that could pass through the sampled points. Such odd shapes would contain frequencies higher than the sinusoid that is assumed. Consult M. Fourier if this point is unfamiliar (see, for example, Chapters 3N and 3L).

[7] We usually state Nyquist's rule as applying to the highest frequency in the input signal (see, e.g., AoE §13.5.1B). It is more accurate – though only rarely different – to state the rule as requiring a sampling rate a little more than twice the signal's *bandwidth*. When signal frequencies begin close to zero, as they do for audio, "highest frequency" and "bandwidth" are the same. But in unusual cases one could, for example, sample a narrow bandwidth at a high frequency using a modest f_{sample}. This might come up, say, in digitizing a radio signal.

[8] In some exceptional cases, aliasing can be harmless or even useful. If the signal that would be aliased is a single frequency, then its alias might be removed with a narrow bandstop filter. And the predictable behavior of aliasing also allows

20N.1 Digital ⇔ analog conversion, generally

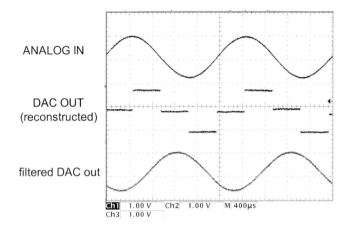

Figure 20N.4 Analog in, to ADC; analog out reconstructed from DAC, steppy waveform smoothed by lowpass filter.

Figure 20N.5 shows a 1 kHz analog input sampled at 1.2 kHz: sampled well below the theoretical minimum "Nyquist" rate of >2 kHz. We have marked the sampling points with small circles, and show the reconstructed waveform (the DAC output), before and after lowpass filtering. The spurious sinusoid in the fourth trace is convincing – but it is a false artifact. (The phases of sampled points, reconstruction and filtered output do not match; this results from delays in the processing that produced these signals.)

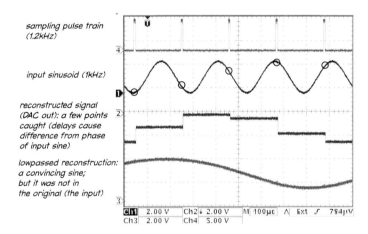

Figure 20N.5 A sinusoid inadequately sampled: Nyquist defied. (Scope settings: 2 V/div except top trace, 5 V/div; 400 μs/div.)

The reconstructed waveform may look funny (and not even sinusoidal). But if one lowpass-filters this funny waveform, one gets a sinusoid, as you can see in Fig. 20N.5. This sinusoid was not in the original signal but is indistinguishable from a genuine input at 200 Hz (the alias appears at $f_{sample} - f_{in} = 1.2\,\text{kHz} - 1\,\text{kHz} = 200\,\text{Hz}$). We will look more closely at such sampling artifacts in Chapter 20S.

To protect against aliasing, you need a good lowpass filter *ahead* of the ADC input – in addition to the one used to smooth the reconstructed signal. This input filter makes sure that no indigestible frequencies ever are fed to the converter. This is a standard feature for ADC systems – although we omit it from Lab 26W, relying instead on the limited bandwidth of the human voice to make the system work. Enough generalities. Let's get back to hardware: let's look at some ways to carry out the conversion, in both directions.

purposeful "folding" of high frequencies to a lower range. This technique can be used to convert a high-frequency radio *carrier* to an "intermediate frequency" (IF) for demodulation. See Aldrich, p. 35.

20N.2 Digital to analog (DAC) methods

A DAC is conceptually simple: All we need is a way to *sum* a binary-scaled set of voltages or currents: if a high input at the MSB should generate an output of 1V, the next bit by itself should generate 0.5 V; both together should generate 1.5 V; and so on. Some methods are pretty self-explanatory; others are not.

20N.2.1 Thermometer DAC

AoE §13.2.1

A voltage divider tapped by switches that are controlled by the digital inputs can do the job.[9] This design has a wonderfully dignified pedigree: it was designed by Lord Kelvin almost two centuries ago, though it's not as musty as that fact might suggest.[10] It is a design now used for digital potentiometers and some small DACs. It requires only a resistive divider, some switches, and some decoding logic.

With good analog switches, a 16-bit divider can do the job – requiring about 65,000 precise resistors (e.g., TI's DAC8564). As the DAC8564's datasheet advertises, the design "minimizes undesired code-to-code transient voltages (glitch)," in contrast to DACs whose monotonicity depends on precise matching of resistor ratios.

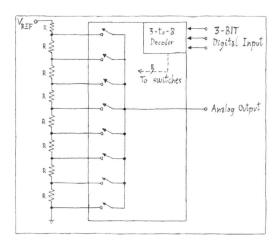

Figure 20N.6 "Thermometer" DAC (after Analog Devices online tutorial).

20N.2.2 An op-amp current-summing circuit

An op-amp current-summing circuit, and resistors of values $R, 2R, 4R \ldots$, can form the DAC's binary-weighted sum. Cozily familiar though it is to those of us comfortable with op-amps, this method rarely is used. It requires the fabrication of a wide range of resistor values to high precision.

Compare AoE Fig. 13.3

20N.2.3 *R–2R* ladder

AoE §13.2.2

Instead, one can get the same result – a binary scaling of currents – with an ingenious circuit called an *R–2R* ladder, a circuit that requires just *two* resistor values. This pair of values is easier to fabricate in a fixed ratio than the *n* different values required for an *n*-bit converter like that in Fig. 20N.7. Those *n* values would need to span a wide range and would require great precision in order to give many bits of resolution.

---

[9] See Aldrich at pp. 148–152; Analog Devices Tutorial at https://LAoE.link/ADI_MT-014.pdf.
[10] This surprising provenance we take from Chapter 3 of Analog Device's useful Data Conversion Handbook. Available at https://LAoE.link/ADI_Conversion_Handbook.html.

20N.2 Digital to analog (DAC) methods

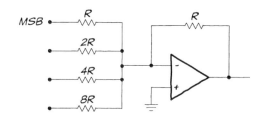

Figure 20N.7 A possible DAC – but hard to fabricate.

How an R–2R ladder divides down a voltage: The ladder is made up of units that look like the one on the left of Fig. 20N.8. Since the unit looks like $2R$, seen from its left side (its input), we can plug in another exactly similar unit in place of its right-hand $2R$ – and so on, extending the chain as far as we like. At the right-hand end of each of the resistors labeled R the voltage is down to half what it was at the left end of the resistor.

Figure 20N.9 contrasts the clever R–$2R$ against a dopey divider that makes a binary-scaled divider by following our usual loading rules. R_{in} for each succeeding stage is $100 \times R_{\text{out}}$ for the stage before to make the division voltages good to 1%. This is pretty mediocre performance (equivalent to fewer than 7 bits of precision); but you'll notice that the resistor values quickly blow up, and with them so does the output impedance at the successive division points. Meanwhile, the smarter R–$2R$ gets the job done with just two R values, and holds R_{out} constant. Whereas *loading* causes errors in the dumb divider, loading achieves the desired division in the clever R–$2R$ divider.

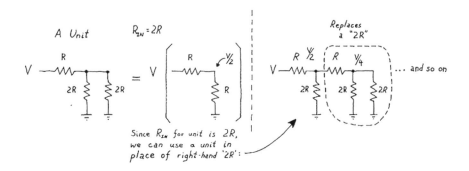

Figure 20N.8 *R–2R ladder.*

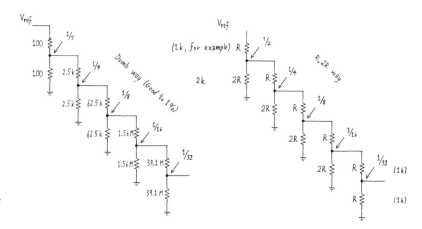

Figure 20N.9 Dumb divider contrasted with the clever R–$2R$ binary divider.

Applying the R–2R ladder: Figure 20N.10 shows two DAC circuits exploiting the R–2R scheme. The left-hand circuit uses such a ladder to source current into the summing junction of an op-amp; the right-hand circuit (the schematic of an IC DAC) omits the op-amp, so the output is a *current*. Both DACs use just *two* resistor values.

AoE §13.2.3

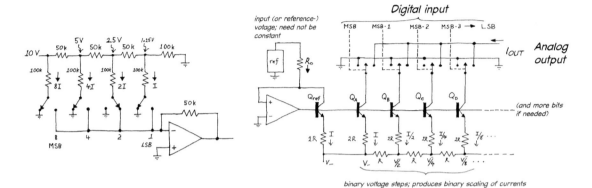

Figure 20N.10 Two DAC designs using R–2R networks: V out, I out.

A detail of Fig. 20N.10 may alarm you at first: the feedback to the op-amp on the right side. But on reflection you'll realize you have seen this oddity before, in the low-dropout regulator of Chapter 11N. In both cases an *inversion* within the feedback loop transforms what looks like positive feedback.

20N.2.4 Switched-capacitor DAC

Capacitors, rather than resistors, can be used to form a DAC's binary-weighted output. IC fabricators favor using capacitors over resistors, especially for CMOS designs. The circuit in Fig. 20N.11 – for a 3-bit DAC – resembles the op-amp summing circuit but is simpler.[11] The capacitive design sums *charges* directly whereas the op-amp summing circuit, which sums currents, requires either scaled current sources or a reference voltage and scaled resistors. (Compare AoE Fig. 13.4.) The capacitive design can leak, but works well in a DAC whose output is used only briefly, as in an SAR ADC (see §20N.3.3).[12]

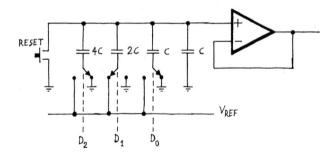

Figure 20N.11 3-bit switched-cap or "charge-scaling" DAC.

Each of the binary-scaled capacitors, after being reset to zero, is connected either to ground or to V_{ref} according to the level of the digital input bit. When all bits are high, V_{out} equals the value of $V_{ref} \times \frac{7}{8}$. In other words "111" would produce a value of 7 V if V_{ref} were 8 V. In Fig. 20N.12 we have sketched the capacitive dividers formed by this and a couple of other input values.

[11] R. Jacob Baker, *CMOS Circuit Design, Layout and Simulation*, IEEE Press, Wiley (1998), p. 806.
[12] See also Analog Devices DAC Tutorial at https://LAoE.link/ADI_MT-015.pdf.

20N.2 Digital to analog (DAC) methods   801

Figure 20N.12 3-bit switched-cap: capacitive dividers formed for three input combinations.

20N.2.5 1-bit DAC; pulse-width modulation (PWM)

AoE §13.2.8

A good DAC is hard to build: Though it is easy to draw a diagram for a DAC, it is difficult to make a DAC that works with good resolution – that is, a DAC that responds properly to a large number of bits. In this Chapter's lab we will use a modest *8-bit* DAC, which implies resolution down to 10 mV given a 2.5 V range. This we can manage, even in our breadboarded circuits.

A low-resolution digital signal – even just a one-bit ON/OFF signal – can achieve high analog resolution when averaged. If this sounds a bit cryptic, recall that you have seen this effect in action, back when you built a PWM circuit in §8L.4. An ON/OFF signal whose duty cycle is varied can produce very high resolution when the pulsing waveform is smoothed by an averaging circuit.

Early personal computers used this method to provide audio. It is also the method used in so-called "Class D" amplifiers, whose outputs are simply switches, usually to a positive or negative voltage. You met one of these amplifiers, the LM4667, in Chapter 12L. Class D amplifiers can achieve very high efficiency and thus are favored in battery powered audio circuits. They also lend themselves to motor control (as you saw in Chapter 8L), where PWM provides easy power modulation.[13]

Figure 20N.13 is a set of scope images showing PWM done with the circuit of §8L.4. Alongside each duty-cycle waveform is an image of an LED driven with that waveform. The LED is driven ON by a logic *low*, so our listing of *duty cycle* in the figure inverts the usual sense of the term.

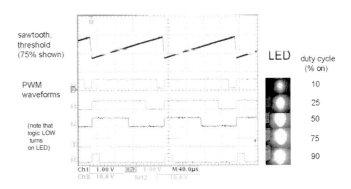

Figure 20N.13 Static PWM (duty-cycle) waveforms, and consequent brightness of driven LED. (Scope settings: sawtooth trace, 1 V/div; five PWM traces, 10 V/div.)

This is a static case, showing a technique well-suited to letting a microcontroller regulate not only

---

[13] Some people believe that PWM provides better torque than can be achieved by varying the voltage to a motor. This myth is disproved in *The Art of Electronics: the X Chapters*, Horowitz and Hill, §9x.5.

802 Analog ↔ Digital; PLL

the brightness of an LED but also the *color* of a bi- or tri-color LED, by varying the relative strength of two or three contributing colors.

PWM that varies with time can produce a waveform. Figure 20N.14 is a hypothetical PWM waveform and the sinusoid that would result if the PWM were averaged by a lowpass filter. This is the method used by Arduino boards for the "AnalogWrite()" output function.[14]

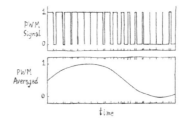

Figure 20N.14 PWM, time-varied, can produce a time-varying waveform.

The fact that a *one-bit* data stream can describe a waveform establishes the foundation for subtler forms of one- and low-bit conversion. In §20N.3.5 below, we will meet the counter-intuitive *delta–sigma* conversion scheme which can use a single bit output stream to provide a high resolution ADC.

20N.3   Analog-to-digital conversion

Going from the continuous world of analog into the discrete categories of digital is harder than going in the other direction and the task has evoked a wide and various set of solutions. Table 20N.1 shows the leading methods for comparison at a glance.

AoE §13.5.2

Table 20N.1 leading analog to digital conversion methods

Method	Speed	Resolution	Applications
"flash" (parallel)	fast ≥5 ns at 10 bits ≥1 ns at 8 bits	low ≤10 bits	storage scope, medical imaging, RF signals
binary search	intermediate ≥1 μs at 20 bits*	quite high ≤20 bits	general purpose widely used
delta–sigma	low intermediate ≥4k samples/sec at 31 bits**	high to very high ≤32 bits	general purpose becoming widely used
dual slope	slow a few samples/sec at 24 bits	high ≤24 bits	DVMs (digital voltmeter) now must compete vs. ΔΣ

* LTC378-20
** LTC244x: 24 bits, no latency; with one-cycle latency, 8k samples/sec. A spectacular 31 bits is offered at 4k samples/sec by TI's ADS1282.

Another representation of alternative ADC methods appears in a good, concise graph in Tietze and Schenk, *Electronic Circuits: Handbook for Design and Applications*, 2nd ed. Springer (2008) §18.12, Fig. 18.68.

---

[14] Some ARM-based Arduino boards also provide a true DAC output on specific pins. See
`https://LAoE.link/Arduino_AnalogWrite.html`.

20N.3 Analog-to-digital conversion

20N.3.1 Open-loop

AoE §13.6

Flash: We will start with the method that is conceptually simplest, though most difficult to fabricate: *flash* (or parallel) conversion. (The design may remind you of the "thermometer" DAC design of §20N.2.1, whose process it runs in reverse.) The comparator outputs provide a thermometer-like result (comparator outputs are high up to the level of the input voltage). That inconvenient code must be compressed by an *encoder* circuit.

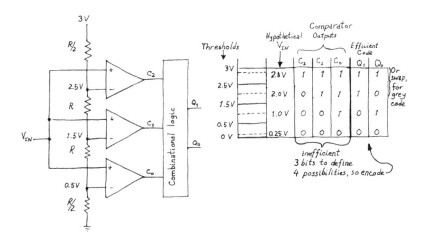

Figure 20N.15
Parallel or "flash" ADC.

This method calls for a lot of analog circuitry to achieve a modest number of bits. A 9-bit converter needs 511 comparators. That may not sound like a lot of parts, if you are accustomed to reading about million-transistor digital parts, and multi-gigabyte memories. But the fact that flash converters now are limited to about 9 or 10 bits underlines the point that *precise, fast analog* circuitry is hard to build, whereas fast digital circuitry is relatively easy.

Each of the outputs of a 10 ns digital memory, say, needs to decide only whether the stored data is closer to a high or to a low value. Each comparator in a 9-bit, 2.5 V-range converter, in contrast, needs to make a decision good to about 5 mV; and it needs to do it fast – in somewhat less than 10-odd ns, for it to run at the same speed as the memory. The comparators' job evidently is much the harder of the two.

The *encoder* is essential. The need for it is not obvious in a tiny example like the one in Fig. 20N.15. That encoder compresses the three comparator output lines to two binary encoded bits: useful, but not a spectacular improvement in efficiency. At 8 bits, the need for the encoder becomes very obvious: rather than take 255 lines from the ADC – the outputs of all comparators – we take the *encoded* version of this information, on 8 lines.

The device looks as if it could operate continuously. Normally, however, it is not used that way. Catching a value when the comparators were in transition could be disastrous: say, partway through a transition from 01 to 10 we could get an output of 00 or 11. Instead, it is operated with a sampling clock, like other ADCs. A sample-and-hold keeps the comparator inputs constant while the comparators settle, and the (digital-) output is caught in a register.[15]

Flash ADCs are rather specialized devices. They are power hungry, low-resolution converters. As standalone parts, they are reserved for the very impatient. But they are used more frequently within "subranging" ADCs that form a higher-resolution answer at somewhat lower speed, by repeatedly applying a flash converter to fractions of the full range.

---

[15] See, e.g., Analog Devices' App. Note 215A at https://LAoE.link/ADI_AN-215A.pdf.

Pipelining: A subranging flash device requires multiple clock cycles, in order to carry out a conversion – but it can run faster than the need for multiple cycles would suggest if it is fed a stream of new data while earlier samples are being processed.

The input is first subjected to a "coarse" 3-bit conversion using a 3-bit flash ADC. Then that ADC value is fed to a DAC and converted to a value representing the top three bits, and this coarse level is subtracted from the input. The result of the subtraction, a "residue," is passed to a second 3-bit flash ADC. The outputs of the two flash ADCs provide a 6-bit answer, and the fact that the two conversions occur one *after* the other mean that if the MSBs from sample N are saved, to be combined with LSBs when the latter are ready, a new MSB conversion, of sample $N + 1$, can be started while the N bits are being completed.

A 16-bit ADC from Linear Technology, for example (LTC2207) achieves the astonishing sampling rate of 105 Msps (megasamples per second): 16 bits converted in about 10 ns. But it pipelines its samples, using five flash ADCs, each time converting a fraction of the total range. Adding some housekeeping time, it needs seven cycles to carry out a single conversion. But this is still an extremely fast conversion – 16 bits in about 65 ns, even if one did not take advantage of pipelining, and simply used the device to do a single conversion. The seven-clock delay of the pipeline is called its "latency."

20N.3.2 Dual-slope

We looked at *single-slope* converter in §17W.2, Fig. 17W.4, on counter applications.[16] The single-slope measured the time a ramp took to reach V_{in}. The dual slope converter is similar: it lets V_{in} determine the size of a current that feeds a capacitor for a fixed time; when that fixed time is up, the capacitor is switched to a fixed discharging current. A counter measures the time for the discharge to take V_{cap} back down to zero.

AoE §13.8.4

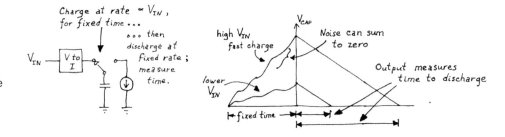

Figure 20N.16 Dual slope converter: charge for fixed time at rate determined by V_{in}; measure time to discharge.

The dual slope differs however in one important respect from the other converters described up till now: it is an *integrating* converter, in contrast with the sampling types. Because of this behavior, variations above and below the average input voltage level tend to get swallowed up, not recorded as the true level. The swallowing-up is perfect if the converter takes in an *integral* number of periods of the interfering noise (the integral of that noise then is zero).

As a result you can get the right answer in the presence of periodic interference. In particular, if you expect periodic interference at 60 Hz and its harmonics (as usually you can expect from United States power lines), you can let a conversion include an integral number of cycles of this signal; then the bumps above and below the average value will cancel. This is the method used in the DVMs we use in our teaching lab.[17]

The second virtue of a dual-slope ADC is its cancellation of DC errors in its components: in order to

---
[16] It also appears in AoE §13.5.2.
[17] Such converters designed for the European market, where the power line frequency is 50 Hz, use a longer integrating time.

know when the circuit has ramped V_{cap} back down to ground, the circuit needs an accurate comparator. To resolve the input voltage to, say, 0.1 mV it might seem that the comparator would need an offset voltage smaller than that. But the dual-slope design eliminates that requirement. The comparator need only return V_{cap} to its starting point, not to *zero* volts. If the starting point was, say, 1.5 mV (supposing this to be the comparator's offset voltage, V_{OS}), a return to that starting point of 1.5 mV introduces no error. The effect of V_{OS} is cancelled.

Another virtue of the dual-slope, particulary compared to the single slope converter of §17W.2, is that the value of the capacitor is not critical. This technique relies on the *ratio* of the fixed charging time to the variable discharge time. Since the same capacitor is used for *both* integrations, its value is not a source of error in the final measurement value.

Dual-slope converters now are getting competition from so-called "delta–sigma" ADCs (see §20N.3.5 below). Like the dual-slope, these are integrating types capable of very high resolution and they can be configured to reject 60 Hz. They would serve well in a DVM.

20N.3.3 Closed-loop

A closed-loop[18] converter, and in particular one that uses a binary search, beats either flash or dual-slope as a general purpose converter. The flash costs too much, and provides only mediocre resolution; the dual-slope is slow.

The closed-loop converter works like a discrete-step op-amp follower: it makes a digital "estimate;" converts that to its *analog* equivalent and feeds that voltage back; a single comparator decides whether that estimate is too high or too low (relative to the analog input of course); the comparator output tells the digital estimator which direction it ought to go as it forms its next, improved estimate. Negative feedback drives the digital estimate close to the input value.

Figure 20N.17 shows contrasting waveforms at the *feedback* DACs of two sorts of feedback ADC. The waveforms show the analog estimates made by the two converters as they strive to drive their estimates close to the input value. (The waveforms show analog voltages reflecting the digital estimates, but let's not forget that it is the *digital* values that we are after, not these analog equivalents. These are, after all, Analog-to-Digital converters.)

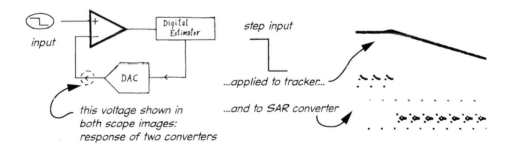

Figure 20N.17 Tracker versus binary-search or successive-approximation converter: trying to follow a step input.

Successive-approximation and tracking: The block diagram in Fig. 20N.18 illustrates the implementation of feedback ADCs using the two forms of *digital estimator*, estimators whose efforts to

---

In principle, one could set the integration time to be an integral multiple of both 50 Hz and 60 Hz periods. Often, an integration time of 0.1s is used. This is 6 periods of US line noise, 5 periods of European line noise. But such a choice could impose unacceptable delay.

[18] This term is not standard, but only our way of contrasting the converters that follow versus the flash and dual-slope types. We have also heard them referred to as feedback converters.

home in on the analog input appear on the right-hand side of Fig. 20N.17. One form is smart; the other is not.

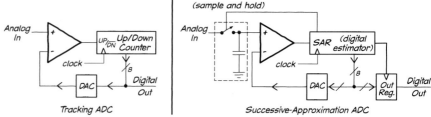

Figure 20N.18 Two closed-loop ADC converters: tracking and binary-search.

Note the presence of a *sample-and-hold* circuit in the block diagram for the binary search type (right-hand circuit of Fig. 20N.18). You met (and built) an S&H in the MOSFET lab, 12L. The tracker, which operates continuously, updating on every clock, does not need a sample-and-hold.

Tracking ADC: The tracking converter – the ADC sketched on the left in Fig. 20N.18 – is dopey: it forms and improves its estimate simply by counting up or down. Although it is not widely used, the tracking ADC persists because of two virtues: first, it always provides the best current answer, unlike the binary-search type which needs multiple clocks to arrive at its answer. Second, it is relatively insensitive to glitches that could cause large errors in the smarter SAR: a spurious clock edge causes an error of only one LSB. So the humble tracker can be the best for really plodding assignments like converting the output of a slowly rotating shaft (a so-called "resolver" application).[19] Its LSB must continually flip back and forth once the tracker has found its best estimate of a constant input. The downside of the tracking converter is that it responds slowly to a step input. It must count up or down, potentially through 2^n counts, until it reaches its new equilibrium value.

Binary-search or "SAR" ADC: By contrast, the successive-approximation estimator (SAR) is clever: it does a binary search starting always at the *midpoint* of the range and asking the comparator which way to go next. As it proceeds, it goes always to the midpoint of the remaining range. That sounds complicated; in fact it is easy for a machine that is by its nature already binary. The comparator tells the binary-search device *bit-by-bit* whether the most recent estimate was too high or too low, relative to the analog input. Unlike the tracking converter, the SAR ADC always takes n clock cycles (where n is the number of bits in the digital output) to convert the input value, even if it jumps full scale.

AoE §13.7

Because of this incremental way of composing its answer, such a converter needs an output register to catch its last, best estimate: its digital estimates look funny until the process is complete so they should not be shown to the outside world. In addition, the analog input must be passed through a *sample-and-hold*, which freezes the input to the converter during the conversion process. We omit this S&H in our lab circuit, in order to keep the circuit simple. We lose certainty in sampling time, and that uncertainty translates to considerable voltage *errors* in our conversions.[20]

Sampling should occur at regular intervals: Any processing of the signal – including the simplest: playback through a DAC and lowpass filter – assumes *periodic* sampling. When the S&H is omitted,

---

[19] See Analog Devices tutorial at https://LAoE.link/ADI_MT-015.pdf.
[20] The converted value will correspond to the input voltage at *some* point during the conversion process. But during that time the input waveform may have changed level by a significant amount, often by far more than the desired ADC resolution of 1/2 LSB.

the converted value equals the input at *some* point in the sampling interval, but one cannot know exactly when. Sampling time thus becomes vague, wandering, in successive samples. We'll see in §20N.4 that sampling at a controlled and regular rate is normally required.[21]

We present the *tracker* only to make the binary-search estimator look good. There are very few applications for which anyone would consider a tracker.

As we suggested earlier, it is hard to make precise analog parts: hard to make, say, a DAC good to many bits. It's easy to make digital parts, and easy to string them together to handle many bits. For example, it would be easy to build an up/down counter or SAR of 100 bits. So it is the *analog* parts – the DAC and comparator – that limit the resolution of the closed-loop converter. They also limit the converter's *speed*, as illustrated in Fig. 20N.19 (cf. Fig. 20L.7).

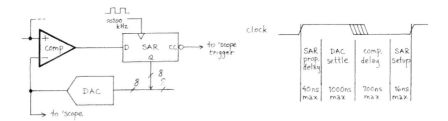

Figure 20N.19 Closed-loop ADC: clock period must allow time for all delays in loop.

20N.3.4 A binary search: example

Here, for anyone not already convinced, is a demonstration of the strength of the *binary search* strategy, used by the successive-approximation register (SAR) that you use in Lab 20L. If someone chooses a number in the range 0 through 255 and tells you whether each of your successive SAR-like estimates is *too low*, you can get to the answer in eight guesses. If you write out the binary equivalents of those guesses (as the "binary value" table does in Fig. 20N.20), you can see how the SAR forms its answer, bit by bit.

It proceeds from MSB to LSB, always setting the next bit *low*. After each guess the comparator tells the SAR whether the current guess is *too low*. If it is, the SAR sets the bit just put *low* back to a *high*, while forcing the next bit low. When the "travelling zero" arrives at the low-order end of the estimator, and has been forced high or left low, the search is over. In the case sketched in Fig. 20N.20 we have arbitrarily assumed that the answer is 156.

Figure 20N.21 is a scope image showing such a search: the SAR drives the DAC output to home in on the analog input. The conversion value differs from one sketched in Fig. 20N.20 in its d3 and d1 values, but the patterns in the two figures are obviously quite similar.

---

[21] We've hedged by using the word "normally," having been chastened by Shannon's observation that "samples can be unevenly spaced," though in that case "the samples must be known very accurately" and "the reconstruction process is also more involved...." (Claude Shannon, op. cit., p. 12.) Not only input levels but also exact timing of the sampling points would need to be known "very accurately." We will steer wide of such subtleties.

20N.3.5 Delta–sigma ADC

AoE §13.9, especially §13.9.4A

The binary-search or successive-approximation converter, long the most versatile and widely-used design, now is challenged by another good conversion strategy, usually called "delta–sigma."[22]

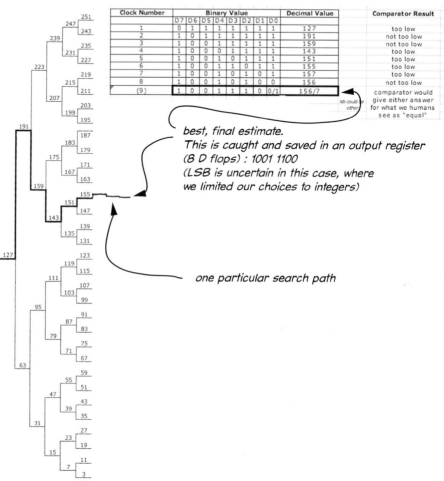

Figure 20N.20 One particular binary search: eight guesses to get 8-bit answer.

Figure 20N.22 is a sketch of the "modulator" that implements this arrangement. We have drawn a triangle around the whole thing to make the point that this is a feedback loop familiar from your experience with op-amps. The feedback signal fed to the "difference" block is a strangely crude signal, banging to the full-scale limits of the input range – either +1 V or −1 V in this hypothetical case. It cannot match *analog in* at any particular time (unless input hits full scale). But the long-term average of the output bitstream *can* match the level of *analog in*.

Figure 20N.22 gives in detail (perhaps more than necessary) an account of how this "modulator" works.

1. The circuit notes the sign of the difference between the analog input and the present "estimate" –

---

[22] Confusingly enough, it sometimes is called "sigma–delta," even when the sequence of operations is not truly interchanged. The labels vary from one manufacturer to another. Historically, sigma–delta may be the more accurate, matching the name applied by the method's inventors. See https://LAoE.link/DeltaSigma.html.

20N.3 Analog-to-digital conversion

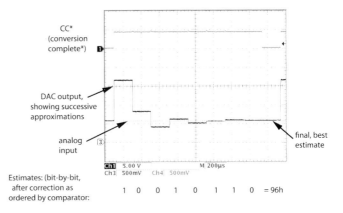

Figure 20N.21 SAR homes in on analog input. (Scope settings: 500 mV/div, except CC (pulse, 5 V/div; 200 μs/div.)

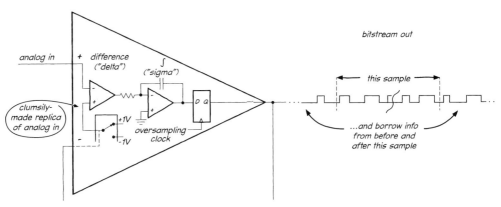

Figure 20N.22 Delta–sigma modulator.

in Fig. 20N.22, an estimate that is just a High or Low extreme, +1 V or −1 V, determined by the one-bit digital output. This difference, appearing at the output of the differential amplifier, is the *delta*. Then...

2. In response, the modulator injects a dollop of charge of appropriate sign and magnitude into an analog integrating circuit (that's the *sigma*), trying to drive the fed-back estimate close to the level of the analog original. The feedback loop attempts to hold the integrator output close to zero. Doing this, the converter produces a bitstream output whose average value matches that of the analog input, over many cycles.

So far this scheme is not surprising. The bitstream output roughly resembles PWM (§20N.2.5). And the loop is just another negative feedback scheme, somewhat resembling the crude *tracker* ADC on page 806, which injects a positive or negative increment into the *digital* summer, the counter, an accumulator that holds the current best estimate. The output of the tracker was formed in a very simple way: it was simply the current multi-bit value to which the Up/Down* counter had been driven. The tracker was very slow, requiring 2^n clocks to find an n-bit digital result, after a full-scale change of the analog input.

The delta–sigma is much smarter than that – and much more surprising. Figure 20N.23 is a block diagram of a delta–sigma ADC, its elements so generic that we don't expect you to get excited by this description – not yet.

Note that the "lowpass filter" in this diagram is a *digital* circuit; its output is a series of n-bit samples,

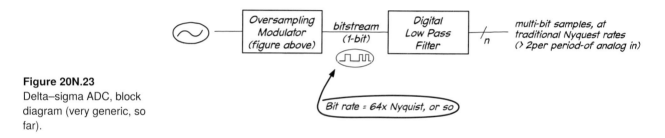

Figure 20N.23
Delta–sigma ADC, block diagram (very generic, so far).

assembled from the fast-flowing stream of single-bit data, produced by the "oversampling modulator," the circuit of Fig. 20N.22. As AoE says, that is where the magic resides.

If a delta–sigma converter used 2^n clocks to get an n-bit result, as a tracker does (worst case) and as a simple PWM bitstream would do, there would be not much to admire or explain in delta–sigma's operation. But practical delta–sigma designs do not do this. Instead, they manage to get high resolution with quite modest oversampling rates, such as 64 (implying a clock rate that is quite manageable). With this limited oversampling they nevertheless can achieve 16 or 24 bits of resolution.

The magic of delta–sigma is extremely counter-intuitive. AoE's great contribution to the literature on this topic is to *admit* that this conversion method resists intuition. The conventional "explanations" of the method abound in glib evasions of explanation – sudden retreats into frequency-domain abstractions[23] – as AoE points out.

Analog Devices' (ADI's) explanation begins, as these delta–sigma clarifiers often do, by acknowledging that everyone else's explanation is hard to follow: "... most commence with a maze of integrals and deteriorate from there. Some engineers... are convinced, from study of a typical published article, that it is too difficult to comprehend easily."[24] To the credit of ADI, the company also provides an interactive time-domain animation of the behavior of a delta–sigma modulator, which one clocks by hand in order to get a sense of the way that the circuit encodes a DC input.[25] This animation does help.

Maxim's application note explaining delta–sigma (which they choose to call "Sigma–Delta") remarks that "Designers often choose a classic SAR ADC instead, because they don't understand the sigma–delta types." "...and no wonder!" was our feeling, when we hit the point in this mostly clear note that says, without explanation, "... the integrator acts as a lowpass filter to the input signal and a highpass filter to the quantization noise."[26] This crucial observation recurs in all discussions of the conversion method – and is explained, at last, in AoE, and below on page 812.[27]

Before we reach this highpass/lowpass issue, let us note a preliminary point in favor of the one-bit converter:[28] the quantization noise is spread over a wide frequency range, and can be reduced in a particular, narrower range of interest. This initial thought is almost a commonplace.

Trading word length for sampling rate: oversampling: Sampling at more than the Nyquist rate (which, as you know, is slightly more than twice the top input frequency) is called *oversampling*. It is possible to convey a given amount of information in shorter "words" (fewer bits per sample) by sending these words at a higher rate.[29] This simple claim certainly is plausible; no magic here (the

---

[23] ...the last refuge of a scoundrel – in the view of some of us abstraction-o-phobes.
[24] ADI application note: https://LAoE.link/ADI_MT-022.pdf.
[25] Kester, Walter, op. cit., and ADI application note: https://LAoE.link/DS_ADC.html.
[26] Maxim Application Note 1870: https://LAoE.link/Maxim_AN1870.pdf.
[27] We also want to acknowledge the wonderfully thorough explanations offered by Nika Aldrich in his book which is rich with waveform images that help to make this difficult topic intelligible.
[28] Some use more than a single bit, and are properly described as "low-bit" converters. Their workings are fundamentally similar to those of the one-bit designs.
[29] See John Watkinson, *The Art of Digital Audio*, 3d ed. Focal Press (2001), p. 253.

20N.3 Analog-to-digital conversion

plot in Fig. 20N.24 describes, for example, how a sub-ranging converter can convert a wide word in multiple cycles at narrower conversions). But if one tries to apply this analysis to an ADC that uses *one*-bit to provide a high-resolution output – say, resolved to 16 bits – one gets a preposterous result. Sixteen bits would require sampling at about 65,000 times the Nyquist rate. That cannot be done, with present technology.

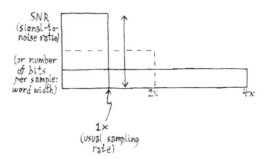

Figure 20N.24 One can trade word-width against sampling rate, holding information rate constant.

...but long words are best for efficient data storage: Although the low-bit strategy can pay in conversion processes, it does *not* pay in data storage. As Watkinson points out,[30] a bit can convey far more information in a longer word: in a 4-bit word, with its 16 possible codes, each bit carries "four pieces of information," as he puts it. In a 16-bit word, with its 65,000-odd codes, each bit carries 4096 "pieces of information." So, for example, no one advocates storing digital audio as 4-bit samples.[31]

Delta–sigma design gets high resolution with quite modest oversampling rates, such as 64 (implying a clock rate that is quite manageable, under 3 MHz, for a 16-bit audio signal). It achieves this by spreading quantization noise over the wider oversampled frequency range, pushing much of that noise into higher frequencies where it can be filtered out.[32]

As the lowest image of Fig. 20N.25 suggests, still greater reduction of noise is possible if the quantization noise can be "shaped" by highpass filtering in the conversion process, then eliminating most of that noise with a lowpass at the output. Some such "shaping" is inherent in the modulator's operation, as we argue below.

The technique is very successful, and now provides the cheapest way to get high-resolution conversion at low frequencies (for example, Analog Devices' AD7748 provides 24-bit results in 20 ms). The method also dominates audio conversion.

How delta–sigma puts quantization noise where it is less harmful: The point that delta–sigma depends on "noise shaping," some of which is inherent in the modulator's operation, is often reiterated in discussions of delta–sigma – as we have complained. The process is sketched in Fig. 20N.25, and is explained in AoE; see also the fine expostion in a web posting by Uwe Beis.[33]

AoE §13.9.5

Here, we try to diagram the effects described in AoE. The delta-modulator sketched in Fig. 20N.26 is an all-analog version that shows an injection of quantization noise (v_{qn}) where the comparator

---

[30] Watkinson, op. cit., p. 254
[31] But we should concede that Sony's Super CD, which stores a single-bit delta–sigma bitstream, stands as a striking exception to this generalization favoring long words for storage. The Super CD never was widely adopted, but persists among audiophiles.
[32] Gosh! Haven't we just uttered the formula that we complained about, above? But we will not stop with this remark. We will try to justify the claim.
[33] https://LAoE.link/DeltaSigma.html.

812 Analog ↔ Digital; PLL

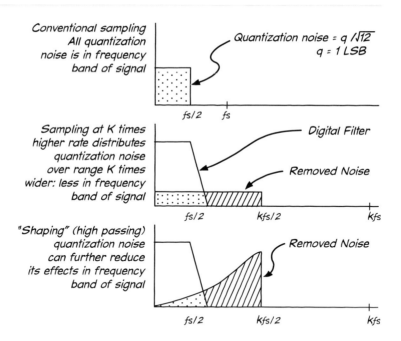

Figure 20N.25 Oversampling spreads quantization noise.

and flip-flop of a digital modulator would appear. The contrast between the loop's treatment of noise (highpassed) versus its treatment of signal (lowpassed) becomes more evident if we redraw the loop for the two cases.

How the noise is highpassed: The quantization noise is injected after the differencing and integration. We have redrawn those two elements as a single triangle – an *op-amp*-like object. That redrawing should not alarm us, because a normal, compensated op-amp is exactly that: a difference amplifier whose gain rolls off as an integrator does.

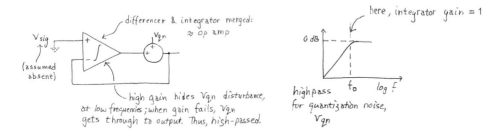

Figure 20N.26 Delta–sigma modulator redrawn to explain highpassing of *quantization noise*.

If you think of the injected quantization noise as just one more "dog" within the feedback loop (mentioned in §6N.9.1), then the "op-amp" hides this dog (is quantization noise its barking?) – but only as long as the amp has sufficient gain to do its feedback magic.

Then, as the gain begins to fall, more and more of the quantization noise survives. Eventually it passes right through to the output, undiminished. In other words it is *highpassed*. Much of the noise normally falls above the range of frequencies that we consider *signal* for the ADC.

How the signal is lowpassed: The modulator, at the same time, is said to *lowpass* the signal. Thus it keeps what we want, while it pushes what we don't want outside the frequency range that interests us. Figure 20N.27 gives the argument that says the modulator can lowpass the signal. Again, we have redrawn differencer and integrator as a single op-amp-like triangle. Notice that we are not claiming that the entire *follower-like* circuit in the figure is itself a filter. Rather, the claim is that the triangle – the modulator circuit fragment that we have likened to an op-amp – implements the lowpass function. We don't normally describe an op-amp that way; but we do know that any ordinary ("compensated") op-amp does, indeed, attenuate higher frequencies. It is a lowpass filter, or an integrator, though overall feedback often hides this fact from our view.

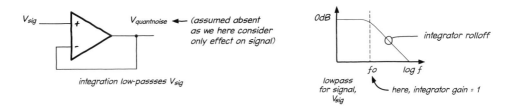

Figure 20N.27 Delta–sigma modulator redrawn to explain lowpassing of *signal*.

The remarkable result: high resolution from a modest number of bits: But even once one understands (or accepts) this point about the shifting of quantization noise, the greater mystery remains (as AoE points out). Delta–sigma offers high resolution (for example, 16 bits) using oversampling of only, say, 64× (that is, 64 times the minimum "Nyquist" rate of two samples per period of the input signal). But that would imply that just 128 sequential bits could encode a value to one part in 64K – a process that ought to take 64K sequential bits. How is this possible?

A note from Analog Devices concedes how puzzling this result is:

For any given input value in a single sampling interval, the data from the 1-bit ADC is virtually meaningless. Only when a large number of samples are averaged, will a meaningful value result. The Σ–Δ modulator is very difficult to analyze in the time domain because of this apparent randomness of the single-bit data output.[34]

Any single 16-bit sample constructed by the delta–sigma ADC, then, uses the digital lowpass of Fig. 20N.23 to tap information from a great many more bits than the nearest 128. The digital filter looks before and even *after* the present digital sample.[35] In other words the bitstream is, in fact, much longer than the 128 bits we announced just above. In fact, the digital filter may take information from *several hundred* or even a thousand successive bits in the stream or "taps".[36] AoE argues, in addition, that bit position matters, so that a 128-bit stream carries much more information than simply its average (which would provide only 7-bit resolution).[37]

It follows that a delta–sigma ADC requires a stream of samples; it cannot do its magic on an isolated sample.

---

[34] Kester, Walter, ADC Architectures III: Sigma–Delta ADC Basics, Tutorial MT-022, p. 6.
[35] This is a peculiarity of digital filters: they can "look ahead in time," having samples available in both directions relative to the "present" sample at the cost of delaying the output.
[36] ADI Tutorial: https://LAoE.link/ADI_MT-022.pdf.
[37] §13.9.4C of AoE says, speaking of a 64-bit stream, "Because the individual bits are weighted differently, there are many more than 64 possible results."

20N.4 Sampling artifacts

20N.4.1 Predicting sampling artifacts in the frequency domain

We have recited Nyquist's rule that one must acquire a little more than two samples per period of an input sinusoid in order to get a full representation of the waveform (assuming we sample multiple cycles): see §20N.1.2 above. We now can present Nyquist's rule in a way that makes it easier to design a sampling system, a way that allows us to predict the artifacts (often called "images") that will be caused by sampling.

In general, if one samples an analog signal of frequency f_{in} (let's assume a single input sinusoid for simplicity), sampled at frequency f_{sample}, the sampling process will produce artifacts at predictable frequencies:

$$f_{artifacts} = (n \times f_{sample}) \pm f_{in} \qquad (20N.1)$$

here n is any integer. We devote Chapter 20S to the subject of sampling artifacts, and we try to explain there why the artifacts appear at the specified frequencies; we will not try to justify equation 20N.1 now.

Figure 20N.28 is a sketch of the first two sets of images for a hypothetical input signal range. The "images" sit above and below f_{sample} and above and below $2 \times f_{sample}$, as if reflected about the sampling rate and its integer-multiples.

In Fig. 20N.29, scope frequency-spectrum images show a particular case, where we applied a single input frequency (rather than the more realistic *range* of frequencies sketched in Fig. 20N.28). This f_{in} is constant at 500 Hz. The sampling rate is reduced progressively from top trace down, and the sampling artifacts appear in these frequency spectra.

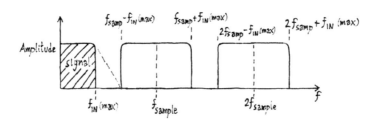

Figure 20N.28 Sampling artifacts sketched: above and below f_{sample} and its multiples.

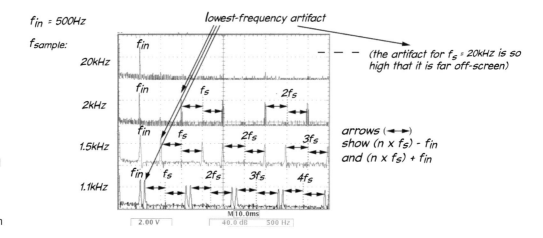

Figure 20N.29 Artifacts introduced by sampling: four sampling rates. (Scope settings: frequency spectrum FFT 500 Hz/div.)

20N.4 Sampling artifacts

In Fig. 20N.29 we have marked the *lowest-frequency artifact*, because it turns out this is the only one that need concern us. We do need to eliminate all the other artifacts as well, as we reconstruct the signal, but a lowpass filter that knocks out the lowest image necessarily takes out the higher artifacts along with the lowest.

So equation 20N.1, predicting endless repetition of false images, is correct, but carries more information than we need. The lowest image – the one at $f_{\text{sample}} - f_{\text{in}}$ is the target. Let's say this boldly, because it is a truth more useful than the more general formula of equation 20N.1. Here is the artifact that we need to worry about – and choose a filter to eliminate:

$$f_{\text{lowest-artifact}} = f_{\text{sample}} - f_{\text{in}} \qquad (20\text{N}.2)$$

As the sampling rate gets stingier, the filtering task becomes more difficult, since the lowest artifact moves closer to the true signal, f_{in}.

The bottom case of Fig. 20N.29, for example, where the lowest image is at 600 Hz and the true signal at 500 Hz, presents a challenge for a filter. The other cases, at higher sampling rates, make it easy to eliminate the lowest image. The original CD audio standard presented a similar challenge (and even a little more difficult): with the sampling rate at 44.1 kHz and audio up to 20 kHz, the lowest image occurred at 24.1 kHz – and the industry found that it could not build so steep a filter without introducing audible distortion.[38] This failure led to the introduction of the "oversampling" strategy that took over digital audio after that initial disappointment. It is also true that such digital filtering soon became cheaper than fashioning excellent analog filters.

The difficulty of filtering out the lowest artifact is sketched in Fig. 20N.30 where the lowest figure mimics the stingy sampling used in the early CD implementation.

The second generation of CD players adopted 2× *oversampling* (§20N.1), achieving an apparent sampling rate of 88 kHz. Later CD players pushed this oversampling higher still. These changes made the task of the analog reconstruction filter manageable.[39]

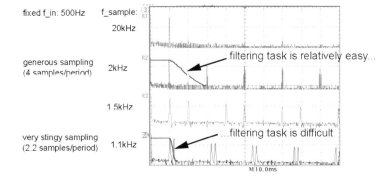

Figure 20N.30 Stingy sampling can make the reconstruction filtering difficult or even impossible. (Scope setting: frequency spectrum FFT at 500 Hz/div.)

---

[38] Well, *audible* to certain golden-eared audio critics, at least. See, for example, this comment on an audiophile site: "... ALL of these '14 bit' players tended to all but remove subtle ambient clues.... The sound was rather 'flat' spatially, the decay of reverb all in the speaker plane, rather than front-to-back.... subsequent Sony players ... showed, for me anyway, a substantial improvement in reproduction of venue acoustics." (Harbeth User Group Forum, post no longer available.) Certainly we are entitled to wonder whether such complaints were well founded.

[39] Oversampling injects zero-valued samples between the original samples during playback. The CD itself is still encoded at a data rate of 44.1 ksps. See Pohlmann, op. cit., p. 94.

20N.4.2 Aliasing is predictable

The frequency-domain view described in §20N.4.1 allows one to predict just where *aliased* artifacts will arise as well. It turns out that frequencies too high for proper conversion will get "folded" to a lower frequency; see Fig. 20N.31.

One can see how this happens by recalling that an image forms at $f_{\text{sample}} - f_{\text{in}}$. If f_{sample} is not more than double f_{in}, then this lowest image must lie below f_{in}. Once this has occurred there is no remedy. A lowpass cannot eliminate the falsehood without also eliminating the true input. This is the effect we saw earlier in a time-domain scope image of aliasing, Fig. 20N.5.

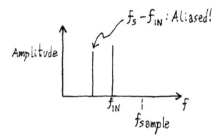

Figure 20N.31 Aliased image where f_{sample} is too low (set at $1.5 \times f_{\text{in}}$).

20N.5 Dither

Here is another counter-intuitive truth: adding some random noise (analog or digital) to a signal can improve the performance of an ADC or DAC. Random noise so used is called "dither."[40] The need for this addition appears only for low-level inputs to an ADC (or outputs of a DAC). The good signal-to-noise ratio of an ADC at full-scale – almost 100 dB, for a 16-bit part – shrinks toward nothing as the signal becomes small. A sinusoid whose amplitude is close to one LSB will evoke a small square wave at the reconstruction DAC, if no dither is added. With this square wave come higher-frequency harmonics, artifacts introduced by sampling.

Dither allows – amazingly enough – resolution well below the level of one LSB.[41] We look more closely at this effect in Chapter 20S, but Fig. 20N.32 gives the scope image indicating the benefits that result from adding noise to a low-level input.

The square wave on the middle trace, left side of Fig. 20N.32, shows the converted value that results from a low-level input that moves between two levels.[42] The bottom trace shows the many (spurious)

---

[40] The word is an archaic one meaning "... to tremble, quake, quiver, thrill. Now also in gen. colloq. use: to vacillate, to act indecisively, to waver between different opinions or courses of action" (Oxford English Dictionary). Either sense describes its function here, pretty well: it makes the input hop back and forth across a threshold.
Nika Aldrich tells a wonderful story, for the origin of dither – a story that we hope is true. British mechanical navigation computers for World War II aircraft were observed to work more accurately when airborne than on the ground. Someone hypothesized that engine vibration was freeing up elements of these machines, pushing cogs, gears, and levers back and forth across mechanical boundaries. Having confirmed this behavior, the British installed vibrator motors in their mechanical navigators: they added random noise, which came to be known as "dither." Aldrich, op. cit. p. 250.
Others tell many similar stories: devices that worked in propeller-driven B29s failed when adopted in jet bombers; gun controllers worked on battleships in Vietnam, jostled by the vibration of firing, but failed on trials in Boston harbor. All very plausible – and hard to confirm.

[41] Pohlmann, op. cit. p. 42, reproducing work of Vanderkooy and Lipshitz. The technique assumes a converter that can handle frequencies well above the frequency of the signal to which the dither is added. The dither frequency can be placed between the max signal frequency and the Nyquist frequency, $f_{\text{sample}}/2$.

[42] To get these images we used a rather large low-level step to simulate LSB. You will notice that this step is not tiny: it is about 250 mV. But the magnitude does not matter for the purpose of this demonstration of dither.

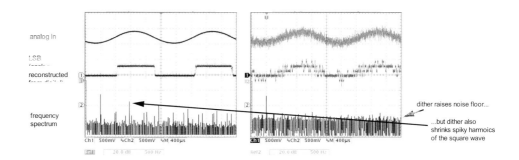

Figure 20N.32 Dither can increase converter resolution. (Scope settings: 500 mV/div; frequency spectrum FFT, 500 Hz/div.)

harmonics in the frequency spectrum of this square wave: strong and audible harmonics added to the original signal by the sampling process.

If one did not notice the frequency of the square wave, one might make the mistake of assuming that the reconstruction lowpass filter at the DAC output could eliminate the harmonics. But that is not so: the harmonics lie well within the audible range, will survive filtering, and will, indeed, sound unpleasant.

The right-hand image in Fig. 20N.32 shows the effect of adding dither: the analog input shows the thickening caused by this added noise. The frequency spectrum of the reconstructed waveform shows this "thickening" as a raised "noise floor." The LSB (a DAC output) in the middle trace now chatters above and below the midpoint level; averaged, this jaggedy waveform would produce a sinusoid close to the original.

Perhaps the raised noise floor distresses you. Probably it need not. It could be low enough to be insignificant in, say, an audio signal. But even if it were audible, this *white noise* would be much less disturbing than the strong patterned artifacts that are the harmonics of the undithered signal. Dither also can be used to randomize low-level noise other than sampling artifacts, masking effects that otherwise would be audible.[43] An excellent discussion of dither appears in Nika Aldrich's book.[44]

20N.6 Phase-locked loop

AoE §13.13.1

A PLL uses feedback to produce a replica or a multiple of an input *frequency*. It is a lot like an op-amp circuit; the difference is that it amplifies not the *voltage* difference between the inputs, but the frequency or *phase* difference. Once the frequencies match, as they do when the circuit is "locked," the remaining error is only a *phase* difference. The phase *error* signal is applied to a VCO[45] in a sense that tends to diminish the phase error, driving it toward zero.[46]

This sounds familiar, doesn't it: like a discussion of an op-amp circuit. The scheme is simple. The only difficulty in making a PLL work lies in designing the *loop filter*. We'll postpone that issue for a little while.

20N.6.1 Phase detectors: simplest, XOR

AoE §13.13.2A

The simplest phase detector is just an XOR gate – a gate that detects *inequality* between its inputs. (We assume digital inputs; sinewave inputs require phase detectors that are different, though conceptually equivalent.)

The phase difference evokes pulses from the XOR output; averaged, these produce a DC level that

The right-hand image of Fig. 20N.32, after the addition of dither, shows small sub-steps smaller than the large step that we

818 Analog ↔ Digital; PLL

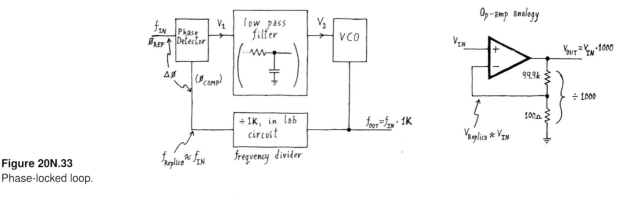

Figure 20N.33
Phase-locked loop.

Figure 20N.34 The simplest phase detector: an XOR gate.

drives the VCO. When the loop locks, there must be some phase difference in order to provide a non-zero feedback signal.

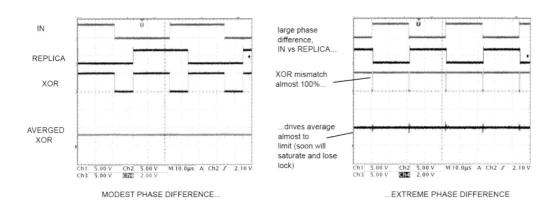

Figure 20N.35 XOR phase detector: phase difference is required to generate VCO feedback signal; phase-difference extremes can lose lock. (Scope settings: 5 V/div on all except bottom trace, showing average: 2 V/div.)

If the phase difference approaches either zero or 180° (as in the right-hand image of Fig. 20N.35), the VCO input signal approaches the limiting value; pushed a little further the PLL loses lock – just as an

---

have labeled LSB. This could not occur using normal dither and is a side-effect of our showing not the true LSB but a square wave larger than our demo DAC's LSB.

[43] Pohlmann, op. cit., p. 39.
[44] Aldrich, op. cit., Chapter 15. Normal dither adds 3 dB to the noise floor, whereas our homebrew example in Fig. 20N.32 adds much more. But we hope our simplified version serves to illustrate the principle.
[45] You met a VCO – Voltage Controlled Oscillator – back in Chapter 13L. There you used a '555 as a rough-and-ready VCO.
[46] Some phase detectors, like the simple XOR described in §20N.6.1, cannot take the phase difference to zero; others, like the edge-sensitive detector of §20N.6.2, can take the error all the way to zero.

20N.6 Phase-locked loop

op-amp loses feedback once its output saturates. A second weakness will appear shortly in §20N.6.2: the XOR detector can let the loop lock when the replica frequency is a multiple of the input.

20N.6.2 Phase detectors: edge-sensitive

The CMOS 74HC4046 that you will meet in the lab includes a fancier phase detector (called "Type II"). It shows two principal virtues: (1) it cannot lock onto multiples of the input frequency; and (2) it locks with zero phase difference between the two waveforms. It also uses less power, by shutting off its drive to the loop filter when it is happily locked. It is inferior to the XOR detector in just one respect: a glitch – a spurious edge – disturbs it much more seriously than a glitch would disturb the XOR detector, which would simply swallow the pulse into its averaging process. It is because of this superior *noise immunity*, important in some applications, that the '4046 PLL offers, as an alternative to its clever "Type II" edge-sensitive phase detector, a simple XOR ("Type I").

The better behavior of the edge-sensitive detector: The scope images in Fig. 20N.36 contrast the responses of the two phase-detectors. The XOR response is already familiar to you.

The left-hand image shows a case, using the XOR as phase detector, where the loop is correctly locked. A phase-difference persists, as we expect, but otherwise the loop is working properly.

This phase difference leads the *edge-sensing* phase detector to generate pulses that would try to drive the phase difference toward zero. (They would succeed if used; they are not applied, however, in this demonstration.) These pulses arise as logic *highs* during the brief time when IN leads REPLICA. These signals, if applied to the VCO, would tell it "Speed up!," driving the phase-difference toward zero. IN and REPLICA would stabilize in-phase.[47]

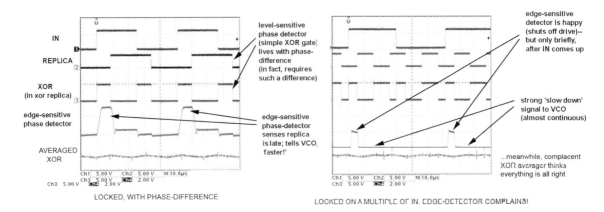

Figure 20N.36 Contrasting responses of 2 phase detectors: XOR needs phase difference, and locks on a multiple of f_{in}. (Scope settings: 5 V/div on all except bottom trace, showing average: 2 V/div.)

The right-hand image shows the edge-detector's response to the case where $f_{replica}$ is a multiple of f_{in}. The XOR loop is satisfied (and locked). The edge-detector is far from satisfied; it delivers a nearly-continuous "Slow down!" signal (a logic low), low except for a short time just after the rising edge of IN (when the detector thinks everything may be OK).

---

[47] When the edge-detector is "happy," not trying to change the VCO frequency, its output is neither High nor Low, but Off (3-stated). AoE points out that this absence of correction signal can cause "hunting," an error corrected in improved versions of the edge-sensitive detector. See AoE §13.13.2A, especially Fig. 13.92.
   In the scope image of Fig. 20N.36 the edge-detector output looks funny: stepping downward rather than sitting at a constant level. This is an artifact caused by loading imposed by the test circuit, and is not significant.

How the edge-sensitive phase detector works: The edge detector in Fig. 20N.37 is sensitive to *edges* rather than levels. It implements a state machine that generates correction signals during the time when one square wave's rising edge has led the other.

When phase difference goes to zero, the edge-sensitive phase detector shuts off its 3-state. At that time, the filter capacitor just holds its voltage, acting like a sample-and-hold, really; not like a filter at all. Figure 20N.38 shows a *flow diagram* for this state machine.

The detector becomes unhappy when it finds one waveform rising before the other. The state machine notices which rising edge comes first. If A – the input signal – comes up before B, for example, the detector says to itself, "Oh, B is slow. Need to crank up the VCO a bit." So it steps into the right-hand block, where it turns on the upper transistor, squirting some charge into the filter capacitor.

Then B comes up; A and B both are up, so this detector (like the XOR) sees equality, and goes back to the middle block, saying, "Everything looks OK for the moment; A and B are behaving the same way." In the middle block, the output stage is *off* (*3-stated*); the capacitor just holds its last voltage.

This pattern tends to force the VCO up, bringing B high a little earlier. But that reduces the time the detector spends in the right-hand block. The machine spends more and more of its time in the middle block – the "Everything's all right" block. This is where it lives when the loop is locked. You will see this on the '4046, when you watch an LED driven by a '4046 pin that means "The output is 3-stated." That is equivalent to "I don't see much phase difference," or simply, "The loop is locked."

Figure 20N.39 is the simple circuit that behaves this way.

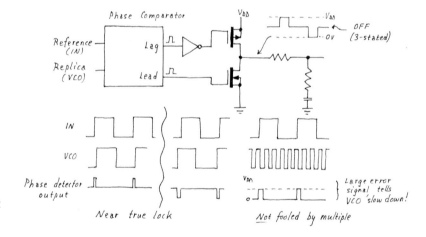

Figure 20N.37 '4046 edge-sensitive phase detector; it's smarter than XOR, and cannot be fooled.

20N.6.3 Applications

The PLL lets one generate a signal that is a precise multiple of an input frequency. Fig. 20N.33 showed a loop used to generate $f_{out}=2^{10} \times f_{in}$. All that's required is a divide-by-1K circuit placed within the feedback loop.[48]

This trick is often used within integrated circuits to drive internal circuitry at a clock rate higher than the off-chip clock frequency. For example, Analog Devices' ADuC848 multiplies the modest oscillator frequency – a standard "watch" frequency of 32.768 kHz[49] – by 384 to get a fast on-chip clock rate of more than 12 MHz (cited in AoE §15.10).

---

[48] In the silly terms of Chapter 6N, this div-by-1K is a dog that induces the feedback circuit to generate "inverse dog," a 1K× multiplication.

[49] If you're wondering why such a strange number, the answer is that this is 2^{15} Hz, so a watch can derive a 1 Hz "tick" just by putting the crystal frequency through a 15-bit binary ripple counter.

20N.6 Phase-locked loop

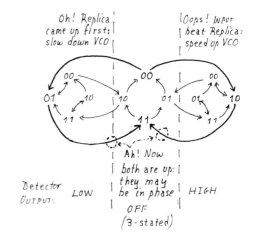

Figure 20N.38 '4046 state diagram (after National Semiconductor data).

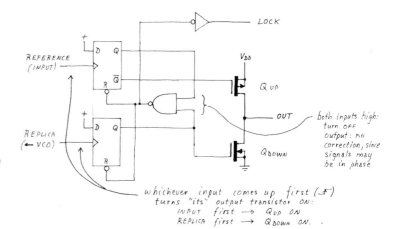

Figure 20N.39 '4046 phase detector: just two flip-flops; the winner of the race turns on "its" output transistor till both inputs go high.

You can use this tactic to create a clock that is a 64× multiple of the *sampling rate* in order to drive the switched-capacitor anti-aliasing and reconstruction filters of a data conversion system. The goal is to let the PLL automatically adjust filter rolloff to fit a sampling rate that we can vary. The filter's f_{3dB} is proportional to its clock rate ($f_{3dB} = f_{clock}/100$). So using the PLL to make the filter clock track the sampling rate allows us to vary the sampling rate freely while the filter automatically adjusts its clock rate as needed.

Figure 20N.40 sketches the scheme. Here are some details of the circuit: the signal labeled "sample" is a narrow pulse, so it is put through a divide-by-two flop to give a 50%-duty-cycle square wave to the PLL (the PLL performs poorly when given a narrow spike as input). The PLL multiplies this by 64, driving the switched-capacitor filter at $32 \times f_{sample}$. The filter's f_{3dB} is $1/100 \times f_{clock}$, so this arrangement puts the filter's f_{3dB} at about 1/3 of f_{sample}. This permits inputs up to about 2/3 of the theoretical maximum that Nyquist permits – 1/2 of f_{sample}.[50]

The PLL also can demodulate an FM signal: see Fig. 20N.41 – just watch the input to the VCO (which is the filtered *error* signal, describing variations in input frequency). This exercise is proposed in §20L.2.2. We could have used a PLL to demodulate the low-frequency FM we used in the group audio project in Chapter 13L – if we had known about PLLs back then.

---

[50] Perhaps we should admit that most students do *not* get around to trying the PLL in this application because they are busy with other issues. Still, it's a rather neat application of the PLL so we hope some of you will try it out.

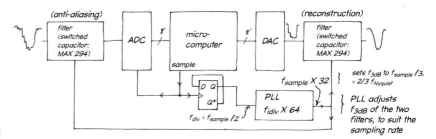

Figure 20N.40 PLL adjusts input and output filters automatically as sampling rate varies.

20N.6.4 Stability issues much like PID's

AoE §13.13.4A

The PLL is vulnerable to oscillations because of an implicit 90° shift within the loop, an *integration* imposed by the phase-detector. An additional simple lowpass filter could impose a total shift of 180° at some frequency, and such a shift could produce oscillations – an endless, restless hunting for *lock*.

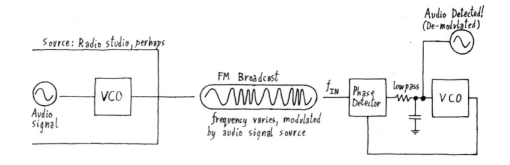

Figure 20N.41 PLL demodulating an FM signal.

This is essentially the same as the hazard you recall from our earlier discussions of op-amp stability and compensation. It is also much like the problem of the motor-control PID lab (10L), which placed an integrator within the feedback loop.

The second resistor in the lowpass is designed to eliminate this hazard by pushing the lowpass phase shift toward zero as frequency rises: see Fig. 20N.42.

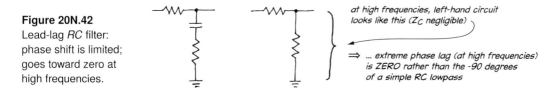

Figure 20N.42 Lead-lag *RC* filter: phase shift is limited; goes toward zero at high frequencies.

Figure 20N.43 shows, on the left, the rollofff of the lead-lag filter: it falls like $1/f$, as an ordinary lowpass would – but then levels off. The effect of this levelling-off upon the PLL is sketched on the right. Rolloff proceeds for a while at a scary -12 dB/octave (like $1/f^2$) – the effect of the implicit integration of the PLL combined with the $-90°$ of the filter. But then the whole loop's gain returns to a safe -6 dB/octave (like $1/f$).

20N.6 Phase-locked loop

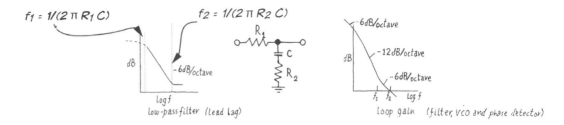

Figure 20N.43 Lead-lag lowpass filter keeps PLL stable by limiting phase lag.

Figure 20N.44 describes the same filter but showing only phase shifts: first (on the left) in the filter alone, and also (on the right) in the PLL loop that includes this filter.

You'll notice that the PLL loop gain goes to the $-180°$ shift that we fear – but makes it back to $-90°$ before closing time, and that's good enough (surprisingly).[51]

Figure 20N.44 Phase shift in lead-lag and entire PLL heads toward trouble – and then backs off.

The phase-detector's integration: The phase-detector performs an implicit integration, because phase (the quantity it measures and applies within the loop) is an integral of frequency. This is an idea that is easy to recite but perhaps harder to get a grip on. A look at units may help to cement the notion that phase is the integral of frequency: if frequency is measured in radians/second, integrating this with respect to time gives radians, a measure of phase.[52]

And in Fig. 20N.45 is a scope image that may help to make this idea less abstract.[53] The images shows the output of the phase-detector as IN and REPLICA run at slightly different frequencies.[54] One signal gradually slides past the other – and the *phase* difference signal shows the ramping we expect an integrator to show when fed a constant difference. This image may convince you that the *integration occurs* – and thus that we face a stability hazard reminiscent of what we met in the PID lab, 10L.

---

[51] You may recall we remarked on the marvel that such a phase lag is permissible back in Lab 10L. In fancy language, this sort of filter makes a "pole-zero network."

[52] It may also help to think of a similar claim that phase-difference is integral of frequency-difference. Imagine that the PLL's f_{in} and $f_{replica}$ differ by some small amount, and this difference remains constant for a while. A scope image of the two signals – a time-domain image – would show one slowly creeping past the other. A phase-difference detector would measure this as a ramping difference (until the difference hit a maximum and rolled over to zero, then started its upward ramp again. . .). This is the effect illustrated in Fig. 20N.45.

[53] An alert student objected to this argument, saying that the integration appeared to be achieved by the *RC*, not by the phase detector. This objection, though plausible, is not correct. It is true that the *ramping* in Fig. 20N.45 is visible as an analog waveform only because of the *RC* that converts the *duty cycle* output of the XOR phase detector to a voltage. But the *integration* carried out by the phase detector is complete before this conversion occurs. The XOR output expresses as *duty cycle* the phase difference between its inputs. This phase difference is an integral of the frequency difference fed to the detector.

[54] This is the *RC*-averaged output of the simple XOR detector.

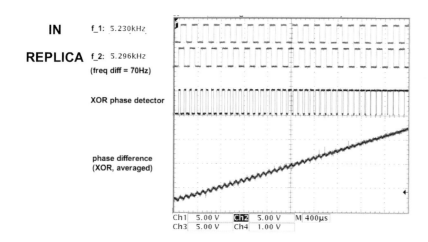

Figure 20N.45 Evidence that phase detector integrates frequency-difference: ramp for constant difference.

20N.7 AoE reading

- Chapter 12: Logic Interfacing:
    - §12.1.1 Logic family chronology – a brief history.
    - §12.1.2 Input and output characteristics.
    - §12.1.3 Interfacing between logic families.
- Chapter 13: Digital Meets Analog:
    - DAC and ADC:
        - §13.2 Digital-to-analog converters (DACs).
        - §13.5 Analog-to-digital converters (ADCs).
        - §13.7 Successive approximation (important to us: this is Lab 20L's method).
        - §13.9 Delta–sigma converters (a difficult topic, bravely explained here).
        - Review of Chapter 13: a good, compact summary of the many ADC and DAC methods.
    - Phase-locked Loops:
        - §13.13 Phase-locked Loops.
        - §13.13.2A The phase detector.
        - §13.13.4A Stability and phase shifts.
        - §13.13.6C FM detection.
        - §13.14: Take just a quick look at §13.14.2 Feedback shift register sequences. (We used this method to generate the dither demonstrated in this chapter.)

20L Lab: Analog ↔ Digital; PLL

This lab presents two devices, both partially digital, that have in common the use of feedback to generate an output related in a useful way to an input signal. The first circuit, an analog-to-digital converter, uses feedback to generate the digital equivalent to an analog input *voltage*. The second circuit, a phase-locked loop (PLL), uses feedback to generate a signal matched in *frequency* to the input signal – or to some multiple or rational fraction of that frequency.

20L.1 Analog-to-digital converter

The A/D conversion method used in this lab, "successive approximation," or "binary search," is probably the method still most widely used, though now competing with delta–sigma designs. It provides a good compromise between speed and low cost. It substitutes some cleverness for the brute force (that is, large amounts of analog circuitry) used in the fastest method, *flash* or parallel conversion.[1]

Note that the converter you will build today with four chips normally would be fabricated on a single chip. We build it up using an essentially obsolete *successive-approximation register* (SAR), so that you will be able to watch the conversion process. In an integrated converter the approximation process is harder to observe because the successive analog *estimates* are not brought out to any pin, though the stream of bits that forms that estimate may be. We also omit, for simplicity, the *sample-and-hold* that normally is included.

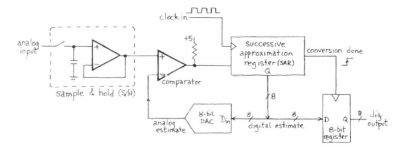

Figure 20L.1 Successive-approximation A/D converter: block diagram.

20L.1.1 D/A converter

The process of converting digital to analog is easier and less interesting than converting in the other direction. In the successive-approximation ADC, as in any closed-loop ADC method, a DAC is necessary to complete the feedback loop: it provides the analog translation of the digital *estimate*,

---

[1] We admit that it is an unusual brute who could put together a flash ADC.

allowing correction and improvement of that digital estimate. As a first step in construction of the ADC we will wire up a DAC, the Analog Devices AD558 of Fig. 20L.2.

This DAC integrates on one chip not only the DAC itself but also an output amplifier and an input latch. The latch is of the *transparent* rather than edge-triggered type, and we will ignore it in this lab, holding the latch in its transparent mode throughout (grounding pins 9 and 10, CS* and CE*, keeps the latch continuously transparent).

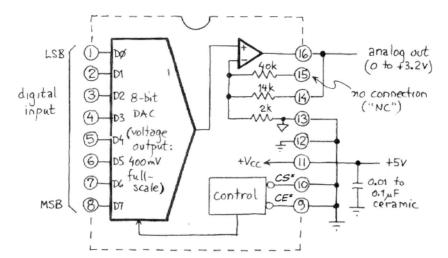

Figure 20L.2 AD558 DAC.

Checkout: (Hurry through this checkout. The fun comes later!)

Confirm that the 558 is working by controlling its two MSBs and its LSB with a DIP switch or simply with wires plugged into ground or +3.3 V. Hold the other five input lines low. (You may find a ribbon cable convenient, though less tidy than a wire-at-a-time; in any case, you will need to feed all eight lines in a few minutes.)

Note the relation between switch settings and V_{out}. Full-scale V_{out} range should be about 0–3.2V. What "weight" (in output voltage swing) should D7 carry? D6?[2]

Note the weight of the LSB. Here, if you look at the output with a scope you are likely to find noise of an amplitude comparable to that of the signal. If the noise is at a very high frequency, however, it may be quite harmless. How does your DVM appear to treat this noise? How do you expect the comparator to treat it when you insert the DAC into the ADC loop?[3]

Figure 20L.3 DAC checkout: voltage weights for particular input bits.

---

[2] D7 should carry the weight of one half full scale; D6, one quarter.
[3] We mean to suggest that very high-frequency noise – radio frequency that one often sees when scope gain is very high – may be too quick to be sensed by this ADC circuit, just as it is too quick to make the DVM respond. Large high-frequency signals, however, can be accidentally rectified by a DVM, and can produce spurious results, as you may possibly have seen back in Lab 9L when the discrete follower sometimes can make a DVM claim to see a voltage beyond the positive supply.

20L.1.2 ADC: watching the conversion process

When you are satisfied that the DAC is working, add the rest of the successive-approximation converter circuit: comparator and SAR along with DAC. If you are using a 74LS503 rather than 74LS502, you must *ground pin 1*, an active-low enable pin; we do not use pin 1 in the '502 circuit, and can leave it open if using a '502.[4]

If you have neither '503 nor '502 but are using custom-programmed logic that emulates the '502 please note that its pinout will not match that of the other two chips.[5] Consult its own datasheet. The comparator output pullup resistor and the Conversion Complete* LED current limiting resistor are connected to 3.3 V to be compatible with either the '502/503 running on 5 V or CMOS programmable logic on 3.3 V.

Connect the eight DAC inputs connected to the SAR Q outputs to the eight breadboard LEDs as well so that you can watch the estimating process. If you have the WebFPGA connected to the logic indicators you will have to temporarily remove it from the breadboard or disconnect it from the logic indicators to avoid damaging the 3.3 V FPGA with the '502/503 5 V outputs. You may instead choose to use the WebFPGA to emulate the '502: see Footnote 5.

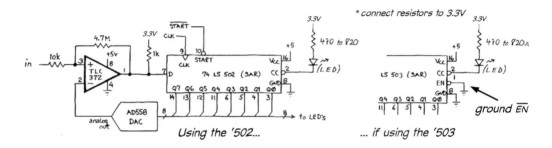

Figure 20L.4 Successive-approximation A/D converter: slow-motion checkout: layout.

Slow motion checkout: Use a debounced switch for Clock and another switch (which may bounce) for Start* (S*). Watch Conversion Complete* (CC*) on an LED.

- You will have to limit the current to this ninth LED using a resistor. We chose the resistor value considering the following points: the LED drops about 2 V when lit and the output pins of the '502/503 LS-TTL or the Lattice iCE40 FPGA can sink up to 8 mA, plenty for a high-efficiency LED like the Everlight HLMP series (which has adequate brightness at 2 mA).

Note: the behavior of S* may strike you as odd:

- First, the SAR is "fully synchronous:" asserting S* by itself has no effect. The SAR ignores S* unless you clock the SAR while asserting S*.
- Second, "Start*" is poorly named. It should be called something like "initialize*," because conversion will not proceed beyond the initial guess until you *release* S*.

---

[4] On the '502 it serves as a synchronized version of the D input; not useful for our purposes. See datasheet at https://LAoE.link/74LS502_Datasheet.pdf.

[5] These parts may be difficult to find. Verilog to create a 74LS502 is available at https://LAoE.link/FPGA/20L_74LS502.v. You can use this code with the WebFPGA; the AD558 has TTL input levels and will work fine with the WebFPGA's 3.3 V CMOS logic outputs.

Ground the analog input and walk the device through a conversion cycle in slow motion.[6] Watch the *digital* estimates on the eight LEDs and the analog equivalents, the *analog* estimates, on DVM or scope.

As you clock slowly through a conversion cycle, the DAC output (showing the analog equivalent of the SAR's digital estimates) should home in on the correct answer, 0 V. Do you recognize the *binary search* pattern in these successive estimates? If you get a digital *1* rather than *0* as your converter's final, best estimate – that is, if the LSB ends up HIGH rather than LOW – make sure that you have grounded the input close to the converter. Then use the scope to look closely at ground and +5 V lines, watching for noise. A digital ... *01* outcome need not shock you, given that an LSB is only about 15 mV, and the comparator's V_{offset} is spec'd at 5 mV (max); a small drop in the ground line added to this V_{offset} could tip the LSB.

Operation at normal speed: Now make three changes.

1. Connect "Conversion Complete*" (CC*) to "Start*" (S*). This lets the converter tell itself to start a new conversion cycle as soon as it finishes carrying out a conversion. (Disconnect the pushbutton that was driving S*, of course.)

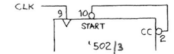

Figure 20L.5 CC* wired to drive start*: SAR starts self.

2. Feed the converter from a pot (2.5k or less: (Why?[7]) rather than from ground: see Fig 20L.6.

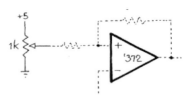

Figure 20L.6 Variable DC analog input.

3. Clock the converter with a 0 to 3.3 V oscillator signal, rather than with the pushbutton. The oscillator built into your breadboard is convenient; use the 3.3 V output of the 74HC4050 level converter. Set $f_{\text{clock}} \approx 100$ kHz.

Watch the DAC output and ADC input on the scope. (If you want a rock-steady picture, trigger the scope on CC*.) Vary the pot setting (analog input voltage), and confirm that the converter homes in on the input value.

Displaying full search "tree": You have watched the converter put together its best estimate, homing in on the input value. If you feed the converter *all possible* input values, you can get a pleasing display as in Fig. 20L.7 that shows the converter trying out every branch of its estimate "tree." Feed the converter an analog input signal from an external function generator: a triangle wave spanning the converter's full input range (0 to about 4 V). Set the frequency of the triangle wave at about 100 Hz. Trigger the scope on CC*.

If necessary, tinker with the frequency and amplitude of the input waveform, until you achieve a

---

[6] Note that the "analog input" is not the non-inverting terminal of the comparator, but one end of the 10k resistor; that resistor is necessary to maintain hysteresis.

[7] You will recognize here our usual concern that R_{source} not mess things up. Here, an over-large pot would mess things up by causing a significant increase in hysteresis.

20L.1 Analog-to-digital converter

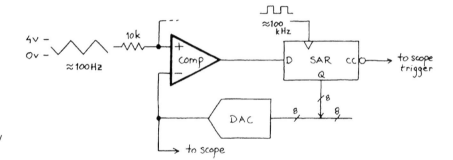

Figure 20L.7 Displaying binary search tree.

display of the entire binary search tree. This can be a lovely display (you may even notice the "quaking aspen" effect). You are privileged to see the binary search in such vivid form – while it remains to most other people only an abstraction of computer science. (An exceptionally noise-free image of the search tree produced by this lab circuit appears as AoE's Fig. 13.33.)

20L.1.3 Speed limit

The ADC completes an 8-bit conversion in nine clock cycles. Evidently, the faster you can clock it the faster it can convert. The faster it can convert, the higher the frequency of the input waveform that you can capture.

How fast can you clock your converter? Figure 20L.8 shows what must happen between clock edges. These numbers suggest a maximum clock frequency of a little less than 600 kHz.

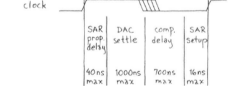

Figure 20L.8 ADC speed limit: what must be accomplished within one clock period.

Feed the converter a DC level and gradually crank up the clock rate as you watch the analog estimates on the scope. This time use the *external* function generator as clock source (the breadboard's top frequency, 100 kHz, is too low). At some clock frequency you will recognize breakdown: the final estimate will change because the clock period will no longer be allowing time for all levels to settle. Chances are, this will happen at a frequency well above the worst-case value of 570 kHz (above 1 MHz, in most cases we have seen.)

20L.1.4 Completing the ADC: latching the digital output

Up to this point we have been looking at the converter's feedback DAC output. Do not let our attention to this *analog* signal distract you from the perhaps obvious fact that the feedback–DAC output is not the *converter's* output. We use an ADC in order to get a *digital* output, of course, and on a practical IC ADC, as we observed at the start of this chapter, the analog estimate is not even brought out to any pin.

We now return our attention to the normal subject of interest: the ADC's *digital* output.

Output register: An integrated ADC normally includes a register to save its output (and incidentally, in the age of microcomputers, such a register routinely includes 3-state outputs, for easy connection to a computer's data bus).

Now we will add an 8-bit register of D flip-flops to complete the ADC. We need to provide a clock pulse, properly timed, that will catch the converter's best estimate and hold it till the next one is ready. Timing concerns make this task more delicate than it looks at first glance.

Conversion Complete* certainly *sounds* like the right signal. It turns out that it is not – not quite. The trailing edge (rise) of CC* comes too late; the other edge (which, inverted, could provide a rising edge) comes too early.

At the beginning (fall) of CC*, the SAR is putting out its initial estimate for the LSB; it has not yet corrected it (set it high) if such correction is necessary. Thus you would lose the LSB data, getting a constant Low, if you somehow used the start of CC* to latch the output.

At the end (rise) of CC*, the SAR is already presenting the first guess of its next cycle (0111 1111).[8]

Figure 20L.9 CC* timing.

What we need is a pulse that ends well away from both edges of CC*: see Fig. 20L.10. A single gate can do the job.

Figure 20L.10 Output register clock needed.

Add this gate and feed its output to the clock of the output register (74HCT574).[9] Let the register's outputs drive the breadboard's eight LEDs. (Don't forget power and ground, not shown in Fig. 20L.11. They are the corner pins, as usual for digital devices.)

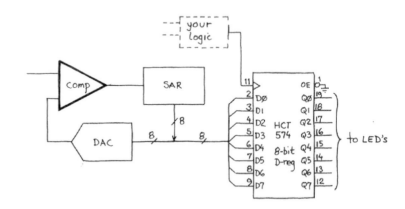

Figure 20L.11 Output register added to SAR ADC (output-register clocking left open for your addition).

Watch the converter's digital output and confirm that it follows the analog input applied from the potentiometer. It may be indecisive, by 1 bit. Is this indecision avoidable?[10]

---

[8] Why? Because the rise of CC* and the initial guess both come in response to the SAR clock, and CC* happens to come up a little *later* (see 74LS502 datasheet: t_{PLH} is slower than t_{PHL}. How's that for fine print?).

[9] A 74HCT574 running on 5 V will work with both the real 74LS502/3 and the 3.3 V WebFPGA emulating the 75LS502. If you are using the WebFPGA, you can also use a 74HC574 running on 3.3 V.

[10] We have nothing subtle in mind; we're just referring, as in §20L.1.1, to the likelihood that noise may be comparable to the value of an LSB, in your breadboarded circuit.

20L.2 Phase-locked loop: frequency multiplier

AoE §13.13.4

We will first apply the a phase-locked loop as if we were using it to generate a multiple of the line frequency, 60 Hz. This example is discussed in detail in AoE §13.13.4 and the circuit in Fig. 20L.12 is the one designed in that discussion, except that we have altered the VCO component values to permit a wider range of operation.

Construct the circuit shown in the figure. The phase detector and VCO are drawn as separate blocks; but note that they are *within one 4046 chip*: you do *not* need two 4046s. The 4040 is a 12-stage ripple counter.[11]

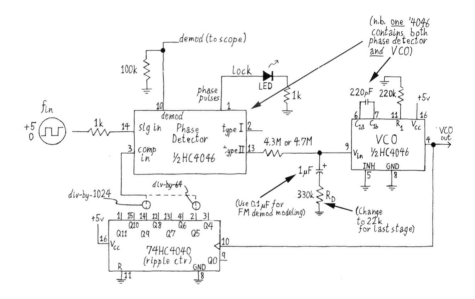

Figure 20L.12 Phase-locked loop frequency multiplier circuit (using 74HC4046 PLL).

20L.2.1 Generating a multiple of line frequency: (Type II detector)

Take the *replica* signal from Q9 of the 4040 (pin 14). (Q9 divides f_{clock} by 2^{10}. In case this seems odd, recall that Q0 divides by 2, not 1.) Set the function generator, which drives the input, to around 60 Hz. Use the scope to compare this *input* signal with the synthesized *replica* at 4046 pin 3. Confirm that the PLL *locks* onto the input frequency within a few seconds. Are the two waveforms in phase? The *lock* LED should light when the loop is locked (a logic *high* at pin 1 indicates that the phase detector output is *3-stated*: that is, the detector is satisfied, rarely seeing reason to correct the *replica* frequency).

See if the replica follows the input as you change frequency slowly; then try teasing the PLL by changing frequency abruptly. You should be able to see a brief *hunting* process before the loop locks again.

The limit of input frequency *capture* and *lock* ranges (identical, for this PLL) is determined by the VCO range: from about 40 kHz to 250 kHz for the specified values of R_1, C_1: 220k, 220 pF. Since the counter in the feedback path divides by 2^{10}, the input frequency range runs from approximately 40 Hz to 250 Hz.

---

[11] The pinout labeling shown follows the usual convention, in which the low-order Q is labeled Q0 on the part. NXP/Philips and others use this convention. National Semiconductor and Fairchild oddly started their Q numbering at *one* (CD74HC4040, on TI site). If you consult a National datasheet you may run into this discrepancy. TI dodges the labeling question by using *letters* rather than numbers: QA, QB,... (SN74HC4040).

The loop filter: The hard part of PLL design comes in designing the lowpass filter to maintain stability. The PLL uses a feedback loop strikingly reminiscent of the loop in the PID motor-control lab, 10L. In the PID exercise, the motor-to-position transducer placed an integration inside the loop. This 90° lag obliged us to use care to avoid injecting another 90° lag at frequencies where loop gain was above unity.

In the PLL, again an integration lies within the loop, and again it threatens us with instability.[12] This time the integration results from the fact that we are sensing a characteristic (phase) that is an integral of the characteristic that our loop means to control (frequency). Here, the integration is harder to localize than in the PID loop where it clearly occurred in the motor-potentiometer unit. In the PLL the integration – the equivalent of the motor-pot combination – is performed by the VCO–phase-detector pair. The detector's output is a signal that measures phase error.[13] For a constant frequency difference, the *phase error* (difference between input signal and the replica generated by the VCO) *ramps*; in other words, it integrates the frequency error: see Fig. 20N.45. This behavior closely resembles what we saw in the motor-pot combination: a constant error signal fed to the motor-pot produced a ramping output voltage: integration, once more.[14]

We need a lowpass filter in the loop to smooth the signal into the VCO – but we cannot afford another 90° lag. What is to be done? The answer turns out to be a simple circuit amendment: use a lowpass whose phase shift at high frequencies goes toward *zero* rather than toward −90°. A lowpass with an extra resistor between cap and ground does the trick, as we argue in §20N.6.4.

If you replace the filter's 330k resistor with 22k, you should find that the loop hunts much longer – overshooting, backtracking.... Watching the *demod* signal at pin 10, which shows the VCO input voltage, may make the hunting and overshooting vivid for you.

By reducing the filter resistor value, you have made a dangerous reduction in the *phase margin* – the safety margin between the phase shift around the loop (at the frequency where the loop gain goes to unity) and the edge of the cliff: the deadly −180° shift. This is a notion you will recall from Lab 9L and from the PID exercise; see, especially Fig. 10N.9.

If you want to live really dangerously, short out the lower resistor; now you have reduced the phase margin to zero, and the circuit may hunt forever. When you have seen this effect, restore the 330k, to restore stability.

What frequency should be present at the VCO output when the loop is locked? Check your prediction by looking with the scope at that point (pin 4 of the 4046). Why is this waveform jittery? (In a world without noise it would *not* be.)[15]

Now look at the output of the phase detector (pin 13 of the 4046). This is the Type II detector, edge-sensitive. You'll notice a string of brief positive going pulses, each decaying away exponentially

---

[12] Here, we restate a point made in the notes, §20N.6.4.

[13] You may be inclined to ask, "Why do this? Why not be more straightforward and measure frequency error directly?" The answer is that we have no way to do that quickly. To measure frequency even indirectly, inferring it from *period*, requires waiting for a full period of the two waveforms. Phase error information is available more promptly: after each of the respective rising edges in the Type II detector.

[14] If you're not exhausted by this discussion, you may want to consider a complication: a *second* integration performed by the *RC*, at least as used in the edge-sensitive "Type II" detector: a constant phase difference produces a ramping output from the phase detector. This second integration is analogous to the *I* used in the PID circuit. The *I* term drives the PID error to zero; the Type II detector, similarly, uses this integration to drive the phase difference to zero.

[15] The jitter arises because the corrections to VCO frequency come only once in every 1024 cycles of the VCO output. Slight wanderings of frequency between these checks go uncorrected. As usual, being college teachers, we are reminded of the fact – probably beneficial to humans, though unsettling in the PLL analogy – that the college checks whether students can do the course work only now and then: at best, weekly; the big test comes only once per term. In the meantime, who knows what goes on in the student's life? No doubt that life would be full of jitter, if one were so rude as to look at it between checkpoints.

when the detector reverts to its 3-stated condition (you may also see smaller negative-going pulses to the extent that the circuit is troubled by noise).

Theory predicts that these correction pulses should vanish in the steady state. But the 10 MΩ load of the scope probe you are using is discharging the filter capacitor fast enough to cause the pulses; the current the scope draws also causes the VCO to be slightly under-driven, causing the VCO to run a little late; a lagging phase difference persists: the loop now *needs* a phase difference to provide the positive pulses that feed the scope's R to ground. If you're feeling energetic, interpose a '358 op-amp follower between the capacitor and the probe. The positive pulses should become narrower – and now the phase difference is reversed: the VCO output comes a little *early*.[16]

20L.2.2 Try FM demodulation (in slow motion)

If you watch the input to the VCO (which is available in buffered form at pin 10), you can see a measure of VCO frequency. That's not very interesting when the input frequency is constant – but becomes interesting when a variation in input frequency carries information, as it does in an FM radio broadcast (or in our analog project, where we sent music across the room with an LED's flashes). We aren't going to pause long enough to do such an exercise, here; but we suggest that you get a glimpse of the way it might work trying a two-minute demonstration.

Replace the 1 μF capacitor, temporarily, with a 0.1 μF part (to make the loop adjust faster), and watch the *demod* signal (pin 10) as you try to vary f_{in} sinusoidally by slowly varying the function generator frequency (or let the function generator apply the frequency modulation if it supports it). You should sweep the scope very slowly – using "roll" mode (perhaps 1 s/div) if you have a digital scope. We hope the demod signal will look like the sinusoid that varies or "modulates" the PLL's input frequency. Too bad we didn't know about PLLs when we did the group audio-LED project. When you've had enough fun with this, restore the 1 μF capacitor.

Type I detector (exclusive-OR): The 74HC4046 includes three phase detectors. AoE describes two of them (Types I and II) in §13.13.2A.[17] The third detector on the HC4046, Type III, is simply a variation on the SR latch, with S and R driven by positive *levels* on the input and replica lines, respectively. We will *ignore* the Type III detector – *two* detectors seem plenty to consider on a first encounter with a PLL.[18] But feel free to check out the Type III if you like: its output appears at pin 15.

The Type I detector output is at pin 2, and the inputs are the same as for the Type II detector; to use the Type I, simply move the wire from pin 13 to pin 2 (and then to 15 for the Type III, if you *must*!). For both Type I and Type III you should be able to see the fluctuation of the VCO frequency over the period of the input, which you can exaggerate by reducing the size of the 1μF loop filter capacitor.

If you make a sudden, large change in the input frequency, you should be able to fool the Type I circuit into locking onto a *harmonic* of the input frequency (a multiple of the input frequency). Such an error would make the circuit useless, so we will normally use the Type II detector. You are likely to make the same choice in most applications.[19]

Note also the phase difference that persists between input and feedback signal in the locked state with the Type I detector. This simple phase detector (like the Type III) *requires* such a phase difference; this difference generates the signal that drives the VCO. If the phase difference ever goes to zero (or to

---

[16] Why? It's a fussy detail, but now, instead of discharging as the scope probe did, the '358's I_{bias} *injects* some current from its PNP input transistor's base. Try a CMOS op-amp to make *that* error go away.
[17] AoE's discussion treats the similar CD4046 rather than the 74HC version.
[18] The Type III detector seems to have been added as a sort of afterthought; early versions of the 4046 did not offer it. Like the simple XOR (Type I) it requires a phase difference between input and replica signal.
[19] The Type I is worth considering is if your input signal is quite noisy and the Type II mistakes the noise for edges.

π), the loop loses feedback: it can no longer correct frequency in both senses as required. When it hits that limit it's like an op-amp feedback loop that fails because the op-amp output hits saturation and thus cannot make a further correction. The Type II detector is altogether classier: it requires no errors to keep the loop locked; it is able to use the capacitor as a *sample-and-hold*, once the loop is locked, rather than as a conventional filter.

When you have finished looking at the behavior (and *mis*behavior) of the Type I detector, revert to the earlier circuit, using the Type II (output at pin 13).

20L.2.3 Expanded lock range: ×64 rather than ×1k

Now let's set up the PLL to expand its lock range: change the *tap* on the 4040 from Q9 to Q5 (pin 2). Now you are generating, at the VCO output, a modest $64\times f_{\text{in}}$. Because we are feeding back a larger part of the output frequency, we need to attenuate more in the loop filter. (We are worrying about the *loop gain* as we did for op-amps.)

Feeding back 1 part in 64, rather than 1 in 1024 – about 16× as much – calls for reducing the fraction preserved by the loop filter proportionately. So we reduce the 330k that was below the 1 μF cap by a factor of 16, to about 22k. The two sets of dividers – analog filter and digital counter – are sketched in Fig. 20L.13 for the two stages: first, where the loop multiplies by 1024; and second, where the loop multiplies by 64. The fraction fed back is held roughly constant (to about 12%) in the two cases.

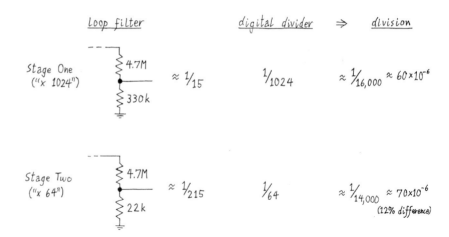

Figure 20L.13 PLL filter divider is adjusted to hold fraction fed back roughly constant in the two stages.

You may notice that we have drawn the "loop filter" as if it were simply a resistive divider. We have done that because at the high frequencies where we anticipate challenges to stability X_C is insignificant relative to the R values.

Over what range of input frequencies does the PLL now remain locked? The range should be significantly wider than before. The '4046 is capable of *capture* and *lock* over a frequency range of about 6:1.[20] We will be content with a range that accepts an input between about 600 Hz and 4 kHz. Capture and lock range are the *same*, for the Type II detector; for the less clever detectors, capture range (the range over which the loop will be able to *achieve* lock) is narrower than lock range (the range over which the loop will *hang on* once it has locked). This ability of the Type II to capture any frequency it can hold seems to follow from its immunity to harmonics: you can't fool the Type II.

---

[20] This is the approximate range as one drives the VCO over its permitted range from 1.1–3.4 V (the datasheet specifies this range with a 4.5 V supply; at 5 V each voltage is presumably about 10% higher).

20S Supplementary Notes: Sampling Rules; Sampling Artifacts

20S.1 What's in this chapter?

Here are reminders of Nyquist's sampling rule and some consequences, with scope images of artifacts that sampling can introduce.

The images in §20S.3, particularly Figs. 20S.3, adequate, and filtered, show four sampling rates applied to a 1 kHz sinusoid – sampling rates that are *lavish, adequate, stingy*, and *inadequate*.[1] But before we look at those images, let's recall the general notions that govern sampling. These notions that will help explain why the frequency spectra look as they do. Finally, in §20S.4, we'll attempt an intuitive explanation of the effects of sampling, particularly, the appearance of images.

20S.2 General notion: sampling produces predictable artifacts in the sampled data

Here, for a start, are some truths about sampling in a tiny nutshell:

- Sampling produces spurious images of the input at $f_{\text{sample}} \pm f_{\text{in}}$ and these images repeat, appearing in pairs above and below each integral multiple of f_{sample}: in general, $f_{\text{image}} = (n \cdot f_{\text{sample}}) \pm f_{\text{in}}$.
- One result of the appearance of these images is expressed as Nyquist's sampling rule: one must sample at a rate slightly more than double the maximum input frequency.[2]

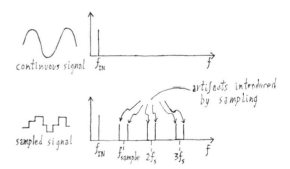

Figure 20S.1 Sampling creates multiple images of the true spectrum at higher frequencies.

It may not be obvious why these spurious images appear. At the end of this chapter we give an argument that we hope will appeal to your intuition.[3]

---

[1] To conserve space here, we omit Sleepy and Grumpy.
[2] This follows from the preceding comment about images because the first (false) image appears at $f_{\text{sample}} - f_{\text{in}}$; to make sure that this image does not overlap with f, one must put f_{sample} above $2 \times f_{\text{in}}$. More on this later.
[3] But here we will leave you with a general argument that they *must* exist given the stepped nature of the sampled signal. Like

Of all these spurious images, only one is troublesome: the image that is lowest in frequency at $f_{\text{sample}} - f_{\text{in}}$. In principle, a lowpass filter can remove this and all other false images – as long as these images lie above the true signal (that is, as long as we have sampled adequately: more on that below). Sometimes, however, this filtering is hard to achieve: it is hard when the spurious image lies close to the signal frequency, as we'll see in one example below.

20S.2.1 Aliasing

Violation of Nyquist's sampling rule – sampling too infrequently – brings on a disastrous error named "aliasing:" this is the generation of artifacts that cannot be disentangled from the true signals.

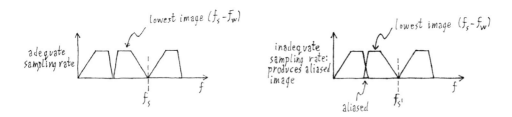

Figure 20S.2 Schematic representation of the problem of aliasing.

We're in trouble if the image lowest in frequency, at $f_{\text{sample}} - f_{\text{in}}$, falls into the frequency range of our signal, the analog input. Once this happens, an output filter cannot save us. The left-hand image in Fig. 20S.2 shows a sampling rate that is just adequate: the lowest part of the lowest image lies slightly above the input signal. So, a good lowpass filter should be able to eliminate the image.

The right-hand image in Fig. 20S.2 shows the mishap called aliasing. It occurs when one does not sample fast enough to satisfy Nyquist: that is, when f_{sample} is not more than double the maximum input frequency. Sample at a rate $> 2 \times f_{\text{in(max)}}$ *or else*! Or else you'll get nonsense: false images mixed in with the truth. Each image is folded from its true frequency (f_{in}) down to an aliased image (at $f_{\text{sample}} - f_{\text{in}}$). When an aliased image appears we can't bomb it with a filter, because it has invaded friendly territory.

20S.2.2 Filtering

CD players reproduce 20 kHz music that was sampled at only about 44 kHz: just 10% above the theoretical minimum rate. They face a challenging problem in filtering because the lowest sampling artifact lies very close to the highest genuine signal. The CD industry learned to solve this problem by faking a higher sampling rate – by interpolating digitally-calculated pseudo-samples between the genuine samples. This trick is called "oversampling." The first time you hear the claim, the assertion that a CD player can "oversample" the music sounds like a fraud. Wasn't the sampling rate fixed at the time the CD was made? Yes, it was. But once you understand what's intended, you discover that "oversampling" is a harmless figure of speech. Some people call the operation "digital filtering" or even "digital interpolating." These phrases sound less magical.[4]

---

the Fourier transform of a square wave, we need an infinite number of higher frequency components with some relation to the input frequencies (both input and sampling) to create a waveform with instantaneous transitions at the sampling frequency.

[4] The digital filtering operation can be more complex than this mere interpolation.

20S.3 Examples: sampling artifacts in time- and frequency-domains

The images in this section come from the test setup used in our class demonstrations in which an *analog* sinusoid was fed to an 8-bit ADC, and that *digital* signal then was fed to a DAC. The reconstructed *analog* signal was then lowpass-filtered.

These are the traces in the scope images:

1. the top scope trace is the input;
2. the second trace is the unfiltered output of the 8-bit DAC that has reconstructed the waveform from the converted digital values;
3. some images include a third trace, which is a filtered version of the reconstructed DAC output; and
4. at the bottom is the frequency spectrum (FFT) of either the raw DAC or the filtered reconstructed output, as labeled.

The four sampling rates illustrate four cases:

- *Lavish* sampling: 100 kHz (extravagantly excessive sampling, in fact).
- *Adequate* sampling: 4 kHz (four samples/period). This is good enough if a good output filter is applied.
- *Stingy* sampling: 2.2 kHz (providing just 10% more than the theoretical minimum of two samples per period). A steep filter (8-pole elliptic–MAX294) is able to knock out the lowest image.
- *Inadequate* sampling: 1.5 kHz, then 1.2 kHz (barely more than one sample per period). This brings on "aliasing."

Lavish sampling rate: Figure 20S.3 illustrates lavish sampling; the sampling rate is so high that the lowest artifact, at 99 kHz, is far out of the picture (far off the right side of the frequency spectrum shown).

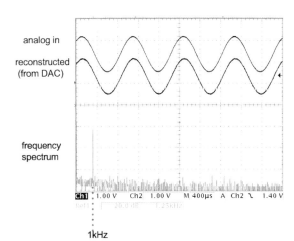

Figure 20S.3 Lavish sampling: 100 samples/period. (Scope settings: 1 V/div; frequency spectrum FFT, 1.25 kHz/div.)

As the reconstructed waveform shows, this way of sampling *works* in one sense: it allows us to reconstruct the analog waveform accurately. But it is a ridiculous way to achieve this result, producing almost fifty times as much digital data as we need to do the job. If this were the way we stored data on a disc, a CD would be at least the size of a large pizza.

Supplementary Notes: Sampling Rules; Sampling Artifacts

Adequate sampling rate: Figure 20S.4 shows more reasonable sampling. The artifacts are far enough above the true signal so that filtering them out does not look difficult – and it isn't.

The image in Fig. 20S.5 shows the effect of the filter applied to the reconstructed waveform: all artifacts are removed from the frequency spectrum – and the steppy quality of the reconstruction disappears from the filtered reconstruction. This smoothing is evident in the third trace in the right-hand image of the figure. The original sinusoid is recovered.

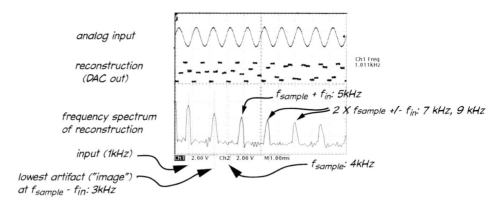

Figure 20S.4 Adequate sampling: artifacts now apparent. (Scope settings: 2 V/div; frequency spectrum FFT, 1.25 kHz/div.)

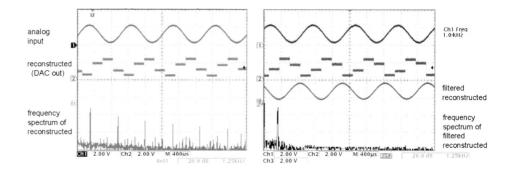

Figure 20S.5 A lowpass filter can remove artifacts introduced by sampling. (Scope settings: 2 V/div; frequency spectrum FFT, 1.25 kHz/div.)

Stingy sampling rate: In general, one aims to sample as stingily as feasible to acquire no extraneous digital samples. The samples must be stored and perhaps transmitted; we want to minimize complexity and cost. Nyquist says we need slightly more than two samples per period. But how much more? The answer depends on the quality of the lowpass filter used to remove sampling artifacts. A good filter, like the MAX294, an 8-pole elliptical active filter that you will use in labs, lets one come quite close to the minimal Nyquist sampling minimum of two samples per period of the analog input.[5]

In Fig. 20S.6, the sampling rate used to acquire the 1 kHz input is just 10% higher than the theoretical minimum of 2 kHz: 2.2 kHz. (Here, we are approximating the stinginess of CD sampling: 44.4 kHz to acquire signals up to 20 kHz: again, about 10% above the theoretical minimum Nyquist rate.) The filter

---

[5] To keep things simple, in this discussion we usually treat the analog input as a single pure sinusoid. Real inputs almost never are so simple. But if the analog input includes many frequencies, one need consider only the highest included frequency – perhaps 20 kHz for music, for example. If the sampling rate is adequate for that frequency, it is adequate for the entire audio input.

20S.3 Examples: sampling artifacts in time- and frequency-domains

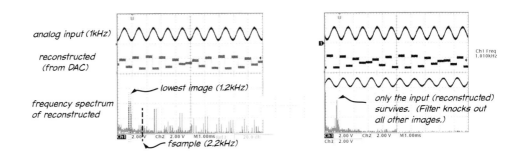

Figure 20S.6 Sampling just 10% above the Nyquist minimum: barely succeeds. (Scope settings: 2 V/div; frequency spectrum FFT, 1.25 kHz/div.)

is barely able to knock out the lowest image, and in fact attenuates the signal slightly as it eliminates that image.

The left-hand scope screen shown in Fig. 20S.6 shows how very close the lowest image lies to the true signal: at 1.2 kHz, just 200 Hz above the signal. This filtering task, though challenging, is not as difficult as the imposed CD audio standard, because CD resolution is 256 times finer than what we are demonstrating with this 8-bit conversion, and therefore calls for greater filter attenuation.[6] Soon after establishing the CD standard, the CD's developers[7] in effect conceded that they had given themselves a task too difficult for analog filters. The industry began, in so-called "second generation" CD players, to use digital "oversampling," interpolating the digitally-calculated samples that we referred to in §20S.2.2.

Inadequate sampling rate: Inadequate sampling – violating Nyquist's requirement that one pick up a little more than two samples per period of the highest frequency of interest[8] – leads to a disaster called "aliasing," as we said in §20S.2.1. When this occurs, an input frequency gets "folded" from its true high-frequency to a spurious low-frequency image.

Aliasing illustrated: Nyquist warns us that we had better sample a 1 kHz signal at better than 2 kHz. If we defy him, we get into trouble. In the case shown in Fig. 20S.7, f_{in}=1 kHz, f_{sample}=1.2 kHz. This sampling rate is far below Nyquist's minimum required rate and produces an *alias*: an artifact at 200 Hz. This is an image at $f_{sample}-f_{in}$. This image lies below the truth, and this fact makes it damaging.

The third trace in Fig. 20S.7 shows a convincing sinusoid at 200 Hz – but it is obviously a phoney.

---

[6] One might fairly protest, however, that a full-amplitude signal at 20 kHz is impossible in music, so the filter attenuation that is required is actually not as great as the contrast between our 8-bit conversion and CD-audio's 16-bit conversion.

[7] The CD standard was developed by Philips, who had introduced the large "laser disc" as a medium for movies, and Sony, who had introduced the portable audio device, the "Walkman" (it was a pocket-sized tape player with headphones).

[8] The phrase, "highest frequency of interest" may bear some explaining.
First, an elementary point: while we illustrate sampling cases in this chapter by applying a single sinusoid as input, real applications normally apply a range of frequencies as input to the ADC. If the highest frequency is sampled adequately, then so necessarily are all lower frequencies, frequencies, as we said in §20N.4.1.
Second – a much subtler point – it is not invariably true that one need sample at better than twice the highest input frequency. Strictly, instead, one must sample at better than twice the frequency *bandwidth* of the signal applied to the ADC. When the input frequency range runs from essentially zero to f_{max}, as in audio applications, the two formulations lead to the same result. But cases do arise where the difference between the two formulations matters. Sometimes one's goal is to sample *a narrow bandwidth at high frequencies*, and in this case the distinction between the two formulations becomes important. If, for example, one needs to convert signals between 1 MHz and 1.1 MHz, one needs a sampling rate of a bit better than 200 kHz, rather than 2.2 MHz. See AoE §13.6.3.

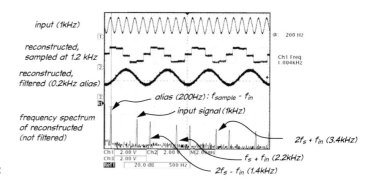

Figure 20S.7 Inadequate sampling rate produces aliasing, an artifact too low to filter out. (Scope settings: 2 V/div, 2 ms/div; frequency spectrum FFT, 500 Hz/div.)

We can see that the only true signal is the 1 kHz sinusoid shown on the top trace, and in the output frequency spectrum. The fact that under-sampling captures the input is quite counterintuitive; it might seem that we are not getting enough information to describe the input fully. But the sampled points, which define the alias, must lie on the 1 kHz original as well; that is where they came from, so 1 kHz is another component of the reconstructed output waveform.[9] Sampling has created the 200 Hz fraud, which of course shows up also in the frequency spectrum. The alias is deadly: a lowpass filter cannot remove it without also eliminating the genuine *signal*.[10]

20S.3.1 Aliasing: details

The filtered signal includes alias... and you can hear it: If this reconstruction is lowpass-filtered to include the genuine signal, as normally it would be after it emerged from a DAC, the filtered result also includes the alias. Figure 20S.8 shows that, in this case, the 200 Hz alias is larger than the true signal at 1 kHz. Together, they sound like a low tone with a harsh, metallic quality (contributed by the 1 kHz): not what a person wants to listen to.

Normally, aliasing is precluded by insertion of a lowpass filter ahead of the ADC. This "anti-aliasing" filter blocks any frequency that cannot be sampled adequately. In these illustrations we omit such a filter because, today, aliasing interests us. Usually it only threatens us.

20S.4 Explanation: the images, intuitively

The rest of this chapter is an attempt to give an intuitive explanation for the fact that sampling introduces artifacts ("images") that appear at $n \cdot f_{\text{sample}} \pm f_{\text{in}}$.

20S.4.1 Sampling resembles amplitude modulation

Radio amplitude modulation – "AM" – provides an analogy that helps.[11] In AM radio, a signal "modulates" the carrier. We will argue in the coming pages for the similarity between such modulation and the sampling of an analog signal. Amplitude modulation – the oldest and simplest way to impose a signal upon a radio "carrier" – multiplies the carrier by the "modulating" signal; see Fig. 20S.9.

---

[9] A point that is surprising, and rarely mentioned in the literature on sampling.
[10] This bad news ought perhaps to be qualified for certain special cases: if an alias is entirely predictable – as it is if the high-frequency input is in a known and narrow frequency range, then perhaps a narrow bandstop filter could be applied. And there are rare cases where the aliasing is useful: a radio carrier frequency can be deliberately aliased, accomplishing what an RF "mixer" ordinarily is used to achieve, but the input signal must be strictly bandwidth limited. See AoE §13.6.3.
[11] Simpson, R., *Introductory Electronics for Scientists and Engineers*, 2nd ed. Allyn & Bacon, (1987), pp. 720–721.

20S.4 Explanation: the images, intuitively

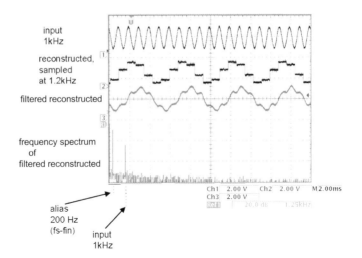

Figure 20S.8 Effect of alias: in this case, second tone forming a chord with the true signal; in the usual case, a range of aliased frequencies would have its frequency spectrum *inverted*. (Scope settings: 2 V/div; frequency spectrum FFT, 1.25 kHz/div.)

Figure 20S.9 AM modulation multiplies modulating frequency times carrier.

$$\text{AM signal} = \underbrace{(1 + \overbrace{m\cos\omega_m t})}_{\text{this preserves the carrier itself}}\underbrace{\cos\omega_c t}_{\text{carrier}}$$

A little trigonometry indicates that the multiplication applied in amplitude-modulation will produce (along with the carrier itself) a sum of the two frequencies, and their difference:

$$\cos\omega_1 \cdot \cos\omega_2 = \frac{1}{2}\cos(\omega_1 + \omega_2) + \frac{1}{2}\cos(\omega_1 - \omega_2)$$

or, more compactly,

$$f_{\text{sidebands}} = f_{\text{carrier}} \pm f_{\text{modulation}}$$

The "$1+\cdots$" in the equation of Fig. 20S.9 appears because (as the figure says) ordinary AM radio deliberately *preserves the carrier* in order to ease the receiver's job. This is convenient, but not essential.

A straight multiplication, without the $1+\cdots$, would produce just the *sidebands*.[12] No *carrier* would appear in the output.

That result is very similar to the effect of our digital sampling, which produces "images" that are sum and difference of input and sampling-rate. Input takes the role of the modulating signal in AM; sampling pulses take the role of carrier. In sum, we argue that our sampling "images" at $f_{\text{sample}} \pm f_{\text{in}}$ are equivalent to the AM sidebands at $f_{\text{carrier}} \pm f_{\text{modulation}}$.

We'll also discover soon that sampling includes an offset term, a sort of "1" like the $1+\cdots$ of AM, in order to preserve the original input that is to be recovered from the digital form; but let's work up to that point gradually.

Figure 20S.10 shows images of AM radio in which a *carrier* is amplitude-modulated by the signal. The sum and difference frequencies appear, in the frequency spectrum of the product, as "sidebands."[13]

---

[12] Some radio modulation schemes do dispense with transmitting the carrier; others suppress one of the two sidebands (single-sideband or "SSB"). Both variations are more power-efficient, but less easy to demodulate.

[13] Simpson, op. cit. p. 120.

Figure 20S.10 Amplitude modulation produces "sidebands" that straddle the carrier frequency. *The AARL Handbook for Radio Amateurs*, 17th ed., (1993), pp. 9–4, 9–5. Images, copyright AARL, used with permission.

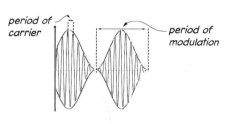
carrier with 100% amplitude modulation

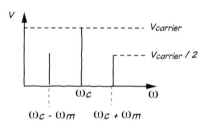
fourier components of 100% modulated carrier

Scope images are more persuasive than sketches, so here are some AM waveforms. Figure 20S.11 first shows an unmodulated carrier (not very interesting), then, to the right, what modulation looks like in the time domain (familiar to you from Lab 3L). Below those scope traces are frequency spectra showing first the carrier alone, then the carrier and sidebands: sum and difference. The sidebands are at $f_{\text{carrier}} \pm f_{\text{signal}}$, where the "signal" in this case is the modulating audio.

The modulation shown in Fig. 20S.11 differs from that sketched in Fig. 20S.10 in that the modulation of the scope image is not a single frequency, but the (more typical) case: a collection of frequencies constituting the signal. In this case, the signal is speech.

In the coming pages we will show some images of AM-like multiplication, done in the lab with an analog multiplier chip. The goal of this examination of the effects of *multiplication* of two waveforms is to try to persuade you that sampling is a very similar process, and thus not much more mysterious than radio's amplitude modulation – a technique that, by now, is an old friend of yours.

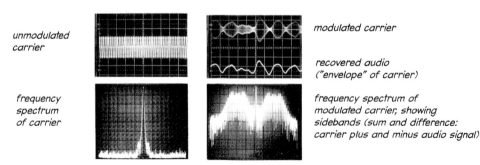

Figure 20S.11 AM radio: "sidebands" straddling carrier resemble conversion "images" that straddle f_{sample}.

20S.4.2 A reminder: spectrum of a sine versus pulse train

More specifically, sampling resembles multiplying an input by a sequence of narrow *pulses* at the sampling rate.[14] So it is worth recalling what the frequency spectrum of such a pulse train looks like.

Sinusoid versus pulse train: The frequency spectrum of a sinusoid (a single frequency) is familiar: one spike. In the left-hand scope image in Fig. 20S.12, such a solitary spike appears at 1 kHz:

In the image to the right in Fig. 20S.12 appears the spectrum of a train of pulses, about $20\,\mu s$ wide, repeating at 1 kHz (the sweep rate is 4× slower in the pulse image). Again we see a spike

---

[14] For the closest analogy to AM radio, we should perhaps state this point backwards: "sampling resembles multiplying a sequence of narrow pulses – whose role resembles the role of the radio *carrier* – by the modulating analog input." We have stated it in the other form, because we usually think of sampling as a process applied to an input, not to the sampler. But the two notions are, of course, equivalent.

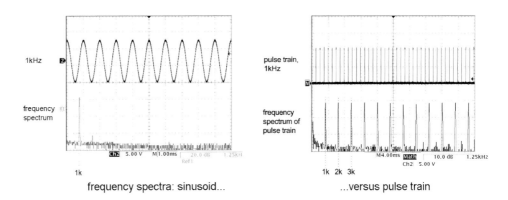

frequency spectra: sinusoid... ...versus pulse train

Figure 20S.12 Frequency spectra contrasted: sine at 1 kHz versus narrow pulse repeating at 1 kHz. (Scope settings: 5 V/div; frequency spectrum FFT, 1.25 kHz/div; sweep rates: left-hand image 1 ms/div, right-hand image 4 ms/div.)

in the frequency spectrum at 1 kHz, as for the sine – but in addition the spectrum shows spikes at 2 kHz, 3 kHz,..., repeating forever. That is, they repeat at integral multiples of f_{in}, where f_{in} is the repetition-rate of the input pulses.

Effect of multiplying a signal by another waveform:

Sine × sine (like AM): You recall that if we multiply two sinusoids we get sum and difference frequencies. Fig. 20S.13 gives two images saying that. One shows a simple case: $f_1=1$ kHz, $f_2=4$ kHz. The difference and sum products appear: 1 kHz and 5 kHz. (Neither contributing signal survives.)

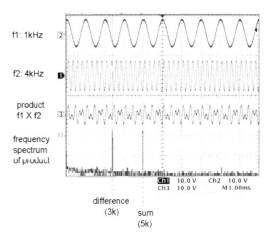

Figure 20S.13 Product of two sinusoids: sum and difference. (Scope settings: 10 V/div, 1 ms/div; frequency spectrum FFT, 1.25 kHz/div.)

Sine × sine with offset: If we multiply a sine at 1 kHz by another sine at 4 kHz – the case we just looked at – but we add a constant to the latter, then we get the sum and difference frequencies, as before, and in addition we get the *original*, the 1 kHz sine. That sounds as if it might be useful, does it not? – given that our goal is to recover an *original* sampled waveform!

This result, in Fig. 20S.14 does, indeed, resemble what we get – and require – in sampling. The

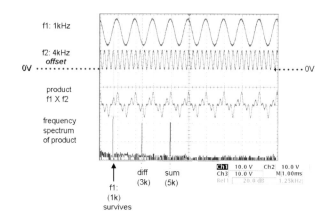

Figure 20S.14 Sine multiplied by sine-with-offset: the result is images – and also the *original*. (Scope settings: 10V /div, 1 ms/div; frequency spectrum FFT, 1.25 kHz/div.)

original is what we're after; the sum and difference frequencies are artifacts, and these we plan to strip away.[15]

Sine × pulse train (more like sampling): We've already shown that in the frequency domain the pulse train looks like sinusoids at $f_{in}, 2f_{in}, 3f_{in}, \ldots$. So you'll not be surprised to see that the spectrum of the product of sine × pulse-train looks like sine × sine endlessly repeated – resulting in repeating differences and sums. The offset applied to the pulse-train makes sure that the original 1 kHz sine also appears.

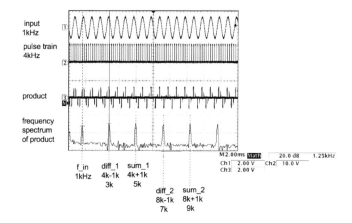

Figure 20S.15 Product: sine × DC-biased pulse train → lots of sum-and-difference pairs – *plus the original sine.* (Scope settings: 2 V/div, 1 ms/div; frequency spectrum FFT, 1.25 kHz/div.)

The survival of the original is surely good news! When we *sample*, this is our *entire* goal: to recover that original signal from among the artifacts. To clear away the artifacts, we need only a good lowpass filter.

To recapitulate: in A/D conversion, the *sampling pulses* play the role of the AM *carrier*; these pulses (whose amplitude is constant) are modulated by the analog input, which plays the role of the *modulating signal* in AM.

The repetition of images around integral multiples of f_{sample} – repetition that does not occur in AM radio – occurs because the sampling pulses are not sinusoidal, but instead are narrow pulses with their many Fourier components: components that keep reappearing at $f_{sample}, 2f_{sample}, 3f_{sample}, \ldots$.

---

[15] We hope you're not rattled by the difference, here, from AM radio. There, it is the *carrier* that is preserved by the offset of the modulating signal. Here it is the modulating signal that is preserved by the offset in the carrier-like pulse train. We are arguing that sampling *resembles* AM, not that the processes are identical.

20S.4 Explanation: the images, intuitively

Readers adept at mathematics may recognize this as simply the convolution of spectra (in the frequency domain) of the product of the sampled waveform with a periodic delta function (in the time domain), a consequence of the convolution theorem.

20S.4.3 The punchline: the sampled data looks a lot like the multiplied data

Figure 20S.16 shows two scope images that support the claim that sampling resembles multiplication by a pulse train. (The sampled waveform has been reconstructed with a DAC, as you can see.)

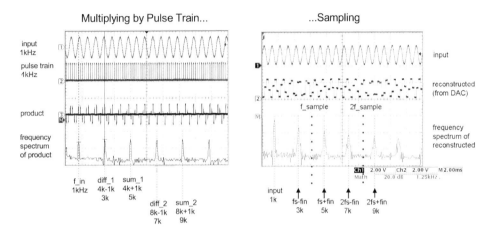

Figure 20S.16 Sampled sine (1 kHz, f_{sample}=4 kHz) versus multiplied sine (1 kHz × pulses at 4 kHz). (Scope settings: 2 V/div, 1 ms/div; frequency spectrum FFT, 1.25 kHz/div.)

The frequency spectra match: the spectrum of the waveform that was *sampled and then reconstructed as analog* matches the spectrum of the *product of sine and pulse train*. This is the result we were pursuing in §20S.4.

20S.4.4 A refinement or correction to the claim

The sampled and reconstructed output (from an output DAC) differs from the multiplied waveforms we have just been discussing in one respect: whereas the sampled output is sustained between *sampled*

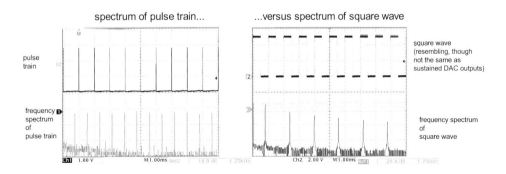

Figure 20S.17 A heuristic argument for the rolloff of sampling images. (Scope settings: left image, 2 V/div, 1 ms/div; frequency spectrum FFT, 1.25 kHz/div; right image same except 2 V/div.)

points, the product (as in Fig. 20S.16) is not sustained, but reverts to zero between multiplying pulses.

The sustaining of the sampled level, as in normal reconstruction of analog from digital, has the effect of lowpass filtering the reconstructed result. The frequency spectrum of the DAC reconstruction shows this lowpass effect: the image amplitudes diminish with frequency, whereas the spectrum of the *multiplied* waveform shows no such fall-off.

Figure 20S.17 attempts an argument by analogy to support this contrast. The figure contrasts the frequency spectra of *pulse train* versus *square wave*. Granted, the square wave is not the same as the sustained "steppy" waveform produced by a DAC; but it resembles it more closely than it resembles the pulsed output. This is only an argument by analogy, trying to make plausible the observed effect; the rolloff of the frequency components of the DAC output.[16]

To compensate for this rolloff – which applies to signals in the frequency band of interest as well as to the artifacts that are to be eliminated, a filter applied to the DAC output should include a slight high-frequency boost in the passband to compensate. Output filters for digital audio devices do include such a boost.

---

[16] Thanks to our colleague Jason Gallicchio for suggesting this heuristic argument – though we can't blame him if you find the argument a stretch.

20W Worked Examples: Analog ↔ Digital

20W.1 ADC

20W.1.1 Size of *error* versus size of *slice*

If a voltage range is sliced into n little slices, each labeled with a digital value, the value will be correct to *1/2* the size of the slice: 1/2 LSB, in the jargon. Perhaps this is obvious to you. If not, try this example.

Example How many bits are required for 0.01% resolution?

Solution 0.01% is one part in 10,000. If a converter spans a 5 V input range, for example, we mean that when we give a digital answer, like "4.411 V," we expect to be wrong by no more than 0.5 mV (1 part in 10,000 of the *full-scale* range: 0.5 mV/5 V).

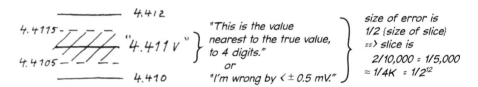

Figure 20W.1 What I mean when I claim to be correct to 0.01%: I say the answer is 4.411 V; I could be wrong by ±0.5 mV.

How many bits does the converter need? We can tolerate slices that are *two* parts in 10,000 wide, or 1/5k. 12 bits give 4K slices (4096), and give an *error* of 1/8K or 0.012%: this does not quite satisfy the specification. One more bit – 13 – would cut that error in half to 0.006%, going well beyond the minimum requirement. So use 13 bits.

20W.1.2 ADC application: digital audio

CDs established a standard that still serves for pretty-good audio: 16-bit samples at 44.1 kHz (about 10% above the theoretical minimum to get 20 kHz music). Let's calculate about how much data had to be stored on a standard CD.[1]

Example How many bytes had to be stored on a 74-minute audio CD?

Solution Two bytes per channel; four bytes per stereo sample. So,

number of bytes = 2 bytes/channel × 2 channel/sample × 44.1k sample/sec × 74 min × 60 sec/min

≈ 783 million bytes ≈ 747M or 747 MB

---

[1] According to legend, the 74 minutes was chosen by Sony to permit recording Beethoven's 9th Symphony: a curious concern, given the likely audience for most CDs. See Ken C. Pohlmann, *Principles of Digital Audio*, 3rd ed., (1995), p. 265.

20W.1.3 A more interesting application; noise and filter considered

Converting a noisy signal: Suppose someone asks you to specify an ADC to transmit some music digitally. The music amplitude variations are to be distinguished to about 50 dB (−50 dB relative to full-scale, "−50 dBFS:" much lower resolution than the CD audio standard), but you are to reproduce the full conventional 20 Hz to 20 kHz frequency range. Here are the details.

Example If the resolution we look for is about 50 dB, between loudest and quietest passages, how many bits do we need to use?

Solution We want the quietest music – and the magnitude of the least significant bit (LSB) – to be 50 dB below full-scale. We can plug that number into the definition of dB:

$$\text{LSB/full-scale} = -50\,\text{dB};\ -50\,\text{dB} = 20\log_{10}\frac{A2}{A1};\ \log\frac{A2}{A1} = -\frac{50}{20};\ \frac{A2}{A1} = 10^{-(5/2)} = \frac{1}{316}.$$

So, we need enough bits to provide about 315 slices. Eight bits allow 256 slices: not enough. So we'll use nine bits, enough for 512 slices.

We could get the same result using a shortcut from the starting *dB* specification. We could use the fact that each additional bit halves the size of a slice, shrinking it by about 6 dB. So, 50 dB requires a little more than eight bits, by this calculation: nine to be safe.

Straightforward case: no noise: Let's suppose the signal is pretty clean: nothing substantial above 20 kHz.

Problem What sampling rate?

Solution The sampling rate must be adequate to acquire the highest frequency of interest, 20 kHz in this case. A naive view of Nyquist's sampling rule would say that we need to sample at 40 kHz to get two samples/period. That answer is wrong.

The sampling rate must take account of the limited slope of any filter that is assigned the task of keeping the highest signal while attenuating the lowest image. Such filtering is sketched in Fig. 20W.2. The filter we use in this problem is one that you met in lab Lab 12L. It is, indeed, steep, as Fig. 20W.3 shows.

The filter will need to attenuate the lowest image to under 1 LSB. The filter's response curve in Fig. 20W.3 shows attenuation of −60 dB at about 20% above the frequency where the filter begins to roll off (the "corner frequency," roughly f_{3dB}). That is sufficient (and we probably can't squeeze a more precise answer from the plot of rolloff). So let's make sure that the lowest image is 20% above the highest frequency of interest. That would put the lowest image at 24 kHz.

What sampling rate puts the lowest image there? Recall the rule that the lowest image appears at $f_{\text{sample}} - f_{\text{in(max)}}$. Plugging in the numbers in our case, we find $f_{\text{sample}} - f_{\text{in(max)}} = 24$ kHz; and $f_{\text{sample}} = 24$ kHz $+ f_{\text{in(max)}} = 44$ kHz. It's not surprising that we should land close to the sampling rate of the audio CD – but it was by chance that we arrived as close as we did.

Note that the MAX294 plot in Fig. 20W.3 shows $f_{3dB}=1$ kHz, but f_{3dB} is adjustable from under 1 Hz to 25 kHz based on the input clock frequency; the plot can be scaled to whatever f_{3dB} you choose.

---

[2] Pohlmann, op.cit. p. 265.

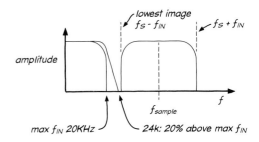

Figure 20W.2 Filter rolloff dictates how close image may come to top input frequency.

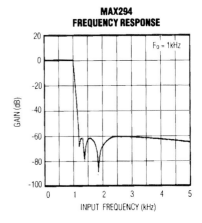

Figure 20W.3 MAX294 frequency response.

A nastier case, noise is added: Now suppose that your converter is fed the strikingly noisy signal shown in Fig. 20W.4, rather than the clean music signal you were led to expect. You may recognize the scope image as a portrait, in time and frequency domains, of music transmitted and received via the infrared audio project on Lab 13L. The signal is polluted by vestiges of the 30 kHz *carrier*.[3]

Let's suppose that our goal is, as before, to convert this and similar signals to digital form, for transmission (by wire) over a longer distance.

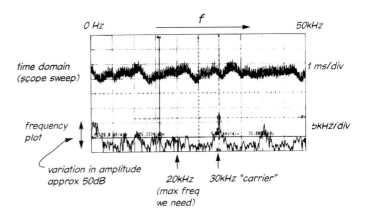

Figure 20W.4 Frequency spectrum of some music.

Problem What happens if you omit an anti-aliasing filter?

Given the sampling rate you prescribed for the "clean" case, qualitatively speaking how would the

---

[3] The alert student will also recognize the underlying music as a phrase from a solo by Miles Davis in *So What?*, on the classic album, *Kind of Blue*. (*Just kidding!*)

30 kHz *carrier* noise appear or sound when your digital transmission was re-converted to analog? What happens to the other large *noise* peak that appears above the carrier frequency, in the frequency spectrum?

Solution *You* wouldn't forget to put an anti-aliasing filter ahead of the ADC; but we did. Assuming a clean signal (or the presence of an anti-aliasing filter), we prescribed a sampling rate of 44 kHz. The 30 kHz carrier frequency shows up as a large peak just right of the center in Fig. 20W.4. This would produce its lowest *image* at $f_{sample} - f_{in(max)} = (44\,\text{kHz} - 30\,\text{kHz}) = 14\,\text{kHz}$. It would be a high but audible whine. The smaller peak at about 40 kHz would produce an image at 4 kHz – even more troublesome. So the solution is to not omit the anti-aliasing filter!

21N Digital Project Lab

Contents

21N.1	**A digital project**		**851**
	21N.1.1	Multiplex the four-digit LED display	851
	21N.1.2	Make some noise	853
	21N.1.3	Use the display to build something fun	853

21N.1 A digital project

The Digital Project Lab is an open-ended two day lab session that gives you the opportunity to design and build something of moderate complexity using the WebFPGA and any of the components and techniques we have discussed in the course so far. Chapter 21L discusses several ideas for projects. You may also want to look at Chapter 17W as well for examples of issues that may arise when using counters.

However, first we would like you to build a four-digit, BCD counter to count from 0000 to 9999 in the WebFPGA. This forms the basis for most of the projects we describe in the lab.[1]

21N.1.1 Multiplex the four-digit LED display

AoE §10.6.2

We would first like you to expand the seven-segment LED display you added in Lab 18L into a full four-digit display (0000 to 9999). To use all four digits of the LED display, you will need to add electronic switches to the anodes of the multiplexed display so that you can turn each digit on or off individually with a digital signal. Figure 21N.1 shows one way to accomplish this.

The hardware: In Lab 18L, you connected the anode of the single seven-segment digit you were using directly to 5 V provide the voltage for the LED V_F and the constant-current TLC5916 display driver output compliance. In order to multiplex the display, you will need to supply +5V to each anode sequentially using a *high side anode switch*. This is a fancy name for a P-channel MOSEFT. "High side" just means it switches the supply voltage to the load rather than connecting the load to ground with an N-channel switch.[2] We use the ZVP3306 ($V_{GS(th)min} = -1.5$ V; $V_{GS(th)max} = -3.5$ V @ 1 mA) for the anode switches, but any small-signal P-channel MOSFET with similar $V_{GS(th)max}$ such as the BSP250P should work.

You will connect the source of each of four MOSFETs to +5 V and the drains to each of the four anodes in the display. To light up the segments in a particular digit, the gate of that MOSFET must be

---

[1] Of course you are welcome to build something other than one of our suggestions but we would like you to try to incorporate the four digit display in some fashion in whatever you come up with.

[2] The display LED V_F is specified as 2.4 V max and the TLC5916 only requires 0.5 V so you could have gotten by with connecting the anode to 3.3 V in Lab 18L. However, here when we add the high-side switch, the MOSFET $R_{DS(on)}$ adds another volt or so drop at max current, increasing the source voltage required to 5 V.

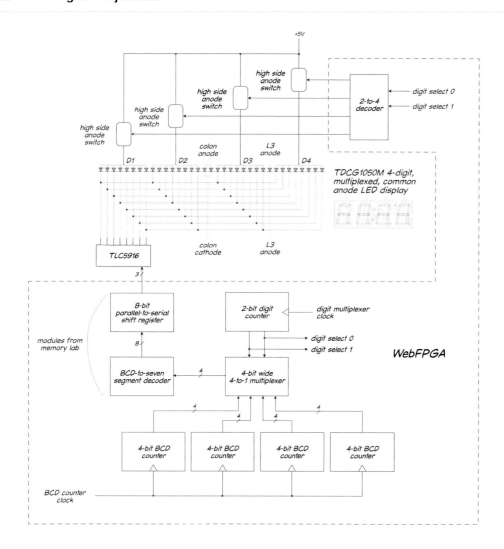

Figure 21N.1 Block diagram of a four digit BCD counter with multiplexed seven-segment display.

pulled down from +5 V by more than 3.5 V; to turn it off, the gate must be within 1.5 V of +5 V.[3] That means the anode control signal at the MOSFET gate must be ≤ 1.5 V to turn on the digit and ≥ 3.5 V to turn it off.

A WebFPGA output can never be higher than 3.3 V, making impossible to ensure the anode switch will turn off if we connected it directly to the MOSFET gate. To be sure we can turn a digit off, we need a level translator to convert the +3.3 V high output of the FPGA to something closer to +5 V. An easy solution is to use an HCT part running on a +5 V supply. HCT logic has TTL input levels; it sees any input ≥2 V as a high, meaning it works fine with 3.3 V CMOS outputs.[4] A 74HCT04 hex inverter has maximum low and minimum high output levels of 0.1 V and 4.4 V respectively on a 4.5 V supply. This guarantees that the anode switches work correctly: see Fig. 21N.2. Using an inverter also makes

---

[3] We say "more than" because at $V_{GS(th)} = -3.5$ V the worst case drain current is -1 mA, too little for our display. However, we will be turning the MOSFET on with $V_{GS} \approx -4.9$ V. The datasheet shows a typical device saturation characteristic of $I_D \geq 60$ mA with $V_{GS} = -4.5$ V and $V_{DS} = -2$ V, more than enough for our display.

[4] HCT logic considers anything less than 0.8 V as a low. The FPGA low output voltage is specified as 0.2 V max at low output currents, so that works as well.

21N.1 A digital project

the WebFPGA signals active-high; a high-level out of the WebFPGA turns on the respective anode switch.

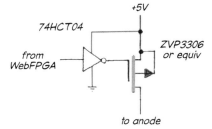

Figure 21N.2 High-side anode switch for multiplexed seven-segment display.

The Verilog code: You will reuse the seven-segment decoder and parallel-to-serial shift register you created in Lab 18L; however, now you need to add a two-bit counter, a multiplexer and a decoder to the WebFPGA to light up all four digits: see Fig. 21N.1. The counter keeps track of the active digit and selects the multiplexer and decoder outputs sequentially. The multiplexer switches the seven-segment decoder input between four 4-bit sources. The 2-to-4 decoder selects one of the four anode drivers.

The source for the multiplexer inputs can be either counters or registers, depending on your final application. You may have to modify your parallel-to-serial converter so it is triggered by the 2-bit counter clock since, even if a digit's value has not changed, the input to the converter now changes with each count.

21N.1.2 Make some noise

Projects often require audio feedback either to provide user feedback, to indicate an error, or to provide an alarm. Any easy way to get a beep tone is to use a self-contained DC buzzer. These devices generate a tone when voltage is applied. They typically run on +5 V and require more current than the FPGA can provide, so you will need to drive them with a MOSFET: see Fig. 21N.3.[5] The buzzer is polarized, so be careful to connect the terminal marked "+" to 5 V.

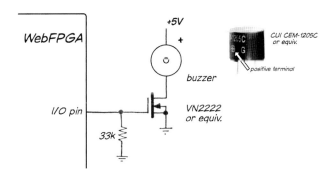

Figure 21N.3 DC buzzer connection to WebFPGA.

21N.1.3 Use the display to build something fun

Once you have a working four-digit display, use it to create something useful, interesting or just fun. We have included some suggested projects in Lab 21L.

---

[5] Any small-signal MOSFET with a $V_{GS(th)} \leq 2.5$ V should work.

21L Lab: Digital Project

This lab is divided into two parts. In the first part everyone will add hardware and Verilog code to implement a four-digit multiplexed seven-segment display. In the second part, you will create something interesting with this display.

21L.1 Multiplex the four-digit LED display

21L.1.1 Build the hardware

In Lab 18L you connected a TLC5916 constant-current LED driver between the WebFPGA and the segment LED cathodes to drive a single digit of the four-digit display. You also wrote Verilog code to convert a hex value into the correct segments to turn on to display that value and sent them to the LED driver with a parallel-to-serial shift register (the hardware and WebFPGA blocks labeled "A" in Fig. 21L.1). Now, add a 74HCT04 and four ZVP3306 P-Channel MOSFETs to your breadboard to allow the WebFPGA to control power to the four digit anodes (be sure to disconnect the anode you used in the earlier lab from +5 V). The 74HCT00 must be powered from the +5 V bus.[1] Connect the inputs of the level shifters to four pins on the FPGA. As with the segment drivers, we connected the WebFPGA to the anode switches in whatever order made the wiring easiest and corrected the order in the pin assignment code later. This new hardware along with the WebFPGA decoder to drive it is labeled "B" in the figure.

21L.1.2 Test the hardware

Modify your single digit display autostart code from step 7 in §18L.2.3 to select one of the four digits to display the hex output. Use two of the available logic switches to select which anode switch to turn on. This should be a minor modification of your working code from the earlier lab. You will need to decode the two-bit input from the logic switches and set one of the four anode outputs high with a 2-to-4 decoder. A decoder in Verilog takes only a single line of code:

```
assign ANODE = 4'b0001 << asel;  // asel is a 2-bit value for the digit to light
```

We will have you multiplex the display in the next section; for now, just select one digit at a time manually to make sure your anode switching hardware works. (You have probably figured out that we believe in testing one thing at a time.) You should note which decoder input value lights which digit so you can order your output correctly later when assigning WebFPGA pins.[2]

---

[1] You must use a 74HCT device to provide the required level shifting but almost any inverting 74HCT gate will work. For example, a quad 74HCT00 or 74HCT02 will work as long as you tie the two inputs of each gate together to create an inverter. You can even use a noninverting HCT gate if you modify the Verilog for active-low anode driver outputs.

[2] Our complete test code is available at https://LAoE.link/FPGA/21L_Anode_Test.v

21L.1 Multiplex the four-digit LED display

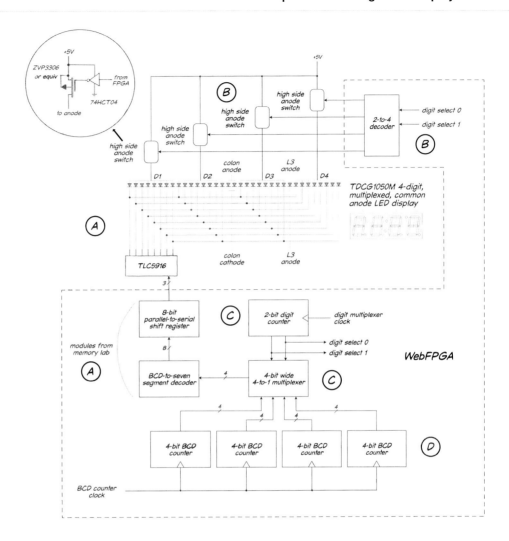

Figure 21L.1 Add high level anode drivers to seven-segment display.

21L.1.3 Multiplex the display

Modify your test code from the previous section to display four multiplexed digits on the display. The displayed value may be hard-coded (we displayed "3.145"). This will require creating a digit multiplexer clock, adding a two-bit counter to sequence through the digits, a 4-bit wide 4-to-1 multiplexer to select one of four hex input values to send to the seven-segment decoder, and possibly a modification to your parallel-to-serial (P2S) converter to output data on every rising edge of the multiplexer clock: see items "C" in Fig. 21L.1.

You will need to refresh each digit fast enough that the user does not see any flicker. Human persistence of vision is on the order of 30 ms, so you will need to run the digit clock at least four times faster (since you have to refresh all four digits in that time). Experiment to see what works best.

21L.1.4 Add BCD counters

Replace the fixed data displayed on the LED display with four cascaded BCD counters. Add a 4-bit BCD counter module to your code. (We modified our 74HC161 counter from Lab 17L.7.1 into a 74HC160.) Instantiate four of these counters and connect them together to form a decimal counter that counts from 0000 to 9999. Connect the output of each counter to the 4-to-1 digit multiplexer. You should see the count displayed on all four LED digits. Here is what our top level module definition looked like:[3]

```
module MUX_COUNTER(
  input WF_CLK,           // used for internal timing
  input CLK,              // clock for four digit BCD counter
  input CLEAR,            // asynchronous active high BCD counter reset
  input CTR_ENABLE,       // active high BCD counter enable
  input LZ_BLANK,         // if high, leading zeros are blanked
  output TLC_SDI,         // TLC5916 outputs
  output TLC_CLK,
  output TLC_LE,
  output [3:0] ANODE      // active high anode switches
);
```

Enhancements: You should now have a working four-digit decimal counter to use in one of the projects that follows (or a project of your choosing). While we don't expect you will want to waste any time now on the following enhancements, you may find some of them useful later depending on what project you choose:

- Leading zero blanking. Normally, decimal numbers are displayed with no leading zeros or with only a single zero ("56" rather than "0056" or "0.1" rather than "000.1"). You may want to add leading blanking to your display. The rules for leading zero blanking are:
  - For integers, only display a single right-justified "0" when the value is zero. Otherwise do not display any leading zeros.
  - For values including a decimal point, display all zeros to the right of the decimal point. Display only a single "0" to the left of the decimal point when the value is less than one.

  This is a bit trickier than it sounds but can be done with clever combinational logic to set the blanking input to the decoder for each digit.

- Decimal point manipulation. If you need to display a fixed decimal point, you will have to arrange your code so the decimal point is selected in the decoder for the proper digit. We added a decimal point flag input to our decoder that turned on the decimal point when asserted. Our code includes a two-bit vector that selects which decimal point to display where 00 is the decimal point of the leftmost digit and 11 turns off all decimal points. (We did not think we ever needed the decimal point to the right of the rightmost digit.) This allows us to select the location of the decimal point dynamically.

- Overflow handing. Our counter displays "----" when the count increments from 9999. It stays this way until the counter is reset. This requires a flag to capture overflow (the falling edge of RCO out of the most significant counter) and a modification to the seven-segment decoder to add a "dash" input that overrides normal numeric display.

---

[3] The LZ_BLANK flag turns leading zero blanking on and off. See "enhancements" following.

21L.2 Some project ideas

21L.2.1 Capacitance meter ("C-meter")

The scheme is this: use a LMC555 timer[4] to generate a pulse whose period is proportional to the capacitor-under-test (CUT) and adjust the BCD counter frequency so the display reads capacitance directly in nanofarads. Your design goal should be to run continuously, displaying the result on the seven-segment display. For example, a 0.1 μF cap might read "100.0", providing a range from 100 pF to 999.9 nF.

We discussed using the '555 as an oscillator (i.e., astable mode) in Chapter 8L. Here we want to use it in monostable mode to generate a single pulse in response to a trigger signal. The circuit we will use is shown in Fig. 21L.2.

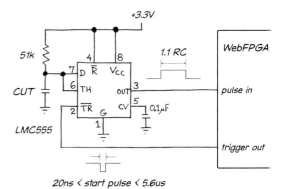

Figure 21L.2 '555 timer configuration to generate a pulse proportional to capacitance.

This circuit outputs a positive pulse with a period of p = 1.1 ∗ RC in response to a negative going trigger pulse. The output pulse is used to enable the decade counters in the WebFPGA which are set up to count to 10,000 with a 1 μF capacitor: see Fig. 21L.3. This allows the meter to display capacitance values between 100 pF and 999.9 nF with the values shown. The multiplexer, decoder and P2S converter show the four output digits on the LED display.

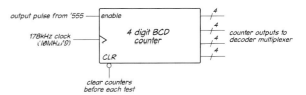

Figure 21L.3 Capacitance meter counter connections.

You will want to design an FSM in the WebFPGA to implement the capacitance meter. For each measurement:

1. The meter should zero the counters.
2. The meter then triggers the LMC555 with a *negative going pulse* having a width of at least 20 ns but less than the shortest possible output pulse (i.e., <5.6 μs).
3. The counters will be enabled by the LMC555 output pulse going high and count the 178 kHz input clock while the LMC555 output pulse remains high.

[4] You must use the CMOS "LMC" version of the timer, not the bipolar "LM" version, since it has to run on 3.3 V in order to interface with the FPGA.

4. When the pulse goes low again, the displays should be updated (and latched) to show the counter value. This will be the capacitance in nF (with the resistor and clock frequency values shown).
5. If you don't latch the counter outputs, the user will see the meter count for about 56 ms at the maximum capacitance.
6. If the value is invalid, the display should show "----".
7. The system should repeat the measurement forever. (Three times per second is a reasonable repetition rate.)

We suggest first designing the Verilog code that takes the four BCD values (plus the location for the decimal point and the signal that the value is invalid) and displays the result independently of the FSM driving the capacitance measurement. This requires modifying the basic system you built in §21L.1.4 to handle a decimal point and the "----" display.

Once you can measure capacitance, you can program the FSM to send the start pulse. You may want to start with manually triggering a measurement with a button input to the FSM and then making it continuous once you get that working.

You can manually change the range by using a 5.1 k resistor to get a 1nF to 9.999 μF range or a 510 k resistor to get a 10 pF to 99.99 nF range or by changing the clock frequency by a decade up (to show smaller values) or down (to show larger values). The latter is easier if you want to use switches to set the range but increases the count interval to over 560 ms for the largest value capacitor, limiting the measurement repetition rate.

If you are feeling ambitious, you can modify your design to autorange by either changing the clock frequency to get the optimal range (1.78 MHz for a 10 pF to 99.99 nF range; 17.8 kHz for a 1 nF to 9.999 μF range) or just increase the width (i.e., number of decades) of the counter and only display the four most significant digits captured by the counter (this is how we implemented autoranging). You will need to move the location of the decimal point in the display if you autorange.

In any event, start with the simplest version and add enhancements once you have everything working.[5]

21L.2.2 Reaction timer

We are somewhat reluctant to suggest this as it is pretty easy given the parts you have already done, but we've promised you multiple times that you will get a chance to complete this project eventually.

The problem: Here are some specifications for the circuit.

- Person A presses a START pushbutton to start the timing. The START pushbutton also turns on a high efficiency green LED (it wants 3 mA at 2 V), to which another person, B, responds by pressing a STOP pushbutton to stop timing.
- Pressing STOP turns off the green LED if it is on.
  - Debounce only where necessary. (You will need to decide if you need to debounce each switch or not.)
- Use the 4-digit BCD counter you wired up in §21L.1.4 to time the duration the green LED is on to one millisecond resolution.
- After the stop button is pressed, display the "reaction time" on the four-digit multiplexed LED display.
- Once either button has been pressed, pressing it again has no effect until the timer is reset.

[5] You can view a video of our c-meter at https://LAoE.link/C-Meter_Demo.html.

- Let a third pushbutton, RESET, reinitialize the machine. When the RESET button is pushed, the machine clears all displays and LEDs and awaits another measurement cycle.
- If person B cheats by hitting the STOP button before A presses START, turn on a RED "cheat" LED.
- If the counter overflows, the display should show "----".
- You should use the 16 MHz clock in the WebFPA (WF_CLK) for timing.

Here are the ingredients we suggest.

You will want to combine the WebFPGA reaction timer control FSM from Lab 19L.1 with the BCD counter and LED display from §21L.1.4 and use the DURATION signal to enable the counter as shown in Fig. 21L.4.

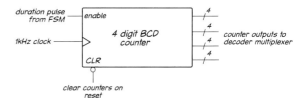

Figure 21L.4 Reaction timer counter connections.

You may still need to modify your counter code to show "----" on counter overflow, but this design is much easier than the C-meter: it runs just once, and the counter stops, allowing everyone to see the final count without any effort to catch and hold that final count in another register.

Wait! That was too easy. Let's make the reaction timer more interesting. Instead of starting the duration pulse and turning on the green LED when the START button is pressed, have the LED light up after a random delay of up to about 8 seconds. Warning, creating a random value in a synchronous system is harder than it sounds. The easiest way is to add a second high speed clock asynchronous with the WebFPGA WF_CLK to increment a counter. As you enter the delay state, you capture that counter's value to use as the delay time.

An alternative is to find some input that is asynchronous to the FSM timing source. In Fig. 21L.5, a 7-bit delay counter is constantly loading its high four bits with the low four bits of the main WF_CLK divider counter as long as it is not in the delay state. The low three bits are set to zero to ensure a minimum half-second delay. When Person A presses the START button, the delay counter stops loading and captures a 7-bit delay value (with the lower three bits always zero). This value is random because Person A is not synchronized with the 16 MHz clock. On the next FSM clock, the counter enters the delay state. Loading is disabled while in this state (and there is no way for Person A to release the button before the FSM enters the delay state less than 1 ms after s/he presses the button, so the captured value does not change).

While in the delay state, the delay counter is enabled every ≈65 ms to increment at about a 15 Hz rate. (Hint: You want to create an RCO signal to enable the delay counter to increment once for every 2^{20} WF_CLK clock cycles. Think about how RCO is generated on the 74HC16X counters.) When the delay counter reaches its terminal count of 7'b1111111, it asserts a delay end flag which tells the FSM to go to the duration start state. (Again, consider the hint.)

Another way to create a random delay is to use a pseudo-random linear feedback shift register (LFSR). This is just a shift register with combinational logic feedback that generates a pseudorandom bit sequence: see Fig. 21L.6.[6] It is pseudorandom because the bit sequence is repeatable but as long as you sample it asynchronously to the shift register clock, the output values will appear to be random.

[6] See AofE §13.14.2: Feedback shift register sequences. Image used with permission.

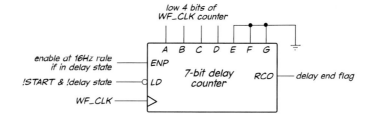

Figure 21L.5 A possible random delay generator.

You must be careful to initialize a LFSR to a non-zero value on a reset; if it ever is cleared to zero, it stays there.

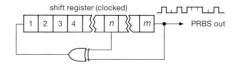

Figure 21L.6 Pseudorandom bit sequence generator.

Here is a Verilog module for a 7-bit LFSR.

```
//////////////////////////////////////////////////////////////////////
// A 7-bit Linear Feedback Shift Register to generate pseudorandom values
// Modified from https://nandland.com/lfsr-linear-feedback-shift-register/
//////////////////////////////////////////////////////////////////////
module LFSR_7BIT   (
  input i_Clk,                // shift register clock
  input i_Seed_DV,            // assert to load Seed Value
  input [6:0] i_Seed_Data,    // seed Value (must be non-zero)
  output [6:0] o_LFSR_Data    // 7-bit pseudo random output
  );

  reg [7:1]      r_LFSR = 7'd25;  // a random seed for the first run
  reg            r_XNOR;

  // Load LFSR with Seed if i_Seed_DV is asserted.
  // Othewise just run LFSR when enabled.

  always @(posedge i_Clk)
    begin
      if (i_Seed_DV == 1'b1)
        r_LFSR <= i_Seed_Data;
      else
        r_LFSR <= {r_LFSR[6:1], r_XNOR};
    end

  // Create Feedback Polynomials.  Based on Application Note:
  // http://www.xilinx.com/support/documentation/application_notes/xapp052.pdf

  always @(*)
    begin
      r_XNOR = r_LFSR[7] ^~ r_LFSR[6];
    end // always @ (*)

  assign o_LFSR_Data = r_LFSR[NUM_BITS:1];

endmodule // LFSR
```

Figure 21L.7 shows how to use the LFSR to generate a random delay after the START button is pressed. It uses a 7-bit version of a 74HC16X style counter. The counter is continually loaded with pseudorandom LFSR values at a ≈15 Hz rate until the START button is pressed. Again, the low three bits are zeroed to ensure a minimum delay of one-half second. When the FSM enters the "delay before duration" state, the counter begins counting up to its terminal count of 7'b1111111. When it reaches the terminal count, it asserts the delay end flag, which causes the FSM to transition to the state which starts the duration pulse and lights the green LED. (If STOP is pressed during the delay, it should still go to the cheat state.) Here, pressing the START button and transitioning to the delay state captures the output of the LFSR to provide a random value.

While this version is easier to implement than the one in Fig. 21L.5, it has two drawbacks. If the synthesizer initializes the LFSR to zero, the reaction timer will require an initial reset before it will work. (The LFSR module initializes the shift register to a non-zero value to try and avoid this.) Also, it uses seven additional flip-flops, although with over 5000 logic cells available in the WebFPGA that should not be an issue for us.

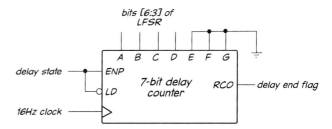

Figure 21L.7 Another random delay generator.

Modify your reaction timer FSM to add a random delay between pressing the START button and when the green LED comes on. This will also make it possible test your reaction time by yourself.

21L.2.3 Digital clock

Use the four digit multiplexed display and the WebFPGA to build a digital clock;[7] see Fig. 21L.8. Use the 16 MHz clock as your timing source. You can use a Logic Switch to select SET/RUN* and pushbuttons to increment or decrement the hour and minute in SET mode.

Since we were not using the decimal point, we connected the colon cathode to the TLC5916 instead and connected the colon anode to one of the digit anodes. That allowed us to control the colon from the WebFPGA. We blink the colon at a 1 Hz rate in run mode and keep it lit constantly in set mode to provide user feedback.

Enhancements include adding an alarm (use a piezo buzzer) or automatically increasing the time set increment/decrement rate if a pushbutton is held for a longer time. You could also use a keypad (see §21L.2.6) to set the time and alarm but that increases the difficulty significantly.

21L.2.4 Digital voltmeter

If you used the WebFPGA to emulate the 74LS502/3 in the previous lab, you have most of the hardware already connected to create a 0 to 2.55 V voltmeter. Build a Successive Approximating Digital-to-Analog converter by adding a LM358 op-amp test voltage source to your breadboard and a connection from pin 15 to pins 14 and 16 on the AD558 to set the range to 2.55 V as shown in Fig. 21L.9.

This voltmeter is designed to measure a voltage between 0 and 2.55 V at the V_{in} input (which is initially connected to a variable voltage divider for testing). The LM358 op-amp is configured to buffer the input signal, which provides high multimeter input impedance. The rest of the circuit is a modified

---

[7] But not a very good digital clock since it loses its time if the power goes out.

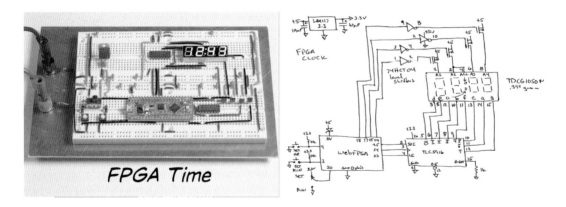

Figure 21L.8 Our FPGA clock.

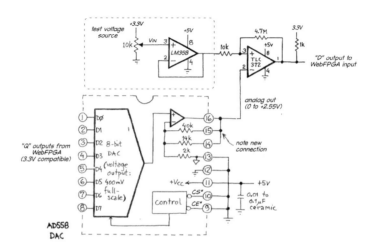

Figure 21L.9 Digital voltmeter hardware.

version of the successive approximation A/D converter from Figure 20L.4 in the data conversion lab. The AD558 DAC is configured to output 0 to 2.55 V for input values of 0x00 to 0xFF. This makes the DAC digital value equal to the voltmeter input.

You will need to program the WebFPGA to replace the 74LS502/3 SAR and to display the final ADC value on three of the four digits on the LED display. You should use an FSM for the SAR.[8] Converting the two-digit hex DAC value to three decimal digits is not easy as easy as you might think. If you get stuck you might want to look at https://LAoE.link/Binary-to-BCD.html. You could also use a brute force method: create a big lookup table to convert the 256 bit possible ADC output values to the digits to be displayed for each ADC value.

This design uses nine WebFPGA pins, one for the comparator output and eight for the DAC inputs, but you can let the pins connected to the logic indicators do double duty by using them for the DAC inputs as well.

Enhancements:

- Reduce the pin count by six by using a 74HCT164 serial-in, parallel-out 8-bit shift register to output the DAC value. This part must be powered by 5 V, but the HCT inputs are compatible with the WebFPGA 3.3 V logic outputs. This adds complexity and makes it a more difficult project.

---

[8] No fair using the structural '502 code we provided in the last lab. Write your own behavioral FSM to implement the SAR.

21L.2 Some project ideas

- Include additional ranges by adding gain and/or attenuation to the LM358 input buffer. This allows you to make the voltmeter autoranging by controlling the gain or attenuation from the WebFPGA with an analog switch or MOSFET. Start with maximum attenuation and if the result does not fill the display, try the next range. Alternatively, add input protection to the LM358 and start with the highest gain and reduce the gain until the four decade WebFPGA counter does not overflow. Remember to adjust the display decimal point as the range changes.

21L.2.5 Bicycle odometer

Use a Hall-effect sensor and magnet to detect the rotation of a bicycle wheel, and translate number of rotations into total distance traveled on the LED display. A digital Hall-effect sensor like the Melexis US5881 has an open-drain output that pulls the output to ground when a magnetic field is detected.[9]

If you mount the magnet on a spoke so it triggers the sensor on each tire rotation, you can count the number of rotations and translate that into miles ridden. A bicycle with 29-inch tires travels 91 inches per rotation, so it takes 696 rotations to travel a mile. You could increment the tenths digit every 70 rotations (a 0.5% error) then roll over to increment the mile at 696 rotations.

Enhancement: Add a trip odometer that can be reset to measure specific trips. Use a button to switch the display between total miles and trip miles (maybe light up the decimal points to show when trip data is being displayed) and use another button to reset the trip odometer.

21L.2.6 Keypad scanner

In Chapter 18N we introduced a 4×4 matrix keypad and discussed how to use a ROM as a lookup table to translate row and column information to key values. This project has you create an FSM in the WebFPGA to scan the keypad and display keys as they are pressed on the LED display. The 4-digit result can be used in several of the projects that follow.

The keypad (Fig. 21L.10) consists of a matrix of SPST switches. There are eight electrical connections to the switch, four row lines and four column lines. When a key is pressed, one column is connected to one row.

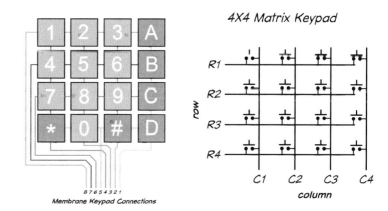

Figure 21L.10 The 4 × 4 keypad.

To use this keypad, you apply a signal to one row and see if the signal is found on any of the four columns. If it is, it means that the switch at the intersection of the driven row and sensed column is

---

[9] Datasheet available at https://LAoE.link/US5661.pdf.

pressed. Normally, you scan the rows sequentially to capture a single key press at the intersection of a row and column.[10]

The keypad requires eight I/O pins, four configured as outputs to drive the rows, and four configured as inputs to sense the columns. Since the columns that do not have any keys pressed are floating, you will need to pull up the WebFPGA column inputs to force them to a default value opposite that of the driving signal.

In operation, one row is driven low at a time while the other three rows are set high. Each of the four columns is normally held high by the pullup resistor. If no column line goes low, then no switch in that row is pressed and the next row should be selected and set low with the others high.[11]

However, if a switch is pressed, the column with that switch will go low when the corresponding row is set low. When a low column is detected, the system should recognize the new key, update the display and then stop scanning until the key is released (be careful not to start scanning again on a key bounce). Figure 21L.11 shows one possible block diagram for the keypad scanner. A "key pressed" flag is asserted while a key is pressed, indicating that the row select and column detector values are valid and may be used to determine which key is pressed. It also signals that scanning should stop until the flag is dissasserted.[12]

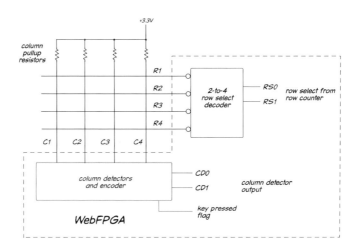

Figure 21L.11 Keypad scanner block diagram.

You should display the value of each newly pressed key in the rightmost digit of the display and move the three low digits to the left. The most significant digit is discarded as each new digit is entered.

Enhancements:

- Use the buzzer to give a short beep each time a key is pressed. Give a longer beep if the key is not used (for example one of the non-numeric keys).

- If you have time you could add simple editing features, such as letting one of the nonnumeric keys reset the display to "0000" and/or letting another delete the rightmost digit, moving the three higher order digits right while setting the most significant digit to zero.

---

[10] If you are wondering how this works if someone presses more than one key at once, remember the doctor's advice: "Don't do that."

[11] You could choose to use pulldown resistors on the columns and drive one row high with the others driven low but we had to choose *something*.

[12] See Chapter 23L §23L.3 for more detail on how to scan the keypad, including a directed graph for the FSM.

21L.2.7 Hotel safe

Combine the membrane keypad and the four digit display to build a hotel safe: see Fig. 21L.12 for operating instructions.[13] You can use an LED to show if the safe is locked/unlocked and a piezoelectric buzzer to warn when an incorrect combination is entered. This is a moderately difficult project.

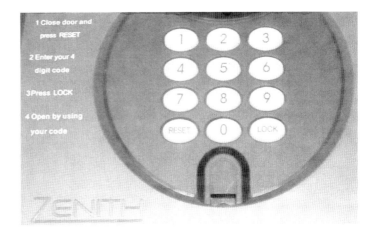

Figure 21L.12 Hotel safe operating instructions.

21L.2.8 Kitchen timer

Use the WebFPGA, display, buzzer, and optionally the keypad to create a kitchen timer. Set the number of minutes to count down using either the logic switches or the keypad, then press a button to start the timer.

You can show the time remaining as MM:SS by using the two left-most digits to show the minutes and the two rightmost for the seconds and turning on the colon. Alternatively, you can just show minutes remaining and blink one or more decimal points at a 1 Hz rate to show the timer is operating. Beep the buzzer when the time remaining reaches zero.

21L.2.9 Function generator

Connect an AD558 DAC to the WebFPGA as shown in Fig. 21L.9.[14] You don't need to include the op-amp or comparator. Use the keypad or logic switches to select sine or triangle wave output and to set its frequency. You can create the triangle wave with an 8-bit up/down counter whose output goes directly to the DAC. For the sine wave, use the output of an 8-bit up counter to address a sine lookup table in ROM, then send the table values to the DAC. Here are the values for a 256 point, 8-bit sine wave:[15]

```
128,131,134,137,140,143,146,149,152,155,158,162,165,167,170,173,
176,179,182,185,188,190,193,196,198,201,203,206,208,211,213,215,
218,220,222,224,226,228,230,232,234,235,237,238,240,241,243,244,
245,246,248,249,250,250,251,252,253,253,254,254,254,255,255,255,
255,255,255,255,254,254,254,253,253,252,251,250,250,249,248,246,
245,244,243,241,240,238,237,235,234,232,230,228,226,224,222,220,
218,215,213,211,208,206,203,201,198,196,193,190,188,185,182,179,
176,173,170,167,165,162,158,155,152,149,146,143,140,137,134,131,
```

---

[13] Image by Bai Timan Lam SAREOYSEE, Creative Commons license. See https://LAoE.link/Hotel_Safe.html.
[14] You may already have this wired if you used the WebFPGA to emulate the 74LS502/3 in the converter lab.
[15] Created by https://LAoE.link/Sine_Gen.html.

```
128,124,121,118,115,112,109,106,103,100,97,93,90,88,85,82,
79,76,73,70,67,65,62,59,57,54,52,49,47,44,42,40,
37,35,33,31,29,27,25,23,21,20,18,17,15,14,12,11,
10,9,7,6,5,5,4,3,2,2,1,1,1,0,0,0,
0,0,0,0,1,1,1,2,2,3,4,5,5,6,7,9,
10,11,12,14,15,17,18,20,21,23,25,27,29,31,33,35,
37,40,42,44,47,49,52,54,57,59,62,65,67,70,73,76,
79,82,85,88,90,93,97,100,103,106,109,112,115,118,121,124
```

21L.2.10 Solutions and code

Here are our solutions to the parts of §21L.1 in case you get stuck (but as always you won't learn how to solve problems with programmable logic unless you try to get your own version working before looking at our answers).

- §21L.1.2: `https://LAoE.link/FPGA/21L_Anode_Test.v`.
- §21L.1.3: `https://LAoE.link/FPGA/21L_Mux_Display.v`.
- §21L.1.4: `https://LAoE.link/FPGA/21L_Mux_Counter.v`.

And here is some useful project code:

- §21L.2.2: 7-bit LFSR module `https://LAoE.link/FPGA/21L_LFSR_7BIT.v`.
- §21L.2.9: sine wave values `https://LAoE.link/FPGA/21L_Sine_Values.v`.

Part VI

Microcontrollers

22N Microcontrollers I: Introduction

Contents

22N.1	**Microcomputer basics**	**869**
	22N.1.1 A little history	869
	22N.1.2 What is assembly language? How does it relate to C?	873
	22N.1.3 Elements of a minimal machine	874
22N.2	**Which microcontroller to use?**	**876**
	22N.2.1 Microprocessor versus microcontroller, again	877
22N.3	**ARM Cortex architecture**	**878**
	22N.3.1 Using the SAMD21 internal peripherals	880
	22N.3.2 Introduction to CMSIS	885
22N.4	**Yikes! Isn't there an easier way?**	**886**
	22N.4.1 Why not just use an Arduino?	886
	22N.4.2 Why not let [insert the name of your favorite AI here] write my code?	888
	22N.4.3 OK, but why can't we just use a wizard?	888
	22N.4.4 We are your wizard	889
22N.5	**The first day of the microcontroller lab**	**889**
22N.6	**AoE reading**	**889**

Why?

Here's today's problem: make a state machine that is fully general and programmable, i.e., transform memory and logic into a computer. Computers are so familiar to you that you may not think of this as much of a challenge. Once upon a time it was.

22N.1   Microcomputer basics

22N.1.1   A little history

Prehistory: before the microprocessor: Yes, there was a time when computers roamed the Earth, but were not based on microprocessors. In the 1930s electromechanical computers were built using relays; some were true "Turing machines," fully programmable. At the same time, the last giant analog computers were growing, unaware that their era was-ending (notably, Vannevar Bush's differential analyzer, built in the late 1920s at MIT). The first fully-electronic computer, using vacuum tubes, was the Colossus, put into operation in 1943 as part of the British cryptanalytic work at Bletchley Park.[1] Because of its application to classified work it remained unknown to the public until the 1970s. In 1946 the first American fully-electronic machine – not secret, and therefore often thought to be the

---
[1] See the summary history at https://LAoE.link/History_of_Colossus.html.

Microcontrollers I: Introduction

first fully-electronic computer – was launched. This was the ENIAC. It was *big* (see Fig. 22N.1), and could run at first only a few hours between disabling tube failures, later – with better tubes and the expedient of never turning the machine off – for about two days between failures. But it was fast, about 1000 times faster than its electromechanical predecessors.

Neither Colossus nor ENIAC was quite the general-purpose machine that we think of as a computer. Both were powerful calculators that had to be set up by arranging physical wiring, using switches and jumper wires. They had not incorporated Turing's conception of a stored program machine.[2]

Figure 22N.1 ENIAC: the first US fully electronic computer (US army photo).

A processor on a circuit board: Transistors relieved the painful difficulty caused by vacuum-tubes' short lifetime. Integrated circuits made digital circuitry much denser and cheaper than it had been. Figure 22N.2 shows a Data General Nova computer's processor board, ca. 1973. The board is about sixteen inches square, and is implemented with small- and medium-scale TTL ICs. Note the afterthoughts implemented with hand-wired jumpers. (This was a production board, not just a prototype.)

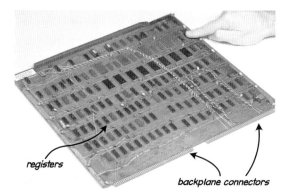

Figure 22N.2 Processor on a PC board: Nova computer, ca. 1973.

The first microprocessors: Intel produced a 4-bit processor called the 4004 in 1970. It was not conceived as a general purpose device, having been designed to implement a calculator for a single customer. Intel gave it more general powers, but no one took much interest. Two years later, Intel produced the first 8-bit microprocessor, the 8008. It had 18 pins, dissipated a watt, cost $400 in 1972 dollars, and was not widely adopted.

In 1974 the Intel 8080 arrived, and this one enjoyed some success, although it was not quite the one-chip processor that one might expect. The 8080 required twenty other chips to implement its interface to the rest of the computer. These early processors found their way into games, notably Pong[3] and Space Invaders. The 8080 also was the brain of the early hobbyist's home computers.

---

[2] Ibid.
[3] Pong was a huge hit - even though its designer scorned it. He said that after the failure of a more complex game, he decided to "build a game so mindless and self-evident that a monkey or its equivalent (a drunk in a bar) could instantly understand it." See M. Malone, *The Microprocessor: A Biography*, Springer-Verlag (1995), p. 137.

22N.1 Microcomputer basics

An issue of *Popular Electronics* from 1975 showing the home-built Altair is said to have convinced Bill Gates and his friend Paul Allen that something was afoot that they ought not to miss. They rushed to write a BASIC interpreter for the 8080, and Gates quit college to start a company to develop this product. Meanwhile, competition was hot among the fancier parts, the *microprocessors*, the brains of full-scale personal computers. In 1976, defectors from Intel moved across the street and formed Zilog, which put out the Z80 (so-called because it was to be the last word in 8080s). As often happens, it did not displace the inferior design.

The following year, Apple launched the Apple II computer, its first mass-produced model, using a 6502 processor made by a small company named MOS Technology. In 1979, Intel's 8086 became the industry leader by getting itself designed into the IBM PC, even though Motorola's new 68000 was a much classier processor (adopted in the underdog Macintosh computers).

Figure 22N.3 shows an 80386 (a descendant of the original Intel 8080 processor) PC motherboard.[4]

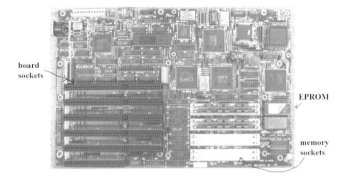

Figure 22N.3 IBM PC motherboard using the Intel 80386/87 processor on a PC board, ca. 1990.

Microcontrollers: Well before anyone imagined a mass market for a "personal" computer, it was not hard to envision uses for a versatile one-chip device that might implement a fancy calculator, or perhaps a game. We have noticed Intel's decision to provide, in the 4004, a part that was more capable than what the single customer's calculator required. And a game like Pong could be made cheaper if it used fewer ICs.

So it is not surprising that the first microcontroller – a self-contained one-chip computer – was born in 1976, just a few years after the first microprocessor. This was again an Intel design: the 8048. Indeed, the surprise for Intel and the world was the rapid spread of the "personal computer." Very few people had envisioned any need for such a gadget. At first it was only a few hardcore hobbyists who thought they needed to own a home computer.

The 8051, Intel's second try at a microcontroller, was born in 1980 and sold in huge volume: 22 million in the first year.[5] Two years later, Motorola brought out the 6805, the best-selling microcontroller of its era (about 5 billion sold by the year 2001). Motorola remained for many years the market leader in controllers, though not in processors, primarily because it dominated the growing automotive market. Later companies now outsell both these original giants in the microcontroller market (for example NXP – a Philips spinoff, Renesas – a Hitachi company, and Microchip outsell Motorola at the time of writing).

Microcontrollers like the 8051 and the 6805 are examples of 8-bit, Complex Instruction Set Computing (CISC) architectures. These processors operate natively on 8-bit data values, and a single machine language instruction can do complex operations such as fetching data from memory, performing logical

---

[4] The motherboard is the printed circuit board containing the microprocessor and into which peripheral cards and memory could be plugged. The early bare PC motherboards like the one in Fig. 22N.3 were computers, but not useful by themselves, as they usually lacked essentials such as a video display circuit or a mass storage device such as a floppy or hard disk drive.

[5] Malone, op. cit.

or arithmetic operations on it, then storing it back to its original memory location. Around the same time, a British computer company, Acorn Computers, began developing a 32-bit Reduced Instruction Set Computing (RISC) architecture processor motivated by research at UC Berkeley.[6] This architecture, which they called ARM for Acorn (later Advanced) RISC Machines, limited the functionality of individual processor instructions to simpler operations that could be executed quicker than the complex instructions on CISC machines while using less power. ARM processors are now used in most smartphones and tablets, as well as many digital televisions and some mobile computers.[7] The Cortex-M0+ microcontroller that you will soon meet in the lab is a 32-bit ARM RISC architecture design.[8]

Of course, microcontrollers keep getting more and more "micro." Here is a comparison of six 8051 microcontrollers. The largest of the packages in Fig. 22N.4 is the original 40-pin, wide-DIP 8051 we used to use in our labs. The PLCC next to it is the same part in the somewhat smaller PLCC package followed by smaller and smaller modern packaging. The tiniest one, on the right, measures 2mm square.[9]

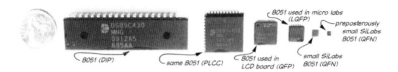

Figure 22N.4 Shrinking microcontrollers: 8051 in six packages.

As packages shrink, so do prices. In quantity 1000 some small controllers now cost 25–30 cents. Some *untested* controllers cost a good deal less than that (we heard a recent estimate of 7 cents) – and by the time you are reading this, no doubt prices will have fallen further.

Recalling what we mean by "a computer": The computers we are accustomed to are Turing machines, state machines that advance, one step at a time, from state to state following a sequence and taking actions at each state. The inputs that call for taking a particular action, and that steer the machine to a particular set of next states, are what we consider to be the machine's *instructions*.

The abstract model proposed by Turing in 1936 was intended to define the limitations of computing, rather than be a design for for a computing machine.[10] Most contemporary computers, including the Cortex-M0+, are also called *von Neumann* designs, meaning that they use a single memory store for both data and instructions. An alternative arrangement, called "Harvard class," places instructions and data in separate stores. This is the scheme employed by the designers of the 8051 microcontroller.[11]

Stored program machine: The processor is a *state machine* (short for "finite state machine" or FSM) within a *state machine*: the computer itself is an FSM. It reads, from memory, one instruction at a time,

---

[6] ARM processors are now available with both 32-bit and 64-bit data width.

[7] While CISC microprocessors have historically dominated the personal computer market, in 2020 Apple Computer began selling Mac systems containing a 64-bit ARM-based RISC processor, the M1 (since updated to more powerful models).

[8] ARM Ltd. does not build its own microcontrollers. Rather, it licenses its technology to manufacturers who integrate ARM's core CPU design with memory and peripherals to create a complete device. The SAMD21 Cortex-M0+ we use in the lab is manufactured by Microchip (previously Atmel).

[9] The parts are, from left: Dallas DS89C430, in DIP and PLCC packages; SiLabs 8051F700 in QFP; SiLabs 8051F410 in LQFP; SiLabs C8051F990 in QFN; and the tiny SiLabs C8051T606-ZM in the 2mm QFN package.

[10] Isn't this oddly reminiscent of Boole's development of his algebra with the goal of making philosophical analysis more rigorous?

[11] There could be some disagreement whether the ARM Cortex-M series is von Neumann or Harvard architecture since it contains features of both. In our opinion, the linear 32-bit address space along with the ability of the SAMD21 we use in the lab to execute code out of either flash memory or static RAM places it squarely in the von Neumann camp.

executes it, then goes back to memory to find out what to do next.[12] The sequence of instructions form a *program* (or "code"), stored in memory.

The smaller-scale state machine that is the *microprocessor* is fed "instructions" – mostly 16-bit binary codes in our case, with a few 32-bit codes thrown in – that tell it which of its several tricks to do (2^{16} or 2^{32} tricks in principle). In a CISC machine, the processor steps through multiple states or steps for a single instruction. In a RISC processor, most instructions execute in a single step. Fig. 22N.5 shows an "instruction decoder" drawn in the form usual for state machines. As it executes each trick it can move data between memory and registers, perform arithmetic or logical operations on operands, turn on 3-states to drive signals here or there, clock input registers to catch signals, and so on.

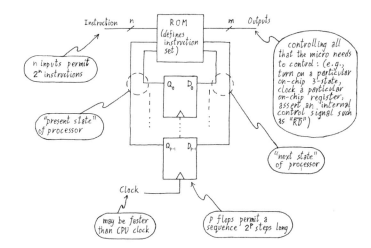

Figure 22N.5
Processor as *state machine*; it is the brains of the larger state machine that is the computer.

22N.1.2 What is assembly language? How does it relate to C?

The instructions stored in a computer's program memory are just ones and zeros known as "machine language." Indeed, in the previous edition of this book, you loaded binary program instructions into the lab computer's memory with a hex keyboard (which was a step up from the toggle switches of the old Data General Nova computer shown in Fig. 22N.7 below). However, coding in machine language is tedious and error prone. Humans are not well suited to recognizing that the two byte hexadecimal instruction 7838 in 8051 machine language loads a constant value into a particular register. Assembly language substitutes quasi-English mnemonics for machine language, so the previous instruction can now be written as "MOV R0, #38h." (The instruction means "copy into scratch-register zero the value 38 hexadecimal," in case you're curious.) Assembly code must be run through a translator program, called an assembler, to convert it into binary form ("object code") before it can be loaded into the computer's instruction memory. The assembler program often runs on a different (and usually more powerful) computer, allowing the programmer to take advantage of editing programs and storage options not available on the target machine.

Another advantage of assembly code is the ability to give symbolic names to program and memory data locations. In machine language, you have to specify the numeric value of an address in memory to store a variable to or to transfer program control to in a loop or branch. In assembly language you can define a variable with a name such as "CurTime DB 0" or a location in program memory with the label "LOOP1:" In addition, the assembler will automatically keep track of the next available memory

---

[12] Most processors use a *pre-fetch* or *pipeline* to get one or more additional instructions from memory, while busy *decoding* the current instruction. The Cortex-M0+ uses a two stage instruction pipeline.

location to assign the variable CurTime to or will readjust the program memory address for the label LOOP1 if you modify the program. It also allows comments, which are ignored by the assembler but are very valuable in explaining what is going on to a human who has to understand and maintain the code.

While a significant improvement over programming in machine language, assembly language has its disadvantages. First of all, it is specific to the CPU architecture used. The examples above are for the 8051 microcontroller. If you switch to a different processor architecture, not only are the instruction mnemonics different but the form of the assembly code, such as the order you write the operands and how you specify directives to the assembler, may change. In addition, the architecture of different CPUs is usually so different that you cannot just translate your code to the new mnemonics and syntax, you have to rethink the entire design of your program. This locks you into a microcontroller architecture unless you are willing to rewrite your code from scratch.

Assembly language asks a lot of the programmer. You need to keep track of and handle data types (bytes, words, characters, etc.) explicitly.[13] Since each assembly language instruction corresponds to a single machine language instruction, you are restricted to the operations available in the CPU's instruction set. Program branching is usually limited to jumps (gotos) and simple loops. If you need a complex math instruction like multiply or would like a branch in the form of a "do ... while" loop, you may find you have to write it yourself. Finally, older microcontrollers like the 8051 often have a convoluted or irregular architecture that requires the programmer to spend time becoming familiar with its quirks and limitations – knowledge that is not useful when transitioning to a new architecture.

Assembly language programming has mostly gone out of style because it costs too much in human programming time.[14] In the programming industry you sometimes hear the statement that a person can produce about ten *lines* of fully debugged code in a day. If this is about right, then surely it pays to produce lines that get more done.

In a higher-level language like C, a single line of code often expresses a more complex operation, one that requires multiple lines of machine code (and of assembly language code). C supports many data types including string and floating point values, and the standard C library included with most compilers provides functions for numeric formatting, math operations, string manipulation, and many other useful operations. C is also portable, meaning a properly written program should compile to microcontrollers with a very different architectures although, as we shall see, that is not much help if you are accessing the hardware features of a particular microcomputer at a low level.[15]

Like assembly code, C code must be run through a translator, called a "compiler," before it can be loaded into a target system's memory. Early compilers translated C to assembly code, which was then run through an assembler to create the object code. Most modern C compilers generate object code directly. Again, we usually "cross-compile" microcontroller C code on a more powerful PC to take advantage of its editing and storage capabilities.

22N.1.3 Elements of a minimal machine

An ordinary *microprocessor* requires a good deal of support in order to do anything useful – in order to form a "microcomputer." It needs memory at very least. And it needs some sort of input/output,

---

[13] While a byte is always eight bits, the size of a word depends on the CPU architecture. On early Intel processors, a word is 16 bits. Otherwise, it is typically the register size of the CPU; 32 bits in the Cortex-M0+ we will be using.

[14] The exceptions are in operating system kernels, interrupt handlers, and device drivers where efficiency and code size are paramount – but even here assembly has often been replaced by C. See, e.g., https://LAoE.link/Linux_Kernel_in_C.html.

[15] If you are unfamiliar with C (or Arduino which is based on a subset of C), §22S.3 offers some suggestions for learning the basics.

22N.1 Microcomputer basics

to act on the outside world. Often additional support ("glue") chips are required to decode address signals, buffer data signals and interface to external circuitry. Figure 22N.6 sketches a block diagram of a generic computer.

Which of these elements are required, which optional? Could one omit RAM? ROM? A disc drive? Keyboard? Display? (Surely we could omit ADC and DAC.) Yes, we could omit any of those elements, though not all.

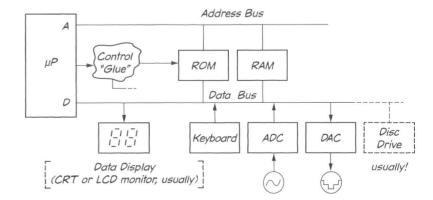

Figure 22N.6 Any computer: which elements are essential?

- RAM: RAM is important for general-purpose computers, but is not required for a dedicated controller.

    In a full-scale computer, RAM can hold program code as well as data, and both code and data normally are loaded from an external disc drive or solid-state storage. But a controller used, for example, to run a vending machine will store its code not in RAM but in ROM. (The machine needs to be able to remember short-term variables; but the amount of memory needed for this purpose may be so small that it can better be called "registers".)

- ROM: ROM allows a computer to run at startup – before any code has been loaded into RAM from an external store. Once upon a time, before ROM, a computer lost all its code in volatile RAM on power-down, and had to be nursed back to life when turned on. Figure 22N.7 shows the front panel of the old Nova computer model that we once used daily.[16] Why all those toggle switches? They were used to enter instruction codes in binary, one word at a time. Put enough in, and you could then run the little boot loader program that read in a larger program from the paper tape.[17]

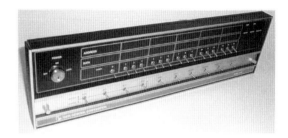

Figure 22N.7 Front panel of a Data General Nova computer (born 1969).

---

[16] One of the authors used this computer to write the 1st edition of AoE, published in 1980. Image source: Wikipedia: http://upload.wikimedia.org/wikipedia/commons/5/58/Nova1200.agr.jpg. Used under Gnu GPL license.

[17] Of course you wanted this initial program to be as short as possible. The paper tape loader program for the Nova 2 was only 13 words long.

"Bootstrap" loading: It's a bit hard to imagine working with such a machine (having paid about $8000 for it): startup was laborious. It's for that reason a computer that simply *ran* when you turned it on seemed almost miraculous. It appeared to be doing the impossible feat of lifting itself up by its own bootstraps. Hence the term "bootstrap loading," now shortened to "booting" of the computer, to mean "starting up," or (more specifically) "loading the operating system."

Figure 22N.8 Miracle of the 1960s: cowboy, like computer, lifts himself by bootstraps.

- I/O: There must be some sort of input/output hardware if the computer is to accomplish anything; but no particular form is required. General-purpose computers use keyboards and visual displays, but a controller regulating a process might conceivably get by with ADC and DAC, and perhaps a few knobs (read by the ADC). A still simpler application – say, control for an electric toothbrush – might need no more than an LED, a transistor switch for the motor, and a pushbutton for input.

22N.2 Which microcontroller to use?

AoE §15.3, 15.10.3

The variety of available microcontrollers is indeed daunting. It might seem that a person would have to put in a month of study before beginning. Usually, though, things are not that bad – largely because most people are not so rational (or is it *compulsive*?) to canvass and evaluate all alternative devices. Instead, most of us ask around, taking the advice of those with experience (or at least working near us).

That makes sense, because it's a big help to be able to ask advice from someone who has used the particular variety that you undertake to use. This sort of conservatism – even among people who take pride in working close to the cutting edge of new technology – accounts for the remarkable fact that the *8051* design that we previously used in these labs is still in use more than forty years after its introduction by Intel in 1980.

In earlier years, the primary reason for this conservatism may have been the desire to take advantage of a large volume of *legacy code* written for the particular processor or controller. This was important when coding was done in assembly language (because a change of processor usually required a complete code rewrite). It is not nearly so important now that coding is done in higher-level languages (for controllers, most often in C). Apart from some C-extensions required for each processor, the higher-level code can be ported to a new processor without the pain of a full rewrite.[18] But it remains convenient to stick with one controller in successive projects. Small differences among controllers do call for learning details that may be pleasant to learn once, but not repeatedly.

---

[18] However, the code controlling peripherals such as timer, serial controllers, etc., often requires extensive rewriting when switching manufacturers, even if using the same CPU core.

22N.2 Which microcontroller to use?

22N.2.1 Microprocessor versus microcontroller, again

The previous edition of this book taught microcontrollers using the "Big Board" in which the venerable 8051 was used as a microprocessor rather than a microcontroller, and was interfaced to external memory and peripherals through a parallel bus on the big, messy circuit board shown in Fig. 22N.9 to form the microcontroller. The system was programmed in machine language (i.e., 1s and 0s) using the hex keypad in the lower right corner of the figure. Software was debugged by single-stepping code while viewing the bus signals in hexadecimal on the LCD display to the left of the keypad. This configuration entailed a lot of wiring, which was good for the soul and developed excellent debugging skills, but was not very representative of modern embedded systems design.

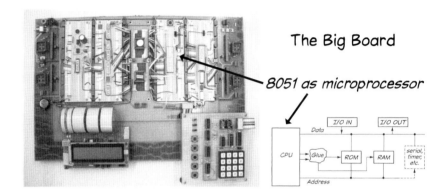

Figure 22N.9
Learning microcontrollers the old way.

Modern microcontrollers contain a plethora of internal peripherals, from simple timers to complex communications controllers, and provide enough memory to handle many, if not most, tasks. They are almost always programmed in a high level language, usually C, using an attached PC for cross-development and debugging. If additional external memory or an unsupported peripheral function is required, it is normally connected to the microcontroller through a serial communications link. Parallel buses use too many pins to be practical.

In the lab, you will meet the SparkFun SAMD21 mini, Fig. 22N.10, a single-chip, self-sufficient microcontroller mounted on a DIP breakout board along with a 3.3 V regulator, several indicator LEDs, connectors, and a small amount of support circuitry.

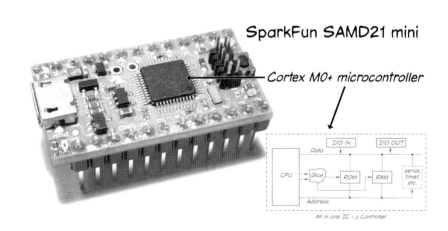

Figure 22N.10
Learning microcontrollers the new way.

The SparkFun board is preprogrammed to work with the Arduino development environment but we will be using it "bare metal," programming it in C using an Integrated Development Environment (IDE), and connecting to it with a hardware debugging tool to allow us to look at the processor's internal memory and registers on an attached PC.[19]

Over the thirty-plus years of its existence, ARM Ltd. has designed a large number of processors for various applications. Their products range from the high-performance Cortex-A series used in smartphones and tablets to the Cortex-M designs intended for low-cost, low-power embedded applications. The SAMD21G18A Cortex-M0+ microcontroller on the Sparkfun SAMD21 mini is at the low-end of the Cortex-M processor product line.[20] Nevertheless, it offers an impressive array of computing resources including:

- a 32-bit Cortex-M0+ CPU core running at up to 48 MHz
- 256k bytes of flash-programmable memory for program storage,
- 32k bytes of static RAM for data storage,
- a Direct Memory Access (DMA) controller,
- a Vectored Interrupt Controller with multiple external and internal interrupt sources,
- a 24-bit SysTick timer used to create a system timing reference,
- 38 I/O pins (18 of which are available at the pins of the SparkFun breakout board),
- eight internal programable clock generators,
- a 12-bit, 14 channel 350ksps ADC with an internal bandgap reference,
- a 10-bit Digital-to-Analog Converter,
- two voltage comparators,
- three simple 16-bit timer/counters (two can be combined into a 32-bit counter),
- three complex counter/timers (one 16-bit and two 24-bit),
- six serial communication units configurable as SPI, I2C, or UART,
- one USB interface,
- a watchdog timer and a real-time clock,
- a peripheral touch controller, and
- an Inter-IC Sound bus for connecting digital audio devices.

All you need to add to the SAMD21 IC to get a working microcontroller is power and bypass.

22N.3 The ARM Cortex architecture

When programming in a higher-level language like C, the details of how to allocate and use registers, make function calls, pass arguments, and return values are mostly hidden from you. However, it is helpful to have an idea of what the internal architecture of the CPU looks like, both to aid in debugging your code and in case you ever want to optimize your program to get the maximum performance possible. You will not write assembly code in the labs but we will look at the low-level code that is generated from your C program to see how the compiler uses the Cortex-M architecture.

While most of the features of the Cortex-M0+ processor are standardized, some functions are optional and may be omitted by a specific manufacturer. For example, the Microchip SAMD21 used on the Sparkfun board includes a SysTick timer (which we will use) and a second internal bus for

---

[19] Actually, since we are programming in C and not assembly language, we are not quite writing the code from scratch. The IDE links in some low level hardware initialization routines before calling our main() function and takes care of setting up the code in the proper order in the flash memory for the ARM architecture.

[20] The higher-end Cortex-M CPUs operate at higher speed and include memory-management units, digital signal processing instructions and floating point hardware, which we will not use and only complicate learning how to use the microcontroller.

Single-Cycle I/O access (which we will not); however, it does not include a Wake-up interrupt controller or the ability to reset all registers. See the datasheet for details.[21]

The ARM Cortex-M architecture is register based. That means that most instructions operate on one of the sixteen 32-bit wide internal registers shown in Fig. 22N.11. Registers and memory may be accessed as word (32-bit), half (16-bit) or byte (8-bit) values.[22] The first thirteen registers are general purpose, and may be used to store data and memory addresses ("pointers"). The convention for C programs on ARM processors allows registers R0-R3 and R12 to be destroyed on function calls. All other registers must be preserved.

The remaining three registers have specific functions. The Stack Pointer points to an area in static RAM used to store and retrieve temporary values without having to devote permanent specific memory locations to the data.[23] The Link Register is used to store the memory address of a program instruction to return to at the end of a subroutine (a "function call" in C). The Program Counter is the address in memory of the next CPU instruction to execute. Program instructions may execute out of either flash memory or static RAM. On a processor reset, the program counter is set to the value stored in location 0x00000004, which holds the address of the first line of program code. The IDE we will use places a startup routine at this address which initializes the microcontroller then calls our `main()` function.

There are several additional special registers which are used to store internal CPU state. These are read or written with special instructions. The Application Program Status Register ("PSR" in Fig. 22N.11) includes a set of flags (i.e., one bit values) that hold the result of the previous calculation such as zero, negative, overflow or carry. It also contains the Q saturation bit, which controls what value results after an arithmetic overflow or underflow. The Exception Mask Registers control the priority of processor interrupts.[24] The Control Register determines which of the two Stack Registers mapped to Register 13 is active and the privilege state which is used to implement access control.

Memory, both program flash and static data memory, as well as the peripheral control registers are accessed via a linear 32-bit internal memory address bus. Although this theoretically gives your code access to over 4 billion memory locations, it is somewhat academic since bus access is limited to the internal resources of the microcontroller – there is no external memory bus and no way to add additional resources to the address space.

In a CISC machine like the 8051, arithmetic and logical instructions can use memory data as operands and can save results directly to memory. In the RISC Cortex-M, all operands are supplied from register values and results are written to a register. Memory values must be loaded into one of the registers before an arithmetic or logical operation and saved back to memory afterwards as a separate step. Registers can be loaded with a value stored as part of the instruction, known as *immediate addressing*, or the value in a register can be used as the address of a location in memory. The data in another register can then be read from that memory location or written to it. This is called *indirect addressing* and is the basis for the pointer data type in the C language.

---

[21] Microchip SAM D21/DA1 Family datasheet (DS40001882G), Table 11-1. Datasheet available at https://LAoE.link/Microchip_ATSAMD21G18_Datasheet.pdf. We recommend you keep the datasheet open whenever you are working with the microcontroller for reference.

[22] However, there are limitations on how memory can be accessed. For example, a word value in memory can only be accessed at an even four byte address (i.e., the two least significant bits of the address must be zero).

[23] Two stack pointers are available but only one is active at a time. The second stack pointer facilitates running a supervisory operating system.

[24] Interrupts are hardware events that cause the CPU to execute a function call. They are useful to respond quickly to a peripheral or external signal or to avoid having to repeatedly check for a rarely occurring event.

880 Microcontrollers I: Introduction

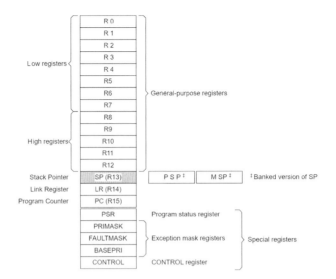

Figure 22N.11 Arm Cortex-M0+ processor core registers.

22N.3.1 Using the SAMD21 internal peripherals

The many internal peripherals of the SAMD21G18A are controlled by additional registers accessed by reading or writing to them at specific addresses in the 2^{32} byte address space. These registers are used to both configure the peripheral as well as to read and write data from/to it. As with any other memory access, you use the CPU instructions that move data between the general purpose registers and the memory address space to use a peripheral. Indeed, there is no difference between reading/writing to memory and to the peripheral registers.[25]

There are far more possible peripheral input and output signals than available pins on the integrated circuit package, so an internal port multiplexer allows you to assign internal peripheral signals to specific external pins. In order to use an internal resource, you must first configure the peripheral, then connect its input and output signals to external pins. You must also configure and connect a clock signal to the peripheral.

The steps required to use most internal peripherals include:

1. Configure any external I/O port pins used as inputs or outputs.
2. Configure the pin multiplexer to connect the peripheral signals to the external pins.
3. Enable the peripheral clock. (On reset, peripheral clocks are disabled to save power.)
4. Connect a clock signal to the peripheral (and configure the clock if not yet set up).
5. Configure and initialize the peripheral.
6. Enable the peripheral.

22N.3.1.1 Configuring Port I/O

However, if all you want to do is use a pin as an input or output, the process is much easier. First you configure the pin as an input or output, then either write one or zero to the register controlling the pin if an output or read the pin state from the input buffer if an input.

---

[25] This method of I/O access is referred to as "memory mapped I/O." An alternative is "port I/O," used by the 8051, which uses separate machine language instructions to access I/O pins. The advantage of port I/O is that it can include instructions to modify individual I/O pins or to apply logical operations to pins, while memory mapped I/O must write a minimum of 8-bits at a time and, in the ARM architecture, can only read and write to memory locations, not perform logical operations. As we shall see shortly, the SAMD21 gets around these limitation with specialized I/O control register hardware.

22N.3 The ARM Cortex architecture

The I/O pins of the SAMD21 are organized into Port Groups of up to 32 pins. These can be configured, set and/or read as a group or individually. The first Port Group is "A" and the second "B." Pins are denoted by their group and number, so the first pin in Port Group A is pin PA00, the last is PA31. These each correspond to a specific physical pin on the integrated circuit.[26] The pins of the IC are connected to the pins of the breakout board as shown in Fig. 22N.12. As you can see, only a subset of the available microcontroller I/O pins are brought out to the breakout board. This means we will have to be careful in assigning peripherals to pins to be able to use all the peripherals in the final lab.[27]

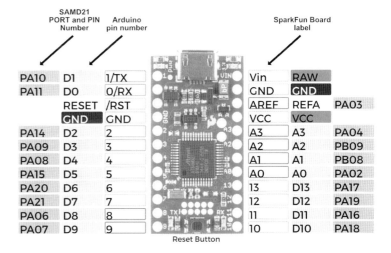

Figure 22N.12 Sparkfun SAMD21 Mini Breakout I/O pin connections.

Any I/O pin on the SAMD21 can be used for input or output. Figure 22N.13 shows the block diagram of an individual external I/O pin ("PAD"). The shaded block labeled "PORT" contains the memory-mapped control registers for pin number "x."

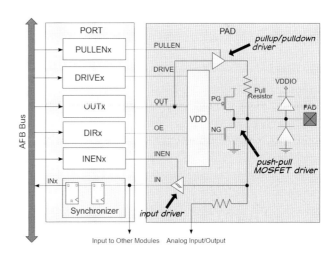

Figure 22N.13 SAMD21 I/O pin configuration.

To configure a pin as an output, the OE control signal must be asserted by writing a 1 to the DIR register. This enables the push-pull MOSFET output driver. An output pin normally can sink

---

[26] See section 5.2.1 in the datasheet. The SparkFun board uses an M0+ in a QFN48 package.
[27] A spreadsheet to keep track of I/O pin use is available at https://LAoE.link/micro1/SAMD_Pin_Spreadsheet.xlsx

2.5 mA and source 2 mA, however if the DRIVE signal is asserted, this is increased to 10 mA and 7 mA respectively.[28] The actual state of the output, high or low, is set by the OUT bit.

For input, the output driver is disabled by setting the DIR register bit to zero which turns off both MOSFET drivers. To allow the state of the input pin to be read onto the internal bus (the "IN" signal), you must enable the input driver by setting the INEN bit high.[29] You can optionally add a pullup or pulldown resistor of $\approx 60\,\text{k}\Omega$ to the input pin by setting the PULLEN control bit high and writing a 1 (for a pullup) or 0 (for a pulldown) to the OUT register.

After a reset, all the input and output drivers are disabled and the pin is in a high impedance state. It cannot be used for digital I/O in this state, but it is still connected to the analog subsystem through a resistor allowing it to be used for output from the internal DAC or input to the ADC and/or comparators.

Table 23-2 in the datasheet summarizes the port control signals for each I/O mode and Section 23.6.3.2 shows a detailed view of each configuration.[30]

22N.3.1.2 Toggling an output pin

It is easier to set up a pin for output than input, so let's first take a look at how you would configure an I/O pin as an output and toggle it from low to high. The Sparkfun SAMD21 has an active high blue LED connected to Arduino pin D13.[31] Looking at figure Fig. 22N.12, you can see that Arduino pin D13 is connected to port pin PA17. We need to set this pin as an output and toggle it from low to high to blink the on-board LED.

The table in Section 23.7 of the datasheet shows the registers used to control the port pins. Each port has 128 addresses in memory assigned to access the port control registers. The first eleven registers are 32 bit wide, while the remaining registers starting with either "PMUX" or "PINCFG" are eight bits wide. The data in the 32-bit registers may be accessed by word, half-word (16 bits) or byte. Note that not all memory locations in the 128 byte block are used in this version of the processor.

Looking at the Peripherals Configuration Summary (datasheet Table 12-1), the Port control registers begin at location 0x41004400 in the 32-bit memory address space. This is the address for the Port A register block. The Port B control register block follows directly 128 bytes later at address 0x41004480.

The first four registers with names starting with "DIR" control the direction (input or output) of each port I/O pin, while the next four starting with "OUT" set the state (high or low) of each pin configured as an output. The first register (the Port A "DIR" register) at offset 0x00 from the port control register base address sets the direction of I/O pins PA00 to PA31 directly. Writing a zero to a bit position sets that I/O pin to an input, while writing a one sets it to an output. So writing 0x00000001 to memory address 0x4100440 sets Pin PA00 as an output and the rest of the pins in Port A to input mode.

So why do we need three more registers to control the I/O direction? The problem is that you will usually not use all 32 pins in a port for a single function. Some pins will control part of your system while other pins control different hardware. As long as you only configure the pin direction once and never change it, writing a 32-bit word to the DIR register works fine. However, if you need to change the direction of a pin later on, or you want to configure I/O pins in different sections of your code, for example if each peripheral initialization routine sets the direction of the pins it uses – a common practice – you would need to read the current direction of the port pins, modify the ones you want to

---

[28] This bit is actually called DRVSTR and is in the pin's PINCFG register. We will deal with the pin configuration register in a later chapter.

[29] Note that when the input driver is enabled, you can read the state of the pin even if it is configured as an output. This can be useful to monitor for shorted I/O pins.

[30] Figure 23-8 shows an output configuration driven through the large value pull resistor rather than the low-impedance MOSFET driver. It is not clear why you would to do this, but it is a possible configuration with the SAMD's port hardware.

[31] The Sparkfun SAMD21 mini comes configured for use with the Arduino development environment. Although we will be programming it directly, the schematic for the board uses the Arduino pin numbers to show internal connections to the microcontroller, so you need to work backwards from the circuit board pin label to determine the port pin number.

change, then write them back to avoid disturbing the state of other pins. In a typical real-time controller application this "read-modify-write" procedure creates a hazard condition. If your code reads the state of the DIR register but then is interrupted by another task that changes the state of a bit in the DIR register before returning to your write operation, you would undo the change made by the interrupting routine.

This is a bit hypothetical if all you are doing is setting the pin's direction, but it becomes a real problem when setting the output state from the foreground program if you are using interrupt-based code. In order to use read-modify-write to change the output state of a single bit in a port, you would need to disable processor interrupts before reading the OUT register, and then reenable them only after writing the updated value. This not only complicates the code (it now takes at least five instructions to change an output bit), but it also increases the latency between when an interrupt occurs and when it is responded to. This is a problem in many real-time applications. Even if you do not have to worry about incorrectly modifying a port register, read-modify-write code is inefficient.

The additional DIRSET, DIRCLR and DIRTGL registers in the SAMD21 allow you to modify individual bits in the DIR register with a single instruction, while not disturbing the state of the remaining pins. Writing a one to a bit position in the DIRCLR register sets the direction of the corresponding I/O pin as an input; the direction of any port pin written with a zero is not changed. Thus, writing 0x00000001 to memory address 0x41004404 (the address of the DIRCLR register for Port A) sets Pin PA00 as an input and leaves the direction of the remaining pins in Port A unchanged.

Similarly, the direction of any I/O pin written with a one bit to the DIRSET register is set as an output while the in/out directions of all other pins are not modified. Finally, the direction of each pin written with a one to the DIRTGL register is toggled; an output is changed to an input and visa versa. Again, the direction of any pin written with a zero is not changed. Writing to a SET, CLR, or TGL register is an "atomic operation;" it cannot be interrupted until the DIR register value has been updated.

The four registers beginning with OUT work similarly, except that they set the value (high or low) of the I/O pins configured as outputs. Here it is even more important not to modify other I/O pins because even pins configured as inputs respond to changes in the OUT register if the pull buffer is enabled. Indeed, it is rare to use the OUT register to set the state of all 32 pins in a port group.[32] Normally, you set individual I/O pins with the OUTSET register, clear pins with the OUTCLR register and toggle individual pins with the OUTTGL register.

Ok, let's see how to blink the blue LED connected to pin PA17 on the Sparkfun board.

We first need to set the direction of pin PA17 to output since all the I/O pins are set as inputs when the code starts after a reset. Once the pin is initialized, we will set it high, then low repeatedly. You use C pointers to write a value to a specific memory address location to access the port control registers.[33] A pointer puts the numeric memory address value into one of the general purpose registers then writes the data in another register to that address via indirect addressing. We first define constants for the three registers used to first set the pin mode and then to blink the LED. Defining numeric values as mnemonic constants documents the code and makes it easier to understand:

```
#include <stdint.h>        // needed for uint32_t data type

// define access to port A registers
#define DIRSETA *((volatile uint32_t *)(0x41004408))
#define OUTCLRA *((volatile uint32_t *)(0x41004414))
#define OUTSETA *((volatile uint32_t *)(0x41004418))
```

[32] However, it may make sense to set eight bits simultaneously with a byte register write to the OUT register if all eight I/O pins are being used as a group.

[33] Indeed, in the ARM RISC architecture, you must use pointers to access any memory location.

The `#include <stdint.h>` in the first line is required to allow us to specify the size of variables explicitly. In C, the number of bytes used for a numeric variable like an integer (`int`) can vary depending on the compiler and target machine. Here we need to tell the compiler to transfer the exact same number of bits as are in the port control register by using the `uint_32t` (unsigned 32-bit integer) data type.

This code snippet defines each of the three Port A registers we will use, DIRSET, OUTSET and OUTCLR, as pointers to 32-bit values to their respective memory addresses. While the assignment looks complicated, it is easy to understand if you break it into parts. The text value after the `#define` is the constant you will use to set the value of the corresponding register. The "A" appended to the register name reminds us that this is a definition for I/O Port Group A. The asterisk after the constant's name signifies that this define is "dereferencing" a pointer. This tells the C compiler that when it sees this constant, it should not use the value in the constant's definition (in this case the hex value at the end of the line) as the data to read or write, but rather it should use the value of the constant as an address into the memory address space to read or write the data value at that address (in this case one of the Port registers). The `int32_t *` is a cast. It tells the C compiler that the hex value following is a pointer to a 32-bit value. Without it, the compiler would not know the size of the data (word, half word or byte) the pointer points to. Finally, the `volatile` keyword warns the compiler that the data at the pointer location can change from time to time. This is necessary to prevent the compiler optimizer from trying to outsmart us and only read the value at the pointer address once and store it somewhere (most likely a register) to eliminate unnecessary memory accesses.

Once we have the port registers defined, we could just write a value with bit 17 set (`0x0002000`) to the DIR and OUT registers to configure and output to PA17, but again it is clearer to define this value with a understandable mnemonic:

```
#define BLUE_LED   (1 << 17)    // bit within register for led (active high)
```

Not only does this make the code much more readable, it also has the advantage that it is easy to change the code if the pin the LED is connected to changes. Rather than search through the program for every instance of 0x00020000 and then make sure it is being used to denote pin PA17, we just have to change the definition of the value for BLUE_LED in one place.

Finally, once the hardware is defined, you blink the LED by first setting pin PA17 as an output, then repeatedly setting it high then low in an endless loop. Since the register constants are all defined as dereferenced pointers, setting one to a value actually changes the data at that memory address, not the value of the pointer itself.

```
int main (void)
{
    DIRSETA = BLUE_LED;        // set PA17 to output

    while (1) {
        OUTSETA = BLUE_LED;            // turn LED on (active high)
        OUTCLRA = BLUE_LED;            // turn LED off
    }
    return 0;
}
```

Of course the blinking is only visible if we single-step the processor. If we let it run at full speed the LED will appear to be on constantly, albeit at less than full brightness.

22N.3.2 Introduction to CMSIS

Whew! That was a lot of work. Wouldn't it be great if we could get someone else to do all that setup defining registers and pins for us? It turns out someone already has – the microcontroller manufacturer.

The Common Microcontroller Software Interface Standard (CMSIS) provides a hardware abstraction layer to ARM Cortex processors.[34] It consists of a set of C program and header files provided by silicon vendors that simplifies accessing resources in Cortex-Microcontrollers and facilitates switching between different models in a Cortex family.

In its most complete implementation, it would be possible to replace one vendor's Cortex-M0+ with another's with almost no code changes. In reality, most manufacturers offer only a subset of the CMSIS components, in particular choosing not to implement the CMSIS specification that supports a standardized peripheral driver interface.[35] Since each manufacturer of a Cortex-M microcontroller designs its own peripherals to attach to the ARM CPU core, that means that code written for one manufacturer's offering is unlikely to work with another's without at least some (and possibly an extensive) rewrite.[36] However, moving between a single manufacturer's products in a single family (such as the many Microchip Cortex-M offerings) is usually not too difficult if you #include the core CMSIS Core files.

The CMSIS files for the SAMD21 family provide defines for all the CPU control and peripheral registers in the microcontroller as well as defining constants for I/O pins and individual register bit functions. They also include a small number of functions to abstract basic tasks such as setting up a system timer and managing interrupts, as well as providing a C interface to some specialized assembly instructions. This leads to much more manageable and readable code. Unfortunately, you will still have to read the datasheet to see how to configure and use the built-in peripherals, you just won't have to define all the pointers and register values required to use them.

We will be using an Integrated Development Environment (IDE) named SEGGER Embedded Studio to write, compile and debug our microcontroller programs.[37] When you create a project in the Segger IDE, it automatically sets a program define to the specific microcontroller part number you select. To use CMSIS, you just have to add the line

```
#include "samd21.h"    // add to use CMSIS
```

to the beginning of your program. This generic include file pulls in all the C and include files for the microcontroller you specified to allow you to use the CMSIS-Core (Cortex-M) features, including the Hardware Abstraction Layer (HAL), system initialization functions, and intrinsic CPU instructions not supported by the C language.

To use a hardware feature of the CPU, you will need to look at the associated include files installed by the IDE to see what are the CMSIS names of the structures, variables and registers used by that feature. In this chapter (and the next) we will be using Port I/O. The Port I/O definitions are contained in two header files, both named port.h (the same named files exist in separate directories so there is no naming clash). One include file defines the individual port registers directly as we did manually above,

---

[34] See https://LAoE.link/CMSIS_Overview.html.

[35] Instead, most microcontroller manufacturers provide some form of configuration "wizard," a software program or web interface that lets you select the processor and peripheral components you want to use. It then generates C code to initialize the system and provides high-level function calls to use the peripherals. For example, Microchip provides the web-based Atmel START tool (https://LAoE.link/Atmel_Start.html) to let you select and configure software modules for its embedded microcontrollers. We do not use these tools in the labs because they generate large amounts of dense code that make it difficult to understand how the microcontroller components work.

[36] For example, the Microchip SAMD21 M0+ provides three registers to let you independently set, clear or toggle each of the 32-bits in the port. TI's microcontroller using the same CPU core provides 32 separate registers, each of which sets one of the 32 bits in the port.

[37] See §22S.2.

while the other defines them as a C structure (which we will use in later labs). Pin definitions and memory map information are contained in two other header files, both named sam21g18a.h, located in the

`<project name>\SAMD21\CMSIS\Device\SAMD21\Include\pio\`

and

`<project name>\SAMD21\CMSIS\Device\SAMD21\Include\`

subdirectories respectively.

The port.h file you want to look at for port output in this lab is in the \Include\instance\ directory and contains a series of pointer definitions for the Port I/O configuration registers. The section of the include file commented as "Register definition for PORT peripheral" defines the pointers for the registers shown in Section 23.7 of the SAMD21 datasheet.[38] The register names ending in "0" refer to the control registers for Port A, while the names ending in "1" are for Port B.

In the previous code, we defined the DIRSET register as:

```
#define DIRSETA *((volatile uint32_t *)(0x41004408))
```

The CMSIS definition for the same register is

```
#define REG_PORT_DIRSET0     (*(RwReg  *)0x41004408U)
```

where RwReg is defined as:

```
typedef volatile        uint32_t RwReg;
```

in one of the samd21g18a.h files.

The advantage of using the CMSIS definitions is that you do not have to know where the register is located in the CPU memory map, and if you change to another processor which has the same register functions located at different memory addresses, your code will still work as long as you select the correct device when you create the project in the IDE and recompile your code.

The other file named `samd21g18a.h` contains definitions for the 32 I/O pins in each port. There are two definitions for each pin, one starting with PIN and one starting with PORT. The PIN_PXNN definition, where "x" is the port letter ("A" or "B" for the M0+) and "NN" is the two digit pin number (00 to 31), is just the pin number from 0 to 31. The PORT_PXNN definition is a 32-bit value with a single "1" in the pin number position. For example, PORT_PA17 is defined as "(1ul << 17)" (which is the binary value 00000000000000100000000000000000).

The PORT definitions for pins are used with the registers that read, write, set, clear, and toggle all 32-bits in a port with a single read or write instruction. You will mostly be using the PORT pin definitions with the M0+, so if you find yourself using a PIN definition (which you will occasionally need – see §23N.4) be sure you understand why.

22N.4 Yikes! Isn't there an easier way?

22N.4.1 Why not just use an Arduino?

Whenever we begin this section of the course, at least one student with eyes glazed over from our discussion of registers, memory maps, ports, pads, peripherals, casts, and pointers asks why don't we just use an Arduino. An Arduino is a microcontroller, usually on a breakout board with additional support circuitry, preprogrammed with code to abstract it to a simpler controller model: see `https://LAoE.link/Ardunio.html`. You write code for an Arduino compatible microcontroller in a

---

[38] The first set of definitions in the include file is for assembly language programs and should be grayed out if you are viewing the file in the Segger IDE. The second set is for C programs and are the ones that you will use.

subset of C which provides a limited set of data types, timing functions, and character, bit, string, and math operators. It supports generic peripherals including pin input and output, data conversion, and serial communication. External pins are abstracted to D<n> for digital I/O pins and A<n> for analog. This means you can substitute one Arduino board for another, even one with a very different processor, with little to no change to your code.

An Arduino user program consists of two functions: (1) an initialization function, `setup()`, which is executed once, followed (2) by a `loop()` function which is executed forever. You use the Arduino IDE to write, compile, and download code (usually using USB) from a PC to your target board. The IDE is available for Windows, Linux, and Mac (both Intel and ARM). The original (Version 1) IDE has no debugging facilities other than a print to console. The new Version 2 IDE includes a debugger allowing breakpoints and single stepping, but is limited to only a subset of Arduino boards using the SAMD architecture.

Our answer to our students (and to you) when asked why they have to learn all this C, hardware, and CMSIS stuff is – if you can do what you want with an Arduino, then you should use an Arduino. It is easy to learn, requires little work to move to a different processor and is a great solution for small/simple microcontroller applications. It is what we teach students in our undergraduate course.

In this book, we are offering you a solution for when your application outgrows the Arduino model. Some of the limitations of Arduino that we circumvent by going directly to the hardware include:

More flexibility using I/O pins: Arduino limits you to setting a pin as an input, an output or an input with a pullup. You cannot set pulldowns and you can only set one pin at a time to a high or low value. You cannot set or read multiple pins simultaneously, nor can you float a pin.

Better input data conversion: Arduino has an `AnalogRead()` function that returns a 10-bit conversion from the microcontroller's ADC. An extension supports higher resolution on a small number of processors. You do not have control of any of the ADCs parameters such as averaging or differential mode and you need to call the function for each sample; there is no way to create an interrupt service routine to generate samples at equal time intervals or at high speed.

Better output data conversion: The default `AnalogWrite()` function provides a pulse-width-modulated digital signal on an output pin. This is low frequency waveform (500 Hz to 1000 Hz) unsuited to reproducing even voice output, much less music. On a limited number of microcontrollers, the internal DAC is supported as well, but the limitations on configuration noted above for the ADC still apply.

No access to internal peripherals: Arduino includes a limited set of timing functions and serial communications. You cannot access a timer counter to program it for PWM as we will do in §25L.1.4 to drive a hobby servo or to generate a sine wave at precise intervals as we will do in §25L.2.3.5 to play a song. Similarly, except for the limited functions included in the Arduino model, you have no access to whatever nifty peripherals are available in the processor you are using. For example, you can use one serial channel for each of the popular serial protocols, SPI, I2C, and UART, but if you need more, you are out of luck.

No way to write real-time applications: A common use of microcontrollers is in real-time systems where responding to external events in a short, precise amount of time is crucial. We do this in the labs by having the occurrence of a hardware event interrupt the processor. We then write a function that is called when the event occurs. While Arduino allows you to attach an interrupt function to an external pin, it does not allow you to do the same with internal devices.

Superloop program model: Finally, the Arduino model forces you to write your application as a single looping function (a "superloop"). This works for small programs/applications but imagine creating a large application, say an oscilloscope, using this model. You would need to monitor user input, sample data input, output waveforms to the display, and update user indicators all in one big loop while maintaining the strict timing necessary to ensure data acquisition fidelity. Such a program is difficult to get right and hard to maintain. Instead, we are going to be using a tiny operating system that allows us to write an application like this as a series of independent tasks with a set of intertask communication channels to pass information between them. The result is code that is robust, easier to debug, and easier to maintain and modify.

This is only a short list of benefits of programming the microcontroller "bare metal" rather than as an Arduino. Of course, the disadvantages are that what we are doing is harder, requires an understanding of the internals of the microcontroller, and may require a significant rewrite if you need to change microcontrollers. In addition, the Arduino world includes software libraries that support a multitude of external peripherals including LCD display, pressure sensors, gas detectors, LIDARs, etc. So, use an Arduino if you can – but keep these lessons in your back pocket in case you outgrow it.

22N.4.2 Why not let [insert the name of your favorite AI here] write my code?

Well, this might work seamlessly someday, but we are not quite there yet. When we first tried to let AI do our work for us a year ago the result was a mess. A more recent (Jan 2024) attempt was much better but not perfect. Asked to write a simple CMSIS program to blink an LED, the resulting code turned on Port I/O in the power manager (unnecessary but not fatal), used software timing loops without any indication of what the CPU clock should be, and #included CMSIS files from a development environment other than the one were were using. Assuming you had already completed the microcontroller chapters in this book, this would have provided a helpful start to the task. But if you had no idea how to code using CMSIS, you likely would have struggled to get the code to compile, much less work properly.

A more complex request to initialize the SAMD21 DAC for a 0 to 1 V output was only partially successful. The code properly enabled the DAC in the power manager and connected the CPU clock to the peripheral, but it connected the DAC reference to the analog supply, not the 1V reference, and oddly tried to set a prescaler that does not exist. It also failed to include write synchronization on registers that require it.[39] Again, helpful if you have the experience to modify the code to work but not the panacea the AI folks would have you believe it is.

22N.4.3 Ok, but why can't we just use a wizard?

Most manufacturers of an ARM microcontroller provide some sort of software "wizard" to automate the generation of code to use their product. Microchip offers their MPLAB Harmony v3 to develop code for their ARM Cortex-based devices: see `https://LAoE.link/Harmony.html`. Wizards are a viable way of creating applications that are able to take advantage of all the features of a particular microcontroller. But in our experience, wizards generate impenetrable code, using lots of very short functions strung together to initialize and use internal devices. They also tie you tightly to a single manufacturer's products. In addition, there can be a steep learning curve to get familiar with a new wizard and the code generated is not portable to other manufacturers' microcontrollers. (To be fair neither is the code we present here.) Finally, the wizard's output may not work with your preferred IDE; typically only a few development environments are supported.

---

[39] Don't worry too much if these details don't mean much to you yet. We will explain it all in Chapter 24N when we initialize the DAC.

Our goal is to show you how to build a working system by using the microcontroller datasheet to understand how the CPU and internal peripherals work, giving you total control in configuring these devices. Once you wade though our low-level, understand-how-the-thing works approach, you may decide never to start coding from scratch again (and we are not sure we would blame you). But we believe that understanding these devices at a hardware level will allow you to better utilize the resources available in a modern microcontroller, even if you decide to take an easier path in future projects.

22N.4.4 We are your wizard

But all is not lost. In earlier versions of this material we asked students to decipher the peripheral sections of the SAMD datasheet and write the initialization code for several of the devices we are using. On reflection, we decided instead to provide the initialization code and leave the fun part for you – writing the application code for the lab exercises. Nevertheless, we still take the time to describe how each peripheral works and why we initialize it as we do, so that you will be able to use any peripheral in the future, even if we do not use it here.

22N.5 The first day of the microcontroller lab

Since the Sparkfun SAMD21 mini breakout board is a complete single-chip microcontroller, you'll have almost nothing to build before you can try out some programs. Hardware-wise, you need only tie the on-board 3.3 V regulator to five volts, add an external LED and current limiting resistor and connect the board to your PC with the J-Link debugging hardware. You will also need to install the Segger Embedded Studio for ARM development environment on your PC.

Since the lab includes next to no hardware work, you will be free to spend your energies on getting used to the development environment and the debugging interface. You will find that you can compare the C code you compile with the assembly code loaded into the SAMD21's flash memory and you can watch the processor's registers change as you single step the code. The first day's program does no more than blink the LED – providing the pleasure that we have admitted everyone seems to want from a controller.

22N.6 AoE reading

Chapter 14 (Computers, Controllers and Data Links)
Introduction through §14.1.2: Some terminology:
- Microprocessors, microcontrollers.
- Computer architecture: CPU and data bus.
- Memory.

§14.2.1 Assembly language and machine language.
§14.3.10 Direct memory access: introduction.

Chapter 15 (Microcontrollers)
§15.1 Introduction.
§15.3 Overview of controller families.
§15.9.1 Software.
§15.10.2 When to use microcontrollers.
§15.10.3 How to select a microcontroller.

Web resources re: the SAMD21

SparkFun SAMD21 Mini Breakout:
- SparkFun SAMD21 Mini Breakout documents:
  `https://LAoE.link/SparkFun_SAMD21_Mini.html`.
- SAMD21G18A datasheet:
  `https://LAoE.link/Microchip_ATSAMD21G18_Datasheet.pdf`.
- Cortex-M0+ Technical Reference Manual w/ instruction set summary:
  `https://LAoE.link/Cortex-M0+_Technical_Reference_Manual.html`.
- SAMD 1x/2x MCU Family Reference:
  `https://LAoE.link/Microchip_Developer_32bit_Help.html`.

Cortex-M0+ Development Tools:
- Segger Embedded Studio:
  `https://LAoE.link/Segger_Embedded_Studio.html`
  `https://LAoE.link/Embedded_Studio_Download.html`. (downloads)
- Segger J-Link EDU Mini:
  `https://LAoE.link/J-Link_Product_Page.html`.
- CMSIS-Core (Cortex-M) Reference:
  `https://LAoE.link/CMSIS-Core_Reference.html`.

22L Lab: Microcontrollers I

22L.1 Introduction to the SparkFun SAMD21 Mini

Modern microcontrollers such as the Cortex-M series require a power source between 1.62 V and 3.63 V, plus a few capacitors, to be usable.[1] The SparkFun breakout board we are using includes a 3.3 V voltage regulator so you can power it from 5 V, along with four LEDs, a USB connector, I/O headers, and a 10-pin SWD (Single Wire Debug) connector: see Fig. 22L.1.[2]

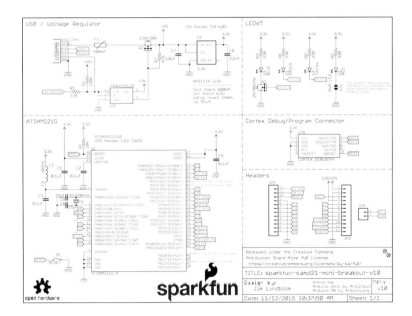

Figure 22L.1
The SparkFun SAMD21 Mini schematic.

The board can be powered either through the USB connector, by connecting a 3.3 V regulated supply to the 3.3 V header pin (labeled vcc on the graphical datasheet) or by connecting a 3.5 V to 6.0 V source to the V_{in} header pin.[3] You are going to power the SparkFun board from the breadboard's +5 V supply through a 1N5817 Schottky diode. The diode isolates the breadboard supply from the five volts on the USB line should you want to connect the USB to a PC. A red LED is lit whenever the board is powered, two additional LEDs (yellow and green) indicate serial communications to and from an

---

[1] See Figure 8-1 in the datasheet.
[2] Original schematic at https://LAoE.link/SparkFun_SAMD21_Mini_Schematic.pdf.
[3] If you are powering the board from +5 V, you can use the 3.3 V "vcc" header pin as an *output* to supply several hundred milliamps to external devices. If you do not have a separate 3.3 V supply, you can power the LED in this lab from the vcc header pin.

attached PC and there is a blue LED that can be turned on and off under program control. The blue user-controlled LED is driven by a N-Channel MOSFET so it is on when Arduino pin D13/SCK is high.

The "Cortex Debug/Program Connector" is a standardized hardware interface used both to program the internal flash memory and to debug program code by single-stepping program instructions and reading and writing the internal registers of the CPU. You will use a software integrated development environment ("IDE") with a hardware interface board, the Segger J-Link, to connect the debug header to a personal computer to write and compile programs, download them to the SAMD21, and then debug your code.

An external 32.768 kHz crystal provides an accurate timing source for real-time applications. The SAMD21G18A has an internal 8 MHz oscillator built in, but it is not as precise as a crystal oscillator.

22L.2 Install the SparkFun SAMD21 Mini

Install the SparkFun mini on your breadboard and connect power and ground as shown in Fig. 22L.2. You should allow room on the left side of the board for the LCD display board, which is three inches wide. You will not be using the USB port, so you do not need to leave space to get access to the connector. The board will be powered from the +5 V breadboard supply. While the microcontroller itself runs on 3.3 V (and *no I/O pin on it should ever be connected to more than* 3.6 V), the VIN pin on the board goes to a 3.3 V voltage regulator that accepts a 3.5 V to 6.0 V input. You should connect the breadboard +5 V supply to the VIN pin through a 1N5817 Schottky diode. This protects the breadboard power supply in the event the USB connector is plugged in at the same time.

Figure 22L.2 SparkFun SAMD21 Mini installation.

Next, install an external high-efficiency green LED near the board with a current limiting resistor from +3.3 V to the LED anode. Connect the cathode of the LED to the pin marked "10." Choose the resistor value to limit the LED current to no more than 2.5 mA. Test your installation by turning on the breadboard. The red "PWR-LED" on the SparkFun board should light.

Turn the breadboard power off and connect the J-Link Mini Edu debug hardware to the SparkFun breakout board with the supplied 10 pin flat ribbon cable. The red stripe on the cable should be near pins 8 and 9 on the SparkFun board and near the words "J-Link" on the debugging board. The pins on the connectors are delicate, so do not force them.

Finally, connect the USB connector on the J-Link to your computer with either the short cable included with the J-Link or a longer standard USB-to-micro-USB cable.

22L.3 Testing your development environment

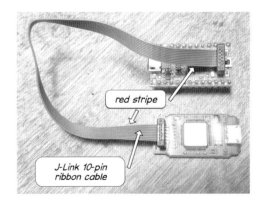

Figure 22L.3 J-Link debugger connection.

22L.3 Testing your development environment

If you have not yet installed the Segger IDE, follow the directions in the Supplementary Notes (Chapter 22S) to download and install it.

22L.3.1 Create a new project

From the IDE home page (if you're not there, open the View menu, and select "Dashboard"), select "Create New" in the Projects panel and choose "A C/C++ executable for Atmel SAMD21." If prompted, select "Create project in a new solution." Give your project a descriptive name, and then press "Next" to open the "Choose common project settings" dialog (Fig. 22L.4).

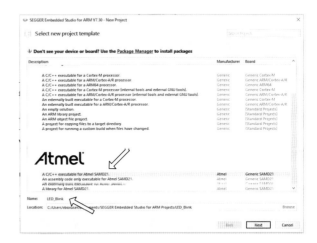

Figure 22L.4 Select a new project template and give it a name.

Set the CPU to the Atmel ATSAMD21G18A by clicking on the three dots on the right of the Target Processor field, then press "Next" (Fig. 22L.5).

Accept the default settings in the "Select files to add to project" and "Select configurations to add to project" dialogs that follow. Press "Finish" to create the project. The project explorer will open. You may need to use the tree view on the left side of the screen to open "Source Files" and double click on "main.c" to view the default main() function.

You should repeat these steps each time you create a new project file.

894 Lab: Microcontrollers I

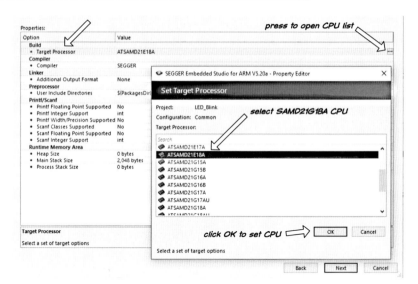

Figure 22L.5 Set the target processor to the Atmel SAMD21G18A.

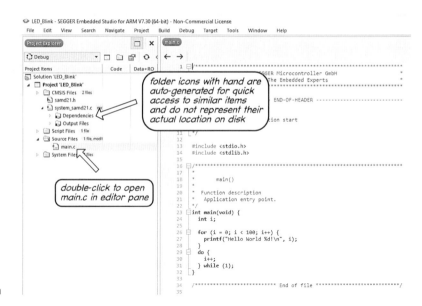

Figure 22L.6 The files in the Project Explorer are displayed hierarchically. Open the main.c file in the editor.

22L.3.2 Blink the on-board blue LED

Delete all of the default main.c code and replace it with the following:[4]

```
/***************************************
 * Learning the Art of Electronics
 * David Abrams 2023-03-19
 * SparkFun SAMD21 Mini Development
 * Environment Test Program
 ***************************************/

#include <stdint.h>        // needed for int32_t data type
```

---

[4] Code available at https://LAoE.link/micro1/22L_Environment_Test.c.

```
// define access to port A registers
#define DIRSETA  *((volatile int32_t *)(0x41004408))
#define OUTCLRA  *((volatile int32_t *)(0x41004414))
#define OUTSETA  *((volatile int32_t *)(0x41004418))
#define BLUE_LED (1 << 17)       // bit within register for led (active high)
                                 // connected to Arduino pin D13 (Port Pin PA17)

int main (void)
{
    DIRSETA = BLUE_LED;       // set PA17 to output

    while (1) {
        OUTSETA = BLUE_LED;   // turn LED on (active high)
        OUTCLRA = BLUE_LED;   // turn LED off
    }
    return 0;
}
```

Build (i.e., compile and link) the test program by pressing F7 or by selecting "Build/Build LED_Blink" from the IDE menu. The code should build with no errors in the "Output" window.

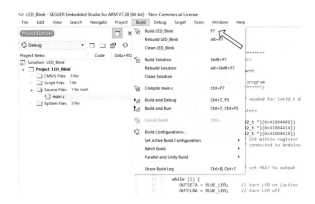

Figure 22L.7 Build the project object file.

Make sure the J-Link is connected to your computer and the SparkFun board is powered on (the red LED should be lit). Press F5 or select "Go" from the IDE Debug menu to start the debugger. If the Disassembly window does not replace the Project Explorer on the left side of the IDE, open the View menu and select "Disassembly."

Single step through the C code. You should see the blue LED on the SparkFun board come on when you execute Line 20 and off when you execute Line 21. If this works properly, you have successfully set up the development environment.

As we have said before, everyone gets a kick out of seeing an LED blink when that blinking means that something you have built is working. Primum lucernam micare.[5]

22L.4 A debugging primer

The Segger Embedded Studio debugger allows you to single step through your program (in either C or through the disassembled code), set breakpoints, view the microcontroller registers, examine variables

---
[5] "First flash the lamp." In fact, LED-love goes deeper than this. An LED may not even need to blink to keep us happy. A glowing LED reassures anyone who plugs in a well-designed battery charger. The chargers that save 50 cents by omitting the LED seem likely to fade from the market. Users like reassurance.

Lab: Microcontrollers I

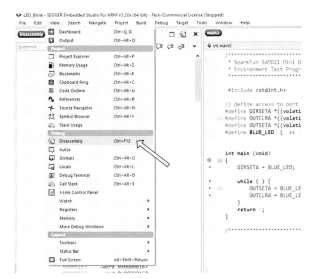

Figure 22L.8 Display the disassembled C code.

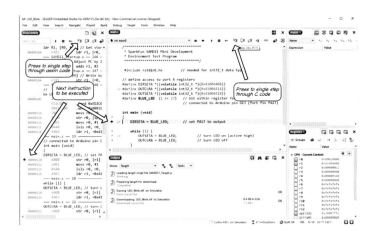

Figure 22L.9 Single step through the code to blink the blue LED.

and memory, and output to a virtual console (by opening the "Output" window) with `printf()` statements.

You can single step through each line of C code (in the C window) or each line of machine code (in the disassembly window). This icon steps *into* function calls. The icon steps *over* function calls. It is especially useful to avoid stepping through library calls such as `printf()`. Nevertheless, if you inadvertently step into a function call, you can run to the end of it with the icon. Finally, the icon runs to the cursor which is useful when you don't want to set a breakpoint just to skip some code.

The yellow arrow and highlighted line indicate the next instruction to be executed in both the disassembly window and the C code. While the next machine language instruction highlight will always be correct, be aware that the highlighted line in the C source window may not be. This often occurs when executing a loop in which a pointer or register was set up by a source line *before* entering the loop and reused in the loop. In that case, the highlight may jump to a source code line before the loop expression (i.e., before the `while()` or `for()`). If you see this behavior you can always look at the disassembly window to see why it is occurring but don't be concerned, the program is executing the loop instructions as you intended.[6]

You can right click on a variable and go to its definition in the source code or display its current value. You can also create a "watch" entry in the watch window to show the current value of a variable when single stepping or when you reach a breakpoint. Registers and watches that are changed when single stepping or a breakpoint is reached display in red text. However, registers, memory, and watches are not updated while running full speed.

You set breakpoints by clicking in the margin to the left of the line of code you want to break at. Breakpoints can be set in the C code or on an assembly instruction. You can then press the Go icon and the program will run full speed until it reaches the break point. You can enable and disable breakpoints by positioning the cursor on the line with the breakpoint and pressing Ctrl-F9. A disabled breakpoint is a displayed as a hollow circle.

The ← icon restarts the debugger at the beginning of main(). The icon ∎ stops the debugger and returns to the editor.

22L.5   Blink the external LED

Once you have a working development environment, you are ready to start using the Cortex-Microcontroller Software Interface Standard (CMSIS) to write programs. CMSIS provides a level of hardware abstraction which frees you from having to create structures and pointers for the features of each individual processor you use. You are going to toggle an I/O output pin again, but this time using CMSIS to avoid having to define the registers and pin values. Rather than blink the on-board LED, you will blink the external high-efficiency LED you installed previously in section 22L.2. The LED is configured active low because a SAMD21 output I/O pin can sink about 25% more current than it can source.

Create a new project following the instructions in section 22L.3.1. In the Segger IDE, Projects are a complete executable existing in a "Solution." A Solution can hold one or more Projects. We recommend creating a new solution for each project to keep your project tree simple. Give this new project a descriptive name like "Blink_External_LED."

Once the project is created, click on main.c in the Project Explorer and replace the default main.c with the following skeleton code.[7] You will use this skeleton (or an updated version you develop from this skeleton) for each program you write in all the microcontroller labs.

```
/***************************************************************************
* Name:
* Create Date:
* Target Device:   Sparkfun SAMD21 Mini Breakout (SAMD21G18A)
* Description:
*
* Revision:
*    Rev 0.01 - File Created
* Additional Comments:
***************************************************************************/

#include "samd21.h"      // add to use CMSIS
#include <stdio.h>
#include <stdlib.h>
```

---

[6] You can see an example of this behavior in our solution to §22L.5.
[7] Code available at https://LAoE.link/micro1/22L_Blink_External_LED_Skeleton.c.

Lab: Microcontrollers I

```
/****************************
 *     main()
 *
 *  Application entry point
 ****************************/

int main (void){

    // Do program setup here

    while (1) {

    // Do program loop here forever

    }
    return 0;
}
/************************** End of file ***************************/
```

If you have done any Arduino programming you will recognize the format of this program. It does any required initialization once (as does the Arduino call to the `void setup()` function) followed by an endless loop (the Arduino `void loop()` function).

The first line of code

```
#include "samd21.h"      // add to use CMSIS
```

is required to use the CMSIS-Core (Cortex-M) features, including the Hardware Abstraction Layer (HAL), system initialization functions, and intrinsic CPU instructions not supported by the C language. To actually bring these files into the project explorer so you can view them, compile the empty main function by selecting "Compile main.c" under the Build menu.[8]

Open the Dependencies folder under `main.c` in the project items tree. You will see a long list of CMSIS files specific to the SAMD21G18 processor. To use a hardware feature of the CPU, you will need to look at the associated include (".h") files to see what are the CMSIS name of the structures, variables, and registers you need for access. In this lab (and the next) you will be using Port I/O. The port I/O definitions are contained in two header files, both named port.h (the same named files exist in separate directories so there is no naming clash, but that is not obvious from the IDE tree listing).[9]

Pin definitions and memory map information are contained in two other header files named `sam21g18a.h` located in the `\Include\pio\` and `\Include\` subdirectories respectively.

Use the SparkFun SAMD21 schematic or the Graphical Datasheet to determine which microcontroller pin is connected to the LED you wired to the SparkFun board. Add the appropriate program setup code to set this I/O pin as an output and then initialize the LED off. Modify the skeleton main program loop (the `while()` loop) to first turn the LED on and then to turn it off. This will allow you to blink the external LED as you single-step through the program.

Compile and run your program. Step through the code to see if it works correctly.

22L.6   Add a delay to permit full speed operation

The program you have only works when single stepped. If you run it full speed, the LED looks like it is always on. Try it. Connect your scope to the LED output I/O pin. What is the frequency of the

---

[8] If you get a *'samd21.h' file not found* error, open "Show Installed Packages" from the Tools menu and make sure you have Version 2.02 of the SAMD21 CPU Support Package installed. Newer versions will not work.

[9] The PORT register definition `include` file resides in the "instance" subdirectory, while the PORT structure definition `include` file is in the "component" subdirectory.

output signal? Does this seem reasonable? Hint: Count the number of instructions in the delay loop in the disassembly window. The Sparkfun board runs at 1 MHz after a reset and most instructions take one clock cycle to execute. This should give you a starting point for the correct delay constant value.[10]

Now add two identical in-line ≈ 500 ms delay routines (i.e., do nothing loops) after each output instruction to get a one second blink rate.

22L.7 Make the delay a function

With these additions, this small program is starting to look quite ugly: the important, central loop is hard to make out because the flow is interrupted by the patches of code that implement the delay. In addition, if you reuse this delay code in many places and need to change it, you must make sure you make the changes everywhere it is used.

We will improve this situation by moving the delay routine into a subroutine, i.e., what is known as a function in C. A subroutine allows you to write code to implement a common function once and use it wherever it is needed. A subroutine works by transferring control from the code that is executing when the subroutine call instruction is encountered, then returning after the subroutine has completed execution. The assembly language instructions that implement subroutine calls are the *call* instruction, which takes as an argument the address of the subroutine to execute, and the *return* instruction which is the last instruction in the subroutine and causes it to return to the instruction after the call instruction.

The call instruction must first save the return address (which is the current value in the PC register (R15) when the call instruction is executed) somewhere so that the return instruction can find it. It then places the address of the subroutine in the program counter to transfer execution to the subroutine code. When the subroutine has finished executing, the return instruction replaces the program counter value with the saved return address, returning control to the calling code.

Move your in-line delay code to a new C function that takes an argument to set the delay in milliseconds:[11]

```
void Delay(uint32_t mSec){}
```

You will want to add new functions such as this one to your code *before* they are called so you do not have to add prototypes to avoid compiler errors. Call this function after each I/O output instruction. Single-step through the assembly code of the function call to see how subroutine (function) calls work in the ARM processor.

In ARM Cortex-M assembly language, the subroutine call instruction is

```
BL <label>      ; branch to subroutine at label
```

where `<label>` is the address of the subroutine to branch (jump) to.[12] This instruction copies the value in the program counter (PC - R15) to the link register (LR - R14), then loads the address of the subroutine into the PC. This causes program execution to transfer to the subroutine.

At the end of the subroutine, the code must execute the return instruction:

```
BX LR           ; branch indirect to location specified by Rm
```

---

[10] When we tried this, our delay loop comprised eight assembly language instructions. That suggested we try an initial delay count of (500 ms/8 μs) = 62,500.

[11] Hints: Use a `#define` for the number of times to loop so it is easy to adjust the delay. Since the CPU runs on a default 1 MHz clock after a reset, you may need a fairly large value to get the desired delay. If your delay does not seem to be working, look at the disassembled code. Some compilers will eliminate any loop that does nothing. You may have to put some "make work" instructions in your delay loop (increment a variable for example) to get the compiler to retain the loop in the output executable. In the next lab we will see how to create a much more accurate Delay() function that does not depend on the microcontroller clock frequency.

[12] In this assembly language, anything after a semicolon to the end of a line is a comment and is ignored. The comments shown here are from the Cortex-M assembly language reference `https://LAoE.link/Cortex-M_Assembly_Language.html`.

Lab: Microcontrollers I

This instruction just copies the contents of the register specified to the program counter.[13]

Notice that the ARM architecture call instruction places the return address for the subroutine in the LR register, and the subroutine moves the contents of the LR register to the program counter (PC) to return to the calling program. What might be a problem with this scheme for subroutine calls? Why do you think the ARM does it this way?[14]

22L.8 Use read-modify-write to toggle the LED

Instead of setting the output pin high then low, use a read-modify-write operation to toggle the output I/O pin. First read the value of the Port A IN register into a variable, then invert *only* the bit the LED is connected to with a bitwise XOR, and finally write the entire thirty-two bits back out to the Port A OUT register.

You will need to add the following line to your setup code to enable the port for reading:[15]

```
PORT->Group[0].PINCFG[PIN_PA18].reg = PORT_PINCFG_INEN;
```

Since you are toggling the LED port pin, you will only need to call the `Delay()` function once each time through the `while()` loop.

Take a look at the assembly language generated by this method of toggling a single output I/O pin. Notice the multiple machine language instructions required.

22L.9 Use a single register write to toggle the LED

Reading the port, changing a bit, then writing it back (Read-Modify-Write) becomes a problem once we start to use interrupts, as we will in future labs.[16] You can avoid these problems if you toggle the LED I/O pin in a single non-interruptible operation. Take a look at the PORT I/O section of the M0+ datasheet (Chapter 23) and find a better way to accomplish this. Hint: The registers with SET, CLR, and TGL in their name implement atomic operations. Compare the assembly language instructions generated using this intrinsic read-modify-write operation with the assembly code generated by the C read-modify-write operation you tested in §22L.8. Be sure your code is properly documented and commented.

A better way to delay? But does it not seem perverse to take a fast processor and then to slow its operation to a crawl by wasting time doing nothing in delay loops? Yes, it is rather perverse – and there is, indeed, a better way to slow execution than by trapping the processor in a loop for hundreds of thousands of cycles. This better way we will explore in the next lab uses the controller's hardware *timers*. These can be loaded once with a delay value, and then told to notify the main program when the delay time has expired. This scheme leaves the processor free to do something useful during the timekeeping process: a much better plan than the delay we demonstrated in §22L.6.

---

[13] "Rm" refers to any of the general purpose registers; for a subroutine return this is normally the link register LR.

[14] This only works for one level of function call. If the function called tries to call another function, the original return address in LR will be overwritten. Most computer architectures use a "stack" in memory to store return addresses in a first-in, last-out fashion to support as many levels of subroutine call as necessary. In the ARM architecture, register to register moves are very fast (taking one clock cycle) but memory access may require multiple clock cycle "wait states." Rather than always using memory for the return address, ARM puts the onus on the subroutine to save the return address to memory if it needs to call another function. This means there is no speed penalty if there is only one level of subroutine call.

[15] We will discuss how this instruction works in the next lab.

[16] See §22N.3.1.2 for an explanation of the problem with read-modify-write operations.

22L.10 Solutions

Here are our solutions to the parts of this lab in case you get stuck. **As always, we suggest you try to get your own version working *before* looking at our answers.**

- §22L.5: https://LAoE.link/micro1/22L_LED_Blink_CMSIS.c.
- §22L.6: https://LAoE.link/micro1/22L_LED_Blink_Delay.c.
- §22L.7: https://LAoE.link/micro1/22L_LED_Blink_Delay_Function.c.
- §22L.8: https://LAoE.link/micro1/22L_LED_Toggle_RMW.c.
- §22L.9: https://LAoE.link/micro1/22L_LED_Toggle_Atomic.c.

22S Supplementary Notes: Microcontrollers I

22S.1 Preparing the SparkFun SAMD21 Mini Breakout

As delivered, the SparkFun mini is missing both the 0.1 inch headers used to connect its inputs and outputs to the breadboard as well as the 2×5, 0.05 inch debugging header used by the J-Link EDU programmer/debugger. Before you can begin the lab, you must obtain and solder these components to the SAMD21 breakout board. Figure 22S.1 lists part numbers for each item from multiple sources.[1]

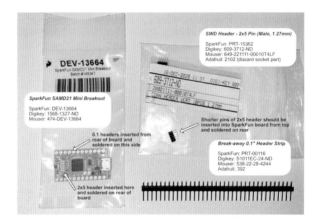

Figure 22S.1 SparkFun SAMD21 Mini and required headers.

Insert the shorter pins of the 2x5 SWD header from the top of the board and solder from the rear. Use care to avoid solder bridges on the closely spaced pins of this connector: see Fig. 22S.2.[2]

Next, cut or break the 0.1 inch header into two 12-pin long strips. These are inserted from the rear of the board with the black insulating material and long pins under the breakout board. Take care to make sure they are perpendicular to the circuit board.[3] These get soldered from the front of the board. Use flux remover or rubbing alcohol to clean off any excess flux from all solder connections.

22S.2 Installing Segger Embedded Studio for ARM

You will use the Segger Embedded Studio for ARM to write, compile, download and debug code on the SparkFun board. The Embedded Studio is an all-in-one solution and is available for Windows, MacOS

---

[1] See online content Chapter 20O (https://LAoE.link/LAoE_Chapter_22O.pdf) for information on alternative SAMD21 boards that are supplied with the debugging header preinstalled and the online parts list for any part number updates.

[2] Use solder wick to remove excess solder if necessary.

[3] Placing the headers in a breadboard strip before soldering helps keeps them vertical, but be careful not to overheat and melt the plastic breadboard.

22S.2 Installing Segger Embedded Studio for ARM

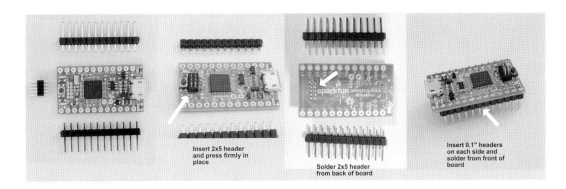

Figure 22S.2 Installing headers on the SparkFun SAMD21 Mini.

(both Intel and ARM) and Linux. The IDE contains a code browser, code editor, C/C++ compiler, C libraries, linker, and a debugger.[4] It is free for noncommercial and educational use.

You will need to download and install the IDE before beginning Lab 22L. Please check the errata page at https://LAoE.link/Installing_EStudio.html before you begin to see if there have been any updates or changes to the procedure below.

22S.2.1 Download the IDE

Download the *Embedded Studio for ARM* software for your operating system from the Segger site (https://LAoE.link/Embedded_Studio_Download.html). Select version 7.30 from the dropdown box before downloading: see Fig. 22S.3.[5] You should also download the Embedded Studio Reference Manual from the same page.

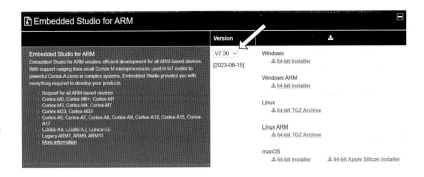

Figure 22S.3 Download version 7.30 of the IDE for your operating system.

22S.2.2 Install the IDE

Install the Embedded Studio software.[6] If using Windows, be sure that the "Install J-Link Device Drivers" checkbox is selected in the Additional Components dialog.[7]

---

[4] It also includes a M0+ simulator so you can test code without the SparkFun board and debugging hardware.
[5] The examples and labs in this book were tested with version 7.30 of Embedded Studio. We will post to the errata page as we have an opportunity to test newer versions.
[6] On MacOS, you may need to go into System Preferences, Security & Privacy general settings to allow the installation of apps from identified developers.
[7] You may not see this dialog on every OS installation. It appears that J-Link support is auto-installed on MacOS.

Figure 22S.4 Be sure to install the J-Link drivers.

When the installation is complete, *unplug the J-Link if connected* and start the program.[8] Answer "yes" if you are asked if you want to create a source directory.

The first time you open the program you need to add the software packages necessary to support the exercises in the labs. From the Tools menu, select "Package Manager." Press "Yes" to download available packages.

Figure 22S.5 Download packages to get CMSIS and SAND21 support.

You want to install:

- CMSIS 5 CMSIS-CORE Support Package;
- CMSIS 5 Documentation Package; and
- Atmel SAMD21 CPU Support Package V2.02 (if Atmel is not listed, try Microchip).

Change the Action column for each item to "Install", and then press "Next" twice to download and install the packages. *You will need to change the default version of the SAMD21 CPU Support Package to V2.02.* Newer versions will not work with the exercises in the labs.[9]

Confirm that the selected packages have been loaded. Press "Finish" to complete the IDE installation.

Finally, you should not update either Embedded Studio or any of the installed packages unless the errata page says we have tested the update and found it compatible with the book material. (*Do not update the SAMD21 support package beyond V2.02.* Microchip has made changes to the CMSIS support in newer versions that do not work with the labs.)

---

[8] The program may hang if the hardware is connected on the initial startup.

[9] Segger has variously listed the SAMD21 under Atmel and Microchip in the Select Packages dialog. In either case, you will need to select V2.02 to get the correct definitions to use with the examples in LAoE.

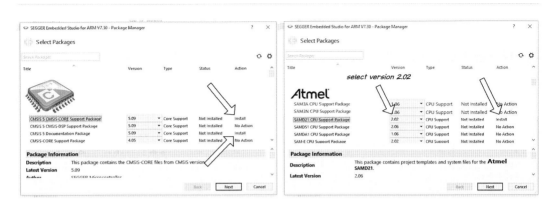

Figure 22S.6 Set these three items to "Install".

Figure 22S.7 Make sure that CMSIS-CORE and SAMD21 V2.02 have been installed.

22S.3 Programming in C

We (the authors) had it easy in the first edition of LAoE. We could pretty much assume none of our readers were familiar Intel 8051 assembly language and just teach it to you. In this edition, we will be programming the ARM Cortex-M0+ in the C programming language. Some of you may already be expert C coders while others have never looked at the language. We have tried to accommodate you all.

The good news is that, with a few exceptions, we will be writing very simple C code. Indeed, if you have programmed an Arduino you already know the syntax, control structures and operators of C. Most of your code will be structured similar to an Arduino program, with an initial setup section followed by an infinite loop for the application code. We will not be using floating point variables or any of the object-oriented features of C++, and we will barely bother with strings other than to send fixed messages to the console or LCD display.

Our exceptions to Arduino-like code are the use of pointers, structures, and unions to access the internal features of the microcontroller. We have endeavored the first time we encounter each of these constructs to explain them in simple terms that we hope will make them manageable. Nevertheless, even if you do not quite grok the concept of these advanced C features, you will not need to use them to create new code, you will merely need to follow our examples to access the registers and bits of the ARM microcontroller.[10]

For those of you unfamiliar with both C and Arduino programming (or if you just need a quick refresher), Stanford University has a nice summary in its CS Education document "Essential C" available at https://LAoE.link/Stanford_C.pdf. An online beginner's tutorial is available at https:

---

[10] If you are unfamiliar with the term "grok," immediately run out and read Heinlein's *Stranger in a Strange Land* (or just look it up in Wikipedia).

`//LAoE.link/C_Tutorial.html`. For an introduction to structures and unions, see the tutorials at `https://LAoE.link/Advanced_C.html`. Multiple videos for beginners available are on YouTube.com, one we like is `https://LAoE.link/Beginners_C_Video.html`.

22O Online Content: Microcontrollers I

22O.1 Alternatives to the SparkFun SAMD21 Mini

Available at `https://LAoE.link/LAoE_Chapter_22O.pdf`.

23N Microcontrollers II: Stacks, Timers and Input

Contents

23N.1	**The computer stack**	**908**
23N.2	**Subroutine (function) calls**	**910**
23N.3	**An improved delay function**	**911**
	23N.3.1 Polling vs. interrupts	911
	23N.3.2 Using the SysTick timer	912
23N.4	**SAMD21 port input**	**913**
	23N.4.1 Using structures to access registers	913
	23N.4.2 A mistake to avoid	916
23N.5	**Intrinsic functions for CPU instructions**	**916**
23N.6	**AoE reading**	**916**

Why?

Our task today is to add to our microcontroller the ability to take in information. We also would like a way to store temporary information efficiently and to not only get accurate delays, but to get delays without tying up the CPU so it can do other work in the meantime. Finally, we take a look at how CMSIS provides access to some unique Cortex-M instructions without forcing us to use assembly language.

23N.1 The computer stack

A common need in computer systems is an efficient way to allocate memory for temporary values. One option is to dedicate a fixed section of data memory to temporary storage, but this is inefficient if only a small fraction of it gets used. On the other hand, if too little memory is reserved the program may crash or overwrite other variables. The traditional method to dealing with this problem is to include hardware in the CPU's design to maintain a program stack.

The stack is an indirectly addressed region of data memory accessed through a dedicated register called the "Stack Pointer." This register is set to point to an unused region of memory at the start of program execution and is designed to provide a Last-In, First-Out data structure supported by machine language instructions to both place data items on the stack and retrieve them quickly. Data values are placed on the stack explicitly by the PUSH instruction and retrieved by the POP instruction or indirectly using the Stack Pointer. Stack data may also be stored and retrieved automatically by the subroutine call and return instructions.[1] The size of the stack grows dynamically, increasing when items are push'ed

---

[1] Subroutines are the low level equivalent of higher level language functions or procedures; independent blocks of code that can be executed ("called") from multiple places in the overall program. Subroutines use machine language instructions that implement a call and a return operation, allowing a program to jump to another address in program memory, execute the code there and then return to the instruction after the subroutine call.

23N.1 The computer stack

onto it and decreasing when items are pop'ed off of it. Stacks can grow either up (increasing memory addresses) or down, depending on the CPU architecture.

In the Cortex-M architecture, the Stack Pointer is register R13.[2] The stack grows downwards, to lower memory addresses, as items are placed on it. After a stack operation, the Stack Pointer points to the last value pushed on the stack, so it is pre-decremented for a push instruction and post-incremented for a pop. Since the stack grows downward, it is normally placed at the top of data memory, and the bottom of data memory is used for static data. On a processor reset, the stack is set to an initial value specified in the object file loaded into flash memory.[3] Stack transfers are always 32-bit wide and occur on word aligned boundaries in memory (i.e., the two least significant address bits are zero). The Stack Pointer is always incremented or decremented in multiples of four bytes.

On the SAMD21G18A, the 32k bytes of static RAM begins at address 0x20000000 and the last byte is at address 0x20007FFF. The Segger development environment locates static data at the bottom of RAM and initializes the Stack Pointer to address 0x20008000. While this points past the top of data memory, the pointer is predecremented by 4 to point to the last word in data memory before the first stack entry is stored. The three panels in Fig. 23N.1 show the operation of the stack for a register value push'ed then pop'ed just after a processor reset.

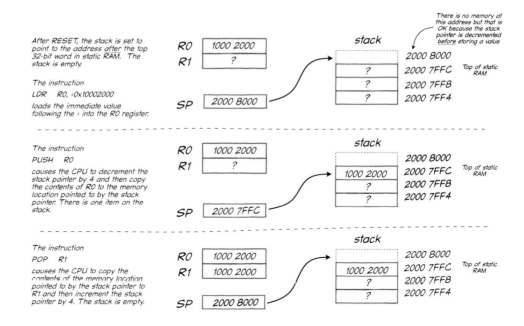

Figure 23N.1 Example of a stack operation immediately after a CPU reset.

In the top panel, the Stack Pointer has been initialized to point to the word after the top of static RAM, 0x20008000, and the value 0x10002000 has been loaded into register R0 by a load immediate data instruction. When a PUSH R0 instruction is executed (center panel), the Stack Pointer is decremented by four so it points to the last word in static RAM (at 0x20007FFC) and the contents of register R0 is

---

[2] There are actually two 32-bit Stack Pointer registers in the Cortex-M0+, but only one is accessible at a time as R13. This allows the creation of a separate stack for application programs when a supervisory program is used.
[3] In the SAMD21, flash memory begins at address 0x00000000. The first two 32-bit words stored in flash memory set up the stack and the start of program execution. After a hardware reset, the word in flash location 0x00000000 is copied to the SP (R13) and the word in flash location 0x00000004 is copied to the PC (R15).

copied to that location. This is all a function of the hardware built into the CPU to execute the PUSH instruction.

In the lower panel, a POP R1 instruction causes the data pointed to by the Stack Pointer to be copied into the specified destination (R1) and the Stack Pointer is incremented by 4 so it points to the last saved word (in this case a non-existent RAM location because the stack is empty). Notice that the POP instruction actually copies data, it does not change what is in memory, but this location will be overwritten on the next stack PUSH. Also, note that the source and destination registers need not be the same. The ARM architecture is unusual in that you can specify multiple registers as arguments to the PUSH and POP instructions to save and restore several registers in a single instruction. However, it is important to always match the number of items push'ed and pop'ed. To do otherwise is an invitation to disaster as the stack is also used to store program addresses to return to at the end of subroutines. If pushes and pops are not matched, it is possible to begin executing code at a random address rather than at the instruction after the subroutine call.

In C (and many other computer languages), the stack is also used to store a function's local variables. When you enter a function, the C compiler generates code to decrement the Stack Pointer by the size of the temporary local storage needed and then, before the return from the subroutine is executed, code is inserted to restore the Stack Pointer back to the value it had on entry. Local memory is reused by the stack once the function exits and the variables cease to exist. That is why the scope of a C local variable is confined to a function and its value is lost when you exit the function.[4]

The stack is also used by interrupts to save the state of the CPU and restore it on the return from the interrupt. Hardware interrupts can occur at any time when interrupts are enabled. Since an interrupt can occur between an arithmetic operation and a check for a zero result, overflow. etc., the interrupt code must save the state of the processor including any general purpose registers it uses as well as any flags it modifies and then must restore them before returning to the interrupted program. This information is typically saved on the stack.

23N.2 Subroutine (function) calls

In most computer architectures, calling a subroutine automatically pushes the value of the Program Counter on the stack, so the return address can be pop'ed into the Program Counter when the subroutine executes a return instruction. This allows subroutines to call other subroutines or even themselves recursively.[5]

The ARM architecture is unusual in that calling a subroutine stores the return address in the link register ("LR," register R14). If the called subroutine calls another subroutine, it is responsible for saving and restoring this return address. Because register moves are faster than memory access, and because subroutines often need to save registers anyway and the ARM supports saving/restoring multiple registers in a single instruction, this design is more efficient than the traditional stack based call/return design.

If you use the debugger to look at the value of the link register after a subroutine call, you will see it is set to an odd address, even though the actual instruction after the call is at an even address. Since machine language instructions on Cortex-M processors always are stored at even addresses, the CPU designers use the LSB of the return address as a flag to indicate that the M0+ processor is using

---

[4] In C, you can preserve a variable's value through function calls by declaring it as `static`. When you do, the compiler allocates storage for the variable in the static area of data memory along with global variables rather than on the stack.

[5] It also guarantees an ugly crash if a subroutine calls itself recursively without executing a return instruction until it runs out of memory. This is known as a stack overflow.

the 16-bit "Thumb" instruction set rather than 32-bit ARM instructions. This bit is cleared before the Program Counter accesses the next instruction on the return from the subroutine. You can see this by stepping through the machine code of a function called by another function. The calling function will push the odd value of LR before transferring control to the called function. After the called function returns, the calling function will return to the code that called it by pop'ing the stored value of LR into the Program Counter. However, the pop'ed value in the PC will be even rather than the odd return value that was on the stack.

23N.3   An improved delay function

There are several problems with creating a program delay by executing a software loop for a fixed number of times. It takes trial and error to figure out the number of loop iterations per time delay and you are limited to delays that are multiples of the time it takes to execute the loop. The delay is a function of the clock speed of the processor and how many clock cycles each instruction takes to execute. That means if you change the clock speed on your processor or change to a different processor in the Cortex-M family, you have to recalculate the constants that set the delay time. Also, the actual delay time can be longer than you expect if background events interrupt the loop execution. Finally, nothing else can get done while the microcontroller executes a "do nothing" loop; rather a waste of good CPU cycles.

A much better design uses the internal microcontroller system clock, which is quite accurate, along with a hardware counter/timer. This allows the program to continue doing useful work while the timer counts the desired delay. The SAMD microcontroller includes a 24-bit SysTick system timer specifically for this purpose – to generate events at timed intervals. The SysTick timer consists of a set of registers including a 24-bit Current Value timer register, a Reload Value register, a Calibration Value register and a Control and Status register.[6] The Current Value register counts down until it reaches a count of zero. This causes a bit, COUNTFLAG, in the Control/Status register to be set and reloads the Current Value register with the value in the Reload Value register. Reading the Control/Status register clears the COUNTFLAG if it is set. The read-only Calibration Value register holds the load value count for a 10ms period and is updated if the system clock frequency changes. If you set the reload register to the value of the 10 ms calibration register divided by 10, the timer will count down to zero, set the COUNTFLAG status bit to 1 to indicate it has reached zero (which means that 1ms has elapsed), then reset the counter to the reload value to time the next one millisecond interval. This all happens automatically once the SysTick registers are initialized.

You could then design an Arduino style `delay(ms)` function to continually read ("poll") the COUNT-FLAG bit and return after the desired number of milliseconds have elapsed. This solves the first problem, having to change the delay calibration value, since the SysTick calibration register ensures that the delay is correct even if the CPU clock speed or the number of clock cycles per instructions changes or the delay function is interrupted. However, no useful work is done while waiting in the delay function. The solution is to use interrupts.

23N.3.1   Polling vs. interrupts

The easiest way to understand the difference between polling and interrupts is to consider inviting your parents over for Thanksgiving dinner at your third-floor flat in the pre-cellphone era. If your doorbell is broken, you must run downstairs every so often to see if your parents have arrived. If you wait too

[6] See https://LAoE.link/SysTick_Timer.html.

long between checks, your poor parents will be huddled freezing on your doorstep. If you check too often, you won't get dinner done on time. This is how polling works. It is wasteful and prone to family discord.

The solution is to get your doorbell fixed. Now you can put all your efforts into preparing a lovely Thanksgiving feast knowing the doorbell will alert you immediately to your parents' arrival. When the doorbell rings, you stop what you are doing, run downstairs to let your folks in, then go back to cooking right where you left off. This is how interrupts work. An interrupt executes a function (answering the doorbell?) on the occurrence of a hardware event.

In the computer world, your program sets up the hardware conditions for an interrupt to occur and you provide a function (the "Interrupt Service Routine" or ISR) to be executed on that condition. The CPU automatically interrupts the program that is executing when the event happens (the "foreground" program), saves the current state of the processor, and then calls your ISR (the "background" program) to handle the interrupt. On the return from the interrupt, the processor state is restored to exactly what it was when the interrupt occurred and the foreground program continues from the next instruction, blissfully unaware that it has been interrupted. Interrupt service routine functions receive no parameters and they cannot return values, i.e., they must be declared "void ISR_Name(void)." However, an ISR can modify variables accessible to the foreground program to let it know the triggering event has occurred or that some new data is available (or that your parents are at the door). Note that since an interrupt can occur at any time, the ISR must be careful not to modify any resource used by any foreground program unless there is an agreement with *all* foreground programs on what can be changed.

23N.3.2 Using the SysTick timer

The SAMD21 SysTick timer hardware includes the ability to generate an interrupt when the count reaches zero. Fortunately, CMSIS supports the SysTick interrupt so you do not have to figure out how to setup the SysTick registers and enable the interrupt to get a constant timing source.[7] To use the SysTick timer, you need to configure the counter with the time interval desired using a CMSIS function:

```
SysTick_Config(uint32_t ticks); //returns 0 - success, 1 - fail
```

This function configures the SysTick timer to generate an interrupt every `ticks` CPU clock cycles.[8] You must also create a SysTick interrupt handler function which will be called when the interrupt occurs:

```
void SysTick_Handler(void)
```

You must use this exact name for your SysTick interrupt handler function.[9] The CMSIS initialization code contains a do-nothing `SysTick_Handler()` function but gives it a "weak" linker attribute. This means that any user function with the same name overrides the default weak function in the final program code.[10]

You can easily calculate the desired ticks value for the configuration call using the CMSIS system variable SYSTEMCORECLOCK which is the frequency of the SysTicks clock in Hz. For example, setting:

```
SysTick_Config(SystemCoreClock / 1000);
```

---

[7] See https://LAoE.link/CMSIS_SysTick_Timer.html.
[8] The SysTick_Config() function only fails if the ticks value is greater than $2^{24} - 1$.
[9] The ARM architecture is unusual in that an ISR is written as a normal C function. Most processor architectures require special instructions in ISRs to save and restore registers and flags and to return from the interrupt, so an ISR cannot be written (or called) as a ordinary function. In ARM, the CPU state and a subset of registers are saved and restored by the CPU interrupt hardware.
[10] The linker would normally give an error if the same function is defined twice in a program.

causes your interrupt handler function to be called once every millisecond.

A typical use of the SysTick interrupt is to maintain a count of time. For example:

```
volatile uint32_t msTicks = 0;  /* Variable to store millisecond ticks */

/* SysTick interrupt Handler. */
void SysTick_Handler(void) {
  msTicks++;
}
```

maintains a count of milliseconds elapsed since the CPU was last reset if the SysTick counter is configured with the value `SystemCoreClock/1000`. The global 32-bit `msTicks` variable holding the count must be declared as volatile since it can be changed by the ISR at any time. The "volatile" keyword tells the C compiler not try to cache the value but rather to get the value anew each time it is used. Note that the 32-bit `msTicks` value overflows every 49.71 days at a 1ms interrupt rate. If this is a problem you can declare the variable as a `uint64_t`.

23N.4   SAMD21 port input

Configuring an I/O pin for input is a bit more complicated than setting it up as an output because you must enable the pin input buffer and [optionally] the internal pullup or pulldown resistor function.[11] This requires accessing the Pin Configuration register for the selected I/O pin.[12]

In the lab, we will connect a normally open pushbutton switch between an I/O pin and ground. Since the pin will either be connected to ground when the button is pressed or floating when it is not, we need a pullup resistor to 3.3 V to set the high value. You could wire an external pullup, but the microcontroller can provide one internally with no extra wiring. We will need to configure the port pin to conform with the settings for "Input with pull-up" as shown in Table 23-2 of the processor datasheet. The direction (DIR) and output (OUT) bits are set using the DIR registers we used to set up the output pin in the first microcontroller lab.[13] However, the pull enable bit (PULLEN) and the input enable bit (INEN) require writing to one of a set of 32 eight-bit Pin Configuration registers (PINCFG), one for each I/O pin in the Port Group.[14]

23N.4.1   Using structures to access registers

Rather than define thirty-two separate pointers to each Pin Configuration register, one of the CMSIS `port.h` files defines a C structure to the entire Port I/O register set. A C structure is a set of variables grouped together that can be accessed as elements of the group, similar to an array. However, unlike an array, the variables do not all have to be of the same size and type. In CMSIS, each I/O Port is defined as a structure named `PortGroup` that replicates the size and order of the 128-byte port configuration register memory space shown in Section 23.7 of the datasheet:

```
typedef struct {
  __IO PORT_DIR_Type         DIR;
  __IO PORT_DIRCLR_Type      DIRCLR;
```

---

[11] See Figures 23-4 and 23-5 in the SAMD21 datasheet.

[12] To be fair, had you wanted to use a pin with high output drive mode enabled you would have had to access the Pin Configuration register when configuring it as an output as well.

[13] Although we will use the DIRCLR register rather than the DIRSET register to set the pin direction to input.

[14] Only four of the eight bits in each SAMD21 Pin Configuration register are used. However, since the designers chose not to pack the configuration bits for two I/O pins in each eight bit register, the remaining bits may be used in other versions of the Cortex-M microcontroller.

```
    __IO  PORT_DIRSET_Type        DIRSET;
    __IO  PORT_DIRTGL_Type        DIRTGL;
    __IO  PORT_OUT_Type           OUT;
    __IO  PORT_OUTCLR_Type        OUTCLR;
    __IO  PORT_OUTSET_Type        OUTSET;
    __IO  PORT_OUTTGL_Type        OUTTGL;
    __I   PORT_IN_Type            IN;
    __IO  PORT_CTRL_Type          CTRL;
    __O   PORT_WRCONFIG_Type      WRCONFIG;
          RoReg8                  Reserved1[0x4];
    __IO  PORT_PMUX_Type          PMUX[16];
    __IO  PORT_PINCFG_Type        PINCFG[32];
          RoReg8                  Reserved2[0x20];
} PortGroup;
```

The first type specifier indicates if the register is read only ("__I"), write only ("__O"), or read/write ("__IO"). The PORT_XXX_Type specifiers are C union definitions for each register. Unions are a way to allow a single variable to be accessed as different types. Here, they are used to allow port register data to be accessed either as a single 8/16/32-bit register value (".reg") or as a set of 8/16/32 individual bits (".bit"). In some definitions, the bits themselves are organized into structures to match the register bit definitions.

For example, the 32 Pin Configuration registers (PINCFG[32]) are defined as:

```
typedef union {
  struct {
    uint8_t  PMUXEN:1;       // Bitfield for peripheral MUX enable
    uint8_t  INEN:1;         // Input enable
    uint8_t  PULLEN:1;       // Pull enable
    uint8_t  :3;             // Reserved
    uint8_t  DRVSTR:1;       // Driver strength
    uint8_t  :1;             // Reserved
  } bit;                     // Struct to access individual bitfields
  uint8_t reg;               // Union, so use reg for WHOLE 8-bit reg
} PORT_PINCFG_Type;
```

This defines a union called PORT_PINCFG_Type that can be accessed either as a byte (uint8_t reg;) or as a set of eight bits (struct{...} bit;).[15] The definition of the bits in the bit structure is a bit confusing, the individual fields are declared as uint8_t, which makes it seem like they are 8 bit values, but the actual size is the value after the colon. So uint8_t PMUXEN:1; is telling the compiler (and you) that PMUXEN is a one bit field (:1) and it is in the least significant bit location of an eight bit structure. The next field (uint8_t INEN:1;) is also one bit wide and is in the next least significant bit position by virtue of its order in the structure definition.

For the SAMD21G18 processor, CMSIS groups two Port Groups (representing Port A and Port B) together as an array into a single Port structure definition representing all the registers to configure and access the microcontroller's two 32-bit I/O ports:

```
typedef struct {
      PortGroup              Group[2];
} Port
```

Finally, PORT is defined in one of the samd21g18a.h files as a pointer to the Port structure located at the address of the actual registers in the 32-bit memory address space:

```
#define PORT          ((Port *)0x41004400UL)
```

---

[15] Don't panic if the rest of this paragraph does not make much sense. This is far more detail than you need to use this method of accessing the SAMD21 registers, as you will see shortly.

To access a port register, you use the arrow operator ("->") to access the Port group from the PORT pointer, then the C structure operator (".") to access the members (i.e., the registers) of that port group. Finally, you also use the dot operator to access the union member in the register. If accessing the entire 8/16/32-bit register value, you append ".reg" to the register specifier. If accessing a bit or group of bits, you append ".bit.<bit field name>" to the register specifier.

So, for example, rather than use the CMSIS defined pointer to the Data Direction Set register for Port A (REG_PORT_DIRSET0) as you did in the first lab, you could access it through the Port structure as:

```
PORT->Group[0].DIRSET.reg
```

where PORT is the pointer to the address of the beginning of the Port registers (0x41004400UL), Group[0] is the group structure of registers for Port A (Group[1] would be Port B) and DIRSET selects the DIRSET register in the structure for Port A. Finally, the .reg allows you to set the register value as a 32-bit unsigned integer value (rather than as individual bits). This means that there are two ways to set Arduino Pin 10 (I/O Pin PA18) as an output using CMSIS:

```
REG_PORT_DIRSET0 = PORT_PA18;      // the method used in uController Lab 1
```

or

```
PORT->Group[0].DIRSET.reg = PORT_PA18;  // register accessed as a structure
```

Both compile to the same machine language code. While the structure access method may seem excessively complicated for most register access, it is the only easy way to access the thirty-two Pin Configuration registers to set the PULLEN and INEN bits (or the bit that controls the output drive strength) for an individual I/O pin. For example, the Pin Configuration register for I/O port pin PA18 is:

```
PORT->Group[0].PINCFG[PIN_PA18].reg
```

Here we use PIN_PA18 and not PORT_PA18 because the I/O port pin number is being used as an index into an array of 32 registers.

The same CMSIS port.h file defines values for the bits in the Pin Configuration register, so to set pin PA18 as an input with the pull enabled you would write:

```
PORT->Group[0].PINCFG[PIN_PA18].reg = PORT_PINCFG_INEN | PORT_PINCFG_PULLEN;
```

where PORT_PINCFG_INEN is defined as 0x02 while PORT_PINCFG_PULLEN is defined as 0x04 in port.h. The "|" operator does a bitwise OR of these two bits into an eight bit value (the Pin Configuration registers are 8 bits wide) with the result of 0x06, setting the other bits to zero.[16]

Alternatively, you could set the bits individually using .bit access to this register:

```
PORT->Group[0].PINCFG[PIN_PA18].bit.INEN = 1;
PORT->Group[0].PINCFG[PIN_PA18].bit.PULLEN = 1;
```

However, this results in two read-modify-write operations. Also, bit access does not change any other bits in the register, even if you need them set to zero, so you would have to clear them individually with additional instructions.

The pull direction, either pulled high or pulled low, is configured by setting the output register for pin PA18 high or low using either the OUT, OUTSET or OUTCLR register using either the pointer definitions or the structure definitions.[17]

---

[16] The other two active bits in the pin configuration register DRIVSTR and PMUXEN should be set to zero for Port input.

[17] You may find it easiest to use the structure definitions only when you must, to access registers organized as arrays, and to use the pointer definitions for individual registers.

23N.4.2 A mistake to avoid

We often see students testing an input button with:

```
if ((REG_PORT_IN0 & PORT_PA18) == 1)   // check to see if pin PA18 is high
```

This conditional always returns false. The result of (REG_PORT_IN0 & PORT_PA18) is either zero or PORT_PA18, never one. To test to see if a port bit (or any bit in a word, half word or byte) is high, use:

```
if (REG_PORT_IN0 & PORT_PA18)    // true if pin PA18 is high
```

or

```
if ((REG_PORT_IN0 & PORT_PA18) != 0)   // true if pin PA18 is high
```

23N.5 Intrinsic functions for CPU instructions

While C does a pretty good job of compiling to the processor's core instructions efficiently, there are some specialized CPU instructions that do not have a C equivalent. For example, the Cortex-M instruction set includes the ROR (Rotate Right) instruction to rotate a 32-bit value right by 1 to 31 bits. C has a shift arithmetic operation but not a rotate, so implementing a rotate in C would require multiple instructions. CMSIS makes the intrinsic CPU Rotate Right instruction available though a function:

```
uint32_t __ROR (uint32_t value, uint32_t shift);
```

Other useful intrinsic functions rotate through the carry bit, reverse the bit order of a word, swap bytes in a 16-bit or 32-bit value, count leading zeros in a 32-bit value and instructions that tell the CPU it can go into a low power mode until an interrupt or some other hardware event occurs. The intrinsic __BKPT(uint_t value) function causes the program to stop if running in the debugger. This is handy to catch execution of unexpected sections of code while developing an application.

For a complete list see the CMSIS documentation.[18]

23N.6 AoE reading

Chapter 14 (Computers, Controllers and Data Links):
  §14.2.2C PUSH and POP instructions.
  §14.3.7 Interrupts.

---

[18] https://LAoE.link/CMSIS_Intrinsic_CPU_Instructions.html.

23L Lab: Microcontrollers II

In this chapter we will begin with an introduction to interrupts, then learn how to configure I/O pins for input, first to read a pushbutton and then to scan a matrix keyboard. We will also look at creating Finite State Machines in the microcontroller.

23L.1 Create a better Delay() function

Replace the software `Delay()` function from the LED blink program in §22L.9 with an accurate delay using the SysTick timer as follows.

In the initialization section of your program, call the CMSIS `SysTick_Config()` function to set the interval between timer interrupts. The global CMSIS variable

`SystemCoreClock`

is set to the frequency of the of the system clock before your `main()` function is called. Use it to initialize the SysTick interrupt

`SysTick_Config(SystemCoreClock/1000);`

to interrupt 1000 times per second to provide 1ms delay precision.

Next create an SysTick interrupt handler to count the number of interrupts. This function must be declared as

`void SysTick_Handler(void)`

so that it will replace the weak do-nothing default handler. All this function needs to do is increment a global tick count variable. The variable holding the count must be declared as `volatile` since it can be changed by the ISR at any time

`volatile uint32_t msTicks = 0;  /* Variable to store millisecond ticks */`

```
/* SysTick interrupt Handler. */
void SysTick_Handler(void) {
  msTicks++;
}
```

Finally, rewrite your previous `Delay(uint_32t mSec)` function to store the current global tick count when Delay() is first entered, then wait until the tick count reaches the starting count plus the passed delay value, then return.[1]

Of course this design still suffers from the problem that nothing else can happen while waiting for the delay period to end, but at least the delay time will be more accurate and will be portable to other CPUs or clock rates. You can improve the power efficiency of your delay function by including the following intrinsic CPU instruction in the loop that waits for the tick count to reach the desired value:

`__WFE();    // call the Wait For Event intrinsic instruction`

---

[1] This function will fail and return immediately if the current tick value plus the delay exceeds 2^{32-1} (about 49.7 days). This is unlikely to be an issue in the LAoE labs but we will eliminate the possibility of failure when we switch to using the timing functions in the Real Time Operating System in the last microcontroller lab.

The `__WFE(void)` instruction does not change any registers or flags, but it tells the CPU it can turn off power to most of the processor until an interrupt or event occurs.[2] You should call this function whenever your code is doing nothing but waiting for an interrupt to occur. There is no reason to continue executing a loop as fast as possible to check if something has happened when the CPU includes hardware to suspend your code's execution to save power until the thing you are waiting for occurs. This is not particularly important in our application where the microcontroller is powered by the breadboard, but could be a significant power saving in a battery operated application.[3]

Modify the Blinky program to use the CMSIS SysTick timer for your Delay() function. Test it by timing a ten second delay with a [cell phone] stopwatch to make sure it works correctly. You should also modify your CMIS skeleton code from the previous lab to include the SysTick interrupt handler and configuration function, since you will be using it for timing in future labs.

23L.2 Add a pushbutton to control the LED

Add a normally open SPST pushbutton switch to your breadboard near the SparkFun SAMD21. Connect one side of the switch to ground and the other to the Arduino 0/RX pin.[4]

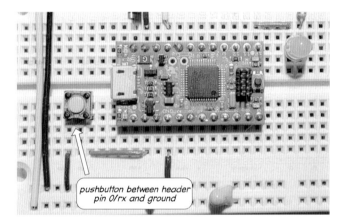

Figure 23L.1 Add a pushbutton to Arduino pin 0/RX.

Use the SparkFun schematic or graphical datasheet to determine which SAMD21G18A I/O pin is connected to the Arduino 0/RX pin.[5] You need to initialize this port pin to conform with the settings for "Input with pull-up" from Table 23-2 in the processor datasheet.

Figure 23L.2 Pin configuration for input with pullup to +3.3 V.

Table 23-2. Pin Configurations Summary

DIR	INEN	PULLEN	OUT	Configuration
0	1	1	1	Input with pull-up

---

[2] Implementation of the power saving function is optional by the microcontroller manufacturer, but this instruction is always supported as a NOP (no operation) by every CPU.

[3] In a future lab we will do the real work in the interrupt function to allow the foreground program to continue running while the background operation is in progress.

[4] We will not be using the TX/RX pins for asynchronous communication in the labs so they are available for general purpose I/O.

[5] The header pin labeled 0/RX is connected to I/O port pin PA11.

23L.2.1 Press the pushbutton to light the LED

Use the direction and output registers for port A to set this port pin as an input (DIRCLR) with the pull direction high (OUTSET). To set the pull enable (PULLEN) and input enable (INEN) bits, use the structure access method to set the pin configuration register for pin PA11 to the bitwise OR of the CMSIS definitions of these bits:[6]

```
PORT->Group[0].PINCFG[PIN_PA11].reg = PORT_PINCFG_INEN | PORT_PINCFG_PULLEN;
```

Edit a copy of your external LED Blinky.c program to blink the external LED when the pushbutton is pressed as follows:[7]

- Configure the I/O pin connected to your pushbutton as an input with the internal pullup enabled.
- Configure the pin with the LED connected to it as an output.
- Write a function IsBtnPressed() which returns an active high Boolean value of one if the button is pressed and zero if not (you will need to #include <stdbool.h> to use Boolean variables).
- Write a function SetLED(<on/off>) which turns the LED off if the parameter <on/off> is zero, or on for any other value.
- The LED should light when the button is pressed and go out when released.

23L.2.2 Press the pushbutton to toggle the LED

Next modify your program so the LED toggles from on-to-off or off-to-on when the button is pressed:

- The LED should only toggle when the button is first pressed, not while it is held down or when it is released.
- Write this as a Finite State Machine (FSM) using the switch statement to evaluate the current state.[8] Hint: Draw a directed graph.
- For each state (i.e., case):
  - Set the output to the correct value. You may not need to change the output in every state.
  - Set the value for the next state based on the current state and the input value. You do not need to explicitly set the state if it does not change.
  - Call Delay() for one FSM state cycle (choose the FSM cycle time to debounce the button).

23L.2.3 Implement the complex blink FSM

Use the microcontroller to implement the complex blink program as a finite state machine.[9] *Use functions to read the pushbutton and turn the LED on and off.*

As a reminder, here is the specification, the directed graph, and the PS/NS table for the FSM:

- Start with the LED off and the button not pressed. (RESET)
- Press the button, the LED comes on.

---

[6] If you are still confused by structure register access, §23S.3 explains how the C compiler interprets this statement.
[7] While it may seem silly to create functions with only a few lines of code to read the pushbutton or to turn on or off the LED, by isolating the hardware from the rest of the program code you allow the program's main() function to be hardware agnostic. This hides active-high/active-low implementation issues and makes it easy to change the hardware if necessary. It also means that once you get the hardware interface functions working correctly, if your program does not do what it is supposed to, you know it is a software issue.
[8] Chapter 23S contains skeleton code for a software FSM.
[9] This is your fifth run-in with Complex Blink. You've already seen it implemented with discrete flip-flops using both logic and memory for the next state computation well as both behaviorally and structurally in the FPGA. We suspect you can do this in your sleep by now.

- Release the button, the LED blinks at a 1 Hz rate.[10]
- Press the button, the LED goes off.
- Release the button, the LED stays off.

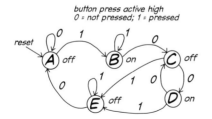

Figure 23L.3 Directed graph and state table for complex blink FSM.

23L.3 Scan a matrix keyboard

Figure 23L.4 below shows the structure of a bare 4×4 membrane switch.[11] This assembly consists of 16 SPST switches labeled 0-9, A-D, # and * in a four-by-four matrix with one terminal of all the switches in a row connected together and the other terminal of the switch connected together with all the other switches in a column.

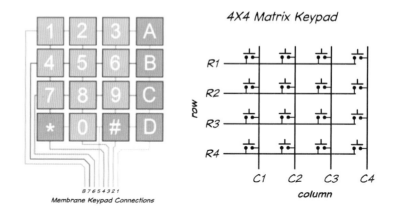

Figure 23L.4 Our 4×4 keypad.

In this lab exercise you will scan the keypad and print the key label character to the IDE debug console when a key is pressed. There are eight electrical connections to the switch, four row lines and four column lines. When a key is pressed, one column is connected to one row.

To use this keypad, you apply a signal to one row and see if the signal is found on any of the four columns.[12] If it is, it means that the switch at the intersection of the driven row and sensed column is

---

[10] This results in an FSM that is slow to recognize button input. If you have time, improve the user response time. Hint: speed up the FSM clock and use a counter to stay in states C and D for the required number of FSM cycles.

[11] A "membrane" switch consists of a sandwich of a plastic sheet with split conductive pads at each position, another sheet with full conductive pads in the identical positions and a spacer sheet between the two with holes over each pad to allow the upper pad to be pressed down and make contact with the split pad completing a circuit. See, e.g., https://LAoE.link//Membrane_Keypad_Construction.html.

[12] This description arbitrarily drives the rows and looks for the driven signal on the columns to see if a key is pressed. However, the switch is symmetrical so you may alternately drive the columns and sense the rows.

pressed. Normally, you scan the rows sequentially to capture a single key press at any intersection of a row and column.[13]

The keypad requires eight I/O pins, four configured as outputs to drive the rows, and four configured as inputs to sense the columns. It is easiest to use Arduino pins 2 through 9 of the SparkFun SAMD21 Mini since that allows the keyboard connector to plug directly into eight consecutive I/O pins without any additional wiring.[14]

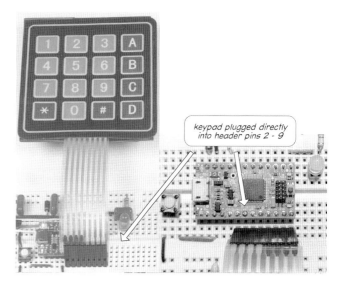

Figure 23L.5 The keypad connects directly to the SparkFun pins labeled 2 through 9.

Since the columns that do not have any keys pressed are floating, you will need to pullup (or pulldown) the column inputs to force them to a default value opposite that of the driving signal. You can use the pull drivers built into the SAMD21 to accomplish this; no external hardware is required.

In operation, one row is driven high at a time while the other three rows are set low. Each of the four columns is normally held low by the I/O port pull settings. If no column line goes high, then no switch is pressed and the next row should be selected and set high with the others low.[15]

However, if a switch is pressed, the column with that switch will go high when the corresponding row is set high. When a high column is detected, the system should recognize the new key, print the value and then stop scanning until the key is released (be careful not to start scanning again on a key bounce).

The two-bit value of the row (R_1 R_0) and the two-bit value of the column (C_1 C_0) identify the key pressed. The valid hex key value ($R_1 R_0 C_1 C_0$) can be decoded with a lookup table into an ASCII character value "0" – "9", "A" – "D", "#" or "*" and printed to the debug console in the Segger IDE with the printf() function.

Write a program to scan the 4×4 matrix keypad and print the key label to the IDE debug output window when a key is pressed.[16] Use SysTick for timing and write your program as an FSM. Your program should only print the character once per press and should not respond to a short (< 10 ms)

---

[13] This scheme only works correctly if a single key is pressed at a time. Pressing multiple keys in a column shorts a high I/O port to a low and could damage the microcontroller. More complex hardware and software can recognize multiple simultaneous key presses. See https://LAoE.link/Keyboard_Matrix.html.

[14] But note that the pins are *only contiguous mechanically*. They do not connect to sequential SAMD21 port I/O pin numbers.

[15] Again, you could choose to drive one row low with the others driven high. If so, the columns would need to be pulled up.

[16] The keypad has a pressure-sensitive adhesive back. We found the keypad is easier to use if you remove the protective liner and bond it to a stiff backing like a square of plastic.

Lab: Microcontrollers II

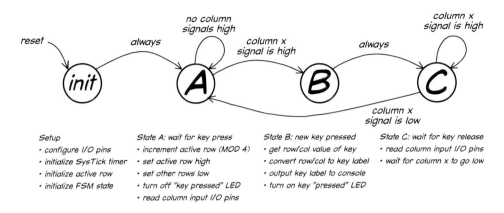

Figure 23L.6 One possible directed graph for the keypad scanner.

bounce on the key. Light the external LED while a key is pressed. Print a message to the console when the key is released. See sample output in Fig. 23L.7.[17]

Hints:

- Write the keypad scanner as an FSM. Fig. 23L.6 shows one possible directed graph.[18]
- Use functions to interface to hardware.
- You are welcome to scan either rows or columns, sensing the other axis, and you can make your scan either active high or active low. (Fig. 23L.6 assumes row scanning and active high signals somewhat arbitrarily.)

Figure 23L.7 An example of the output from the keypad scanner test program.

- Since the SAMD I/O port numbers associated with the contiguous Arduino pins 2 through 9 are not sequential, consider using a lookup table (i.e., an array) to access the four driven lines sequentially.
- A lookup table (i.e., an array) is also a good way to convert the row/column value of a key to an ASCII character (the key label) to print to the debug console.
- You can use `printf()` to send strings to the debugger output window. Make sure the Debug Terminal window is open when you test your program.
- A sample FSM microcontroller skeleton `main.c` is provided in Chapter 23S. In the keypad scanner initialization (i.e., setup) code:

---

[17] There is short demonstration video of how the keypad scanner should work at https://LAoE.link/Keypad_Demonstration.html.
[18] The INIT state corresponds to the setup() function in an Arduino program, while the remaining states compose the loop() function.

- Initialize a 1ms SysTick counter.
- Set up the LED and row driver I/O pins as outputs.
- Start with all row drivers low and the LED off.
- Set up the column I/O pins as inputs with pulldown.

23L.4 Solutions

Here are our solutions to the parts of this lab (break glass in case of desperation).

- §23L.1: `https://LAoE.link/micro2/23L_Systick_Delay.c`
- §23L.2.1: `https://LAoE.link/micro2/23L_Pushbutton_LED_On.c`
- §23L.2.2: `https://LAoE.link/micro2/23L_Pushbutton_Toggle_FSM.c`
- §23L.2.3: `https://LAoE.link/micro2/23L_Complex_Blink_FSM.c`
- §23L.2.3: Here is a version that runs the FSM clock at 20 Hz for a better user experience: `https://LAoE.link/micro2/23L_Complex_Blink_FSM_Fast_Resp.c`
- §23L.3: `https://LAoE.link/micro2/23L_Matrix_Keypad_Scan.c`

23S Supplementary Notes: Creating Robust, Readable and Maintainable Code

23S.1 Write functions to isolate hardware from your application

Using functions to interface to hardware not only makes your code clearer, it also makes it easier to maintain. Consider the LED and switch connected to the Arduino I/O pins shown in Fig. 23S.1 below.

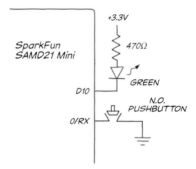

Figure 23S.1 Two simple I/O devices connected to the SparkFun SAMD21.

To use these devices, you first need to initialize I/O pin PA18 (Arduino pin D10) as output and I/O pin PA11 (Arduino pin 0/RX) as input with an internal pullup to 3.3 V. You probably also want to set the LED on or off to begin with. In your main code, you will determine if the button is pressed by reading the Port A IN register and testing bit 11 for zero (from the schematic the button is wired active low). To turn the LED on you need to set pin 18 of Port A low, since the LED is wired active low as well (because a SAMD21 I/O pin can sink a bit more current than it can source).

Here is a program to use this design with the CPU specific hardware interface instructions interwoven into the application code:

```
/*****************************************************************
* Target Device:  Sparkfun SAMD21 Mini Breakout (SAMD21G18A)
* Description: Example of SparkFun SAMD21 inline hardware I/O
*****************************************************************/

#include "samd21.h"    // add to use CMSIS
#include <stdio.h>
#include <stdlib.h>

/****************************
*       main()
****************************/

int main (void){

  // set up Arduino pin 10 for green LED
  REG_PORT_DIRSET0 = PORT_PA18;        // set output for LED on Arduino Pin 10
  REG_PORT_OUTSET0 = PORT_PA18;        // start with green LED off (active low)
```

23S.1 Write functions to isolate hardware from your application

```
  // set up Arduino pin 0/RX for pushbutton
  REG_PORT_DIRCLR0 = PORT_PA11;           // set button pin to input
  // enable pin input and pull drver
  PORT->Group[0].PINCFG[PIN_PA11].reg = PORT_PINCFG_INEN | PORT_PINCFG_PULLEN;
  REG_PORT_OUTSET0 = PORT_PA11;           // set for high pullup

    while (1) {
      if ((REG_PORT_IN0 & PORT_PA11) == 0) {  // if button pressed
          REG_PORT_OUTCLR0 = PORT_PA18;       // turn green LED ON
      } else {
          REG_PORT_OUTSET0 = PORT_PA18;
      }
    }
    return 0;
}
```

Someone looking at this code would need the schematic in Fig. 23S.1 to see what is connected to the Arduino pins, the schematic for the SparkFun SAMD21 to see what M0+ pins the Arduino pins correspond to and the SAMD21+ datasheet to understand what the register manipulations in your code do.[1]

A much better practice is to abstract (i.e., isolate) the hardware from the functional code with understandable, active high functions:[2]

```
/****************************************************************************
* Learning the Art of Electronics
* David Abrams
* 23S_Isolated_Hardware.c mru 2023-03-19
* Target Device:  Sparkfun SAMD21 Mini Breakout (SAMD21G18A)
* Description: Example of SparkFun SAMD21 with functions to isolate hardware
****************************************************************************/

#include "samd21.h"      // add to use CMSIS
#include <stdio.h>
#include <stdlib.h>
#include <stdbool.h>     // adds boolean data type

/*** function to initialize I/O ports ***/
void InitPortIO(void){
  // set up Arduino pin 10 for green LED
  REG_PORT_DIRSET0 = PORT_PA18;           // set output for LED on Arduino Pin 10
  REG_PORT_OUTSET0 = PORT_PA18;           // start with green LED off (active low)

  // set up Arduino pin 0/RX for pushbutton
  REG_PORT_DIRCLR0 = PORT_PA11;           // set button pin to input
  // enable pin input and pull drver
  PORT->Group[0].PINCFG[PIN_PA11].reg = PORT_PINCFG_INEN | PORT_PINCFG_PULLEN;
  REG_PORT_OUTSET0 = PORT_PA11;           // set high pullup as button active low
}

/*** function to return active high button pressed state ***/
bool ButtonState(void) {
  return ((REG_PORT_IN0 & PORT_PA11) == 0);  // return true if button pin low
  }

/*** function to set LED state, active high ON ***/
void SetLEDOn(bool LEDState) {
```

---

[1] That someone might be you. You will be surprised how quickly code that seemed obvious when you wrote it becomes incomprehensible when you look at it again a few weeks later.

[2] Code available at https://LAoE.link/micro2/23S_Isolated_Hardware.c

```
        if (LEDState == true) {
            REG_PORT_OUTCLR0 = PORT_PA18;   // turn green LED ON (active low)
        } else {
            REG_PORT_OUTSET0 = PORT_PA18;
        }
}

/****************************
 *      main()
 ****************************/

int main (void){

  InitPortIO();        // set I/O ports for LED and switch; LED is off

  while (1) {
    if (ButtonState() == true) {
        SetLEDOn(true);
    } else {
        SetLEDOn(false);
    }
  }
  return 0;
}
```

Not only is the functional code in main() *much* easier to read and understand, once the hardware interface functions are tested and verified, you no longer need to worry about how the external devices are connected or controlled if you are only concerned about the application. Someone reading main() never needs to look at any schematic or even the CPU datasheet. But the real benefit comes if you ever need to make changes to the hardware.

Imagine you decide to switch to a different M-series CPU or even to a different vendor's processor. If you are lucky, the former may not need many changes to the code if CMSIS works as it should, but the latter could require a significant rewrite to all the I/O code. With the hardware interface isolated from the functional code, you can be sure that once you modify the interface functions to work with the new hardware, the program will work as it did previously, since you do not have to change any application code to make the hardware change. For example, if an external device changes from active high to active low, it only affects the interface function; the application remains unchanged.

By isolating hardware manipulation as much as you can, you not only make your code much easier to understand – you also make it easier to debug. If a hardware interface function does not do what you expect, you know where to look for the problem. In a larger application, it makes sense to isolate the hardware-dependent functions in their own file so switching processors just requires compiling and linking in a different source file.

You will see an example of this design technique in the sample programs accompanying the Real Time Operating System (RTOS) we use in the last microcontroller lab. A single application-code file is accompanied by dozens of hardware interface files for many different breakout/evaluation boards using many different CPUs from many different manufacturers. Each hardware dependent file implements the same set of interface functions. This allows the demo programs to work unchanged on all the different platforms.

23S.2 Write your application as a finite state machine

The Finite State Machine design technique we used earlier with discrete and programable logic offers many of the same benefits when used to write microcomputer code. It provides a systematic way to think about the design of a complex system with outputs that respond to inputs based on a current state. This describes many controller applications from simple blinking lights to the complex musical jukebox we will build in the last microcontroller lab.

If the state transition diagram directed graph properly implements your system and you have created the Next State/Present State table correctly, creating the code from the NS/PS table is purely mechanical. For each state, set the outputs, read the inputs and set the next state according to the table. In addition, debugging is simplified because only the code in the current state case/break block can contribute to an erroneous output or incorrect next state. This localizes the code you need to look at if your application does not work properly.

The software FSM replaces the hardware flip-flops with a state variable and the FSM clock with a fixed delay.[3] States are implemented as a C switch statement very similar to what we did in the FPGA with the case control structure. For each FSM cycle, modify the generic template to set the outputs for the current state, evaluate the inputs, and then set the state variable to the next state based on the current inputs and current state.[4]

```
/*************************************************************************
* Learning the Art of Electronics
* David Abrams
* 23S_FSM_Skeleton.c mru 2023-03-19
* Description: Generic template for microcontroller finite state machine
*************************************************************************/
#include "samd21.h"      // add to use CMSIS
#include <stdio.h>
#include <stdlib.h>
#include <stdbool.h>     // adds "bool" data type

#define FSM_CLK <millisec per FSM cycle>    // set your FSM clock period here

// define FSM state constants
// replace "STATE<n>" with descriptive names that convey what each state represents
#define STATE1 1
#define STATE2 2
#define STATE3 3

// create 1ms system clock for timing
volatile uint32_t msTicks = 0;    // number of ms since program start

void SysTick_Handler(void)   {    // SysTick interrupt Handler
  msTicks++;                      // counts ms since program start
}

/*******************************
*      main()
*******************************/
int main (void)
{
   int      state = STATE1;       // set initial state
   uint32_t endTicks;
```

---

[3] This generic template does not make use of the wasted cycles during the clock delay but once we transition to the RTOS, other parts of our application will be able to run during the FSM delay time.
[4] Code available at https://LAoE.link/micro2/23S_FSM_Skeleton.c.

```
/************** Initialization *************/
// Configure SysTick to generate an interrupt every 1ms for FSM timing
SysTick_Config(SystemCoreClock/1000);
// complete any other one-time initialization here

/******* Finite State Machine loop *********/
while (1) {
  endTicks = msTicks + FSM_CLK;   // calculate FSM cycle end time
  switch (state) {
    case STATE1:
        <SET OUTPUT>
        <READ INPUT>
        <SET NEXT STATE>
      break;

    case STATE2:
      break;

    case STATE3:
      break;

    default:
      state = <SET DEFAULT STATE>;
      break;
  }

  // delay one state cycle
  while (msTicks < endTicks)  {   // wait for end of FSM clock period
     __WFE ();                    // tell CPU ok to power down till interrupt
  }
 }
 return 0;
}
```

The FSM clock is implemented by saving the current value of `msTicks` at the start of each cycle, then waiting until `msTicks` equals this starting value plus the period of the FSM clock in milliseconds at the end of the cycle. As long as the execution time of each switch case is less that the FSM clock period, this guarantees an accuracy to within one millisecond.[5] However, if the `case` code for a state takes longer to execute than the FSM period, the actual FSM clock cycle time for that state will be the longer of the two. The SysTick timer can be set to interrupt at a higher frequency if you need a faster FSM clock for better timing resolution.

There are several operational differences between this software FSM and the hardware machines we built earlier out of flip-flops and programable logic. The hardware FSM evaluates the next state at the end of the clock cycle, while the software FSM sets the next state *at the beginning* of each state cycle. This simplifies the template by using a common `while()` delay loop after the switch statement. This is usually not an issue – it adds a one cycle delay to state changes – however it can be eliminated by calling a delay function in each state after setting the output and only reading the inputs and setting the next state just before the `break;` statement. This complicates the code and is unnecessary in most applications.

In the software FSM, you do not have to worry about stuck states even if the number of states is not a power of two. The `default:` case takes care of any unexpected value of the `state` variable.

---

[5] The FSM delay short-cycles if the program has been running for 49.71 days and msTicks + FSM_CLK > 2^{32-1} due to 32-bit variable overflow. If this is an issue, SysTick interrupts can be counted in a 64-bit variable. For a more elegant solution, see https://LAoE.link/millis_overflow.html.

23S.2.1 A Software FSM example

Figure 23S.2 below shows a directed graph and PS/NS table for an odd sequence, divide-by-three counter similar to the one we looked at in Chapter 19N, albeit with the addition of a synchronous active high reset input.

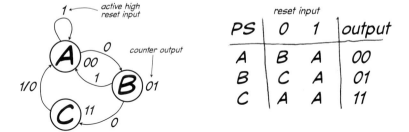

Figure 23S.2 Divide-by-3 counter with odd sequence and reset.

The software FSM loop to implement this counter would look like this:[6]

```
/******* Finite State Machine loop *********/
while (1) {
  endTicks = msTicks + FSM_CLK;   // calculate FSM cycle end time

  switch (state) {
    case OUTPUT00:
        SetOut(0b00);
        if (DoReset() == true) {
          state = OUTPUT00;     // not really necessary since no state change
        } else {
          state = OUTPUT01;
        }
      break;

    case OUTPUT01:
        SetOut(0b01);
        if (DoReset() == true) {
          state = OUTPUT00;
        } else {
          state = OUTPUT11;
        }
      break;

    case OUTPUT11:
      SetOut(0b11);
      state = OUTPUT00;
      break;

    default:
      state = OUTPUT00;
      break;
  }

  // delay one state cycle
  while (msTicks < endTicks) {   // wait for end of FSM clock period
     __WFE ();                    // tell CPU ok to power down till interrupt
  }
}
```

[6] You may assume the hardware interface functions *call*'ed operate as you would expect.

This is just a mechanical translation of the PS/NS table, similar to what we did in the FPGA behaviorial implementation. If you wanted an active low reset, all you would need to do is change the `DoReset()` function to return `true` when the reset line was low rather than high. This code, the application FSM implementation, would not change.

23S.3 A deeper dive into the CMSIS structure format

The notes describe how to access all of the SAMD21 registers using the CMSIS structure definitions. For example, you set the value of the PINCFG register for Pin PA18 to enable the input buffer and a pull (up or down depending on the OUT value) using the following statement:

```
PORT->Group[0].PINCFG[PIN_PA18].reg = PORT_PINCFG_INEN | PORT_PINCFG_PULLEN;
```

This may look daunting, but if you take it step-by-step it is not too hard to understand.[7] "PORT" is defined by CMSIS as a pointer to the beginning of the PORT register address space in the microcomputer. A pointer is just a variable that contains a memory address. Defining a variable as a pointer tells the compiler that the variable should be used to access the contents of the memory address the variable value is pointing to. The pointer is said to be dereferenced when we use it to access the contents of memory.

In the samd21g18a header file PORT is defined as:

```
#define PORT            ((Port    *)0x41004400UL)
```

This defines PORT as an unsigned long integer equal to 0x41004400. The "(Port *)" type definition tells the compiler that the variable should be used as a pointer to a Port structure, which is a defined in the port.h header and corresponds to the physical memory structure of the Port I/O registers described in Sec. 23.7 of the datasheet. Each Port on the SAMD21 occupies 0x80 sequential bytes of the memory address space (not all locations are used) and there are two contiguous sets of Port registers in the SAMD21 Cortex-M0+, the first for Port Group A (I/O pins A00 to A31) and the second for Port Group B (I/O pins B00 to B31). So access to the pin configuration register for pin PA18 starts with a pointer to memory address 0x41004400.

The "->" following it is the C structure pointer operator and is the way you select a particular member of the structure pointed to by a pointer.

The next item, Group[0] selects Port A from an array of (two on the SAMD21G18) PortGroup structures and is added to the address of PORT.[8] It has a value of 0x0 since PORT already points to the registers for the first port.[9] So the value of "PORT->Group[0]" is still 0x41004400.

The "." is the structure member operator. It is used to select a particular member of a structure. Here it selects the 8-bit pin configuration register structure for pin PA18 in PortGroup structure. The PINCFG array starts at offset 0x40 in the PortGroup. The register for pin PA18 is at an offset of 18 bytes (0x12) from the start of the array. That means that value of PINCFG[PIN_PA18] is 0x40 + 0x12 = 0x52 from the beginning of the PortGroup. Again, it gets added to what came before it to get the absolute offset of the register in the memory address space. So PORT->Group[0].PINCFG[PIN_PA18] has a value of

0x41004400 + 0x00000000 + 0x00000052 = 0x41004452

---

[7] OTOH, you don't really need to understand it to use the structure definitions, so go take a nap if you prefer.
[8] The PortGroup structure orders the individual Port I/O register structures into the 0x80 byte long Port I/O address space.
[9] You would use Group[1] to access Port B pins. The value of "Group[1]" is 0x80, the offset from the PORT pointer to the second set of Port registers.

23S.3 A deeper dive into the CMSIS structure format

The "." following this address specifier is the union member operator.[10] It tells the compiler the pointer will access the "reg" union definition (not the "bit" definition). In this case it tells the compiler to write eight bits to the address pointed to by the dereferenced pointer. This is necessary because there are two size definitions for the PINCFG register; one as a byte and the other as a set of eight bits. The compiler cannot guess which union definition you want to use, you have to tell it.

This register (i.e, memory location) is set to the value

"PORT_PINCFG_INEN | PORT_PINCFG_PULLEN"

which ORs two bits, one that enables reading the state of the I/O pin (INEN = 0x03) and one that allows you to set a pullup or pulldown on the pin (PULLEN = 0x04) respectively.

23S.3.1 What does this look like in the debugger?

The actual ARM disassembly code created by this statement is only three lines.

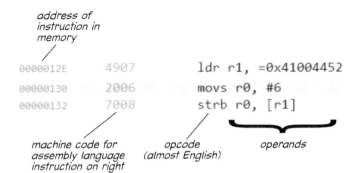

Figure 23S.3 Assembly code to set Pin PA18 PINCFG register.

The *opcode* (operation code) is more-or-less English for what you want the processor to do. The *operands* are the data you are doing it to. There can be zero, one, two or three operands following an opcode in ARM assembly.

The first instruction, "ldr," loads a register with a value from memory. The "=" indicates the 32-bit value following is stored in a nearby table, not as part of the instruction itself.[11] The 4-byte machine language translation of this instruction (shown to the left) contains the relative offset ("07") in words from the current Program Counter to the value in the table to load into the destination register (r1). This is known as *relative addressing*.[12]

The next instruction, "movs," is a data move instruction. The destination of the move is the left operand (r0) and the source is on the right (#6). The hash/pound sign, "#," invokes a very simple addressing mode called *"immediate addressing."* It means "use this value itself, not the thing stored at this address." In this case, the result is that register r0 is loaded with the value 0x6

(PORT_PINCFG_INEN | PORT_PINCFG_PULLEN).

This instruction can only load immediate values from 0 to 255. If you needed to load larger values, you would use the ldr instruction. The "s" on the end of the instruction is optional, it tells the CPU to update the condition codes (zero, negative, etc.) after the move.

The final instruction, "strb," stores the value in a register to the memory location specified by the

---

[10] The operator for accessing members of a union and for accessing members of a structure are the same.
[11] This is actually a pseudo-instruction that generates the most efficient machine code to load the register with a constant. If it is possible to store the constant in the instruction (usually when the constant value is fairly small), it will, otherwise it creates a pc-relative instruction and stores the value in a nearby table.
[12] When you look at the table in memory, you may see it proceeded by a do nothing NOP instruction. This is because the table must start on a word boundary, while Thumb instructions are on half-word (16-bit) boundaries.

second operand. The brackets ("[]") around r1 indicate *indirect addressing*. Rather than copying the contents of register r0 to register r1, the brackets indicate the source should be copied to the memory location whose address is in register r1. Here, that means the CPU writes the value 0x6 (setting the INEN and PULLEN bits) to memory location 0x41004452 (the PINCFG register for pin PA18). The "b" (byte) on the end of the instruction is a size specifier to tell the CPU to only write eight bits to memory.[13]

---

[13] The "str" instruction by itself without the "b" transfers a word (32 bits), while "strh" transfers a half-word (16 bits). The ldr instruction uses the same suffixes to specify the data size to read from memory.

24N Microcontrollers III: Using Internal Peripherals

Contents

24N.1	**Initialize the DAC**	**934**
	24N.1.1 Configure the peripheral I/O pin[s]	934
	24N.1.2 Configure the peripheral's core clock	936
	24N.1.3 Turn on power to the peripheral	937
	24N.1.4 Finally (!) configure the DAC	937
24N.2	**Write the DAC output function**	**939**
24N.3	**Lab preview**	**939**
24N.4	**AoE reading**	**939**

Our [peripheral] cup runneth over

Early microcontrollers typically provided only a few built-in peripheral devices in addition to the CPU and memory that made them stand-alone devices. For example, the original version of the 8051 developed by Intel in 1980 included only two 16-bit timer/counters and a serial port (UART), in addition to the basic CPU, memory and I/O ports.[1] In contrast, the SAMD21G18A we are using has an abundance of peripheral devices, including a USB interface, three simple timer counters and three complex timer counters, a 12-bit ADC with a 14-channel input analog multiplexer, two analog comparators, a 10-bit DAC, six serial interface units, each of which can be configured for UART, SPI, or I2C communications, as well as a specialized I2S digital audio sound link controller, a real-time counter with alarms to keep track of time, a watchdog timer, the SysTick timer we introduced in the last chapter, and a touch controller for 12x10 touch input. To support this plethora of devices, the SAMD21 has an interrupt controller, I/O pin multiplexers, multiple clock generators, a power controller, a Direct Memory Access (DMA) controller, an event system to allow communication between peripherals without CPU intervention, as well as on-chip voltage references and brownout detectors.

As you can imagine, taking advantage of this huge selection of devices comes with significant complexity. Where the 8051 required only two 8-bit registers to configure and use the serial port, the SAMD21 has almost a dozen registers up to 32-bits wide devoted to each serial interface unit, plus it also requires initialization of many more to select and configure the I/O pins, the peripheral's clock, and to enable power to the peripheral. Nevertheless, learning to use the internal peripherals of modern microcontrollers is the key to building useful and powerful systems with a minimum of external components.

---

[1] Newer versions of the 8051, such as the one from Silabs previously used in this book, include many additional peripheral devices similar to those found in modern microcontrollers.

24N.1 Initialize the DAC

In this chapter we will get our feet wet by setting up the Digital-to-Analog Converter (DAC) to output a synthetic sine wave to the breadboard speaker. Control of the DAC is relatively simple compared to some of the other devices, requiring writing to only a few registers if all we are trying to accomplish is simple analog output. However, we still must configure the I/O pin used for output, connect a clock to the peripheral, and enable power to the DAC to turn it on. These steps are required no matter which peripheral device we are using.[2] We normally bundle these configuration steps into a single peripheral initialization function, for example `DAC_Init()`, called from the setup section of `main()`.

24N.1.1 Configure the peripheral I/O pin[s]

While there are some peripheral devices that do not need to communicate with the outside world, many devices require one or more I/O pins to interface with external signals.[3] With only the limited peripherals available on the 8051, it was feasible for the CPU designers to assign dedicated I/O pins to the UART and timer/counter. Modern microcontrollers, in contrast, typically have many more peripheral inputs and outputs than available I/O pins. Indeed, the number of peripherals you can take advantage in a given application is often limited by the number of physical I/O pins available.[4] The version of the Cortex-M0+ we are using comes in a 48 pin package, 38 of which are available for I/O. However, only 19 of these are brought out on the SparkFun Mini breakout board. Each I/O pin can be configured for Port I/O (as we did in the first two microcontroller labs) or connected through a Peripheral Multiplexer to one of the internal peripheral devices. However, rather than allowing any I/O pin to be connected to any peripheral device with a crossbar switch as in some microcontroller designs, the SAMD21 has a limited set of eight multiplexer configurations (A – H) which can be selected on a pin-by pin basis.[5] This further restricts the pin/device combinations available, and it may require careful planning to take full advantage of all the peripherals needed for a project.

In this chapter's lab we will only need to connect the DAC output to a single I/O pin. Take a look at Table 7-1 in the datasheet listing the port multiplexing options for the SAMD2xG microcontroller. The eight multiplexer function choices, labeled A to H, are each devoted to different groups of peripherals. Function B handles all of the analog peripherals as well as the touch controller. Looking at the DAC column of function B, we can see that the DAC voltage output (VOUT) can only be connected to a single pin, PA02 (Arduino Pin A0). Therefore, we have no choice except to devote this pin to the DAC if we want to use its output externally. That means we will not be able to use pin PA02 as an external interrupt pin (EXTINT[2]), an analog input to the ADC (AIN[0]), with the touch interface (Y[0]), or for output from timer/counter for control 3 (the other choices in the row for pin PA02).[6]

This pin is defined in CMSIS header file `samd21g18a.h` both as a generic pin and port and as the DAC output:[7]

---

[2] In the next chapter we will add the additional step of configuring the interrupt controller and writing the ISR function so we do not have to poll the device to see when it is ready for more data.

[3] For example, the DAC can be used to provide a reference voltage to the internal analog comparators, in which case no external pins are required.

[4] An updated Excel spreadsheet tracking our SparkFun SAMD21 I/O pin use is available at
https://LAoE.link/micro3/SAMD_Pin_Use.xlsx.

[5] See Chapter 7, *I/O Multiplexing and Considerations*, in the datasheet.

[6] SparkFun has an expanded graphical datasheet that shows which peripheral devices can be connected to each header pin. It is a useful planning tool to get maximum use of available pins. See
https://LAoE.link/SparkFun_SAMD21_Graphical_Datasheet.pdf.

[7] This header file contains additional PA02 definitions for its use as an ADC analog input, an external interrupt, and with the touch controller as well.

24N.1 Initialize the DAC

```
#define PIN_PA02                2
#define PORT_PA02               (1ul << 2)

#define PIN_PA02B_DAC_VOUT      2L    // DAC signal: VOUT on PA02 mux B
#define MUX_PA02B_DAC_VOUT      1L    // 0 = MuxA; 1=MuxB, etc
#define PORT_PA02B_DAC_VOUT     (1ul << 2)
```

Both sets of PIN and PORT defines resolve to the same numeric values and both will generate the same machine code. However, if you use PIN_PA02 in your code, someone reading it only knows which I/O pin you are using. A much better practice is to use the second set of definitions to make the code self-documenting. Now anyone reading it knows not only that you are using external pin PA02, but that you are using it to output the DAC voltage using Peripheral Multiplexer function B.

A pin is multiplexed to a peripheral by setting the enable bit, PMUXEN, in the PINCFG register for the pin (as discussed in the previous chapter, there are 32 pin configuration registers per port, one for each pin in the port group) to the value PORT_PINCFG_PMUXEN and selecting the desired Peripheral Multiplexer function in the appropriate PMUX register for the pin. This is a bit more difficult because there are only 16 Peripheral Multiplexer registers, PMUX0 through PMUX15, one for every two port pins. These registers are eight bits wide and the low four bits select the multiplexer function for the even number pin while the upper four bits select the function for the odd pin.

We index into the Pin Multiplexer array using the CMSIS structure definitions:

```
// select correct one of 16 registers
PORT->Group[0].PMUX[PIN_PA02B_DAC_VOUT >> 1].reg
```

and set it with the macros defined in the port.h header file:

```
PORT_PMUX_PMUXE(<even pin mux function>) | PORT_PMUX_PMUXO(<odd pin mux function>)
```

If you are using only one of the two pins assigned to this register for peripheral I/O, you can eliminate the unused even or odd pin mux function macro in the assignment.[8]

In addition to selecting and enabling Peripheral Multiplexer function B for I/O pin PA02, we must configure the DAC output pin for analog output as shown in Fig. 24N.1. This disables the output driver, the pull driver, and the input buffer, leaving the input pin connected only to the analog subsystem.

Figure 24N.1 Pin configuration for analog input or output.

Table 23-2. Pin Configurations Summary

DIR	INEN	PULLEN	OUT	Configuration
0	0	0	X	Reset or analog I/O: all digital disabled

In summary, to configure PA02 as the DAC analog output, we must:

- Set the DAC pin direction to zero by writing to the DIRCLR0 register.[9]
- Set the PINCFG register for this pin to DRVSTR = 0, PULLEN = 0, INEN = 0 and PMUXEN = 1.
- Set the low nibble for the even pin multiplexer value in the the PA02/PA03 PMUX register to Multiplexer B (MUX_PA02B_DAC_VOUT).[10]

---

[8] You must always write all eight bits of the Peripheral Multiplexer register at once. If you are multiplexing both I/O pins controlled by the register, you must either configure them together or use a read-modify-write operation. However, if you are not multiplexing the other pin and the PMUXEN bit in the PINCFG register for that pin is not set, then the other nibble's value is ignored.

[9] Theoretically, we could skip this step if we know the CPU has just been reset, since the default value for the data direction register is 0x0000. Nevertheless, it is good programming practice to either reset the peripheral in your initialization code if you are going to rely on default values, or explicitly set up the configuration you require. If you choose to rely on the default value, be sure to document it in a comment.

[10] While CMSIS provides much of the support to get this initialization correct, you are still responsible for figuring out whether the I/O pin is even or odd and using the correct PORT_PMUX_PMUX[E/O] macro.

You should always use the CMSIS register and bit defines, both to make your code easier to understand and to ensure portability with other microcontrollers by the same manufacturer.

24N.1.2 Configure the peripheral's core clock

Every built-in peripheral device requires two clocks. The interface clock, which always runs at the CPU clock speed, is used to communicate with the CPU and other devices over the internal peripheral bus. This clock is enabled in the Power Manager. See §24N.1.3.

The core clock is the timing source for the internal logic of the peripheral device. The SAMD21's Generic Clock Controller (GCLK), shown in Fig. 24N.2, contains clock generators that are used to create the core clock signals. Each peripheral can be connected to one of the eight clock generators (GCLKGEN0 to GCLKGEN7).[11] Selecting a slower core clock for a peripheral provides lower power operation while a faster clock gives better performance.

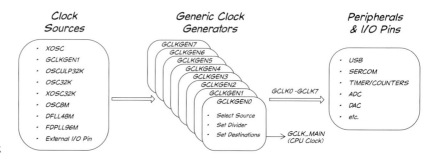

Figure 24N.2 SAMD21 Generic Clock Controller block diagram.

Each clock generator can use one of several internal or external oscillators as sources. A clock generator can be set to divide its input source by a selectable division prescaler to provide a wide range of frequencies. Configuring a clock generator is a somewhat complex operation because the clock controller uses a single set of registers to configure each of the clock generators indirectly. The clock generator registers are defined in the GCLK.H header files.

After a CPU reset, The SAMD21 Clock Generator 0 (GCLKGEN0) is connected to the internal 8 MHz oscillator (OSC8M) prescaled by eight to provide a 1 MHz clock. The output of this clock generator is always used as the main CPU clock. Clock Generator 2 (GCLKGEN2) is connected to the internal low power 32,768 Hz oscillator (OSCULP32K).

For this lab, we will connect the GCLKGEN0 output to the DAC. A clock generator is configured with the Generic Clock Generator Control (GENCTRL) register by specifying a source and setting the prescaler value. However, since this clock generator is already initialized to 1 MHz after a reset, we just need to configure the generic clock multiplexer using the Generic Clock Control (CLKCTRL) register. To connect Generic Clock 0 to the DAC clock, write the bitwise OR of the source generic clock (GCLK_CLKCTRL_GEN_GCLK0), the destination peripheral (GCLK_CLKCTRL_ID_DAC) and the clock enable bit (GCLK_CLKCTRL_CLKEN) to the Generic Clock Control register.[12]

This register is the first we have used that requires synchronization. In this case, write synchronization

---

[11] The datasheet conflictingly states there are eight clock generators in one paragraph and nine elsewhere. CMSIS only provides defines for eight, which is more than we need.

[12] You can only connect one generic clock source to one peripheral clock at a time. The source clock and destination peripheral are *numeric* values. Trying to OR together multiple source or destination values makes no sense. For peripherals like the analog comparator that require two generic clocks, you must set the Generic Clock Control register twice, once for each clock connection.

is required, which means we must wait until the SYNCBUSY bit in the generic clock STATUS register is cleared after writing to the CLKCTRL register.[13]

The CLKCTRL and STATUS registers may be accessed as register pointers at:

```
REG_GCLK_CLKCTRL
REG_GCLK_STATUS                 // test bit GCLK_STATUS_SYNCBUSY
```

respectively, or using structures:

```
GCLK->CLKCTRL.reg               // 16-bit register access
GCLK->STATUS.bit.SYNCBUSY       // single bit read access
```

To keep the bus from stalling, we should:

- Write to the register requiring synchronization (this sets the SYNCBUSY bit high)
- Then wait until the SYNCBUSY bit in the status register is clear in a while() loop:

```
GCLK->CLKCTRL.reg =
    GCLK_CLKCTRL_ID_DAC | GCLK_CLKCTRL_CLKEN | GCLK_CLKCTRL_GEN_GCLK0;
while(GCLK->STATUS.bit.SYNCBUSY);                //Wait for bus clock sync
```

24N.1.3 Turn on power to the peripheral

To conserve power, the SAMD21 does not enable the interface clock to many of the on-chip peripherals after a processor reset. Table 12-1 in the datasheet summarizes the configuration of each peripheral including its base memory address, the interrupt line it uses, if it is initially powered ("Enabled at Reset"), etc. The I/O port (PORT) clock is enabled at reset, which is why we were able to use Port I/O immediately in the previous microcontroller labs. However, the DAC interface clock is disabled and cannot be used until it is enabled in the Power Manager (PM).

Luckily, turning on the DAC clock is a fairly easy task, requiring only a single register access. The bit that controls the DAC clock is in the APBCMASK register defined in the pm.h header files.[14] You must set the PM_APBCMASK_DAC bit in this register using a read-modify-write operation to make sure you do not inadvertently disable any other peripherals in use. You can use either the register pointer or structure definitions:

```
REG_PM_APBCMASK |= PM_APBCMASK_DAC;   // explicit RMW
or
PM->APBCMASK.bit.DAC_ = 1;  // implied RMW
```

24N.1.4 Finally (!) configure the DAC

Once the DAC is powered by enabling the interface clock, connected to an internal generic core clock and its output multiplexed to an external I/O pin, it is time to configure the DAC itself. The DAC is a relatively simple peripheral if all we want to do is output a digital value as an analog voltage.[15]

The DAC converts a 10-bit DATA value between 0x000 and 0x3ff to a voltage:

$$V_{\text{out}} = \frac{\text{DATA}}{0\text{x3FF}} * V_{\text{ref}}$$

---

[13] If you fail to properly synchronize peripheral register access, your code will still work but the CPU will stall waiting for the bus to be available again. This can result in a delay in responding to interrupts. See chapter 14 and section 15.6.6 in the datasheet.

[14] There are three 32-bit peripheral mask registers to control the dozens of peripherals in the SAMD microcontroller. See Table 16.7 in the datasheet to determine which one to use to turn on a specific peripheral.

[15] It becomes more complicated if you want to use DMA, double buffering, interrupts, and/or events. See Chapter 35 in the datasheet.

The reference voltage, V_{ref}, may be set to the power supply (3.3 V), an internal 1.0 V reference, or an external reference connected to an I/O pin. In Chapter 24L, we will use the internal reference to get a 0 to 1 V output range. There are only two registers plus the status register necessary to set up the DAC for simple voltage output. However, both of these registers require write synchronization. One register, control register A (CTRLA), contains bit flags to reset the DAC and to enable it. The other, control register B (CTRLB), selects whether the 10-bit DAC value is written right-justified or left-justified to the DATA out register. Other bits in control register B select the DAC reference and enable the output buffer, as well as several other functions that we will not use.

Like many of the internal peripherals, the DAC can only be configured when it is disabled, i.e., when the CTRLA.ENABLE bit in control register A is not set. We will explicitly reset the DAC before we start configuring it, which will ensure that this bit is cleared.

To configure the DAC, you should:

- Wait for the both the `DAC_CTRLA_SWRST` and the `DAC_STATUS_SYNCBUSY` bits to be clear (low).
- Set the CTRLA register to `DAC_CTRLA_SWRST` to initiate a software reset.
- Wait for the both the `DAC_CTRLA_SWRST` and the `DAC_STATUS_SYNCBUSY` bits to be clear again.
- Select the 1V internal reference, right-justification and enable the output driver in the CTRLB register.
- Wait for the `DAC_STATUS_SYNCBUSY` bit to be clear.
- Set the `DAC_CTRLA_ENABLE` bit in the CTRLA register to enable the DAC.
- Wait for the `DAC_STATUS_SYNCBUSY` bit to be clear.

As a general rule of thumb, it is good practice to initiate a software reset of an internal peripheral before configuring it, unless it is a device used by several peripherals like the Power Manager or Generic Clock Controller. This sets the device to a known state and allows you to use the default values where applicable. To reset the DAC, first wait for the SYNCBUSY and SWRST bits to be clear (not strictly necessary but a good practice in case some other code just reset the peripheral) before setting the `DAC_CTRLA_SWRST` bit in control register A.[16]

After setting the software reset bit, we must wait until it and SYNCBUSY are clear before further set up of the DAC:

```
/* Wait for reset and busy synchronization */
while(DAC->STATUS.bit.SYNCBUSY || DAC->CTRLA.bit.SWRST);

REG_DAC_CTRLA = DAC_CTRLA_SWRST; // Initiate a software reset of DAC

/* Wait for reset and busy synchronization */
while(DAC->STATUS.bit.SYNCBUSY || DAC->CTRLA.bit.SWRST);
```

Once the DAC is reset, we must set the voltage reference used for full scale, the data justification, enable the output I/O pin driver and then enable the DAC. This requires writing to both control registers, each of which requires a synchronization after the write. Note the comment documenting that a default value is being used.

```
// Set reference to 1V supply for 0 - 1V output and enable output driver
// Right justified data is the default after reset so not set explicitly
REG_DAC_CTRLB = DAC_CTRLB_EOEN | DAC_CTRLB_REFSEL_INT1V;
while(DAC->STATUS.bit.SYNCBUSY);    // wait for SYNCBUSY synchronization

REG_DAC_CTRLA = DAC_CTRLA_ENABLE;   // turn on the DAC
while(DAC->STATUS.bit.SYNCBUSY);    // wait for the enable to finish
```

[16] Since you are resetting the peripheral, it is permissible to do a full register write with the other bits zero.

At this point the DAC initialization is complete and it is ready to accept data.

24N.2 Write the DAC output function

Once the DAC is initialized, we need a function to send data to the DAC. This function should take an unsigned 16-bit value as its argument and write it to the DAC DATA register (with synchronization). The DATA register is a write-only, 16-bit register to which the digital value of the analog voltage is written. However, only the low 10 bits written to this register set the DAC output voltage. The analog output is updated immediately whenever a new value is written to the DATA register.[17]

You access the DAC DATA register as:
```
REG_DAC_DATA    // pointer definition
```
or
```
DAC->DATA.reg   // structure definition
```

24N.3 Lab preview

In the third microcontroller lab, you will first add a comparator to the DAC output to build an ADC and then an audio amplifier so you can listen to its output on the breadboard speaker. Initially, you will send data to the DAC from your `main()` program, using calls to `Delay()` between output updates (we will replace the Delay function with a timer interrupt in the following lab). You will also increase the CPU clock speed to get a faster DAC update rate.

24N.4 AoE reading

Chapter 15 (Microcontrollers):
  §15.1 Introduction.
  §15.2 Design example 1.
  §15.10.2 When to use microcontrollers.
  §15.10.3 How to select a microcontroller.

---

[17] The second write-only register, DATABUF, is only used when double buffering the DAC. Double buffering is a technique that allows you to write the next DAC value before it is needed and have it transferred automatically to the DATA register at a specific time or when a specified event occurs. This frees the foreground program from having to deliver data to the DAC at precise timing intervals. In the next chapter, we will use a timer peripheral and interrupts to accomplish the same precise timing while freeing the foreground program from updating the DAC entirely.

24L Lab: Microcontrollers III

In this chapter we will use the internal DAC peripheral first to create an analog-to-digital converter, then to output a synthetic sine wave.

24L.1   DAC skeleton code

The skeleton code below initializes the DAC, sets up the SysTick timer to provide a 1ms interrupt for the Delay() function and then outputs a sawtooth waveform on Arduino pin A0.[1] Take a look at the DAC_Init() function. After configuring the output I/O pin, it connects a Generic Clock to the DAC, turns on the bus clock in the Power Manager to enable the peripheral, and initializes the DAC for one volt full scale, right justified data. The first three steps are common to initializing most peripherals.[2] The function

```
DAC_OutData(uint16_t data)
```

sends a 10-bit value (right justified) to the DAC. The DAC conversion rate is 350 ksps, implying $\approx 3\,\mu s$ settling time.

The main() function calls the hardware initialization functions and then outputs a sawtooth analog waveform by counting repeatedly from 0x000 to 0x3ff in a tight loop as fast as possible and sending the count value to the DAC.

```
/****************************************************************************
* Company: Learning the Art of Electronics
* Engineer: David Abrams
* Create Date: 2022-07-08
* Module Name:  main.c (24L_DAC_Skeleton.c)
* Project Name:   24L
* Target Device:  Sparkfun SAMD21 Micro
* Description: Skeleton code to set up DAC for output. Initially with
*              default CPU clock, then with CPU clock set to 48MHz.
****************************************************************************/
#include "samd21.h"      // add to use CMSIS
#include <stdio.h>
#include <stdlib.h>
#include <string.h>
//#include "ClockSysInit48M.h"

/* 10-bit, 128 element sampled sine wave array generated in MATLAB */
uint16_t sine[] = {
512, 537, 562, 587, 612, 637, 661, 685, 709, 732, 754, 776, 798, 818, 838,
857, 875, 893, 909, 925, 939, 952, 965, 976, 986, 995, 1002, 1009, 1014,
1018, 1021, 1023, 1023, 1022, 1020, 1016, 1012, 1006, 999, 990, 981, 970,
959, 946, 932, 917, 901, 884, 866, 848, 828, 808, 787, 765, 743, 720, 697,
673, 649, 624, 600, 575, 549, 524, 499, 474, 448, 423, 399, 374, 350, 326,
303, 280, 258, 236, 215, 195, 175, 157, 139, 122, 106, 91, 77, 64, 53, 42,
33, 24, 17, 11, 7, 3, 1, 0, 0, 2, 5, 9, 14, 21, 28, 37, 47, 58, 71, 84, 98,
114, 130, 148, 166, 185, 205, 225, 247, 269, 291, 314, 338, 362, 386, 411,
```

---

[1] Code available at https://LAoE.link/micro3/24L_DAC_Skeleton.c.
[2] The exceptions are peripherals that do not use any I/O pins and peripherals that are already enabled after a reset (see datasheet Table 12-1).

24L.1 DAC skeleton code

```c
    436, 461, 486, 511
};

#define SAMPLES sizeof(sine) / sizeof(sine[0])  // number of samples in sine array above

/***********************************************************************
 *              System Timer interrupt and Delay function
 ***********************************************************************/

// use 1ms system clock for timing
volatile uint32_t msTicks = 0;

void SysTick_Handler(void){           // SysTick interrupt Handler - interrupts every 1ms
    msTicks++;
}

// Function to delay for n milliseconds
void Delay (int DelayMSec){
  uint32_t endTicks;

  endTicks = msTicks + DelayMSec;

  while (msTicks < endTicks) {        // wait for DelayMSec SysTick interrupts
      __WFE ();                       // tell CPU it can power-down until next event/interrupt
  }
}

/***********************************************************************
 *   Initialize PORT I/O pin PB09 (Arduino A2) for comparator input
 ***********************************************************************/
#define PORT_COMPIN    PORT_PB09
#define PIN_COMPIN     (PIN_PB09 & 0x1f) // CMSIS defines this as 0x29

void PORT_Init(void){
    // Set comparator pin as an input
    REG_PORT_DIRCLR1 = PORT_COMPIN;

    // Configure comparator pin with DRVSTR = 0, PULLEN = 1, INEN = 1 and PMUXEN = 0
    PORT->Group[1].PINCFG[PIN_COMPIN].reg = PORT_PINCFG_PULLEN | PORT_PINCFG_INEN;
    REG_PORT_OUTSET1 = PORT_COMPIN;     // set pull high (TLC372 has open-drain output)
}

/***********************************************************************
 *                    DAC Support Functions
 ***********************************************************************/

void DAC_Init(void){

// Only pin PA02 (Arduino pin A0) can be used for DAC output.
// Peripheral multiplexer Function B is used for all analog functions.

    /*************** PORT I/O Initialization  *************/
    /* Set up DAC output pin as analog I/O (see table 22-1) */
    /*****************************************************/

    // DIR bit for an analog pin must be set to 0
    REG_PORT_DIRCLR0 = PORT_PA02B_DAC_VOUT;

    // Configure the DAC output pin with DRVSTR = 0, PULLEN = 0, INEN = 0 and PMUXEN = 1
    PORT->Group[0].PINCFG[PIN_PA02B_DAC_VOUT].reg = PORT_PINCFG_PMUXEN;  // enables pin mux

    // set pin multiplexer configuration B for even pin PA02
    PORT->Group[0].PMUX[PIN_PA02B_DAC_VOUT >> 1].reg = PORT_PMUX_PMUXE(MUX_PA02B_DAC_VOUT);

    /******** GENERIC CLOCK CONTROLLER Initialization ********/
    /* Connect GCLK 0 to DAC and enable DAC clock.           */
    /* GCLK 0 is enabled and initially set to 1MHz.          */
    /*********************************************************/

    GCLK->CLKCTRL.reg = GCLK_CLKCTRL_ID_DAC | GCLK_CLKCTRL_CLKEN | GCLK_CLKCTRL_GEN_GCLK0;
    while(GCLK->STATUS.bit.SYNCBUSY);     // requires write synchronization

    /************** POWER MANAGER Initialization  *************/
    /* Enable the DAC bus clock in Power Manager              */
    /**********************************************************/
```

Lab: Microcontrollers III

```c
    PM->APBCMASK.reg |= PM_APBCMASK_DAC;

    /****************** DAC Initialization *********************/
    /* Setup DAC for 10-bit, right justified, 0 to 1V output  */
    /***********************************************************/

    // Wait for reset and busy synchronization
    while(DAC->STATUS.bit.SYNCBUSY || DAC->CTRLA.bit.SWRST)

    // Perform a software reset
    REG_DAC_CTRLA = DAC_CTRLA_SWRST;

    // Wait for reset and busy synchronization
    while(DAC->STATUS.bit.SYNCBUSY || DAC->CTRLA.bit.SWRST);

    // Set reference to internal 1V ref for 0 - 1V output.  Enable output driver
    // (right adjusted data is the default after reset so not explicitly set)
    REG_DAC_CTRLB = DAC_CTRLB_EOEN | DAC_CTRLB_REFSEL_INT1V;

    // enable the DAC
    REG_DAC_CTRLA = DAC_CTRLA_ENABLE;      // turn on the DAC

    // wait for the enable to finish
    while(DAC->STATUS.bit.SYNCBUSY);       // wait for the enable to finish
}

//****************************************************
// function to send one 10-bit sample out the DAC port

void DAC_OutData(uint16_t data){
    REG_DAC_DATA  = data;                  // write value to DAC
    while(DAC->STATUS.bit.SYNCBUSY);       // wait for write to finish
}


/*********************************************
 *             utility functions
 *********************************************/

// function to print DAC value as a three
// digit voltage to the debug console

void PrintVoltage(uint16_t DAC_Val) {
    int voltage = (DAC_Val * 1000)/1023;
    printf("Input = 0.");
    if (voltage < 100) {
        printf("0");
    }
    if (voltage < 10) {
        printf("0");
    }
    printf("%d Volts\n", voltage);
}

/*********************************************
 * main(): setup DAC, then output forever
 *********************************************/
int main (void){

    // system initializations
    DAC_Init();                                  // DAC to 1V FS, right justified
    SysTick_Config(SystemCoreClock/1000);        // SysTick to interrupt every 1ms

    while(1) {
      for (int i = 0; i <= 0x3ff; i++) {        // create sawtooth DAC output
          DAC_OutData(i);
      }
    }

  return 0;
}
/*************************** End of file ****************************/
```

24L.2 Test the SAMD21 DAC

24L.2.1 See if the DAC functions work

Create a new project in the IDE and replace the default `main()` with the skeleton code. Connect your oscilloscope to the DAC output (trigger on the falling edge) and see if the code works properly. You should see the voltage rise from zero volts to one volt, and then drop back to zero. Calculate how long it takes to output one sample by dividing the period of the sawtooth wave by 1024.[3]

24L.2.2 Use the DAC to build an ADC

Add a TLC372 single-supply comparator to your breadboard as shown in Fig. 24L.1.[4] The non-inverting input to the comparator should be connected to the DAC output, the inverting input is connected to a 0–1 V adjustable voltage divider. Connect your scope and multimeter as shown.

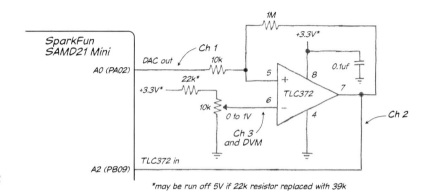

Figure 24L.1 TLC372 comparator for ADC.

24L.2.2.1 Test a simple but slow version

The output of the TLC372 is low when the input from the voltage divider is greater than the DAC output and high when the DAC is greater. We can use this to determine the divider voltage by capturing the DAC value that trips the comparator.

Replace the small `while()` loop in `main()` with the following code to implement a single slope ADC:[5]

```
// Implement a simple ADC converter
PORT_Init();                        // enable Arduino pin A2 as input
while(1) {
  int done = 0;                     // reset done flag before next run
  for (int i = 0; i <= 0x3ff; i++){
    DAC_OutData(i);                 // create sawtooth output on A0
    if ((REG_PORT_IN1 & PORT_COMPIN) && done == 0){
      PrintVoltage(i);              // capture DAC value that matches input
      done = 1;                     // set flag to only print once per cycle
    }
  }
}
```

---

[3] It took 35.4 μs per sample when we tried it.
[4] The comparator and input test voltage divider may alternatively be powered from 5 V as noted.
[5] Code available at https://LAoE.link/micro3/24L_Simple_ADC_Loop.c

This is just the sawtooth code in the skeleton with a check of the TLC372 output each time we increment the DAC voltage. When the comparator goes high, a helper function, `PrintVoltage()`, scales the 0 to 1023 DAC value to 0 to 1 V and prints the value to the debug console. The done flag makes sure we only print the output value once per conversion cycle. How long does each conversion take?[6]

24L.2.2.2 Convert this into a tracking converter.

Modify the code in §24L.2.2.1 to make a tracking converter. This is one where the DAC value is incremented if the DAC output is lower than the input voltage or decremented if it is higher. Ensure that the DAC value does not exceed 0x3ff or go below zero. It would not make sense to print out the value on each cycle, so think about how to print out the DAC value every nth time (every three thousand conversions worked for us).

Once you have it working, try replacing the voltage divider with a slow 0 to 1 V sine wave at about 1 Hz. (Be sure to check the function generator output with the scope *before* connecting it to the TLC372 to avoid damaging the comparator.) Now increase the sine frequency. What problem do you see with the tracking converter?[7]

24L.2.2.3 Design a faster ADC

Try using the DAC and the comparator to create a successive approximation ADC. To refresh your memory, in this type of converter you initialize the result to zero. You then set the most significant bit to one and compare the DAC output to the input voltage. If the DAC output is less than the input voltage, you save this bit as a one in the final value. If the DAC output is greater than the input, you set this bit to zero in the result. You repeat this for each bit, shifting your test bit right until you get to the least significant bit. Once you have tested all ten bits, print the result to the console.[8] See Fig. 24L.2 for a suggested flowchart of the converter code and an example of the DAC output.

Test your successive approximation ADC. How would you compare it with the simple ramp-based converter and the tracking converter in speed and code complexity? (Clearly, the hardware requirements are the same for all three.)

24L.3 Add an audio amp, so you can listen

LM386 amp: Let's feed the output of the DAC to an audio amp capable of driving an 8 Ω speaker. (The SAMD21 I/O pin driver doesn't come close to being able to do that job by itself.) The LM386 audio amp is designed for just this task.

Add the power amplifier shown in Fig. 24L.3 to your breadboard near the speaker.[9] The amplifier is a fairly simple single-chip device that only requires power (+5 V/ground) and a few components to operate. You will also need a potentiometer for volume control; you can either use a small trimmer near the output of the DAC or the 10k pot at the bottom of the breadboard. The latter will require running longer wires from the microcontroller DAC output pin to the potentiometer then back up to the amplifier.

---

[6] We measured 52.5 ms, about 45% longer than the sawtooth alone. The longer time per point is due to the extra code to check if the comparator output is high.

[7] The tracking converter works fine when the input moves slowly, but it gives erroneous output value when the input changes so quickly it cannot keep up. Our converter only worked up to about 5 Hz.

[8] It is a good idea to set the DAC to zero at the end of each measurement. This makes it easier to trigger the oscilloscope on the beginning of the next cycle when the DAC is set to its mid-scale value.

[9] The audio amp will remain on the breadboard until the final lab. You can remove the comparator circuit (we won't be using it again) or leave it if you would like to play with it further. It won't interfere with the audio tests that follow.

24L.3 Add an audio amp, so you can listen

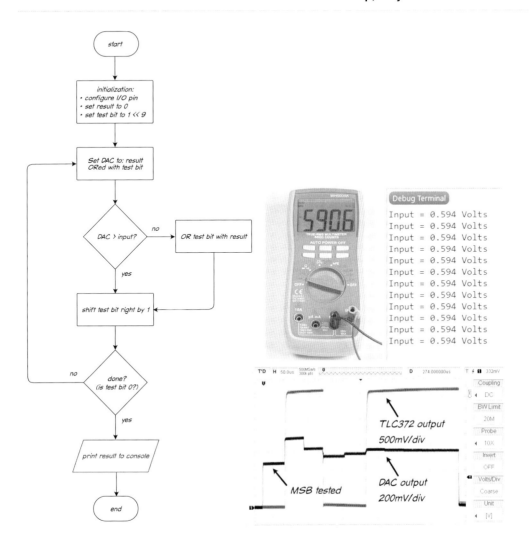

Figure 24L.2 Flowchart and output for successive approximation ADC.

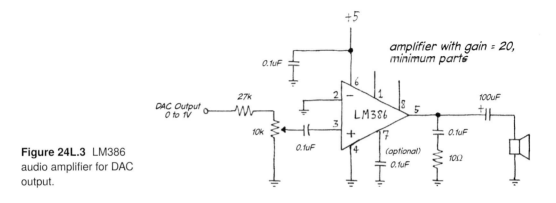

Figure 24L.3 LM386 audio amplifier for DAC output.

General truth: we need to block the DC bias of these single-supply parts: Because the signal source from the SAMD21 DAC is unipolar (centered on 2.5 V) and so is the output of the internally-biased LM386, you need *twice* to insert blocking capacitors to knock out this DC bias.

The LM386 likes a small input signal symmetric about ground – strange to tell – even though it uses only a positive supply.[10] So, you need the 0.1 μF blocking capacitor to feed the input. The input nominal input resistance of the '386 is 50k, giving an f_c of 32 Hz. The 27k resistor and the 10k pot divide the 0–1 V input signal down to 0 to 0.25 V so the gain of 20 in the amplifier (the minimum possible) provides a maximum output of 0 to 5 V.[11]

The blocking cap between the output of the '386 and the speaker is very large because this capacitor forms a highpass filter with the very low 8 Ω resistance of the speaker. Be sure to observe the correct polarity on this capacitor.

The LM386 is predisposed to parasitic oscillations. Don't forget to *decouple* the '386 power supplies, using both a ceramic (0.1 μF) and a larger tantalum (at least 1 μF), close to the +5 V supply pin of the IC. The 0.1 μF capacitor and 10 Ω resistor on the amplifier output is a "snubber" circuit to prevent oscillation driving the inductive speaker load. You may want to try adding a 0.1 μF capacitor from bypass pin 7 to ground on the '386 if you get excessive noise on the speaker with the input off.

Once you have the amplifier connected to the DAC output, you will be able to hear the signals you generate in the next part of the lab.

24L.3.1 Now output a sine wave

Restore the original skeleton `while(1)` loop and change the value sent to the `DAC_OutData()` function from the count index *i*, to the value in the sine array: `sine[i]`. This array holds a set of values computed to generate a sinusoidal waveform. You will need to change your counter to count only up to a maximum of 0x7f since there are only 128 values in the array to output.

What is the frequency of the sine wave? How long does it now take to output each sample point? Do you expect this to be longer or shorter than when you output the sawtooth signal?[12]

24L.3.2 We need to giddyap here!

One way to speed up the time per point is to move the code in the function `DAC_OutData()` that updates the DAC DATA register into the `main()` `while` loop to avoid the function call. This is known as "inline" code, and it saves time because it avoids the overhead of the function call. Try it. What is the sine frequency now? Time per point? What does this tell you about C function call overhead on this processor?[13]

Set a breakpoint at the beginning of the `for()` loop and run until it is hit. Look at the assembly code of the `for` loop in the Disassembly window. Count the number of instructions in the loop. (The easiest way to do this is to single step through the assembly instructions until you get back to where you started, counting instructions. Single step to the branch ("b") instruction then count the number of instructions until you get back to it.) Assume each instruction takes one CPU cycle (a not unreasonable assumption on an ARM CPU) and calculate the CPU instruction cycle time. Is the SAMD21 running at its maximum clock speed of 48 MHz?[14]

---

[10] It understands a signal as much as 0.4 V below its negative supply, which ordinarily is *ground*.
[11] If the input capacitor seems in an odd place (after the divider), it is because we are going to replace the mechanical pot with an electronic one that does not like negative voltages in a later lab.
[12] Since we have added the extra step of a table lookup (i.e, the index into the sine[] array), you would expect it to take longer to output each value and that is what we saw. We measured 44 μs per point for a sine wave of about 178 Hz.
[13] We got a 246 Hz sine wave or 32 μs per point, a 38% improvement. So the function call overhead is definitely significant.
[14] We counted 12 instructions which suggests about an execution time of ≈ 2 μs per instruction. This is definitely not the maximum CPU speed.

24L.3.3 We need to giddyap a whole lot more!

You probably discovered that the CPU is running much slower than its maximum clock frequency. It is going to be pretty difficult to play any decent music using the DAC if we are limited to no more than ≈ 250 Hz. In order to get the highest possible DAC output update rate, you will want to run the SAMD21 at its maximum 48 MHz CPU clock rate.

Copy `ClockSysInit48M.c` and `ClockSysInit48M.h` from Chapter 24W into separate files and place them in the project directory containing `main.c`.[15] (You do not need the function `EnableGCLK0IO()`. It is only used in the worked example.)

Uncomment `#include "ClockSysInit48M.h"` in the list of include files at the top of `main.c` and add the following line to the initializations at the beginning of your `main()` function *before* the call to `SysTick_Config()`.[16]

```
ClockSysInit48M();         // set CPU clock to 48MHz
```

Finally, add the new C file to the list of project Source Files.[17]

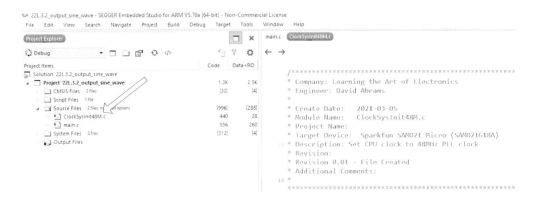

Figure 24L.4 Source file addition to run at 48 MHz CPU clock.

Now recompile and try the DAC output program again. What is the frequency of the sine wave? What is your calculated time per point? What is your new estimate of the instruction cycle time? What is the ratio of the previous cycle time to this one?[18]

24L.3.4 Whoa, Nelly! Lets slow it down a bit.

Now that you can get a reasonably fast sine wave out of the DAC, slow it down by calling the `Delay()` function after each point. Try the minimum delay of 1 ms. What frequency sine wave do you expect to see? What frequency do you actually measure? Next, keeping the call to the Delay() function, replace the direct DAC DATA register load in the while loop with the original call to the `DAC_OutData()` function. Does the sine frequency change? Why do you think you get the result you do?[19]

---

[15] Code available at https://LAoE.link/ClockSysInit48M.c and https://LAoE.link/ClockSysInit48M.h.

[16] The function that increases the CPU clock also updates the value of the `SystemCoreClock` variable from 1,000,000 to 48,000,000. If you increased the clock speed after the call to `SysTick_Config()`, your `Delay()` function would run 48 times too fast.

[17] You will want to update your lab skeleton code with these additions as well, since we will be running the SAMD21 at its maximum CPU speed from now on.

[18] Our sine frequency was 12.4 kHz with inline output to the DAC. (It was too high to hear on the definitely not high-fidelity breadboard speaker.) This implies a time of 630 ns per point or an instruction time of about 39 ns. This is about 50 times faster than with the default 1 MHz CPU clock.

[19] With the minimum 1 ms delay, the output sine wave is 7.81 Hz or a period of 128 ms. There is no change between inline code and calling the `DAC_OutData()` function. This is not a surprise. The overhead to call the function is so short (we

24L.4 Conclusion

It does not appear that Delay() is a reasonable way to control the frequency of the sine wave. You could try going back to a delay based on a for loop in place of the function using the SysTick timer, but that would not give you accurate timing and you would have to calibrate for each build (debug vs production) of the code. Another problem with the current design is that the main program loop must constantly update the DAC at precise intervals to keep the frequency constant and the harmonic distortion low. This makes it difficult for the microcontroller to do another task, such as scanning for user input or updating a display.

A better solution is to use one of the timer/counter peripherals to trigger an interrupt to update the DAC at a fixed rate based on the stable CPU clock without any foreground program intervention.[20] That is what we will do in the next lab, so leave the audio amplifier connected to the microcontroller DAC output.

24L.5 Solutions

- §24L.2.2: https://LAoE.link/micro3/24L_DAC_Comparator_ADCs.c.
- §24L.3: https://LAoE.link/micro3/24L_Test_DAC_Sine.c.

---

measured about 11 μs from §24L.3.2) compared to the 1 ms delay that it does not affect the frequency by a measurable amount. In both cases the output is $f_{out} = 1/(128 * 1\,\text{ms}) = 7.8\,\text{Hz}$.

[20] To be fair, you could also increase the resolution of the Delay() function by using a smaller divider in the call to SysTick_Config(), but using the timer and interrupts allows the CPU to do other work while the DAC gets updated in the background.

24S Supplementary Notes: Waveform Processing

You now have a working DAC available in your microcontroller. We are going to use the built-in ADC to allow us to digitize analog signals as well. Once you've got these peripherals available, it's fun to try altering waveforms, fun to see the result on a scope and fun to hear the result.

A controller like the Cortex-M0+ is not particularly well suited to signal processing. A Digital Signal Processor (DSP) is the device designed to do such a job. It includes fast, wide multipliers and accumulators, and usually speeds its processing with "pipelining."[1] Nevertheless, it is useful and fun to try to compose some tiny signal-processing programs as you try out the ADC and DAC.

24S.1 Add the hardware

You already have the DAC output connected to the LM386 power amplifier. All you need to add is a connection for the input signal. Almost any I/O pin on the SAMD21 can be used as an analog input: see Table 7-1, column B in the datasheet. We will use Arduino pin A3 (PA04) with the simple input protection circuit shown in Fig. 24S.1.

We have left out an input anti-aliasing filter and an output reconstruction filter in the interests of simplicity. Without the former, we will be able to show the effects of aliasing by increasing the input frequency to greater than one-half the sampling frequency. The lack of a reconstruction filter means the output will be stepped, not smooth.

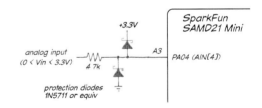

Figure 24S.1 Analog input for waveform processing.

24S.2 Test your waveform processing system

The skeleton code in §24S.4 includes initialization functions for the two peripherals as well as functions to acquire a 10-bit value from the ADC (`ReadADC()`) and output a 10-bit value to the DAC (`DAC_OutData()`). The input and output ranges are identical at 0 to 3.3 V.[2] The `main()` function sets the

---

[1] Hardware for efficient DSP is included in many modern FPGAs as well. Even the relatively small Lattice iCE40 used in the WebFPGA includes eight hardware multiply and accumulate blocks designed for efficient DSP operations.

[2] This means the speaker volume will be much louder than with a 1 V DAC output. Be sure you turn the volume down and increase it slowly.

CPU clock to 48 MHz, initializes the DAC and ADC, then enters an infinite loop reading the analog input and writing it to the analog output.

You can choose to operate on the offset binary value out of the ADC (0x0000 is the lowest value, 0x03ff the highest, and 0x0200 the midpoint) or you can convert the ADC value to 2's-complement so that the midpoint is 0x0000 with higher values positive and lower values negative. If you use the latter you have to convert back to offset binary before writing the value to the DAC. The skeleton includes code to convert from offset binary to 2's-complement and back again.

Create a new project, add the 48 MHz files and replace the default `main.c` with the code below.

Connect your function generator to the analog input and set it for a 1 kHz sine wave with a peak-to-peak amplitude of about 2 V centered around a DC voltage of 1.65 V. Make sure the generator output does not exceed the input range of the ADC: 0 to 3.3 V. The DAC output should look like the input since the default processing loop just samples the input and copies the value returned by `ReadADC()` to the output.

Increase the input sine frequency to 25 kHz and capture a single sweep of the input. You should now be able to see the individual samples and estimate the acquisition rate of the system.[3]

24S.3 Some suggested lab exercises

24S.3.1 Is Nyquist right?

Your ears should answer this question (but it is pretty obvious on the scope as well). To make it easier to see, slow down the sample rate to 1 kHz by adding a 1ms delay to the acquisition loop. Set the input sine wave to 100 Hz and uncomment the "`Delay(1)`" in the code. Listen to the DAC output. It won't sound like a pure sine wave because the lack of a reconstruction filter lets the high frequency components of the sample steps through to the speaker.

Now increase the input frequency past the Nyquist rate of 500 Hz. You should hear the output frequency *decrease* as the input frequency increases. Indeed, the output frequency above the Nyquist rate is 1 kHz minus the input frequency: see Fig. 24S.2.

Have some fun with this setup. It's not every day you get to confirm these sampling notions, which often seem rather abstract and arcane.

24S.3.2 Modifying waveforms

Here's a chance to write some of your own code to alter waveforms caught by the ADC. We'll help you with the first one. But first, comment out the delay in the acquisition loop to go back to the higher speed sample rate.

Full-wave rectify: Let an input sine input (which must lie within the ADC's 0 to 3.3 V input range) generate a full-wave-rectified output. "Rectify" about the midpoint of the voltage range of ADC and DAC, as in Figure 24S.3.

Full-wave rectify code: The sample is tested by checking to see if it is in the bottom half of the range.

```
if (sample < 0x0200){        // if below midpoint
   sample = ~sample;         // flip all the bits
}
```

---
[3] We measured 10 μs per sample for a conversion rate of 100 kHz.

24S.3 Some suggested lab exercises

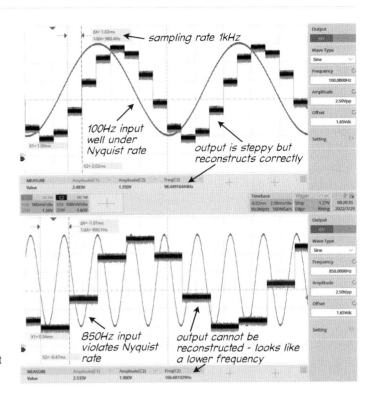

Figure 24S.2 It looks like Nyquist was right!

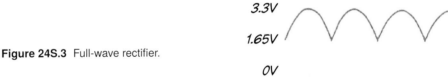

Figure 24S.3 Full-wave rectifier.

Note that the midpoint of the 10-bit ADC range is the value 0x0200h. The MSB flips as the input signal crosses this midpoint. Uncomment the sample code to try the offset-binary full-wave converter.

ADC data formats: offset-binary versus 2's-complement: The code above handles the "rectify" operation in offset-binary format, the format used by the SAMD21 ADC. You may feel like trying the alternative 2's complement format so that the midpoint of the ADC range is zero and values below the midpoint are negative. To go between the two formats, flip the MSB and sign extend the 2's-complement version to 16-bits.[4] The skeleton code converts the offset binary ADC sample to a signed 16-bit variable "ssample" for you. Feel free to do your processing with either.

Figure 24S.4 Comparison of 10-bit values: offset binary vs. 2's-complement.

```
0x3FF  11 1111 1111  ←—— TOP   ——→  0x1FF  01 1111 1111
0x200  10 0000 0000  ←—— MID   ——→  0x000  00 0000 0000
0x000  00 0000 0000  ←—BOTTOM——→   0x200  10 0000 0000
       offset binary                      2's complement
```

The offset binary program transforms 0x01ff to 0x0200, and leaves 0x0200 alone; thus two original values map to one output value (two "zeroes"). Two's-complement does not do this: zero would be untouched (as 0x0200 is in the program above); minus one (0xffff) would be transformed not to zero

---

[4] The sign extension is required in order to have the C compiler handle negative values properly.

but to plus one (0x0001). Not only does this seem cleaner, you can let the compiler deal with negative values (i.e., voltages below the midpoint).

If you do change to 2's-complement in software, you must flip the MSB and [optionally] clear the high bits in ssample to convert it back to 10-bit offset binary before sending it out the DAC.[5] Uncomment the second example to try out the 2's-complement version of the full wave rectifier.

Half-wave rectify: Here's a challenge: write code that substitutes for a resistor and *one* diode. Again, in Fig. 24S.5 we mean to rectify about the midpoint voltage – corresponding to 0x0200. (If you prefer, you can use the 2's-complement ssample variable and treat negative values as signals below the midpoint.) Certainly this code can be very similar to the full-wave program.

Figure 24S.5 Half-wave rectification centered about waveform midpoint.

Lowpass filter: It turns out that a program to simulate a lowpass filter is straightforward, though, once again, the performance of this primitive version compares poorly against that of 50-cents' worth of analog parts. Proper digital signal processing, done usually with dedicated processors or with large arrays of logic, like FPGAs, can produce extremely good filters. Such devices do what we're doing, but in a fancier and smarter way.

Let the processor average the current sample with the previous average, and output the result. Give the most recent sample a weight equal to the previous average. This filter, shown in Fig. 24S.6, is called an *Infinite Impulse Response* (IIR) filter because, using feedback, its response to an impulse never totally dies away. (We'll see the contrasting case in a moment.)

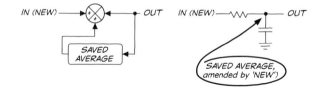

Figure 24S.6 IIR low pass filter block diagram compared to RC analog filter.

Write code to implement the IIR lowpass filter in the skeleton program's while(1) loop. Test your routine by feeding it a square wave of about 2.5 kHz or lower. Does the shape look roughly like

$$V_{\text{out}} = V_{\text{in}} \left(1 - e^{-t/RC}\right)?$$

If you're not sure, does it climb fast at first then slowly? Does the waveform travel in each step just halfway to its destination as shown in Fig. 24S.7 (about like Zeno's hare)?[6] The answer to all these questions should be yes.

Test the circuit's treatment of a sinewave. Do you find the usual lowpass filter phase shift effects? (You will see a constant delay; this is an artifact of the digital processing caused mostly by the ADC conversion time, and this is different from the phase shift that characterizes the analog lowpass filter.) Find your filter's $f_{3\text{dB}}$ point.

If you know the sampling rate, you know the rate at which the averaging of old and new occurs,

---

[5] You do not need the sign extension converting back to offset binary, but zeroing the high six bits makes the variable value look less confusing in the debugger.

[6] See https://LAoE.link/Zenos_hare.html.

24S.3 Some suggested lab exercises

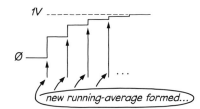

Figure 24S.7 IIR filter's response to step input.

and the time between "steps." Since we know that at every step the output should go halfway to the destination (from where it was to V_{in}), and we learned back in Lab 8L that the LMC555 waveform goes halfway in $0.7RC$, can you infer the effective "RC" of this filter, at a given sampling rate? To try the idea, suppose that the averaging steps come at 10 microsecond intervals, and see if you can calculate the "virtual RC."[7]

Tune the filter? To change f_{3dB} you can change the sampling rate (easy, but crude) or change the relative weights given to *new_sample* versus *running_average*. Changing those weights requires using two *multiply* operations, but the Cortex-M0+ includes a hardware integer multiply so, if we use integer values for the weight variable, it should not significantly impact the sample rate compared to a simple one-bit, divide-by-two right shift.[8]

Create a variable to control the weight of the new sample:

`uint8_t weight = 0x80;  // between 0 and 0xff`

A weight of 0x80 is equivalent to a simple average (well almost, $\frac{127}{2}$ would give exactly equal weights to the sample and previous average).

To calculate the new average:

- multiply the new sample by the weight;
- multiply the current average by (0xff - weight);
- add the results together and shift right by 8 bits to get the new average; then
- send the average to the DAC.

Use of multiply makes the filter more versatile – but also introduces round-off errors. Assuming that you're obliged to attenuate new, then old before adding the two, you'll find (if your program resembles the one we wrote) that heavy attenuation of the *new* sample causes the final value to fall short of the large input values, on both positive and negative excursions (error noticeable with weight value set to 020h – about one part in eight). Rounding the most significant byte of the average rather than just truncating (i.e., shifting right by 8) helps. You may find a more ingenious solution

An illustration of the tunable lowpass: Figure 24S.8 is a scope image – actually several, superimposed – showing response to a step, with the weight applied to the new sample set to four alternative values. This program applies to the new sample attenuation values of 1/2, 1/4, 1/8 and 1/16. The exponential shape is most evident for the easiest case, 1/2 (multiplier value of 080h).

FIR filter: Finite Impulse Response (FIR) describes a filter that doesn't remember *all* past inputs (as an analog filter would do) but instead knows about a few samples – usually, but not always, *past*

---

[7] Not hard, we think you'll agree. Time for each step is 10 μs, so RC is about 14 μs. That implies a f_{3dB} of 11.3 kHz, which is about what we measured (and it looked like 45° of phase shift as well).

[8] We measured a one microsecond (i.e., 10%) increase in sample time with the additional multiplies.

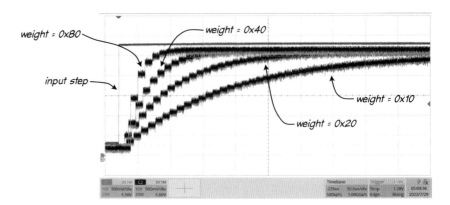

Figure 24S.8 Response of lowpass to step input with multiply used to vary weight of new sample relative to running average.

samples.[9] It lets those flow in and out of a "pipeline," and multiplies these by differing weights to form a weighted average that is the output.

For example, we could put out an *average* of newest sample and *one-ago sample*, as in Fig. 24S.9. This would be a not-very-conservative lowpass.

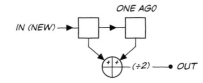

Figure 24S.9 FIR filter: lowpass.

Or we could put out the *difference* between the newest sample and *one-ago sample*, as in Fig. 24S.10. This would be a differentiator or highpass, of a very rudimentary sort. A longer pipeline would let us make a more refined filter.

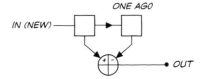

Figure 24S.10 FIR filter: highpass.

Reverb: This is lots of fun to listen to and play with. If you use a microphone as input, you can put yourself in a pipe or a barrel or a hall or a canyon, as you vary echo time and the quickness of decay.

Reverb needs a "circular buffer" – a table in RAM that forms a pipeline for as long as the echo time desired. Each time a sample comes in, we combine it (add it) with an attenuated version of an old running-accumulation value – pulled from a place DELAY ago in the pipeline – and then we play this sum and store it away. If input falls silent, for example, we'll hear the old running-accumulation value gradually die away.

The circular buffer is no more than a table with moving pointer, and pointer reinitialization when it hits the end of the table. The algorithm that pulls out the "old" value also needs to be smart enough to "wrap around" the end of the table. Apart from that, it's straightforward.

---

[9] How can it know "future" samples? By letting us process or operate on a sample that is not the most recent, but rather is one surrounded by later as well as prior samples.

24S.4 Skeleton code for waveform processing

Code to repeatedly read a 10-bit sample from Arduino pin A3 using the ADC and write it to the DAC.[10]

```c
/***************************************************************************
* Company: Learning the Art of Electronics
* Engineer: David Abrams
*
* Create Date:   2022-07-14
* Module Name:   main.c (24S.Waveform_Processing.c)
* Project Name:
* Target Device: Sparkfun SAMD21 Micro
* Description: Signal processing using ADC and DAC
***************************************************************************/

#include "samd21.h"      // add to use CMSIS
#include "ClockSysInit48M.h"
#include <stdio.h>
#include <stdlib.h>
#include <string.h>
#include <stdbool.h>

/***********************************************************************
*              Set up System Timer and Delay function
***********************************************************************/
// use 1ms system clock for timing

volatile uint32_t msTicks = 0;

void SysTick_Handler(void)  {         // SysTick interrupts every millisecond
    msTicks++;
}

// Function to delay for n milliseconds
void Delay (uint32_t DelayMSec)  {
  uint32_t endTicks;

  endTicks = msTicks + DelayMSec;

  while (msTicks < endTicks)  {       // wait for DelayMSec SysTick interrupts
      __WFE ();                       // tell CPU ok to power-down until interrupt
  }
}

/***********************************************************************
*              ADC Support Functions
***********************************************************************/

// We are using Arduino pin A3 (PA04) for the signal input (must be >0V and <3.3V).
// We will acquire 10 bits of data for the 10-bit DAC.  The ADC clock is 1MHz.

void ADC_Init(void) {

   /**************** PORT I/O Initialization  **************
    * Set up I/O pin as analog input according to table 23-2
    *
    * PIN_PA04B_ADC_AIN4 is analog input (Arduino pin A3)
    * PORT_PA04B_ADC_AIN4
    ********************************************************/

   // DIR bit for an analog pin must be set to 0
   REG_PORT_DIRCLR0 = PORT_PA04B_ADC_AIN4;

   // Configure the analog inputs with DRVSTR = 0, PULLEN = 0, INEN = 0 and PMUXEN = 1 (Table 23-2)
   PORT->Group[0].PINCFG[PIN_PA04B_ADC_AIN4].reg = PORT_PINCFG_PMUXEN;    // enable mux to map to ADC input

   // Enable pin multiplexer configuration B (use RMW in case other pin already set)
   PORT->Group[0].PMUX[PIN_PA04B_ADC_AIN4 >> 1].reg |= PORT_PMUX_PMUXE_B; // mux even pin number to B

   /************* POWER MANAGER Initialization  ************/
   /* ADC peripheral is already enabled after reset          */
```

---

[10] Code available at https://LAoE.link/micro3/24S_Waveform_Processing.c

Supplementary Notes: Waveform Processing

```c
/********************************************************/
/********* GENERIC CLOCK CONTROLLER Initialization *********/
/* Clock 3 is enabled and set to 8MHz by ClockSysInit48M() */
/********************************************************/

   GCLK->CLKCTRL.reg = GCLK_CLKCTRL_ID_ADC | GCLK_CLKCTRL_CLKEN | GCLK_CLKCTRL_GEN_GCLK3;
   while (GCLK->STATUS.bit.SYNCBUSY);

   /***************** AC Initialization ************************************/
   /* Configure ADC for 8-bit, GCLK3/8 (1MHz) manually triggered conversion */
   /************************************************************************/

   ADC->CTRLA.bit.ENABLE = 0;              // disable ADC before reset (33.6.2.2)
   while(ADC->STATUS.bit.SYNCBUSY);        // wait for disable to finish

   /* Perform a software reset of ADC */
   REG_ADC_CTRLA = ADC_CTRLA_SWRST;
   while(ADC->STATUS.bit.SYNCBUSY || ADC->CTRLA.bit.SWRST);

   /* set reference to 3.3V/2 = 1.65V to give 0 to 3.3V scale (input will be divided by 2) */
   REG_ADC_REFCTRL = ADC_REFCTRL_REFSEL_INTVCC1;

   /* Set clock prescaler (to get 1MHz) , 10-bit, single triggered */
   /* conversion mode, right justified, single ended              */
   REG_ADC_CTRLB = ADC_CTRLB_PRESCALER_DIV8 | ADC_CTRLB_RESSEL_10BIT;
   while(ADC->STATUS.bit.SYNCBUSY);

   /* extend sample time a bit to charge ADC input capacitor in case source imedance high */
   REG_ADC_SAMPCTRL = 2;

   /* divide input by 2 so matches ref, use AIN[4], ground as reference */
   REG_ADC_INPUTCTRL = ADC_INPUTCTRL_MUXPOS_PIN4 | ADC_INPUTCTRL_MUXNEG_IOGND
   | ADC_INPUTCTRL_GAIN_DIV2;
   while(ADC->STATUS.bit.SYNCBUSY);

  /* Enable ADC peripheral */
   REG_ADC_CTRLA = ADC_CTRLA_ENABLE;       // turn on ADC (ok to set other bits to 0)
   while(ADC->STATUS.bit.SYNCBUSY);        // wait for the ADC enable to finish
}

/* function to return 10-bit value (0 to 1023) from AIN[4] */

uint16_t ReadADC(void) {
    REG_ADC_INTFLAG = ADC_INTFLAG_RESRDY;               // clear result ready flag

    REG_ADC_SWTRIG = ADC_SWTRIG_START;                  // start a conversion
    while(ADC->STATUS.bit.SYNCBUSY);

    while(ADC->INTFLAG.bit.RESRDY == 0) {               // wait for conversion done
      __WFE();
      }
    return REG_ADC_RESULT;                              // clears ready flag
}


/******************************************************************
*              DAC Support Functions
 ******************************************************************/

void DAC_Init(void){

// Only pin PA02 (Arduino pin A0) can be used for DAC output.
// Peripheral multiplexer Function B is used for all analog functions.

    /*************** PORT I/O Initialization *************/
    /* Set up DAC output pin as analog I/O (see table 22-1) */
    /*****************************************************/

    // DIR bit for an analog pin must be set to 0
    REG_PORT_DIRCLR0 = PORT_PA02B_DAC_VOUT;

    // Configure the DAC output pin with DRVSTR = 0, PULLEN = 0, INEN = 0 and PMUXEN = 1
    PORT->Group[0].PINCFG[PIN_PA02B_DAC_VOUT].reg = PORT_PINCFG_PMUXEN;   // enable pin mux
```

24S.4 Skeleton code for waveform processing

```c
    // set pin multiplexer configuration B for even pin PA02
    PORT->Group[0].PMUX[PIN_PA02B_DAC_VOUT >> 1].reg = PORT_PMUX_PMUXE(MUX_PA02B_DAC_VOUT);

    /******** GENERIC CLOCK CONTROLLER Initialization ********/
    /* Connect GCLK 0 to DAC and enable DAC clock.           */
    /* GCLK 0 is enabled and initially set to 1MHz.          */
    /*********************************************************/

    GCLK->CLKCTRL.reg = GCLK_CLKCTRL_ID_DAC | GCLK_CLKCTRL_CLKEN | GCLK_CLKCTRL_GEN_GCLK0;
    while(GCLK->STATUS.bit.SYNCBUSY);        // requires write synchronization

    /************** POWER MANAGER Initialization ************/
    /* Enable the DAC bus clock in Power Manager            */
    /********************************************************/

    PM->APBCMASK.reg |= PM_APBCMASK_DAC;

    /***************** DAC Initialization *********************/
    /* Setup DAC for 10-bit, right justified, 0 to 3.3V output */
    /***********************************************************/

    // Wait for reset and busy synchronization
    while(DAC->STATUS.bit.SYNCBUSY || DAC->CTRLA.bit.SWRST)

    // Perform a software reset
    REG_DAC_CTRLA = DAC_CTRLA_SWRST;

    // Wait for reset and busy synchronization
    while(DAC->STATUS.bit.SYNCBUSY || DAC->CTRLA.bit.SWRST);

    // Set reference to analog 3.3V supply for 0 - 3.3V output.  Enable output driver
    // (right adjusted data is the default after reset so not explicitly set)
    REG_DAC_CTRLB = DAC_CTRLB_EOEN | DAC_CTRLB_REFSEL_AVCC;

    // wait for SYNCBUSY synchronization
    while(DAC->STATUS.bit.SYNCBUSY);        // wait for SYNCBUSY synchronization

    // enable the DAC
    REG_DAC_CTRLA = DAC_CTRLA_ENABLE;     // turn on the DAC

    // wait for the enable to finish
    while(DAC->STATUS.bit.SYNCBUSY);       // wait for the enable to finish
}

//*****************************************************
// function to send one 10-bit sample to the DAC port

void DAC_OutData(uint16_t data){
REG_DAC_DATA = data; // write value to DAC
}

/******************************
*           main()
******************************/
int main (void)
{
uint16_t sample;                    // 10-bit offset binary ADC/ADC value
int16_t ssample;                    // signed two's- complement version of sample

 // system initializations
 ClockSysInit48M();                             // set CPU clock to 48MHz
 SysTick_Config(SystemCoreClock/1000); // configure SysTick for Delay() timer
 ADC_Init();                                    // used to sample voice input
 DAC_Init();                                    // outputs audio to LM386

// read the ADC and send data to DAC forever

  while (1) {
    sample = ReadADC();

    ssample = (sample) ^ 0x0200;        // convert to 2's- complement
    if (ssample & 0x0200) {             // and sign extend to 16-bits
      ssample = ssample | 0xfe00;
    }
```

Supplementary Notes: Waveform Processing

```
    // waveform processing goes here

    // example - offset binary full wave rectifier
    //if (sample < 0x0200){
    //   sample = ~sample;
    //}


    //Delay(1);                          // to slow down acquisition to see aliasing

    // to convert 2's-complement back to offset binary
    //sample = (ssample ^ 0x0200) & 0x03ff;

    DAC_OutData(sample);
  }                                    // end while loop                              // end while loop
  return 0;
}
/*************************** End of file ***************************/
```

24S.5 Solutions

- §24S.3: `https://LAoE.link/micro3/24S_Waveform_Processing_Sol.c`

24W Worked Examples: Speeding Up the SAMD21 CPU Clock

24W.1 Why does the SAMD21 run so slowly after a reset?

After a microcontroller reset, the SAMD21 internal 8 MHz oscillator is initialized with a prescaler of eight to configure the OSC8M output as a 1 MHz clock. Generic Clock 0 (GCLK0) is enabled and connected to OSC8M as its source.[1] Since the SAMD21 CPU is always clocked by the output of Generic Clock 0, the processor will run at a relatively slow 1 MHz unless changed by the programmer's initialization code.[2]

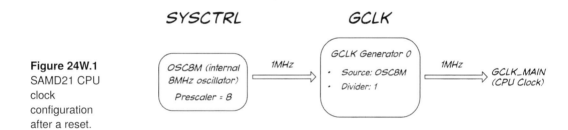

Figure 24W.1 SAMD21 CPU clock configuration after a reset.

The power used in a CMOS circuit is directly proportional to clock frequency. Current only flows into or out of a CMOS gate to charge or discharge the input gate-to-source capacitance when the input switches from low to high or visa-versa. In the SAMD21, power supply current is a linear function of the CPU clock frequency. Table 37-8 in the datasheet shows that typical supply current in an endless `while(1)` loop is $(60*\text{freq}(\text{MHz})+136)$ μA. That means that at a supply voltage of 3.3 V, the processor dissipates about 647 μW when clocked at 1 MHz. At the maximum CPU clock speed of 48 MHz, the power consumption would be about 3 mW, still a small amount but significant if running on batteries. Additional power is consumed as peripherals are enabled. It appears that the CPU designers felt lower power consumption should be the default over speedier processing.

24W.2 Let's look at the CPU clock (literally!)

To see changes to the CPU clock as we change it first from 1 MHz to 8 MHz, and then to 48 MHz, add this function to the §23L.1 CMSIS blinky program `main.c` and call it as part of your initialization code to connect the CPU clock to Arduino pin D2.[3]

```
/*************************************************
 *      EnableGCLK0IO()
```

---

[1] See §14.8, §17.8.9 and §15.8.4 in the SAMD21 datasheet.
[2] The output of GCLK0 is sent as GCLK_MAIN to the Power Controller which generates the CPU clock, CLK_CPU.
[3] Code available at https://LAoE.link/micro3/24W_EnableGCLK0IO.c.

Worked Examples: Speeding Up the SAMD21 CPU Clock

```
 *
 * Enables the output of Generic Clock
 * Generator 0 on I/O Pin PA14 (Arduino pin D2)
 ************************************************/

void EnableGCLK0IO(void) {
    uint32_t rmwValue;                  // read-modify-write value of GCLK0
    uint8_t* REG_GCLK_GENCTRL_ID8;      // ptr to 8-bit address field of GENCTRL

    // Configure up I/O pin PA14 for GCLK0 output
    REG_PORT_DIRSET0 = PORT_PA14H_GCLK_IO0;
    PORT->Group[0].PINCFG[PIN_PA14H_GCLK_IO0].reg = PORT_PINCFG_PMUXEN;
    PORT->Group[0].PMUX[PIN_PA14H_GCLK_IO0 >> 1].reg =
                                    PORT_PMUX_PMUXE(MUX_PA14H_GCLK_IO0);

    // To read the value of a generic clock control configuration, you have
    // to first write the ID of the clock generator to the low byte of the
    // indirect access GENCTRL register. Since the CMSIS register definitons
    // access the register as 32-bit we need a pointer to just the 8-bit ID
    // to set it to the generator we want to access without disturbing the
    // existing configuration.

    REG_GCLK_GENCTRL_ID8 = (uint8_t*) &REG_GCLK_GENCTRL;
    *REG_GCLK_GENCTRL_ID8 = 0x00;       // select access to GCLK Gen 0

    rmwValue = REG_GCLK_GENCTRL;        // enable I/O output to existing config
    rmwValue |= GCLK_GENCTRL_OE;
    REG_GCLK_GENCTRL = rmwValue;
}
```

You should also add

```
printf("SystemCoreClock is set to %d.", SystemCoreClock);
```

after the call to EnableGCLK0IO(void) to display what CMSIS thinks the current system clock is on the debug console.

Connect your oscilloscope to the pin marked "2" on the SparkFun SAMD Mini and run your Blinky program.[4] The LED should blink at a 1 Hz rate, the console should show SystemCoreClock set to "1000000" and the 'scope should show a 1 MHz square wave.

24W.3 Speeding up the SAMD21

24W.3.1 Increasing the CPU clock to 8 MHz is easy

The CPU initializes the 8 MHz oscillator prescaler to a value of 8 (by setting PRESC[1:0] bits in the System Controller osc8m register to a value of 0x3.[5] All you need to do to get an 8 MHz CPU clock is to reset the prescaler to a value of 1 (PRESC[1:0] = 0) and update the value of the global variable SystemCoreClock.[6]

Add the following function to your Blinky program and call it as part of your initialization.[7] Be sure to call this function *before* printing out the CPU clock frequency.

---
[4] Be sure any bandwidth limiter on the scope channel you are using is disabled.
[5] The 8 MHz clock is divided by 2^{PRESC} where PRESC is a two bit value from 0 to 3.
[6] CMSIS includes a function, SystemCoreClockUpdate(void), which is supposed to examine the microcontroller clock registers and calculate the current core clock. However, the version included with Segger Embedded Studio CMSIS (in file *system_samd21.h*) always sets SystemCoreClock to 1000000, so you should never call it.
[7] Code available at https://LAoE.link/micro3/24W_ClockSysInit8M.c.

24W.3 Speeding up the SAMD21

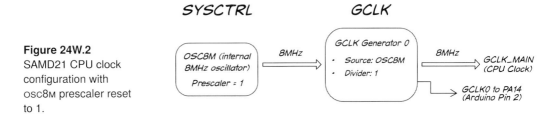

Figure 24W.2 SAMD21 CPU clock configuration with OSC8M prescaler reset to 1.

```
/*************************************
 *   ClockSysInit8M()
 *
 * Modify prescaler value of OSC8M
 * to produce 8MHz output
 *************************************/
void ClockSysInit8M(void) {
  SYSCTRL->OSC8M.bit.PRESC = 0;     // prescale by 1
  SYSCTRL->OSC8M.bit.ONDEMAND = 0 ; // oscillator is always on if enabled
  SystemCoreClock = 8000000;        // update global variable with CPU freq
}
```

The LED should still blink at a 1 Hz rate, but now the console and oscilloscope should show an 8 MHz CPU clock. This demonstrates the benefit of using the SysTick timer and using the value of SystemCoreClock to calculate the delay time. Even though the CPU is now executing code eight time faster, the LED blink rate has not changed. Had you used a software delay loop, you would need to change the delay count to a value eight times larger.

24W.3.2 Getting to 48 MHz is lot more work

Although the SAMD21 is specified for a CPU clock speed up to 48 MHz, the fastest internal oscillator is the 8 MHz OSC8M and the maximum speed of an external oscillator connected to the XOSC pin is 30 MHz. However, the microcontroller includes a Digital Frequency Locked Loop (DFLL48M) which can track other clock sources and provide a higher frequency output at a multiple of an input reference frequency. The DFLL can operate in open-loop or closed-loop mode. The latter uses a low frequency, high accuracy reference clock. We will operate in closed-loop mode, using the external 32.768 kHz crystal oscillator (XOSC32K) as the reference clock and connecting the DFLL48M output to Generic Clock Generator 0 (GCLK0) which will output the 48 MHz clock as GCLK_MAIN. We will also move the 8 MHz oscillator (with a prescaler of 1) to GCLK3 for use as a peripheral clock.

The steps to set up the 48 MHz CPU clock are shown in Fig. 24W.3. The actual procedure is pretty complicated, so feel free to skip the rest of this section.[8]

To create the 48 MHz CPU clock initialization function[9]

```
void ClockSysInit48(void)
```

begin by setting up the following definitions:

```
// Constants for Clock Generators
#define GENERIC_CLOCK_GENERATOR_MAIN       (0u) /* Initialized at reset for 1MHz */
#define GENERIC_CLOCK_GENERATOR_XOSC32K    (1u)
```

---
[8] However, be sure to follow the directions in Sec. 24W.3.3 to run the CPU at 48 MHz to show that the 1 Hz blink is not affected.
[9] The code that follows has been modified from the SAM D21 DFLL48M 48 MHz Initialization Example at https://LAoE.link/Microchip_48MHz_App_Note.html. The complete modified version is available at https://LAoE.link/ClockSysInit48M.c.

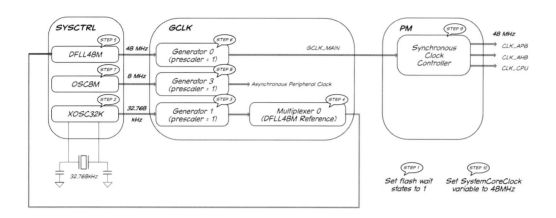

Figure 24W.3 Steps to set the SAMD21 CPU clock to 48 MHz.

```
#define GENERIC_CLOCK_GENERATOR_OSCULP32K (2u) /* Initialized at reset for WDT */
#define GENERIC_CLOCK_GENERATOR_OSC8M     (3u)
// Constants for Clock Multiplexers
#define GENERIC_CLOCK_MULTIPLEXER_DFLL48M (0u)

// Constants for DFLL48M
#define MAIN_CLK_FREQ (48000000u)
#define EXT_32K_CLK_FREQ (32768u)
```

24W.3.2.1 Step 1: Flash don't fail me now

SAMD21 programs are stored in the internal flash memory, which normally operates with no wait states.[10] Above a CPU clock frequency of 24 MHz, the SAMD21 nonvolatile memory requires one wait state to operate reliably.[11] We need to enable the extra wait state *before* switching the CPU clock to 48 MHz. The number of Read Wait States (RWS) for the flash memory is set in the Nonvolatile Memory Controller (NVMCTRL):

```
    NVMCTRL->CTRLB.bit.RWS = 1;      /* 1 wait state required @ 3.3V & 48MHz */
```

24W.3.2.2 Step 2: Turn on the XOSC32K reference oscillator

Although the SparkFun SAMD Mini breakout includes an external 32.768 kHz crystal connected to the microcontroller XOUT32 and XIN32 pins, the XOSC32K crystal oscillator is not enabled by default after a reset. We need to enable it in the System Controller before it can be used as a source for a generic clock generator (whose output will be used as the DFLL reference clock).

```
// Configure SYSCTRL->XOSC32K settings
SYSCTRL_XOSC32K_Type sysctrl_xosc32k = {
  .bit.WRTLOCK = 0,          /* XOSC32K configuration is not locked */
  .bit.STARTUP = 0x4,        /* 500ms start-up time */
  .bit.ONDEMAND = 0,         /* Osc. is always running when enabled */
  .bit.RUNSTDBY = 0,         /* Osc. is disabled in standby sleep mode */
  .bit.AAMPEN = 0,           /* Disable automatic amplitude control */
  .bit.EN32K = 1,            /* 32kHz output enabled for crystal */
  .bit.XTALEN = 1            /* Crystal connected to XIN32/XOUT32 */
};
```

---

[10] Wait states are additional clock cycles inserted by the CPU when accessing slower memory to ensure the memory output is valid before the data is latched.

[11] See Table 37-42 in the datasheet.

```c
// Write these settings
SYSCTRL->XOSC32K.reg = sysctrl_xosc32k.reg;
// Enable the Oscillator - Separate step per datasheet recommendation (sec 17.6.3)
SYSCTRL->XOSC32K.bit.ENABLE = 1;

// Wait for XOSC32K to stabilize
while(!SYSCTRL->PCLKSR.bit.XOSC32KRDY);
```

24W.3.2.3 Step 3: Make XOSC32K the source for Generic Clock Generator 1

In order to connect a source to a peripheral (in this case the DFLL reference input), we need to first create a Generic Clock using that source. Since GCLK0 must be the 48 MHz output of the DFLL in order to be used as the CPU clock, we connect the external 32.768 kHz oscillator to the next unused Generic Clock Generator, GCLK1.

```c
// Set the Generic Clock Generator 1 output divider to 1
// Configure GCLK->GENDIV settings
GCLK_GENDIV_Type gclk1_gendiv = {
  .bit.DIV = 1,                                 /* Set output division factor to 1 */
  .bit.ID = GENERIC_CLOCK_GENERATOR_XOSC32K    /*   for Generator 1 */
};
// Write these settings
GCLK->GENDIV.reg = gclk1_gendiv.reg;

// Configure Generic Clock Generator 1 with XOSC32K as source
GCLK_GENCTRL_Type gclk1_genctrl = {
  .bit.RUNSTDBY = 0,   /* Ignored if OE not 1 */
  .bit.DIVSEL = 0,     /* Use GENDIV.DIV value to divide the generator */
  .bit.OE = 0,         /* Disable generator output to GCLK_IO[1] */
  .bit.OOV = 0,        /* GCLK_IO[1] output value when generator is off */
  .bit.IDC = 1,        /* Generator duty cycle is 50/50 */
  .bit.GENEN = 1,      /* Enable the generator */
  .bit.SRC = 0x05,     /* Generator source: XOSC32K output */
  .bit.ID = GENERIC_CLOCK_GENERATOR_XOSC32K        /* Generator ID: 1 */
};
// Write these settings
GCLK->GENCTRL.reg = gclk1_genctrl.reg;
// GENCTRL is Write-Synchronized...so wait for write to complete
while(GCLK->STATUS.bit.SYNCBUSY);
```

24W.3.2.4 Step 4: Connect GCLK1 to the DFLL reference input

We now can use the Generic Clock Multiplexer to connect the output of Generic Clock 1 (the external 32.768k oscillator) to the reference input of the DFLL48M digital frequency locked loop.

```c
    GCLK_CLKCTRL_Type gclk_clkctrl = {
        .bit.WRTLOCK = 0,     /* Generic Clock is not locked from subsequent writes */
        .bit.CLKEN = 1,       /* Enable the Generic Clock */
        .bit.GEN = GENERIC_CLOCK_GENERATOR_XOSC32K,     /* GCLK1 is the XOSC32k source */
        .bit.ID = 0x00        /* Generic Clock Multiplexer 0 is DFLL48M Reference */
    };
    // Write these settings
    GCLK->CLKCTRL.reg = gclk_clkctrl.reg;
```

24W.3.2.5 Step 5: Enable the DFLL 48 MHz clock source

With the reference oscillator connected, we can configure and enable the 48 MHz DFLL. This is first done in open-loop mode, then switched to closed-loop mode to use the reference oscillator.[12]

---

[12] This operation is not well documented in the SAMD21 datasheet. The code here is copied directly from the Microchip application note.

964 Worked Examples: Speeding Up the SAMD21 CPU Clock

```c
// Enable the DFLL48M in open loop mode. Without this step, attempts to go
// into closed loop mode at 48 MHz will result in Processor Reset (you'll
// be at the Reset_Handler in startup_samd21.c). PCLKSR.DFLLRDY must
// be one before writing to the DFLL Control register. Note that the
// DFLLRDY bit represents status of register synchronization - NOT clock
// stability (see Data Sheet 17.6.14 Synchronization for detail).

uint32_t tempDFLL48CalibrationCoarse;   /* temp for DFLL48 coarse calibration value */

while(!SYSCTRL->PCLKSR.bit.DFLLRDY);
SYSCTRL->DFLLCTRL.reg = (uint16_t)(SYSCTRL_DFLLCTRL_ENABLE);
while(!SYSCTRL->PCLKSR.bit.DFLLRDY);

// Set up the Multiplier, Coarse and Fine steps
SYSCTRL_DFLLMUL_Type sysctrl_dfllmul = {
  .bit.CSTEP = 31,    /* Coarse step - use half of the max value (63) */
  .bit.FSTEP = 511,   /* Fine step - use half of the max value (1023) */
  .bit.MUL = 1465     /* MAIN_CLK_FREQ (48MHz)/EXT_32K_CLK_FREQ (32768 Hz) */
};
// Write these settings
SYSCTRL->DFLLMUL.reg = sysctrl_dfllmul.reg;
// Wait for synchronization
while(!SYSCTRL->PCLKSR.bit.DFLLRDY);

// To reduce lock time, load factory calibrated values into DFLLVAL
// (cf. Data Sheet 17.6.7.1) Location of value is defined in Data
// Sheet Table 10-8. NVM Software Calibration Area Mapping

// Get factory calibrated value for "DFLL48M COARSE CAL" from NVM Software
// Calibration Area
tempDFLL48CalibrationCoarse = *(uint32_t*)FUSES_DFLL48M_COARSE_CAL_ADDR;
tempDFLL48CalibrationCoarse &= FUSES_DFLL48M_COARSE_CAL_Msk;
tempDFLL48CalibrationCoarse =
    tempDFLL48CalibrationCoarse>>FUSES_DFLL48M_COARSE_CAL_Pos;
// Write the coarse calibration value
SYSCTRL->DFLLVAL.bit.COARSE = tempDFLL48CalibrationCoarse;
// Switch DFLL48M to Closed Loop mode and enable WAITLOCK
while(!SYSCTRL->PCLKSR.bit.DFLLRDY);
SYSCTRL->DFLLCTRL.reg |=
    (uint16_t) (SYSCTRL_DFLLCTRL_MODE | SYSCTRL_DFLLCTRL_WAITLOCK);

// Wait for the frequency to lock
 while (!SYSCTRL->PCLKSR.bit.DFLLLCKC || !SYSCTRL->PCLKSR.bit.DFLLLCKF);
```

24W.3.2.6   Step 6: Make DFLL48 the source for Generic Clock Generator 0

Now that Digital Frequency Locked Loop is running, switch the CLKGEN0 source to DFLL48 to run the CPU at 48 MHz. The GCLK0 divider is initialized at reset to 1, so we do not have to set it again.

```c
GCLK_GENCTRL_Type gclk_genctrl0 = {
  .bit.RUNSTDBY = 0,   /* Generic Clock Generator is stopped in stdby */
  .bit.DIVSEL =  0,    /* Use GENDIV.DIV value to divide the generator */
  .bit.OE = 0,         /* Do not enable generator output to output pin */
  .bit.OOV = 0,        /* GCLK_IO[0] output value when generator is off */
  .bit.IDC = 1,        /* Generator duty cycle is 50% */
  .bit.GENEN = 1,      /* Enable the generator */
  .bit.SRC = 0x07,     /* Generator source: DFLL48M output */
  .bit.ID = GENERIC_CLOCK_GENERATOR_MAIN          /* Generator ID: 0 */
};
GCLK->GENCTRL.reg = gclk_genctrl0.reg;
// GENCTRL is Write-Synchronized...so wait for write to complete
while(GCLK->STATUS.bit.SYNCBUSY);
```

24W.3.2.7 Step 7: Reset the OSC8M prescaler to 1

We want to use the 8 MHz clock for slower peripherals that do not need to run at 48 MHz. Since, after a hardware reset, the prescaler for this oscillator is initialized to divide the 8 MHz by eight to create a 1 MHz clock, we need to reset the prescaler to one as we did in §24W.3.1 to create an 8 MHz Generic Clock.

```
SYSCTRL->OSC8M.bit.PRESC = 0;         /* Prescale by 1 */
SYSCTRL->OSC8M.bit.ONDEMAND = 0 ;     /* Oscillator is always on if enabled */
```

24W.3.2.8 Step 8: Connect the 8 MHz clock to Clock Generator 3

Clock Generator 2 is initialized to the internal very low power 32,768 kHz oscillator (OSCULP32K) after a microcontroller reset. That leaves GCLKGEN3 as the next unused clock generator. We will connect it to OSC8M (now running at 8 MHz) to use as a generic clock for devices that do not need to run at the high-speed 48 MHz clock.

```
GCLK_GENDIV_Type gclk3_gendiv = {
  .bit.DIV = 1,                                  /* Set output division factor = 1 */
  .bit.ID = GENERIC_CLOCK_GENERATOR_OSC8M        /* Apply division factor to Generator 3 */
};
// Write these settings
GCLK->GENDIV.reg = gclk3_gendiv.reg;

// Configure Generic Clock Generator 3 with OSC8M as source
GCLK_GENCTRL_Type gclk3_genctrl = {
  .bit.RUNSTDBY = 0,   /* Generic Clock Generator is stopped in stdby */
  .bit.DIVSEL = 0,     /* Use GENDIV.DIV value to divide the generator */
  .bit.OE = 0,         /* Disable generator output to GCLK_IO[1] */
  .bit.OOV = 0,        /* GCLK_IO[2] output value when generator is off */
  .bit.IDC = 1,        /* Generator duty cycle is 50% */
  .bit.GENEN = 1,      /* Enable the generator */
  .bit.SRC = 0x06,     /* Generator source: OSC8M output */
  .bit.ID = GENERIC_CLOCK_GENERATOR_OSC8M        /* Generator ID: 3 */
};
// Write these settings
GCLK->GENCTRL.reg = gclk3_genctrl.reg;
// GENCTRL is Write-Synchronized...so wait for write to complete
while(GCLK->STATUS.bit.SYNCBUSY);
```

24W.3.2.9 Step 9: Set the synchronous bus clock dividers to 1

The next step is to make sure both the CPU clock and the synchronous system and peripheral buses are running at full speed by setting their respective bus prescalers to 1.[13]

```
PM->CPUSEL.reg  = PM_CPUSEL_CPUDIV_DIV1 ;
PM->APBASEL.reg = PM_APBASEL_APBADIV_DIV1_Val ;
PM->APBBSEL.reg = PM_APBBSEL_APBBDIV_DIV1_Val ;
PM->APBCSEL.reg = PM_APBCSEL_APBCDIV_DIV1_Val ;
```

24W.3.2.10 Step 10: Update the SystemCoreClock

The final step is to update the global SystemCoreClock variable to the new CPU frequency so we can calculate the proper value for the SysTickTimer interrupt.[14]

```
SystemCoreClock = MAIN_CLK_FREQ;
```

This completes the `ClockSysInit48()` function. You can either include the function in your `main.c` source file or create a separate source file along with a header file containing the function prototype.

---

[13] This is likely unnecessary since the prescalers are set to divide-by-1 on a microcontroller reset.
[14] See Footnote 6.

If the latter, you will need to #include the header file in main.c and add both the source and header files to the project directory.[15]

24W.3.3 Try your Blinky program at 48 MHz

Update your Blinky program to run at 48 MHz by removing the call to function ClockSysInit8M() added in §24W.3.1 and replacing it with a call to ClockSysInit48M(). If you have not already followed the directions in §24L.3.3 to add the 48 MHz function header and source files to your project, do so now.

As with the 8 MHz test, the LED should still blink at a 1 Hz rate, but now the console and oscilloscope should show a 48 MHz CPU clock. (If you have a lower bandwidth oscilloscope, for example 100 MHz, or you have enabled the bandwidth limiter on the scope channel this signal will look sinusoidal rather than square. Why?[16])

---

[15] The function header file, ClockSysInit48M.h, contains the single line "void ClockSysInit48M(void);", the prototype for the function. If you add the function code to main.c, you will not need the header file.

[16] Because the third and higher odd harmonics of the 48 MHz square are attenuated by the limited bandwidth of the oscilloscope, so you are only seeing the fundamental sine wave.

25N Microcontrollers IV: Timers & Interrupts

Contents

25N.1	**The SAMD21 Timer/Counter peripherals**	**967**
	25N.1.1 Start by selecting the output pin	968
	25N.1.2 Turn on power to the TC4 peripheral	968
	25N.1.3 Set up the TC4 clock	968
	25N.1.4 Configure the Timer/Counter to toggle the I/O pin	969
25N.2	**Interrupts**	**970**
	25N.2.1 Why interrupts?	970
	25N.2.2 SAMD21 Interrupts	971
25N.3	**Lab preview**	**975**
25N.4	**AoE reading**	**975**

In the previous lab, you configured the SAMD21's internal 10-bit Digital-to-Analog converter to output an analog voltage to an I/O pin and then used the DAC to synthesize a 128 point sine wave. While this worked, updating the DAC in a loop did not provide precise control over the frequency of the output signal and the process of sending data values to the DAC consumed all the processor CPU cycles. This design also did not ensure that samples were output at constant time intervals.

In this chapter, we will correct all these deficiencies by using an internal timer peripheral to control how often the DAC is updated, giving precise control over the sine wave frequency, and will use the timer's interrupt to update the DAC in a background process, leaving the main foreground loop free for user input or other processing tasks.

25N.1 The SAMD21 Timer/Counter peripherals

The SAMD21G18A has three 16-bit Timer Counter ("TC") peripherals. Each TC can either count clock pulses in timer mode or external events in counter mode. The TCs can be configured as 8-, 16- or 32-bit wide (the latter by combining two of the TCs into a single counter). Each TC has two Waveform Output signals (`WO[0]` and `WO[1]`) of which none, one or both can be connected via the peripheral multiplexer to an output pin. By default, counters count up to their maximum value (termed "`MAX`" in the datasheet) then wrap to zero; however, they may also be configured to count up from zero to a preset compare value (the "`TOP`" value) or to count down from a preset load value to zero and then, in each case, toggle an output pin for frequency generation or pulse-width modulation. They can trigger an interrupt on overflow or a compare match. The TCs also support an automatic capture mode used to measure the period or pulse-width of an external signal.[1]

---

[1] The microcontroller also includes three Timer Counters for Control ("TCC"). Timer Counters for Control support all the functions of the simpler TC, but add more complex pulse-width modulation and sophisticated timing for motor control and other power control operations. We do not use the TCCs in this lab because we do not need their enhanced functionality and they are more complex to configure.

968 Microcontrollers IV: Timers & Interrupts

You will be using one of the TCs in 16-bit timer mode counting the 48 MHz clock and resetting on a compare match. This will allow timing events with 20.9 ns precision, and theoretically allow you to update the DAC at a rate from 48 MHz down to 732 Hz under program control by changing the compare value.[2]

25N.1.1 Start by selecting the output pin

It might seem that we do not need any I/O pins in this application, since all the operations are internal. One of the timer/counters will count up until it reaches the value stored in the TC compare register. The counter will then generate an interrupt and reset to zero. The interrupt will trigger a call to the Interrupt Service Routine function which will write the next synthetic data point to the DAC output register. None of this requires external I/O. However, it is convenient to initially connect the timer output to an I/O pin so that we can use an oscilloscope to see if the timer is doing what we expect while testing the timer configuration code. We will also use this output in several of the experiments in Chapter 25L.

Look at datasheet Table 7-1, column E to see the port multiplexing options for the SAMD21G TC output. The SAMD21G has three Timer/Counters, TC3–TC5, and each TC can be connected to two separate sets of output pins.[3] This gives the illusion that we have quite a bit of latitude in selecting output pins for the timer/counters. However, only two of the timers, TC3 and TC4, are available on header pins. Looking at the SparkFun SAMD21 Mini graphical datasheet, we see that only the outputs for TC4 do not conflict with the pins used for the keypad and the LED.[4]

This means we are limited to using TC4 and choosing either PB08 (Arduino pin A1) or PB09 (Arduino pin A2). Choosing WO[0] connects TC waveform output 0 to PB08 and puts the square wave output next to the DAC output on Arduino pin A0, making it easy to find with the scope.[5]

25N.1.2 Turn on power to the TC4 peripheral

Like the DAC, the SAMD21 does not enable the timer/counters after a reset, in order to conserve power. (See Table 12-1 in the datasheet.) Since TC4 is initially disabled, it must be enabled in the Power Manager before it can be used. We set the bit in the APBCMASK register that enables the TC4 with a RMW operation to avoid inadvertently disabling any other enabled peripherals.

25N.1.3 Set up the TC4 clock

We also need to connect a generic clock to timer/counter 4 to be used as the timing source. We will connect the timer/counter to the 48 MHz CPU clock on GCLKGEN0. That will allow us to adjust the period of the DAC sine wave in $\approx 2.7\,\mu s$ increments.[6]

---

[2] "Theoretically" because calling the interrupt function and writing data to the DAC takes many 20.9 ns CPU cycles, so the maximum frequency at which the program works properly will be significantly lower. In addition, the DAC is only rated for a maximum of 350 ksps.

[3] Table 7-1 in the datasheet appears to show five TCs available, but the configuration summary in the SAMD21 summary datasheet (Document 40001884A) states that three are available in the G version. In any case, the CMSIS definitions only provide access to TC3–TC5.

[4] The WO[1] output of TC3 is available on pin PA19 (Arduino pin 12) but we will need that pin to connect to the LCD display in the next lab.

[5] We admit this is pretty arbitrary, so go ahead and use PB09 if you prefer.

[6] To select a sine frequency, we will set the number of 48 MHz cycles to count before updating the DAC to the next sample. Since there are 128 samples per period, the period change by increasing the timer count by one is $\frac{1}{45\,\text{MHz}} * 128 = 2.67\,\mu s$.

25N.1.4 Configure the Timer/Counter to toggle the I/O pin

Once TC4 is powered and connected to both an internal clock and an external I/O pin, we can configure the timer/counter itself. Initially, we will just set up TC4 to output a square wave on Arduino Pin A1 so we can test it with the oscilloscope. Later, we will also enable an interrupt on a timer compare which will output data to the DAC in the ISR.[7]

An earlier version of the datasheet provided a succinct explanation of how to configure the Timer/Counter for waveform generation:

> When one of the compare/capture channels is used in compare mode, the TC can be used for waveform generation. Upon a match between the counter and the value in one or more of the Compare/Capture Value registers (CCx), one or more output pins on the device can be set to toggle.[8]

As with the other peripherals, two separate CMSIS include files allow you to configure the TC either as individual registers (`tc4.h`) or as a structure of joined register structures (`tc.h`). The defines for the count and capture registers in `tc4.h` end in "8," "16," or "32" depending on the counter resolution selected. The registers we need to configure are:

```
REG_TC4_CTRLA           // to reset TC, configure size, mode, prescaler and enable
REG_TC4_COUNT16_CC0     // to set the count up compare value for WO0
REG_TC4_STATUS          // to write synchronize two previous registers
```

If using the structure access method, the register name is prefixed with "COUNT8", "COUNT16" or "COUNT32" to indicate the TC resolution:

```
TC4->COUNT16.CTRLA.reg              // to reset TC, configure size, mode, prescaler and enable
TC4->COUNT16.CC[0].reg              // to set the count up compare value for WO0
TC4->COUNT16.STATUS.bit.SYNCBUSY    // status bit for write synchronization
```

To configure the Timer/Counter in waveform generation mode to output a square wave on a Channel 0 output pin we:

- Disable TC4 by writing a '0' to the ENABLE bit in the CTRLA register. This step may be skipped if you are sure the TC has not been previously enabled. The CTRLA register requires write synchronization so you should wait for the SYNCBUSY bit in the STATUS register to go low after clearing the ENABLE bit.
- Reset the TC4 peripheral then wait for both the SWRST bit in the CTRLA register and the SYNCBUSY bit in the STATUS register to go low (write/reset synchronization).
- Use the CTRLA register to set 16-bit mode, the prescaler to divide by 1, the waveform operation to "Match frequency" mode which counts up to the compare register value, and also set the ENABLE bit to enable the timer/counter. Wait for the SYNCBUSY bit to go low before writing to another register.
- Set the 16-bit Channel 0 Compare/Capture Value register (CC0) to an initial value of 239. This will divide the 48 MHz clock by 480 to give an output square wave of 100 kHz. (Why do you think we set the compare value to 239 instead of 240 or 480?[9]) This register requires write synchronization.

In the lab, you should see Arduino Pin A1 output a 100 kHz square wave as soon as this initialization is complete.

---

[7] You should always write and test your code incrementally. Like building hardware, it is much easier to test and troubleshoot a small functional block than try to figure out why a circuit or program, not tested until complete, does not work.

[8] Atmel–42181G–SAM–D21_Datasheet–09/2015, p. 613.

[9] Each time the count matches the value in the Compare Value register, the Counter/Timer resets the count to zero and toggles the output pin. You load a value of 239 instead of 240 because the counter has a synchronous clear. On the 239th clock pulse the match occurs and clear is asserted. However, it only resets back to zero on the *next* (240th) clock rising edge. Since the output is toggled, it takes two resets for a single output cycle for a total divider of 480.

25N.2 Interrupts

We first encountered interrupts in §23N.3.1 where we contrasted them with the alternative, polling. Interrupts are a special type of function call invoked not under program control, but by a hardware event. This hardware event can be external, such as a transition on an I/O pin, or an internal event in a peripheral such as the ADC having new data available after completing a conversion or a Timer/Counter reaching the value of a compare register. Unlike a normal function call, an interrupt can occur at any time to break the foreground program flow.[10]

25N.2.1 Why interrupts?

The name certainly suggests the answer: interrupt when you're in a hurry. Certainly there's a sense in which this is true, and another in which it is not.

When you need a prompt response: Interrupts free the programmer from having to continually *poll* a device (i.e., expressly read its status under program control) to see if it is ready or if a specific event has occurred. Replacing polling with interrupts is particularly valuable when the event occurs rarely but must be responded to quickly, such as saving important data to non-volatile storage if power to the microcontroller unexpected goes down (it won't do for the computer to say, in effect, "The sky is falling? Well, I'll get to it once I finish calculating pi to 20 places.").[11] Interrupts are also useful when timing is important, for example to send synthetic sine data values to the DAC at precise intervals to minimize harmonic distortion in the reconstructed analog signal.

To free the CPU from mundane tasks: Interrupts are also useful to leave the main program free to do something useful while waiting for the interrupting event. For example, rather than sitting in a software do-nothing loop waiting to send each character in a line of text to a slow serial interface, the data can be placed in a FIFO buffer and fed to the serial peripheral by an ISR set to interrupt on an empty output buffer condition. In this case the ISR does little, just reading the next character from the FIFO and writing it to the serial peripheral's DATA output buffer. When the FIFO is empty, the ISR disables its own interrupt. Here, the speed with which the ISR responds is not very important, but once the CPU has loaded the text string into the FIFO it is free to continue with other computing tasks.[12]

Latency: When an interrupt occurs, the microcontroller completes any single-cycle instruction being executed, saves the state of the CPU and then transfers execution to the ISR.[13] Although an interrupt provides a faster response than if you had to wait until the program got around to polling, there is a lag between the event calling for action and the response. This "interrupt latency" can be troublesome particularly if it is variable (known as latency "jitter"), depending upon what the processor is doing when the interrupt request occurs. The Cortex-M0+ is unusual in that its interrupt latency is deterministic and fixed at 15 CPU clock cycles (assuming zero-wait state memory). In addition, ARM processor hardware automatically pushes the registers normally destroyed by compiled C functions, R0–R3, R12, on the stack as well as the link register (LR), PSR, and PC when an interrupt occurs and restores them when the ISR returns. This not only allows interrupt service routines to be written as normal

---

[10] Here, "foreground" refers to normal program execution while "background" refers to the Interrupt Service Routine (ISR) code executed when the interrupt occurs.
[11] On the SAMD21, the 3.3 V Brownout Detector ("BOD33") monitors the supply voltage and can reset or interrupt the processor when a specific threshold voltage is reached.
[12] You will get an opportunity to create this type of serial output system in Chapter 26W.
[13] Multicycle instructions are abandoned and restarted from the beginning on the return from the interrupt. See Chapter 25S for a detailed look at how the processor responds to an interrupt.

C functions, but also frees the ISR from saving and restoring these registers, simplifying coding and reducing the time spent in the background program.[14]

You have already seen one type of interrupt, the SysTick timer interrupt which called, at regular intervals, the `SysTick_Handler()` ISR. That interrupt was relatively easy to use, since it was supported directly by CMSIS. Now we are going to look at the steps to configure an interrupt for an I/O pin or an internal peripheral device. In this lab you will configure the Timer/Counter TC4 to send data to the DAC at regular intervals, but the steps to set up the interrupt are similar for other devices.

25N.2.2 SAMD21 interrupts

In the SAMD21G18A, there are 28 maskable interrupt sources available from internal peripherals plus one non-maskable interrupt ("NMI"). Each peripheral can generate an interrupt when one or more conditions or events occur. Maskable interrupts can be enabled and disabled under program control. The non-maskable interrupt will still interrupt the CPU even when maskable interrupts are globally disabled.

One of the internal peripherals, the External Interrupt Controller ("EIC"), supports up to 16 software maskable external interrupts (EXTINT0 to EXTINT15) and the non-maskable interrupt. External interrupts can be configured to trigger on the rising edge, the falling edge, or both edges, or on a high or low level of an I/O pin (although edge-triggering is the norm and level-triggered interrupts generally should be avoided). Each external interrupt including the NMI is connected to an I/O pin selected with Peripheral Multiplexer function A. The block diagram of the SAMD21 interrupt structure is shown in Fig. 25N.1.

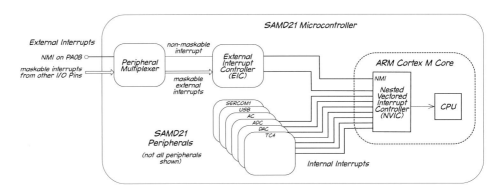

Figure 25N.1 SAMD21 interrupt structure.

Since an interrupt can occur at any time and must handled by the microcontroller hardware, the CPU has to know where each interrupt service routine is located in program memory. A table of pointers to each ISR, known as the Interrupt Vector Table ("IVT"), is kept at a fixed location in memory. In ARM microcontrollers, the IVT starts at the beginning of memory at location 0x00000004, just after the 32-bit value at memory address 0x00000000 containing the initial stack pointer. The first location in the IVT (at 0x00000004) contains the starting address of the code to run after reset. After a reset, the CPU copies the values in the first two memory locations to the Stack Pointer and Program Counter registers respectively, to begin execution of the user program. When an interrupt occurs, the CPU calculates the offset of the ISR in the interrupt vector table based on the interrupt source and jumps to the function pointed to by the entry in the IVT.

Since ISRs receive no parameters and can return no values, they are of the form:

---

[14] ARM processors can also "chain" interrupts, so if multiple interrupts are pending, the CPU will not restore the processor state before executing the next ISR, reducing the latency between chained interrupts to six CPU cycles.

```
void <Peripheral Name>_IRQHandler(void)
```
CMSIS defines default interrupt handlers for all internal peripherals in the system file
```
ATSAMD21G18A_Vectors.s
```
which is included in every Segger IDE project. These default handlers are declared "weak" and do nothing other than enter an endless loop if called, effectively halting normal program execution. The IVT is initialized by CMSIS with these default handlers for each peripheral. However, the weak linker declaration allows any of the default handlers to be replaced by a custom handler with the same name in your program. This is how you will replace the default handler in the lab with one to output sequential data to the DAC on each TC4 compare match.

Interrupt Service Routines should be kept short and should only access variables and resources in coordination with the foreground program. You should never include a delay, wait or call printf() or similar library call in an ISR.[15] The goal is to handle the triggering event as expeditiously as possible and then return control to the interrupted program. When an interrupting event or external signal edge occurs, it sets an internal interrupt request flip-flop which can be read in a bit of the peripheral's INTFLAG register. This flag usually must be cleared by writing to that bit in the INTFLAG register before the device or pin can interrupt again, although some interrupts are cleared by other means such as reloading an empty buffer.[16] Typically, the interrupt flag is cleared as the last step in the ISR before returning to the foreground program.

25N.2.2.1 Enabling interrupts

After a processor power-on reset, all interrupts are disabled by default. This makes sense because normally one needs to initialize the machine before interrupts can be handled correctly.[17] When you're ready to let the processor accept interrupts, you *enable* them. You do this in three steps:

1. Set the interrupt priority for a specific peripheral device and enable that device's interrupt in the Nested Vector Interrupt Controller (NVIC).
2. Configure and enable the interrupt source(s) in the peripheral device.
3. Write a custom Interrupt Service Routine to replace the default CMSIS handler.

1. Configure the Nested Vector Interrupt Controller: The NVIC is the part of the ARM Cortex-M0+ core that is responsible for handling interrupt requests from the peripherals included in the microcontroller. It prioritizes interrupts and allows individual interrupts to be masked (i.e., ignored) or enabled.[18] If a higher priority interrupt occurs during execution of a lower priority ISR, the lower priority interrupt handler will be suspended while the higher priority ISR is executed. The priority register in the NVIC is eight bits wide, but the Cortex-M0+ only supports two bits of interrupt prioritization, providing four priority levels. ARM uses a reversed priority numbering scheme where zero is the highest priority and larger values indicate decreasing priority.

The steps to configure a peripheral to interrupt in the NVIC are:

- Disable the peripheral's interrupt prior to configuring it.

---

[15] Since an interrupt can occur at any time, it is possible to interrupt while the foreground program is executing its own call to printf(). The resulting output to the console would certainly make no sense.

[16] Check the SAMD21 datasheet to see if the interrupt is cleared automatically or if you must do it programmatically.

[17] One arguable exception is the Watchdog Timer (WDT). This timer forces a complete reset of the CPU if it is not repeatedly cleared by software within a predefined time-out period. Its purpose is to restart the microcontroller if the program code hangs or fails. It can be used in mission-critical applications to attempt to recover from runaway or deadlocked code or a recoverable hardware malfunction. The SAMD21 has a small amount of non-volatile memory (NVM) that includes a bit to automatically enable the WDT after a reset. As received from the factory, this bit is disabled but, if enabled by the user, all program code must make sure to continually clear the WDT within the programmed time period.

[18] The Cortex-M0+ also includes a global interrupt enable/disable flag in the CPU PRIMASK register. When interrupts are disabled with the CPSID I assembly instruction, only the non-maskable interrupt can interrupt the processor. Interrupts are re-enabled with the CPSIE I instruction.

- Clear any interrupts pending for that peripheral.
- Set the priority of interrupts from the peripheral.
- Enable the peripheral to interrupt.

CMSIS provides functions for all these steps. Whenever you want to enable interrupts from an internal peripheral (including external interrupts from the EIC), you should execute the following four CMSIS function calls in order:[19]

```
void NVIC_DisableIRQ(IRQn_Type IRQn)
void NVIC_ClearPendingIRQ(IRQn_Type IRQn)
void NVIC_SetPriority(IRQn_Type IRQn, uint32_t priority)
void NVIC_EnableIRQ(IRQn_Type IRQn)
```

where "IRQn" is the interrupt number defined in the CMSIS samd21g18a.h header file. For TC4, it is "TC4_IRQn."

The CMSIS documentation for the NVIC_SetPriority(uint_32t priority) function states that it "[s]upports 0 to 192 priority levels." However, these are mapped to one of the four levels supported in the SAMD21G18 by shifting the priority parameter left by six bits and AND'ing the result with 0xC0. This means that only the two least significant bits in the parameter passed to the function are used to set the interrupt priority. Thus a priority value of zero is the highest priority while a value of "3" is the lowest (as is any value of (priority MODULO 4) equal to 3). Priority should be assigned based on the importance of ensuring a minimum delay in beginning and completing the interrupt service request. For example, we want the DAC to be updated on as regular intervals as possible to minimize harmonic distortion, so you should assign it a priority of zero. The ISR will then always be called within the minimum latency of the CPU unless there is already an ISR with a priority of zero executing (and you can eliminate that possibility by assigning all other interrupts to a lower priority). On the other hand, in the next lab you will use an ISR to send characters to an LCD display to unload the processor from handing a simple but time-consuming serial communication task. Since the human reading the information displayed on the LCD is unlikely to notice a delay of microseconds or even milliseconds, it makes sense to assign this interrupt the lowest priority of 3.

2. Configure the peripheral interrupt: Each SAMD21 peripheral has one or more events or conditions that can generate an interrupt. When the event or condition occurs, it sets an internal interrupt request flip-flop in the peripheral's INTFLAG register.[20]

For example, each Timer/Counter has five possible sources of interrupts. A TC in 16-bit mode can interrupt on counter overflow (OVF), when a TC error occurs (ERR), when the synchronization status clears (SYNCRDY), or when a match compare occurs on compare register 0 or 1 (MC0 and MC1). Each of these events sets a bit in the INTFLAG register.[21]

If interrupts have been enabled in the peripheral for one or more of these conditions, an interrupt request is sent to the NVIC when the interrupt flag is set. Since there is only one NVIC interrupt request per peripheral, all the enabled interrupt flags for a particular peripheral are OR'ed together to generate the NVIC request. The ISR for the peripheral must read the INTFLAG register to see which

---

[19] The NMI is non-maskable in the Nested Vectored Interrupt Controller so none of these steps are required to use it. However, it is only triggered if the PA_08 peripheral multiplexer is enabled and set to Mux A and a non-zero NMISENSE value in register NMICTRL is set in the External Interrupt Controller. The EIC itself is not required to be enabled; however, a GCLK must be connected for edge detection. If none of this makes sense – don't worry about it. You are unlikely to need the NMI in simple microcontroller applications.

[20] The one exception is an external interrupt set to interrupt on a level not an edge. In that case, the actual external signal value is reflected in the INTFLAG register.

[21] Even if you are not enabling a particular interrupt or even any interrupts from a peripheral, you can still read the INTFLAG register to check the status of these bits for polled operation.

event or events triggered the interrupt, then respond, handle, and clear each one in turn, before another interrupt request can be generated.

How you clear an interrupt depends on the peripheral and the specific event that caused it to be set. In some cases, the interrupt flag is reset by some other action that naturally clears it. For example, you did not have to explicitly clear the SysTick interrupt in your `SysTick_Handler()` ISR because the interrupt was asserted when the counter reached a count of zero and cleared when the counter was reloaded with the countdown value on the next CPU clock edge.

In other cases you must clear the interrupt by explicitly writing to the INTFLAG register. In this lab you will enable an interrupt on a TC comparison match on channel 0 and then write sequential sine values to the DAC in the ISR. *Reading* the INTFLAG register returns the value of each interrupt flag; however, *writing* to the same register works like a "CLR" register for the Match or Compare Channel flags. Your TC4 ISR must explicitly clear the compare match flag bit by writing a one to the MC0 bit position in the INTFLAG register.[22]

In general, if there is no action that would naturally clear an interrupt flag, such as reading data after a "Receive Complete" interrupt or reloading a register after a "Data Register Empty" interrupt, you have to manually clear it by writing to the INTFLAG register. In the TC comparison match example above, you do not need to read or write any registers after a match occurs, so you must clear the *Match or Compare Channel 0* interrupt flag (MC0) manually.[23]

You should always look at the datasheet description for each interrupt flag in the INTFLAG register you are using to determine whether you have to explicitly clear the interrupt flag. If you fail to clear the interrupt flag after an interrupt, you cannot receive another interrupt from that peripheral since most interrupts are edge-triggered and an interrupt request to the NVIC is generated only when the INTFLAG bit goes from low to high.

There is one exception to the edge-triggered interrupt rule. The External Interrupt Controller allows an interrupt to be generated not only on the rising or falling edge (or both) of an external pin, but also on either a high or low level of that pin. Nevertheless, you should try to avoid level-sensitive interrupts; unless you change the level of the interrupt line before exiting the ISR, the interrupt will reoccur immediately. Therefore, your ISR must ensure the state of the *external pin* is toggled before exiting to use level-sensitive interrupts. In addition, unlike an edge-triggered event, the level change is not captured in an internal flip-flop; rather the INTFLAG bit for a level-sensitive external interrupt reflects a real-time comparison of the I/O pin state with the interrupt condition. This means a level-sensitive interrupt could be lost if the level changes before the ISR is called. This might happen if the level change changes while the CPU is busy responding to higher priority interrupts, clearing the interrupt bit in the INTFLAG register.

You select which INTFLAG bits generate an interrupt request to the NVIC by setting an event bit in the peripheral's INTENSET register or disabling it in the INTENCLR register. These work like all the other "SET" and "CLR" registers, a high bit causes the interrupt to be enabled or disabled respectively, while a zero bit has no effect and the interrupt's enable status is unchanged.

In this lab, the only condition you care about is when the count matches the value in the Compare Match register 0. To enable the interrupt, set the MC0 interrupt enable bit to one using the INTENSET register:

```
REG_TC4_INTENSET = TC_INTENSET_MC0;
```
or
```
TC4->COUNT16.INTENSET = TC_INTENSET_MC0;
```

---

[22] That is why the full name of the TC INTFLAG register in the datasheet is "Interrupt Flag Status and Clear." See 30.8.10.
[23] However, if the interrupt occurs on a *capture* event, the MC0 bit is cleared when you read the capture value.

3. Write a custom interrupt handler: The final step in handling an interrupt is to write an interrupt service routine function to deal with the condition or event that triggered the interrupt. This is much simpler on an ARM microcontroller than most other architectures because an ISR is just a normal function; you do not need to save the processor state or use a special return instruction (or, in C, tell the compiler that the function is an ISR). Your ISR function should:

- Check each possible condition that could have caused the interrupt.
- Handle that condition.
- Clear the corresponding bit in the peripheral's INTFLAG register unless cleared automatically (as with the SysTick interrupt).
- Repeat for each set interrupt flag which for which an interrupt is enabled.

The first step in handing an interrupt is to determine what caused the interrupt. Since all enabled interrupt flags are OR'ed together to generate the request to the NVIC, you will need to read the INTFLAG register to see which flags are set for the enabled interrupts. However, if only a single interrupt flag is enabled to interrupt you can skip this step. This is the case for the Timer/Counter match interrupt in this lab; the ISR is only called when the count equals the match register 0 value.

Handling the interrupt condition depends on the peripheral and what the condition signifies. In the case of the TC4 interrupt, your interrupt handler needs to look up data points in the 128 point `sine[]` array and write sequential values to the DAC DATA register on each interrupt. This should be similar to the code in `main()` from Chapter 24L used to update the DAC. The array index can be a local variable within the ISR function, but you will want to declare it as `static` to ensure the compiler maintains the value through successive invocations of the TC4 ISR.

Finally, you must clear the interrupt by clearing the MC0 flag in the TC4 INTFLAG register before exiting the ISR. If you do not clear the interrupt flag, the ISR will not be called again when the next match compare occurs. This register is accessed as:

```
REG_TC4_INTFLAG
```
or
```
TC4->COUNT16.INTFLAG
```

You may *read* this register using the structure ".bit" convention, but you should always write it as a full eight bit register value since, as previously discussed, it acts like a "CLR" register when written to, allowing you to clear individual interrupt flag bits in an atomic operation.

25N.3 Lab preview

In Lab 25L, we will first configure Timer/Counter 4 to output a square wave on I/O pin A1 at an adjustable frequency. We will then set up TC4 to generate an interrupt when a match occurs and update the DAC output to generate a controllable frequency sine wave requiring only a tiny faction of the CPU's available computing cycles. Finally, we will use this sine generation system to play simple tunes on the breadboard's speaker.

25N.4 AoE reading

§14.3.7 Interrupts.
§14.3.8 Interrupt handling.
§14.3.9 Interrupts in general.

25L Lab: Microcontrollers IV

In this chapter you will configure the Timer/Counter peripheral to interrupt the CPU at a constant rate to output a sampled sine wave from the DAC. On the way, you will land on the Moon.

25L.1 Use Timer/Counter 4 to generate a square wave

25L.1.1 Test the timer/count square wave output

Compile and run the skeleton code in §25L.3.[1] You will need to include `ClockSysInit48M.c` and `ClockSysInit48M.h` in your project.

Look at the output of TC4 on Arduino pin A1 with your oscilloscope. You should see a 100 kHz square wave, the result of toggling the output pin every $\frac{240}{48\,\text{MHz}}$ µs. Notice that the foreground program is an empty `while(1)` loop. Once it has been configured, the Timer/Counter creates the output square wave with no additional CPU support required.

25L.1.2 Make the square wave frequency adjustable

Complete the skeleton function

```
void SetTC4Div(uint16_t freq_divider)
```

This function should set Compare/Capture register 0 to the argument value. This register requires write synchronization.

Call this new function in your `main()` initialization to change the Timer/Counter output to a 750 Hz square wave. Can you get this exact frequency? How about 900 Hz?[2]

What is the highest and lowest frequency you can produce with this system assuming a 48 MHz CPU clock? Confirm your calculations experimentally.[3]

25L.1.3 Use the TC4 interrupt to update the compare count

The TC4 ISR function, `TC4_IRQHandler()`, is called each time the counter value equals the value in the channel 0 compare/capture match register (cc0). Replace the do-nothing Timer/Counter interrupt handler in the skeleton code with the following function[4]

```
uint16_t hightime = 959;
```

---

[1] Code available at `https://LAoE.link/micro4/25L_TC_Skeleton.c`.
[2] You can get exactly 750 Hz but not 900 Hz. Only frequencies that are a subset of the factors of 24 MHz ($3 * 5^6 * 2^9$) are exactly realizable.
[3] The output frequency is $f = \frac{48\,\text{MHz}}{2*(n+1)} = \frac{24\,\text{MHz}}{n+1}$ where $0 \le n \le 2^{16} - 1$. So you can create outputs from 24 MHz down to about 366.2 Hz; however, the spacing between possible frequencies is variable, larger at higher frequencies and smaller at the lower end.
[4] Code available at `https://LAoE.link/micro4/25L_TC4_Handler.c`.

25L.1 Use Timer/Counter 4 to generate a square wave

```
uint16_t lowtime = 3839;

void TC4_IRQHandler (void) {
    if (REG_PORT_IN1 & PORT_PB08E_TC4_WO0) {   // if output pin is high
        TC4->COUNT16.CC[0].reg = hightime;     // count this many CPU clock cycles
        while(TC4->COUNT16.STATUS.bit.SYNCBUSY);
    } else {
        TC4->COUNT16.CC[0].reg = lowtime;      // else if low count this many
        while(TC4->COUNT16.STATUS.bit.SYNCBUSY);
    }
    TC4->COUNT16.INTFLAG.reg = TC_INTFLAG_MC0; // clear the interrupt
}
```

Your while(1) main loop should still be empty.

The conditional if statement looks at the TC4 output pin (Arduino pin A1) and sets the compare/match value for the next count based on whether it is high or low. The last line clears the interrupt flag so the next interrupt will be recognized. (Since this register operates as a "SET" register when written to, you should never use a RMW operation.)

What do you expect the output to look like? Frequency? Duty cycle? Test to see if you are right.[5]

In the foreground program, set the values of the hightime and lowtime global variables to output a 50 kHz waveform with a 60% duty cycle (i.e., the output should be high for 60% of the period).[6]

25L.1.4   It's time to land on the Moon.

Now that you can control the frequency and duty cycle of the timer output, you can use it to drive a hobby servo. These are described in more detail later in §26N.9, but they consist of a geared motor that provides an angular position over 90 degrees of rotation in response to a pulse width change from 1 ms to 2 ms. (Some servos provide control over 180 degrees of position. That's fine, but you don't want a continuous rotation servo.)

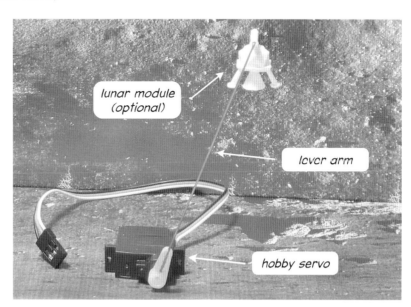

Figure 25L.1 Lunar lander hardware.

---

[5] The ISR changes the time for the high and low counts. The output high time is $(959 + 1)/48\,\text{MHz} = 20\,\mu\text{s}$ and the low time is $(3839 + 1)/48\,\text{MHz} = 80\,\mu\text{s}$. The result is a 10 kHz, 20% duty cycle output.

[6] 50 kHz is a 20 μs period. That means you want the total count to be $48\,\text{MHz} * 20\,\mu\text{s} = 960$. To get the duty cycle, 60% of that should be hightime (576 - 1 = 575) and 40% should be lowtime (384 - 1 = 383).

Hobby servos come various sizes. Any size will work for this lab; Fig. 25L.1 shows a micro-servo. The lever arm is made of music wire or a paper clip. The lander module is optional, but files to 3D print it are available at https://LAoE.link/LEM_3D_Files.html. The servo connects to +5 V, ground, and the Timer/Counter output on Arduino pin A1: see Fig. 25L.2.

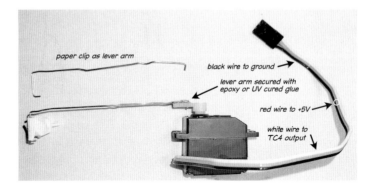

Figure 25L.2 Lever arm construction and electrical connections.

Write a new while(1) main loop as a finite state machine to alternately land on the Moon and return to orbit each time you press the pushbutton connected to Arduino pin 0. If the lander is in orbit, pressing the button should cause it to *slowly* descend to the lunar surface. (If you just set the servo position to the correct angle, it will descend too fast and crash.) If the lander is on the Moon, pressing the button should cause it blast off and slowly return to orbit.[7]

Some hints for the Lunar Lander loop:

- You will have to experiment with the servo to get the correct pulse widths for the orbit and landing angles. You can use a pulse generator to determine these values (be sure the output pulse voltage is 0 to 3.3 V). The pulse width for orbit may be longer or shorter than the surface value depending on the geometry you chose. You may have to reindex the attachment arm on the servo output splined shaft to get a geometry you like.
- One possible directed graph for the Lunar Lander is shown in Fig. 25L.3.
- The FSM cycle time must be short enough that the user does not perceive a delay in response but does not have to be shorter than the servo pulse period since you can only change the pulse width once per period. A reasonable value is 50 ms to 100 ms.
- You do not have to worry about button bounce if the FSM cycle time is longer than the bounce time.
- Use the ReadButton() function to get the state of the pushbutton.
- You must wait for the pushbutton to be released before looking to see if it has been pressed again. Otherwise holding the button down will cause the lander to oscillate between orbit and the surface.
- The frequency of the servo control pulse is not very important but 50 Hz is a typical value. That means you will need to comment out the initialization call to ClockSysInit48M() so TC4 counts at 1 MHz, not 48 MHz, to make it possible to get a 50 Hz pulse frequency with a 16-bit TC match count. This results in the TC4 match count register value equalling the pulse width in microseconds.
- You can let the C compiler compute numeric values for you. For example, if you want to descend in 3 seconds with an FSM cycle time of FSM_CLK milliseconds, the increment to add to hightime each FSM cycle in State C is:

---

[7] A video demonstrating the desired behavior is available at https://LAoE.link/Lunar_Lander_Demo.html. If you would like to start with a simpler project, you can design the FSM to descend down to the surface when the button is pressed and ascend to orbit when it is released.

25L.2 Use the TC4 interrupt to update the DAC

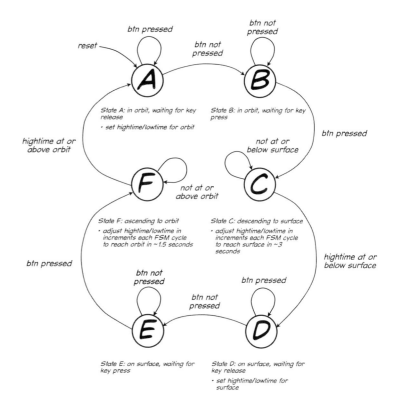

Figure 25L.3 One possible directed graph for the Lunar Lander FSM.

```
#define DN_INC  (SURFACE_PULSEW - ORBIT_PULSEW)*FSM_CLK/3000
```

where `SURFACE_PULSEW` and `ORBIT_PULSEW` are the TC4 count values in microseconds for the pulse time at those positions and "3000" is the desired transition time in milliseconds. Be sure to adjust `lowtime` anytime you change `hightime` to keep the frequency at 50 Hz.

- The test for reaching orbit or the surface in Fig. 25L.3 is an inequality so it works even if the increment does not divide evenly into the delta between the orbit and surface counts. That is also the reason that States A and D set the TC to the correct orbit and surface values respectively after the ascent and descent.

25L.2 Use the TC4 interrupt to update the DAC

Now let's use the Timer/Counter to implement a DAC sine wave output requiring no foreground code other than loading a value for the desired frequency into the TC match count register.

25L.2.1 Copy your DAC initialization function from the last lab

Uncomment the call to `ClockSysInit48M()` to restore the 48 MHz system clock and replace the empty `DAC_Init()` skeleton function with your working code from Lab 24L. It should not require any changes since you have already tested and debugged it.

25L.2.2 Write a new interrupt service routine

Replace the Timer/Counter interrupt handler from §25L.1.3 with one that sends sequential sine values to the DAC. Your interrupt handler must maintain an index into the `sine[]` array and output the next value to the DAC on each execution of the ISR. This index may be local to the interrupt service routine, but must be declared `static` to make sure it is maintained between invocations of the function.[8] To keep the index in the valid range of 0 to 127, AND it with 0x7f each time you increment it. Before returning from the interrupt, be sure to clear the interrupt flag by writing a one to the MC0 bit in the TC4 INTFLAG register.

Try to make your interrupt service code as fast as possible since its execution time limits the highest frequency sine wave you will be able to produce. If your ISR takes longer to execute than the Timer/Counter interrupt period, your program will appear to hang as it continually reenters the ISR each time it exits from it. You can speed up your ISR by eliminating the test for SYNCBUSY after writing the next value of the sine array to the DAC as long as you do not try to read or write to the same peripheral in the next line of code.[9]

25L.2.3 Test your interrupt-driven DAC

The skeleton code contains an empty `while(1)` loop in `main()`. Modify this main loop as necessary to complete the following tests. In each test you will want to watch the DAC output on your oscilloscope at Arduino pin A0 and listen to it on the breadboard speaker.

25L.2.3.1 Output a 1 kHz sine wave

Creating a steady tone merely requires calling the `SetTC4Div()` function once with the correct divider. The tone will be maintained continuously without additional CPU intervention. The frequency of the output sine wave is $f = \frac{48\,\text{MHz}}{(cc0+1)*128} = \frac{375{,}000}{cc0+1}$.

The divider to produce a specific frequency can be calculated as $cc0 = \frac{375{,}000}{f} - 1$ where cc0 is the value set with the `SetTC4Div()` function. The C compiler can calculate the divider necessary for a given frequency at compile time. For example, to output a 125 Hz tone set the compare/match register to:

```
#define F125HZ (375000/125) - 1 // constants must begin with a letter
```

25L.2.3.2 Create a European police car siren

Modify `main()` to switch between two frequencies, 413 Hz and 550 Hz every 350 ms. This simulates a European police siren. Here, finally, we give the CPU something (albeit not much) to do in the main loop.

25L.2.3.3 Sweep a sine wave

Now repeatedly sweep the sine wave from around 6150 Hz down to 375 Hz in about 9 seconds (about 10 ms per frequency). If your program appears to lock up at the higher frequencies, you will need to either lower the upper frequency of the sweep or modify your ISR to execute in less time (see §25L.2.2).

---

[8] While it might seem that declaring "`static i = 0;`" inside the interrupt service routine would cause the compiler to initialize the variable each time the ISR function is called, C only initializes static variables once when they are created, no matter whether they are global or local.

[9] If you do not include write synchronization where the datasheet indicates it is required, your program will still work correctly. However, if you try to read or write to the same peripheral before SYNCBUSY is cleared, the processor waits until the peripheral is ready and interrupt latency is no longer deterministic.

25L.2.3.4 Create an American police car siren

An American "wail" siren repeatedly increases in frequency from about 375 Hz to 1500 Hz in 2.5 to 3.5 seconds then decreases back down to 375 Hz in the same amount of time. Create a main loop that simulates an American siren.

25L.2.3.5 Play a tune

The supplied `PlaySong()` function takes a pointer to a Song structure as its argument and plays a song through the speaker by changing the Timer/Counter divider to produce notes for set periods of time. The Song structure consists of two arrays, one for the notes and one for their duration, along with a count of the number of notes, the name of the song (which we will use in the last microcontroller lab) and a tempo flag. To play the Song structure for *Mary had a Little Lamb* call the function as follows:

`PlaySong(&Mary);`

The C "&" operator returns a pointer to its argument, the Mary Song structure in this example. The song includes a pause at the end so you can call it repeatedly if you want.

25L.2.3.6 Write a tune

The Song data structure is explained in more detail in online Chapter 25O.[10] Try creating and playing a song. If you are feeling ambitious, create an original song. Otherwise, create the Song structure for the music to Frère Jacques shown in Fig. 25L.4.[11] If you prefer something shorter (and more dramatic) try this: `http://LAoE.link/Beethovens_Fifth.html`.

Figure 25L.4 A simple tune to encode and play as a Song structure.

25L.2.3.7 (Optional) Make a piano

If you have time, turn your timer/DAC tone generator into a piano. Copy your key scanning code from Lab 23L into the Timer/DAC skeleton code and plug the 4 × 4 keypad back into your breadboard if you removed it. Modify your FSM so it loads a timer note value (G6_NOTE, F6_NOTE, etc.) into the TC compare register when you press a key and loads the QUIET timer value when you release it rather than sending messages to the console.

25L.3 Timer/DAC skeleton code

The skeleton code below configures TC4 to output a square wave on Arduino pin A1.[12] You will need to copy your tested `DAC_Init()` function from the third microcontroller lab into this code skeleton where noted. You will add the interrupt handler once you have tested the basic Timer/Counter waveform generation output. See Chapter 25N for an explanation of the code that configures TC4 for waveform generation mode.

---

[10] Available at `https://LAoE.link/LAoE_Chapter_25O.pdf`.
[11] Mysid, CC0 (public domain), via Wikimedia Commons: see `https://LAoE.link/Frere_Jacques.html`.
[12] Code available at `https://LAoE.link/micro4/25L_TC_Skeleton.c`.

Lab: Microcontrollers IV

```c
/**************************************************************************
* Company: Learning the Art of Electronics
* Engineer: David Abrams
*
* Create Date:    2021-02-25
* Module Name:    main.c (Micro 4 - DAC out with Timer/Counter Interrupt)
* Project Name:   25L_TC_Skeleton.c
* Target Device:  Sparkfun SAMD21 Micro
* Description: Set up DAC for output with CPU clock set to 48MHz PLL clock.
* Use timer ISR to output synthetic sine wave
* Revision: 2021-04-20 clean up for lab; add PlaySong() function
*           2022-05-26 rename SetTC4Div()
**************************************************************************/
#include "samd21.h"        // add to use CMSIS
#include "ClockSysInit48M.h"
#include <stdio.h>
#include <stdlib.h>
#include <stdbool.h>

/* 10-bit, 128 element sampled sine wave array generated in MATLAB */
uint16_t sine[] = {
512, 537, 562, 587, 612, 637, 661, 685, 709, 732, 754, 776, 798, 818, 838,
857, 875, 893, 909, 925, 939, 952, 965, 976, 986, 995, 1002, 1009, 1014,
1018, 1021, 1023, 1023, 1022, 1020, 1016, 1012, 1006, 999, 990, 981, 970,
959, 946, 932, 917, 901, 884, 866, 848, 828, 808, 787, 765, 743, 720, 697,
673, 649, 624, 600, 575, 549, 524, 499, 474, 448, 423, 399, 374, 350, 326,
303, 280, 258, 236, 215, 195, 175, 157, 139, 122, 106, 91, 77, 64, 53, 42,
33, 24, 17, 11, 7, 3, 1, 0, 0, 2, 5, 9, 14, 21, 28, 37, 47, 58, 71, 84, 98,
114, 130, 148, 166, 185, 205, 225, 247, 269, 291, 314, 338, 362, 386, 411,
436, 461, 486, 511
};

#define SAMPLES sizeof(sine) / sizeof(sine[0])  // number of samples in sine array above

// data for song demonstration

// note definitions (48MHz/128 = 375,000)
#define G6_NOTE ((375000/1568)-1)    // G6
#define F6_NOTE ((375000/1397)-1)    // F6
#define E6_NOTE ((375000/1318)-1)    // E6
#define D6_NOTE ((375000/1175)-1)    // D6
#define C6_NOTE ((375000/1040)-1)    // C6
#define B5_NOTE ((375000/988)-1)     // B5
#define A5_NOTE ((375000/880)-1)     // A5
#define G5_NOTE ((375000/783)-1)     // G5
#define F5_NOTE ((375000/698)-1)     // F5
#define E5_NOTE ((375000/659)-1)     // E5
#define D5_NOTE ((375000/587)-1)     // D5
#define C5_NOTE ((375000/523)-1)     // C5
#define B4_NOTE ((375000/494)-1)     // B4
#define QUIET   0xffff                // 5Hz - effectively silent note

// note duration definitions
#define BRK  50             // time between notes (ms)
#define QTN  350            // time for quarter note (ms)
#define QTNH (QTN * 1.5)    // quarter note held
#define HFN  (QTN * 2)      // half note is twice as long as quarter note
#define HFNH (QTN * 3)      // half note held
#define WHN  (QTN * 4)      // whole note
#define ETN  (QTN / 2)      // eighth note
#define ETNH (ETN * 1.5)    // eighth note held

// data for Mary Had a Little Lamb
const uint16_t mary_notes[] = {E5_NOTE, D5_NOTE, C5_NOTE, D5_NOTE, E5_NOTE,
                               E5_NOTE, E5_NOTE, D5_NOTE, D5_NOTE, D5_NOTE,
                               E5_NOTE, G5_NOTE, G5_NOTE, E5_NOTE, D5_NOTE,
                               C5_NOTE, D5_NOTE, E5_NOTE, E5_NOTE, E5_NOTE,
                               D5_NOTE, D5_NOTE, E5_NOTE, D5_NOTE, C5_NOTE, QUIET};

const uint32_t mary_htime[] = {QTN, QTN, QTN, QTN, QTN, QTN, HFN, QTN, QTN, HFN,
                               QTN, QTN, HFN, QTN, QTN, QTN, QTN, QTN, HFN,
                               QTN, QTN, QTN, QTN, WHN, 1000};

#define MARY_NUM_NOTES sizeof(mary_notes) / sizeof(mary_notes[0])

struct Song {
```

25L.3 Timer/DAC skeleton code

```c
    const uint16_t* notes;
    const uint32_t* htime;
    const uint32_t num_notes;
    const char name[16];    // only room for 8 characters on 2 x 16 LCD display
    const int tempo;        // 0 = normal; 1 = 50% slower; 2 = half speed
};

struct Song Mary = {.notes = &mary_notes, .htime = &mary_htime,
                    .num_notes = MARY_NUM_NOTES, .name = "Mary", .tempo = 0};

/************************************************************************
 *              Set up System Timer and Delay function
 ************************************************************************/
// use 1ms system clock for timing

volatile uint32_t msTicks = 0;

void SysTick_Handler(void)  {       // SysTick interrupts every millisecond
    msTicks++;
}

// Function to delay for n milliseconds
void Delay (uint32_t DelayMSec)  {
  uint32_t endTicks;

  endTicks = msTicks + DelayMSec;

  while (msTicks < endTicks)  {     // wait for DelayMSec SysTick interrupts
      __WFE ();                     // tell CPU ok to power-down until interrupt
  }
}

/************************************************************************
 *               DAC Support Functions
 ************************************************************************/

void DAC_Init(void){
    // insert DAC_Init() function from previous lab
}

/************************************************************************
 *               Timer/Counter Support Functions
 ************************************************************************/

// Pin definitions from samd21g18a.h for reference
//   #define PIN_PB08E_TC4_WO0    40L        / PB08 (TC4:0 - Arduino pin A1 on mux E)
//   ***** you will need to bitwise AND this value with 0x1f to get valid pin number
//   #define PORT_PB08E_TC4_WO0   (1ul << 8)

// Pin multiplexer definitions in port.h
//   #define PORT_PMUX_PMUXE_E    (PORT_PMUX_PMUXE_E_Val << PORT_PMUX_PMUXE_Pos)

void TC4_Init(void){

    /*************** PORT I/O Initialization  *************/
    /* Set up Arduino Port A1 (PB_08) as output pin       */
    /* unnecessary except for debugging                   */
    /*****************************************************/

    // Set pin PB08 as a TC4 WO0 output
    REG_PORT_DIRSET1 = PORT_PB08E_TC4_WO0;           // port group 1 (PB pins)

    // Configure the TC4 output pin with DRVSTR = 0, PULLEN = 0, INEN = 1 and PMUXEN = 1
    PORT->Group[1].PINCFG[(PIN_PB08E_TC4_WO0 & 0x1f)].reg = PORT_PINCFG_PMUXEN | PORT_PINCFG_INEN;

    // Enable pin multiplexer configuration E (even pin)
    PORT->Group[1].PMUX[(PIN_PB08E_TC4_WO0 & 0x1f) >> 1].reg = PORT_PMUX_PMUXE_E;

    /****************** POWER MANAGER Initialization  ****************/
    /* On reset power to Timer Counters are disabled, must be enabled */
    /*****************************************************************/

    PM->APBCMASK.reg |= PM_APBCMASK_TC4;    // enable the TC4 bus clock

    /******** GENERIC CLOCK CONTROLLER Initialization ********/
    /* Connect Clock 0 to TC4 and enable TC4 clock.          */
```

```c
    /* Clock 0 is enabled and set to 48MHz                   */
    /*********************************************************/

    GCLK->CLKCTRL.reg = GCLK_CLKCTRL_ID_TC4_TC5 | GCLK_CLKCTRL_CLKEN | GCLK_CLKCTRL_GEN_GCLK0;
    while(GCLK->STATUS.bit.SYNCBUSY);         // requires write synchronization

    /***************** TC4 Initialization **********************/
    /*    Setup TC4 for 16-bit waveform generation operation   */
    /***********************************************************/

    // Reset the TC before configuring it
    TC4->COUNT16.CTRLA.reg = TC_CTRLA_SWRST;
    while(TC4->COUNT16.CTRLA.bit.SWRST);        // wait for reset synchronization
    while(TC4->COUNT16.STATUS.bit.SYNCBUSY);    // wait for sync synchronization

    // configure for 16 bit mode, waveform generation, no prescaler and run in standby mode
    TC4->COUNT16.CTRLA.reg =
           TC_CTRLA_MODE_COUNT16 | TC_CTRLA_WAVEGEN_MFRQ | TC_CTRLA_PRESCALER_DIV1 | TC_CTRLA_ENABLE;
    while(TC4->COUNT16.STATUS.bit.SYNCBUSY);    // wait for enable bit synchronization

    // configure initial compare value to 239 (100Khz timer square wave)
    TC4->COUNT16.CC[0].reg = 239;
    while(TC4->COUNT16.STATUS.bit.SYNCBUSY);    // wait for sync synchronization

    /*******************************************************************************
     * Enable TC4 interrupt in NVIC and configure TC4 to interrupt on CC0 match
     *******************************************************************************/

    // Configure and enable the NVIC interrupt for TC4 w/ high priority
    NVIC_DisableIRQ(TC4_IRQn);
    NVIC_ClearPendingIRQ(TC4_IRQn);
    NVIC_SetPriority(TC4_IRQn, 0);   // 0 is highest priority; 3 lowest
    NVIC_EnableIRQ(TC4_IRQn);

    // Enable the TC4 interrupt request on the MC0 (match/compare 0) bit
    TC4->COUNT16.INTENSET.reg = TC_INTENSET_MC0;
}

/******* TC4 Interrupt Service Routine (not done) *******/
void TC4_IRQHandler (void) {
  TC4->COUNT16.INTFLAG.reg = TC_INTFLAG_MC0;     //clear the interrupt so that it will run again
}

/****** Set timer interrupt frequency (not done) ******/
// Argument is actual value for timer counter
// divider NOT the frequency desired

void SetTC4Div(uint16_t freq_divider) {
}

/******************************************
 *           utility functions
 ******************************************/

/** enable pushbutton on Arduino Pin 0 as input with pullup **/
void PORT_Init(void) {
  REG_PORT_DIRCLR0 = PORT_PA11;
  // Configure the button input pin with DRVSTR = 0, PULLEN = 1, INEN = 1 and PMUXEN = 0
  PORT->Group[0].PINCFG[(PIN_PA11)].reg = PORT_PINCFG_PULLEN | PORT_PINCFG_INEN;
  REG_PORT_OUTSET0 = PORT_PA11;
}

/** function to return true if pushbutton pressed **/
bool ReadButton(void) {
  return (REG_PORT_IN0 & PORT_PA11) == 0;   // true if button pressed (active low)
}

/** function to play a song on DAC **/
void PlaySong(struct Song *song) {
        printf("Playing: %s\n", song->name);
        for (int i = 0; i < song->num_notes; i++) {
             SetTC4Div(song->notes[i]);
             Delay(song->htime[i]);
             if (song->tempo == 1) {
                Delay(song->htime[i] >> 1);
             } else if (song->tempo == 2) {
```

```
                    Delay(song->htime[i]);
            }
            SetTC4Div(QUIET);       // 50ms between notes
            Delay(BRK);
    }
}
/*******************************
 *          main()
 *******************************/
int main (void)
{
    // system initializations
    ClockSysInit48M();                       // set CPU clock to 48MHz
    SysTick_Config(SystemCoreClock/1000);    // Configure SysTick for Delay() timer
    DAC_Init();                              // 1V full scale, 10-bit right justified
    TC4_Init();                              // initialize Timer/Counter 4 for DAC output timing
    PORT_Init();                             // enable pushbutton on Arduino pin 0

    // main loop - DAC and Timer/Counter Interrupt tests.
    while(1) {
    }

    return 0;
}
/************************* End of file *************************/
```

25L.4 Conclusion

Replacing the call to Delay() you used in the last lab with the Timer/Counter and an ISR offloads the work of updating the DAC at regular intervals, thus freeing the CPU for other tasks. We still waste most of the CPU's cycles in the PlaySong() calls to Delay() but we will fix that when we look at using a Real Time Operating System in the last microcontroller lab. But before that, we need to add some user output in the form of an alphanumeric Liquid Crystal Display ("LCD") which will require us to configure yet another peripheral, the Serial Communications Interface in the next chapter.

25L.5 Solutions

- https://LAoE.link/micro4/25L_DAC_Part_2_Sol.c
- https://LAoE.link/micro4/25L_Piano.c

25S Supplementary Notes: A Detailed Look at SAMD21 Interrupt Handling

So what really happens when a program loads? When an interrupt event occurs? Well, a lot!

Lets take a look at what happens when we run a simple program that uses the SysTick Timer interrupt to update a counter every second.[1]

25S.1   The test program

```
/*******************************************************************
* Company: Learning the Art of Electronics
* Engineer: David Abrams
* Create Date:   2022-06-24
* Module Name:   main.c (25S_Interrupt_Demo.c interrupt demonstration)
* Target Device:  Sparkfun SAMD21 Micro
* Description: Shows how interrupts work using SysTick timer
*******************************************************************/
#include "samd21.h"          // add to use CMSIS
#include <stdio.h>
#include <stdlib.h>

volatile uint32_t msTicks = 0;     // count of seconds since program start

void SysTick_Handler(void)  {    // called on each SysTick interrupt
  msTicks++;                     // increment count every second
}

int main (void)
{
   unsigned int i = 0;

   SysTick_Config(SystemCoreClock);    // Configure SysTick to generate interrupts

   while(1) {                          // do some math (forever)
     i = i + 3;
     i = i - 1;
   }

   return 0;
}
```

The CMSIS function `SysTick_Config(SystemCoreClock)` is called once to configure the SysTick timer to count down at the CPU clock frequency from a starting value of `SystemCoreClock` (the CPU frequency in Hz) and generate an interrupt when the counter reaches zero. The counter is then automatically reloaded with the value `SystemCoreClock`. The result is that an interrupt is generated every second.

When the interrupt occurs, the foreground program is interrupted and the `SysTick_Handler()` function (the background program) executes. This function increments a counter and exits.

The foreground program does some simple math on the local variable "i." It adds three to i, then subtracts one from i. (It's really not important what the foreground program does – we just need a foreground program for the SysTick timer to interrupt.)

---

[1] Available at https://LAoE.link/micro4/25S_Interrupt_Demo.c.

25S.2 Let's take a look at the foreground program

The test program is compiled and loaded into flash memory by the Segger IDE.

After a reset, the CPU copies the first two words in flash memory, starting at address 0x00000000, to the Stack Pointer and the Program Counter, respectively. These values are set by the programming environment, in our case the IDE. Here is what that looks like when we load the test program in the debugger:

Figure 25S.1 Startup stack and pc values.

The Stack Pointer is set to the [non-existent] location four bytes past the top of read/write memory.[2]

While it might seem like the starting Program Counter value should be the address of our `main()` function (which is at 0x124), it actually points to a Segger initialization routine, `SAMD21_Startup.s`, at address 0x24E.[3] This code calls a series of initialization functions (for example, one sets the value of `SystemCoreClock` – unfortunately to a fixed value of 1,000,000) before finally calling our `main()` function.[4]

Figure 25S.2 shows the "bl" (branch to subroutine) instruction at address 0xDA that calls our `main()` function. While our main() code is expected to never return, the call to it is followed by an infinite loop at address 0xDE just in case. ("`b 0x000000DE`" is a jump to itself.)

Figure 25S.2 Segger call to our main() function.

Figure 25S.3 shows the state of the CPU after the branch to subroutine `main()` call executes. The

---

[2] Since the Stack Pointer is predecremented before storing values in a push operation, it would waste a word of memory if it was initialized to point to the actual top of memory.

[3] The value at location 0x00000004 (0x24F) is odd to let the CPU know the program uses the shorter 16-bit "Thumb" instructions. The LSB is cleared when this value is copied to the Program Counter.

[4] You can run this program without a SAMD21 and the J-Link tool. It works fine in the Segger IDE debugger using microcontroller simulation. Set a breakpoint in the disassembly window at the initial PC value and restart the debugger to single step through the startup code.

Supplementary Notes: A Detailed Look at SAMD21 Interrupt Handling

Link Register (lr), which holds the return address in a subroutine call, points to the infinite loop code we just looked at just in case we try to return from main.[5]

The Program Counter now points to the beginning of our test program at 0x124. The stack is empty.

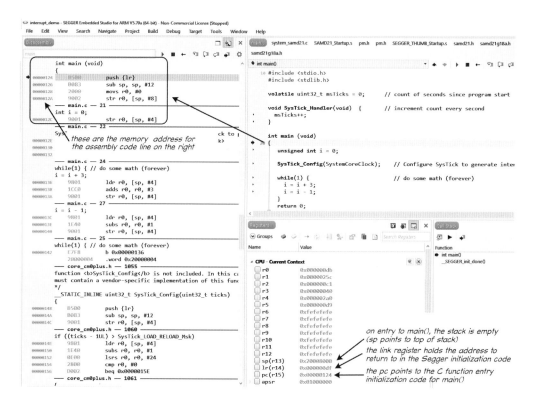

Figure 25S.3 CPU state at start of main().

The first four assembly instructions of `main()` are the C function entry code inserted automatically by the compiler. The first line saves the return address of the caller (`SEGGER_THUMB_Startup.s`) by pushing the value of lr on the stack. The next line allocates three words of local variable memory by decrementing the Stack Pointer by 12 bytes.[6] After these two lines execute, the stack pointer will have been decremented by 16 (four words – the saved lr value and the three local 4-byte variables). The word at [sp + 4] is the local variable *i*.[7]

Figure 25S.4 shows the endless `while(1)` arithmetic loop code. Each line of C code compiles to three assembly instructions followed by an unconditional branch back to the start of the loop at address 0x136. For each calculation, the code copies the value of variable *i* from the stack into register r0, does the arithmetic calculation, then stores the value of r0 back into the local variable on the stack. This means r0 always contains the current value of *i*.[8]

You can step through this code to see it update the value of the variable *i*, then jump back to the top of the loop to do it again.

---

[5] Again the value in the Link Register is odd to indicate the use of the Thumb instructions set. The actual return code is at 0xDE.

[6] While our main function never returns, if it did the compiler would first restore the Stack Pointer by executing the instruction "add sp, sp, #12." It would then pop the saved return address off the stack into the Program Counter to return to the caller.

[7] It is unclear what the other two variables at [sp] and [sp + 8] are for, they are not used by the function.

[8] To be fair, this is compiled with the "debug" build configuration selected. It is possible the compiler is smarter when the "release" configuration is selected.

25S.2 Let's take a look at the foreground program

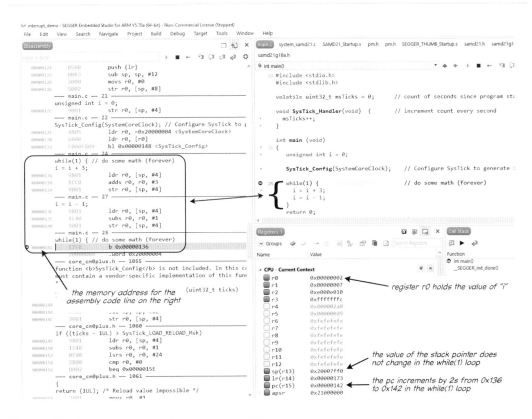

Figure 25S.4 CPU state in the foreground while(1) loop.

25S.2.1 What happens when an interrupt occurs?

Figure 25S.5 diagrams the CPU's response to an interrupt request. Interrupts are a subset of the broader group of ARM exceptions, which are defined as "a condition that changes the normal flow of control in a program."[9] An exception can be generated not only by a peripheral interrupt, but also by a CPU core fault, i.e., an abnormal event such as an attempt to execute an undefined instruction.

25S.2.2 Step 1: The interrupt request is captured as "pending."

In general, interrupts are asynchronous to the SAMD21 CPU clock.[10] That means that interrupts can occur at any time with respect to the CPU's operation. Normally, an interrupting event is captured by a flip-flop and causes the NVIC to assert an interrupt pending signal to the CPU.[11]

Most CPU architectures only check for pending interrupts between instructions, which can lead to interrupt "jitter," a variation in the time to respond to the interrupt. ARM microcontrollers have a fixed interrupt latency (15 CPU cycles for the Cortex-M0+ if there are no memory wait states). This is possible because the majority of ARM instructions take only one CPU cycle to execute, so they complete before a pending ISR is called. To guarantee fixed latency, instructions that take more

---

[9] The following discussion assumes only a single interrupt is asserted, no other ISR is in process and the triggered ISR completes before any higher priority exception occurs. For a more complete discussion of ARM Cortex-M exceptions, see https://LAoE.link/ARM_Cortex-M_Exception_Handling.html.

[10] That is not true for the SysTick timer interrupt since the counter is decremented by the CPU clock, but it is true for many, if not most, other hardware interrupt sources.

[11] Level-sensitive interrupts are an exception. These only cause an interrupt if they are still asserted when the CPU is ready to handle them.

Supplementary Notes: A Detailed Look at SAMD21 Interrupt Handling

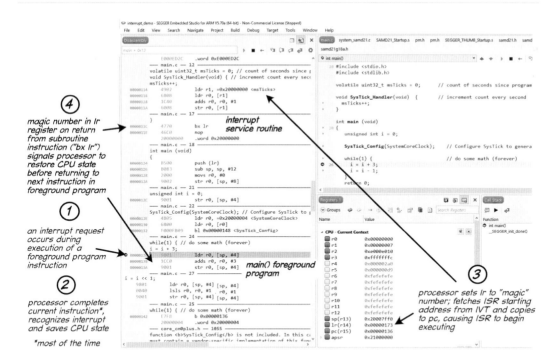

Figure 25S.5 Interrupt handling in the ARM Cortex-M processor.

than one CPU cycle are either stopped, then continued ("interruptible-continuable" instructions) or are abandoned and then restarted when the ISR returns.

25S.2.3 Step 2: The processor saves the current CPU state.

To respond to the pending interrupt, the ARM processor hardware saves all registers that could be changed by the ISR on the stack. These include the Program Counter, the Application Program Status Register (aspr) containing the condition codes of the last operation, registers R0-R3 and R12 (the standard registers used but not saved/restored by C function calls), and the Link Register. As long as the Interrupt Service Routine conforms to the ARM Architecture Procedure Call Standard (which the IDE C compiler does), restoring these registers will put the CPU back in the same state it was before the ISR was executed.

25S.2.4 Step 3: The processor calls the ISR.

To call the correct Interrupt Service Routine function, the processor must know where it is located in program memory. It looks up the address of the ISR corresponding to the pending exception in the Interrupt Vector Table (IVT), stored in flash memory.[12] The IVT is a table of 32-bit addresses, one for each possible exception source. Each exception is numbered, starting at 1, and the vectors (i.e., function pointer) to their respective ISRs start at address 0x0000004 in flash memory.[13]

---

[12] In some versions of the processor, the IVT can be moved to R/W memory after the program starts if the programmer needs to dynamically change the vector table.

[13] But wait! Isn't that the location where the initial Program Counter value is stored? It is indeed. The first entry in the IVT is for a processor reset exception, which certainly meets the definition of "a condition that changes the normal flow of control in a program."

25S.2 Let's take a look at the foreground program

The SysTick timer is exception number 15, so its vector (the address of the ISR function) is stored at location 4 * 0xF = 0x3C. The value at this memory location (Fig. 25S.6) tells the NVIC that the `SysTick_Handler()` starts at address 0x114.

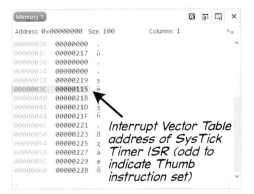

Figure 25S.6 Interrupt Vector Table entry 15 (at 0x3C) holds the address of SysTick Timer ISR.

The CPU can start execution of the ISR by just copying the IVT table entry to the Program Counter. However, the ISR is written as an ordinary function, so when it is done it will return to the calling program by copying the contents of the Link Register to the Program Counter. The problem is that here there is no calling program – the contents of the Link Register are indeterminate when an exception occurs. The CPU could copy the current Program Counter to the Link Register before jumping to the ISR, that would cause the foreground program to pick up right where it was when it was interrupted. That sounds like what we want but if it did this, the CPU state would still be determined by the ISR, not restored to what it was when the exception occurred. To work properly, the CPU state needs to be restored to exactly what it was when the foreground program was interrupted.

In many computer architectures, this problem is handled by a requiring an ISR to end with a special return from interrupt instruction. (On the 8051 you use RETI instead of the normal RET return from subroutine instruction.) On ARM processors, the CPU loads the Link Register with a "magic" number called the EXC_RETURN value before jumping to the ISR.[14] When this value is copied into the Program Counter at the end of the ISR to return from the subroutine, it triggers special Return From Exception processing that restores the saved registers before returning to the foreground program. This is transparent to the ISR, which is why it can be written as a normal C function.[15]

Figure 25S.7 shows the CPU registers after the loading the IVT vector for the SysTick interrupt into the Program Counter. The Stack Pointer is decreased by eight words (32 bytes) due to the CPU saving the state before the jump (compare to Fig. 25S.5). The Program Counter holds the value 0x114 (the IVT value with the LSB cleared – the address of `SysTick_Handler()`) and the Link Register holds the EXC_RETURN value, an address into the system portion of the memory address space.

25S.2.5 Step 4: The CPU state is restored on the return from the ISR.

The SysTick interrupt ISR function increments the `msTicks` value by loading a pointer to the location of the variable (in normal R/W memory, not the stack, since it is global) into register r1, copying the current value of `msTicks` into register r0, adding one to register r0, and storing the updated `msTicks`

---

[14] In computer design and programming, "magic" numbers are data values that have special meaning. For example, in a null-terminated string, a zero value indicates the end of the string rather than a character.

[15] The Cortex-M reserves the top of the memory for system functions. There are actually several EXC_RETURN values in this address space; the five low order bits are used to restore some additional internal state information in the CPU. See https://LAoE.link/ARM_EXC_RETURN_register.html.

Supplementary Notes: A Detailed Look at SAMD21 Interrupt Handling

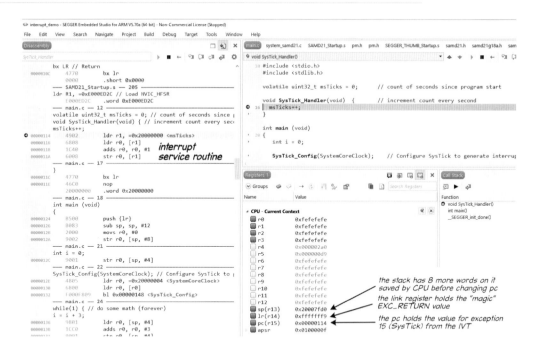

Figure 25S.7 CPU registers after jump to beginning of ISR.

value back to R/W memory at the address in r1. The `adds` instruction also modifies the arithmetic flags in the aspr register. If you single-step through the ISR return instruction, "`bx lr`" (branch to the program address in lr), registers r0–r3, r12, lr and the apsr are pop'ed off the stack, restoring the values these registers held before the interrupt (in particular note that r0 again holds the value of the variable i) and the Program Counter is set to an instruction in the foreground program.[16]

25S.3 Conclusion

From a C programmer's point of view, interrupt handling is relatively painless on the ARM Cortex-M. You write your handler as an ordinary function and, as long as you name it correctly, it will replace the default handler. Just be sure to use static variables for values that must be maintained between invocations of the ISR. You can use either global variables (which are always static) or local variables declared as `static` if they are not used outside the ISR. If your global variables are read or written by the foreground program, you should declare them as `volatile` to make sure the current value is always fetched from memory. You can even call your ISR from your foreground program should you find a reason to do so.

---

[16] It is not possible to know which foreground program instruction was executing when the interrupt occurred (other than inferring it from where the ISR returns to) because the IDE turns off interrupts when single-stepping. This means you have to run at full speed for a breakpoint to be recognized in the ISR.

25W Worked Examples: Using SAMD Interrupts

25W.1 An interrupt-driven pulse measurement system

Suppose we want to measure the width of a short, positive-going pulse that occurs infrequently using the hardware shown in Fig. 25W.1.

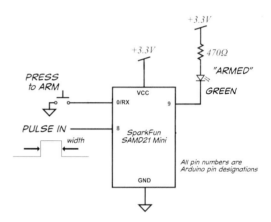

Figure 25W.1 Interrupt-driven pulse width measurement hardware.

The system should work as follows:

- When the SPST "PRESS TO ARM" pushbutton is pressed, the green ARMED LED should light and the system waits for the rising edge of the next pulse in.
- On the rising edge of the PULSE IN signal, the ARMED LED should go out and the system should start measuring the pulse high time.
- On the falling edge of the PULSE IN signal, the debug console should display the pulse high time in milliseconds.
- When the system is not armed, it ignores the PULSE IN signal and waits for the PRESS TO ARM button to be pressed to rearm the system to measure the next pulse. This is the reset state.

This is a fairly simple system that could be implemented by just polling the PULSE IN signal, but where's the fun in that? Let's capture the edges of the input pulse with an interrupt and use the ISR to save the starting time on the rising edge and then calculate the pulse width on the falling edge. The foreground program will be responsible for monitoring the ARM button and handling output, including turning the LED on and off and printing the pulse width to the debug console.

The foreground program should be written as an FSM. Since the background program is doing the measurement, the foreground FSM can run at a leisurely clock rate; it just needs to be fast enough to not seem slow responding to user input. A twenty to fifty millisecond FSM cycle time should be more than adequate to meet this requirement. The background ISR will respond to the edges of the pulse in less than 20 μs, even at the default 1 MHz CPU clock rate, so the resolution of the measurement will be limited by the 1 ms period of the SysTick timer. One possible directed graph for the FSM is shown in Fig 25W.2. State D ensures we do not re-arm if the button is still pressed when the measurement is complete.[1]

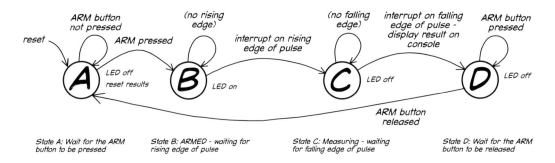

Figure 25W.2 Directed graph for the pulse measurement FSM.

25W.1.1 Start with your generic FSM template

In addition to the normal header files, we want to include stdbool.h to add a boolean true/false data type.[2] We also need to define the FSM clock period. Since the timing is done in the background by the interrupt service routine, the speed or accuracy of the FSM clock is not very important. While we could just use the Delay() function we created in the second microcontroller lab, it is a good idea to get in the habit of using the more accurate "loop start plus FSM period" version shown in the FSM skeleton code in Chapter 23S.

```
/*************************************************************************
* Company: Learning the Art of Electronics
* Engineer: David Abrams
*
* Create Date:   2022-06-02
* Module Name:   main.c (interrupt-driven pulse measurement)
* Project Name:  25W_Interrupt_Pulse_Measurement
* Target Device: Sparkfun SAMD21 Micro
* Description: Implements an FSM to time pulse signals and display width on
*              the IDE debug console. Uses the EIC to generate an interrupt
*              on the rising and falling edge of the pulse to capture
*              the starting and ending msTicks count to calculate the width.
* Revision:
* Revision 0.01 - File Created
* Additional Comments:
*************************************************************************/
#include "samd21.h"           // add to use CMSIS
#include <stdio.h>
#include <stdlib.h>
#include <stdbool.h>          // adds "bool" data type

#define  FSM_CLK 50           // FSM period in ms
#define  ON    true           // for LED abstraction function
#define  OFF   false

/*********************************************************
 *          Set up 1ms SysTick Timer                     *
 *********************************************************/

volatile uint32_t msTicks = 0;   // count of millisecond since program start (rolls over in 49 days)

void SysTick_Handler(void) {     // SysTick interrupt Handler - interrupts every millisecond
  msTicks++;
}
```

[1] Since all four states are entered multiple times, the FSM writes the pulse width to the console once during the *transition* from state C to D. This avoids an extra one-cycle state to output the measurement result.

[2] Complete code for this example is available at https://LAoE.link/micro4/25W_Interrupt_Pulse_Measurement.c.

25W.1.2 Initialize SAMD Port I/O pins

The next step is to configure the Port I/O pins. We must translate the Arduino pin numbers shown on the schematic in Fig. 8L.7 to SAMD pin numbers using the SparkFun SAMD21 mini schematic or graphical datasheet. You could use the generic CMSIS PORT and PIN defines from samd21g18a.h directly, but it is better to use them to define more descriptive names that describe their function in the circuit. This eliminates the need to look at the schematic to understand the code. Multiplexed pins are configured in their respective peripheral initialization code, so only the pins used for normal Port I/O are initialized in Init_IO() – the active low SPST pushbutton switch on pin PA11 and the active low LED on pin PA07. The pushbutton pin needs to be configured as an input with a pullup and the LED pin as an output.

```
/************************************************************************
 *        Define pulse timer I/O Connections                            *
 ************************************************************************/
#define PORT_ARM_BAR_PA11    PORT_PA11
#define PIN_ARM_BAR_PA11     PIN_PA11
#define PORT_ARMED_LED_PA07  PORT_PA07

/************************************************************************
 * Init_IO() - Configure non-interrupt pulse timer I/O pins             *
 ************************************************************************/
void IO_Init(void){
    REG_PORT_DIRSET0 = PORT_ARMED_LED_PA07;
    REG_PORT_OUTSET0 = PORT_ARMED_LED_PA07;    // set LED as output and start off

    REG_PORT_DIRCLR0 = PORT_ARM_BAR_PA11;
    PORT->Group[0].PINCFG[PIN_ARM_BAR_PA11].reg = PORT_PINCFG_INEN | PORT_PINCFG_PULLEN;
    REG_PORT_OUTSET0 = PORT_ARM_BAR_PA11;      // set button pin as pullup
}
```

Since we want the pulse connected to external Arduino I/O pin 8 to generate an interrupt, we need to figure out how to use the External Interrupt Controller peripheral. This is the most difficult part of this problem. However, like all the SAMD21 peripherals we have used so far, we must:

1. PORT: Configure I/O pins (if needed) as inputs or outputs and enable the I/O multiplexer, then select the correct multiplexer.
2. PM: Make sure the peripheral is enabled in the Power Manager.
3. GCLK: Connect the peripheral to a generic clock. If the 48 MHz CPU clock is configured, use the 8 MHz clock on GCLK3, otherwise use the default 1 MHz CPU clock on GCLK0.
4. Configure the peripheral. Here we need to consult the datasheet section for the peripheral we are using. Typically, there will be many registers and lots of choices but the process is usually:
   - Reset the peripheral, then wait for the reset and sync bits to be clear.
   - Set the peripheral mode in the CTRLx register. Wait for the sync bit to clear if the datasheet says write synchronization is required.
   - If using interrupts, use the INTSET register to enable the condition you want to interrupt on, and then configure the peripheral interrupt in the NVIC.
   - Enable the peripheral and wait for the sync bit to clear.
   - If using interrupts, write the Interrupt Service Routine. Be sure to clear the interrupt (usually by writing to the INTFLAG register) before exiting the ISR.

25W.1.3 Configure the External Interrupt Controller

25W.1.3.1 Select and initialize the EIC I/O pins

The only pin we are using with the EIC is Arduino Pin 8 which is SAMD pin PA06. Looking at Table 7-1 in the datasheet, PA06 is connected to EIC EXTINT[6] using multiplexer setting A.

Table 7-1. PORT Function Multiplexing for SAM D21 A/B/C/D Variant Devices and SAM DA1 A/B Variant Devices

Pin[1]			I/O Pin	Supply	A	B[2][3]				C	D	E	F	G	H
SAMD2xE	SAMD2xG	SAMD2xJ			EIC	REF	ADC	AC	PTC	DAC	SERCOM[2][3]	SERCOM-ALT	TC[4] /TCC	TCC	COM	AC/ GCLK
7	11	15	PA06	VDDANA	EXTINT[6]		AIN[6]	AIN[2]	Y[4]			SERCOM0/ PAD[2]	TCC1/WO[0]	TCC3/ WO[4]		

Figure 25W.3 EIC interrupt and multiplexer configuration for interrupt on PA06.

We need to set this pin as an input, with the input buffer and multiplexer enabled, and then set the pin multiplexer to A at the beginning of the `EIC_Init()` function.[3] We can use the CMSIS port/pin definition for the interrupt as it describes the peripheral, the pin and the interrupt.

```
/***********************************************************************
* EIC_Init() - External Interrupt Controller (EIC) initialization
* Configure to interrupt on the rising and falling edge of
* the pulse signal on Arduino pin 8
***********************************************************************/
void EIC_Init(void)
{
    /*************** PORT I/O Initialization **************/
    /* Set up Arduino Pin 8 (SAMD pin PA06) as an external */
    /* interrupt pin. Set as input and enable port input   */
    /* and multiplexer then set multiplexer A.             */
    /******************************************************/

    REG_PORT_DIRCLR0 = PORT_PA06A_EIC_EXTINT6;

    // enable I/O mux: with DRVSTR = 0, PULLEN = 0, INEN = 1 and PMUXEN = 1
    PORT->Group[0].PINCFG[PIN_PA06A_EIC_EXTINT6].reg = PORT_PINCFG_INEN | PORT_PINCFG_PMUXEN;

    // Enable pin multiplexer configuration A (external interrupt)
    PORT->Group[0].PMUX[PIN_PA06A_EIC_EXTINT6 >> 1].reg = PORT_PMUX_PMUXE_A;   // mux even pin number to A
```

No need to enable the EIC in the Power Manager

Unlike many of the internal SAMD peripherals, power to the EIC is enabled after a processor reset so you can skip turning it on the Power Manager. See Table 12-1 and the reset default values in section 16.8.8 of the datasheet.[4]

25W.1.3.2 Connect the EIC to a Generic Clock

If the CPU clock is set to 48 MHz, connect the EIC to the 8 MHz clock on GCLK3; otherwise, we will use the 1 MHz CPU clock on GCLK0.

```
/******** GENERIC CLOCK CONTROLLER Initialization *******/
/* Connect EIC to 1MHz CPU clock GCLK0                  */
/********************************************************/
GCLK->CLKCTRL.reg = GCLK_CLKCTRL_ID_EIC | GCLK_CLKCTRL_CLKEN | GCLK_CLKCTRL_GEN_GCLK0;
```

25W.1.3.3 Configure the EIC peripheral

The previous configuration steps are common to most internal peripherals, only the bit definitions change depending on the peripheral. Now we need to configure the EIC itself. This requires looking at

---

[3] Normally, you would not need to enable the input buffer to use a pin as an interrupt source, but in this application we will need to read the state of the pin in the ISR to determine which edge of the pulse caused the interrupt.

[4] Indeed, all the peripherals controlled by the APBAMASK register in the Power Manager are enabled by default. These include, among others, the Real Time Clock, the Watchdog Timer, the Generic Clock Controller, and the Power Manager itself.

the EIC register definitions and figuring out which ones are relevant to our application. It is helpful to first skim the datasheet description of the peripheral to get an idea of how it works. The block diagram in Fig. 21-1 of the datasheet shows that each external interrupt pin goes through a filter to an edge or level detector which can then either wake up the microcontroller from low power mode, generate an interrupt, or generate an event.[5] We are not going to need the filter, since our input is a clean digital pulse, nor do we need to wake up the device or generate an event. We do need to set the edge detector for both the rising and falling edge since we want an interrupt to occur on either edge.

Section 21.6.2 describes the basic operation of the EIC. Noteworthy is the order of steps in section 21.6.2.1; the EIC registers should be configured *before* enabling the EIC. This is true of most peripherals.

Looking at the register summary in Table 21.7, we will need to use the CTRL and STATUS registers to reset and enable the device. Only the CTRL register requires write synchronization: see §21.6.9.

Since we are not using the NMI, waking up the CPU or generating events, we do not need the NMICTRL, NMIFLAG, EVCTRL or WAKEUP registers. We will need to use the INTENSET register to enable the EXTINT6 interrupt but, since we will never need to disable the interrupt, we won't need the INTENCLR register.

We won't need to read the INTFLAG register to see which external interrupt invoked the call to the ISR function (because we are only enabling one possible external interrupt source), but we will need to write to it to clear the interrupt. See §28.8.8 "Writing a one to [the EXTINTx] bit clears the External Interrupt x flag."

Finally, we only need to write to one of the two CONFIG registers, the one that configures EXTINT6. This means we only have to deal with five of the eleven EIC registers.

Reset the peripheral

A good first step in configuring any peripheral (other than devices that control many other peripherals such as the Power Manager or Generic Clock Manager) is to reset it. This ensures the peripheral is disabled and all registers reset to their default values. So, start by writing a 1 to the SWRST bit of the CTRL register and then wait for the SWRST bit and the SYNCBUSY bit in the STATUS register to both go low before continuing:[6]

```
/***** EXTERNAL INTERRUPT CONTROLLER Initialization *****/
/* Reset EIC peripheral and configure to interrupt       */
/* on the rising and falling edge of duration signal     */
/*********************************************************/

/* reset the EIC before configuring it*/
EIC->CTRL.reg = EIC_CTRL_SWRST;
while(EIC->CTRL.reg & EIC_CTRL_SWRST);
while(EIC->STATUS.reg & EIC_STATUS_SYNCBUSY);
```

Configure the interrupt conditions

Since it takes one bit to select interrupt filtering and three bits to set the sense of the interrupt for each external interrupt source, it requires two 32-bit configuration registers to configure the 16 possible external interrupts. CONFIG0 configures external interrupts 0 to 7, while CONFIG1 configures interrupts 8 to 15. After a reset, all 16 external interrupts are set to filtering off and no interrupt detection (i.e, both CTRL registers are set to zero). See section 21.8.10 in the datasheet. We can set the sense of external interrupt 6 to interrupt on both edges by setting CONFIG0 to the CMSIS value `EIC_CONFIG_SENSE6_BOTH`.

---

[5] The interrupt filter is a digital function that samples the external pin on three successive clock cycles and only generates an interrupt if the pin is high for at least two of the three samples. See Table 21-1.

[6] In general you set up the EIC once in your program initialization code for all external interrupt sources, so it is ok to reset it first. If you needed to change the EIC once it it is set up, you would have to forego the reset and set every register necessary without relying on default values or changing any of the existing settings.

This clears the EXTINT6 filter bit and disables all other interrupts controlled by this register (interrupts 0–5 and 7). This is ok, since we are only using a single external interrupt. If we needed to configure more than one external interrupt, we would need to write the OR'ed sense and filter values for all the interrupts being used at once or else use a RMW to configure a single external interrupt without modifying the others.[7]

```
/* configure EXTINT6 to interrupt on both edges of duration signal */
EIC->CONFIG[0].reg = EIC_CONFIG_SENSE6_BOTH;
```

Enable the peripheral interrupt in the NVIC

The Nested Vector Interrupt Controller is part of the core CPU that enables/disables the interrupt request signal from each internal SAMD peripheral and sets its priority. Even if a peripheral is asserting an interrupt request, it will not be recognized if it is not first enabled in the NVIC. Every peripheral interrupt is configured with the same four CMSIS calls; the only thing that changes is the interrupt source and the priority. Here, the source is EIC_IRQn, the interrupt request from the EIC, and we set the priority to zero, the highest of the four possible values.[8]

```
/* enable the system wide interrupt from the EIC in NVIC */
/* set high priority to get accurate timing */
NVIC_DisableIRQ(EIC_IRQn);
NVIC_ClearPendingIRQ(EIC_IRQn);
NVIC_SetPriority(EIC_IRQn, 0);
NVIC_EnableIRQ(EIC_IRQn);
```

Enable the EXTINT6 interrupt in the EIC

Each of the 16 external sources can be individually enabled and disabled in the EIC. The EIC peripheral itself must be enabled, a specific external interrupt must be enabled in the EIC, and EIC peripheral interrupts must be enabled in the NVIC for an external interrupt service request to be acknowledged by the CPU. Individual interrupts are enabled by writing a 1 to their respective bit positions in the INTENSET register and cleared by writing a 1 bit to the INTENCLR register. These registers, like all SET and CLR registers, perform an atomic RMW operation on the register value. Here we need to enable the EXTINT6 interrupt, which is connected to the pulse input on I/O pin PA06.

```
/* enable the interrupt on the pulse pin in EIC */
EIC->INTENSET.reg = EIC_INTENSET_EXTINT6;
```

Enable the EIC

Finally, enable the EIC by writing a one to the ENABLE bit in the CONFIG register. It is acceptable to set the entire register to EIC_CTRL_ENABLE, since the only other valid bit in the register is SWRST which we want to remain zero. The ENABLE bit requires write synchronization by waiting until the STATUS bit goes to zero.

```
/* enable the EIC peripheral */
EIC->CTRL.reg = EIC_CTRL_ENABLE;
while (EIC->STATUS.reg & EIC_STATUS_SYNCBUSY);
}   // end EIC_Init()
```

25W.1.4 Write the ISR and its support functions

The Interrupt Service Routine function (EIC_IRQHandler()) is called on both the rising and falling edge of the input pulse. The ISR must time the pulse width using the 1ms SysTick timer counter.

---

[7] Note that ORing a sense value into the current register value only works if we know that the bits being ORed to are zero. Otherwise you need to read the register, clear the bits you want zero with an AND operation and set the bits you want one with an OR before writing the value back.

[8] Given that the resolution of the timer is 1ms, one thousand CPU cycles at 1 MHz, any priority would likely work fine in this application.

25W.1 An interrupt-driven pulse measurement system   999

On the first rising edge it should save the value of `msTicks`. On the next falling edge interrupt, it can calculate the pulse width by subtracting the saved starting value from the current msTicks value. Since it is possible for the system to be armed while the pulse is high, the ISR should ignore a falling edge interrupt that occurs before the first rising edge. It should also ignore all subsequent rising and falling edges after the first measurement is complete.

We do not have to read the INTFLAG register to see which external pin caused the interrupt because only one external interrupt (EXTINT6) is enabled. If the ISR gets called, it must be due to the pulse edge interrupt on PA06. However, we do have to read the I/O pin level in the ISR using the Port IN register to see which edge triggered the interrupt. If the pin is high, the rising edge was the culprit, low means it was the falling edge.

The ISR and its support functions use two static variables to keep track of the interrupt function status, `start_time` and `pulse_width`. Both are initialized to zero to arm the pulse measurement. When the first rising edge on the pulse input triggers an interrupt, `start_time` is set to the current msTicks value. Falling edge interrupts are ignored while `start_time` is zero. The first falling edge interrupt triggered after `start_time` is set to a nonzero value tells the ISR to subtract `start_time` from the current value of msTicks to calculate the pulse width and store it in `pulse_width`. Once `pulse_width` holds the pulse in milliseconds (i.e., is nonzero), further interrupts on either edge are ignored. However, every execution of the ISR must clear the interrupt by writing a one to the EXTINT6 bit position in the INTFLAG register before returning.

Three support functions are called by the foreground program to provide communication with the ISR. `ResetPulseTimer()` sets both `start_time` and `pulse_width` to zero to arm the timer for the next pulse measurement. `GetISRStatus()` returns a value to the foreground program indicating the current measurement state: ARM'ed and waiting for a rising edge if `start_time` = 0, measuring and waiting for a falling edge if `start_time` is not zero but `pulse_width` = 0 or measurement complete if `pulse_width` is not zero. Finally, `GetPulseWidth()` returns the width of the last pulse measured in milliseconds. It returns zero if there is no valid measured pulse width yet.

```
/*************************************************************
 *              ISR and support functions                    *
 *************************************************************/

// These two variables are global (i.e., static) but are only used locally by
// the ISR and its support functions. They are declared volatile so they will
// not be cached by the compiler.

volatile uint32_t start_time;
volatile uint32_t pulse_width;

/*************** EIC Interrupt Service Routine ******************/
// The ISR must be named properly to replace the CMSIS default handler.
void EIC_IRQHandler (void) {

    if (REG_PORT_IN0 & PORT_PA06A_EIC_EXTINT6) {    // nonzero if rising edge
        if (start_time == 0) {                       // if start time is 0, first rising edge - ignore others
            start_time = msTicks;
        }
    }
    else if (pulse_width == 0 && start_time != 0) {  // save width if first falling edge
        pulse_width = msTicks - start_time;          //    after first rising edge
    }
    EIC->INTFLAG.reg = EIC_INTFLAG_EXTINT6;          // always clear the interrupt
}

/******************* ISR Support Functions *********************/
// function to reset timer by zeroing start time and pulse width
void ResetPulseTimer(void) {
    start_time = 0;   // 0 means waiting for pulse
    pulse_width = 0;  // when non zero, measurement done
}
```

```
// function to return ISR status
// returns:
#define ISR_ARM   -1  // waiting for pulse rising edge
#define ISR_MEAS   0  // pulse is high (being measured)
#define ISR_DONE   1  // if pulse width measurement done

int GetISRStatus(void) {
  if (start_time == 0) {
    return ISR_ARM;              // waiting for start of pulse
  } else if (pulse_width == 0){
    return ISR_MEAS;             // timing pulse
  } else {
    return ISR_DONE;             // pulse width available
  }
}

// function to return pulse width (returns 0 if not done)
uint32_t GetPulseWidth(void) {
    return pulse_width;
}
```

25W.1.5 Write the hardware abstraction functions

Hardware should always be isolated from the main application program by abstraction functions. These serve two purposes. First, they make all operations active-high logic, even if the hardware is active-low. Second, by isolating the hardware from the application, they make it easier to change hardware without modifying (and possibly breaking) the program's functionality.

```
/**********************************************************************
 *              Hardware abstraction functions                        *
 **********************************************************************/

// function to return state of ARM button (true = pressed)
bool IsARM(void) {
    return (!(REG_PORT_IN0 & PORT_ARM_BAR_PA11));
}

// function to set ARM LED "ON" (true) or "OFF" (false)
void SetLED(bool OnOff) {
  if (OnOff) {
    REG_PORT_OUTCLR0 = PORT_ARMED_LED_PA07;
  } else {
    REG_PORT_OUTSET0 = PORT_ARMED_LED_PA07;
  }
}
```

25W.1.6 Write the program's application code as an FSM

Because all of the communication with the system's hardware is encapsulated in the interface functions above, the application in main() is relatively short and straightforward. It mimics the Arduino program structure of a setup function called once and a loop function called forever.

Here, setup consists of function calls to initialize the 1ms SysTick timer, Port I/O and the EIC interrupt. The program then enters an infinite while() loop that implements the FSM depicted in the directed graph of Fig. 25W.2. The advantage of this design is that once the FSM properly implements the application, the hardware can be changed without modifying the code in main(). For example, you could replace the Microchip SAMD21 with a microcontroller from a different manufacturer. You would probably have to rewrite IO_Init() and EIC_Init(), since each ARM licensee designs their own peripherals, but the application would still run correctly once the new hardware interface code was debugged.

```
/**********************************************************************
 *                          main()                                    *
 **********************************************************************/
```

25W.1 An interrupt-driven pulse measurement system

```c
// FSM State Definitions
#define WAIT_ARM   1    // waiting for ARM button to be pressed (reset state)
#define ARMED      2    // waiting for pulse to go high
#define WAIT_DONE  3    // waiting for pulse to go low
#define WAIT_RES   4    // wait for ARM button to go low

int main (void)
{
  uint32_t endTicks = 0;
  int state = WAIT_RES;

  // like Arduino setup() - runs once
  SysTick_Config(SystemCoreClock / 1000);     // SysTick generates an interrupt every millisecond
  IO_Init();              // configure pulse timer signal I/O pins
  EIC_Init();             // set up interrupt on rising and falling edge of pulse signal

  // like Arduino loop() - runs forever
  while(1) {
    endTicks = msTicks + FSM_CLK;           // calculate FSM period end time

    // state machine to handle pulse timer signals
    switch (state) {
        case WAIT_ARM:          // wait for ARM button to be pressed
          SetLED(OFF);
          ResetPulseTimer();
          if (IsARM()) {
             state = ARMED;
          }
          break;

        case ARMED:             // LED on while waiting for rising edge of pulse
          SetLED(ON);
          if (GetISRStatus() != ISR_ARM) {   // pulse started or done
             state = WAIT_DONE;
          }
          break;

        case WAIT_DONE:         // wait for end of pulse
          SetLED(OFF);
          if (GetISRStatus() == ISR_DONE) {
             printf("Pulse Width: %d ms\n", GetPulseWidth());
             state =  WAIT_RES;
          }
          break;

        case WAIT_RES:          // wait for ARM button to be released
          SetLED(OFF);
          if (!IsARM()) {
             state = WAIT_ARM;
          }
          break;

        default:
           state = WAIT_ARM;
    }                                   // end of switch() loop

      //delay one state cycle
      while (msTicks < endTicks){       // wait for end of FSM clock period
         __WFE ();                      // tell CPU ok to power down till interrupt
      }
  }                                     // end of while(1) loop
   return 0;
}                                       // end of main()
/************************* End of file *************************/
```

25W.1.7 Test your system

Be sure to test your final design for correct operation to meet the specification.[9] Figure 25W.4 shows the function generator set up to output a 125 ms pulse (1 Hz frequency with 12.5% duty cycle), the scope measuring the pulse width at 125 ms, and our program also measuring it at 125 ms. We used an Excel spreadsheet to graph a series of tests comparing the scope's pulse width measurement with our program's pulse width.

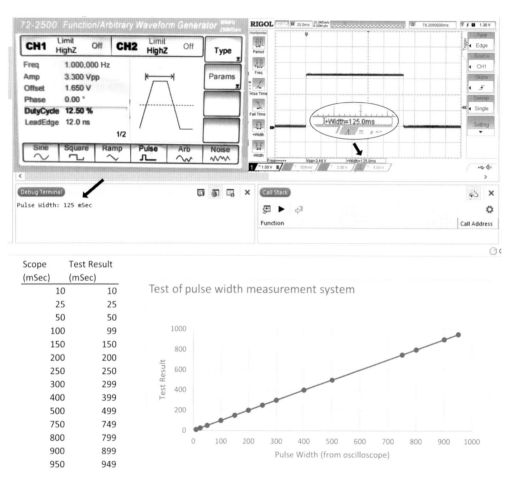

Scope (mSec)	Test Result (mSec)
10	10
25	25
50	50
100	99
150	150
200	200
250	250
300	299
400	399
500	499
750	749
800	799
900	899
950	949

Figure 25W.4 Testing the pulse measurement program.

---

[9] The initial ISR we tested did not work properly if the pulse was high when the ARM button was pressed, as it tried to calculate the pulse width on the next falling edge (which occurred before a start time had been saved). To fix this bug, we modified the ISR to ignore falling edge interrupts if a rising edge had not been first detected, by doing nothing on a falling edge interrupt if the start time is zero.

25O Online Content: Creating PlaySong() data

25O.1 What is PlaySong?

Available at `https://LAoE.link/LAoE_Chapter_25O.pdf`.

26N Microcontrollers V: Serial Communication

Contents

26N.1	**Some serial buses**	**1004**
	26N.1.1 UART/RS232	1005
	26N.1.2 Universal Serial Bus (USB)	1007
	26N.1.3 Serial Peripheral Interface	1008
	26N.1.4 I^2C bus	1010
26N.2	**The SAMD21 SERCOM peripheral**	**1010**
26N.3	**Other serial protocols...**	**1011**
	26N.3.1 Infrared remote control	1011
	26N.3.2 Hobby servos	1011
	26N.3.3 One-wire	1012
	26N.3.4 JTAG	1012
	26N.3.5 Single Wire Debug interface	1012
	26N.3.6 Firewire: "IEEE 1394" and Thunderbolt	1013
	26N.3.7 SATA	1013
	26N.3.8 Ethernet: 10–100–1000 Mb/s	1013
26N.4	**AoE Readings**	**1013**

Why?

In the previous chapters we used several of the built-in peripherals in the SparkFun SAMD21 Mini including the DAC, the Timer/Counter and the EIC (in a worked example). While modern microcontrollers like the SAMD21 have an impressive selection of internal devices, many systems incorporating a microcontroller require peripherals not available internally or may need to communicate with some external computer or system. To handle access to external devices and systems, most microcontrollers support some form of external communications.

26N.1 Some serial buses

While early devices, such as the Intel 8051 previously used in this book, supported an external parallel bus for memory mapped I/O devices, modern microcontrollers rely almost exclusively on serial data communications protocols. The most common serial standards are the Universal Serial Bus ("USB"), Serial Peripheral Interface ("SPI"), Inter-Integrated Circuit ("I^2C" – verbalized as "I-Squared-C" or "I-Two-C") and the Universal Asynchronous Receiver Transmitter ("UART," "USART", "RS-232," or "Serial"). The ARM Cortex-M0+ supports all of these protocols.

In microcontrollers, serial communication is appealing because of controllers' chronic shortage of pins.¹ Serial buses have become fast enough so that even parts like FPGAs that might earlier have run many parallel lines now often include *serializer/deserializer* blocks ("SerDes").²

Let's sketch a few of the common serial protocols.

26N.1.1 UART/RS232

AoE §14.7.8

The oldest of the standard serial protocols is relatively complex, though slow: it is usually called RS232 (or, UART).³ It is the protocol used by the now-obsolete "COM" ports on computers. Most microcontrollers include an RS232 port; the 8051 does, as do most (all?) ARM microcontrollers.⁴ These ports are useful for communicating with a full-scale computer (a PC, usually); but are not much used for communication with *peripherals* (serial RAM, serial ADCs, real-time clocks, serial sensors, for example). You may, however, run across older external devices that still use RS232 for communications and control.

RS232 requires intelligence in the receiver because it uses just one line (plus ground), or two for bidirectional use.⁵ The lack of any separate synchronizing signal – a clock or strobe – means first that the receiver must be set up to expect approximately the correct transmission bit rate (called the "baud" rate); and second – and more subtly – that the receiver must also *synchronize* itself with the transmission in order to sample bit levels at appropriate times. A *start* bit kicks the receiver into life. The receiver then samples at a time that its clock tells it is the midpoint in each bit period. A *stop* bit terminates the byte transmission.⁶

The absence of a separate clock line is not unique to RS232. USB, Firewire, SATA, and PCIe share this characteristic. But the lack of a clock does complicate the hardware needed to communicate, in contrast to the simple SPI for example (see §26N.1.3).

If only seven bits of data are sent (enough for the standard ASCII⁷ character set) then one error-checking "PARITY" bit was often included to describe the transmitted data: the extra bit was set so the total number of one bits in the data was either even or odd depending on whether even or odd parity was selected. Usually, a UART will send eight bits of data. So a typical UART transmission uses ten bits for seven or eight bits of data.

RS232 is a relatively slow protocol, promising 19.2 kbps over 50 feet, with higher speeds (up to about 460kb/s) possible over shorter distances.

The top trace of Fig. 26N.1 shows an RS232 transmission from a PC to a Dallas DS89C430 microcontroller (which accepts downloaded code in this format). The second trace is the microcontroller's

---

[1] Another way to put this point is to say that people using controllers are chronically stingy: they could have lots of pins if they were willing to pay for them. But low cost is much of the appeal of microcontrollers.

[2] See, e.g., http://LAoE.link/SerDes_Devices.pdf, re SerDes generally

[3] This acronym stands for a full name that is quite a mouthful: "Universal Asynchronous Receiver/Transmitter." Even RS232 has a history: "Recommended Standard #232," as AoE reminds us in §14.7.8. A related protocol is "Universal Synchronous and Asynchronous Receiver/Transmitter" (USART), which adds synchronous operation.

[4] Since each licensee of the ARM architecture decides what peripherals to include in their offerings, we are hedging our bets a bit here.

[5] The standard connector for many years included *25 pins*, later reduced to 9 by IBM to allow both a serial and parallel port connector to fit on a PC add-in card. This worked because most of the pins on the 25-pin RS232 connector were not being used.

[6] A student asked how a start bit can be distinguished from an ordinary data bit if no dead time is required between byte transmissions. It's a good question, to which the answer is that a loss of synchronization soon would be detectable because a stop bit would be lacking. At that point the receiver could ask for a re-transmission. If you wonder "couldn't a non-synchronized bit pattern by chance deliver what looked like a correct stop bit?" the answer is Yes – but this is a rare event so that soon, probably on the next byte, the error would be detected. Error detection and re-transmission then will restore synchronization. See AoE, §14.7.8.

[7] ASCII: "American Standard Code for Information Interchange," a binary encoding standard for text. See https://LAoE.link/ASCII_Codes.html.

response, which it issues as soon as it has seen the initial ASCII character for "Carriage Return" (CR).[8] In the original specification, RS232 achieved high noise immunity by use of wide voltage swings (as wide as ±15 V). Nowadays, it is more common to see only the standard 5 V logic swings shown in the figure below.

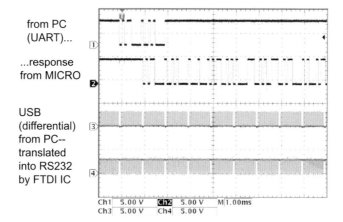

Figure 26N.1 Micro loader signal exchange using UART and USB serial protocols.

If we scrutinize the bit stream in the top trace: see Fig. 26N.2, we can just make out the pattern – which ought to be CR followed by line-feed (LF). The hexadecimal values for these ASCII characters are 0D, 0A.

The micro's response is a longer declaration – the second line of Fig. 26N.1 – announcing that this is a Dallas part speaking: see Fig. 26N.3 showing the response on the screen of the computer linked to the microcontroller.

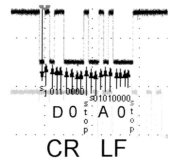

Figure 26N.2 Two RS232 characters decoded from the bit stream (time increases to the right so the carriage return, 0x0D, is sent first).

Figure 26N.3 Controller's text response sent back to PC as RS232 serial stream of characters.

```
DS89C420 LOADER VERSION 1.0 COPYRIGHT (C) 2000 DALLAS SEMICONDUCTOR
>
```

A UART transmission from a SiLabs C80512F410 microcontroller to a PC is displayed at increasing scope sweep rates in Fig. 26N.4. In this figure, the smart oscilloscope decodes the 8-bit data value as ASCII characters.[9] At the highest sweep rate (shown in the lowest of the three traces) you may be able

---

[8] How's that for an anachronism? "Carriage" refers to a typewriter's mechanism that moves the paper to print sequential characters because the typebars all impact in the same center location.

[9] This is a Tektronix MSO2014, using a RS232 decoder module.

26N.1 Some serial buses

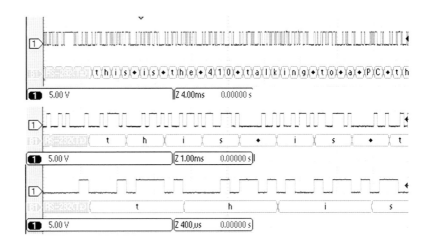

Figure 26N.4 A decoded UART transmission at three sweep rates.

to figure out the binary code that was sent. It is hard though. First, the scope has been set to invert the logic levels to make a graphical high correspond to a logic high. Second, the 8 bits of data are padded with two extra bits: a Start bit (graphically, a low in the figure) and a Stop bit (graphically, a high).

The data is sent LSB first, so we are obliged to read the clusters of four bits "backwards" (MSB on the right). We have labeled the stream with hex values just above the bit stream in Fig. 26N.5.[10] The value for "t" is 74h, as it should be. See if you can make out the 68h for the character "h." This scope finds the decoding easier than we do.

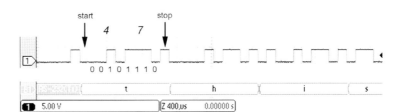

Figure 26N.5 UART detail, showing ASCII "74h" for character 't'.

26N.1.2 Universal Serial Bus (USB)

AoE §14.7.13

COM ports (and printer ports as well) have disappeared from ordinary computers, replaced by USB as the standard external peripheral interface bus.[11] USB is a much more complex protocol. The transmission gets its good noise immunity by using differential signaling in place of the UART approach of very wide voltage swings.

USB is not a true *bus*, in the sense that one cannot tie multiple devices to a shared set of USB lines; but with some additional electronics (a "hub" that fans from one to many, and many to one), one can achieve the equivalent result. USB ("full speed") achieves 12 Mbits/s; the newer USB2 ("high speed") betters that rate 40-fold, to 480 Mb/s, and USB3 ("SuperSpeed") specifies 5–10 Gb/s, max.[12]

Figure 26N.6 is a detail of a few of the bursts of USB activity that looks like no more than fuzz in Fig. 26N.1.

The lower two lines show the differential signaling. The signals are complements – except for a

---

[10] The ASCII code is only seven bits, but the eighth bit, the MSB, here is sent as zero.
[11] However, if you must connect to an RS232 device, USB-to-RS232 adapters are readily available.
[12] The USB4 specification, released in 2019, promises speeds of up to 40 Gbs.

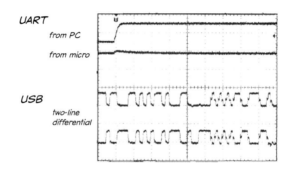

Figure 26N.6 Detail of USB differential signals.

couple of hiccups: one near the middle of the screen, one at far right. These are used to signal bus states such as "device reset."

USB can't stop talking: it is notably noisy, and this noise can be a nuisance if, for example, USB is feeding a circuit that includes analog functions.

One additional benefit of USB is that it can supply power as well as transfer data. Originally limited to a fixed 5 V at a maximum of 500 mA, the latest USB Power Delivery (USB-PD) specification allows devices to request higher voltages and currents up to 100 W from capable hosts.[13]

The SAMD21 contains a single USB interface module compatible with the USB 2.1 specification. We don't use it but it is used to program and control the SparkFun SAMD21 Mini when it is used as an Arduino compatible microcontroller.

26N.1.3 Serial Peripheral Interface

AoE §14.7.1

Microcontroller peripherals need a serial interface simpler than USB and faster than RS232, and several protocols are available. The one you will meet in this chapter's lab is probably both the simplest and the most widely used. It is called the Serial Peripheral Interface (SPI) bus or protocol. An SPI communications system consists of a controller responsible for initiating and timing the communications data stream and one or more devices that receive and respond to the controller's signals.

SPI was created by Motorola in the early 1980s. It uses just three lines (a clock and a data line in each direction) plus ground. However, if more than one device is to be controlled, SPI requires one chip select line per peripheral.[14] This requirement of a line per peripheral makes it not quite a true *bus* like Philips' more complex I^{2}C which uses a single clock and data line (plus ground) to control multiple peripherals. See §26N.1.4. (But USB also is not a true bus, either; so we can't be too shocked at SPI's inelegance.)

SPI scheme: SPI sends a clock along with data that is to be sampled on one of the edges of the clock (exactly *which* edge is determined by the particular flavor or "mode" of SPI). Thus the scheme is "synchronous," in contrast to RS232. In addition, you must add one [usually] active low CHIP_SELECT line for each peripheral, since no *address* is transmitted.

This synchronous scheme makes SPI very easy to transmit and to receive. The number of bits

---

[13] This makes it practical to run relatively high current notebook computers from USB-C power adapters. In January 2020, the European Parliament voted to recommend a single universal charger for most portable devices based on USB-C to reduce electronic waste.

[14] Different manufacturers use various names for the four SPI signal lines. The controller data out line was original called MOSI but you will also see it labeled SDO, DO, DOUT, and COPI. The data in line may be called MISO, SDI, DI, DIN, or CIPO. The chip select may be some variant of SEL, CE, CS, or SS and is normally active low. The clock line is usually labeled with a variant of CLK or CK, but we have seen devices labeling it as SC for serial clock.

transmitted is flexible. The data is transmitted MSB first for the devices used in our labs but may be reversed in other applications. No start or stop bit, nor parity or checksum complicates the scheme (or protects it from error). The data rate depends on the hardware and can exceed 10 MHz. The SAMD21 can output SPI data at a prodigious rate, up to the CPU clock speed divided by 16, but the maximum transfer speed is dependent on the capability of both the Mentor and the Student.[15] We were unable to run the LCD display used in this chapter's lab reliably at rates higher than 50 kHz; the slower device wins.

A SPI Mentor connection to multiple Student devices along with the timing of a one byte device communication with Student 1 is shown in Fig. 26N.7.

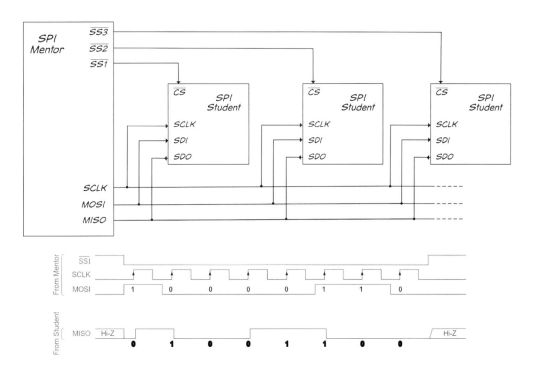

Figure 26N.7 SPI pseudo-bus configuration with sample transfer timing.

SPI's virtue is that it asks so little of a peripheral: only a shift-register is needed. The Mentor device need only...

- Select a device (if more than one), by asserting its private CHIP_SELECT line.[16]
- Put out a bit level on the transmit line, which goes by the name MOSI ("Mentor Out Student In").
- Then provide a clock edge (some peripherals demand rising edge, others falling).
- Meanwhile, the Mentor can receive data on the MISO line ("Mentor In Student Out"), and can do this even as data flows out (this is "full-duplex" communication).

---

[15] The original literature for SPI uses the disfavored terms "master" and "slave" for the device providing the clock and the devices receiving the clock respectively. Here, we will use the terms "mentor" and "student" to replace the original terms. These words have the advantage of conveying the correct relationships while maintaining the standard abbreviations for the protocol signals. (The SAMD21 datasheet uses the terms "Host" and "Client.") You will, however, continue to find the original terms used in many datasheets.

[16] The Mentor must assert only one Student select line at time, since all Student devices drive the MISO line with 3-state drivers. The CHIP_SELECT on commercial peripheral devices are usually active low, but there is no reason they have to be. If you design both the Mentor and Student device, you may make the chip select either active high or active low.

26N.1.4 I²C bus

> AoE §14.7.2

I²C is a fancier controller-to-peripheral scheme, as we have said. It is a true bus, using two-lines (signal and clock, plus ground), and sends address as well as data. This eliminates the need for SPI's individual CHIP_SELECT lines. It also requires a good deal of cleverness from both sender and receiver. Device address must be extracted from the serial stream. It's neat – but elaborate to code.[17]

I²C is a low to medium speed protocol, limited by the use of a resistive pullup bus and open-collector (or open-drain) drivers. The original specification max'ed out at only 100 kb/s but later added Fast mode (400 kbs). The protocol pushed the max speed to 3.4 Mb/s with High-speed mode, albeit by requiring an active pullup on the controller and a lower maximum bus capacitance.[18]

I²C uses a 7-bit addressing scheme, ostensibly allowing up to 112 devices after subtracting reserved addresses.[19] Although this may seem adequate for an embedded system, each peripheral connected to the bus must have a unique address. I²C peripheral devices normally have their address hardwired in silicon during manufacture. Some manufacturers only provide a fixed I²C address, such as the NXP PCF8563 (0xA3 for read operations, 0xA2 for write). Many devices do allow the user to choose an alternative address by grounding a spare pin. Silicon Lab's Si5338 clock synthesizer lets you select between I²C address 0xE0 or 0xE1 with a pin labeled "I2C_LSB." However, if you need to use more than two of these devices in your design, you are out of luck (unless you can order in high enough quantity that the manufacturer will provide a version with a custom address). Occasionally, manufacturers are more generous; Analog Devices' AD7294 multifunction monitor provides the option to program all five least significant bits of the I²C address, allowing up to 32 devices on a single bus.

Controllers from Phillips (the inventor) provide built-in I²C, and so do many microcontrollers and peripherals. Linking such hardware-equipped devices is easy; doing it all in code is not easy. In contrast, you can implement SPI using port I/O instructions ("bit banging") with only a small amount of code.

In the lab we do not use the elegant but complex I²C protocol, even with the SAMD21 which implements I²C in hardware, preferring the simplicity of SPI.

26N.2 The SAMD21 SERCOM peripheral

As previously discussed, each ARM licensee designs their own peripherals around a common CPU core. Many, if not most, manufacturers provide separate subsystems for each protocol. For example, STM's STM32G081xB Cortex-M0+ microcontroller includes four USARTs (plus one low power UART), two I²C and two SPI interfaces. NXP's LPC112x provides three UARTs, two SPIs and one I²C.

Atmel (now Microchip) took a different approach in the SAMD21G18A. Rather than guessing how many instances of each protocol users would need, it includes six SERCOM peripherals, each of which can be configured for either the UART/USART, I²C or SPI protocol.[20] While this has the advantage of flexibility, it also adds complexity. Because it is the easiest of the three to set up, in this chapter's lab we will first program the SERCOM to communicate with a digital potentiometer using the SPI

---

[17] For Philips' detailed description and promotion of this format, see https://LAoE.link/Philips_I2C_Manual.pdf.
[18] The MIPI Alliance released a specification for a successor to I²C in 2018 called MIPI I3C which supports higher data rates while being backward compatible with I²C. See https://LAoE.link/I3C.html.
[19] One of the reserved addresses enables an expanded 10-bit address space; however, it appears that most commercial devices only support the 7-bit scheme.
[20] The USB interface is a separate peripheral.

protocol, then to communicate with an alphanumeric display. Once you have your feet wet with SPI, we hope you will be able to use the datasheet to program the other two protocols on your own.

26N.3 Other serial protocols...

Here we mention some other serial formats in somewhat less detail.

26N.3.1 Infrared remote control

The *remote* that controls many household devices uses a scheme much like the UART's but transmits data over an IR link. The receiver and its remote need to share a transmission rate. A $\overline{\text{START}}$ pulse synchronizes the receiver, which then samples the bit level at regular intervals. The remote transmitter sends not quite binary levels, as a UART does, but instead a series of bursts of IR from a light emitting diode, typically at a frequency of 36 kHz to 40 kHz. A burst of IR is a one; an absence of an IR burst is a zero. A signal-conditioning circuit *integrates* the magnitude of these bursts and passes that result to a comparator. The comparator puts out the binary levels shown on the lower three traces of Fig. 26N.8.[21] The program that decodes this transmission first looks for the $\overline{\text{START}}$ pulse, then samples four successive bits to get the codes for keys 0 through 9 in this example.[22]

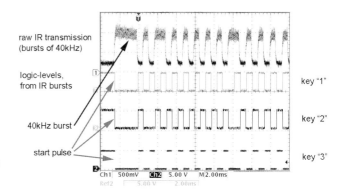

Figure 26N.8 Infrared remote signals: bursts of 40 kHz, and conversion of this to logic levels.

26N.3.2 Hobby servos

Hobby servomotors, like the one in Fig. 26N.9 and which we used in the last lab's Lunar Lander, originally were used in radio control models but now are now widely available for use in robotics or other motion control applications. This ingenious actuator uses a single-wire serial control signal to move the rotor to an *angular position* determined by the width of the pulse fed to it.[23] The motor has three leads: red, white and black. Red gets +5 V, black goes to ground, white takes the logic-level pulse from your computer's I/O port.

Within the servomotor is a one-shot. Every time an input pulse arrives, the motor is driven one way or the other for the duration of the mismatch between the two pulses: *input* pulse versus *one-shot* pulse. Since the one-shot pulse width is adjusted by motor position (the motor turns a pot rigged as variable

---

[21] The figure shows transmitted codes from *three* successive key presses; only the bottom trace matches the sequence of bursts shown in the top trace (in inverted form).
[22] Actual IR protocols tend to be much more complicated, with multiple fields and duplicated data included for error checking.
[23] Typically, a pulse width varying from 1 ms to 2 ms moves the rotor over its full range. Continuous rotation servomotors are also available; these translate the input pulse width into direction and speed of rotation.

Figure 26N.9 A hobby servomotor.

resistor), the one-shot pulse width adjusts to match input pulse width. If input outlasts one-shot, the motor runs in one direction during the mismatch time; if one-shot outlasts input, the motor is driven the other way during the mismatch. Full adjustment and consequent homing of the motor usually occurs over many pulse repetitions. The scheme is slow but neat, and very simple to drive. This is a perfect application for one of the SAMD21's Timer/Counters configured in pulse-width-modulation mode. Typically, pulses are output at 50 Hz but the actual rate is not very important, the rotor position is determined by the pulse high time and the system response is limited by the mechanical system.

Servomotors are available in several sizes, degrees of rotation and torques. For an introduction to hobby servos, see `https://LAoE.link/Hobby_Servo_Tutorial.html`.

26N.3.3 One-wire

AoE §14.7.3

Really? *one* wire? Well, that's a bit of puffery: it's one wire plus ground; but still impressive. All devices are tied to one line that is normally high (and not-so-incidentally *powering* them as well). The controlling device sends data and address information – encoded as longer or shorter low pulses. A peripheral that recognizes its address responds (pulling the line low; internal capacitors keep each part alive during the low pulse). You may have seen these around in stand-alone form: as mysterious stainless-steel buttons on walls: a security guard, making his or her rounds, touches the button with a device that reads the button's identifying code, proving that they passed that way (and when).

26N.3.4 JTAG

AoE §14.7.4

Joint Test Action Group[24] is a serial scheme, originally invented to test printed-circuits by testing all the installed ICs. The circuits are daisy-chained, a test signal is routed through all of them, the result is read back. JTAG's use has been widely extended into both the programming of parts and in the debugging of circuits, such as microcontrollers (otherwise painfully opaque). A processor or controller, for example, can be designed with JTAG-readable internal registers so that it can be examined and even single-stepped while installed in a circuit. JTAG uses five pins plus ground on the target device.

26N.3.5 Single Wire Debug interface

ARM microcontrollers use a two-pin Serial Wire Debug (SWD) programming and debugging interface. In the SAMD21 the two connections are on PA30 (swdclk – called swclk in some parts of the datasheet) and PA31 (swdio). SWDIO is a bidirectional data line and SWDCLK provides the clock. This is the connection used by the J-Link device to program the flash memory in the microcontroller and to allow you to single-step and debug it in the Segger IDE.

26N.3.6 Firewire: "IEEE 1394" and Thunderbolt

AoE §14.7.14

Firewire, introduced by Apple but now dropped from its computers in favor of Thunderbolt, was less widely used than USB; 1394a achieves same nominal speed as USB2, about 480 Mb/s, but in practice is the faster of the two, coming closer than USB to using its full capacity. It also is fully duplex.

26N.3.7 SATA

AoE §14.7.6

Serial-ATA connects your PC to mass storage devices like hard drives, optical drives and solid-state drives. It replaced an earlier parallel mass storage standard called IDE. It originally transferred at 1.5 Gb/s but the current third-generation version runs at a transfer rate of 6 Gb/s.

26N.3.8 Ethernet: 10–100–1000 Mb/s

AoE §14.7.16

This is what ties your PC to computer networks and the Internet, for example – when you aren't using a wireless link. Modern Ethernet is no longer a bus architecture, it operates point-to-point using switches to forward traffic to its destination.

26N.4 Readings

AoE Chapter 14 (Computers, Controllers and Data Links):
§14.7 Serial Buses and Data Links.
Microchip SAM D21/DA1 Family Datasheet:
`https://LAoE.link/Microchip_ATSAMD21G18_Datasheet.pdf`.
SERCOM: Chapter 25.
SERCOM USART: Chapter 26.
SERCOM SPI: Chapter 27.
SERCOM I2C: Chapter 28.
USB: Chapter 32.

---

[24] Named after the industry group that set up this standard.

26L Lab: Microcontrollers V

In this lab you will use one of the internal SAMD21G18A SERCOM peripherals to send information to several external devices using the SPI protocol.

26L.1 Using the SAMD21 SERCOM peripheral

We are going to configure the SERCOM with the simplest of the three protocols, SPI, requiring only a clock and data line along with an optional device select line.[1] You implemented an SPI interface (perhaps without knowing it) when you connected the WebFPGA to the TLC5916 constant current driver on the first day of the Digital Project Lab. Now we will program the SAMD SERCOM interface explicitly to operate as an SPI Mentor device to control two external peripheral devices, an alphanumeric LCD display and a digital potentiometer (digipot).

For both devices, you only need to send commands and data *to* the peripherals, no data is returned. This means we just need the MOSI and SCLK lines, the MISO line is not used.[2] We will create the SPI Chip Select signal two ways, automatically using a SERCOM option for the LCD display and explicitly using Port I/O for the digipot.[3] The chip select signal also provides a convenient trigger to view the SPI output signals on the oscilloscope.

26L.2 Selecting the SparkFun SAMD21 Port I/O pins

Selecting the SparkFun SAMD21 Mini breakout board pins for the three SPI output signals, MOSI, SCLK (called "SCK" on the digipot, the LCD display and in the SAMD21 datasheet) and $\overline{\text{SS}}$ ("$\overline{\text{CS}}$" on the digipot and "/CS" on the display), is more difficult here than in the DAC lab. Unlike the DAC, where there was only a single choice for output, here there are three signals we need to assign to I/O pins. These are connected to the available pins based on the choice of six SERCOM units, two possible pin multiplexer configurations (C and D) and four choices for assigning the three SPI signals to three I/O pins. Contributing to the difficulty is that only some of the possible pins are available on the breakout board and a number of these are already used for the DAC, the keypad, the external LED and the pushbutton.

The SPI Select signal can be either controlled automatically by the SERCOM hardware or manually

---

[1] Various device datasheets use different terms for the device select signal including chip select ($\overline{\text{CS}}$) and student or SPI select ($\overline{\text{SS}}$). They all refer to the signal that, when asserted, indicates to a particular device that it should receive data from the mentor device on the MOSI line and output data to the mentor device on the MISO line on the clock edges. If there is only a single student device connected to the mentor, it may be possible to just wire the device ($\overline{\text{CS}}$) low.

[2] However, the worked example in Chapter 26W shows how to use both SPI input and output to connect a 32k RAM to the microcontroller to create a [rather limited] voice memo recorder.

[3] The digipot requires the ($\overline{\text{CS}}$) to remain low for a full 16-bit data word. Since the SERCOM is an 8-bit device, there is no way to keep ($\overline{\text{CS}}$) asserted through two output bytes if we used the automatic option.

26L.2 Selecting the SparkFun SAMD21 Port I/O pins

using Port I/O. The disadvantage of automatic hardware control is it limits the choice of which SERCOM unit we can use because then all three pins are handled by the SERCOM rather than only two. That means we must find three free I/O pins that are supported by a particular SERCOM. The advantage is simplicity, we do not have to assert the $\overline{ss}$ signal under program control before sending a byte, then disassert it afterwards. This makes interrupt-driven SPI communications much simpler.[4] We will write our SERCOM initialization routine to allow either option (manual or automatic chip select).

The SERCOM input and output signals are assigned to four internal "PAD"s (labeled PAD[0] to PAD[3]) and there is a configuration setting in the SERCOM that maps the PAD[]s to the actual SPI signals. A good way to figure out which settings to use is to list the I/O pins still free and see which SERCOM they can be multiplexed to. Since we want to add the LCD to what we have built so far, let's take a look at what pins are already in use to see which ones are still available: see Fig. 26L.1.

Figure 26L.1 SAMD21 pin use before SERCOM assignment.

Next, look at Table 7-1 in the datasheet showing the possible I/O pin multiplexer assignments to see what SERCOM and which PAD[] can be connected to the I/O pins that have not yet been assigned. We need to find at least one of the six SERCOM units in column C or D that has three different PAD[]s (for MOSI, SCLK and SS) connected to free pins. Here are all the available unused I/O pins and their Mux C and D SERCOM assignments:

I/O Pin	SERCOM Connection
PA10	SERCOM0/PAD[2] (Mux C) or SERCOM2/PAD[2] (Mux D)
PA03	none
PA04	SERCOM0/PAD[0] (Mux D)
PB09	SERCOM4/PAD[1] (Mux D)
PB08	SERCOM4/PAD[0] (Mux D) (currently used for testing TC4 but could be used)
PA17	SERCOM1/PAD[1] (Mux C) or SERCOM3/PAD[1] (Mux D)
PA19	SERCOM1/PAD[3] (Mux C) or SERCOM3/PAD[3] (Mux D)
PA16	SERCOM1/PAD[0] (Mux C) or SERCOM3/PAD[0] (Mux D)

We cannot use SERCOM0 or SERCOM4 because neither connects to at least three free pins.[5] That means we must use either SERCOM1 or SERCOM3 connected to pins PA16, PA17 and PA19 (Arduino pins 11, 13 and 12 respectively). Since none of the SERCOM units is in use, choosing SERCOM1 and MUX C is a reasonable choice.

However, before we can wire an SPI device to the microcontroller, we need to figure out which of the three SPI signals we are using (MOSI, SCK and $\overline{ss}$) connects to which I/O pin. This is configured by the DOPO (Data Out Pinout) field in the SERCOM CTRLA register. We must be able to connect PAD[0],

---
[4] See Chapter 26W for an example of interrupt-driven SPI output.
[5] However, had we chosen manual SPI Select control, SERCOM4 would work using pins PB08 and PB09.

PAD[1] and PAD[3] to the SPI output signals. Looking at the table of DOPO[1:0] options in register CTRLA of the SERCOM SPI chapter of the datasheet (Sec. 27.8.1), the only choice which includes all three of these PADs is when DOPO is set to 0x3. In that configuration, PAD[0] (I/O Pin PA16) is connected to MOSI (called DO here); PAD[3] (I/O Pin PA19) is connected to Serial Clock; and PAD[1] (I/O Pin PA17) is connected to SPI Select.

Let's add these connections to our pin assignment diagram: see Fig. 26L.2.

Figure 26L.2 SAMD21 pin use after SERCOM assignment.

26L.3 The SparkFun SerLCD display

SparkFun sells a family of alphanumeric liquid crystal displays ("LCD") that include an 8-bit AVR microcontroller to allow them to accept input in all three serial protocols: RS232, SPI and I²C. The interface to use is controlled by which interface pins you connect to.[6] They call these devices SerLCD; two models provide two lines of 16 characters while the third has four lines of 20 characters. While any of the three will work in this lab, we will be using the LCD-16396 16 × 2 SerLCD.[7] This display shows dark letters on a light background and includes a red-green-blue (RGB) LED backlight that can be set to produce a wide variety of background colors.

Figure 26L.3 SparkFun 16 × 2 SerLCD with RGB backlight.

The SPI protocol only describes the transfer of raw bytes between devices. The actual meaning of the data transferred depends on the device being controlled. The SerLCD display uses a simple data protocol, with data sequences beginning with the vertical bar character, "|" (124 decimal or 0x7C hex), indicating a control sequence. A sequence beginning with decimal 254 is a cursor-positioning

---

[6] It appears the microcontroller monitors all three inputs simultaneously.
[7] See https://LAoE.link/LCD-16396_SerLCD.html.

command. Any other string is just printed to the display sequentially from the current cursor position. Some useful control sequences (in decimal) include:

Command	Command Sequence (decimal)	Notes
Clear Display	124, 45	Cursor set to row 0, col 0 (upper left)
Set Backlight Color	124, 43, <red>, <grn>, <blu>	0 (off) to 100 (brightest) for each color
Set Contrast	124, 24, <contrast>	0 (darkest) to 100 (lightest): try 5 to 10
Set Cursor Position	254, 128, <rowcol>	rowcol = row * 64 + col

26L.3.1 Add the SerLCD to your breadboard

Add the LCD display to your breadboard and connect it to the microcontroller as shown in Fig. 26L.4. While the display is a 3.3 V device (and no I/O pin should ever be connected to a voltage greater than 3.3 V), it includes a voltage regulator to allow it to be powered from a 3.3 V to 9 V source. You should power it by connecting the 5 V supply to the pin labeled the "RAW" and the pin labeled "-" to ground as shown in the figure.

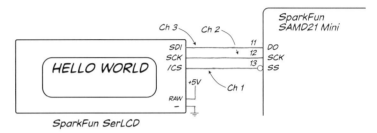

Figure 26L.4 SAMD21 to SerLCD connections with scope connections labeled for debugging.

26L.4 Program the SERCOM for SPI communications

Programming the SERCOM peripheral requires the usual steps of configuring the Port I/O pins, setting the peripheral multiplexer, enabling power to the device in the Power Manager, choosing and connecting a Generic Clock, then finally configuring the peripheral itself. The SPI test code consists of an initialization routine and a set of functions to send byte values out the SPI port. The generic SPI send function, SPI_SendData(), sends a specified number of bytes from an array to the display. Several helper functions use this generic function to send a null terminated string, clear the display, set the contrast, the cursor position and the backlight color. There is also a helper function we will use to control a digital potentiometer in the last part of the lab.

The initialization function, SPI_Init(), is a bit different from our previous peripheral initialization routines in that it takes a boolean argument, AutoCS, which selects between automatic chip select handled by the SERCOM peripheral (AutoCS == true) and manual control of chip select using Port I/O when false. We use the former with the LCD display and the latter with the digipot to allow us to send 16-bit commands.

26L.4.1 Initialize the SAMD21 Port I/O pins

We set all three Port I/O pins as outputs and multiplexed to function C. If the AutoCS parameter is false, we do not enable the multiplexer in the PINCFG register for the chip select pin. This results in the multiplexer setting for that pin being ignored and the pin operates as a normal Port I/O output using the OUT, OUTSET and OUTCLR registers.

26L.4.2 Enable power to the SERCOM

Power to all the SERCOM units is disabled on reset. See Table 12-1 in the datasheet. We enable power by setting the SERCOM1 bit in the APBCMASK register using a RMW operation.

26L.4.3 Connect a Generic Clock to the SERCOM

As you did setting up the DAC and the TC, you need to connect a clock to the SERCOM peripheral. This not only clocks the internal SERCOM peripheral logic, it is also used to derive the SPI SCLK clock (called the BAUD rate) as well.[8] In order to make debugging easier, we want SCLK to be slow enough to be easily viewed on any oscilloscope. The largest BAUD divider is 0xff, making the 48 MHz CPU clock (Generic Clock 0) a poor choice. However, after a reset, Generic Clock 2 is set to 32,768 Hz which allows a BAUD rate of about 5 kHz to be set. This is fast enough to update the display imperceptibly while being easy to debug with the scope.

26L.4.4 Configure the SERCOM for SPI communications

Once the SERCOM is powered, connected to an internal clock, and the external I/O pins have been configured and multiplexed to the peripheral, it is time to configure the SERCOM unit for SPI communications. The datasheet contains two relevant sections; a general SERCOM section (Ch. 25) which describes how to configure the peripheral to USART, I2C or SPI, and a section detailing the use of the SERCOM in SPI communications (Ch. 27). As with the other peripherals you have used, resetting the SERCOM before configuring it allows you start from a known default state.

To set the SERCOM for SPI operation, the MODE field in register CTRLA must be set to "SPI in host operation" after the reset. See datasheet Table 25-1.

The clock rate of the SPI data output is a function of the Generic Clock connected to the SERCOM divided by the eight-bit value in the BAUD register. Table 25-2 shows how to set the value of BAUD to get a particular output frequency, f_{baud}. The SPI is a synchronous protocol, so we use the last row of the table to calculate the value of the BAUD register. We let the compiler calculate a value for BAUD to get a clock frequency of about 5 kHz.[9]

The MSSEN bit in the Control B register enables auto chip select. We always set this bit but it doesn't do anything unless the pin multiplexer is enabled when AutoCS is true.

To configure the SERCOM for SPI communications:

- Set the SWRST bit in CTRLA to one to reset the SERCOM peripheral.
- Wait for the ENABLE and SWRST bits in the SYNCBUSY register to be clear.
- Configure the CTRLA register:
  - We use the default values to send the MSB first and transfer on the rising edge of the clock.
  - We ignore the CPHA, DIPO and FORM fields since they control input data, which we are not using.
  - The DOPO field is set to match the pin assignments in Fig. 26L.2.
  - MODE must be set to host (i.e., Mentor) operation.
  - Other bits can be left at their default [zero] value.
- Configure the CTRLB register:
  - Disable the receiver.

---

[8] Baud is a measure of the number of symbols per second sent on a communications channel. For a binary transmission system like SPI, this is equivalent to bits per second.

[9] "About" because the divider must be an integer.

- AMODE is unused.
- Enable hardware chip select by setting the MSSEN bit (but the bit has no effect if AutoCS is false and the pin multiplexer is not enabled).
- Set the character size to 8 bits.
- Wait for write synchronization on the CTRLB bit in the SYNCBUSY register.
• Set the BAUD register to the BAUD value calculated by the compiler.
• Finally, enable the SERCOM by setting the ENABLE bit in the CTRLA register and waiting for write synchronization on the ENABLE bit in the SYNCBUSY register.

At this point the SERCOM is ready to transmit data via the SPI protocol.

26L.5 LCD display tests

Create a new project and replace the default main.c file with the test code in §26L.10. You will need to add the 48 MHz .c and .h files to the project as well. The main() function initializes the 48 MHz CPU clock, the SysTick interrupt and the SERCOM. It then sends sequential values 0x00 to 0xff to the LCD display over a 76 second period repeatedly.

26L.5.1 Basic SPI byte output test

Use your scope to look at the three SPI signals, triggering on the falling edge of CS. If your oscilloscope supports protocol decoding, add a hex display of the SPI output. You should see the MOSI output signal counting from 0x00 to 0xff with the MSB on the left of the screen and the LSB on the right: see Fig. 26L.5.

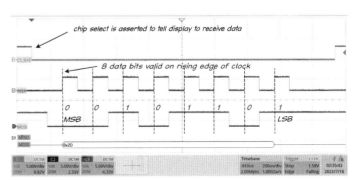

Figure 26L.5 SPI output of value 0x2D to LCD display.

26L.5.2 LCD control and output test

Comment out the first while(1) loop to enable the LCD display tests. The second while loop prints a fixed string to the display, then uses the cursor-positioning commands to update a count at the end of the string every second. Each time though the loop, the backlight color is changed. The display takes a bit of time to initialize when it is powered on before it accepts commands. That is the reason for the delay after the call to SPI_Init().

Adjust the defined value for CONTRAST as necessary to get the best display. Set your scope triggering mode to "Normal" and reduce the horizontal sweep rate to see the full SPI output command sent to the display.

26L.6 The MCP41010 digital potentiometer

SPI peripheral: a digital potentiometer We propose that you now try using the SPI bus to control a *digital potentiometer* with the three SPI lines. This digital potentiometer, the Microchip MCP41010, is a 10kΩ pot whose slider can be placed anywhere between one end and the other, in 256 steps controlled by a byte value sent to it rather than with a mechanical connection.

Connect the digital pot to your microcontroller Add the digipot to your breadboard and connect it to the microcontroller and test equipment as shown in Fig. 26L.6. You should leave the LCD display connected to the same SPI lines as you will need it in Chapter 27L.[10]

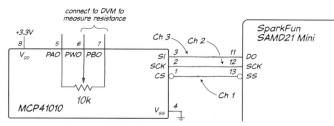

Figure 26L.6 SPI digipot wiring.

26L.7 SPI digipot test program

Replace the main() function in your LCD test program with the code below.[11] This program initializes the SERCOM to use manual chip select (so we can send 16-bit values in one transaction) then sends 16-bit commands to the digipot to set the value from 0x00 to 0xff sequentially over about 25 seconds. Watch the SPI signal lines on the scope and read the resistance between the pot wiper (PWO) and the "B " end terminal (PBO) on your multimeter.

```
/*******************************
 *          main()
 *******************************/
int main (void)
{
  // system initializations
  ClockSysInit48M();                  // set CPU clock to 48MHz
  SysTick_Config(SystemCoreClock/1000); // Configure SysTick for Delay() timer
  SPI_Init(false);                    // need manual CS for digital pot

/******* 23L.2.3.1 test digipot as variable resistor ********/
  while(1) {
    for (int i = 0; i <= 0xff; i++) {
      Digipot_SetValue(i);
      Delay(100);
    }
  }
  return 0;
}
```

You should see two bytes sent to the digipot on each iteration of the `for` loop with a single assertion of chip select. The first byte, 0x11, is a command telling the digipot to set pot 0 (some versions of the device contain a second potentiometer, pot 1) to the value of the next byte: see Fig. 26L.7.

---
[10] The LCD will display whatever data you send to the digital pot. Enjoy the show.
[11] Code available at https://LAoE.link/micro5/26L_SPI_Digipot_Main.c.

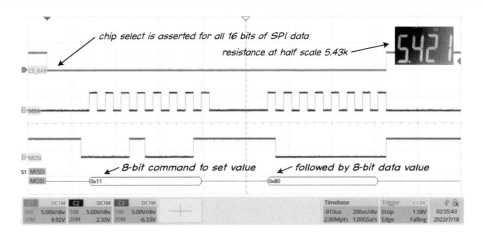

Figure 26L.7 SPI output to set digipot value to mid range.

Graph the output at several values from zero to full scale. How linear is the pot? How accurate is it? Does it meet the specifications on the spec sheet? Here is what we measured at zero, half-scale and full scale:

digipot value	resistance measured
0x00	85 Ω
0x80	5.42k
0xff	10.65k

26L.7.1 Control the DAC output volume

Temporarily replace the 10k manual trimpot that controls the DAC output volume in Fig. 24L.3 with the MCP41010.[12] Be sure the output increases as the value sent to the digipot increases.

Copy the DAC and Timer/Count code from the previous lab the new project. You want the version of TC4_IRQHandler that outputs to the DAC. Be sure to put the copied code *before* main() so you don't have to create prototypes. You also need to copy the sine[] array and the SAMPLES define into the new code. Finally, replace main() with one that includes initializations for the DAC and timer:[13]

```
/*********************************************************************
* DAC Support Functions
*********************************************************************/
void DAC_Init(void){
// insert DAC_Init() function from previous lab
}
/*********************************************************************
* Timer/Counter Support Functions
*********************************************************************/
void TC4_Init(void){
// insert TC4_Init() function from previous lab
}

void TC4_IRQHandler (void) {
  static int i = 0;

  REG_DAC_DATA = sine[(i & 0x7f)];            // write value to DAC
  i = i + 1;
  TC4->COUNT16.INTFLAG.reg = TC_INTFLAG_MC0;  // writing a 1 to the MC0 bit clears the interrupt
}
```

---

[12] You will put the manual potentiometer back for the next lab, so it is easiest to pull the trimpot and run wires over from the digipot.

[13] Complete code available at https://LAoE.link/micro5/26L_Digipot_Volume.c.

```
/****** Set timer interrupt frequency ******/
// Argument is actual value for timer counter
// divider NOT the DAC frequency desired

void SetTC4Div(uint16_t freq_divider) {
    TC4->COUNT16.CC[0].reg = freq_divider;
    while(TC4->COUNT16.STATUS.bit.SYNCBUSY);    // wait for sync synchronization
}

/*******************************
 *          main()
 *******************************/
int main (void)
{
  // system initializations
  ClockSysInit48M();                  // set CPU clock to 48MHz
  SysTick_Config(SystemCoreClock/1000); // Configure SysTick for Delay() timer
  SPI_Init(false);                    // need manual CS for digital pot
  DAC_Init();                         // 1V full scale, 10-bit right justified
  TC4_Init();                         // initialize timer/counter 4 for DAC output timing

  SetTC4Div(375000/1000);             // set DAC to 1kHz output


/******* test digipot as variable voltage divider ********/
  while(1) {
    for (int i = 0; i <= 0xff; i++) {
      Digipot_SetValue(i);
      Delay(10);
    }
    for (int i = 0xff; i >=0; i--) {
      Digipot_SetValue(i);
      Delay(10);
    }
  }

  return 0;
}
```

Test this program. You should hear a 1 kHz tone go from almost off to maximum volume and then back down to almost off again repeatedly over a few seconds.

26L.8 Build a theremin

Let's use the digipot, DAC and Timer/Counter to build an electronic musical instrument called a theremin, after its inventor.[14] The theremin is unique in that it is played without physical contact with the instrument.[15] In the original design, the user varied the frequency of radio frequency oscillators by moving his or her hands over metal antennas, giving an eerie sound. You may have heard the theremin, possibly without realizing it, on the soundtrack for horror and science fiction movies.

In our design we will control the frequency and volume of the output by placing our hands between a light source and photosensitive resistors that change resistance depending on the amount of light striking them.[16] We will use the SAMD21's internal ADC and input multiplexer to measure the two sensor voltages, the Timer/Counter and DAC to output a sine wave at the frequency set by the amount of light striking one sensor, and the digipot to control the amplitude set by the light striking the other as you move your hands above the LDRs.[17] The block diagram of our theremin is shown in Fig. 26L.8.

---

[14] Léon Theremin. See https://LAoE.link/Theremin.html.
[15] See https://LAoE.link/Theremin_theremin_demo.html.
[16] Photosensitive resistors are also referred to as "Light Dependent Resistors" (LDRs) or "CdS cells" (for Cadmium-Sulfide).
[17] See demonstration video at https://LAoE.link/Our_theremin_demo.html.

26L.8 Build a theremin

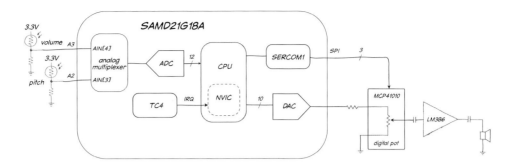

Figure 26L.8 SAMD21 Theremin block diagram.

26L.8.1 The sensor

Photoresistive sensors are inexpensive devices whose resistance decreases with increasing light intensity. The dark resistance may be as high as several megohms, dropping to a few kiloohms or less in bright light. In this lab we use two LDRs in separate voltage dividers to create output voltages inversely related to their illumination: see Fig. 26L.9.[18]

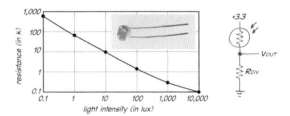

Figure 26L.9 LDR typical response curve and application circuit.

26L.8.2 Add the theremin hardware

To select the divider resistor, R_{div}, mount two LDRs on your breadboard in locations that will make it easy to move one hand over each.[19] Connect your multimeter to one LDR at a time and record the resistance with your hand blocking the room light about 18 inches above the sensor and then with your hand an inch or so above it. Calculate $R_{\text{div}} \approx \sqrt{R_{\text{far}} * R_{\text{near}}}$. Repeat for the other sensor. Add the resistive dividers to the microcontroller as shown in Fig. 26L.10.[20]

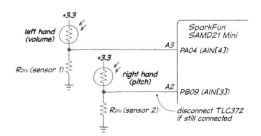

Figure 26L.10 LDR connections to SAMD21.

---

[18] For a nice tutorial on using LDRs see https://LAoE.link/LDR_tutorial.html.
[19] Almost any LDR will work. We tested the NTE 01-LDR2, the GL5528 and the GL5537. The PDV-P8001 and Adafruit's Product ID 161 CdS photoresistor should work as well, but we have not tested them.
[20] If you still have the TLC372 connected to Arduino pin A2, you must either remove it or disconnect it from the pin in order to sense the divider voltage.

26L.8.3 Test the LDR dividers

Add the following ADC initialization and data acquisition functions to the code from §26L.7.1, and replace main() with this one that initializes the ADC and prints out the voltage values of the LDR voltage dividers.[21]

```
/*********************************************************************
 *                  ADC Support Functions
 *********************************************************************/

// We are using two input pins for the two light sensors,
// one on Arduino pin A2 (PB09) and one on Arduino pin A3 (PA04).

void ADC_Init(void) {

    /***************** PORT I/O Initialization ***************/
    /* Set up I/O pins as analog inputs according to table 23-2 */
    /*********************************************************/

    //  PIN_PB09B_ADC_AIN3 must be ANDed with 0x1f
    //  PORT_PB09B_ADC_AIN3 is sensor 1 (pin A2)
    //  MUX_PB09B_ADC_AIN3

    //  PIN_PA04B_ADC_AIN4 is sensor 2 (pin A3)
    //  PORT_PA04B_ADC_AIN4
    //  MUX_PA04B_ADC_AIN4

    // DIR bit for an analog pin must be set to 0
    REG_PORT_DIRCLR0 = PORT_PA04B_ADC_AIN4;
    REG_PORT_DIRCLR1 = PORT_PB09B_ADC_AIN3;

    // Configure the analog inputs with DRVSTR = 0, PULLEN = 0, INEN = 0 and PMUXEN = 1 (Table 23-2)
    PORT->Group[0].PINCFG[PIN_PA04B_ADC_AIN4].reg = PORT_PINCFG_PMUXEN;
    PORT->Group[1].PINCFG[PIN_PB09B_ADC_AIN3 & 0x1f].reg = PORT_PINCFG_PMUXEN;

    // Enable pin multiplexer configuration B (use RMW in case other pin already set)
    PORT->Group[0].PMUX[PIN_PA04B_ADC_AIN4 >> 1].reg |= PORT_PMUX_PMUXE_B;  // mux even pin number to B
    PORT->Group[1].PMUX[(PIN_PB09B_ADC_AIN3 & 0x1f) >> 1].reg |= PORT_PMUX_PMUXO_B; // mux odd

    /************ POWER MANAGER Initialization ***********/
    /* ADC peripheral is already enabled after reset     */
    /*****************************************************/

    /******** GENERIC CLOCK CONTROLLER Initialization ********/
    /* Clock 3 is enabled and set to 8MHz by ClockSysInit48M() */
    /*********************************************************/

    // connect 8MHz to ADC
    GCLK->CLKCTRL.reg = GCLK_CLKCTRL_ID_ADC | GCLK_CLKCTRL_CLKEN | GCLK_CLKCTRL_GEN_GCLK3;
    while (GCLK->STATUS.bit.SYNCBUSY);

    /***************** AC Initialization *********************/
    /* Configure ADC for 16-bit, 32 sample averaged conversion, */
    /* GCLK3/8 (1MHz) manual conversion                      */
    /*********************************************************/

    ADC->CTRLA.bit.ENABLE = 0;              // disable ADC before reset (33.6.2.2)
    while(ADC->STATUS.bit.SYNCBUSY);        // wait for disable to finish

    /* Perform a software reset of analog to digital converter */
    REG_ADC_CTRLA = ADC_CTRLA_SWRST;
    while(ADC->STATUS.bit.SYNCBUSY || ADC->CTRLA.bit.SWRST);

    /* set reference to 3.3V/2 = 1.65V to give 0 to 3.3V scale */
    REG_ADC_REFCTRL = ADC_REFCTRL_REFSEL_INTVCC1;  // input will be divided by 2 as well

    /* Set clock prescaler (to get 1MHz or less), 16-bit for averaging,
       single conversion mode, right justified, single ended */
    REG_ADC_CTRLB = ADC_CTRLB_PRESCALER_DIV8 | ADC_CTRLB_RESSEL_16BIT;
    while(ADC->STATUS.bit.SYNCBUSY);

    /* average 32 samples  to reduce noise, right shift to 12 bits */
```

---

[21] Complete code available at https://LAoE.link/micro5/26L_ADC_LDR_Test.c.

```
    REG_ADC_AVGCTRL = ADC_AVGCTRL_ADJRES(4) | ADC_AVGCTRL_SAMPLENUM_32;

    /* extend sample time a bit */
    REG_ADC_SAMPCTRL = 16;

    /* divide input by 2 so matches ref, start with AIN[4] */
    REG_ADC_INPUTCTRL = ADC_INPUTCTRL_MUXPOS_PIN4 | ADC_INPUTCTRL_MUXNEG_IOGND
      | ADC_INPUTCTRL_GAIN_DIV2;
    while(ADC->STATUS.bit.SYNCBUSY);

    /* Enable ADC peripheral */
    REG_ADC_CTRLA = ADC_CTRLA_ENABLE;          // turn on ADC (ok to set other bits to 0)
    while(ADC->STATUS.bit.SYNCBUSY);           // wait for the ADC enable to finish
}

/* function to return 12-bit value 0 to 4095 for selected input */

uint16_t ReadADC(uint32_t AInPin) {
    // set input pin and gain
    REG_ADC_INPUTCTRL = AInPin | ADC_INPUTCTRL_MUXNEG_IOGND | ADC_INPUTCTRL_GAIN_DIV2;
    while(ADC->STATUS.bit.SYNCBUSY);

    REG_ADC_INTFLAG = ADC_INTFLAG_RESRDY;                 // clear result ready flag

    REG_ADC_SWTRIG = ADC_SWTRIG_START;                    // start a conversion
    while(ADC->STATUS.bit.SYNCBUSY);

    while(ADC->INTFLAG.bit.RESRDY == 0) {                 // wait for conversion done
      __WFE();
      }
    return REG_ADC_RESULT;                                // clears ready flag
}

/*******************************
 *          main()
 *******************************/
int main (void)
{
  // system initializations
  ClockSysInit48M();                     // set CPU clock to 48MHz
  SysTick_Config(SystemCoreClock/1000);  // Configure SysTick for Delay() timer
  SPI_Init(false);                       // need manual CS for digital pot
  DAC_Init();                            // 1V full scale, 10-bit right justified
  TC4_Init();                            // initialize timer/counter 4 for DAC output timing
  ADC_Init();                            // initialize ADC for 8 bit conversion

  SetTC4Div(375000/1000);                // set DAC to 1kHz output

/******* demonstration of ADC reading LDR dividers ********/
  uint16_t A2, A3;
  while(1) {
    A2 = ReadADC(ADC_INPUTCTRL_MUXPOS_PIN3);
    printf("Arduino Pin A2 voltage %d mV; ", (A2 * 3300) / 4096);
    A3 = ReadADC(ADC_INPUTCTRL_MUXPOS_PIN4);
    printf("Arduino Pin A3 voltage %d mV\n", (A3 * 3300) / 4096);
    Delay(500);
  }

  return 0;
}
/*************************** End of file ***************************/
```

The function `ADC_Init()` sets the two pins connected to the LDR dividers as analog inputs, connects an 8 MHz clock to the ADC, configures the ADC to convert at a 1 MHz rate, averages 32 samples, and computes a 12-bit result by shifting the sum right by 5 bits.[22] The ADC is set to use a $\frac{3.3\,\text{V}}{2}$ reference but the function also sets the input gain to $\frac{1}{2}$. This results in an ADC COUNT between 0 and 4095 corresponding to an input of $V_\text{in} = \frac{\text{COUNT}}{4096} * 3.3\,\text{V}$.

---

[22] Averaging is technique used to reduce the effect of any random noise at the analog input on the digitized result by adding a number of ADC samples together. If we take n samples and average them, we reduce the effect of input noise by a factor of sqrt(n). Of course it takes at least n times as long to collect the result, but that is not an issue here where we can acquire the 32 averaged samples in about 500 μs and only need to sense the [relatively slow] movement of a human hand.

The function `ReadADC()` returns the sampled ADC value of one of the analog multiplexer inputs. The CMSIS header file ADC.H contains a list of available multiplexer input choices for the function argument `AInPin`.[23] The volume LDR is connected to `ADC_INPUTCTRL_MUXPOS_PIN3` (MUXPOS AIN3, SAMD21 pin PA04, Arduino pin A3). The pitch LDR is connected to `ADC_INPUTCTRL_MUXPOS_PIN4` (MUXPOS AIN4, SAMD21 pin PB09, Arduino pin A3).

You might also want to try using the ADC to read the internal 1.1 V bandgap reference voltage:

```
REG_SYSCTRL_VREF |= SYSCTRL_VREF_BGOUTEN;   // enable bandgap to ADC
printf("Bandgap voltage %d mV\n", (ReadADC(ADC_INPUTCTRL_MUXPOS_BANDGAP)* 3300) / 4096);
```

and the 3.3 V supply voltage (divided by four):

```
printf("Vcc voltage %d mV\n", (ReadADC(ADC_INPUTCTRL_MUXPOS_SCALEDIOVCC) * 4 * 3300) / 4096);
```

26L.8.4 Now turn all this into a theremin

You now have the Timer/Counter generating an interrupt at a fixed rate which updates the DAC to output a sinewave at a frequency controllable by the `SetTC4Div()` function. You can set the volume of the sine wave using the `Digipot_SetValue()` function, which communicates with the digital potentiometer over SPI signals generated by the SERCOM peripheral. You can read the LDR divider values with the ADC by calling the `ReadADC()` function. Modify the demonstration `while(1)` loop to implement a theremin:

- As you move your hand closer to the pitch LDR, the pitch should increase (smaller number in the timer counter match register).
- As you move your hand closer to the volume LDR, the volume should decrease (smaller number sent to the digipot).
- You will need to use the test program to see what the upper and lower limits of the ADC output are as you move your hands over the LDRs. Use these values in constructing your theremin loop. (It is simpler to just use the raw ADC values than the scaled voltage calculation.)
- If you set the TC interrupt frequency too high, it will appear that the system has crashed (i.e., you have no response to user input). Actually, it is just executing the ISR over and over without ever returning to the foreground program. So it is a good idea to limit the value written to the TC to avoid this problem (and set a maximum frequency for the sine wave).
- You may also want to set a floor on LDR volume input. If the voltage goes below that just set the volume to zero.
- `Digipot_SetValue()` takes an 8-bit argument and `SetTC4Div()` a 16-bit value. You will calculate the arguments for these functions from the 12-bit ADC value. Be careful about overflow and scaling to avoid discontinuities in operation.

26L.9 Cleanup

In preparation for the last lab, you will need to remove the digipot and reinstall the manual trimpot to control DAC output volume. Make sure the LCD still functions correctly with SPI commands and plug in the 16-key keypad if you removed it. We will not be using the TLC372 comparator or the LDRs, so feel free to remove them as well.

---

[23] See also the values for the MUXPOS[4:0] field in the INPUTCTRL register, datasheet section 33.8.8.

26L.10 SPI test code

Code available at https://LAoE.link/micro5/26L_SPI_LCD_Test.c

```c
/***************************************************************************
 * Company: Learning the Art of Electronics
 * Engineer: David Abrams
 * Create Date:    2022-07-14
 * Module Name:    main.c (26L_SPI_LCD_Test.c)
 * Target Device:  Sparkfun SAMD21 Micro
 * Description: Set up SERCOM to control LCD display and MCP41010 digital pot
 ***************************************************************************/

#include "samd21.h"       // add to use CMSIS
#include "ClockSysInit48M.h"
#include <stdio.h>
#include <stdlib.h>
#include <string.h>
#include <stdbool.h>

#define CONTRAST 5      // LCD contrast setting - 0 (darkest) to 100 (lightest)

/***********************************************************************
 *              Set up System Timer and Delay function
 ***********************************************************************/
// use 1ms system clock for timing

volatile uint32_t msTicks = 0;

void SysTick_Handler(void)   {      // SysTick interrupts every millisecond
    msTicks++;
}

// Function to delay for n milliseconds
void Delay (uint32_t DelayMSec)  {
  uint32_t endTicks;

  endTicks = msTicks + DelayMSec;

  while (msTicks < endTicks)  {    // wait for DelayMSec SysTick interrupts
      __WFE ();                    // tell CPU ok to power-down until interrupt
  }
}

/***********************************************************************
 *                   SPI Support Functions
 ***********************************************************************/

// Define pins to use for SPI communications (MISO is not needed b/c there is no input data)
// Here we use SERCOM1 mapping C and pinout DOPO 0x3.  I chose to use my own definitions
// that documented what each pin was used for rather than the CMSIS SERCOM defines.

#define SPI_SERCOM_DOPO_GROUP    3               // DOPO value to get the pin mapping below

// Pad[1] Arduino pin 13 on SparkFun board is student select
#define PIN_SS     PIN_PA17
#define PORT_SS    PORT_PA17

// Pad[3] Arduino pin 12 on SparkFun board is SCK
#define PIN_SCK    PIN_PA19
#define PORT_SCK   PORT_PA19

// Pad[0] Arduino pin 11 on SparkFun board is MOSI
#define PIN_MOSI   PIN_PA16
#define PORT_MOSI  PORT_PA16

// The SPI clock frequency is set by an 8-bit BAUD register
// That limits the slowest clock to the reference clock divided by 255
// We use the 32,768 real-time clock oscillator (Generic Clock 2)
// to allow for an approx 5kHz output easy to see on the oscilloscope.

#define SPI_SERCOM_CLKCTRL_GEN   GCLK_CLKCTRL_GEN_GCLK2
#define SYS_CLK_FREQ       32768        // clock generator 2 is set up for 32.768Hz clock at reset
#define SPI_BAUD           5000         // SPI can go much faster but this is easy to look at on scope
```

Lab: Microcontrollers V

```c
#define SPI_CHSIZE_8BIT     0       // SPI data size

// The active low Chip Select (CS*) on Arduino pin 13 can be either automatic (AutoCS true)
// or manual.  SPI_SendData() uses the same value of AutoCS that is used when you call SPI_Init().
// The difference is that manual CS keeps CS asserted until all SPI_SendData() bytes are sent.
// This is necessary to control the digital potentiometer which requires two bytes sent as'
// a 16-bit value.

bool Auto_CS;               // static used by SPI_SendData()

void SPI_Init(bool AutoCS){

    Auto_CS = AutoCS;       // make CS mode visible to SPI_SendData

    /***************** PORT I/O Initialization *************/
    /* Set up PORT_SS, PORT_MOSI and PORT_SCK as outputs,   */
    /* multiplexed to peripheral function C (but PORT_SS    */
    /* mux not enabled unless AutoCS true).                 */
    /*******************************************************/
    REG_PORT_DIRSET0 = PORT_MOSI | PORT_SCK | PORT_SS; // set as outputs
    REG_PORT_OUTCLR0 = PORT_MOSI | PORT_SCK;           // start with MOSI and CLK low
    REG_PORT_OUTSET0 = PORT_SS;                        // and student select high (not selected)

    // set mux C on PA16, PA17 and PA19
    PORT->Group[0].PMUX[PIN_PA16 >> 1].reg = PORT_PMUX_PMUXE_C | PORT_PMUX_PMUXO_C;
    PORT->Group[0].PMUX[PIN_PA19 >> 1].reg = PORT_PMUX_PMUXO_C;

    PORT->Group[0].PINCFG[PIN_MOSI].reg = PORT_PINCFG_PMUXEN;   // enable the mux so PA16 and
    PORT->Group[0].PINCFG[PIN_SCK].reg  = PORT_PINCFG_PMUXEN;   // and PA19 connect to the SERCOM
    if (AutoCS){
       PORT->Group[0].PINCFG[PIN_SS].reg   = PORT_PINCFG_PMUXEN; // don't enable mux on PA17 if manual cs
    }

    /*************** POWER MANAGER Initialization ***************/
    /* On reset, power to SERCOM1 is disabled, must be enabled  */
    /************************************************************/

    PM->APBCMASK.reg |= PM_APBCMASK_SERCOM1;      //Enable the SERCOM 1 under the Power Manager w/ RMW

    /******** GENERIC CLOCK CONTROLLER Initialization ********/
    /* Connect Clock 2 to DAC and enable SERCOM1 clock.       */
    /* Clock 2 is enabled and set to 32.768kHz                */
    /*********************************************************/

    GCLK->CLKCTRL.reg = GCLK_CLKCTRL_ID(SERCOM1_GCLK_ID_CORE)
      | GCLK_CLKCTRL_CLKEN | SPI_SERCOM_CLKCTRL_GEN;
    while(GCLK->STATUS.bit.SYNCBUSY);         // requires write synchronization

    /***************** SERCOM1 Initialization *********************/
    /* Setup SERCOM for SPI Mentor, DOPO 3, auto SS and 8-bit data */
    /**************************************************************/

    /* Perform a software reset and wait for it to complete */
    SERCOM1->SPI.CTRLA.bit.SWRST = 1;
    while(SERCOM1->SPI.SYNCBUSY.bit.SWRST || SERCOM1->SPI.CTRLA.bit.SWRST);

    /* Set SERCOM1 mode as SPI Mentor and DOPO to PAD[0,3,1] for MOSI, SCK and SS */
    SERCOM1->SPI.CTRLA.reg = SERCOM_SPI_CTRLA_MODE_SPI_MASTER
            | SERCOM_SPI_CTRLA_DOPO(SPI_SERCOM_DOPO_GROUP);

    /* Select auto SS control and 8 data bits in control register B *
     * but auto SS ignored if mux not enabled (AutoCS false)        */
    SERCOM1->SPI.CTRLB.reg = SERCOM_SPI_CTRLB_MSSEN | SERCOM_SPI_CTRLB_CHSIZE(SPI_CHSIZE_8BIT);
    while(SERCOM1->SPI.SYNCBUSY.bit.CTRLB);                        //Write synchronize

    // SPI clock rate calculation according to table 25-2 for synchronous operation
    uint16_t BAUD_REG = ((float) SYS_CLK_FREQ / (float)(2 * SPI_BAUD)) - 1;   //Calculate BAUD register value
    SERCOM1->SPI.BAUD.reg = SERCOM_SPI_BAUD_BAUD(BAUD_REG);        //Set the SPI baud rate

    SERCOM1->SPI.CTRLA.reg |= SERCOM_SPI_CTRLA_ENABLE;             //Enable the SERCOM1 SPI
    while(SERCOM1->SPI.SYNCBUSY.bit.ENABLE);                       //Wait for the enable to finish
}

/*******************************************/
/*           SPI Output Functions          */
/*******************************************/
```

26L.10 SPI test code

```c
/********** SPI_SendData() *****************/
/*                                          */
/* Send a fixed number of bytes out the SPI */
/* port. Does not return until all bytes    */
/* have been sent.  Returns number of bytes */
/* sent.                                    */
/*                                          */
/* If manual chip select (AutoCS false),    */
/* CS is asserted until all bytes sent.     */
/********************************************/

int SPI_SendData(uint8_t *data, int count)
{
  if (!Auto_CS){
    REG_PORT_OUTCLR0 = PORT_SS;                    // if manual, assert CS
  }
  for (int i = 0; i < count; i++) {
    while(SERCOM1->SPI.INTFLAG.bit.DRE == 0);      // wait until Data Register Empty from last send
    SERCOM1->SPI.DATA.reg = data[i];               // and send next byte
  }
  if (!Auto_CS){
    while(SERCOM1->SPI.INTFLAG.bit.TXC == 0);      // wait for last byte to be sent
    REG_PORT_OUTSET0 = PORT_SS;                    // disassert CS
  }
  return count;                                    // and return when done
}

/*************************************************/
// Send bytes from a null terminated string out
// the SPI port. Returns number of bytes sent.

int SPI_PrintData(char *outstr) {
  return SPI_SendData((uint8_t*) outstr, strlen(outstr));
}

/*************************************************
 * Function to set digital potentiometer value *
 * value = 0x00 to 0xff                        *
 *************************************************/

int Digipot_SetValue(uint8_t value) {
  uint8_t DPCmd[] = {0x11, 0};      // command to set value of potentiometer followed by value byte

  DPCmd[1] = value;
  return SPI_SendData(DPCmd, sizeof(DPCmd));
}

/***********************************************************************
 *                     LCD Support Functions
 ***********************************************************************/

void LCD_Clear(void) {
  uint8_t ClrCmd[] = {124, 45};      // command to clear display

  SPI_SendData(ClrCmd, sizeof(ClrCmd));
}

void LCD_SetBacklight(uint8_t red, uint8_t grn, uint8_t blu) {
  uint8_t BLCmd[] = {124, 43, 0, 0, 0};

  BLCmd[2] = red;              // set brightness to 0 to 100 for each color
  BLCmd[3] = grn;
  BLCmd[4] = blu;
  SPI_SendData(BLCmd, sizeof(BLCmd));
}

void LCD_SetContrast(uint8_t ctrst) {  // contrast = 0 (darkest) to 100 (lightest)
  uint8_t CtCmd[] = {124, 24, 0};      // 5 is a good starting value

  CtCmd[2] = ctrst;
  SPI_SendData(CtCmd, sizeof(CtCmd));
}

/* function to set cursor position */
/* ROW = 1 or 2                    */
```

```c
/* COL = 1 to 16                  */
#define MAXCOLIDX 15

void LCD_SetCursor(uint16_t row, uint16_t col) {
  uint8_t SCCmd[] = {254, 0};

  if (row <= 1)
      row = 0;           // 0 or 1 means first row
  else
      row = 64;          // otherwise second row

  if (--col > MAXCOLIDX) col = 0;   // 0 to 15 allowed for 2x16 display

  SCCmd[1] = (128 + row + col);
  SPI_SendData(SCCmd, sizeof(SCCmd));
}

/******************************
 *           main()
 ******************************/
int main (void)
{
  uint32_t counter = 0;
  uint32_t returnCode;
  uint8_t k = 0;
  char lcdbuf[20];      // buffer for strings to send to LCD display

  // system initializations
  ClockSysInit48M();                      // set CPU clock to 48MHz
  SysTick_Config(SystemCoreClock/1000);   // configure SysTick for Delay() timer
  SPI_Init(true);                         // Display uses automatic chip select
  Delay(1000);                            // but does not respond to commands immediately after power up

/*************************************************************
    Use the following test code to test your SPI routines
    using the oscilloscope to look at the SCLK and MOSI
    signals (trigger on SS). One you have SPI working,
    delete (or comment out) this while() loop to send
    real data to the LCD display.
*************************************************************/
while(1) {
  SPI_SendData(&k, 1);  // 0x00 to 0xff one byte at a time out SPI port
  k = (k + 1) & 0xff;
  Delay(100);
}

// Use SPI to send data to the LCD display and change the backlight color every second

  LCD_Clear();
  LCD_SetBacklight(100, 100, 100);          // start with brightest white backlight
  LCD_SetContrast(CONTRAST);

  SPI_PrintData("Hello World!    Counter: ");

  while (1) {                               // cycle through backlight at zero, 50%, 100% each color
    switch (k++) {
    case 0:
        LCD_SetBacklight(50, 0, 0);   // red, green, blue - each 50%, then 100%
        break;
    case 1:
        LCD_SetBacklight(100, 0, 0);
        break;
    case 2:
        LCD_SetBacklight(0, 50, 0);
        break;
    case 3:
        LCD_SetBacklight(0, 100, 0);
        break;
    case 4:
        LCD_SetBacklight(0, 0, 50);
        break;
    case 5:
        LCD_SetBacklight(0, 0, 100);
        break;
```

```
            case 6:
                LCD_SetBacklight(50, 50, 50);
                break;
        case 7:
                LCD_SetBacklight(100, 100, 100);
                k = 0;
                break;
        default:
                k = 0;
        }

        //Change cursor position to 2nd row (first row is 1), column 10 (first col is 1)
        LCD_SetCursor(2, 10);

        sprintf(lcdbuf, "%d", counter++);      // and display the cycle count there
        SPI_PrintData(lcdbuf);
        Delay(1000);                            // delay one second until next iteration
        }

    return 0;
}
/*************************** End of file ****************************/
```

26L.11 Conclusion

While the theremin used a number of the internal peripherals in the SAMD21, the application loop itself was fairly simple. However, a simple project can often grow into a larger one, and trying to shoehorn new code into an Arduino loop type application may not only be difficult, it is easy to add bugs to a working system when modifying existing code. In the next chapter we will look at a better way to build complex systems using a Real Time Operating System (RTOS).

26L.12 Solutions

- §26L.8.4: https://LAoE.link/micro5/26L_Working_Theremin.c

26W Worked Examples: Interrupt-Driven Serial I/O and SPI Off-chip RAM

26W.1 An interrupt-driven version of SPI_SendData()

Serial data input and output are classic applications where interrupt-driven I/O makes sense. Rather than sit in a loop wasting CPU cycles waiting for each byte to be sent or received, an ISR can load a new byte into the output register each time the previous byte has been sent, or store each new byte in a buffer as they are received. Let's take a look at how we can make the SPI output function from Lab 26L more efficient by using interrupts.

26W.1.1 Why?

The main loop of the function that we used in the lab to send data out the SPI port looked like this:

```
for (int i = 0; i < count; i++) {
   while(SERCOM1->SPI.INTFLAG.bit.DRE == 0);   // wait until Data Register Empty from last send
   SERCOM1->SPI.DATA.reg = data[i];            // and send next byte
}
```

The program waits for the Data Register Empty (DRE) bit in the INTFLAG register to be set, indicating the SERCOM is ready for new data to transmit. It then loads the next byte in the DATA register and waits until that byte is sent.[1] It repeats this until all bytes to send are complete.

Even at 5 kHz, it can take several milliseconds for SPI_SendData() to return. For example, a 16-character line would take more than 25 ms to send. Since nothing else (other than interrupt service routines) can happen while SPI_SendData() is waiting for DRE to go high, this would waste over one million cycles at a 48 MHz CPU clock.

26W.1.2 How?

A much better design (and one that is typical for serial communications) is to send the data in a background process. There is no change in the foreground program's call to SPI_SendData() but, instead of waiting, the function buffers the data to be sent and returns immediately, allowing the CPU to continue foreground activity.

To convert the lab SERCOM1 code from polled to interrupt-driven:

- Using SERCOM1_IRQn as the interrupt number, enable SERCOM1 interrupts in the NVIC at the end of the SPI_Init() function. We do not have to set a high priority for this interrupt since the SPI will work fine if it is delayed a few hundred microseconds by higher priority interrupts; it will just take a bit longer to send all the data to the display.

---

[1] The SERCOM DATA register address actually accesses two registers. A byte *written* to this address is saved in the data output register which is then sent out by the hardware on the serial MOSI communication line. *Reading* from this address returns the value received in the data input register from the MISO data stream.

- Modify `SPI_SendData()` so that it copies the data to send into a FIFO, enables the DRE interrupt (i.e., sets the SERCOM_SPI_INTENSET_DRE bit in the INTENSET register) to start the background task sending data, and then returns.[2] (If the FIFO is full the function will have to wait until there is room for all the new data before returning. We will make the FIFO buffer large enough that this never happens.)
- Add an interrupt service routine `void SERCOM1_IRQHandler(void)` to be called each time the SPI DATA register is empty and the interrupt is enabled.
- If there is still data to send when the interrupt occurs, the ISR copies the next byte to the SPI DATA register (which clears the interrupt).
- If there is no more data to send, the ISR should disable the interrupt by writing to the INTENCLR register.

With short output strings, you will likely not notice any difference between the polled version and the interrupt-driven version. As an indication that interrupts are working properly, the demonstration code (§26W.1.3) turns on the external LED when `SPI_SendData()` is called and turns it off when the ISR disables interrupts.

26W.1.3 The code for interrupt-driven SPI output

`SPI_Init()` is identical to the non-interrupt version with the addition of four lines at the end to enable interrupts in the NVIC and one line to initialize (i.e., clear) the FIFO. The interrupt should not be enabled in the SERCOM at initialization, as there is no data to output until `SPI_SendData()` is called.

The five functions beginning with "FIFO_" implement a small FIFO to save the data strings to send out the SPI port. Since the display is only 32 characters max, a 64-byte FIFO should be more than adequate.

`SPI_SendData()` stores the string to send in the FIFO and enables the DRE interrupt (in case it is not already enabled) to start data being sent. If the FIFO is full, it must wait until there is enough room for the new data, but with 64 bytes available this is extremely unlikely.[3] It also turns on the LED connected to Arduino pin 10 to indicate data is being sent. (This is purely to highlight that the system is working, you can remove it once you're convinced.) To keep this example simple, the value of AutoCS is ignored and automatic chip select is always used.[4]

The ISR writes the next byte in the FIFO to the SERCOM output register. If the FIFO is empty, it just turns the interrupt (and the LED) off.

To test the interrupt driven SPI driver, modify your LCD display test code from §26L.5 by replacing the SPI Support Functions with following code.[5] The LCD display should work the same as it did with the polled code. The only difference is that the external green LED should flash once per second to show the ISR is doing all the work of sending data to the display.

```
/***********************************************************
*          SPI Support Functions                           *
***********************************************************/
/*********************************************/
/*        Create SPI output FIFO             */
/* Filled by SPI_SendData() and emptied      */
/* by DRE interrupt service routine.         */
/*********************************************/
```

[2] A FIFO buffer – First In, First Out – always returns the oldest item when read.
[3] Since the display is intended to convey information to a human, it would not make sense to display LCD messages more quickly than a user can read them.
[4] The complication with manual chip select is that it requires handling a second interrupt source, Transmit Complete (TXC), to turn off CS in the ISR after the last byte is sent.
[5] Code available at https://LAoE.link/micro5/26W_SPI_Interrupt_Driver.c.

```c
#define FIFO_SZ 64   // should be plenty for 2x16 display
volatile uint8_t fifo[FIFO_SZ];
volatile int fifo_start;  // index of next empty location
volatile int fifo_end;    // index of last data stored

void FIFO_Reset(void){   // empty fifo
  fifo_start = 0;
  fifo_end = 0;
}

bool Is_FIFO_full(void){
 if (((fifo_end + 1) % FIFO_SZ) == fifo_start)  // no room!
   return true;
 else
   return false;
}

bool Is_FIFO_empty(void){
 if (fifo_end  == fifo_start)
   return true;
 else
   return false;
}

bool FIFO_push(uint8_t data){
  if (Is_FIFO_full()){
    return false;
  } else {
    fifo[fifo_end] = data;
    fifo_end = (fifo_end + 1) % FIFO_SZ;
    return true;
  }
}

bool FIFO_pop(uint8_t *data){
  if (Is_FIFO_empty()){
    return false;
  } else {
    *data = fifo[fifo_start];
    fifo_start = (fifo_start + 1) % FIFO_SZ;
    return true;
  }
}

// optional LED indication of interrupt output

#define PORT_LED_D10 PORT_PA18;

// Define pins to use for SPI communications (MISO is not needed b/c there is no input data)
// Here we use SERCOM1 mapping C and pinout DOPO 0x3. I chose to use my own definitions
// that documented what each pin was used for rather than the CMSIS SERCOM defines.

#define SPI_SERCOM_DOPO_GROUP    3            // DOPO value to get the pin mapping below

// Pad[1] Arduino pin 13 on SparkFun board is student select
#define PIN_SS     PIN_PA17
#define PORT_SS    PORT_PA17

// Pad[3] Arduino pin 12 on SparkFun board is SCK
#define PIN_SCK    PIN_PA19
#define PORT_SCK   PORT_PA19

// Pad[0] Arduino pin 11 on SparkFun board is MOSI
#define PIN_MOSI   PIN_PA16
#define PORT_MOSI  PORT_PA16

// The SPI clock frequency is set by an 8-bit BAUD register
// That limits the slowest clock to the reference clock divided by 255
// We use the 32,768 real-time clock oscillator (Generic Clock 2)
// to allow for an approx 5kHz output easy to see on the oscilloscope.

#define SPI_SERCOM_CLKCTRL_GEN  GCLK_CLKCTRL_GEN_GCLK2
#define SYS_CLK_FREQ       32768       // clock generator 2 is set up for 32.768Hz clock at reset
#define SPI_BAUD           5000        // SPI can go much faster but this is easy to look at on scope
```

26W.1 An interrupt-driven version of SPI_SendData()

```c
#define SPI_CHSIZE_8BIT     0       // SPI data size

// The active low Chip Select (CS*) on Arduino pin 13 is set to be automatic to keep
// this example simple.  Manual CS requires two interrupt sources since we cannot
// de-assert CS until the TXC bit is set.  I left the function argument
// so you do not have to change the main() function call; however, it is ignored.

void SPI_Init(bool AutoCS){

    /**************** PORT I/O Initialization **************/
    /* Set up PORT_SS, PORT_MOSI and PORT_SCK as outputs,   */
    /* multiplexed to peripheral function C.                */
    /*******************************************************/
    REG_PORT_DIRSET0 = PORT_MOSI | PORT_SCK | PORT_SS | PORT_LED_D10;   // set as outputs
    REG_PORT_OUTCLR0 = PORT_MOSI | PORT_SCK;                // start with MOSI and CLK low
    REG_PORT_OUTSET0 = PORT_LED_D10;                        // LED off
    REG_PORT_OUTSET0 = PORT_SS;                             // and student select high (not selected)

    PORT->Group[0].PMUX[PIN_PA16 >> 1].reg = PORT_PMUX_PMUXE_C | PORT_PMUX_PMUXO_C;  // set mux C on PA16,
    PORT->Group[0].PMUX[PIN_PA19 >> 1].reg = PORT_PMUX_PMUXO_C;                     // PA17 and PA19

    PORT->Group[0].PINCFG[PIN_MOSI].reg = PORT_PINCFG_PMUXEN;    // enable the mux so PA16 and
    PORT->Group[0].PINCFG[PIN_SCK].reg  = PORT_PINCFG_PMUXEN;    // and PA19 connect to the SERCOM
    PORT->Group[0].PINCFG[PIN_SS].reg   = PORT_PINCFG_PMUXEN;    // mux on PA17 for auto cs


    /**************** POWER MANAGER Initialization **************/
    /* On reset, power to SERCOM1 is disabled, must be enabled  */
    /************************************************************/

    PM->APBCMASK.reg |= PM_APBCMASK_SERCOM1;    //Enable the SERCOM 1 under the Power Manager w/ RMW

    /******** GENERIC CLOCK CONTROLLER Initialization ********/
    /* Connect Clock 2 to DAC and enable SERCOM1 clock.      */
    /* Clock 2 is enabled and set to 32.768kHz               */
    /*********************************************************/

    GCLK->CLKCTRL.reg = GCLK_CLKCTRL_ID(SERCOM1_GCLK_ID_CORE) | GCLK_CLKCTRL_CLKEN
                      | SPI_SERCOM_CLKCTRL_GEN;
    while(GCLK->STATUS.bit.SYNCBUSY);           // requires write synchronization

    /**************** SERCOM1 Initialization ********************/
    /* Setup SERCOM for SPI Mentor, DOPO 3, auto SS and 8-bit data */
    /************************************************************/

    /* Perform a software reset and wait for it to complete */
    SERCOM1->SPI.CTRLA.bit.SWRST = 1;
    while(SERCOM1->SPI.SYNCBUSY.bit.SWRST || SERCOM1->SPI.CTRLA.bit.SWRST);

    /* Set SERCOM1 mode as SPI Mentor and DOPO to PAD[0,3,1] for MOSI, SCK and SS */
    SERCOM1->SPI.CTRLA.reg = SERCOM_SPI_CTRLA_MODE_SPI_MASTER
                | SERCOM_SPI_CTRLA_DOPO(SPI_SERCOM_DOPO_GROUP);

    /* Select auto SS control and 8 data bits in control register B */
    SERCOM1->SPI.CTRLB.reg = SERCOM_SPI_CTRLB_MSSEN | SERCOM_SPI_CTRLB_CHSIZE(SPI_CHSIZE_8BIT);
    while(SERCOM1->SPI.SYNCBUSY.bit.CTRLB);                  //Write synchronize

    // SPI clock rate calculation according to table 25-2 for synchronous operation
    uint16_t BAUD_REG = ((float) SYS_CLK_FREQ / (float)(2 * SPI_BAUD)) - 1;   //Calculate BAUD register value
    SERCOM1->SPI.BAUD.reg = SERCOM_SPI_BAUD_BAUD(BAUD_REG);  //Set the SPI baud rate

    SERCOM1->SPI.CTRLA.reg |= SERCOM_SPI_CTRLA_ENABLE;       //Enable the SERCOM1 SPI
    while(SERCOM1->SPI.SYNCBUSY.bit.ENABLE);                 //Wait for the enable to finish

    /******** NVIC Initialization (new) ********/
    /* Setup NVIC for SERCOM1 Interrupt.       */
    /* It will be enabled in the SERCOM later  */
    /* when there is data to send.             */
    /*******************************************/

    NVIC_DisableIRQ(SERCOM1_IRQn);
    NVIC_ClearPendingIRQ(SERCOM1_IRQn);
    NVIC_SetPriority(SERCOM1_IRQn, 50);         // does not require high priority
    NVIC_EnableIRQ(SERCOM1_IRQn);

    FIFO_Reset();   // start with nothing to send
```

```
}
/**********************************************/
/*          SPI Output Functions              */
/**********************************************/

/********** SPI_SendData() *****************/
/*                                           */
/* Queues a fixed number of bytes in the     */
/* Transmit FIFO and enables the DRE         */
/* interrupt in case not already enabled.    */
/**********************************************/

int SPI_SendData(uint8_t *data, int count)
{
  for (int i = 0; i < count; i++) {
      while (Is_FIFO_full()) {__WFE();}          // wait if no room in FIFO
      FIFO_push(data[i]);
  }
  SERCOM1->SPI.INTENSET.reg = SERCOM_SPI_INTENSET_DRE;   // make sure DRE interrupt enabled
  REG_PORT_OUTCLR0 = PORT_LED_D10;                        // turn on LED for visual indication of output
  return count;
}

/**********************************************/
// Send bytes from a null terminated string to
// the SPI port.

int SPI_PrintData(char *outstr) {
  return SPI_SendData((uint8_t*) outstr, strlen(outstr));
}

/*********** SERCOM1 ISR *********************/
/* Outputs bytes from FIFO until empty.      */
/* When FIFO is empty, disables DRE interrupt. */
/**********************************************/
void SERCOM1_IRQHandler(void) {
    uint8_t txdata;
    if (FIFO_pop(&txdata)) {                    // if FIFO not empty
      SERCOM1->SPI.DATA.reg = txdata;           // send next byte
    }
    else {
      SERCOM1->SPI.INTENCLR.reg = SERCOM_SPI_INTENCLR_DRE;   // else disable interrupt if nothing more to
      REG_PORT_OUTSET0 = PORT_LED_D10;                       // send and turn off LED to give visual indication of operation
    }
}
```

26W.2   Adding more RAM

Let's add more memory to the SparkFun SAMD21 using an external RAM chip. The Microchip 23K256 is a 32k × 8 serial SRAM that transfers data via SPI at up to 20 MHz. Adding this to the SAMD21 doubles the total RAM available. While this memory cannot be used directly by the C compiler, it can be used to store data or tables by writing functions to access it. The 32K SPI RAM that we use is capable of storing or reading a single byte at a specified address. But for building and reading tables it provides a neat alternative: it can store a sequence of bytes at successive addresses using an internal counter, automatically incremented upon each data write. The playback works the same way. Both multi-byte operations – write and read – require only an initial address specification. After that, the RAM is pretty much autonomous. This speeds up data transfer since we no longer have to provide a command and a 16-bit address for each byte read or written.

26W.2.1 The serial access method

The serial RAM available in the Microchip device has some brains within it, brains that allow it to communicate despite its paucity of pins: just eight. Like other SPI peripherals (including even the simple SPI Digipot used in Lab 26L) this serial access part needs to be told what to do with values fed to it. It therefore understands a small collection of *commands*:

Instruction Name	Description
READ	Read data from memory starting from an address
WRITE	Write data to memory starting from an address
RDSR	Read STATUS register
WRSR	Write STATUS register

The STATUS register is a bit of a misnomer since this register is also used to *set* the operating mode, in addition to reading the current mode:

Mode	Description
BYTE OPERATION	Read or write one byte at a given address
PAGE OPERATION	Read or write sequentially within a 32 byte page starting at a given address
SEQUENTIAL OPERATION	Read or write sequentially starting at a given address

You must set the desired mode in the STATUS register before reading or writing to the memory.[6] Once the mode is set, you do not have to write to the status register again unless you want to use a different mode. In each of the three modes, you first write a command (READ or WRITE) followed by a 16-bit address.[7] You then either write a data byte if a write command or read a byte if a read. In the two sequential modes, you can continue to read or write bytes as long as you keep chip select asserted without having to send another command or address. Data is written to or read from sequential addresses until you terminate the command by dissasserting chip select (which is why we cannot use the SERCOM's automatic chip select since it always deselects $\overline{CS}$ after every eight bit transfer).

26W.2.1.1 An illustration: single-byte write and readback

Before sending data we must tell the SPI RAM how we want it to handle the data that we present. In Fig. 26W.1, the "status write" command, 0x01, tells the RAM "what's coming in the next byte will be the new mode." The byte that follows selects the mode: here, "single" byte read/write mode, 0x01. These two bytes are written as a single 16-bit value with $\overline{CS}$ asserted continuously.

Once we've told the RAM to take a single byte, we can issue a data write command (0x02), followed by the 16-bit address and data (a total of 32 bits). Figure 26W.2 shows the writing of a single byte to an arbitrary address, then its readback. The microcontroller writes the data value 0xAB into location 0x2445, then reads it back, as you can see in the figure.

The very first item in each frame – whose beginning is flagged by the assertion of $\overline{CS}$ – is a *command*, as you would expect: *write* (02h), followed by *read* (03h).

26W.2.1.2 An illustration: multi-byte write and readback

When the goal is to store data in a table and then to read it back, the RAM's ability to handle a long sequence of data bytes is just what we need. In Fig. 26W.3 the controller sends three bytes of data (AAh, BBh, CCh) to three addresses beginning at the start of RAM (address zero).

The successive writes occur quickly, thanks to the automatic incrementing of address. In contrast to the single-byte action of §26W.2.1.1, here there is no need to specify the successive addresses (or repeat the command), once the start address has been defined.

---

[6] But if you don't, the default is BYTE OPERATION.
[7] Only fifteen bits are required to address 32K but both commands require a full 16 bits with the MSB always set low.

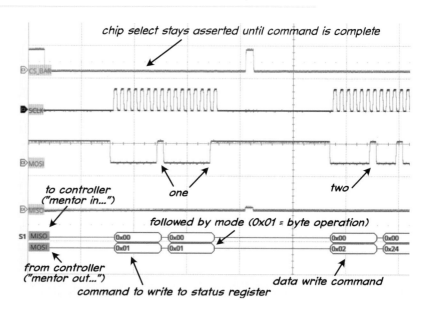

Figure 26W.1 First send the SPI RAM a command to tell it which of its "modes" to use. This only has to be done once unless you want to change modes.

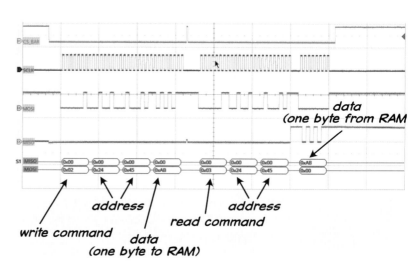

Figure 26W.2 SPI single byte write and readback using Microchip 32K serial RAM.

26W.2.2 Add the SPI RAM to your microcomputer

This is a more complicated SPI device to use than the display or digipot as you must read as well as write data to it. Rather than connect to the exiting SERCOM1 SPI bus we will use a second SERCOM peripheral and four other I/O pins for the RAM. This is necessary to allow the RAM to run faster than the 5 kHz LCD bus speed.[8]

Connect the Microchip 23A256 SPI RAM to the SparkFun SAMD21 as shown in Fig. 26W.4. We are using some of the pins used by the 4-by-4 keypad; however, as long as you do not press a key the keypad will not interfere with the RAM. You will need to remove the RAM chip when you want to use the keypad again. If you do not have a separate 3.3 V supply, power the RAM from the SAMD21Mini "vcc" header pin.

---

[8] We have a Goldilocks problem. The LCD display thinks the RAM clock is too fast and the RAM finds the display clock painfully slow for getting anything useful done.

26W.2 Adding more RAM

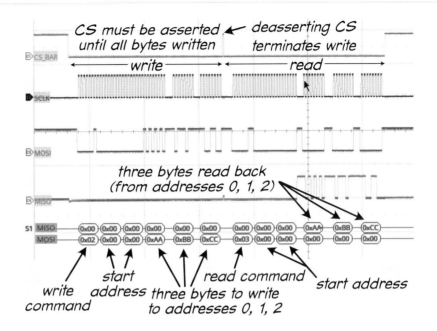

Figure 26W.3 SPI RAM takes and delivers three successive bytes, with only the start address specified.

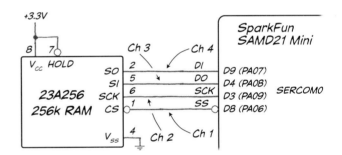

Figure 26W.4 Connect the SPI RAM to the microcontroller with scope channels as shown.

26W.2.3 Code to demonstrate the SPI RAM

To try out the SPI RAM, replace main() in the LCD display test program (either the polled version in §24L.5 or the interrupt version above in 26W.1.3) with the code below.[9] The pushbutton connected to Arduino pin 0/RX alternately switches the RAM between byte operation and sequential operation. (Hold the button down to send the mode switching command repeatedly to capture it on your scope.) Unlike the LCD display, the SPI RAM chip select must be handled manually using Port I/O so that we can keep it asserted during multibyte operations.

```
/****************************************************************
 *          SPI RAM Support Functions (using SERCOM0)
 ****************************************************************/

#define PORT_RAM_CS   PORT_PA06   // not mux'ed; handled by Port I/O
#define PORT_RAM_DI   PORT_PA07   // mux D
#define PORT_RAM_DO   PORT_PA08   // mux C
#define PORT_RAM_SCK  PORT_PA09   // mux C
```

---

[9] Complete code available at https://LAoE.link/micro5/26W_SPI_RAM_Demo.c.

Worked Examples: Interrupt-Driven Serial I/O and SPI Off-chip RAM

```c
void SPIRAM_Init(void){
    /*************** PORT I/O Initialization *************/
    /* Set up DO, SS* and SCK as outputs, DI as an input   */

    REG_PORT_DIRSET0 = PORT_RAM_DO | PORT_RAM_SCK | PORT_RAM_CS;  // set DO, SS* and SCK as outputs
    REG_PORT_DIRCLR0 = PORT_RAM_DI;                               // set DI as an input
    REG_PORT_OUTCLR0 = PORT_RAM_DO | PORT_RAM_SCK;                // start with MOSI and CLK low
    REG_PORT_OUTSET0 = PORT_RAM_CS;                               // and student select high (not selected)

    PORT->Group[0].PMUX[PIN_PA08 >> 1].reg = PORT_PMUX_PMUXE_C | PORT_PMUX_PMUXO_C; // set mux C on PA08,
    PORT->Group[0].PMUX[PIN_PA07 >> 1].reg |= PORT_PMUX_PMUXO_D;                    // PA09 data in is on mux D

    // enable multiplexer on pins controlled by SERCOM
    PORT->Group[0].PINCFG[PIN_PA07].reg = PORT_PINCFG_PMUXEN | PORT_PINCFG_INEN;
    PORT->Group[0].PINCFG[PIN_PA08].reg = PORT_PINCFG_PMUXEN;
    PORT->Group[0].PINCFG[PIN_PA09].reg = PORT_PINCFG_PMUXEN;

    /*************** POWER MANAGER Initialization **************/
    /* On reset, power to SERCOM0 is disabled, must be enabled    */
    /***********************************************************/

    PM->APBCMASK.reg |= PM_APBCMASK_SERCOM0;     //Enable SERCOM 0 w/ RMW

    /******** GENERIC CLOCK CONTROLLER Initialization *********/
    /* Connect Clock 3 (8MHz) to DAC and enable SERCOM0 clock */
    /**********************************************************/

    GCLK->CLKCTRL.reg = GCLK_CLKCTRL_ID(SERCOM0_GCLK_ID_CORE) | GCLK_CLKCTRL_CLKEN
                        | GCLK_CLKCTRL_GEN_GCLK3;
    while(GCLK->STATUS.bit.SYNCBUSY);        // requires write synchronization

    /*************** SERCOM0 Initialization *******************/
    /* Setup SERCOM for SPI Mentor, DOPO 0, DIPO 3 and 8-bit data */
    /**********************************************************/

    /* Perform a software reset and wait for it to complete */
    SERCOM0->SPI.CTRLA.bit.SWRST = 1;
    while(SERCOM0->SPI.SYNCBUSY.bit.SWRST || SERCOM0->SPI.CTRLA.bit.SWRST);

    /* Set SERCOM0 mode as SPI Mentor (host) and DOPO to PAD[0,1,2] for MOSI, SCK and SS */
    SERCOM0->SPI.CTRLA.reg = SERCOM_SPI_CTRLA_MODE_SPI_MASTER
            | SERCOM_SPI_CTRLA_DOPO(0x0) | SERCOM_SPI_CTRLA_DIPO(0x3);

    // set baud rate to 8MHz clock divided by 8
    SERCOM0->SPI.BAUD.reg = SERCOM_SPI_BAUD_BAUD(3);           //Set the SPI baud rate to 1MHz

    SERCOM0->SPI.CTRLA.reg |= SERCOM_SPI_CTRLA_ENABLE;          //Enable the SERCOM0 SPI
    while(SERCOM0->SPI.SYNCBUSY.bit.ENABLE | SERCOM0->SPI.SYNCBUSY.bit.CTRLB);  //Wait for enables to finish
}

/*******************************************/
/*          SPIRAM Output Functions        */
/*******************************************/

/*********** SPIRAM_SendData() **************
 * Send a fixed number of bytes out the SPI *
 * port. Does not return until all bytes    *
 * have been sent.  Returns number of bytes *
 * sent.                                    *
 * RAM CS is controlled by caller so we can *
 * use multi byte writes .                  *
 ********************************************/

int SPIRAM_SendData(uint8_t *data, int count){
  for (int i = 0; i < count; i++) {
    while(SERCOM0->SPI.INTFLAG.bit.DRE == 0);     // wait until Data Register Empty from last send
    SERCOM0->SPI.DATA.reg = data[i];              // and send next byte
  }
  while(SERCOM0->SPI.INTFLAG.bit.TXC == 0);       // wait for last byte to be sent
  return count;                                    // and return when done
}

int SPIRAM_SendByte(uint8_t data){
  while(SERCOM0->SPI.INTFLAG.bit.DRE == 0);       // function to send a single byte (for efficiency)
  SERCOM0->SPI.DATA.reg = data;
  while(SERCOM0->SPI.INTFLAG.bit.TXC == 0);
```

26W.2 Adding more RAM

```c
    return 1;
}

/*********** SPIRAM_GetByte() ****************
 * Receives a single byte from the SPI RAM.  *
 * RAM CS is controlled by caller so         *
 * can use multi byte reads.  We enable the  *
 * receiver before and disable it after to   *
 * avoid buffer overflow.  Setting CTRLB as  *
 * a reg is ok b/c we want all other bits 0. *
 ********************************************/
uint8_t SPIRAM_GetByte(void){
  uint8_t data;

  while(SERCOM0->SPI.INTFLAG.bit.TXC == 0);        // wait for last transmission to complete

  SERCOM0->SPI.CTRLB.reg = SERCOM_SPI_CTRLB_RXEN;  // enable receiver
  while(SERCOM0->SPI.SYNCBUSY.bit.CTRLB);

  SERCOM0->SPI.DATA.reg = 0;                       // send something to start recieve
  while(SERCOM0->SPI.INTFLAG.bit.RXC == 0);        // wait until Data Register receives something
  data = SERCOM0->SPI.DATA.reg;

  SERCOM0->SPI.CTRLB.reg = 0;                      // disable receiver (flushes buffer)

  return (uint8_t) data;
}

/******* manual CS functions ******/
void EnableRAMCS() {
  REG_PORT_OUTCLR0 = PORT_RAM_CS;
}

void DisableRAMCS() {
  REG_PORT_OUTSET0 = PORT_RAM_CS;
}

/*******************************************
 *  SPI RAM functions
 *  these use the generic read/write above
 *******************************************/
// defines for SPIRAM_SetMode "mode" argument

#define RAM_BYTE_MODE 0x00
#define RAM_PAGE_MODE 0x01
#define RAM_SEQ_MODE  0x02

int SPIRAM_SetMode(int mode){
  static uint8_t WRSRCmd[] = {0x01, 0x00};        // command to write RAM status register
  static uint8_t ModeTable[] = {0x01, 0x81, 0x41}; // mode commands see RAM datasheet Table 2-2

  if (mode < 0 || mode > RAM_SEQ_MODE ){
    return -1;
  } else {
    WRSRCmd[1] = ModeTable[mode];
    EnableRAMCS();
    SPIRAM_SendData(WRSRCmd, sizeof(WRSRCmd));
    DisableRAMCS();
    return 0;
  }
}

uint8_t SPIRAM_GetMode(void){
  static uint8_t WRSRCmd[] = {0x05};              // command to read RAM status register
  uint8_t status;

  EnableRAMCS();
  SPIRAM_SendData(WRSRCmd, sizeof(WRSRCmd));
  status = SPIRAM_GetByte();
  DisableRAMCS();
  return status;
}

// The following two function DO NOT assert chip select
// in order to support sequential mode.  You must assert
// CS before calling either function then dissassert it
```

```c
// when all bytes have been read/written.
int SPIRAM_WriteByte(uint16_t address, uint8_t data){
  static uint8_t WRByte[] = {0x02, 0x00, 0x00, 0x00};     // command to write byte to RAM

  if (address > 0x7fff){   // only 32k bytes in RAM
    return -1;
  } else {

    WRByte[1] = (uint8_t) (address >> 8);
    WRByte[2] = (uint8_t) (address & 0xff);;
    WRByte[3] = data;

    SPIRAM_SendData(WRByte, sizeof(WRByte));
    return 0;
  }
}

uint8_t SPIRAM_ReadByte(uint16_t address){
  static uint8_t RDByte[] = {0x03, 0x00, 0x00};      // command to read byte to RAM
  uint8_t data;

  if (address > 0x7fff){   // only 32k bytes in RAM
    return 0;              // no error code
  } else {

    RDByte[1] = (uint8_t) (address >> 8);
    RDByte[2] = (uint8_t) (address & 0xff);;

    SPIRAM_SendData(RDByte, sizeof(RDByte));  // send command and address
    data = SPIRAM_GetByte();
    return data;
  }
}

/*****************************
 *    pushbutton functions    *
 *****************************/

#define PORT_PB_P0   PORT_PA11  // active low pushbutton connected to Sparkfun pin 0/RX

void PORT_Init(void){
   // set up pushbutton pin as input with an internal pullup resistor
   REG_PORT_DIRCLR0 = PORT_PB_P0;         // set button pin to input
   PORT->Group[0].PINCFG[PIN_PA11].reg = PORT_PINCFG_INEN | PORT_PINCFG_PULLEN; // enable button input w/ pull
   REG_PORT_OUTSET0 = PORT_PB_P0;         // set pull voltage for button input pin hig
}

bool Is_Btn_pressed(void) {
  return ((REG_PORT_IN0 & PORT_PB_P0) == 0);   // active low
}

/*****************************
 *            main()          *
 *****************************/
// FSM defines

#define IN_BYTE_MODE    0
#define SET_SEQ_MODE    1
#define IN_SEQ_MODE     2
#define SET_BYTE_MODE   3

int main (void)
{
  int state = SET_BYTE_MODE;         // start in setting byte mode
  char lcdbuf[20];                   // buffer for strings to send to LCD display
  uint8_t databuf[10];               // buffer for data sent/received from RAM

  // system initializations
  ClockSysInit48M();                 // set CPU clock to 48MHz
  SysTick_Config(SystemCoreClock/1000); // configure SysTick for Delay() timer
  SPIRAM_Init();
  SPI_Init(true);            // Display uses automatic chip select
  PORT_Init();
  Delay(1000);                       // but does not respond to commands immediately after power up

// Use SERCOM1 SPI to send data to the LCD display
```

```c
    LCD_Clear();
    LCD_SetBacklight(100, 100, 100);        // start with brightest white backlight
    LCD_SetContrast(CONTRAST);

    while (1) {
      switch (state) {
        case IN_BYTE_MODE:                  // demonstrate single byte R/W
          EnableRAMCS();
          SPIRAM_WriteByte(0x2445, 0xab);
          DisableRAMCS();

          EnableRAMCS();
          SPIRAM_ReadByte(0x2445);
          DisableRAMCS();

          if (Is_Btn_pressed()) {
            LCD_Clear();
            SPI_PrintData("Setting Seq ModeRelease for next");
            state = SET_SEQ_MODE;
          }
          break;

        case SET_SEQ_MODE:                  // set sequential mode
          SPIRAM_SetMode(RAM_SEQ_MODE);
          SPIRAM_GetMode();

          if (!Is_Btn_pressed()) {
            LCD_Clear();
            SPI_PrintData("SPI RAM Seq R/W Prs btn for next");
            state = IN_SEQ_MODE;
          }
          break;

        case IN_SEQ_MODE:                   // demonstrate sequential mode
          EnableRAMCS();
          SPIRAM_WriteByte(0x0000, 0xaa);
          SPIRAM_SendByte(0xbb);
          SPIRAM_SendByte(0xcc);
          DisableRAMCS();

          EnableRAMCS();
          SPIRAM_ReadByte(0x0000);
          SPIRAM_GetByte();
          SPIRAM_GetByte();
          DisableRAMCS();

          if (Is_Btn_pressed()) {
            LCD_Clear();
            SPI_PrintData("Settng Byte ModeRelease for next");
            state = SET_BYTE_MODE;
          }
          break;

        case SET_BYTE_MODE:                 // set byte mode
          SPIRAM_SetMode(RAM_BYTE_MODE);
          SPIRAM_GetMode();

          if (!Is_Btn_pressed()) {
            LCD_Clear();
            SPI_PrintData("SPI RAM Byte R/WPrs btn for next");
            state = IN_BYTE_MODE;
          }
          break;

        default:
          state = SET_BYTE_MODE;
          break;
      }
      Delay(200);                           // state cycle
    }
    return 0;
}
/*************************** End of file ****************************/
```

26W.3 Make a voice recorder

We can use the extra memory in the Microchip SPI RAM to build a simple voice memo recorder. Impressively, this application uses all the peripherals we have discussed so far: the ADC to record voice from a microphone, the DAC to output the recorded samples to the speaker amplifier, the timer/counter interrupt to sample the ADC voice signal and output the samples to the DAC at a constant rate, SERCOM1 to output user messages to the LCD display and SERCOM0 to send and receive data to/from the SPI RAM: see Fig. 26W.5.

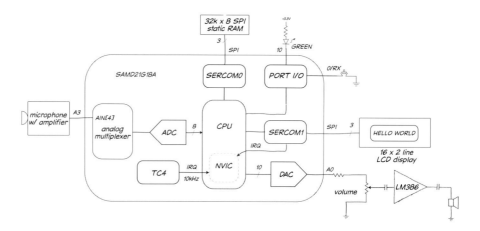

Figure 26W.5 Voice recorder block diagram.

The memo recorder uses the pushbutton connected to Arduino pin 0/RX for user input, the LED connected to Arduino pin 10 for status and the LCD display for instructional messages. Pressing and holding the pushbutton for one second lights the LED and begins recording. The systems records as long as you hold the pushbutton down or until it runs out of memory, which ever occurs first. When recording ends, the LED goes off. The ADC samples the input at 10 kHz, so the 32k × 8 SPI RAM provides about 3.3 seconds of recording time.

Once the RAM contains a recording, the LED blinks at a one-second rate. Pressing and releasing the pushbutton in less than one second starts playback of the recording. Pressing the pushbutton during playback deletes the current recording.[10]

26W.3.1 Build the hardware

The only hardware you need to add to your build (assuming you already installed the SPI RAM in Fig. 26W.4) is an amplified microphone: see Fig. 26W.6.[11] All the 3.3 V devices can be powered from the "vcc" header pin on the SAMD21 Mini if you do not have a 3.3 V supply available.

(We are cheating a bit in our analog system to keep the build simple. In general you should use a better antialiasing filter on the ADC input and a good (e.g., MAX294) reconstruction filter on the DAC output. Nevertheless, the single pole, 2.8 kHz lowpass on the input and the dreadful high-frequency response of the breadboard speaker on the output are adequate to provide an intelligible voice recorder, albeit not high-fidelity. However, feel free to remedy these deficiencies if you wish.)

---

[10] See demonstration video at https://LAoE.link/26W_Voice_Recorder_demo.html.
[11] Any amplified microphone with a zero to 3.3 V max output will do fine. We are using the Adafruit 1062 but you could also use the SparkFun offering at https://LAoE.link/SparkFun_Microphone_Breakout.html. Or you could pull out the microphone amplifier that you built back in Lab §7L.5, if you modify it to work on 3.3 V.

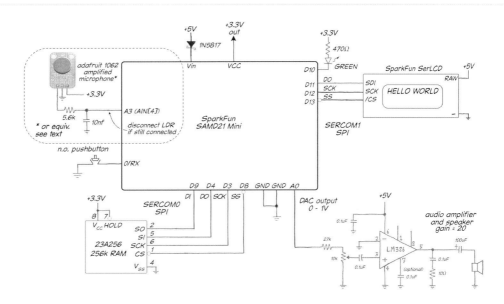

Figure 26W.6 Add a microphone to use the SPI RAM as a voice recorder.

26W.3.2 Download and run the voice recorder code

The code for the voice recorder is too long to print here. Download it from https://LAoE.link/micro5/26W_Voice_Recorder.c.

Except for the TC4 interrupt handler and the new application code, most of it consists of identical or slightly modified code from previous labs. There is no change to the SysTick timer, DAC, LCD SPI (the interrupt version from §26W.1.3 is included but the polled version should work) and the SPI RAM code from §26W.4. The TC4 initialization code is identical to Lab 25L.3 TC4_Init() except for the removal of the code that outputs the timer/counter square wave on Arduino pin A1.[12] The ADC code is similar to the LDR acquisition code from §26L.8.3 with a few changes: ADC_Init() sets the ADC for an 8-bit conversion to match the RAM width, the sample is the average of four conversions to reduce noise, the clock is increased to 1 MHz to minimize the acquisition time, and ReadADC() no longer takes a multiplexer selection as an argument, it always samples the microphone input on AIN[4].

26W.3.2.1 The new TC4 interrupt handler

The big changes are to the TC4 interrupt service routine, TC4_IRQHandler(), which does all the data acquisition heavy lifting, and to main(), which communicates with the ISR to implement the voice recorder. The TC4 ISR uses three global variables to communicate with the FSM; a counter sampctr() which counts the number of samples written to the SPI RAM during recording or read during playback, a counter sampcnt() which is set to the number of samples stored in the RAM when recording ends, and a flag used by the foreground program to tell the ISR to abort recording.

The TC4 ISR is always called at a 10 kHz rate. When not recording or playing back, sampctr() is zero and the ISR just clears the interrupt and returns, taking about 800 ns to execute: see Fig. 26W.7.[13]

The application FSM starts recording or playback by initiating sequential RAM access in the foreground, sending a read or write command to the RAM along with the starting address (always 0x0000, the beginning of RAM). It then sets the sampctr() to 1, which is the flag for the ISR to record or playback. If sampcnt() is zero, there is no saved data and the ISR starts recording, incrementing

---

[12] We will be using pin A1 to measure the ISR execution time.
[13] To measure interrupt execution time, we had the ISR set Arduino pin A1 high on entry and clear it just before it returns. That is the signal shown on channel 4 of the scope trace.

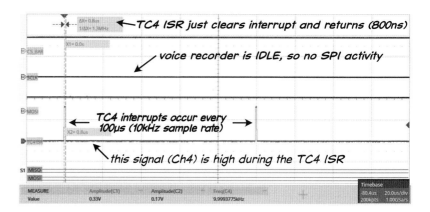

Figure 26W.7 TC4 interrupt service execution time while voice recorder is idle.

sampctr() with each sample. If sampcnt() is non-zero, the ISR plays back that number of bytes, using sampctr() to keep track of the number of bytes it has sent to the DAC.

If the stoprec flag is set by the foreground program during recording, the ISR stops recording and sets sampcnt() equal to the number of bytes saved (the value in sampctr()). It does the same thing if it runs out of memory. When done recording or playing back, the ISR sets sampctr() to zero, which is a flag to the foreground program that the ISR is idle.

It is important to realize that communication between the foreground and background is tricky because the foreground can be interrupted at any time. It is easy for subtle bugs to creep in that only occur rarely when the interrupt happens at the wrong time. Here, the foreground program should be careful modifying more than one global variable at a time, since there is the possibility that it could be interrupted between operations if two were required. Also, since sampctr() is the action flag to the ISR, it should always be set last, after the other two variables are modified, to avoid the ISR responding to a foreground request while the variables are in a partially initialized state.

```
/******* TC4 Interrupt Service Routine for DAC out *******
 * these two variables tell ISR what to do
 * foreground program (FSM) sets:
 *   | sampctr     | sampcnt    | on interrupt do
 *   ---------------------------------------------------
 *   | 0           | don't care | do nothing (idle)
 *   | > 0         | 0          | record: read ADC, write byte to RAM (stop on 32767 or flag)
 *   | > 0         | > 0        | playback # of sampcnts
 ***********************************************************/

volatile bool stoprec = false; // flag to ISR to stop recording when true
volatile int  sampctr = 0;     // -1 means record, 0 means off, 1 to 32767 is playing
volatile int  sampcnt = 0;     // number of stored samples in RAM

void TC4_IRQHandler (void) {
uint8_t samp;

  if (sampctr > 0 && sampcnt > 0) {      // play back sampcnt-1 samples (FSM read first byte)
    samp = SPIRAM_GetByte();
    REG_DAC_DATA = (samp << 2);          // convert 8-bit sample to 10-bit DAC value
    sampctr++;                           // increment count of samples output to DAC
    if (sampctr >= sampcnt) {            // see if done (foreground sets sampcnt to 0 to erase)
      sampctr = 0;                       // turn off playback
      REG_DAC_DATA = 0x80;
      DisableRAMCS();                    // end sequential RAM read
    }
  } else if (sampctr > 0 && sampcnt == 0) { // record (FSM wrote first byte to RAM)
    samp = ReadADC();
    SPIRAM_SendByte(samp);               // save voice sample
    sampctr++;                           // increment count of samples saved to RAM
    if (sampctr == 32767 || stoprec) {   // if done (out of memory or foreground flag)
      sampcnt = sampctr;                 // save length of recording
```

```
        sampctr = 0;                          // turn off recording
        DisableRAMCS();                       // end sequential RAM write
        Set_LED(OFF);
      }
    }
    TC4->COUNT16.INTFLAG.reg = TC_INTFLAG_MC0;  // clear the interrupt
}
```

This ISR is unusual in that is does a lot, more than the interrupt handlers we have previously seen. We must make sure that the ISR completes in the time available, $100\,\mu s$ for the 10 kHz sampling rate. Figures 26W.8 and 26W.9 show interrupt handler during playback and record respectively. Playback requires reading a byte from the RAM (taking a bit over $2\,\mu s$) and writing it to the DAC output register.[14] This is a fairly quick operation, taking about 6.4 μs. Data acquisition takes longer. The ISR during record takes about 32 μs. (It took 20 μs when only two samples were averaged and 56 μs to average eight so each sample takes about about 6 μs plus an additional 8 μs or so of overhead.) This still leaves plenty of time for the LCD SPI interrupt service handler and the foreground program to execute.

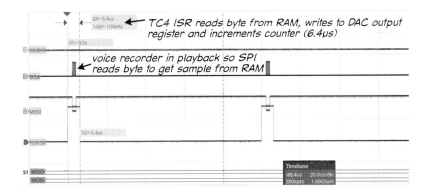

Figure 26W.8 TC4 interrupt service execution time during playback.

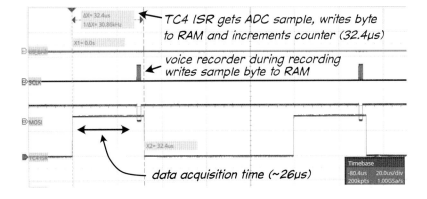

Figure 26W.9 TC4 interrupt service execution time while recording.

---

[14] This is why we had to run the RAM SPI on a different SERCOM than the LCD. The LCD SPI would have taken longer than the 10 μs available at a 10 kHz sample rate.

26W.3.2.2 The foreground application program

The application program that implements the voice recorder is written as an FSM. It runs fairly slowly (10 Hz state cycle) since all the time-sensitive sampling is done by the TC4 ISR. One hundred milliseconds is fast enough to not seem unresponsive to the user while also ensuring that switch bounce is not a problem. A number of the states are transitory to put information on the LCD display. If we tried to do this in a multicycle state, like IDLE, the display would flicker.[15]

A local variable, `statectr`, allows timing using state cycles. Note that times are `#define()`'ed as milliseconds divided by the FSM cycle time. This allows the cycle time to be changed without having to change the other constants.

The voice recorder is a good example of a single-threaded, moderately complicated application. It is helped by having much of the work done in the timer/counter interrupt handler, allowing the timing requirements of the foreground application to be fairly lax. Nonetheless, as the application becomes more complex, maintaining a single-threaded application bug-free becomes more difficult. In the next chapter, we will look at a solution – a Real Time Operating System that allows applications to be multi-threaded. This will allow us to parse the application into a number of separate tasks that intercommunicate to coordinate the desired functionality, rather than forcing us to pack all the functions of the application into a single main loop.

26W.4 Solutions

- §26W.1: `https://LAoE.link/micro5/26W_SPI_Interrupt_Driver_Complete.c`.

---

[15] In the worked example of §25W.1.6 we output to the display in the transition from one state to another to ensure we only displayed the information once. You can also set a flag to indicate the LCD has been updated to avoid flashing. All these techniques are fine and you should choose the one that makes the most sense to you.

27N Microcontrollers VI: Using an RTOS

Contents

27N.1	Real-time operating systems		1049
27N.2	Segger embOS		1050
	27N.2.1	Introduction and features	1050
	27N.2.2	Multitasking operating systems	1052
	27N.2.3	Anatomy of an embOS program	1054
	27N.2.4	The hardware independent application program	1054
	27N.2.5	The hardware specific board support package (BSP.c)	1056
	27N.2.6	Resource usage	1058
27N.3	Readings		1059

27N.1 Real-time operating systems

All the programs we have created so far follow the Arduino model of a set of initialization functions that execute once (akin to Arduino `setup()`) followed by a `while(1)` loop that executes forever (like the Arduino `loop()` function). All the application code, the input/output functions and the user interface, each with its own timing requirements, must be shoehorned into a single execution "superloop."[1] As programs grow and become more complicated, maintaining the Arduino single-threaded structure becomes more difficult and less efficient.

AoE §15.9.2

One solution to the problem of building reliable and maintainable complex embedded system applications (in addition to writing them as finite state machines) is to use a Real Time Operating System. An RTOS is an operating system intended for real-time applications in which multiple tasks or processes can run with well-defined timing requirements. "Real Time" means the operating system provides a fast, deterministic response to hardware events. Unlike a user-oriented operating system like Windows or macOS, an RTOS typically does not include user interface and storage I/O functions, rather it concentrates on making real-time embedded systems easier to create and operate efficiently. The application programmer creates separate tasks (functions) for each operation and assigns a priority to them in the RTOS. The RTOS schedules each task according to its priority. Higher priority tasks run before lower priority ones. Tasks with the same priority run "round robin," each getting a fixed slice of time to run before it is suspended and the next task of equal priority given a chance to execute.

Many, if not most, tasks will have a point in their programming where they need to delay for a fixed amount of time or wait for I/O or access to a resource such as a display or serial device. An RTOS provides efficient APIs for delays, input, and resource access, and also allows tasks to communicate with one another. Rather than sitting in a loop wasting CPU cycles as we did with much of our code

---

[1] However, as the Voice Memo Recorder in the §26W.3 worked example demonstrated, it may be possible to use interrupt service to move some of the more stringent timing issues out of the main program loop.

while waiting for a flag to clear or a count to reach a specified value, the RTOS APIs assign CPU time to some other task that is ready to run while making the waiting task dormant.

You could use global variables for intertask communication, as we did with interrupt handlers, but that would still require a task to poll the variable to see if the information or resource was available. Instead, a task signals the operating system when it has nothing to do until necessary information becomes available or an event occurs. The RTOS then blocks the task and it does not run until whatever resource it is waiting for becomes available. Since even high priority tasks normally do not need 100% of the CPU time, this allows lower priority tasks to run and gives the appearance that all tasks are running simultaneously. Unlike the simple do-nothing `Delay()` routine you created in the second microcontroller lab using the SysTick timer, the RTOS delay function makes the caller dormant during the delay time and allows other tasks to run. This makes for much more efficient use of the microcontroller.

Another major benefit of using an RTOS is it makes complex applications much simpler to write and easier to debug. Since each task operates as a stand-alone function, interactions between functions are limited to the OS communications mechanisms provided.[2] Like the FSM, this makes it easier to isolate problems when a system does not work as intended. Even if you have only a single task, an RTOS can provide useful resources including timers with microsecond precision, idle sleep-mode support for power saving, and easy expansibility into a multitasking application.

27N.2 Segger embOS

There are real-time operating systems available for almost every microcomputer architecture, some are open source while others are proprietary.[3] We will be using *embOS* from Segger, a real-time, priority-controlled multitasking system. Although proprietary, it is available free of charge for evaluation and non-commercial use. Segger's embOS is lightweight (i.e., uses only a small amount of the microcontroller's program and data memory) and easy to use, with a rich set of APIs and clear documentation. It also integrates nicely with CMSIS, so you will find that it takes little effort to integrate the code you have already written into the RTOS. That will allow you to spend most of your lab time creating the application code to implement your project's functionality.

27N.2.1 Introduction and features

While we will not be able to cover all the features of embOS, we will use a representative selection of its APIs to demonstrate typical RTOS operations. In particular, in Chapter 27L you will be using:

- Tasks,
- Software Timers,
- Mutexes,
- Mailboxes, and
- Task Events,

to build a jukebox that plays children's lullabies.[4] Figure 27N.1 is the system block diagram for the jukebox. The shaded blocks take advantage of embOS features; the SPI Driver and the Timer Interrupt

---

[2] This removes a common source of errors and/or crashes by eliminating global variables, troublesome because they can be modified from anywhere in the code and are difficult to police.

[3] See https://LAoE.link/RTOS_List.html.

[4] While we are discussing specific embOS features here, these resources are common to most real-time operating systems, although possibly with different names. See, e.g., https://LAoE.link/CMSIS-RTOS2.html.

and DAC blocks are just copies of the code from the earlier labs, unchanged, that interface to the onboard peripherals.

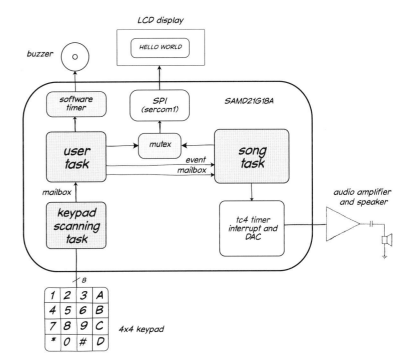

Figure 27N.1 Lullaby jukebox block diagram.

Tasks: A Task is a function that runs forever in a `while` loop.[5] Tasks that share memory are called *threads*.[6] Tasks can be in one of four states: *not existing* before they are created or after they terminate, *waiting* for an event to move them to the ready state, *ready* to run, or *running*. Only one task can be running at a time. The operating system maintains the state of each task in a Task Control Block (TCB) structure and schedules a waiting task to run when it is ready and no higher priority task is ready to run. It time-slices a ready task with other tasks of the same priority that are also in the ready state. From the task's point of view, it executes as though it owned the CPU. You normally design tasks to be simple and to handle one particular aspect of the system you are building. In the jukebox, we will create three tasks: 1) a user interface and jukebox control task; 2) a keypad-scanning task; and 3) a song-playing task. Each of the `main()` loops we have written in previous labs can be turned into a task with little or no changes.

Software Timers: In many cases, creating a task is overkill if all you need to do is to complete a simple operation after a specific amount of time has expired. Software Timers operate like a one-shot monostable, allowing you to execute a function call after a set time delay. In the jukebox, we use software timers to create the keypad feedback beep when a key is pressed. Upon receiving a keypress notification, the user-control task turns on the buzzer and starts a software timer. When the programmed

---

[5] A task can choose to terminate itself if it is no longer needed; however, all the jukebox tasks run forever (or until you turn off the power).

[6] *Processes* are similar to threads but have their own memory address space. This requires a microcontroller with a memory management unit, something lacking on the Cortex-M0+.

time delay ends, the timer calls a function to turn off the buzzer.[7] Software timers use a mere 20 bytes of R/W memory compared to over 800 bytes for a task and its associated stack.

Mutexes: Mutexes allow multiple tasks to share a non-interruptible resource. In the jukebox, both the user task and the play task each need to update separate lines of the LCD display. However, the display can only be used by a single task at a time. This is because the information sent to the LCD consists of combined commands and data that must be sent as an atomic unit. If the RTOS switches to another task while a lower priority task is in the middle of sending command/data to the LCD and the new task then writes to the LCD, the result will likely corrupt the information displayed since the new task's SPI data would be misinterpreted as part of the earlier task's command/data string.

A mutex (short for "mutual exclusion object") is a token associated with a non-sharable resource that is claimed by a task to make sure the operating systems stops any other task (even a higher priority task) from accessing that resource until the first task has finished using it. Once the task is done with the resource, it must release the token so other tasks can use the resource. While one task has claimed a Mutex on a resource, any other task that needs the resource will be blocked and cannot run. This means a task should not hold onto a Mutex any longer than it needs to.

Mailboxes: Mailboxes allow tasks to send short, fixed-length messages to another task.[8] While mailboxes can be used for multi-character messages, the jukebox only needs single character messaging. The keypad-scanning task sends a single byte to the user-control task mailbox each time a new key is pressed, and the user task sends single byte song numbers to the play task mailbox when the user selects a song and presses the enter key. The embOS mailbox system has special, efficient one-byte versions of its `put` and `get` APIs that we will take advantage of in the jukebox intertask communication. One useful feature of mailbox messaging is that the RTOS queues messages to a task in a FIFO so the play task can fetch each song to be played sequentially.

Task Events: Finally, Task Events allow tasks to send Boolean flags to another task. This is convenient to convey true/false state such as if pushbutton is pressed or a signal has changed state. In the jukebox, we need a way to allow the user to abort playing all songs in the queue by pressing a key. You cannot just put a stop command in the play mailbox, because the play task would not see it until it had finished playing all the songs in the mailbox queue. In addition, the play task only checks the song mailbox after playing a complete song. Even if the sender clears the mailbox first, the play task would not check for the stop message until the end of the song. That means you need an out-of-song-mailbox channel that the play task can check frequently during song play to stop playing songs. Since stop is a binary event, we can can use a Task Event flag as a convenient way to signal the play task to stop playing. Up to 32 Task Events are available on the Cortex-M0+ and require no additional resources, instead using a word allocated in the Task Control Block for each task.

27N.2.2 Multitasking operating systems

Multitasking operating systems give the appearance that multiple tasks (functions) are executing simultaneously. Each task behaves as though it has exclusive use of the CPU and, unless it has need to, has no knowledge of any other task running concurrently. The operating system manages this magic by allocating CPU cycles among the running tasks. In embOS, each task has its own stack. To switch

---

[7] In the jukebox, the actual `Beep()` function coding is a bit more complicated to allow it to produce one or three beeps passed to the function as an argument.

[8] Although we won't need them for the jukebox, embOS provides *Queues* to send longer, variable length intertask messages.

tasks, embOS pushes the CPU registers of the task being suspended to that task's stack, then stores the value of the stack pointer in that task's TCB. To wake the next task, it copies the saved stack pointer of the task being activated from its TCB to the SP register and then loads the CPU registers from the values stored on its stack when it was last suspended. Tasks switching can be either *cooperative* or *preemptive*; embOS supports both types of task switching.

Cooperative task switching: In cooperative switching, a task gives up use of the CPU when it has nothing to do, for example when it is waiting for input or needs to delay for a fixed time. Many embOS APIs come in two flavors, ones that always return immediately and ones that do not return until some data is available ("blocked" APIs).[9] For example, in the mailbox messaging system a task can request the next queued message with the non-blocked char OS_MAILBOX_Get() API call. This function returns zero if there is a message (the message itself is returned in a separate memory buffer) or non-zero if there is no message. If the calling function has other processing to do, even if no message is available, this API is appropriate.

However, in a well designed RTOS application the calling function may have a single purpose – to handle messages in the mailbox. In a non-RTOS environment, the function would keep calling the Get function until a message is available, wasting CPU cycles. In embOS, it can call the blocked void OS_MAILBOX_GetBlocked() API. This call does not return until a message is ready (hence the void return value). If no message is available, embOS suspends the calling task and wakes up the next ready task of the same or lower priority. (A task with higher priority would have triggered a preemptive task switch as soon as it became *ready*.)

The embOS system has Blocked versions of most API calls that return a discretionary value. If you would normally put the API call in a loop that does nothing until the call returns a value, you should always use the Blocked version to cooperatively give up the CPU to another task. Some APIs, such as OS_TASK_Delay(), are always blocked so the function name does not explicitly include the word.

Preemptive task switching: In a preemptive task switch, the OS suspends the active task due to an external event. This can occur if a higher priority task becomes ready to run; for example, if its blocked API call can return. A task may also be preemptively suspended if there is a task of equal priority and the time slice allocated to the running task in the round-robin scheduler expires. This allows all tasks of equal priority to get an opportunity to execute, even if they do not cooperatively yield the CPU.

Interrupts: As long as an ISR does not need to use any OS APIs, interrupts in embOS work the same as they do without an RTOS, unaware that the RTOS even exists. Indeed, we will use unchanged the timer/counter interrupt handler that updates the DAC with the next sine sample in the jukebox running under embOS. Segger calls this a "zero latency interrupt" because the OS adds no additional delay to the normal 15 cycle Cortex-M0+ interrupt latency. Interrupts in the top half of Cortex-M0+ priority levels are considered zero latency interrupts.

If an ISR requires access to RTOS resources, for example a mailbox, you will need to use "embOS interrupts." This entails calling the ISR entry function OS_INT_Enter() when you enter the ISR and the OS_INT_Leave() function when you exit. Interrupts in the bottom half of the priority levels (CMSIS levels 2 and 3) are considered embOS interrupts.

In either case the usual rules for interrupt service routines apply. ISRs must preserve all registers (this is handled by the compiler when writing in C) and interrupt handlers must be short and return

---

[9] There are also some "Timed" API calls that are blocked but return after a maximum delay time even if the event has not occurred.

quickly. That means you should never call a blocked API function or a delay function from interrupt service.[10]

27N.2.3 Anatomy of an embOS program

An embOS RTOS program consists of [at least] two source files: a hardware independent application and a hardware board support package, BSP.c.[11] The board support package connects the application code to a particular piece of hardware. The advantage of partitioning the application this way is that it makes porting your application to different hardware relatively easy, you only need to change BSP.c if you change CPUs or external I/O connections.[12]

A typical Board Support Package for simple port I/O contains descriptive defines for I/O pin names, port initialization code and functions to toggle port output pins and read input pins. Any hardware dependent function intended to be called externally by the application program has a "BSP_" prefix and is prototyped in the BSP.h file.

27N.2.4 The hardware independent application program

Let's take a look at the Segger sample application (header removed for brevity):

```
/*
File    : OS_StartLEDBlink.c
Purpose : embOS sample program running two simple tasks, each toggling
          an LED of the target hardware (as configured in BSP.c).
*/

#include "RTOS.h"
#include "BSP.h"

static OS_STACKPTR int StackHP[128], StackLP[128];  // Task stacks
static OS_TASK         TCBHP, TCBLP;                // Task control blocks

static void HPTask(void) {
  while (1) {
    BSP_ToggleLED(0);
    OS_TASK_Delay(50);
  }
}

static void LPTask(void) {
  while (1) {
    BSP_ToggleLED(1);
    OS_TASK_Delay(200);
  }
}

/**********************************************************************
*
*       main()
*/
int main(void) {
  OS_Init();      // Initialize embOS
  OS_InitHW();    // Initialize required hardware
  BSP_Init();     // Initialize LED ports
  OS_TASK_CREATE(&TCBHP, "HP Task", 100, HPTask, StackHP);
  OS_TASK_CREATE(&TCBLP, "LP Task",  50, LPTask, StackLP);
  OS_Start();     // Start embOS
  return 0;
}
```

---
[10] The embOS user guide includes columns for each API indicating if a particular API can be called from interrupt service.
[11] There are a number of other source code programs that are included in a newly created RTOS project, but these two are the ones that you must supply to create your application.
[12] The Segger sample RTOS application includes board support for over 150 ARM evaluation boards from more than a dozen different manufacturers, all of which use the same OS_StartLEDBlink.c application program but each with its own custom board support package.

The two `#include` files define constants, structures, and function prototypes for their corresponding source files.[13] `RTOS.h` defines constants, structures and prototypes for the embOS APIs while `BSP.h` provides the same information for the hardware interface functions.[14]

Tasks, like any function, require a stack to store local variables and temporary values. In addition, when embOS switches tasks it must save the CPU registers of the task being deactivated so it can restore them when the task is reactivated. Each task has its own stack. When a task is created, it must provide a pointer to an area of memory to be used for its stack. The size required is dependent on how heavily the task function uses the stack, but Segger recommends at least 512 bytes to start.[15] Debug builds of embOS applications call a function `OS_Error()` in file `OS_Error.c` if the stack overflows. This function enters an endless loop, but you can place a breakpoint in it to view the error code if it is ever called.

The sample application code

```
static OS_STACKPTR int StackHP[128], StackLP[128];
```

defines two 512 byte arrays to be used as the stacks for the sample tasks.

The next line defines the task control block structures. Most embOS features, including tasks, mailboxes, mutexes, etc., require a control structure used by the OS to keep track of the identity and status of the item. A pointer to this structure is passed as an argument to embOS when the feature is created. The sample program creates two tasks, so it needs to define two `OS_TASK` structures. While the line

```
static OS_TASK TCBHP, TCBLP;
```

allocates memory for these structures, the TCBs are not initialized until the tasks are created.

The two functions following the TCB declaration implement the two sample tasks. In this example, the two tasks run independently, with no knowledge of the other (i.e., there is no communication between them). Each task toggles an LED and then calls the embOS API function to delay for a set period of time to blink the LED; a 50 ms delay (10 Hz blink rate) for the task that will be assigned the higher priority (`HPTask`) and a 200 ms delay (2.5 Hz blink rate) for the low priority task (`LPTask`). Task functions normally never return, although they can request termination through an API.

Functions prefixed with `BSP_` indicate hardware functions in the board support package. This is merely a convention, but like the `OS_` convention, it makes it easier to understand what each function does and where to find it. When a task calls the `OS_TASK_Delay()` function, it is deactivated and receives no CPU time until the delay has expired. Once the delay time has elapsed, embOS sets the task state to waiting and returns to it from the delay API call when it has priority or a time slice. Since each task uses only a few cycles to toggle the LED and call the delay function, the vast majority of the CPU cycles in this example end up given to a default embOS `OS_Idle()` function. On the Cortex-M0+, this function calls the `__WFI` instruction to enter sleep mode until some hardware event occurs (in this case a timer interrupt).

The remainder of the application program, the `main()` function, initializes the system, and then starts the RTOS. The first two lines are required for embOS to run:

`OS_Init()` initializes the embOS kernel and must be called before any other API function is called.

---

[13] A C function prototype tells the compiler about a function's inputs and outputs so that you can call the function in your code before it is defined. A function prototype is just a function definition with the body replaced by a semicolon. They are usually collected in a header file so that you can `#include` them to use the functions in another source code file. We have avoided function prototypes to this point by always creating our functions before we use them. That is no longer possible now that we are splitting the application between two files.

[14] While you need a different board support package source file (`BSP.c`) for each board, only a single board support package header file (`BSP.h`) is required since you write the same interface functions for all target hardware.

[15] See §5.1 in "CPU & Compiler specifics for Cortex-M using Embedded Studio," Segger Document UM1063 at https://LAoE.link/embOS-MPU.pdf.

1056   Microcontrollers VI: Using an RTOS

OS_InitHW() initializes the hardware timer used by the OS to time slice tasks, create accurate delays and maintain an elapsed time variable. embOS uses the SysTick timer for this, so you should never attempt to use or modify it once the RTOS has been started.

The call to the user-written BSP_Init() function does any necessary hardware initialization specific to the microcontroller and hardware it is connected to. Here, it only needs to set the I/O pins connected to the LEDs as outputs and turn the LEDs off to start. In the jukebox, we will call a series of initialization functions to set up the TC4 counter to interrupt when the programmed count is reached, configure the DAC for 0 to 1 V output, initialize the keypad pins, four as outputs and four as inputs with pulldowns, set the buzzer pin as an output, and configure SERCOM1 for SPI output to the display. Most of these initializations will be either identical to or only slightly modified from what we did in the earlier labs (although we will rename them to meet the BSP_ convention).

The next two lines create the two sample tasks. The task creation function is passed a pointer to the allocated task control block, a name for the task (used by the Segger embOSView debugger), the task priority (higher values indicate higher priority, the opposite of the NVIC), a pointer to the task function, and a pointer to the task stack.[16]

On a 32-bit processor like the Cortex-M0+ the priority parameter can be any value between one and $2^{32} - 1$. The actual value is unimportant, only the relative values count. In this example, assigning the HPTask task a priority of 2 and the LPTask task a priority of 1 would have the same effect as the values 100 and 50.

In more complicated programs you would create any other resources you need at this point in main(), such as mailboxes, software timers, mutexes, etc., with the appropriate embOS API calls as well.

Once the initializations are complete, the call to OS_Start() starts embOS, which begins activating tasks, highest priority first and time slicing same priority tasks when it is their turn to run. This call never returns.

27N.2.5   The hardware specific board support package (BSP.c)

This is the SparkFun SAMD21 Mini Board Support Package for the Segger sample application (some comments removed for brevity):

```
/*********************************************************************
File    : BSP.c
Purpose : BSP for SparkFun SAMD21 Mini Breakout board.
*/

#include "BSP.h"

#define PORTA_BASE_ADDR    (0x41004400u)
#define PORTA_DIRSET       (*(volatile unsigned int*)(PORTA_BASE_ADDR + 0x08u))
#define PORTA_OUTCLR       (*(volatile unsigned int*)(PORTA_BASE_ADDR + 0x14u))
#define PORTA_OUTSET       (*(volatile unsigned int*)(PORTA_BASE_ADDR + 0x18u))
#define PORTA_OUTTGL       (*(volatile unsigned int*)(PORTA_BASE_ADDR + 0x1Cu))

#define PORTB_BASE_ADDR    (0x41004480u)
#define PORTB_DIRSET       (*(volatile unsigned int*)(PORTB_BASE_ADDR + 0x08u))
#define PORTB_OUTCLR       (*(volatile unsigned int*)(PORTB_BASE_ADDR + 0x14u))
#define PORTB_OUTSET       (*(volatile unsigned int*)(PORTB_BASE_ADDR + 0x18u))
#define PORTB_OUTTGL       (*(volatile unsigned int*)(PORTB_BASE_ADDR + 0x1Cu))

#define LED0_BIT_MASK      (1u << 27)  // Green LED0 connected to PA27, low active
#define LED1_BIT_MASK      (1u <<  3)  // Green LED1 connected to PB03, low active

/*********************************************************************
```

---

[16] An API function in all caps such as this one indicates the function is a macro, i.e., that is there is an actual API function OS_TASK_Create() that takes more parameters and that this one uses default values for the missing values. For example, OS_TASK_CREATE() determines the size of the task stack using the C SIZEOF operator and uses a TimeSlice parameter value of 2 ms. The full API call allows these values to be specified separately.

27N.2 Segger embOS

```c
*       BSP_Init()
*/
void BSP_Init(void) {
  PORTA_DIRSET = LED0_BIT_MASK;
  PORTA_OUTSET = LED0_BIT_MASK;
  PORTB_DIRSET = LED1_BIT_MASK;
  PORTB_OUTSET = LED1_BIT_MASK;
}
/**********************************************************************
*       BSP_SetLED()
*/
void BSP_SetLED(int Index) {
  if (Index == 0) {
    PORTA_OUTCLR = LED0_BIT_MASK;
  } else if (Index == 1) {
    PORTB_OUTCLR = LED1_BIT_MASK;
  }
}
/**********************************************************************
*       BSP_ClrLED()
*/
void BSP_ClrLED(int Index) {
  if (Index == 0) {
    PORTA_OUTSET = LED0_BIT_MASK;
  } else if (Index == 1) {
    PORTB_OUTSET = LED1_BIT_MASK;
  }
}
/**********************************************************************
*       BSP_ToggleLED()
*/
void BSP_ToggleLED(int Index) {
  if (Index == 0) {
    PORTA_OUTTGL = LED0_BIT_MASK;
  } else if (Index == 1) {
    PORTB_OUTTGL = LED1_BIT_MASK;
  }
}
```

This should look familiar, it is quite similar to the code we used in the first microcontroller lab before we added CMSIS to our projects. The sample board support package is using the yellow (the code comment is incorrect) TX_LED LED on pin PB03 and the green TX_LED LED on pin PA27 to demonstrate embOS's multitasking. The initialization function first defines memory-mapped addresses for the Port I/O registers, then sets the pins connected to the LEDs as outputs and initializes them high (active low LEDs off). The three hardware interface functions that follow use the CLR, SET and TGL registers to turn the selected LED on, off, and toggle its state, respectively.

There is only one sample application program (`OS_StartLEDBlink.c`) but a different board support program `BSP.c` for each target board.[17]

For comparison, here is the `BSP_ToggleLED()` function from the `BSP.c` for an evaluation board using the Texas Instruments MSP432 Cortex-M4 microcontroller:

```c
/*
File    : BSP.c
Purpose : BSP for TI MP432 Launch Pad
*/

#define LED0_BIT    (0)
#define LED1_BIT    (1)
#define LED2_BIT    (2)

#define P1OUT       (*(volatile unsigned char*)(0x40004C02u))
#define P2OUT       (*(volatile unsigned char*)(0x40004C03u))
```

---

[17] You need a different board support package not only for each different microcontroller, but also for each board since the LEDs are external to the microcontroller and may be connected to different I/O pins on each one.

```
void BSP_ToggleLED(int Index) {
  if (Index == 0) {
    P1OUT ^= (1u << LED0_BIT);  // Toggle P1.0 (LED1)
  } else if (Index == 1) {
    P2OUT ^= (1u << LED0_BIT);  // Toggle P2.0 (RGBLED_RED)
  } else if (Index == 2) {
    P2OUT ^= (1u << LED1_BIT);  // Toggle P2.1 (RGBLED_GREEN)
  } else if (Index == 3) {
    P2OUT ^= (1u << LED2_BIT);  // Toggle P2.2 (RGBLED_BLUE)
  }
}
```

This code does not use an atomic toggle operation; rather the toggle is accomplished with an XOR of the I/O pin connected to the selected LED. Since the hardware differences are isolated in the BSP.c interface file, the Segger sample program works with no change to the RTOS application code.

27N.2.6 Resource usage

Like most real-time operating systems, embOS uses very little RAM and ROM. The embOS kernel uses ≈160 bytes of the former and ≈2100 bytes flash memory. Tasks, Software Timers, and Mutexes use 308 bytes (plus stack space), 20 bytes, and 16 bytes of RAM respectively per resource. Mailboxes require 24 bytes plus the RAM for the FIFO buffer, whose size is set by the programmer and is dependent on the system requirements. Task events require no additional memory (a word in the task control block is devoted to the event flags).

Since embOS provides timer and memory management APIs as well as automatic low-power support while the CPU is idle, it may be resource-positive to use it even with some single task (i.e., superloop) applications. Even if slightly resource-negative, the benefits of splitting application operations into separate tasks and the ability to set task priorities make writing applications in the RTOS worthwhile.

The Segger two-task embOS sample application above uses 3 KB of flash and 3.2 KB of RAM in a Release build.[18] For comparison, the same program written as a superloop using the SysTick timer for the delay uses 720 bytes of flash and 2 KB of RAM, also in a Release build.[19] However, changing the LED blink rate in the embOS version is easy, you just set the delay for each task to one-half the period of the desired blink frequency. Changing the LED blink rate in one task has no effect on the other. For the superloop, you must calculate the least common denominator of the two delays and check to see when the slower LED loop count MOD denominator is reached to toggle that LED:

```
/* embOS sample program rewritten as a superloop  */
/* uses same same BSP.c as embOS version          */

#include "samd21.h"              // add to use CMSIS
#include "BSP.h"                 // MemoryInit and INTERWORK commented out

volatile uint32_t msTicks = 0;   // number of millisecond since program start (rolls over in 49 days)

void SysTick_Handler(void) {     // SysTick interrupt Handler - interrupts every millisecond
  msTicks++;
}

void Delay(uint32_t mSec){
  uint32_t EndCnt;

  EndCnt = msTicks + mSec;
  while (msTicks < EndCnt) {
    __WFE();                     // tells CPU it can power down until interrupt ocurs
  }
}
```

---

[18] By default, we have been using the Debug build configuration, which uses some additional RAM and ROM to add debugging and error checking to your executable. Once you have a properly working program, you would normally build a Release version, which would use fewer resources and run slightly faster.

[19] Since this is a very simple application, the RTOS resource overhead is large compared to the application code. In an actual application, the opposite is likely to be true.

```
}
/******************/
int main(void) {
  int i = 0;

  SysTick_Config(SystemCoreClock/1000);   // sets up 1000 interrupts per second
  BSP_Init();                             // initialize LED pins as outputs

  while (1){

    BSP_ToggleLED(0);          // toggle 50ms LED

    if ((i % 0x4) == 0) {      // toggle 200ms LED every four loops
      BSP_ToggleLED(1);
    }

    Delay(50);                 // common denominator: one-half period of faster LED
    i++;
  }
}
```

In the worst case, there will be no common denominator and you will have to check the MOD result for both LEDs every millisecond. In addition, while not relevant in this example, you cannot set separate priorities for the two LED blink tasks in the superloop. Finally, adding another operation to the superloop can be painful and error prone. In the lab you will add several more tasks to the Segger sample program. As you do, consider how you would do the same thing in the superloop version above.

27N.3 Readings

AoE Chapter 15 (Micocontrollers) – design examples:
embOS User Guide & Reference Manual at https://LAoE.link/embOS_manual.pdf.
embOS Specifics for Cortex-M at https://LAoE.link/embOS-MPU.pdf.

27L Lab: Microcontrollers VI

In the previous microcontroller labs you controlled the I/O ports directly to scan a matrix keyboard, programmed a timer to interrupt at regular intervals to output a sine wave at a specific frequency, and initiated SPI communications with a LCD display to show text messages. In this lab you are going to integrate these elements into a jukebox that plays children's lullabies using an RTOS.

27L.1 Complete the hardware build

This lab will use the external peripherals from the previous five microcontroller labs. However, you need to add one more item to your breadboard, a keypad audio feedback buzzer: see Fig. 27L.1.

Figure 27L.1 A small 5 V buzzer used to provide keypad feedback.

Figure 27L.2 is the complete schematic for this lab's hardware.[1] The only external devices that require 3.3 V are the green LED and the optional IR receiver (used in the worked example). If you have only a 5 V supply available, you can use the 3.3 V output on the SAMD21 Mini marked "vcc" to power these devices.

Add the buzzer and MOSFET driver to your breadboard as shown.[2] The buzzer is polarized, so be careful to connect the terminal marked "+" to 5 V. The MOSFET is necessary both because the buzzer requires 5 V and because its 35 mA current consumption is much higher than the I/O pins of the SAMD21 microcontroller can sink. The 33k resistor ensures that the buzzer is off when the I/O pin is high impedance (floating).[3]

Pin assignments (both Arduino and SAMD21) for the build are shown in Fig. 27L.3.

27L.2 Install embOS and updated sample code

If you have not yet done so, follow the directions in Chapter 27S to install and test embOS in Segger Embedded Studio.

---

[1] The schematic includes the optional IR receiver used in the Chapter 27W worked example.
[2] Any small signal N-Channel MOSFET with $V_{GS(th)} \leq 2.6$ V, such as the VN10, should work.
[3] I/O pins are configured as inputs after a microcontroller reset. Without the pulldown resistor you would have to set the AREF pin as an output or enable an internal pulldown in every program to avoid the possibility of the buzzer screeching at you continually.

27L.2 Install embOS and updated sample code 1061

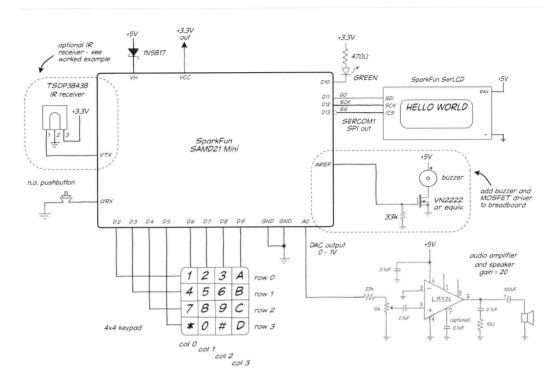

Figure 27L.2 Hardware schematic for Lab 27L (Arduino pin names shown).

Figure 27L.3 SAMD21 pin use for the final microcontroller lab.

Once you have the basic embOS environment working, modify the sample BSP.h, BSP.c and OS_StartLEDBlink.c files with updated versions as follows (Fig. 27L.4).

Add the following includes, defines, and function prototypes to BSP.h.[4] Insert the following header includes after "#define BSP_H":

```
#ifndef bool
  #include "stdbool.h"    // to use boolean data type
#endif

#ifndef uint32_t
```

[4] The updated BSP.h is available at https://LAoE.link/micro6/BSP.h.

1062   Lab: Microcontrollers VI

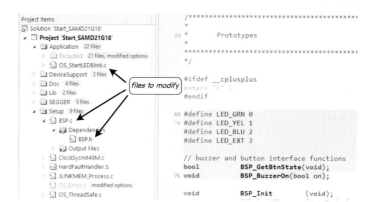

Figure 27L.4 embOS files to modify before starting lab.

```
    #include "stdint.h"    // to use std data types
#endif
```

Insert the following definitions and prototypes before "`void BSP_Init (void);`"

```
// index of three onboard LEDs and added external LED on Arduino Port 10 for LED functions
#define LED_GRN 0
#define LED_YEL 1
#define LED_BLU 2
#define LED_EXT 3

// buzzer and button interface functions
bool        BSP_GetBtnState(void);
void        BSP_BuzzerOn(bool on);
```

Replace `BSP.c` with the file: `https://LAoE.link/micro6/BSP.c`. This version uses CMSIS in place of the Segger direct register defines, adds support for the on-board blue user LED, the external LED on Arduino pin D10, and adds functions to get the state of the pushbutton on pin 0/RX and turn on and off the buzzer you just added.

Finally, replace the two tasks in `OS_StartLEDBlink.c` with the following functions:[5]

```
static void HPTask(void) {
  while (1) {
    BSP_ToggleLED(LED_GRN);
    OS_TASK_Delay(200);
  }
}
static void LPTask(void) {
  while (1) {
    BSP_ToggleLED(LED_YEL);
    OS_TASK_Delay(500);
  }
}
```

The only changes from the Segger sample code tasks are the use of self-documenting defines in place of LED index numbers, and slowing down both blink rates (yellow to 1 Hz, green to 2.5 Hz) to make the timing easier to verify visually.

27L.3   Get familiar with the RTOS

Open the sample embOS application by clicking on the "Start_SAMD21G18.emProject" file in the 📁 `Start ▸ BoardSupport ▸ Microchip ▸ SAMD21G18_SAMD21_Mini_Breakout` directory.

---
[5] The full updated file is available at `https://LAoE.link/micro6/27L_OS_StartLEDBlink.c`.

This will open the Segger IDE and load the demonstration project. If it is not already shown, click on `OS_StartLEDBlink.c` in the Application folder to display the file in the editor.

Compile and run the demonstration program. Each task blinks its respective LED as though it owns the CPU. In this example, there is no need for either task to communicate with the other. This shows how easy it is to port existing code to the RTOS. You just move the `main()` loop of your standalone program into a new task function. You should, however, isolate hardware and CPU dependent code into functions in `BSP.c`. Existing code can still use peripherals and interrupts without change, as long as no other task needs to access the same hardware.

Experiment with task priority: The two sample tasks are created with different priorities; the "H[igh]P[riority]Task" with a priority of 100 and the low priority task with a priority of 50. For tasks like these, where the task consists of a few dozen instructions at most followed by hundreds of milliseconds not requiring any CPU time while waiting for a delay to expire, it really does not matter that one task (the low priority one) would not run if the high priority task never yielded for a delay or while waiting for a resource or input. However, you can verify that the higher priority task gets first dibs on CPU time by setting breakpoints at the first instruction of each task (on the calls to `BSP_Toggle_LED()`) and restarting the program. The breakpoint for the high priority task will always occur first and the low priority breakpoint will occur afterward.

Try commenting out the call to `OS_TASK_Delay()` in the high priority task and replacing it with a software delay:

```
for (int i=0; i < 750000; i++);
```

Rebuild the project. Again, set breakpoints on the LED toggle instruction in each task. Now the low priority task never gets called since the high priority task never gives up control of the CPU. Disable the breakpoint in the high priority task to verify this. The green LED will blink at about 2.5 Hz, but the yellow LED stays off and the `LPTask()` breakpoint never is reached. This is because the high priority task is never yielding execution by calling `OS_TASK_Delay()` or any other blocked API function. This illustrates why it is important to use the embOS APIs and not software loops for delays if you want lower priority tasks to run.

However, if both tasks have the *same* priority, embOS will use preemptive task switching to run both tasks round-robin even though the HPTask never gives up the CPU. Temporarily change the HPTask priority to 50 in the call to `OS_TASK_CREATE()` to match the priority of the LPTask and run the program again. Now both tasks execute as expected. The software timing loop on the HPTask will be off by a bit, but not by much since the LPTask only runs for a few CPU cycles before it returns control to the OS with the call to `OS_TASK_Delay()`.

Add another task: Restore the demo code to the original version, using `OS_TASK_Delay()` in both tasks and different priorities for each. Add a new task, `BTNTask()`, and allocate a stack for it. Modify `main()` to start this new task.[6] You can choose any priority since the other two tasks use very little CPU time. Code this new task function to toggle the external LED (LED_EXT in `BSP.h`) when you press the pushbutton, following the comments below:

```
static void BTNTask(void) {
  while (1) {
    // if button not pressed, wait OS_TASK_Delay(<delay in ms>)
    // else button pressed - toggle external LED
    //    and then wait for button to be released calling OS_TASK_Delay()
  }
}
```

---

[6] Be careful. A common error is copying and modifying an existing `OS_TASK_CREATE()` call in `main()` but forgetting to change to the new task control block or the new stack. With two tasks trying to use the same memory, a crash is guaranteed.

Be sure to call the BSP functions `BSP_GetBtnState()` and `BSP_ToggleLED()` to read the pushbutton state and toggle the external LED, respectively, to keep the application itself hardware agnostic. Remember also that the BSP functions are active high. Set the delay time long enough to debounce the pushbutton and short enough that the user does not experience a lag between pressing the button and seeing the LED toggle; 100 ms is a reasonable value. You do not have to include any port initialization code as it is already included in the call to `BSP_Init()`.[7]

Run your new application and make sure it works as you expect. All three tasks should execute as though each had full ownership of the microcontroller. While you could have accomplished this in a single application loop, it would have required you to adjust the time delay to the shortest common denominator (the pushbutton debounce delay) and count multiple delays to get the correct LED blink delays. If you later decided to change either of the blink rates or the debounce time, you would have had to recompute and change your counters. Alternatively, you could program multiple Timer/Counters and toggle the LEDs in separate interrupt service routines, but hopefully you agree that the effort would be much greater than letting the RTOS do the work for you.

With an RTOS, you can change the blink rate of one LED independently of the timing of the other tasks. Go ahead and change the blink rate of the low priority LED so it is no longer a multiple of the high priority LED's rate. Try a toggle delay of 333 [ms] to give a ≈ 1.5 Hz blink rate. With the RTOS, this requires changing a single value in one task. A superloop would require much more extensive changes with a correspondingly higher chance of introducing bugs in your code.

Add a software timer: It would be nice to have some audio feedback on the button press (in case you have a visually impaired user). Let's beep the buzzer for BEEPTIME milliseconds when the button is pressed. You could do this in the button task, but if BEEPTIME is not the same as the debounce time you have to manage two separate delays. You could also create a new Task to beep the buzzer, but that would be overkill for such a simple function; particularly since a task requires additional memory for its stack and the task control block.[8] An easy way to accomplish a beep is to turn the buzzer on in the button task, then start a timer to turn it off when the delay elapses. This is what a Software Timer does.

Create a Software Timer object by adding

`static OS_TIMER Timer_BeepDly;`

to the top of `main.c` just after declaring the Task Control Blocks. Add a define for BEEPTIME as well so you can adjust it to a pleasing length. Fifty milliseconds (`#define BEEPTIME 50`) is a good starting value.

Next, write the callback function:

`static void BuzzerOff(void)`

Place it before the button task (`TaskBTN()`) so you don't have to create a prototype. This function should just use the `BSP_BuzzerOn()` function to turn off the buzzer.

Now create the software timer in your `main()` function by calling the timer create macro before the call to `OS_Start()`:

`OS_TIMER_CREATE(&Timer_BeepDly, BuzzerOff, BEEPTIME);`

The first argument is a pointer to the software timer structure, while the second is the callback function. The third argument is the timer delay in milliseconds.[9] The macro both creates the timer and starts it. (You must start a Software Timer once before you can restart it. We will *restart* it in

---

[7] You can also reuse your FSM loop from the second microcontroller lab as long as you use the BSP functions and change the state delay clock to call `OS_TASK_Delay()`.

[8] A Task requires approximately 308 bytes of RAM for the TCB plus 512 bytes for the stack; a Software Timer requires 20 bytes.

[9] While you won't need to do it in this lab, embOS allows you to change the delay time after you have created the timer if necessary.

`TaskBTN()` when the button is pressed, so we need to start it once before that task executes.) Since the buzzer is already off, calling the callback function to turn it off here has no effect.

To provide the button pressed feedback tone, modify your button task to turn on the buzzer using the `BSP_BuzzerOn()` function when the button is initially pressed and then restart the timer with:

```
OS_TIMER_Restart(&Timer_BeepDly);
```

After BEEPTIME milliseconds, the timer will call `BuzzerOff()` which turns off the buzzer to end the beep tone. The timer will then turn itself off and stay dormant until it is restarted on the next button press.

Go ahead and add this to your button toggle program and see how it works. Feel free to experiment with the beep delay to find a length that you like.

What we need here is a way to communicate:[10] So far, each of the tasks we have created has been independent, with no connection to one another. In most RTOS applications you will want to isolate functionality in individual tasks, even those that need to communicate with each other, to achieve the desired application performance. As an exercise, let's separate the button sensing task from the external LED toggling and put the latter in its own task called `TGLTask`. But once you do, you will need a way for the button task to tell the toggle task it should toggle.

The embOS system provides three ways (in order of increasing functionality and complexity) for a task to send a message to another task: Task Events, Mailboxes and Queues. Since the button task only needs to tell the toggle task to toggle the LED, all you need here is the simple Boolean messaging provided by Task Events. Task Events are easy to implement, require no memory, and involve only a few API function calls. Like the other messaging APIs, Task Events provide blocked versions so a task can remain dormant while waiting for an event to occur.

To create a Task Event, you normally first declare the event object:[11]

```
static OS_TASKEVENT BtnEvent;
```

On ARM processors, embOS Task Events provide the ability to signal that up to 32 different events (one per bit in an `int`) have occurred. You need to choose a bit position in the `int` to signal the pushbutton has been pressed and then give it a name. Here you only have a single event so you might as well use bit 0:

```
#define EVENT_BTN (1u << 0)    // event bit for pushbutton pressed
```

To signal to the toggle task that the event has occurred, the sending task (`TaskBTN()`) should call:

```
OS_TASKEVENT_Set(&TCBTGL, EVENT_BTN);    // tell toggle task button pressed
```

The first parameter is a pointer to the task control block of the task to receive the event, and the second is the bit mask of the event(s) bits to set. You can set multiple events in a single `Set` call if you want by OR'ing the event bits to set in the second parameter.

The receiving task has to choose between two types of API calls to get the event flags: blocking or non-blocking. The non-blocking versions return the bit flags immediately, even if none of the event bits are set, i.e., none of the events have occurred since the event flags were last cleared. The blocking calls do not return until one or more event flags in the event mask passed in the `Get` call (discussed below) are set. If the calling function would normally do nothing but poll the event flags waiting until one of them is set, the blocking function allows other tasks to use the CPU cycles that the receiving task would have wasted polling for the event. Tasks should always take advantage of embOS Blocked

---

[10] With apologies to *Cool Hand Luke*. See https://LAoE.link/Comm_Failure.html.
[11] However, the event bits themselves are already included in the task's header when the task is created. The `OS_TASKEVENT` object is just an `int` used to save the return value from an event `Get` API call to see which event bits are set. If only a single event is possible or you do not care which event caused a blocked call to return, you do not need to create or use this object.

API calls to allow the operating system to efficiently allocate processing time to other tasks that are ready to run if they can do nothing until the call returns with whatever the task is waiting for.

In this example, the toggle task has nothing to do until the button task tells it that the button has been pressed. Therefore, it should use the

```
BtnEvent = OS_TASKEVENT_GetBlocked(EVENT_BTN);
```

API call to see if it needs to toggle the LED. The event mask argument to the function is the bitwise OR of the bits flags to wait for, and the blocked function will always return with at least one of the bits in the argument set. This means you only have to check the return value if the argument had more than one bit set and you need to know which event occurred. If you do, you can AND the return value with a bit mask to see which event bits are set. If the task is only waiting on a single event bit or does not care which event occured, it can just respond to whatever it was waiting for. In this problem, when you return from the blocking get call, you know the button has been pressed and you can just toggle the LED and then wait for the next event without examining the return value.[12]

Modify your program so the button-checking task and the external LED-toggling task are separate (i.e., create a new TGLTask) and add a Task Event to allow the button-sensing task to signal the toggling task to toggle. Finally, add a blocking get event call to the toggling task to respond to the button event. You can have either the button task or the toggle task beep the buzzer (but to my mind it is the button task's responsibility). Remember to create and start your new task in main().[13]

After this modification, the program should operate exactly as it did previously when the button-sensing and LED-toggling commands were combined in a single task. However, you now have two independent tasks, one responsible for monitoring the pushbutton (and beeping the buzzer) and one for toggling the LED, talking to each other to coordinate the application's functionality. The toggling task is almost always dormant, waking up only for a few instruction cycles to change the state of the LED when the button pressed event OS_TASKEVENT_GetBlocked() API call returns.[14]

27L.4   Build a lullaby jukebox

Let's use embOS to build a real application, a jukebox that plays children's lullabies. We will use the keypad for input, the LCD display for user feedback and status and the DAC to output the tunes: see Fig. 27L.5.

The jukebox stores up to ten songs in the frequency/length format we used in Lab 25L to test the timer/DAC interrupt system. A user can select a song by pressing one of the numeric keys on the keypad. The [shortened] name of the song is shown on the top line of the display after a "Select: " prompt (note space after colon). To play a selected song, the user presses the enter key ('#'), which places the song in the play queue. As in a real jukebox, a user can select and enter songs, even while a song is playing, and they will play sequentially in the order entered. While a song is playing its name is shown on the bottom line of the display as "Playing " (again, a space is included at the end of the string), and the display backlight cycles through various colors like an old fashioned Wurlitzer: see

---

[12] One "gotcha" to watch out for is that the blocked version of the API get call clears all events after it returns the bits of the event that occurred, even ones not specified in the event mask argument. The non-blocking call does not. If you call the non-blocking OS_TASKEVENT_Get() API and it returns with any bits set, those bits will continue to be set until you call a blocked version or clear the events yourself. There is also a blocking OS_TASKEVENT_GetSingleBlocked() API which operates similarly to the OS_TASKEVENT_GetBlocked() call but only clears the bits that were specified in the event mask.

[13] Since there is only one event, you do not need to create an event object or save the Get call return value unless you want practice for a future application.

[14] A solution is available at https://LAoE.link/micro6/27L_Task_Event.c. However, we *strongly* suggest you attempt a solution on your own. You cannot learn to program without doing it yourself, and debugging code is an essential skill.

27L.4 Build a lullaby jukebox

Figure 27L.5 The complete Lullaby Jukebox.

https://LAoE.link/Wurlitzer.html. The '*' key stops the music and empties the queue. User feedback is provided by a buzzer that beeps once when a valid key is entered and three times when an unused or invalid key is pressed.[15] See https://LAoE.link/Jukebox_Demo.html for a video demonstration of the jukebox.

You could try to create this application by scanning the keyboard, handling user input, and sending notes to the DAC all in one large Arduino-style superloop. The difficulty with this approach is the differing time requirements of these program elements. The keyboard scanner needs to operate slowly enough to debounce the keypad while the song loop needs to output notes at precise intervals to keep similar notes (i.e., all quarter notes) the same length. This may be possible (but not necessarily easy) in a simple application like this, but as the complexity of the system grows, it becomes more difficult to manage multiple tasks, each with its own timing requirement. Instead we will create separate tasks for each of these operations in embOS, making each task responsible only for its own timing issues. Intertask Mailboxes and Event Tasks will allow the tasks to communicate with one another to implement the jukebox. See the system block diagram in Fig. 27N.1.

27L.4.1 Create the jukebox project

You will modify the sample embOS application to create the Jukebox project. Make the following changes to the sample application you used in to familiarize yourself with the RTOS in §27L.3:

- Replace BSP.h in directory ▭Start▸Inc
  with https://LAoE.link/jukebox/BSP.h.
- Copy https://LAoE.link/jukebox/songs.h to directory ▭Start▸Inc.
- Replace BSP.c in directory
  ▭Start▸BoardSupport▸Microchip▸SAMD21G18_SAMD21_Mini_Breakout▸Setup
  with https://LAoE.link/jukebox/BSP.c.
- Copy https://LAoE.link/jukebox/OS_StartJukebox.c to directory
  ▭Start▸BoardSupport▸Microchip▸SAMD21G18_SAMD21_Mini_Breakout▸Application.

These new versions of BSP.h and BSP.c support the peripherals needed for the jukebox including the DAC, Timer/Counter, and SERCOM for SPI output to the LCD. In each case the code is copied

---

[15] In the basic design, the letter keys 'A'–'D' are not used. Some suggested enhancements are included at the end of the lab to take advantage of these extra keys.

from the earlier labs with only a few changes other than renaming the function with a "BSP_" prefix.[16] The embOS system supports interrupt code that has no knowledge it is running under an RTOS, so we do not need to make any changes to the timer and SPI interrupt service routines. Indeed, if the interrupt priority is set to one of the two highest Cortex-M0+ priorities, the ISR will have priority over the RTOS scheduler.[17]

There is a BSP function to read a keypad column:

`int BSP_ReadKeypadCol(int row);`

You pass this function the row to read (0 to 3) and the return value is −1 if no key is pressed in that row, or the column number (0 to 3) of the column with the key pressed. You will use this function to write the keypad-scanning task.[18]

The include file `songs.h` contains the data structures for the songs available on the jukebox. If you want to add a song, create a new `Song` structure and add a pointer to it in the `Songs` array in this file. See online Chapter 25O for an example.

`OS_StartJukebox.c` is a skeleton program for the Jukebox application. Much of the program has been written for you, but you will be asked to complete the keypad-scanning task and the user interface and program control task.

Once you have copied these files to your development environment, rebuild the last version of the sample application you tested in §27L.3 (the LED button toggle with Task Event) and make sure it still works with the updated version of `BSP.c`.

Now replace the sample application with the jukebox skeleton program. In Segger Embedded Studio, right-click on `OS_StartLEDBlink.c` and select "Remove" to remove the sample application as `main()`. Then right-click on the "Application" folder and select "Add Existing File." Select `OS_StartJukebox.c` to add it to the project and make it the new `main()` application. Build and run the skeleton code. You should see a welcome message on the display after which the jukebox plays a short intro tune.

27L.4.2 Let's take a look at the jukebox application code

`OS_StartJukebox.c` begins with the usual header `#includes`, then defines several textual prompts. In a professional application you would put all the user text in a separate header file to make it easy to create versions in multiple languages. You could then use a compile-time define to select the language version to create.

27L.4.2.1 Create the RTOS data structures

Following the prompt definitions, we create the data structures for the embOS features we are using, Tasks, Mailboxes, and Mutexes, and create stacks for the three tasks. We are also using one Task Event, but it only requires declaring a single word `PlayEvents` to save the event flags returned by the `OS_TASKEVENT` API call. We then define the bit EVENT_PLAYSTOP we use to signal the Play task to abort playing the current song. Finally, we create the data structures for the two Software Timers used to create the keypad beep tones and define the length in milliseconds of the beep ON time (SHTBEEP) and the delay between multiple beeps (BEEPDLY) used by the timers.

---

[16] These changes include setting the initial compare value for TC4 to 0xffff to start with the speaker quiet. The SERCOM1 interrupt no longer turns on the external LED when active so you can use the LED for other purposes (see the optional enhancements described in §27L.4.4).
[17] See https://LAoE.link/embOS_interrupts.html.
[18] BSP.c also contains several interface functions used for the remote-control addition described in the worked example of Chapter 27W. These are unused in this lab and you can ignore any compiler warnings that suggest declaring them `static`.

27L.4.2.2 Share the LCD display safely

Next, we deal with resource sharing in a multitasking environment. Multitasking operating systems use Mutexes and Semaphores to handle resource contention, allocation, and synchronization. Semaphores are the more complicated mechanisms, supporting resource synchronization. Mutexes merely keep one task from accessing a shared resource until another task releases it. As explained in the notes, output to the LCD should not be interrupted to avoid display corruption. That means that any task sending a command/data text string to the display must be allowed to send the complete character string to the SPI FIFO before any other task can do the same. In this application, the User task uses the top line of the display while the Play task uses the bottom line. Since the Play task has higher priority than the User task, it could interrupt User SPI output when it starts playing a new tune. We use a Mutex wrapper around the BSP LCD functions to make sure each task sends its complete command/data string to the display before the other is granted access.

Using a mutex to protect the LCD display from corruption is very easy. The Play and User tasks need to use two of the BSP LCD function calls. Rather than call them directly, we have created two new function calls that first lock the display with a OS_MUTEX_LockBlocked() API call, call the BSP function to store the LCD command/data string in the SPI output FIFO, and then call OS_MUTEX_Unlock() to release the display to the other task. Because the lock call is a blocked function, a higher priority task calling it to get access to the display will be suspended if there is already lower priority task holding the mutex. When the lower priority task releases it, the higher priority task is restarted. Figure 27L.6 shows an example of a higher priority task attempting to access the display when a lower priority task has not completed its output to the LCD. Since the Play task has higher priority than the User task, it will interrupt the User task, even if the User task has locked the display. However if the Play task tries to output to the display, it will be suspended and the User task allowed to finish sending its string to the FIFO before the Play task is reactivated.[19]

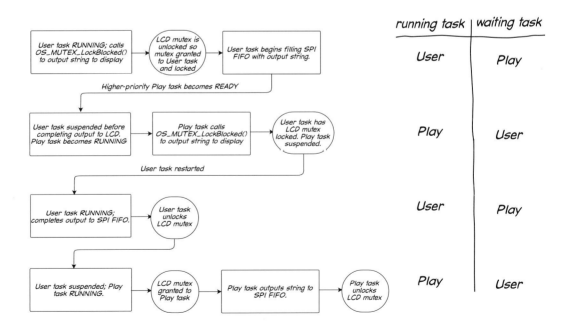

Figure 27L.6 LCD mutex ensures output to display is not corrupted.

---

[19] Because the Play task has higher priority, it will never be interrupted by the User task whether it it owns the mutex or not.

27L.4.2.3 The play task

The Play task code has been provided for you. This task fetches song indexes to play from the play mailbox and plays them using the timer/counter and DAC. The infinite task loop turns off the audio output (`BSP_SetNote(QUIET)`), sets the backlight back to bright white, and puts the "Stopped" status message on the bottom line of the LCD while waiting to play a song. It then calls the blocked mailbox get API call. If there is nothing to play, the task is suspended until there is.

When a song index is returned from the mailbox, the Play task first makes sure the index is valid (less than the constant NUMSONGS) and plays the corresponding song if it is. To play a song, it puts the name of the song on the lower line of the LCD after the "Playing " status message, resets the stop Task Event flag, and enters a play loop almost identical to the `PlaySong()` function used to play a tune in Lab §25L.2.3.5. The important differences are we use `OS_TASK_Delay()` for the note duration and rest delays to allow lower priority tasks to run and we check the stop event flag between notes to see if we should abort playing. If the stop flag is set, we stop playing and go back to waiting for a song to play. (The Play queue mailbox is cleared by the User task when it sets the abort event flag.) This is an example of a case where you must call the non-blocking version of an embOS API since, if the event flag is not set, we need to return immediately to play the next note. This task also calls a helper function to change the LCD backlight color after each note to simulate a Wurlitzer jukebox's colored lights.

27L.4.2.4 The keypad-scanning task

This one you must write. Since you have to separate the hardware Port I/O access from the key scan logic, you will have to rewrite your §23L.3 keyscan code slightly. When you detect a key is pressed you should still translate it into the ASCII character corresponding to that shown on the keypad ('1', '2', 'A', etc.); however, instead of displaying the character on the debug console, you will put it in the keyboard mailbox for use by the User task (see comment in skeleton code). Once a keypress is detected, the task should wait until the key is released before continuing to scan. If no key is pressed (i.e., the BSP function returns −1), the key-scanning task should delay long enough to debounce the keypad and then try the next row. Be sure to use the embOS delay function to allow other tasks to run during the debounce time.

This task is well suited to a FSM running at the debounce delay rate. The skeleton user task reads data from the MBKey mailbox and prints the value to the debug terminal. This allows you to get your keypad-scanning task working before writing the user task.

27L.4.2.5 The Beep() function

Again, this one has been written for you. You implemented a single beep previously using a Software Timer in §27L.3. Here we have two timers to allow multiple beeps. One timer creates the beep and the other sets the delay between beeps. A beep counter variable counts down the number of times the beep timer callback function is called. Each callback function restarts the other timer until the beep count goes to zero.

When you call the `Beep()` function, it saves the passed beep count argument in a global beep-count variable (although it is only used locally by `Beep()` and the two callback functions), turns on the buzzer and restarts the beep-delay timer. When the timer expires, the callback turns off the buzzer and decrements the beep count. If the count is zero, it is done. Otherwise it restarts the quiet-delay timer. At the end of this delay, the callback turns the buzzer on and restarts the beep-delay timer again. The result is the buzzer beeps the number of times passed to `Beep()`.

27L.4.2.6 The user application task

The User task (`UsrTask()`) creates the application. It is this task that turns this collection of hardware and software into a jukebox. In the reference design described, it fetches keypad characters from the keypad-scanning mailbox, beeps to provide audible feedback, and then determines how to handle the key. It either: a) shows the name of the song attached to a numeric key on the top line of the LCD; b) adds the song index to the play mailbox queue when the enter ('#') key is pressed; or c) stops the play task if it is playing a song and clears the play queue when the stop playing key ('*') is pressed. Pressing enter or stop also clears the selected song name from the display.

If the enter key is pressed when no song is selected, it just gives an error signal (three beeps). If an unused song key is pressed, it gives the error signal and clears any selected song name. If an unused key ('A' through 'D') key is pressed it gives the error signal but does not clear the display.

You should write the task loop as finite state machine following the directions in the skeleton comments.[20] Note that the user FSM does not require a time-based delay, since it sits idle in a blocked mailbox get call and only is activated when a key is pressed.

Some hints for a professional design:

- Create one state per function, not per key. This means a given state may handle multiple keys (case statement matches fall through any case statements following, so you can stack them to execute the same code block). For example, it makes sense to have a single state handle keys '1' through '0'. Note that the '0' key selects song index 10.[21] The basic demonstration code uses only four states (Enter, Song Key, Stop, and Unused Key).
- Try to make your code easy to update. Rather than hard coding the currently unused song keys to a separate error state, have a single "song key pressed" state and compare newkey - '0' (this converts the ASCII characters '0'–'9' to a digit 0 to 9) to the NUM_SONGS define and give an error only if the key is unused. That way, if you add a new song to songs.h all you have to do is recompile the code to activate it on the next available song key.
- Implement one state and debug it before working on the next state. (This is beauty of using an FSM. Each state is independent and only executes when specific keys are pressed.) A good way to start is to implement the unused-error beep as the default. Then every key will give an error.
- Next implement the Enter key. You can temporarily put your favorite song index in the play mailbox until you implement the song keys. The enter key should clear the selected song title on the top line of the display. (The demonstration code replaces the song title with an "In Queue" status message on the top line for a half second.) When no song is selected, the Enter key should give an error beep.
- The Song keys are the most complicated state. Pay attention to the specification. Pressing a valid song key should display the song title after the "Select: " prompt on the top line of the display. Pressing an unused key should clear any song title on the top line and give an error beep. Pressing the Enter or the Abort key should clear the top line song title as well.
- Don't forget to output a single beep every time a valid key is pressed.

27L.4.2.7 The main() initialization function

The Jukebox main() function is just a supercharged version of the one used in §27L.3. The main() function in an embOS program is solely an initialization function. It is executed once to initialize the operating system and hardware and to create the tasks, timers, mailboxes, mutexes, etc., used by the application. It then starts embOS before control is passed to the RTOS scheduler.

As in the sample application, we start by calling the embOS hardware and software initialization functions OS_Init() and OS_InitHW(). These must be called before any other OS function to initialize the RTOS internal variables and set up the hardware timer. This is followed by the board specific

---

[20] You may also want to watch the jukebox video demonstration again at https://LAoE.link/Jukebox_Demo.html to see how the user interface operates.

[21] Song index 0 is the startup Intro song. It is played on entry to the User task but cannot be selected from the keypad.

hardware initialization calls to set up the SAMD21 Mini Port I/O and the SERCOM SPI initialization so startup messages can be displayed on the LCD. If you try to send messages to the LCD immediately after turning on the power, they get garbled, so there is a one second delay before accessing the display. Note that calling `OS_TASK_Delay()` before the OS is started is explicitly allowed by the documentation.

Once SPI output is available, the display is cleared, the contrast and backlight set, and a welcome message displayed for a short time. These calls do not need a mutex because no tasks are running and the operating system has not yet been started. Since the tasks have not been started, there is no possibility of the LCD functions calls being interrupted. After the LCD message, the DAC and timer peripherals are initialized to start sending a sine wave to the speaker.[22]

After the hardware initializations, we create all the embOS objects we will be using including the two timers, the two mailboxes, the mutex used to share the LCD, and the three tasks. With all the initializations complete, `main()` starts embOS and never returns.[23]

27L.4.3 Complete the jukebox code

Create a working Lullaby Jukebox by completing the code for the keyboard scanning (`KSTask()`) and user (`UsrTask()`) tasks. Write and debug small sections of code at a time, following the hints in §27L.4.2.6. Test your program's functionality against the written specification and the demonstration video (but feel free to modify the user interface if you don't like they way we did it).

Debugging tip: Before you start debugging your code, set a breakpoint on the first line of the function `OS_Error()` in setup file `OS_Error.c`: see Fig. 27L.7. The operating system checks for a number of common errors and calls this function if any of them occur. If the execution stops at this breakpoint, the "Locals" window will display the error code. For example, the break shown in Fig. 27L.8 occurred because the stack overflowed.[24] You could then try increasing the size of each of task stack and re-running the code to see which one is causing the problem.

27L.4.4 Optional jukebox enhancements

It seems a shame to waste keys 'A' though 'D.' Here are some ideas to add enhancements to the jukebox and utilize the extra keys. Feel free to come up with your own feature improvements as well.

Add a "Skip Current Song" key: Make the 'A' key stop playing the current song and skip to the next one in the queue. This is really easy; you only have to add the key to the User task. The function is identical to the stop function but you don't clear the mailbox queue. The Play task will see the stop event flag, stop playing the current song, then get the next song to play out of the play mailbox. You will want to add a short (400 ms) quiet break in the Play task after aborting the currently playing song to separate it from the next song in the queue.

Add a "Repeat" key: Make the 'B' key put the Play task into repeat mode. When a user presses 'B' while a song is playing, have the Play task turn on the external green LED (remember to use the BSP functions) and play the current song repeatedly. If the 'B' key is pressed again, the Play task should turn off the LED and fetch the next song in the queue when it is done playing the current one. You will have to add a new Event Flag and modify both the User and the Play tasks to implement this.

---

[22] The `BSP_TC4Init()` function initializes the timer to generate an inaudible sine wave but even if it didn't, the Play task turns off the sound once it is started if its mailbox queue is empty.

[23] Like Charlie on the MTA. See `https://LAoE.link/Charlie_MTA.html`.

[24] A list of error codes and a short description of each one is available at the top of file `OS_Error.c`.

27L.4 Build a lullaby jukebox

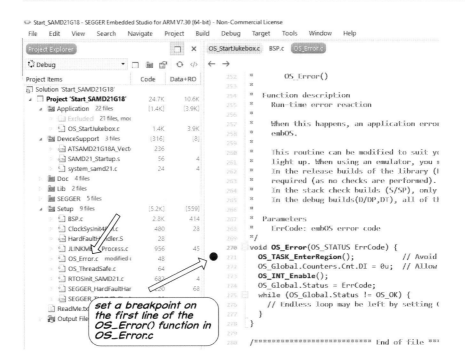

Figure 27L.7 Set up a breakpoint on OS_Error() to catch errors while debugging.

Figure 27L.8 If the OS_Error breakpoint occurs, an error code will appear in the Locals window. Here OS_ERR_TASK_STACK suggests a task stack that is too small.

Add a "Pause" key: Make the 'C' key pause the currently playing song. When a user presses 'C' while a song is playing, have the Play task stop fetching and playing notes until the pause key is pressed again. You should replace the "Playing " status message with "Paused ". You will have to add yet another a new Event Flag and modify both the User and the Play tasks to implement this. (You will also have to decide how you want the other new functions to work when in Paused mode.)

Support more than 10 songs: This is a difficult enhancement primarily because it complicates the user interface.[25] This enhancement will require a major rewrite of the user task. You will need to replace the "Select: " prompt on the top line of the display with an editable song number and add a

---

[25] It certainly is not a resource issue. Our debug build with all the previous enhancements, IR remote control support and eight songs uses 27k (10%) of program memory and 8k (25%) of RAM.

backspace key to edit song numbers larger than '0' (song 10). In addition, you will not be able to give an error for an invalid song number until the user presses enter, since only then do you know the user has finished editing the selection. You should put "None" as the song title for intermediate edits that do not select a valid song.

27L.5   Testing and debugging an embOS program

You can use the IDE debugger to debug an embOS program just as you would any other program. However, Segger has a special tool called embOSView that allows you to analyze the target application as an RTOS program with an understanding of tasks, task stacks, priority, status, etc. The embOSView tool is included in the embOS download and documentation is in the embOS manual.

27L.6   Conclusion

You could probably implement the Lullaby Jukebox as a superloop, but it would be messy to combine all the jukebox functionality into a single program loop. Using embOS and breaking the program into separate tasks simplifies both programming the application and debugging it. If we haven't convinced you yet, take a look at the worked example in Chapter 27W where we add an infrared remote control to the jukebox without changing a single line of code in the existing tasks.

27L.7   Solutions

- §27L.3: `https://LAoE.link/micro6/27L_Task_Event.c`.
- §27L.4: `https://LAoE.link/micro6/27L_OS_StartJukebox_final.c`.

27S Supplementary Notes: Installing embOS

27S.1 Installing Segger embOS

You will need to download, install, and modify the Segger embOS according to the following instructions before beginning Lab 27L. Please check the errata page at
https://LAoE.link/Installing_embOS.html
before you begin to see if there have been any updates or changes to the procedure.

27S.1.1 Download embOS

Download the zip file for the "Segger embOS for Cortex-M and SEGGER Embedded Studio" from the Segger site: https://LAoE.link/embOS_download.html, see Fig. 27S.1. This is a large file, so be sure you have room on your hard drive to both download it and install it. You will need about 1.2 GB of free space. (We will pare this down significantly when we modify it to work with just the SparkFun SAMD21 Mini breakout board.)

Figure 27S.1 Version of embOS to download.

27S.1.2 Install embOS

Unzip the embOS file into a new directory in your "SEGGER Embedded Studio for ARM Projects" directory.[1] The unzipped directory will contain release notes, the embOS manuals, the embOS debugger application (embOSView), and the 🗁 Start directory containing the embOS sample application. The Start directory contains subdirectories for the embOS object libraries and include files common to all hardware, and a "🗁 BoardSupport" directory containing directories for the many different hardware supported by the Segger sampler application.

Removing unneeded board support files: You can reduce the space requirements to only 95 MB (from over a gigabyte) by removing all the board support directories except the one for Microchip, the manufacturer of the SAMD21G18A: see Fig. 27S.2.

---

[1] You can place the embOS files anywhere but this will group them with your other Segger projects.

Supplementary Notes: Installing embOS

Figure 27S.2 Directory tree of embOS before removing unneeded board support.

You can further reduce the storage requirements by half, to about 52MB, by deleting all the Microchip board support directories except the one for the SAMD21 Mini Breakout: see Fig. 27S.3.

Figure 27S.3 Directory tree of embOS after removing all board support except Microchip.

27S.1.3 Update the interrupt vector table file

Replace the file
samd21g18a_Vector.s in directory
🗀 Start ▸ BoardSupport ▸ Microchip ▸ SAMD21G18_SAMD21_Mini_Breakout ▸ DeviceSupport
with the file https://LAoE.link/micro6/samd21g18a_Vectors.s.
If this directory already contains a file named ATSAMD21G18A_Vectors.s, you can skip this step.

27S.1.4 Test your basic embOS installation

Load the embOS sample project into the IDE by closing any active project and opening the "Start_SAMD21G18.emProject" project file from the
🗀 Start ▸ BoardSupport ▸ Microchip ▸ SAMD21G18_SAMD21_Mini_Breakout
directory: see Fig. 27S.4.

Build and run the project. The green Tx LED should blink rapidly (10 Hz), while the yellow Rx LED blinks at a slower rate (2.5 Hz). Each LED is controlled by a separate task, i.e., a function that runs as though it has exclusive control of the CPU.

27S.1 Installing Segger embOS   1077

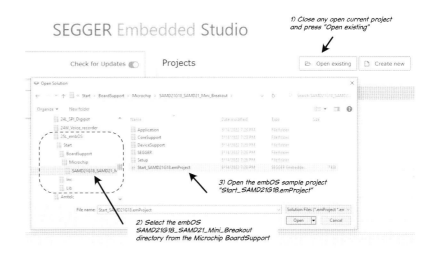

Figure 27S.4 Load the embOS sample application project.

27S.1.5 Modify embOS to run at full CPU speed.

Make the following changes to the embOS sample application to enable the SAMD21 to run at 48 MHz.[2]

- Copy ClockSysInit48M.h to directory 📁Start▸Inc.
- Copy ClockSysInit48M.c to directory
  📁Start▸BoardSupport▸Microchip▸SAMD21G18_SAMD21_Mini_Breakout▸Setup.
- In the IDE Project Explorer, right click on the *Setup* folder and select "Add Existing File…" to add ClockSysInit48M.c to the project: see Fig. 27S.5.
- Edit the file system_samd21.c in the DeviceSupport folder as follows:[3]
  - Add #include "ClockSysInit48M.h" to the file.
  - Replace "(7998000)" with "(48000000)" in the line #define __SYSTEM_CLOCK (7998000).
  - Remove (or comment out) the line SYSCTRL_OSC8M &= ~SYSCTRL_OSC8M_PRESC_MASK; in function SystemInit() and replace it with ClockSysInit48M();.

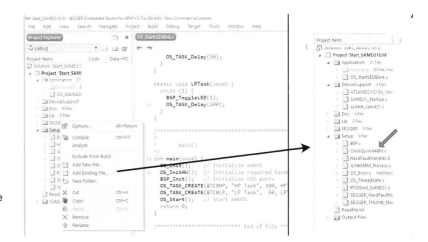

Figure 27S.5 Add the 48 MHz initialization function to project.

---
[2] Stock embOS increases the default CPU clock from 1 MHz to 8 MHz, but we need to run the microcontroller at full speed to get the DAC output sine wave range needed for the PlaySong() function in the jukebox application.
[3] You may download the modified file at https://LAoE.link/micro6/system_samd21.c.

Supplementary Notes: Installing embOS

Recompile the sample application and make sure it works as it did before. Add `SystemCoreClock` to your Watch list to be sure the CPU clock is 48 MHz; it should show 48000000 (or 0x02dc6c00) when you pause execution.

Figure 27S.6 Check to make sure the CPU is running at 48 MHz.

27W Worked Examples: Adding IR Remote Control to the Jukebox

The advantage of using a real-time operating system (embOS in our case) becomes abundantly clear when you need to add features or modifications to an existing program. The natural partitioning of an RTOS application into individual tasks with limited communication channels makes modifying existing code not only much more contained, but also much less likely to introduce new bugs into working code.

27W.1 Remote control

In this example, we will add infrared remote control to the Lullaby Jukebox from Chapter 27L to control the jukebox in parallel with the keypad. While the BSP.c code to decode the IR signals is relatively complicated, adding remote control to the application requires no changes to any of the existing tasks, only the addition of a new task to feed IR commands into the keypad MBKey mailbox.[1] While a mailbox allows only a single task to read from it, it is fine if multiple tasks write to it. The user task does not know or care where the key values come from, so feeding it from both the keypad and remote control works transparently.

27W.2 The hardware

A consumer IR remote control puts out bursts of a carrier at 30 kHz to 56 kHz depending on the remote. A three-terminal receiver with an integrated IR photodiode and detector circuit translates those bursts into logic levels. The device is shown is Fig. 27W.1.

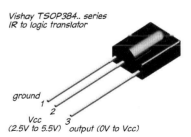

Figure 27W.1 IR Receiver Module.

Figure 27W.2 shows the ≈40 kHz oscillation that a Sony TV remote sends; this signal is integrated by the receiver IC, then sent to a comparator included on the IC to give a logic-level output as shown

---

[1] When we were developing this example, it took some time to figure out the BSP code to measure the IR signal pulse width and generate IR codes, but once the BSP functions worked, integrating IR into the application took only about fifteen minutes.

in the figure, where the bottom trace shows logic levels corresponding to the carrier bursts. The output is active low: high when the carrier is off and low when it is on.

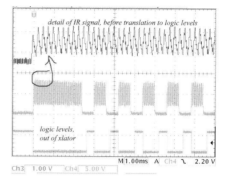

Figure 27W.2 Detail of IR remote transmission: bursts of 40 kHz oscillations.

The IR remote signal is robust because it need only sense whether a burst of carrier is present or not; it does not need to detect each carrier transition as our audio project did back in Lab 13L. In addition, most IR signaling protocols include error checking, often sending the command twice, once normally followed by the command inverted.

Add a Vishay TSOP38438 IR Receiver to your breadboard as shown in the schematic of Fig. 27L.2 with the output connected to Arduino pin 1/TX (SAMD21 pin PA10).[2]

27W.3   Choosing an IR transmitter

Since we can assign the output signals of the IR remote to any ASCII keypad value we want in software, our goal in choosing a remote control was to find one with an easy-to-decode IR output format. We have a variety of IR remote controls in our lab so we tested several of them to see which made the task easiest. We connected an oscilloscope to the output of the IR receiver and sent signals from various remotes until we found one that looked easy to decode. We chose a small remote control intended for Microsoft's Media Center that put out a single code for each button press: see Fig. 27W.3.[3]

Figure 27W.3 Media center remotes tested for use with Lullaby Jukebox. We ended up using the one on the right.

---

[2] The TSOP38438 expects a carrier frequency of 38 kHz. If you have a specific remote control you want to use with the jukebox that operates on a different carrier, there are five versions in the TSOP384XX series designed for carrier frequencies from 30 kHz to 56 kHz.

[3] Most of the universal remote controls we tried put out codes with more bits, which seemed more difficult to decode.

The command received from this remote when a key was pressed appeared to be in the popular NEC IR format: see Fig. 27W.4.[4]

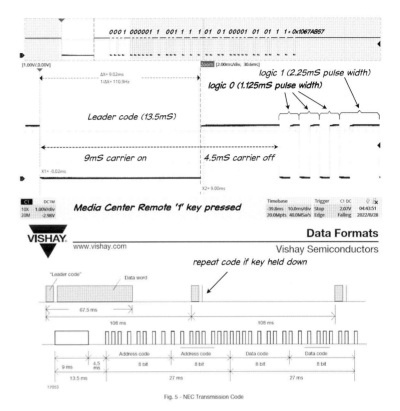

Figure 27W.4
Demodulated output of Media Center Remote compared to NEC format.

27W.4 Decoding the demodulated IR signal

Each key press of our media center remote generated an IR output signal beginning with an NEC format *Leader code* – 9ms of carrier followed by a pause of 4.5 ms. The leader code was followed by a series of 32 data bits in Pulse Distance Coding format. Each PDC data bit consists of carrier on for 600 μs followed by carrier off for $\approx$ 525 μs for a zero (a pulse width of 1.125 ms) or $\approx$ 1.65 ms for a one (a pulse width of 2.25 ms). There is a final carrier on pulse of 600 μs to mark the end of the last bit. In Fig. 27W.4, pressing the '1' key on the remote results in an output of 0x1067A857.

If a key is held down, the remote sends a repeat code consisting of the Leader code followed by a single 600 μs carrier on pulse every $\approx$ 100 ms after the initial key code until the key is released: see Fig. 27W.5.

Once we understood the IR code format, we needed a program to capture the remote's key codes (since decoding the scope trace by hand is tiresome). We wrote a stand-alone IR decoder based on the pulse measurement program in the worked example code from Chapter 25W.[5] That problem measured

---

[4] See https://LAoE.link/IR_Formats.pdf for a description of the two most common IR formats, NEC and RC5.
[5] We initially thought we could use the automatic pulse width and period measurement (PPW) mode built into the SAMD21's timer/counters to time the period for each pulse, but the datasheet was pretty impenetrable on how to set it up. We were finally able to get it working (mostly because someone from Sweden who posts as "Lajon" had uploaded sample code to a discussion forum). It required setting up a 1 MHz GCLK, configuring the External Interrupt Controller to

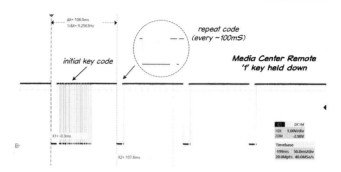

Figure 27W.5 Demodulated output of Media Center Remote with key pressed and held.

the *width* of a positive-going pulse by programming the External Interrupt Controller (EIC) to interrupt on both the rising and falling edge of the input signal to measure the pulse width. The ISR started a timer on the rising edge and stopped it on the falling edge. The main changes to measure the IR pulse *period* are to increase the SysTick timer interrupt to increment every microsecond rather than every millisecond to get sub-millisecond time resolution of the NEC pulses, and to interrupt only on the falling edge of the input signal to get period rather than width. We use the falling edge because the output of the TSOP38438 is inverted, making the falling edge the start of each pulse. We also used the BSP_ naming convention for any function we knew would be called by the jukebox application when we integrated the code into embOS.

The complete decoder program, which prints the remote's 32 data bits to the debug console in hex format, is available at https://LAoE.link/micro6/27W_IR_Remote_Decode.c. The program begins by initializing the CPU clock to 48 MHz and configures the SysTick timer to interrupt every microsecond. As in our earlier code, the SysTick timer ISR just increments a count but now it does so with 1000 times greater precision. Next, the call to BSP_EICInit() configures the EIC to interrupt on the falling edge of the PA10 signal. See the detailed explanation in §25W.1.3 to understand the steps to configure this interrupt. After initialization, main() enters a tight loop waiting for IR codes to be received, printing them to the console.

The interrupt service routine, EIC_IRQHandler, is a two-state FSM (using if statements rather than a switch) to measure the number of microseconds since the last time the interrupt occurred. The two states are: (1) waiting for the Leader pulse marking the start of key press data (WAIT_START); and (2) decoding the key press data into 1s and 0s (PULSE_CAP). Timing is maintained in 64-bit variables to avoid overflow between key presses.[6] We convert to 32-bit values for the actual bit calculations. Based on our measurements over multiple iterations, it appears that the pulse widths are very consistent with only a few percent change in length. Our max/min pulse width #define values for the length of a Leader pulse, a one pulse and a zero pulse allow for up to ±12% variation in the period timing.

The interrupt finite state machine stays in the WAIT_START state until a Leader pulse is detected. As long as it is in this state the global done flag and ircode variables stay set to zero. When a Leader pulse is received, the FSM transitions to the PULSE_CAP state to capture the next 32 pulse widths as 1s or 0s and shift them into the ircode global variable.[7] If any of the pulses do not meet the specification for the correct pulse width of a zero or one, the FSM is reset to the WAIT_START state.

recognize a high signal on pin PA10, enabling the Event System, configuring the TC in an obscure way, and writing an ISR. We decided that was too complicated to teach and reworked it into the EIC-based decoder described here.

[6] We don't actually expect anyone to wait 585 years between key presses, but a 32-bit variable would overflow and could miss the next key press if someone put the remote down for 1.2 hours. This would occur in the rare case where the overflow count was seen as a false Leader pulse.

[7] Although we assigned the long pulses a 1 and the short ones a 0, it would not matter if we decoded them the other way around since we just use whatever number we get to identify a unique remote key.

27W.4 Decoding the demodulated IR signal

This ensures that 600 μs repeat codes are ignored. After 32 data bits, the done flag is set and a helper function, `BSP_GetIRCode(void)` returns valid IR codes to `main()` to print on the console. IR codes are not buffered so the foreground program must check for the 32-bit `ircode` value at least every 100 ms before the next repeat code could arrive and overwrite the buffer.

27W.4.1 Examining the decoded data

`IR_Remote_Decode` is a useful tool for exploring IR remotes. (You may need to modify the timing constants and NUMBITS in the interrupt handler for your remote based on an examination of the output signal on the scope.) We modified it to output both binary and hex and used it to capture the twelve key codes we planned to use in the jukebox (the '1' through '0' keys, the enter key and the stop key). Finally, we put the data in an Excel spreadsheet to make some sense of it: see Fig. 27W.6.

																																		32-bit value	Third Byte
Key 0	0	0	0	1	0	0	0	0	0	1	1	0	0	1	1	1	1	0	0	0	0	0	1	0	0	1	1	1	1	1	0	1	0x1067827d	0x82
Key 1	0	0	0	1	0	0	0	0	0	1	1	0	0	1	1	1	1	0	1	0	1	0	0	0	0	1	0	1	0	1	1	1	0x1067a857	0xa8
Key 2	0	0	0	1	0	0	0	0	0	1	1	0	0	1	1	1	0	0	1	0	1	0	0	0	1	1	0	1	0	1	1	1	0x106728d7	0x28
Key 3	0	0	0	1	0	0	0	0	0	1	1	0	0	1	1	1	0	1	1	0	1	0	0	0	1	0	0	1	0	1	1	1	0x10676897	0x68
Key 4	0	0	0	1	0	0	0	0	0	1	1	0	0	1	1	1	1	0	0	1	1	0	0	0	0	1	1	0	0	1	1	1	0x10679867	0x98
Key 5	0	0	0	1	0	0	0	0	0	1	1	0	0	1	1	1	0	0	0	1	1	0	0	0	1	1	1	0	0	1	1	1	0x10671877	0x18
Key 6	0	0	0	1	0	0	0	0	0	1	1	0	0	1	1	1	0	1	0	1	1	0	0	0	1	0	1	0	0	1	1	1	0x106758a7	0x58
Key 7	0	0	0	1	0	0	0	0	0	1	1	0	0	1	1	1	1	0	1	1	1	0	0	0	0	1	0	0	0	1	1	1	0x1067b867	0xb8
Key 8	0	0	0	1	0	0	0	0	0	1	1	0	0	1	1	1	0	0	1	1	1	0	0	0	1	1	0	0	0	1	1	1	0x106738c7	0x38
Key 9	0	0	0	1	0	0	0	0	0	1	1	0	0	1	1	1	0	1	1	1	1	0	0	0	1	0	0	0	0	1	1	1	0x10677887	0x78
Enter	0	0	0	1	0	0	0	0	0	1	1	0	0	1	1	1	0	0	1	1	0	0	0	0	1	1	0	0	1	1	1	1	0x106730cf	0x30
Stop	0	0	0	1	0	0	0	0	0	1	1	0	0	1	1	1	1	1	0	0	0	0	0	1	0	0	1	1	1	1	0	1	0x1067c23d	0xc2

<----------- 8 Bits -----------> <----------- 8 Bits inverted ----------->

Data Code

Figure 27W.6 Demodulated and decoded output of IR remote keys '0' to '9'. enter and stop.

The first 16 bits are always 0x1067 and never change. However, the third byte and the fourth byte are the bitwise inverse of each other and are unique for each key. We decided to use the third byte to look up the proper ASCII character to send to the user task.

27W.4.2 Integrating the remote into the jukebox

To add remote control to the jukebox, we need to move the hardware interface functions from the IR test program into `BSP.c`, add prototypes for them in `BSP.h`, and add a task in `OS_StartJukebox.c` to poll for IR codes every 50 ms or so and, if one is available, look up the correct ASCII character and put it in the keypad mailbox to send to the user task.[8]

Changes to BSP.c: The `BSP.c` and `BSP.h` you downloaded in §27L.4.1 already have the IR support functions included:

```
void        BSP_EICInit(void);
uint32_t    BSP_GetIRCode(void);
```

The `BSP.c` function `BSP_EICInit()` is unchanged from the code in the stand-alone program. However, the EIC interrupt service routine needed a few changes to use in the embOS environment. Since embOS uses the SysTick timer for its own timing, the ISR cannot use it to time the IR pulses. Luckily, embOS

---

[8] Since the EIC interrupt routine is RTOS aware, we could also have had the ISR translate IR codes into keypad characters directly and put them in the user mailbox to avoid making *any* changes to the application code. This is allowed because API Table 9.2 in the embOS User guide shows `OS_MAILBOX_Put1()` may be called from interrupt service. Alternatively, if we wanted to keep the ISR as short as possible we could have added a new mailbox to send IR codes from the interrupt routine to the new task to eliminate polling. Instead of the OS Delay call, the foreground task would call the blocked Mailbox Get API and only be given CPU time when a IR command was available.

provides a rich set of timing API functions that we can use to replace the usTick count in the stand-alone program. In particular, the API call OS_TIME_Getus64() returns the current system time in microseconds as a 64-bit value. We use this in place of the SysTick timer interrupt count in the jukebox.

However, to use embOS APIs in an interrupt service routine, the interrupt must: (1) be low priority (i.e., in the lowest half of the interrupt priority levels); and (2) must call OS_INT_Enter() at the beginning of the ISR before anything else and call OS_INT_Leave() just before returning. The first condition was already met in the stand-alone program, NVIC priority was set to two.[9] The stand-alone program also included commented-out embOS entry and exit calls; in BSP.c these have been enabled.

The BSP function BSP_GetIRCode() returns the most recent IR code or 0 if there is none. There is only a single byte buffer for codes, so you need to poll often enough that they are not overwritten. Since we only have to be faster than the human pressing the remote buttons, polling every 50 ms is fine. The buffer is cleared by the call to BSP_GetIRCode() so you will only receive an IR code once for each button press.

Changes to OS_StartJukebox.c: Obviously we need to include the call to BSP_EICInit() in main() to initialize the external interrupt controller for the IR receiver interrupt. We then created a new task to check for IR commands, IRTask(), and declared a task control block (TCBIR) and stack (StackIR[256]) for it as well.

Finally, we need to write the task to check for a new IR command, translate it into a keypad character, and put it in the keypad mailbox. The user task will respond to a keypad character sent from the IR task in the same way it responds to characters sent from the actual keypad. (Indeed, the user task has no way to determine the source of any mailbox message.) We wrote the IR task as a simple FSM. While a lookup or hash table might have been more efficient, the human instigated IR commands are so infrequent that we would not see any change in performance at all. This task spends most of its time suspended while in the call to OS_TASK_Delay(), allowing other tasks to run.[10]

```
/*****************************************************
*    IRTask() - Task to send IR charcters to user task
*
* Polls the EIC ISR received ir codes and when
* one becomes available puts corresponding
* ASCII charcter in key mailbox.
*
* The remote control used sends four bytes
* the third byte is unique for each key as follows:
* Key 0     0x82
* Key 1     0xa8
* Key 2     0x28
* Key 3     0x68
* Key 4     0x98
* Key 5     0x18
* Key 6     ox58
* Key 7     oxb8
* Key 8     0x38
* Key 9     0x78
* Enter     0x30
* Stop      0xc2
*
* The exhanced commands are assigned as follows
* >>| (skip)    0xe2
* REPEAT        0xf8
* || (pause)    0x42
*****************************************************/
```

---

[9] CMSIS on the Cortex-M0+ supports four interrupt priority levels; 0 is the highest and 3 the lowest.

[10] The complete IR enabled jukebox application is available at
https://LAoE.link/micro6/27W_OS_StartJukebox_IR.c. This version includes all the optional jukebox enhancements from §27L.4.4 as well, except for support for more than 10 songs.

27W.4 Decoding the demodulated IR signal

```
static void IRTask(void) {
  uint32_t ircode;
  uint8_t  irkey;
  char     curkey;

  while(1) {
    if ((ircode = BSP_GetIRCode()) != 0){
      ircode = ircode >> 8;
      irkey = ircode & 0x00ff;
      switch (irkey) {
        case 0xa8:
          curkey = '1';
          break;
        case 0x28:
          curkey = '2';
          break;
        case 0x68:
          curkey = '3';
          break;
        case 0x98:
          curkey = '4';
          break;
        case 0x18:
          curkey = '5';
          break;
        case 0x58:
          curkey = '6';
          break;
        case 0xb8:
          curkey = '7';
          break;
        case 0x38:
          curkey = '8';
          break;
        case 0x78:
          curkey = '9';
          break;
        case 0x82:
          curkey = '0';
          break;
        case 0xc2:
          curkey = '*';
          break;
        case 0x30:
          curkey = '#';
          break;
// enhanced commands
        case 0xe2:
          curkey = 'A';   // skip rest of current song
          break;
        case 0xf8:
          curkey = 'B';   // repeat/play current song toggle
          break;
        case 0x42:
          curkey = 'C';   // pause/continue toggle
          break;

        default:
          curkey = 'X';   // causes user tasks to error beep

      }
      if (curkey != 0) {
        OS_MAILBOX_Put1(&_MBKey, &curkey);    // put new key in mailbox buffer
        curkey = 0;
      }
    }
    OS_TASK_Delay(50);
  }
}
```

If you want to add an IR remote to your jukebox, you will need to modify the `irkey` computation to select the appropriate field from your remote's command code and change the switch statement compare values to match your codes. You may also need to modify the EIC interrupt service routine in `BSP.c` to match your remote's output format.

Summary: Because we used an RTOS, we were able to add a significant feature to our application without changing *any* existing function other than adding the new task structures and initializations to `main()`. All other modifications to the jukebox to add IR remote control were new, separate functions. This makes adding the feature not only easier, it lessens the chance of breaking existing code.

28N Project Possibilities: Toys in the Attic

Contents

28N.1	**Projects: an invitation and a caution**		**1088**
28N.2	**Some pretty projects**		**1088**
	28N.2.1	Laser character display	1088
	28N.2.2	Sound source detector	1090
	28N.2.3	Computer-driven Etch-a-Sketch	1091
	28N.2.4	A giant Etch-a-Sketch	1091
	28N.2.5	An attempted insect	1091
28N.3	**Some other memorable projects**		**1093**
	28N.3.1	X–Y scope displays	1093
	28N.3.2	Complications and improvements	1095
	28N.3.3	Size control; zoom	1095
	28N.3.4	True vector drawing: position-relative	1097
	28N.3.5	Animation	1097
	28N.3.6	Computed drawing	1097
	28N.3.7	Laser X–Y display	1098
	28N.3.8	Charmingly simple LED display	1099
	28N.3.9	Vehicles	1099
	28N.3.10	A neat application for the independent-drive steerable chassis	1100
	28N.3.11	A computer-steered toy car	1101
	28N.3.12	A perennial heartbreaker: inverted pendulum	1102
28N.4	**Games**		**1105**
	28N.4.1	Pong on a scope screen	1105
	28N.4.2	Pacman on a scope screen	1106
	28N.4.3	Asteroids on a scope screen	1106
	28N.4.4	Other great games	1106
28N.5	**Sensors, actuators, gadgets**		**1107**
28N.6	**Stepper-motor drive**		**1107**
28N.7	**Project ideas**		**1109**
	28N.7.1	Nifty new and untested projects for your inspiration	1109
28N.8	**And many examples are shown in AoE**		**1110**
28N.9	**Now go forth**		**1110**

Project Possibilities: Toys in the Attic

28N.1 Projects: an invitation and a caution

In our own version of this course, only a minority of the busy students choose to do projects. But a project can be heaps of fun. To help you conceive of one, here is some information on gadgets and ideas that might inspire a project builder, along with sketches of some great projects of yester-term.

General advice:

- Try your ideas on someone else, early in the process. Someone with experience can steer you away from the impossibly grandiose; less often, she might tell you that your invention is too modest to be called a "project."
- Choose a project appropriate to the little computer, not a project that's better done on a full-scale computer. That means, among other points, keep the programming modest. It also may mean for example, *making a motor spin* rather than doing a computation and then displaying a number.
- Try to include some hardware additions along with your invented software. This point is partly a corollary of the preceding point, but also reflects a second aim: the project often works to help you firm up your grip on ideas, by letting you do some designing. It can serve as a sort of review. A broader review is better than a pure exercise in programming. (And this book is all about hardware.)
- Make your project *incremental* rather than all-or-nothing. For example, the people who built the computer-controlled RC car (§28N.3.11) started out proposing a game in which the computer would make one car chase a second car that was controlled by a human. We persuaded them to try this in stages:
    - first, see if the computer could control the car (pretty easy);
    - second, see if their ultrasonic ears could get significant information from the *delay* times between the car's ultrasonic squeak and the reception at the computer's three "ears;"and
    - third, build the link between the little computer and the PC.
- You get the idea. In fact, these students were able to get most of their original scheme to work. But tasks nearly always take longer than one expects, and we want you to feel satisfied with a modest project, rather than frustrated by a grand one.

28N.2 Some pretty projects

Later in these chapter, beginning at §28N.3, we mention some memorable past projects from our course. The accompanying online Chapter 28O, suggests some hardware that we offer to builders – along with the much richer possibilities of hardware available from hobby and robotics sites.[1] In this preliminary section, in hopes of exciting ambitions, we offer a glimpse of some of the more photogenic gizmos produced by the energetic minority who found time for a project.

28N.2.1 Laser character display

This student persevered despite our advice to try something less ambitious. What he proposed seemed much too complicated: to spin a set of differently inclined mirrors, while switching a laser on and off appropriately to produce a projection of characters on a screen.

We gasped when he brought the thing in. We also felt uneasy for his beautiful machine, noting that it was mounted with *plasticene*. Even the aiming of the laser was accomplished by pushing the plasticene blob this way or that.

---

[1] Available at `https://LAoE.link/LAoE_Chapter_28O.pdf`.

28N.2 Some pretty projects

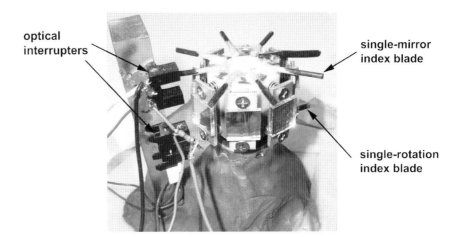

Figure 28N.1 Detail of spinning-mirrors structure in laser display.

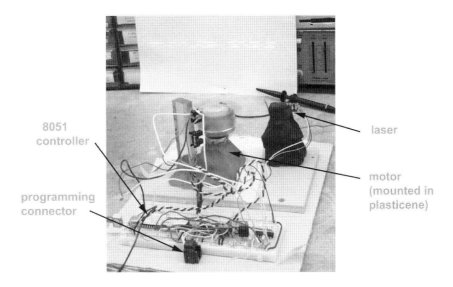

Figure 28N.2 Laser display; showing projection on a sheet of paper (incoherent image in this photo).

Figure 28N.1 shows the prettily fabricated set of mirrors, each with its indexing toothpick. One other toothpick serves to index the start of a full rotation. These toothpicks break a light beam in two opto-interrupters read by the controller; these are the machine's only inputs. The set-screws above and below each mirror allowed hand-adjustment of each mirror's tilt. Software permitted still-finer tuning, to make up for small imperfections in the left–right positioning of each mirror. Figure 28N.2 shows the machine in action. The text is illegible in this photo – but was perfectly legible in life. It spelled out a long message, beginning with the name of the course and finishing with the name of the author, Mr. Uscinski. He deserves to be remembered for this machine![2]

---

[2] If you are not already impressed this, along with most of the project described in this section, used the old Intel 8051 microcomputer – either the very limited "Big Board" version or the newer single-chip Silabs version, programmed in assembly language.

28N.2.2 Sound source detector

28N.2.2.1 Differential sound source compass

This project, done with the big-board version of the computer rather than with a single-chip implementation as in the other cases, came together with surprising speed: in one long afternoon. It survives partly because its builders, Sian Kleindienst and Kristoffer Gauksheim, took the trouble to mount it on a big piece of foamcore.

It displays the compass heading of a loud sound (usually a handclap). The computer is fed three binary signals, taken from three microphone amplifiers and comparators. The microphones are separated by a few centimeters. The computer gets an approximate direction by noting simply the *order* in which the comparators fire. It gets a more detailed reading in one 60° pie-slice – implemented in just one segment of the circle because the student designers decided they should call it a day and move on to revising for exams. Their circuit gets that detailed reading by noting time delay among the three microphones.

Figure 28N.3 shows the three microphone boards, and the computer board reading "320," the compass heading of a recent handclap. Figure 28N.4 shows the circuit with a scope displaying the staggered arrival of the handclap that produced a particular compass reading.

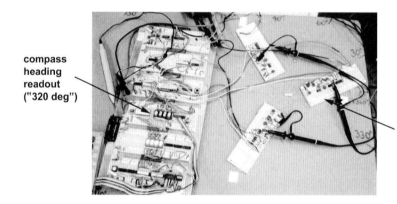

Figure 28N.3 Sound source detector project.

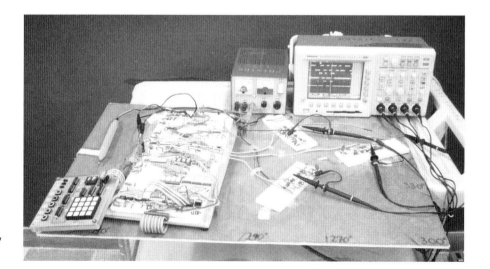

Figure 28N.4 Sound source detector showing scope display of microphone-comparator inputs.

28N.2.2.2 The Clap 'n Snap

Two students in our undergraduate class (Ali Forelli and Mary Carmack) took the differential sound source measurement concept one step further. They used an Arduino Pro Micro microcontroller to calculate the time difference between sounds received by two microphones and triangulated to control a digital camera mounted on a hobby servo. When a user clapped, the camera swiveled and took their picture: see Fig. 28N.5.

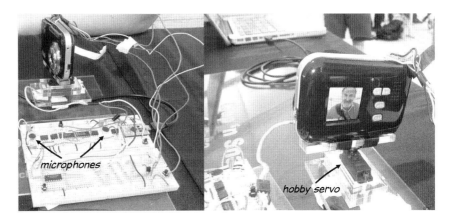

Figure 28N.5 The Clap 'n Snap with an example of its microphone-comparator selfie output.

28N.2.3 Computer-driven Etch-a-Sketch

This project implemented a notion less original than the ones above: let two stepper motors turn the knobs of a child's Etch-a-Sketch display. The mechanical part of the task was a pesky challenge, because the toy's knobs' resistance to turning is considerable, and the shafts wobble a bit. The gadget did, however, laboriously spell out "P 1 2 3." That doesn't sound like much of an achievement, perhaps. But having witnessed the struggles that came before, as the designers implemented the stepper driver in software, including the refinement of half-step, we were thoroughly impressed when we saw our course number emerge on the screen (not shown in Fig. 28N.6, unfortunately).

28N.2.4 A giant Etch-a-Sketch

A somewhat similar drawing machine was fashioned, again using two stepper motors, from a massive X–Y table that we rescued from the trash. The resolution of this machine is fantastically fine – but this project applied it to writing a short message in the manner of the toy Etch-a-Sketch.

A later student added a solenoid that permitted the pen to lift. It would be nice to see this excellent X–Y table used to route copper for a printed circuit board layout. But the software for such a thing would be way out of line for the purposes of our students' projects.

28N.2.5 An attempted insect

Not all projects succeed, unsurprisingly. But failures after intelligent effort please us too. Figure 28N.8 shows one: a nicely fabricated six-legged insect. Each leg is driven by a servo-motor (the sort that takes an angular position determined by the width of a pulse). This insect carries its battery, a flat Polaroid 6 V type, in its belly. The tin can on top regulates that to 5 V.

The designers discovered, when they set their machine to walking, that insects need knees – or some other scheme that lets a foot rise and fall from the ground. The creature of Fig. 28N.8 was unable to

Project Possibilities: Toys in the Attic

Figure 28N.6 Etch-a-Sketch driven by stepper motors.

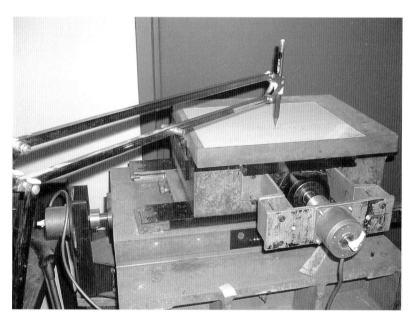

Figure 28N.7 Big X–Y table driven in the manner of the Etch-a-Sketch.

walk, as designed; it could only raise and lower its body, in the manner of someone curtseying. It could not proceed forward or backward.

The designers of the Radio Shack insect of Fig. 28N.9 understood that the creature needed to be able to lift some of its feet while they moved forward, then to lower them to the ground as they moved back.

28N.3 Some other memorable projects

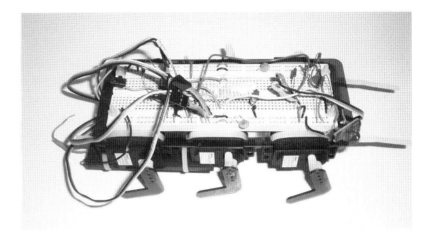

Figure 28N.8 A six-servo insect in need of further evolution.

Figure 28N.9 Insect legs better designed: these can lift from the ground as they reach forward.

eccentric leg mount allows middle leg to rise and fall, thus lifting front and rear legs from the ground as middle leg moves

The middle leg is an eccentric rod that periodically lifts the fore and rear legs from the ground. While lifted, these legs move forward, then they regain contact as they move backward relative to the insect's body. But if the creature of Fig. 28N.8 was not a successful insect, it still was a handsome one, and nicely constructed. Next time, no doubt, it will acquire knees.

28N.3 Some other memorable projects

Most of the memorable projects that we have seen do not look like much so we will rely on descriptions more than on photographs in this section.

28N.3.1 X–Y scope displays

This is the quickest and easiest way to draw a picture on a scope screen. The projects described used analog CRTs, not digital scopes with their LCD screens.[3]

Using a CRT, if you use an RC slow-down circuit (basically, a lowpass filter) to join the points that your program puts out, you achieve a quick and easy "vector" display, good at drawing lines, including diagonals, whereas the more usual raster-scan (TV-like) display does lines laboriously with a succession of dots – and incidentally makes ugly lines whenever the slope approaches vertical or horizontal (an effect called "aliasing").

Hardware: You will need to give your computer a second DAC (*two* more if you implement the hardware "Zoom": see §28N.3.3). The complicating wrinkles suggested in the later subsections require a bit more hardware. Most of the work, however, lies in the programming, and in building the data tables your program is to send out. The payoff comes when you get to put whatever you choose on the scope screen.

---

[3] Although our attempt looked acceptable on our digital scope. See https://LAoE.link/XY_Display.html.

Project Possibilities: Toys in the Attic

Preliminary note – DAC types: In the lab we have three sorts of DAC:

- Parallel bus "Microprocessor compatible," single supply: the AD7569 (A/D, DAC in one package); and the AD558, an 8-bit single-supply part (you used this DAC in Lab 20L, where it provided feedback for the successive-approximation A/D). Avoid the AD558 in any application where it is fed directly from a data bus, because of its unacceptably-long setup time.
- Parallel "Multiplying" ("MDAC"). These offer an output that is the *product* of its digital input and an analog input. An example of this type is the Analog Devices DAC8229 containing dual 8-bit DACs.
- Serial bus parts suited for connecting to microcontrollers lacking an external parallel bus, such as the ARM Cortex-M0+. We have had success with the MCP48X2 series from Microchip, an 8-pin DIP containing two 8-, 10- or 12-bit single-supply DACs, along with an internal reference. The MCP49X2 series are selectable between an internal or external reference and use a 14-pin DIP package. Using the external reference makes it a multiplying DAC, since the full scale output is proportional to the reference voltage. We exploit the external reference in the Zoom feature described more fully in §28N.3.3.

The advantage of the parallel bus parts is they can easily be used without a microcontroller; for example, converting the output of a 74LS469 counter to an analog voltage.

If the hardware method appeals to you, you should decide at the outset whether you mean to zoom, in order to choose the appropriate DAC.

The dot array hardware: For a 256×256 array of dots, the dual 8-bit MCP4802 SPI DAC is sufficient. For a 1024 × 1024 array use the MCP4812 or for a 4096 × 4096 array use the MC4822.[4] All three devices are pin compatible and take a 4-bit command followed by a 12-bit value, so there is no change to the hardware interface when switching to higher resolution; you just output more significant digits in the 12-bit value field.[5] Chip select is asserted for the full 16-bit data word so it must be controlled by Port I/O, not the SAMD21 SERCOM Auto-CS mode. See the SPI RAM code in Chapter 26W for a similar scheme (although the MCP48X2 is write-only so you do not need to receive data back to the microcontroller).

The $\overline{\text{LDAC}}$ signal controls a set of transparent latches for the dual DAC. It should be set high while you download the X and Y values into each DAC, then brought low to update both analog outputs simultaneously.

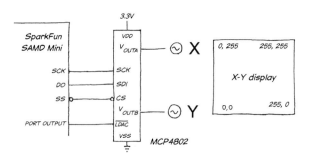

Figure 28N.10 Simple X–Y display; scope screen map.

The program should fetch a byte (or half-word if you are using the '4812 or '4822) at a time from successive locations in a table of X–Y display values, add the 4-bit command to set the proper DAC,

---

[4] However, it is unlikely you will be able to see any difference on your display scope with more bits.
[5] The low two bit are ignored in the '4812 and the low four bits are ignored in the '4802.

and then send the 16-bit result to the SERCOM output function. The output voltage range on the MCP48X2 is fixed at 0 V to ≈ 2.05 V for full scale. If you include a call to your delay subroutine, you will be able to tinker with the drawing rate.

28N.3.2 Complications and improvements

Connect the dots: As you learned in elementary school, connecting a few dots can give a coherent picture. So, with relatively little programming effort (few coordinates to list in your data table), you can draw straight-line pictures. To connect the dots, add a lowpass filter at the output of each DAC. Try RC of a few microseconds (e.g., $R=2.2$ k, $C=0.01$ μF). But you will have to experiment; the visual effect will vary with the drawing-update frequency, as well as with the filter's RC. The filter slows the movement of the output voltages, of course, so that the movement of the scope trace becomes visible. You will notice that this new scheme obliges you to pay attention to the sequence in which your program puts out these dots.

If you enter a large number of data points, you will begin to see flicker in the display, even if your program includes no delay routine. The screen needs to be refreshed about 30 times/second in order not to flicker annoyingly. You can calculate the number of dots you can get away with – or you can just try a big table and examine the resulting display for flicker.

28N.3.3 Size control; zoom

You can *zoom* in software but if you want to make your programming task a little easier, and like the neat feature of a picture symmetrical about zero as its size changes, then consider the zoom hardware described just below.

Zooming, you will recall, is the operation that can exploit the multiplying capability of certain DACs. You use a dual multiplying DAC for the X and Y outputs and use the regular internal microcontroller DAC to vary the reference voltage input of the X and Y DAC to control the size of the image.

Zoom using a dual MDAC:

- The Microchip MCP49X2 series is a dual SPI DAC using an external reference. This device comes in 8-bit ('4902), 10-bit ('4912) and 12-bit ('4922) versions. Since the output voltage is $V_{out} = V_{ref} \cdot \frac{\text{digital\_input}}{2^N}$, varying the reference voltage linearly scales the output to zoom the image.

Zoom using three DACs: You will need to swap the MCP4802 for a MCP4902. The pinout is different because this multiplying DAC requires two external reference inputs necessitating a 14-pin package. You need another DAC (either the DAC internal to the microcontroller or the MCP4802 you just removed) and connect its output to both reference inputs of the MCP4902. This allows the computer to scale the X–Y outputs by varying the '4902's reference voltage.

Figure 28N.11 is a surprisingly subtle image drawn with similar hardware. This pyramid appears to show cleverly gradated shading. In fact, the image shows nothing fancier than a diminishing square. The square was drawn as usual by defining four points, then connecting the dots by slowing each DAC output with an RC.

The apparent shading in Fig. 28N.11 results from the CRT beam's gradual slowing as it moves from source to destination. Since the movement of the dot's position slows exponentially (the rate approaching zero as the dot approaches its target), the trace grows progressively brighter.

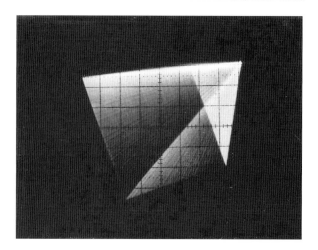

Figure 28N.11 Receding square drawn with scaled X–Y DACs.

Centering the zoomed image: A circuit refinement is proposed below to center the image on the CRT regardless of *size*. Without this circuit addition, a change of size also moves or translates the image: since X and Y are always positive, the visual effect will be as if a figure that grows were moving up and to the right on the screen, as well as toward you. If you prefer to make your figures "approach" head-on, then you should make the modification shown below: an op-amp is added to the output of each of the DAC's (X and Y); the op-amp allows you to center the coordinate output at the mid-value of the largest peak-to-peak value, 1.65 V for a maximum 3.3 V '4902 reference input.

In Fig. 28N.12's circuit, half the MDAC reference voltage is summed with the output voltage, centering that voltage as the scaling voltage varies.[6]

Notes on the offset adjust circuit in Fig. 28N.12:

- The resistor values are not critical as long as you maintain a 2:1 ratio. Keep them between 10k and 200k to minimize loading and op-amp errors.
- The voltage, V_{OFFSET}, connected to the 150k resistor at the op-amp positive input sets the no-signal DC output voltage to $\frac{V_{OFFSET}}{2}$. With V_{OFFSET} = 3.3 V as shown, the mean output voltage is 1.65 V allowing the internal SAMD21 to zoom the DAC outputs from 0 V to 3.3 V on single supply operation.
- If you prefer the output to be centered around zero volts, set V_{OFFSET} to zero volts. You will need to run the op-amp on split supplies (±3.3 V or ±5 V).
- You can run the op-amp from 3.3 V by replacing the LM358 with a LMC6482. Alternatively, the LM358 will run on 3.3 V if you set the V_{REF} DAC output range to 0 to 1 V and set V_{OFFSET} to 1 V as well.
- This circuit will work with the 8-bit MCP4902, the 10-bit MCP4912 or the MCP4922 dual SPI DAC; they all use the same pinout and the DAC value is left justified for all three. The sample code uses 8-bit XY coordinates in order to fit each data point in an ARM 16-bit half-word. When the 8-bit point value is loaded into the DAC, the low four bits in the 12-bit data field are all set to zero.
- Sample code using SERCOM0 to communicate with the external DAC is available at https://LAoE.link/28N_XY_Display.c. The Excel spreadsheet we used to create the points[] array is available at https://LAoE.link/28N_LAoE_Vectors.xlsx. A video of an advanced version of this program in action is available at https://LAoE.link/28N_XY_Display.html.

---

[6] Thanks to D. Durlach for this nifty concept.

28N.3 Some other memorable projects

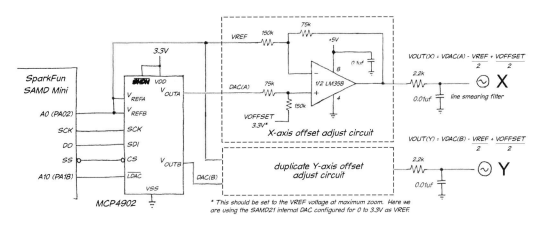

Figure 28N.12 Op-amp added to DAC to scale V_{out} symmetrically about $\frac{V_{\text{ref(max)}}}{2}$.

28N.3.4 True vector drawing: position-relative

Your drawing table need not store absolute screen locations. Instead, it can store vectors: direction and length relative to present screen location.[7] The relative vectors suggested in Fig. 28N.13 are listed in the manner of compass headings: ESE = East–South–East. This scheme would work with a 4-bit direction specification (16 directions). The remaining four bits could define magnitude; that magnitude could be used with the multiply instruction to stretch the unit-length direction vector (approximate unit lengths will do!).

Figure 28N.13 Drawing with true vectors: *relative* movement.

This way of defining a figure allows one to rotate it without much difficulty. Some rotations are described in §28N.3.6. The programming is a good deal harder than for an absolute X–Y figure.

28N.3.5 Animation

If you enter the coordinates for two or more similar but different pictures in memory, and then draw these pictures in quick succession (changing the picture, say, every 10th of a second) then you can animate a drawing. Evidently, you will need a large table to animate even a simple figure, so start modestly. You will want to use the follow-the-dots scheme to minimize the number of data points required to draw a single image. Figure 28N.14 is a crude example, showing first a hand-drawn sketch of a creature with three leg positions, then a scope image showing an implementation that was a little more ambitious. The creature seems to have morphed from perhaps horse to terrier, and the implementation apparently allowed the dog's head to bob, and its tail to wag.[8]

28N.3.6 Computed drawing

The tedious loading of values into memory demanded by the table-reading schemes we have suggested probably has made you yearn to turn over to the machine the job of determining what points it should

---
[7] Thanks to Scott Lee for proposing and demonstrating this arrangement.
[8] Our video demonstration link above includes an animation of the crude horse/dog.

1098 Project Possibilities: Toys in the Attic

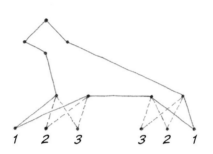

Figure 28N.14 Animation using vector drawing: a sketch and an implementation.

draw. Of course you can do this. You could write a program that would draw a rectangle, say, by incrementing the X register for some steps while holding Y fixed, then incrementing the Y register while holding X fixed, then decrementing the X register..., and so on.

Two excessively experienced programmers have used the X–Y hardware to draw cubes, which they then could rotate about any of three axes.[9] Figure 28N.15 showing multiple cubes is a multiple exposure showing the cube rotating over time. This sort of task requires too much programming sophistication – and too much code – for ordinary mortals.

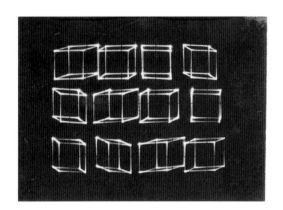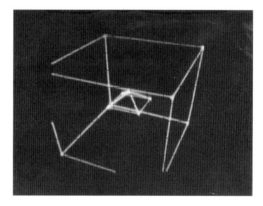

Figure 28N.15 Computed drawing: two versions of cube that can be rotated about its axes: Shumaker/Gingold.

28N.3.7 Laser X–Y display

A mirror that can be tilted on each of two orthogonal axes can steer a laser beam to project an image on a screen (or ceiling). The first students to try this built an admirable structure using two audio speakers, shown in Fig. 28N.16.

The massive mechanism of Fig. 28N.16 *worked*, sort of: the mirror did tilt in response to signals from two DACs; the projected laser beam thus could be placed here or there on the ceiling under program control. But as a way to draw an image it was a sad failure. The mechanism was much too massive and therefore much too slow.

---

[9] Grant Shumaker did this first (1986), and did it entirely in assembly language, entered by hand. This is a feat roughly comparable to climbing one of the middle-sized Himalayas without oxygen. David Gingold did it again, this time putting shapes within cubes; David used oxygen (in the form of a C compiler), as noted in §28N.4.3 in a discussion of David's remarkable asteroids game.

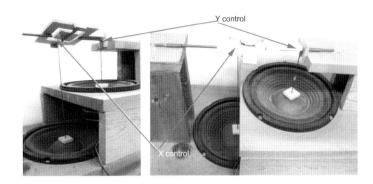

Figure 28N.16 X–Y laser mirror assembly – sadly oversized.

A later student made the scheme work by radically reducing the scale of the mechanism. Instead of eight-inch woofers he used the tiny speakers in a set of headphones, and to these he attached a tiny shard of mirror.

28N.3.8 Charmingly simple LED display

You may have seen persistence of vision displays made by spinning or oscillating sticks of LEDs; some can be attached to a bicycle wheel; others attach to fan blades, or to an oscillating rod. A student wanted to make such a thing – and at first was stalled by the difficulty of getting electrical signals to the object that was spinning.[10] To rig brushes for sliding contacts – one for each LED – seemed problematic. (One can buy such wiring harnesses, but we didn't have one at the time.) He wisely shied away from committing much effort to fabrication.

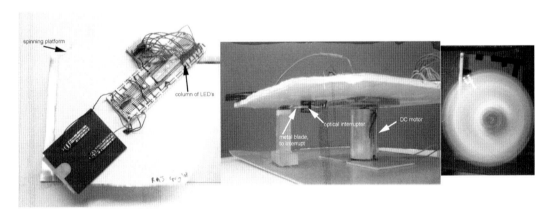

Figure 28N.17 Spinning display: computer and battery spin, along with display.

His solution was to spin *everything*: not just the display LEDs but also the computer and its power supply (a battery). His display didn't have much to say: it counted out the digits 0 through 9. But it *worked*, and that was gratifying to see.

28N.3.9 Vehicles

Tractors with independent drive for two wheels: Two motors mounted end to end, and each with a rubber-tired wheel on its shaft form a tractor that can be steered, in the manner of a bulldozer or tank, by driving the two wheels independently. It has the useful ability to pivot in place.

---

[10] This was Ryan Jamiolkowski, 2008.

Figure 28N.18 Tractor/tanks: steered by independent drive of the two rear wheels.

The version on the right in Fig. 28N.18 is a commercial base. This model is no longer available, but similar chassis are.[11] Students have used it to make a line tracer and an odd sort of dancing machine that would learn what steps to dance by scanning a nearby radio-frequency "tag."

The home-built version on the left in Fig. 28N.18 was applied by a couple of ambitious students (Mike Pahre and Danny Vanderryn) to blunder its way through a maze. Its only sensor was its nose-bump switch, but this was enough to allow the tractor, after laboriously finding its way to the far side of a maze built with boxes and books, to make the return trip without going down any blind alleys. It cruised calmly out, avoiding all the dead-ends it had found on the way in. When the tractor had completed its graceful exit from the maze it then did a victory dance in the manner of a football player who has just scored a touchdown. We thought it was entitled to feel pleased with itself.

That was a challenging project, and putting together the tractor itself requires some machine-shop skills. Because the motors draw a lot of current, the tractor was fed through an umbilical cable, which carried logic signals as well as power for the motors.

The commercial tractor (really just a chassis), in contrast, uses low-power servo motors that have been modified for continuous rotation. See §28O.3. These motors run happily from a flat polaroid battery or a 3-cell AAA pack carried by the little vehicle.[12] But lacking the stepper-motor drive, which allows easy repeatability of a travel path, this tractor would need some feedback and recording scheme to mimic the stepper-tractor's feat of maze mapping.

28N.3.10 A neat application for the independent-drive steerable chassis

Two students (Lucas Kocia and Farrell Helbling) took advantage of the ready-made steerable chassis to make a neat car with a little bit of intelligence. The idea was simple and they implemented it in an afternoon: a car that is smart enough not to roll over the edge of a cliff (well, really the edge of a table).

They gave their car a long snout, and on it put an optical ranging device – a clever Sharp infrared

---

[11] Parallax offers a circular base, SEN-13285 (`https://LAoE.link/Robot_Base.html`). Pololu offers a similar round chassis plate with cutouts for wheels (Pololu 5" Robot Chassis RRC04A) (`https://LAoE.link/Robot_Chassis.html`). For either base one would need to select appropriate wheels and then improvise servo motor mounts.

[12] In the days before digital cameras, if you wanted instant photography you bought a Polaroid camera whose disposable film pack included an integral battery. This made for a plentiful supply of overdesigned, flat 6.8 V batteries capable of supplying amps of current for a short time: see `https://LAoE.link/Polapulse.html`. Alas, nowadays you will have to make do with alkaline or rechargeable lithium-ion batteries for your project.

28N.3 Some other memorable projects

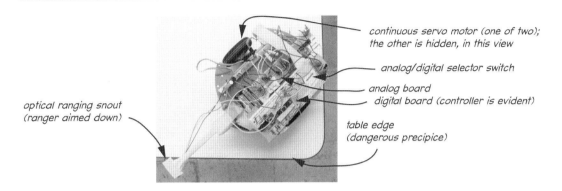

Figure 28N.19 Smart anti-suicide car.

module with range of 4–30 cm – pointing down.[13] The ranger thus can sense when the snout protrudes over the edge of a table.

The ranging device uses *synchronous detection* to ignore ambient light. It applies a square wave Enable to its IR emitter. While the emitter is On, the integrator accumulates the reflected light plus ambient. When the emitter is Off, the receiver *inverts* its signal – ambient only – before feeding it to the integrator. Thus ambient light cancels itself.

The two students showed us this car flirting with disaster but always turning back from the cliff – and then startled us by revealing that they had done it with pure analog methods (apart from the arguably digital use of a '555 as a one-shot). Probably we looked disappointed to hear that they had not taken advantage of the microcontroller that they had met in the course. A day later they invited us to watch the car again do its suicide-avoidance routine. But this time they had done the task using the little SiLabs controller. A switch selected between the two modes, *analog* versus *digital*. We were charmed: what a neat way to demonstrate the wide range of skills that we hope students would have picked up by the end of this course.

28N.3.11 A computer-steered toy car

Two students (Jonathan Wolff and George Marcus) proposed a project that we thought wildly excessive: their little computer was to steer an *RC*-controlled car, chasing a car steered by a human. (We mentioned their ambitions in §28N.1.) "Please," we pleaded with them, "couldn't you start by seeing if the computer can just steer a car?" Reluctantly, they consented – looking as if they considered this a dull and trivial task. We had ruined their game.

Many days later, they demonstrated such a design, and it had turned out to be a very challenging task. The "chase me" game never arrived. Their design entailed several substantial tasks:

- The hardest part of their project lay in letting the computer locate the little car. Their solution was to put ultrasonic "squeakers" on the car, then to listen for the squeak with three ultrasonic detectors. Their computer would command a squeak;[14] the time delays to the three squeak detectors would, in principle, indicate the car's position, through triangulation.
- *Triangulation*?! Trigonometry on the little computer, in assembly language? They recognized that this sounded painful, and decided to turn this computation-heavy work over to a desktop PC.

---
[13] Sharp GP2Y0A41SK0F, available from Sparkfun, which also offers a longer-range version: see https://LAoE.link/Proximity_Sensor.html.
[14] Sad to say, under pressure of time, they sneaked a wire to the little car, rather than command the squeak through a radio signal.

To that end, they built a little-computer-to-PC interface (a job that would have been a respectable project in itself). The little computer shipped the raw numbers – time delays; PC responded by shipping back coordinates of the car's location.

* Given these coordinates, the little computer figured out how to get the car from where it was to where it ought to be.
* Finally, the little computer "drove" the car to its destination. To do this, it controlled switches that were designed to be controlled by a human pressing pushbuttons. Implementing such switches was not hard. But steering the car was. They had bought the cheapest car offered by Radio Shack, and this car could not be told simply "turn left." In order to turn left, it would have to turn the front wheels *right*, and then *back up*. Only then could it drive forward to "go left."

Remarkably, they made all this work. Figure 28N.20 shows the little car, now much battered, and also a snapshot of the two designers looking at their project with odd detachment. (Shouldn't they be dancing with excitement?) In the figure, one can make out two of the three ultrasonic detectors attached to breadboards placed on two stools.

Figure 28N.20 Little *RC* car. Triangulation located its ultrasonic squeak.

28N.3.12 A perennial heartbreaker: inverted pendulum

This is a problem routinely assigned in *controls* courses. It is also very difficult, done seat-of-the-pants in the style of this course. People have come close to success, but real success has eluded everyone, so far.[15]

Fearless pioneer tried two dimensions: The first person to try attacked the problem in a style that made success extra-unlikely, and yet made an impressive approach to doing the job. He had not

---

[15] Examples of others' successful implementations appear on YouTube: see, e.g.,
https://LAoE.link/Inverted_Pendulum_1.html (stick on printer-like base); or
https://LAoE.link/Inverted_Pendulum_2.html (flying inverted pendulum). Search for "inverted pendulum" for many more.

28N.3 Some other memorable projects

studied controls and probably didn't know how hard a task he had set himself. If he had, would he have undertaken it in *two* rather than the usual *one* dimension?

He used simple and entirely analog methods to attempt to keep a pencil balanced on its point. The choice of a *pencil* rather than the more usual long rod made his job harder still. Nevertheless, his primitive machine was able to keep a pencil balanced on its point for a few seconds, by moving a little platform about on a smooth base.

This entirely analog design shone an LED on the upright pencil while a pair of photodetectors watched to see if the pencil leaned away from the vertical. The photodetectors fed a differential amplifier, which drove a DC motor whose response tended to force the pencil upright. The motor did this by spinning a pulley on which a string was wound, and the string would draw the balancing-block one way or the other on the base platform.

Two such LED-photodetector-pair units took care of watching for X and Y lean. The negative feedback in this arrangement made the mechanism work quite well even though the construction was crude in the extreme: sewing thread, wooden pulleys, a composition-board base, and odds and ends of adhesive tape were its materials.[16]

The mechanism worked pretty well until the little platform reached one of the edges of the base. There, the pencil would fall. What the whole design lacked was any awareness of the position of the moving block upon the base – and the slowness of its response sets it up as vulnerable to instabilities like those you saw with the PID motor control. Later efforts were more sophisticated, but not more charming than Paul Titcomb's.

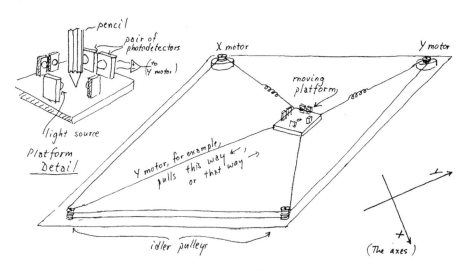

Figure 28N.21 A wonderful partial success: pencil balancer.

One-dimensional inverted pendulum: After this initial attempt, everyone tried the task in the more manageable form, in *one* dimension, using the guts of an old printer to move the base of a long stick.

---

[16] This balancer was the work of the inspired tinkerer *extraordinaire*, Paul Titcomb. Another of Paul's amazing projects deserves mention here; it may inspire someone to a similar effort. Paul made a drawing-mimicker using his lab computer. Paul rigged a 2-jointed arm that held a pen. He placed a potentiometer at each joint, and as he guided the arm, drawing a picture by hand, he had the computer read the joint-potentiometers at regular time-intervals. Then, when he had finished a drawing, Paul would ask the computer to play back the motions it had sensed, by driving a pair of stepper motors on the joints of a similar arm also holding a pencil. The machine never quite worked: its drawings showed a rather severe palsy – probably an effect of its mechanical crudity and the fact that (unlike the pencil balancer) it operated *open loop* – without the error-forgiving magic of feedback. It was an impressive gadget nevertheless, and the arm and its last, shaky drawings may still languish at the back of our lab awaiting someone to perfect the scheme.

Project Possibilities: Toys in the Attic

We began to collect discarded printers, and fitted one with a second potentiometer that indicated the *position* of the moving base in addition to the *tilt* information given by a first pot. This information provided a warning when adjustment range was close to finished.

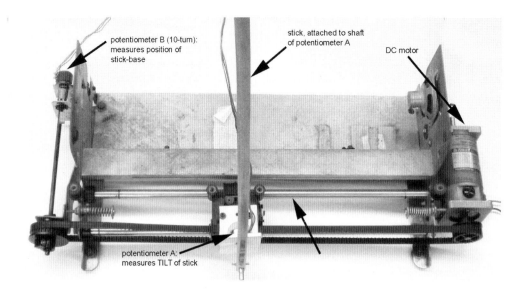

Figure 28N.22 Inverted pendulum hardware: the guts of an old dot-matrix printer drive.

The results have been mixed. One student whose computer control of this setup failed wrote a thorough and somewhat formula-rich report concluding that he had failed because the motor's response was too slow to allow it to catch the falling stick.[17] He proposed that the job could not be done till we got a heftier motor.

A term or so later, two students (Justin Albert and Partha Saha) came closer to success using purely analog methods. These students were sophisticated enough to be well aware of the inherent stability problems, and tried summing various sorts of error signal, making a PID loop that resembled the one you built in Lab 10L. Their circuit worked well enough that we asked them to show it off to the new crop of students on opening day. In September, they pulled out the machine that had worked in May – and were unable to get it to work again. This is a sad tale. Is it an argument against analog circuits trimmed with multiple potentiometers – an argument in favor of digital methods?

A recent effort used a digital implementation of a PID loop. It showed admirable intelligence, shimmying the base in its efforts to keep the stick vertical. After a few seconds, it would lose the struggle, unable to provide sufficient corrections. It was clearly trying, though. The tantalizing *near* success of the machine is visible in the photos of Fig. 28N.23: first, with anxiety and apprehension, the designers launch the thing;[18] then they show their delight as the machine makes its frantic small adjustments, keeping the ruler upright for several seconds. We omit the sadder conclusion in which the ruler falls.

---

[17] Is it significant that this fellow was an MIT grad?
[18] These are Emily Russell and Lusann Wang (2012). Their collaborator, Tom Dimiduk, is not shown. They used an Arduino controller rather than our then-standard 8051, because Tom was familiar with the device.

Figure 28N.23 Two students test their brainchild. Their expressions show apprehension and hope.

28N.4 Games

28N.4.1 Pong on a scope screen

Students Ashlyn Frahm and John Alex Keszler created a Pong game using only analog electronics and discrete digital devices – no microcomputer! (Perhaps even more impressive, both were freshmen with no prior electronics experience at the time.) The walls, ball and paddles were implemented with separate analog circuits and multiplexed to the scope X and Y inputs with 8-to-1 analog multiplexers. They used linear potentiometers for the paddles. Window detector comparators sensed collision with either the walls or the paddles. 74HC161 counters kept score and the entire game ran on single-supply +5 V. The build was frightening but worked: see Fig. 28N.24.

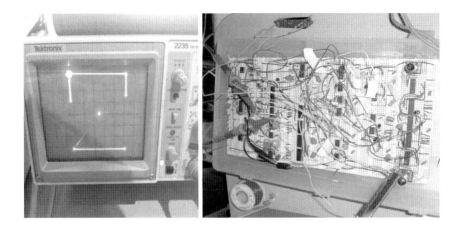

Figure 28N.24 An X–Y scope Pong game.

28N.4.2 Pacman on a scope screen

This was insanely excessive: it required a home-made "video board" to refresh the scope screen; the processor was not quick enough to redraw the full screen, over and over. Instead, the designer set up RAM that was written initially with the screen layout, and then re-written rapidly using hardware counters.

The computer's display duties thus were reduced to writing only *changes*: Pacman's mouth opened and closed, and he moved, as monsters moved too.

Figure 28N.25 Two Pacman screenshots.

We wish we could show it to you in action – but during one summer somebody pulled the battery plug, and we've been too lazy to re-enter the code so far. We could not bring ourselves, however, to destroy the hardware; so, maybe someday....

28N.4.3 Asteroids on a scope screen

Pacman used a raster-scanned display, with a mostly-fixed display defining the maze. Asteroids used vector drawing like that described earlier in §28N.3.6.[19]

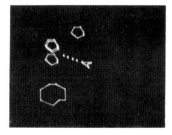

Figure 28N.26 Asteroids on a scope screen.

The game included the gravitational field of the "real" asteroids game. It was a marvel to watch (and this is another of those pieces of hardware that we keep around, unable to bring ourselves to tear it apart).

28N.4.4 Other great games

Other wonderful games have appeared, but we lack photos of these; we'll have to settle for words.

---

[19] This was David Gingold's project, so hugely ambitious that we have to note that he got credit for it in two courses; the other was one in computer animation. He put together not only the code but also hardware that would draw lines of constant brightness, and having compiled the C code on a full-scale computer, he needed to pump it into his little home-made computer. This need arose on a weekend, so rather than ask us for help in finding a UART IC he wrote a serial interface in software – as if he needed a challenge to make his project a little harder. His serial link worked, as did the game.

Stephane's *virtual-bubble* game. Stephane Ryder figured out how to decode the signals from a Nintendo *Power-Glove* – a neat toy that can detect position and orientation of a glove, and also finger closings. He used these signals to let his computer implement a game. The computer would issue beeps at a rate that increased as the glove came nearer to the 3D coordinates of a "virtual-object" – just some arbitrary X,Y,Z point in the air. When the glove was at the target point, one could close the hand and then *move* the target point, depositing it wherever one opened the hand. Dazzling!

Nomeer & Clay's punch-out interface. These students[20] replaced the keypad of a Nintendo boxing game with an array of photosensors set in a square that looks like an empty picture frame. Behind the frame they put a TV set showing the boxers. To punch the adversary the user punches air in the picture frame. The nine segments of the frame defined functions earlier assigned to keys on the keypad: uppercut, body-blow, etc. It was not a complicated project – but fun.

Scope maze game. Computer handled scope display, but the patterns displayed were stored as state machines in PALs. So, one could change the maze (a set of rooms: rectangles with passages to other rectangles) by swapping PALs. (Meredith Trauner & Robby Klein).

Tetris on an LED array. (Kurt Shelton): 8051 running from flash. An awful lot of coding – but it did work.

28N.5   Sensors, actuators, gadgets

We've moved this section to an online chapter so we can try to keep it current:
`https://LAoE.link/LAoE_Chapter_28O.pdf`.
Please let us know if you find any information on parts or suppliers in it that needs to be updated.

28N.6   Stepper-motor drive

Generally: A stepper motor contains two coils surrounding a permanent-magnet rotor; the rotor looks like a gear, and its "teeth" like to line up with one or the other of the slightly offset coils, depending on how the current flows in the coils. A DC current through these coils holds the rotor fixed. A reversal of current in either coil moves the rotor one "step" clockwise or counter-clockwise (a coarse stepper may move tens of degrees in a step; a moderately fine stepper may move 1.875°: 200 steps/revolution). The sequence in which the currents are reversed (a gray code) determines the direction of rotation: see Fig. 28N.28.[21]

Integrated stepper driver: Integrated stepper-motor driver chips can make driving a stepper extremely easy: the chip usually is just a bidirectional counter/shift-register, capable of sinking and sourcing more current than an ordinary logic gate. Fancier ICs, like the Allegro A3967SLB[22] can provide finer control: so-called "microstepping."[23] The A3967SLB permits 1/8 steps. This part also includes driver transistors capable of more than that expected in a modest 24-pin package: 750 mA

---

[20] Noah Helman, Sameer Bhalotra and Clay Scott.
[21] See also the helpful animation at `http://LAoE.link//How_Steppers_Work.html`.
[22] This surface mount part is available in a handy breakout board from Sparkfun: ROB-12779, "EasyDriver - Stepper Motor Driver."
[23] Microstepping – which can be pushed to extremes such as 1/256 step – works by energizing *both* windings at graduated relative currents. The price one pays for this result is drastic reduction in torque so one should not rely on microstepping for extreme resolution. See `http://LAoE.link/Microstepping.html`.

at up to 30 V (though not both at the same time). This is not enough for a large stepper motor, but is impressive, indeed.

Such drivers reduce the controller's job to determining *direction* of rotation and rate of stepping (an low-to-high edge on the STEP input pin advances the motor one increment).

The controller can drive the stepper: If you don't want to be so fancy, and so lazy, you can use a few pins of the controller to drive power transistors wired to the motor. A power MOSFETs with $V_{GS(Th)} \leq 2.6$ V will do the job: the IRLZ34 used in Lab 12L (55 V, 30 A max) would do.[24] Fig. 28N.27 shows the scheme. Integrated stepper-motor drivers normally include current-limiting. The home-brew scheme shown in Fig. 28N.27 lacks such limiting, so it is up to you to select a motor supply *voltage* that will not overheat the motor. It can get somewhat hot to the touch without damage.

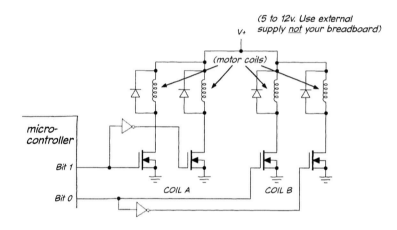

Figure 28N.27 Stepper driver hardware using two PORT I/O bits.

One way to generate such successive patterns is to load a value into a register and rotate that value, feeding two adjacent bits to the motor's two coil-drivers: see Fig. 28N.28.[25]

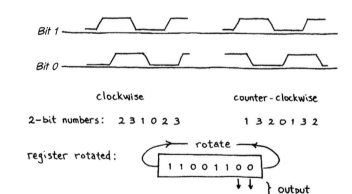

Figure 28N.28 Rotated register can produce the 2-bit drive pattern for stepper.

---

[24] The snubber diodes shown in Figure28N.27 will protect the transistors from voltage spikes on transistor turn-off. These should be power diodes such as 1N400x (the "x" specifies max reverse voltage, not critical in this circuit). Small-signal diodes like our usual 1N4148/914 cannot handle the current.

[25] Here the CMSIS _ROR intrinsic instruction would come handy. However, this machine level instruction callable from C rotates a 32-bit value so you will have to duplicate the register values in Fig. 28N.28 four times. There is no ROL instruction. To rotate left by 1 bit, rotate right by 31 bits.

Using the stepper motor: You will dream up your own ways to use a motor, but here are a few uses students have tried.

Crane: Two motors, each with a spool on its shaft to wind cord, can rest on a tabletop with the load suspended from both cords. The two motors can work together to lift, move, then lower the load. This is easy to do crudely but would be challenging if one wanted to let the load move vertically, then horizontally, then vertically again. A small electromagnet could let the machine pick up and drop an iron load (a washer, perhaps). (A small spool of wire-wrap wire is handy for an instant home-made coil; put an iron bolt through its core.) (One student, Dylan Jones, built such an electromagnet, hung it from a stepper-driven "winch," and mounted the whole thing on the tractor like that in Fig. 28N.18.)

Drawing machine: As we said back in §28N.2.3, two motors can drive the X and Y knobs of a child's sketching toy: see Fig .28N.29.

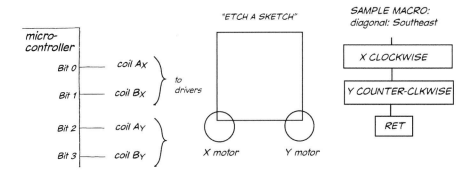

Figure 28N.29 Two stepper motors can drive an X–Y drawing toy using four PORT I/O pins.

28N.7 Project ideas

28N.7.1 Nifty new and untested projects for your inspiration

Speech recognition: You certainly don't want to try it the conventional, full-scale way: transforming to the frequency domain and then comparing against a template. That is a huge programming task, ill-suited to this little machine and our tools. (But see the cheap trick suggested at the end of §28O.5 – use someone's IC that does the hard work for you.) We want a rough-and-ready method.

One strategy – used, for example, on an old Radio Shack chip that could distinguish "Stop" from "Go" for toys – is to look at just *zero-crossings*: in other words, find the dominant frequency (this method might detect the "Sssss" in "Stop"). Count zero-crossings in hardware or software, and check against a few range definitions: "≥ 5 kHz$\Longrightarrow$Sssss," and so on.

A little more refined: use a pair of steep filters, highpass, lowpass, with little overlap; then compare amplitudes (time-averaged, we suppose) above and below this frequency boundary. Bell Labs experimented with this method in the early '50s, placing the boundary around 900 Hz, and found they could distinguish the 10 digits pretty well.

Audio encryption: Here's an idea we saw proposed in a trade magazine: simple encryption of audio – say, for a cordless phone – by inverting the frequency spectrum of the speech.

Do this by exploiting a violation of Nyquist's rule: multiply the speech data (in digital form) by a large square wave at a frequency just slightly above the top signal frequency. The result is an inverted frequency spectrum (because of aliasing or folding). De-crypt by multiplying by the same frequency. The "multiplication" is simpler than it sounds: just flip the MSB of the data, at the multiplication rate (apply a 1, then a zero, then a 1... to an XOR with the MSB of data). This also can be done with analog methods.

Music effects: chorus, etc.? You may dimly recall a phase-shifter, back in Lab 7L, that could vary phase by varying a resistance to ground.[26] If you use a digipot like the one you met in Lab 23L you can make a computer-controlled phase shifter.

An alternative way to get the same result might use a *multiplying DAC* – which can be described as a digitally-controlled attenuator; output is the product of the analog and digital inputs.

Combined with a summing circuit, in principle either shifter can make interesting rock-and-roll sounds – like those vaguely mentioned in Lab 7L. For some other sound effects circuits, try `https://LAoE.link/Effects.html`.

28N.8 And many examples are shown in AoE

AoE §15.8

We have suggested just a few possible peripherals for your controller in Chapter 24N. AoE suggests many more.

28N.9 Now go forth

We hope that you'll be moved to build a *project* now that you've made it through the book. But our larger wish of course is that you'll take what you've learned of electronics and apply it in many settings. And we hope too that you'll continue to find electronics intriguing and fun. The customer in Fig. 28N.30 goes a bit beyond our ideal: we do want your devices to be useful. But we admire his enthusiasm. Probably all of us geeks have in our basements some gizmos like the one he's buying – cool things that we just had to have.

Figure 28N.30 A model of enthusiasm for electronics – though maybe not for its utility. www.cartoonstock.com

---

[26] This was the second of two phase-shifter designs.

28O Online Content: Toys in the Attic Sensors, Actuators, and Gadgets

28O.1 A few good sources of sensors, actuators, and other devices

28O.2 A source of small parts for mechanical linkages

28O.3 "Servo" motors

28O.4 Nitinol muscle wires

28O.5 Transducers

28O.6 Displays

28O.7 Interface devices

Available at https://LAoE.link/LAoE_Chapter_28O.pdf.

A Appendix: Debugging Circuits

Figure A.1 Threatening your circuit rarely works.
www.cartoonstock.com

You will minimize the time you spend debugging by taking your time and *building circuits carefully and neatly*. Break your circuit into self-contained blocks and build each block individually. Test and debug a block before moving on to the next one. This has two benefits over building your circuit all at once: 1) If you build an entire circuit and it does not work, you have no idea where to start looking for the problem; and 2) if you have more than one problem you don't know when you have fixed one of them since the circuit as a whole still does not work.

Nevertheless, many times your circuits will not work as you expect. It is best to take a methodical approach to debugging. Start with the most likely explanation (the power is off, you connected something incorrectly, you used an incorrect part, the LED is in backwards, etc.) rather than assuming the less likely (the op-amp is bad, the microcontroller is running code incorrectly). Here are some useful troubleshooting steps to take when you have a problem.

If your circuit is not working: REALLY, REALLY IMPORTANT! Make sure that you have an accurate schematic of what you built with all parts labeled and pin numbers annotated! It is near impossible to troubleshoot if you do not have adequate documentation.

A.1 A debugging checklist

1. Is the power supply on and is power getting to your circuit?
2. Use an oscilloscope to check your power supply buses and ensure that the voltages are at the correct value and don't have odd transients (a multimeter won't work for this).
3. Do your power supply buses have bypass capacitors?
4. Check integrated circuits and make sure that they are correctly powered and connected to ground.
5. Use a scope to check the value of the supply voltages at the supply pin of the individual component - sometimes wires can slip out, a pin can be bent or the wire can be stripped too short. These problems can be very difficult to debug because everything *looks* ok.
6. Make sure that your probes are switched to the 10× mode. You can use the square wave calibration output of the oscilloscope to verify that your scope is measuring things properly.
7. Measure caps/resistors to make sure that they are the correct value (it can get hard to read stripes/codes) and that polarized capacitors are correctly oriented. You must remove components from the breadboard to measure their value with the multimeter.
8. If things are still not working, get a second set of eyes on it! Even explaining what you are doing out loud to someone who has no idea what you are talking about often helps find the problem. (Odd but true - it is surprising how often explaining your problem out loud makes *you* realize that something doesn't make sense.) You can also try building the circuit on a separate part of your breadboard - it is unlikely that you will make the same non-conceptual mistake twice.

Every time you make a change, start at the top of this checklist again!

A.2 David's Axiom

If you cannot explain why a change to your circuit fixes your problem or only explains some of the symptoms, you have not fixed the problem.

Sometimes, after hours of frustrating debugging, a change makes the problem go away but you cannot explain why the change should have fixed the problem. From years of experience we can assure you that not only is the problem *not* fixed, but it will reoccur at the worst possible time. Resist the temptation to not worry about why it is working and move on.

Persevere. At some point you will actually fix the problem and it will be crystal clear what the actual problem was and why the fix corrects all the symptoms. At that point, you have the problem solved.

B Appendix: Pinouts

B.1 Analog

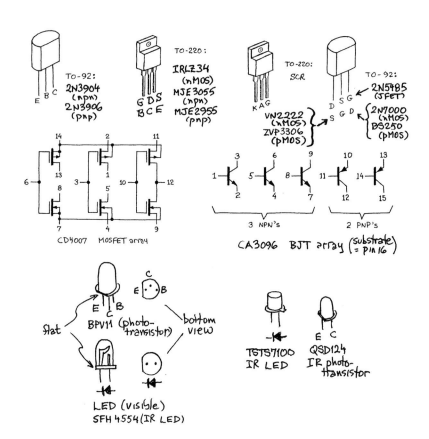

Figure B.1
Transistors and LEDs.

B.1 Analog

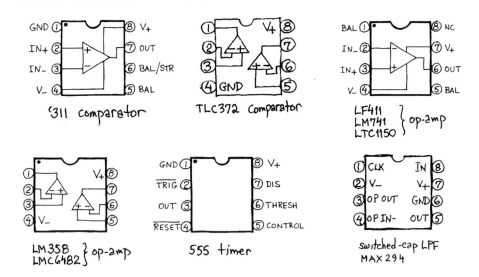

Figure B.2 Linear ICs: op-amps, comparators, etc.

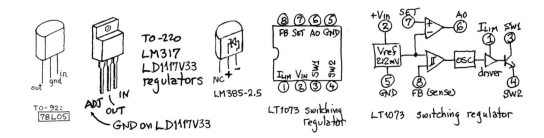

Figure B.3 Voltage regulators.

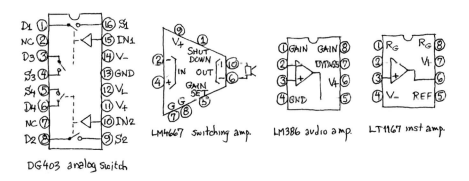

Figure B.4 Miscellaneous analog ICs.

B.2 Digital

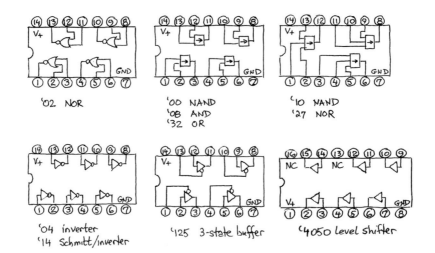

Figure B.5 Gates.

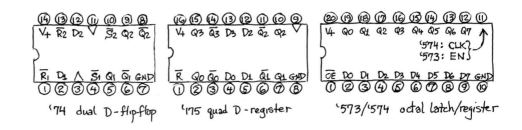

Figure B.6 Registers.

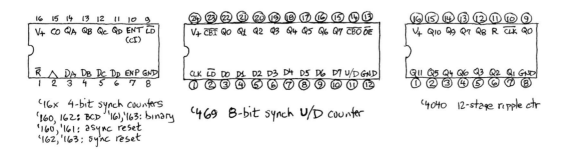

Figure B.7 Counters.

B.2 Digital

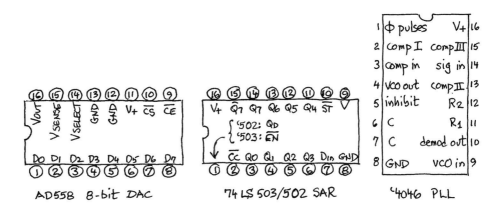

Figure B.8 Converters: PLL, DAC, and SAR.

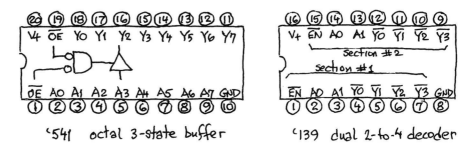

Figure B.9 Three-state buffer; decoder.

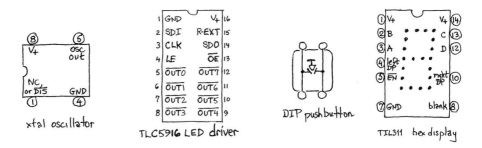

Figure B.10 Digital miscellany.

1118   **Appendix: Pinouts**

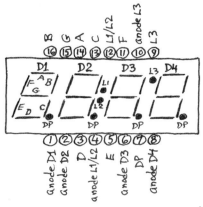

Figure B.11 Seven-Segment LED display.

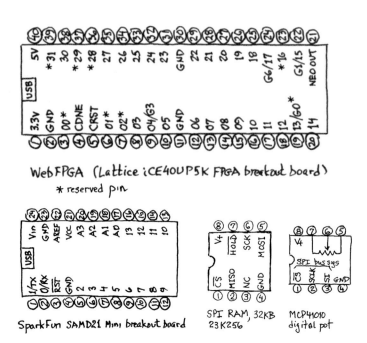

Figure B.12
Microcontrollers,
FPGA, SPI RAM, and
digipot.

C Appendix: Transmission Lines

Contents

C.1	**A topic we have dodged till now**		**1119**
	C.1.1	The impedance rules we have applied so far, when *A* drives *B*	1120
C.2	**A new case: transmission line**		**1120**
	C.2.1	Considering frequency (of sinusoids)	1121
	C.2.2	Considering risetime	1121
C.3	**Reflections**		**1122**
	C.3.1	Pulses rather than sinusoids	1122
	C.3.2	Improper termination 1: end open	1122
	C.3.3	Improper termination 2: end shorted	1123
C.4	**But why do we care about reflections?**		**1124**
	C.4.1	Remedies: termination	1125
	C.4.2	Classic remedy: terminate the line in matched resistance	1125
	C.4.3	Series termination: include a driving resistor matched to the line's characteristic impedance	1126
C.5	**Transmission-line effects for sinusoidal signals**		**1127**
	C.5.1	Familiar: lowpass behavior	1128
	C.5.2	Novel: transmission line effects	1128
	C.5.3	Partial impedance mismatch	1129

C.1 A topic we have dodged till now

Because, in our labs, we have not encountered the high frequencies that call for consideration of *transmission line* behaviors, you might leave this course unaware of this way of looking at impedance questions. You will need this way when you encounter frequencies significantly higher than the highest we have seen in this course. We have seen signals of only a few MHz, with perhaps a single exception: the fast parasitic oscillation of the discrete follower of Lab 9L. That frequency did come close to 100 MHz for some of you. In addition, you clocked your microcontroller at 48 MHz, but you were not obliged to do any handling of this clock; you just had to confirm its frequency on your scope.

We should correct our claim that we have not met high frequencies in the course, by taking account of the *fast edges* that we did, indeed, meet in the digital parts of our work. The steep edges of digital signals include very high frequencies – 100 MHz and above. But we did not ask these digital signals to drive lines long enough to cause difficulties.

At worst, for example, such edges driving a ten-inch breadboard trace evokes some ugly junk, as shown in Fig. C.10 on page 1127 – but probably not enough to cause false clocking. In contrast, a longer line driven by such an edge can produce really troublesome junk. See, for example, the large swings at the far end of the eight feet of coaxial cable driven by such an edge, in Fig. H.7 in *The Art of Electronics*, Appendix H. There we see a sort of ringing, at amplitudes easily large enough to cause false clocking.

AoE Fig. H.7

C.1.1 The impedance rules we have applied so far, when A drives B

The most common case – voltage source at modest frequencies: On the very first day of the course we began to drum away at a theme concerning the recurrent case in which circuit fragment *A* drives another fragment, *B*. We have said, over and over, that we like to make sure that *A*'s Z_{out} is low relative to *B*'s Z_{in}. And we elevated to the status of a *rule of thumb* the notion that this relation should be a ratio of 10 to 1:

$$Z_{out\_A} \leq \frac{1}{10}(Z_{in\_B})$$

This design rule remains valuable. Don't worry: we are not about to tell you that transmission line rules make this rule obsolete.

A less common case – current source at modest frequencies: Now and then we have met signals that are *currents* rather than voltages. A *current source*, like the photodiode of Lab 6L (drawn in Fig. C.1 with a smaller *R*), has tastes opposite to those of a voltage source.

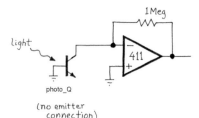

Figure C.1 Current source as signal calls for *low* R_{in}.

The photodiode in Fig. C.1 likes to drive a load whose R_{in} is *low*, and its favorite load is (of all things!) a short circuit. The op-amp current-to-voltage converter of Fig. C.1 (sometimes called by the fancy name "transimpedance amplifier") comes close to providing this ideal R_{in}. At this *virtual ground* R_{in} is very low; the golden rules would put it at *zero*.[1] (We doubt that you need to be reminded that R_{in} for this circuit is *not* 1M!)

C.2 A new case: transmission line

We have worked with coaxial cables throughout this course (we often call these BNC cables, referring to their connectors). They usually have worked for us quite transparently. Occasionally – if you found yourself using not a function generator but a circuit with an appreciable R_{source} to drive a cable – you may have become aware of the cable's substantial capacitance (about 30 pF per foot). That's as complicated as things have got so far.

But the behavior of the familiar coax cable becomes unfamiliar when we push on to frequencies higher than what we have been using in this course, while driving fairly long cables. (The effect is not limited to cables, but these are easiest to discuss because of their standardized characteristics.)

Specifically, we get new results in cases that can be described in terms using either *frequency* or *time*. We can describe in two ways the cases where we must treat a signal path as a *transmission line*:

AoE §H.1

---

[1] You're more sophisticated than that though. You recall that the actual $R_{in} = R_{feedback}/A$, where *A* is the open-loop gain. So, at a frequency where *A* was 10k, for example, R_{in} would be 1M/10k = 100 Ω.

1. frequency: when the line length[2] is, to quote AoE, "a significant fraction... of the wavelength of the highest frequency"; and
2. time: when "the round-trip propagation time is a significant fraction of the signal risetimes."

C.2.1 Considering frequency (of sinusoids)

Let's try applying the first of these criteria – the one that speaks of *frequency* – to the signals we have met in this course. Let us see whether we can confirm that our experience makes sense: we noticed no transmission line effects.

At what cable length should we begin to worry about transmission-line effects?[3] We are applying a rule of thumb that says we'll worry if cable length – the "electrical length," taking account of cable dielectric – is about *1/10* of the wavelength of the applied sinusoid.

Frequency (Mhz)	Wavelength, λ, (meters)	λ physical length (meters coax)	Xmssn-line threshold length (meters; cable = $\lambda/10$)
1	300	200	20
3	100	66	6.6
10	30	20	2
30	10	6.6	0.66
100	3	2	0.2

The "physical length" of one wavelength in the *cable*, at a given frequency, is less than in free space because the signal travels more slowly in the dielectric. So, difficulties appear at shorter cable lengths than one might expect.

But the table makes clear how we managed to steer clear of transmission-line effects in this course: we used function generators which top out between 3 MHz and 30 MHz. And on the rare day when we set the generator to a few MHz, we did not run sinusoids down a 7-meter cable.

C.2.2 Considering risetime

So much for sinusoids (except for one more glance in §C.5). But steep digital waveforms can raise transmission-line problems, even when the repetition rate of the waveform is low. The signals we met in our course did sometimes show the potentially troublesome combination of fast edges with fairly long lines. We did not worry about these effects because the result of these reflections was only to make our logic signals somewhat ugly, but not so ugly as to produce spurious crossings of logic thresholds. We admire digital circuitry for its ability to ignore a good deal of ugliness.

We did not drive long coaxial cables with logic signals. The fastest logic clock you might have encountered was the 4 MHz clock driven over a few inches of #22 wire used with the SPI RAM in Chapter 26W.

We will look at such bused waveforms in §C.4.3. But first, let's note some rules of thumb that can relate *edge rates* (or "risetimes") to signal path lengths, so as to learn when we ought to worry about transmission-line issues.

---

[2] Strictly, as AoE says, it is the "electrical length" that matters: electrical length takes into account the slowing of signal propagation in the cable dielectric. – and see §C.2.1.
[3] The table assumes signal speed is 2/3 the speed of light, about right for cable with solid polyethylene dielectric.

C.2.2.1 Relating risetime to path length

The cause of trouble It's easiest to see what causes trouble by describing what sort of reflections do *not* cause trouble: those from short lines that return to the driving end while the driver is still in transition. The reflection may distort the edge somewhat, but typically is harmless. So where risetime is well over the round-trip signal propagation time, we can ignore transmission line effects.[4]

Rules of thumb that tell one when to worry can be formulated in several ways. Here is a very direct rule for estimating line length where transmission line effects should begin. The rule describes the length of a printed circuit trace whose reflection would arrive just at the completion of the transmitting edge.

Begin treating a line as transmission line at length...

$$\text{three inches} \times t_r \text{ (in nanoseconds)}^5$$

This rule is based on the argument that such a length allows a reflection to return during the transition time. A risetime of 1 ns, for example, allows time for about 6 inches of travel. (Wave velocity on FR4 printed circuit is assumed to be about one half the free space velocity, which is approximately one foot per nanosecond.) A round-trip on a three-inch length of PC trace should take 1 ns. Hence the rule. A risetime of 3.5 ns (close to 74HC driver edge rates[6]) would put the transmission line "critical length" around 10 inches.

A less direct rule of thumb translates risetime to the highest frequency present in the waveform:

$$\text{Bandwidth (that is, max sine frequency)} \approx 0.35/t_r \text{ where } t_r \text{ is risetime}$$

So, for example, a t_r of 3.5 ns translates to 100 MHz. The table of §C.2.1 suggests, then, that for the 3.5 ns edge, transmission line effects would begin at around 0.2 m, or about 8 inches for coaxial cable or 6 inches for a PC trace. This rule is more conservative than the simpler risetime rule, "three inches $\times t_r$," which put the boundary at about ten inches. But the critical-length values from the two rules are not far apart.

C.3 Reflections

C.3.1 Pulses rather than sinusoids

AoE §H.1.2

The behaviors of pulses on transmission lines are easier to understand than the behavior of sinusoids, so let's start with these. We'll reach sinusoids later in §C.5. When everything is done right, as we'll see in §C.4.1, nothing remarkable happens when a pulse travels down even a long cable. The interesting – and troublesome – behaviors, called "reflections," result from improper *termination*. You may have seen equivalent effects demonstrated for waves on a taut rope.[7] We'll begin with the extremes of wrong terminations of a long cable to which we'll apply a pulse.

C.3.2 Improper termination 1: end open

If the end of the transmission line is left open, which in electrical terms is equivalent to driving a load whose impedance is much larger than the transmission line's impedance, a pulse is reflected at

---

[4] See https://LAoE.link/Critical_Length.pdf.
[5] Rule proposed for FR4 printed circuit. *Ibid.* A nicely done informal explanation is available for download at https://LAoE.link/What_is_Critical_Length.pdf.
[6] See Philips user's guide: https://LAoE.link/HCMOS_Specs.pdf, indicating 4 ns risetime for bus driver IC's, 6 ns for ordinary 74HC. More recent HC parts may show faster edges.
[7] A good animation and film clip of pulses: https://LAoE.link/Wave_Reflections.html. Less good is Wikipedia's treatment: https://LAoE.link/Pulse_Physics.html.

the end. The pulse is of the same polarity as the incident pulse (thus reaching double the level of the input pulse), and travels back to the input.[8] Here is an example. In Fig. C.2, the pulse is narrow enough so that the returning pulse arrives well after the extinction of the initial pulse. The two pulses at the driving end – the original and the reflection – lie about 32 ns apart. This is about what we would predict for a 3 m cable.[9]

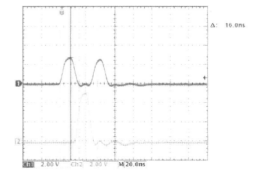

Figure C.2 Reflected pulse appears non-inverted one round-trip propagation time after the incident pulse when end is *open* (3 m cable). (Scope settings: 2 V/div; 20 ns/div.)

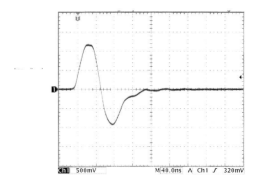

Figure C.3 Reflection from shorted end comes back *inverted* (3 m cable). (Scope settings: 500 mV/div; 40 ns/div.)

C.3.3 Improper termination 2: end shorted

At the other extreme of mismatch, the far end is *shorted* rather than open. In this case the pulse is reflected, but returns *inverted*. Fig. C.3 shows the effect for a pulse sent down a 3 meter cable.

Step input, end shorted: A curious variation on this circuit can be used to generate a narrow pulse from a step input. A first thought, if one considers what the transmit end would look like with the far end shorted, might be "You'll get nothing if R_{out} is 50 Ω, since we're looking at a divider between 50 Ω and 0 Ω." That's almost correct: you *do* get nothing in the steady state. But it takes some time for the *transmit* end to discover that the end is shorted. Meanwhile, the voltage looks like the driving voltage (or, more exactly, like $V_{source}/2$, since R_{out} forms a divider with the cable's characteristic impedance of 50 Ω). Only when the reflected wave returns does the voltage drop to zero. The pulse width equals the round-trip delay time. In Fig. C.4, the pulse begins to fall perhaps 30 to 40 ns after the rise to full amplitude. The calculated round-trip time is about that, as we argued in §C.3.2.

---

[8] This "reflecting" behavior, like that of the case of *shorted* end, result from conservation of energy. For the open end, current is zero, for the shorted end voltage is zero. The induced reflection dissipates energy in the line and driving source impedance. See https://LAoE.link/Reflections.pdf.

[9] The round-trip length is 6 m or 20 feet. Wave velocity in this coax is 0.66 what it is in free air, so about 0.66 feet/ns, and round-trip time therefore is about 30 ns.

1124   Appendix: Transmission Lines

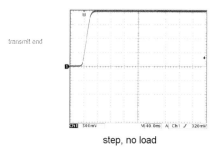
step, no load

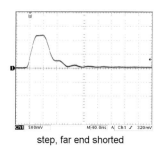
step, far end shorted

Figure C.4 Step input into shorted cable produces a narrow pulse (3.5 m cable). (Scope settings: 500 mV/div; 40 ns/div.)

C.4   But why do we care about reflections?

One can make a plausible argument that these reflections may be harmless. Send a pulse down a long coaxial cable from a logic gate (low output impedance) to another logic gate (high input impedance). What happens?

This case, with a logic gate rather than function generator driving the cable, differs from the "...end open" case of §C.3.2 in that the signal source is not matched to the cable as the function generator was (50-ohm output impedance is built into the generator, as in your lab instruments). That $50\,\Omega$ source impedance swallowed the reflected waveform, so only one reflection occurred. The low output impedance of the logic gate should, instead, produce a new reflection (and inversion), which travels anew, and will get reflected. In short, it looks as if things will get complicated and messy.

Indeed they do. Figure C.5 shows the waveform into an unterminated 6-foot coax driven by a collection of 7HCT logic gates, paralleled, for extra current).[10] Ugly!

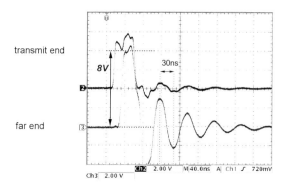

Figure C.5 74HCT logic driving 6-foot coaxial cable, unterminated. (Scope settings: 2V/div; 40ns/div.)

Not only is the waveform ugly, but the 8 V peak at the far end of the cable could damage a 5 V logic gate. The oscillations look, at first glance, like resonances that we have seen produced by stray L and C. You may have seen such junk if you forgot to attach a scope probe's ground lead, for example. But the oscillations are not that. Instead, they result from reflections that bounce off the far end, then off the transmitting end, and so on, banging back and forth. When the pulse reflects from the low-impedance at the transmitting end, it is *inverted*, and that would account for the negative-going excursion at the far end, a swing that does occur about one round-trip time (30 ns) after the initial positive pulse reaches that far end. The very best evidence that this ugliness results from reflections is the fact that proper termination cleans things up (see §C.4.3, Fig. C.7).

[10] We used all eight gates of a 74HCT541, for nominal current of about 50 mA, enough to drive a $50\,\Omega$ termination to legal logic HIGH.

C.4.1 Remedies: termination

AoE §H.1.2A

AoE §12.10.2B

The reflection effects described in §C.3.2 and §C.3.3 are intriguing. But they are, of course, undesirable – except for the short-pulse generator of Fig. C.4, which could be useful. For fast signals, proper *termination* will clean up signals that otherwise might be distorted by reflections.

C.4.2 Classic remedy: terminate the line in matched resistance

If we "terminate" the cable – that is, join signal line to ground or another voltage source through a resistor – with a resistance equal to the *characteristic impedance* of the cable, then the cable appears

AoE §H.1.1

to continue forever. The wave or pulse never hits a discontinuity.

The "characteristic impedance" of a cable describes a surprising effect of the capacitance and inductance that inhere in its construction. Two conductors, side-by-side, running off to infinity (or properly "terminated") will look to the source driving it like a *resistive* load. This is surprising, surely, because the construction of the cable presents both inductance per unit length (in the conductors) and capacitance per unit length (in the capacitance that results from the proximity of the two conductors). In addition, there will be some resistance per unit length, and some leakage through the capacitor's dielectric; but these effects usually are small enough to be neglected.

More particularly, ignoring these small losses, Z_0, the characteristic impedance, simplifies to

$$Z_0 = \sqrt{\frac{L}{C}} \tag{C.1}$$

where L and C are inductance and capacitance per unit length.

For the coaxial cables we use in lab – the ones we call "BNC cables" – that characteristic impedance is 50 ohms.[11]

If we terminate the 3.5 m coaxial cable that produced a double pulse back in Fig. C.2, no returning pulse occurs. In Fig. C.6 we show the two cases, unterminated and terminated.

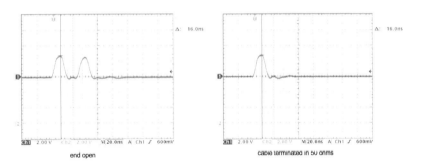

Figure C.6 Open-ended versus terminated cable (10 feet): termination eliminates the reflection. (Scope settings: 2 V/div; 20 ns/div.)

The same remedy – a 50 Ω resistor between the signal line and ground at the far end of the cable – cleans up the pulse sent by the logic gates of Fig. C.5, as seen in Fig. C.7.

This termination does clean things up – but it calls for a lot of current from the signal source: as much as 100 mA, if the termination goes to ground and is driven to 5 V.[12]

---

[11] This is the standard impedance for radiofrequency use. The standard for video applications is $Z_0 = 75\,\Omega$.

[12] The current would be less by half if we used a termination *divider* to one-half of the supply voltage, making its $R_{\text{Thev}} = 50\Omega$. This is a bit more trouble to wire. DC current could be eliminated if the termination were AC-coupled, using a blocking capacitor.

Appendix: Transmission Lines

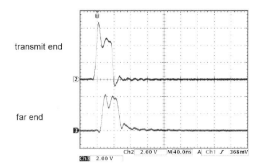

Figure C.7 50 Ω termination cleans up 74HC-to-coax drive. (Scope settings: 2 V/div; 40 ns/div.)

C.4.3 Series termination: include a driving resistor matched to the line's characteristic impedance

AoE §H.5

A less perfect but more practical protection is simply to install a series resistor on the transmitting end, matched to the cable's characteristic impedance. This is so standard on function generators that you may not even think of this as a "remedy." But that series resistance driving the cable helps a lot.

It does not prevent reflections from the far end (as we have seen in §C.3.2), but it does prevent a *second* reflection at the driving end, and prevents the later rebounds from both ends. It swallows up any reflected wave, preventing the bang-bang nonsense visible in Fig. C.5. It also attenuates the driving signal so that the level that reaches the far end of the line is full-size, not *double* size.

For logic gates, this simple scheme is preferable to the classic termination at the far end, which

Figure C.8 Series termination at the transmit end will swallow up reflections.

would call for large quiescent currents. Figure C.9 shows the cleanup provided by insertion of a 47 Ω resistor at the output of the '541 logic driver as it drives a 6-foot cable.

Driving PC board traces rather than coaxial cable: For the similar case of driving printed circuit traces, inserting a series resistance at the driving end again helps a lot. Figure C.10 shows the improvement where logic gates drive a 10-inch bus trace. The left-hand figure shows direct drive; the right-hand screen shows the cleanup provided by driving through 100 Ω. The driver is 74HCT541.[13]

Use an inherently-terminated logic family: LVDS, the differential digital logic you met back in Chapter 14N, solves the reflection problem elegantly. At first glance, the fast transitions of LVDS might seem troublesome – as low as about 0.25 ns.[14] But a glance at the typical LVDS wiring in Fig. C.11 may reassure you.[15]

The characteristic impedance of the printed circuit traces is around 100Ω, and yes, that resistor at the receiver sure looks like a terminating resistor matched to the line's impedance. Indeed it is, and it is not an afterthought. The 100Ω is not just desirable as termination: it is also required in order to generate the signal at the receiver. The transmitter's current, sourced through one wire, sunk through the other, generates the *difference* signal of 350mV.

---

[13] The drive is again the set of eight gates paralleled for high current capability used for the coaxial cable conventionally terminated in Fig. C.7.
[14] See https://LAoE.link/LVDS_Data_Handbook.pdf.
[15] Compare National/TI LVDS User's Guide: https://LAoE.link/LVDS_Guide.pdf, Fig. 1-1.

C.5 Transmission-line effects for sinusoidal signals

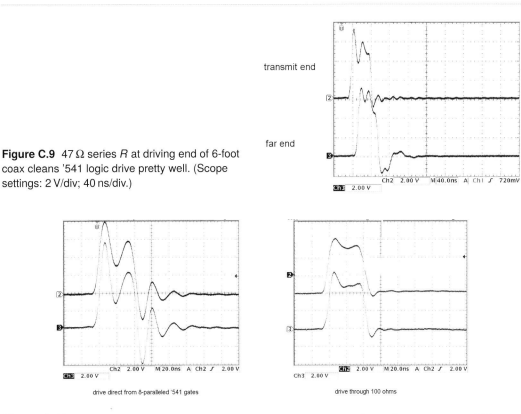

Figure C.9 47 Ω series R at driving end of 6-foot coax cleans '541 logic drive pretty well. (Scope settings: 2 V/div; 40 ns/div.)

drive direct from 8-paralleled '541 gates

drive through 100 ohms

Figure C.10 Matched drive impedance cleans up 10-inch printed circuit drive. (Scope settings: 2 V/div; 20 ns/div.)

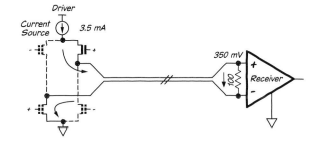

Figure C.11 LVDS transmitter and receiver wiring.

LVDS scope image: Figure C.12 shows differential signals, small but pretty clean, as they arrive at the far end of a 10-inch printed circuit trace.

C.5 Transmission-line effects for sinusoidal signals

Where the signal applied to a transmission line is sinusoidal rather than a pulse, the effects are even farther out of the range of what we saw in lab. We did apply digital signals with fast edges, but never applied high frequency sinusoids. If we had, we would have found results surprisingly different from what we saw at lower frequencies.

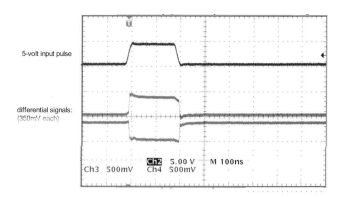

Figure C.12 LVDS inherently-terminated signal line produces pretty clean signals on 10-inch PC trace. (Scope settings: 5 V/div (input), 500 mV/div (diff signals); 100 ns/div.)

C.5.1 Familiar: lowpass behavior

At lowish frequencies, a coaxial cable behaves like a low-valued capacitor to ground. If we use our function generators whose R_{out} is 50 Ω to drive a long cable – say, the 3.5 m cable we used in most of the pulse tests above – we usually see no attenuation. This is what we are accustomed to in lab.

But we anticipate that at some high frequency we would begin to see a *lowpass* effect from the RC of 50 Ω driving the cable's 30 pF/foot of stray capacitance. If we calculate this effect for 3.5 m of RG58 coax, we get an f_{3dB} of ≈9 MHz. This relatively high frequency explains why we have not noticed any such effect in our labs: our function generators do not run that high – or if they do, we've not used them at such a frequency. And we certainly never drove 10 feet or so of coax at high frequencies.

C.5.2 Novel: transmission line effects

If we try driving this 3.5 m coax with a sinusoid, pushing the frequency beyond what have tried in lab, we get a new behavior that results from transmission line effects. Specifically, as frequency climbs, approaching one where the cable length is 1/4 wavelength, amplitude at the transmitting end falls – looking superficially like the behavior of a lowpass. Amplitude goes right down to zero when the cable length is exactly a quarter wavelength. This occurs at about 16 MHz. And if frequency climbs further, amplitude begins to rise, going all the way to *full* amplitude when cable length is 1/2 wavelength (32 MHz). This behavior is brand new. What's going on? Reflections from the unterminated end of the cable are causing this result.

Figure C.13 shows a manually swept frequency that begins at 1 MHz and moves up to about 110 MHz. The scope is watching the *transmitting* end of the unterminated cable. The "nulls" are obvious, the first at about 16 MHz.

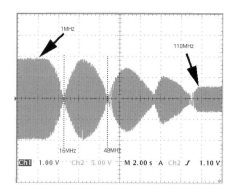

Figure C.13 Manually swept frequency shows several *nulls* where reflected waveform cancels the driving waveform. (Scope settings: 5 V/div, 2 s/div.)

When the cable length is 1/4 wavelength, the reflected waveform arrives back at the transmit end at a time when the driving signal has reached 1/2 its period. The reflected wave is then an inverted version of the driving signal. The two cancel. When the round-trip time is a full wavelength there is no interference, and again (at 32 MHz) we see full amplitude.

Let's try the numbers to see if the *nulls* of Fig. C.13 make sense. Round-trip time for the 3.5 m cable – about 11.5 feet – is the time to travel 23 feet. In the cable, where signal velocity is about 2/3 wave velocity in free space, 23 feet takes about 23×1.5 ns ≈ 35 ns. A null should appear when this time is a half period, making the full period around 70ns. That period corresponds to a null frequency of 1/70 ns=0.015 GHz=15 MHz: about what we observed in Fig. C.13. Seems to make sense.

Since the cause of the problem is the same as the cause of the pulse reflections that we saw earlier, the remedies are the same as well: matching impedances of source and load to the cable.

C.5.3 Partial impedance mismatch

We have looked at only the extreme mismatch cases: far end open and far end shorted. Less extreme mismatches produce, as you might guess, *partial* reflections. These are also undesirable. We will not illustrate them here.

D Appendix: Scope Advice

Contents

D.1	What we don't intend to tell you	**1130**
D.2	What we'd like to tell you	**1130**
	D.2.1 Triggering	1130
	D.2.2 Aliasing on a digital scope	1133

D.1   What we don't intend to tell you

We don't believe we can tell you how to use a scope by writing a lot of words. You'll learn by trying it – and, if you're lucky, by having someone with experience look over your shoulder as you try. But there are a few points that may be worth writing out and trying to illustrate.

D.2   What we'd like to tell you

> *AoE Appendix O: The Oscilloscope*

D.2.1   Triggering

Triggering is the hard element in scope use. *Triggering* is jargon for the scope's mechanism for synchronizing its successive *sweeps* from left to right across the screen. This synchronization is necessary in order make the image coherent rather than a muddled overlaying of traces. On successive sweeps, the trace had better redraw the waveform in the same left-right position as the previous sweep. If it doesn't, the traces seem to wander horizontally, as in Fig. D.1. The visual effect is thoroughly annoying.

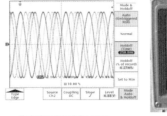

Figure D.1 Bad triggering produces muddled overlaying of images on scope screen.

Choose the appropriate signal to trigger on (reject all the duds) – starting with the most general issues...:

Don't start by adjusting LEVEL   Novices often start by twiddling the trigger *LEVEL* knob hoping

D.2 What we'd like to tell you

to see the waveform stabilize. That is the wrong way to begin. Figures D.2 and D.3 illustrate this for analog and digital scopes, respectively.

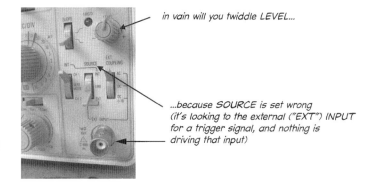

Figure D.2 Don't start by fussing with the LEVEL knob. First, set SOURCE appropriately.

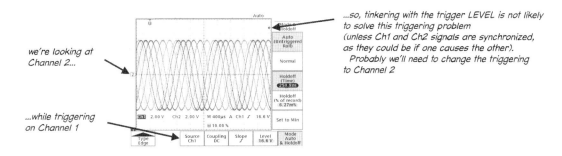

Figure D.3 LEVEL won't solve this problem either (digital scope version).

Instead, start with the most general issue: INT/LINE/EXT Before fine-tuning can work (as by adjustment of LEVEL) you must tell the scope which sort of signal to trigger on. The most general selections are those labeled "INT/LINE/EXT" on our scopes.

- INT means "internal," and refers (despite the strange name) to the scope's INPUT channels. Nearly always this is the correct choice
- LINE triggers the scope on the power supply of the scope. This is good for stabilizing any waveform synchronized with the 60 Hz "line" voltage (the supply that comes out of a wall socket).[1] If you see low-frequency noise or wobble, and wonder if it's caused by the 60 Hz noise that pervades indoor settings, LINE trigger usually answers that question: if LINE stabilizes the wobble, then, yes, the wobble is caused by power-supply noise.
LINE, then, can be useful – but not often.
- EXT means "EXTERNAL," and takes its signal from a separate BNC input somewhere on the scope.[2] This is useful when the usual trigger source – one of the input channels (INT) – is marred by noise that can upset triggering. We also use it when the function generator is in sweep mode to trigger at the start of the frequency sweep and all the normal input channels are being used for other signals.[3]

---

[1] 50 Hz in much of the rest of the world.
[2] Usually on the front of older analog scopes; in the Siglent SDS2104X Plus and Tektronix TDS3014 digital scopes, the connector is on the back.
[3] Most function generators that support sweeping include a "trig" or "sync" output for this purpose.

1132 **Appendix: Scope Advice**

Figure D.4 EXTERNAL trigger is useful when an input signal is muddled by noise.

...then choose the particular internal trigger source: You will nearly always trigger on an input channel rather than LINE or EXT. You will need to choose *which* channel – and there is a trap lurking, here.

Beware the evil "VERTICAL MODE": The analog scope shows two choices: Ch 1 and Ch 2. Between them – tempting the wishy-washy liberal – is a compromise: VERT MODE ("vertical," whatever that might mean).

Figure D.5 The evil VERTICAL mode tempts the unwary.

This mode triggers on the two channels *independently* – and as a result *loses the relative phase or timing between the two channels*.

This mode is named differently on our old digital scopes (TDS2014), and you would have to hunt through menus to find it – so it is very unlikely to lead you astray.[4] The TDS2014 calls this triggering mode "Alternating" – not to be confused with the display mode of analog scopes that is called "ALT"![5]

But if you use this mode, with analog or digital scope, you will get a display that *falsely* displays

---
[4] Our new Siglent digital scopes don't appear to include this confusing feature.
[5] You probably recall the contrast between ALT and CHOP display modes for an analog scope: ALT draws one trace for a full sweep (say, Ch 1), then draws another trace (say, Ch 2). CHOP, in contrast, jumps the beam rapidly back and forth between the two channels on a single screen sweep. You normally use ALT for faster sweep speeds and CHOP for slower ones.

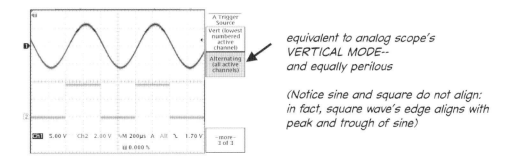

Figure D.6 VERTICAL mode (called "alternating…" on the digital scope) falsifies relative timing among channels.

relative timing of the input channels. Figure D.6 for example, shows the function generator's MAIN output and SYNC output, falsely displayed by this mode, on the digital scope:

We know that the timing display in Fig. D.6 is false because the MAIN output (a sinusoid here) and the TRIG_SYNC output (a square wave) are locked in phase by the design of the function generator, the TRIG edges coinciding with peak and trough of the MAIN output. That relation is not shown here. The scope is misrepresenting the relative timing of the two signals.

Why, you ask, do scope designers provide this nasty trap? As you would expect, there is indeed a case for which VERT mode is useful. That is the very rare situation when signals on the several channels are not synchronized. In this case, VERT does give a stable display for all channels at the expense of losing information concerning relative timing. We have to be smart enough to remember that relative timing is lost, or VERT mode will mislead us. In the entire set of labs that we do in this course, you are unlikely to want this mode even once.

"AUTO" versus "NORM": Once you have told the scope which signal to use as trigger, you still face a choice, labeled "AUTO" versus "NORM." AUTO displays traces whether or not you have set the trigger LEVEL properly; NORM does not: it shows you nothing if you have not fed the scope a good trigger signal. Since a dark screen – or frozen screen, in the case of a digital scope, which holds the last image it caught – is even worse than a muddled one, we advise you to think of NORM as short for *ABNORMAL*. We use it only in exceptional cases.

Figure D.7 shows the contrasting results of AUTO and TRIG choices for a case where the LEVEL has been set too high (above the highest level of the Ch 2 waveform chosen as trigger source; the arrow shows this trigger level, in both left- and right-hand images). The left-hand image, using AUTO, shows a stuttering display; the NORM display shows us nothing. Usually we prefer something to nothing.

The case for NORM: NORM becomes useful (rather than just frustrating) when we want to display a waveform whose trigger event is infrequent. AUTO gets impatient and sweeps when it decides enough time has passed; NORM waits patiently for the next occurrence of the trigger event. You will recognize the cases that call for NORM when you meet them. But expect these to be exceptional.

D.2.2 Aliasing on a digital scope

AoE Appendix O.2

Digital scopes are pretty nearly irresistible, but they are vulnerable to one misbehavior that is quite baffling to someone who has not seen it before (and sometimes puzzling to someone who *has* seen it before, as we can attest). This is the effect called "aliasing," the generation of a false image when the scope does not sample the input waveform often enough.

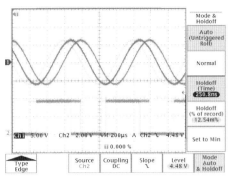
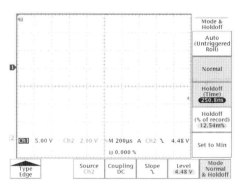

AUTO trigger, with LEVEL set too high...   ...*NORMAL trigger, with LEVEL set too high*

Figure D.7 AUTO and NORMAL trigger results when LEVEL is mistakenly set higher than the input waveform.

The topic of sampling, and the artifacts it can introduce, is treated at some length in Chapters 18N and 18S. If you are reading this appendix early in the course, it is too early to work at a full understanding of sampling effects. We will settle here for showing you an example, and then adding a few words on why the aliasing looks as it does. If you return to this appendix after reading Chapter 18N we hope the points sketched here will look familiar.

Aliasing occurs when we violate the standard sampling rule (often referred to as Nyquist's or Shannon's sampling theorem). This rule says that one needs somewhat more than two samples per period of an input to get enough information about that signal. If one samples less often, one doesn't just miss some data. One gets *disinformation*: a pseudo signal called an "alias."

For example, a sinusoid at 5 kHz should be sampled at more than 10 kHz. Sampling creates a lowest frequency artifact at $f_{sample} - f_{in}$.[6] So a 5 kHz signal sampled at 7.5 kHz would create an artifact at 2.5 kHz. In the example illustrated by Fig. D.8 we will see a similar effect.

Aliasing can produce a strange low-frequency image on a digital scope: To demonstrate this hazard, we fed a sinewave of just under 5 kHz to a digital scope[7] that was sweeping rather slowly (200ms/division). Figure D.8 is what the scope showed us.

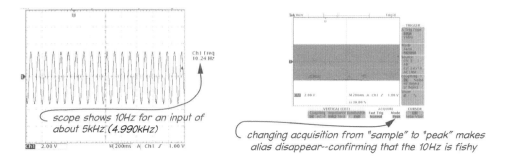

Figure D.8 A digital oscilloscope can show an unexpected low-frequency when sweeping too slowly to resolve the input signal.

---

[6] More generally, sampling produces "images" at $n \times f_{sample} \pm f_{in}$. But usually it is only the image of lowest frequency that concerns us.
[7] A Tektronix TDS3014.

Ten hertz, when we fed the scope about 5 kHz (5 kHz less 10 Hz: 4,990 Hz – a detail that turns out to be important)?[8] Is the scope crazy?

No, it's not crazy, it's just stuck with being a sampling machine. This scope collects 10,000 data points as it fills a screen. So, when sweeping slowly at 200 ms/division, a trip across the scope screen takes 2 seconds. So the 10k data points must have been picked up (sampled) at 5 kHz. This is much too low a rate to get an honest image of a signal coming in at close to 5 kHz. The result is the spurious 10 Hz signal that the scope shows us. (This alias occurs at the frequency $f_{\text{sample}} - f_{\text{in}} = 5\,\text{kHz} - 4.990\,\text{kHz} = 10\,\text{Hz}$.)

The right-hand scope image in Fig. D.8 shows what happens when (confronted by the weirdness of a 10 Hz output) we change scope acquisition from "sample" (the usual setting) to "peak," which catches highest and lowest levels, acquiring samples at a high rate despite the slow sweep rate.

We don't want to scare you with this example of aliasing. You won't see this often and it isn't really likely to fool you (you're not likely to believe that 5 kHz looks like 10 Hz, despite what the scope says). But it's useful to have heard about the possibility of aliasing. When your digital scope suddenly talks nonsense – especially if it's surprising low-frequency nonsense – we hope you'll remember this possible cause.

---

[8] The nearness of f_{in} to f_{sample} is what produces the striking low-frequency sinusoid in Fig. D.8, rather than just a puzzling messy display. But other input frequencies would have similar effects. An input close to any multiple of 5 kHz evoke the same sort of strangeness.

O Online Appendices

We've moved several chapters and appendices online; in some cases because they are rarely (or possibly never) used and in others to allow us to update them when necessary. An example of the former is the appendix on creating Verilog testbenches. We suspect many, if not most, of our readers will have no interest in the process. The parts list fits squarely in the second category. Electronics parts are occasionally obsoleted and we would like to provide information on substitutes, workarounds or alternative sources when possible. As with our redirected links, if you find something in the online chapters that is incorrect or does not work, let us know by email at `BadLink@LAoE.link` and we will endeavor to correct it.

All the online chapters are available from `https://laoe.link/Online_Chapters.html`. The additional appendices available online are:

Appendix E: Creating Verilog Testbenches – `https://laoe.link/Testbenches.pdf`.

Appendix F: Outfitting Your Lab – `https://laoe.link/Equipment.pdf`.

Appendix G: Where to Go to Buy Electronic Goodies – `https://laoe.link/Sources.pdf`.

Appendix H: Parts Lists
　　Analog Labs Parts List – `https://laoe.link/Analog_Parts.html`.
　　Digital Labs Parts List – `https://laoe.link/Digital_Parts.html`.

The parts lists are also downloadable in Excel and PDF format at `https://laoe.link/Online_Chapters.html`.

If you have an update for the parts lists or have trouble finding a part, please let us know at `parts@LAoE.link`.

Index

'311 (lab), 369
'555, 373–375
#344 lamp (lab), 375
#47 lamp (lab), 183
1N5294 current-limiting diode, 374, 487
1N5294 current-limiting diode (lab), 223, 372
1N5711 Schottky diode (lab), 143
2N3904 (lab), 177
2N3906 (lab), 289
4046 PLL, 819
74HC00 (lab):, 592
74HC14 (lab), 371
74HC161, 693
74HC175 (lab), 673
74HC4040 (lab), 831
74HC4046 (lab), 831
74HC574 (lab), 829
74HC74 (lab), 668
74LS00 (lab), 592
74LS469 (lab), 701
74LS503 (lab), 827
78L05 regulator (lab), 497
8051 microcontroller, 871

A in feedback loop: open-loop gain, 429
AB in feedback: loop gain, 429, 431
AC/DC (scope input select), 37
active filter, 386
   Bessel, 419
   Bessel (lab), 406
   Butterworth, 419
   Butterworth (lab), 406
   Chebyshev, 419
   Chebyshev (lab), 406
   transient response, 420
   VCVS, 386
active filter (lab), 405–406
active pullup gate output: CMOS (lab):, 598
active rectifier, 299
active-low
   . . . in Verilog, 631, 633
   explained, 573
   why prevalent, 580
AD558 (lab), 825
AD630
   integrated modulator, 456
ADC, see analog ↔ digital conversion
ADC (lab), 827–830
ADC & DAC to micro
   modify waveforms
      lowpass filter, 952–953
      rectify: full-wave, 950
      rectify: half-wave, 952
   number formats, 951
adder (digital)
   worked example, 613–614
AGC (automatic gain control) lab, 543
aliasing, 796, 836, 839–840
ALU
   design with gates, 617
   worked example, 617
AM radio (lab), 143–144
ammeter (lab), 32
analog
   contrasted with digital, 564
analog ↔ digital conversion
   ADC designs, 802–813
      . . . compared, 802
      "flash" or parallel, 803
      binary search (SAR), 806–808
      delta–sigma, 808–813
      dual-slope, 804
      how delta–sigma highpasses quantization noise, 811–813
      how delta–sigma lowpasses signal, 813
      oversampling, 810
      pipelining, 804
      single-slope, 716
      tracking, 806
   aliasing, 816
   DAC designs, 798–802
      current summing R:2R, 798
      PWM, 801
      switched-capacitor, 800
      thermometer, 798
   dither, 816
   quantization error, 795
   R:2R ladder, 798–800
   resolution, 794
   sample-and-hold, 806
   sampling artifacts, 814–816
   sampling rate, 795–797
      aliasing, 796
analog switch
   application
      chopper (lab), 534
      flying capacitor (lab), 536

Index

sample-and-hold (lab), 535
switched-capacitor filter (lab), 536–538
AND
  as "if", 578
  as Pass/Block* function, 577
arduino, 336
ASIC, 620
assembly language, 873
assertion level logic notation, 573–634
`assign`, *see* Verilog
audio amplifier (LM386), 944

B in feedback loop: fraction fed back, 429
ballast resistor (in BJT circuits), 513
bandpass filter (worked example), 107
Bessel filter, 419
beta, 166
bias current, 306
biasing
  divider design, 188
biasing BJT circuits
  details, 171
  why required, 169
binary numbers
  BCD (binary coded decimal), 570
  defined, 564
  hexadecimal notation, 569
  number codes, 568
  signed (two's-complement), 568
  two's-complement (worked example), 611–613
  unsigned, 568
binary search (lab), 825
binary search ADC, 806–808
binary search display (lab), 828
bipolar transistor, *see* transistor (bipolar)
bit defined, 564
Black, Harold
  formalizing feedback, 267
  memoir, 263
bleeder resistor, 151
blocking capacitor, 78
blocking vs non-blocking assignments, *see* Verilog
Boole, George, 570–572
bootstrap (op-amp input), 430
breadboard (use of), 25
BS250P (lab), 289, 496, 854
bullet timer (worked example), 722
bus, *see* digital logic
Butterworth filter, 419

CA3096 array (lab), 219
CA3600 MOSFET array (lab), 597
cache memory, 733
capacitor
  blocking, 78
  decoupling, 79
  dynamic description, 56
  exponential RC charging, 58
  hydraulic analogy, 57
  polarized, 89
  reading value, 89
  static description, 56
  tolerance codes, 93
carrier
  in lock-in amplifier, 453
carrier frequency
  group audio project, 554
carry, *see* counters
CC* (conversion complete*) (lab), 827
CD4007 MOSFET array (lab), 597
Chebyshev filter, 419
chopper op-amp, 349
CISC (Complex Instruction Set Computers), *see* microcontroller
clamp, diode (lab), *see* diode
clipping
  differential amplifier, 237
  emitter follower, 181
clock (synchronous digital circuits), 655, 656
  gating clocks, 684
CMRR
  in op-amp (worked example), 258
  op-amp from array (lab), 229
CMSIS, *see also* microcontroller programming; SAMD21
  input/output (I/O)
    port groups, 914
  intrinsic functions, 916
  introduction, 885
  NVIC functions, 973
  pointer definitions, 886
  structure definitions, 913, 930
  SysTick timer, 912
combinational logic, *see* digital logic
common-emitter amplifier
  . . . modest gain, 165
  bypassed-emitter, 207
  distortion, 203
    remedy: R_E, 205
common-emitter amplifier (lab), 182–183
common-mode amplifier (discrete)
  common-mode gain, 215
common-mode gain
  measuring (lab), 223
Compact Disk (CD) standard, 847
comparator, 352–358
  . . . contrasted with op-amp, 352
  digital
    magnitude (design problem), 615, 648
    Verilog (worked examples), 648–649
    worked example, 614–615
  hysteresis, 356–358
    AC, 358
    how much?, 357
    Woody Allen, 357
  Schmitt trigger, 356
comparator (lab), 369–370
compensation capacitor
  op-amp (lab), 228
compliance (voltage), *see* current source, current mirror
counters, 685–689
  74HC16X series, 693
  8-bit (lab), 701–703
  applications

bullet timer (worked example), 722
    stopwatch (lab), 706
BCD, 692
carry, 690
carry-out glitch (lab), 704–705
carry-out: synchronous, 714
cascading counters, 690
clear (reset) input, 691
divide-by-n, 692
divide-by-n (lab), 703
load input, 692
overflow (worked example), 725
ripple counter, 685
ripple counter (lab), 699
synchronous counter, 687–689
synchronous counter (lab), 700
up/down, 692
Verilog counters (behaviorial), 695
Verilog counters (lab), 707
worked examples), 714
counters (lab), 699–700
crossover distortion
    demonstration (lab), 181
    op-amp remedy, 277
crossover frequency, 344
current limit
    op-amp output, 260
current meter "burden", 28
current mirror, 210, 239–247
  as load for op-amp first stage, 258
  output voltage compliance, 239
current mirror (lab), 232
current source
    op-amp, 275
    output voltage compliance, 165
      defined, 253
    single-transistor, 164
current source (lab), 181–182
current-hogging, 512
current-to-voltage converter (op-amp), 276

D flip-flop, *see* flip-flops, D type
D flip-flop (lab), 668–669
DAC, *see* analog ↔ digital conversion
DAC (lab), 825–826
Darlington transistor pair
    defined, 253
    switch applied (worked example), 381, 383
debouncer, 653, 661–662
debouncer (lab), 671–673
debugging
    digital circuits, 709
    folded pin, 709
    hints and tips (appendix), 1112
      checklist, 1113
      David's Axiom, 1113
    oscillation frequency as clue, 712
    power through I/O pin, 710
decibel
    two definitions, 44
    use in filter applications, 68

decoder
    2-to-4, 609
decompensated op-amp, 416
decoupling capacitor, 79
decoupling capacitor (lab), 186
decoupling power supplies (lab), 186, 280
delta–sigma ADC, *see* analog ↔ digital conversion
demodulation
    in group audio project, 555
    lock-in amplifier, 454
DeMorgan's theorem, 572
DeMorgan, Augustus, 570
depletion mode, 547
DG403 (lab), 532
DG403 analog switch, 518
difference amplifier, 198, *see* differential amplifier
    op-amp version, 318
difference amplifier (discrete), 211–215
    ...evolves into op-amp, 216
    design, 213–215
    differential gain, 214, 215
differential amplifier, 198
    clipping, 237
    op-amp version, 318
differential amplifier (discrete), 211–215
    ...evolves into op-amp, 216
    design, 213–215
    differential gain, 214
differential amplifier (lab), 219–224
differential gain
    measuring (lab), 223
differentiator
    op-amp, 318–320
    passive RC, 60
differentiator (lab), 83–84
Digikey, 506
digital
    alternatives to binary, 566
    binary, 564
    contrasted with analog, 564
    glossary, 601
    resolution, 565
    why, 565
digital logic
    as functions, 577
    buses, 729
    combinational logic, 570
    combinational vs sequential, 651
    dangers
      gating clocks, 684
      glitches, *see* glitches (digital hazard)
      hazard, 684
      metastability, 660, 685
      race, 682
    ground rules, 587
    minimization, 576
    redundant cover, 684
    sequential logic, 651
    sum-of-products, 576
    synchronous vs asynchronous, 657
    three-state, 729

Index

diode
  truth table, 571
  circuits, 122
  clamp, 123
  clamp (lab), 145
  current-limiting, 223, 487, 541
  diode drop, 123
  diode test (DVM), 29
  I versus V (lab), 32
  Schottky, 126
  zener, *see* zener diode
DIP package
  breadboarding, 220
discrete transistor (defined), 160
distortion
  differential amplifier, 237–239
    symmetric, 238
dropout voltage (lab), 497
dual-slope ADC, *see* analog ↔ digital conversion
dynamic RAM, *see* memory
dynamic resistance, 126

E12 resistor values, 41
Early effect, 241–244
  calculating R_{out}, 243
  emitter resistors as remedy, 245
  Wilson mirror, 244
Ebers–Moll
  I_S, 201
  r_e, 202
  I versus V curve, 201
  model, 200
  reconciling with $I_C = \beta \times I_B$, 210
edge triggering
  advantages, 658
  how implemented, 660
edge versus level-sensitive inputs, 657
edge-triggering, 655
electret microphone (lab), 341
electronic justice, 205
emitter follower, 165
  . . . as impedance changer, 167
  clipping, 172, 181
  input impedance, 170
  push–pull, 173
emitter follower (lab), 178–180
EPROM, *see* memory
Ethernet, *see* serial buses
exponential
  charge and discharge, 58
    estimating, 59
  function, 58

f_{3dB}
  placing (worked example), 105
$f_{crossover}$, 344
FB73-110 ferrite bead (lab), 408
feedback
  . . . versus open-loop, 268
  effect on R_{out}, 270
  examples without op-amps, 269
  generally, in electronics, 267
  positive, 352
  quantitative view
    A, 429
    B, 429
    benefit from phase shift, 431–432
    loop gain, 429, 431
  the golden rules, 269
feedthrough, analog switch (lab), 533
ferrite bead (lab), 408
FET (field effect transistor)
  introduction, 160
  saturation region, 550
  similarities to bipolar transistors, 509
  structure, 546
  virtues, 511
filter
  software
    FIR, 953
finite state machine, *see* state machine
FIR filter, 953
Firewire, *see* serial buses
flag (as signal), *see* flip-flops
flash ADC, *see* analog ↔ digital conversion
flip-flops, 651
  as edge-sensitive "flag", 677, 720
  clock, 655, 656
  counters, 685
  cross-coupled latch, 652
  D type, 655, 656
  D type (lab), 668–669
  debouncer, 653, 661–662
  debouncer (lab), 671–673
  divide-by-two (lab), 669–670
  edge versus level-sensitive inputs, 657
  edge-triggering, 655
  jam (asynchronous) clear, 657
  level-sensitive latch, 657
  master–slave, 656
  one-shot, synchronous (lab), 674–675
  pathology
    slow clock edge, 681
    slow clock edge (lab), 670
  pathology: slow clock edge (lab), 670
  propagation delay (lab), 670
  reset (clear) and set inputs, 657
  setup time, 660, 688
  shift register, 662
  shift-register (lab), 673–674
  SR (set-reset) (lab), 668
  SR (set-reset) latch, 652, 663
  sync set, async clear, 677
  sync set, sync clear, 678
  synchronous do-this/do-that (generalized), 678
  toggle (T flip-flop), 670, 687
    built from D, 688
  toggle (T flip-flop), (lab), 670
  transparent latch, 654, 657
  tricks, 677–678
  Verilog, *see* Verilog
flip-flops (lab), 668
float

function generator common (lab), 221
floating logic inputs (lab), 593–595
FM demodulator
　PLL (lab), 833
　group audio project, 555
FM using '555, *see* "oscillator..."
folded pin, *see* debugging
follower (op-amp), 270
Ford, Henry, 570
Fourier (lab), 137
Fourier series
　square wave, 121
FPGA, *see* PLD; WebFPGA; Verilog
frequency compensation (op-amp), 412–416
frequency domain (lab), 85
FSM, *see* state machine
full-wave rectifier, 124
fuse
　duration of overload, 130
　slow-blow, 130
glitches (digital hazard), 682
　asynchronous carry-out, 714
　redundant cover as remedy, 684
golden rules (feedback), 269
ground
　defined, 6
　two senses, 20
ground loop, 423–424
ground noise, 423
grounded-emitter amplifier, 200, 202–204
h_{fe}, 166
hazard (digital), *see* digital logic, dangers
HDL (hardware description language), 624
heat sink
　gasket, 506
　thermal analysis (worked example), 505
highpass filter (lab), 86–87
hobby servos, *see* serial buses
HX711, 336
hysteresis (lab), 370
IC current source, 486
　JFET type, 487
　LT3092, 486
　REF200, 486
ICM7555 (lab), 373
IGBT, 513
impedance
　RC circuit, 72
　　worst-case, 72
　...in a direction, 171
　...output, 15
　capacitor, 63
　relations: a rule of thumb, 21
INA149 difference amplifier, 294
inductor, 116
input protection clamp (lab), 594
instantiation, *see* Verilog
instrumentation amp (lab)
　using ADC, 336
　using Arduino, 336

integrator
　op-amp, 313–318
　passive *RC*, 61
integrator (lab), 84
interfacing among logic families
　CMOS can behave like TTL, 604
　low-voltage (<5V), 604–605
　TTL to CMOS (5V), 603–604
interrupts, *see also* SAMD21
　interrupt service routine (ISR), 912
　latency, 970
　uses, 970
　vs polling, 911
inverting amplifier (op-amp), 272
IRL2505PBF (lab), 529
IRLIB9343 (lab), 496
IRLZ34 (lab), 529
JFET (junction field effect transistor), 509, 546
　as variable resistance, 548
　as variable resistance (lab), 539
　linear region, 548
JTAG, *see* serial buses
KiBi (kilobyte), 734
Kirchhoff's Laws, 9
lab equipment (online appendix), 1136
latency, *see* interrupts
LC circuit, 117
LCD alphanumeric display, 1016
lead-lag filter, 822–823
LF411 (lab), 280
line noise, 422–423
little r_e, 202
LM311, 352
LM311 (lab), 369
LM317, 483
LM358 (lab), 340
LM385–2.5 (lab), 495
LM386, 944
LM4667 (lab), 538
LM4667 switching amplifier, 514
LM50 temperature sensor, 380
LM555, LMC555, 361
LM723 (lab), 495
LM741 (lab), 280
LM78xx, 482
LMC6482 (lab), 496
LMC7555 (lab), 373
load input, *see* counters
loading, 15
lock-in amplifier
　notes, 453
logic families, 578
　TTL and CMOS circuitry, 579
logic gates
　active pullup, 584
　input threshold, 579
　logic functions (lab), 595
　LVDS (low voltage differential signaling), 582
　noise immunity
　　differential transmission, 582

Index

hysteresis, 564, 581
noise margin, 579
open collector, 584
open drain, 584
push–pull, 584
speed versus power consumption, 585
three-state, 584
  where needed, 729
totem pole, 584
tri-state, 584
universal gate, 572
wired OR, 584
logic probe (lab), 588
loop gain, 429, 431
lowpass filter (lab), 85–86
LT1073 regulator (lab), 501
LT1215 (lab), 410
LT3092, 486
LTC1150 (lab), 327
LUT (look up table), 622
LVDS, *see* logic gates
master–slave flip-flop, *see* flip-flops
MAX294
  frequency response, 848
memory
  as data organization, 730
  EPROM, 732
  PROM, 732
  RAM
    new technology, 733
    pseudo-static, 732
    SPI serial (worked example), 1036
    static vs dynamic, 732
    synchronous SRAM, 732
    types, 732
    vs ROM, 731
  ROM
    to implement state machines, 765, 780
    varieties: EEPROM, 732
    varieties: EPROM, 732
    varieties: Flash, 732
    vs RAM, 731
  ROM (lab), 743–752
  shadow RAM, 733
  specifications, 733–735
  types, 731–733
  Verilog
    using FPGA blocks, 754
    using logic cells, 736
  Verilog (lab), 742
  WOM (write-only memory), 732
meter movement (analog), 30
microcomputer
  ...as state machine, 872
  bootstrap startup, 876
  elements of a computer, 872
    ...minimal, 874
  history
    microcontroller, 871
    microprocessor, 869

microcontroller
  ...contrasted with microprocessor, 877
microprocessor
  ...contrasted with microcontroller, 877
microcontroller
  ...contrasted with microprocessor, 877
  8051, 871
  architecture
    ARM, 871
    CISC vs RISC, 871
  choosing a controller, 876
  read-modify-write vs atomic operations, 882
  SAMD21, *see* SAMD21
microcontroller programming, *see also* CMSIS; SAMD21
  as state machine, 927
  counter/timer (lab), 976
  foreground vs background processing, 912
  function calls, 910
  in C, 905
  interrupt handler, 975
  interrupts, 911
  interrupts (lab), 976
  pointers, 884
  RTOS (lab), 1060
  RTOS (real time operating system), 1049
  RTOS (worked example), 1083
  state machine (lab), 919
  structures, 913
  subroutines, 910
  switch statement, 927
  unions, 914
  waveform processing, 949
  writing robust code, 924
microprocessor
  ...contrasted with microcontroller, 877
Miller effect, 403, 513
MJE2955 (lab), 410
MJE3055 (lab), 183, 410
modulation
  lock-in square wave reference, 456
module, *see* Verilog
MOSFET, 160, 546
  $R_{DS(on)}$, 512
  analog switch, 517–528
    application: multiplexer, 520
    application: sample-and-hold, 521–528
    application: switched cap filter, 520
    charge injection, 522, 526
    CMOS, 518
    imperfections, 518
  body diode, 511, 517
  CMOS
    logic gate, 517
  IGBT, 513
  input capacitance, 513
  logic gate, 516
  power switch, 512
    application: switching amplifier, 514
    paralleling, 512
  symbols, 510
multimeter (lab), 30–32

Index

multiplexed display (lab), 851–856
multiplexer
   decoder included, 607
   implemented with gates, 609
   implemented with three-states, 610
   implemented with transmission gates, 609
multiplexing, 606
multiplier
   worked example, 616
NAND latch (lab), 668
negative feedback
   in ordinary usage, 266–267
noise
   "line", 422–423
   ground, 425
   ground loop, 423–424
   ground noise, 423
   parasitic oscillation, 426–428
   power supply (lab), 185–186
   RF pickup, 422, 425
non-inverting amplifier (op-amp), 271
NRE (Non-Recurrent Engineering costs), 620
Nyquist sampling theorem, 795

offset binary, 951
offset current (I_{OS}), 308
offset voltage (V_{OS}), 304
Ohm's law
   hydraulic analogy, 5
   why it holds, 7
one-shot, synchronous (lab), 674–675
open-collector (lab), 589
operational amplifier
   … designed from discrete transistors, 216
   AC amplifier, 323
   active rectifier, 299
   amplifier, inverting (lab), 283
   amplifier, non-inverting (lab), 283
   amplifier, summing (lab), 284
   as comparator (lab), 368
   chopper, 349
   current source, 275
   current source (lab), 289–290
   current-to-voltage converter, 276
   current-to-voltage converter (lab), 287–289
   detailed schematic ('411), 428
   difference amplifier, 318
   differential amplifier, 318
   differentiator, 318–320
   differentiator (lab), 328–329
   feedback generalized, 277
   finite gain
      effect of phase shift, 434
   follower, 270
   follower (lab), 281
   frequency compensation, 412–416
      … imposes integration, 412–413
      uncompensated and decompensated op-amps, 416
   golden rules
      conditional application, 299
      when they apply, 274
   imperfections, 302
      balancing paths for I_{bias}, 306
      bias current (I_{bias}), 306
      DC, 303
      dynamic, 302
      gain rolloff (GBW product), 309
      input and output voltage range, 310
      noise, 310
      offset current (I_{offset}), 308
      offset voltage, 304
      output-current limit, 310
      slew rate, 309
   instrumentation amplifier, 320
   instrumentation amplifier (lab), 330–340
   integrator, 313–318
   integrator (lab), 325–328
      offset trim, 326
      self-trimming op-amp, 327
   integrator tests op-amp, 317
   inverting amplifier, 272
   made from transistor array (lab), 224–229
   non-inverting amplifier, 271
   open-loop (lab), 281
   phase shifter (lab), 284–285
   photodetector (lab), 287–289
   power booster, 276
   push–pull buffer (lab), 286–287
   rail-to-rail, 312
   single-supply AC amplifier (lab), 340–341
   slew rate (lab), 330
   source of name, 269
   specifications, ordinary and premium, 312
   stability
      split feedback (worked example), 436
   summing amplifier, 273
   transresistance amplifier, 276
   virtual ground, 273
OR as Set/Pass* function, 578
OS-CON capacitor (lab), 502
oscillator, 359–367
   FM using '555 (lab), 374
   IC: '555, 361
   IC: recent, 363
   relaxation (IC inverter), 360
   relaxation (op-amp), 359
   sinusoid: Wien bridge, 363–367
   triangle (lab), 374
oscillator (lab)
   IC oscillator: '555, 373–375
   op-amp relaxation, 371
   PWM motor drive, 372
   RC: IC Schmitt trigger, 371–373
   sinusoid: Wien bridge, 375–376
oscilloscope
   lab: first view, 35
   sweeping frequencies, 86, 97
   triggering, 35–36
   X–Y mode, 86, 98
oscilloscope advice (appendix)
   triggering, 1130–1133
      "auto" vs "norm", 1133

vertical mode, 1132
oscilloscope probe
  compensation, 113, 116
output current limit
  op-amp (lab), 282
output impedance, 15
oversampling, *see* analog ↔ digital conversion

parasitic oscillation, 426–428
  discrete follower, 400–403
  discrete follower (lab), 406–408
  general remedies, 403
  generally, 388
  op-amp
    capacitive load as problem, 394–398
    capacitive load: remedy 1, feed back less, 396
    capacitive load: remedy 2, move feedback, 397
    capacitive load: remedy 3, split feedback, 397
    capacitive load: remedy 4, specialized driver, 397
    frequency compensation as remedy, 392–393
    gain rolloff as remedy, 393
    generally, 388–391
    phase lag as cause, 390
    stability criteria, generally, 398–400
  operational amplifier (lab), 408–411
    remedy: snubber, 411
    remedy: split feedback, 411
  remedy: base resistor (lab), 408
  remedy: ferrite bead (lab), 408
parasitic oscillation (lab)
  comparator, 369
parts lists (online appendix), 1136
parts sources (online appendix), 1136
phase margin, 415
phase shift
  hiding from it, 64
  rules of thumb, 73
phase splitter, 198–199
phase splitter (worked example), 191
phase-locked loop, 817–823
  applications, 820–821
  FM demodulation (lab), 833
  loop filter (lab), 832
  phase detector
    edge-sensitive, 819–820
    edge-sensitive compared to XOR, 819
    XOR, 817
    XOR (lab), 833
  stability, 822–823
    lead-lag filter, 822–823
phase-locked loop (lab), 831–834
phasor diagram, 74
PID
  controller design, 442
  D: derivative of error
    calculating (lab), 466
  D: derivative of error (lab), 465–467
  effect of derivative, 446–452
  formal description, 443
  I: integral of error (lab), 467–469
  integral, 452

motor control loop, 440
P: proportional to error (lab), 464–465
phase margin, 445
sample applications, 440
setting D gain, 449
the problem, 439
pinouts (appendix), 1114
PLD (programmable logic device)
  bitstream, 623
  FPGA (field programmable gate array), 621
  history, 621
  logic cell, 622
  PAL (programmable array logic), 621
PLL, *see* phase-locked loop
positive feedback
  hysteresis, 356–358
  oscillators, 359–367
potentiometer, 12
  as variable resistor, 13
  construction, 13
  digital (SPI), 1020
power, 4, 8
  $P = IV$ justified, 9
power dissipation
  variable resistor, 502
  choosing a heat sink, 505
  in resistor, 8
power supply
  filter capacitor, 128
  fuse, 129
  split, 124
  transformer current rating, 129
  transformer voltage, 127
  unregulated, 127–130
probe
  compensation, 116
  oscilloscope, 113
programmable logic device, *see* PLD
programming, *see* microcontroller programming
projects
  digital project lab, 851
    capacitance meter, 857
    digital clock, 861
    digital voltmeter, 861
    hotel safe, 865
    kitchen timer, 865
    reaction timer, 858
  examples
    Etch-a-Sketch, 1091
    game: asteroids on scope, 1106
    game: pacman on scope, 1106
    game: pong on scope, 1105
    inverted pendulum, 1102–1104
    laser character display, 1088–1089
    laser X–Y display, 1098
    scope X–Y display, 1093–1098
    sound source detector, 1090
    spinning LED display, 1099
    vehicles, 1099–1102
    X–Y table, 1091
  general advice, 1088

Index

hardware
    stepper motor drive, 1107–1109
labs
    lullaby jukebox, 1066
    lullaby jukebox IR (worked example), 1079
    lunar lander, 977
    theremin, 1022
    voice recorder (worked example), 1044
PROM, *see* memory
propagation delay (lab), 670
pulse-width modulation (PWM)
    DC motor control, 372
    used as DAC, 801
    using microcontroller, 977
push–pull follower
    ... simple, 166
    as op-amp output stage, 226
    first lab version (low current), 181
    high current, 410

Q (quality factor), 119–121, 136–138
QAM (quadrature amplitude modulation), 566
quiescent defined, 187

R:2R ladder, 798–800
r_e
    ... deriving, 216
R_{on}
    analog switch (lab), 533
$R_{DS(on)}$
    MOSFET (lab), 532
race (digital hazard), 682
radio
    AM demodulation, 131, 134
    AM modulator (lab), 229
    AM receiver, 130–134
radio (lab), *see* AM radio
rail-to-rail op-amp, 312
RAM, *see* memory
ramp waveform, 57
RC circuit
    impedance, 72
        worst-case, 72
    time constant, 59
reactance
    capacitor, 63
    defined, 63
reaction timer (lab), 858
rectifier
    full-wave, 124
    half-wave, 123
rectifier, full-wave (lab), 142
rectifier, half-wave (lab), 141
REF200 current-source IC, 486, 506
reflections, *see* transmission lines
relaxation, *see* oscillator
relay, 529
resistance, 7
    dynamic, 8
    parallel, 11
    shortcuts, 11
resistor

"E12" values, 40
"ten percent" values, 40
carbon composition, 38
color code, 39
fabrication, 7
metal-film, 38
power, 41
tolerance, 39
resonance
    RLC circuit, 117–122
resonant circuit (lab), 136–138
reverb, 954
ringing, 121
    LC circuit, 146
ringing (lab), 138
ripple counter, *see* counters
RISC (Reduced Instruction Set Computers), *see* microcontroller
RLC (lab), *see* resonant circuit
RLC circuit, 117–122
    resonance, 118
    ringing, 121
ROM, *see* memory
Romeo and Juliet, 578, 603
RS232, *see* serial buses
RTOS, *see* microcontroller programming
SAMD21, *see also* CMSIS; microcontroller programming
    adding external RAM (worked example), 1036
    architecture, 878
    CMSIS, *see* CMSIS
    CPU clock speed, 947, 959
    debugging code, 895
    IDE (integrated development environment), 902
    internal peripherals
        counter/timer, 911, 967
        DAC, 934
        DAC (lab), 940
        external interrupt controller (EIC), 971
        generally, 880
        initialization, 934, 967, 1014
        port I/O, 880
        SERCOM (serial communications), 1014
    interrupt handling explained, 986
    interrupts, *see also* interrupts
        architecture, 971
        interrupt vector table (IVT), 971
        IR using EIC (worked example), 1081
        serial I/O (worked example), 1032
    interrupts (lab), 917
    interrupts (worked example), 993
    intrinsic instructions, 916
    introduction, 877
    introduction (lab), 891
    nested vector interrupt controller (NVIC), 972
    port I/O
        port input, 913
        port input (lab), 917
        port output, 880
        port output (lab), 891
    read-modify-write vs atomic operations, 882

registers, 878
    link register (LR), 910
    stack pointer (SP), 909
    setup, 902
    software development, 893
    SysTick timer, 911
    SysTick timer (lab), 917
sample-and-hold, 806
    as analog switch application, 521
    charge injection, 522, 526
    errors, 523
    used with ADC, 806
sampling
    aliasing, 836, 839–840
    analogy with amplitude modulation, 840–845
    artifacts of sampling, described in frequency domain, 840
    filtering artifacts, 836
    sampling rate, 795
    seen as multiplication, 843–845
SAR, *see* binary search ADC
SAR ADC (lab), 825
SATA, *see* serial buses
saturation region (FET), 550
saturation voltage (bipolar transistor)
    effect upon R_{in}, 208
    lab, 184
Schmitt trigger, 356
Schmitt trigger (lab), 370
Schottky diode, *see* diode
sensitivity list, *see* Verilog
sequential circuit, *see* digital logic
SerDes (serializer/deserializer), 1004
serial buses, 1004–1013
    I^2C, 1010
    "one-wire", 1012
    COM port, 1005
    Ethernet, 1013
    Firewire, 1013
    hobby servo, 978, 1011
    IR remote, 1011
    IR remote (worked example), 1079
    JTAG, 1012
    SATA, 1013
    SerDes, 1004
    SPI, 1008–1010
    SPI (lab), 1014
    SWD (Singe Wire Debug), 1012
    Thunderbolt, 1013
    UART & USART (RS232), 1005
    USB, 1007
SET*, RESET* (lab), 669
setup time (flip-flop), 660, 688
Shannon, Claude, *see* Nyquist sampling theorem
shift register, 662
shift-register (lab), 673–674
sideband
    lock-in amplifier, 453
single-slope ADC (worked example), 716
slow clock edge (lab), 670
slow-blow fuse, 130, 151
SPI, *see* serial buses

LCD display, 1016
SPI RAM, 1037
split feedback (lab), 411
split feedback (worked example), 436
split power supply, 124
SR flip-flop (lab), 668
SR latch, 652
SR latch (lab), 668
SRAM, *see* memory
stack
    for temporary storage, 908
    PUSH and POP, 909
state machine, 759–770
    design steps, 761
    directed graph, 762
    examples
        complex blink, 776
        two-bit counter, 760
    implemenation
        discrete logic, 764
        microcontroller, 786
        ROM, 765, 780
        Verilog, 766
    Moore vs Mealy, 760
    PS/NS table, 762
    state transition diagram, 762
state machine (lab), 771
static RAM, *see* memory
sticky flag, 677, 678
stopwatch (lab), 706
summing circuit, 273
SWD, *see* serial buses
sweeping frequencies (lab), 85
sweeping function generator frequencies, 97
    analog generator
        analog scope: timed display, 101
        analog scope: XY, 98–101
        digital scope, 101
        artifacts caused by fast sweep, 100, 102
    digital generator
        digital scope, 102
switch
    bipolar transistor, 174
    mechanical, 653
    MOSFET, 509, 546
switch debouncer, 653, 661–662
switch, bipolar transistor (lab), 183–185
switched-capacitor DAC, 800
switching amplifier, 514
switching amplifier (lab), 538
SYNC_OUT (function generator output), 36
synchronizer
    on carry-out (worked example), 726
synchronizer (lab), 673
synchronous
    meaning, 663, 687
synchronous carry, *see* counters
    for overflow (worked example), 726
synchronous circuit
    advantages, 658
synchronous counter, *see* counters, 689

synchronous counter (lab), 700
synchronous SRAM, *see* memory

T flip-flop, *see* flip-flops
T network, 316
temperature effects
   bipolar transistor, 206
   current mirror, 240
temperature instability (BJT), 205–209
   remedy
      R_E, 206
      compensation with second transistor, 209
      explicit feedback, 209
temperature stability
   diff-amp (worked example), 257
thermal calculation: MOSFET (lab), 530
thermal runaway
   op-amp push–pull, 260
thermally conductive gasket, 506
Thevenin
   model, 15
   resistance, 16
      shortcut, 16
      shortcut: justifying, 16
   voltage, 15
Thevenin model (lab), 29–30
three-state, *see* logic gates
three-state (lab), 598
Thunderbolt, *see* serial buses
time constant, 59
time constant (lab), 82–83
time-domain (lab), 82
TIP110 Darlington transistor pair, 381
TLC372, 380, 943
TLC5916, 744
toggle flip-flop, *see* flip-flops
tolerance (resistor), 39
tolerance codes, capacitor, 93
tracking ADC, *see* analog ↔ digital conversion
transistor (bipolar)
   . . . why they are hard, 198
   beta, 166–168
   biasing, 169
   biasing (worked example), 187
   common-emitter amplifier
      bypassed-emitter, 207
   defined, 160
   Early effect, 241–244
      calculating R_{out}, 243
      emitter resistors as remedy, 245
      limit to amplifier gain, 247
      remedies, 244–246
      Wilson mirror, 244
   Ebers–Moll model, 200
   effect of saturation upon R_{in}, 208
   first model, 163
   four topologies, 175
   grounded-emitter amplifier, 202–204
   impedance at collector, 175
   phase splitter, 198
   simple model, 163

   symbol, 162
   temperature effects
      mirror, 240
   temperature instability, 205–209
transistor pinout (lab)
   by experiment, 177
transistor summary
   Early Effect, 252
   impedances, common-emitter amp, 251
   impedances, diff-amp, 252
   Miller Effect, 253
   switch, 252
transistor switch
   bipolar, 174
transmission gate, 509
transmission lines, 23
transmission lines (appendix), 1119
   LVDS, 1126
   reflections, 1122
   series termination, 1126
transparent latch, 654
transresistance amplifier, 276
tri-state, *see* logic gates
triangle waveform, 57
truth table, 571
TTL (function generator)
   syncing output, 36
two's-complement, *see* binary numbers, 611

UART, USART, *see* serial buses
uncompensated op-amp, 416
universal gate, *see* logic gates
USB, *see* serial buses

variable resistor, 13
VCO
   group audio project, 554
VCO (voltage controlled oscillator)
   group audio project, 553
VCVS (voltage-controlled voltage-source) filter, 386, 418
VCVS filter (lab), 405–406
Verilog, 624–680
   `always` block, 664
   `assign`, 625
   bit-select, 626
   blocking vs non-blocking assignments, 665, 679–680
   `case` expression, 736
   comparators (worked examples), 648
   concatenation, 626
   conditional statements, 665
   constants (literals), 625
   counters, 695
   finite state machines, 766, 781, 783
   flip-flops, 663
   hierarchical design, 629
   instantiation, 629–630
   loops, 740
   macro primitives, 754
   module, 624
   operators, 626
   parameters, 739
   part-select, 626

RAM (lab), 742
reg declaration, 665
replication, 626
ROM (lab), 743–752
scalars, 626
sensitivity list, 664
signals, 626
structural vs. behavioral design, 628
taming active-lows, 631
testbenches (online appendix), 1136
vectors, 626
verilog
   testbenches (online appendix), 1136
VHDL, 623, 633
virtual ground, 273
VN2222 (lab), 704, 853, 1060
volatility (of memory), 733
voltage divider, 12
voltage divider (lab), 29
voltage regulator
   78xx (fixed), 482
   crowbar (lab), 500
   crowbar circuit, 487
   current limit, 480
   dropout voltage, 481
   evolving a design , 477–480
   IC regulator
      adjustable, 483
      fixed, 482
   LM317 (adjustable), 483–484
   low-dropout regulator, 481
   stabilization, 479
   switching, 488–493
      configurations: boost, buck, invert, 490
      efficiency, 490
      inverting (lab), 504
      step-down ("buck") (lab), 503
      step-up ("boost") (lab), 501
      web applications for design, 492
      Williams, Jim, 493
   switching (lab), 500–504

thermal calculation (lab), 497
thermal design, 484
three-terminal, adjustable (lab), 499
three-terminal, fixed (lab), 497–499
voltage reference IC, 478
zener, 478
voltage, defined, 6
VP01 (lab), 496
VP0106 (lab), 289
waveform
   ramp, 57
   triangle, 57
WebFPGA
   description, 641
   development environment, 641
   introduction (lab), 636
   using open-drain I/O, 645
Wheatstone bridge, 321
Wien bridge, 363
Wien bridge oscillator (lab), 375–376
Wilson current mirror, 244
wired OR, 584
Woody Allen, 357
worst-case impedance, 72
XOR as Invert/Pass* function, 578
Z_C, 64
zener diode
   compared to IC reference, 494
   description, 126
   in overvoltage protection circuit, 500
   in voltage regulator, 478
   VI curve (lab), 140
ZVP3306 (lab), 289, 496, 854